AF598604

Encyclopedia of Applied Physics

SPONSORS

AMERICAN INSTITUTE OF PHYSICS
DEUTSCHE PHYSIKALISCHE GESELLSCHAFT
JAPAN SOCIETY OF APPLIED PHYSICS
PHYSICAL SOCIETY OF JAPAN

DEUTSCHE
PHYSIKALISCHE
GESELLSCHAFT

JAPAN SOCIETY
OF APPLIED PHYSICS

PHYSICAL SOCIETY
OF JAPAN

ENCYCLOPEDIA OF APPLIED PHYSICS

VOLUME 18
Servo and Control Devices to Sonic Noise

Edited by
GEORGE L. TRIGG

Associate Editors
EDUARDO S. VERA
WALTER GREULICH

Managing Editor
EDMUND H. IMMERGUT

Assistant Managing Editor
CHRISTOPHER THOMAS MORAN

George L. Trigg
275 Beaver Dam Road
Brookhaven, New York 11719

Walter Greulich
Ziegeleiweg 30
D-69488 Birkenau
Federal Republic of Germany

Edmund H. Immergut
Christopher Thomas Moran
2 Sidney Place
Brooklyn, New York 11201

Eduardo S. Vera
Science and Technology Information Center (ICT)
University of Chile
Beaucheff 850
Santiago, Chile

Library of Congress Cataloging-in-Publication Data

Encyclopedia of applied physics.
(Revised for Vol. 18)

"Sponsors, American Institute of Physics . . . [et al]."
Includes bibliographical references and index.
1. Physics—Encyclopedias. 2. Engineering—Encyclopedias. I. Trigg, George L. II. Vera, Eduardo S. III. Greulich, Walter. IV. American Institute of Physics.
QC5.E543 1991 530′.05 91-8738
ISBN 1-56081-077-7 (VCH Publishers)

British Library Cataloguing in Publication Data

Encyclopedia of applied physics.
Vol. 18
I. Trigg, George L. (George Lockwood) II. Vera, Eduardo S. III. Greulich, Walter
621
ISBN 1-56081-058-0 (set)
ISBN 1-56081-077-7 (vol. 18)

Printed in the United States of America.

ISBN 3-527-28140-1 (Volume 18) VCH Verlagsgesellschaft mbH
ISBN 3-527-26841-3 (set) VCH Verlagsgesellschaft mbH

Printing History:
10 9 8 7 6 5 4 3 2 1

Published jointly by:

VCH Publishers, Inc.
333 7th Avenue
New York, NY 10001

VCH Verlagsgesellschaft mbH
P.O. Box 10 11 61
69451 Weinheim
Federal Republic of Germany

VCH Publishers (UK) Ltd.
8 Wellington Court
Cambridge CB1 1HZ
United Kingdom

ADVISORY BOARD

EDITORIAL CONSULTANTS

MAIN ENTRIES

The subject matter in the *Encyclopedia of Applied Physics* is presented in approximately 500 individual articles, arranged alphabetically. The topics can be classified into 20 sections, similar to the AIP Physics and Astronomy Classification Scheme (PACS):

01 General Aspects: Mathematical, Computational, and Information Techniques
02 Measurement Science, General Devices and/or Methods
03 Nuclear and Elementary Particle Physics
04 Atomic and Molecular Physics
05 Electricity and Magnetism
06 Optics (classical and quantum)
07 Acoustics
08 Thermodynamics and Properties of Gases
09 Fluids and Plasma Physics
10 Condensed Matter A: Structure and Mechanical Properties
11 Condensed Matter B: Thermal, Acoustic, and Quantum Properties
12 Condensed Matter C: Electronic Properties
13 Condensed Matter D: Magnetic Properties
14 Condensed Matter E: Dielectrical and Optical Properties
15 Condensed Matter F: Surfaces and Interfaces
16 Materials Science
17 Physical Chemistry
18 Energy Research and Environmental Physics
19 Biophysics and Medical Physics
20 Geophysics, Meteorology, Space Physics, and Aeronautics

Each article has been assigned a code number consisting of two digits which denotes the section, and a letter which gives the type of article. There are six types: A = Devices, Equipment; B = Materials; C = Methods, Processes; D = Phenomena, Effects; E = Scientific or Technological Fields; F = Institutions, Companies, Societies and other organizations.

CONTRIBUTORS

Berni J. Alder, University of California, Lawrence Livermore National Laboratory, P.O. Box 808, Livermore, CA 94550
Simulation By Molecular Dynamics

Maurizio Arienzo, United Technologies Research Center, 411 Silver Lane, MS 129-7, East Hartford, CT 06108
Silicon, Polycrystalline

Janick F. Artiola, Soil, Water, and Environmental Science Department, University of Arizona, 429 Shantz, Tucson, AZ 85721
Soil Pollution

Herbert J. Bernstein, Bernstein + Sons, 5 Brewster Lane, Bellport, NY 11713-2803
Software Engineering

R. Brückner, Fritz-Müller-Str. 57, D-82467 Garmisch-Partenkirchen, Germany
Silicon Dioxide

Mark L. Brusseau, Soil, Water, and Environmental Science Department, University of Arizona, 429 Shantz, Tucson, AZ 85721
Soil Pollution

I. Dostrovsky, Department of Environmental Sciences and Energy Research, Weizmann Institute of Science, 76100 Rehovot, Israel
Solar Energy

Monika Fieberg, Chimet SpA, Badia al Pino (Arezzo), Italy
Silver, Gold, and Other Noble Metals

A. V. Granato, Department of Physics, University of Illinois at Urbana-Champaign, 1110 W. Green Street, Urbana, IL 61801-3080
Solids, Acoustical Properties of

Russell J. Hemley, Geophysical Laboratory and Center for High-Pressure Research, Carnegie Institution of Washington, 5251 Broad Branch Road NW, Washington, D.C. 20015
Solids, Static High-Pressure Effects in

Atsushi Hiraiwa, Semiconductor and Integrated Circuits Division, Hitachi, Ltd., Kodaira, Tokyo 187, Japan
Si–SiO_2 Interface, Electronic Properties

Robert Hołyst, Institute of Physical Chemistry, Polish Academy of Sciences, Kazpraka 44/52, 01/224 Warsaw, Poland
Solubility and Mixing in Fluids

Y. Horie, Box 7908 CE, North Carolina State University, Raleigh, NC 27695
Shock Waves

Joseph L. Horner, Rome Laboratory, Hanscom Air Force Base, MA 01731
Signal Processing, Optical

Bahram Javidi, Department of Electrical and Systems Engineering, University of Connecticut U-157, 260 Glenbrook Road, Storrs, CT 06269-2157
Signal Processing, Optical

Anthony J. Ladd, University of California, Lawrence Livermore National Laboratory, P.O. Box 808, Livermore, CA 94550
Simulation By Molecular Dynamics

George C. Maling, Jr., Noise Control Foundation, P.O. Box 2880, Arlington Br., Poughkeepsie, NY 12603
Sonic Noise

Ho-kwang Mao, Geophysical Laboratory and Center for High-Pressure Research, Carnegie Institution of Washington, 5251 Broad Branch Road NW, Washington, D.C. 20015
Solids, Static High-Pressure Effects in

Christoph E. Nebel, Walter Schottky Institut, Technische Universität München, Am Coulombwall, D-85748 Garching, Germany
Silicon, Amorphous

W. J. Nellis, Lawrence Livermore National Laboratory, P.O. Box 808, L-299, University of California, Livermore, CA 94550
Solids, Dynamic High-Pressure Effects in

Wendy A. Orr-Arienzo, MICRUS, Hopewell Junction, NY 12533
Silicon, Polycrystalline

Sokrates T. Pantelides, Department of Physics and Astronomy, Vanderbilt University, Nashville, TN 37235
Silicon, Crystalline

Rob Phillips, Division of Engineering, Box D, Brown University, Providence, RI 02912
Solids, Crystalline—Mechanical Properties

Hermann Renner, Degussa A.G. (retired), D-63403 Hanau, Germany
Silver, Gold, and Other Noble Metals

F. L. Riley, School of Materials, University of Leeds, Leeds LS2 9JT, England
Silicon Nitride

Akira B. Sawaoka, RELM-Tokyo Institute of Technology, 4959 Nagatsuta, Midori, Yokohama 2278, Japan
Shock Waves

Guenther Schlamp, Stettinerstr. 25, D-61449 Steinbach, Germany
Silver, Gold, and Other Noble Metals

M. A. Shea, Space Physics Division, Geophysics Directorate, Phillips Laboratory, Hanscom Air Force Base, Bedford, MA 01731-3010
Solar Radiation

D. F. Smart, Space Physics Division, Geophysics Directorate, Phillips Laboratory, Hanscom Air Force Base, Bedford, MA 01731-3010
Solar Radiation

Martin Stutzmann, Walter Schottky Institut, Technische Universität München, Am Coulombwall, D-85747 Garching, Germany
Silicon, Amorphous

R. Tenne, Department of Environmental Sciences and Energy Research, Weizmann Institute of Science, 76100 Rehovot, Israel
Solar Energy

James P. Trevelyan, Department of Mechanical and Materials Engineering, The University of Western Australia, Nedlands 6907, Western Australia, Australia
Servo and Control Devices

Charles Wert, Department of Materials Science and Engineering, University of Illinois at Urbana-Champaign, 1304 W. Green Street, Urbana, IL 61801
Solubility and Segregation in Alloys

A. Yogev, Department of Environmental Sciences and Energy Research, Weizmann Institute of Science, 76100 Rehovot, Israel
Solar Energy

D. J. Young, School of Materials Science and Engineering, The University of New South Wales, Sydney 2052, Australia
Solids, Diffusion in

Markus Zahn, Massachusetts Institute of Technology, Room 10-174, Cambridge, MA 02139
Solid, Liquid, and Gaseous Electrical Insulation

SERVO AND CONTROL DEVICES

JAMES P. TREVELYAN, *Department of Mechanical and Materials Engineering, The University of Western Australia, Nedlands, Western Australia*

INTRODUCTION

To understand how servo and control devices work, we need to think about the machines or processes they regulate. We can then think of a *control system** that comprises the control device and the regulated machine or process. When we drive a car, we form a control system comprising ourselves (controller) and the car on the road (process).

*Important technical terms are introduced *in italics*.

Control engineering is the technology of control systems; it is a pervasive technology. It surrounds us and we rely on it for many of our basic needs. Control systems keep electric generators turning at exactly the right speed to provide a constant frequency of alternating current in our electricity supply. Hydraulic and electrical control systems keep aircraft and transport systems running. Other systems ensure the continuity of our water supply, control the temperature and environment inside our buildings, and regulate the production processes that produce

3-527-28140-1/96/$5.00 + .50

the material goods we use in our everyday lives.

Yet control is even more pervasive than this, for our own bodies contain control systems. They maintain our internal and chemical environment at precisely the conditions needed to maintain our intricate biochemistry. Other systems keep us balancing upright, allow us to walk where we want to and to follow rapidly moving objects with our eyes, and even keep us breathing when we forget to.

James Lovelock in his now classic work on the Gaia theory has shown how biological control systems can be seen to regulate the temperature and physical composition of the Earth's environment. His work has created considerable controversy, as he claims that this degree of control is the essence of life itself. However, the controversy revolves around the meaning of the word "life" and the facts are not in dispute (Lovelock, 1987).

It is surprising, therefore, that some relatively simple concepts can be applied to understand the behavior of all of these diverse systems.

1. INTRODUCTION TO CONTROL SYSTEMS

1.1 History of Control Engineering

Much of our knowledge of control systems was developed during and immediately after the Second World War in the 1940s. Stimulated by the need for accurate and effective weapons, the advent of radar and navigation devices opened the possibility of automatic control for guns, missiles, and aircraft. The classical control theory of linear systems was essentially complete by the early 1950s. A *linear system* is one for which the response to two simultaneous inputs is the sum of the responses to each input presented separately. Few real control systems behave exactly like this, but linear theory can give a useful explanation for many systems. In the years since, control theory has been modified and extended to cope with more complex systems and some nonlinear systems. However, a comprehensive theory covering nonlinear systems in the way that classical control covers linear systems has so far been elusive.

In the 1960s control theorists turned their attention to other applications. As the study of economics and business became more and more based on quantitative methods, control theory seemed to become more and more applicable. Governments started to use these ideas to control their economies. Unfortunately, control theorists knew little about the human values intrinsic to government and economic activity, and neither politicians nor economists cared much for the systematic analysis and mathematics that provide the basis for control theory. We are still encumbered with the consequences of these misunderstandings in the way our governments attempt to control economies. However, in their attempts to deal with the uncertainties of economic systems, control theorists developed some remarkably useful tools for systems with random variability.

Since the 1970s, computers have been used to extend greatly the performance and reliability of simple control systems. Since computers are so reliable, complex control systems can now be built. However, in the 1990s there is a big gap between control theory and application. Complex control systems are feasible in theory, but are seldom used in practice, particularly where safety is critical. We do not yet understand how to design complex control systems that are absolutely safe. Therefore, designers often aim to simplify control systems, at the expense of performance, so that safety can be assured.

2. THINKING ABOUT CONTROL SYSTEMS—THE BASIC PRINCIPLES (DEFINING STANDARD TERMS)

Most electric ovens have a control knob to set the internal temperature. The control device is often a temperature-sensitive (thermostat) switch that simply turns the heating element on if the temperature is too low and off when the temperature is too high (Fig. 1).

Many controllers work on a similar principle. The "switch on" temperature is a little less than the "switch off" temperature. If this difference is small, given constant heating and cooling rates, the power will need to be switched on and off rapidly. Usually the difference is a few degrees so the power is switched every minute or so. The power will

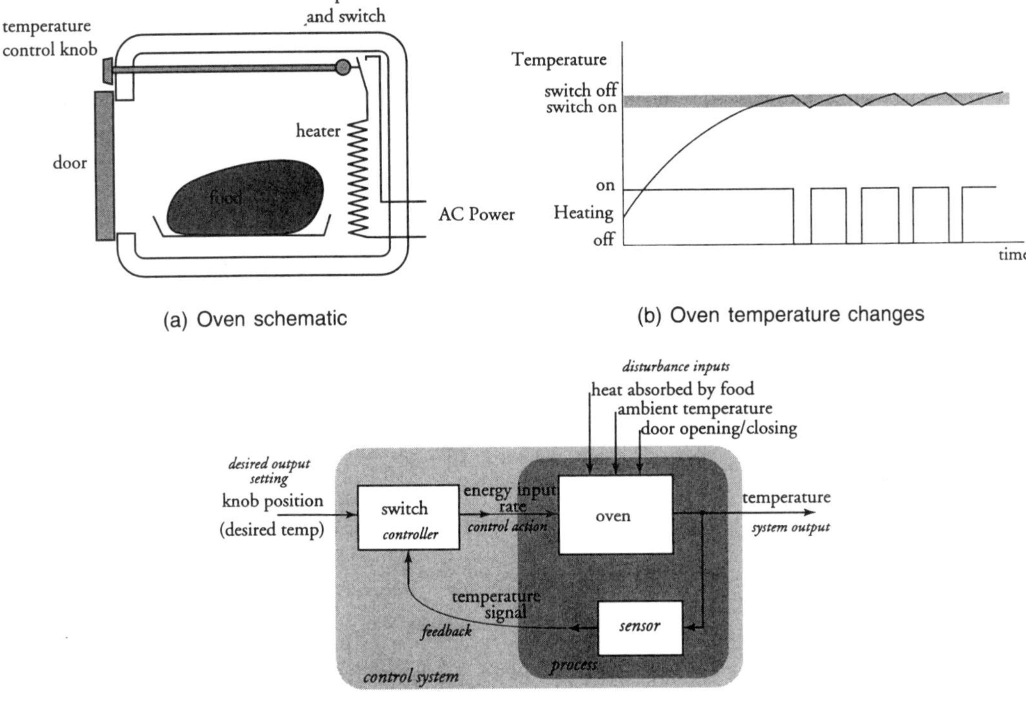

(a) Oven schematic

(b) Oven temperature changes

(c) Block diagram of control system

FIG. 1. An electric oven: a common example of a control system. The graph (top right) shows how the temperature will change with time after the oven is switched on. The block diagram (lower) represents the action of the control system in a systematic format explained in the text.

be on for a greater proportion of the time if a high temperature has been set, since more heat is lost through the walls.

This is a simple and common example of a control device, but it illustrates some important principles.

1. In any control system, we must define which system property we need to regulate. In this case it is temperature. We might need to regulate more than one property. Each property we want to regulate is usually called an *output state variable,* or *system output.*
2. Some machines require only two, or a small number of, possible system-output values. These control systems are often called *discrete control systems.*
3. We must be able to observe or measure the system output. This is known as the *observability* principle. Sometimes, we can only take indirect measurements of the system output. For example, the temperature of materials heated to extremely high temperature might only be observable by measuring heat transferred to nearby objects. In other situations, measurements of the system output might include large random errors.
4. There must be some way of changing the state of the machine or process in order to change the system output. In this example, we change the oven state by adjusting the electric power to the heating element. This is called the *control action* (again, there may be more than one control action). This is a fundamental principle known as *controllability.* A process may not always be controllable. For example, if the temperature outside the oven is higher than the set temperature, the oven temperature cannot be controlled by the control action (electric heating).
5. There are usually several factors other than the control actions that affect the system output. In the case of the oven, heat absorbed by the food being cooked,

heat lost through the door opening or through the wall insulation, and the outside temperature all affect the system output, and cannot be controlled. These are called *disturbance inputs.*

As in the oven example, many controllers adjust the control action in response to measurements of the system output. This is known as *feedback control.* The oven controller adjusts the control action (electrical heating) in response to the difference between the system-output measurement (internal temperature sensed by the thermostat switch) and the *desired output* setting (position of oven temperature knob). A stable control system must use negative feedback; thus the control action must be decreased if the system output is too high.

We can visualize a control system with an abstract diagram that shows the process, the controller, and the feedback sensor as separate blocks. The links between them are abstract variables and do not necessarily represent material or energy flow, as a process flow diagram might portray. From this we can see that a feedback-control system forms a loop. Thus feedback control is also known as *closed-loop control.*

Most control systems use feedback. However, this is not necessarily the case for all control systems. Some control systems use *open-loop control* where there is no feedback during the operation of the control system. There is no need to measure the system state while the system is operating. However, such a system can only be designed if we know the process characteristics accurately, and the disturbance inputs either are known or affect the system output only slightly. Open-loop control systems can be simpler and more stable, but feedback control is often necessary to achieve the required accuracy.

2.1 Subtle Effects of Feedback Control—Is the Air Conditioning Colder in Summer?

Feedback-control systems sometimes behave contrary to common sense. They require very careful thinking and systematic methods of analysis. One example will illustrate this.

Many of us have experienced the frustration of feeling so cold inside an air-conditioned building in summer that we need extra clothes or even a heater to feel comfortable. Why is this so? Common sense says that it must be warmer on the inside if the weather is hotter on the outside, and if you feel cold it is just because you have become adjusted to the hot weather outside.

The air conditioner pumps cold air into the living space until the measured air temperature reaches the desired level set by the thermostat knob. The air enters the living space and absorbs heat coming through the walls and ceiling, and heat given off by people in the building. Even with air circulation, the temperature varies from one part of the building to another. The temperature sensor, located high on the wall, is in one of the warmer spots. Therefore, the temperature at the sensor is slightly higher than the temperature where the people are, say 1° in winter.

In summer, with more heat coming through the walls, the temperature difference between the center of the room and the sensor on the wall is greater than in winter—the sensor may be 5° warmer in summer. Therefore, when the room has been cooled so that the measured temperature is correct, the actual temperature at the center of the room will be 4° lower than in winter. So your senses may not be deceiving you.

We can learn from this that the *sensor location* is of critical importance. This lesson will be reinforced later.

2.2 Self-Filling Water Tank

Almost every toilet has a self-filling water tank that uses a simple control device to regulate the level of water stored in the tank (see Fig. 2). The water inlet valve is operated by a lever with a float on the end. As the water level rises, lifting the float, the lever turns and gradually closes the valve. Eventually, the water level rises high enough to close the valve and no more water enters the tank.

We can predict the behavior of this kind of tank by making some simple assumptions about the valve. Given the water level h and the full level h_{max}, assume that water inflow is proportional to the drop in level below h_{max} until it is fully open at h_1 below h_{max}. Thus,

$$q_{in} = q_{max} \quad \text{if} \quad h < h_1$$

(valve fully open), and

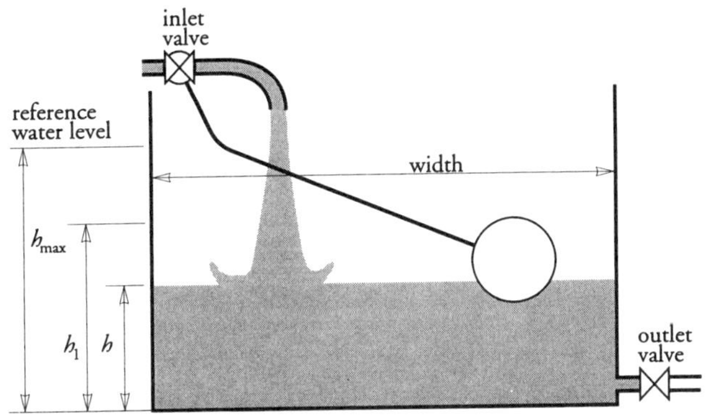

FIG. 2. Self-filling water tank used for toilets and animal drinking troughs (refer to text).

$$q_{in} = k(h_{max} - h) \quad \text{if} \quad h \geq h_1$$

(valve part open), where

$$k = q_{max}/(h_{max} - h_1).$$

Knowing the surface area A of the water in the tank (assume vertical sides so that the area is constant), calculate the level change over a short time interval δt by assuming the net inflow does not change:

$$\delta h \approx \delta t(q_{in} - q_{out})/A.$$

Given an initial water level h_0 and the outflow q_{out} as a function of time, we can repeat these computations to calculate how the water level in the tank changes at each time step. A spreadsheet is convenient for this and a drawing package can graph the results. Excel calculated the results drawn by Illustrator in Fig. 3 (Microsoft, 1994; Adobe, 1994). A sample of an Excel spreadsheet for simulation is in the Appendix.

Initially the tank is empty, and for a while no water leaves the tank. It fills at a constant rate until the level reaches h_1 and then at a decreasing rate until the tank is full. If all the water is flushed from the tank, the level drops again. Later we see what happens in the case of a leak—the steady-state level is less than h_{max}. A control system like this exhibits a *transient* response, when the level is changing with time, and later a *steady-state* response when the level is effectively constant.

Calculating the behavior of the tank like this is known as *simulation*. The equations represent a mathematical model of the tank and, in effect, we have solved a differential equation that governs its behavior.

We can build a block diagram of the tank using the model (Fig. 4). The system output that we want to control is the water level in the tank. The desired output is represented by the arrangement of the float on the lever. If we bend the float lever, we will change the desired output. This method is often adopted by plumbers or farmers who maintain these tanks. The process is the tank itself and, perversely, the outflow of water is a disturbance

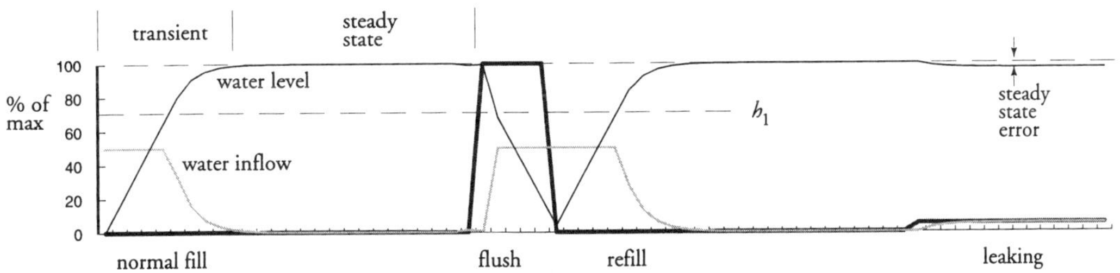

FIG. 3. Graph of tank behavior, predicted by simulation. The graph shows how the water inflow, outflow, and level vary with time. Each is expressed as a percentage of maximum level or flow rate.

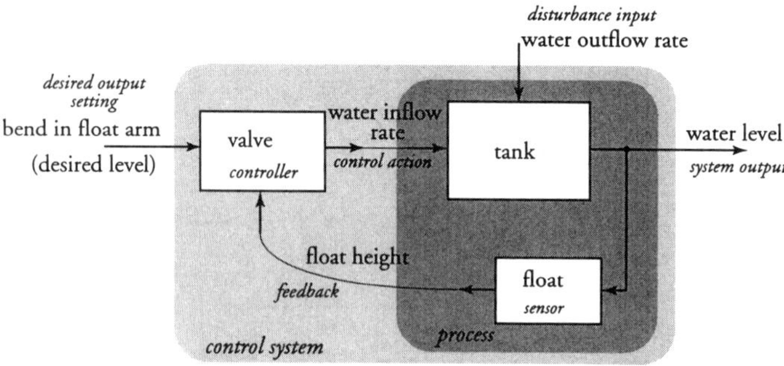

FIG. 4. Control block diagram of self-filling water tank in same systematic format as Fig. 1(c). Notice that the system output is not the water from the tank, as a process flow diagram might indicate. The system output is what we want to control, i.e., the water level.

input. This is because the rate of outflow is set outside the system, but it has an effect on the tank (process). The feedback sensor is the float, and the lever and control valve can be considered as the controller. The control action, therefore, is the water inflow rate.

Most introductory texts show how systems can be modeled using block diagrams and differential equations, and then how the solutions of these equations can be characterized in general terms (see any control textbook such as D'Azzo and Houpis, 1981). Linear control theory stems from this approach, and not only predicts the response of a linear system, but also shows how to design a controller to provide the desired response.

However, most real control systems are nonlinear, including this simple example when the water level is less than h_1 or greater than h_{max}. Practical performance limits must be taken into account, and the most effective way to do this is simulation as shown above. Later we will see when linear analysis can be helpful and when simulation is needed.

2.3 Hydraulic Actuator

Hydraulic servo-actuators are used extensively in aircraft to move the flight controls because they are fast, light, powerful, and, under the right conditions, 100% reliable. When the correct seals are used, friction is low and the oil does not leak. Similar actuators move robots in response to instructions from a control computer; the sheep-shearing robot is a good example (Trevelyan, 1992). A simple actuator can develop controlled forces of up to several tonnes in a few thousandths of a second. The position of a load weighing several hundred kilograms can be changed within a few tens of milliseconds.

The diagram (Fig. 5) shows the basic principles. A voltage proportional to the required position of the piston is supplied to an amplifier. Another voltage is generated by the potentiometer—this voltage is proportional to the actual position of the piston. The difference between the two voltages (which is proportional to the error in the piston position) is amplified and a current proportional to the voltage difference is supplied to the servo valve. The servo valve directs oil flow to either end of the cylinder to move the piston. The oil flows at a rate proportional to the current in the valve coils—thus the piston moves toward the required position at a velocity that is proportional to the error between the required position and its current position.

The servo valve is the key to understanding this impressive device, for it uses an electric current of approximately 10 mA to regulate an oil flow of up to 120 liters per minute at a pressure of up to 35 MPa. (The high-pressure oil is supplied from a separate pump unit.) Typical high-performance servo valves respond in 2 to 4 ms.

If we were to look more closely at a servo valve, we would find that it contains a control system of its own. This control system causes the valve spool to move in proportion to the electric current, thus influencing the oil flow. Again, the amplifier contains a controller to regulate the electric current so that it is proportional to the voltage difference. Thus one control system might consist of several control systems in its component parts. Typical costs are shown in Fig. 5.

Parts of the valve must be built to extremely fine tolerances—most notably the

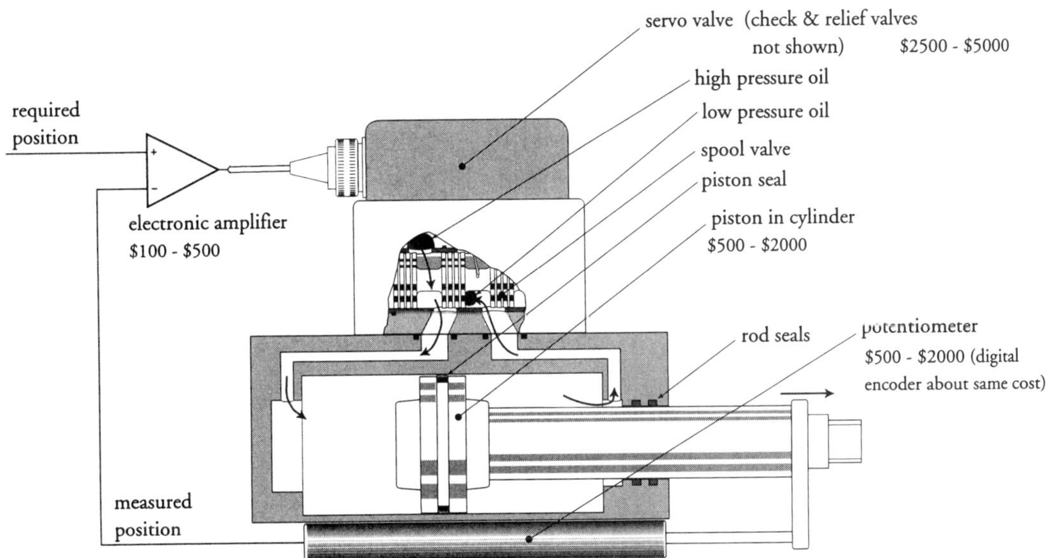

FIG. 5. Hydraulic servo-actuator. High-pressure oil supplied by a pump (not shown) is directed to one end of the cylinder by the servo valve, and the oil from the other end returns through the same valve to the oil-storage tank. A linear potentiometer measures the piston position, but there are many alternatives such as electromagnetic sensors (LVDT), magnetostrictive sensors, and digital optical encoders. Refer to text for further explanation.

clearance between the spool and the valve liner insert. The mating surfaces are highly polished, and most valves incorporate some form of filtration to minimize the chance of contaminating particles in the oil causing damage to the surfaces. The slightest damage can cause a dramatic failure of a hydraulic control system; see Sec. 6.1. However, by following appropriate design approaches, one can eliminate contamination and achieve very high reliability.

2.4 Electric Motor Positioner

Small electric motors are widely used for moving parts of a machine, such as the lens in an autofocus camera, which is moved to keep the image sharply focused. Electric motors, using digital encoders for position and speed feedback, are used for some of the most precise position controllers in computer-controlled instruments, machine tools, and robots (Fig. 6). Electric drives are popular because they are extremely reliable, repeatable, and precise and are easy to maintain. The power-to-weight ratio of hydraulic actuators can be 10 times greater, but they are less accurate unless expensive noncontact sealing arrangements are used.

The most direct way to control an electric motor exploits the relationship between current, force, and magnetic field strength. In a constant field, the sideways force is proportional to the electric current in a wire. Thus, the current in a permanent-magnet motor (a common type used for servo applications) generates a proportional torque.

If we use a transistor to regulate current directly, considerable heat is liberated at the transistor. Designing transistors that can withstand the effects of this heat is difficult, and so most motor controllers use a *switching controller* (or more fully, a pulse-width–modulated switching controller) (Bradley *et al.,* 1991). An electronic semiconductor switch does not dissipate much heat at all. A dc power source is rapidly switched electronically from one direction to the opposite in response to pulses (Fig. 7). If the "on" time is the same as the "off" time, the average current is zero. If the "on" time is longer than the "off" time, then the average current will be positive in the direction shown; otherwise it will be negative. Because the inductance of the motor is large, the current takes some time to change. Thus, switching the supply at a frequency of 15–40 kHz provides effective control of current without dissipat-

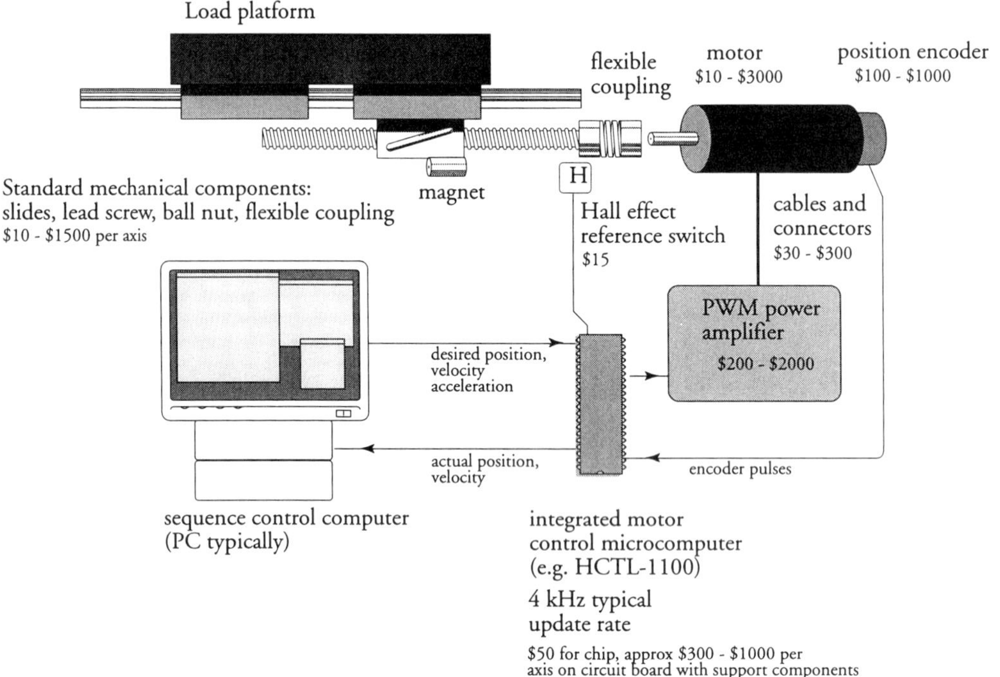

FIG. 6. Typical electric servo positioner built from standard components as used in robots and machine tools.

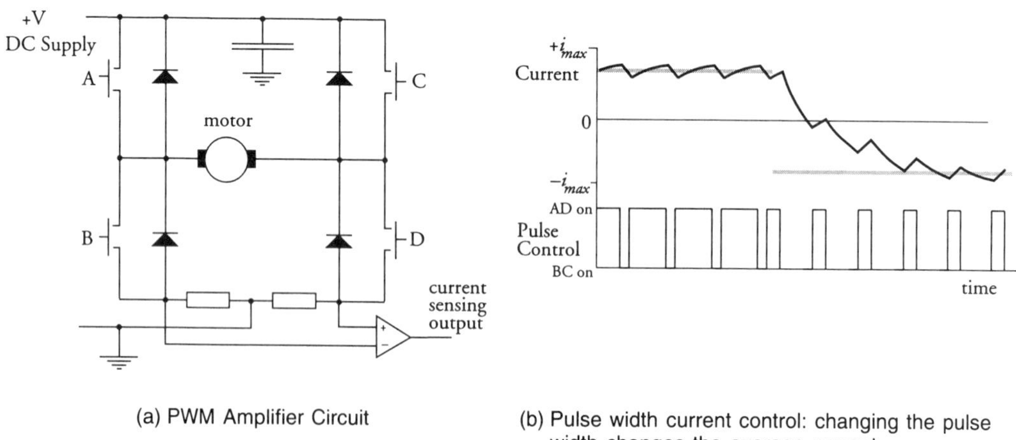

(a) PWM Amplifier Circuit

(b) Pulse width current control: changing the pulse width changes the average current.

FIG. 7. Pulse-width–modulated power amplifier for a permanent-magnet electric motor. Electronic switches alternate the voltage applied to the motor. We control the average motor current by changing the relative duration of the forward and reverse voltage pulses. The graph shows a typical current variation with time. The pulse frequency is typically 15 kHz or greater. The average current is measured and used as a feedback signal for current control. Diodes protect the electronic switches from induced voltage surges. This circuit configuration is often called "H bridge" because the layout resembles a letter H.

ing heat in the switching transistors. The average current can be measured electronically, and then used to adjust the pulse width, thus the term *pulse-width modulation* (PWM).

Notice how there is close similarity between pulse-width modulation and the simple switching thermostat control of an electric oven. The essential difference is that the frequency of pulse-width modulation is usually constant. We achieve apparently continuously variable control of current (in this instance) by using simple on/off control at a "microscopic" level.

The most accurate and convenient means of measuring the position of a motor shaft is an optical encoder (Fig. 8). In its most common form, it provides two pulse streams as the shaft turns (usually from 100 up to 20 000 pulses or more per revolution). One pulse stream leads the other by one-quarter of a cycle (phase quadrature), and so the direction of motion can be detected. By first moving the load to a reference position (usually a reference proximity switch) and then counting the pulses, allowing for direction changes, we can measure position to within (at least) one-quarter of a pulse period. Absolute encoders are available at greater cost for comparable resolution and are increasingly used in machine tools and robots because there is no need to move to a reference position first. Resolvers are also used and provide comparable accuracy.

A single microcomputer can be used to receive the desired motor position and/or speed from a master control computer, count the pulses, and produce the pulse train used to switch the power supply.

The cost of electric servo systems ranges from a few dollars (for radio-control models) to thousands of dollars, depending on speed of response, torque, power, and accuracy. A typical accurate instrumentation-positioning servo might cost between $2000 and $10 000. An instrumentation servo can be built for less if skilled electronics and mechanical technicians are available to help, but 2–6 months full-time effort should be allowed to build one from components without extensive prior experience. Unfortunately, few common standards have been adopted by component suppliers and so linking them together is not easy.

3. CONTROL-SYSTEM PERFORMANCE

3.1 Bandwidth

One of the most important concepts to understand is *bandwidth,* or, in more detailed terms, the *frequency response* of a con-

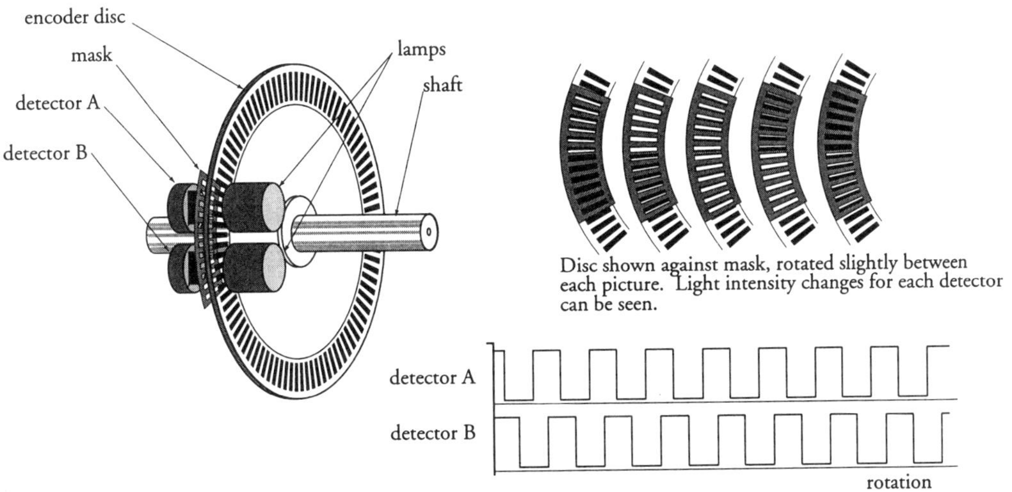

FIG. 8. Incremental optical encoder of a type often used for measuring the motor shaft position and speed in low-cost servos. The two detectors produce pulses as the shaft rotates and the light intensity changes. The mask has two sets of slots arranged such that pulses from the detectors are 90° out of phase (see sequence at top left). The detectors incorporate threshold logic to convert light intensity changes to an on/off electrical signal. Many encoders also provide an additional detector to provide a reference pulse once per revolution.

trol system, process, sensor, or controller. One way to visualize this is to imagine what happens if you drive a car over a road with regular undulations of a given height and spacing (say 3-cm height, 30-cm spacing). If you drive regularly on unsealed roads, you may have encountered this yourself.

Driving very slowly, the car rides up and down slowly and follows the undulations.

Driving fast, the wheels follow the undulations but the springs absorb the wheel motion and the car body hardly moves at all (but you will feel the throbbing of the wheels going up and down).

At some speed between these extremes, the car bounces *in resonance* with the undulations, and may bounce as much as two or three times the undulation height if the shock absorbers are not working well.

The frequency response of a process defines the *amplitude ratio* (ratio of output amplitude to input) and *phase angle* for a range of input frequencies. A frequency response is usually plotted logarithmically. For a typical control system, the frequency response resembles the graph in Fig. 9, which shows the idealized linear response predicted by theory, and a typical measured response. Frequency-response measurements can be made with specialized equipment such as a frequency analyzer, commonly used for experimental studies of vibrations or audio equipment (D'Azzo and Houpis, 1981).

A typical control system responds with a unit amplitude response and zero response lag (i.e., the output almost exactly follows the desired value) up to a certain frequency limit, beyond which the system responds less and less. The phase angle (negative, meaning a response lag) rapidly increases too. The bandwidth, ω_{max}, is usually interpreted as the range of frequencies for which the response is no less than 70% of the input and the phase lag is less than 90°.

The *response time* of a system is the time needed for the system output to travel the first 63% of the desired change and is roughly equivalent to $1/\omega_{max}$ where ω_{max} is in radians/second (2π radians per second = 1 hertz).

Thinking about bandwidth is important because it is difficult to design a control system for which the response time of the system is significantly less than its components.

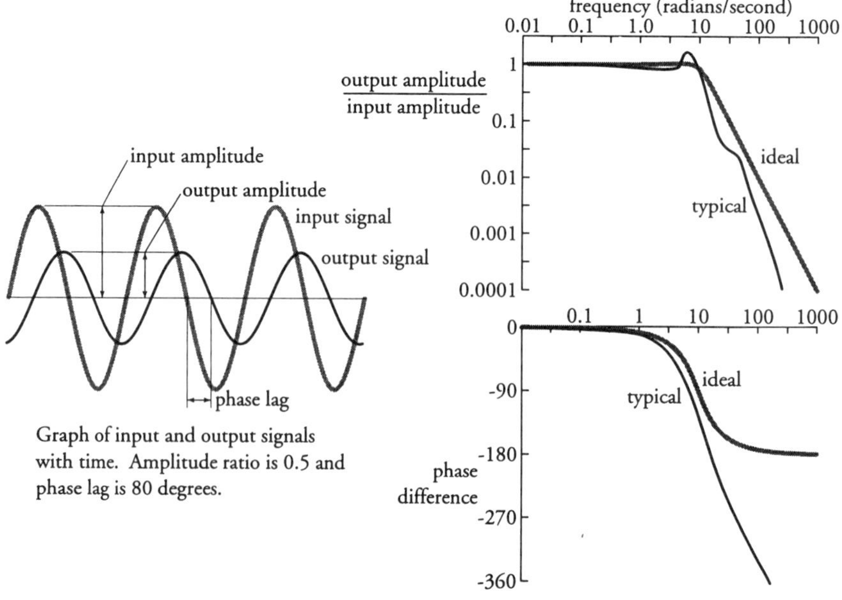

FIG. 9. Graph of typical control-system frequency response. The phase of a real system lags more than theory predicts because of factors such as computation delay in the controller (which is typically a small microprocessor), transport delays in fluids or even solid members where motion is limited to the speed of sound, mechanical backlash, and friction.

This is only possible if the characteristics of the components can be accurately predicted, either from theory or experimental studies.

Take, for example, the control of a space-robot manipulator. Factory robots are built with rigid links that do not deform significantly under load. Therefore, the position of the end of the robot can be controlled by controlling the relative positions of the links, typically using electric servos to drive rotary joints to known angles. Rigid links are usually heavy, and so robots carry loads that are small compared with their weight. Robots with flexible links may be much lighter, and very large robots for use in space cannot be made rigid. To control a flexible robot, we must compensate for deflection under load and eliminate oscillations.

A large space robot might oscillate at a natural frequency of one cycle every 10 or 20 seconds when carrying a large load (0.3–0.6 rad/s). This implies that it will be difficult to achieve a control bandwidth of 0.03–0.06 rad/s and therefore a realistic response time might be 30 s. Faster control of flexible robots has only been achieved under idealized laboratory conditions, and requires accurate modeling of the arm and intricate computations.

The sheep-shearing robot provides a useful illustration of the importance of bandwidth. This gives us an insight into a simple performance measure that predicts the maximum speed for tracking an unknown surface using a sensor on the robot's tool.

For a robot to follow an unknown surface, a robot has to guide its end effector over the surface using information from one or more sensors. Figure 10 shows some simulations of surface following with a position-control system. The robot end effector has a distance sensor, and approaches the surface at a constant speed, traversing the surface toward the right. The gray horizontal line shows the desired surface tracking distance δ. If we are not to hit the surface, this represents the maximum trajectory error allowed. This arrangement performs best if the sensed displacement error is limited to $\pm\delta$. For stable behavior the approach velocity must be limited.

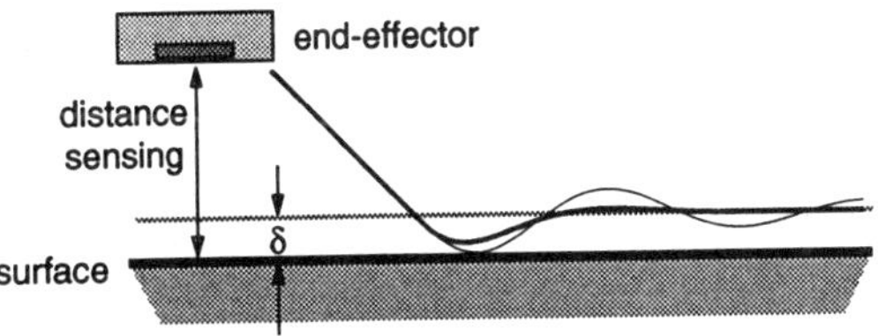

FIG. 10. A robot approaches an unknown surface using a sensor mounted on its tool or end effector to measure the distance error. The maximum approach speed must be limited to ensure that the robot does not hit the surface.

A remarkably simple relationship predicts the minimum bandwidth required of the control system, ω (rad/s):

$$\omega \geq v/\delta,$$

where v is the maximum approach speed perpendicular to the surface. This simple measure can be applied in many different contexts where sensor outputs modify planned robot movements.

3.2 Transient Response—Simulating a dc Servo Drive

We have seen how control systems exhibit two different types of behavior after a change in the desired output setting. First there is a transient while the process is adjusted to the new setting, and then the system settles to a new steady state. Clearly, the transient should be as quick as possible and *stable:* There will be little or no tendency for the output to overshoot or oscillate about the desired value.

Not all processes are easy to control and, as we saw earlier, correct analysis of a feedback-control system requires great care in thinking. It helps to follow the basic principles outlined in Sec. 2. A dc motor servo is a useful example.

Let us take, for example, a typical small rare-earth permanent-magnet motor. These specifications are based on a small dc servo motor (Maxon, 1994): 10 W rating, 9 V nominal voltage, 585 rpm/V back EMF, 16.3 mN m/A torque constant, max continuous current 1.5 A, 1.2 Ω coil resistance, rotor inertia 9.38 g cm^2 ($\times 10^{-7}$ to convert to kg m^2), mechanical time constant 4.24 ms, inductance 0.12 mH.

For a given current input i (amperes) the motor produces torque $\tau = 0.0163i$ N m. The inertia of the motor is I kg m^2 (1 kg m^2 $= 10^7$ gm cm^2) and the load equivalent inertia is I_L, and so the acceleration is $\alpha =$

$(0.0163i - \tau_L)/(I + I_L)$, where τ_L is the load torque (friction, steady load, etc.).

The model of the motor can be set up in a sequence of three steps that calculate the behavior over a small time interval δt. The calculation is repeated over enough time steps to calculate the response of the control system. At each given time step (j),

$$\text{shaft acceleration } \alpha_j = (0.0163i - \tau_L)/(I + I_L); \quad (1)$$

$$\text{shaft speed } \quad \omega_j \approx \omega_{j-1} + \alpha_j \delta t; \quad (2)$$

$$\text{shaft position } \quad \theta_j \approx \theta_{j-1} + \omega_j \delta t. \quad (3)$$

Step (2) above calculates the change in shaft speed over the time interval, $\alpha_j \delta t$, and adds this to the previous value of shaft speed to obtain the new speed. Step (3) does the same for shaft position. These calculations are not exact but good enough for the purpose here. More precise methods are only useful if the parameter values are known exactly, which is seldom the case.

There are some excellent simulation and design packages for control systems (such as MATLAB with a Control Toolkit, or Matrix-X) but an exposition of their use is beyond the scope of this article. A spreadsheet, once again, is quite sufficient to explore some stability problems.

3.3 Proportional Control

The most common controller arrangement is simply to use the difference between the system output and desired setting, or error, multiplied by a fixed constant called *gain*. The desired shaft position is θ_d and we arrange our simple controller to determine the value of motor current by

$$\text{current } \quad i_j = k(\theta_d - \theta_{j-1}). \quad (4)$$

Notice the negative sign for θ_j—this is the negative-feedback term. The term $\theta_d - \theta_{j-1}$ is the position error. The gain is k. Notice that the gain is also a unit conversion from radians to amperes.

If, for the sake of this example, we set the initial position and speed of the motor at position zero, the gain 5.0, the desired position 0.5, and the load torque 0.002, we obtain the response shown in Fig. 11. Notice how the shaft position oscillates in an unstable manner. Usually we can improve the system stability by decreasing the gain. However, this type of control system is still unstable, but with a decreased frequency. Not all control systems behave like; this many are quite stable with proportional control.

There is one important detail. The length of the time step, δt, needs to be chosen with some care. It should be short enough to represent the shortest response time in the system by about 20 steps. If there are large changes in each variable from one step to the next, reduce the time step. Check that the results do not change much if the time-step value is altered.

3.4 PID Controller

The behavior of the proportional motor-control system above can be modified by using a different controller. The most common form of process controller used industrially is a PID (proportional + integral + deriva-

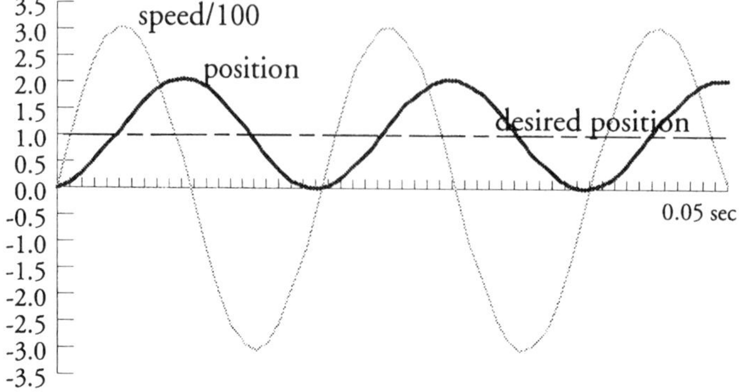

FIG. 11. Simulated response to a step change in desired position with a proportional controller.

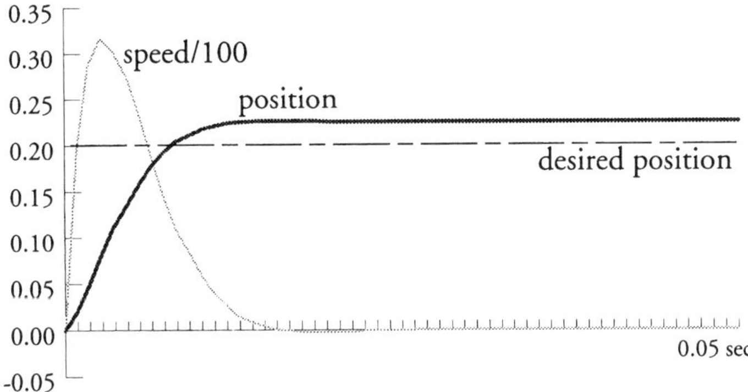

FIG. 12. Proportional + derivative control for an ideal system. Notice the steady-state error remaining after the transient.

tive) controller that combines three forms of the error variable used above:

1. a proportional component (as used above),
2. the derivative of the error, and
3. the integral of the error.

To do this, step (4) above is rewritten as two steps: integrate the error,

$$e_j = e_{j-1} + \delta t(\theta_d - \theta_{j-1}); \tag{4a}$$

$$i_j = k_p(\theta_d - \theta_{j-1}) \quad \text{(proportional term)}$$

$$k_d(\omega_d - \omega_{j-1}) \quad \text{(derivative term)}$$

$$k_i e_j \quad \text{(integral term).} \tag{4b}$$

Then, the gains k_p, k_d, and k_i are adjusted to give the best result, most often by trial and error.

In practice the magnitude of e_j, the integrated error, is limited to a range $\pm e_{\max}$ to prevent *integral windup,* which can occur during sudden large disturbance or setting changes. Windup limiting:

$$\text{if} \quad e_j > e_{\max} \quad \text{then} \quad e_j = e_{\max};$$
$$\text{if} \quad e_j < -e_{\max} \quad \text{then} \quad e_j = -e_{\max}. \tag{4c}$$

The usual procedure for adjusting the gains goes like this. First, the proportional gain is increased with $k_d = 0$ and $k_i = 0$ until the process responds fast enough (Fig. 11). Second, if unwanted oscillations occur, k_d is increased to help reduce them (0.025 in Fig. 12). (The effect of introducing the derivative term by using k_d has, in this case, an effect similar to adding mechanical damping.) When the quickest stable response is obtained, k_i is used to remove unwanted steady-state error remaining after the transient response has died away ($k_d = 0.03$, $k_i = 350$ in Fig. 13). (In this case the steady-state error is due to the load torque; Sec. 6.2 describes other causes.)

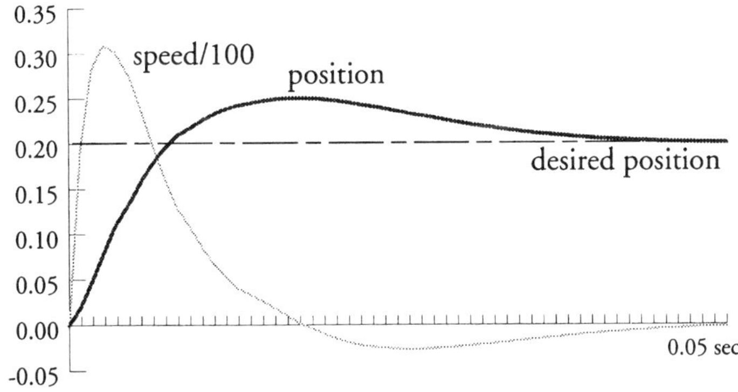

FIG. 13. Proportional + derivative + integral control. The steady-state error has been removed, but the transient lasts longer. In practice, many such simulation studies would be performed using control-system inputs representative of the actual application. Friction, time delays, power limits, sensor resolution, and other practical limitations need to be taken into account.

3.5 Limits of PID Control—When to Use Other Methods

We have seen how PID control works for a simple motor controller. PID control has been successfully applied to many processes and machines, but it has its limitations.

1. It is difficult to control nonlinear systems in which the gains need to change. For example, some positioning servos perform much better using error-squared control. Here, the control signal is proportional to the error squared, and thus the gain is higher when the error is large. When the error is small, the gain decreases, so that stability for final approach to the desired position is better.

   If the inertia of a motor shaft changes when the load changes, then a PID controller with fixed gains may go into unstable oscillations.
2. If the dynamics of the process are more complex, then a PID controller may not provide a satisfactory stable response. For example, if a motor drives a load on a long shaft that can twist by a significant amount, the simple model [steps (1)–(3) above] may not be accurate. A more complex model of the dynamic behavior will be needed, and a more complex controller may be necessary to control the torsional vibrations of the load on the long, compliant shaft.

   Advanced control techniques are available to design such a controller. These include cascade controllers, where extra feedback loops are used to stabilize parts of the system, making the whole system easier to control, and adaptive controllers that automatically adjust themselves to changing dynamics.

3.6 Fuzzy-Logic Control

No discussion of servos would be complete today without some mention of fuzzy logic. Fuzzy logic has become very popular with designers of industrial control systems, and many commercial design packages and hardware systems are available.

Fuzzy logic does not perform as well as most other control techniques can, but it is much easier to use for industrial designers. Also, a fuzzy-logic controller can incorporate the expertise of skilled process operators in the form of explicit rules (Johnston, 1994).

Each input to the controller is given a number of possible states. For example, a positioning error might be classified as follows: above 5 mm, "large positive"; 5 to 1 mm, "medium positive"; 1 to 0.1 mm, "small positive"; 0.1 to -0.1 mm, "zero"; -0.1 to -1 mm, "small negative"; -1 to -5 mm, "medium negative"; below -5 mm, "large negative." The controller is programmed with a particular rule for each possible combination of input states. Thus with three inputs and seven states each, 7^3 rules are needed.

Although the input classification might seem like coarse resolution, the control action from the controller is usually a nonlinear continuous function of the inputs with apparently infinite resolution. The controller is also programmed with "membership functions" so that a particular input value, say 3.5 mm, can be classified as "large positive," "medium positive," "small positive," etc. with different "degrees of membership" between 0 and 1. In this case, "medium positive" would have the highest membership value. Then the results of the rules for all input states with nonzero membership values are combined to give some optimum weighted output value.

Fuzzy-logic controllers have provided excellent results with processes that behave in a simple manner. Fuzzy control often beats PID control because the nonlinear nature of real processes and controllers can be taken into account. However, more conventional control techniques are far superior to fuzzy control for processes with complex dynamic behavior (such as a flexible robot arm), *provided* the behavior is observable and can be accurately predicted.

3.7 Self-Tuning and Adaptive Controllers

To control some processes, we need to adjust the gains of a simple PID controller to keep the response stable and eliminate unwanted errors. Several techniques have been developed to do this automatically. A number of companies offer self-tuning controllers, which adjust their gains automatically in response to changing conditions. Broadly speaking, there are two types of controllers:

1. gain-scheduling controllers, in which a table of gain values for particular conditions has been programmed in advance;
2. true adaptive controllers, which adjust the gains automatically on line to maintain optimum performance. This technique works well when relatively slow changes in gains are sufficient. This type of controller needs to be able to observe the system behavior for a little while before attempting to change the gain settings, and it is useful when gain scheduling is impractical (see Borrie, 1986).

3.8 Advanced Control Techniques

So far, we have looked entirely at single-output control systems. Many industrial process and machine controllers must work with several controlled outputs. For example, an aircraft autopilot must control aircraft speed, height, attitude (angle to the ground), and compass heading. Separate controllers cannot be used because all these aspects of aircraft behavior are "tightly coupled" together. Changing height, for example, requires an attitude change and engine adjustment, and will affect speed.

Advanced control techniques have been extensively developed for this and several other applications. Standard textbooks on control used for university engineering courses provide the theoretical background on these methods. However, most advanced control methods need to be applied with great care, and it is essential to seek advice from people with application experience (Borrie, 1986, discusses several different approaches).

4. COMPUTERS—FROM PERFORMANCE PREDICTION TO CONTROL

In Sec. 3.3, we predicted the performance of a control system by careful analysis and simulation. Step (4) represented the action of the controller.

In many, if not most, control systems today, the controller is a microcomputer. Computers have replaced analog electronic circuits used widely until the 1970s, because they are more reliable, more versatile, and easier to maintain.

The program in the control computer performs the same calculations as shown in step (4). Thus, in the case of a motor controller, the shaft position and velocity would be taken from input registers set by encoder pulse-counting logic, and the current setting would be sent to an output register, or converted to a PWM pulse train by the same microcomputer. Microcomputers have been specially designed for many typical applications. Hewlett-Packard, for example, produces the HCTL-1100 integrated circuit, which counts encoder pulses and controls a variety of electric motors or hydraulic actuators.

Many control and instrumentation systems are based on IBM personal computers or clones (PCs). Specialist companies supply interfaces to allow sensors to input feedback signals and the computer outputs to operate the process directly.

There are some special practical issues that must be considered. The most fundamental is the computation delay time.

4.1 Computation Delay Time

Computer control is based on the notion that the computer will regularly input process measurements and recalculate the control action to regulate the process. These operations are repeated indefinitely. The elapsed time between measurement at the feedback sensor and delivery of the newly calculated control-action signal to the process is the *computation delay time.*

Looking at steps (4a), (4b), and (4c) (Sec. 3.4 above), we can see that the computations are very simple indeed, and a modest computer could easily perform these in a few microseconds. We might therefore conclude that computation delay is not a significant issue. However, there are many other factors that lengthen the delay time.

First, the feedback sensor may itself incorporate a microcomputer. Often known as "smart" sensors, they are used because they are usually self-calibrating, and diagnose their own fault conditions. However there will be some computation delay because of this.

If the sensor signal is analog, it must be converted to digital form (A/D converter). To do this it is sampled, and then converted. The delay between sampling and arrival of the converted value in the computer memory must be taken into account.

If the sensor signal is transmitted to the control computer in digital form, then there will be data-transmission delay. Digital-data networks are increasingly used in control systems to reduce the cost and complexity of wiring. However, transmission delays can be much longer than the data-transmission rate (bits/second) might suggest, and most suppliers do not specify transmission delay times.

Similar delays may occur in outputting the computed control-action signal, particularly if it must be transmitted through a network again.

Next, the control computer will most likely need to check the validity of input signals to make sure that fault conditions are detected promptly. Usually this is more complex than just checking to make sure a signal is within a given range. Fault conditions are not easy to detect, especially if they are to be detected in time to take useful safety precautions, with sufficient reliability to reduce false alarms to an acceptably low level. While specialized "real-time" expert systems are being developed to do this, we do not yet understand how to design such systems in a straightforward manner.

Experience suggests that delays resulting from all these factors can far exceed the time required to perform the control calculations. Further, it is difficult if not impossible, in most cases, to determine the actual delay times from hardware-manufacturer data sheets. If delay times are stated, they are typically for only part of the measurement process.

For example, an interface data sheet may specify the maximum conversion rate for analog signals as 20 kHz. However, it may be difficult to achieve a computation delay of less than 1 ms in practice using DOS, even though the control calculations might take a few microseconds. Using a faster processor to reduce input delay may not help as much as expected.

All this assumes that the software running in the control computer has complete control over the computer's operating system.

Modern computer systems are making increasing use of networks to link computers together, and of graphical user interfaces (GUIs) such as Windows or OSF Motif. However, networks and GUIs can take over a computer so that a process-control program running in the computer may suspend operations for 100 ms or more. Delays such as this usually make direct control impractical using this type of operating system. However, Labview is an example of GUI software that can be used for direct control of processes with slower response times.

Because of this, specialized real-time operating systems are available for process control. Examples are QNX and Lynx for PCs and VX-Works for Unix computers. These operating systems have restricted features for general use, but ensure that delays caused by the operating system are minimized. Even so, total delays of less than 100 μs are difficult to achieve.

An interesting example is the transputer chip introduced for high-speed parallel computing in the 1980s. While each chip is fast and apparently ideal for control, communication across an array of chips incurs a series of cumulative delays that dominates the total computation delay.

One paradox of modern computing is that computer-speed increases have had only a marginal effect on response time. One reason for this is that input/output (I/O) operations take their timing from a standardized I/O bus compatible with both slower and faster machines. Thus delays will correspond more to the speed of the slowest (and probably the oldest) computer in a given range. Many I/O devices for PCs synchronize their timing with the original XT model introduced in the early 1980s. Another reason is that operating systems are designed to handle more and more options, networks, GUIs, etc., and so they consume an ever-increasing proportion of the computing power of the central processor.

As a rule of thumb, the computation delay in a feedback-control system should be less than 1/20 of the desired response time (Fig. 14). Thus, if a system-response time of 20 ms is adequate (approximately 10-Hz bandwidth), a PC computer could serve as controller. Otherwise specialized hardware and/or operating systems should be considered essential.

As a second rule of thumb, the computation delay should be measured with representative equipment and software, rather than being estimated from performance

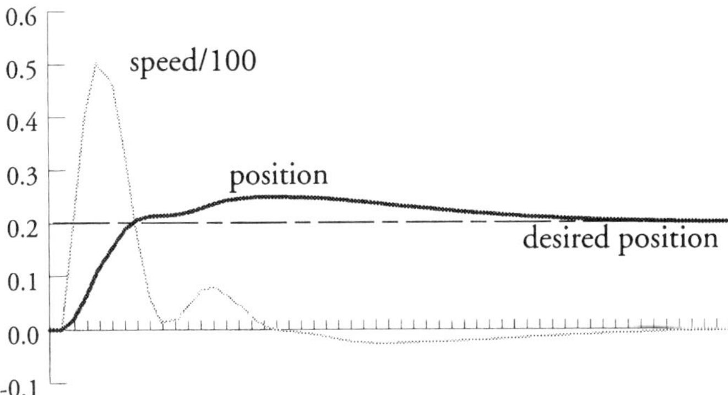

FIG. 14. Simulated behavior of the same PID control system as the preceding graphs, but with a 1-ms time delay in the controller. The reduction in stability is apparent from oscillation in the speed trace. A delay of 2 ms is sufficient to cause unstable behavior.

specifications. Measurement is not difficult. Apply a known-frequency signal to the sensor and observe the response lag in the control action.

The sheep-shearing robot was controlled by a computer capable of 3 million instructions per second (MIPS), which is modest by today's standards. With special software and hardware, the total delay was reduced to about 900 μs for the fastest actuator in the robot. Of this, the operating system contributed about 100 μs, control took about 300 μs, and the rest was needed for communication delays, A/D sampling, scaling, and fault detection.

Typical dc motors have response times of between 4 and 20 ms; this is the time needed for the motor shaft to accelerate to 63% of maximum speed with the rated voltage, assuming no load, and a zero-resistance power supply. Thus the computation delay of a motor controller should be less than 200 μs.

4.2 Sensor Resolution

The next factor in using a computer for control is the resolution of the input signal. An analog signal is said to have infinite resolution, which means that tiny signal changes can, in theory, be measured. However, when an analog signal is converted to digital form, the level must be represented by a finite number of bits, typically 8, 10, 12, 14, or 16. 8-bit resolution provides a maximum of 1 part in 256. However, a typical sensor might operate within a limited voltage range, say, 1 to 6 V. An A/D converter providing 8 bits of resolution usually has a range of -10 to $+10$ V, and so only 64 different values of sensor output can be represented between $+1$ and $+6$ V: The useful resolution is only 1 part in 64. For this reason, we aim to provide as much A/D resolution as possible within price, speed, size, and power restrictions. The useful resolution limits the accuracy that a control system can achieve, and if it is poor, it may affect stability.

Digital position encoders present a different resolution problem. While the resolution is theoretically limited by the number of pulses on the encoder, in practice the maximum pulse rate that can be reliably transmitted and counted may be the limiting factor. As we have seen, velocity feedback is essential for stable control. Velocity can be obtained from the pulse frequency, but at small velocities this information loses resolution because the time between pulses becomes longer. In some important situations, the loss of velocity information at low speed can seriously affect stability.

Digital encoders are often preferred because they are reliable, accurate, and repeatable. Resolution can be a problem, particularly if, for example, an encoder is being used to measure a relatively small position or angle change. If the resolution is poor, such that accurate positioning to within 1 encoder count is essential, a special design may be needed. A normal control system is likely to oscillate between one count and the next.

4.3 Output Resolution

In the same way that resolution of the input signal is a problem, output resolution can also be a problem, but the requirements are usually less stringent. For example, in the case of the electric motor, resolution of the output may not be fine enough to generate a current (and hence torque) to balance the steady load exactly. Once again, the effects can be explored by simulation (Fig. 15).

In this instance, mechanical friction can be helpful. If the mechanical-friction torque corresponds to a range of three or more possible output values (and hence motor-current values), output resolution should not cause oscillation problems.

In Excel, the effects of resolution can easily be computed. Given a quantity in, say, cell F7 that is to be measured to a resolution of x, the measurement can be computed by the formula

INT(F7/x + 0.5) * x.

5. SOME IMPORTANT LIMITATIONS

5.1 Steady Torque and Thermal Load on a Motor

The calculations shown so far are an approximate model of an ideal control system and electric motor. Real systems have important limitations that must be taken into account. In an ideal system the gains can be increased indefinitely to speed up the response of the motor. However, in doing so, the current i will far exceed what a practical power supply can provide and what a real motor could absorb without serious damage through heating, mechanical distortion of the windings, or demagnetization. The motor used as a case study can take a current of as much as 15 A for a brief instant, but the maximum continuous current is only 1.5 A.

The maximum continuous current is limited by the thermal capacity of the motor, particularly the rotor. Brushless dc motors, ac motors, and switched-reluctance motors can provide much more power for a given size and weight, but require more complex controllers (Kenjo, 1991).

If the controller can supply the peak current (implying that the dc power voltage is considerably higher than the nominal rating of the motor), then the motor may be damaged by a sustained high current that may occur during testing, or through a fault condition. The motor must be protected by a special circuit breaker to prevent damage.

Figure 16 illustrates a typical problem that must be considered in designing an electric servo. A motor is required to position a load vertically, and carry the weight of the load. This is a simplified version of an important problem in designing robot manipulators with several motors at the joints in the arm.

The first problem is that the load has inertia, which increases the apparent inertia of the motor shaft, slowing the response. However, in most applications, there is a gear reduction head on the motor so that the winding shaft rotates more slowly than the motor. The effective inertia is reduced by the square of the gear ratio, and so the motor inertia is usually the dominant component.

A more significant problem is the steady load torque experienced by the motor at rest. This must be balanced by a motor current unless the gear ratio is so high that friction in the mechanism can resist the weight of the load. This is unlikely if the load is to have a fast response. A steady current liberates heat in the windings through resistance

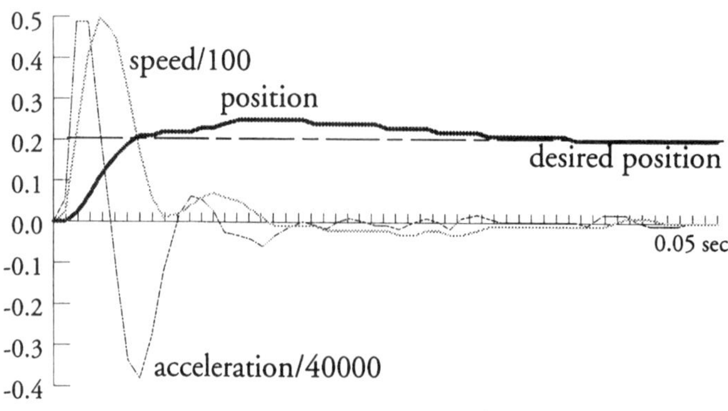

FIG. 15. Simulated PID controller from previous graphs, with 0.01 unit position resolution and 0.02 mA output resolution. The erratic acceleration trace shows how limited resolution introduces noise into a system, which will cause audible vibrations and noise in the servo. A simple simulation such as this is sufficient to confirm that noise and stability may be a problem. Experimental investigation would be necessary to measure the levels of noise and effects on accuracy.

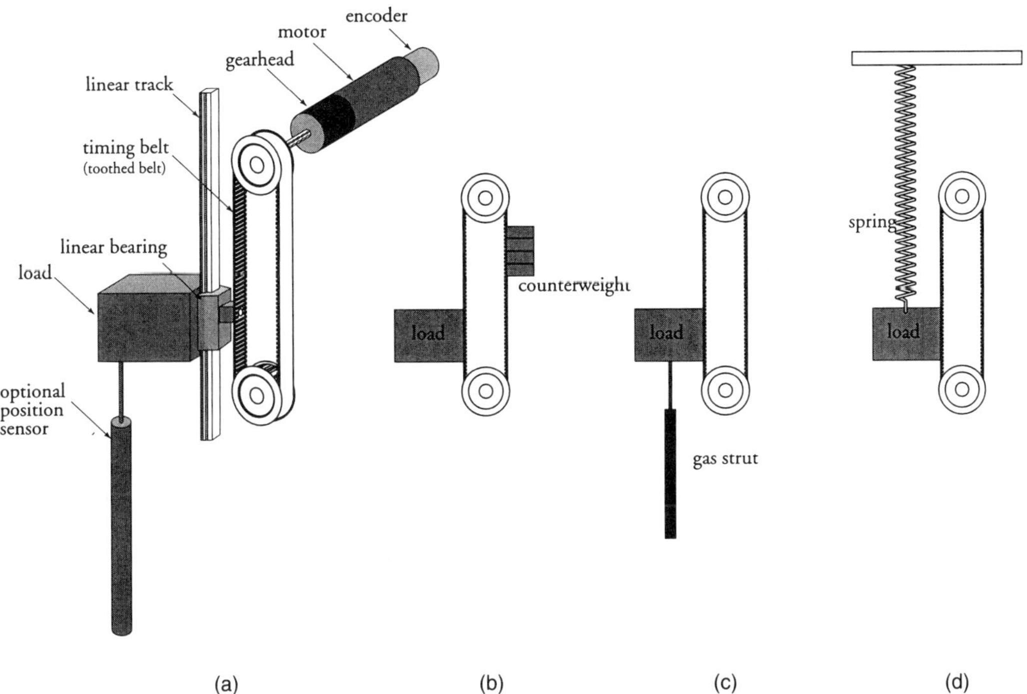

FIG. 16. (a) Typical position servo arrangement where load must be moved vertically against gravity force. Servo limitations influence important mechanical design choices. (b) Counterweight balances load weight. (c) Gas strut balances load weight. (d) Tension spring balances load weight. A weight, as in (b), can provide a constant counterbalance. This works well unless the added inertia slows the response too much. Further, the load mass may vary during operation so that a weight may not always balance the load.

effects, and the steady current that can be supplied to a typical motor is much less than the peak current (see Sec. 3.2). Therefore, most designers would provide some form of counterbalance in this instance (Bradley *et al.*, 1991).

A spring or gas strut (often used to balance the weight of the rear door of a car) provides balance without added mass. However, the force is often dependent on position and, in the case of a gas strut, speed.

Thus, the rate at which heat can be removed from a motor is often a major performance limitation, and indirectly affects several aspects of mechanical design. If the motor is close to heat-sensitive components (such as electronics or plastic materials), thermal loadings may have to be kept well below the maximum specification for the motor. Higher temperatures may change the motor characteristics enough to affect control performance.

The maximum continuous torque is limited by the heat dissipated in the motor windings. ac motors provide a higher power output for a given size, but require more complex controllers.

5.2 Speed Limitations

As the speed of a permanent-magnet motor increases, the windings generate a reverse voltage referred to as *back EMF* proportional to shaft speed. This opposes current flow, and for the motor in Sec. 3.2 is 1 V for each 585 rpm. Therefore, since the dc power-supply voltage is limited, the maximum current will be limited as the motor speed increases, and hence the maximum forward acceleration. The maximum reverse acceleration will increase at the same time, however, provided the power supply is designed to take the increased current. Braking with excessive reverse current can cause mechanical damage to some motors.

5.3 Power-Supply Limitations

Power supply can be a limitation in both electric and hydraulic systems. In the former, it is usually most important for battery-driven systems. Here there are two considerations. First, the overall drain on the battery needs to be minimized, and second, the supply current and voltage will drop as the battery loses charge. This may affect stability of the servo. A good computer control system can monitor battery charge and make allowances.

A hydraulic system is seldom self-contained, and so energy conservation is seldom so important. In fact, a hydraulic servo is very energy inefficient as the energy consumed on each stroke of the piston is independent of the load (power used is supply pressure × piston velocity × piston area). Further, servo valves require a constant oil flow to operate, called a pilot flow, typically 1%–3% of full rated flow. So the major consideration in a hydraulic system is pressure loss in the pipes and fittings between the pump and the servo valve, and then between the valve and the cylinder.

Note that the characteristics of both electric motors and hydraulic actuators change with temperature.

5.4 Sensor Location

Normally the feedback sensor is placed where the most direct output measurement can be obtained. This helps to minimize accuracy problems discussed in Sec. 2.1. If we look again at Fig. 16, we see that there is a complex mechanical transmission between the motor and the sensor if we use the load position sensor. This will introduce some compliance, stiction, and backlash between the motor and the sensor. Backlash is caused by clearance between adjacent parts: The motor must turn through some angle before the load moves. The term *stiction* describes a friction effect that is common in mechanical systems. When the friction coefficient decreases as sliding speed increases, motion tends to be intermittent or jerky and the moving parts seem "stuck" between motions. This effect is referred to as stiction and, like backlash, affects the ultimate accuracy that can be achieved.

If a slow response is sufficient, the control system may be stable. However, if a fast response is needed, these mechanical effects may destabilize control. For this reason, it is common to place the position encoder on the motor shaft, and use a stiff mechanical transmission. Figure 16 shows a timing belt being used. In Fig. 6, a lead screw and recirculating ball nut are shown. These components provide a stiffer and more precise mechanical transmission, but cost more and add inertia.

A good designer will place the feedback sensor and actuator (or motor) as close to the point of action as possible, and for the fastest response, the sensor and motor should be colocated. If the encoder is on the motor shaft, a reduction gearhead or lead screw helps to increase the effective position resolution.

Two sensors could be used if the mechanical transmission cannot be made rigid: one on the motor shaft and another at the load. Alternatively, a tachometer could be used at the motor and used to control motor speed with one controller, and an encoder could be used at the load for position control. However, more complex control is needed.

6. FAULTS IN CONTROL SYSTEMS—UNEXPECTED BEHAVIOR

When control systems work well for a while, they tend to be taken for granted. Failure is often marked by a sudden and dramatic change in behavior rather than a gradual deterioration seen in other machinery. This often occurs because the feedback that compensates for process variation can also mask progressive failure of a system component. The failure then only becomes evident when the controller can no longer compensate for this and the process variation combined.

While each type of control system (mechanical, electronic, hydraulic, etc.) has characteristic failures, there are failures common to all in principle.

6.1 Runaway

One of the most alarming and dangerous failures occurs when a position controller drives its load to one end of its range at maximum speed, possibly causing mechanical damage.

This type of failure emphasizes the need to cushion the load at each end, no matter how reliable the controller may be. Hydraulic cylinders are provided with optional hydraulic cushions as part of their design. Electric drives require limit switches to disable the motor if it is driving the load toward the end, and these must be placed far enough from the ends to allow the motor to run down and stop. With large loads or inertias, rubber cushions or shock absorbers are essential.

Possible explanations for runaway are the following:

1. Feedback-sensor failure. For example, if the mechanical coupling to the position sensor fails, then the measured position will not change, even if the motor shaft rotates. When a new position is requested, the motor turns to drive the load toward the desired position. However, the feedback signal does not change, and so the controller drives the motor faster and faster until the load reaches the end of its allowed motion . . . *Bang!* No external stimulus is needed—small natural perturbations can start a runaway some time after the component failure occurs. Apart from mechanical failure, a break in a sensor wire, a faulty connector, or failure of the sensor itself can cause similar effects. If the failure is interpreted by the controller as a large position change, runaway will occur immediately. Otherwise, it may happen later.

   A controller can be designed to shut off power automatically if there is no detected motion when the motor power level exceeds a given level. Usually this is detected by monitoring the error signal: Failure is presumed if the error exceeds a predetermined limit.
2. Amplifier failure. Failure of the electronic switches or equivalent devices in an electrical amplifier (Fig. 7), or of the valve in a hydraulic or pneumatic controller (Fig. 5), can cause similar effects.

   Once again, a controller can be designed to shut off power when the error limit is exceeded. Note that the power-shutoff mechanism or circuit must be independent of the amplifier. Hydraulic systems usually include an emergency shutoff valve that closes off the high-pressure supply to the servo valves. Some hydraulic servos also include check valves between the servo valve and the cylinder. These valves are held open during normal operation, but close if there is a power or controller failure. Electric and pneumatic servos may need a brake to stop motion safely.
3. Mechanical failure. If the load is not directly driven by the motor, failure of the transmission or counterbalance can also cause runaway.

Experience suggests that wiring faults form a large proportion of all failures. Particularly with robots, wires and connectors are subjected to repeated flexure and vibration, and will fail with intermittent faults eventually.

Normal flexible connecting wire consists of seven copper strands. Special cables are made to withstand repeated flexure, and each wire may consist of 100 or more extremely fine strands in specially chosen sleeving. Stress-relieving entry glands should be used where free cables enter rigid boxes or connectors, and where possible, mechanical "cable-track" chains should be used to constrain moving wires and prevent accidental kinking.

6.2 Persistent, Steady Error

1. Feedback-sensor failure. For example, if the mechanical coupling to the position-measurement encoder has slipped, then the measured position will be wrong. Thus, the motor turns to drive the load toward the desired position as measured by the sensor. If the error is large, this can cause a collision with the end stops. (Check the measured position when the load is at a known position. If the encoder has not slipped, check that the encoder mounting is secure.)
2. In the case of an incremental encoder, false pulses may be generated by electromagnetic interference (EMI). The interference may come from sources such as flickering fluorescent tubes or other motors starting or stopping. It can also come from the switching pulses of the servo motor—this may be the hardest to detect

and confirm. Absolute encoders help to prevent this problem. (Run a length of wire alongside the sensor wire and measure the EMI picked up by the wire using an oscilloscope. Compare the observed level with the voltage generated by the sensor.)

EMI often causes obscure faults that can be transient and difficult to observe. It is best to take basic precautions:

a. Amplify weak sensor signals close to their source, and where possible use transmission methods that are resistant to EMI such as a current loop, differential line drivers/receivers, or fiber optics.

b. Place dc power-supply regulators close to all sensitive circuits. dc isolation can be achieved by using a high-frequency ac power supply and converting to dc power at each sensor electronics unit. A high-frequency, low-power supply can be very compact.

c. Shield cables (sensor, motor, and power cables), and pass the shield drain wire through connectors back to a reliable common earth point for the whole system.

d. Use shielded twisted-pair cable designed for signal transmission, and if necessary, also for repeated flexure.

e. Use electronic filters to prevent EMI from entering boxes along wires or power supplies.

f. If possible, avoid running sensor cables parallel or alongside power cables.

Note that some EMI precautions (such as line filters or line drivers) can attenuate pulses from an incremental encoder if the pulse frequency is too high. Thus pulses may be lost during rapid motion. This can also result in a steady persistent error.

3. Mechanical overload—the load on the motor is too great for the controller to compensate. Try removing or reducing the load.

4. An excessive offset in amplifier or servo valve can result in a steady error. (An offset exists if the amplifier or servo-valve input is nonzero when the motor is not moving.)

6.3 Unstable Transient Response

After some time in operation, the control system may become less stable, with more overshoot after large position changes and more oscillation.

1. If the motor drive power is reduced (or hydraulic pressure is reduced) the maximum drive power is reduced. This may cause unstable behavior. (Measure maximum torque/force/acceleration and speed available.)

2. Change in load inertia or stiffness. Usually, increased inertia causes more overshoot. However, if a controller is "tuned" to the characteristics of a particular load, any change in load may reduce the controller performance. (Test controller with different loads.)

3. Often the friction in mechanical systems decreases with use, and this can reduce the energy dissipated by friction. If a bearing or lubrication fails, friction can increase with age and this can also cause instability. (Measure friction changes.)

4. Controller settings (such as gain) may have been accidentally changed following maintenance. Controllers should be designed such that gain settings do not have to be changed for maintenance or testing procedures. (Always record gain settings.)

5. Backlash (mechanical slack movement) will occur in joints and bearings of mechanisms subject to wear. As backlash increases, unstable behavior of the controller may increase. Earlier we noted the importance of choosing an appropriate sensor location. If the backlash occurs between the actuator and the sensor, it will almost certainly cause instability. Otherwise, it will cause steady errors. However, even when the sensor is mounted on the actuator, backlash can cause unstable behavior if it occurs between the actuator and the load, because it changes the apparent load characteristics. (Make provision for mechanical wear—provide adjustment for backlash, use wear-resistant components, keep contamination away from joints and seals, and measure backlash.)

7. DISCRETE CONTROL SYSTEMS

So far we have talked about control systems with variables that change more or less continuously. For example, the potentiometer of the hydraulic actuator provides a voltage output that varies continuously with the position of the piston. The servo valve generates an oil flow to the cylinder that is continuously variable.

There is another class of control systems where these variables have a small number of possible values. The simplest of these are on/off control systems. Figure 17 shows an on/off hydraulic actuator controlled with a four-way valve. Although the valve spool is very similar to that of a servo valve, its operation is quite different. If neither the "A" solenoid nor the "B" solenoid is activated electrically, the spool remains at the center position and no oil can flow to the cylinder. If the "A" solenoid is activated, the spool moves to the left, causing the piston also to move to the left at a constant speed.

The position of the piston is sensed with microswitches, which provide a signal when the piston has reached either end of its travel. Thus we can only be sure about the position of the actuator when the piston has reached the position of either microswitch.

There are countless examples of on/off control systems because they are often much simpler and cheaper than servo systems. Most automated factory machines use this kind of control. Pneumatic machines of this type are in widespread use around the world. Several manufacturers supply a vast range of standardized components (such as Festo, SMC, Parker, ABB, Siemens, Texas Instruments, and Rexroth, to name just a few). The controller (usually a microcomputer) is usually known as a programmed logic controller (PLC).

While discrete control systems operate quite differently from continuous control systems the same basic principles apply. The system must be observable—the system state we wish to control must be able to be measured. The system must also be controllable. Even though the system input is either on or off, it must still affect the system state in a predictable manner for the control system to work.

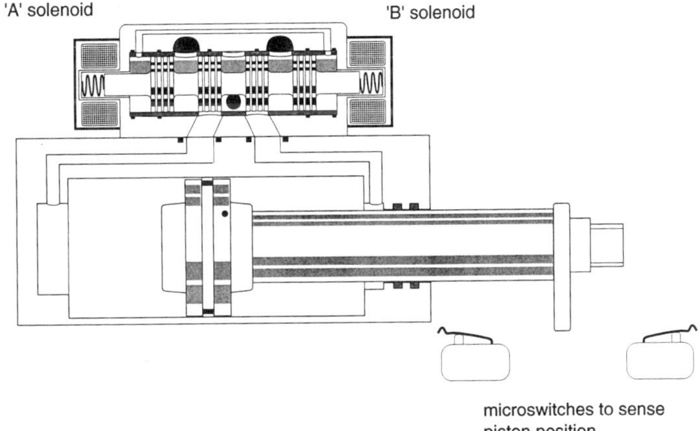

FIG. 17. A simple on/off hydraulic actuator for positioning—refer to text.

APPENDIX

Excel Spreadsheet Formulas for dc Servo Simulation (Proportional Control, Fig. 11):

A	B	C	D	E
1) Time step	0.001		Desired posn	1
2) Gain	5		Inertia	9.38
3)				
4) Time	Current	Accn	Speed	Shaft
5)				posn
6) 0			0	0
7) =A6 + B1	= B2* (E1 − E6)	= (0.0163* B7 + 0.002)/(E2* 0.0000001)	= D6 + C7* B1	= E6 + D7* B1
8) = A7 + B1	= B2* (E1 − E7)	= (0.0163* B8 + 0.002)/(E2* 0.0000001)	= D7 + C8* B1	= E7 + D8* B1
9) = A8 + B1	= B2* (E1 − E8)	= (0.0163* B9 + 0.002)/(E2* 0.0000001)	= D8 + C9* B1	= E8 + D9* B1
10) = A9 + B1	= B2* (E1 − E9)	= (0.0163* B10 + 0.002)/(E2 0.0000001)	= D9 + C10* B1	= E9 + D10* B1
11) = A10 + B1	= B2* (E1 − E10)	= (0.0163* B11 + 0.002)/(E2* 0.0000001)	= D10 + C11* B1	= E10 + D11* B1

Row 7 represents the formulas in steps 1–4, and is copied down as far as necessary to obtain the desired number of results.

Time step	0.0010		Desired posn	1.000
Gain	5.000		Inertia	9.380
Time	Current	Accn	Speed	Shaft posn
0.0000			0.000	0.000
0.0010	5.000	89019.190	89.019	0.089
0.0020	4.555	81284.580	170.304	0.259
0.0030	3.703	66487.397	236.791	0.496
0.0040	2.519	45913.325	282.704	0.779
0.0050	1.106	21349.981	304.054	1.083

Plot columns D and E against time (shaft speed and position).

ACKNOWLEDGMENTS

The author would like to acknowledge the contributions of Michael Ong, who worked with the author as a research officer on the robotic sheep-shearing project and more recently developing low-cost servo-system components. Michael explored many of the practical aspects of servo systems and not only mastered them, but recorded his designs and calculations in great detail. He has not published his reports. However, many of the insights presented here stem from his work. David Elford and Jan Baranski, colleagues in the sheep-shearing project, also have contributed useful insights and design details.

GLOSSARY

Actuator (Sec. 2.3): A device that moves in response to an external command signal (Figs. 5, 6).

Adaptive Control (Sec. 3.7): Control system in which the controller behavior is adjusted automatically as the system properties change with time.

Amplitude Ratio (Sec. 3.1): Part of a frequency response, the ratio of amplitude of output to the amplitude of the input signal (Fig. 9).

Back EMF (Sec. 5.2): Reverse voltage generated in a spinning electric motor.

Backlash (Sec. 5.4): Clearance between adjacent mechanical parts that transmit motion.

Bandwidth (Sec. 3.1): Range of frequencies over which a device operates in a certain manner.

Closed-Loop Control (Sec. 2): A control system that adjusts the control action in response to measurements of the system output (Fig. 1).

Compliance (Sec. 5.4): Inverse of stiffness; the position of a mechanical transmission with compliance will be affected by the applied force.

Computation Delay Time (Sec. 4.1): Time between measurement and change in computer output signal, needed to transmit measurement to computer, process the measurement, and transmit the calculated output signal to the process.

Control Action (Sec. 2): Means of changing the state of a machine or process in order to change the system output (Fig. 1).

Control Engineering (Sec. 1.1): Technology of control systems.

Control System (Sec. 1.1): A control device and a regulated machine or process connected together (Fig. 1).

Controllability (Sec. 2): Degree to which the state of a machine or process can be changed.

Derivative Control (Sec. 3.4): Control system in which the control action is proportional to the rate of change of error (difference between the measured system output and desired output).

Desired Output Setting (Sec. 2): Control-system input that defines the desired value of the control-system output (Fig. 1).

Discrete Control Systems (Secs. 2, 7): Control systems in which variables have a finite number of possible values, usually a small number, often just two possible values or states (Fig. 16).

Disturbance Inputs (Sec. 2): Factors other than the control actions that affect the system output (Fig. 1).

Electromagnetic Interference (Sec. 6.2): False signals caused by wires acting as radio aerials, which pick up local electrical fields radiated from other wires.

Encoder (Sec. 2.4): An optical/electronic sensor for measuring position of an actuator or machine (Fig. 8).

Excel (Sec. 2.2): Software by Microsoft for manipulating spreadsheets.

Feedback Control (Sec. 2): A control system that adjusts the control action in response to measurements of the system output (Fig. 1).

Frequency Response (Sec. 3.1): Response of a device to a time-changing input that varies with a given frequency (Fig. 9).

Gain (Sec. 3.3): Factor by which an error, or other signal, is multiplied in a control system. Usually constant, but in some systems can be varied automatically.

Graphical User Interface (Sec. 4.1): Means of human–computer interaction by using a pictorial or graphical display and possibly a device (such as a mouse) to designate a point on the display, as well as a keyboard.

Illustrator (Sec. 2.2): Software by Adobe for creating illustrations and graphs.

Integral Control (Sec. 3.4): Control system in which the control action is proportional to the accumulated sum (or integral) of the error (difference between the measured system output and desired output).

Integral Windup (Sec. 3.4): An undesirable by-product of using integral control when large changes in the desired output setting occur; the error integrator accumulates a large unwanted offset during the transient response period. To avoid this, the integrator output is limited to a certain range.

Linear System (Sec. 1.1): One for which the response to two simultaneous inputs is the sum of the responses to each input presented separately.

Matrix-X (Sec. 3.2): Software package used for control-system simulation and design.

MATLAB (Sec. 3.2): Software package used for control-system simulation and design.

Observability (Sec. 2): The degree to which certain properties or states of a system can be measured.

On/Off Control Systems (Secs. 2, 7): Control systems in which variables have just two possible values or states (Figs. 1, 16).

Open-Loop Control (Sec. 2): Control system that adjusts the control action without measuring the system output; absence of feedback.

Output State Variable (Sec. 2): Property of control system that needs to be regulated (Fig. 1).

Phase Angle (Sec. 3.1): Part of a frequency response; the degree to which a system output either lags behind the system input, or leads ahead of it (Fig. 9).

PID Controller (Sec. 3.4): Controller that uses combined proportional, integral, and derivative control.

Potentiometer (Sec. 2.3): An electrical sensor often used to measure the position of an actuator (Fig. 5).

Programmed Logic Controller (Sec. 7): Small computer that can be used as a controller for a process, typically used for on/off or discrete control systems.

Proportional Control (Sec. 3.3): Control system in which the control action is proportional to the error (difference between the measured system output and desired output).

Pulse-Width–Modulation (PWM) Amplifier (Sec. 2.4): Common electronic-circuit arrangement for regulating current in an electric motor (Fig. 7).

Real-Time Operating System (Sec. 4.1): A computer-operating system designed to keep computation-delay time within specified limits.

Resolution (Sec. 4.2): Least detectable change in a signal.

Resonance (Sec. 3.1): Peak in amplitude-ratio part of frequency response (Fig. 9).

Response Time (Sec. 3.1): Time for the system output to reach the first 63% of the desired change.

Self-Tuning Control (Sec. 3.7): Control system in which the controller settings (such as proportional gain) are adjusted automatically as the system properties change with time.

Servo (Sec. 2.3): A device that changes its setting automatically in response to an external command.

Servo Valve (Sec. 2.3): An electromechanical valve that regulates the flow of fluid in a hydraulic control system (Fig. 5).

Simulation (Sec. 2.2): Calculation of behavior from mathematical equations that represent the properties of a system (Fig. 3).

Stability (Sec. 3.2): The degree to which the transient response disappears quickly.

Steady-State Response (Sec. 2.2): Response of control system after transient responses have stopped when the process or machine state no longer changes with time (Fig. 3).

Stiction (Sec. 5.4): A friction effect, common in mechanical systems, which occurs when the friction coefficient decreases as sliding speed increases so that motion tends to be intermittent or jerky and the moving parts seem "stuck" between motions.

Switching Controller (Sec. 2.4): Common electronic-circuit arrangement for regulating current in an electric motor (Fig. 7).

System Output (Sec. 2): Property of control system that needs to be regulated (Fig. 1).

Thermal Capacity (Sec. 5.1): Ability of electric motor to dissipate excess heat.

Transient Response (Sec. 2.2): Response of control system that changes with time but that stops after some time (Fig. 3).

Works Cited

Adobe (1994), *Adobe Illustrator Users Guide,* Adobe Systems, Inc.

Borrie, J. A. (1986), *Modern Control Systems: A Manual of Design Methods,* Englewood Cliffs, NJ: Prentice-Hall.

Bradley, D. A., Dawson, D., Burd, N. C., Loader, A. J. (1991), *Mechatronics—Electronics in Products and Processes,* London: Chapman and Hall.

D'Azzo, J. J., Houpis, C. H. (1981), *Linear Control System Analysis and Design,* New York: McGraw-Hill.

Johnston, R. (1994), "Fuzzy Logic Control," *GEC J. Res.* **11** (2), 99–109.

Kenjo, Tak (1991), *Electric Motors and Their Controls,* Oxford: Oxford Univ. Press.

Lovelock, J. (1987), *The Practical Science of Planetary Medicine,* London: Pan Books.

Maxon (1994), *Maxon Motor Catalog,* 1994/1995 edition, Sacshseln, Switzerland: Interelectric AG, p. 79, model RE 025 055 34EAA200A.

Microsoft (1994), *Excel Spreadsheet Software Package User Manual Version 4. for Windows,* Microsoft, Inc.

Trevelyan, J. P. (1992), *Robots for Shearing Sheep,* Oxford: Oxford Univ. Press.

Further Reading

Almost all of the thousands of learned papers published each year on control and servo systems focus exclusively on mathematical control theory, and so information on practical servo systems can be hard to find. The following is a small selection of books, articles, and catalogs to which the reader could progress for more detail. Many commercial catalogs have good application notes from which useful information can be gleaned.

Control Theory

Borrie, J. A. (1986), *Modern Control Systems: A Manual of Design Methods,* Englewood Cliffs, NJ: Prentice-Hall.

D'Azzo, J. J., Houpis, C. H. (1981), *Linear Control System Analysis and Design,* New York: McGraw-Hill.

Electric Motors

Bradley, D. A., Dawson, D., Burd, N. C., Loader, A. J. (1991), *Mechatronics—Electronics in Products and Processes,* London: Chapman and Hall.

Katebi, M. R., McIntyre, J., Lee, T., Grimble, M. J. (1993), "Integrated Process and Control Design for Fast Coordinate Measurement machine," *Mechatronics* **3,** 335–342.

Kenjo, Tak (1991), *Electric Motors and Their Controls,* Oxford: Oxford Univ. Press.

Maxon (1994), *Maxon Motor Catalog,* 1994/1995 edition, Sacshseln, Switzerland: Interelectric AG, p. 79, model RE 025 055 34EAA200A.

Yoshiaki, Hachisuka (1991), "Compact and High Torque Accelerators for Locomotion Robot," in: *Proceedings of the International Symposium on Advanced Robotics Technology (ISART),* Tokyo: Japan Industrial Robot Association, pp. 343–368.

Hydraulic and Pneumatic Systems

Boulet, B., Daneshmend, L., Hayward, V., Nemri, C. (1991), "System Identification and Modelling of a High Performance Hydraulic Actuator," in: *Proceedings of Second International Symposium on Experimental Robotics,* Berlin: Springer.

Daley, S. (1990), "Self Tuning Control of a Hydraulic Test Rig Using a Personal Computer," *GEC J. Res.* **7** (3), 157–167.

Dransfield, P. (1990), "PC Controlled Air Powered Servodrives—Their Response Potential," in: *Proceedings of the 3rd International Fluid Power Workshop,* Boston: IFS.

Jacobsen, S. C., Iverson, E. K., Knutti, D. F., Johnson, R. T., Biggers, K. B. (1986), "Design of the Utah/MIT Dextrous Hand," in: *Proceedings of the IEEE International Conference on Robotics and Automation, San Francisco,* New York: IEEE, pp. 1520–1532.

Trevelyan, J. P. (1992), *Robots for Shearing Sheep. Shear Magic,* Oxford: Oxford Univ. Press.

Uusijarvi, R. W., Torngren, E. (1994), "Introducing Distributed Control in Mobile Machines Based on Hydraulic Actuators," *Mechatronics* **4,** 139–157.

There are relatively few references on modern hydraulic servo systems. Good information is available from leading suppliers such as Moog, Parker, and Rexroth through local distributors.

Mechanical Design of Servo Systems and Transmission Elements

Rivin, E. I. (1988), *Mechanical Design of Robots,* New York: McGraw-Hill.

Stiffler, A. K. (1992), *Design with Microprocessors for Mechanical Engineers,* New York: McGraw-Hill.

Several bearing manufacturers such as THK supply specialized components such as linear tracks, recirculating ball bearings, and ball nuts.

Robotics

McKerrow, P. J. (1991), *Introduction to Robotics,* Reading, MA: Addison-Wesley.

Sharkey, P. M., Murray, D. W., Vanevelde, S., Reid, I. D., McLauchlan, P. F. (1993), "Modular Head/Eye Platform for Real-Time Reactive Vision," *Mechatronics* **3,** 517–535.

Trevelyan, J. P. (1992), *Robots for Shearing Sheep. Shear Magic,* Oxford: Oxford Univ. Press.

Trevelyan, J. P. (in press), "Robot Force Control without Stability Problems," in: *Proceedings of 3rd International Symposium on Experimental Robotics, Tokyo,* Berlin: Springer.

Fuzzy Control

Johnston, R. (1994), "Fuzzy Logic Control," *GEC J. Res.* **11,** 99–109. A good general introduction.

Other

Lovelock, J. (1987), *The Practical Science of Planetary Medicine,* London: Pan Books. Gives some interesting comments on the relevance of feedback systems to environmental science.

Catalogs

Major suppliers include Sankyo Denki, Escap, ABB, and GEC/Alsthom.

Motorola, National Semiconductor, and Hewlett-Packard manufacture a range of motion-control and servo components, including encoders and control microprocessors. Most semiconductor companies supply switching circuits in various capacities.

SHOCK WAVES

Y. HORIE, *North Carolina State University, Raleigh, North Carolina, U.S.A.*

AKIRA B. SAWAOKA, *Tokyo Institute of Technology, Nagatsuta, Yokohama, Japan*

INTRODUCTION

In physical science, the term "shock wave" refers to a narrow region of compression (wave front) moving at supersonic speed across which the pressure, velocity, and other related quantities of the medium change in a nearly discontinuous manner. The term is also used to signify the entire wave including the wave's trailing expansion phase, such as the boom from a supersonic aircraft, or the blast wave from an explosion.

Shock waves arise from many sources: a sudden release of various forms of energy in a limited space; supersonic motion of a body in a fluid medium; or the impact of gases, liquids, and solids at high speed. Physically, the formation of a shock wave is attributed to the fact that in most materials the sound speed increases with increasing pressure. Thus, when a finite-amplitude wave is created in a medium, its profile becomes steeper during propagation and develops into a near discontinuity as a result of a high-pressure regime overtaking a low-pressure regime. The maximum steepness of the wave front is limited by the dissipative processes of viscosity and heat conduction.

The phenomena of shock waves are truly ubiquitous. They occur on scales ranging from atomic to cosmic. Electrons moving in a tiny semiconductor device may cause such a wave, as may a massive star collapsing under gravity at the end of its evolution. The shock wave from the latter is the basis of the supernova explosion that is thought to be the origin of most heavy elements in the galaxies. Life, then, is born of seeds scattered by cosmic shock waves (Glass, 1974).

Shock waves are man-made as well as natural. A sonic boom is a conical shock wave from the passage of a supersonic aircraft through the atmosphere. Shock waves are often invisible without the assistance of a special diagnostic device, but the devastating damage they can cause, e.g., from the detonation of chemical or nuclear explosives, is often witnessed. However, even nuclear explosions are minute when compared with the shock waves generated by the impact of large meteorites with planets. Such an impact on Earth is thought to have caused the demise of dinosaurs many millions of years ago. The recent comet impact on Jupiter provided a spectacular astronomical event readily observable on Earth.

Chemical explosives are routinely used in mining and civil engineering for excavation

3-527-28140-1/96/$5.00 + .50

and demolition. The almost instantaneous reaction of explosives is indispensable for the deployment of air bags that often save lives in auto accidents. High-pressure shock waves are used to synthesize diamond and other hard materials for applications in grinding and polishing. Shock-synthesized diamond powders are said to have unique polishing properties that do not occur with naturally made diamond. Their lack of sharp cutting edges avoids the creation of deep grooves. A focused shock wave from an electric discharge or a small amount of explosives has been used to fracture gallstones to facilitate their removal for several years now. Shock waves associated with supersonic flight are to be avoided because of their disturbing qualities to humans and animals, but, in the case of reentry vehicles, shock waves help to increase drag forces; e.g., by increasing the gas density, they aid in deceleration through the Earth's atmosphere.

There are also many exciting examples of potential applications that are currently under investigation. For instance, scramjets (supersonic combustion ramjet) are being considered for the proposed hypersonic National Space Plane that can carry passengers from New York to Tokyo in less than three hours. Laser fusion, one of the several techniques of obtaining a practically inexhaustible source of energy, uses an imploding shock to heat and compress a target with the aim of initiating and controlling the thermonuclear-fusion reaction therein.

Obviously, shock waves are not limited to industrial applications. They have been used to develop our understanding of materials response to extreme conditions of high pressure and temperature (see EQUATIONS OF STATE; SOLIDS, DYNAMIC HIGH-PRESSURE EFFECTS IN) and also to test the laws of the physical world. There have been many fundamental contributions in the studies of gas dynamics, chemical kinetics of gases at high temperatures, high-pressure and high-temperature equations of state and other properties of solids, and the planetary interiors. The study of cosmic shock-wave phenomena has been a key to understanding of the creation of the universe and the evolution of stars, the interstellar medium, our solar system, and even the Earth. This article only touches a very small portion of truly ubiquitous shock-wave phenomena.

1. FUNDAMENTAL FEATURES OF SHOCK WAVES

Theorizing about shock waves dates as far back as Rankine's memoir (Rankine, 1870). Mathematical formulation varies from a simple physical approach in one space dimension to the approach based on weak solutions of hyperbolic differential equations in multiple spatial dimensions. In this article, discussion will be limited to an idealized, one-dimensional shock wave in a perfect fluid. This is a simple model, but it describes basic shock-wave relationships. A more general derivation and the inclusion of such effects as chemical reactions, ionization, and electromagnetic fields may be found in references provided at the end of the article.

As shown schematically in Fig. 1, a pair of vertical lines represents a narrow region of a plane shock wave propagating from left to right in a perfect fluid across which the flow variables change almost suddenly. The conservation laws of mass, momentum, and energy applied across this near discontinuity yield (Courant and Friedrichs, 1948)

$$\rho_1 u_1 = \rho_0 u_0, \tag{1}$$

$$p_1 + \rho_1 u_1^2 = p_0 + \rho_0 u_0^2, \tag{2}$$

$$\begin{aligned} \rho_1 u_1 (E_1 + \tfrac{1}{2} u_1^2) + p_1 u_1 \\ = \rho_0 u_0 (E_0 + \tfrac{1}{2} u_0^2) + p_0 u_0, \end{aligned} \tag{3}$$

where p is pressure, ρ is density, E is specific internal energy, and u is flow velocity. The subscripts 0 and 1 denote the values of variables upstream and downstream of the discontinuity, respectively. These equations are known to hold in the normal direction for even an arbitrary discontinuous surface.

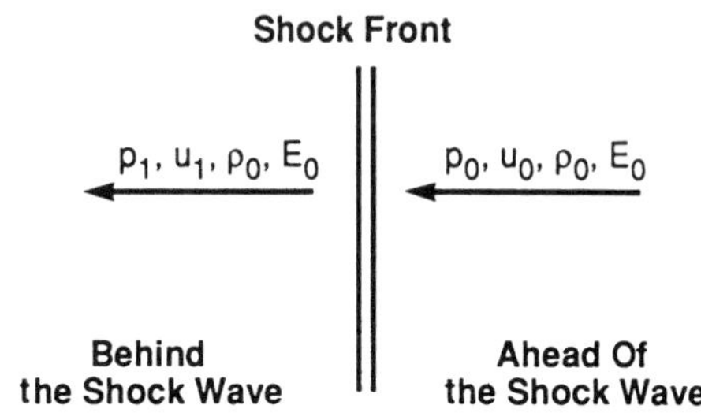

FIG. 1. A schematic diagram of a normal plane shock.

By rearranging variables in Eqs. (1)–(3), one can transform them into the more useful form

$$m^2 = (p_1 - p_0)/(V_0 - V_1), \tag{4}$$

$$u_1 = u_0 \pm (p_1 - p_0)/m, \tag{5}$$

$$E_1 - E_0 = \tfrac{1}{2}(p_1 + p_0)(V_0 - V_1), \tag{6}$$

where $m = \rho_0 u_0 = \rho_1 u_1$ and $V = 1/\rho$. The last equation is known as the Hugoniot equation. In many applications, enthalpy is a more useful variable than internal energy. Then, the Hugoniot equation becomes

$$H_1 - H_0 = \tfrac{1}{2}(p_1 - p_0)(V_0 + V_1). \tag{7}$$

The speed of the shock, D, is

$$D = V_0[(p_1 - p_0)/(V_0 - V_1)]^{1/2} + u_0. \tag{8}$$

The locus of final states is called the Hugoniot curve (or shock adiabat) and represents the states that can be reached from the initial state (p_0,V_0) by a single shock. It does not represent the successive states of compression such as an isentropic compression curve (see Fig. 2). Since the shock adiabat involves two parameters, the initial state and the shock strength, two successive shocks cannot, in general, be effected by a single shock wave. Equations (4) and (5) have useful geometric meaning in the p-V plane. For instance, $-m^2$ is the slope of the chord called the Rayleigh line, drawn from (V_0,p_0) to (V_1,p_1). Then, the shock in the internal energy is the area under this Rayleigh line (see Fig. 2).

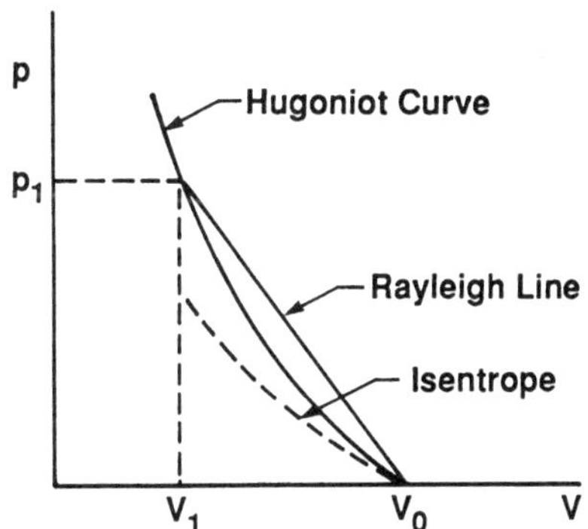

FIG. 2. Hugoniot curve in the p-V plane.

There are four variables in the shock equations, and so a fourth equation is required to determine the Hugoniot curve. This is typically accomplished by hypothesizing an equation of state such as the ideal-gas law or by introducing experimentally obtained data on the D-u relationship.

Consider the ideal-gas law

$$E = pV/(\gamma - 1), \tag{9}$$

where γ is the ratio of the specific heats at constant pressure and at constant volume. Then, it may be found that

$$\frac{V_1}{V_0} = \frac{(\gamma + 1)p_0 + (\gamma - 1)p_1}{(\gamma - 1)p_0 + (\gamma + 1)p_1}, \tag{10}$$

$$\frac{T_1}{T_0} = \left(\frac{p_1}{p_0}\right)\frac{(\gamma + 1)p_0 + (\gamma - 1)p_1}{(\gamma - 1)p_0 + (\gamma + 1)p_1}. \tag{11}$$

This shows that an arbitrarily strong shock cannot compress the gas by a factor greater than $(\gamma + 1)/(\gamma - 1)$. For a monatomic gas of $\gamma = 5/3$, this factor is 4, a large density change mentioned previously regarding the increase of drag forces on reentry vehicles.

A linear D-u relationship is known to hold for a variety of condensed matter over a wide range of pressures, even up to a megabar range for some materials; $D = C_0 + gu_p$ where C_0 is comparable to the bulk sound speed at ambient pressures, u_p is the particle velocity of shock running into an initially stationary state $(u_0 - u_1)$, and g is a constant. Then, the Hugoniot in the p-V plane is given by

$$p_1 = \frac{\rho_0 C_0(1 - V_1/V_0)}{[1 - g(1 - V_1/V_0)]^2}. \tag{12}$$

Computations of other thermodynamic variables such as temperature require mechanical and thermal equations of state. For instance,

$$p(V,E) = p_i(V) + b(E - E_i), \tag{13}$$

$$E = E_i + C_V(T - T_i), \tag{14}$$

where p_i and E_i are pressure and internal energy on the isotherm, C_v is the specific heat at constant volume, and b is a constant. Then, the pressure and temperature on the Hugoniot are given by

$$p_1 = \frac{p_i - b(E_i - E_0)}{1 - b(V_0 - V_1)/2}, \tag{15}$$

$$T_1 = T_0 + \frac{p_1 - p_i}{bC_V}, \tag{16}$$

where $p_0 = 0$.

For a normal fluid for which

$$(\partial^2 p/\partial \rho^2)_S > 0,$$

where S is the entropy, it is known (Zel'dovich and Raizer, 1965) that this entropy change is of the third order in terms of the shock strength $p_1 - p_0$ or $V_0 - V_1$. This underlies the important fact that the motion of a shock wave, even in an ideal fluid, is an irreversible process and leaves a permanent change in the medium after the wave has passed. In fact, this condition is used to select the physically admissible solutions from among those that satisfy the shock conditions. Since the entropy increases only in compression for normal materials, an expansion shock wave does not exist in such a medium. A typical exception is a medium that involves a phase transition or a chemical reaction. Other noteworthy consequences of the entropy increase are that

1. the shock is compressive ($p_1 > p_0$ for $V_0 > V$), and
2. the shock wave is supersonic relative to the upstream material, but the downstream flow is subsonic.

Since a true discontinuity is physically impossible in a real material, the shock transition occurs in a thin layer (shock structure). But the entropy change across the shock does not depend on the mechanisms of entropy production. These mechanisms control only the structure and the thickness. In a gas, this dissipation is due to the collision of atoms and molecules. On the macroscopic level, if high-temperature effects such as ionization can be ignored, the collisional process may be approximated, model-wise, by viscosity and heat conduction. Then, the structure of gasdynamic shock can be adequately described by the Navier–Stokes equations (Hayes, 1960).

The Navier–Stokes equations have proven useful, not only for investigating numerous inert dynamic problems, but also for problems that involve complicating effects such as dissociation, chemical reactions, and ionization (an extreme example of which is the compression of targets in laser fusion). With any of these processes, the shock conditions require no modification; the only complications that require change are the equations of state or the details of the structure (Hayes, 1960).

If temperatures exceed a few tens of thousands of degrees, there will be an additional complication resulting from radiative heat transfer. Since the mean free paths of light are much greater than those of gas particles, the radiative transfer controls the thickness of the shock structure. This requires the modification of the basic conservation equations. Furthermore, if an external magnetic field exists, then shock waves must be reconstructed on the basis of the equations of magnetohydrodynamics. However, for many technological applications, shock waves with an infinitely thin discontinuity provide an adequate first-order description.

The notion of viscosity and heat conduction has also been used to represent the macroscopic shock structure of solids, but viscous and conductive behaviors require different types of descriptions because of their underlying heterogeneous and anisotropic properties of shear strength (Graham, 1993). Likewise, multiphase heterogeneous systems such as gas–solid mixtures will also require new formulations (Hetsroni, 1982).

There are even collisionless shocks in space plasma where the structure is no longer due to collisions, but is created by electromagnetic interactions. The resulting shock thickness is known to stretch over a distance of many thousands of kilometers. The Earth's bow shock, one of the best examples, is caused by the supersonic flow of ionic particles (solar wind) past Earth's magnetosphere (Stone and Tsurutani, 1985; see also MAGNETOSPHERE).

2. ELEMENTARY WAVE INTERACTIONS

In a physical world, a shock wave is rarely steady or solitary. It interacts with other types of waves as well as material bodies. Without support, a shock wave attenuates quickly and "degenerates" into waves dispersed over a wider region through decompression. In this section, elementary examples of plane shock-wave interactions in an ideal compressible fluid are considered. First, the relationships for propagation of continuous disturbances, referred to as simple waves, are developed.

2.1 Characteristic Equations and Riemann Invariants

The continuum time-dependent equations for conservation of mass, momentum, and energy can be written in one space dimension as

$$\frac{d\rho}{dt} + \rho\frac{\partial u}{\partial x} = 0, \tag{17}$$

$$\rho\frac{du}{dt} + \frac{\partial p}{\partial x} = 0, \tag{18}$$

$$\frac{dE}{dt} + p\frac{dV}{dt} = 0 \quad \text{or} \quad \frac{dS}{dt} = 0, \tag{19}$$

where t = time, x = Eulerian space coordinate, S = entropy, and

$$\frac{d}{dt} = \frac{\partial}{\partial t} + u\frac{\partial}{\partial x}. \tag{20}$$

Since $\rho = \rho(p,S)$ for an ideal fluid, Eqs. (17) and (18) can be transformed into the characteristic form (Lieberman and Velikovich, 1986)

$$\left[\frac{\partial p}{\partial t} + (u \pm C)\frac{\partial p}{\partial x}\right] \pm \rho C\left[\frac{\partial u}{\partial t} + (u \pm C)\frac{\partial u}{\partial x}\right] = 0, \tag{21}$$

where

$$C = (\partial p/\partial \rho)_s^{1/2}. \tag{22}$$

Recalling the physical meaning of the total time derivative along a particle path, Eq. (21) can be reduced to

$$\rho C\frac{du}{dt} \pm \frac{dp}{dt} = 0, \quad \text{along}\ \frac{dx}{dt} = u \pm C. \tag{23}$$

The directions specified by dx/dt in the x-t plane are called characteristics ($C_\pm$ respectively) and define a system of new coordinates. Physically, they represent the local propagation paths of all analytic disturbances (simple waves). Equation (19) is also the characteristic equation (C_0) along particle motion:

$$\frac{dS}{dt} = 0, \quad \text{along}\ \frac{dx}{dt} = u. \tag{24}$$

Since the sound speed is a function of density alone, Eq. (23) can be further simplified by functions called Riemann invariants,

$$I_\pm = u \pm \int \frac{dp}{\rho C}. \tag{25}$$

The dynamic equations become

$$I_\pm = \text{constant}, \quad \text{along}\ \frac{dx}{dt} = u \pm C. \tag{26}$$

The flow is now completely described by two variables: u and p (or any one of the other variables: ρ or C) along the characteristics. The Riemann invariants provide the conservation conditions along grids of the new coordinate systems.

For illustration, consider a small disturbance propagating in one direction. On setting $u = u_0 + \Delta u$, $p = p_0 + \Delta p$, it may be seen that (Zel'dovich and Raizer, 1965)

$$I_\pm = \Delta u \pm \Delta p/\rho_0 C_0 = \text{const}. \tag{27}$$

If the perturbation is a forward-facing wave (a simple wave moving to the right), then I_- is constant since no disturbance is propagating in the C_- direction. Likewise, if it is backward facing, moving to the left, then I_+ is constant. In general, a disturbance originating at some point p in the space-time coordinate (see Fig. 3) can only influence events that lie in the region demarked between the C_+ and C_- through p advancing in time. Likewise, the state at p can be affected by a certain region of events in the past (domain of dependence).

It may now be seen that if the speed of acoustic waves increases with pressure, a continuous, finite pressure wave propagating in one direction steepens its profile with increasing time because of converging C_+

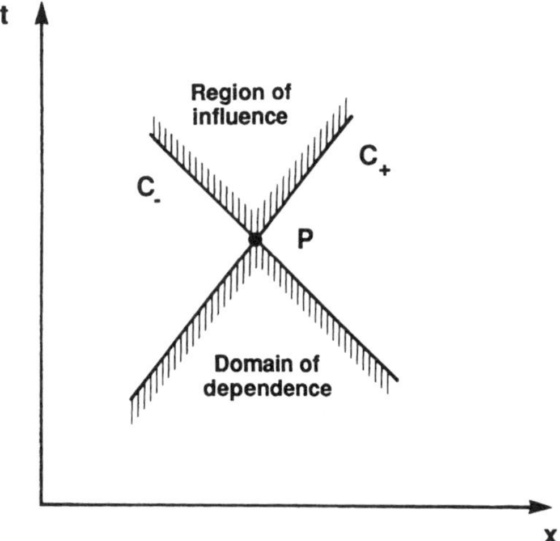

FIG. 3. Region of influence and domain of dependence defined by $C_{\pm}$ characteristics.

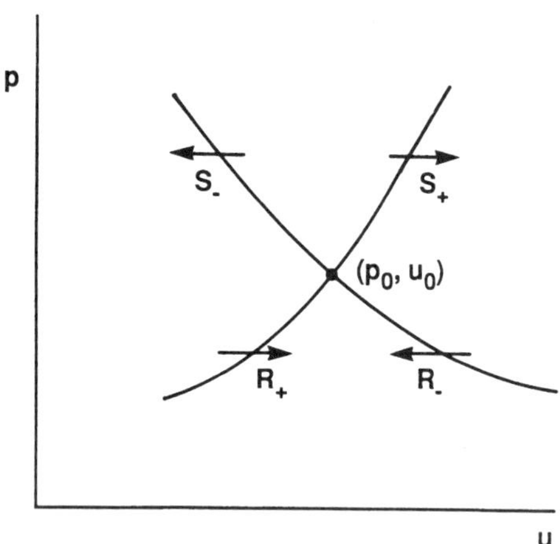

FIG. 4. Four types of wave transition in the *p-u* plane.

grids and that after some time it develops into a situation in which the two characteristic curves intersect. This is physically impossible because the flow field becomes multiple-valued. The result is the formation of a discontinuous shock wave. It may also be seen that diverging expansion waves (rarefaction waves) do not develop the non-uniqueness problem.

Generalization of the characteristics of problems involving more than two independent variables is found in advanced texts (e.g., Courant and Friedrichs, 1948). Here, note only that the characteristic grids in two dimensions become multidimensional surfaces (e.g., a cone in two space variables) and integration of the governing equations still involves partial differential equations. Therefore, engineering applications of these equations are limited in comparison with numerical approaches. However, their solid foundation in mathematics makes them an indispensable tool, not only in mathematical analysis of wave propagation, but also in the numerical solution of engineering problems.

2.2 Examples of Wave Interactions

The essential point of the previous discussion is that the change of an arbitrary state created by wave interactions is connected by specific paths described by Eqs. (5) and (25) in the *p-u* plane as illustrated in Fig. 4. In this figure, the labels *S* and *R* indicate a shock and a simple expansion wave originating at the initial state (p_0, u_0). The signs + and − designate forward- and backward-facing waves, respectively. To illustrate the use of these fundamental relationships, two applications (Duvall, 1971; Zel'dovich and Raizer, 1965) will be considered.

Figure 5 sketches a solution to the Riemann problem, where at an initial time, a semi-infinite region of a high-pressure gas on the left-hand side is in contact with a low-pressure gas on the other side (an idealized shock-tube problem). At subsequent times, a uniform shock travels to the right (a forward-facing shock S_+) and a family of expansion waves travel to the left (backward-facing, simple expansion waves R_-). When viewed in the *p-u* plane, the continuity of p and u for the state between the separating waves requires that this state must be given by the intersection of S_+ originating at (p_0, $u_0 = 0$) and R_- at (p_1, $u = 0$).

A special application of the above solution is a plane shock induced by explosive detonation. If the detonation process is assumed to be an adiabatic expansion of an ideal gas from a high-pressure state p_1 to an ambient pressure p_0, then the velocity of the products is given by (Schall, 1971)

$$u = \int_{p_0}^{p_1} \frac{dp}{\rho C} = \frac{2C_1}{\gamma - 1}\left[1 - \left(\frac{p_0}{p_1}\right)^{(\gamma-1)/2\gamma}\right], \qquad (28)$$

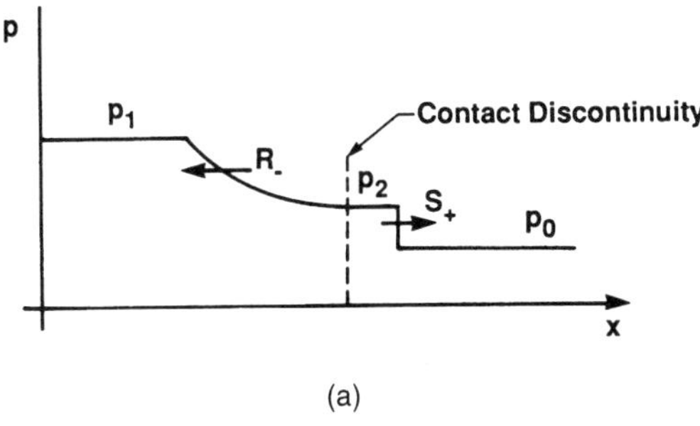

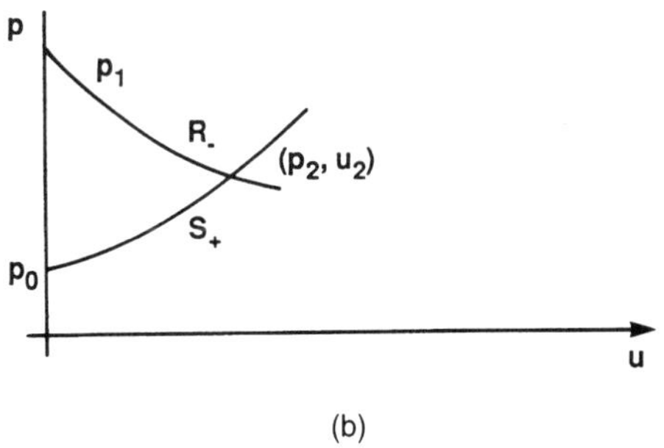

FIG. 5. Schematic diagrams of wave interactions in a shock tube: (a) a spatial wave profile and (b) wave transition in the p-u plane.

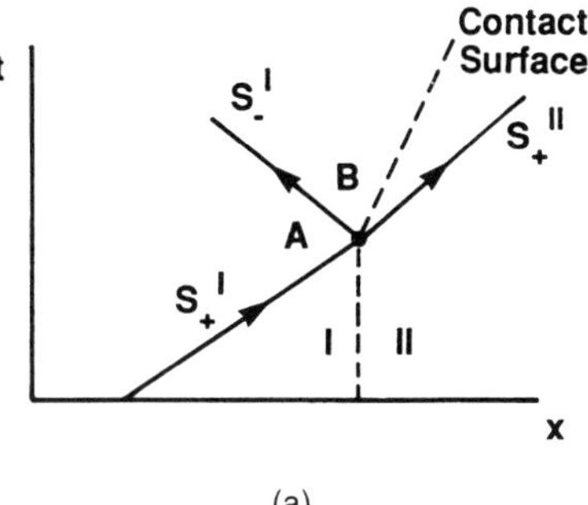

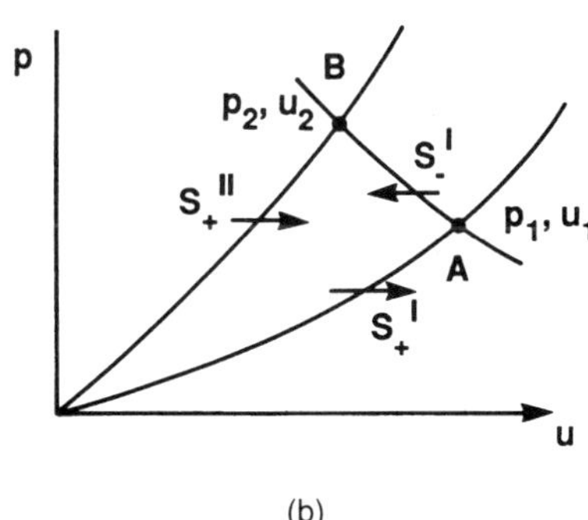

FIG. 6. Schematics of shock-wave transmission through a material interface: (a) shock trajectories in the x-t plane and (b) shock transition in the p-u plane.

where C_1 is the sound speed of the detonation product. If this shock interacts with an inert medium like a metal plate without impacting it, then the strength of the shock transmitted in the medium can be located using the same procedure as previously given. If the gas impacts with a finite velocity, then it belongs to the type considered next.

The second case is concerned with the transmission of a shock through a material interface. See Fig. 6. The incident shock S_+^I propagates from left (material I) to right (material II). It may be assumed, without loss of generality, that impedances satisfy $\rho_{02}D_2 > \rho_{01}D_1$. Then, both the transmitted and reflected waves are shock waves. Again the continuity of p and u on the material interface indicates that the state separating the two waves is found at B, the intersection of S_+^I and S_-^I. The latter originates at A, the state for the incident shock. If the second material is a rigid wall, it can be shown that

$$\frac{p_2 - p_0}{p_1 - p_0} = \frac{\rho_1 D_{21}' + \rho_0 D_1}{\rho_0 D_1} \underset{p1 \to p0}{\to} 2, \qquad (29)$$

where D_1 = velocity of the incident shock, D_{21}' = the velocity of the reflected shock relative to the material ahead, and $p_B > p_A$. It may be seen that if the magnitudes of shock impedances are reversed, then B is related to A by R_-.

3. VARIOUS PROBLEMS

As briefly mentioned in the Introduction, there are many interesting problems regarding shock waves. The following is a short list of such topics that are beyond the scope of the present article. They are complex because

1. the mathematics of shock-wave propagation is nonlinear;
2. the shock wave and materials response are interdependent—the shock wave affects materials behavior and vice versa; and

3. the wave motion occurs in multiple dimensions.

In the real world, the normal shock waves described previously are a special case of a more general family of oblique waves in which the mass flow across the shock front has a component parallel to the discontinuity. These shock waves are obviously two- and three-dimensional in nature. The discussions of these oblique waves are found in most textbooks of gas dynamics. Nevertheless, so far as the shock conditions are concerned, the obliquity does not change the shock conditions in the normal direction. For plane waves, the change required is the condition that the tangential component of the flow velocity is preserved. One often-illustrated oblique shock is the supersonic flight of a wedge- or cone-shaped object. Many of these shocks are illustrated in the collections of Van Dyke (1982) and Glass (1974).

When temperatures are high enough, shock waves inevitably induce chemical reactions in their wake. They occur not only in gases, but also in solids and even nuclear substances (nucleosynthesis on a cosmological scale). If the amount of heat is large enough and it is deposited fast enough, the reactions can lead to a unique phenomenon called detonation. This mode of combustion is distinguished from ordinary combustion in that the reacted materials move in the same direction as that of the propagating reaction wave (Fickett and Davis, 1979).

Converging and diverging shock waves are difficult problems to solve even in one dimension. They are fundamentally transient problems and require either a numerical method or a special analytical technique such as the similarity method (Sedov, 1959).

Beyond the simple condition mentioned earlier, the question of stability is a difficult issue because it could arise from geometry, or material properties, or both (Whitham, 1974).

4. EXPERIMENTAL AND DIAGNOSTIC TECHNIQUES

In some areas such as astrophysics, it is difficult to conduct experiments of observed phenomena under controlled conditions. But in gas dynamics and aerodynamics, shock tubes, wind tunnels, or combinations of both have been used to study shock-wave phenomena for many decades. According to Anderson (1990), Vieille used the former in 1899 and Busemann conducted experiments with the latter in the early 1930s.

A wind tunnel consists essentially of a channel through which a uniform stream of gas is driven at controlled velocities. For subsonic tests, the system is typically driven by a fan or propeller, but it requires a tank of compressed gas and a properly shaped nozzle to obtain supersonic streams at a high Mach number (u/C) of 4 or 5. Obviously, the latter are run only for short periods of time.

Tests in wind tunnels typically involve a scaled model of a real object and require a careful matching of the Mach and the Reynolds numbers (the ratio of momentum to viscous forces). As a result, a hypersonic tunnel such as the one at the NASA Langley Research Center is required to employ cryogenic nitrogen gas. Induced shock temperatures will reach 20 000 K and above.

In shock tubes, shock waves are produced in a closed tube by a sudden deposition of energy. Some of the most common sources of energy are compressed gas, electric discharge, explosives, and combustion. Each covers a different range of loading conditions. No single facility can cover the entire range of pressures and temperatures required to study diverse topics ranging from a simple shock in an inert gas, to weak shocks in automobile exhaust pipes, to shocks involving chemical reactions and ionizations (Green, 1964).

The scope and breadth of shock-tube and shock-tunnel applications may be seen by the following chapter titles excerpted from the proceedings of the 17th International Symposium on Shock Waves and Shock Tubes, *Current Topics in Shock Waves* (Kim, 1990). The chapter titles include shock formation; focusing and implosion; shock reflection and diffraction; turbulence, ionization, and plasma interaction; chemical kinetics, pyrolysis, and soot formation; ignition of detonation and combustion; particle entrainment and shock propagation through particle suspension; boundary layers; and blast simulation.

A variety of diagnostic techniques have

been developed to study the properties of shock waves. The measurements typically monitor the shock velocity and other quantities such as density, pressure, and temperature, and chemical compositions. Generally, they are classified as optical and nonoptical techniques. The optical techniques include Schlieren and shadow photography, interferometry, spectroscopy, and x-ray absorption. Particularly noteworthy is the rapid development in laser technology and its applications to spectroscopic measurements. Examples of nonoptical techniques are piezoelectric and ionization gauges.

Techniques for producing shocks in condensed matter are numerous. They include chemical explosives, high-speed impact of projectiles, and the deposition of energy by laser or charged-particle irradiation. When compared with gases, shock-induced changes of temperature and compression are, in general, small with the obvious exceptions of porous and powder materials, though pressures as high as several terapascals can be achieved in these materials.

Diagnostic techniques for solids, again, include optical and nonoptical techniques; however, with opaque solids, optical techniques are limited to surface measurements such as interferometric velocity measurements, luminescence, and surface temperatures. There exist not only stress gauges such as manganin and quartz gauges, but also electromagnetic gauges for particle-velocity measurements. The measurements of other physical properties such as electrical, magnetic, and optical properties have also provided valuable information, though the measurement of chemical compositions is limited at the present time to post-shock samples as exemplified in diamond synthesis.

5. NUMERICAL TECHNIQUES

It may not be an overstatement to say that, in the technical and industrial applications of shock waves, the single most important development over the last several decades is that of numerical calculations based on finite differences. The progress has been truly explosive. The solution of fluid-dynamic problems with shocks is no longer restricted to specialists, but it is becoming (if it is not already) an integral part of routine design tasks for engineers.

What is also significant is that with the advent of fast workstations, access to supercomputers is no longer critical to solving problems. For instance, the calculation of shock-free airfoils described in the following requires many thousands of cells and equations, but can be approached by anyone with a workstation and an idea. This has leveled the playing field and created a robust community of diverse ideas. For a historical review of the subject, with special focus on computational fluid dynamics, refer to two excellent articles by MacCormack (1985, 1993).

Before the advent of fast digital computers, the method of characteristics was a reasonably accurate means of numerically integrating hyperbolic systems of partial differential equations. For instance, the first plan of a supersonic wind tunnel was designed by a numerical method based on the characteristic equations by Prandtl and Busemann [a pedagogical account is found in Anderson (1990)]. But it is a cumbersome method because the characteristic net deforms as the flow changes (see Sec. 2.1) and may introduce inordinate numerical errors. In contrast, finite-difference techniques are more straightforward than the method of characteristics and are less prone to errors associated with grids. Hence, most contemporary approaches to numerical solutions of shock problems are based on finite-difference methods.

6. APPLICATIONS

There are no distinct boundaries between the scientific and technological applications of shock waves, but there is a useful distinction. In the scientific application, shock waves alone are the object of research and this research, in turn, provides information about the source or the medium in which shock waves propagate. Astronomical and interstellar shock waves are typical examples. Shock-wave research in condensed matter, motivated mostly by weapons development, has provided an understanding of shock-compression behavior under extreme conditions of high pressures, high temperatures, and high rates of loading.

In industrial and technological applications, shock waves or the force of shock waves are used either constructively or destructively, or both. Explosive metal forming and demolition in civil engineering are typical examples of the constructive/destructive use of shock waves. Metal forming exploits the force of high-pressure shock waves to do some useful work that may not be possible by other means. Shock waves may also form the metal at a much faster rate or more efficiently than other means. Scramjets, as described in the following, are another example. Extracorporeal lithotripsy is an example in which the destructive power of shock waves is used for a constructive purpose.

The subjects of lithotripsy, aerodynamics, diamond synthesis, and inertial-confinement fusion are treated in detail in other articles in the *Encyclopedia*.

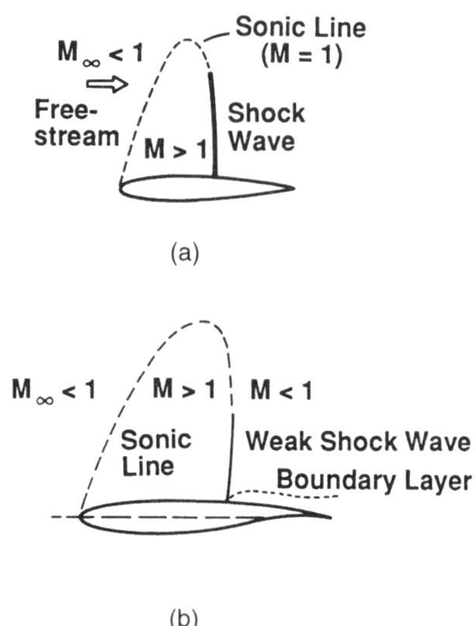

FIG. 7. A sketch of flow pattern near airfoils: (a) a section with a strong shock and (b) a section with no shock or a weak shock.

6.1 Aerodynamics and Propulsion

It may not be entirely wrong to state that the study of shock waves in the early days was motivated by the desire to fly at supersonic speed. The literature is too vast to conduct even the review of a single topical area. Therefore, the following examples of shock-free airfoil and supersonic combustion are meant to convey just the flavor of current interest.

A shock-free airfoil is not an application of shock waves in the strictest sense, but it is directly tied to shock-wave research because the goal is to eliminate shocks that occur even in subsonic flight as a result of the acceleration of air flow above the wing to generate lift (see Fig. 7) (see AERODYNAMICS). Such airfoils reduce the drag generated by shocks (about 5% of the lift force) to near zero. In practice, however, the complete elimination of shock waves is found to be very difficult because any small perturbation in a flow results in the appearance of relatively weak shocks. The design of such airfoils is a prime example of the application of numerical techniques through computer power in solving aerodynamic problems.

The concept of air-breathing supersonic combustion (scramjets) is an old idea that dates back to the 1960s (Waltrup *et al.*, 1976). It is proposed for both jets at high Mach numbers (>6) and booster rockets. Recent revival of interest has been triggered by President Reagan's proposal (1985) for hypersonic planes such as the National Aerospace Plane that will fly from New York to Tokyo in less than three hours. These planes are proposed to cruise at a speed of about 8 km/s in an Earth orbit.

There are numerous advanced ideas of scramjets, but there is not yet an engine in practical production. Many technical problems need to be overcome (Billing, 1986). One such issue is the problem of slow mixing in supersonic flow. This requires a long combustion section that is counter to the enhanced thrust from supersonic combustion. The gas dynamics of shock formation from the release of chemical energy at the entrance of the combustion section is another problem. Nevertheless, technical arguments supporting the use of scramjets for hypersonic flight are very strong. Scramjet designers today have many advantages over the pioneers of the 1960s. They have a critical tool, the computer, for large-scale computations of complex reacting flow from shocks. It is not unrealistic to expect to see a prototype space plane in the near future.

Connected with scramjets, there is also a revival of the concept of a supersonic combustion ramjet-in-tube to accelerate a projectile to a very high speed. In this scheme, a projectile flies through a premixed fuel–oxidizer mixture and uses energy through its supersonic combustion (Hertzberg *et al.*, 1988).

6.2 Materials Synthesis

Excluding military applications, the early study of shock (detonation) waves was carried out on solids. In the 18th century, there were attempts to exploit the high-pressure shock waves generated by impacting rifle bullets on graphite powder in a steel cavity to manufacture diamond through phase transition. Although the actual discovery and subsequent commercialization of shock-synthesized diamond did not take place until the early 1960s, present understanding of the shock synthesis of diamond shows that diamond powders probably were produced in the early experiments but were unidentifiable because of their microscopic size.

Because of the very short duration of the shock pulse (on the order of microseconds), the product powders are small, ranging typically from 0.01 μm to several microns. However, these particles are aggregates of microcrystallites and have a blocky shape with thousands of microcrystallite cutting edges that have unique cutting and polishing properties. As a result, they are used in special superprecision applications in the microelectronics industry. In Russia, shock-synthesized diamond powders are being used as an additive lubricant in oil as well as a raw material for manufacturing hard compounds.

Diamond powder is not the only substance that is commercially manufactured by the shock-compression technique. Wurtzite-type and zincblende-type boron nitride have also been produced on a commercial scale and used for the manufacture of cutting and grinding tools.

6.3 Medical Applications

The use of explosive shock waves is common in demolishing structures, breaking boulders, and removing earth. Now the same destructive power is used for noninvasive crushing of bladder renal calculi. The technique is named extracorporeal shock-wave lithotripsy (ESWL) (Chaussy *et al.*, 1980). Since the first clinical application of one such device in 1980, its effectiveness has been recognized worldwide (20 000 cases were reported in 1985) and ESWL is now considered the preferred treatment for urinary stones.

Figure 8 is a schematic of such a device. Box S represents the generator of focusing shock waves. The coil represents the high-voltage source or the detonator of a microexplosion. Shock waves are generated at the first focus of an ellipsoidal cavity and are refocused at the second focus point where a target stone is located (Takayama, 1993). The mechanisms of stone disintegration are not yet well understood, but are thought to be caused by the cavitation of water on the stone's surface, hammering from water jet, and compressive and tensile waves in stones. Efforts to understand the mechanisms better are hampered by the fact that there is no adequate device to measure high pressures in living tissues. Existing gauges are made inoperable because of ionic, chemical, and electromagnetic activities occurring in living tissues.

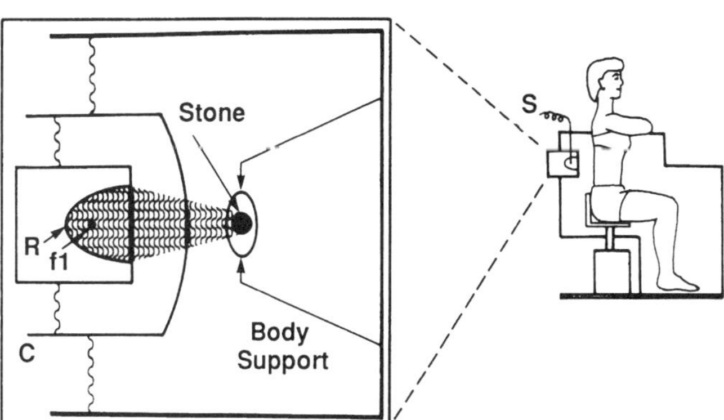

FIG. 8. A sketch of ESWL: R = reflector, C = a wave-coupling device, f1 = shock-wave source located at first focal point [after Takayama (1993)].

There are many new studies of focused shock waves in medical applications. One application that has already passed the stage of clinical tests is extracorporeal shock-wave osteoclasis (ESWO). In this technique, shock waves are used for the treatment of pseudoarthosis and delayed (and controlled) union of the fractures by creating microfissures and bony fragments (Schlecbergur and Sengo, 1992). Another application is the shock destruction of cancer cells.

6.4 Metalworking

The use of high-intensity shock waves for a variety of metal processes such as forming, cutting, embossing, and joining is a mature subject; nevertheless, new applications are far from being exhausted. The earliest patent on the expansion of bicycle tubes with low explosives dates back to 1898. High explosives are probably the most common source of energy today, but a high-voltage discharge and a magnetic field are also used for forming.

Techniques of explosive forming and welding do not require sophisticated equipment. In forming, an explosive is detonated and its energy (in the form of a shock wave) is transmitted through a fluid medium, commonly water, to a workpiece that is typically held on a female die. In the 1960s, large structural elements with a diameter as large as 3–5 m were fabricated by the explosive technique. Since then, for a variety of reasons such as noise, precision, and cost, the technology has been on the wane except for the following few bright areas:

1. the cladding of corrosion-resistant titanium on steel plate, and
2. the consolidation of amorphous metal.

The latter is still in its infancy, however, because of the problem of cracking during the expansion of shock waves. Nevertheless, the coating of thin amorphous film or powder is known to have passed the experimental stage (Murr, 1988).

An interesting new trend in metal forming is the redirection of the technology from mass production to the production of unique pieces that require precision. One example is the fabrication of denture plates made of pure Ti (Kani, 1977). These plates have several advantages over existing plastic pieces: thermal conductivity, thickness, and dimensional accuracy. Shock hardening of pure Ti helps to develop subsequent dimensional stability. Another area of experimentation is the creation of artwork. Again, the aim is not the mass production of ornamental pieces as in the past, but a unique artwork that is akin to painting and demands a close collaboration between the artist and the engineer in charge of production. An example is shown in Plate 1.

6.5 Inertial-Confinement Fusion

The role of shock waves in the quest for fusion power through inertial confinement has elements that are analogous to those in metal forming and propulsion. A small fuel pellet is heated and compressed by laser or ion beams irradiated uniformly over the pellet surface for a few nanoseconds (see Fig. 9).

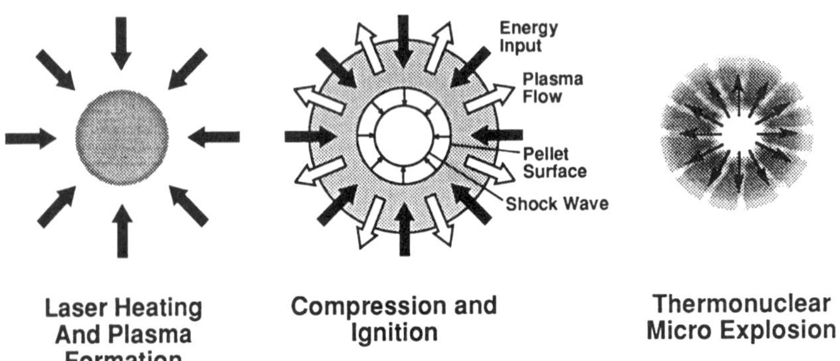

FIG. 9. A sketch of laser fusion.

A spherically heated outer layer becomes an ablating plasma, moves outward, exerts through recoil a pressure of 10^8 to 10^9 bars on the surface, and creates a radially converging, strong shock wave as well as an imploding mass motion. When the core reaches thermonuclear conditions (on the order of 2×10^5 kg/m^3 and 100 million degrees centigrade for deuterium–tritium fuel), it ignites and burns. If there are enough fusion reactions and sufficient inertial force to hold the compressed pellet, the reactions propagate outward, releasing much more energy, and produce an overall positive energy gain. Someday microscopic imploding shock waves may bring us abundant, clean energy through man-made, controlled nuclear reactions.

On a cosmic scale, shock waves generated in the final throes of collapsing, massive stars (supernovas) are known to play a crucial role in the creation of elements heavier than helium, the origin of life, in some sense, in the universe. In type I supernova having solar masses of 1.44 to about 8, the shock wave is driven by explosive nuclear reactions (Prange and Girvin, 1990). But in type II, a supernova with a solar mass greater than 8, the shock wave is said to be generated by the impact of an outer layer, collapsing as a result of gravity, on a core that has become a nuclear matter whose density is of the order of 2×10^7 kg/m^3! Type II is epitomized by the recent case of the 1987A supernova that captivated the attention of not only astronomers and physicists, but also laypeople. The scale of a supernova event is beyond any human experience—e.g., the gravitational energy release of over 10^{46} joules in less than a tenth of second and the initial temperature of 200 billion degrees at the center. Study of such a shock is not limited to the interior evolution of supernovas. As the shock bursts out of a supernova, it provides a large mass of hot gas and interacts with the surrounding media. Thus, the shock waves become a rich source of x rays and infrared emissions and provide a valuable means of investigating interstellar media.

7. CONCLUSION

Shock waves differ from other mechanical waves in that

1. they move at supersonic speed,
2. their amplitudes are determined by the energy of the wave, and
3. entropy increases across their near discontinuous wave fronts.

Technological applications are mostly based on the high-pressure and high-temperature effects of shocks under high rates of loading as exemplified by scramjets, inertial fusion, and diamond synthesis. Scientific applications, due to the ubiquitous nature of shock phenomena on microscopic to cosmic scales, have contributed to the understanding of such diverse subjects as solid state, interstellar structure, and supernova.

GLOSSARY

Detonation: A supersonic, self-sustained mode of combustion. The reaction region is preceded by a shock wave.

Hugoniot Equations: Equations that express the changes of flow properties across shock waves, based on the conservation of mass, momentum, and energy.

Laser Fusion: An approach to liberation of nuclear energy by heating and compressing a small pellet by the irradiation of laser light.

Mach Number: A dimensionless quantity (M) given by u/C. If $M > 5$, the motion is said to be hypersonic.

Navier–Stokes Equations: Partial differential equations governing the motion of a viscous, heat-conducting fluid.

Ram Jet: A compressorless jet engine. Oxidizer air in an engine duct is compressed by the forward motion of the aircraft, mixed with fuel, and burned. The exhaust from the rear of the engine produces thrust.

Rarefaction Wave: An expansion wave across which the pressure decreases.

Reynolds Number: A dimensionless parameter (Re) given by the ratio of inertial forces to viscous forces.

Schlieren (and Shadow) Photography: An optical technique for photographing density gradient (of disturbances) in fluid motion under a special lighting arrangement.

Scramjet: An air-breathing engine in which combustion occurs at supersonic air velocity.

Solar Wind: Motion of ionized particles from the sun's corona with velocities of 300–1000 km/s.

Supernova: A massive star that suddenly explodes into great brilliance by a factor of 10^{10}.

Works Cited

Anderson, A. D., Jr. (1990), *Modern Compressible Flow,* New York: McGraw-Hill, Chaps. 9 and 11.

Billing, F. S. (1986), *Ramjets with Supersonic Combustion,* AGARD LS-136, Neuilly-Sur-Seine, France: North Atlantic Treaty Organization, pp. 8.1–8.29.

Chaussy, C., Brendel, W., Schmiedt, E. (1980), *Lancet* **2,** 1265–1268.

Courant, R., Friedrichs, K. O. (1948), *Supersonic Flow and Shock Waves,* New York: Interscience.

Duvall, G. E. (1971), in: P. Caldirola and H. Knoeppel (Eds.), *Physics of High Energy Density,* New York: Academic, pp. 7–47.

Fickett, W., Davis, W. C. (1979), *Detonation,* Berkeley: Univ. of California Press.

Glass, I. I. (1974), *Shock Waves & Man,* Toronto: Univ. of Toronto Press.

Graham, R. A. (1993), *Solids under High Pressure Shock Compression,* New York: Springer.

Green, E. F. (1964), *Chemical Reactions in Shock Waves,* New York: Academic.

Hayes, W. D. (1960), *Gasdynamic Discontinuities,* Princeton: Princeton Univ. Press.

Hertzberg, A., Bruckner, A. P., Bogdanoff, D. W. (1988), *AIAA J.* **26,** 195–203.

Hetsroni, G. (1982), *Handbook of Multiphase Systems,* New York: McGraw-Hill.

Kani, K. (1977), in *Proceedings of 6th International Conference on High Energy Rate Fabrication, Essen, West Germany,* Denver: Denver University Press, pp. 3.8–3.815.

Kim, Y. W. (1990), *Current Topics in Shock Waves,* New York: American Institute of Physics.

Lieberman, M. A., Velikovich, A. L. (1986), *Physics of Shock Waves in Gases and Plasmas,* New York: Springer-Verlag, Chap. 1.

MacCormack, R. W. (1985), AIAA Paper 85-0032, New York: American Institute of Aeronautics and Astronautics.

MacCormack, R. W. (1993), AIAA Paper 93-3291-CP, New York: American Institute of Aeronautics and Astronautics.

Murr, L. E. (1988), *Shock Waves for Industrial Applications,* Philadelphia: Noyes.

Prange, R. E., Girvin, S. M. (1990), *Supernovae,* New York: Springer.

Rankine, W. J. M. (1870), *Trans. R. Soc. (London),* 160–277.

Schall, R. (1971), in: P. Caldirola, H. Kneppel (Eds.), *Physics of High Energy Density,* New York: Academic, pp. 239–240.

Schlecbergur, R., Sengo, T. (1992), *Arch. Orthop. Trauma Surg.* **111,** 224–227.

Sedov, L. L. (1959), *Similarity and Dimensional Methods in Mechanics,* New York: Academic.

Stone, R. G., Tsurutani, B. T. (1985), *Collisionless Shocks in the Heliosphere: a Tutorial Review,* Washington, DC: American Geophysical Union.

Takayama, K. (1993), *Jpn. J. Appl. Phys.* **32,** 2192–2198.

Van Dyke, M. D. (1982), *An Album of Fluid Motion,* Stanford, CA: Parabolic Press.

Waltrup, P. J., Anderson, G. Y. Stull, F. D. (1976), in *Proceedings of 3rd International Symposium on Air Breathing Engines,* pp. 835–861.

Whitham, G. B. (1974), *Linear and Nonlinear Waves,* New York: Wiley.

Zel'dovich, Ia. B., Raizer, Yu. P. (1965), *Physics of Shock Waves and High Temperature Hydrodynamics,* New York: Academic.

Further Reading

Journals/Proceedings

Shock Waves; The Journal of Fluid Mechanics; Physics of Fluids; Combustion, Explosion, and Shock Waves (tr. of Fiz. Goreniya Vzryva); International Journal of Impact Engineering.

International Symposium on Shock Waves and Shock Tubes (Biannual since 1957), *Symposium International on Detonation.*

Mathematics

Lax, P. D. (1973), *Hyperbolic Systems of Conservation Laws and the Mathematical Theory of Shock Waves,* Philadelphia: SIAM

Le Veque, R. J. (1992), *Numerical Methods for Conservation Laws,* Boston: Birkhauser.

Liberman, M. A., Velikovich, A. L. (1986), *Physics of Shock Waves in Gases and Plasmas,* New York: Springer.

Menikoff, R., Plohr, B. J. (1989), *Rev. Mod. Phys.* **61,** 75–130.

Oran, E. S., Boris, J. P. (1987), *Numerical Simulation of Reactive Flow,* New York: Elsevier.

Potter, D. (1973), *Computational Physics,* New York: Wiley.

Richtmyer, R. D., Morton, K. W. (1967), *Difference Methods for Initial-Valued Problems,* New York: Wiley.

Scramjets

Avery, W. H., Dugger, G. L. (1964), *Astronaut. Aeronaut.* **2,** 42–47.

Northam, G. B., Anderson, G. Y. (1986), AIAA

Paper 86-0159, New York: American Institute of Aeronautics and Astronautics.

Medical Applications

Kandel, L. B., Harrison, L. H., McCullough, D. L. (1987), *State of the Art Extracorporeal Shock Wave Lithotripsy,* New York: Futura.

Inertial-Confinement Fusion

Verlarde, G., Ronen, Y., Martinez-Val, J. M. (Eds.) (1993), *Nuclear Fusion by Inertial Confinement,* Boca Raton, FL: CRC Press.

Yamanaka, C. (1991), *Introduction to Laser Fusion,* New York: Harwood Academic Publishers.

Collisionless Shocks

Stone, R. G., Tsurutani, B. T. (1985), *Collisionless Shocks in the Heliosphere: Review of Current Research,* Washington, DC: American Geophysical Union.

Supernovae and Interstellar Shock Waves

Bethe, H. (1986), in: S. L. Shapiro, S. A. Teukolsky (Eds.), *Highlights of Modern Astrophysics,* New York: Wiley, Chap 3.

McKee, C. F., Draine, R. T. (1991), *Science* **252,** 397–403.

Ostriker, J. P., McKee, C. F. (1988), *Rev. Mod. Phys.* **60,** 1–68.

Material Synthesis and Metal Working

Crossland, B. (1982), *Explosive Welding of Metals and Its Applications,* Oxford: Clarendon.

Horie, Y., Sawaoka, A. B. (1993), *Shock Compression Chemistry of Materials,* Tokyo: KTK Scientific Publishers.

Sawaoka, A. B. (Ed.) (1993), *Shock Waves in Materials Science,* Tokyo: Springer.

Plate 1. An explosively formed art work by Yasuko Sawaoka (by permission of the artist).

Si–SiO$_2$ INTERFACE, ELECTRONIC PROPERTIES

ATSUSHI HIRAIWA, *Central Research Laboratory, Hitachi, Ltd., Kodaira, Tokyo, Japan*

INTRODUCTION

Information technology today owes much to the progress of large-scale integration (LSI) technology. This has been achieved by scaling down and integrating devices. The high reliability and high production yield indispensable to the widespread use of LSI, on the other hand, are due to the excellent properties of the SiO$_2$–Si interface. This is because most current LSI uses metal–oxide–semiconductor (MOS) transistors as switching devices.

Although the idea of MOS transistors had

3-527-28140-1/96/$5.00 + .50

been proposed in the 1920s (Lilienfeld, 1926), premature technology prevented the device operation. The study of the SiO_2/Si interface started in the 1940s in search of the origin of the shielding layer that obstructed the switching operation of MOS transistors, and during these studies bipolar transistor action was discovered in 1947 (Bardeen and Brattain, 1948). This discovery shifted the research activities from the interface to the Si substrate.

During the development of bipolar transistors, anomalous properties of junctions emerged as reliability problems. As researchers realized that these anomalies were not a sole result of bulk properties, the study of the interface was reactivated in the 1950s in an effort to passivate the Si surface. Thus the development of a high-quality SiO_2 surface led to a successful implementation of MOS transistors in 1960 (Kahng and Atalla, 1960).

In the early MOS transistors, the most important issue was mobile charges in the SiO_2 because these charges (which were later attributed to Na contamination) degraded the devices as a result of operation. The 1964 invention of MOS stabilization treatment using P_2O_5 solved this problem. In the 1960s, the interface trap density was reduced to the low 10^{10}-cm^{-2} range by annealing the MOS transistors in an atmosphere containing hydrogen. Because such a low trap level causes negligible device degradation, since the mid-1970s interest in the interface focused not on the characteristics of the as-made interface, but on its degradation after fabrication. One kind of degradation is induced by the device operation and is known as the hot-carrier effect (Ning *et al.*, 1975). Another kind is due to exposure to ionizing radiation (see, for example, Ma and Dressendorfer, 1989). The interest in this problem followed the failure of the telecommunication satellite Telstar 1, which passed through the Van Allen belt in 1962. Although the Telstar failure was in a bipolar transistor, MOS transistors were also found to be vulnerable to such radiation. These two kinds of postfabrication degradation spurred the investigation of carrier traps at the SiO_2–Si interface as well as in bulk SiO_2. For more details of the history, refer to the publication by Nicollian and Brews (1982).

From a practical viewpoint the hot-carrier effect is still a crucial issue in device development. Furthermore, a better SiO_2–Si interface has recently proved essential for making low-power dynamic random-access memories (DRAMs), and for making nonvolatile memories of flash type. These devices require the interface to be controlled far more strictly and will thus stimulate new activities in this field.

1. PREPARATION OF SiO_2

1.1 Thermal Oxidation

A high-quality SiO_2–Si interface is usually made by oxidizing a Si substrate in an atmosphere containing oxygen at a temperature ranging from 800 to 1100 °C. A steam ambient is widely used in the oxidation both of thin oxides for gate isolation (10–20-nm thick) and of thick oxides for device isolation (about 500-nm thick). This is because the steam oxide grows faster than the dry oxide. Another reason is the empirically superior breakdown tolerance of the steam oxide. The oxidation is carried out in a high-purity quartz tube oriented vertically in order to avoid the influx of open air by convection. Figure 1 shows an example of the kind of furnaces used industrially. Before the Si substrate is oxidized, it is cleaned to reduce particulate and metallic contamination to or below the levels of 0.1 particle/cm^2 and 1×10^{10} atoms/cm^2.

The oxide growth at the SiO_2–Si interface is due to the diffusion of oxidizing molecules, such as O_2 and H_2O, through the oxide (see Grove, 1967). This means that the interface is continuously renewed during the oxidation and that the properties of the interface are determined mostly by the final step of the oxidation. In the case of SOI (silicon-on-insulator) substrates, the top Si is oxidized at the interfaces with both the surface and buried oxide. The oxide that grows at the top-Si/buried-oxide interface increases the thickness of the buried oxide, as shown in Fig. 2.

1.2 Formation of SiO_2 Other Than by Oxidation

1.2.1 Chemical Vapor Deposition (CVD) A SiO_2 film can also be formed by the thermal reaction of monosilane (SiH_4) or

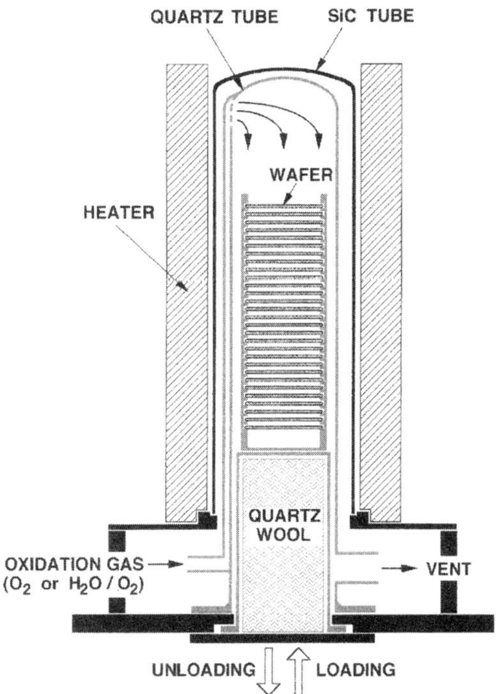

FIG. 1. Vertical oxidation furnace.

dichlorosilane (SiH_2Cl_2) with oxidizing gas in a reactor kept at 10–30 Pa to get a uniform and conformal film. This is called chemical vapor deposition [see CVD (CHEMICAL VAPOR DEPOSITION)]. Various combinations of reactant gases are used depending on the deposition temperature, and one of the criteria considered in selecting the gases is their thermal stability. A mixture of SiH_2Cl_2 and N_2O is used in the deposition at temperatures over 850 °C, below which SiH_4 is used instead of SiH_2Cl_2. Deposition at a temperature below 550 °C is done using a gas mixture of SiH_4 and O_2 to increase the growth rate. It is usually at atmospheric pressure.

The CVD oxides applied to poly-Si gate MOS capacitors have an interface quality comparable to that of thermal oxides. MOS transistors fabricated on polycrystalline Si films in static random access memory (SRAM) use a CVD oxide as the gate oxide, because the thermal oxide formed on poly-Si films has poor insulation characteristics due to the crystal defects.

1.2.2 Liquid-Phase Deposition (LPD) This is an extreme case of low-temperature preparation developed by Nagayama *et al.* (1988). Oxide film is deposited in a silica-supersaturated solution prepared by dissolving silica powder in hydrofluorosilicic acid (H_2SiF_6). The deposition proceeds, even at room temperature, by the reaction

$$H_2SiF_6 + 2H_2O \Leftrightarrow 6HF + SiO_2$$

and is facilitated by continuously adding boric acid, which pushes this reaction toward SiO_2 formation by consuming the hydrofluoric acid:

$$H_3BO_3 + 4HF \Leftrightarrow BF_4^- + H_3O^+ + 2H_2O.$$

The LPD oxide capacitor has an interface

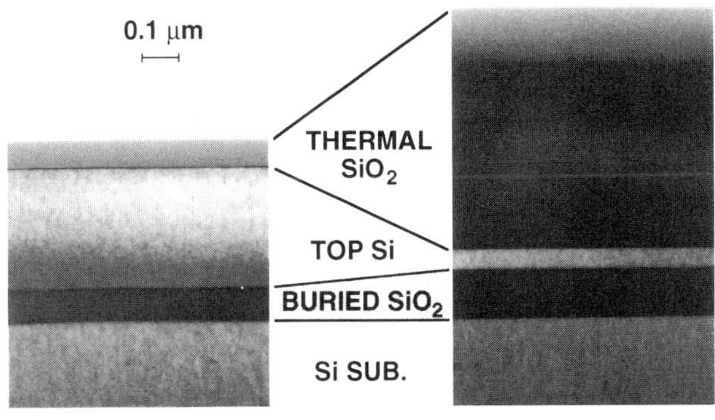

FIG. 2. Cross-sectional TEM images of SOI (silicon-on-insulator) wafers. The thickness of the buried oxide increased by 40 nm after oxidation in a steam ambient at 1350 °C. After Nakashima *et al.* (1994); © 1994 IEEE.

trap density of about 1×10^{11} cm^{-2} eV^{-1} at the middle of the Si band gap, one order of magnitude greater than that of the thermal oxide. For repeated gate-voltage scans, the capacitor exhibits different capacitance-vs-voltage curves, indicating the existence of bulk-oxide traps (Yeh *et al.,* 1994).

2. CHARACTERIZATION OF THE SiO_2–Si INTERFACE

2.1 Morphology

High-resolution transmission electron microscope (TEM) images of the SiO_2–Si interface indicate the existence of an undulation with an amplitude corresponding to one or two atomic layers of Si. We should note, however, that the actual interface is not necessarily as flat as it seems because TEM images are formed by electronic diffraction and scattering through the samples. This means that for any structure to be detected it must have sufficient extent along the direction of electron-beam incidence. Theoretical analyses of TEM imaging indicate that the actual roughness corresponds to more than four atomic layers (Goodnick *et al.,* 1985; Ohdomari *et al.,* 1986). The results of atomic force microscope (AFM) surface profiling are consistent with the larger undulation.

The SiO_2 is mostly amorphous even near the interface, but there is some evidence that a crystalline phase exists in the thermal oxide. Rochet *et al.* (1989) observed epitaxial crystalline growth of SiO_2 along the $[1\bar{1}0]$ ledges on a vicinal (001) Si surface by using reflection high-energy electron diffraction (RHEED) together with TEM. Takahashi *et al.* (1993) also observed the crystalline phase on a (001) Si surface by using x-ray crystal truncation rod (CTR) scattering analysis, and they concluded that the crystalline phase not only exists at the interface but that it also makes up a few percent of the volume of the bulk SiO_2.

2.2 Bonding Structure

Grunthaner and Grunthaner (1986) and Himpsel *et al.* (1988) have used x-ray photoelectron spectroscopy (XPS) to investigate the chemical bonding structure of the SiO_2–Si interface, and an electron spectroscopy for chemical analysis (ESCA) spectrometer is widely used in XPS. The x ray of the Al $K\alpha$ line (1486 eV) is monochromatized and used to irradiate the samples in a chamber kept at a pressure of 1×10^{-7} Pa or less. The energy spectrum of photoelectrons from the Si $2p$ core level is measured with a multichannel analyzer. The sensitivity with which the interfacial Si atoms are detected can be improved by using photons with an energy of about 130 eV radiated from a synchrotron. This is because the escape depth of the photoelectrons is about 3 Å in Si at this photon energy, whereas the escape depth in the conventional spectrometer is about 15 Å. In the case of 130-eV photon irradiation, only a small fraction of the photoelectrons from the Si substrate are detected because of the shorter escape depth. This suppression of the Si signal makes the signal from the interface more visible. The XPS measurement is limited to interfaces covered by an oxide whose thickness is less than 10 times the escape depth.

Figure 3 shows an example of the XPS analysis. The spectrum consists of a number of peaks. These peaks correspond to the shifted core-level binding energies that reflect the valence charge redistribution induced by the change in the local atomic arrangements. This redistribution is known as chemical shift. The peaks are associated with the intermediately oxidized Si atoms—Si^+ ($Si_3{\equiv}Si{-}O$), Si^{2+} ($Si_2{=}Si{=}O_2$), and Si^{3+} ($Si{-}Si{\equiv}O_3$)—that form the transition region at the SiO_2–Si interface.

The total amount of the intermediately oxidized Si atoms indicates that the interfacial transition region extends for about one (Grunthaner and Grunthaner, 1986) or two (Himpsel *et al.,* 1988) layers of Si. As shown in Fig. 3, for each of the Si surface orientations the intensity of Si^{3+} is far greater than expected from the ideal image of the local atomic geometry at the interface. This greater intensity of Si^{3+} also suggests that the interface is rougher than is observed by TEM.

2.3 Electrical Characterization

2.3.1 Theoretical Evaluation of Interface Traps Carrier traps at the SiO_2–Si interface exchange charges with the Si surface,

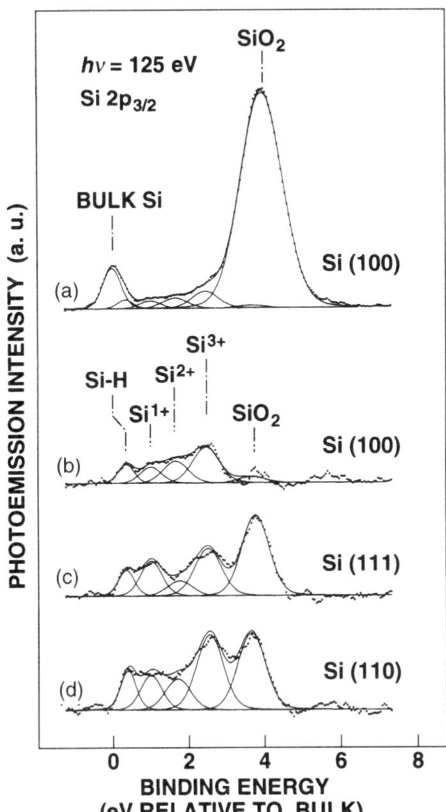

FIG. 3. Photoelectron spectra of a 2-nm-thick layer of oxide on Si. (a) The original spectrum on (100) Si. (b)–(d) The spectra after the removal of the bulk-Si and SiO_2 components. (After Niwano *et al.*, 1991.)

and this leads to an important property: Their charges change depending on the energy level relative to the Fermi potential. This property results from the fact that the trap energy level is located in the Si band gap. The interface traps are classified into two groups according to their charging characteristics. One consists of acceptor-type traps that are negatively charged when the energy level is below the Fermi level and are otherwise neutral. The other group consists of donor-type traps that are neutral when the level is below the Fermi level and are otherwise positive.

Sakurai and Sugano (1981) theoretically investigated the trap states assuming the Bethe lattice structure and obtained the following results, which are consistent with experiments. A perfect interface has no states in the Si band gap even if the angle of adjacent Si–O–Si bonds is varied. A dangling bond of oxygen has no band-gap state either. It is the dangling bond of Si backbonded to three other Si atoms that becomes a trap located at the middle of the Si band gap. Besides this bond, extended Si–Si bonds (with a bonding orbital) between $Si_3{\equiv}Si\cdot$ and $\cdot Si{\equiv}O_3$ produce trap states in the lower half of the Si band gap. They are donorlike because of the bonding orbitals. They might be regarded as oxygen vacancies. On the other hand, weak Si–O bonds (with an antibonding orbital) between $Si_3{\equiv}Si\cdot$ and $\cdot O{-}Si$ produce trap states in the upper half of the band gap. They are acceptorlike because of the antibonding orbitals. As shown in Fig. 4, the energy levels of both types approach the midgap when the bond length increases. Assuming a practical distribution of bond lengths, Sakurai and Sugano reproduced the U-shaped distribution of trap states that is

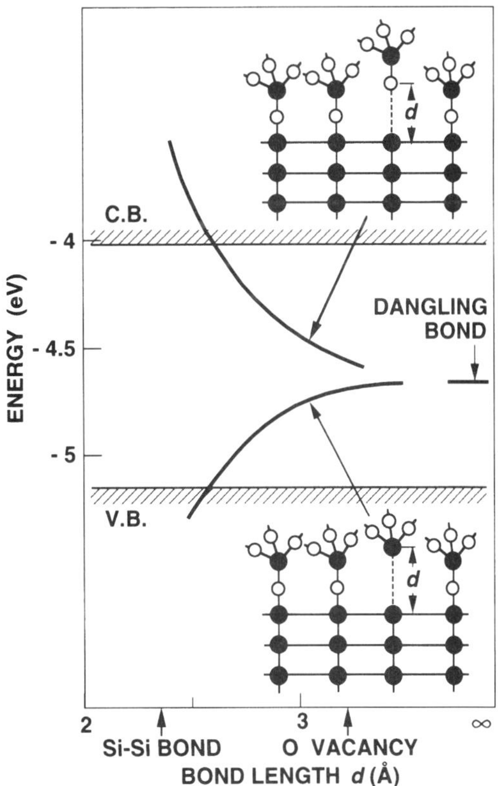

FIG. 4. Theoretical energy levels of Si–O and Si–Si weak bonds at the SiO_2–Si interface. Insets show the atomic configuration model used in the calculation. (After Sakurai and Sugano, 1981.)

widely observed. They also showed by calculation that the trap levels of both types go into either the valence band or the conduction band using H, Cl, F, or OH to terminate the dangling bonds.

2.3.2 Measurement of Interface Traps

2.3.2.1 Quasistatic Method A voltage applied to an electrode formed on a SiO$_2$–Si structure controls the potentials (relative to the Fermi potential) of interface-trap levels and changes the interface-trap charges (see Sec. 2.3.1). This change is detected by various capacitance methods and helps us characterize the traps. The quasistatic method developed by Berglund (1966) is the most popular among the capacitance methods because of its simplicity. In this method a low-frequency small modulation is superimposed on the voltage applied to the electrode (called gate after the naming in MOS transistors; see Sec. 3.1). The interface-trap and Si surface charges follow the slow modulation, and the entire MOS capacitance C_{LF} is given by $C_{LF} \equiv dQ_T/dV_G$, where Q_T stands for the areal density of total charges and V_G is the gate voltage. From the charge neutrality requirement and Maxwell's equation we get $Q_T = -(Q_{it} + Q_s + Q_{ox})$ and $Q_T = C_{ox}V_{ox}$, where Q_{it} is the areal density of interface-trap charges, Q_s is the areal density of Si surface charges, Q_{ox} is the areal density of bulk-oxide charges, C_{ox} is the oxide capacitance, and V_{ox} is the voltage across the oxide layer. We should take care that Q_{ox} is not a simple integration of oxide charge density $\rho(x)$, but a moment with respect to the gate given by

$$Q_{ox} = \frac{1}{d}\int_0^d x\rho(x)\,dx, \tag{1}$$

where d is the thickness of the oxide layer (Fig. 5). The voltage V_{ox} is given by $V_{ox} = (V_G - \psi_s) - (V_{G0} - \psi_{s0})$, where ψ_s is the band bending at the Si surface, and V_{G0} and ψ_{s0} are the values of V_G and ψ_s when $V_{ox} = 0$. Differentiating the above equations with respect to ψ_s, we get $C_{LF} = C_{ox}(C_S + C_{it})/[C_{ox} + (C_S + C_{it})]$, where $C_{it}(\psi_s) \equiv -dQ_{it}/d\psi_s$ and $C_s(\psi_s) \equiv -dQ_s/d\psi_s$ are the interface-trap capacitance and the Si surface capacitance. The interface-trap capacitance is related to the trap level density D_{it} (cm^{-2} eV^{-1}) at the Fermi level ϕ_s through the equation $D_{it}(\phi_s) = C_{it}(\psi_s)/q^2$, where q is the electronic charge. The Fermi level ϕ_s is equal to $\phi_s = \psi_B - \psi_s$, where ψ_B is the bulk Fermi level. Both ϕ_s and ψ_B are measured from the intrinsic level E_i ($\psi_B > 0$ for p-type Si).

The capacitance meter fails to provide accurate results in low-frequency modulation measurements because the signals are noisy.

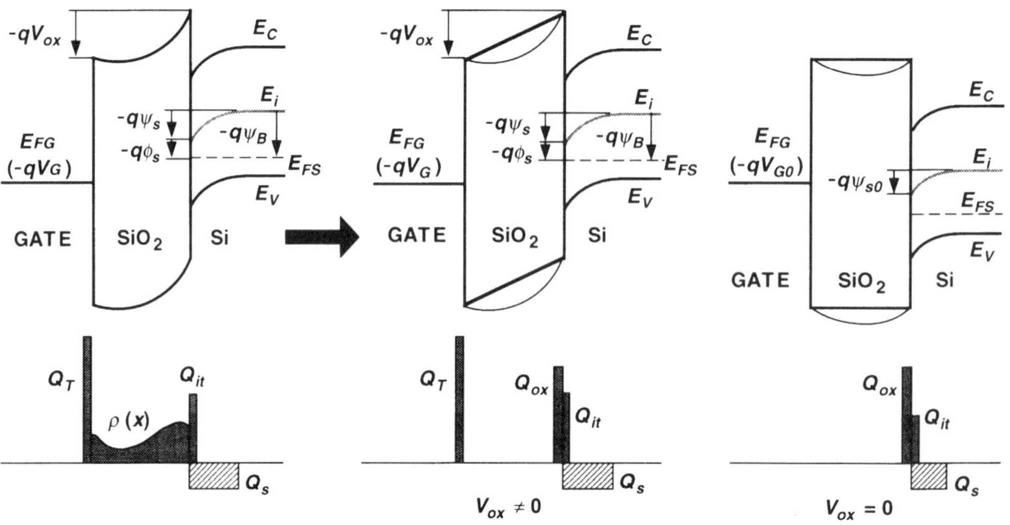

FIG. 5. Energy-band diagrams of MOS capacitors with bulk-oxide charges.

Accordingly, C_{LF} is usually obtained by measuring the displacement current $I_G = C_{LF}dV_G/dt$ induced by varying the gate voltage slowly, from which this method is named *quasistatic.*

The Si surface capacitance is theoretically calculated as (see Sze, 1981; Nicollian and Brews, 1982)

$$C_s(\psi_s) = \left|\frac{\epsilon_s}{2L_{Di}} \frac{[e^{\beta\psi_B}(1 - e^{-\beta\psi_s}) + e^{-\beta\psi_B}(e^{\beta\psi_s} - 1)]}{F(\beta\psi_s,\beta\psi_B)}\right| \quad \left(\beta \equiv \frac{q}{kT}\right), \quad (2)$$

where ϵ_s is the dielectric constant of Si, $L_{Di} = \sqrt{\epsilon_s kT/2q^2 n_i}$ is the intrinsic Debye length, n_i is the intrinsic carrier density of Si, k is Boltzmann's constant, T is temperature, and $F(\beta\psi_s,\beta\psi_B)$ is the normalized electric field at the Si surface defined by

$$F(\beta\psi_s,\beta\psi_B) = [e^{\beta\psi_B}(e^{-\beta\psi_s} + \psi_s - 1) + e^{-\beta\psi_B}(e^{\beta\psi_s} - \psi_s - 1)]^{1/2}.$$

Thus we can get the interface-trap density by measuring C_{LF} and using the following equation:

$$D_{it}(\phi_s) = \frac{1}{q^2}\left[\frac{C_{ox}C_{LF}(V_G)}{C_{ox} - C_{LF}(V_G)} - C_s(\psi_s)\right]. \quad (3)$$

Figure 6 shows examples of quasistatic capacitance curves of MOS capacitors fabricated on p-type Si. The samples, under a positive gate bias (4.3 V), were irradiated with x rays from a Mo tube. As the x-ray dosage increases, the capacitance curve is distorted by the interface-trap buildup (see Sec. 4.1.1).

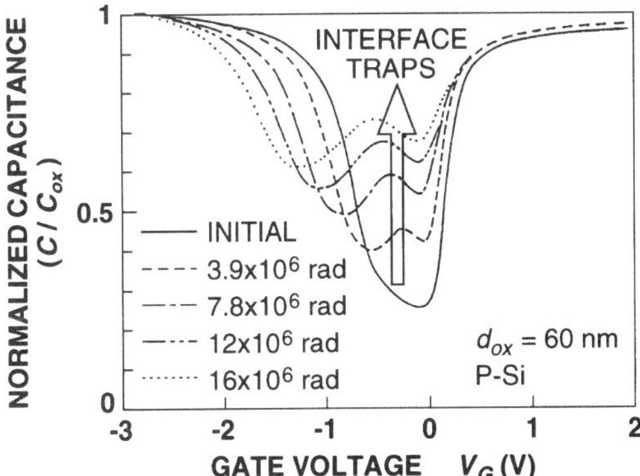

FIG. 6. Quasistatic capacitance curves of MOS capacitors under a positive bias V_G = 4.3 V and irradiated with x rays from a Mo-cathode tube (50 keV). (After Watanabe *et al.*, 1988.)

Berglund (1966) developed an experimental method to get the bandbending ψ_s as a function of V_G. He differentiated the equations $Q_T = C_{ox}V_{ox}$ and $V_{ox} = (V_G - \psi_s) - (V_{G0} - \psi_{s0})$ with respect to V_G and obtained

$$\psi_s = \int_{V_{G1}}^{V_G}\left[1 - \frac{C_{LF}(V)}{C_{ox}}\right]dV + \psi_s|_{V=V_{G1}}, \quad (4)$$

where V_{G1} is the initial gate voltage. If we start from the strong accumulation or inversion, $\psi_s|_{V=V_{G1}}$ is approximately equal to $(E_V - E_{FS})/q$ or $(E_C - E_{FS})/q$ (where E_{FS} is the Fermi level in the bulk Si) in p-type Si. By changing the gate voltage, we can control ψ_S and get the interface-trap density as a function of the trap level in the Si band gap.

2.3.2.2 High-Low–Frequency Capacitance Method When using the quasistatic method, we need to know the Si surface capacitance as a function of the gate voltage, and this capacitance is usually calculated theoretically using Eq. (2). This requires us to know the dopant distribution in the Si substrate, and this knowledge is not always easy to get. Castagné and Vapaille (1971) therefore developed the high-low–frequency method, in which the surface capacitance is obtained experimentally from a high-frequency capacitance measurement in the following way. The interface-trap level shifts relative to the Fermi level as the gate voltage changes. Those traps that crossed the Fermi level exchange carriers with a Si conduction or valence band near the interface. The exchange of carriers has a noninfinitesimal time constant τ given by $\tau = 1/(\sigma_n v_{th} n_s + \sigma_p v_{th} p_s)$, where σ_n and σ_p are, respectively, the electron and hole capture cross sections; n_s and p_s are the electron and hole concentrations near the interface; and v_{th} is the carrier thermal velocity (Iunovich, 1958). Because of this time constant, the traps respond to the gate-voltage modulation like a resistor R_{it} and a capacitor C_{it} in series: $\tau = R_{it}C_{it}$.

Because of the resistive element R_{it} of the interface traps, the entire circuit responds to the high-frequency (about 1 MHz) modulation as if there were no interface traps. This is an ideal case (no traps), and the impedance of the entire MOS system is given by a capacitance $C_{HF} = C_{ox}C_S/(C_{ox} + C_S)$. In the lower limit of the modulation frequency, we can regard the interface traps as a pure capacitance, and the entire system is reduced to the capacitor C_{LF}. Because C_S is the same for these two cases if the gate voltage is the same, the capacitance associated with the interface-trap charges is given by

$$C_{it} = \left(\frac{1}{C_{LF}} - \frac{1}{C_{ox}}\right)^{-2} - \left(\frac{1}{C_{HF}} - \frac{1}{C_{ox}}\right)^{-1}. \qquad (5)$$

Figure 7 shows the high-frequency capacitance curves of the same samples as in Fig. 6. The negative shift of the *CV* curves is due to positive charges produced by the x-ray irradiation (see Sec. 4.1.1). Slight distortions caused by the interface traps are also seen in the curves. Figure 8 shows the interface-trap level distribution in the Si band gap derived from Figs. 6 and 7 by using the method of this section.

2.3.2.3 Charge Pumping Brugler and Jespers (1969) developed the charge-pumping method as a probe of SiO$_2$–Si interface traps. As shown in Fig. 9, the source and drain of MOS transistors are connected together and reversely biased with respect to the substrate. A pulse applied to the gate drives the surface between inversion and accumulation. The interface traps in *p*-type Si are filled with carriers during the inversion

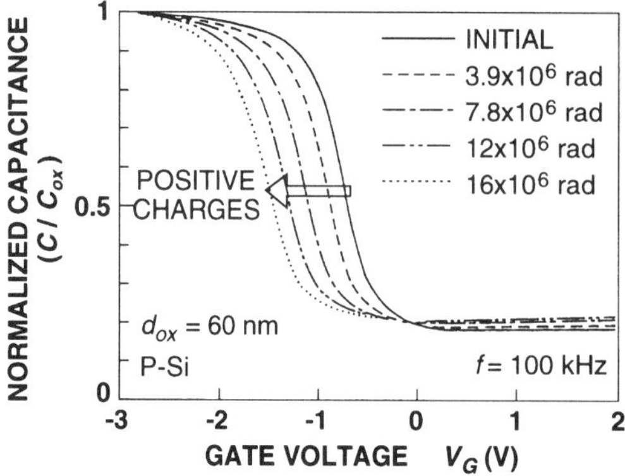

FIG. 7. High-frequency capacitance curves of the same samples as in Fig. 6. Measured at 100 kHz. (After Watanabe, unpublished.)

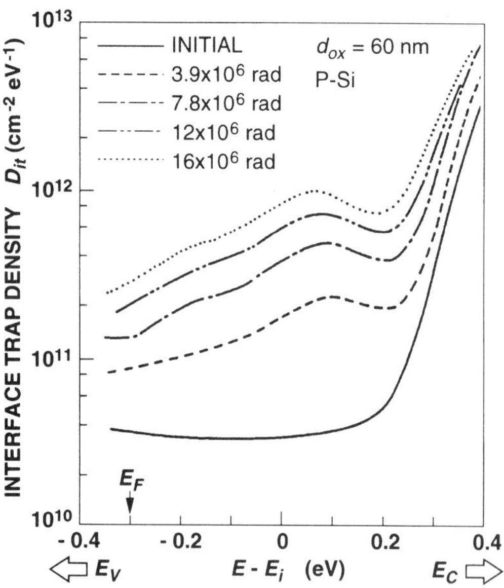

FIG. 8. Interface-trap density of the same samples as in Figs. 6 and 7. (After Watanabe *et al.*, 1988.)

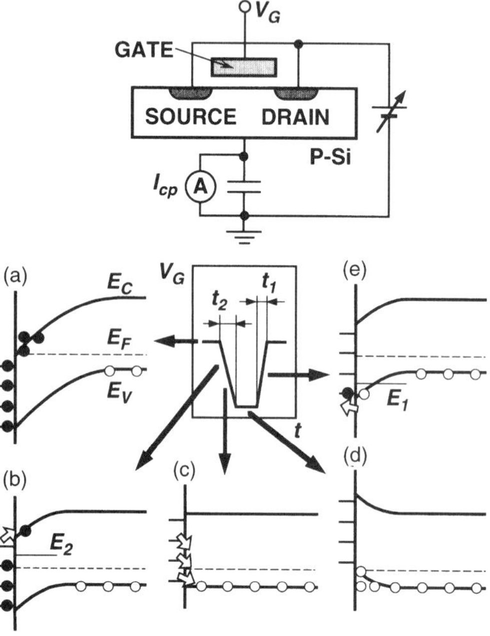

FIG. 9. Experimental setup and band diagrams illustrating the charge-pumping method.

period [Fig. 9(a)]. When the surface turns from inversion to accumulation, the inversion-layer carriers flow into the source and drain first. Most of the trapped carriers subsequently recombine with the majority carriers in the substrate [Fig. 9(c)]. They produce a dc component of the substrate current given by

$$I_{cp} = qS_G f_{cp} \int_{E1}^{E2} D_{it} dE_{it}, \tag{6}$$

where S_G is the gate area, f_{cp} is the pulse frequency, and D_{it} is the interface-trap density (cm^{-2} eV^{-1}). Some of the trapped carriers ($E > E_2$) are emitted into the conduction band (the edge of which is E_C) and do not contribute to the substrate current [Fig. 9(b)]. On the other hand, some traps ($E < E_1$) capture electrons from the valence band [the edge of which is E_V, Fig. 9(e)] and reduce the substrate current. These electron-emission and -capture processes ($E_2 \rightarrow E_C$, $E_V \rightarrow E_1$) are suppressed by keeping the transition times t_2 (inv. → acc.) and t_1 (acc. → inv.) short. When t_2 is too short, however, some of the inversion-layer carriers do not flow into the source or drain but instead recombine with the substrate majority carriers. This effect produces an additional substrate current ("geometric component") that has nothing to do with the interface traps. The use of short-channel MOS transistors (less than 1 μm) is effective in avoiding this effect.

Charge pumping is a direct method of measuring the interface traps. It is applicable to MOS transistors with a gate area as small as 1 μm^2. The capacitance method, in contrast, requires a gate area of at least 10^4 μm^2 when the commercially available capacitance meter is used. The basic charge-pumping method [Eq. (6)] provides only the average value of D_{it}, and many subsequent studies have tried to extract the distribution of D_{it} by changing the high or low levels of the gate pulse that controls the values of E_2 and E_1 (Haddara and Christoloveanu, 1986).

2.3.2.4 Electron-Spin Resonance The microscopic nature of the interface traps was revealed by electron-spin-resonance (ESR) measurements (see ELECTRON PARAMAGNETIC RESONANCE). ESR detects electronic spins associated with dangling bonds, and from the resonance absorption line we can get information about their structure. Dangling bonds, named P_b centers, which have a resonance g value of 2.008, are found at the SiO_2–Si interface. Poindexter *et al.* (1984) showed experimentally that the P_b center is a dangling bond of an interfacial Si atom backbonded to three other Si atoms, just as predicted by Sakurai and Sugano (1981). The experiments using unannealed samples (Poindexter *et al.*, 1984) and radiation-damaged samples (Lenahan *et al.*, 1981) showed that some of the interface traps indeed are P_b centers.

Not all of the interface traps are P_b centers, however. Cartier *et al.* (1993) reported a case in which only 10% of the generated interface traps were P_b centers. This result is another support for the extended bond model of interface traps (see Sec. 2.3.1).

3. THE SiO_2–Si INTERFACE AND DEVICE PERFORMANCE

This section reviews the ways in which the SiO_2–Si interface affects the performance of MOS transistors, which are the most important components of current LSI. There are two ways for the interface traps to degrade the performance of MOS transistors: by degrading switching performance and by increasing junction leakage current.

3.1 Operational Principle of MOS Transistors

First we briefly review the basic operational principle of MOS transistors, shown schematically in Fig. 10. The details are given in the textbook by Sze (1981). The MOS transistors are composed of three electrodes. Two of them, called source and drain, are heavily doped n-type regions in the lightly doped p-type Si substrate. The source is usually used as a voltage reference, and the electrical connection between the source and drain is controlled by a third electrode (gate) that is placed above the Si substrate and insulated from it by SiO_2 (gate oxide). A positive bias applied to the gate induces a two-dimensional electron layer, called an inversion layer, in the surface region of the Si. Thus electrons driven by the positive bias applied to the drain flow from the source to the drain. This electron flow is called drain current. Without the positive

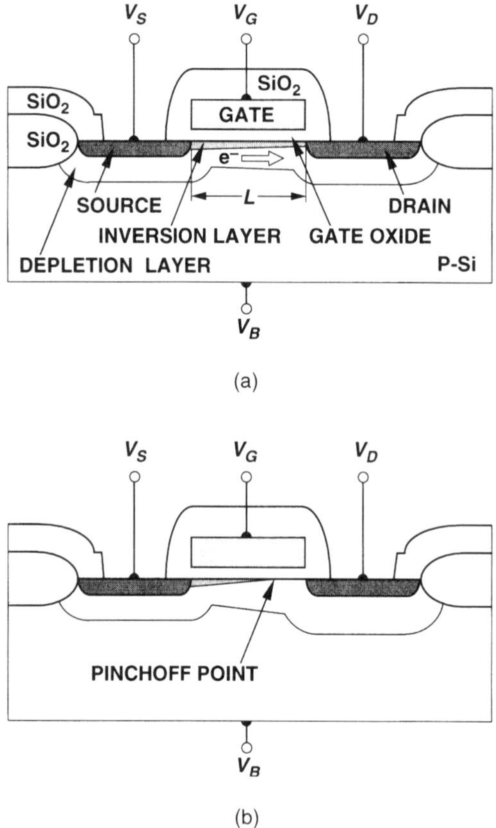

FIG. 10. Schematic cross sections of MOS transistors operated in (a) linear and (b) saturation regions.

gate bias, the drain is reversely biased to the substrate and electrically isolated from the source. A MOS transistor of this type is called an n-channel transistor, because the conduction layer (channel) is fed by negative carriers (electrons). We treat only n-channel MOS transistors in any detail in this article.

The speed of the LSI circuit is determined by two factors, one of which is the speed with which the transistor itself is driven into the "on" state. Before the formation of the inversion layer, a space-charge layer is formed below the gate, and most of the interface traps are filled by carriers. Because of these processes only a small number of carriers are induced in the Si surface during the buildup period of gate voltage. The drain current I_D depends exponentially on the gate voltage in this period, and we can use a subthreshold swing S, which is defined as the gate-voltage swing necessary to increase the drain current by a factor of 10, as a measure of the switching speed of MOS transistors. In the presence of interface traps, S is given by

$$S \equiv \frac{dV_G}{d\log_{10}I_D} = S_0\frac{1 + (C_D + C_{it})/C_{ox}}{1 + C_D/C_{ox}}, \qquad (7)$$

where $S_0 \approx (kT/q)\ \ln 10(1 + C_D/C_{ox})$ is the subthreshold swing in the absence of interface traps and C_D is the depletion-layer capacitance, which is equal to ϵ_s/x_D (where x_D is the width of the depletion layer). Equation (7) shows that the traps have an effect of adding a capacitance C_{it} to the system in parallel with the depletion capacitance, thus retarding the formation of the inversion layer. The smaller the subthreshold swing is, the faster the MOS transistor is driven into the "on" state.

The other factor determining the circuit speed is the magnitude of the drain current. The signal of component circuits (such as inverters) is output after charging or discharging stray or intentionally prepared capacitors through MOS transistors. Accordingly, the magnitude of current that the transistor can drive (drain current) determines the circuit speed. The drain current depends on the applied voltages in a different way for gate voltages $V_G - V_T > V_D$ (linear region) and $V_G - V_T < V_D$ (saturation region). Here V_D is the drain voltage and V_T is the minimum gate voltage for formation of the inversion layer (called threshold voltage). In the linear region, an almost uniform inversion layer is formed, and the carrier drifts along the channel driven by the field $F_C = V_D/L$ (where L is the channel length). If μ is the carrier mobility, the drift velocity v is expressed as $v = \mu F_C$. The inversion charge density is equal to the oxide capacitance multiplied by the potential difference between the gate and the inversion layer. If we put qN_I as the charge density near the source, the charge density near the drain is equal to $qN_I - C_{ox}V_D$ because the potential of the inversion layer is lower than that near the source by the amount V_D. Thus the average charge density is given by $qN_I - C_{ox}V_D/2$, and the drain current I_D is given by

$$I_D = W(qN_I - C_{ox}V_D/2)\mu V_D/L, \qquad (8)$$

where W is the channel width and q is the

electron charge. For a small value of V_D, this equation is reduced to a linear function of V_D.

In the saturation region, on the other hand, both N_I and v vary along the conduction path. In the lowest-order approximation, $N_I(x)$ and $v(x)$ may be replaced by the averages $N_I/2$ (N_I is the carrier density at the source) and v. The factor $\frac{1}{2}$ in the average carrier density is due to the density being N_I at the source and about zero at the pinchoff point, where the surface potential is $V_G - V_T$. Thus the drain current is expressed as

$$I_D = \tfrac{1}{2}WqN_Iv.$$

The drift velocity is again expressed as $v = \mu F_C$, where F_C is the average field, which is equal not to V_D/L (V_D is the drain voltage), but to $(V_G - V_T)/L$. This is because the drain current remains constant even when the drain voltage V_D increases beyond the pinch-off voltage $V_P = V_G - V_T$. The excess voltage $V_D - V_P$ is imposed on the depleted region near the drain and has little effect on the enhancement of the current. Thus we get

$$I_D = W\mu(qN_I)^2/2LC_{ox}, \tag{9}$$

where $qN_I = C_{ox}(V_G - V_T)$. We note that the value given by this equation does not depend on the value of V_D and is equal to the maximum of Eq. (8) (at $V_D = qN_I/C_{ox} = V_G - V_T$). The term *saturation* comes from these results.

From a practical point of view, the saturation region is much more important than the linear region. In order to show this we briefly review the operational principle of an inverter, which is one of the elementary components of an LSI circuit. The inverter consists of an MOS transistor and a load connected in series (Fig. 11). The signal (high or low voltage) is input to the gate of the transistor, and the inverted signal is output to the drain. Now we consider the case where the input voltage turns from low to high. When the input voltage is low, the drain voltage is high because the load is connected to the power supply, and no current flows through the transistor. During the buildup period of the input voltage, drain current begins to flow, and the transistor is in the saturation region because the gate voltage is not high enough. The drain current discharges stray capacitors connected to the drain and decreases the drain voltage. The operational state of the transistor then turns into the linear region. Accordingly, the switching time is mostly determined by the saturation drain current.

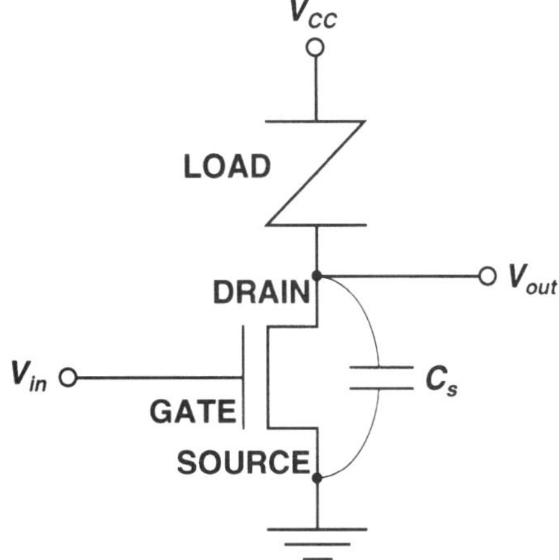

FIG. 11. Inverter.

The mobility is physically defined only for low fields where the velocity is proportional to the field. In practical applications, however, the mobility in the nonlinear region is extensively used for the analysis of device characteristics. In this case we regard the mobility as defined by the equation $\mu = v/F_C$. Then μ is a function of F_C, expressed by Thornber (1980) as

$$\mu = \mu_0\left[1 + \frac{(\mu_0F_C/u_l)^2}{G + \mu_0F_c/u_l} + \left(\frac{\mu_0F_c}{v_{sat}}\right)^2\right]^{-1/2} \tag{10}$$

where μ_0 is the mobility at low fields, u_l is the longitudinal acoustic phonon velocity, G is a fitting parameter, and v_{sat} is the saturation velocity. It is easily seen that $\mu \approx \mu_0$ for $F_C \ll u_l/\mu_0$, whereas $\mu \approx v_{sat}/F_C$ for $F_C \gg u_l/\mu_0$. Typical values of the parameters used in the simulation are listed in Table 1.

3.2 Degradation of MOS Transistor Performance

3.2.1 Increase in Subthreshold Swing

The subthreshold swing S in current MOS transistors is from 70 to 80 mV, and Eq. (7)

Table 1. Parameters of mobility. After Yamaguchi (1983).

	u_l (cm/s)	G	v_{sat} (cm/s)	μ_0 (cm^2/Vs)
Electron	4.9×10^6	8.8	1.04×10^7	1400
Hole	2.93×10^6	1.6	1.0×10^7	480

shows that S increases appreciably when the interface-trap capacitance C_{it} becomes comparable to the depletion-layer capacitance $C_D \approx 3 \times 10^{-8}$ F/cm^2. The trap density of as-fabricated transistors is little more than 10^{10} cm^{-2} eV^{-1} and corresponds to $C_{it} \approx 5 \times 10^{-9}$ F/cm^2, which is an order of magnitude less than C_D. Thus, the generation of interface traps after fabrication is the dominant factor that increases the subthreshold swing.

3.2.2 Reduction of Drain Current (1): Loss of Surface Carriers In n-channel MOS transistors under operation, the Fermi level is close to the conduction-band edge in the channel. Accordingly, the acceptor-type interface traps (areal density $N_{it,C}$) are negatively charged and reduce the inversion carrier density from N_I to $N_I - N_{it,C}$, leading to I_D degradation as seen From Eq. (9). Similar phenomena are observed in p-channel MOS transistors where the Fermi level approaches the valence-band edge, and donor-type interface traps that are positively charged reduce N_I. Because the inversion carrier density in current devices is about 8×10^{12} cm^{-2}, the interface traps of as-fabricated devices (<5 $\times$ 10^{10} cm^{-2}) have a negligible effect. Device operation, however, generates trap levels beyond the 4×10^{11}-cm^{-2} level that cause a 10% reduction of the drain current.

3.2.3 Reduction of Drain Current (2): Decrease in Mobility The interface traps reduce the low-field mobility μ_0 in Eq. (10) through Coulomb scattering, phonon scattering, and surface-roughness scattering. Assuming Matthiessen's rule, we can express the mobility as

$$\frac{1}{\mu_0} = \frac{1}{\mu_C} + \frac{1}{\mu_{ph}} + \frac{1}{\mu_{SR}},$$

where μ_C, μ_{ph}, and μ_{SR} are the mobilities determined by the three scattering processes.

3.2.3.1 Screened Coulomb Scattering This process limits the mobility when the inversion carrier density is low. The scattering of carriers is caused by charged centers: space charges (ionized dopants) in the depletion layer (qN_{depl}), charges in the gate oxide (qN_{ox}), and interface charges ($qN_{it,C}$). It should be noted that only the charged interface traps ($N_{it,C}$), not the interface traps as a whole (N_{it}), contribute to this process. Sah *et al.* (1972) expressed the results of their theoretical and experimental analysis as follows:

$$\mu_C \propto TN_I^{\alpha}/N_C \qquad (\alpha \approx 1.2\text{–}1.4), \tag{11}$$

where T is the temperature and N_C is the total density of scattering centers (given by $N_C = N_{depl} + N_{ox} + N_{it,C}$). The term N_I^{α} (missing in their original report) represents the screening effect by the carriers themselves, which effect was grounded theoretically by Stern and Howard (1967). When N_I is large, the mobility is less sensitive to N_I and is not approximated by such a simple formula.

By reducing N_I, the interface traps reduce the mobility, as seen from Eq. (11). They further reduce the mobility by increasing N_C. In as-fabricated practical devices, both the oxide and interface charges are little more than 10^{10} cm^{-2}, and their contribution to N_C is negligible when compared to that of the depletion-layer charges, $N_{depl} \approx 10^{12}$ cm^{-2}. Thus, the reduction of mobility by Coulomb scattering is mostly caused by the decrease in the number of inversion carriers.

3.2.3.2 Phonon Scattering This scattering is caused mostly by acoustic phonons. Kawaji (1969) considered the two-dimensional nature of the inversion carriers and obtained the result

$$\mu_{ph} = \frac{\hbar^3 q\rho c_l^2 \langle z \rangle}{\Xi^2 k_B T m_t^2} \propto \frac{1}{TF_{eff}^{1/3}}, \tag{12}$$

where $\hbar$ is Planck's constant divided by 2π, ρ is the mass density, c_l is the longitudinal sound velocity, $\langle z \rangle$ is the effective thickness of the inversion layer (proportional to $F_{eff}^{-1/3}$; Kawaji, 1969), Ξ is the deformation potential (12 eV; Ezawa *et al.*, 1971), m_t is an electron effective mass parallel to the interface, and $F_{eff} = q(N_{depl} + \eta_{ph}N_I)/\epsilon_S$ and corresponds to the average field in the inversion layer. The value of η_{ph} was reported as $\frac{11}{32}$.

In this process, the interface traps increase the mobility by reducing F_{eff}. This effect is small, however, and as seen from Eqs. (8) and (9), the drain current itself decreases because of the reduction of inversion carrier density.

3.2.3.3 Surface-Roughness Scattering This process becomes dominant when the inversion carrier density is high, and it generally is the most important process because the device is usually operated under such a condition. The SiO_2–Si interface is not completely flat. The potential barrier that confines the inversion carriers outside SiO_2 fluctuates because of this roughness and scatters the inversion carriers in the same way as in Coulomb scattering. The variation ϕ of the barrier height is given as $\phi \approx F_{eff}\Delta$, where F_{eff} is the effective field given by $F_{eff} = q(N_{depl} + \eta_{SR}N_I)/\epsilon_S$, and Δ is the average displacement of the interface from flatness. Matsumoto and Uemura (1974) and Ando (1977) theoretically obtained

$$\mu_{SR} = \hbar^3\left\{F_{eff}^2 q m_t^2 \Lambda^2 \Delta^2 \int_0^{\pi} d\theta(1 - \cos\theta) \times \exp[-\tfrac{1}{2}k_F^2\Lambda^2(1 - \cos\theta)]\right\}^{-1}, \quad (13)$$

and $\eta_{SR} = \frac{1}{2}$, where k_F is the wave number of the Fermi level and Λ is the correlation length of spatial variation of the interface undulation and is equal to $\frac{1}{3}$ for Gaussian-correlated undulations. Matsumoto and Uemura (1974) obtained $\Delta \approx 5$ Å and $\Lambda \approx 15$ Å for (100) electrons at 1.7 K.

In the surface-roughness scattering, the interface affects the mobility through the asperity Δ of the interface. The interface traps increase the mobility through the decrease in the effective field similarly to the way they do in phonon scattering.

Recent experiments by Takagi *et al.* (1990) give us a good view of the mobility as a function of inversion carrier density covering the phonon and surface-roughness scattering (Fig. 12). In the region of a moderate field and at room temperature, the mobility changes in proportion to $1/TF_{eff}^{1/3}$. As a matter of course we cannot observe this component at temperatures below 100 K. On the other hand, mobilities are less sensitive to the effective field than they are to the theoretical F_{eff}^{-2} for a higher field at high temperatures. This discrepancy is ascribed to the phonon scattering.

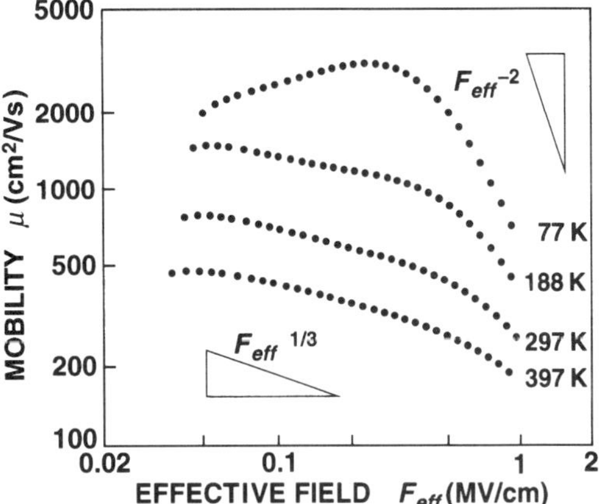

FIG. 12. Electron mobility in the inversion layer on (100) Si. Acceptor concentration of the substrate is 3.9×10^{15} cm^{-3}, and F_{eff} is defined by setting $\eta = 1/2$. (After Takagi *et al.*, 1990.)

3.3 Excessive Junction Leakage Current

3.3.1 Requirements of Low-Power DRAMs A *p-n* junction is formed between the source (or drain) and Si substrate of MOS transistors and is reversely biased. Leakage current, driven by the reverse bias, flows through the junction and causes LSI failures when it is generated excessively. DRAM is one of the LSIs that are sensitive to the leakage current. It stores one-bit information in a unit (called memory cell) that consists of a MOS transistor and a capacitor (Fig. 13). The capacitor consists of a pair of electrodes separated by an insulator and fabricated extending over the MOS transistor to reduce the memory-cell area in most DRAMs of 4 Mbits or more. When the logical "1" state is stored in a cell, there are more positive charges in the lower capacitor electrode (storage electrode) than in the case when the "0" state is stored. The potential of the plate electrode is kept constant. The MOS transistor switches the charging and discharging of the capacitor. The storage electrode is electrically isolated from the Si substrate by the *p-n* junction. Because some leakage current is inevitable in the junction, the stored charges are lost in the course of time. In DRAMs the charges are read from the capacitor before they are lost, and the full charges are again stored in the capacitor. This operation for data retention (called refreshing)

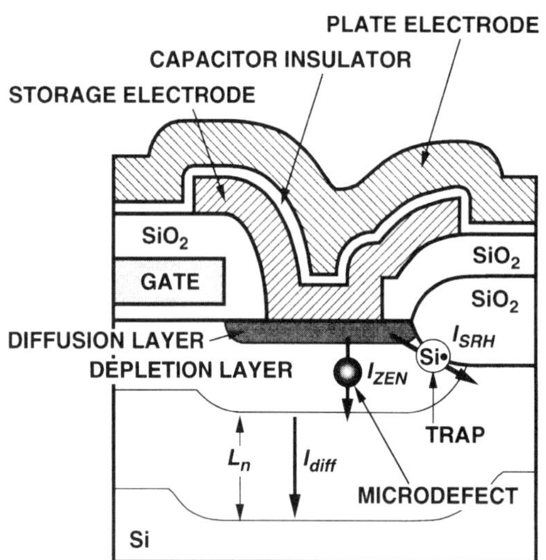

FIG. 13. Schematic cross section showing possible junction leakage paths in a DRAM cell. The cell is of a stacked type used in most DRAMs of 4 Mbits or more.

consumes much power. Accordingly, to meet the market need for low–dissipation-power DRAMs, it is important to make the refreshing interval as long as possible, namely, to reduce the junction leakage current. Figure 13 shows typical paths of the junction leakage: SRH (Shockley–Read–Hall) process, local Zener effect, and diffusion. Although only the SRH process is related to the SiO$_2$–Si interface (Kircher, 1975), we look over all three of these paths.

3.3.2 Shockley–Read–Hall Process A space-charge layer (called depletion layer) exists between the neutral n-type and p-type regions that form the junction. An electric field is generated by the space charge and sweeps the carriers (electrons and holes) in the depletion layer toward the neutral regions. The carriers thus swept lead to the leakage current and are generated by thermal excitation of Si valence electrons into the conduction band. This excitation has an activation energy of the Si band gap (E_g = 1.12 eV) and leads to only a negligible leakage current near room temperature. Carrier traps have energy levels in the Si band gap (see Sec. 2.3.1) and assist the thermal carrier generation by serving as stepping stones to the excitation. This process was investigated in detail by Shockley and Read (1952) and Hall (1952), and is called the SRH process. The leakage current produced by this process is expressed as

$$I_{\mathrm{SRH}} = \int_{depl} qN_t G_{\mathrm{SRH}}\, dV, \tag{14}$$

where N_t is the trap density and the integral should be carried out over the depletion region. G_{SRH} is the carrier generation rate given by

$$G_{\mathrm{SRH}} = \sigma_p \sigma_n v_{th} n_i \left[\sigma_n \exp\left(\frac{E_t - E_i}{kT}\right) + \sigma_p \exp\left(\frac{E_i - E_t}{kT}\right)\right]^{-1}, \tag{15}$$

where σ_n and σ_p are the electron and hole capture cross sections, n_i is the intrinsic carrier density, E_t is the trap energy level, and E_i is the Fermi level of nondoped Si (close to the middle of the Si band gap). According to this equation, most of the leakage current is caused by the traps close to E_i. For these traps the SRH current has almost the same temperature dependence as n_i, which has an activation energy of $E_g/2$.

The electric field in the present LSI appreciably enhances this SRH current (Hurkx *et al.,* 1992). In the absence of the field, the trapped electrons can be emitted over the entire trap depth $E_C - E_t$ only by thermal excitation. Under an electric field, they can be emitted by a combination of thermal excitation over a part of the trap depth and tunneling through the remaining potential barrier along the field. Hole emission is similar, and these processes enhance the carrier generation rate as follows:

$$G_{\mathrm{SRH}}(F) = G_{\mathrm{SRH}}\left[1 + 2\sqrt{3\pi}\,\frac{F}{F_\Gamma} \exp\left(\frac{F}{F_\Gamma}\right)^2\right], \tag{16}$$

where F is the field, $F_\Gamma = [24m^*(kT)^3]^{1/2}/q\hbar$, and m^* is the effective mass of carriers. Hurkx *et al.* (1992) experimentally obtained $m^* = 0.25\ m_0$, where m_0 is the electron mass.

Experiments using junctions of different areas and edge lengths show that there is a leakage-current component that is proportional to the edge length and has an activation energy close to $E_g/2$. This component is attributed to the SHR current generated by the traps at the interface between the Si substrate and the SiO_2 that is formed surrounding the devices for electric isolation. As the device size is reduced, the edge component dominates the areal component because the edge length is scaled by a constant K while the area is scaled by K^2 ($K < 1$). The contribution from the bulk traps in Si is negligibly small because their density in the current LSI is less than 10^{12} cm^{-3}.

3.3.3 Other Current Generation Processes

3.3.3.1 Local Zener Effect The local Zener effect is the mechanism of anomalous current generation in defective junctions (Ohyu and Hiraiwa, 1992). Crystalline microdefects, such as oxide and metallic precipitates, locally enhance the electric field in the depletion layer around them because the dielectric constant of a defect differs from that of the crystalline Si. Circular metallic precipitates enhance the field by a factor of as much as 3; oxides, by a factor of as much as 1.4. The enhanced electric field causes band-to-band (valence band to conduction band) tunneling of electrons, which is known as the Zener effect (Zener, 1934). The current generated is expressed as

$$I_{\text{Zener}} = AV \exp(-B/\xi F), \tag{17}$$

where V is the volume of the microdefect, F is the external field at the microdefect, and ξ is the field enhancement factor. B is a physical constant that depends on the Si band gap and tunneling effective mass and is experimentally given as 1.9×10^7 V/cm (Hurkx *et al.*, 1992). The local Zener current largely depends on the electric field at the defect and only slightly depends on temperature through the term $E_g^{2/3}$ involved in B.

3.3.3.2 Diffusion The diffusion current is fed by minority carriers (electrons in the case of n^+-p^- junction) that are generated in the bulk Si (p^- in the case of n^+-p^- junction) within a diffusion length L_n from the depletion-layer edge and flow into the depletion layer. It is given by Shockley (1949) as

$$I_{diff} = S_j q \frac{L_n}{\tau_n} \frac{n_i^2}{N_A}, \tag{18}$$

where S_j is the area of the depletion layer, N_A is the dopant concentration in the bulk Si, and τ_n is the minority-carrier lifetime in the bulk Si within L_n from the depletion-layer edge, given by $\tau_n = 1/(\sigma_n v_{th} N_t)$.

The diffusion current as a function of temperature T is mostly determined by the term $n_i^2 \propto \exp(-E_g/kT)$ of Eq. (18) because the terms L_n and τ_n change slowly with T. Accordingly, the activation energy of the diffusion current is almost equal to that of n_i^2 ($E_g = 1.12$ eV). The magnitude of this component is proportional to the area of the junction and becomes larger than that of the SRH-process current at temperatures above 100°C.

4. GENERATION AND PASSIVATION OF INTERFACE TRAPS

4.1 Physical Origins

Interface traps are produced and annihilated by various processes during LSI fabrication and device operation. The processes are classified into two groups: physical and chemical. The former is driven by energetic beams. From a quantum mechanical point of view, the energy is carried by particles that are classified according to their mass as follows; photon (no mass), electron or hole, and atom or molecule. The results depend not only on the kind of particles but also on their energy, and the results are described separately for the respective particle and energy.

4.1.1 Photon Beams

4.1.1.1 Below 8 eV Photons of this energy range excite electrons in Si and inject them into SiO_2 (internal photoemission, IPE), thus causing a negative oxide charging similar to that caused by electron injection (Williams, 1965). Emptying of the trapped carriers is also observed. Current–voltage measurements of the IPE electrons through the oxide (called the photo I-V method) are effective for investigating the charge distribution in the film (Berglund and Powell, 1971; DiMaria, 1976). The amount of IPE electrons is mostly limited by the local field near the electron-emitting electrode. Because the field is affected by the oxide charge, we can derive the charge distribution from the difference between the gate voltages that correspond to the same IPE current before and after the operation of interest. Letting ΔV_G^+ and ΔV_G^- be the respective voltage shifts for the electron emissions from the Si (positive V_G) and the gate (negative V_G), we can express the actual charge change ΔQ and the center $\bar{x}$ (centroid) of the charge distribution averaged by the moment by

$$\Delta Q = \frac{\epsilon_S}{d}(\Delta V_G^- - \Delta V_G^+), \tag{19}$$

and

$$\bar{x} = d\Delta V_G^+/(\Delta V_G^+ - \Delta V_G^-). \tag{20}$$

The charge change ΔQ and centroid $\bar{x}$ are related to the difference of distributions $\Delta\rho(x)$ (equal to $\rho_{\text{after}} - \rho_{\text{before}}$) by the equations

$$\Delta Q = \int_0^d \Delta\rho(x)\,dx,$$

and

$$\bar{x} = \frac{1}{\Delta Q}\int_0^d x\Delta\rho(x)\,dx.$$

4.1.1.2 From 8 eV to 470 keV Vacuum ultraviolet light (VUV, 5 eV to 1 keV), x rays (100 eV to 100 keV), and some γ rays (>10 keV) are classified here. VUV photons are produced by the plasma used for depositing films and for etching the selectively masked films to produce LSI patterns. Gamma rays are emitted from nuclear reactors and from the outer walls of artificial satellites irradiated by cosmic rays (beta rays) through the process of bremsstrahlung.

The most fundamental process involved is electron–hole pair generation in the bulk SiO$_2$. The holes thus generated are trapped at oxide defects (presumably oxygen vacancies), producing paramagnetic defects associated with positive charges (Lenahan and Dressendorf, 1984):

$$O_3{\equiv}Si{-}Si{\equiv}O_3 + h^+ \rightarrow O_3{\equiv}Si\cdot + {}^+Si{\equiv}O_3. \tag{21}$$

These defects are called E'_γ centers (g = 2.0004) and are shown schematically in Fig. 14(b). When they result from a hard x ray that penetrates the entire film, both the E'_γ centers and positive charges are located close to the SiO$_2$–Si interface (Lenahan and Dressendorfer, 1984). This supports the hypothesis that the hole-trapping defects are O vacancies whose density is enhanced by strain (Ohmameuda *et al.*, 1991). A soft x ray, on the other hand, produces the paramagnetic defects only in the penetrated surface region of the film (Yokogawa *et al.*, 1990). In this case the defects show up only after an extended x-ray irradiation, suggesting that there are fewer O vacancies in the region far from the SiO$_2$–Si interface. The

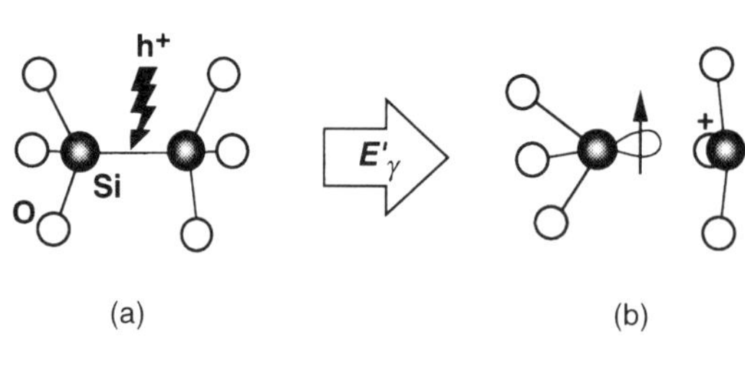

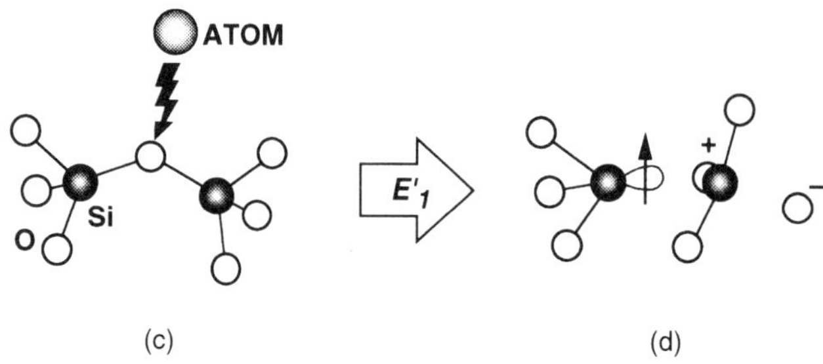

FIG. 14. Schematic illustrations of (a), (c) the precursors and (b), (d) the ESR-active states of the E' centers in SiO$_2$. (b) corresponds to the E'_γ center, and (d) to the E'_1 center. (After Warren *et al.*, 1992.)

positive charges caused the negative shift of the high-frequency *CV* curves shown in Fig. 7. The positive charging increases almost linearly with the photon dosage, and more charges were produced for the positive gate bias. This positive bias effect is ascribed to the enhanced drifting of holes to the interface.

Interface-trap generation is closely connected with E'_γ-center formation. Atomic hydrogen is released following electron trapping at water-related traps (O_3Si–H in "dry" oxide and O_3Si–OH in wet oxide) by the reaction

$$O_3{\equiv}Si{-}H + e^- \rightarrow O_3{\equiv}Si{:}^- + H^0$$

or

$$O_3{\equiv}Si{-}OH + e^- \rightarrow O_3{\equiv}Si{-}O{:}^- + H^0. \quad (22)$$

We should notice that a nominally dry oxide contains an appreciable amount of water (from 10^{17} to 10^{20} H atoms/cm^3 in the form of ≡Si–OH or ≡Si–H) as reported by Revesz (1979). The atomic hydrogen H^0 dimerizes by the reaction

$$H^0 + H^0 \rightarrow H_2 + 4.5\ \text{eV} \quad (23)$$

and forms hydrogen molecules that diffuse to the interface. The H_2 molecule reacts with the E'_γ centers, presumably releasing H^0 again:

$$O_3{\equiv}Si\cdot + H_2 \rightarrow O_3{\equiv}Si{-}H + H^0. \quad (24)$$

Reactions (22) to (24) lead to hydrogen migration from the bulk SiO_2 to the interface, which was demonstrated by Gale *et al.* (1983) in the case of avalanche injection of electrons.

The atomic hydrogen finally produces the interface traps. Considering that more interface traps are generated under a positive bias, an electronic involvement by the reaction

$$Si_3{\equiv}Si{-}H + H^0 + e^- \rightarrow Si_3{\equiv}Si{:}^- + H_2 \quad (25)$$

is likely in the interface-trap formation (Griscom, 1985). Some hole involvement other than reaction (21), however, cannot be ruled out (see Sec. 4.1.2.1).

In the case of soft (nonpenetrating) x rays, on the other hand, the atomic hydrogen seems to be supplied to the interface both by the direct reaction (22) throughout the oxide and by reactions (21) to (24) in the surface region penetrated by the x rays. Winokur and Sokoloski (1976) reported a similar interface-trap buildup for both hard and soft x rays.

The number of interface traps thus produced is comparable to the number of positive charges. As shown in Fig. 8, the trap distribution has a peak 0.1–0.3 eV above the midgap.

For photons of energies close to the resonance absorption, the diffusion of excitons also seems responsible for the interface-trap generation (Weinberg and Rubloff, 1978). At higher energies up to 50 keV, the photoelectric effect is predominant, and electrons are scattered away from the bound atoms, completely absorbing the photons. For photons with energies larger than 50 keV, though, Compton scattering is the predominant effect. In this process, photons only lose part of their energy in the first scattering event and continuing scattering until all of it is absorbed. Thus the resulting phenomenon is essentially the same as in the energy range below 50 keV. The electron scattering is followed by emission of Auger electrons or characteristic x rays.

4.1.1.3 Above 470 keV In this region, at least a part of the electrons scattered by the Compton effect subsequently displace the atoms in SiO_2. For energies greater than 1 MeV, electron–positron pair generation takes place, although its frequency is much less than that of the Compton effect. Before the phenomena resulting from atomic displacement or pair generation become dominant, interface degradation caused by electron–hole pair generation occurs. Accordingly, from the practical point of view, the device degradation mechanism is the same as that with photons of lower energies.

4.1.2 Electron and Hole Flux

4.1.2.1 Below 10 eV The second group of physical origins that affect the interface is electrons and holes. The drift electrons of MOS transistors (or holes in *p*-MOS transistors) are confined outside of the SiO_2 because of the SiO$_2$–Si potential barrier: 3.1 eV (3.8 eV for holes). Energetic electrons and holes (called hot carriers) are produced by impact ionization near the drain region of the Si substrate, where a large electric field exists (see Sec. 3.1). These carriers pass the barrier, are trapped in the SiO$_2$, and generate interface traps. Additionally, in flash memories where information is stored as negative charges in an electrically isolated gate, electrons are forced to flow through the oxide by a high field.

To study this trapping phenomenon, Nicollian *et al.* (1969) developed the avalanche-injection method. They repeatedly applied a large-amplitude ac voltage to the gate of MOS capacitors. This voltage capacitively induces a deep depletion layer accompanied by a high field in the Si substrate, causing impact ionization and subsequent injection (driven by the dc component of the gate voltage) of hot carriers into the oxide. This method has advantages over other techniques, such as high-field stressing (HFS): Only electrons or holes are injected into the film, and the field applied to the film may be suppressed below 3 MV/cm.

The bulk trapping and interface-trap generation depends on the kind of carriers (holes or electrons) and on the way they are introduced. As shown in Fig. 15, avalanche-injected holes produce numbers of positive charges that increase monotonically with increase in the number of holes injected into the SiO$_2$. The efficiency of positive-charge generation is independent of the magnitude of the applied field (Schwerin *et al.*, 1990). Postoxidation annealing (POA) enhances the positive charging, and this enhancement is ascribed to an increase in the number of O vacancies. The number of interface traps is also increased by hole injection, and there is a linear relationship between the number of traps generated and the number of positive charges. Different values of the ratio of interface traps to holes are reported: 1 (Al gate) and 0.4 (poly-Si gate) by Thanh *et al.* (1986) and 0.05 (poly-Si gate) by Schwerin *et al.* (1990). The interface traps are mostly located from 0.1 to 0.3 eV above the midgap in the Si band gap. We cannot understand this interface-trap generation in the framework of the mechanism described in Sec. 4.1.1.2 because the negligible electron fluence involved means that atomic or molecular hy-

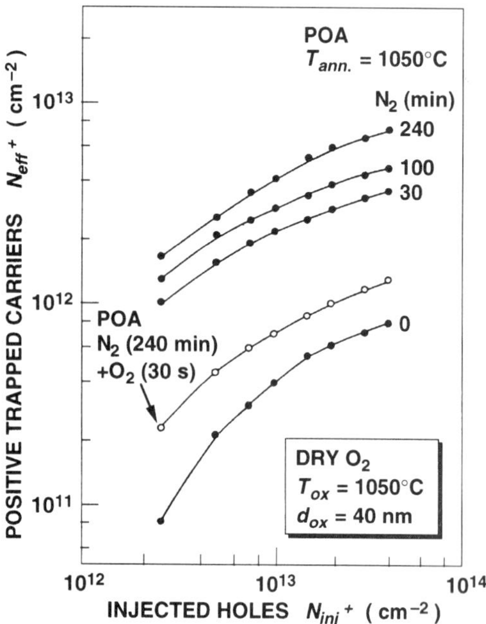

FIG. 15. Density of positive trapped carriers as a function of avalanche-injected hole fluence. The samples were treated by POA in N$_2$ for different periods. Reprinted from Thanh *et al.* (1986), with kind permission from Elsevier Science Ltd.

drogen is not supplied to the interface. The two-step model by Lai (1983) seems more likely to apply here. A strained Si–O–Si bond first captures a hole, which breaks the Si–O bond, and the oxygen atom thus freed moves to a position of lower strain. The trapped hole eventually captures an electron, restoring the charge neutrality, but the oxygen atom does not return to its original position. This leads to a dangling bond or a weak bond, resulting in an interface trap as depicted by Sakurai and Sugano (1981). This process might be at least partly responsible for the interface traps generated by photon irradiation.

In the case of avalanche injection of electrons, negative charges first build up and then decrease with extended injection (Fig. 16). This is the so-called turnaround effect. The fluence of electron injection necessary to produce an appreciable charging is about three orders of magnitude larger than that necessary when holes are injected. The efficiency of the electron trapping is appreciably enhanced by oxide fields above 4 MV/cm (Heyns *et al.*, 1989). This is in contrast to

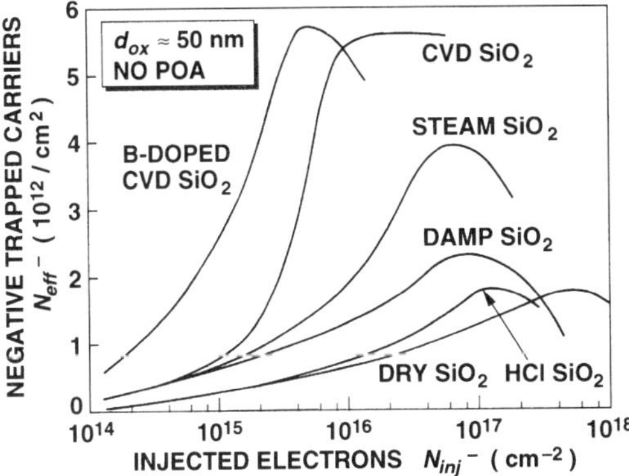

FIG. 16. Density of negative trapped carriers as a function of avalanche-injected electron fluence. Oxidation was performed at 1000 °C in the ambient of dry O_2, dry O_2 + 3% HCl (HCl SiO_2), dry O_2 + 3% H_2O (DAMP SiO_2), or steam. CVD oxide films were deposited using SiH_4 + O_2 at 800 °C. B-doped films were doped with 5 wt % B_2O_3. (After Gdula, 1976.)

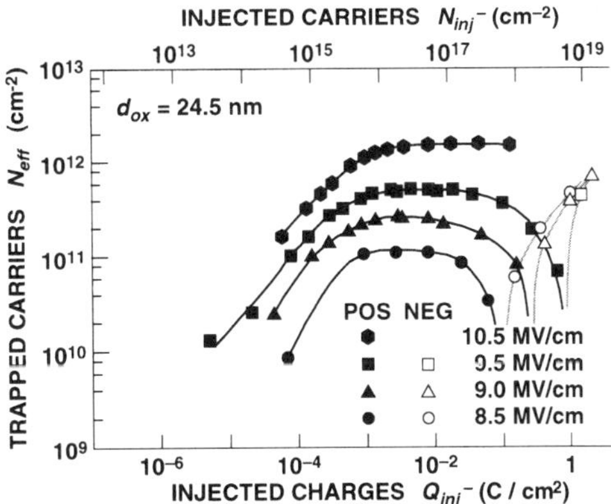

FIG. 17. Density of trapped carriers as a function of injected electron fluence. The electrons were introduced into SiO_2 driven by a high field. Closed symbols are for positive net charge, and open symbols are for negative net charge. (After DiMaria *et al.*, 1993.)

what is observed in hole injection where the trapping efficiency was not enhanced (Schwerin *et al.*, 1990). A field of 4 MV/cm in the oxide produces hot electrons with the average energy larger than the 2.3-eV threshold for hydrogen release (DiMaria and Stasiak, 1989). The negative charging is more remarkable in wet oxides. From this we can see that the water-related species, such as Si–H and Si–OH, are possible electron traps. The decreased negative charging after an extended injection is due to the buildup of positive charges that are presumably injected from the electrode. Here again the relationship between the generated interface traps and negative charges is almost linear. In poly-Si gates the ratios of traps to charges are from 0.2 to 1 (Thanh *et al.*, 1986; Heyns *et al.*, 1989). The interface-trap distribution reaches a peak from 0.1 to 0.3 eV above the Si midgap for fields above 4 MV/cm, below which a broad distribution in the Si band gap is reported.

In the case of high-field stressing, most of the current that flows through the SiO_2 is fed by electrons, regardless of the biasing polarity. Nonetheless, the net charge is initially positive and turns negative only after an extended stressing (Fig. 17). This is due to the trapping of holes that are generated by band-to-band ionization in the oxide. The electron fluence where the positive charging begins to saturate agrees well with that of the appreciable negative charging shown in Fig. 16. This coincidence indicates that the saturation of positive charging is due to charge compensation by the buildup of negative charges. Higher fields increase the fraction of holes in the current and enhance the efficiency of the positive charging. DiMaria *et al.* (1993) reported that the number of interface traps is about 1/10 of that of the positive charges in the early stage of trap generation.

4.1.2.2 Above 10 eV Plasma etching for LSI pattern formation is a source of electrons with energies from 10 eV to a few hundred electron volts. The thick oxide for device isolation is exposed to this irradiation. Electron-beam (EB) lithography is another source of high-energy electron irradiation (20–30 keV). Because the oxide is covered by an EB-sensitive polymer film during the irradiation, the interface degradation in this process is mostly caused by the low-energy secondary electrons produced by the incident electrons.

Electrons with energies from 10 eV to a few hundred electron volts excite electron–hole (e-h) pairs that generate interface traps and bulk positive charges in the same way that photons do, and the interface-trap distribution in the Si band gap again peaks from 0.1 to 0.3 eV above the Si midgap. Yunogami and Mizutani (1993) reported the following processes specific to the electronic irradiation: plasmons first produced at the surface or in the bulk of SiO_2 decay, resulting in e-h pair generation. All the e-h pairs are not produced by plasmons, but oscillations specific to the plasmon energies are clearly observed when the number of generated interface traps or oxide charges is plotted as a function of the incident electron energy.

Electrons with energies greater than a few hundred kiloelectron volts displace the constituent atoms of the oxide, resulting in phenomena similar to the atomic displacement in Sec. 4.1.3.

4.1.3 Atomic and Molecular Beams Atoms introduced by ion implantation displace (elastic scattering) and ionize (inelastic scattering) the atoms in SiO_2. The elastic scattering produces paramagnetic (ESR active) defects called E_1' centers. These defects are the products of O recoil from the network $O_3{\equiv}Si{-}O{-}Si{\equiv}O_3$, and consist of ESR-active $O_3{\equiv}Si\cdot$ and positively charged $^+Si{\equiv}O_3$ [Fig. 14(d)]. The positive charges that are observed in the implanted SiO_2 even after a thermal annealing might be due to these E_1' centers. The inelastic scattering, on the other hand, produces electron–hole pairs that cause charging and trap generation in a way similar to the way photon, electron, and hole injection does.

The ion implantation also produces interface traps, and the number of traps remaining after 900 °C annealing is determined solely by the energy dissipated in the nuclear stopping events (Fig. 18). This figure clearly shows that the interface traps are produced by the displacement of host (SiO_2) atoms by nuclear collisions with the impinging and recoiled atoms. We also found that the implanted oxygen atoms restore the broken bonds at the interface to the state before the implantation (Hiraiwa, to be published). But because this repairing process was not observed when Ar, Ne, and C ions were implanted, it seems to be a chemical process.

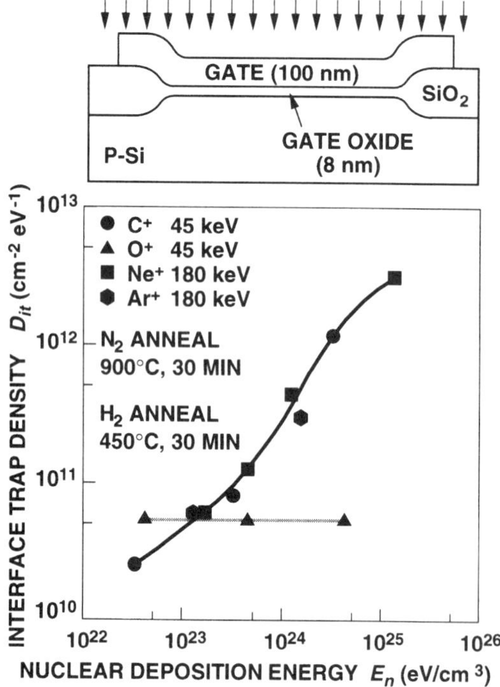

FIG. 18. Midgap interface-trap density of ion-implanted SiO_2 after N_2 annealing at 900 °C. The samples received the ion implantation after the formation of phosphorus-doped Si film as a gate. The ion dose is normalized by the total energy that the ions dissipated through a nuclear stopping process. (After Hiraiwa, to be published.)

4.2 Chemical Origins

4.2.1 Thermal Annealing The results of thermal annealing are somewhat controversial, probably because the purity of the annealing ambient and of the sample itself is not always controlled. The traditional furnace usually contains not a small amount of moisture (as impurity) from the environment, and the source-gas purity is easily degraded if the piping is not prepared and operated carefully. And because even a trace amount of moisture (10 ppm) passivates the interface traps, we should take special care to ensure the purity of the ambient when we study annealing. Figure 19 shows that in a high-purity nitrogen annealing, the number of P_b centers (some of the interface traps, see Sec. 2.3.2.4) increases during annealing times shorter than 5 min. This increase is attributed to the desorption of hydrogen atoms from Si–H bonds that are easily broken above 500 °C by the reaction

$$Si{-}H \rightarrow Si\cdot + H^0. \tag{26}$$

Brower *et al.* (1990) obtained the rate equation as

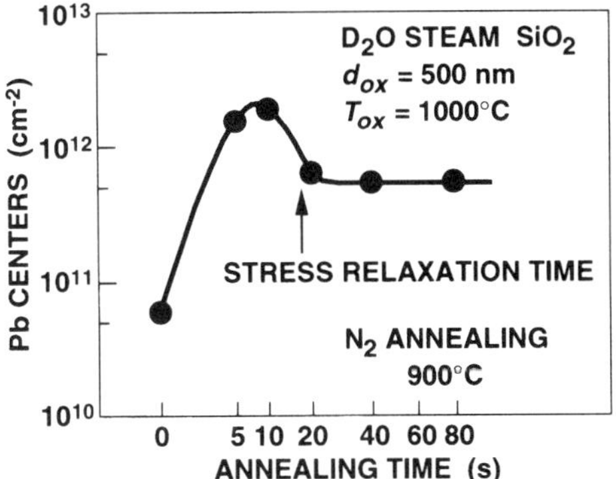

FIG. 19. Density of P_b centers as a function of annealing time. After Ohji *et al.* (1990); © 1990 IEEE.

$$\frac{d[\text{Si–H}]}{dt} = -k_d[\text{Si–H}], \tag{27}$$

where

$$k_d = (1.2 \times 10^{12}\ \text{s}^{-1}) \exp\left(-\frac{2.56\ \text{eV}}{kT}\right).$$

The number of P_b centers decreases during an extended annealing, showing thermal passivation in the absence of hydrogen-related species. As a result of this thermal passivation effect, a plot of the number of P_b centers as a function of annealing temperature has a peak (near 800 °C in Fig. 20). Montillo and Balk (1971) reported the direct thermal passivation of the interface traps, but some involvement of hydrogen (or deuterium) supplied from the wet or nominally dry oxide itself cannot be ruled out.

Not only hydrogen desorption but also the thermal decomposition of SiO$_2$ seems to contribute to the generation of interface traps. This decomposition is enhanced by the catalytic action of metallic impurities, such as Cu, W, and Pt. In extreme cases, portions of contaminated oxide films disappear during high-temperature ultrahigh-vacuum annealing (Tromp *et al.*, 1985). The decomposition proceeds at the SiO$_2$–Si interface by the reaction Si + SiO$_2$ → 2 SiO ↑. The feasibility of interface-trap generation by this process is supported by the result of an experiment in which one Cu atom present during the oxidation produced one interface trap on the average (Hiraiwa and Itoga, 1994).

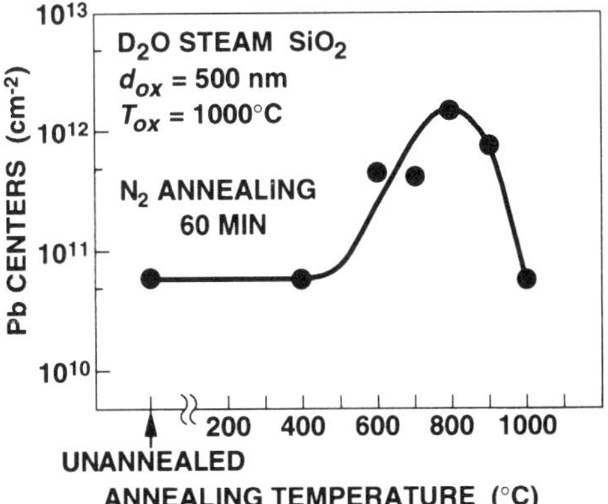

FIG. 20. Density of P_b centers as a function of annealing temperature. Reproduced from Ohji *et al.* (1990); © 1990 IEEE.

4.2.2 Hydrogen Annealing In LSI processes the interface traps are electrically passivated by annealing in the final step of device fabrication. This annealing must be done below 450 °C, however, because the Al wirings are already formed, and we cannot expect thermal passivation to occur at these low temperatures (Sec. 4.2.1). Thus, we usually terminate the interfacial dangling bonds by exposing the samples to a hydrogen-containing ambient.

The results of hydrogen annealing depend on whether a gate electrode is formed or not. Figure 21 shows the case without the gate, which is a kind of postoxidation annealing. According to Brower and Myers (1990) again, the hydrogen molecules passivate the interface traps by the reaction

$$\text{Si}\cdot + \text{H}_2 \rightarrow \text{Si–H} + \text{H}^0, \tag{28}$$

where

$$\frac{d[\text{Si}\cdot]}{dt} = -k_f[\text{Si}\cdot][\text{H}_2],$$

$$k_f = (1.94 \times 10^{-6}\ \text{cm}^3\,\text{s}^{-1}) \exp\left(-\frac{1.66\ \text{eV}}{kT}\right). \tag{29}$$

We should take care that reactions (26) and (28) are reversible. In the case of H$_2$ annealing, H^0 is so scarce that reverse reactions are not likely. In ample atomic hydrogen, on the other hand, reverse reactions take place. Thus, we observe both passivation and de-

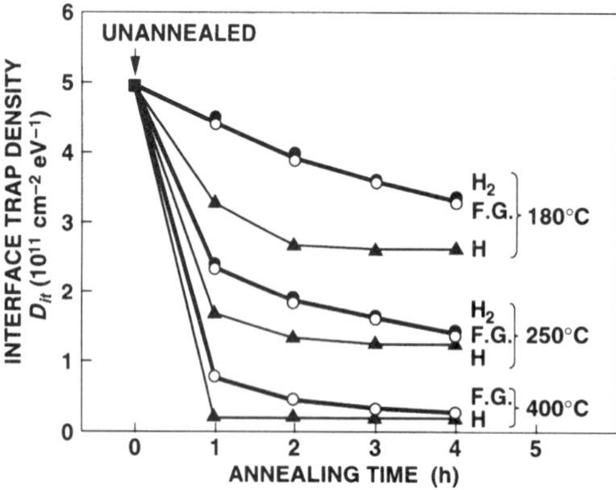

FIG. 21. Interface-trap density at midgap as a function of annealing time at various temperatures in various ambients: H_2, molecular hydrogen; F. G., forming gas (10% H_2 in N_2); and H, atomic hydrogen produced by a high-voltage discharge. (After Thanh and Balk, 1988.)

passivation of the interface traps depending on the temperature and initially passivated state (Thanh and Balk, 1988). The result is more interface traps at lower temperatures, which in turn leads to more interface degradation occurring as a result of radiation or carrier injection (where the atomic hydrogen is released from the bulk SiO$_2$, as described in Sec. 4.1) because the interface is kept at low temperatures, except during annealings.

The results of postmetallization annealing differ from those of POA because of the presence of the gate. For both polycrystalline Si (poly-Si) and metals such as Al, even annealing in an ambient without any hydrogen-containing molecules reduces the number of interface traps. This reduction is ascribed to hydrogen atoms bound to crystalline defects (such as grain boundaries and dislocations) in the poly-Si gate. In an Al gate, the number of interface traps is reduced much more than in a poly-Si gate, possibly because the Al electrode reduces Si–OH that was incorporated into SiO$_2$ during oxidation or adsorbed onto the film before Al metallization (Reed and Plummer, 1988).

Fluorine atoms also terminate the interface traps. They are much larger than hydrogen atoms and require a higher temperature to reach the SiO$_2$–Si interface. Because the energy of the Si–F bond is greater than that of the Si–H bond, the F passivation is more stable during thermal treatments. Accordingly, the introduction of F atoms before oxidation is also effective in passivating the interface traps. The F passivation also makes the interface more resistant to the effects of radiation and hot-carrier injection (Nishioka *et al.*, 1989). We should take care, however, that we do not increase the number of traps and decrease the dielectric constant of the oxide by introducing too many F atoms.

5. BREAKDOWN OF SiO$_2$

Destructive breakdown of the SiO$_2$ film occurs when a large electric field is applied to MOS capacitors. The breakdown field differs among capacitors (Yamabe and Taniguchi, 1985). When the capacitor area is smaller, we observe more breakdown events at the maximum field (greater than 10 MV/cm). From these results many researchers consider that SiO$_2$ has its own (intrinsic) dielectric strength and that the breakdown is triggered by some structural defects in the SiO$_2$.

Even if the SiO$_2$ field is smaller, the breakdown occurs when the field is applied for an extended period. In the estimation of breakdown reliability, we usually accelerate the breakdown by increasing the field far beyond the daily-use level because we cannot wait for 10 years (certified lifetime of the devices). To predict the daily-use reliability from the accelerated testing, an understanding of the breakdown mechanism is indispensable. There is little doubt about the view that the intrinsic breakdown is caused by the current that is induced by the field and flows through the SiO$_2$. Although most of the current is carried by electrons, it seems that holes are deeply involved in the breakdown event. This is because the total amount of holes that pass through the SiO$_2$ until breakdown occurs is constant independent of the applied field, whereas the electron fluence varies (Chen *et al.*, 1986). The holes are considered to be generated in the oxide by impact ionization or to be injected into the oxide from the anode by tunneling with possible assistance of excitation by electrons (incident from SiO$_2$). DiMaria *et al.* (1993), on the other hand, reported that the breakdown is correlated with the bulk-SiO$_2$ and SiO$_2$–Si-interface trapping. In any case, however, the fundamental processes of the intrinsic breakdown are still beyond our understanding.

From a practical point of view, the intrinsic dielectric strength is enough for MOS transistor operation because the SiO_2 field is no more than 4 MV/cm. Thus, most of the electronics engineers are concerned about how to reduce the oxide defects. There are many studies that have investigated oxide-defect formation in relation to MOS capacitor processing conditions. Through these studies Si-crystal defects (such as oxide precipitates in Czochralski-grown Si) are found to cause the breakdown (Itsumi *et al.*, 1992). It is also reported that metallic contamination degrades the SiO_2 by lowering the potential barrier of SiO_2 (Takiyama *et al.*, 1994). However, it is still unclear what properties make the oxide defects different from the intrinsic region.

6. SUMMARY

The transition from bulk Si to SiO_2 seems to be sharp, extending for only one or two atomic layers, but it is not as flat as it appears in TEM images: the actual morphology of the interface presumably has undulations corresponding to more than four atomic layers. Defects at the interface act as carrier traps, some of which are paramagnetic and are detected as P_b centers by ESR measurement. The P_b center is ascribed to a dangling bond of an interfacial Si atom that is back-bonded to three other Si atoms. These centers, however, form only some of the interface traps, and the rest of the traps have not been identified.

The properties of the SiO_2–Si interface are crucial to MOS transistor performance that is characterized by the subthreshold swing and the drain current. The increase in subthreshold swing is caused by the interface traps and reduces the switching speed of the MOS transistor itself. The drain current, on the other hand, determines the switching time of an LSI circuit after the onset of the "on" state of MOS transistors. The interfacial undulation reduces the drain current because surface-roughness scattering reduces the carrier mobility. The interface traps, as well as bulk-oxide traps, reduce the drain current mostly by reducing the number of surface carriers. On the other hand, the interface traps generate a current by the SRH process in the depleted region of junctions, and this leads to an excessive junction leakage current that causes data-retention errors of DRAMs.

The interface traps are generated by ionizing radiation and carrier injection, and thermal annealing not only passivates but also generates the traps. Many researchers consider the generation kinetics in the following way, although other explanations cannot be ruled out. Some of the interface traps are generated by the release of terminating hydrogen from Si–H bonds. This proceeds either thermally ($Si\text{–}H \rightarrow Si\cdot + H^0$) or by a reaction with atomic hydrogen ($Si_3{\equiv}Si\text{–}H + H^0 \rightarrow Si_3{\equiv}Si\cdot + H_2$). One of the three processes that produce the atomic hydrogen is the electron trapping at water-related traps, such as Si–H and Si–OH bonds. Another is the thermal decomposition of these bonds. The other is the cracking of hydrogen molecules at oxide defects known as E' centers. The hydrogen molecules are supplied by the dimerization of atomic hydrogen or by diffusion from the ambient. The E' centers are formed by hole trapping at oxygen vacancies that are concentrated where the SiO_2 network is strained, such as in the interfacial region. Interface traps might also be generated by a two-step process in which a hole is first trapped at the strained Si–O–Si bond and then captures an electron, thus leaving a weak or dangling bond. The electrons and holes thus involved in the trap generation are injected from the Si substrate or electrode. They are also generated in pairs by ionizing radiation, or band-to-band ionization under a high field. The passivation of the interface traps proceeds by reactions that are the reverse of those generating the traps, and these passivation reactions are facilitated by high temperature and by ambient atmospheres containing hydrogen molecules. The annealing of the traps at high temperatures is also observed even when hydrogen is not present.

Destructive breakdown of SiO_2 occurs when a high field is applied. Even a lower field causes breakdown after an extended application. The breakdown is caused by the current that is induced in the SiO_2 by the field. Although most of the current is carried by electrons, it seems that holes are deeply involved in the breakdown. The correlation of the breakdown with bulk-SiO_2 and SiO_2–Si-interface trapping is also reported. Structural defects existing in the SiO_2 trigger the breakdown and reduce the lifetime. The physical reality of the defects, however, is still unclear.

Many problems concerning the SiO_2–Si interface remain unsolved despite many studies. The increasing number of damage processes, and the growing need for a better interface (such as is required for low-power DRAMs), will spur further studies of the SiO_2–Si interface.

ACKNOWLEDGMENTS

The author acknowledges valuable suggestions on MOS transistor operation from Shin'ichiro Kimura and Ryo Nagai. The section on junction leakage current owes much to discussions with Kiyonori Ohyu, and many thanks are owed to Kikuo Watanabe, Masataka Kato, Takashi Yunogami, Yuzuru Ohji, and Ken'etsu Yokogawa for their discussions of the oxide traps.

GLOSSARY

Avalanche Injection: One of the methods to introduce carriers from the Si substrate into the SiO_2 film. The carriers are produced by impact ionization in the Si depletion layer, induced by a large-amplitude ac gate voltage. They are subsequently driven into the oxide by the dc component of the gate voltage. The field across the oxide can be kept below 3 MV/cm.

***CV* Method:** A method to quantify bulk oxide charges and interface traps by measuring the capacitance of MOS capacitors as a function of the gate voltage.

ESR: Electron-spin resonance measurement. From the resonance absorption line, we can get information about the structure of dangling bonds.

Interface Trap: Carrier trap that exists at the SiO_2–Si interface and exchanges a charge with the Si surface. Some are ascribed to the dangling bond of an interfacial Si back-bonded to three other Si atoms.

LSI: Large-scale integration or large-scale–integrated circuit. Currently more than 30 million components can be integrated in a single chip.

MOS: A structure consisting of stacked metal, oxide, and semiconductor layers (substrate). In the strict sense of the word, most present devices do not have the MOS structure, but they do have the SOS (semiconductor-oxide-semiconductor) structure because they have polycrystalline silicon gates heavily doped—with P in *n*-channel devices or with B in some *p*-channel devices.

SRH Process: The process that restores the carrier density to its equilibrium value, assisted by traps that are energetically located in the Si band gap. This process generates carriers and causes an excess leakage current in a reversely biased junction. Named after Shockley, Read, and Hall, who formulated the process.

Works Cited

Ando, T. (1977), *J. Phys. Soc. Jpn.* **43,** 1616–1626.

Bardeen, J., Brattain, W. H. (1948), *Phys. Rev.* **74,** 230.

Berglund, C. N. (1966), *IEEE Trans. Electron Devices* **ED-13,** 701–705.

Berglund, C. N., Powell, R. J. (1971), *J. Appl. Phys.* **42,** 573–579.

Brower, K. L., Myers, S. M. (1990), *Appl. Phys. Lett.* **55,** 162–164.

Brugler, J. S., Jespers, P. G. A. (1969), *IEEE Trans. Electron Devices* **ED-16,** 297–302.

Cartier, E., Stathis, J. H., Buchanan, D. A. (1993), *Appl. Phys. Lett.* **63,** 1510–1512.

Castagné, R., Vapaille, A. (1971), *Surface Sci.* **28,** 157–193.

Chen, I.C., Holland, S., Young, K. K., Chang, C., Hu, C. (1986), *Appl. Phys. Lett.* **49,** 669–671.

DiMaria, D. J. (1976), *J. Appl. Phys.* **47,** 4073–4077.

DiMaria, D. J., Stasiak, J. W. (1989), *J. Appl. Phys.* **65,** 2342–2356.

DiMaria, D. J., Cartier, E., Arnold, D. (1993), *J. Appl. Phys.* **73,** 3367–3384.

Ezawa, H., Kawaji, S., Nakamura, K. (1971), *Jpn. J. Appl. Phys.* **13,** 126–155.

Gale, R., Feigl, J., Magnee, C. W., Young, D. R. (1983), *J. Appl. Phys.* **54,** 6938–6942.

Gdula, R. A. (1976), *J. Electrochem. Soc.* **123,** 42–47.

Goodnick, S. M., Ferry, D. K., Wilmsen, C. W., Liliental, Z., Fathy, D., Krivanek, O. L. (1985), *Phys. Rev. B* **32,** 8171–8186.

Griscom, D. L. (1985), *J. Appl. Phys.* **58,** 2524–2533.

Grunthaner, F. J., Grunthaner, P. J. (1986), *Materials Science Reports 1*, Amsterdam: North Holland, pp. 65–160.

Haddara, H., Christoloveanu, S. (1986), *Solid-State Electron.* **29,** 767–772.

Hall, R. N. (1952), *Phys. Rev.* **87,** 387.

Heyns, M. M., Krishna Rao, D., De Keersmaecker, R. F. (1989), *Appl. Surf. Sci.* **39,** 327–338.

Himpsel, F. J., McFeely, F. R., Taleb-Ibrahimi, A., Yarmoff, J. A., Hollinger, G. (1988), *Phys. Rev. B* **38,** 6084–6096.

Hiraiwa, A., Itoga, T. (1994), *IEEE Trans. Semicond. Manufact.* **7,** 60–67.

Hurkx, G. A. M., Graaff, H. C., Kloosterman, W. J., Knuvers, M. P. G. (1992), *IEEE Trans. Electron Devices* **ED-39,** 2090–2098.

Itsumi, M., Nakajima, O., Shiono, N. (1992), *J. Appl. Phys.* **72,** 2185–2191.

Iunovich, A. E. (1958), *Sov. Phys. Tech. Phys.* **3,** 646–650.

Kahng, D., Atalla, M. M. (1960), *IRE-IEEE Solid-State Device Research Conference, Carnegie Institute of Technology, Pittsburgh, PA,* New York: IEEE.

Kawaji, S. (1969), *J. Phys. Soc. Jpn.* **27,** 908–910.

Kircher, C. J. (1975), *J. Appl. Phys.* **46,** 2167–2173.

Lai, S. K. (1983), *J. Appl. Phys.* **54,** 2540–2546.

Lenahan, P. M., Brower, K. L., Dressendorfer, P. V., Johnson, W. C. (1981), *IEEE Trans. Nucl. Sci.* **NS-28,** 4105–4106.

Lenahan, P. M., Dressendorfer, P. V. (1984), *J. Appl. Phys.* **55,** 3495–3499.

Lilienfeld, J. E. (1926), U.S. Patent 1,745,175 (filed in 1926), 1,877,140 (filed in 1928), 1,911,018 (filed in 1928).

Ma, T. P., Dressendorfer, P. V. (Eds.) (1989), *Ionization Radiation Effects in MOS Devices and Circuits,* New York: Wiley.

Matsumoto, Y., Uemura, Y. (1974), *Jpn. J. Appl. Phys.* **Suppl. 2,** Pt. 2, 367–370.

Montillo, F., Balk, P. (1971), *J. Electrochem. Soc.* **118,** 1464–1468.

Nagayama, H., Honda, H., Kawahara, H. (1988), *J. Electrochem. Soc.* **135,** 2013–2016.

Nakashima, S., Katayama, T., Miyamura, Y., Matsuzaki, A., Imai, M., Izumi, K., Ohwada, N. (1994), "Thickness Increment of Buried Oxide in a SIMOX Wafer by High-Temperature Oxidation," *Proceedings of the 1994 IEEE International SOI Conference, Nantucket Island, Massachusetts,* New York: IEEE, pp. 71–72.

Nicollian, E. H., Brews, J. R. (1982), *MOS (Metal Oxide Semiconductor) Physics and Technology,* New York: Wiley.

Nicollian, E. H., Goetzberger, A., Berglund, C. N. (1969), *Appl. Phys. Lett.* **15,** 174–177.

Ning, T. H., Osburn, C. M., Yu, H. N. (1975), *Appl. Phys. Lett.* **26,** 248–250.

Nishioka, Y., Ohyu, K., Ohji, Y., Natsuaki, N., Mukai, K., Ma, T. P. (1989), *J. Appl. Phys.* **66,** 3909–3912.

Niwano, M., Katakura, H., Takeda, Y., Takakuwa, Y., Miyamoto, N., Hiraiwa, A., Yagi, K. (1991), *J. Vac. Sci. Technol.* **A9,** 195–200.

Ohdomari, I., Mihara, T., Kai, K. (1986), *J. Appl. Phys.* **60,** 3900–3904.

Ohji, Y., Nishioka, Y., Yokogawa, K., Mukai, K., Qiu, Q., Arai, E., Sugano, T. (1990), "Effects of Minute Impurities (H, OH, F) on SiO_2/Si Interface as Investigated by Nuclear Resonant Reaction and Electron Spin Resonance," *IEEE Trans. Electron Devices* **ED-37,** 1635–1642.

Ohmameuda, T., Miki, H., Asada, K., Sugano, T., Ohji, Y. (1991), *Jpn. J. Appl. Phys.* **30,** L1993–L1995.

Ohyu, K., Hiraiwa, A. (1992), *Ext. Abst. 1992 Int. Conf. Solid State Devices and Materials,* Tsukuba: The Japan Society of Applied Physics, pp. 73–75.

Poindexter, E. H., Gerardi, G. J., Rueckel, M.-E., Caplan, P. J., Johnson, N. M., Biegelsen, D. K. (1984), *J. Appl. Phys.* **56,** 2844–2849.

Reed, M. L., Plummer, J. D. (1988), *J. Appl. Phys.* **63,** 5776–5793.

Revesz, A. G. (1979), *J. Electrochem. Soc.* **126,** 122–130.

Rochet, F., Froment, M., D'Anterroches, C., Roulet, H., Dufour, G. (1989), *Philos. Mag. B,* **59,** 339–363.

Sah, C. T., Ning, T. H., Tschopp, L. L. (1972), *Surf. Sci.* **32,** 561–575.

Sakurai, T., Sugano, T. (1981), *J. Appl. Phys.* **52,** 2889–2896.

Schwerin, A. V., Heyns, M. M., Weber, W. (1990), *J. Appl. Phys.* **67,** 7595–7601.

Shockley, W. (1949), *Bell Syst. Tech. J.* **28,** 435–489.

Shockley, W., Read, W. T. (1952), *Phys. Rev.* **87,** 835–842.

Stern, F., Howard, W. E. (1967), *Phys. Rev.* **163,** 816–835.

Sze, S. M. (1981), *Physics of Semiconductor Devices,* 2nd ed., New York: Wiley, Chaps. 1, 2, 7, and 8.

Takagi, S., Iwase, M., Toriumi, A. (1990), *Ext. Abst. 22nd Conf. Solid State Devices and Materials,* Sendai: The Japan Society of Applied Physics, pp. 275–278.

Takahashi, I., Shimura, T., Harada, J. (1993), *J. Phys.: Condens. Matter* **5,** 6525–6536.

Takiyama, M., Ohtsuka, S., Sakon, T., Tachimori, M. (1994), *IEICE Trans. Electron.* **E77-C,** 464–472.

Thanh, L. Do, Aslam, M., Balk, P. (1986), "Defect Structure and Generation of Interface States in MOS Structure," *Solid-State Electron.* **29,** 829–840.

Thanh, L. Do, Balk, P. (1988), *J. Electrochem. Soc.* **135,** 1797–1801.

Thornber, K. K. (1980), *J. Appl. Phys.* **51,** 2127–2136.

Tromp, R., Rubloff, G. W., Balk, P., LeGoues, F. K. (1985), *Phys. Rev. Lett.* **55,** 2332–2335.

Warren, W. L., Poindexter, E. H., Offenberg, M., Muller-Warmuth, W. (1992), *J. Electrochem. Soc.* **139,** 872–880.

Watanabe, K., Kato, M., Nagata, M., Okabe, T. (1988), *Electron. Commun. Jpn.* **71,** P. 2, 83–92.

Weinberg, Z. A., Rubloff, G. W. (1978), *Appl. Phys. Lett.* **32,** 184–186.

Williams, R. (1965), *Phys. Rev.* **140,** A569–575.

Winokur, P. S., Sokoloski, M. M. (1976), *Appl. Phys. Lett.* **28,** 627–630.

Yamabe, K., Taniguchi, K. (1985), *IEEE Trans. Electron Devices* **ED-32,** 423–428.

Yamaguchi, K. (1983), *IEEE Trans. Electron Devices* **ED-30,** 658–663.

Yeh, C. F., Lin, S. S., Yang, T. Z., Chen, C. L., Yang, Y. C. (1994), *IEEE Trans. Electron Devices* **ED-41,** 173–179.

Yokogawa, K., Yajima, Y., Mizutani, T., Nishimatsu, S., Suzuki, K. (1990), *Jpn. J. Appl. Phys.* **29,** 2265–2268.

Yunogami, T., Mizutani, T. (1993), *J. Appl. Phys.* **73,** 8184–8188.

Zener, C. (1934), *Proc. Roy. Soc.* **A145,** 523–529.

Further Reading

Ando, T., Fowler, A. B., Stern, F. (1982), *Rev. Mod. Phys.* **54,** 437–672.

Grove, A. S. (1967), *Physics and Technology of Semiconductor Devices,* New York: Wiley.

Schroder, D. K. (1990), *Semiconductor Material and Device Characterization,* New York: Wiley.

SIGNAL PROCESSING, ACOUSTICAL

See UNDERWATER AND ATMOSPHERIC ACOUSTIC SIGNAL PROCESSING

SIGNAL PROCESSING, OPTICAL

BAHRAM JAVIDI, *Department of Electrical and Systems Engineering, University of Connecticut, Storrs, Connecticut, U.S.A.*

JOSEPH L. HORNER, *Rome Laboratory, Hanscom Air Force Base, Massachusetts, U.S.A.*

INTRODUCTION

One of the most important applications of optics and optoelectronics is in information technology. Processing of information with optics (Goodman, 1968; Guenther, 1990; McAulay, 1991; Javidi and Horner, 1994) offers many advantages and capabilities, including high speed, large-volume data handling, compactness, low power consumption, and ruggedness. Electronic machines are mostly serial processors while optical systems are inherently parallel processing systems (Lohmann and Brown, 1966; Flannery and Horner, 1989). When one is dealing with large volumes of data, this feature offers a powerful alternative to electronic computing. Presently, compact custom-made optical hardware can process two-dimensional arrays of data of up to a million pixels at kilohertz frame rates, which is substantially higher than their electronics counterpart.

In the past, most applications of optical-processing systems have been for military hardware because of high cost and performance demands. With recent advances in optical devices and components (Saleh, 1991) such as optics memory and display, optical processing systems have become

3-527-28140-1/96/$5.00 + .50

more attractive for commercial applications. However, the progress in the field still seems to be of an evolutionary nature. Progress is needed in the development of reliable active optical devices and materials to realize low-cost optical systems.

This article presents the fundamentals of coherent optical signal processing. We discuss optical information-processing techniques, optical pattern recognition, optical neural computing, and materials and devices for optical processing. Signal-processing algorithms and architectures in optics as well as basic hardware concepts such as the fundamentals of optical spatial light modulators are covered. Each section begins with a review of the basic concepts and follows with a discussion and examples of recent advances in the field. A complete bibliography on the fundamentals of each topic is also included to assist the reader.

1. FOURIER-TRANSFORM PROPERTY OF LENSES

An attractive feature of an optical system is that the Fourier transform of an image can be generated in the space domain (O'Neill, 1963). As a result, various types of filtering operations for information processing can be implemented. It is well known that the complex amplitude of the light distribution in the back focal plane of a lens is the two-dimensional complex Fourier transform of the transmittance in the front focal plane of the lens (Goodman, 1968). For convenience the lens is modeled as a thin spherical lens. By a thin lens, it is meant that a ray of light entering at coordinates (x,y) on one face emerges at approximately the same coordinates on the opposite face. In other words, there is negligible translation of light within the lens volume. Therefore, a thin lens simply introduces a phase shift (delay) to an incident wave front by an amount proportional to both the thickness of the lens at each point and the index of refraction of the lens material n (for the wavelength λ). The lens transmittance function $t_1(x,y)$ can be expressed as (Goodman, 1968; Saleh, 1991)

$$t_1(x,y) = \exp\left[j\frac{2\pi}{\lambda}n\Delta_0\right]\exp\left[-j\frac{\pi}{\lambda f}(x^2+y^2)\right], \tag{1}$$

where Δ_0 and f are the maximum thickness and the focal length of the lens, respectively. Equation (1) is valid only under the paraxial approximation, i.e., only the portions of the wave front that are near the lens axis are delayed in phase according to Eq. (1). The complex amplitude of the light distribution in the back focal plane of a lens [see Fig. 1(a)], $E(\alpha,\beta;\lambda)$, in terms of that in the front plane, $e(x,y;\lambda)$, is

$$E(\alpha,\beta;\lambda) = K\iint e(x,y;\lambda) \times \exp\left[-j\frac{2\pi}{\lambda f}(\alpha x+\beta y)\right]dxdy, \tag{2}$$

where a time-dependent factor $\exp[j\omega t]$ has been dropped for convenience; K is a constant under coherent illumination; (x,y) are the coordinates in the input plane, which is the front focal plane of the lens; (α,β) are the coordinates in the Fourier plane, which is the back focal plane of the lens; and λ is the wavelength of the illuminating light. This Fourier-transform relationship is the fundamental relationship in the analysis of the optical processors. If we place a second lens L2 behind the Fourier plane as shown in Fig. 1(b), the light distribution at the back focal plane of L2 is the double Fourier transform of the input field. Here, we have assumed that lenses L1 and L2 have the same focal lengths. Different image-processing operations can be achieved by placing a spatial filter at the Fourier plane. For example, by placing an opaque spot at the origin of the Fourier plane, we can block the low or zero spatial frequencies of the input signal, thus generating a high-pass filtered version of the input field. Similarly, for more sophisticated types of image processing, a complex spatial filter $F(\alpha,\beta)$, where $F(\alpha,\beta)$ is the Fourier transform of $f(x,y)$, can be inserted at the Fourier plane, resulting in a light field of $F(\alpha,\beta)E(\alpha,\beta)$ leaving the filter plane. Therefore, at the output plane P3, we obtain the Fourier transform $F(\alpha,\beta)E(\alpha,\beta)$, which is equivalent to the convolution of the input signal $e(x,y)$ with the filter function $f(x,y)$. By proper choice of the spatial filter, numerous signal-processing operations can be performed, and since it is two dimensional, it lends itself naturally to image processing.

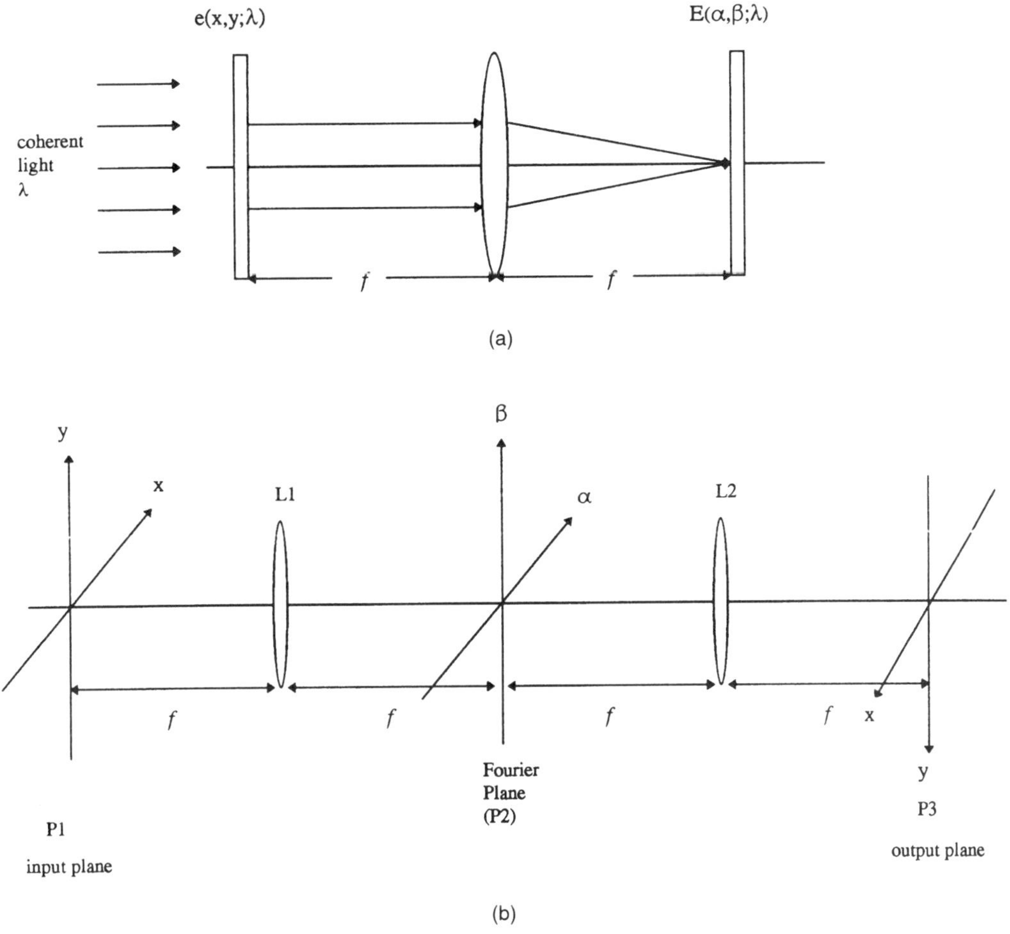

FIG. 1. (a) The Fourier-transform property of a thin lens. (b) A 4*f* optical system performs the Fourier transform twice. Plane (α,β) is the Fourier plane, where a filter function can be inserted to execute different image-processing operations.

2. FILTERS

2.1 Correlation and Matched Filtering

One of the early achievements in the field was the invention of the optical matched filter correlator by VanderLugt (1964). The matched filter, developed during World War II for extracting radar returns from noisy background and detector noise, gives the optimum theoretical response of a system when the signal is corrupted by additive overlapping Gaussian. In the derivation of the matched filter, optimum is defined as maximizing the signal-to-noise ratio (SNR), which is defined as the ratio of the output signal peak to the root mean square of the output noise. The definition of "optimum" and the fact that the noise overlaps or blankets the target or the signal is very important (Javidi and Wang, 1992). If different criteria are used, the matched filter is no longer optimum. This is discussed in more detail later.

The procedure to synthesize an optical matched spatial filter by a holographic technique is shown in Fig. 2. It turns out that the matched filter designed for detecting a specific image or target in the presence of white noise (Javidi and Horner, 1994) is just the target itself, in the spatial domain. In the Fourier domain it is equal to the complex conjugate of the Fourier transform of the target. It is possible to produce the matched spatial filter of a reference signal $s(x,y)$ at the

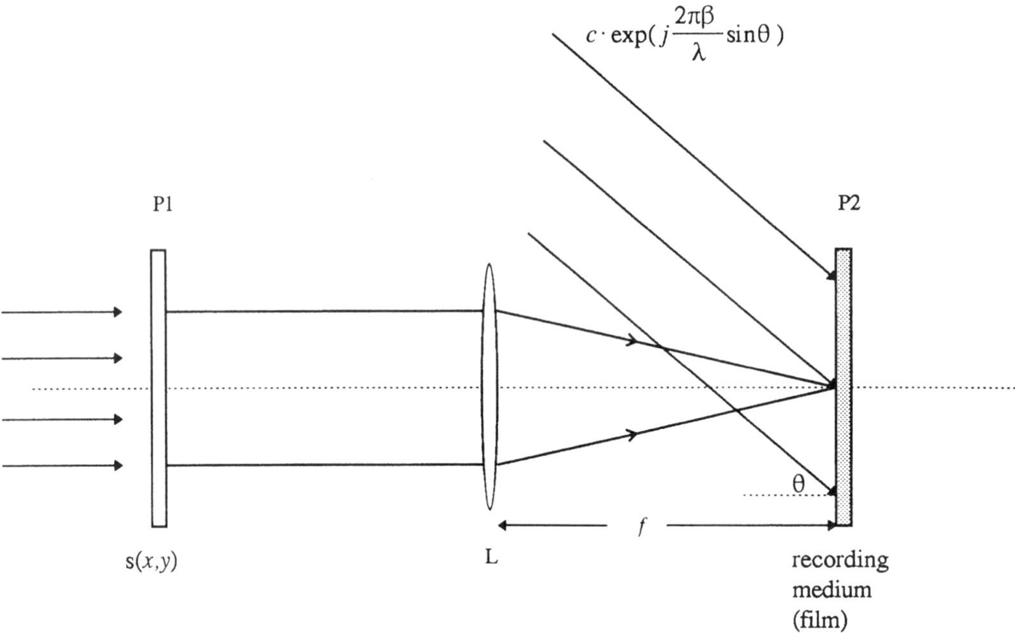

FIG. 2. An interferometric technique for synthesizing an optical spatial filter.

filter plane P2. Referring to Fig. 2, the reference signal $s(x,y)$ is inserted at the input plane P1. The light distribution $S(\alpha,\beta)$ at the filter plane P2 is the Fourier transform of the input reference signal $s(x,y)$:

$$S(\alpha,\beta;\lambda) = K\iint s(x,y) \times \exp\left[-j\frac{2\pi}{\lambda f}(\alpha x + \beta y)\right] dxdy, \quad (3)$$

where (x,y) and (α,β) are the coordinates in planes P1 and P2, respectively; f is the focal length of the transform lens L; and K is a complex constant. A plane-wave reference beam of uniform amplitude incident on the plane P2 at an angle θ with respect to the optical axis can be written as

$$R(\alpha,\beta) = c\exp\left(-j\frac{2\pi}{\lambda}\sin\theta\cdot\beta\right), \quad (4)$$

where $\sin\theta/\lambda$ is the spatial frequency corresponding to the plane wave with wavelength λ (Goodman, 1968). To produce the matched filter at the Fourier plane for detecting the reference signal $s(x,y)$, a conventional holographic technique is used to record the interference patterns of the reference-signal transform $S(\alpha,\beta)$ with the reference beam $R(\alpha,\beta)$. This can be done by placing a detector such as a high-resolution photographic film at plane P2. The total intensity distribution at the filter plane $I(\alpha,\beta)$ is obtained as

$$I(\alpha,\beta) = |R(\alpha,\beta) + S(\alpha,\beta)|^2. \quad (5)$$

The film is then developed to produce a transmittance function $H(\alpha,\beta)$.

For simplicity, we assume that $H(\alpha,\beta)$ is proportional to the intensity of the distribution $I(\alpha,\beta)$. Under this condition, the filter transmittance function can be written as

$$H(\alpha,\beta) = c^2 + |S(\alpha,\beta)|^2 + cS(\alpha,\beta)\exp\left(j2\pi\frac{\sin\theta}{\lambda}\beta\right) + cS^*(\alpha,\beta)\exp\left(-j2\pi\frac{\sin\theta}{\lambda}\beta\right). \quad (6)$$

The first and second terms of Eq. (6) represent the on-axis terms, since they are not modulated by a spatial carrier term. The last term in Eq. (6) is the desired matched spatial filter of $s(x,y)$, which is proportional to the complex conjugate of the reference signal's spectrum $S^*(\alpha,\beta)$. The third term produces the convolution term.

Referring to Fig. 1(b), if the matched spatial filter described above is placed at the Fourier plane, and an arbitrary signal $g(x,y)$ is inserted at the input plane, then the complex amplitude of the light leaving the filter plane is the product of the filter's transmittance function and the input signal's spectrum; i.e.,

$$t(\alpha,\beta) = G(\alpha,\beta)S^*(\alpha,\beta)\exp\left(-j\frac{2\pi\beta}{\lambda}\sin\theta\right), \tag{7}$$

where $t(\alpha,\beta)$ is the light leaving plane P2, and $G(\alpha,\beta)$ is the Fourier transform of the input signal $g(x,y)$. For simplicity, we have considered only the matched spatial filter term of the filter's transmittance function, because the spatial frequency $\sin\theta/\lambda$ spatially separates the correlation term from the other terms at the output plane P3. Plane P2 is located at the front focal plane of lens L2, as shown in Fig. 1(b). Therefore, the light pattern in plane P3 is proportional to the Fourier transform of $t(\alpha,\beta)$ and is given by

$$r_c(x,y) = \iint g(\xi,\eta) \times S^*(\xi - x, \eta - y + f\sin\theta)\, d\xi d\eta, \tag{8}$$

where $r_c(x,y)$ is the light distribution at the output plane.

We note that $r_c(x,y)$ is the cross-correlation between the input signal $g(x,y)$ and the reference signal $s(x,y)$ that is diffracted in the vicinity of $(0, f\sin\theta)$ at the output plane. If the input signal is equivalent to the reference signal $s(x,y)$, then the autocorrelation of the reference signal is obtained at the output plane.

2.2 Spatial Filter Development

The holographic matched filter discussed above was a breakthrough in its time, but being described by a complex mathematical function, it is difficult to implement on real-time devices like spatial light modulators. In 1984 the phase-only filter (Horner and Gianino, 1984) was introduced. This approach makes use of the fact that when we look at the phase and amplitude information of a signal in the Fourier transform plane, the phase information is vastly more important than the amplitude information. Interesting examples of reconstructing two-dimensional signals (images) from the phase information were shown by Oppenheim and Lim (1981). This concept was first applied to the matched filter (Horner and Gianino, 1984) with very good results. That is, if we represent the Fourier transform of a signal $r(x,y)$ by

$$\mathcal{F}(r(x,y)) = R(\alpha,\beta)\exp[j\phi_R(\alpha,\beta)], \tag{9}$$

where $\mathcal{F}(.)$ represents the Fourier transform and $R(\alpha,\beta)$ and $\phi_R(\alpha,\beta)$ are the amplitude and the phase of the Fourier transform of the signal $r(x,y)$, the transmittance function of the matched filter for this signal is

$$H(\alpha,\beta) = R(\alpha,\beta)\exp[-j\phi_R(\alpha,\beta)], \tag{10}$$

and that of the phase-only filter (POF) is

$$H_{\mathrm{POF}}(\alpha,\beta) = \exp[-j\phi_R(\alpha,\beta)]. \tag{11}$$

The advantage of the POF as compared with the matched filter is that it produces a very narrow correlation peak. This means that if there are multiple targets in the input, their correlation signals will not overlap and cause false alarms. Also, in an optical system, the POF has the advantage of passing all the light incident on it, since it is a pure phase element. The holographic implementation of the matched filter typically puts only a fraction of a percent of the light into the correlation plane (Horner, 1982). Figure 3 shows a comparison of the matched filter, the phase-only filter, and the further developed filter called the binary phase-only filter, which we discuss next.

2.3 Binary Phase-Only Filters (BPOFs)

In the POF, the amplitude information was discarded, leaving only the phase information. A further simplification is possible by forcing the phase function to assume one of two values, 0 or π (Psaltis *et al.*, 1984; Horner and Leger, 1985). If we think of each regime, phase and amplitude, as being represented by a one-byte word, the BPOF rep-

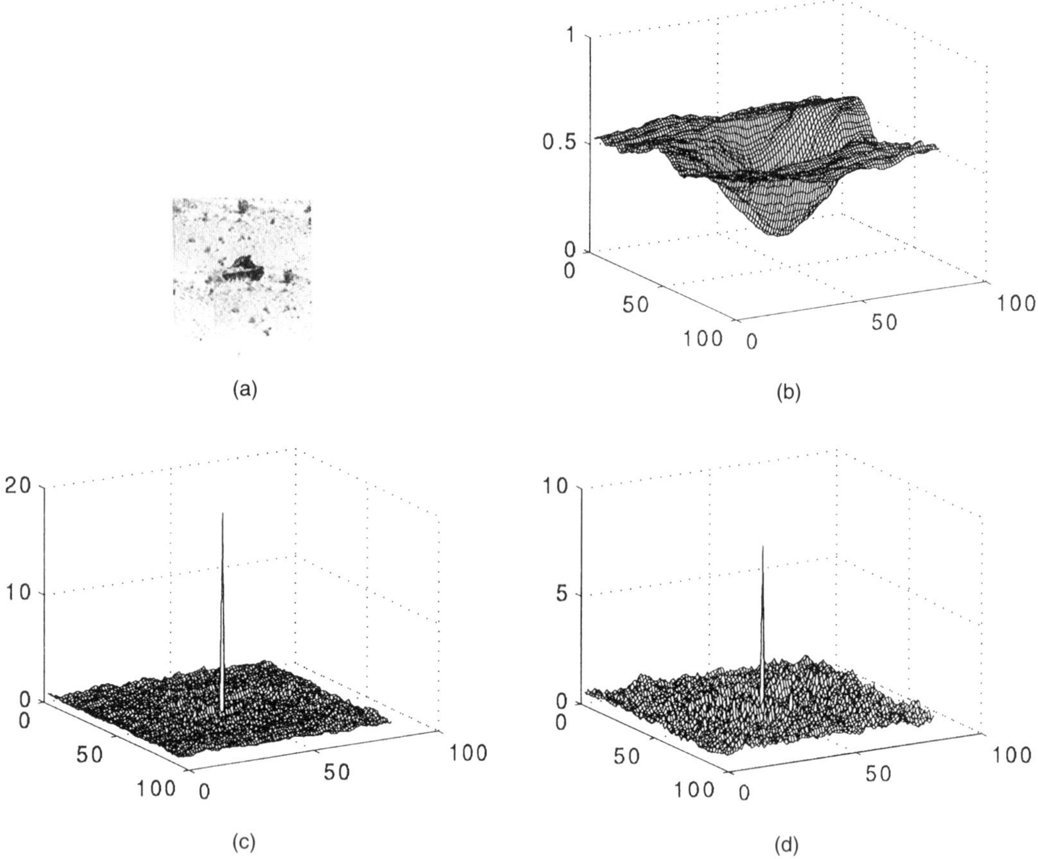

FIG. 3. The comparison of the classical matched filter, the phase-only matched filter, and the binary phase-only filter. A tank image is the target to be detected. (a) The input scene. (b) Correlation output of the classical matched filter. (c) Correlation output of the POF. (d) Correlation output of the BPOF.

resents a reduction of 94% in the information of the signal target. It is surprising that the filter works at all, yet, as Fig. 3 shows, it works almost as well as the POF, except that the noise background is a little higher. That is the trade-off. Using a nonlinear Fourier-transform method of analysis, it was shown that the BPOF could be represented as a sum of infinite harmonic terms, the first term of which is just the term for the POF (Javidi, 1990; Horner *et al.*, 1992). That is,

$$\begin{aligned} H_{\mathrm{BPOF}}(\alpha,\beta) = {} & k_1 \exp[-j\phi_R(\alpha,\beta)] \\ & + k_2 \exp[j\phi_R(\alpha,\beta)] \\ & + \textstyle\sum \text{other harmonics}, \end{aligned} \tag{12}$$

where k_1 and k_2 are constants. The second term is the POF for an inverted-target signal. Therefore, the BPOF cannot distinguish the difference between an object and its inverted version. Whether or not this is a problem depends on the application. For example, if we are trying to find tanks in a battlefield, this would not be a disadvantage. In practice the BPOF, like the POF, can be implemented in a real system in real time on a single phase-modulating spatial light modulator (SLM), which is now commercially available (Lu and Saleh, 1990).

2.4 Optimum Filtering for Spatially Nonoverlapping Target and Scene Noise

The matched filtering (VanderLugt, 1964; Turin, 1960) and its variations (Flannery and Horner, 1989; Horner and Gianino, 1984; Horner and Leger, 1985; Javidi, 1990) are well-established techniques for optical pattern-recognition applications. The matched filter function is derived to optimize the conventional SNR defined in Sec. 1, under the condition that the additive input noise is overlapping with the target and is, at least in the wide sense, stationary. In many pattern-recognition applications, however, the input scene noise does not overlap the target (it is then sometimes called disjoint noise). This means that the target is in the foreground and blocks the scene noise. For this class of problems, the matched filter and the optimum filter derived under the assumption of overlapping input target and scene noise may not perform well (Javidi and Wang, 1992).

Now, we consider that an input image $s(t)$ (one-dimensional notation is used for simplicity) is real and contains a target $r(t)$ in the presence of random noise that includes nonoverlapping scene noise and distortion noise on the target (the second and third terms in the following equation, respectively):

$$s(t,\tau) = r(t - \tau) + n_b(t)[w_0(t) - w_r(t - \tau)] + n_r(t)w_r(t - \tau),$$

where $w_0(t)$ and $w_r(t)$ are window functions of the input image and the target, respectively. The window $w_0(t)$ is defined to be 0 outside the input image and 1 inside. The target window $w_r(t)$ is defined to be 1 wherever the target exists and 0 otherwise. The quantity τ is a random variable that indicates the random location of the target in the input scene, and $n_b(t)$ and $n_r(t)$ are noise for modeling the background noise and the distortion on the target, respectively.

One solution for this problem is using multiple-hypothesis testing to design an optimum receiver for the disjoint input target and scene noise (Javidi *et al.*, 1993). It is shown that for a noise-free target, the optimum receiver is similar to a correlator normalized by the input scene energy within the target window. In addition, given that the target is noise free and the scene noise probability-density function is bounded, then the actual scene noise statistics becomes irrelevant to the detection process.

Another solution is the optimum-filter approach (Javidi and Wang, 1994) for detecting targets in spatially disjoint scene noise. The filter is designed by maximizing a new performance metric, peak-to-output energy (POE), that is defined as the ratio of the square of the expected value of the output signal at the target location to the expected value of the average output signal energy. So the filter produces a sharp output signal at the target location with a low output-noise floor.

We provide a test result of the optimum filter to show its performance. Three target tanks and two objects (a car and a vehicle) are embedded in white Gaussian-distributed background noise with mean of $m_b = 0.4$ and standard deviation of $\sigma_b = 0.3$ [Fig. 4(a)]. Target tank 1 is identical to the reference tank used in the filter design. Target tank 2 is rotated by 4°. Target tank 3 is scaled up by 10%. The target-distortion noise parameters are chosen as mean $m_r = 0$ and standard deviation $\sigma_r = 0.2$ in the filter design. The filter output is plotted in Fig. 4(b) and compared with the output of the conventional matched filter in Fig. 4(c). Here the peak-to-sidelobe ratio (PSR) is defined as the correlation peak intensity to the largest noise peak intensity at the output plane.

3. CORRELATIONS

3.1 Classical Joint-Transform Correlators

Another architecture for correlation of objects is the joint-transform correlator (JTC) (Goodman, 1968; Weaver and Goodman, 1966), as shown in Fig. 5. Here the reference function $r(x,y)$ and an unknown input object $s(x,y)$ are presented together in the input plane, and their combined or joint Fourier transform is produced in the focal plane behind the first lens. If the joint Fourier transform is recorded on a square-law or energy detector such as photographic film and a second Fourier transform is taken, a correlation of the two objects can be realized. The main advantage of the joint-transform correlator is that both the input signal and the reference signal are Fourier-transformed simultaneously, and the interference between the transforms is achieved in one single step. Thus, the need for making filters is eliminated. The input images can be displayed on a spatial light modulator (SLM) for real-time operation. A disadvantage of the JTC is that

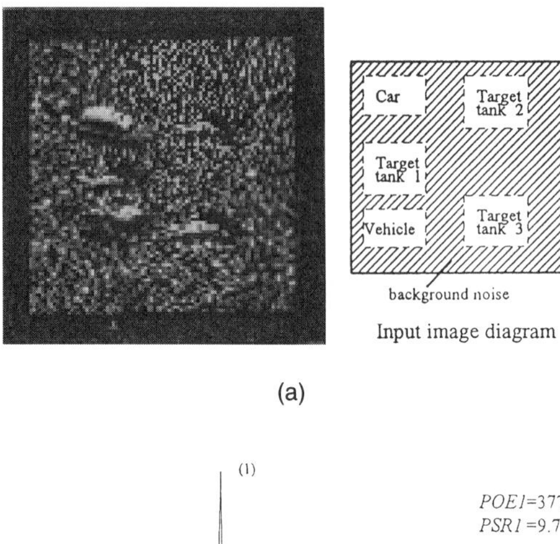

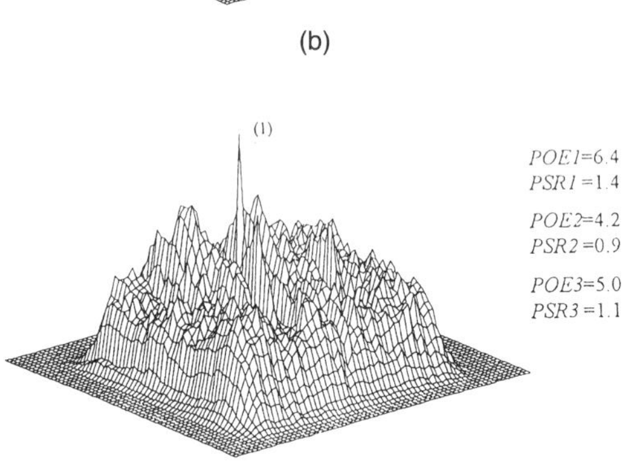

FIG. 4. The performance of the optimum filter for spatially disjoint target and scene noise (Javidi and Wang, 1994). (a) The input scene where three target tanks and two objects (a car and a vehicle) are embedded in white Gaussian-distributed background noise with mean of $m_b = 0.4$ and standard deviation of $\sigma_b = 0.3$. Target tank 1 is identical to the reference tank used in the filter design. Target tank 2 is rotated by 4°. Target tank 3 is scaled up by 10%. The target distortion noise parameters are chosen as $m_r = 0$ and $\sigma_r = 0.2$ in the filter design. (b) Correlation output of the classical matched filter. (c) Correlation output of the optimum filter.

the space–bandwidth product (equivalent to the total number of image pixel elements) available in the input plane must be shared by the input and reference signals as well as a space between these signals to ensure separation of correlation patterns from the zero-frequency or dc terms at the output plane.

The implementation of the joint-transform correlator using a spatial light modulator is shown in Fig. 6 (Javidi *et al.*, 1991). Plane P1 is the input plane that contains the reference signal $r(x + x_0, y)$ and the input signal $s(x - x_0, y)$. The amplitude of the light distribution at the back focal plane of the transform lens FTL1 is the sum of the Fourier transforms of the two input image functions; i.e.,

$$\begin{aligned} I(\alpha,\beta) &= S(\alpha,\beta)\exp[j\phi_S(\alpha,\beta)]\exp(-jx_0\alpha) \\ &\quad + R(\alpha,\beta)\exp[j\phi_R(\alpha,\beta)]\exp(jx_0\alpha), \end{aligned} \tag{13}$$

where (α,β) are the spatial-frequency coordinates and $S(\alpha,\beta)\exp[j\phi_S(\alpha,\beta)]$ and $R(\alpha,\beta) \times \exp[j\phi_R(\alpha,\beta)]$ correspond to the Fourier transforms of the input and reference signals $s(x,y)$ and $r(x,y)$, respectively. The Fourier-transform intensity distribution at plane P2 can be written as

$$\begin{aligned} E(\alpha,\beta) &= |I(\alpha,\beta)|^2 = S^2(\alpha,\beta) + R^2(\alpha,\beta) \\ &\quad + S(\alpha,\beta)\exp[j\phi_S(\alpha,\beta)]R(\alpha,\beta) \\ &\quad \times \exp[-j\phi_R(\alpha,\beta)]\exp(-j2x_0\alpha) \\ &\quad + S(\alpha,\beta)\exp[-j\phi_S(\alpha,\beta)]R(\alpha,\beta) \\ &\quad \times \exp[j\phi_R(\alpha,\beta)]\exp(j2x_0\alpha), \end{aligned} \tag{14}$$

where we assume that $E(\alpha,\beta)$ has been normalized.

For the linear or classical joint-transform correlator, the inverse Fourier transform (or the Fourier transform with coordinates reversed) of Eq. (14) will produce the following correlation signals at the output plane:

$$\begin{aligned} h(x',y') &= R_{11}(x',y') + R_{22}(x',y') \\ &\quad + R_{12}(x' - 2x_0, y') \\ &\quad + R_{21}(x' + 2x_0, y'), \end{aligned} \tag{15}$$

where

$$\begin{aligned} R_{21}(x',y') &= R_{12}(-x',-y') \\ &= \iint s(\xi,\zeta)r(\xi - x', \zeta - y')\,d\xi d\zeta, \end{aligned}$$

$$R_{11}(x',y') = \iint s(\xi,\zeta)s(\xi - x', \zeta - y')\,d\xi d\zeta,$$

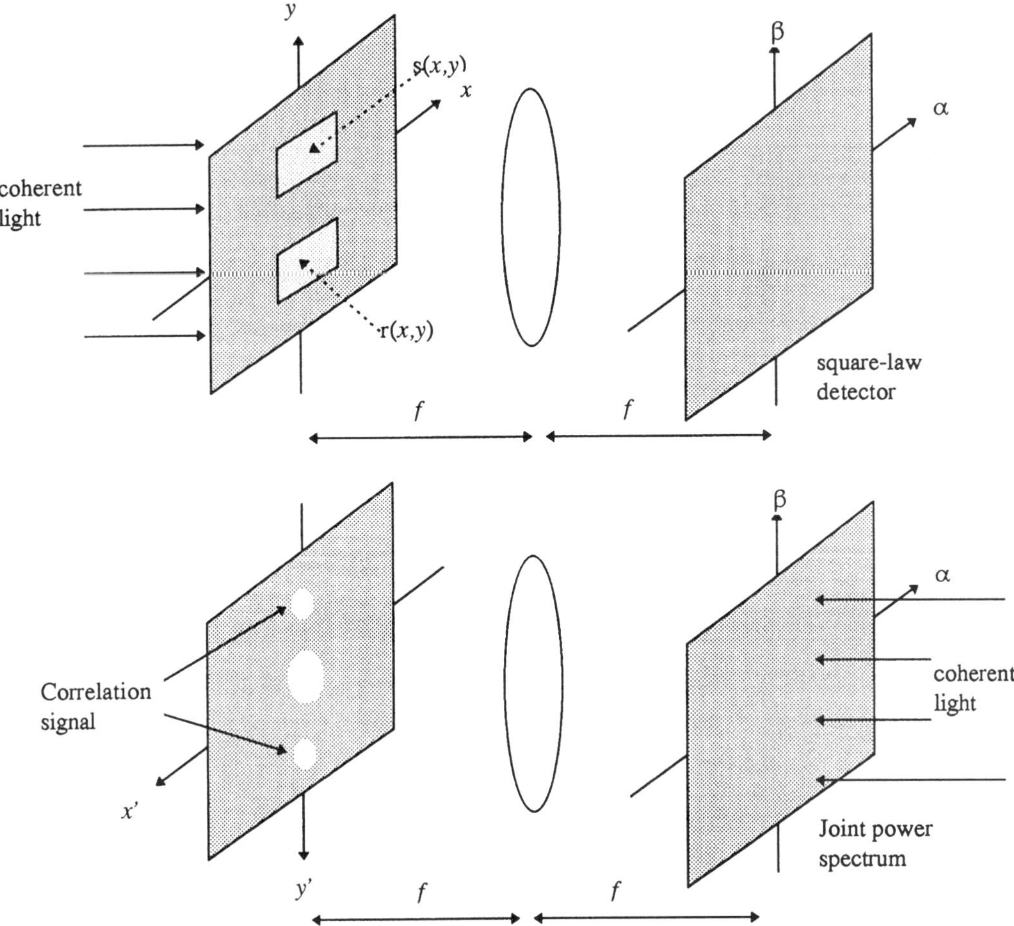

FIG. 5. Joint-transform correlator—an architecture for correlation of objects.

$$R_{22}(x',y') = \iint r(\xi,\zeta)r(\xi - x', \zeta - y')\, d\xi d\zeta.$$

The first two terms of Eq. (15) are autocorrelation terms. The terms of interest are the third and fourth terms, which are the cross-correlations of the reference signal with the input signal. Here ξ and ζ are just dummy variables.

3.2 Nonlinear Joint-Transform Correlators

More recently, nonlinearities were introduced into the JTC. The binary JTC is obtained by binarizing the joint power spectrum into two values (Javidi and Kuo, 1988). It has been shown that in terms of discrimination a binary joint-transform correlator has superior performance compared with that of the conventional linear JTC. Later, the binary JTC was generalized to nonlinear JTCs (Javidi and Kuo, 1988; Javidi, 1989; Javidi and Horner, 1989a, 1989b), and thus a family of correlators and kth-power–law nonlinear JTCs, including the conventional JTC for $k = 1$ and the binary JTC for $k = 0$, were introduced. Here, k represents the severity of the nonlinearity of the transformation of $\mathrm{sgn}(E_m)|E_m|^k$, where E_m is the modified joint power spectrum ($E_m = E - S^2 - R^2$) and sgn(.) is the signum function; $k = 1$ corresponds to the linear case, and $k = 0$, to the binary case. It was also shown that JTCs with $k = 1$ correspond to classical matched filters, and JTC's with $k = 0$ correspond to inverse filters. Theoretical and experimental

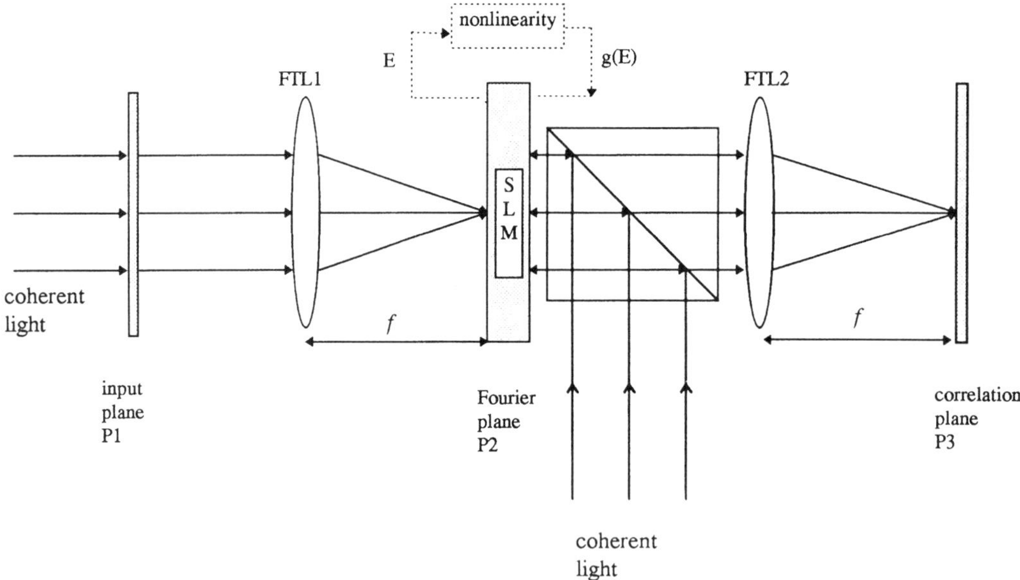

FIG. 6. Nonlinear joint-transform correlator (Javidi and Horner, 1994).

studies have shown that the nonlinear joint-transform correlator can produce very good correlation performance. The nonlinear JTC can use the nonlinearity of a nonlinear device such as a SLM at the Fourier plane to alter the Fourier-transform interference intensity. It has been shown that compared with the classical correlator, the compression type of nonlinear joint-transform correlator ($k < 1$) provides higher peak intensity, larger peak-to-sidelobe ratio, narrower correlation width, and better correlation sensitivity.

In Fig. 6, the joint-transform correlator is implemented with use of a liquid-crystal light valve (LCLV) as an optically addressed SLM at the Fourier plane. The Fourier-transform interference pattern, Eq. (14), is displayed at the input of the LCLV to obtain the intensity of the Fourier-transform interference. For the nonlinear JTC, the LCLV nonlinearly transforms the joint power spectrum according to the nonlinear characteristics of the device. The transformed joint power spectrum can be considered as the output of a nonlinear system. The nonlinear characteristic of the nonlinear device is denoted by $g(E)$, where E is the Fourier transform of the interference intensity. An expression for the nonlinearly transformed interference intensity can be obtained by an approach similar to the analysis of nonlinear systems using the transform method of communication theory (Horner *et al.*, 1992; Javidi, 1989). Let the Fourier transform of the nonlinear characteristic of the nonlinear device be defined by

$$G(\omega) = \int_{-\infty}^{\infty} g(E) \exp(-j\omega E)\, dE. \tag{16}$$

The output of the nonlinear device can be written as

$$g(E) = \sum_{\nu=0}^{\infty} H_\nu[R(\alpha,\beta),S(\alpha,\beta)] \times \cos[2\nu x_0\alpha + \nu\phi_S(\alpha,\beta) - \nu\phi_R(\alpha,\beta)], \tag{17}$$

where

$$H_\nu[R(\alpha,\beta),S(\alpha,\beta)] = \frac{\epsilon_\nu}{2\pi} j^\nu \int G(\omega) \exp\{j\omega[R^2(\alpha,\beta) + S^2(\alpha,\beta)]\} \times J_\nu(2\omega R(\alpha,\beta)S(\alpha,\beta))\, d\omega.$$

Here,

$$\epsilon_\nu = \begin{cases} 1, & \nu = 0, \\ 2, & \nu > 0 \end{cases},$$

and J_ν is a Bessel function of the first kind, order ν.

It can be seen from Eq. (17) that for $\nu = 1$, the nonlinear system has preserved the phase of the cross-correlation term $\phi_S(\alpha,\beta) - \phi_R(\alpha,\beta)$, and only the amplitude is affected. This explains the good correlation properties of the first-order correlation signal at the output plane. For a kth-power–law nonlinearity, i.e.,

$$g(E) = |E|^k \operatorname{sgn}(E - \nu_T),$$

where $\nu_T = R^2 + S^2$, Eq. (17) becomes

$$g_k(E) = \sum_{\nu=1}^{\infty} \frac{\epsilon_\nu \Gamma(k+1)[R(\alpha,\beta)S(\alpha,\beta)]^k}{\Gamma(1 - \frac{1}{2}(\nu - k))\Gamma(1 + \frac{1}{2}(\nu + k))} \times \cos[2\nu x_0\alpha + \nu\phi_S(\alpha,\beta) - \nu\phi_R(\alpha,\beta)], \quad (18)$$

where Γ is the gamma function, and k is the severity of the nonlinearity; $k = 1$ corresponds to a linear device, and $k = 0$ corresponds to a hard-clipping nonlinearity.

It can be seen from the above equation that each harmonic term is phase modulated by ν times the phase difference of the input-signal and the reference-signal Fourier transforms, and the higher-order correlation signals are diffracted to $2\nu x_0$. The correct phase information of the joint power spectrum is obtained for the first-order harmonic term ($\nu = 1$); i.e.,

$$g_{1k}(E) = \frac{2\Gamma(k+1)[R(\alpha,\beta)S(\alpha,\beta)]^k}{\Gamma(1 - \frac{1}{2}(\nu - k))\Gamma(1 + \frac{1}{2}(\nu + k))} \times \cos[2x_0\alpha + \phi_S(\alpha,\beta) - \phi_R(\alpha,\beta)]. \quad (19)$$

It is noted that varying the severity of the nonlinearity k will produce correlation signals with different characteristics. For highly nonlinear transformations (small k), the high spatial frequencies are emphasized, and the correlation becomes more sensitive in discrimination.

4. SPATIAL LIGHT MODULATORS (SLMs)

Spatial light modulators are very important optical components of optical information-processing systems. They are used in optical computing systems, programmable optical interconnects, optical neural nets, and optical pattern-recognition systems. The SLM input is either a time-dependent electrical signal or a light distribution such as an image. SLMs with electrical or optical input are called electrically addressed or optically addressed SLMs, respectively. The SLM modulates the amplitude and/or phase, or polarization, of the read-out light beam as a function of the input signal, which can be an optical image or an electrical signal. Figure 7 shows an example of an optically addressed SLM. The writing light $A_i(x,y)$ is incident on the input of the SLM. In general, $A_i(x,y)$ is a two-dimensional spatially varying amplitude distribution, imaged from a display monitor onto the SLM. The output light distribution $A_0(x,y)$ is a function of the input light amplitude $A_i(x,y)$. The details of the component pieces are discussed in Sec. 4.4 (Liquid-Crystal Light Valve).

Various SLMs differ in addressing methods and the modulating materials used. In Fig. 8, the input light is converted to an electric field distribution by a photoconductor. The electric field is related to the input light intensity (optically addressed SLM). The electric field can also be directly applied by use of transparent conductive electrodes (electrically addressed SLM). This electric field modifies the properties of the electro-optic or modulating material. For example, it may change the optical refractive index of the modulating material. The read-out light beam is modulated by the modulating element and reflected back to create the read-out image. Some modulating properties are electro-optic effect, molecular alignment by electric field that exists in liquid crystals, photorefractive effect, electrostatic deformation, and acousto-optic effect. The electro-optic effect, which is the change in index of refraction of the medium as a function of the applied electric field, is used in a number of SLMs, such as the Pockels read-out optical modulator. The SLMs can also operate in the transmissive mode. A SLM operating in complex amplitude mode has an output

$$A_0(x,y) = \hat{A}_0(x,y) \exp[jP_0(x,y)], \quad (20)$$

where $P_0(.)$ represents the phase. The phase-only SLM output is $\exp[jP_0(x,y)]$, with $\hat{A}_0(x,y)$

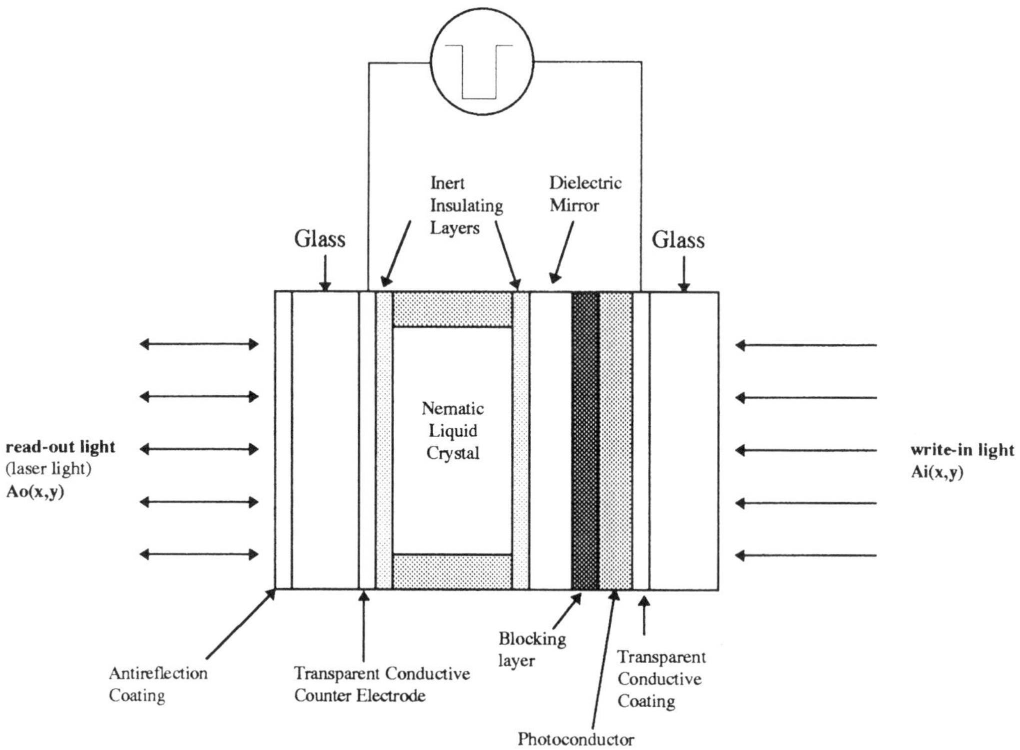

FIG. 7. Structure of the Hughes liquid-crystal light valve.

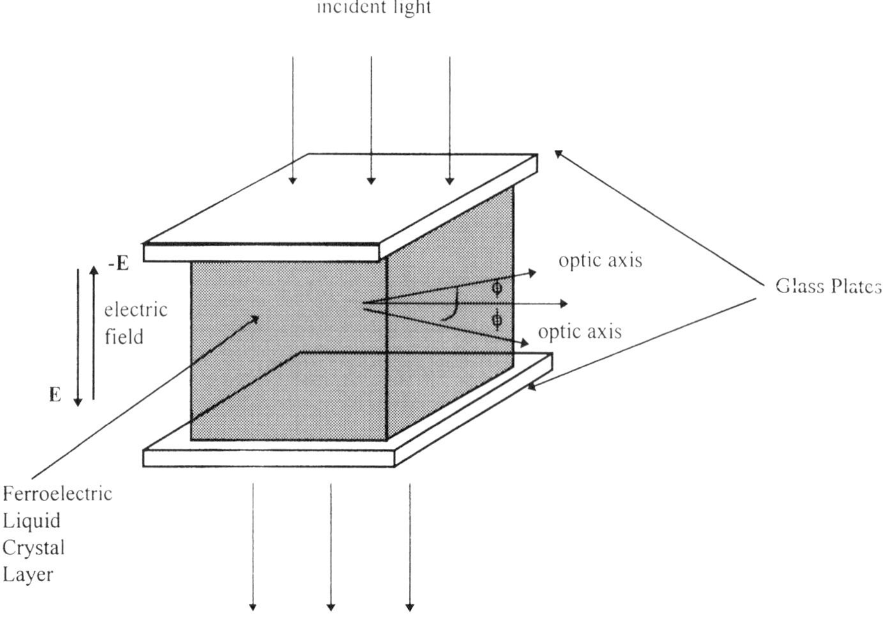

FIG. 8. Ferroelectric liquid-crystal device.

a constant. The amplitude-only SLM output is $\hat{A}_0(x,y)$, with $\exp[j(P_0(x,y)]$ being constant.

4.1 Capabilities of SLMs

Some of the capabilities of SLMs are listed here:

1. *Image storage.* This is useful for optical memory, optical database/knowledge-base processors, optical pattern recognition, and neural networks.
2. *Incoherent to coherent light conversion.* This is used for converting real scenes illuminated under natural or other incoherent light into a coherent image. The real scene is imaged onto a SLM which is read out by coherent light. The coherent image can then be processed by optical signal-processing techniques.
3. *Real-time holography.* The interference generated between the object beam and the reference beam can be positioned on the SLM and the holographic pattern therefore displayed on the SLM. However, the space–bandwidth product (SBP) of the typical SLM limits the effectiveness of this method.
4. *Fourier-plane spatial filtering.* Spatial filters can be displayed on SLMs in the Fourier plane. In this case, the input image Fourier transformed by a lens is multiplied by the filter on the Fourier-plane SLM. An additional Fourier transformation using a lens will produce the convolution between the input image and the impulse response of the filter. This can be used in optical spatial filtering, pattern recognition, and neural nets. If the filter is generated electronically by a computer, a SLM with electrical-to-optical conversion is used to display the filter.
5. *Nonlinear transformation and thresholding.* SLMs can be used to transform an image nonlinearly or binarize an image. This property is also useful for logic operations and switching in digital optical computing. In information processing, nonlinear characteristics of the SLM can be used in nonlinear filtering and nonlinear signal processing. Contrast reversal can be performed using SLMs.
6. *Image amplification.* SLMs are used in projection systems for image amplification. An object is imaged onto the SLM. It is read out from the SLM to provide a brighter image.

4.2 SLM Performance Considerations

There are many considerations in designing and using a SLM for optical processing. Frame rate determines how fast an image written on the SLM can be updated. Spatial resolution is a measure of how fine an image displayed on the SLM can be. Space–bandwidth product is a measure of the number of pixels available. Dynamic range is the number of gray levels that can be represented by a pixel. Contrast ratio, a measure of the ability of an SLM to block the light, is the ratio of the maximum and minimum output light levels. Flatness of the mirrors or windows of the SLM to a fraction of a wavelength of the light is important for optical processing where phase information is critical. Nonlinear input–output characteristics of the SLM are often considered for the specific image-processing applications. The exposure sensitivity and read-out light efficiency define the light budget of the system. SLMs also have write-in and read-out wavelength range, and electrical driving-signal power requirements.

4.3 Photorefractive Materials and Devices

Photorefractive materials can store optical images using variations in the index of refraction through the electro-optic effect. A photorefractive material, upon exposure to a light beam or an image, generates free charges, such as electrons or holes. These charges are redistributed and fall into trapping sites, thereby creating a charge distribution. This space-charge field produces a spatially dependent electric field that changes the index of refraction of the material through the electro-optic effect. The variations in the index of refraction result in the refraction or diffraction of light. The image stored in the photorefractive material can be read out by an optical beam. For a 1D signal with no applied field, the change in the index of refraction $\Delta n(x)$ as a function of the input intensity $I(x)$ is

$$\Delta n(x) = -K \Delta I(x)/I(x), \qquad (21)$$

where K is a constant dependent on the electro-optic coefficient, refractive index of the material, temperature, and electron mobility. Examples of photorefractive materials are lithium niobate ($LiNbO_3$), barium titanate ($BaTiO_3$), and bismuth silicon oxide (BSO). Photorefractive materials are used in optical storage and memory, real-time optical information processing, neural networks, holography, distortion compensation, and phase conjugation.

Photorefractive devices can be used to generate spatial filters or holograms in real time and have been used for an optical lock-in amplifier/detector. An image $I(x,y)$ is spatially mixed with a reference beam, and their interference intensity is recorded in a photorefractive device. The interference intensity produces changes in the refractive index $\Delta n(x,y)$ that are stored in the form of a volume phase hologram. When the device is illuminated by the reference wave, the object beam $I(x,y)$ is reconstructed. For spatial filtering, the Fourier transform of $I(x,y)$ is stored in the device as a filter function.

An interesting property of photorefractive crystals is real-time holographic wave mixing. The interference between a reference beam and an object beam produces a grating that couples them, and energy can be exchanged. Each wave reflects from the grating and can add to the other beam. As a result, one wave may be amplified and take energy away from the other beam. The first complete theoretical treatment of this was given by Kogelnick (1969). Table 1 shows a number of available SLMs and their characteristics (optically addressed). In the following subsections, we discuss some of the SLMs in detail.

4.4 Liquid-Crystal Light Valves

A liquid-crystal light valve (LCLV) is a device commonly used as a spatial light modulator to convert an incoherent light signal to a coherent light signal for optical-processing applications. Figure 7 shows an example of a multilayer sandwich structure of the LCLV (Hughes H-4060). The device consists of a nematic liquid-crystal layer, a photoconductor (CdS or silicon), a dielectric mirror that provides the high reflectivity required, and a light-blocking layer used to isolate optically the reading light from the writing light. The optical birefringence of the liquid-crystal layer is electrically controlled by the voltage applied on the layer through a pair of transparent conductive electrodes. The operation of the LCLV can be explained by a hybrid field-effect mode (Bleta *et al.,* 1972). The voltage applied on the device is primarily divided between the photoconductor and the liquid-crystal layer. With changes of the voltage drop across the photoconductor, the voltage drop across the crystal layer changes. The liquid crystal used is a nematic type with positive dielectric anisotropy and homogeneously aligned with 45° twist. In the "off" state (no writing light), the voltage drop on the liquid-crystal layer is very small. The linearly polarized reading light will be reflected from the dielectric mirror and the polarization state of the reflected light leaving the LCLV is unchanged. In the "on" state, the resistance of the photoconductor decreases proportionally to the light intensity of the writing light. The voltage across the liquid-crystal layer then increases coordinately. This causes the liquid-crystal molecular orientation to tilt toward the homeotropic alignment. The polarization ori-

Table 1. A number of available optically addressed SLMs and their characteristics.

	Microchannel SLM	PROM	LC	Deformable mirror
Modulating element	$LiNbO_3$ crystal	$Bi_{12}SiO_{20}$	CdS; Si; photoconductor	Micromirror
Write light	X ray to 1 mm	Blue	Visible	UV to long IR
Read-out	Blue to 4 mm	Red	Visible	
Sensitivity	2–30 mJ/cm^2	10 mJ/cm^2	1–5 mJ/cm^2	
Spatial resolution	10 line pairs/mm	5 line pairs/mm	20–40 line pairs/mm	50 line pairs/mm
Frame rate	5–60 Hz	10–30 Hz	40 Hz (nematic) MHz (Ferroelectric)	4 kHz
Operating voltage	3 kV	4 kV	5–10 V	Low, <12 V
Storage time	Days	1 h in dark		
Contrast ratio	1000:1	1000:1	40:1 (nematic)	40:1

entation of the reading light exiting the liquid crystal will rotate by a certain angle from its original polarization direction. The degree of the rotation of the polarization is proportional to the voltage drop across the liquid-crystal layer and, therefore, proportional to the intensity of the writing light. Therefore the LCLV can modulate the polarization direction of the coherent reading light proportionally to the intensity of the writing light. By passing the reading light reflected from the LCLV through an analyzer, the modulation on the polarization orientation can be easily converted to an amplitude modulation. When the writing-light intensity has spatial variation, different portions of the photoconductor generate different voltage drops across the different portions of the liquid-crystal layer. As a result, spatial variation of the polarization of the reading light is achieved.

The input–output characteristic of a LCLV can be controlled electronically by changing the frequency, wave form, and voltage of the power supply of the LCLV (Efron *et al.*, 1985; Javidi *et al.*, 1991; Warde and Fisher, 1987; Javidi and Zhang, 1992). Some other LCLVs, such as a wire-grid-mirror LCLV (Efron *et al.*, 1989; Javidi *et al.*, 1994), have interesting and useful nonlinear input–output characteristics. The resolution of the LCLV is better than 40 line pairs/mm (a wire-grid-mirror LCLV has a lower resolution about 10 line pairs/mm because of the wire-grid structure of the mirror). The frame speed of the LCLV can vary from several frames per second to a few hundred frames per second. The speed of the LCLV depends on the response time of the twisted liquid crystal and the photoconductors. Usually, a silicon photoconductor is faster than a CdS, and a binary LCLV is faster than a gray-scale LCLV. The contrast of the LCLV is better than 100:1.

4.5 Liquid-Crystal Television Displays (LCTVs)

Liquid-crystal devices are widely used in small television sets, television projectors, and lap-top computers. These displays have been used in the optical signal-processing community for the last several years because of their low cost and commercial availability (Young, 1986). The liquid-crystal displays used in LCTVs were not originally designed for coherent optical systems. Their optical quality is not ideal for a coherent system mainly because of the phase variation of materials and the surfaces of the devices, which are less critical in an incoherent optical system than a coherent one. However, recent experiments show that the LCTV is still a good device in the applications where cost is an important factor and an electrically addressed device is needed.

The liquid-crystal display in a LCTV consists of a 90° twisted liquid-crystal layer sandwiched between two polarizers that have parallel polarization direction. The electric field can be applied to the liquid crystal through two transparent conductive electrodes on the two sides of the liquid-crystal layer. The transparent electrodes are pixelated and can be electrically addressed. When there is no electric field applied, the orientation of the input light is rotated by 90° from one side of the liquid-crystal layer to the other side, and the result is no light passing through because the two polarizers are parallel. When an electric field is applied, the twist and the tilt of the liquid-crystal molecules are altered depending on the voltage across the liquid-crystal layer. As a result, a fraction of the light passing through the liquid-crystal layer will retain the same polarization as the input light and, therefore, will pass through the second polarizer. The fraction of the light that passes through the display is proportional to the voltage applied to the liquid-crystal layer.

Phase modulation and complex modulation can also be obtained by properly setting up the polarization axes of the polarizer and analyzer (Kirsch *et al.*, 1992).

Liquid-crystal displays used in projector-type LCTVs usually have more pixels (300 × 300 or more) and smaller pixel size than those in conventional LCTVs. The contrast of the display is about 10:1.

4.6 Ferroelectric Liquid-Crystal Spatial Light Modulators

The surface-stabilized ferroelectric liquid-crystal (FLC) device (Johnson and Moddel, 1989) contains a FLC layer between two glass plates. The FLC has two optical axis orientations, which are parallel to the glass surface and have the tilt angles $\phi = 22.5°$

and $\phi = -22.5°$ corresponding to the sign of the electric field applied across the FLC layer, as shown in Fig. 8. Since the FLCs have a strong interaction with light (the material birefringence is typically 0.1 to 0.2), with proper choice of the thickness of the FLC layer it functions like an electrically switchable half-wave plate for a specific wavelength of the incident light. When the device is placed between a crossed polarizer and analyzer, and has one of the optical axes of the material aligned with the polarizer, the incident light will not change its polarization on passing through the FLC. Thus, the light will be extinguished by the analyzer at the output. If the electric field is reversed, the optical axis of the material is also rotated by $2\phi = 45°$. This results in a 90° rotation of the polarization of the incident light. Thus, the light passing through FLC is transmitted by the analyzer.

Ferroelectric liquid-crystal spatial light modulators are binary devices and have high speed. They have small pixel size and can be fabricated with very high resolution.

4.7 Deformable-Mirror Spatial Light Modulators

A deformable-mirror spatial light modulator consists of a number of movable mechanical mirror elements that can be electrically addressed by an underlying array of MOS transistors. The materials used for a deformable-mirror spatial light modulator can be categorized as elastomers, cantilever beams, and membranes. The elastomer requires high address voltage. It is not good for high-density integration. The cantilever-beam approach does not require high voltage (<12 V) and can be easily fabricated on silicon (Hornbeck, 1983). It was invented by Texas Instruments Corp.

Each deformable mirror element consists of a small mirror that is supported by the polysilicon lines that also serve as the base for transistor gates. The structure of a mirror element is shown in Fig. 9. There is an air gap between the mirror and the silicon substrate. When the addressed gate is on, the mirror is deflected toward the silicon substrate according to the voltage drop across the air gap. If the mirror is undeflected and illuminated by light, it acts just as a conventional mirror. If the mirror is deflected, the

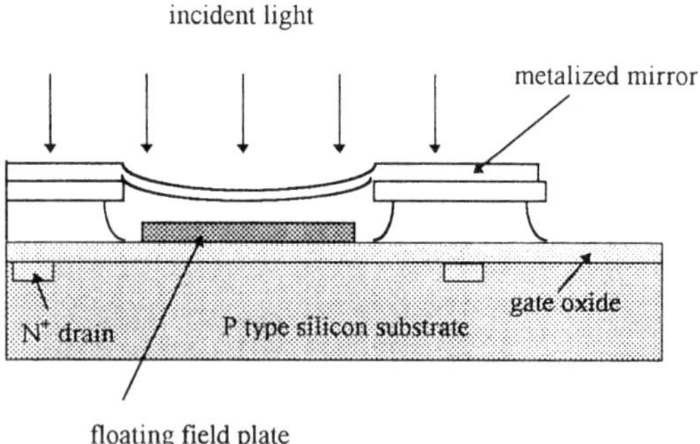

FIG. 9. Deformable-mirror element.

direction of the reflected light will be different from the direction when the mirror is undeflected. If the mirror is adjusted to reflect the light along the optical axis of the system when the mirror is in the "on" state, and off the optical axis of the optical system when the mirror is in the "off" state, the light can be turned "on" and "off" by using an aperture function to block out the off-axis light.

The deformable-mirror device has small pixel size, about $15 \times 15\ \mu m^2$, and can be fabricated as large as $\sim 10^6$ pixels in a two-dimensional array. It has a high contrast. Since it is a mirror-type modulator, it works at a wide range of wavelengths and has an almost pure phase-modulating characteristic. Texas Instruments Corp. is using it in their high-definition TV (HDTV) system with good results.

5. ACOUSTO-OPTIC SIGNAL PROCESSING

So far we have been discussing two-dimensional optical processors, which use spatial light modulators to impress the information of the light beam (see MODULATORS AND DEMODULATORS, OPTICAL). Another class of systems exist which are basically one dimensional and use ultrasound or acousto-optical principles to perform their signal processing.

We will first consider the basic acousto-optic (AO) modulator, the device that impresses an electrical signal on a beam of light. This is shown in Fig. 10. The transducer, which launches an acoustic wave into the medium above it in the figure, is typi-

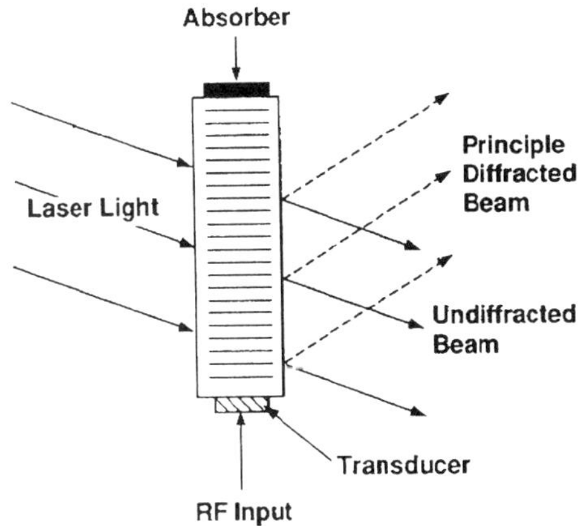

FIG. 10. Acoustic-optic modulator.

cally a piezoelectric crystal. The acoustic wave induces compression and rarifaction into the medium, thus modulating its index of refraction. Hence, the medium becomes in effect a moving phase grating that diffracts the light incident on the device from the left. The upper end of the device usually contains an absorbent material to quench the incident wave and prevent reflections and standing waves that could impair the operation of the device. Note that the device is inherently one dimensional, in contrast to a SLM, which is two dimensional. The number of distinct signal elements (so-called space–bandwidth product) in currently available devices is about 2000.

The AO modulator can be used in a system to correlate two electrical signals. There are two basic configurations for this; the time-integrating (TI) correlator and the space-integrating (SI) correlator. We will discuss the TI correlator first, shown in Fig. 11. One input signal modulates the light-emitting diode (LED) shown in the far left side of the figure. The other electrical signal is fed into the AO modulator as shown. The resultant multiplication of the two signals emerging from the right side of the AO cell is

$$s_1(t - x/v)^*s_2(t), \tag{22}$$

where x is measured from the central optical axis and v is the velocity of the acoustic wave in the AO cell. The lens L_1 takes a spatial Fourier transform of this in order to provide a means of blocking the zero-order undiffracted light from the AO cell. The second lens, L_2, takes a second Fourier transform and restores the spatial signal minus the dc. The light is detected by a detector, such as a CCD detector, and if it is integrated for a time period T, the desired correlation is achieved,

$$\int_0^T s_1(t - x/v)s_2(t)\,dt. \tag{23}$$

The other architecture commonly used is the space-integrating system, shown in Fig. 12. A LED, through a collimating lens, illuminates the AO cell as before. In this case the two signals are fed in to opposite ends of the AO cell. This results in the two propagating waves passing each other and in effect doing the shifting required in the correlation operation. The light again goes through a Fourier-transform lens where the undiffracted zero-order light is blocked to improve performance by cutting down on stray light. Lens L_2 retransforms the light and focuses it down on to the detector. The instantaneous detector output gives a temporal profile of the correlation signal. If just the value of the correlation peak is desired, the detector can be fed to an electronic peak detecting circuit.

There are many variations of the basic schemes we have discussed here, and the reader is referred for these to Berg and Lee (1983), Korpel (1988), and Xu and Stroud

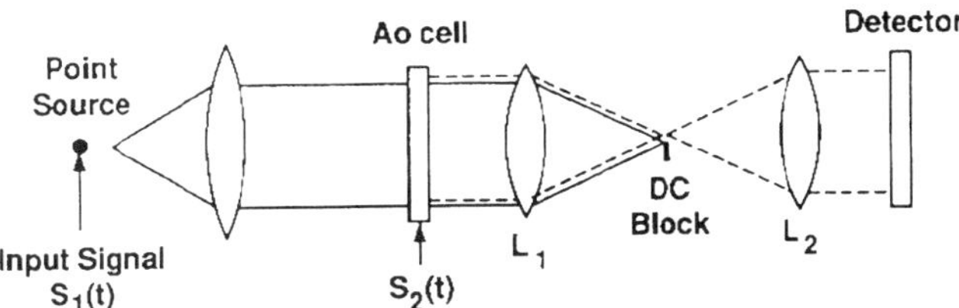

FIG. 11. Time-integrating correlator using an acoustic-optic modulator.

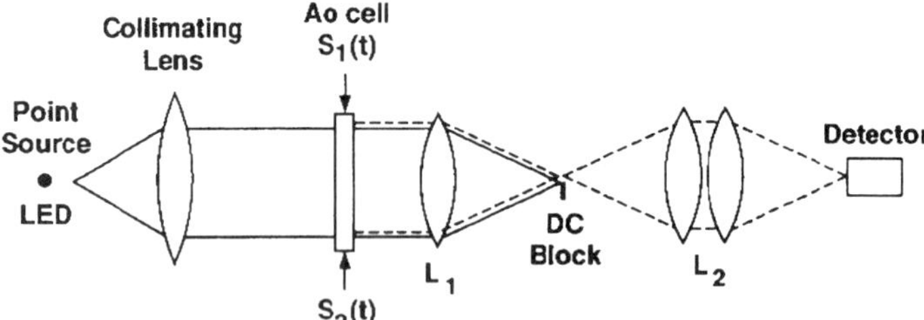

FIG. 12. Space-integrating correlator using an acoustic-optic modulator.

(1992). So far we have been limited to one-dimensional signals. A clever architecture was proposed by Psaltis (1984) that effectively makes the AO correlator a two-dimensional system capable of correlating images and, therefore, pattern recognition and target tracking. An impressive functional system using this scheme has been built and demonstrated by a group at Sandia National Laboratories (Molley and Stalker, 1990).

Comparing the two architectures, the space-integrating systems have single-element detectors, good dynamic range and unlimited time-delay ranges, and wide bandwidths. The time-integrating systems have limited time-delay ranges, require a large detector, i.e., many pixels, but generally have a larger processing gain.

The simplest application of the technology is the so-called channelizer. It consists of a light source, a Fourier-transform lens, and an AO cell. It is simply an optical analog spectrum analyzer. It takes an rf input, from a receiver for instance, and separates it into the various component parts, optically.

6. OPTICAL INTERCONNECTIONS

Interconnections are a major communication problem in high-speed, high-density, VLSI digital and electronic circuits. Optical interconnections (*q.v.*) are intended to complement and/or replace conventional electrical interconnections, such as wires, cables, or connectors on integrated circuits. Optical interconnections have definite advantages over electronic interconnections in terms of speed and density, and can remedy these problems. Free-space optical interconnections can be three dimensional, and beams can pass through one another without cross talk. In contrast, electronic interconnections are planar and cannot intersect without adequate insulation and some degree of cross talk. Thus, optics can offer advantages in terms of the number of interconnections in a given area. Time delays are also a problem in VLSI circuits. Photons travel at the speed of light, and the velocity is independent of the number of interconnections. On the other hand, in electrical transmission, delay times increase with the number of fan outs. In addition, the coupling in electronic interconnections is affected by the bandwidth of the data carried by the connection. This is not the case in optics.

Optical interconnections can be realized by a variety of methods and components, including free-space optics, optical fibers, optical wave guides, optical holograms, and holographic devices. In free-space optics, simple optical components (mirrors, lenses, and gratings) can be used to realize the interconnections by mapping points between the object and image plane. The motivation for using free-space optics for interconnections is the high space–bandwidth–product handling capability of optics. For example, a million points in an area of a square millimeter in the object plane are imaged/connected to a millon points in the image plane using a lens. An electronic counterpart would require a million connections in a square millimeter with special care taken to ensure that the connections are properly insulated to avoid short circuits. Optical components may be used for mapping input beams to fan in, fan out, shift, shuffle, etc. For fan in, a lens focuses the input beam onto its focal point. For fan out, a source placed at the focal point of a lens is simply collimated. Shuffle and shift of beams are accomplished by holographic and diffractive optical elements. In general, an optical mask is designed using computer-generated holograms (Dallas, 1971) to perform the interconnections. This mask may be considered as containing a large number of phase gratings. Each grating has a different spatial frequency, which deflects or tilts the beam at a given angle. Thus, arbitrary optical interconnection maps can be created. Computer-generated holograms may have very high spatial frequencies, up to thousands of lines per millimeter. A phase grating has a complex amplitude transmittance $\exp[-j2\pi(f_x x + f_y y)]$, where f_x and f_y are the spatial frequencies in the x and y directions. When a light beam illuminates the grating, it is modulated by the grating, which causes the beam to tilt by an angle proportional to the grating spatial frequency. By changes in the grating frequency, the tilt angle of the illuminating beam is controlled. Any arbitrary interconnection pattern may be obtained by a computer-generated hologram that contains a number of grating segments with different spatial frequencies. These grating segments tilt the illuminating beam to create the desired interconnection.

Dynamic or reconfigurable interconnections may be obtained by writing the grating or the hologram onto a spatial light modulator. Currently, the available spatial light modulators have a limited space–bandwidth product, which limits the number of interconnection points that can be achieved. Reconfigurable interconnections have been accomplished by writing the gratings into photorefractive materials.

7. OPTICAL MEMORY

In this section, we briefly discuss the significance of optical memory and its application to optical information processing and computing. Optical storage is currently being used in many commercial applications for audio and video recordings, electronic photography, and computer memories. Optical memory is compact, easily removable, and inexpensive. The current storage density of an optical disk is one bit/μm^2. With the development of short-wavelength semiconductor lasers, the storage density is likely to increase. In general it is read-only memory.

The true potential of optical memory is its parallel access capability that may prove advantageous over magnetic storage. By illuminating a disk by a broad optical beam, many stored bits are accessed in parallel. As shown in Fig. 13, the illuminated bits on the

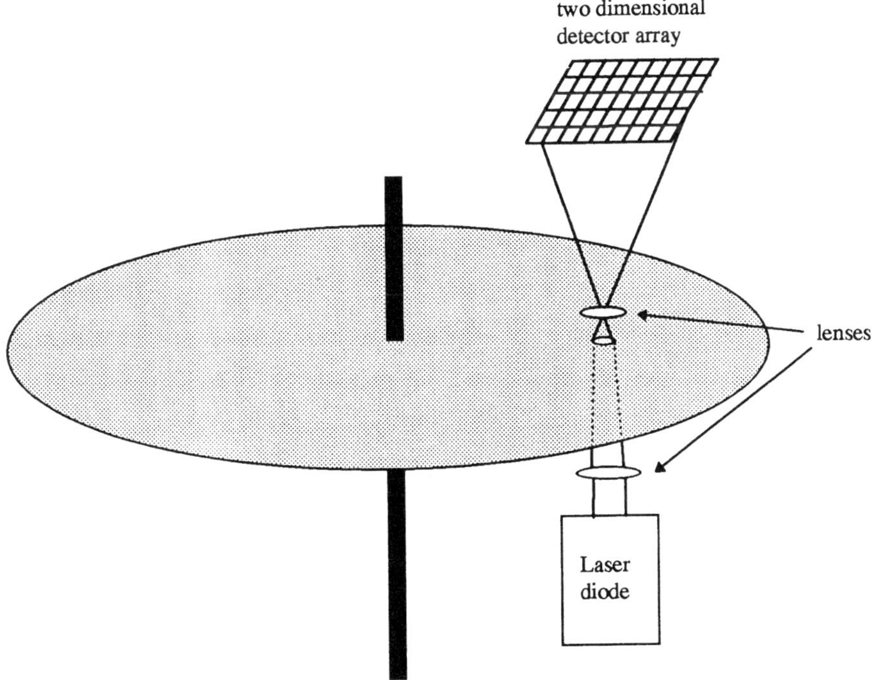

FIG. 13. Optic memory accessible in parallel.

disk are imaged onto a detector array and converted into electronic signals to be utilized by the computer. Given the commercially available optical components and devices, 100 000 channels can be accessed in parallel. The access time to a block of data is a few tens of milliseconds. At present, most optical disks have been of the write-once, read-many (worm) type. The challenge is to develop an erasable—read/write many—optical disk. New materials are now making this possible.

The optical disk can be used to store binary data as 2D blocks for analog image processing. Computer-generated holographic techniques can be utilized to store images in a binary format. The Fourier transform of the data is written in the holographic format, which has the advantage of robustness to misalignments between the disk and the detector array. The data are retrieved by illuminating the hologram on the optical disk, which reconstructs the stored images. The parallel-access optical memory has numerous applications in image processing, database management, and networks. It provides the capability to access a large volume of data rapidly. Parallel-access optical memory is attractive in pattern recognition and associative memory by recording the weights or reference patterns. The pattern to be inspected/searched is displayed on a spatial light modulator to obtain the product between the input pattern and the weights. The product is imaged on a detector, normalized according to the intensity of the input pattern, and is maximum when the input pattern matches the data illuminated on the disk. This process is iterated by rotating the disk and illuminating various portions of the disk to search the entire data to obtain the maximum output. This system is attractive because it provides processing of a large volume of stored data in one disk revolution. The disadvantage of holographic encoding is that it is inefficient in its use of space–bandwidth product: only about one third of the space–bandwidth product is used for the actual information.

Additional improvements in the storage capacity of optical memory can be accomplished by utilizing thick-medium optical disks. The data are stored holographically by interfering the data (such as images) and a plane-wave reference beam. The image is retrieved by reilluminating the hologram on the optical memory with a similar reference beam. In a thick-medium hologram, the image is reconstructed only if the Bragg condition is satisfied, which means that the angle of the reconstruction beam should be equal to the angle of the reference beam used to synthesize the hologram. This property allows the storage of multiple images in the same hologram using angular multiplexing, and retrieving or accessing the images using the proper angle of illumination. The 3D disk is mechanically rotated to scan the entire volume disk. The recording materials are similar to those used in photoreflective recording, such as lithium niobate, and can provide up to 10^{12} pixels for a 5-cm-radius disk. Using acoustic-optic deflectors, the access time to each stored datum in a hologram is 10 μs, which makes the total access time 10 s. Rapid advances have taken place in photorefractive holographic materials such as the commercially available photopolymers and in optoelectronic devices, and the need for large–storage-capacity memory has stimulated much interest in research and development of optical memory. Parallel access and fast data-transfer rates seem to be the key to likely success of optical memory.

The applications remain diverse, ranging from pattern recognition, neural networks, and image processing to database management. The volume optical disk or the 3D disk provides additional storage capacity and makes optical storage a very attractive alternative to its nonoptical counterparts.

8. NEURAL NETWORKS

8.1 Overview of Neural Networks

Artificial neural networks (*q.v.*) (NN), which are also referred to as neuromorphic systems, parallel-distributed processing models, and connectionist machines, are intended to provide humanlike performance by mimicking biological neural systems. They are used in image processing, signal processing, and pattern recognition. They are characterized by massive interconnection of simple computational elements, or nodes, called neurons. Neurons are nonlinear and typically analog, and can have a slow response, typically several hundred hertz. A neuron produces an output by nonlinearly transforming a sum of N inputs as shown in Fig. 14(a), where $f(.)$ represents the nonlinear characteristic of the neuron and w_i is the weight of the interconnection. Three types of neuron nonlinearity are shown in Fig. 14(b). Neural nets provide many computational benefits. The information is stored in the interconnections (w_i). Training or learning changes the interconnection weights.

Because of their large degree of parallelism and massive interconnection capability, NN provide fault tolerance, since damaging a few nodes will not affect the overall performance significantly. Neural networks do not require complete knowledge of the statistical models of the signals to be processed, and instead they utilize available training data. Neural systems are best for problems with no clear algorithmic solutions.

Neural nets are characterized by the network topology, neuron input/output characteristics, and learning rules. Learning plays an important role in the performance of the NN. The ability to adapt the weights is essential in applications such as pattern recognition where the underlying statistics are not available, and the new inputs are continuously changing. Conventional statistical techniques are not adaptive. For classification, a neural-network algorithm is used to compute matching values between the input and the stored data and then to select the class that generates the minimum value. A probabilistic model is used to compute the likelihood or probability that the input belongs to a certain class. If Gaussian distribution is used, simple solutions can be obtained.

Neural-net classifiers may outperform conventional statistical techniques when the underlying distribution of data is generated by nonlinear processes and is strongly non-Gaussian. Neural-net classifiers contain more than one stage. The output of the first stage exhibits the degree of matching between the input and the weights stored in the network.

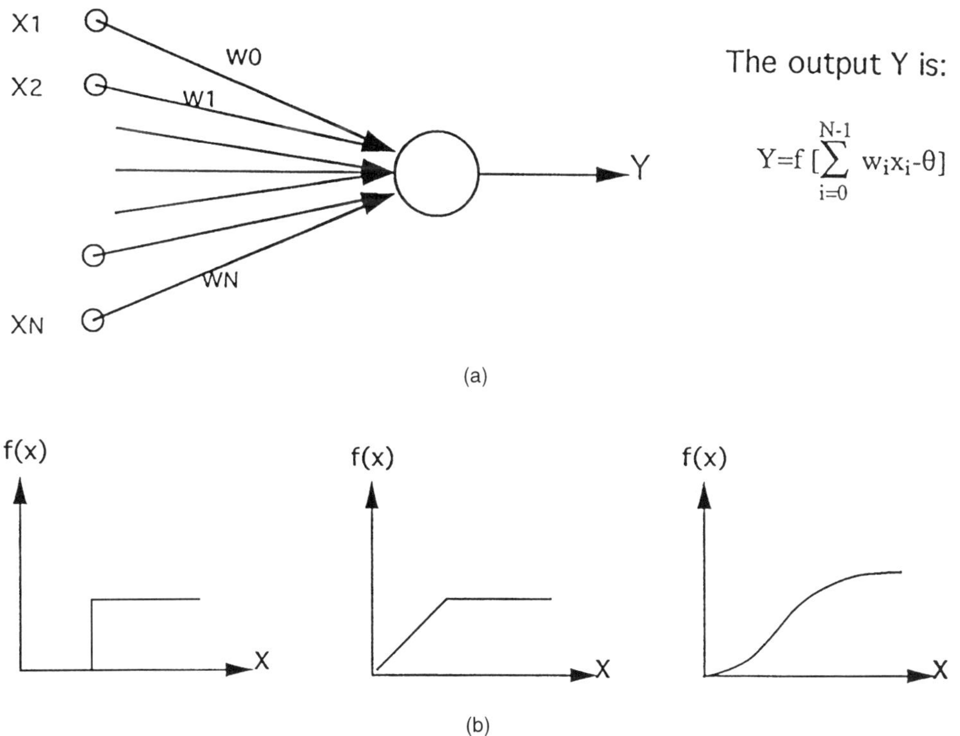

FIG. 14. (a) A neuron produces an output (Y) by nonlinearly transforming a sum of N inputs (X_1–X_N). (b) Three types of neuron nonlinearity.

The maximum of these values is enhanced, and the outputs are forwarded to the second stage. This provides a strong output corresponding to the most likely class. If supervision is provided, this information can be used to adapt the weights of the network using a learning rule, which will improve the performance of the system by reducing the probability of error.

Neural-net–based classifiers are used in three important applications. The first is to identify the class of an input pattern when the input is partially obscured or distorted. This has applications in pattern recognition and classification. The second is in vector quantization and data compression, which are useful in signal and image processing. They are used to reduce the amount of data to be processed without losing the important information. The third is signal and image processing.

8.2 Perceptron

The perceptron learning rule can be implemented in both single-layer and multilayer networks. Figure 15(a) is a single-layer perceptron with a single output that classi-

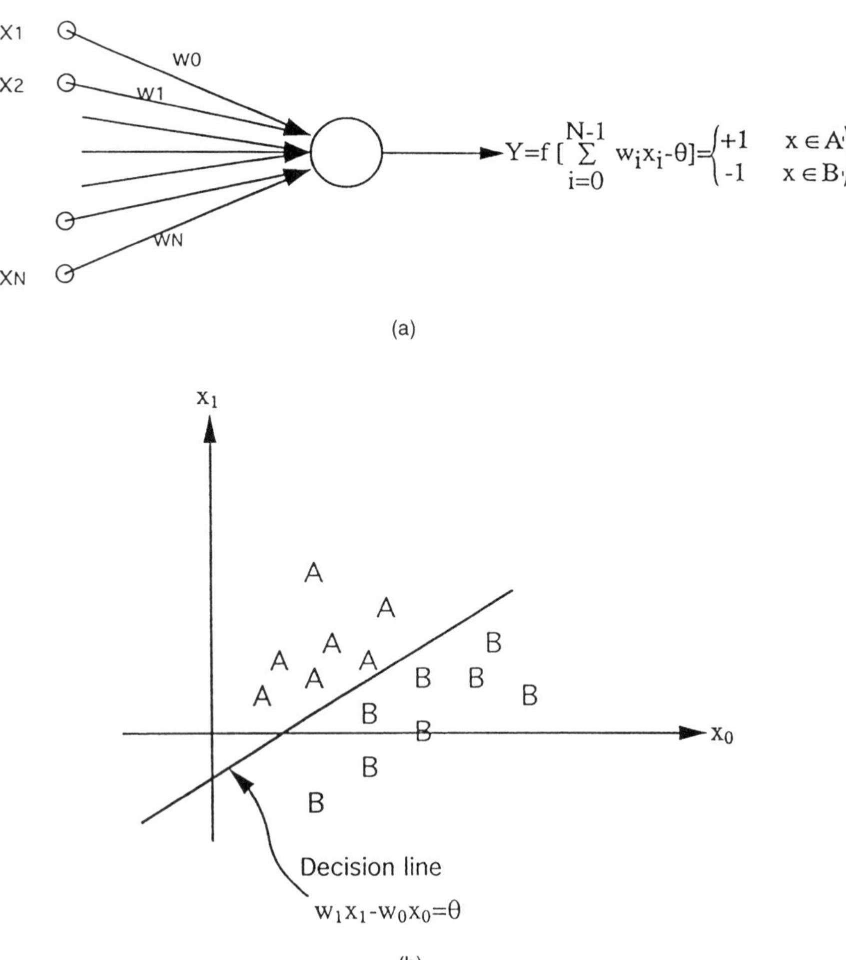

FIG. 15. (a) A single-layer perceptron network with a single output. (b) The decision boundary for the network that separates two classes, A and B. The input vector has two elements, and the decision boundary is a line.

fies an input into two classes, A and B. The decision boundary for this network is a hyperplane that divides the space representation of the input. For example, if the input vector has only two elements, the decision boundary is a line, as shown in Fig. 15(b). The output is the inner product of the inputs and the weights adjusted by a threshold and binarized. An output of +1 represents class A, and −1 represents class B. The connection weights can be adopted using the following learning rule:

$$w_i(t + 1) = w_i(t) + \alpha[d(t) - y(t)]x(t), \qquad (24)$$

with

$$d(t) = \begin{cases} +1, x \in \mathrm{A}, \\ -1, x \in \mathrm{B}, \end{cases} \qquad (25)$$

where $0 < \alpha < 1$ is a positive gain, $d(t)$ is the desired response, $x(t)$ is the input, $y(t)$ is the actual response, and $w_i(t)$ is the weight. If the inputs from the two classes are separable such that they are on opposite sides of a hyperplane, then the perceptron classifier works successfully by placing the decision boundary between the two classes. In this algorithm, the statistical properties of the input are not known. Instead, the procedure focuses on the error between the ideal output and the actual output, which represents the overlap between the different classes. Thus, the algorithm works well when the input statistics are non-Gaussian, generated by nonlinear processes.

8.3 Multilayer Perceptron

When the classes cannot be represented by hyperplane decision boundaries, and are separated by complex decision surfaces, a multilayer perceptron is needed. A multilayer perceptron is a feed-forward network that consists of an input layer, an output layer, and as many hidden layers as needed. A two-layer perceptron is shown in Fig. 16. The nonlinearities used within the nodes of the multilayer perceptron provide the capability to generate the complex decision boundaries.

The back-propagation algorithm is used to train the multilayer perceptron. It is an iterative algorithm designed to minimize the mean square error (MSE) between the desired output and the actual output of a multilayer network. Thus, for each training input x, a desired output d is specified, and continuous differentiable nonlinearities are used in the network. The actual outputs y_i are calculated using the input, weights, and nonlinearities. The weights are adapted to minimize the mean square difference

$$\mathrm{MSE} = n^{-1} \sum_{i=0}^{n-1} (d - y_i)^2. \qquad (26)$$

The number of nodes, the number of hidden layers, and the thresholds need to be set. This poses a practical difficulty. There is no general way to resolve these issues, and the solutions are application dependent. Thresholds can be set using training.

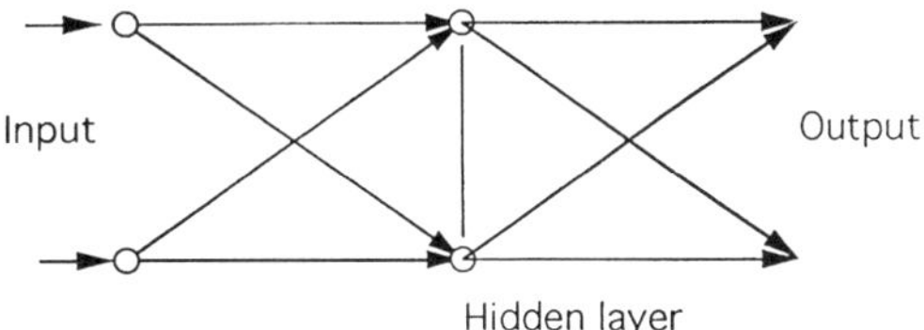

FIG. 16. A two-layer perceptron network.

8.4 Associative Memory

An associative-memory processor stores signals or patterns in the memory (Lippmann, 1987). It is capable of producing an output that is a reproduction of the stored input pattern in response to an input that is a partially obscured or distorted version of one of the stored patterns. For one-dimensional signals, the patterns are stored in a matrix W. The output of the associative memory is the input vector multiplied by the matrix W followed by nonlinear transformations. The associative processors include autoassociative memory and heteroassociative memory. In autoassociative memory, the output recollection is the same as the stored pattern associated with the input. If x is one of the stored patterns, and x' is a distorted version of x, then the output is x. An important property of associative memory is fault tolerance, which means being able to produce the original pattern in response to an input that is a noisy or partially obscured version of a stored pattern. There are also applications to content-addressable memory. When the associative memory is used for pattern recognition, the output is compared to the stored signals to determine if there is a match. The heteroassociative memory produces outputs that are arbitrarily associated to a given input x.

The Hopfield associative memory, which is used with binary inputs such as black and white images or binary data, does not perform as well when continuous gray-scale input values are used. The network contains N nodes with hard limiting nonlinearity $f[.]$ and binary outputs. The output is fed back to the input. The weights t_{ij} are fixed using the M associative signals x^k:

$$t_{ij} = \begin{cases} \sum_{k=0}^{M-1} x_i^k x_j^k, & i \neq j, \\ 0, \; i = j; \end{cases} \tag{27}$$

$0 \leq i, j \leq M - 1$. The output at the time t is μ_i and at time $t + 1$ is

$$\mu_i(t + 1) = f\left[\sum_{i=0}^{n-1} t_{ij}\mu_i\right]. \tag{28}$$

The input is an unknown pattern x' and $\mu_i(0) = x'$. The process is repeated until the output remains unchanged. The output is forced to match one of the patterns x^k stored in the weight. The network converges to a correct solution if the output is the correct version of the distorted input. Graded nonlinearities improve the performance of the Hopfield network. The Hopfield network has many major limitations. First, the storage capacity is limited. If many patterns are stored, the network may converge to a false memory pattern different from all stored patterns, which will produce a no-match output when used as a classifier. This problem can be remedied if the patterns are generated randomly and the number of classes M is less than 0.15 times the number of the input nodes N. The second limitation is that the network may not converge to a correct solution if the stored patterns are too similar to one another. In this case, the stored patterns are considered to be unstable. This problem can be remedied by orthogonalizing the patterns before storing them in the network (Lippmann, 1987).

8.5 Vector-Matrix Architectures

Psaltis and Farhat (Psaltis and Farhat, 1985; Farhat *et al.*, 1985) proposed an optical implementation of the Hopfield (1982) network, as shown in Fig. 17. The Hopfield network is a single-layer network with feedback. It can be used as an autoassociative memory. In this network, neurons accept the input and present the output, and each neuron is connected to all other neurons via the interconnection weights. The weights form a matrix that is called the memory matrix. The matrix is generated by summing all the outer products of the exemplars, which are used to represent the pattern to be retrieved or recognized, in the learning process (an outer product of two vectors is calculated by the matrix multiplication of one column vector with transpose of the second column vector). The diagonal elements of the matrix are set equal to zero to prevent the connection of each node to itself. In the recall process, when an unknown input vector is presented to the network, the output is obtained by performing the matrix multiplication of the input vector and the memory matrix, which is a summation of the stored exemplars weighted by the inner product between the input and the corresponding stored exemplars. The iterations are repeated until the output converges to a stable state, which is one of the stored exemplars that has the fewest different bits from the input. The Hopfield network has limited storage capacity, and only binary patterns can be stored and recalled.

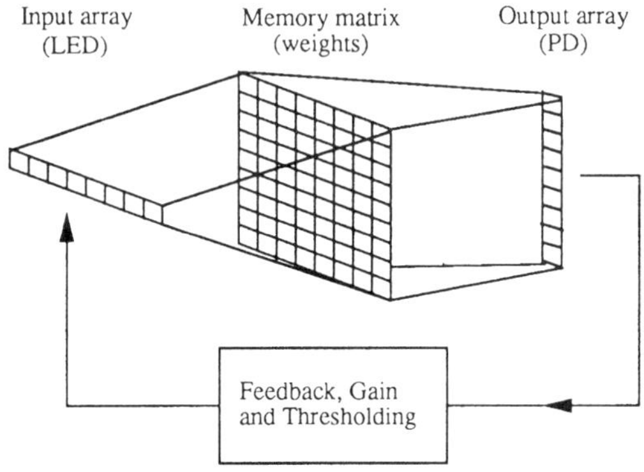

FIG. 17. Scheme of an optical vector-matrix multiplier with nonlinear feedback.

In the architecture shown in Fig. 17, a LED array is used to represent the input array, a photodiode (PD) array is used to represent the output array, and a programmable spatial light modulator is used to store the weights. An anamorphic lens is used between the input and the weight mask to perform the multiplication of the input and the memory matrix, and another anamorphic lens is used between the weight mask and the output array to carry out the summation of the multiplication results in the row direction and to generate the output. When an unknown input is imposed on to the network, the product of the input and the memory matrix is obtained at the output array, and the output is fed back optically to the input through thresholding and gain. Thus, the Hopfield network is implemented by this architecture.

A number of novel optical implementations for autoassociative modules providing versatile adaptive learning capabilities were proposed by Fisher *et al.* (1987). These modules can be configured into larger multiple-module architectures.

Paek *et al.* (1992) described mathematically how all the features of the Hopfield model and higher-order models of neural networks can be derived automatically without any biological assumptions. They also derived the computational circuits for implementing these neural-network models.

8.6 Optical-Correlator Neural Nets

Optical correlators are inherently two-dimensional systems. They can perform some operations on two-dimensional images. As noted in Sec. 8.5, Psaltis and Farhat (1985) proposed using optical correlators to implement the Hopfield model for two-dimensional images. An autoassociative content-addressable memory using the optical correlator is shown in Fig. 18. (Hsu *et al.*, 1990) Originally, the Hopfield network was based on one-dimensional vectors, and outer products of the vectors were used. To implement the outer product between two-dimensional images, spatial-frequency multiplexing is introduced by using two optical correlators. The first correlator is used to obtain the

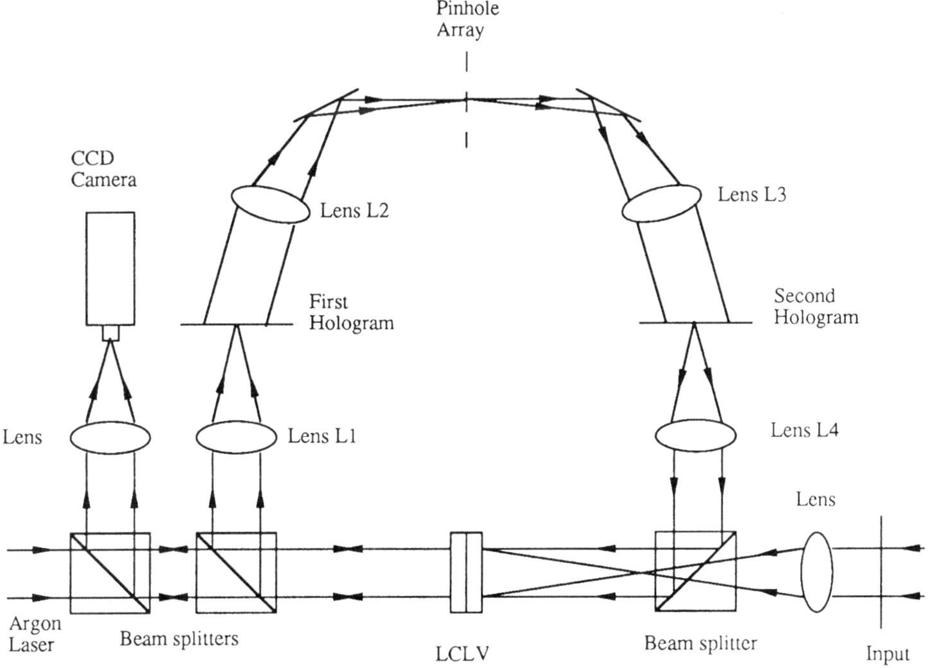

FIG. 18. Schematic diagram of a double-correlator neural network with optical loop.

cross-correlations between the input image and the stored images. The 2D images are stored holographically by means of the Fourier-transform hologram. The hologram is recorded on a thermoplastic plate. Each image is recorded with a different reference beam angle, so-called angular multiplexing. In the retrieval process, the input is imaged onto a liquid-crystal light valve (LCLV) (Sec. 4.4). An argon laser is used to read out the input image from the LCLV. Lens L1 is used to perform the Fourier transform of the input image. On the back focal plane of lens L1, the Fourier transform of the input image is multiplied by the transmittance of the hologram, which records the complex conjugate of the Fourier transforms of the stored images. The product is then passed through another Fourier-transform lens L2. On the focal plane of lens L2, a pinhole array is placed to sample the correlation signals between the input image and the stored images recorded on the hologram. The separation between the pinholes is determined by the focal length of L2 and the reference beam angles used for recording the images on the hologram. In the second correlator architecture, the output light of the pinhole array illuminates another hologram that is similar to the first one. The output at the back focal plane of lens L4 is a weighted superposition of the images stored in the second hologram, with the weights proportional to the correlations of the input and the images stored in the first hologram. By doing this, the outer-product model with associative property is realized for two-dimensional images. The same LCLV is used to obtain the output and feed it back into the first correlator to form the iteration loop. The iterations are repeated until the output is stable. A facial image retrieval test with four people's images stored on the hologram and with partial and distorted images as inputs was shown (Hsu *et al.*, 1990).

Paek *et al.* (1990) proposed a holographic memory, shown in Fig. 19. A photorefractive crystal, lithium niobate ($LiNbO_3$), is used to store multiple holograms. An argon laser with $\lambda = 514.5$ nm is used to record the holograms. A low-threshold electronically pumped vertical-cavity surface-emitting microlaser diode array (VCSEL) with $\lambda = 960$ nm is used to retrieve the images stored in the holograms. The angular separation of the diodes is 0.04° to ensure the angular separation for selective read out. The system has a 10^4-bits/s access speed.

Khoury *et al.* (1992) proposed using the binary phase-only filter (Horner and Gianino, 1984) to realize an associative memory. A 2*f* correlator architecture (Fielding *et al.*, 1990) was used instead of a 4*f* correlator. Because the binary phase-only filter provides sharp autocorrelation, no feedback or nonlinearity was employed. A mirror was used in the correlation plane to reflect the correlation output back to the binary phase-only filter to restore the stored image.

8.7 Hologram-Based Neural Nets

Owechko *et al.* (1987) proposed an autoassociative memory, shown in Fig. 20. This network uses the outer-product neural-network model. The desired images are stored in Fourier-transform holograms with different reference beams in different angles. A

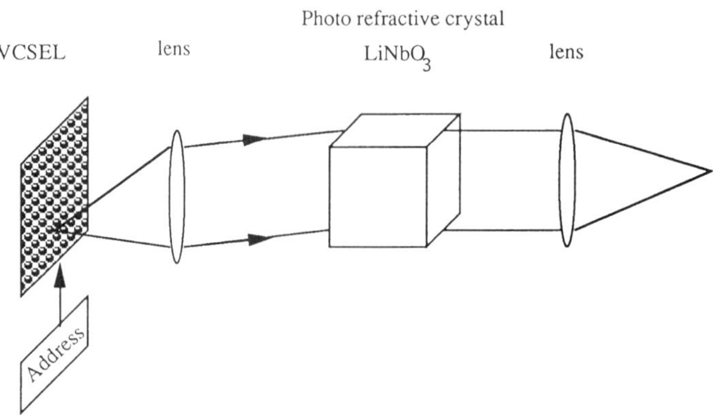

FIG. 19. A holographic memory system.

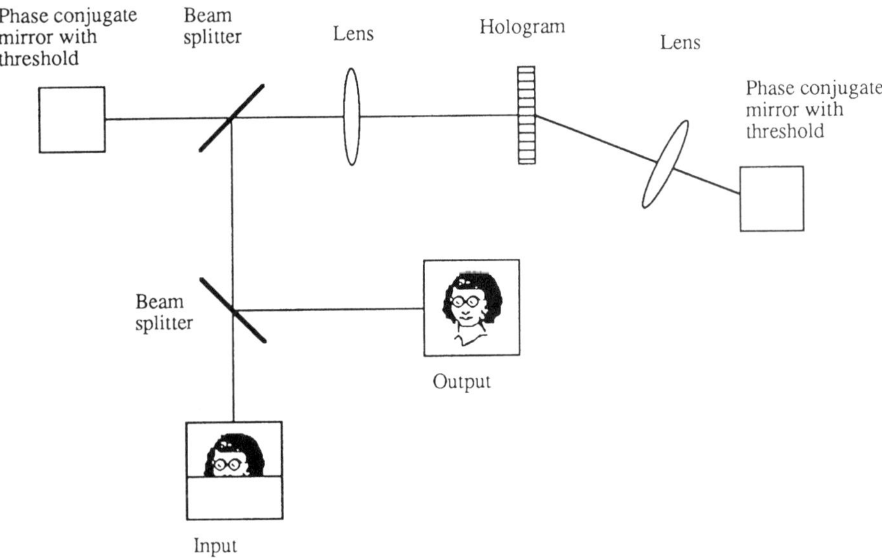

FIG. 20. Schematic diagram of a nonlinear holographic associative memory.

thermoplastic film is used to record the hologram. An input is Fourier transformed by the first lens, and the Fourier transform of the input is multiplied by the hologram that records the conjugate of the Fourier transforms of the stored images. The correlations between the input and the stored images are obtained on the focal back plane of the second lens. A phase-conjugate mirror is used to reflect the correlation signal with a conjugated phase back to the hologram. This way, the hologram acts as a memory that, with the illumination of the correlation signals, generates a weighted superposition of the stored images. This newly formed image is then reflected back by another phase-conjugate mirror as a new input. Thus, the two phase-conjugate mirrors provide a resonator cavity for feedback, and all the stored images oscillate inside the cavity. With the threshold and the nonlinear reflectivity of the phase-conjugate mirror, the system will converge to the strongest correlated stored image while the other images will vanish.

A single-image experiment is performed by storing a portrait in the hologram and retrieving it from a partial version of the original image.

Paek and Lehmen (1989) realized a holographic associative memory that is capable of identifying individual words and inserting word breaks into a concatenated word string. The architecture is quite similar to the two-correlator architecture described above, except that electronics was used at the correlation plane to find the correlation peaks and stretch them in the horizontal direction. The stretched correlation peaks were forwarded to the second correlator to restore the desired words with proper space between them.

8.8 Hughes Programmable Multilayer Optical Neural Net

In a hybrid optoelectronic system (Owechko and Soffer, 1988), shown in Fig. 21, two liquid-crystal light valves are used to form the oscillation cavity, and a photorefractive crystal is used to store the volume phase holograms. It is a modification of the system described in the previous section (Owechko *et al.,* 1987). Computers are used to realize the feedback loops. It is claimed that 7×10^5 neurons and 7×10^7 interconnections can be achieved with this system. The neuron update rate can be 10^7 neurons/s, and the data rate can be 2×10^9 interconnects/s.

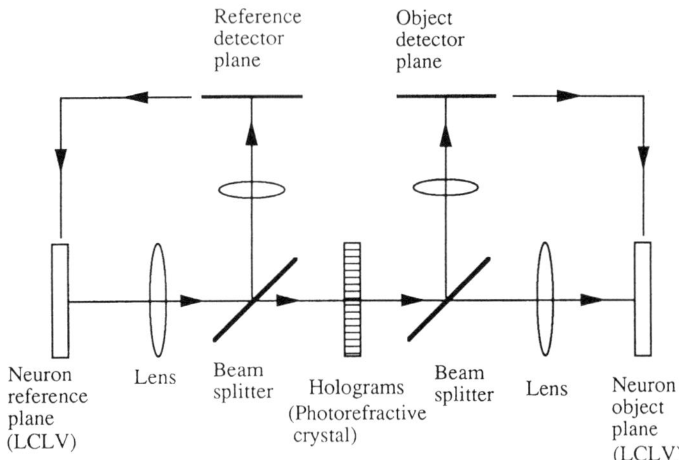

FIG. 21. Schematic diagram of a programmable optical neural network.

8.9 Learning in Multilayer Neural Nets

A handwritten-character recognition system was built by Psaltis and Quio (1990). It was realized by a multilayer optical neural network, shown in Fig. 22. A rotating mirror was used to change the reference beam in 26 different directions, two $LiNbO_3$ photorefractive crystals were used to represent the input layer and the hidden layer, and a CCD camera was used to represent the output layer. The learning method proposed by Kanerva (1986) was used to train the system. The weights are initially assigned random values and are updated with the new inputs. Each input character has 100 pixels, the hidden layer contains 10^5 units, and the output layer has 26 units that represent one of the 26 letters of the alphabet. In tests, 104 patterns were used to train the system, and a test set

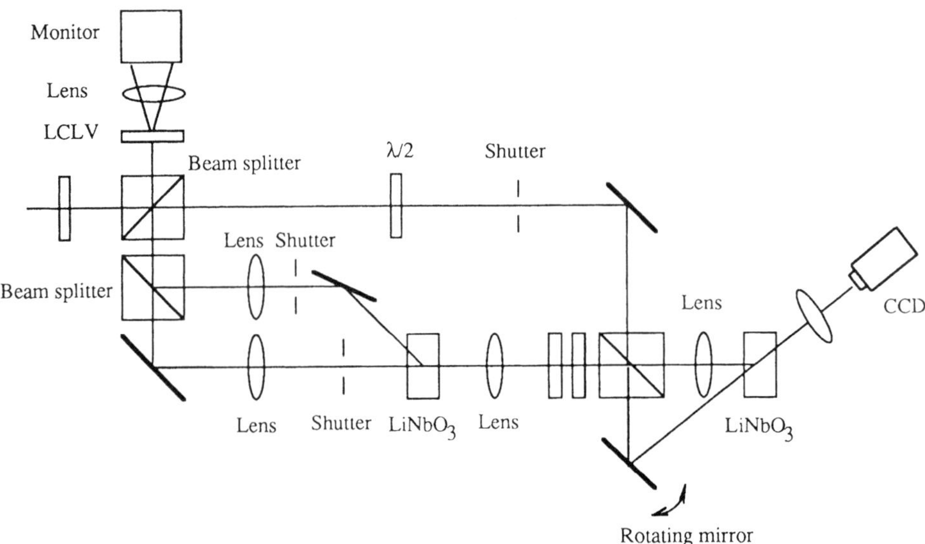

FIG. 22. A two-layer optical network for character recognition.

of 520 patterns was used. The error probability was 44%. It is claimed that the system has a rate of 10^{12} multiplications/s.

9. CONCLUSIONS

We have tried to present a balanced and comprehensive overview of the field of optical signal processing/computing as it has evolved from its beginnings in the early 1960s to today. It has broadened considerably in the intervening four decades—starting out with image enhancement using optical high-pass and low-pass filtering and pattern recognition with the optical correlator. Within optics it is a very large field. For example, the Optical Society of America publishes a separate monthly research journal on information processing alone. Many of the innovative algorithms developed in the context of optical processing are also implementable on a digital computer and perform extremely well compared with various algorithms developed by the digital community. Improvements are occurring rapidly in the basic building block for optical computing and processing, namely, the spatial light modulator. Nearly half a dozen companies are pursuing this technology, which stimulates new advances in applications and materials such as liquid crystals. New areas of application are opening all the time, such as the use of this technology for law enforcement, security, and anticounterfeiting. Although government-funded research is undergoing a transition at this time, with the end of the cold war, we feel the field has a bright and continuing future.

ACKNOWLEDGMENT

We acknowledge the support of Rome Laboratory at Hanscom Air Force Base in preparing this article.

Works Cited

Berg, N., Lee, J., (1983), *Acousto-Optic Signal Processing,* New York: Marcel Dekker.

Bleha, W. P., Lipton, T., Wiener-Avnear, E., Grinberg, J., Reif, P. G., Casasent, D., Brown, H. B., Markevitch, B. V. (1978), "Application of the Liquid Crystal Light Valve to Real Time Optical Data Processing," *Opt. Eng.* **17,** 371.

Dallas, W. J. (1971), "Computer-Generated Hologram," in: B. R. Frieder (Ed.), *The Computer in Optical Research,* Topics in Applied Physics, Vol. 41, Berlin: Springer.

Efron, U., Grinberg, J., Braatz, P. O., Little, M. J., Reif, P. G., Schwartz, R. N. (1985), "The Silicon Liquid-Crystal Light Valve," *J. Appl. Phys.* **57,** 1356.

Efron, U., Au, A., Bak, C. S., Goodwin, N. W., Reif, P. G., Garvin, H. L., Byles, W., Owechko, Y., Welkowsky, M. S. (1989), in: B. Javidi (Ed.), *Optical Information Processing Systems and Architectures,* SPIE Proceedings No. 1151, Bellingham, WA: SPIE, p. 591.

Farhat, N. H., Psaltis, D., Prata, A., Paek, E. (1985), "Optical Implementation of the Hopfield Model," *Appl. Opt.* **24,** 1469.

Fielding, K. H., Horner, J. L., Makekeau, C. (1990), "Modified two-focal length optical correlator," *Appl. Opt.* **29,** 4332.

Fisher, A., Lippincott, W., Lee, J. (1987), "Optical Implementations of Associative Networks with Versatile Adaptive Learning Capabilities," *Appl. Opt.* **26,** 5039.

Flannery, D. L., Horner, J. L. (1989), "Fourier Optical Signal Processors," *Proc. IEEE* **77,** 1511.

Goodman, J. W. (1968), *Introduction to Fourier Optics,* New York: McGraw-Hill.

Guenther, R. D. (1990), *Modern Optics,* New York: Wiley.

Hopfield, J. J. (1982), "Neural Networks and Physical System with Emergent Collective Computational Abilities," *Proc. Natl. Acad. Sci. USA* **79,** 2554.

Hornbeck, L. J. (1983), "128 × 128 Deformable Mirror Device," *IEEE Trans. Electron and Devices* **ED30,** 539.

Horner, J. L. (1982), "Light Utilization in Optical Correlators," *Appl. Opt.* **21,** 4511.

Horner, J. L., Gianino, D. (1984), "Phase-Only Matched Filtering," *Appl. Opt.* **23,** 812.

Horner, J. L., Leger, J. R. (1985), "Pattern Recognition with Binary Phase-Only Filters," *Appl. Opt.* **24,** 609.10.

Horner, J. L., Javidi, B., Wang, J. (1992), "Analysis of the Binary Phase-Only Filter," *Opt. Commun.* **91,** 189.

Hsu, K.-Y., Li, H.-Y., Psaltis, D. (1990), "Holographic Implementation of a Fully Connected Neural Network, *Proc. IEEE* **78,** 1637.

Javidi, B. (1989), "Nonlinear Joint Power Spectrum Based Optical Correlation," *Appl. Opt.* **28,** 2358.

Javidi, B. (1990), "Generalization of the Linear Matched Filter Concept to Nonlinear Matched Filters," *Appl. Opt.* **29,** 1215.

Javidi, B., Horner, J. L. (1989a), "Single Spatial Light Modulator Joint Transform Correlator," *Appl. Opt.* **28,** 1027.

Javidi, B., Horner, J. L. (1989b), "Multifunction Nonlinear Signal Processor: Deconvolution and Correlation," *Opt. Eng.* **28,** 837.

Javidi, B., Horner, J. L. (1994), *Real Time Optical Information Processing,* Boston: Academic.

Javidi, B., Kuo, C. (1988), "Joint Transform Image Correlation Using a Binary Spatial Light Modulation at the Fourier Plane," *Appl. Opt.* **27,** 663.

Javidi, B., Wang, J. (1992), "Limitation of the Classic Definition of the Correlation Signal-to-Noise Ratio in Optical Pattern Recognition with Disjoint Signal and Scene Noise," *Appl. Opt.* **31,** 6826.

Javidi, B., Wang, J. (1994), "Design of Filters to Detect a Noisy Target in Non-overlapping Background Noise," *J. Opt. Soc. Am.* **11,** 2604.

Javidi, B., Zhang, G. (1992), "Experiments on Nonlinearly Transform Matched Filters," *Opt. Eng.* **31,** 934.

Javidi, B., Tang, Q., Gregory, D., Hudson, T. (1991), "Experiments on Nonlinear Joint Transform Correlators Using an Optically Addressed SLM in the Fourier Plane," *Appl. Opt.* **30,** 1772.

Javidi, B., Refregier, P., Willett, P. K. (1993), "Design of an Optimum Receiver for Pattern Recognition with Spatially Disjoint Signal and Scene Noise," *Opt. Lett.* **18,** 1160.

Javidi, B., Zhang, G., Fazlollahi, A. H., Efron, U. (1994), "Application of a Wire-Grid-Mirror Liquid-Crystal Light Valve in a Nonlinear Joint Transform Correlator," *Appl. Opt.* **33,** 2834.

Johnson, K. M., Moddel, G. (1989), "Motivations for Using Ferroelectric Liquid Crystal Spatial Light Modulators in Neurocomputing," *Appl. Opt.* **28,** 4888.

Kanerva, P. (1986), in: J. S. Denker (Ed.) *Neural Networks for Computing,* New York: American Institute of Physics.

Khoury, J., Kane, J. S., Hemmer, P., Woods, C. (1992), "Binary Phase-Only Filter Associative Memory," *Appl. Opt.* **31,** 1818.

Kirsch, J. C., Gregory, D. A., Thie, M. W., Jones, B. K. (1992), "Modulation Characteristics of the Epson Liquid Crystal Television," *Opt. Eng.* **31,** 963.

Kogelnick, H. (1969), "Coupled Wave Theory for Thick Hologram Gratings," *Bell Syst. Tech. J.* **48,** 2909.

Korpel, A. (1988), *Acousto-Optics,* New York: Marcel Dekker.

Lohmann, A. U., Brown, B. R. (1966), "Complex Spatial Filtering with Binary Masks," *Appl. Opt.* **5,** 967.

Lippmann, R. P. (1987), "An Introduction to Computing with Neural Nets," *IEEE ASSP Mag.* 4.

Lu, K., Saleh, B. (1990), "Theory and Design of the Liquid Crystal TV as an Optical Spatial Phase Modulator," *Opt. Eng.* **29,** 240.

McAulay, A. D. (1991), *Optical Computer Architecture,* New York: Wiley.

Molley, P. A., Stalker, K. T. (1990), "Acousto-Optic Signal Processing for Real-Time Image Recognition," *Opt. Eng.* **29,** 1073.

O'Neill, E. L. (1963), *Introduction to Statistical Optics,* Reading, MA: Addison-Wesley.

Oppenheim, A., and Lim, J. (1981), "The Importance of Phase in Signals," *Proc. IEEE* **69,** 529.

Owechko, Y., Soffer, B. H. (1988), "Programmable Multi-layer Optical Neural Networks with Asymmetrical Interconnection Weights," *Proc. IEEE International Conference on Neural Networks,* p. 385.

Owechko, Y., Dunning, G. J., Marom, E., Soffer, B. H. (1987), "Holographic Associative Memory with Nonlinearities in the Correlation Domain," *Appl. Opt.* **26,** 1900.

Paek, E., Lehmen, A. (1989), "Real-time Holographic Associative Memory for Identifying Words in a Continuous Letter String," *Opt. Eng.* **28,** 519.

Paek, E. G., Wullert, J. R., II, Jain, M., Lehmen, A. V., Scherer, A., Harbison, J., Florez, L. T., Yoo, H. J., Martin, R., Jewell, J. L., Lee, Y. H. (1990), "Compact and Ultrafast Holographic Memory Using a Surface-Emitting Microlaser Diode Array," *Opt. Lett.* **15,** 341.

Paek, E., Liao, P., Gharavi, H. (1992), "Derivation of Neural Network Models and Their Computational Circuits for Associative Memory," *Opt. Eng.* **31,** 986.

Psaltis, D. (1984), "Two Dimensional Optical Processing Using One Dimensional Input Devices," *Proc. IEEE* **72,** 962.

Psaltis, D., Farhat, N. H. (1985), "Optical Information Processing Based on an Associative-Memory Model of Neural Nets with Thresholding and Feedback, *Opt. Lett.* **10,** 98.

Psaltis, D., Qiuo, Y. (1990), "Optical Neural Networks," *Optics and Photonics News,* p. 17.

Psaltis, D., Paek, E., Vankatesh, S. (1984), "Optical Image Correlation with Binary Spatial Light Modulator," *Opt. Eng.* **23,** 698.

Saleh, B. E. A. (1991), *Fundamentals of Photonics,* New York: Wiley.

Turin, J. L. (1960), "An Introduction to Matched Filters," *IRE Trans. Inf. Theory* **IT-6,** 311.

VanderLugt, A. (1964), "Spatial Detection by Computer Filtering," *IEEE Trans. Inf. Theory* **IT-10,** 139.

Warde, C., Fisher, A. (1987), in: J. L. Horner (Ed.), *Optical Signal Processing,* New York: Academic.

Weaver, C. S., Goodman, J. W. (1966), "Technique for Optically Convolving Two Functions," *Appl. Opt.* **5,** 1248.

Xu, J., Stroud, R. (1992), *Acoustic-Optic Devices: Principles, Design, and Applications,* New York: Wiley.

Young, M. (1986), "Low-Cost LCD Video Display for Optical Processing," *Appl. Opt.* **25,** 1024.

SILICON DIOXIDE

R. Brückner, *Institut für Nichtmetallische Werkstoffe, Technische Universität Berlin, Garmisch-Partenkirchen, Germany.*

INTRODUCTION

Silicon dioxide is a very simple chemical compound that is widely spread over the earth's crust in the form of quartz, the most stable modification at normal temperature and pressure conditions. It serves as a basic raw material for silicate glasses and silicate ceramics. As simple as this compound is on the one hand, its crystallographic modifications and their structures are manyfold and interesting on the other. However, the most interesting version with respect to properties and physical applications is the noncrystalline state of SiO_2, that is, silica glass and silica gel. Therefore, in this chapter the many varieties of silica glass will be discussed preferably.

1. THE VARIOUS PHASES OF SILICON DIOXIDE

1.1 Phases at Atmospheric Pressure, Properties

At normal atmospheric pressure, the following modifications of silica exist (Fig. 1): α-quartz is the most stable low-temperature

3-527-28140-1/96/$5.00 + .50

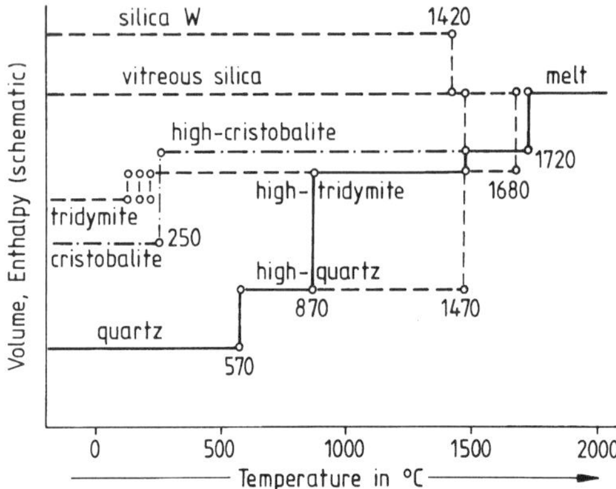

FIG. 1. Silica phases at atmospheric pressure (semi-schematic).

trigonal phase. At 573 °C the structure changes to the high-temperature hexagonal one, and at 1470 °C this phase converts to cubic high-cristobalite that has a melting temperature of T_m = 1713 °C. Upon cooling, a very viscous melt from above T_m undercooling is obtained very easily without crystallization that is frozen to a glass below about 1200 °C (Hetherington *et al.*, 1964), the so-called glass transition temperature *Tg*.

In impure SiO_2, with impurities such as alkali oxide, hexagonal high tridymite occurs at 867 °C from quartz, and then at 1470 °C it converts into the cubic high cristobalite. Because of impurities, the orthorhombic low tridymite has numerous variations, called tridymite M with inversions I to III, S with I to VI, and tridymite U. In a strict sense tridymite does not belong to the pure SiO_2 system (Flörke, 1956a). All low–high inversions of quartz, cristobalite, and tridymite are reversible displacive transformations that take place rapidly, whereas the conversions to another phase, e.g., quartz to cristobalite or cristobalite to tridymite and vice versa, are reconstructive transformations with very slow rates. These transformation rates are one reason, among others, that SiO_2 is a very good glass former.

Another low-pressure modification of SiO_2 occurs at high temperatures under reducing, water-vapor-free conditions and under conditions at which oxidation of SiO occurs: the so-called silica W that crystallizes in long needles with a chain structure such as (Sosman, 1965)

$$\rangle Si\langle{}^{O}_{O}\rangle Si\langle{}^{O}_{O}\rangle Si\langle. \qquad (1)$$

This modification is unstable and will change in the presence of water vapor.

On the basis of density, the following series comprise approximately the low-pressure low-temperature modifications, in addition to the diagram of Fig. 1: quartz, 2.65; cristobalite, 2.33; tridymite, 2.46; fused silica, 2.20; and silica W, 1.96 g cm^{-3}. The refractive indices of these modifications change parallel to the changes in densities.

1.2 High-Pressure Phases, Properties

Keatite is created hydrothermally at about 380–540 °C and at 88–130 MPa. The density is about 2.50 g cm^{-3}, and its thermal expansion is negative. The densest phase of silica (4.35 g cm^{-3}) is stishovite, which is produced approximately at 16 GPa and at 1200–1400 °C. It is the only modification of SiO_2 with a sixfold coordination of Si^{4+} (Sosman, 1965).

Coesite, the second heaviest phase of silica (density 3.01 g cm^{-3}), is formed at temperatures of 300–1700 °C and under pressures of 1.5–4.0 GPa.

1.3 Morphology and Main Applications

Although silicon dioxide exists in a large number of modifications and is able to maintain the various fundamental phases at normal pressure and room temperature, even the high-pressure modifications, only a few of them are suitable for applications. The reason is that the crystals of many of the modifications have only been prepared so far in sizes <1 mm in their maximum dimension (Sosman, 1965). Quartz and fused silica are the great exceptions. Cristobalite is of importance in the high-temperature furnace industry (Jebsen-Marwedel and Brückner, 1980).

1.3.1 Quartz and Cristobalite The appearance of quartz in nature is manyfold. Relatively large crystals are rock quartz, smoky quartz, rose quartz, amethyst, mild quartz, cat's eye, tiger's eye, etc. Fine crystalline quartz are chalcedon, onyx, sardonyx, jade, agate, etc. Quartz crystals as well-developed large pieces can be found in nature as rock crystals that were grown up hydrothermally from solutions. Also the artificially produced quartz crystals can be obtained as large crystals and are produced via the hydrothermal route. Usually the crystals are of excellent transparency that may be reduced by γ radiation (smoky rock crystals) or that will be reduced selectively resulting in amethyst or in rose quartz. While the smoky rock crystals can be decolored by heating up to about 250–300 °C where the paramagnetic color centers or the structural defects are regenerated (Sosman, 1965; Griscom, 1986), the violet or purple and red color of the amethyst and rose quartz, respectively, cannot be removed because they are produced by iron, manganese, and titanium impurities. Other remarkable properties of quartz are the birefringence with an optical indicatrix of a rotational ellipsoid with the two refractive indices $n_\omega = 1.544$ and $n_\epsilon = 1.553$; the rotary power for plane-polarized light along the *c* axis of the hexagonal low-temperature modification in a dextro- or levo-oriented sense depends on the twisted structure of the SiO_4 tetrahedra along the *c* axis. This property is less pronounced in quartz crystals with lower quality due to twinning. Amethyst has an "alternating dextro- and levostructure with such a remarkable—never artificially reached—symmetry of twinning that no rotary power for plane-polarized light exists in it" (Sosman, 1965).

One of the applications of the optical rotary power of quartz is an optical element, called half-shadow plate, for dextro- and levocircularly polarized light. Well known are the $\lambda/10$, $\lambda/20$, and $\lambda/4$ plates for optical compensators in order to measure birefringence of other substances or the so-called quartz wedges for continuously produced birefringence and, in conjunction with crossed Nicol prisms, to get optimum contrast [Nurmarski method for the phase-contrast microscope (see OPTICAL MICROSCOPY)].

Another very important application of well-developed quartz crystals is based on the piezoelectric effect. This property is the basis of ultrasonic sources usually in the megahertz range for longitudinal and transversal waves and is also the basis of steering elements in chronometry (for more details see QUARTZ).

Cristobalite can be obtained only as small crystals usually by a reconstructive transformation at high temperatures (~1300–1650 °C) from quartz or from fused silica. The main reason for the small crystal sizes is the displacive high–low (cubic-tetragonal) transformation with a volume contraction of ≤4% at temperatures between 180 and 270 °C, which depends on impurities and on the temperature of the reconstructive transformation. The lower the latter temperature and the larger the impurity concentration, the smaller are the displacive transformation temperatures and the volume changes. The reason is the increasing disorder of the structure (Flörke, 1956b). The main application of cristobalite is the use as refractory bricks, particularly in glass industry furnaces, particularly as a light material (ρ = 2.33 g cm^{-3}) for the furnace ceiling (Jebsen-Marwedel and Brückner, 1980).

1.3.2 Vitreous Silica, Silica Gel, Silica-Glass Fibers These varieties are the glassy versions of SiO_2 and respectively the amorphous versions of SiO_2 gels. The glassy version is also called vitreous silica, silica glass, fused quartz, or quartz glass. It can be obtained in two ways: first of all by melting (see Fig. 1) and recently by the sol-gel and the colloid methods; likewise, in principle, the silica glass fibers. Because of the noncrystalline condition, these SiO_2 varieties can be formed into all geometries that are possible by the usual glass-forming principles. This formability is the main reason for the larger range of and potential for applications as compared to the crystalline versions of SiO_2 (Sec. 1.3.1).

The properties that are unique for the noncrystalline versions of SiO_2 are the low coefficient of thermal expansion (~5 × 10^{-7} K^{-1}), the high glass transition temperature (~1100–1200 °C), the high tensile strength of the fibers (~10 GPa), the excellent elastic constants, the low optical absorption (losses <0.5 dB/km), and the excellent chemical stability.

These excellent properties, which are useful for many important applications (e.g., the use of SiO_2 crucibles in semiconductor production, the use in optical waveguides, and the use in microelectronic elements on the basis of silicon, see Sec. 2.6), are possible only through special preparation techniques that have lead to many kinds of fused silica. These may be divided into a few groups or types, listed in Table 1. The different impu-

Table 1. Characteristic types of vitreous silica developed over decades for special properties and applications and from different sources.

Type	Method of manufacture	Typical impurities in ppm				Characteristic optical and other properties, remarks
		Al	Na	OH	Cl	
I	Electric melting of quartz crystals in crucibles under vacuum	10–100	4	≤5	Not detected (n.d.)	Good transmittance in the near infrared, absorption band at 247 nm
II	H_2/O_2 flame fusion of quartz crystals	<30	1	150–300	(n.d.)	Good UV transmittance
III	Hydrolysis of $SiCl_4$ (or other silanes) in H_2/O_2 flame, deposition and direct consolidation in the same step	⩽1	⩽1	1000	<50	Excellent UV transmittance, absorption bands at 1.4, 2.2, and 2.7 μm
IIIa	Hydrolysis of $SiCl_4$ (or other silanes) in H_2/O_2 flame, deposition of soot particles to form a porous intermediate body, consolidtaion to full density in a separate step; no dehydration	⩽1	⩽1	200	<50	Excellent UV transmittance, better IR transmittance than type III
IIIb	As IIIa wtih dehydration of intermediate soot body by Cl treatment	⩽1	⩽1	<0.1	1000	Excellent transmittance from UV to IR
IV	Oxidation of $SiCl_4$ in water-vapor–free plasma flame	⩽1	⩽1	<5	<50	Purest type so far
Va	Electric melting of very pure pegmatitic quartz in Mo or W crucibles in H_2-containing atmosphere as melted	10–20	~1	100	(n.d.)	OH absorption at 2.7 μm and small SiH IR-band, large absorption around 245-nm; high gas content caused by physically dissolved H_2
Vb	As Va but out-gassed 10 h at 1080 °C in vacuum or air	10–20	~1	<1–15	(n.d.)	Absorption at 247 nm higher viscosity than types II–IV; volume-wise the most importnat type at present, nearly H_2 free
VI	Sintering of colloidal SiO_2 glass powder preforms after chloride reaction purification	⩽1	⩽1	⩽1	100–500	Fabricated on pilot scale so far
VII	Sol-gel route of Si-alkoxide, drying and sintering	<1	<1	<1	(n.d.)	Fabricated on laboratory scale so far

rities of the various silica-glass types effect the various properties and will be discussed in Sec. 2. The principal preparation techniques are the electrical or flame (H_2/O_2 or plasma) melting of (rock) quartz crystals, the hydrolysis and pyrolysis of $SiCl_4$, the sol-gel route starting from Si-alkoxide solution, and sintering of a silica gel or sintering together with colloidal amorphous SiO_2 powder (Clasen, 1987; Hench, 1993). The advantage of the sol-gel route is the high purity of the silica glass (even without OH^- and Cl^-; Hench, 1993). The advantage of the colloidal particle sintering technique is the high cleaning effect from impurities such as metal ions and water and the possibility of the introduction of specific ions for coloring and of colloidal coloring metals (Clasen, 1993).

The preparation of silica gel can be done in an additional way to the sol-gel route.

Starting material is the so-called water glass, an alkali silicate glass with ≤50 wt % Na_2O. The Na_2O content is leached out by water; the water is exchanged by alcohol or by fluid CO_2, which is transformed to a temperature and pressure above the critical point of CO_2 where the surface tension cannot contract the porous silica gel. In this manner silica gels with a large porosity can be obtained in large pieces without shrinking cracks. These are the so-called aerogels with densities as small as 0.005 g cm^{-3}. The main properties of these are their ultrasonic properties (see Sec. 3) and excellent thermal isolation.

Silica-glass fibers have much higher tensile strength than the bulk silica glass. Additionally, in the case of types IIIb, IV, and Vb silica glasses (Table 1), application is usual as optical waveguides with high optical transparency or low optical losses in the near-infrared part of the spectrum. The production is possible usually via a so-called preform and pulling out the heated end to a fiber of the desired diameter. In the case of simple fibers for reinforcing purposes of composite materials and thermal isolating, the preform consists only of a silica-glass rod of about 8 mm in diameter. If d_r is the diameter of the rod, d_f that of the fiber, and V_f the drawing velocity of the fiber, the velocity V_r of the rod results from the relation

$$V_r{:}V_f = d_f^2{:}d_r^2.$$

The drawing temperatures are well above 2000 °C. The silica-glass fibers as optical waveguides, on the other hand, are built up as step-index or gradient-index fibers, where the refractive index of the core is larger than that of the slad. More details are given in Sec. 2.4.2.

1.4 The System Si–O, Bond Nature and Structural Aspects

The total system Si–O is not well known, except for the two extreme components, Si and SiO_2. The meta composition, SiO, is stable only at high temperatures and decomposes into SiO + SiO_2 and Si + SiO at somewhat lower temperatures and into Si + SiO_2 at still lower temperatures (Sosman, 1965). Silicon monoxide, SiO, is formed as one of the gaseous products when silica is heated above 1500 °C or is in contact with silicon, carbon, hydrogen, or hydrocarbon at or above temperatures of about 1000 °C. Figure 2 shows the partial pressures of the decomposition of SiO_2 in a neutral atmosphere between 1500 and 4000 K (Fanderlik, 1991).

The building unit of crystalline and vitreous silicon dioxide is the SiO_4 tetrahedron, or better, the $SiO_{4/2}$ tetrahedron. The only exception is stishovite with its SiO_6 octahedron. The SiO_4 tetrahedron establishes the short-range order and is based on the excited state of the covalent part of the silicon atom forming the sp^3 hybride: $1s^2\ 2s^2\ 2p^6\ 3s^1\ 3p^3$. This arrangement corresponds to four valencies with σ bonds directed toward the four corners of the regular tetrahedron occupied by four oxygen atoms. However, the Si–O bond is, to a relatively large extent (~40%), also ionic or heteropolar with respect to the Pauling scale of electronegativity data. This mixed bonding character was established among others by determination of the electron density of low-temperature quartz by means of x-ray techniques (Brill *et al.*, 1942).

The three-dimensional linkage of adjacent $SiO_{4/2}$ tetrahedra at the oxygen corners determines the long-range order of the different crystalline silica modifications. Thus, the Si–

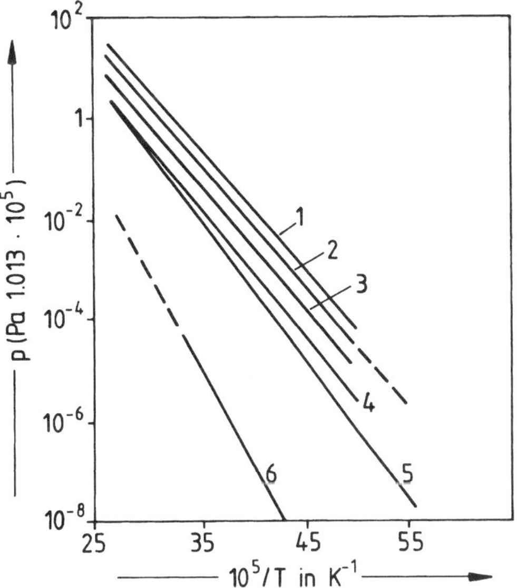

FIG. 2. Partial pressure of the reaction products of SiO_2 in the range of 1500–4000 K in neutral atmosphere. 1 = p_{total}, 2 = p_{SiO}, 3 = p_{O_2}, 4 = p_O, 5 = p_{SiO_2}, 6 = $p_{Si_2O_2}$ (reproduced with kind permission of Elsevier Science from Fanderlik, 1991).

O–Si bond angle is 144° for the trigonal low quartz, 146° for the hexagonal high quartz, 148° for the tetragonal low cristobalite, and 152° for the cubic high cristobalite. In contrast to the crystalline modifications, this bond angle has a wide distribution in vitreous silica: 120° to 180° with a maximum frequency at 145°. The reason for this broad distribution is the absence of long-range order in vitreous silica that is responsible for the large flexibility of the silica-glass network.

While the structures of the crystalline phases of SiO_2 are well determined by x-ray and neutron diffraction methods, the investigations and discussions on silica-glass structure and glass structures in general are still going on. There is the random-network hypothesis of Zachariasen (1932), Warren *et al.* (1936), and Warren (1991) on the one hand and the crystallite hypotheses of Lebedev (1921) and Valenkov and Porai-Koshits (1936) on the other. The discussions had been very controversial until Lebedev (1940) himself wrote that there is no significant difference between the two hypotheses because one can find ordered regions in ideal disordered systems, and it is possible to produce such small crystals that they are practically no longer crystals (paracrystals, x-ray amorphous). From the overwhelming amount of literature about the structure of silica glasses, the following references may be cited: Wright and Leadbetter (1976), Porai-Koshits (1977), Konnert *et al.* (1982), Zarzycki (1982), Wright *et al.* (1982), and Hosemann *et al.* (1986). Because of the uncertainties of the diffraction methods when applied to amorphous structures, many other structural analysis methods had been applied such as NMR (nuclear magnetic resonance), EPR (electron paramagnetic resonance), Raman and infrared spectroscopy, XPS (x-ray photoelectron spectroscopy), and STEM (scanning transmission electron microscopy). Representatively for those investigations may be cited the monograph of Wong and Angell (1976).

It is beyond the scope of this article to go into more detail of the numerous, partly very selective, and sometimes controversial interpretations and discussions. With respect to the silica glasses, the structural concept of a nearly complete continuous three-dimensional network of $SiO_{4/2}$ tetrahedra with various selective short-range orders, with the lack of long-range order, and with a not yet exactly known intermediate-range order appears to be generally accepted, as may be concluded also from the excellent physical properties of the various silica glasses (Table 1) described in the following paragraphs of this contribution.

The longer-range order of the crystalline modifications as well as the short-range order of the silica glass can be changed by irradiation with energetic neutrons (1 MeV, $>10^{19}$ n/cm^2). Thus, quartz loses its rotary power and its piezoelectric ability and reaches a density of 2.26 g cm^{-3}. When exposed to similar doses, vitreous silica, on the other hand, has an increase of density from 2.20 to 2.26 g cm^{-3}; i.e., the same final density value as the irradiated crystalline modifications (Primak, 1958). In the structure of this disordered version of SiO_2, there is a decrease in the Si–Si distances and a shift of the Si–O–Si bond-angle distribution to smaller angles. From the view of the vitreous silica structure, the disordered version shows also a shift to more crystallinity, which is evident from increase in the density and refractive index and also from increases of thermal conductivity at low temperature as compared with an unirradiated silica glass (Cohen, 1958; Evans *et al.*, 1982; see also Sec. 2.3.2).

2. PROPERTIES OF QUARTZ AND PRIMARILY OF VITREOUS SILICA, SPECIAL APPLICATIONS

A typical behavior of the vitreous state is a sensitive dependence of the properties on the mechanical and thermal prehistory in a defined manner. Vitreous silica (and other glasses to a certain degree, too) can be densified permanently even at room temperature up to about 15% of specific volume or density after application of high pressure (various authors; see Brückner, 1970/71, pp. 148–152). The density of a glass is also dependent on the cooling rates from temperatures above T_g to at least 100 K below T_g. This behavior of glasses is a consequence of the lack of long-range order, and this lack affects other glass properties such as viscosity, refractive index, density, and glass-fiber properties and, in the case of silica glasses, also the thermal expansion. Therefore, it is very important to know the thermal and mechanical prehistory of any glass, here of the silica glass; for instance: the smaller the cooling rate, the smaller is the density of vitreous silica. These effects are the consequence of the glass transition range, characterized by the glass transition temperature T_g (or by the fictive temperature T_f; see Sec. 2.1.1) at which the structure of the melt will be frozen in and at which the coefficient of thermal expansion changes its amount reversibly if the cooling and heating rate are equal. Therefore, they are not observed in crystals with or without displacive and reconstructive transformations.

2.1 The Vitreous State of Glasses and the Vitreous and Amorphous State of Silicon Dioxide

The vitreous state of usual, multicomponent glasses as compared with their crystalline state is shown in Fig. 3. If crystallization is prevented by fast enough cooling, the structure of the melt (above or below T_m, the melting point) is maintained by freezing to a solid body, called glass, at a fictive temperature $T_f > T_g$. If the cooling rate is low enough, the metastable (undercooled) glass melt freezes in at a lower temperature than T_g (at $T_f < T_g$) and at a smaller specific volume with a less open structure that is determined by the metastable equilibrium curve (solid-dashed straight line in Fig. 3). The points of intersection between the two straight lines with the lower temperature coefficient and the metastable straight line with the larger temperature coefficient indicate the fictive temperatures T_f, which depend strongly on the cooling (or quenching) rate. Therefore, the term "fictive temperature" is a very practical measure for the thermal prehistory. The concept of fictive temperature in glass science (Tool, 1946) is that temperature T_f to which the structural state corresponds after rapid cooling or rapid heating, where "rapid" means a rate $v = |\Delta T/\Delta t| > |\Delta T/\tau|$, with τ being the relaxation time of the structural rearrangement or that time during which a property (volume, enthalpy, etc.) relaxes to the e^{-1} amount of the deviation from the metastable equilibrium curve (see arrows from the curves of fast cooling or that of fast heating to the metastable equilibrium curve at a temperature $T = T_{meas.}$ in Fig. 3, where $T_{meas.}$ means the temperature at which a certain property is measured, and is time dependent). Thus, many properties of glasses depend on T_f. In the case of fibers, very high T_f values are reached. The so-called glass transition temperature T_g is obtained by cooling and heating a glass with a rate of $v = |dT/dt| = 2$ to 5 K/min. In this case $T_g = T_f$ by definition.

Silica glass is unusual in that the metastable equilibrium curve does not have a positive temperature coefficient as in the case of all known glasses, but it does have a negative one between 1000 and about 1500 °C that

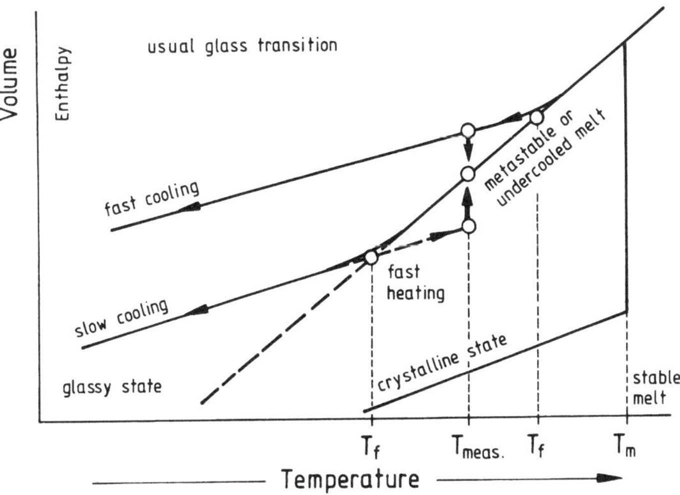

FIG. 3. Enthalpy and/or specific volume of multicomponent silicate glasses versus temperature for two different cooling and heating rates and fictive temperatures, $T_f \gtrless T_g$. T_g can be between T_{meas} and T_f or at T_{meas}. Comparison with the crystalline state (schematic). The partly solid and dashed straight line is the metastable equilibrium curve for a usual (multicomponent) oxide glass.

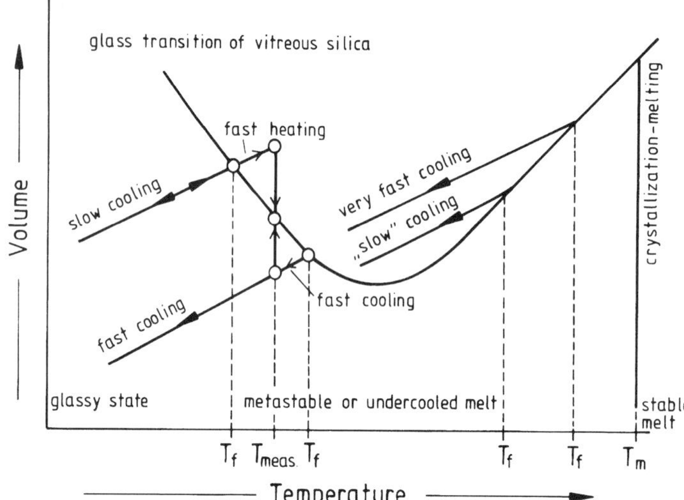

FIG. 4. Specific volume of vitreous silica versus temperature in the metastable and glassy range for various fictive temperatures (schematic), different cooling and heating rates, and fictive temperatures, $T_f \gtrless T_{meas}$. T_g can be between T_{meas} and T_f or at T_{meas}. The arrows indicate the route of relaxation to the metastable equilibrium at T_{meas} for samples with two different fictive temperatures T_f (schematic).

changes to a positive one with increasing OH^- content (Figs. 4 and 5) and/or network modifier concentration (e.g., alkali oxide) (Brückner, 1970/71). This change means that the metastable range of hydroxyl-free and -poor silica glasses have a negative temperature coefficient at least partly in the temperature range of 1000–1500 °C. There is also a negative coefficient in the temperature range below 140 or 210 K, depending on both thermal history and hydroxyl content (see Sec. 2.3.1).

The amorphous state of silica gel (see Secs. 1.3.2 and 3) is due to its low-temperature treatment (far below T_g) and due to its preparation technique (Fricke, 1988). If silica gel (or aerogel) is heated to the T_g range, a sintering process with a large volume contraction leads to vitreous silica. Thin films of SiO_2 (or $2SiO + O_2$) produced by vaporization on a cool target are initially amorphous, not glassy or vitreous, because, similar to the gel state, they do not have the characteristic T_g, but they will be transformed also to the vitreous state by heating to the T_g range or beyond it in neutral (or oxidizing, O_2) atmosphere. In this way the glassy state is more stable than the amorphous state, although the glassy state itself is thermodynamically unstable and only preserved by the large viscosity (see Sec. 2.3.4).

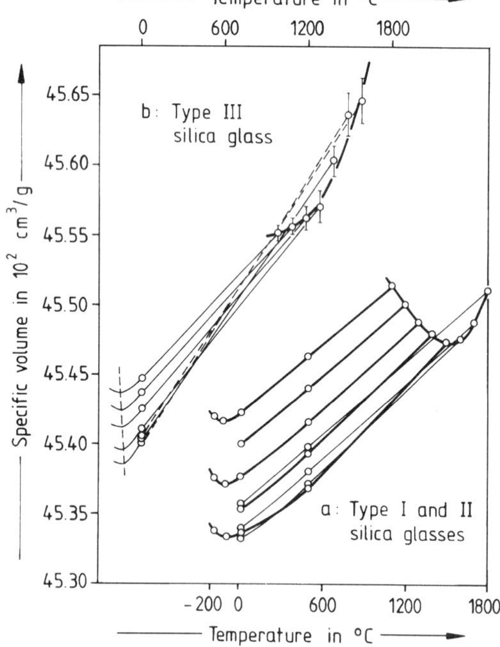

FIG. 5. Volume–temperature behavior of hydroxyl-poor (types I, II, and IV) and hydroxyl-rich (type III) silica glasses in the metastable range (thick curves). The more or less straight lines (thin curves) are mean thermal volume expansion curves of the glasses with different fictive temperatures. Upper and lower temperature scales belong to type-III and type -I, -II, and -IV silica glasses, respectively (Brückner, 1970/71).

2.2 Mechanical Properties

2.2.1 Elastic Constants and Internal Friction The elastic moduli also have unusual properties: the temperature coefficients of the elastic moduli and of the longitudinal elastic strain (of fibers and rods) are positive. The elastic moduli may be expressed by experimentally determined relations: $E = 73.3 \times (1 + 5.75\epsilon)$ GPa, $G = 32.1(1 + 3.06\epsilon)$ GPa, where E is Young's modulus, G is the rigidity modulus, and ϵ is the strain. The consequence is the positive pressure coefficient of the compressibility, β_m (p = compressive stress):

$$\beta_m = a(t) + b(T)p = \frac{1}{V_0}\frac{V_0 - V_p}{p - p_0} = 1/K = \frac{3(1 - 2\nu)}{E} \tag{3}$$

with

$$a(T) = (26.43 - 0.0025T) \times 10^{-12}\ \text{Pa}^{-1}, \tag{3a}$$

$$b(T) = (43.6 - 0.080T) \times 10^{-17}\ \text{Pa}^{-1}, \tag{3b}$$

where ν is Poisson's ratio and T is in °C. These data are valid in the temperature range of 22–260 °C and in the pressure range of 0–300 MPa (see Brückner, 1970/71). Thus, the compressibility increases with pressure, whereas those of quartz and of most other glasses and substances decrease with pressure. Above 350 MPa the compressibility has a normal dependence on pressure for fused silica.

The elastic properties measured at room temperature at acoustic frequencies are a function of the thermal history (see Sec. 2.1) because the density depends also on it. As will be shown in Sec. 2.3.1, the density increases with the fictive temperature in the temperature range 1000–1550 °C before a normal negative temperature coefficient is obtained (see Brückner, 1970/71). In this way the ultrasonic shear velocity, the shear modulus, and the compressibility decrease with increasing density (increasing fictive temperature), whereas the dilatational velocity, the Lamé parameter, and Poisson's ratio increase with increasing density. Increasing hydroxyl and chloride content makes vitreous silica "softer" (lower moduli, lower viscosity).

The deviation of Young's modulus from linear positive temperature dependence at 800–1000 °C is connected to an increase of internal friction (see Brückner, 1970/71). Structurally, these friction losses are related to the initial breakdown of the polymer network from a rigid glassy state to an undercooled liquid metastable state. As in the case of other properties, the hydroxyl content influences the internal friction coefficient; the hydroxyl-rich silica glasses (e.g., type III) have a steep increase in internal friction at lower temperatures than the hydroxyl-free or -poor silica glasses (types I, II, IV, to VII), a behavior that is closely connected to the viscosity (see Sec. 2.3.4).

At low temperatures (<60 K) and high frequencies (ultrasonic longitudinal waves in the kilohertz and megahertz range), another internal-friction peak is observed together with a minimum of rigidity due to a relaxation mechanism (see Brückner, 1970/71). It is remarkable that this kind of absorption does not occur in the crystalline quartz structure. The low activation energy of 4.2 kJ/mole indicates that the losses may be due to Si–O–Si bond deformations of those oxygen atoms having alternative metastable equilibrium positions of equal energies. Again the maximum of the attenuation of the intensity of ultrasonic waves is shifted by the hydroxyl content: from 43.5 K for hydroxyl-poor types, I and II, to 47.5 K for type-III fused silica.

2.2.2 Strength and Fracture, Fracture Toughness and Fatigue Strength is a statistically influenced and surface-dependent property for an ideal elastic brittle body such as silica glass. Small mechanical damage causes a large weakening. Therefore, bulk material with surface cracks has strengths ranging from 60 to 100 MPa. The highest observed tensile strengths reported for silica fibers are about 10–15 GPa, close to the theoretical strength of about 30 GPa (see Brückner, 1970/71) and for which the microcracks (so-called Griffith cracks) are on the order of magnitude of nanometers. In addition to a much smaller statistical distribution of cracks within the fibers, the elongation and thinning of the cracks during the fiber-drawing process might explain the large difference of the strength between the fiber and bulk material.

On account of brittle fracture, the fracture toughness of silica glass, measured by the critical stress-intensity factor, is 0.79 MPa $m^{1/2}$ at 300 K.

The static fatigue behavior of silica-glass fibers (20–40-μm diameters) was studied extensively. While in vacuo at 78 K no fatigue was observed during a time of days, only half of the strength at 78 K was observed in vacuo at room temperature with a decrease from about 900 to 700 MPa during a month. In air, at room temperature, a decrease from 600 to about 300 MPa during a year was measured (Proctor *et al.*, 1967; see Brückner, 1970/71).

2.3 Thermal Properties

2.3.1 Density and Thermal Expansion

One of the most interesting anomalies of silica glasses as compared with multicomponent glasses (see GLASSES) is the density or specific volume and its thermal behavior. As shown in Fig. 4 (Sec. 2.1), the metastable equilibrium curve has partly (between 1000 and 1500 °C) a negative temperature coefficient. This coefficient is a consequence of the decrease of specific volume within the metastable range, which has a minimum at about 1500–1550 °C for the hydroxyl-poor or hydroxyl-free silica glasses and which is not the case for the silica glass with 0.12% OH^- content (type III) (Fig. 5). The low-temperature minimum around −100 °C is still present; it is only shifted to lower temperatures. Both minima are influenced also by the fictive temperature; thus, the minimum temperatures for the hydroxyl-poor silica glass is shifted from −133 to −95 °C, and that of the hydroxyl-rich version is shifted from −120 to −57 °C (see also Fig. 7) with increasing fictive temperature.

The thermal expansion of silica glasses is also dependent on the fictive temperature and on hydroxyl content. Figure 6 shows thermal expansion data for type-II silica-glass samples with differing fictive temperatures. Hydroxyl-poor samples with a fictive temperature corresponding to the minimum of the specific volume at 1550 °C and hydroxyl-rich silica glasses at 1460 °C have the highest thermal expansion (curve 4); those of still higher or lower fictive temperatures have lower thermal expansion (e.g., curve 3). These effects are consistent with the Grüneisen relation:

$$\beta c_v/\alpha V = \text{constant}, \qquad (4)$$

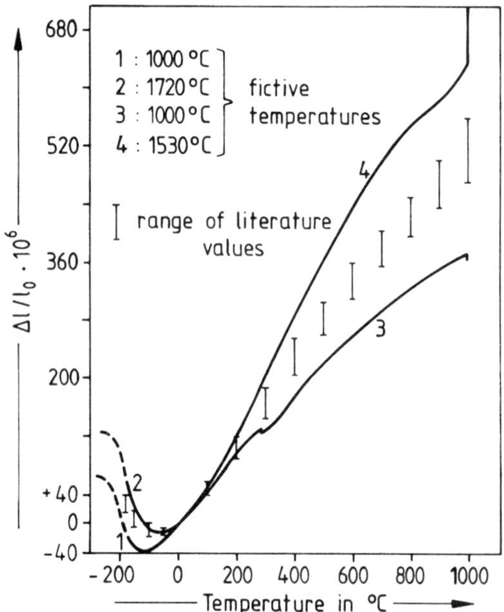

FIG. 6. Thermal expansion below T_g of type-II silica-glass samples with various fictive temperatures versus temperature (Brückner, 1970/71).

where β = compressibility, c_v = specific heat, α = coefficient of thermal expansion, and V = specific volume. If V has a minimum, α should have a maximum, and if β has a minimum, c_v should have a maximum. These effects are at least qualitatively the case for fused silica when the fictive temperature or when the temperature itself is changed from low (1000 °C) to high (1800 °C) (Brückner, 1970/71).

The low-temperature minimum is shifted to higher temperatures with increasing fictive temperature, i.e., it is independent of the high-temperature minimum. There is a linear relationship between the minimum temperature and fictive temperature (Fig. 7). Again a difference between the groups of hydroxyl-poor and hydroxyl-rich silica glasses is found (Brückner, 1970/71).

It is interesting that the crystalline high-temperature modifications, quartz, and crystobalite have a change from positive to negative thermal expansion between 800 and 1000 °C, indicating that structural fragments or preordered regions exist in the vitreous state that are responsible for the high-temperature minimum of silica glass. The low-

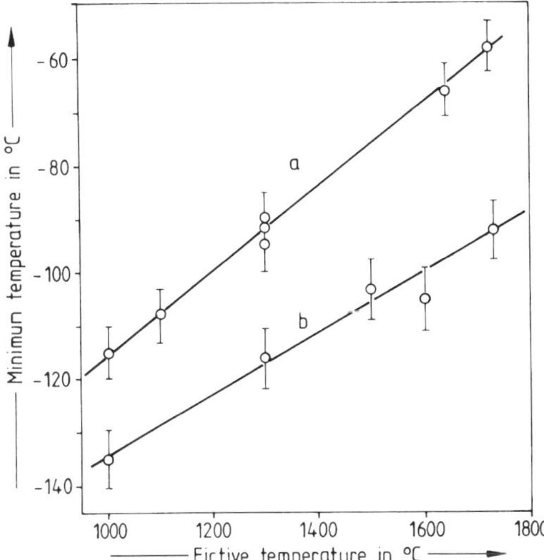

FIG. 7. Shift of the low-temperature minimum with fictive temperature for type-II (curve a) and type-III (curve b) vitreous silica (Brückner, 1970/71).

temperature minimum on the other hand is not found for quartz and crystobalite, indicating that it results from randomly disordered regions in which the preordered regions are embedded. Its shift with fictive temperature may be caused by structural stresses produced by quenching and freezing-in the structure at the fictive temperatures (Brückner, 1970/71). The effect of negative thermal expansion of vitreous silica, as well as that of high cristobalite and high quartz, may be understood qualitatively with special emphasis to the transversal vibrations of the oxygen atoms and with the concept that the major contribution to the thermal behavior is produced by the transversal oxygen vibrations.

2.3.2 Heat Capacity and Heat Conduction According to the classical expression for the temperature dependence of the specific heat,

$$c_v = 3RE(x) \tag{5}$$

in which E is the Einstein function,

$$E(x) = x^2e^x/(e^x - 1)^2, \tag{6}$$

with $x = h\nu/kT = \Phi/T$, where h = Planck's constant, ν = frequency of proper oscillation, k = Boltzmann's constant, R = gas constant, and Φ = the Debye temperature, the specific heat of vitreous silica was calculated, and a relatively good agreement with experimental values was obtained (see Brückner, 1970/71). If the following three Debye temperatures are chosen: Φ_S = 1100 K for each of the three silicon vibrations, Φ_T = 370 K for each of the three transversal oxygen vibration modes, and Φ_L − 1220 K for each of the two longitudinal oxygen oscillations, the Φ_i correspond to vibrations in the IR range of 23×10^{12}, 7.7×10^{12}, and 25.4×10^{12} Hz.

The deviations at low temperatures <100 K are only seemingly small when plotted on an approximate scale; however, they contain an excess of specific heat not only for silica glasses but for all glassy materials, even for metallic glasses, an effect that does not exist in the corresponding crystalline phases, e.g., quartz where the T^3 law is valid. Hunklinger (1982) pointed out that the excess heat is a consequence of relaxation processes that "originates from the modulation of the thermodynamic equilibrium of a subsystem in a sample by the strain and stress field of the sound waves" (phonons). With the introduction of only two possible configurational states, the double-well potential, and of resonant transition between the energies of a two-level system (TLS) for phonons, models have been developed to explain the large difference, a factor of 1000, for the specific heat between the SiO_2 glass and the crystalline state (quartz) at 25 mK. Low-energy excitations (phonons) interact strongly with the disordered network structure. Since the relaxation time of these interactions is in the range of 1 μs up to 10^4 s, the specific heat depends on the duration of a measurement, as was shown by various authors (see Hunklinger, 1982). It is suggested that the nature of these low-energy excitations, which also determine the ultrasonic properties of glasses, is based on clusters of atoms that undergo structure relaxation even at the lowest temperatures by tunneling between the states of the two-level system and/or by rotation of SiO_4 tetrahedra.

Although no exact explanation can be given so far for the excess specific heat at temperatures below 20 K, especially below 5 K, it is important that, in a qualitative way (inconsistent with lattice dynamics), a model is given with the two-level concept in which the Si–O–Si bending vibrations can be made responsible for all the low-temperature specific heat. All of the frequencies required to account for this excess heat capacity are to be found in the region of optical modes of very low frequencies with a continuum extending from 560 cm^{-1} down to 8 cm^{-1} as can be shown directly by the Raman spectrum and the Brillouin scattering spectra of fused silica (see Hunklinger, 1982).

A characteristic difference between the vitreous and crystalline state is very evident through the temperature dependence of heat conduction (Fig. 8). According to the regular atomic arrangements in crystal lattices, the crystals have large values of the mean free path for the phonons at low temperatures and therefore large values of heat conduction. The lack of periodicity and symmetry of a lattice in the glass network causes large interchange energies between thermoelastic waves in such a way that the mean free path, Λ, is significantly shorter. It is assumed that in case of glasses, Λ is independent of temperature, as is also the average sound velocity v; thus, the heat conduction,

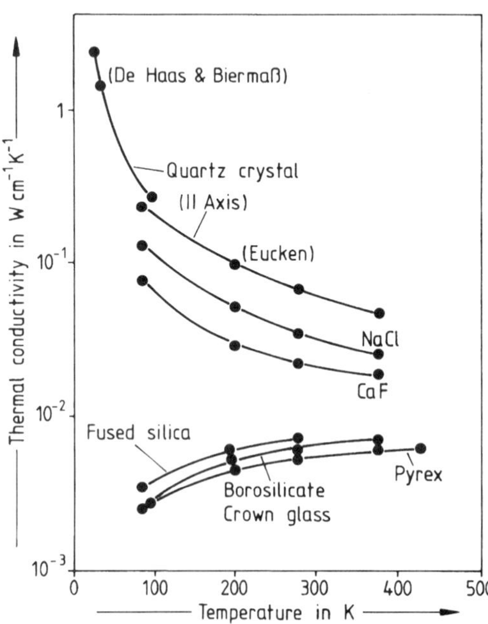

FIG. 8. Heat conduction of crystalline and glassy substances, particularly that of quartz and silica glass, versus temperature (gathered from various authors; see Brückner, 1970/71).

$$\lambda_c = cv\Lambda/3 \tag{7}$$

is predominantly proportional to the specific heat c at most temperatures. For temperatures well above room temperature, the radiation heat conductivity (photon conduction),

$$\lambda_r = 16\sigma n^2T^3/3\epsilon, \tag{8}$$

is superimposed, where σ is the Stefan-Boltzmann constant, n is the refractive index, ϵ is the optical absorption coefficient, and T is the absolute temperature.

The proportionality between λ_c and c is valid to temperatures as low as liquid oxygen temperature (Hunklinger and Schickfus, 1981). At lower temperatures the conductivity decreases less than that expected for a much longer mean free path for the longitudinal than for the transversal waves caused by a small fraction of three-phonon processes at lower frequencies and temperatures. This deficiency leads to completely uncoupled longitudinal waves with lengths very much longer than those of the transversal waves, and this decoupling produces an additional conductivity (curve I in Fig. 9) proportional to T in the range 0–5 K in addition to curve II, resulting by superposition to the experimental curve III. Above 5 K a partial coupling causes a decrease of curve I that leads to the "knee" of curve III.

Besides thermal fluctuations, defects and voids with low-frequency mechanical resonances may have spin–lattice interactions and relaxations and, hence, unusual scattering of phonons that influence the short mean free path, especially those of the transversal phonons. Therefore, a dependence of thermal conductivity on thermal prehistory and on hydroxyl-poor and hydroxyl-rich silica glasses is to be expected.

2.3.3 Diffusion The gas permeation and the mobility of network-forming (oxygen, silicon) and network-modifying (alkali ions, OH^- groups) species are important in applications of vitreous silica as, e.g., high-temperature recipients. Figure 10 gives an impression of the large range of diffusivities of various atomic species as a function of temperature. Helium and hydrogen have high diffusivities due to their small sizes and due to the open network structure of vitre-

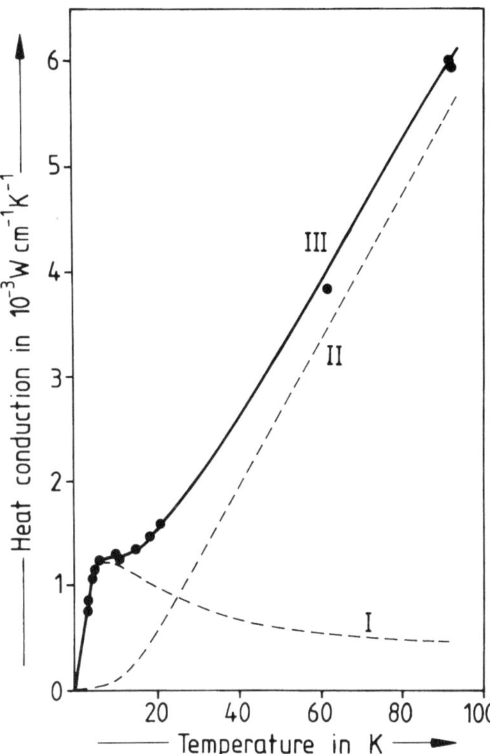

FIG. 9. Heat conduction of vitreous silica at low temperature. Curves I and II, calculated curves; curve III, sum of curves I and II. Symbols: measured values (gathered from Kittel, 1949, and Berman, 1951; see Brückner, 1970/71).

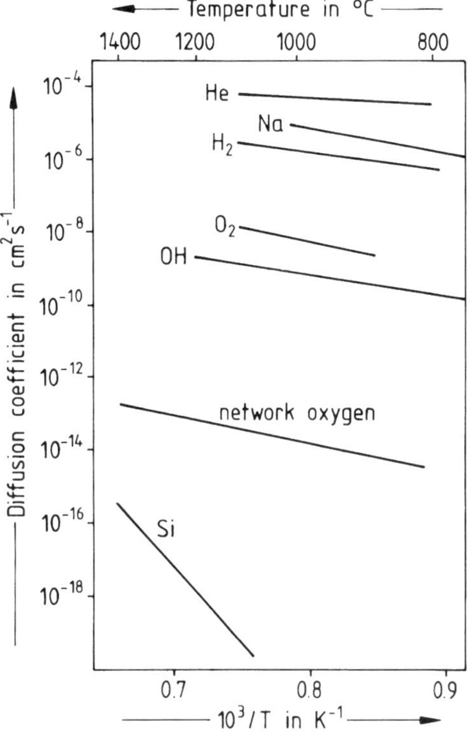

FIG. 10. Diffusion coefficient of various network-forming and network-modifying species in vitreous silica versus reciprocal absolute temperature (from Schaeffer, 1984).

ous silica. The diffusion coefficients and the activation energies of gases depend on the diameter of their ions, atoms, and molecules. Increasing activation energies are obtained in the series He $\rightarrow$ H_2 $\rightarrow$ Ne $\rightarrow$ O_2 $\rightarrow$ N_2 $\rightarrow$ Ar and decreasing values for the permeabilities. Structurally incorporated or interacting species usually have low diffusivities depending on bonding forces by which they are attached to the glass network. Network oxygen (not O_2) and silicon have the smallest diffusivites. Typical examples for interacting diffusion processes are for H_2O, which diffuses via an interaction with the silica network by forming SiOH groups, hydrogen, and network oxygen diffusion. Diffusion of O_2 plays an important role in the oxidation processes of silicon-containing materials producing noncrystalline silica layers (see Sec. 2.5.4). Schaeffer (1984) showed that the parabolic growth of SiO_2 layers ($d \propto t^{1/2}$) is connected to a parabolic rate constant that is, by a factor of 2, larger than the diffusion coefficient of network oxygen (see also Lamkin *et al.*, 1992).

In addition to the complex diffusivity of oxygen, the interaction diffusion process of hydrogen is also complex. To understand the behavior of the OH^- groups in vitreous silica preparation, conditions of SiO_2 glasses (see Table 1) must be known since they affect diffusion. In the case of type-Va vitreous silica, it may be noted that chemically bonded hydrogen can be set free at elevated temperatures from the so-called unstable OH^- groups. This reaction was originally written

$$2SiO_2 + H_2 \rightleftarrows \equiv SiOH + SiH \quad (9)$$

or, better formulated,

$$\equiv Si{-}O{-}Si \equiv + H_2 \rightleftarrows \equiv SiOH + \equiv SiH. \quad (10)$$

In this reaction diffusion model, the amount

of easily and totally extractable hydrogen below T_g increases, together with physically dissolved H_2 (Stone, 1987; Witzke, 1990). It explains the so-called diffusion anomaly of hydrogen (Lee, 1964; Ogawa *et al.*, 1977) in vitreous silica. On the other hand, the interacting diffusion process of Si–OH groups results in, originally written (Stone, 1987; Witzke, 1990),

$$2SiO_2 + H_2O \rightleftarrows 2 \equiv SiOH \quad (11)$$

or, better formulated,

$$\equiv Si–O–Si\equiv + H_2O \rightleftarrows 2 \equiv Si–OH, \quad (12)$$

the so-called stable OH groups, which cannot be easily and totally removed by thermal treatment below T_g.

A typical example for a diffusion process of structurally incorporated species is sodium ions, which have the highest diffusivity with the exception of helium (Fig. 10). The diffusivity depends not only on ionic radius and bonding force but also on the network structure. This is obvious from Fig. 11, which shows that there are large differences particularly parallel and perpendicular to the *c*-axis directions in quartz as well as in the hydroxyl-free (e.g., type IV) and hydroxyl-rich (e.g., type III) silica glasses. While the difference between the two orientations in quartz is a pure steric effect, the difference in the two silica-glass types is mainly a consequence of the different structure of the two silica-glass versions and of the interaction between Na^+ and OH^- in type III.

The diffusivity of cations in vitreous silica decreases usually with increasing charge. The ratio of the different diffusion coefficients in type-I silica glasses at 1000 °C is $D_{Na^+}:D_{Ca^{2+}}:D_{Al^{3+}}:D_{O^{2-}} = 1:2.5 \times 10^{-3}:1.3 \times 10^{-8}:1.1 \times 10^{-9}$ on the basis of an exchange of these ions with the impurities of the type-I silica glass.

A very special effect was observed with liquid and gaseous helium in contact with vitreous silica. While ^4He can be absorbed by vitreous silica at temperatures below 4.2 K, ^3He is not absorbed below 4.2 K (see Brückner, 1970/71). Thus, vitreous silica may be useful in separating ^3He from ^4He by selective absorption (isotope sieve).

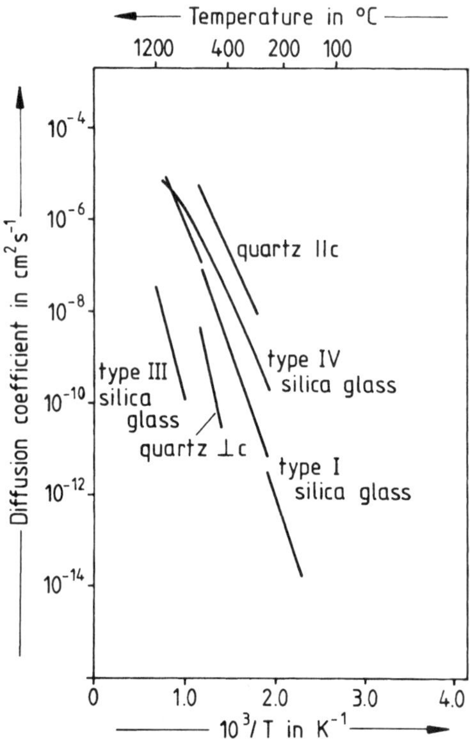

FIG. 11. Diffusivity of sodium in quartz and in various types of vitreous silica versus reciprocal absolute temperature (from Frischat, 1969, in Brückner, 1970/71).

2.3.4 Viscosity The stable and undercooled melt of silica is a high-viscosity fluid with the largest temperature range for a large viscosity range ($\Delta\eta \sim 10^{12}–10^3$ Pa s) and the highest glass transition temperature, $T_g \approx 1200$ °C, among all oxide glass compositions at all. Even at the melting temperature of cristobalite ($T_g = 1713$ °C), the viscosity is still above 10^6 Pa s. This range of temperatures is one of the reasons for the good glass formability of SiO_2 and for the long duration to reach the metastable structural equilibrium for viscosities above 10^9 Pa s (below about 1300 °C) and for various other properties such as density, refractive index, and dielectric properties. On the other hand, this viscosity-temperature range is also the reason for the fact that nearly all properties depend on thermal history, as shown in Sec. 2.1. The hydroxyl content plays an important part, too.

The influence of thermal history on viscosity is seen in Fig. 12 where the low-tem-

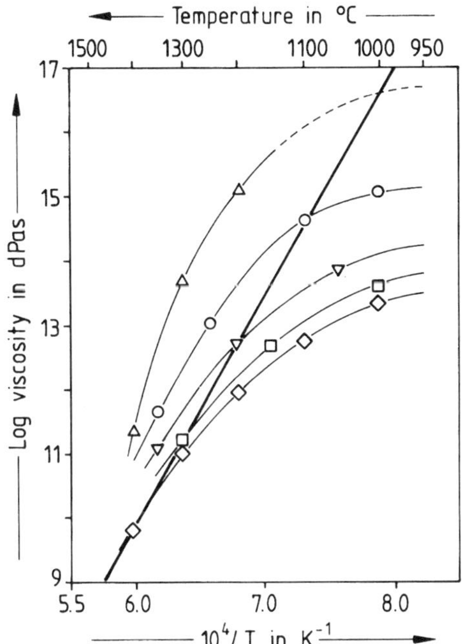

FIG. 12. Viscosity of vitreous silica, type I, versus reciprocal of absolute temperature with different thermal histories (reproduced with kind permission of The Society of Glass Technology from Hetherington *et al.*, 1964). Solid straight line is the metastable equilibrium curve; fictive temperatures: △ 1000 °C, O 1100 °C, ▽ 1200 °C, □ 1250 °C, ◇ 1400 °C.

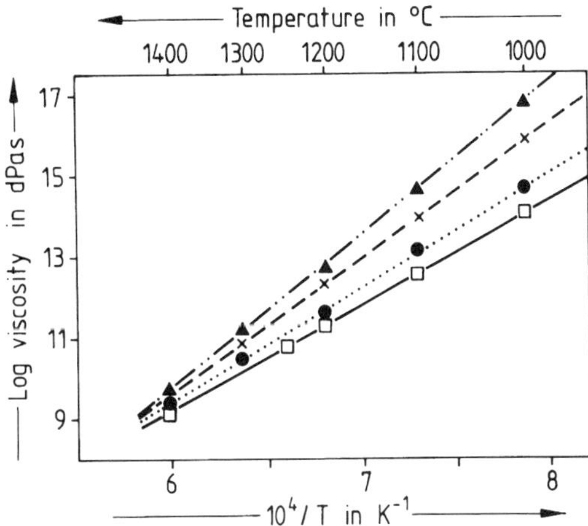

FIG. 13. Equilibrium viscosities of silica glasses with various water content in the high-viscosity temperature range (reproduced with kind permission of The Society of Glass Technology from Hetherington *et al.*, 1964). Curves from above: ▲ = type I, X = type II with 0.027 wt % OH, ● = type II with 0.04 wt % OH, □ = type III with 0.12 wt % OH.

perature viscosities of type-I silica-glass samples were measured with five different fictive temperatures between 1000 and 1400 °C. Those viscosities that have been determined from samples with fictive temperatures below the metastable equilibrium curve (the straight line in Fig. 12) are larger than the equilibrium viscosities, those that have been determined from samples with fictive temperatures beyond the metastable equilibrium curve are smaller than the equilibrium viscosities. Finally, fictive temperatures that agree with those of the metastable equilibrium curve have equal viscosities, because larger fictive temperatures are connected to a glass network structure of lower viscosities than those of the corresponding (lower) temperatures of the equilibrium curve and vice versa.

The influence of hydroxyl content is clearly seen from Fig. 13. Increasing (OH^-) content decreases viscosities by several orders of magnitude by the creation of open bonds or of nonbridging oxygens in the SiO_2 network: ≡Si–OH HO–Si≡. Similar to that is the creation of open bonds by Cl^-, e.g., in type-IV silica glass: ≡Si–Cl Cl–Si≡.

The whole temperature range for the hydroxyl-poor silica glasses may be represented by the Vogel-Fulcher-Tammann equation (Brückner, 1970/71):

$$\log \eta = A + B/(T - T_0) = -2.49 + 15\,004/(T - 253). \qquad (13)$$

The activation energy of viscous flow, Q, obtained from the Boltzmann equation

$$\ln \eta = \ln \eta_0 + Q/kT, \qquad (14)$$

is 670 J/mol at high and 511 J/mol at low viscosities; thus, Q is altered only by a factor of 1.3.

2.4 Optical Properties

Silica in the form of glass and quartz is very useful as optical material because of its high transmittance of visible, ultraviolet, and near-infrared light and, in addition, for quartz because of the birefringence and rotary power for plane polarized light.

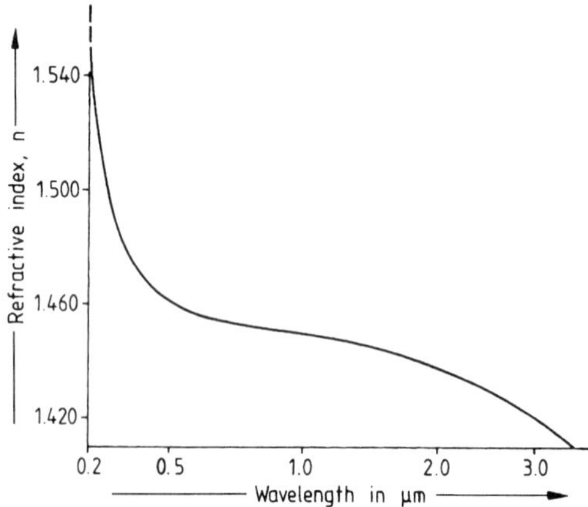

FIG. 14. Refractive index of type-II silica glass versus wavelength (Heraeus prospect).

2.4.1 Refractive Index, Reflectance, Transmittance, Luminescence, Light Scattering Silica glass is optically isotropic with a refractive index of 1.4589 at 789.3 nm wavelength. The dispersion curve is shown in Fig. 14. Like the density, the refractive index depends on thermal history and shows a maximum at a fictive temperature of about 1550 °C (Brückner, 1970/71). Increasing hydroxyl content decreases the refractive index and density. In contrast to the glass, the crystalline silica modifications are birefringent; thus, low-quartz has refractive indices of 1.5442 for the ordinary and 1.5533 for the extraordinary ray at 589.3 nm. The optical activity of quartz, due to its helical structure, corresponds to a rotary power of 21.7° at 589.3 nm for a 1-mm-thick sample.

Vitreous silica becomes also birefringent under stress and has a stress-optical coefficient of 3.4×10^{-12} Pa^{-1}.

The optical transmittance of electromagnetic waves is limited at high frequencies by the ultraviolet band edge on account of interactions with the electrons of silica (electron intertransfer between oxygen and silicon or impurity cations). Increasing the electron bonding energy shifts the UV edge to higher frequencies. Therefore, the most "perfect" silica glasses, regarded as especially "pure," such as types III–VI, have the smallest absorbance between the infrared bands and the valence-band to conduction-band transitions (Fig. 15). Because the absorption edge in the UV is affected by impurities (for example, Fe^{3+} absorbs at 230 nm), and by some intrinsic defects of silica (SiO_{2-x}, $x \sim 10^{-4.5}$, type I, absorbs at 247 nm) and chemical disorder in the silica structure (see Sec. 2.5.3), absorption in the UV region may not correspond to the band-to-band transition. While nonstoichiometry causes an absorption band at 247 nm, impurities or doping of vitreous silica, e.g., with titanium or cerium oxide, can cause tailor-made shifts of the UV edge to longer wavelengths. Such shifts are important for applications, among others for special UV lamps (Witzke *et al.*, 1994).

The transmittance in the visible range is usually >99% and reduced only by reflectivity, which is relatively low because the refractive index is low, and it can be reduced by optical interference (λ/4 –) layers.

By incorporation of transition-metal or rare-earth metal oxides, colored silica can be produced (Shultz, 1974; Clasen, 1993), simi-

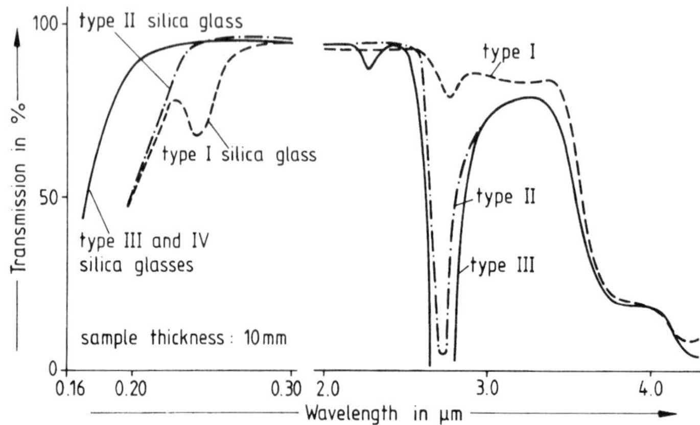

FIG. 15. Spectral transmittance of various types of silica glasses.

lar to silicate glasses, as a result of electron intratransfer within the 3*d* or 4*f* orbitals of the cations.

When natural rock quartz and silica glasses of types I and II are irradiated by x rays, γ rays, or neutrons, they become brown, smoky quartz, and their spectra have three absorption bands near 220, 300, and 550 nm due to the metal impurities (Al, Na; see Table 1) (see Sec. 2.5.2). Irradiated silica glasses of types III and IV as well as types VI and VII do not have color centers absorbing in the visible range but only a shift of the (weak) 220-nm absorption to 215 nm, which is possibly due to a radiation-induced electron transition of an oxygen vacancy into the conduction band (Weeks and Sonder, 1963). The corresponding electron transition of a nonradiated bridging oxygen is due to an absorption below 180 nm. The electronic states of the defects induced by radiation can be removed at temperatures below and around 300 °C by emission of photons of various energies (thermoluminescence bands at about 300, 500, and 600 nm) where traps are involved of which the deepest lie at 1.1 eV (Lippingcott *et al.*, 1958).

The optical transmission in the near infrared also depends on impurities; e.g., Fe^{2+} absorbs at 1.1 μm and the hydroxyl groups at 2.7 μm (Si–O–H) with overtones and combination vibrations at 2.4, 2.2, and 1.4 μm. In the far infrared the Si–O oscillations produce strong and broad absorption bands at 9.1 μm (asymmetric valency vibration), at 21 μm (transversal vibration), and at 12.5 μm in the Raman spectrum (symmetric pumping vibration). The force constant was calculated as $K_{Si-O} = 4.0$ N cm^{-1} (Zarzycki and Naudin, 1960).

Natural quartz and silica glasses of types I and II have luminescence bands at 280 nm (4.4 eV) and 396 nm (3.1 eV) when excited by UV light of wavelengths smaller than 280 nm—e.g., by Hg resonance lamps (254 nm). These luminescence bands are connected to the UV-absorption bands at about 205 nm (6 eV) and 247 nm (5.1 eV), respectively. The origin of the luminescence is not yet adequately understood, but it is assumed that both reduction of SiO_2 to SiO_{2-x} ($x \sim 10^{-4.5}$) and impurities, such as aluminum, germanium, and sodium, produce trapping sites for electrons and holes. Particularly, the luminescence band at 396 nm, connected to the absorption band at 205 nm and to alkali impurities, is assumed to be produced by charge-trapping centers of nonbridging oxygen atoms, while the luminescence band at 280 nm, connected to the absorption band at 247 nm, is produced by vacant oxygen sites (Lell *et al.*, 1966; Stroud *et al.*, 1965).

Very pure silica glasses of types III and IV do not show luminescence and probably also not types VI and VII.

The situation is changed if photoluminescence is considered. Photoluminescence is the result of photoexcitation of electrons in a sample that was previously treated by defect-center–producing or ionizing radiation. Again, the excitation (absorption) energy has to be larger than the energy of emission of the defect center because of nonradiative transitions. The emission spectra of irradiated silica glasses are complex and show bands (Friebele, 1991) at 650 nm (1.9 eV), 440 nm (2.8 eV), 290 nm (4.3 eV), and 185 nm (6.7 eV). The exact position, intensity, and half-width of the bands depend on the type of irradiation, the temperature, and the type of silica glass. Although all bands are influenced by impurities, inclusively OH^- (Wang *et al.*, 1988), they are all related to intrinsic defects. Roughly, the following attributions of these four bands may be given: the 1.9-eV luminescence has an excitation band centered near 4.8 eV and is probably associated with a nonbridging oxygen center of OH^- in high-OH^--containing silica glass, but it seems possible that it also may be several bands arising from different defect centers. The 2.8-eV emission band is common to all quartz and silica-glass types, with larger intensity for quartz than for the glasses. The 4.3-eV emission band is present only in oxygen-deficient and in neutron-irradiated silica samples. The decay of this band correlates well with the time derivative of the decay of the transient E' absorption center at 230 nm, indicating that the emission results from hole–electron pair recombination at the sites of oxygen vacancies. The 6.7-eV emission band was observed in high-purity silica glasses and quartz during 2.5-MeV electron bombardment (radioluminescence), but its origin has not been investigated intensively enough (Treadaway *et al.*, 1975).

Besides the normal luminescence, photoluminescence, and radioluminescence, thermal luminescence is of interest in order to get indications about the origin of the various emission bands observed during heating samples, during or after irradiation, or from annealing at various temperatures. Those investigations are related to optical absorption bands due to radiation and to paramagnetic centers (see Sec. 2.5.3, particularly Fig. 23). These can also be identified by annealing (thermal bleaching or annihilation) at different temperatures between 100 and 800 K (Friebele, 1991; Griscom, 1986). Thus, as an example, the intense blue luminescence during thermal bleaching results from the annihilation rate of the E' centers near 215 nm, unless connected with the restoring of bridging oxygen vacancies with absorption band near 215 nm. It is also obtained during fracture of silica glass under ultrahigh-vacuum condition (triboluminescence). Interesting is the observation that pulsed electron irradiation produces transient absorption centers near 215 nm accompanied by a blue emission with a similar time decay characteristic (see Sigel, 1973/74).

Silica glass for optical purposes has a relatively low light scattering. If, however, optical fibers are taken into consideration, the light scattering as well as relatively low absorption play an important role. This problem will be discussed in more detail in the following section.

2.4.2 Optical Waveguides Optical fibers as waveguides are widely and commercially used as transmittance medium in modern communication technologies. Several hundred times as many channels can be transmitted through one glass fiber with a diameter of 120 μm as can be transmitted through a copper coaxial tube with a diameter of 6 mm.

Silica-glass fibers have a dominant role, particularly in long-distance communication links. The fibers are either a step-index monomode or gradient-index fiber. For short-distance communication links, usually the less expensive multicomponent silicate-glass fibers are used. The production of a step-index monomode fiber starts with a thick-walled silica-glass tube whose interior surface is covered by a chemically or physically vapor-deposited layer of additives giving a higher refractive index than that of the silica glass, e.g., GeO_2 or TiO_2. This prepreg or preform is drawn and collapsed to a fiber with a core diameter of about 1–4 μm and a coating wall thickness of about 60 μm. The criteria that only one mode of a monochromatic wave can be transported is given by the relation between the difference of the two refractive indices, n_i = index of the core glass and n_0 = index of the coating glass, and the maximum diameter D_{max} of the core:

$$D_{max} = 2.403\lambda/\pi(n_i^2 - n_0^2)^{1/2}; \tag{15}$$

e.g., if the difference $n_i - n_0$ is 1.0%, D_{max} has to be 4 μm at a wavelength of the light, λ = 850 nm. For more details, see FIBER OPTICS.

The production of a gradient-index fiber, for which the refractive index has a parabolic profile, starts usually with a silica-glass tube with a diameter of 10–15 mm. The interior surface is covered with about 100 or more layers produced by chemical or physical vapor decomposition, each layer changing in chemical composition in such a way that a gradually increased refractive index is produced layer by layer. This prepreg is collapsed by drawing out the composite in a reduced pressure to a rodlike preform with a diameter of about 4–6 mm. Through this collapse the parabolic profile of the refractive index is preformed. The preform is then drawn to the final optical fiber with an outer diameter of about 120 μm.

The vapor deposition process is based on the deposition of the chlorides of Si, together with other chlorides such as $GeCl_4$, BCl_3, $TiCl_4$, etc., giving successively increasing refractive indices as oxides or better as silicates with increasing concentrations of the added components. The modes of the monochromatic optical wave from a laser and the refractive indices of core and coating are shown in Fig. 16 for the different types of optical fibers. In the monomode step index and in the gradient-index fibers, broadening of the light signal over long distances is small, while in the multimode step-index fiber, large broadening of the signal takes place because of different time delays of the various modes.

The most important advantages of the optical fibers based on silica glass are the high transmission or the low optical losses (less than 1 dB/km at 850 and 1300 nm) due to the low impurity concentration of the syn-

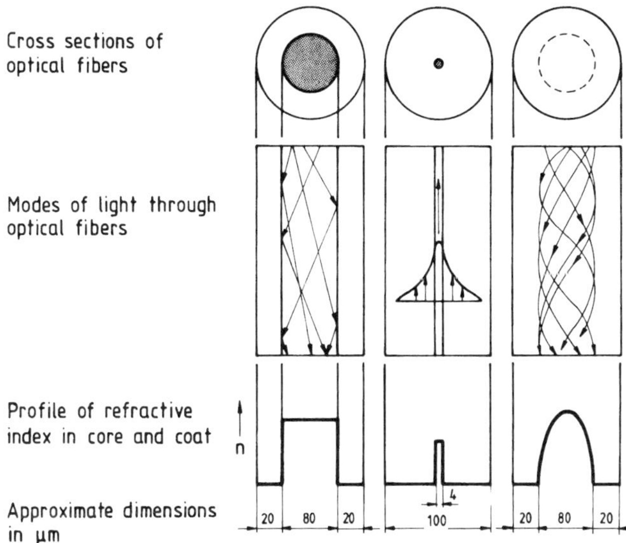

FIG. 16. Three types of optical glass fibers (schematic): above, cross section; middle, longitudinal section; and below, profile of refractive index. Step-index or multimode (left), single-mode (middle), and gradient-index fibers (right). n = refractive index; the numbers at the bottom are approximate dimensions in μm. Arrows indicate the propagation of the modes. The single-mode fiber carries along a relative large amount of intensity of the wave within the slad.

thetically produced silica glasses (types IV–VI). The limit of losses has been achieved. It is determined by the Rayleigh light scattering and phonon absorption (Fig. 17). Although this limit is more favorable for halogenide and chalcogenide glasses, a chemical purity approaching that of vitreous silica is not reached yet.

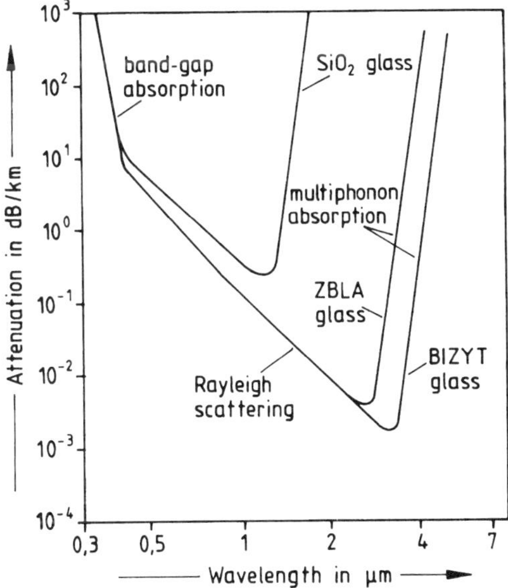

FIG. 17. The attenuation–wavelength window of minimum light losses due to Rayleigh scattering and phonon absorption of silica-glass fibers as compared with zirconium fluoride glasses (from Lucas and Adam, 1989). ZBLA = Zr–Ba–La–Al–fluoride, BIZYT = Ba–In–Zn–Y–Th–fluoride.

Another great advantage of silica-glass fibers is their large tensile strengths and large Young's modulus (see Sec. 2.2.1) as compared to multicomponent, halide, and chalcogenide glass fibers. If fibers are drawn under very clean conditions (CO_2-laser mirror furnace) and coated on line with polymers in order to prevent surface microcracks, average strengths of 5100 MPa for 4-cm proof length and about 4690 MPa for 20-m proof length (Kalish *et al.*, 1979) are measured.

A further advantage is the high chemical resistivity of the silica-glass fibers in contrast to fibers from other systems, particularly of halide and chalcogenide fibers.

2.4.3 Nonlinear Properties of Optical Silica-Glass Fibers; Optical Fiber Sensors

Optical glass fibers are not only passive or linear media for which an increase of power output increases proportionally with the input but are also active ones. In some materials, nonlinear selective power-dependent effects may occur with large optical intensities such as optical gain and frequency conversion (Stolen, 1979; Hill *et al.*, 1979). The main point is that such nonlinear effects depend on intensity and on the interaction length. On account of the low losses in small-core optical fibers and therefore of the high intensities transmitted over lengths of more than 1 km, large intensity amplification of 10^5–10^8 is possible. This enhancement decreases the threshold power for nonlinear processes that makes the effects useful in fiber-optical amplifiers, oscillators, and modulators. Thus, many optical sources used for the measurement of fiber losses and pulse dispersion may be based in future on fibers themselves.

Stimulated Raman scattering (SRS) and stimulated Brillouin scattering (SBS), which are the lowest threshold nonlinear processes in glass fibers, are observed as a result of the long interaction length available in optical glass fibers. Figure 18 shows an example of the Stokes and anti-Stokes scattered intensities for fused silica at 0 and 300 K. Also, optically induced birefringence, multiphoton effects, and two-step absorption from small impurity concentrations have been observed. Phase-matched four-photon mixing, self-phase modulation, and nonlinear absorption have been detected in silica-glass fibers. While these nonlinear effects are of interest as devices with optical glass fibers for tunable Raman oscillators and optical Kerr effect, they have to be also of particular concern for communication fibers because these processes can be detrimental.

Various fiber parameters can serve as sensors for the detection and localization of insertion losses by faults, defects, and connectors using the optical time-domain reflectometer (OTDR). These parameters are attenuation, bandwidth, local losses, mode mixing length, and, under certain conditions, resolution of scattering losses and absorption by attenuation measurements (Rourke *et al.*, 1979). The OTDR method is a nondestructive testing technique and is based on a measurement of the length dependence of optical fiber attenuation. The backscattered power from a probe pulse at a certain position in the fiber is detected by the OTDR. A pulse of light from the source is coupled into the fiber. If the propagating pulse encounters one or several absorption and backscattering sites, which may be dielectric discontinuities that result in Fresnel reflection, point defect sites, and Rayleigh scattering sites within the fiber, each site converts a portion of the forward traveling pulse into a reverse traveling wave that is directed toward the photodetector. Local alterations produced by external physical or chemical influences on the fiber properties can be detected and localized.

Many devices of optical fiber sensors are based on coherently illuminated pairs of fibers, a so-called probe fiber as the active sensor for measuring the parameters of interest such as temperature, pressure, etc., and a shielded reference fiber. Any external influence can change the phase of light in the probe by alteration of the fiber length, the core diameter, or the refractive index. This change of phase can be measured very sensitively in a two-arm fiber interferometer. A silica-glass fiber, for instance, as a pressure sensor has a pressure sensitivity of -4.1×10^{-5} rad Pa^{-1} m^{-1}. An interesting

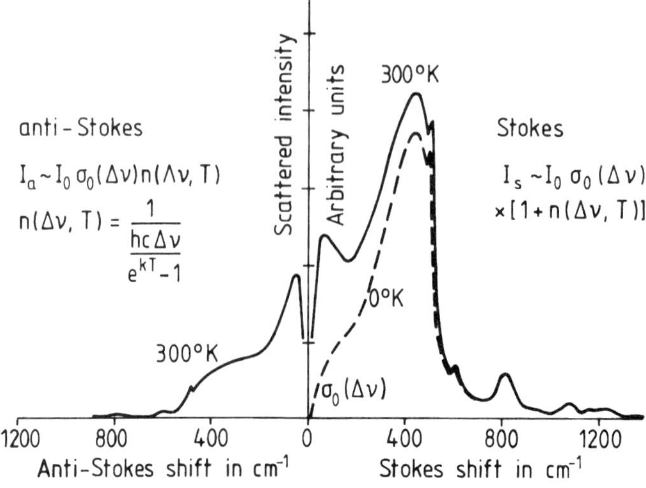

FIG. 18. Stimulated Raman scattering. Stokes and anti-Stokes scattered intensities for vitreous silica at 0 and 300 K; $n(\Delta\nu)$ is the phonon population factor, h is Planck's contant, k is Boltzmann's constant, T is absolute temperature, c is velocity of light, and $\Delta\nu$ is frequency shift in cm^{-1} (from Stolen, 1979, copyright American Telephone and Telegraph Company; used by permission).

application of a pressure-sensing fiber is the hydrophone. The gyroscope is another type of sensor based on fiber optics and on the classical Sanac "ring" interferometer in which the conventional air paths are replaced by single-mode fibers.

2.5 Electrical and Electronic Properties, Structural Defects

2.5.1 Conduction and Losses Quartz and silica glasses are among the best electrically insulating substances even at relatively high temperatures. Also, their surface electrical conductivity is low since hydrated film layers do not contain alkali ions as in the case of conventional silicate glasses.

The electrical properties of vitreous silica are usually attributed to the presence of impurities, especially of sodium ions. This fact holds for dc fields and for low-frequency ac fields. At high frequencies, other relaxation losses occur. In the case of type-III silica glass with reduced OH^- concentration (400 ppm), conductivities of the order of 10^{-12} Ω^{-1} cm^{-1} at 350 °C were measured.

The dielectric constant ϵ and dielectric loss, tan δ, show a typical dispersion and absorption in the temperature range 450–760 °C (Fig. 19; Owen and Douglas, 1959). The larger the hydroxyl content of the silica glasses, the larger are ϵ and the maximum of tan δ. The maximum shifts to higher temperatures with increasing frequency, indicating that a relaxation process is responsible for the shift. Therefore, the results in Fig. 19 are plotted in reduced scales $(\epsilon - \epsilon_\infty)/(\epsilon_S - \epsilon_\infty)$ or $\tan \delta_{ac}/\tan \delta_{ac}^{max}$ versus the ratio of frequencies f/f_{max}, where $\tan \delta_{ac}^{max}$ is the maximum of $\tan \delta_{ac}$ and f_{max} is the frequency at which $\tan \delta_{ac}^{max}$ occurs. Figure 19 indicates that the distribution of relaxation times of these migration losses, caused by the migration of the mobile impurity ions (Na^+, H^+, OH^-, etc.) in silica glasses, is less than in alkali silicate glasses.

Another type of dielectric loss in vitreous silica occurs at frequencies 10^4–10^8 Hz (depending on temperature). The origin of this loss is attributed to bendings and distortions of nonbridging oxygen bonds and to deformations of the positions of bridging oxygens (see Bruckner, 1970/71). Therefore, they are called deformation losses and are based on

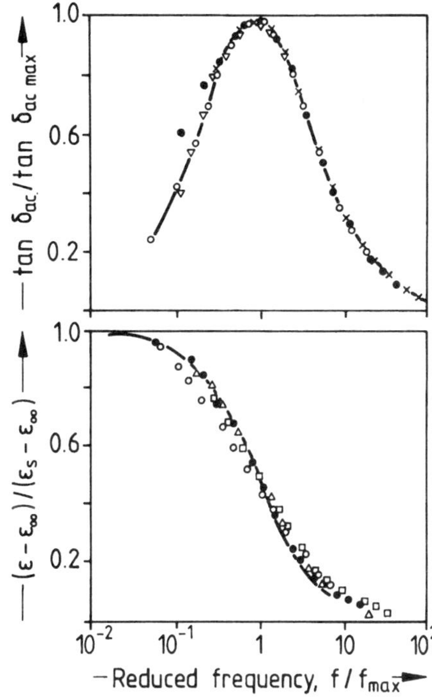

FIG. 19. Dispersion and absorption curves of silica glass versus frequency (reduced variable) in soft electrical fields due to relaxation of mobile impurities (reproduced with kind permission of The Society of Glass Technology from Owen and Douglas, 1959).

the relaxation of these deformations with a much lower relaxation time and a much lower activation energy than those of the migration losses in the low-frequency range.

The transport of electrons in vitreous silica may be measured particularly at high field strengths (e.g., Ferry, 1978). This transport is of special interest because of the properties of vitreous silica as a dielectric medium for use in metal-oxide-silicon (MOS) semiconductor devices. With respect to the transport properties, the primary attention has usually centered on the breakdown phenomena of vitreous SiO_2 at high electric fields. The drift velocity of hot electrons as a function of the applied electric field begins to saturate at fields of the order of 8×10^5 V/cm. Scattering due to polar optical modes of 0.153 and 0.063 eV is found and a "saturated" velocity exists at about 1.9×10^7 cm/s. An impact ionization is found to occur at about 10^7 V/cm.

In contrast to the high velocities of the

(hot) electrons, the velocity of the holes is found to be much lower when hole–electron pairs are produced by photo- or radiation-induced transient hole conductivity. Electrons that escape initial recombination depending on the applied electric field across the vitreous silica sample are rapidly swept out, and most of the holes remain behind near their points of generation and cause negative threshold voltage shifts in a MOS device. During a period of time the holes are transported relatively slowly, depending on temperature and electric field (10^{-6} s at room temperature, 10^{-3} s at cryogenic temperature), through the wet-grown, 96-nm-thick SiO_2 layer on ⟨100⟩ Si and are removed at the interface (Si under positive gate bias). This hole transport is highly dispersive and takes place over many decades of time (Fig. 20). It is assumed that polaron hopping between randomly distributed localized sites is the microscopic charge-transfer mechanism. The average hopping distance in the direction of the electric field is found to be 0.9 ± 0.2 nm for fields in the range of 1–6 MV/cm (McLean *et al.*, 1978).

2.5.2 Electronic Structure and Properties; Density of States Another approach to properties and structure of a material is the density of states, $N(E)$, of the electronic energy of a material or system, which can be calculated for a band from the band structure. The total density of states is the sum of the states of all bands. The band structure is the energy spectrum of an electron in an ideal crystal. The description of the energy states $E_\nu(\mathbf{k})$ for nonmetals usually consists of filled states (the valence band) separated from empty states (the conduction band) by an energy (the band-gap energy). These band gaps are large for insulators, as is the case for SiO_2 for which it is about 10 eV. The energy bands have the symmetry of the point group of a crystal. Most important for the properties of a solid are those bands that are near the Fermi energy. In the case of large band-gap (nonmetallic) crystals, the Fermi level is usually assumed to be approximately at the middle of the gap (Kittel, 1949). Figure 21 shows the results of calculations of band structures based on the extended tight-binding method for low-temperature quartz and cristobalite. Their broad energy gaps are very similar. From these band structures the densities of state are calculated and are shown in Fig. 22 for the two crystalline SiO_2 modifications quartz and cristobalite. A handicap exists for the noncrystalline vitreous silica because of its lack of long-range-order structure. Therefore, an approximation of the Zachariasen–Warren network model has been developed that has no rings of bonds but an infinite surface: the Bethe lattice, from which the density of states has been calculated for vitreous silica [Fig. 22(c)] that is very similar to those of the crystalline modifications and relatively slightly averaged in the energy distribution (Chadi *et al.*, 1978).

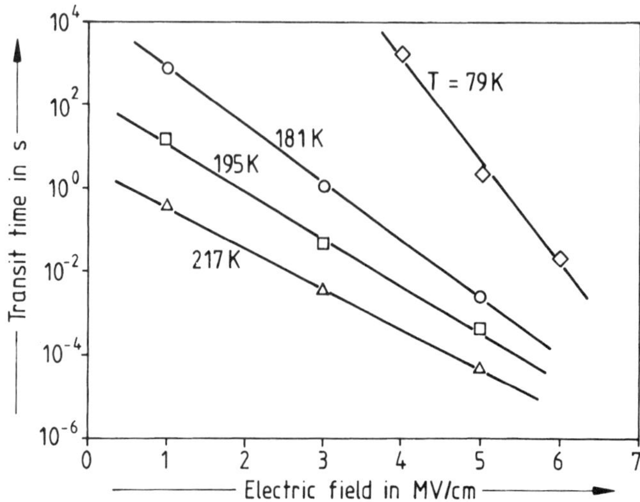

FIG. 20. Hole transit time, $t_{1/2}$, as a measure of the average inverse mobility versus electrical field in the oxide. $t_{1/2}$ is the time at which half recovery occurs (from McLean *et al.*, 1978).

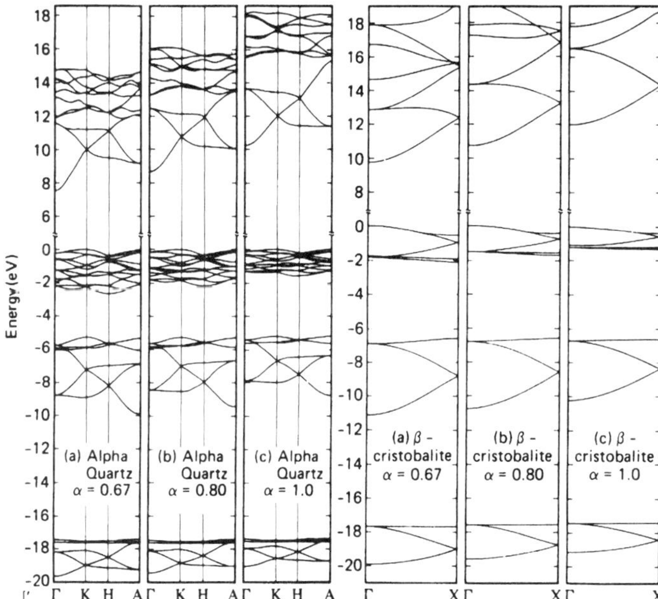

FIG. 21. Energy-band structure of (a) quartz and (b) crystobalite for different values of the exchange parameter α (from Batra, 1978).

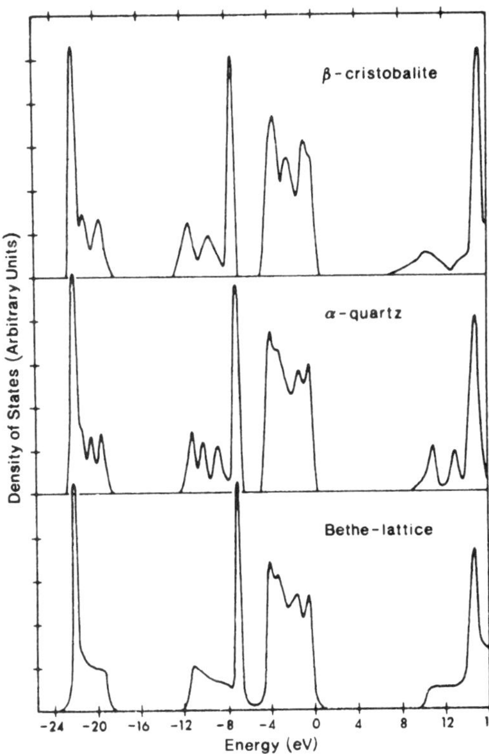

FIG. 22. Density of states for crystobalite, quartz, and vitreous silica (Bethe lattice). The valence-band maximum of quartz has been chosen for zero of energy (from Chadi *et al.*, 1978).

2.5.3 Structural Defects, Paramagnetic and Color Centers Numerous properties of quartz and silica glass are strongly influenced by the presence of defects. There are various defects localized in the structure of SiO_2. A "perfect" structure may be a continuous crystalline lattice (quartz) or a continuous network (silica glass) of $SiO_{4/2}$ tetrahedra joined together at the corners. All kinds of local deviations from a "perfect" structure will be called structural defects, including impurities, dopings, vacancies, radiation damage, and the so-called point defects. These defects may be charged or uncharged, diamagnetic or paramagnetic, and many of them may be associated with optical absorption bands. While the interpretation of optical data is connected with uncertainties, electron paramagnetic resonance (EPR) spectroscopy gives the most detailed data for interpretation. Since the application of EPR implies the presence of unpaired electrons, it is usually necessary to produce such paramagnetic centers by radiation in order to activate precursor sites prior to trapping of electrons and holes at the defect centers of interest. Various impurities such as Fe^{3+}, Cu^{2+}, and Ti^{3+} are paramagnetic.

In natural quartz there are frequently aluminum centers as impurities, which are responsible for the absorption and brown/black color when irradiated (smoky quartz). Due to the two different Si–O distances (0.160 and 0.162 nm) in quartz at exact crystallographic positions that determine the dextro- and levo-oriented arrangement of the SiO_4-tetrahedra, the exact local position of the paramagnetic center within the unit cell can be determined if angle-dependent studies of single crystals relative to the magnetic field of the EPR spectrometer are made and compared with the known point-group symmetries of the crystal. In such a way and by x-ray or γ-ray irradiation at room temperature and cryogenic temperatures, it could be shown that Al is coordinated by four oxygens forming an AlO_4-tetrahedron and that the paramagnetic center is $[Al,e^+/M^+]^{1+}$ with the precursor $[Al^-/M^+]^\circ$, where the hole is trapped at the one oxygen with the small distance and the alkali ion (M^+) between the two oxygens with the longer distance to Si. Many other impurities in quartz have been studied in this way, such as Ge^{4+}, Ti^{3+}, H^0, and Fe^{3+}, the last one being partly responsible for the amethyst color (Griscom, 1978).

If pure quartz is irradiated by neutrons, paramagnetic centers of a type other than those produced by impurities occur that are defect centers termed E'_1, E'_2, and E'_4. They are correlated with optical bands at 5.8 eV. These paramagnetic states are due to unpaired electrons at oxygen vacancies that belong to two differently and asymmetrically relaxed silicon atoms (Griscom, 1978; Weeks, 1994). In neutron, electron, γ-ray, and x-ray irradiated silica glass, only the E'_1 paramagnetic center has been observed (Weeks, 1994).

Intrinsic and extrinsic defect centers play an important role also in pure silica glass (types III and IV) and in silica-glass fibers. Three general intrinsic defect types are known to date: the E' center at a bridging oxygen vacancy (≡Si·, where the point denotes an unpaired spin). Thus, starting from the neutral situation of an oxygen vacancy ≡Si–Si≡, the creation of an $E'_1 = E'_\gamma$ center by hole trapping can be expressed:

$$\equiv\text{Si–Si}\equiv \xrightarrow{h\nu} \equiv\text{Si}\cdot\ \text{Si}\equiv + e^-, \quad (16)$$

with the spin at the site of the oxygen vacancy (Weeks, 1956; Griscom, 1986; Weeks and Nelson, 1960). The second paramagnetic center is nearly the same one with the only difference being that the unpaired spin is at the site away from the vacancy toward an interstitial position and a radiolytically produced hydrogen atom at the second silicon atom with a preexisting oxygen atom (Weeks, 1994):

$$\equiv\text{i–Si}\equiv + \text{H}^\circ \xrightarrow{h\nu} \equiv\text{Si}\cdot\ \text{H–Si}\equiv. \quad (17)$$

In another model the strained bond ≡Si–Si≡ is not attacked by a hydrogen, and the radiolytically displaced oxygen is separated far away, thus, the following reaction is assumed: $\equiv\text{Si–Si}\equiv \xrightarrow{h\nu} \equiv\text{Si}\cdot\equiv\text{Si} + e^-$, where the unpaired spin is at the site away from the oxygen vacancy. In both models for E'_β, the threefold-bonded silicons with no unpaired electrons, ≡Si, are electrically charged with e^+; the sites of the trapped electrons, e^-, are still not known. The third defect center is connected to the formation of peroxide radicals by hole trapping on peroxide linkages. The creation of it is assumed by a reaction such as (Griscom, 1984, 1986)

$$\equiv\text{Si–O–O–Si}\equiv \xrightarrow{h\nu} \equiv\text{Si–O–O}\cdot\text{Si}\equiv + e^- \quad (18)$$

being a hypothetical one (Weeks, 1994).

All these kinds of defects were analyzed by EPR-hyperfine interactions of the unpaired spin with ^{29}Si and ^{17}O isotopes, respectively, and additionally by isochronical radiation and anneal kinetics. In this way Griscom (1986, 1984) has found that subtle but definite differences in the g values seem to correlate with the defect production and/or annealing kinetics, and he has found on this basis an additional very short-living E' center (E'_α) that anneals at 100 K in the order of minutes to hours. With respect to the high OH^--containing type-III silica glasses, there exists a second oxygen-associated hole center that is related not to a bridging but to a nonbridging oxygen, and therefore it is not an intrinsic but an extrinsic defect center produced by the radiolysis of the hydroxyl group:

$$\equiv\text{Si–OH} \xrightarrow{h\nu} \equiv\text{Si–O}\cdot + \text{H}^\circ. \quad (19)$$

Its lifetime is relatively short at 300 K as compared with those of the E'_β and E'_γ and with that of the peroxide radical (Fig. 23). The annealing mechanisms are connected to the diffusion of various species and observed in silica-glass types III and IV.

Other extrinsic defects than those based on OH^- groups are associated with Cl^-, particularly in type-IV silica glass. There are those based on dopants such as Ge and P in the core of optical fibers (Griscom, 1988). In type I and II silica glasses, the Al and Na centers as natural impurities from rock crystals are centers that are very similar to those in quartz.

2.5.4 SiO_2 Films in Microelectronic Devices This section will consider the key developments and novel aspects of technology associated with the use of silica films in solid-state silicon-based devices and circuits. Investigations of the physical and electrical properties in thin dielectric SiO_2 glass films are based on a perspective other than those in studying bulk SiO_2 glass. This other perspective is connected to the various demands on dielectric properties of the thin SiO_2 layers on Si substrates in contrast to bulk silica glass. Two examples may be given for a first impression:

1. Electrical fields across the layers are in the MV/cm range, but those across bulk samples are in the kV/cm range. This difference is due to the relatively small thickness of SiO_2 in films (10–100 nm) compared with typical bulk samples that are about a millimeter thick.
2. Dielectric losses in bulk glasses are measured at sodium concentrations in the tenth-percent range and larger, while instabilities in device properties are observed at concentrations of one part per million and less.

Therefore, any kinds of impurities, defects, or charges have consequences in the quality and reliability of the numerous MOS-based systems. With respect to the principal functions, to the large variety, and to the technology of integrated solid-state circuits, the reader is referred to the articles SI-SIO$_2$ INTERFACE, ELECTRONIC PROPERTIES OF and HETEROSTRUCTURES, SEMICONDUCTOR in this encyclopedia and to Sze (1983, 1991), Widmann *et al.*, (1988), and Aitken and Irene (1985). Only the preparation techniques, functions, and properties of the pure and doped SiO_2 films in contact with the heavily doped Si substrates will be described here.

The most usual preparation techniques are the direct oxidation of silicon and chemical vapor deposition (CVD). Variations are physical vapor deposition (PVC), high-pressure oxidation, and plasma oxidation.

The process of thermal oxidation is of the diffusion-reaction type and leads to a linear-parabolic equation for the thickness growth of the oxide layer with linear and parabolic rate constants, k_l and k_p, respectively:

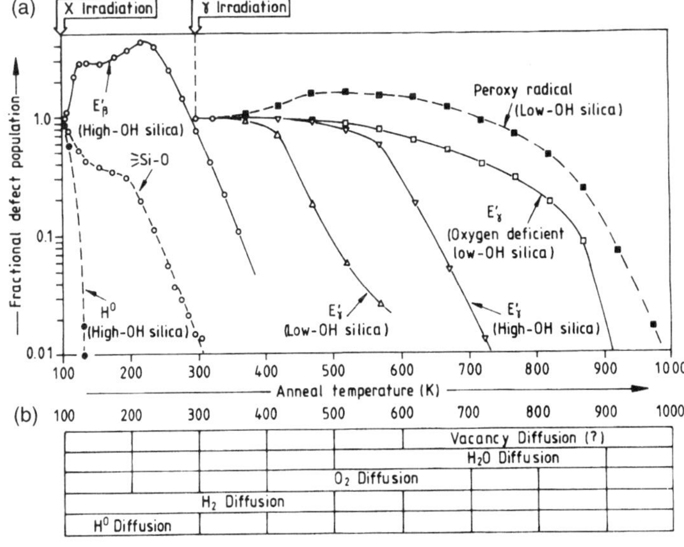

FIG. 23. (a) Isothermal anneal curves for radiation-induced defect centers in high-purity type-III (1200 ppm OH) and type-IV (<5 ppm OH) silica glasses. (b) Effective temperature ranges for various diffusion-limited anneal mechanisms, predicated on the 5–10-min anneal times employed in the experiment (with kind permission of Materials Research Society, from Griscom, 1986).

$$k_l^{-1}(l - l_0) + k_p^{-1}(l^2 - l_0^2) = t - t_0, \tag{20}$$

where l and t are the oxide thickness and time of oxidation, respectively, and l_0 and t_0 are the offsets on the l and t axes. For oxidation in pure H_2O, l_0 and t_0 are both zero, whereas for the oxidation in dry O_2 there exists an initial oxidation regime that is not described by the linear-parabolic law but for which l_0 and t_0 are valid (see Aitken and Irene, 1985). The silicon interface is currently removed during oxidation, and the oxide layer is fully oxidized because the oxidant is moving through the oxide. The electrically active region, the Si–SiO_2 interface, is affected by the oxidation process, and any impurity present in the oxidation ambient or defects caused by oxidation will be located at this important interface. Usual temperatures for oxidation are between 780 and 1200 °C. Lower-temperature oxidation results in a higher density of SiO_2. High-pressure oxidation accelerates the oxidation rate; e.g., for 10 atm steam, a temperature of about 900 °C is comparable in oxidation rate with 1 atm steam at 1200 °C.

Both CVD and PVD are usual techniques for masking and insolation. However, CVD-SiO_2 films do not possess the same electrical quality in terms of fixed and mobile charges, charge trapping, interface states, and dielectric integrity as compared with thermally oxidized SiO_2 layers. However, there are several advantages to the CVD process. The CVD leaves the Si–SiO_2 interface intact, and high deposition rates at lower temperatures, as compared with thermal oxidation of Si, are usual, e.g., 10 nm/min SiO_2 at 750 °C with CVD for which in O_2 1200 °C is needed.

In the field of microelectronics it is well known that differently prepared films may yield considerably different electrical behavior. Therefore, a current physicochemical characterization besides that of the electrical properties is necessary. Quantitative compositional analysis is available from electron microprobe, nuclear backscattering, secondary ion mass spectroscopy (SIMS), and Auger and x-ray photoelectron spectroscopy (XPS). Scanning and transmission electron microscopy (SEM and TEM) techniques are tools for determining the morphology of SiO_2 films and Si–SiO_2 interfaces. EPR has also been used to determine interface states.

The knowledge and preparation of the SiO_2 films produced by growth, by etching, or by secondary passivation with Si_3N_4 films is very important, particularly that at the gate (base) of MOSFETs (FET: field effect transistor). The most common mechanical method uses a sensitive stylus, and optical measurements are usually made by visible-light interference or by ellipsometry.

The principal functions of SiO_2 thin films in MOSFET devices are as follows:

1. primary passivation to reduce the number of electrically active surface defects of the various doped semiconductors (Si) (defects influence the dielectric properties);
2. insulation (secondary passivation) of conducting layers from each other or to insulate individual devices on a common silicon substrate—in this case, phosphosilicate glass films act as an impurity getter, and Si_3N_4 layers act as protection against corrosion, further oxidation, and mechanical damage;
3. masking against dopants in the silicon surface;
4. diffusion source for doping the silicon interface; and
5. increase in the breakdown voltage.

Although the dielectric phenomena and explorations of dielectric constant and dielectric breakdown in thin SiO_2 glass films are similar to those for bulk glass (see Sec. 2.5.1), the passivation of the silicon surface by SiO_2 and the Si–SiO_2 interface are unique phenomena. The MOS-based devices therefore are defined and described by the dielectric and passivating properties of the thin glass films—that means, particularly, by the capacitance-voltage (C-V) characteristics. A MOS capacitor and its equivalent circuit [Fig. 24(a)] has usually a C-V curve, shown in Fig. 24(b) for low and high frequencies. In this diagram a small-signal ac capacitance is measured while a dc bias is applied to the gate. The capacitance of this device depends on the properties of the glass layer and the semiconductor. The C-V data contain information about the dielectric constant, film thickness, interfacial quality, and doping type and concentration of the semiconductor. For more information, see Grove (1967) and Sze (1991). In principle the capacitance of the semiconductor arises from the space-charge layers that are formed by different potentials at the Si/SiO_2 interface due to dif-

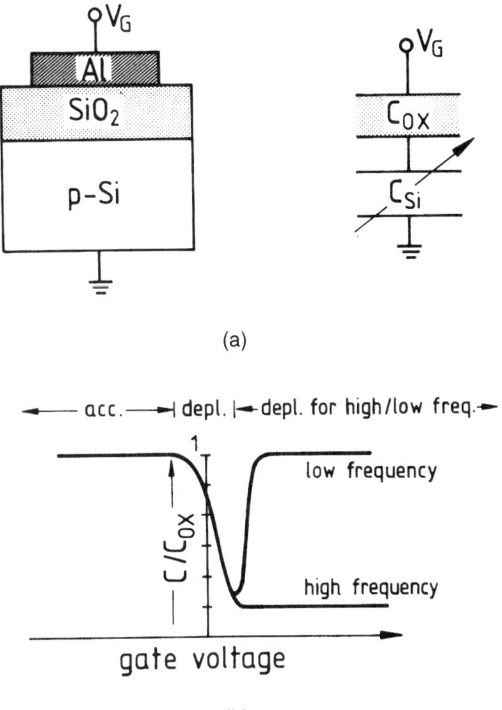

FIG. 24. *C-V* characteristic of an Al–SiO_2–Si device: (a) physical setup of a MOS capacitor and its equivalent circuit, (b) the capacitance-voltage curve measured at high and low frequencies (reproduced with kind permission of Academic Press Ltd., after Aitken and Irene, 1985).

ferent dc bias conditions. Three space-charge conditions in the *p*-type Si, marked along the voltage axis, are produced: At negatively biased voltage the majority of the holes are drawn to the interface, forming a thin dense layer that makes the interface look metallic (accumulation of the mobile majority carriers at the interface, and only the capacitance of the SiO_2 is measured). At smaller negative and positive voltages, the majority carriers are driven away from the interface (depletion range) and the finite capacitance is inversely proportional to the depletion width. This depletion width is increased, and the capacitance is decreased, as voltage increases until any additional positive bias at the gate does not increase the charge in the depletion layer but rather supplies additional minority carriers to the inversion layer. Depletion and inversion layers have quite different frequency responses because the depletion charge is a majority-carrier effect and responds rapidly to the changes in the applied field, and the inversion layer is a minority-carrier effect with a slow response to the field due to the fact that time is needed for the thermal generation of these carriers. Thus, at high frequencies (beyond 10 Hz) only the depletion layer and the steplike high-frequency curve are obtained.

From the preceding description of the MOS *C-V* properties, it becomes clear that not only the quality of the doped semiconductor but also that of the SiO_2 film and Si–SiO_2 interface are of significant importance. The quality of passivation of the thin SiO_2 films determines the quality of the intrinsic electrical properties of the Si–SiO_2 interface and its stability. Thus, any kind of electrically active defects at the interface and in the SiO_2 layer influences the dielectrical properties of the device negatively. These defects fall into two categories: interface traps (also called surface states) and charges, the latter being fixed or mobile.

With respect to the SiO_2 layers, different types of defects have to be considered and their numbers reduced in order to produce high-quality SiO_2 films and Si–SiO_2 interfaces. Since electrically active defects are traps, able to capture carriers, one has to know the various kinds of traps in SiO_2 and their concentrations. DiMario (1978) has given a review on electron and hole traps in thermal SiO_2. The most important traps are sodium-connected traps. There is first the 2.4-eV trap, which is a Coulombic attractive site for electrons in the SiO_2 band gap that reduces the well-known total band gap of about 9 eV. This trapping site is possibly related to sodium in an immobile configuration, particularly produced in steam-grown SiO_2. In contrast to that there are usually mobile traps that are connected to Na^+ ions that are immobilized at 77 K near the Si–SiO_2 interface. In as-grown SiO_2, traps were observed caused by water incorporation into the film during deposition. Trapped hole sites near the Si–SiO_2 interface that trap electrons into energetically deep sites can be injected into SiO_2 films or may be generated during processing conditions (growth, temperature, and annealing). There are hole traps in an as-grown silica film that are introduced into the SiO_2 valence band through excitation by far UV light, x rays, or high electric fields and injections from the Si substrate across the SiO_2 band gap. Deep hole traps near the Si–SiO_2 interface capture holes permanently at room temperature. Hole traps are sensitive to processing conditions and can be increased by annealing above 950 °C. It is believed that they are caused by impurities introduced into the SiO_2 glass and/or by strained or broken bonds due to stresses at the interface. Even the flow of current across the Si–SiO_2 interface can create new traps (interface traps). This creation is observed when capacitors have been stressed by hot-electron injections, by exposure to ionizing radiation, and by high electric fields.

Besides the just-described defects, there is a series of intentionally produced traps with differing properties produced by introduction of tungsten (W traps), by x-ray radiation, and by ion implantation of Al, As, and P, for example. Usually the last ones produce such extensive damage that any carrier injected into the SiO_2 layer is trapped. For more information about these special kinds of traps, see DiMario (1978).

2.6 Applications of Vitreous Silica

The fields of application of vitreous silica are determined by the properties of the various types, which again depend on quality, i.e., on the kinds and concentrations of impurities (see Table 1). Therefore, it will be suitable to list the fields of application of the seven types of vitreous silica, which naturally have relatively large regions of overlap.

Type I is used for IR optics in optical applications. The priority of applications is tubing for lamps of different kinds, particularly for IR lamps that are used for drying applications in the field of coatings. Generally, it is used for nonoptical applications, e.g., for semiconductor-processing equipment, and as a material for structural elements.

Type II is suitable for general and precision optics, for tubing elements of optical fiber preforms, for tubing elements of high-quality discharge lamps, and for semiconductor-processing equipment.

Types III and IIIa were developed for a series of modern applications such as precision (UV) optics, high-power laser optics, mask substrates of optical microlithography, substrates for optoelectronics, and, last but not least, core material for speciality fiber preforms, e.g., PCS.

Type IIIb is useful for optical telecommunication fibers as preforms with doped and undoped core and as material for cladding and jacketing.

Type IV is a nearly hydroxyl-free vitreous silica with highest purity and serves as standard material for scientific research and is used in precision optics and in fluorine-doped cladding for speciality fiber preforms.

Type Va is a precursor for type Vb silica glass.

Type Vb has applications in semiconductor-processing equipment and as special lamp tubing, e.g., metal-halide, mercury, and other discharge lamps as well as metal-halide incandescent lamps.

Type VI is fabricated on a pilot scale so far and not yet on an industrial scale. It appears useful for experimental studies on purification and doping of the porous precursor material, on sintering, and on preparation techniques on the basis of colloidal SiO_2 particles as raw material.

Type VII is a material prepared on a laboratory scale and is a typical laboratory material. It is based on the sol-gel technique (silicon alkoxide); it has mainly experimental applications in production of coatings and thin films on substrates.

Additional applications due to typical properties of vitreous silica and quartz are given in Secs. 1.3.1 and 1.3.2. Special applications are less connected to the typical properties but rather to the material SiO_2 as a part of a system, e.g., as optical waveguides, sensors, etc., as pointed out in Secs. 2.4.2 and 2.4.3.

Not included in the seven types of vitreous silica are the thermally produced thin films of SiO_2 on metallic Si surfaces by oxidation, particularly as a passivation element in semiconducting devices (Sec. 2.5.4). The reason is that the quality of these thin layers depends strongly on preparation conditions such as furnace atmosphere, impurities, and dopings of the Si substrate. In this way this group of silica glasses is not a class in the sense of the other types, and, additionally, it is strongly connected to the metallic Si substrate with special preparation-dependent properties of the interface Si/SiO_2.

3. SILICA GEL, AEROGELS

During the last decade, a pure SiO_2 product with remarkable properties due to its high porosity has been given attention. Silica gels with densities between 0.3 and 0.005 g cm^{-3}, called aerogels because they contain more air than amorphous SiO_2, have been prepared. There are usually two routes of production. In the first route, water glass (sodium silicate) plus HCl is mixed. After gelation the sodium is washed several times with water until a pure SiO_2 skeleton is obtained. The water in this aqua gel has to be washed out by methanol and is dried under supercritical conditions (above 240 °C and above a pressure of 80 bars) in order to prevent shrinking and the collapse of the porous gel by capillary forces. The second route starts from a mixture of tetramethoxysilane, $Si(OCH_3)_4$, and water plus an acid as catalyzer. After formation of hydroxyl, $Si(OH)_4$, and release of CH_3OH and separation of water, Si–O–Si bondings are formed into a very fine wet network of SiO_2 particles (gel). This product has to be dried supercritically.

By virtue of the high porosity and small pore sizes of 1–100 nm (comparable with the mean-free-path length of the air molecules of 70 nm!), these gels have excellent heat isolation for windows and solar energy collectors (Fricke, 1988). Heat conductivities are obtained between 0.005 and 0.010 W m^{-1} K^{-1} (for comparison, styropore has conductivity of 0.026). Aerogels are used as Cherenkov sensors in high-energy physics in order to record relativistic protons, muons, and pions. On account of their nanometer structure, they have very low ultrasonic velocities (about 100 m s^{-1}, which is 50 times smaller than that for bulk silica glass). At a density of 0.005 g cm^{-3} the velocity is 20 m s^{-1}. Such velocities provide material for acoustic antireflection layers because the product of the small densities and sonic velocities has a low acoustic impedance. Therefore, a high acoustic intensity (by a factor of 1000) is obtained in air if a thin layer of aerogel is glued on to a usual ultrasonic piezo source because the impedance in air is also very low. Applications are seen for measuring distances to the rear of cars or in roboter systems, for instance.

4. SUMMARY

In this article silicon dioxide is shown to be a chemically simple composition, but a very complicated one with respect to its crystallographic modifications and properties, especially quartz, cristobalite, and fused silica. Particularly the noncrystalline version of it, the vitreous (glassy) silica or silica glass, is of great interest in science and of great importance in practical applications due to its excellent physical, chemical, and thermal properties and due to its good workability. It has applications in numerous devices, in electronics, in microelectronics, in optics, in sensors, in the modern communication field (optical fibers), for the industrial fabrication of special lamps, and in semiconductors as melting crucibles with highest purity. The survey includes the crystalline, vitreous, and amorphous states of silicon dioxide; its mechanical, thermal, optical, electrical, and electronic properties; and also the various anomalies of vitreous silica as compared with the properties of the large family of the multicomponent silicate glasses. Additionally, some discussion is given to silica gels and aerogels.

ACKNOWLEDGMENTS

The author is grateful to Dr. Englisch and Dr. Witzke, Heraues Quarzschmelze, for fruitful discussions and supplementations and to Prof. Dr. R. Weeks for linguistic and complementary revision of the manuscript.

GLOSSARY

Aerogel: High-porous silica gel, amorphous SiO_2.

Cristobalite: Modification of SiO_2, stable above 1470 °C.

Extrinsic Defects: Defects produced by impurities or nonstoichiometric local composition.

Intrinsic Defects: Structural defects.

Keatite, Coesite, Stishovite: High-pressure, high-temperature modifications of SiO_2.

Quartz: Most frequent and thermodynamically stable modification of SiO_2 on Earth.

Silica Gel: Amorphous SiO_2.

Silica W: Fibrous SiO_2, stable above 1000 °C in water-free atmosphere.

Silicon Dioxide, Silica: Chemical compound SiO_2.

Thermal SiO_2: Glassy SiO_2 films grown on Si surfaces at elevated temperatures (passivation films).

Trapping Sites: Electron or hole traps.

Tridymite: Modification of SiO_2, stabilized by alkali oxide.

Vitreous Silica, Fused Silica, Silica Glass: Glassy modification of SiO_2.

Works Cited

Aitken, J. M., Irene, E. A. (1985), in: M. Tomozawa, R. H. Doremus (Eds.), *Treatise on Material Science and Technology,* Vol. 26, Orlando: Academic, pp. 1–56.

Batra, J. P. (1978), in: S. T. Pantelides (Ed.) *The Physics of SiO_2 and Its Interfaces,* New York: Pergamon, pp. 65–69.

Brill, R. W., Hermann, G., Peters, C. (1939), *Naturwissenschaften* **40,** 676–677.

Brill, R. W., Herman, G., Peters, C. (1942), *Ann. Phys. (Leipzig)* **41,** 233–244.

Brückner, R. (1970/71), *J. Non-Cryst. Solids* **5,** 123–216.

Chadi, D. J., Laughlin, R. B., Joannopoulos, J. D. (1978), in: S. T. Pantelides (Ed.), *The Physics of SiO_2 and Its Interfaces,* New York: Pergamon, pp. 55–59.

Clasen, R. (1987), *Glastechn. Ber.* **60,** 125–132.

Clasen, R. (1993), *Glastechn. Ber.* **66,** 299–304.

Cohen, A. F. (1958), *J. Appl. Phys.* **29,** 591–598.

DiMario, J. G. (1978), in: S. T. Pantelides (Ed.), *The Physics of SiO_2 and Its Interfaces,* New York: Pergamon, pp. 160–178.

Evans, K. M., Gaskell, P. H., Nex, C. M. M. (1982), in: P. H. Gaskell, I. M. Parker, E. A. Davis (Eds.), *The Structure of Non-Crystalline Materials,* Proceedings of the Second International Conference, London: Taylor & Francis.

Fanderlik, J. (1991), *Silica Glass and Its Application,* Amsterdam: Elsevier.

Ferry, D. K. (1978), in: S. T. Pantelides (Ed.), *The Physics of SiO_2 and Its Interfaces,* New York: Pergamon, pp. 29–34.

Flörke, O. W. (1956a), *Naturwissenschaften* **43,** 419–420.

Flörke, O. W. (1956b), *Ber. Dtsch. Keram. Ges.* **33,** 319–321.

Fricke, J. (1988), *Sci. Am.* **258** (5), 92–97.

Friebele, E. J. (1991) in: D. R. Uhlmann, N. J. Kreidl (Eds.), *Optical Properties of Glass,* Westerville, OH: American Ceramic Society, pp. 205–262.

Frischat, G. H. (1969), *Phys. Status Solidi* **35,** K47–K49; *Glastechn. Ber.* **42,** 351–359.

Griscom, D. L. (1978), in: S. T. Pantelides (Ed.) *The Physics of SiO_2 and Its Interfaces,* New York: Pergamon, pp. 232–252.

Griscom, D. L. (1984), *Nucl. Instrum. Methods* **B1,** 481–488.

Griscom, D. L. (1986), in: F. L. Galeener, D. L. Griscom, M. J. Weber (Eds.), *Defects in Glasses,* Pittsburgh: Materials Research Society, pp. 213–221.

Griscom, D. L. (1988), in: R.A.B. Devine (Ed.), *The Physics and Technology of Amorphous SiO_2,* New York: Plenum Press, pp. 125–134.

Grove, A. S. (1967), *Physics and Technology of Semiconductor Devices,* New York: Wiley.

Hench, L. L. (1993), in: A. K. Varshneya, D. F. Bickford, P. P. Bihuniak (Eds.), *Advances in Fusion and Processing of Glass, Ceram. Trans. Vol. 29,* Westerville, OH: American Ceramic Society, pp. 591–597.

Hetherington, G., Jack, K. H., Kennedy, J. C. (1964), *Phys. Chem. Glasses* **5,** 130–136.

Hill, K. O., Kawasaki, B. S., Johnson, D. C., Fujii, Y. (1979), in: B. Bendow, S. S. Mitra (Eds.), *Fibre Optics,* New York: Plenum Press, pp. 211–240.

Hosemann, R., Hentschel, M. P., Schmeißer, U., Brückner, R. (1986), *J. Non-Cryst. Solids* **83,** 223–234.

Hunklinger, S. (1982), *J. Phys. (Paris)* **C9,** suppl. au no. 12, Tome 43, 461–474.

Hunklinger, S., Schickfus, M. von (1981), in: W. A. Phillips (Ed.), *Amorphous Solids,* Topics in Current Physics Vol. 24, Berlin: Springer-Verlag, pp. 81–104.

Jebsen-Marwedel, H., Brückner, R. (1980), *Glastechnische Fabrikationsfehler,* Berlin: Springer-Verlag, pp. 358–384.

Kalish, D., Key, L. L., Kurkjian, C. R., Tariyal, B. K., Wang, T. T. (1979), in: S. E. Miller, A. G. Chynoweth (Eds.), *Optical Fibre Telecommunications,* New York: Academic, pp. 401–433.

Kittel, C. (1949), *Introduction to Solid State Physics,* New York: Wiley.

Konnert, J. H., D'Antonio, P., Karle, J. (1982), *J. Non-Cryst. Solids* **53,** 135–141.

Lamkin, M. A., Riley, F. L., Fordham, R. J. (1992), *J. Europ. Ceram. Soc.* **10,** 347–367.

Lee, R. W. (1964), *Phys. Chem. Glasses* **5,** 35–42.

Lebedev, A. A. (1921), *Arb. Staatl. Opt. Inst., Leningrad* **2** (10).

Lebedev, A. A. (1940), *Bull. Acad. Sci. USSR, Phys. Ser.* **4,** 584–587.

Lell, E., Kreidl, N. J., Hensler, J. R. (1966), *Prog. Ceram. Sci.* **4,** 1–38.

Lippingcott, E. R., van Valkenburg, A., Weir, Ch. E., Bunting, E. N. (1958), *J. Res. Natl. Bur. Std.,* 61–72.

McLean, F. B., Boesch, H. E., McGarrity, J. M. (1978), in: S.T. Pantelides (Ed.), *The Physics of SiO_2 and Its Interfaces,* New York: Pergamon, pp. 19–23.

Lucas, J., Adam, J.-L. (1989), *Glastechn. Ber.* **62,** 422–440.

Ogawa, K., Morikawa, T., Imamura, H., Akai, Y. (1977), *J. Light Vis. Env.* **1,** 16–21.

Owen, A. E., Douglas, R. W. (1959), *J. Soc. Glass Technol.* **43,** 159–165.

Primak, W. (1958), *Phys. Rev.* **110,** 1240–1251.

Porai-Koshits, E. A. (1977), *J. Non-Cryst. Solids* **25,** 87–128.

Rourke, M. D., Jensen, S. M., Barnoski, M. K. (1979), in: B. Bendow, S.S. Mitra (Eds.), *Fibre Optics,* New York: Plenum Press, pp. 255–268.

Schaeffer, H. (1984), *J. Non-Cryst. Solids* **67,** 19–33.

Schultz, P. C. (1974), *J. Am. Ceram. Soc.* **57,** 309–313.

Sigel, G. H. (1973/74), *J. Non-Cryst. Solids* **13,** 372–398.

Sosman, R. B. (1965), *The Phases of Silica,* New Brunswick, NJ: Rutgers Univ. Press.

Stolen, R. H. (1979), in: S. E. Miller, A. G. Chynoweth (Eds.), *Optical Fibre Telecommunications,* New York: Academic, pp. 125–150.

Stone, J. (1987), *J. Lightwave Technol.* **5,** 712–733.

Stroud, J. S., Schreures, W. H., Tucker, R. F. (1965), *Proceedings of the 7th International Congress on Glass,* Bruxelles, 42.1–42.18.

Tool, A. Q. (1946), *J. Am. Ceram. Soc.* **29,** 240–253.

Treadaway, M. J., Passenheim, B. C., Kitterer, B. D. (1975), *IEEE Trans. Nucl. Sci.* **NS-22,** 2253–2258.

Valenkov, N., Porai-Koshits, E. A. (1936), *Nature* **137,** 273–274.

Wang, P. W., Haglund, R. F., *et al.* (1988), *J. Non-Cryst. Solids* **102,** 288–294.

Warren, B. E., Krutter, H., Morningstar, O. (1936), *J. Am. Ceram. Soc.* **19,** 202–206.

Warren, B. E. (1941), *J. Am. Ceram. Soc.* **24,** 256–261.

Weeks, R. A. (1956), *J. Appl. Phys.* **27,** 1376–1381.

Weeks, R. A. (1994), *J. Non-Cryst. Solids* **179,** 1–9.

Weeks, R. A., Nelson, C. (1960), *J. Am. Ceram. Soc.* **43,** 399–402.

Weeks, R. A., Sonder, E. (1963), in: *LOW Symposium on Paramagnetic Resonance,* Vol. 2, New York: Academic, pp. 869–879.

Witzke, H.-D. (1990), *Glastechn. Ber.* **63K,** 333–341.

Witzke, H.-D., Takke, R. St. Thomas, Englisch, W. (1994), *Glastechn. Ber., Glass Sci. Technol.* **67c,** 346–349.

Wong, J., Angell, C. A. (1976), *Glass Structure by Spectroscopy,* New York: Marcel Dekker.

Wright, A. C., Leadbetter, A. J. (1976), *Phys. Chem. Glasses* **17,** 122–145.

Wright, A. C., *et al.* (1982), *J. Non-Cryst. Solids* **49,** 63–102.

Zachariasen, W. H. (1932), *J. Am. Chem. Soc.* **54,** 3841–3851.

Zarzycki, J. (1982), *J. Non-Cryst. Solids* **52,** 31–43.

Zarzycki, J., Naudin, F. (1960), *Verres Réfractaires* **14,** 113–123.

Further Reading

Fricke, J., Emmerling, A. (1992), in: R. Reisfeld, C. K. Jörgensen (Eds.), *Structure and Bonding,* Vol. 77, Berlin: Springer-Verlag, pp. 37–87.

Phillips, W. A. (Ed.) (1981), *Amorphous Solids, Low Temperature Properties,* Topics in Current Physics Vol. 24, Berlin: Springer-Verlag.

Sze, S. M. (Ed.) (1983), *VLSJ Technology,* New York: McGraw-Hill.

Sze, S. M. (Ed.) (1991), *Physics of Semiconductor Devices,* New York: Wiley.

Walrafen, G. E., Revesz, A. G. (Eds.) (1986), *Structure and Bonding in Noncrystalline Solids,* New York: Plenum Press.

Widmann, D., Mader, H., Friedrich, H. (1988), *Technologie hochintegrierter Schaltungen, Halbleiterelektronik,* Vol. 19, Berlin: Springer-Verlag.

SILICON NITRIDE

F. L. Riley, *School of Materials, University of Leeds, Leeds, United Kingdom*

INTRODUCTION

The element silicon forms only one nitride that is stable at normal temperatures. This is a solid and has two crystalline phases, designated α and β. The stoichiometric composition is Si_3N_4, but departures from this stoichiometry can occur in the amorphous, noncrystalline form. Silicon nitride does not occur naturally in the oxidizing environment of the earth's crust (though trace occurrence in minerals of meteoritic origin has been reported); it can thus be regarded as a man-made material.

Although silicon nitride has been known to chemists for over a century, its potential as a polycrystalline ceramic material only became apparent in the late 1950s to early 1960s, when considerable development work was initiated on the production and characterization of nonoxide compounds of low molecular weight and strong interatomic bonding, and thus likely to have useful electrical and structural properties. The possibility of using silicon nitride as a component to withstand temperatures of 1300 °C and higher in a new range of high-efficiency internal-combustion engines was quickly appreciated and was to lead to intensive efforts over the next 25 years to develop high-strength silicon nitride ceramics for a wide range of applications at room and high temperature. Much of this work has been reported in conference proceedings, of which Carlsson *et al.* (1992) and Hoffmann *et al.* (1994) are good examples.

This development work in the ceramic area, involving the production of increasingly

3-527-28140-1/96/$5.00 + .50

complex chemical and crystallographic systems, has led to the realization that silicon nitride, strictly, should be regarded as a very large class of materials. Some members of the class are relatively simple [pure chemical vapor deposition (CVD) silicon nitride, for example]; others (those prepared with oxide-sintering additives) can be complex solid solutions involving other metallic elements and oxygen. Well-recognized examples of this type are the "sialon" group of materials, based on two- or more-component subsystems. Full discussion of these complex systems is beyond the scope of a short overview article on silicon nitride, which will be restricted to single-component materials and, with only brief excursions, where essential, into the field of silicon nitride–metal oxide systems. A detailed overview of the sialon materials has been provided by Ekstrøm and Nygren (1992).

Simultaneously, though on a smaller scale and less visibly, the uses of silicon nitride have been explored in a range of applications in electronics. This work led in particular to improved understanding of the production and electrical and electronic properties of amorphous silicon nitride, usually in thin (~100 nm)-film form. The use of silicon nitride in microelectronic devices was first reported in 1964, when it was discovered that thin films, because of their dense structures, had better masking properties than silicon dioxide (silica) against metals and, in particular, against the diffusion of sodium. It was also established that the films had the capability of remembering and storing for long times electrons and holes, and thus acting as "read only" memory (ROM) devices. Subsequently the effectiveness of silicon nitride as a mask against oxygen and water vapor at temperatures up to ~1000 °C allowed the development of the local silicon oxidation technique; silicon nitride thin films now play a key role in silicon– and gallium-arsenide–based microelectronic technology. Because of their mechanical hardness and radiation resistance, silicon nitride thin films are also used as final protection layers to finished devices. Silicon nitride is now being considered as an alternative or additional gate or node dielectric in advanced metal-oxide-semiconductor (MOS) devices such as random-access memory arrays. A full treatment of the application of silicon nitride in the electronics area is given by Belyi *et al.* (1988).

The purpose of this article is to introduce silicon nitride and its properties and to indicate the nature of the large volume of work dedicated to realizing its considerable potential as a structural material on a very wide range of scale. At the present time there is no single-authored textbook dedicated to silicon nitride; important volumes of other types dealing wholly or in part with silicon nitride are listed under the "Works Cited" and "Further Reading" headings. The Further Reading section gives selected volumes the contributions to which are also well referenced; both sets provide an introductory guide to the very extensive literature (over 25 000 publications) dealing wholly or in part with silicon nitride materials. Numerical data quoted in this review are taken from these volumes. A comprehensive survey and compilation of factual and numerical data on silicon nitride is in the process of being constructed (Gmelin Institut für Anorganische Chemie, 1991, 1995, 1996).

1. CRYSTAL CHEMISTRY

This section deals with the structural arrangement of atoms within the silicon nitride crystal, and the differences between the two crystalline phases, α- and β-Si_3N_4.

1.1 Chemical Bonding

The silicon–nitrogen single bond is strongly covalent; estimates of ionicity are in the range 11–30%. This entails considerable localization of the electrons and directed bonding. Silicon forms four sp^3 hybrid bonds directed to the corners of a tetrahedron with 109.5° angles, and nitrogen, three trigonal sp^2 hybrids at 120° angles in a plane. The basic structural units can thus be regarded as SiN_4 tetrahedra, joined together in a three-dimensional network by sharing (N) corners; each N is thus common to three tetrahedra. Figure 1 shows schematically the arrangement of the SiN_4 tetrahedra in the puckered sheets that form the basis for the α and β structures of Si_3N_4. The marked tetrahedra are at height $\frac{1}{2}$; the basis for the unit cell is also shaded. Using normal covalent radii for Si and N, the Si–N bond length can be cal-

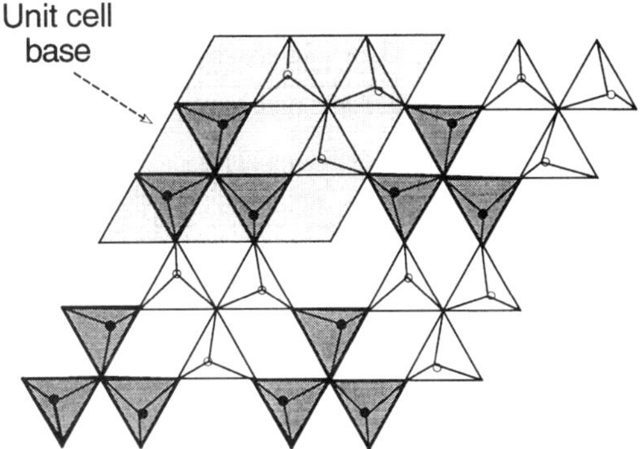

FIG. 1. Schematic illustration of the arrangement of the SiN_4 tetrahedra in the puckered sheets making up the Si_3N_4 structure. The shaded tetrahedra are at a height of $\frac{1}{2}$; the basis for the unit cell is also shaded (Courtesy D. P. Thompson).

culated to be 187 pm, which is much larger than those seen experimentally in silicon nitride (170–177 pm). It is assumed that the Si 3d orbitals make a significant contribution to p^{π}-d^{π} interactions. The mean Si–N bond energy is calculated to be 330 kJ mol^{-1}.

1.2 Crystal Structures

The SiN_4 tetrahedra link to form puckered six-membered rings, which in turn delineate large (and normally empty) voids; these linked rings can be regarded as "AB" and "CD" layers, in which the CD layer is an inverted AB, with protruding Si and N bonds. In the idealized silicon nitride structures, the layers can be stacked in two ways: AB layers stacked vertically on top of each other give the β structure, . . .ABAB. . . (space group $P6_3$); stacking of an AB layer alternately with an offset CD layer obtained by the operation of a c-glide plane gives the α structure, . . .ABCDABCD. . . (designated space group $P31_c$). In α-Si_3N_4 the bottom half of the unit cell is the same as in β-Si_3N_4, and the unit-cell dimension in the c axis is approximately twice that of the β-Si_3N_4 unit cell. The unit cells of α-Si_3N_4 and β-Si_3N_4 are shown schematically in Figs. 2 and 3. These arrangements lead to two rather open, and low-density, crystal structures that, while they contain marked differences, are also in many respects closely similar. The hexagonal β unit cell containing Si_6N_8 has continuous channel-like voids, aligned parallel to the c axis: in the α structure these are replaced by large closed cages. The β structure is almost strain free and has an average Si–N bond length of 173.6 pm. In contrast, the α structure in departing from the idealized structure has a marked degree of strain and a slightly lower single-crystal density (α-Si_3N_4 ~ 3.167–3.171, β-Si_3N_4 ~ 3.192 Mg m^{-3}). For this reason it is generally assumed that the β structure is thermodynamically the more stable. However, practical experience is that under clean conditions and particularly at relatively low temperatures (<1300 °C), the α phase is kinetically the more readily formed, both as a result of gas-phase reactions and by the lower-temperature recrystallization of amorphous silicon nitride.

These observations would be in accordance with the view that the α phase is the low-temperature modification, and the β phase is the high-temperature modification. The α- to β-phase reconstructive transformation of secondary coordination occurs fairly readily at high temperature (>1600 °C), particularly in the presence of a liquid solvent; the reverse process has, however, never been reliably observed.

There has been considerable controversy over the nature of the α-Si_3N_4 phase, particularly with regard to whether it should be considered to be the nitride low-temperature phase or, alternatively, as an oxynitride of low oxygen content and fairly precisely defined stoichiometry. This view translates to that of the β phase being formed under "low" oxygen potentials ($\leq \sim 10^{-20}$ bar) and the α phase under "high" oxygen potentials. Uncertainty was sustained by observations that α-Si_3N_4 could be shown to contain significant amounts of structural oxygen and had a wider range of experimental unit-cell dimensions (a = 774.91–775.72 pm; c = 561.64–562.21 pm), and thus densities, than the β phase (a = 760.3 pm; c = 290.6 pm). The generally accepted current view is that both are true nitrides but that the strained α structure is more readily able to accommodate the small amounts of oxygen inevitably present in formation environments; oxygen seems not to be necessary for the stability of the structure.

1.3 Amorphous Silicon Nitride

The use of chemical vapor deposition techniques at relatively low temperatures (<1000 °C) to form thin films of silicon ni-

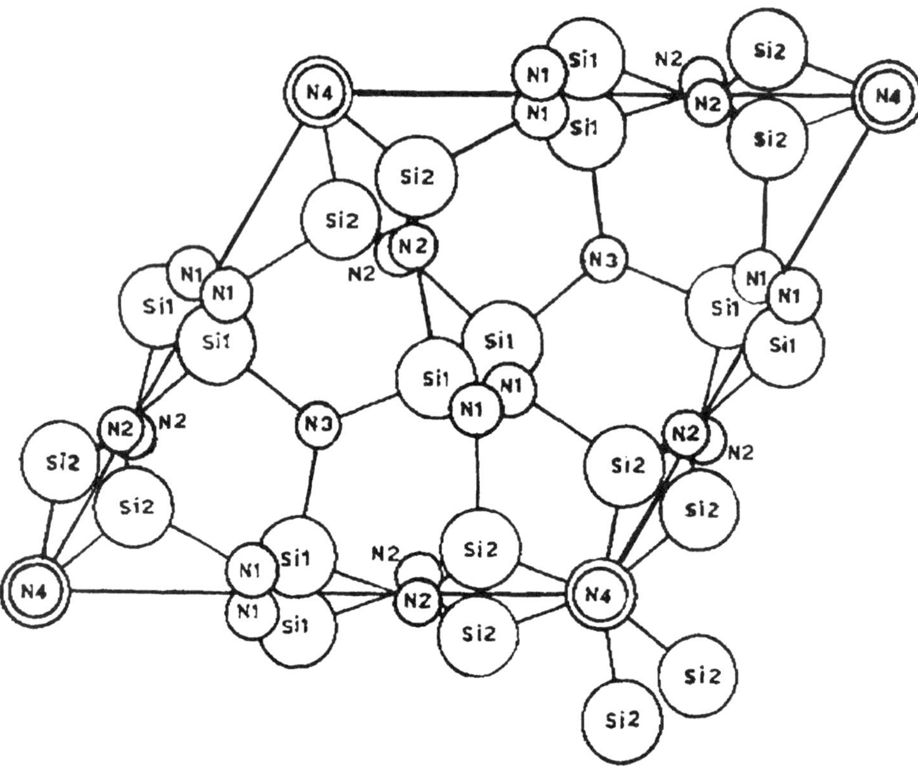

FIG. 2. The unit cell of α-Si_3N_4 projected along the *c* axis (courtesy S. Wild).

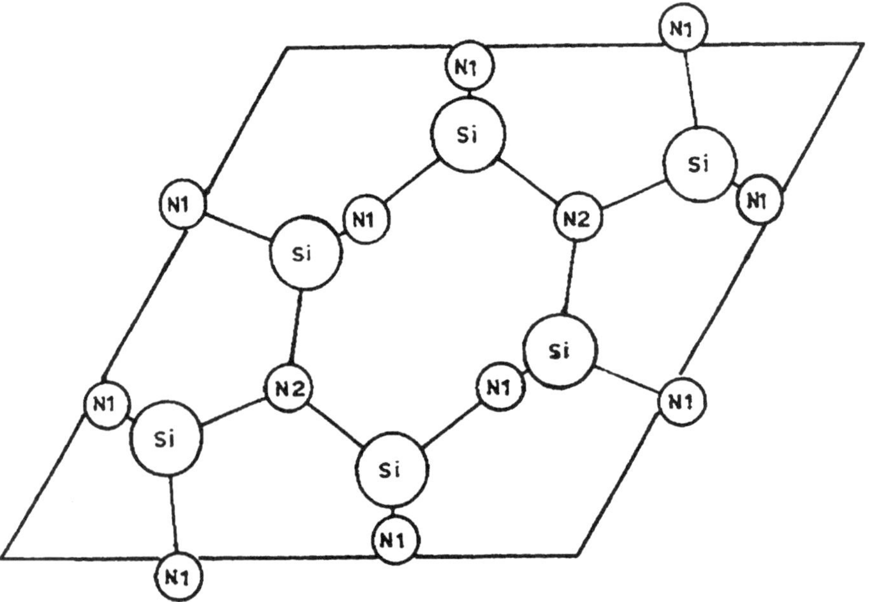

FIG. 3. The unit cell of β-Si_3N_4 projected along the *c* axis (courtesy S. Wild).

tride invariably gives a noncrystalline, amorphous material. Because of the presence of hydrogen as a carrier gas, or in the ammonia (NH_3) and silane (SiH_4) reactant gases, hydrogen is always incorporated into this form of silicon nitride. Strictly its composition should be represented by $Si_xN_yH_z$, in which the N/Si ratio ranges from 0.5 to 1.6 (stoichiometric = 1.33), depending on the synthesis conditions. The amount of hydrogen incorporated is proportional to the hydrogen concentration in the reaction environment and inversely proportional to the deposition temperature: its concentration can range from <1 to >30 at.%. Hydrogen forms chemical bonds with the Si and N (Si–H and N–H groups), with a distribution between the two types depending on deposition conditions. Annealing in vacuum at temperatures above ~900 °C readily leads to complete loss of the hydrogen. The presence of hydrogen in amorphous silicon nitride is of fundamental importance for the chemical, physical, and electrophysical properties of the thin silicon nitride films essential for applications in microelectronics technology, and it significantly affects the properties of the film, such as etch rate, film density, and diffusion-barrier characteristics.

Oxygen is also often present in amorphous silicon nitride as an impurity element, as it is in α-Si_3N_4, and this too can have a very great effect on film properties. However, amorphous CVD films normally contain <0.5% of oxygen.

2. PRODUCTION METHODS

This section provides an introduction to the methods used to produce silicon nitride in its forms of technical importance: powders (the precursors to silicon nitride ceramics), amorphous thin film, and polycrystalline (and often multiphase) ceramic.

2.1 General

Because most ceramic production processes use powders as the starting point, much attention has been directed to the preparation of high-purity and fine–particle-size silicon nitride powders. In order to appreciate better the objectives in this respect, it has to be realized that it is impossible to sinter even submicrometer particle-size silicon nitride powder to full density using conventional ceramic conditions. There are two major reasons for this: the very low intrinsic diffusivities of Si and N in the Si_3N_4 lattice, even at temperatures above 1700 °C, and the volatility of silicon nitride provided by the decomposition

$$Si_3N_4 = 3Si_{(v)} + 2N_{2(g)}. \qquad (1)$$

The nitrogen molecule is a very stable species, and silicon nitride undergoes appreciable decomposition to silicon and nitrogen gas at temperatures above 1200 °C; this process provides an alternative surface-energy lowering mechanism to that required (the movement of material from particle–particle contact points into the surrounding void space) to permit the approach of particle–particle centers and densification. Silicon nitride powder must therefore be sintered in the presence of small amounts of a wetting liquid solvent of carefully controlled composition, to provide the low-conductivity pathway necessary for silicon nitride transport and particle fusion. Suitable liquids have been found to be the metal silicates, normally used at the 5- to 10-vol.% level and giving liquids at temperatures (depending on composition and quantity) from ~1200 °C upward. These silicates are formed *in situ* by reaction of added metal oxide with the silicon dioxide film naturally present on the silicon nitride particles, and providing 3–5% by weight of SiO_2; for this reason the "oxygen" content of a silicon nitride powder is an important aspect of its specification.

Silicon nitride powder suitable for sintering to a high-density and high-strength component will thus have high chemical purity (>99.5%), low oxygen content (<2%), and a specific surface area of ~10–20 $m^2\ g^{-1}$ (corresponding to mean particle sizes in the region of 200–100 nm). It will also consist of >95% α phase, which generally gives a higher mechanical strength sintered product, for reasons connected with the α-phase to β-phase transformation that takes place during sintering, with associated important microstructural developments (see Sec. 2.3.2).

2.2 Powders

Four routes are used for the commercial production of high-grade powders. These rely on the ready availability of materials such as high-purity silicon, silicon dioxide, and silicon tetrachloride. High-quality commercial silicon nitride powders are widely available at prices currently in the region of ($50 to $100)/kg.

2.2.1 Silicon Nitridation The earliest commercial synthesis of silicon nitride used the direct nitridation of micron-dimension, loosely packed, silicon powder by reaction with nitrogen or ammonia gas at temperatures in the range of 1250–1450 °C. This reaction can be summarized by the simple equation

$$3Si_{(s)} + 2N_{2(g)} = Si_3N_{4(s)}, \quad \Delta H_{1700\ K} = -724\ kJ\ mol^{-1}, \qquad (2)$$

though this disguises a complex reality. Because the process is strongly exothermic, conditions must be carefully controlled to avoid a runaway reaction and premature fusion of the silicon powder particles (melting point, 1409 °C), which cannot then be fully nitrided. For this reason the process must be very slow (of the order of 20–50 h depending on the furnace loading), even with computer interlinking of reaction rate and furnace temperature control. The product powder tends to be strongly agglomerated and requires extensive wet milling to give the required submicrometer particles, coupled with subsequent chemical purification to remove metallic contamination and excess surface silicon dioxide.

2.2.2 Carbothermal Nitridation of Silicon Dioxide Silicon dioxide is not reactive toward nitrogen but can be converted to silicon nitride in the presence of carbon at temperatures using nitrogen or nitrogen/hydrogen mixtures, in the temperature range 1200–1450 °C:

$$3SiO_{2(s)} + 2N_{2(g)} + 6C = Si_3N_{4(s)} + 6CO_{(g)}. \qquad (3)$$

This is also a slow reaction, and several hours can be required for completion, depending on temperature and gas flow rate: too high a temperature (>1460 °C with 1 bar pressure of nitrogen) favors the competing reaction of C with SiO_2 to form silicon carbide. Very fine amorphous silicon dioxide powder is used; the silicon nitride powder product is also fine, with a high α-phase content and of high chemical purity, with the exception of some contamination by residual carbon (which can with care be reduced to <0.5%).

2.2.3 Diimide Decomposition Silicon tetrachloride and liquid ammonia at −40 °C react rapidly to form the intermediate silicon diimide, which after removal of the ammonium chloride by sublimation or solution at 900 °C, or by washing with liquid ammonia, is decomposed by heating at ~1300 °C to yield crystalline silicon nitride. This complex series of steps, which may be summarized as

$$SiCl_{4(l)} + 6NH_{3(l)} = Si(NH)_{2(s)} + 4NH_4Cl_{(s)}, \qquad (4)$$

$$3Si(NH)_{2(s)} = Si_3N_{4(s)} + 2NH_{3(g)}, \qquad (5)$$

leads to a product of very high-purity α-Si_3N_4, with well-controlled submicrometer particles and morphology; there may be some chlorine contamination. The route is, however, expensive.

2.3 Bulk Forms

Silicon nitride ceramic is widely produced with two distinctive types, porous reaction-bonded silicon nitride and dense sintered or hot-pressed silicon nitride.

2.3.1 Reaction-Bonded Silicon Nitride Work in the 1950s on the nitridation of micron-size silicon powders led to the discovery that strong components of dimensions exactly those of the compacted silicon starting powder could be produced on heating at 1250–1450 °C under nitrogen or ammonia gas. This porous form of ceramic was termed "reaction bonded," or "reaction sintered," and was, historically, the first ceramic form of the material to be produced. The microstructure consisted of well-bonded silicon nitride microcrystals, with interconnected voids filled with material of a fine, whisker morphology.

Continuing detailed investigations showed that the silicon was nitrided through vapor-phase reactions involving $Si_{(g)}$, and probably also the gaseous silicon monoxide ($SiO_{(g)}$) generated from contaminant oxygen:

$$2Si_{(s)} + O_{2(g)} = 2SiO_{(g)} \qquad (6)$$

$$3SiO_{(g)} + 2N_{2(g)} = Si_3N_{4(c)} + \tfrac{3}{2}O_{2(g)}. \quad (7)$$

Nucleation and growth of silicon nitride takes place in the interparticle void spaces on silicon and silicon nitride surfaces, with the result that the increase in molar volume of ~22% is accommodated entirely within the interparticle void spaces of the compacted silicon powder; at the low temperatures used, there can be no sintering shrinkage of the product. Because compacted powders normally contain ~40–50% of void space, the reaction-bonded silicon nitride product contains 18–28% of porosity, which then has a dominant influence on the properties of the material. Figure 4 illustrates the characteristic appearance of a polished section of reaction-bonded silicon nitride, in which porosity shows as the dark phase.

2.3.2 Sintered Silicon Nitride A fully dense form of polycrystalline silicon nitride was first produced in 1961 by hot-pressing silicon nitride powder in the presence of magnesium oxide (generating liquid magnesium silicate at temperatures of ~1550 °C). This method is cumbersome, slow, and restrictive in terms of the shapes that can readily be produced by it. In 1975 a technique for the more versatile production of pressureless sintered silicon nitride was developed, based on the recognition that the main problem with obtaining dense sintered material was the volatility and dissociation of silicon nitride. This could be suppressed by immersing the component in a bed of powder of the same composition, or by using high pressures (>10 bar) of nitrogen gas. Both techniques have as objective the attainment of a thermodynamically equilibrated environment for the sintering silicon nitride powder, suppressing the tendency for dissociation into $N_{2(g)}$ and $Si_{(v)}$ [reaction (1)].

FIG. 4. Typical microstructures in two grades of porous reaction-bonded silicon nitride, shown in polished section. The dark regions are interconnected porosity. (Courtesy R. G. Stephen.)

Liquid-forming systems developed for hot pressing prove to be equally effective as sintering aids, though increased quantities are needed. Oxides such as MgO and Y_2O_3 at the 3–15 mol% level have been widely used as liquid-forming densification aids, and many other liquid-forming systems have since been investigated. Al_2O_3 is also a widely used additive, often in conjunction with Y_2O_3 although because a significant proportion of the Al and O now enter the Si_3N_4 crystal lattice, the product phase should strictly be regarded as a sialon. The additive oxides initially react with the nanometer-thickness oxide film naturally present on the Si_3N_4 particle surfaces at the ~3% level to form viscous silicate liquids in which silicon nitride has at high temperature (>1700 °C) significant solubility and mobility. Normal liquid-phase sintering occurs by dissolution and recrystallization of the silicon nitride, and fully dense materials can be obtained in 1–2 h at temperatures in the range 1700–1850 °C, provided effective measures are taken to control Si_3N_4 evaporation.

During the sintering process the ~100-nm-dimension α-Si_3N_4 particles convert to the more stable β phase. A major objective of the sintering process is to ensure the development of a microstructure of fine (1–10 μm) β-Si_3N_4 grains with fibrous, high–aspect-ratio, morphology shown in Fig. 5, which is of importance for the mechanical properties of the material. This microstructure is normally achieved by the use of a high–α-phase starting powder containing a low concentration of β-Si_3N_4 nuclei, and by prolonging sintering times to allow full development of the β-Si_3N_4 grains. A standard powder-bed sintering schedule for a 10–15 $m^2\ g^{-1}$ silicon nitride powder containing 5% Al_2O_3 and 5% Y_2O_3 by weight would consist of heating to 1000 °C under vacuum with a 1-h hold, followed by heating under 1 bar nitrogen at 1700 °C for 1 h, with a final anneal at 1800 °C for 2 h. Alternatively, the nitrogen pressure could be increased to 10 bar and

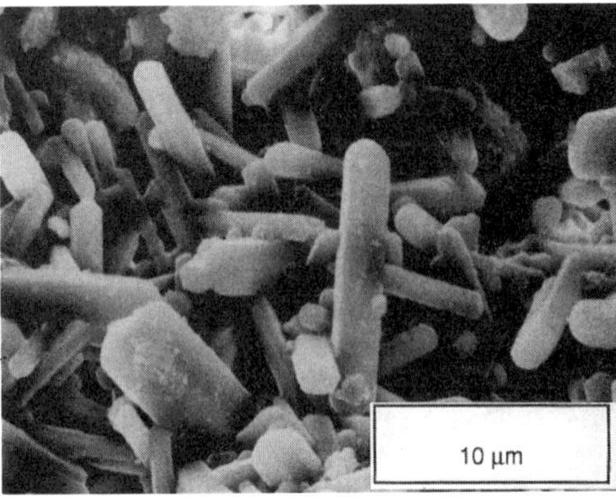

FIG. 5. The characteristic high–aspect-ratio β-Si_3N_4 grains in a high strength-sintered silicon nitride, delineated through removal by chemical etching of the inter-grain glass phase (courtesy P. Andrews).

heating continued at 1700 °C for a further 2 h. The powder in which the component is immersed is composed of a coarser grade of silicon nitride powder and contains ~50% of inert boron nitride powder to prevent excessive sintering of the bed.

During cooling the intergranular liquid forms secondary intergranular phases, which may be glassy or crystalline. Generally the silicate glass is the continuous phase, and it thus controls to a very large extent the high-temperature properties of the material. There is also a very thin (~1–2 nm) film of amorphous silicate material at all β-Si_3N_4 grain faces, thus blocking the continuity of the crystalline lattice. Crystallization of the intergranular glass would be highly desirable, but it is difficult to achieve fully at the grain boundaries. Densities of materials of this type are normally in the range 3.20–3.35 Mg m^{-3}, depending on the additive composition and quantity.

2.3.3 Hot-Pressed Silicon Nitride The uniaxial hot pressing of silicon nitride powder using graphite dies and pressures up to 30 MPa was the first method by which fully dense silicon nitride was obtained, but it has now generally been replaced by the more versatile pressureless sintering process. However, even with careful control of the sintering environment, it remains difficult consistently to produce 100% dense silicon nitride, completely free from voids. For this reason a high-pressure treatment may be carried out, in order to eliminate completely the last traces of porosity.

A well-established procedure for the production of small components is hot-isostatic pressing. This consists of the encapsulation of the component in a glass sheath (or presintering under normal conditions to the closed porosity stage, corresponding to >90% density) and heating in a cold-wall furnace at temperatures up to 1800 °C using nitrogen at pressures up to 400 MPa as the pressurizing fluid. Unlike uniaxial die pressing, there are no restrictions on component shape. A further major advantage of this process is that, because of the much higher pressures available, smaller volumes of liquid-forming additives can be used, with consequent benefits for the high-temperature mechanical properties of the material.

2.4 Thin Films

With the early realization that silicon nitride powder was very difficult to sinter to a fully dense material (although porous silicon nitride components could be readily produced by the nitridation of silicon powder), a natural step was to investigate the possibility of using CVD methods to produce dense silicon nitride in thin layer or sheet form. The CVD process utilizes the vapor-phase reaction at high temperature of volatile or gaseous precursor components, with nucleation and growth of the solid product phase occurring on any convenient surface or substrate. In the case of silicon nitride, silicon tetrachloride and ammonia are cheap and readily available materials, reacting at temperatures >700 °C:

$$3SiCl_{4(g)} + 4NH_{3(g)} = Si_3N_{4(s)} + 12HCl_{(g)}. \quad (8)$$

Silane, SiH_4, is an alternative and more reactive gaseous silicon compound:

$$3SiH_{4(g)} + 4NH_{3(g)} = Si_3N_{4(s)} + 12H_{2(g)}. \quad (9)$$

Carrier nitrogen gas is used. The reaction temperature and pressure determine the type and morphology of the product, with reactant pressures normally being of the order of $\sim 10^{-3}$–10^{-1} bar. At temperatures below ~1200 °C, the product is dense amorphous silicon nitride. This may be crystallized by heating at higher temperatures, or by prolonged heating at 1000 °C, though shrinkage and cracking is likely to occur. At higher deposition temperatures, films of dense, high-purity, polycrystalline α-silicon nitride are formed. Reaction rates under these conditions can be of the order of many μm h^{-1}, and while films of up to 100-μm thickness are easily developed, it is difficult to build up thicker stress- and crack-free layers by this method, and the technique has limited applicability in the context of silicon nitride component production. Prolonged annealing of α-silicon nitride films at temperatures >1500 °C permits conversion to the β form; this is a process useful for the production of very high-purity β-Si_3N_4, but otherwise, at present, it is only of academic interest.

Very thin, amorphous, silicon nitride films can be obtained by the direct nitridation of silicon surfaces by ammonia at temperatures up to 950 °C. This reaction is, however, self-limiting to thicknesses of the order of 10 nm because of the very low diffusivity of nitrogen in silicon nitride.

Chemical vapor deposition is the most common deposition method used in the electronics industry; this can be carried out at atmospheric pressures (APCVD) and at pressures between 10^{-4} and 10^{-2} bar (LPCVD). Until 1976 the standard deposition reaction used silane and ammonia at 1 bar pressure in a cold-wall reactor, and with temperatures between 700 and 900 °C. The SiH_4/NH_3 ratio was >150, and deposition rates were >10 nm min^{-1}. A typical, now commonly used, hot-wall low-pressure LPCVD reactor uses dichlorosilane ($SiCl_2H_2$) and ammonia at temperatures between 700 and 800 °C, and with pressures between 2×10^{-4} and 1×10^{-3} bar:

$$3SiCl_2H_{2(g)} + 4NH_{3(g)} = Si_3N_{4(s)} + 6HCl_{(g)} + 6H_{2(g)}. \quad (10)$$

Growth rates are ~10 nm min^{-1}, and a hundred and more silicon wafers can be coated simultaneously and uniformly. Figure 6 shows schematically a horizontal hot-wall LPCVD reactor, in which silicon slices are mounted vertically and with a spacing corresponding to the molecular mean free path, to improve film quality.

This technique is now routinely coupled with complex molecular excitation schemes such as plasma enhancement (PECVD), laser enhancement (LECVD), photo (UV) enhancement (PHCVD), and microwave and electron cyclotron resonance (ECR). When silicon nitride coatings are used with devices containing aluminum or gold metallization, lower deposition temperatures, between 200 and 350 °C, are needed; to achieve this, plasma enhancement is used, in which a low-pressure (10^{-4} to 4×10^{-3} bar) glow discharge provides the reaction energy, with power supplies in the 30-kHz to 13-MHz range, and effective temperatures as high as 10^5 K. Growth rates are usually ~30 nm min^{-1}.

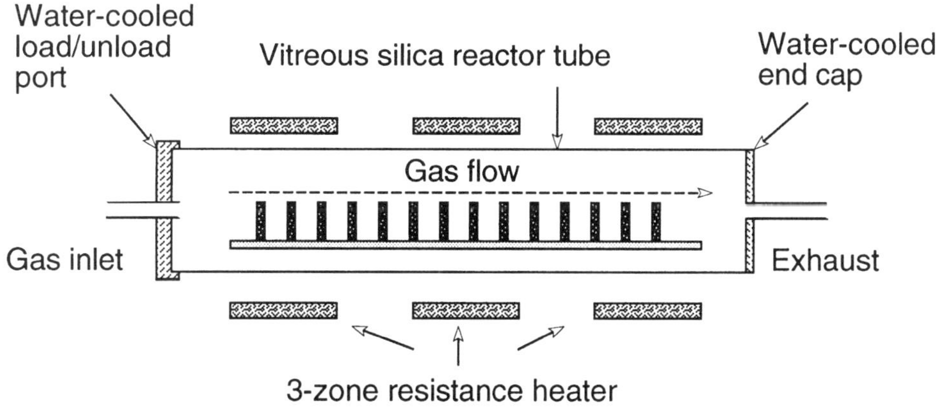

FIG. 6. Schematic diagram of a hot-wall LPCVD reactor.

3. PROPERTIES OF MATERIALS

The materials requirements for any one successful application are generally several in number. Silicon nitride has a very wide range of applications, and the range of properties of interest is therefore correspondingly broad.

3.1 Physical

3.1.1 Density Because of the open rather than "close-packed" crystal structures of α- and β-Si_3N_4, the density of pure crystalline silicon nitride is relatively low at ~3.17–3.19 Mg m^{-3}. Polycrystalline ceramics densified with the aid of heavier metal oxide systems (Y_2O_3, for example) can have densities higher than 3.2 Mg m^{-3}, because of the presence of the metal silicate intergranular phases. Amorphous silicon nitride has densities in the lower range from 2.4 to 3.1 Mg m^{-3} depending on production conditions: the low-pressure (and low-temperature) plasma technique yields the lowest densities and highest extents of hydrogen incorporation.

3.1.2 Diffusivity The intrinsic mobility of both silicon and nitrogen atoms in the Si_3N_4 lattice is very low and provides an additional impediment to the sintering of silicon nitride powder. Values for Si and N self-diffusion coefficients in Si_3N_4 have been difficult to determine, partly because of the difficulty of producing large enough high-purity specimens, and also because of the absence of convenient isotopes of Si and N. The most reliable data are for the tracer diffusion of ^{30}Si in α-Si_3N_4, for which $D = 10^{-15}\ m^2\ s^{-1}$ at 1500 °C. The extrapolated value for the ^{15}N tracer-diffusion coefficient at 1500 °C is $10^{-19}\ m^2\ s^{-1}$, indicating that in this temperature region Si is the more mobile species.

3.1.3 Electrical The electrical properties of silicon nitride depend strongly on its purity and, in the case of sintered materials, particularly on the presence of intergranular glass. Pure silicon nitride is an electrical insulator with a band gap of between 3.85 eV (experimentally for amorphous silicon nitride) and 6.49 eV (theoretically for β-Si_3N_4). The dielectric constant of pure CVD α-Si_3N_4 depends on its density, ranging at 35 GHz from 6.77 (2.64 Mg m^{-3}) to 7.68 (3.15 Mg m^{-3}). This is in close agreement with room-temperature values measured for dense hot-pressed polycrystalline silicon nitrides containing 1% MgO and 8% Y_2O_3, which gave values of 8.32 and 8.25, suggesting that the behavior is intrinsic in silicon nitride. The value rises by a factor of ~1.1 over the temperature range 25–1400 °C. The loss tangent, in contrast, is almost entirely controlled by the intergranular glass, which contains mobile or polarizable metallic ions. For pure CVD α-Si_3N_4, it is $\sim 2 \times 10^{-4}$ at room temperature, rising to 3×10^{-2} at 1400 °C. Values for commercial dense silicon nitrides containing glass are ~10× larger. Amorphous silicon nitride films have dielectric constants in the range 6–7 for LPCVD material and 6–9 for PECVD materials.

The room-temperature dc resistivity of pure silicon nitride is $\sim 10^{14}\ \Omega$ m, though PECVD films of variable composition have a wide spread of 10^4 to $10^{11}\ \Omega$ m. In the case of impure materials with intergranular glass, this value drops rapidly with increasing temperature. At 400 °C the reported dc resistivity of silicon nitride hot-pressed with 5% MgO is $\sim 4 \times 10^7\ \Omega$ m; at 1000 °C the value is $\sim 10^3\ \Omega$ m. With larger amounts of additive oxide, the values fall even further. The ac resistivity becomes less frequency dependent with increasing temperature and approaches dc values at ~1000 °C. The dielectric strength of CVD silicon nitride is of the order of $10^9\ V\ m^{-1}$.

The electrophysical properties of amorphous CVD silicon nitride have been studied in detail. The properties of nonstoichiometric silicon nitride vary over wide ranges, probably because of the important role of trapped hydrogen in controlling the transport properties of silicon nitride. Charges in near-electrode regions of silicon nitride are caused by the electrons and holes injected, and the electrophysical properties of silicon nitride can only be studied effectively together with the properties of the contacts in metal-insulator-semiconductor structures. The trap depth varies from ~2.5 eV (the middle of the forbidden zone) to ~1.2 eV, depending on the fabrication method and hydrogen content. It is considered that electrons and holes have essentially the same importance for charge transport.

3.1.4 Optical Pure silicon nitride is colorless, though polycrystalline ceramic forms normally range from light gray to almost black, presumably because of the presence of traces of metallic impurity. Absorptions in the infrared occur at 830–835 cm^{-1} and 870–880 cm^{-1} (the strongest two bands), 920 cm^{-1}, and 1015–1020 cm^{-1} (weak bands). The refractive index of amorphous silicon nitride is in the range 1.8–2.5 and is sensitive to purity: low values indicate oxygen contamination; values higher than 2.0 indicate excess silicon.

3.1.5 Thermal The thermal expansivity of crystalline silicon nitride is very low, even for ceramic materials. The thermal expansion coefficient is a function of temperature, but a mean value for pure β-Si_3N_4 of 3.4 MK^{-1} over the temperature range 40–1200 °C is reported. Commercial polycrystalline materials containing intergranular glass, and possibly substitutional aluminum and oxygen, often include slightly lower values with a reported range of 2–4 MK^{-1}. The low expansivity is a major reason for the exceptionally good resistance to thermal shock of silicon-nitride–based materials, and the basis for many of their high-temperature structural applications.

The thermal diffusivity of dense sintered silicon nitride is mainly controlled by the intergranular phases, and by the glass in particular. Occurrence of the α-Si_3N_4 to β-Si_3N_4 phase transformation is seen to increase thermal diffusivity, possibly because of the higher intrinsic diffusivity of the β phase. Room-temperature values typically are in the range 10–25 $m^2\ s^{-1}$. Measured thermal conductivities at room temperature are 15–50 W $m^{-1}\ K^{-1}$ for dense material and 4–30 W $m^{-1}\ K^{-1}$ for porous reaction-bonded materials. Quoted specific heat values are in the range 700–1000 J $kg^{-1}\ K^{-1}$ with a mean of 780 J $kg^{-1}\ K^{-1}$ and independent of material type.

3.2 Mechanical

Silicon nitride is intrinsically hard and strong, and ceramic forms have been developed largely because of their good mechanical properties, particularly at high (>1000 °C) temperature. The strength of a good-quality sintered silicon nitride is almost constant up to 1200 °C. At higher temperatures strength begins to fall because of grain-boundary sliding aided by the intergranular glass film. This feature is enhanced by a good resistance to thermal shock, allowing its use for components that may be subjected to rapid temperature changes. Quenching from temperatures in the range 400–700 °C is needed to cause significant loss of strength during a standard water quench test.

3.2.1 Creep The slow deformation under load at high temperature is an important aspect controlling the high-temperature applications of silicon nitride materials. Most creep measurements are made at temperatures above 1100 °C and under loads typically of ~100 MPa. The creep of dense silicon nitride is largely controlled by grain-boundary sliding through the intergranular glass. This leads to grain-boundary separation, cavity formation, and cracking. Cavity nucleation occurs mainly at multiple grain edges. The process is complicated in air by simultaneous oxidation, during which intergranular cations diffuse to the outer surface, leading to a hardening effect and reduced creep rates. The volume fraction and composition of the intergranular phase have an important bearing on creep rates, and crystallization of the intergranular glass gives significant reductions in creep rates. The creep behavior of porous reaction-bonded silicon nitride is adversely affected by internal oxidation, which generates silicate glass with incorporation of trace metallic impurities in the silicon powder.

3.2.2 Fracture Toughness Most measurements of fracture toughness made on sintered polyphase silicon nitride using the diamond-indentation or notched-beam techniques show that fracture toughness is strongly influenced by the microstructure, and by grain aspect ratio in particular. In the case of dense β-silicon nitride materials with a fibrous grain structure, K_{1c} values as high as 12 MPa $m^{1/2}$ have been reported. Normal sintered materials will have values in the region of 5–7 MPa $m^{1/2}$. Measurements made using the Vickers indentation method on isotropic single-crystal α-silicon nitride give much smaller values of 1.9–2.9 MPa $m^{1/2}$, indicating that crack bridging and grain pullout are the primary mechanisms for toughening in sintered silicon nitride. The porous reaction-bonded form has a lower fracture toughness; values in the range 1.5–2.8 MPa $m^{1/2}$ have been reported.

3.2.3 Hardness Silicon nitride has an indentation hardness similar to that of silicon carbide (carborundum). Measurements made on single crystals of CVD α-silicon nitride show an average Knoop hardness value of 23 GPa, and the Vickers values are in the range ~28 to ~30 GPa for the 1010, 0001, and 1120 faces. These values are significantly higher than those measured on sintered polyphase silicon nitride materials, which contain a softer intergranular glass.

3.2.4 Strength Like all ceramics, silicon nitride fails at low temperature by brittle fracture: because there is no plastic flow (except at very high temperature), the stress-strain curve is virtually linear to the fracture point, and stress-concentrating defects are significant factors determining the strength of a component. The strength (σ) of a brittle material is often expressed in terms of the Griffith relationship

$$\sigma = (1/Y)[2E\gamma_s/c]^{1/2}, \tag{11}$$

where E is Young's modulus; γ_s is the fracture surface energy; c is the dimension of the critical defect and often equated to the size of a microstructural defect such as a pore, large grain, or surface scratch; and Y is a constant related to flaw geometry. The term $[2E\gamma_s]^{1/2}$ is generally condensed to a critical fracture-toughness term, K_{1c}. Strength is normally measured by breaking bars in bend; tensile strength is experimentally much more difficult to measure, and values are rarely quoted (as a rough rule of thumb, the corresponding tensile strength would be 0.6 times a bend strength value).

The strength of the porous but otherwise single-phase reaction-bonded silicon nitride is controlled by the void fraction (indicated by density) and the size of the largest void. Most commercial materials have densities in the range 2.3–2.5 Mg m^{-3}, corresponding to void fractions of 0.72–0.80, and bend strengths in the range 150–250 MPa. Maximum void sizes are normally in the range 100–200 μm, though with care in the preparation and packing of the silicon powder before nitriding, they can be reduced to 50–20 μm, allowing strengths of 400–500 MPa to be obtained.

The strengths of dense, sintered silicon nitrides are much higher; room-temperature bend strength values are typically around 700 MPa, though values above 1000 MPa are routinely obtained. This is the result of the 1- to 10-μm-dimension β-Si_3N_4 grains and absence of large voids, coupled with the high fracture toughness given by the fibrous microstructure. Quoted strength values must, however, be treated with caution because of their sensitivity to sample size and to the quality of the surface finish.

The high-temperature strength of these materials is largely controlled by the intergranular glass. There is very little glass in reaction-bonded silicon nitride, and the strength remains almost unchanged up to 1400 °C. Sintered silicon nitride materials, in contrast, may contain 15% or more by volume of glass, and although they may retain good strength to 1200 °C, above this point their strength falls rapidly as a result of grain-boundary sliding and cavitation. The thickness of the intergranular amorphous film at the grain faces is ~1 nm and independent of liquid volume though controlled by the liquid composition. The bulk of the glass phase is concentrated at grain edges and grain corners. An important feature, therefore, is the use of liquid-forming sintering additives that either give very high-viscosity glasses, or can be crystallized by a post-densification annealing process. For this reason much development work has been carried out into silicon nitride materials sintered with additions of Al_2O_3 and Y_2O_3, which subsequently allow the crystallization by a high-temperature anneal of the intergranular yttrium aluminosilicate glass to α'- and β'-sialon solid-solution phases.

3.2.5 Young's Modulus This is a function of density (or void fraction), and to a lesser extent of composition. Porous reaction-bonded materials (of variable void fraction) usually have Young's-modulus values in the range 120–220 GPa. Dense sintered silicon nitride has a value in the region of 300–330 GPa, depending on the volume of lower-modulus glass present.

3.3 Chemical

Because most of the applications foreseen for the ceramic forms of silicon nitride involve high temperatures, oxidation and corrosion behavior have been studied in detail. High-temperature reactions with metals are less well studied. The properties of thin films of amorphous films of silicon nitride have been examined from the point of view of their effectiveness as barriers to the diffusion of metals, and oxygen, at temperatures in the region of 1000 °C.

3.3.1 High-Temperature Oxidation and Corrosion Silicon nitride is thermodynamically unstable with respect to oxidation to silicon oxynitride and silicon dioxide:

$$2Si_3N_{4(s)} + \tfrac{3}{2}O_{2(g)} = 3Si_2N_2O_{(s)} + N_{2(g)}, \quad (12)$$

$$Si_3N_{4(s)} + 3O_{2(g)} = 3SiO_{2(s)} + 2N_{2(g)}. \quad (13)$$

Oxide films of molecular thickness are always present on silicon nitride surfaces, and because of the low mobility of oxygen in them, they provide a barrier against continuing oxidation. Within thin films the film stoichiometry is continuously variable, from Si_3N_4 at the interior to SiO_2 at the surface. At temperatures above ~1000 °C, oxygen mobility becomes significant, and there is continuing oxidation, normally following approximately parabolic kinetics. However, oxidation rates are still relatively slow, and components can be used for many thousands of hours at this temperature. The high-temperature oxidative and corrosive degradation of silicon nitride materials has been studied extensively: the reviews in Nickel (1994) provide a good introduction to this subject.

The key factor under these conditions is the very effective diffusion barrier provided by both silicon oxynitride and pure silicon dioxide toward oxygen transport to the underlying silicon nitride. However, silicon dioxide reacts readily at high temperature with most metal oxides to form silicates of much higher conductivity toward oxygen; the protectiveness of the silicon dioxide film is thereby lost. Metal oxides, such as sodium oxide, potassium oxide, or vanadium pentoxide, are present in significant concentrations in many combustion environments and will accelerate oxidation by reacting with the protective silicon dioxide film. In sintered silicon nitrides, a process that can be regarded as autocorrosion occurs, through the diffusion of metal ions from the intergranular glass to the surface silicon dioxide, with the same effect. The composition and quantity of the continuous intergranular glass phase thus become of significance for the high-temperature oxidation resistance of sintered silicon nitride.

The oxidation behavior of reaction-bonded silicon nitride is influenced by similar factors, and also by the accessible void space, which permits internal oxidation to occur. Oxidation rates become slower at temperatures >~1200 °C, because of the development of a sealing outer barrier layer of silicon dioxide.

High-purity CVD films of silicon nitride are also effective barriers to the diffusion of oxygen and water vapor at temperatures of interest for the production of microelectronic devices (700–900 °C). This is likely to be the result of the formation of a very thin outer layer of silicon oxynitride or silicon dioxide; the nitride film is thus to this extent sacrificial.

The ability of thin, CVD- or plasma-deposited amorphous films of silicon nitride to act as diffusion barriers to dopants or impurities such as B, Al, Ga, In, and As is of great importance in the production of Si- and GaAs-based devices. Sodium penetrates less than 5 nm in 22 h at 600 °C, for example. Film stoichiometry is of importance for its effectiveness; small amounts of oxygen allow readier diffusion of Ga.

3.3.2 Molten-Metal Attack At high temperatures silicon nitride has a good resistance to attack by many molten metals, such as iron. It is particularly effective in the presence of molten aluminum, which reacts readily with many oxide refractory materials by abstraction of oxygen. In the case of the nitride it is likely that a thin barrier film of AlN is developed.

4. APPLICATIONS

This section deals with the applications of silicon nitride, treated in terms of the two important types of material—amorphous thin film and the bulk polycrystalline ceramic.

4.1 Thin Film

Extensive use is made of 100- to 200-nm thin films of amorphous silicon nitride in the semiconductor device industries. This is because the production of an integrated circuit uses multiple stages to give precisely controlled doped layers making up the individual components, as well as the deposition of other dielectric or metallic layers. Because of the high stress levels attained in CVD silicon nitride films applied to a silicon substrate, they are normally only deposited on top of a 5- to 50-nm pad or buffer layer of silicon dioxide. This serves to prevent peeling and cracking of the silicon nitride layer.

4.1.1 Masking Layers for Silicon Local Oxidation Because the oxidation rate of silicon nitride is <0.1 that of silicon, silicon nitride films are extensively used in most silicon device processes to allow the local oxidation of silicon. The surface of a silicon wafer is selectively protected by a 100- to 400-nm LPCVD or PECVD layer of silicon nitride, and the oxide layer is thermally grown in the exposed areas. At the end of the oxidation cycle, the nitride film is selectively removed by hot phosphoric acid. Figure 7 shows schematically the steps in the production of a simple recessed-oxide (ROX) structure, prepared on a silicon surface using a silicon nitride film as an oxidation mask.

4.1.2 Diffusion Barriers Amorphous silicon nitride has a higher density than amorphous silicon dioxide ($\sim$2.2 Mg m^{-3}) and is a better barrier to impurity-metal diffusion. It is therefore often used in very large-scale integration (VLSI) production lines to prevent the cross penetration of dopants during thermal processing stage—for example, where a lightly doped circuit component layer must be grown over a more heavily doped substrate. To prevent dopant escaping from the back surface of the substrate to the front, the back of the substrate is sealed by protective CVD silicon nitride film. The CVD silicon nitride films are also used to prevent the diffusion of elements such as boron from heavily doped devices already grown in a silicon substrate to neighboring regions. In this situation the film can have an additional function of providing a dielectric layer to form a capacitor.

Silicon nitride overcoating films are used to seal devices during fabrication against alkali-ion (sodium, for example) contamination arising from some encapsulants and to improve the resistance of the chips to mechanical damage before final bonding or encapsulation operations. The PECVD silicon nitride is often used between metallizing wiring levels as protection against mechanical damage.

Silicon nitride films are equally important in GaAs device fabrication, because of the tendency of GaAs to lose As by evaporation at temperatures >650 °C. A stoichiometric PECVD silicon nitride film deposited at 250–350 °C is an effective barrier against both As and Ga diffusion and maintains GaAs stoichiometry during thermal treatment (as well as improving the mechanical strength of the device). Films are readily removed after completion of processing by dilute hydrofluoric acid or a plasma etch.

4.1.3 Silicon Nitride Dielectrics Silicon nitride is a component of a useful gate dielectric in specific metal–insulator–semiconductor (MIS) memory device applications. Individually erasable cells are pro-

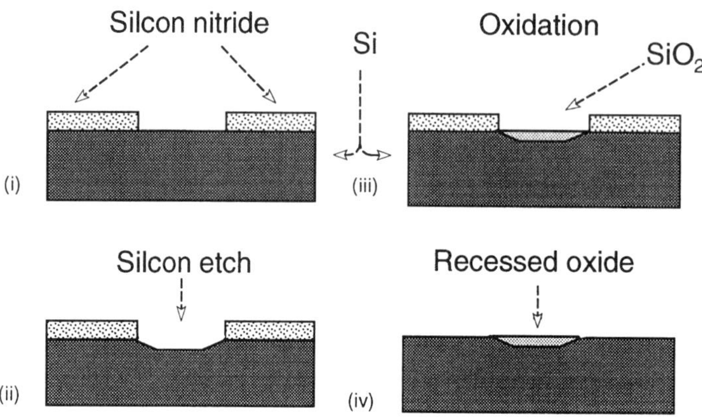

FIG. 7. Steps in the production of a recessed oxide pattern on a silicon surface, by localized oxidation using a silicon nitride mask.

duced with a dual dielectric control gate consisting of a silicon dioxide (~2 nm) and a silicon nitride (~50 nm) film, in metal nitride oxide silicon (MNOS) devices. Charge stored in traps at the oxide–nitride interface is retained when power is removed. The write/erase properties depend on the properties of the interfaces, and on those of the CVD silicon nitride, and are thus very sensitive to processing conditions.

The dielectric constant of silicon nitride is ~2× that of silicon dioxide and allows the production of a capacitor of the same value, but of half the thickness, as one made from silicon dioxide. It is thus an important material when very thin (<10 nm) dielectric films are required. Silicon nitride is used in very thin (~10 nm) gate dielectrics as a dual dielectric to provide increased breakdown field strength in field-effect transistors, and in dual-dielectric storage capacitors in multimegabit random-access memory arrays. Composite, (NO) LPCVD silicon nitride (4 to 8 nm)–silicon dioxide films with low leakage and a high resistance to boron penetration are also successfully used as gate insulators in insulated-gate field-effect transistors (IGFETs).

4.2 Bulk Ceramic

Dense silicon nitride has properties that distinguish it from other technical ceramics: its major features are very good resistance to thermal shock (because of a low thermal expansion coefficient), high strength and wear resistance, and a high fracture toughness giving a relatively low defect sensitivity (arising from a fine grain and fibrous microstructure). It also has the hardness of silicon carbide and is thermally stable under nitrogen or air to very high temperatures. Porous reaction-bonded silicon nitride has the same good thermal properties and a complex shape fabricability. As a structural material, silicon nitride has a number of established applications, most of which are generally on a small scale. Popper (1994) has provided a summary overview. Despite the many demonstrated viable applications, it has still to realize its full potential as a mass-produced high-temperature structural ceramic material.

4.2.1 Cutting-Tool Inserts Cermets such as cobalt-bonded tungsten carbide perform well at cutting speeds of <100 surface m min^{-1} but are thermally degraded at higher speeds. Dense sintered silicon nitride materials (including sialon-based materials) will cut cast iron, hard steel, and nickel-based alloys with surface speeds up to 25× those obtainable with tungsten carbide. Extensive use is made of silicon nitride based cutting tool inserts for machining cast iron in the automotive industry and for machining nickel-based superalloys for the aero industries.

4.2.2 Internal-Combustion Engines Much of the early work carried out on silicon nitride in the 1970s and 1980s was directed at the introduction of dense and porous silicon nitride hot-zone components into a new range of advanced gas-turbine and reciprocating engines, with projected significantly higher operating temperatures and therefore efficiencies. As a result of many intensive research programs, a large number of test engines were developed and satisfactorily evaluated. However, major stumbling blocks to the widespread incorporation of silicon nitride components into standard engines have been the cost of the raw materials and, probably more importantly, the technological problems associated with the mass production of high-grade ceramic engine components with a very high degree of reproducibility and long-term reliability. Reproducibility and reliability are not problems unique to silicon nitride ceramics: they have their roots in the most fundamental aspects of ceramic powder blending and sintering. Nonetheless, a number of engines make routine use of dense sintered silicon nitride components such as diesel glow plugs and turbocharger rotors, where stresses and temperatures are more modest and the consequences of failure less catastrophic. Current development work concerns the introduction of silicon nitride components where improved hot wear resistance is required, such as exhaust valves, bucket tappets, and rocker-arm pads. Figure 8 shows examples of engine components made from sintered silicon nitride. A useful feature of silicon nitride is its very low density, low inertia, and high modulus compared with metallic materials, which make possible the design of valve

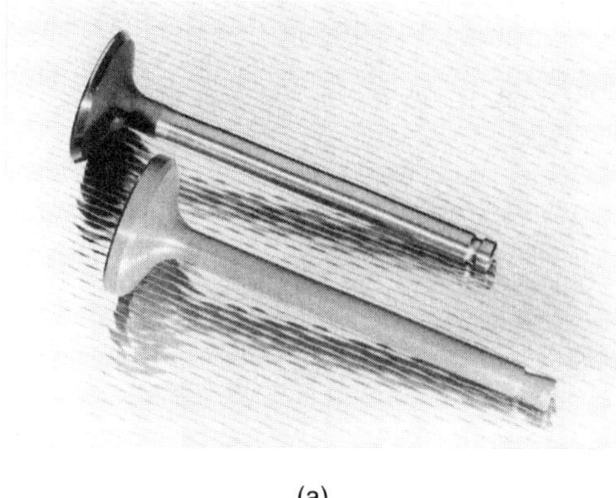

(a)

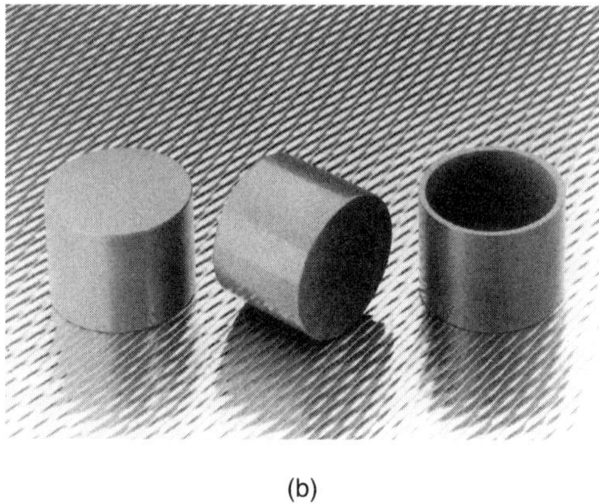

(b)

FIG. 8. Examples of sintered silicon nitride components: (a) silicon nitride (bottom) and metal (top) engine valves; (b) automobile bucket tappets. (Courtesy T&N Technology Ltd.)

springs for use in very high-frequency applications without resonance being set up.

4.2.3 Bearings The wear resistance, high stiffness, and low density of dense silicon nitride have led it to be considered for a high-temperature roller and ball bearing material. These features are well suited to combat the high Hertzian and centrifugal stresses induced in high-speed systems working at well above room temperature. Rolling contact fatigue tests have shown that silicon nitride can have a much longer life than conventional higher-density steel and cermet bearings (Horton, 1991).

4.2.4 Industrial Lower-density silicon nitride has been found to be useful under lower stress conditions, where refractoriness, chemical stability, and a good resistance to thermal shock are the main requirements. There is a steady demand for thermocouple sheath materials for use in the hot-metal processing areas and for components involved in the handling of molten reactive metals such as aluminum. Reaction-bonded silicon nitride is ideal for such components, in that fairly complex shapes can readily be fabricated in compacted silicon powder, and then nitrided. Threaded inert-gas–welding torch nozzles also provide a consistent market for this material. On a smaller scale, specialist kiln furniture of low thermal mass and good thermal shock resistance has been satisfactorily produced in reaction-bonded silicon nitride. Figure 9 shows examples of reaction-bonded silicon nitride components.

Dense silicon nitride has been evaluated as a coil and disc spring material for use in chemical plant in oxidizing or corrosive environments, at temperatures (>500 °C) where normal metallic springs would not be viable.

5. SUMMARY

This article has attempted to introduce the reader to some of the fundamental aspects of silicon nitride and to show its interest and challenge. The material has many usable properties in both the electrical and the structural areas. For this reason it has established a very wide spectrum of applications on the micro and macro scales, as thin films of amorphous material, and as large polycrystalline and multiphase components of carefully tailored microstructure. Many prob-

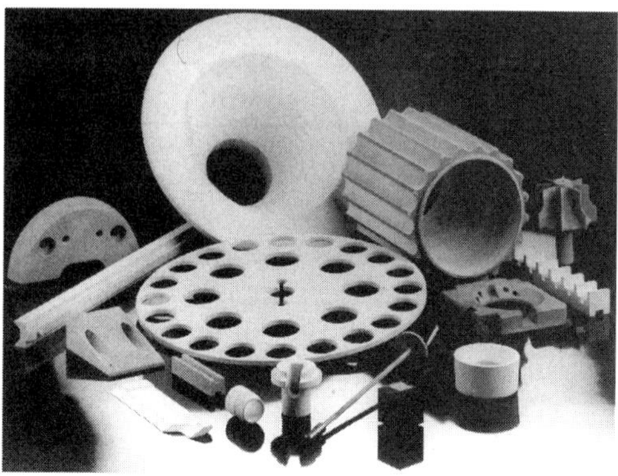

FIG. 9. Examples of reaction-bonded silicon nitride components (courtesy T&N Technology Ltd.).

lems have been overcome; there are many areas still requiring exploration in order to develop a full understanding of the structure-property relationships in these materials. The volume of published literature concerning silicon nitride is vast and generally not easy to plot a course through. The short list below provides no more than a brief introduction to the subject.

Works Cited

Belyi, V. I., Vasilyeva, L. L., Ginovker, A. S., Gritsenko, V. A., Repinsky, S. M., Sinitsa, S. P., Smirnova, T. P., Edelman, F. L. (1988), *Silicon Nitride in Electronics,* Materials Science Monographs, Vol. 34, Amsterdam: Elsevier.

Carlsson, R., Johansson, T., Kahlman, L., (Eds.) (1992), *4th International Symposium on Ceramic Materials and Components for Engines,* London: Elsevier Applied Science.

Ekstrøm, T., Nygren, M. (1992), "SiAlON Ceramics" *J. Am. Ceram. Soc.* **75** (2), 259–276.

Gmelin Institut für Anorganische Chemie (1991), *Gmelin Handbook of Inorganic and Organometallic Chemistry,* 8th ed., Silicon Supplement Vol. B5c: *Silicon Nitride in Microelectronics and Solar Cells,* Berlin: Springer Verlag.

Gmelin Institut für Anorganische Chemie (1995), *Gmelin Handbook of Inorganic and Organometallic Chemistry,* 8th ed., Silicon Supplement Vol. B5d1 and B5d2: *Silicon Nitride: Electrochemical Behavior, Colloidal Chemistry and Chemical Reactions,* Berlin: Springer Verlag.

Gmelin Institut für Anorganische Chemie (1996), *Gmelin Handbook of Inorganic and Organometallic Chemistry,* 8th ed., Silicon Supplement Vol. 5b1, *Silicon Nitride: Mechanical and Thermal Properties,* Berlin: Springer Verlag.

Hoffmann, M. J., Becher, P. F., Petzow, G. (Eds.) (1994), *Silicon Nitride '93,* Aedermannsdorf, Switzerland: Trans Tech Publications.

Horton, S. A. (1991), in: F. L. Riley (Ed.), *3rd European Symposium on Engineering Ceramics,* London: Elsevier Applied Science, pp. 35–50.

Nickel, K. G. (1994), *Corrosion of Advanced Ceramics: Measurement and Modelling,* Dordrecht: Kluwer Academic Publishers.

Popper, P. (1994), in: M. J. Hoffmann, P. F. Becher, G. Petzor (Eds.), *Silicon Nitride '93,* Aedermannsdorf, Switzerland: Trans Tech Publications, pp. 719–723.

Further Reading

Bonnell, D. A., Tien, T. Y. (Eds.) (1989), *Preparation and Properties of Silicon Nitride Based Materials,* Materials Science Forum Vol. 47, Aedermannsdorf: Trans Tech Publications.

Chen I.-W., Becher P. F., Mitomo M., Petzow G., Yen T.-S. (Eds.) (1992), *Silicon Nitride Ceramics,* Materials Research Society Symposium Proceedings Vol. 287, Pittsburgh: Materials Research Society.

Hoffmann, M. J., Petzow, G. (Eds.) (1994), *Tailoring of Mechanical Properties of* Si_3N_4 *Ceramics,* NATO Advanced Study Institutes, Series E: Applied Sciences, No. 276, Dordrecht: Kluwer Academic Publishers.

Riley, F. L. (Ed.) (1977), *Nitrogen Ceramics,* NATO Advanced Study Institutes, Series E: Applied Sciences, No. 23, Leyden: Noordhoff.

Riley, F. L. (Ed.) (1983), *Progress in Nitrogen Ceramics,* NATO Advanced Study Institutes, Series E: Applied Sciences, No. 65, The Hague: Martinus Nijhoff.

SILICIDES

See TRANSITION METALS, SILICIDES AND NITRIDES

SILICON, AMORPHOUS

MARTIN STUTZMANN AND CHRISTOPH E. NEBEL, *Walter Schottky Institut, Technische Universität München, Garching, Germany*

INTRODUCTION

Amorphous silicon (*a*-Si), the disordered analog of crystalline silicon, is a prominent member of the large family of glassy or amorphous semiconductors that have been studied extensively during the last 50 years. Amorphous semiconductors are interesting materials in the field of applied physics because they can be deposited on many different substrate materials at relatively low temperatures while maintaining their basic semiconducting behavior. In addition, some specific properties of disordered semiconductors (bistable switching, radiation hardness, lack of wave-vector conservation, large-area and thin-film capability, chemical metastability, etc.) are particular to the amorphous state and have no direct analogy in crystalline semiconductors.

Particularly, amorphous silicon has captured most of the attention in the field of amorphous semiconductors since it was discovered around 1970 that *a*-Si deposited by low-temperature plasma-enhanced chemical vapor deposition (PECVD) of silane (SiH_4) has a very low defect density (approximately 10^{15} cm^{-3}) and can be doped both n and p type, qualitatively similar to crystalline Si (Spear, 1977). As a direct consequence of these discoveries, hydrogen-containing ("hydrogenated") amorphous Si (*a*-Si:H) has been employed in a number of electronic devices: thin-film solar cells and photodetectors using Schottky diodes or *p-i-n* junctions, thin-film field-effect transistors for display or printing applications, *a*-Si:H xerographic drums, particle detectors, and many others.

Whereas all of these devices worked in principle, it was discovered early on that their ultimate performance was seriously limited by the tendency of *a*-Si:H to create additional deep defects spontaneously under conditions of strong electronic excitation. The best known example is the so-called "Staebler–Wronski effect," which manifests itself via the creation of metastable Si-dangling-bond defects under strong illumination (Staebler and Wronski, 1980). Thus, while prior to 1980 most of the investigations dealing with amorphous Si were aimed at a better understanding of the structural, dynamical, and electronic properties of *a*-Si or intrinsic *a*-Si:H, since 1980 more applied issues and questions have dominated the field: modeling and optimization of specific devices and deposition processes, the microscopic nature and energy distribution of structural defects, doping, impurities, or structural and electronic stability.

During the last decade, amorphous silicon has also made a continuously growing contribution to the market of commercial elec-

3-527-28140-1/96/$5.00 + .50

tronic devices. Large-area thin-film solar cells with stable conversion efficiencies of 10% or more now enter the mass production stage and will replace the limited applications of low-efficiency cells widely used in solar-powered calculators during the 1980s. Amorphous-silicon–based active-matrix flat displays are currently commercialized in high-end portable computers and present a strongly growing market for the near future. Amorphous silicon technology today is mainly attractive whenever large-area devices have to be combined with standard micrometer-lithography techniques. Besides flat-panel displays and thin-film solar cells, amorphous silicon therefore provides a promising technology for future applications, such as large-area image scanners, high-connectivity neural networks, or three-dimensional integration.

1. PREPARATION AND STRUCTURAL PROPERTIES

1.1 Deposition Methods

Amorphous silicon can be prepared by a number of different techniques, including thermal evaporation from a solid source, sputtering of crystalline Si targets (dc, rf), and decomposition of Si–H compound gases (silane, SiH_4, or disilane, Si_2H_6, etc.) via thermal, optical, or glow-discharge excitation. State-of-the-art *a*-Si:H samples today are mainly prepared by using a plasma-enhanced chemical vapor deposition (PECVD) system schematically outlined in Fig. 1. Reactive and carrier gases (SiH_4, H_2, PH_3, B_2H_6, etc.) enter the reactor via a gas manifold equipped with mass-flow controllers. The gases are dissociated in a parallel-plate capacitor that is part of a critically tuned rf-resonance circuit. One of the two capacitor plates is heated (≅200–400 °C) and acts as the substrate holder. The reactor is operated under constant pressure and flow rate using a throttled pumping system. Most systems use the industrial rf at 13.56 MHz for plasma excitation, but good results have also been obtained for reactors operating from dc up to 2.45 GHz. Thermal decomposition in hot-wall reactors or using hot filaments has also been applied successfully to *a*-Si:H deposition, especially for films with lower hydrogen concentration. Attempts to deposit *a*-Si:H via optical excitation of reactive gases (photo-CVD using CO_2^- or excimer lasers) have not met wide acceptance mainly because of window contamination problems and low deposition rates.

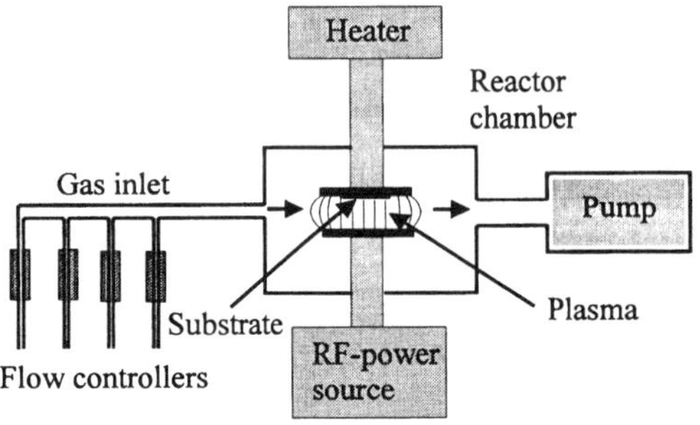

FIG. 1. Schematic diagram of a plasma-enhanced chemical vapor deposition reactor used for the preparation of amorphous-silicon thin films.

1.2 Atomic Structure and Structural Stability

Amorphous Si or hydrogenated amorphous Si prepared at temperatures below 500 °C is amorphous in the sense that neither x-ray scattering, Raman scattering, nor electron microscopy is capable of detecting significant fractions of ordered regions with diameters exceeding 20–30 Å. Instead, the amorphous structure is well described as a continuous random network with fourfold tetrahedral coordination of most atoms, but with no structural order extending beyond second-nearest-neighbor distances. Pure amorphous silicon contains a large number of vacancies and small voids, causing a density deficit of several percent with respect to crystalline Si, as well as a large number of electronic defects that make the material unsuitable for semiconductor devices. Moreover, the reactivity of elemental Si with oxygen leads to a rapid oxidation of internal and external surfaces. The chemical stability of *a*-Si is much improved if reactive atomic hydrogen is present during deposition or diffused into the sample afterwards. H atoms cause a passivation of internal surfaces and electronic defects by the formation of Si–H bonds. This reduces the density of electronic defects (Si dangling bonds) down to about 10^{16} cm^{-3} or less. Device-quality *a*-Si:H films prepared by PECVD of silane typically contain 5–10 at.% hydrogen.

Figure 2 presents a schematic view of the a-Si:H network, indicating several characteristic hydrogen-bonding units. In addition to the tetrahedrally coordinated disordered Si lattice, various hydrogen-related local structures can be identified: isolated Si–H bonds or Si–H_2 bridges, short fragments of polysilane $(Si–H_2)_n$, interstitial H_2 molecules, hydrogenated vacancies and microvoids, and other more hypothetical bonding configurations, such as the mobile H_2^* defect thought to be responsible for metastability phenomena of a-Si:H networks (see Sec. 2.5).

More quantitative information concerning the structural properties of a-Si and a-Si:H can be obtained from the radial distribution function, $D(r)$, which is determined from x-ray or electron-scattering experiments. $D(r)$ describes the probability of finding a network atom (only Si, since H does not contribute to the scattering efficiency to first order) at distance r from a given origin atom. In crystalline semiconductors, $D(r)$ consists of a series of discrete peaks with decreasing spacing corresponding to the various shells of atoms around the origin. In amorphous silicon, because of the long-range disorder, only the first two or three coordination shells can be identified (see Fig. 3). For large values of r, $D(r)$ approaches quickly the homogeneous-medium limit, $D(r) = 4\pi r^2\rho$, where ρ is the average macroscopic density. As shown in Fig. 3, the strength, position, and width of the radial distribution peaks contain information about the number of atoms, the average values, and the variation of bond lengths and bond angles. The first peak in a-Si is basically identical to that of c-Si, except for a small broadening due to bond-length variations ($\Delta\alpha/\alpha \approx 1\%$) and a lower integrated intensity reflecting the smaller density of a-Si. Similar arguments hold for the second peak (broadened by bond-angle disorder, $\Delta\varphi \approx 10°$) and the third peak (broadened by variation of the dihedral angle, $\Delta\psi$).

Hydrogenated amorphous Si is structurally stable over more than 20 years when stored at room temperature. In contrast, amorphous Si without hydrogen quickly oxidizes in air similarly to the crystalline Si surface. All forms of amorphous Si crystallize in an exothermic reaction upon heating to about 650 °C. Highly strained thin amorphous Si films sometimes exhibit so-called

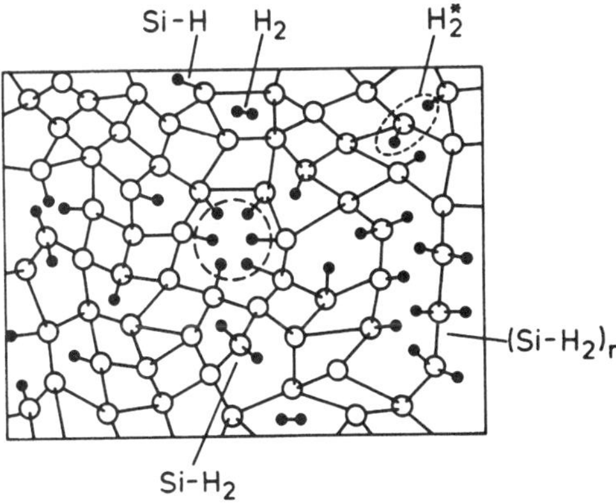

FIG. 2. Different hydrogen-related bonding configurations in hydrogenated amorphous silicon (a-Si:H). See text for details.

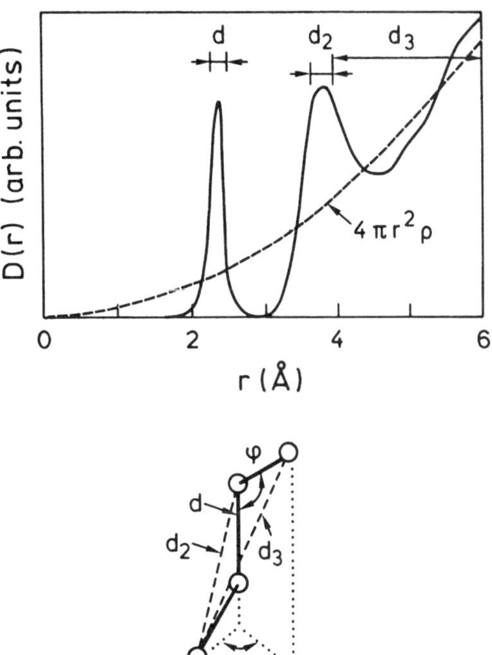

FIG. 3. Experimental radial distribution function $D(r)$ of amorphous silicon. The dashed line shows the homogeneous limit $4\pi r^2\rho$, where ρ is the macroscopic density. d, d_2, and d_3 indicate nearest-, second-nearest-, and third-nearest-neighbor distances according to the schematic model in the lower part of the figure. φ and ψ are the bonding and the dihedral angles, respectively.

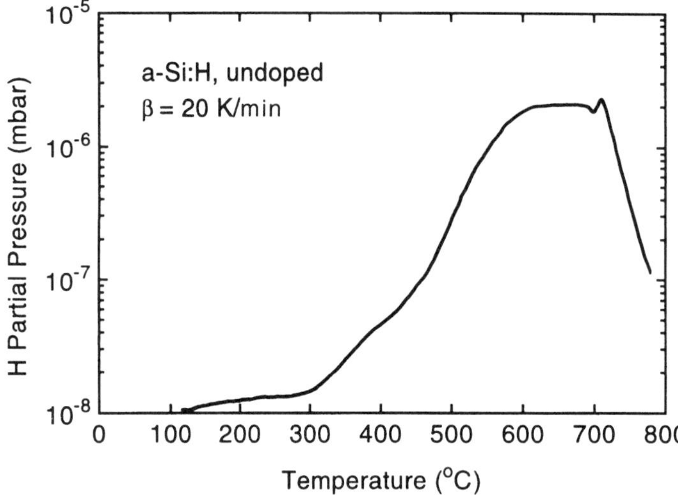

FIG. 4. Hydrogen effusion spectrum of undoped *a*-SiH. The sample was heated with a linear temperature ramp of 20 K/min in a constantly pumped vacuum system. The small peak at ≈700 °C is due to hydrogen release during crystallization.

explosive recrystallization even at lower temperatures when suitably activated. Prior to crystallization, *a*-Si:H looses the bonded hydrogen at temperatures between 300 and 550 °C. Weakly bound hydrogen (SiH_2 bonds, clustered H) is desorbed between 300 and 400 °C, whereas strongly bound H (isolated Si–H bonds) leaves the amorphous network between 500 and 600 °C (Fig. 4) (Beyer, 1985).

1.3 Vibrational Modes

The vibrational modes of amorphous silicon can be separated into network vibrations with energies below 520 cm^{-1} and hydrogen vibrations with energies above 600 cm^{-1}. Si-network vibrations fall into four bands corresponding to the transverse (T) and longitudinal (L) acoustic (A) or optical (O) phonons of the crystalline Si lattice (Lannin, 1984). Since the wave vector **k** of the phonons is no longer a good quantum number in an amorphous semiconductor (missing translational symmetry), only the density of phonon states can be used to describe the vibrational properties of the disordered network. The relation between the phonon density of states (DOS) and the phonon dispersion curves of crystalline Si is shown in Fig. 5. The DOS of lattice vibrations in *a*-Si can then be viewed as a disorder-broadened version of the crystalline DOS. Experimentally, the phonon DOS can be measured by inelastic neutron scattering (upper curve of Fig. 6),

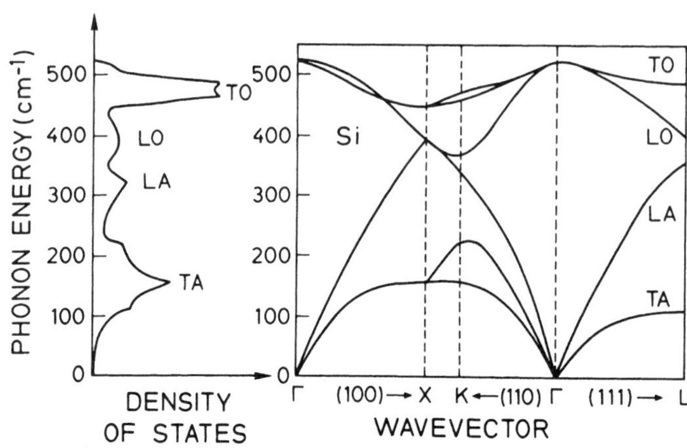

FIG. 5. Phonon dispersion curves of crystalline silicon and the corresponding phonon density of states. TO, LO, LA, and TA denote transverse and longitudinal optical and acoustic phonons.

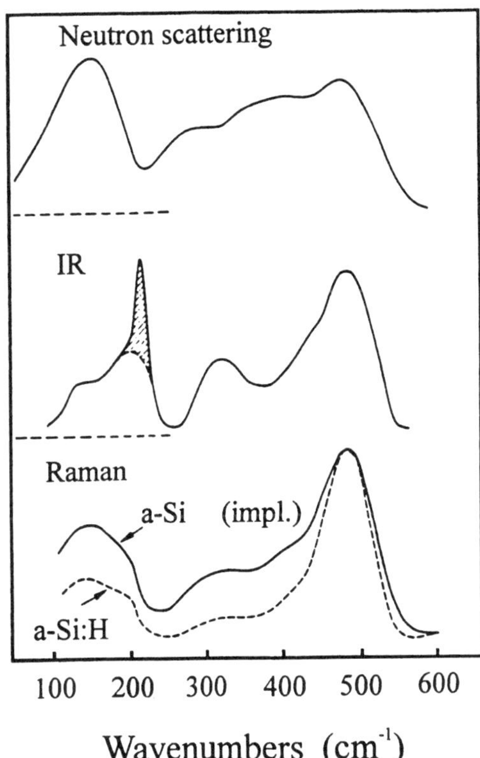

FIG. 6. Neutron scattering, infrared absorption (IR), and Raman spectra of amorphous Si. The shaded peak in the IR spectrum is due to a hydrogen-related mode (Si–H resonant mode). The two different Raman spectra correspond to device quality *a*-Si:H and amorphous Si produced by implantation.

by infrared (IR) absorption (middle curve), or by Raman scattering (lower trace). In the latter two techniques, corrections for the oscillator strengths and the scattering cross section of the different modes, respectively, have to be taken into account. Of practical importance is the use of Raman scattering as a convenient way to analyze the amount of bonding disorder and nanocrystalline admixture. Empirically, the mean deviation $\Delta\varphi$ of the bond angle from the tetrahedral value $\varphi = 109.5°$ is related to the full width at half maximum (FWHM), Γ, of the TO-Raman peak at 480 cm^{-1} via the relation $\Gamma^2 = \Gamma_0^2 + (A\Delta\varphi)^2$, with $\Gamma_0 = 33$ cm^{-1} and $A = 7$ cm^{-1}/deg. The presence of nanocrystallites, on the other hand, can be easily detected via a sharp Raman peak at approximately 520 cm^{-1}, corresponding to the zone-center TO/LO-phonon energy (cf. Fig. 5).

The Si–H bonds in *a*-Si:H give rise to various local vibrational modes (LVM), which can be divided into bond-bending and bond-stretching vibrations. For the assignment of the various modes, see Table 1.

IR absorption spectra of hydrogen LVM are of considerable importance because they allow a fast assessment of the total hydrogen content and the quality of *a*-Si:H films (Cardona, 1983). Dense, homogeneous *a*-Si:H films only show the bending and stretching vibrations of isolated Si–H bonds at 630 and 2000 cm^{-1}, respectively. In contrast, *a*-Si:H samples with inferior electronic quality commonly exhibit Si–H_2- or Si–H_3-related IR features as well as a significant fraction of bond-stretching vibrations at 2100 cm^{-1}. The total hydrogen content can be estimated from the integral of the Si–H bending absorption at 630 cm^{-1} according to the relation

$$[H] \approx (1.6 \times 10^{19}\ \text{cm}^{-2}) \int \frac{\alpha(\omega)}{\omega}\, d\omega,$$

where α is the absorption coefficient in units cm^{-1}.

2. ELECTRONIC PROPERTIES

The discussion of electronic properties of amorphous Si has to take into account three principal features introduced by the characteristic disorder of the amorphous network. A first feature is the preservation of the short-range order. Thus, a similar overall electronic density of states as in crystalline Si is to be expected [Fig. 7(a)].

Second, the long-range disorder introduced by bond-length and bond-angle variations causes localization of electronic states at the band edges of the conduction and valence bands in which the electron wave func-

Table 1.

Mode energy	Bonding configuration	Mode
630	Si–H, Si–H_2	Bend, wag
875	Si–H_2	Bend ("scissors")
2000	Si–H	Stretch
2090	Si–H_2	Stretch
2100	Si–H + void	Stretch

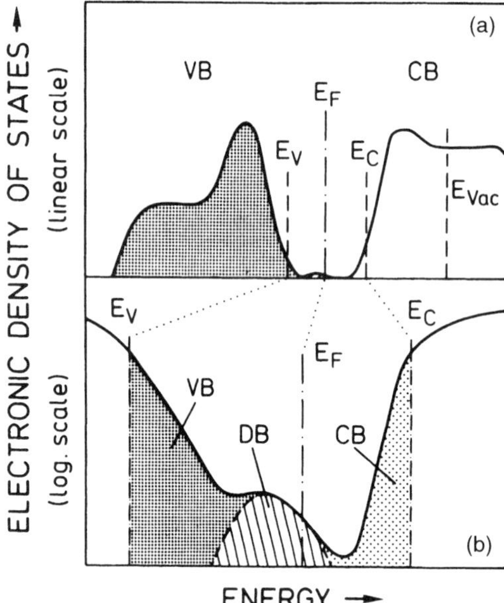

FIG. 7. (a) Main features of the electronic density of states of amorphous silicon. (b) Details of the localized states in the band gap on a logarithmic scale. E_V, E_C, and E_F indicate the valence- and conduction-band mobility edges and the Fermi level. VB, DB, and CB are the valance-band tail, the dangling-bond defect band, and the conduction-band tail.

tion is confined to a small volume. The density of localized states at the conduction- and valence-band edge decreases exponentially toward midgap. These "band-tail states" [VB and CB, Fig. 7(b)] control the transport properties because conduction takes place predominantly at the band edges. The transition energy separating extended from localized states is called the "mobility edge" (E_V and E_C in Fig. 7) because here the carrier mobility drastically decreases at $T = 0$ (Mott, 1987).

Third, deeper in the gap, electronic states due to defects control the recombination of electrons and holes. In *a*-Si:H, the main deep defect is the silicon dangling bond (DB, a threefold-coordinated Si atom), which can be charged positively (D^+) or negatively (D^-), or can be neutral (D^0).

In crystalline semiconductors the wave functions of the electronic states in the bands are solutions of the Schrödinger equation compatible with the periodic lattice potential (Bloch waves). In amorphous Si, however, the lattice potential is not periodic, so that the wave vector **k** is no longer a good quantum number. Instead, inelastic scattering of electrons occurs after a mean free path of only 10–15 Å, strongly decreasing the carrier mobility compared to crystalline semiconductors. A schematic view of delocalized versus localized electronic wave functions in *a*-Si:H is shown in Fig. 8. The detailed effects of disorder on optical absorption and electronic transport are discussed in the following two subsections.

2.1 Optical Absorption

Optical transitions in crystalline semiconductors are subject to the momentum conservation rule $\Delta k = 0$ because of the negligible wave vector of a low-energy photon. Well-known consequences are the strong optical transitions (absorption and luminescence) at the band edge of direct semiconductors and the weak, phonon-assisted transitions of indirect semiconductors, such as silicon. In amorphous Si the missing long-range periodicity destroys the concept of indirect or direct optical transitions. Instead, the dependence of the absorption coefficient α on photon energy $\hbar\omega$ is determined mainly by the combined density of occupied states in the valence band and empty states in the conduction band (Ley, 1984):

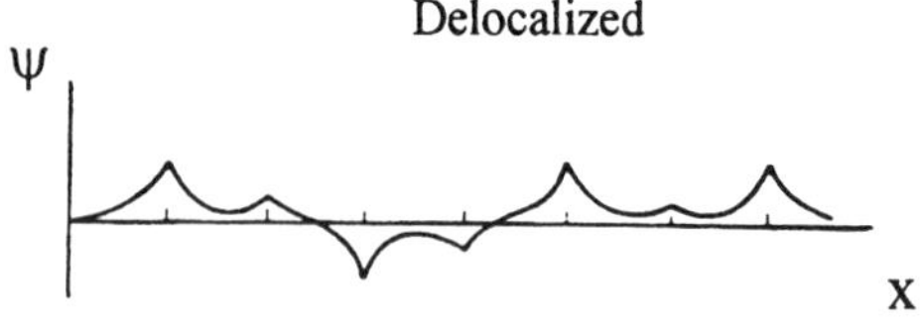

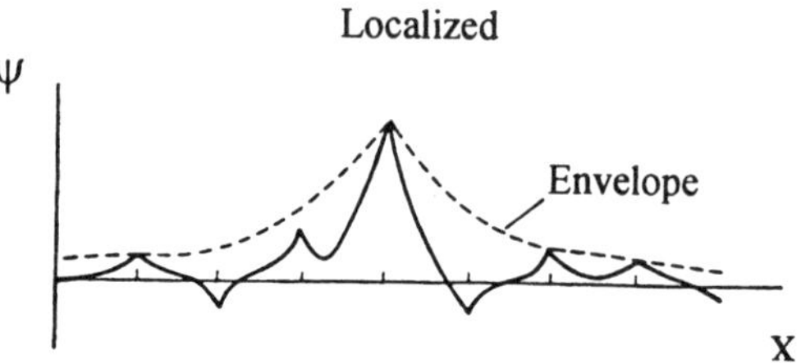

FIG. 8. Schematic view of delocalized and localized electronic wave functions in an amorphous network.

$$\alpha(h\omega) = \left(\frac{2\pi e}{m\omega}\right)^2 \frac{\omega}{hcV} \int M^2(E)N_{VB}(E) \times N_{CB}(E + h\omega)dE,$$

where V is the absorbing sample volume, $N(E)$ denotes the electronic density of states at energy E, and M is the optical transition-matrix element. In Fig. 9, experimental absorption spectra for *a*-Si and *a*-Si:H are compared with those of crystalline and microcrystalline Si. All forms of Si exhibit a strong absorption coefficient of $\alpha \approx 10^6$ cm^{-1} (corresponding to an absorption length of ≈100 Å) for photon energies above 3.2 eV. Significant differences only exist below the first direct optical transition of *c*-Si at 3.2 eV. In this energy range, the absorption of *c*-Si decreases rapidly by two to three orders as a result of the indirect, phonon-assisted nature of the transitions. The absorption finally vanishes almost completely below the indirect band gap (≈1.1 eV at 300 K) in high-quality crystals.

In contrast, device-quality *a*-Si:H and pure *a*-Si maintain a high level of absorption ($\alpha \approx 10^5$ cm^{-1}) in the visible spectral range because of the structural disorder (loss of **k**-vector conservation). As a direct consequence, amorphous silicon photoreceptors (solar cells, photodiodes) usually can be built with film thicknesses of 1 μm or less for visible light, whereas crystalline Si devices have to be at least one order of magnitude thicker

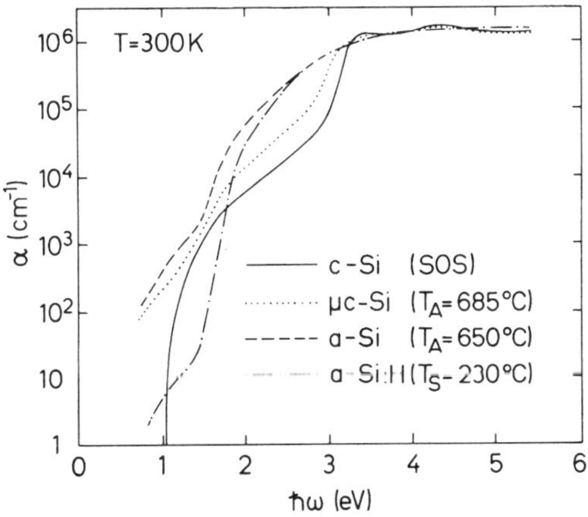

FIG. 9. Linear absorption coefficient, α, of crystalline silicon (*c*-Si), microcrystalline silicon (μ*c*-Si), amorphous silicon (*a*-Si), and hydrogenated amorphous silicon (*a*-Si:H).

to ensure a high quantum efficiency. In the near-infrared energy region below 1.5 eV, hydrogenated amorphous silicon according to Fig. 9 has a lower absorption coefficient than *c*-Si, because of the opening of the optical band gap by the incorporation of hydrogen. As a rule of thumb, one finds for the optical band gap E_{04} (energy at which $\alpha = 10^4$ cm^{-1}), as a function of hydrogen content,

$$E_{04}(H) \approx 1.67 \text{ eV} + (0.02 \text{ eV})[H],$$

where [H] is the hydrogen concentration in at.%. Another quantity commonly used for the description of the optical band gap in disordered semiconductors is the Tauc gap, which is obtained by plotting $\sqrt{\alpha h\omega}$ as a function of the photon energy $\hbar\omega$ (Tauc, 1974). In *a*-Si:H, one finds $E_{Tauc} \approx E_{04} - 0.2$ eV. The difference between *a*-Si and *a*-Si:H, as far as the optical gap E_{04} is concerned, can be easily seen in Fig. 9.

A characteristic feature of optical absorption in all amorphous semiconductors in general and amorphous silicon in particular is the existence of an exponential absorption tail ("Urbach tail") below E_{04}, in which the absorption coefficient, α, follows the empirical relation $\alpha \propto \exp[\hbar\omega/E_0]$, where E_0 (the Urbach slope parameter) is a material parameter. Device-quality *a*-Si:H has a rather steep Urbach tail with $E_0 \approx 50$ meV, whereas unhydrogenated *a*-Si has a much shallower tail ($E_0 \approx 100$ meV, see Fig. 9).

At energies around $\hbar\omega \approx 1$ eV, the absorption spectra of amorphous Si are dominated by optical excitation related to defect levels ("dangling bonds") in the gap (see also Fig. 7). This defect absorption appears as a low-energy shoulder in the spectra of Fig. 8. The defect density N and the absorption coefficient at $\hbar\omega = 1$ eV are found to be proportional to each other (see also Sec. 2.3).

2.2 Transport and Doping

The most prominent parameter to describe transport in *a*-Si:H is the electrical conductivity σ_D in the dark. σ_D is a macroscopic quantity that represents an average behavior of the propagating carriers. The interpretation of the experimental data therefore has to take into account all phenomena related to transport in disordered semiconductors like carrier scattering, tunneling rates, and trapping and thermal re-emission processes as well as the distribution of and the thermal occupation of states. In general, the conductivity σ_D is given by the product of carrier density n, carrier mobility μ, and the elementary charge:

$$\sigma_D = ne\mu.$$

In systems like amorphous Si with a distribution of possible conduction paths, it is useful to write the macroscopic conductivity as a summation over different conduction channels (Overhof and Thomas, 1989):

$$\sigma_D = \int N(E)e\mu(E)f(E,T)\, dE,$$

where $f(E,T)$ is the Fermi occupation function and $N(E)$ is the density-of-states distribution. The integral in general contains contributions from electrons and holes. When conductivity takes place far from the Fermi energy E_F and is dominated by a single type of carrier, then activated conduction is observed:

$$\sigma_D = \int N(E)e\mu(E) \exp\left[-\frac{E - E_F}{kT}\right] dE.$$

The dominant transport path is determined by the density of states, the carrier mobility, and the Boltzmann factor. For samples with a high density of localized defect states or at very low temperatures, conduction occurs by hopping in the distribution of localized states close to E_F. However, in state-of-the-art *a*-Si:H, conduction takes place predominantly in extended states above the mobility edge E_C of the conduction band, and in boron-doped *a*-Si:H below the mobility edge E_V of the valence band. Conduction in extended states is by thermal activation of electrons and holes from the Fermi level to the mobility edge. This leads for n-type conductivity to

$$\sigma_D = \sigma_{min} \exp\left[-\frac{E_C - E_F}{kT}\right],$$

where σ_{min} is the minimum metallic conductivity:

$$\sigma_{min} = N_C e\mu_C kT,$$

with μ_C the carrier mobility at the mobility edge.

For transport in delocalized band states above the mobility edge, the most important parameter is this conductivity prefactor, σ_{min}. In the limit of degenerate doping, σ_{min} would be approximated by

$$\sigma_{min} = en\mu = e(kTN_C)\left(\frac{1}{6}\frac{e}{kT}r_i^2\nu\right) \approx 200\ (\Omega\ \mathrm{cm})^{-1},$$

where $N_C \approx 3 \times 10^{21}\ \mathrm{cm}^{-3}\ \mathrm{eV}^{-1}$ is the density of states at the conduction-band mobility edge, $r_i \approx 5$ Å is the inelastic scattering length, and $\nu \approx 10^{14}\ \mathrm{s}^{-1}$ is the hopping frequency of carriers. This value is also corroborated by experiments performed on highly phosphorus-doped *a*-Si:H samples, where the Fermi level is close to the conduction-band mobility edge, E_C. A different relation for the minimum metallic conductivity was first proposed by Mott (1987) as $\sigma_{min} = 0.03\ e^2/\hbar r_i$, which for $r_i \approx 5$ Å leads to the same value of 200 (Δ cm)$^{-1}$ determined above.

For transport in the nondegenerate case, where carriers are thermally or optically excited above the mobility edges, the most obvious mechanisms by which the prefactor σ_{min} could be different from the expected value of 200 (Ω cm)$^{-1}$ are thermally induced shifts of the mobility edge E_C, of the optical gap E_{gap}, and of the Fermi level (statistical shift). The statistical shift of the Fermi level and the shift of the mobility edge are dependent on the specific density-of-states distribution and cannot be predicted in the general case, whereas the shift of the optical gap is known experimentally to be

$$\gamma_G = -4.4 \times 10^{-4}\ \mathrm{eV/K}.$$

There is still a discussion going on whether the concept of a sharp mobility edge is a reasonable assumption. However, most of the existing experimental data are analyzed on the basis of this concept. From a theoretical and experimental point of view, hopping conduction becomes significant whenever the density of states at the Fermi level exceeds about $10^{18}\ \mathrm{cm}^{-3}$ and the temperature falls below a critical value.

Small amounts of phosphine or diborane added to silane in the glow-discharge deposition process cause variations in conductivity ranging from 10^{-13} to 10^{-2} S/cm (Spear, 1977). As an example, Fig. 10 shows the variation of the dark conductivity of *a*-Si:H upon doping with phosphorus. Around room temperature, σ_D is thermally activated with an activation energy between 0.8 eV (undoped) and 0.15 eV (highest doping level). The variations in the slope at high temperatures are discussed in more detail below. The discovery that amorphous silicon can be doped at all was surprising because of the theoretical prediction that substitutional doping of an amorphous semiconductor should be impossible. This prediction was based on the $8 - N$ rule (where N is the number of valence electrons per atom), saying that all atoms in a random amorphous network are able to achieve bonding configurations with their optimal chemical valency. Accordingly, both boron and phosphorus are expected to be threefold coordinated and electronically inactive. In crystalline semiconductors, on the other hand, the periodic lattice forces the impurity to have the same coordination as the network atoms, so that the phosphorus and boron atoms are incorporated with fourfold coordination and become electronically active.

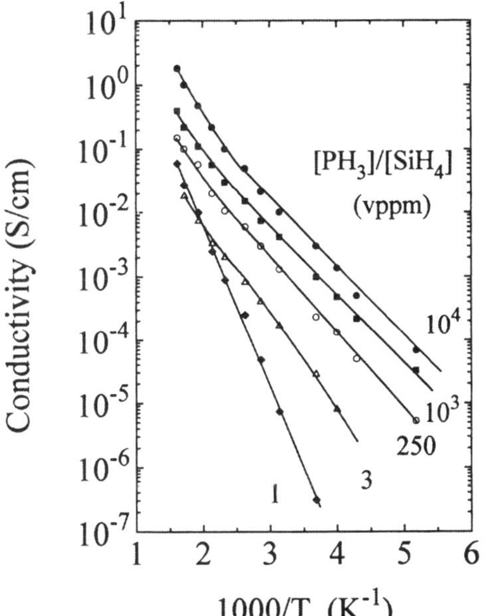

FIG. 10. Dark conductivity of *n*-type amorphous silicon as a function of temperature for different gas-phase doping levels with phosphorus.

Figure 11 summarizes further experimental results related to doping in *a*-Si:H. The variation of the Fermi-level position relative to the mobility edges E_V and E_C as a function of dopant gas concentration is shown in 11(d). As the Fermi level passes through the middle of the gap, the type of conduction changes from *p* type (E_F close to E_V) to *n* type (E_F close to E_C), as evidenced by the sign reversal of the thermopower in 11(c). The movement of the Fermi level through various regions of the localized states distribution (cf. Fig. 7) upon doping can also be traced via the changes of the spin-resonance *g* value in 11(b): for heavy boron doping, the

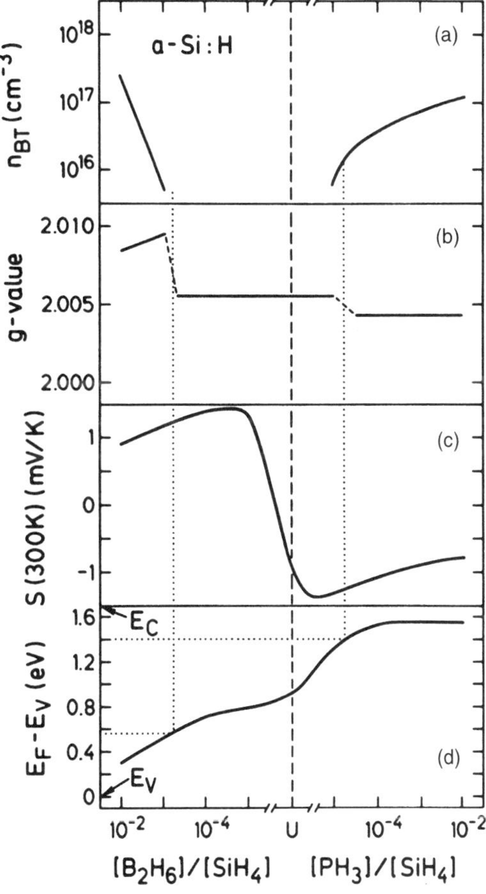

FIG. 11. Dependence of (a) the shallow carrier density in the band tails, (b) the electron-spin resonance *g* value, (c) the thermoelectric power, and (d) the Fermi-level position relative to the valence-band mobility edge in amorphous silicon as a function of doping with diborane and phosphine.

Fermi level lies in the valence-band tail (VB), and spin resonance is due to trapped holes (g = 2.008–2.01). In weakly doped and undoped a-Si:H, E_F is situated in the dangling-bond defect band (DB, g = 2.0055). Finally, in heavily phosphorus-doped samples, E_F shifts into the conduction-band tail (CB), and the spin-resonance signal is due to trapped electrons (g = 2.0044). In the limit of strong doping, electron spin resonance can also be used to measure the density n_{BT} of shallow carriers trapped in band-tail states, which increases exponentially as the conductivity activation energy decreases [11(a)].

The mobility μ of carriers is a further important transport parameter, especially for many device applications. It contains information about the transport mechanism and the density-of-states distribution. For insulating thin films such as a-Si:H, the time-of-flight technique has been the most successful way to determine μ (Tiedje, 1984), mainly because Hall experiments in amorphous materials are not very well understood. To perform a time-of-flight experiment, the a-Si:H layer is deposited between two blocking electrodes consisting of either n- and p-type layers or two Schottky barriers. Carriers are generated near one of the electrodes by a short laser flash and separated in an applied electric field. One type of charge is collected immediately by the nearby electrode, and the carriers with opposite sign cross the sample to the far electrode. The average motion of carriers is recorded via a fast digitizing oscilloscope. As a result of the fast trapping of carriers in the band tails (within femtoseconds or picoseconds), the deduced drift mobility μ_D represents an average over free and trapped carriers. The free mobility in transport states, μ_C, and the experimentally determined drift mobility are related by

$$\mu_D = \mu_C/(1 + f_{\text{trap}}),$$

where the trapping-time fraction f_{trap} for a single trap level is given by

$$f_{\text{trap}} = \frac{N_T}{N_C kT} \exp\left(\frac{E_T}{kT}\right).$$

Here N_T is the volume trap density, E_T the energy gap between the mobility edge and the trap level, and N_C the density of states at the mobility edge. Thus, the drift mobility in a-Si:H strongly depends on the density-of-states distribution close to the mobility edges and on the temperature. For high temperatures the drift mobility approaches the free-carrier mobility because the density of states in the band is higher than in the band tail. It is also important to note that in a-Si:H the tail states are distributed in energy (exponential tail), so that a simple one-trap model cannot be applied. More realistically, time-of-flight experiments have to be discussed on the basis of the multiple-trapping model and are commonly interpreted to evaluate the distribution of localized states at the conduction- and valence-band edges.

Experimental results for the electron and hole mobilities as functions of temperature are shown in Fig. 12. The general features of electron and hole mobilities are comparable, but electrons have significantly higher mobilities than holes. The extrapolations of the high-temperature mobilities indicate that the extended state mobility μ_C of electrons is 10–20 $cm^2/V \cdot s$, and that of holes is 0.5–5 $cm^2/V \cdot s$. Toward lower temperatures the drift mobility decreases strongly. Below T = 250 K for electrons and 410 K for holes, μ_D exhibits thermally activated behavior with activation energies dependent on the applied electric field. This is the dispersive transport regime where the transit of the carriers across a thin film is electric-field and thickness dependent (Nebel and Street, 1993). The activation energies of electron mobilities vary from 200 to 90 meV and of hole mobil-

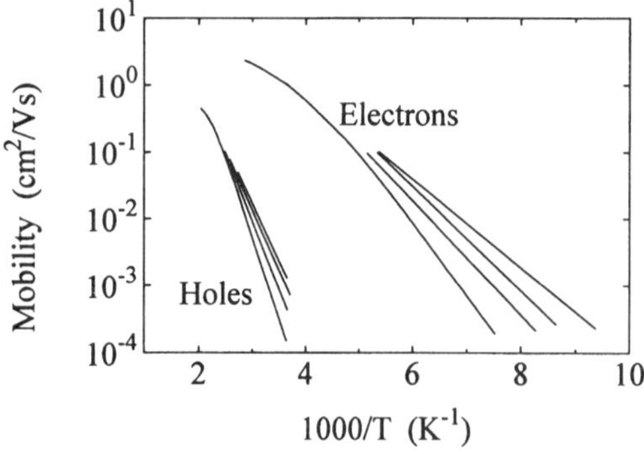

FIG. 12. Temperature dependence of the electron and hole drift mobility for different electric fields. (For a given temperature, the mobility increases with increasing electric field.)

ities from 500 to 350 meV, indicating that on the average holes are more deeply trapped than electrons. This is due to the broader valence-band tail, which decreases exponentially with a characteristic energy of about 45–50 meV (Urbach slope) compared with about 22–25 meV for the conduction-band tail.

The dispersive nature of transport is caused by the distribution of localized states over a broad energy range. Carrier interaction with these states is dominated by successive trapping and release events, which cause the center of the charge distribution to thermalize deeper and deeper in energy during the transit from the front to the back contact. The average trap depth of carriers, as they cross the sample, is given by the demarcation energy:

$$E_D = kT \ln(\nu_0 t),$$

where ν_0 is the attempt frequency for thermal emission from the traps ($\cong 10^{12}$ s^{-1}) and t is the time measured relative to the moment of carrier generation. Calculations show that the activation energy of the drift mobility represents reasonably well the demarcation energy at the moment when carriers reach the back contact.

As a third fundamental transport phenomenon, we mention the Hall effect in a-Si:H. The Hall effect has not contributed to the understanding of transport and doping because of its "double-sign anomaly," meaning that the Hall field has the wrong sign compared to the case of electron or hole transport in crystalline semiconductors. The sign anomaly is characteristic for amorphous semiconductors and seems to be linked to the small scattering distance in these materials. Figure 13 summarizes the Hall mobilities measured in intrinsic and phosphorus- and boron-doped a-Si:H using a conventional Van de Pauw contact configuration. Note that the Hall mobilities are significantly different from the drift mobilities deduced from time-of-flight experiments (Fig. 12).

Although amorphous Si can be doped substitutionally with group-V or group-III atoms just like crystalline Si, considerable differences exist as far as the doping efficiency, and effects of doping on the electronic density of states are concerned. As shown in Fig. 14, the total doping efficiency (number of

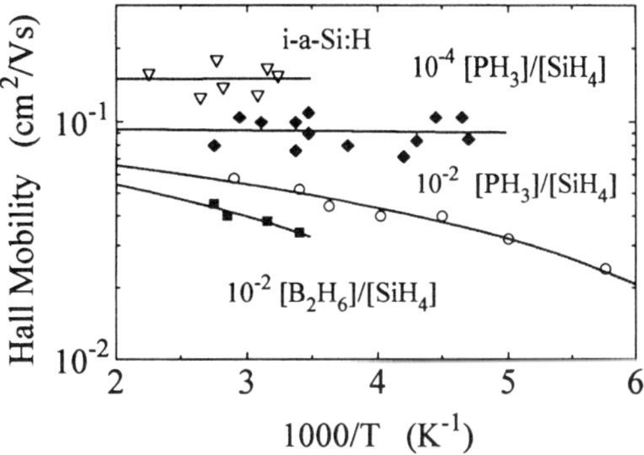

FIG. 13. Temperature dependence of the Hall mobility in intrinsic, n-type, and p-type a-Si:H.

charges added to the material per incorporated dopant atom) decreases like the square root of the dopant gas concentration in the deposition plasma, from unit doping efficiency in the parts per million range down to a doping efficiency of less than 1% for the highest dopant concentrations. At the same time, the density of deep dangling-bond defects increases, from less than 10^{16} cm^{-3} in undoped a-Si:H to more than 10^{18} cm^{-3} in heavily doped samples (see Fig. 15). Both observations can be qualitatively explained by

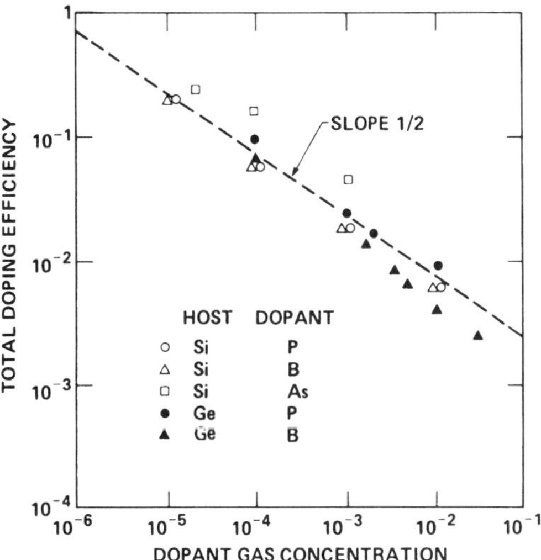

FIG. 14. Total doping efficiency versus gas-phase dopant concentration for different dopant-host systems. The dashed line indicates the expected behavior according to the autocompensation model.

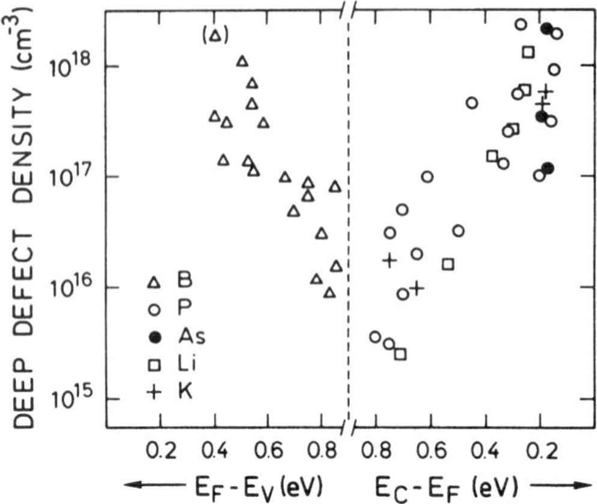

FIG. 15. Dependence of the deep defect density in amorphous silicon on the Fermi-level position for different dopants.

an autocompensation reaction of the form (Street, 1985)

$$Si_4^0 + P_3^0 \Leftrightarrow Si_3^- + P_4^+.$$

Here Si_4^0 denotes a fourfold-coordinated network atom, P_3^0 an inactive, threefold-coordinated phosphorus atom, Si_3^- a negatively charged, doubly occupied dangling-bond defect below the Fermi level, and P_4^+ an ionized substitutional donor state. Such an autocompensation reaction is believed to occur during and even after deposition. Assuming complete thermal equilibration and applying the law of mass action to the autocompensation reaction, one obtains

$$[P_4^+] = [Si_3^-] \propto [P_3^0]^{0.5},$$

where [ ] denotes a solid phase concentration, and we have used the approximations $[P_3^0] \gg [P_4^+]$ and $[Si_4^0] \approx$ const. Thus, the dangling-bond density should increase as the square root of the phosphorus concentration, whereas the doping efficiency, $[P_4^+]/([P_3^0] + [P_4^+])$, should decrease as $1/[P_3^0]^{0.5}$, in accordance with Figs. 14 and 15.

The autocompensation effectively keeps the Fermi level in doped *a*-Si:H between the deep defect band and the shallow donors, so that *a*-Si:H cannot be doped degenerately. Indeed, the number of shallow charge carriers, n_{BT} in Fig. 11(a), is always much smaller than the deep defect density in Fig. 15 for the same doping level. Further details of this phenomenon are discussed in Sec. 2.5.

As in crystalline Si, un-ionized neutral donors in *a*-Si:H are paramagnetic and can be observed in electron spin resonance at low temperatures. Figure 16 shows the corresponding ESR spectra of P_4^0 and As_4^0 levels in *n*-type *a*-Si:H. The multiple lines are due to ^{31}P and ^{75}As hyperfine interaction. Analyzing the hyperfine splitting and the occupancy (i.e., the resonance intensity) as a function of temperature allows one to estimate the localization radius (10 Å) and the position of the donor level relative to the conduction-band mobility edge (≈100–150 meV). Compared to crystalline Si, donor states in *a*-Si:H are thus more localized and energetically deeper (Stutzmann *et al.,* 1987).

2.3 Defects

We have already mentioned in conjunction with Fig. 7 that localized states in *a*-Si:H can be divided into three different

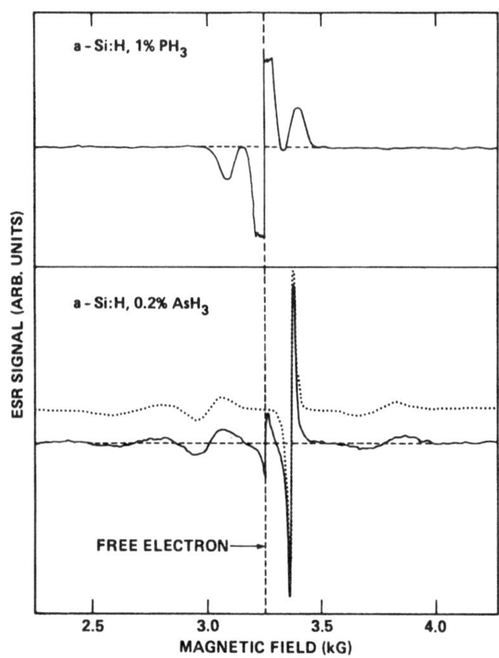

FIG. 16. Electron-spin resonance signal of neutral donors (phosphorus and arsenic) in amorphous silicon. The spin-resonance signal close to the free-electron position (dashed line) is due to electrons in conduction-band tail states present in all *n*-type samples.

types, according to their energetic position in the gap and also via the different g values in spin resonance [cf. Fig. 11(b)]. Microscopically, these three different types of localized states are commonly assigned to bonding and antibonding orbitals of weak or distorted Si–Si bonds (valence-band and conduction-band tail states, respectively) and to threefold-coordinated Si atoms, where one of the four sp^3-hybrid orbitals remains without a suitable bonding partner (deep dangling-bond defects).

Experimental evidence for this assignment only exists for the deep defects. In their neutral charge state, dangling bonds are occupied by a single electron and are, therefore, paramagnetic. The corresponding electron-spin-resonance signal has a characteristic g value of 2.0055 and is shown as the sharp central resonance in Fig. 17. The assignment of this signal to dangling-bond defects is corroborated by the existence of two weak hyperfine lines due to defect orbitals localized on ^{29}Si atoms (natural abundance 4.7%, nuclear spin $I = \frac{1}{2}$) (Stutzmann and Biegelsen, 1988). This hyperfine structure can be seen more clearly in ^{29}Si-enriched samples, where almost all Si atoms carry a nuclear spin and thus produce a hyperfine splitting (lower part of Fig. 17). A detailed analysis of this hyperfine interaction allows one to estimate the degree of localization of the defect orbital and the projected 3s and 3p contributions of the Si atomic orbitals. These estimates (60% localization on the central atom, 10% s, and 90% p character) agree very well with what is to be expected for the relaxed configuration of threefold-coordinated Si. An alternative model for the deep defect in a-Si:H, namely, a fivefold-coordinated Si state ("floating bond," Si_5), has not been widely accepted.

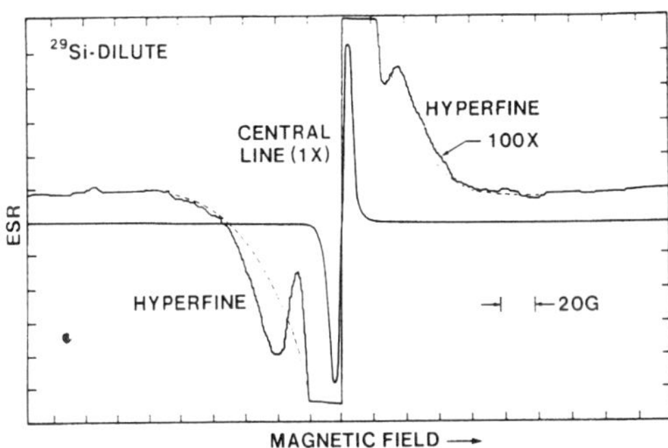

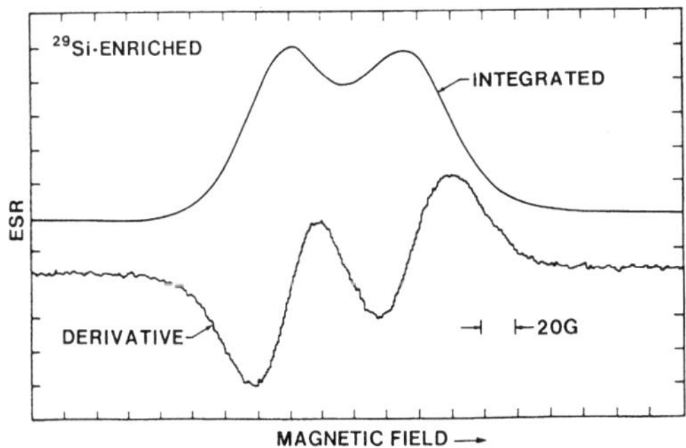

FIG. 17. Hyperfine structure of the dangling-bond spin-resonance signal in intrinsic amorphous silicon with the natural isotope concentration of ^{29}Si (upper part) and in ^{29}Si-enriched material.

An important property of localized states in general is their effective correlation energy U_{eff}, which determines the relative energetic position of the different charge states of a given defect. In Fig. 18 we describe the case of an amphoteric defect with three different charge states in the gap and a positive correlation energy. The occupation factor $g(E)$ versus energy is shown in 18(a) for $T = 0$ and $T > 0$. At $T = 0$, states above the Fermi energy E_F are empty ($g = 0$), states between E_F and $E_F - U$ are occupied by one electron ($g = 1$), and states below $E_F - U$ are doubly occupied ($g = 2$). In Figs. 18(b)–18(d) we show the resulting energy distributions of empty, singly, and doubly occupied states for a band of defect levels centered at an energy E_D and having a width W. Note that this defect band describes the energy distribution for the $g = 0$ to $g = 1$ transition of individual defects. As indicated in 18(d), the electron–electron repulsion, which causes a positive correlation energy, shifts the doubly occupied levels up in energy by the amount U. It is commonly accepted that dangling-bond defects in amorphous Si have a positive correlation energy of the order of 0.2–0.4 eV. Negative correlation energies occur when the change in charge state is accompanied by a strong exothermic structural relaxation, which pulls the $1 \rightarrow 2$ transition energetically below the $0 \rightarrow 1$ transition. In this case, the singly occupied states are not stable in thermal equilibrium but decay spontaneously into pairs of empty and dou-

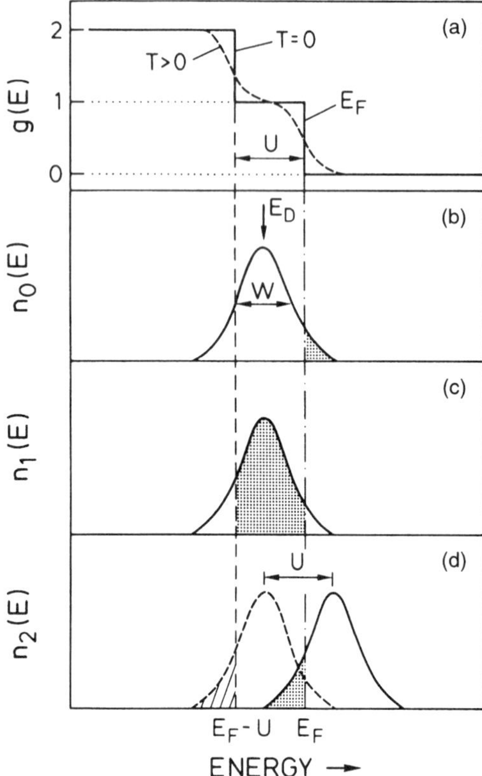

FIG. 18. Electron occupation of a defect band with a positive correlation energy U.

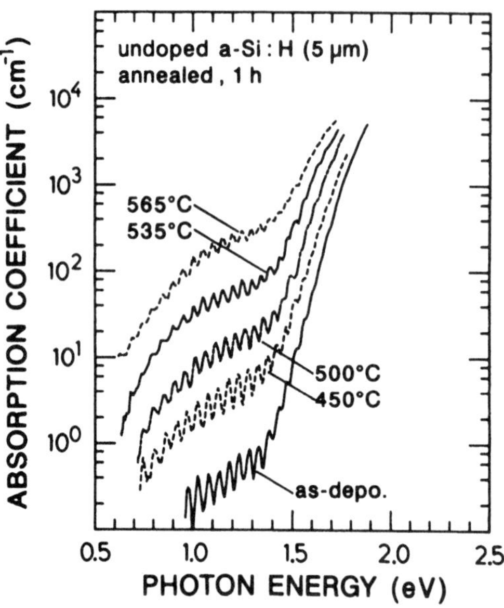

FIG. 19. Sub-band-gap absorption in amorphous silicon with different defect densities produced by thermal annealing.

bly occupied defects. This situation occurs, for example, in chalcogenide glasses.

For all applications of amorphous Si, the density and energy distribution of deep dangling-bond defects are important material parameters. Probably the most reliable experimental method to address this problem is the measurement of the subgap absorption by photothermal deflection spectroscopy (PDS) or the constant photocurrent method (CPM), which can be used for amorphous Si irrespective of doping concentration, film thickness, etc. Typical subgap absorption spectra for undoped *a*-Si:H with different thermal histories are summarized in Fig. 19. As-deposited material exhibits an exponentially decaying Urbach tail (cf. Fig. 8) due to excitation of carriers from the valence-band tail into the conduction band. This absorption tail is followed by the dangling-bond–related defect absorption for photon energies $\hbar\omega \leq 1.4$ eV. Shape and energy position of this absorption band are compatible with a broad (FWHM ≈ 0.3 eV) distribution of defect levels close to the middle of the mobility gap. The absorption coefficient α at a fixed energy position within the defect absorption is proportional to the density of defects, which in the case of Fig. 19 is altered by thermal dissociation of Si–H bonds via high-temperature annealing. As a rule of thumb one has

$$N\,(\mathrm{cm}^{-3}) \approx 10^{16}\alpha(1.3\ \mathrm{eV})(\mathrm{cm}^{-1})$$

as the relation between the defect density in units cm^{-3} and the absorption coefficient α at 1.3 eV in units cm^{-1}. The strict linearity between subgap absorption and deep defect density is demonstrated by comparing subgap absorption with other experimental techniques, such as ESR. Figure 20 shows a plot of $\alpha(1.3$ eV) versus the dangling-bond–related ESR spin density N_S for undoped *a*-Si:H in which the defect density has been changed by light soaking (cf. Sec. 2.5) or by thermal annealing, as in Fig. 19. Data such as those in Fig. 20 suggest that in undoped *a*-Si:H indeed the majority of defects are in their neutral, paramagnetic charge state and have a correlation energy U much larger than kT at room temperature (see Fig. 18).

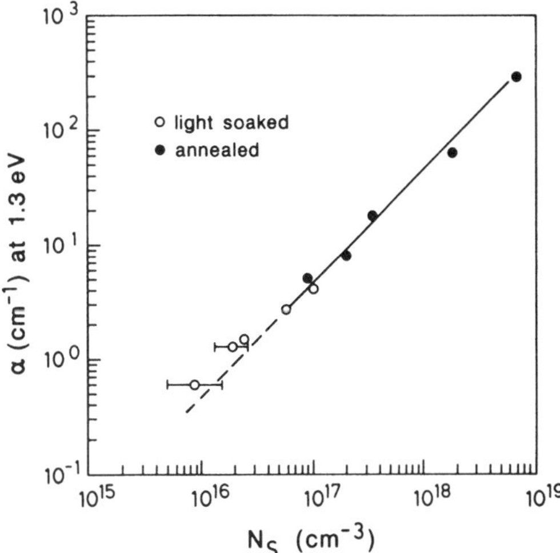

FIG. 20. Sub-band-gap absorption coefficient at a photon energy of 1.3 eV versus spin-resonance defect density in undoped samples. Different defect densities were produced by light-soaking (open symbols) or thermal annealing (closed symbols).

2.4 Recombination

The fate of charge carriers in *a*-Si:H following the creation—e.g., by optical absorption or injection—is another central question for applications of this material. The recombination of excess carriers in *a*-Si:H is schematically displayed in Fig. 21, which summarizes the basic features of a very complex phenomenon. The model has been developed combining data from a variety of experiments, most importantly photoconductivity, photoluminescence (PL), electron spin resonance, spin-dependent recombination, charge-collection measurements, etc. Electron–hole recombination takes place via three basic steps, each of which is sensitively dependent on the density and distribution of gap states, the temperature, the carrier density, and the applied electric field:

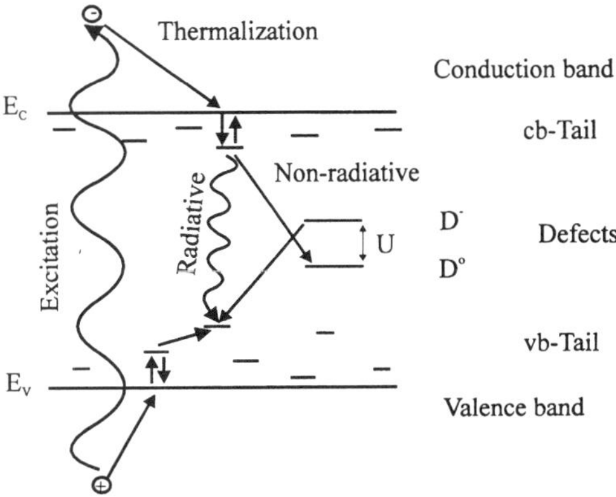

FIG. 21. Fundamental processes involved in optical excitation and recombination.

1. After photogeneration, electrons and holes undergo rapid inelastic scattering (thermalization) in the conduction and valence bands on a time scale typically shorter than 10^{-12} s. Following thermalization, carriers are trapped in the shallow tail states of the respective bands.
2. Further energy dissipation is dominated by thermalization within the band tails. Tunneling transitions (hopping down) are dominant at temperatures lower than 100 K. At higher temperatures multiple thermal excitation and retrapping governs the thermalization process. Typical time scales for this process are between 10^{-10} and 10^{-6} s.
3. Finally, carriers recombine either radiatively or nonradiatively. Radiative recombination proceeds by transitions of electrons from the conduction-band tail to trapped holes in the valence-band tail, causing a broad luminescence emission. The *a*-Si:H luminescence band is featureless, with a peak at 1.3–1.4 eV and a half-width of 0.25–0.35 eV. The luminescence quantum efficiency is practically 100% at low temperatures but decreases rapidly at higher temperatures and also as a function of defect density above 10^{17} cm^{-3}. The radiative recombination is mainly geminate (electron and hole stay together as a pair during their lifetime) when there is insufficient thermal energy to dissociate an electron–hole pair, which is bound by Coulomb interaction. The pair may separate by about 50 Å during thermalization, but when the Coulomb attraction is strong enough, the particles eventually diffuse back together and recombine. The transition from geminate to nongeminate recombination occurs when different electron–hole pairs start to overlap at sufficiently high generation rates or when the pairs diffuse apart at high temperatures or in high electric fields.

With increasing temperature, nonradiative recombination becomes significant because of thermal excitation of trapped carriers, which enhances the mobility. The enhanced mobility gives rise to faster trapping at defects, which is the primary reason for the thermal quenching of the luminescence. In high-quality *a*-Si:H with a low defect density ($<10^{16}$ cm^{-3}), the nonradiative recombination via defects begins to influence the lifetime at temperatures higher than 100 K. At room temperature the luminescence intensity is reduced by about four orders of magnitude, and the recombination is dominated by direct capture of electrons and holes into defects. Whether this capture of carriers by defects occurs via transport in the extended band states or, in contrast, as a diffusive hopping transition depends on the density and distribution of gap states.

Nonradiative recombination of photoexcited carriers in *a*-Si:H above 100 K via dangling-bond defects as the main recombination center has been investigated in great detail by spin-dependent photoconductivity (SDPC). In this technique one makes use of the fact that spin resonance of paramagnetic levels generally enhances electronic transition rates via these levels. SDPC spectra of device-quality, undoped *a*-Si:H taken at 150 K are shown in Fig. 22. The complex SDPC response can be separated into two underlying basic processes with slightly different characteristic time constants. In the first step an excess electron localized in the conduction-band tail (spin g value, g = 2.0044, see Fig. 11) is trapped by a neutral dangling-bond defect (g = 2.0055), charging the latter negatively. This transition is seen in SDPC as a combined resonance at the mean g value, g = 0.5(2.0055 + 2.0044) = 2.0050. In order to complete the recombination, a hole trapped in the valence-band tail (g = 2.01) has to recombine with the negatively charged defect. At 150 K, transport of holes in *a*-Si:H occurs by sequential hopping between tail states. This hopping motion is again spin dependent and gives rise to the second constituent resonance in Fig. 22 at g = 2.01.

Whereas spin-dependent photoconductivity provides a qualitative picture of the recombination process, other methods can provide more quantitative parameters for specific recombination steps. For example,

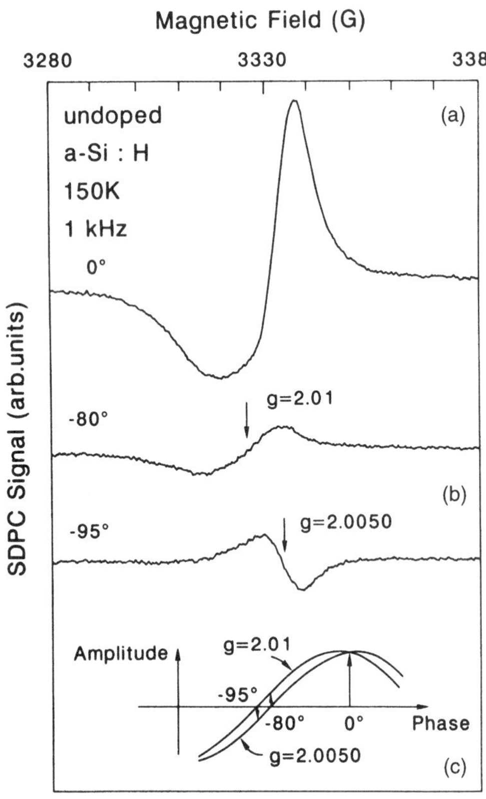

FIG. 22. Spin-dependent photoconductivity response of undoped amorphous silicon at 150 K. (b), (c) Decomposition of the signal in (a) into two constituents with a different phase shift relative to the magnetic-field modulation used for detection.

the capture cross section of dangling bonds in *a*-Si:H has been measured by a combination of ESR and charge-collection experiments in the temperature regime where the interaction between carriers and defects is dominated by direct capture (ballistic capture). Charge-collection experiments evaluate the $\mu\tau$ products that are obtained by measuring the collected charge as function of applied electric field in a time-of-flight experiment. The collected charge Q versus the applied electric field F is given by the Hecht formula as

$$\frac{Q(F)}{Q_0} = \frac{\mu_D \tau F}{d}\left[1 - \exp\left(-\frac{d}{\mu_D \tau F}\right)\right],$$

where τ is the deep trapping lifetime, Q_0 the total generated charge, and d the layer thickness. Figure 23 shows some results for in-

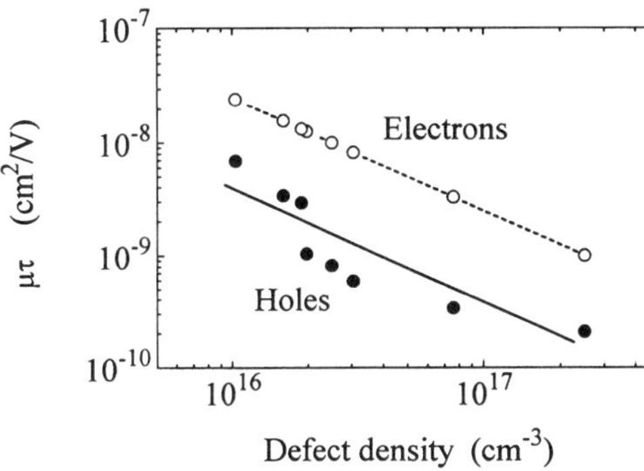

FIG. 23. Mobility-lifetime products of electrons and holes in undoped *a*-Si:H as a function of the defect density.

trinsic *a*-Si:H with different defect densities determined by ESR. The $\mu_D\tau$ products of electrons and holes are inversely proportional to the defect density. The hole $\mu_D\tau$ product is approximately a factor of 7 smaller than the electron $\mu_D\tau$ product, reflecting the smaller drift mobility of holes in *a*-Si:H. The capture cross section Σ_C of the defects for ballistic capture can be calculated as

$$\Sigma_C = \frac{el}{6kT}\frac{l}{\mu_D\tau N_D},$$

where e is the elementary charge, l the mean free path (assumed to be ≅10 Å for electrons and ≅5 Å for holes), k the Boltzmann factor, T the temperature, and N_D the defect density. In intrinsic *a*-Si:H, the dominant defect (Si dangling bond) is neutral and can capture either an electron or a hole. This is the reason for the linear decrease of both electron and hole $\mu_D\tau$ products as the defect density is increased (see Fig. 23). In doped *a*-Si:H, the defects are charged and can only trap carriers of opposite charge. This reduces the minority-carrier $\mu_D\tau$ product significantly for electrons in *p*-type material and holes in *n*-type *a*-Si:H. Charged defects have larger cross sections than neutral defects on account of the long-range attractive Coulomb interaction with excess carriers. The trapping parameters for the four transitions and the related capture cross sections for ballistic capture are summarized in Table 2 (Street, 1991).

The photogeneration of carriers enhances the conductivity of *a*-Si:H by many orders of magnitude, dependent on illumination power and defect density. Examples are given in Fig. 24, where the illumination power has been kept constant at 50 mW/cm² while the defect density has been varied over three orders of magnitude by electron bombardment. The almost linear dependence of σ_{ph} on the inverse defect density $1/N_D$ is in agreement with the role of dangling bonds as the main recombination center.

The photoconductivity in intrinsic *a*-Si:H is governed over the entire temperature range by electron transport. At low temperatures, $T < 150$ K, the weaker localization of electrons and the steeper conduction-band tail prevents freeze-out of electrons in deep traps. At temperatures higher than 150 K, electrons dominate transport as a conse-

Table 2.

Transitions	$\mu_D\tau N_0$ (V cm)$^{-1}$	Σ_C (cm^2)
$e \to D^0$	2.5×10^8	2.7×10^{-15}
$e \to D^+$	5×10^7	1.3×10^{-14}
$h \to D^0$	4×10^7	8×10^{-15}
$h \to D^-$	1.5×10^7	2×10^{-14}

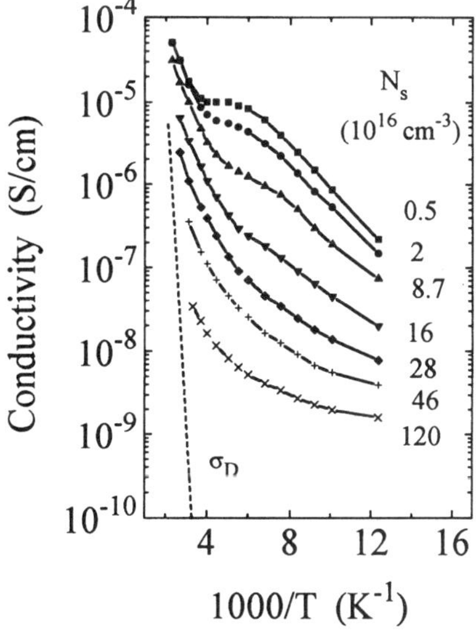

FIG. 24. Dark conductivity (σ_D) and photoconductivity in undoped *a*-Si:H with different defect concentrations produced by electron bombardement. (After Dersch *et al.*, 1983.)

quence of their propagation in delocalized states near or in the conduction band. The mobility of electrons is significantly larger than that of holes (see Fig. 12), which incidentally is also the reason for the n-type conduction in intrinsic and slightly compensated a-Si:H. In contrast, the broader valence-band tail causes holes to be more immobile, especially at low temperatures where they get trapped in the deep valence-band tail states ("save traps"). Here the density of traps is too small for rapid tunneling.

Of particular interest in Fig. 24 is also the specific temperature dependence of the photoconductivity at a given defect density. In the low-defect material, σ_{ph} is singly activated at low temperatures ($T < 150$ K, regime I), shows a shoulder or even a maximum at ≈ 170 K (regime II), and finally increases steeply toward higher temperatures (regime III). In regime I all holes are frozen out in the deep valence-band tail states (save hole traps). Recombination is limited by tunneling transitions of holes to the defects. The thermal activation is due to the increase in electron mobility (see Fig. 12). In regime II (170 K $< T <$ 250 K) holes become mobile via thermal excitation out of their deep traps and start to interact with the defects more efficiently. This causes a decrease in the effective electron lifetime and a corresponding decrease of σ_{ph}. Finally, in regime III ($T >$ 250 K) the increase in σ_{ph} is due to thermal excitation of electrons into shallow traps close to the conduction-band mobility edge. Electrons become very mobile and cause the drastic increase in σ_{ph}. With increasing defect density, σ_{ph} decreases, and the specific minimum associated with the hole freeze-out vanishes. For $N_D \geq 10^{17}$ cm^{-3}, σ_{ph} is governed by defect recombination via tunneling capture.

Figure 25 shows the absorption, photoluminescence, and photoluminescence excitation spectra of intrinsic a-Si:H. The luminescence is most effectively excited with homogeneously absorbed light ($\hbar\omega \approx E_{04}$). The observed radiative recombination occurs between electrons in the CB tail and holes in the VB tail. The photoluminescence intensity distribution is shifted from the Urbach edge by about 0.4 eV and is centered around 1.3–1.4 eV. The FWHM of the photoluminescence band is 0.25–0.3 eV (Street, 1984). The origin of the luminescence bandwidth and of the energy shift is attributed to

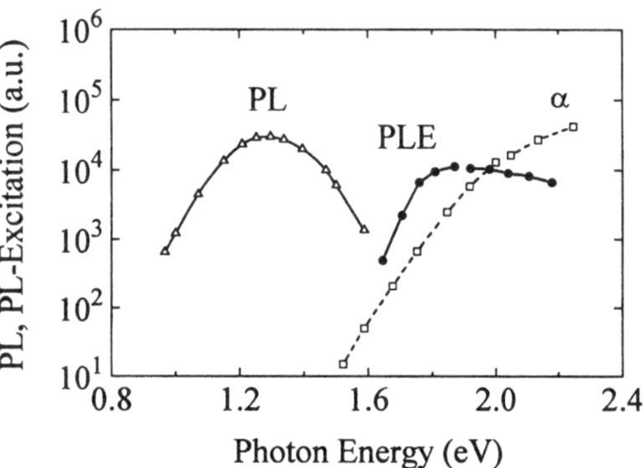

FIG. 25. Photoluminescence (PL), photoluminescence excitation (PLE), and absorption spectra (α) of amorphous silicon at low temperature.

1. disorder broadening due to the spread in energy of band-tail states and
2. dynamic broadening by phonons.

The identification of the relative contributions of disorder and phonon broadening in the luminescence spectrum is rather complex and not clear up to now. The phonon model explains the photoluminescence data reasonably well, indicating that disorder broadening contributes less. On the other hand, the disorder model also gives reasonable agreement, although the interpretation of the low-energy luminescence tail would result in an unrealistically broad valence-band tail.

The effect of defects on the luminescence intensity is shown in Fig. 26. The intensity drops drastically for defect densities larger than 10^{17} cm^{-3}, indicating a change of the recombination mechanism to a nonradiative process. At 10 K it is reasonable to assume that both radiative and nonradiative recombination are governed by tunneling transitions rather than by thermal excitation. The nonradiative tunneling rate is given by

$$R_{nr} = \nu_0 \exp[-2r_c/\alpha],$$

where r_c is the critical tunneling distance for the onset of nonradiative recombination, ν_0 the attempt-to-tunnel frequency ($\approx 10^{12}$ s^{-1}), and α the localization length ($\approx$10–12 Å). If an electron is closer to a defect than to the nearest hole, nonradiative recombination via defects is more likely. Equating R_{nr} with the observed radiative recombination rate, $1/\tau_r \approx 10^3$ s^{-1}, gives a critical distance $r_c \cong 10\alpha$,

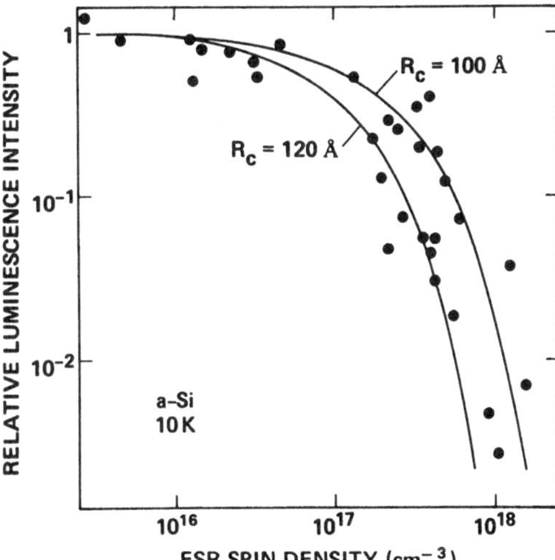

FIG. 26. Defect-density dependence of the luminescence intensity at 10 K. R_C is the critical radius for capture by defects discussed in the text. (After Street *et al.*, 1978.)

which is 100–120 Å. The functional dependence of the luminescence efficiency η_{PL} on defect density N can be written as

$$\eta_{PL} = \exp[-4\pi r_c^3 N/3].$$

According to Fig. 26, this relation also gives critical radii r_c between 100 and 120 Å. In *a*-Si:H with a high defect density, a second PL band at ≈0.9 eV is observed, which is ascribed to radiative recombination between an electron in the CB tail and a dangling-bond defect.

Figure 27 shows the thermal quenching of the luminescence. Above T = 50 K the luminescence intensity drops significantly, and the photoconductivity rises over orders of magnitude. Obviously, the thermally increased carrier mobility (predominantly of electrons) quenches the luminescence and results in an effective separation of electron–hole pairs, thus giving rise to the increase in photoconductivity. At high temperatures thermal excitation is much faster than tunneling transitions, and, therefore, the interaction with defects and nonradiative recombination increases.

The competition between recombination via close electron–hole pairs (radiative tail-to-tail transitions of excitonic pairs) and via

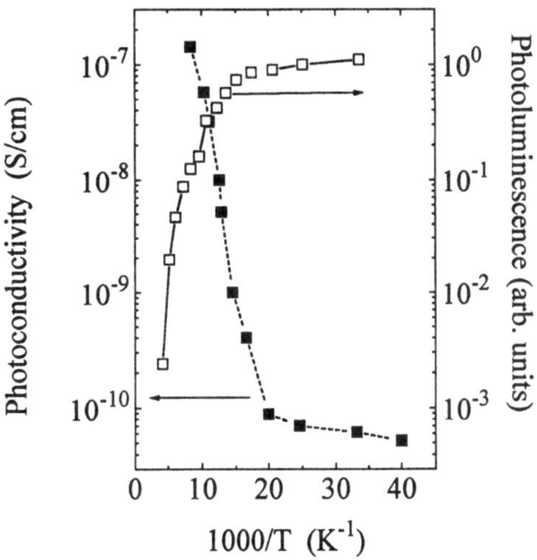

FIG. 27. Photoconductivity (left) and photoluminescence intensity (right) as a function of temperature.

defects (nonradiative recombination) can also be seen in spin-dependent recombination (optically detected magnetic resonance or spin-dependent photoconductivity). As an example, Fig. 28 shows the changes of the SDPC spectra of undoped *a*-Si:H at an intermediate temperature of 175 K as a function of defect density, which in this case is increased by extended exposure to intensive light (light-induced degradation; see next section). At 175 K, radiative recombination and electron–hole pair dissociation followed by nonradiative defect recombination according

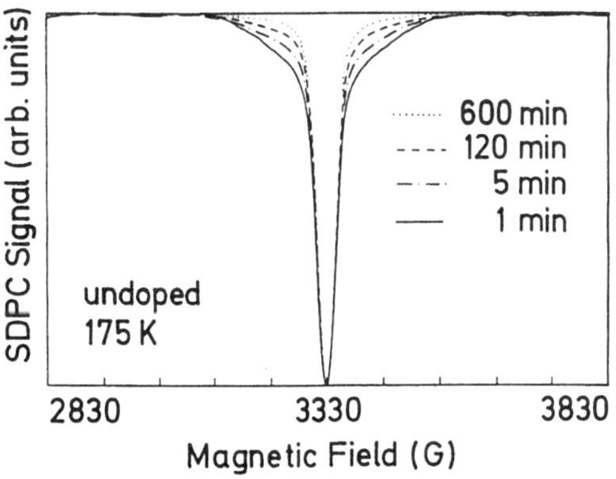

FIG. 28. SDPC signal of *a*-Si:H at 175 K after exposure to intense light for different durations. Note the gradual disappearance of the broad background line due to close electron–hole pairs.

to Fig. 27 both occur with similar probability in device-quality a-Si:H. In this case the SDPC spectrum consists of a broad and a narrow quenching line. The narrow line is the same as discussed in Fig. 22, except for the fact that in Fig. 28 we show the integrated SDPC spectrum and not the first derivative spectrum obtained by magnetic-field modulation. The broad quenching line, which can be seen as a pronounced tail to the narrow line, on the other hand, is due to excitonic pairs. The spin–spin interaction between the spatially correlated electrons and holes causes a broadening of the spin resonance to a FWHM of about 200 G, compared to ≈10 G for distant spin pairs. The main observation in Fig. 28 is that with increasing defect density (i.e., increasing illumination time), the relative contribution of the broad signal to the overall SDPC response decreases, showing directly the increase of the defect-related recombination path (Brandt and Stutzmann, 1991).

Another interesting phenomenon is the influence of high electric fields on carrier propagation and recombination. The photoluminescence, for example, is significantly quenched at fields larger than 5×10^4 V/cm. Field effects have also been detected in dark-conductivity experiments on intrinsic and doped a-Si:H. Time-of-flight measurements show that carrier mobilities are strongly enhanced in the presence of high fields. On the basis of Monte Carlo calculations, Marianer and Shklovskii (1992) have introduced an effective temperature, T_{eff}, which takes into account electric field effects on transport and recombination in a-Si:H:

$$\frac{T_{\text{eff}}^2}{T^2} = 1 + \left\{\frac{0.67e\alpha F}{kT}\right\}^2,$$

where e is the elementary charge, α the localization length, F the electric field, k the Boltzmann factor, and T the temperature. Applying this equation to experimental data results in reasonably good fits with correct values for the localization length α of electrons (≈7–8 Å) and of holes (≈4 Å) in the respective band-tail states.

2.5 Metastability

Of great importance for applications of amorphous silicon is the stability of this material under strong optical or electronic excitation (exposure to intense light or high electric fields, current stressing, thermal cycling). Unfortunately, both undoped and doped a-Si:H show a variety of instability and metastability phenomena, which, at present, impose severe limitations on the performance of a-Si:H–based electronic devices.

The best known example of such a metastability is the light-induced degradation (metastable defect formation) of a-Si:H, also known as the Staebler–Wronski effect (Staebler and Wronski, 1980). The main experimental observations associated with this effect are summarized in Fig. 29. Upon illumination of undoped a-Si:H with intense, strongly absorbed light ($\hbar\omega \geq 2$ eV, $I \geq 100$ mW/cm^2), both the photoconductivity σ_{ph} and the dark conductivity σ_D decrease by at least one order of magnitude within several hours or days [Fig. 29(a)]. This change in conductivity is metastable in the sense that the effect of prolonged illumination can be completely reversed by annealing at ≈200 °C in the dark for 1 h. This defines two distinct states of an undoped a-Si:H sample, namely, the annealed, stable state A and the metastable state B obtained after long light exposure. According to Fig. 29, these two states differ not only by their photoconductivities and dark conductivities but also by the conductivity activation energy and the recombination kinetics. As shown in Fig. 29(b), the decrease in the dark conductivity of a-Si:H upon light soaking is due to a change in the Fermi-level position from $E_C - E_F \approx 0.7$ eV in state A to $E_C - E_F \approx 0.85$ eV in state B. Moreover, the dependence of σ_{ph} on the light intensity I in Fig. 29(c) indicates a transition from a bimolecular recombination in state A ($\sigma_{ph} \propto I^{1/2}$, recombination via direct electron–hole annihilation) to monomolecular statistics in the light-soaked state, B ($\sigma_{ph} \propto I$, recombination via defects acting as recombination centers).

Direct evidence for the creation of metastable defects (Si dangling bonds) upon light soaking of undoped a-Si:H comes from spin-resonance measurements. As an example, Fig. 30 shows the increase of the number of dangling bonds as a function of illumination time for an a-Si:H sample exposed to 300 mW/cm^2 of monochromatic light ($\hbar\omega \approx 2$ eV) at 40 °C.

The total sample volume in this case is about 3 μm × 0.5 cm^2, so that 10^{12} dangling

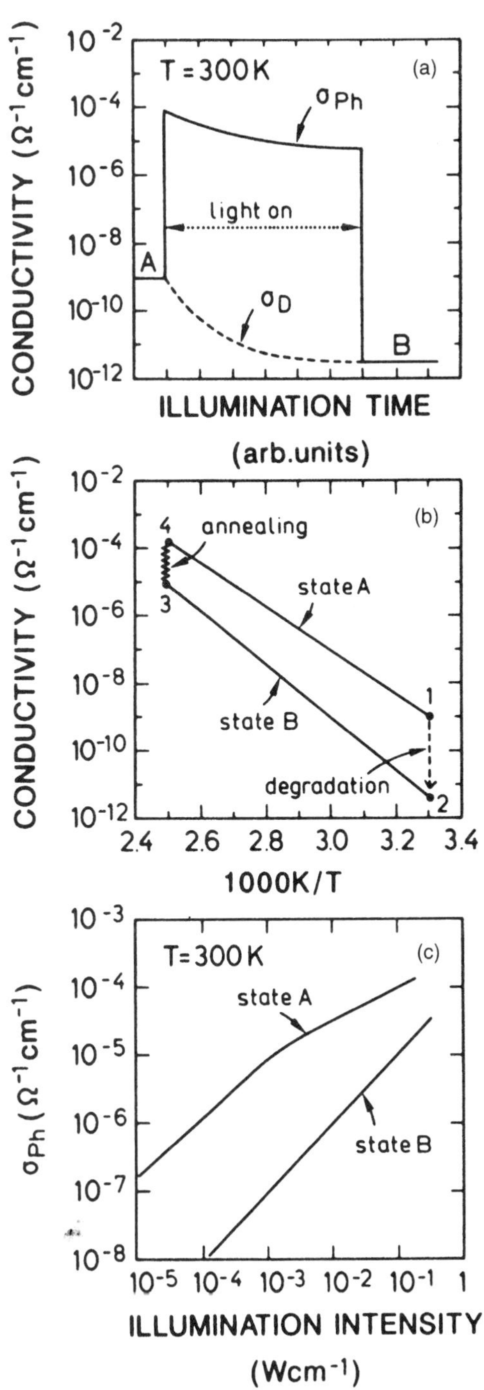

FIG. 29. Main effects of light-induced degradation on dark conductivity and photoconductivity in undoped amorphous silicon (Staebler–Wronski effect).

bonds correspond to a defect density of about 10^{16} cm^{-3}. The main features in Fig. 30, following the initial stage *I*, are a sublinear increase of the number of defects with a characteristic $t^{1/3}$ dependence on illumination time (II) and a subsequent saturation of the effect at a defect density of about 2 × 10^{17} cm^{-3} (III).

The dependence of spin density on illumination time as observed in Fig. 30 can be explained via the following rate equation:

$$\frac{dN}{dt} = c_{sw}\frac{G^2}{N^2} - (N - N_0)\nu_0 \exp\left[-\frac{E_A}{kT}\right] - (N - N_0)B\frac{G}{N}.$$

In this equation, G is the generation rate of photoexcited electrons and holes; N is the total defect density; N_0 is the initial (stable) defect density; and c_{sw}, ν_0, E_A, and B are material parameters. The first term on the right-hand side describes the creation of new metastable defects by nonradiative tail-to-tail recombination of electrons and holes, where both the electron and hole densities in intrinsic *a*-Si:H at 300 K are proportional to G/N (monomolecular limit). c_{sw} is a susceptibility parameter of *a*-Si:H for metastable defect formation and is about 10^{-15} cm^3 s^{-1}. Taking into account only this defect creation term, one obtains from the above equation

$$N^3(t) - N_0^3 = 3c_{sw}tG^2,$$

which quantitatively accounts for parts I and II of Fig. 30 (Stutzmann, 1992).

The negative terms in the rate equation for dN/dt describe the thermal and the light-induced annealing of metastable defects, respectively. Thermal annealing occurs with an activation energy $E_A \approx 1.1$ eV. The prefactor ν_0 is about 10^{10}–10^{12} s^{-1}. Light-induced annealing is only important for high optical generation rates, G, and low temperatures ($T \le 300$ K). The steady state in Sec. III of Fig. 30 is due to a balance between defect creation and annealing. The value of ≈2 × 10^{17} cm^{-3} for the saturated defect density occurs under standard illumination conditions. Saturation values in excess of 10^{18} cm^{-3} can be obtained by illumination with high-intensity short laser pulses and active sample-temperature stabilization at 300 K.

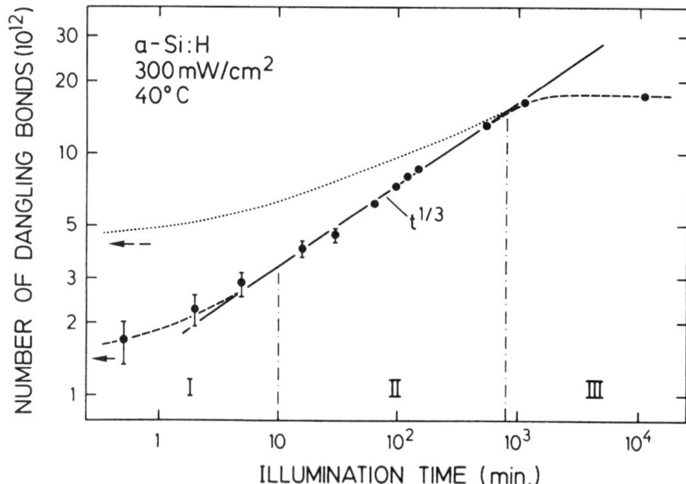

FIG. 30. Evolution of the dangling-bond defect density (areal density) in *a*-Si:H during illumination with white light of intensity 300mW/cm². Note the $t^{1/3}$ dependence for intermediate times and the subsequent saturation.

Light-induced metastable defect creation in *a*-Si:H is the major problem for applications of this material in photosensors and solar cells. The metastable defects act as additional recombination centers that decrease the photoconductivity (cf. Fig. 24) and, as a consequence, e.g., the efficiency of a solar cell using *a*-Si:H as the absorber material (see Sec. 4.1).

The microscopic origin of light-induced defect creation in *a*-Si:H is not yet clear. Both intrinsic (involving Si and H atoms only) and extrinsic mechanisms have been discussed. Figure 31 summarizes the most basic intrinsic relaxation processes that could give rise to metastable dangling-bond formation:

1. dissociation of a weak Si–Si bond due to electron–hole recombination;
2. disproportionation of tetrahedral Si atoms into a dangling bond and fivefold-coordinated atom ("floating bond");
3. metastable creation of neutral dangling bonds, D^0, by electronic charge exchange between charged defect pairs, D^+ and D^-;
4. Si–Si bond breaking stabilized by hydrogen bond switching;
5. creation of two metastable dangling bonds by photodissociation of a H_2^* complex;
6. photoinduced Si–H bond dissociation followed by migration and retrapping at an internal surface (e.g., in hydrogen-terminated voids).

An extrinsic origin of light-induced metastability has been suggested in connection with N, C, or O atoms, since only these impurities are present in device-quality *a*-Si:H with sufficient concentrations (10^{17}–10^{18} cm^{-3}, 10–100 ppm). Considerable efforts have been made to minimize light-induced degradation by producing ultrapure *a*-Si:H samples, but below a critical impurity content of typically 10^{20} cm^{-3}, no positive effect has been observed. As an example, Fig. 32 summarizes the dangling-bond densities in the annealed state, *A*, and after light soaking, *B*, as a function of controlled O or N doping of ultrapure *a*-Si:H. Below 10^{20} cm^{-3} neither O nor N contamination affects the degree of light-induced metastability. A similar observation holds for contamination with C. On the other hand, above a critical concentration of about 10^{20} cm^{-3}, all impurities tend to increase both the stable and the metastable defect density in *a*-Si:H, probably through an overall effect on the bonding properties of Si and H network atoms.

A second type of metastability with practical importance is observed in *a*-Si:H intentionally doped with electronically active shallow donors and acceptors (P, B, As, . . .). As already discussed in Sec. 2.2 in some detail, substitutional dopants are mostly incorporated in the growing amorphous film with threefold coordination (P_3,B_3). Partial thermodynamic equilibrium within the film during growth determines the fraction of fourfold-coordinated, electrically active dopants (P_4,B_4), which varies with doping level and is

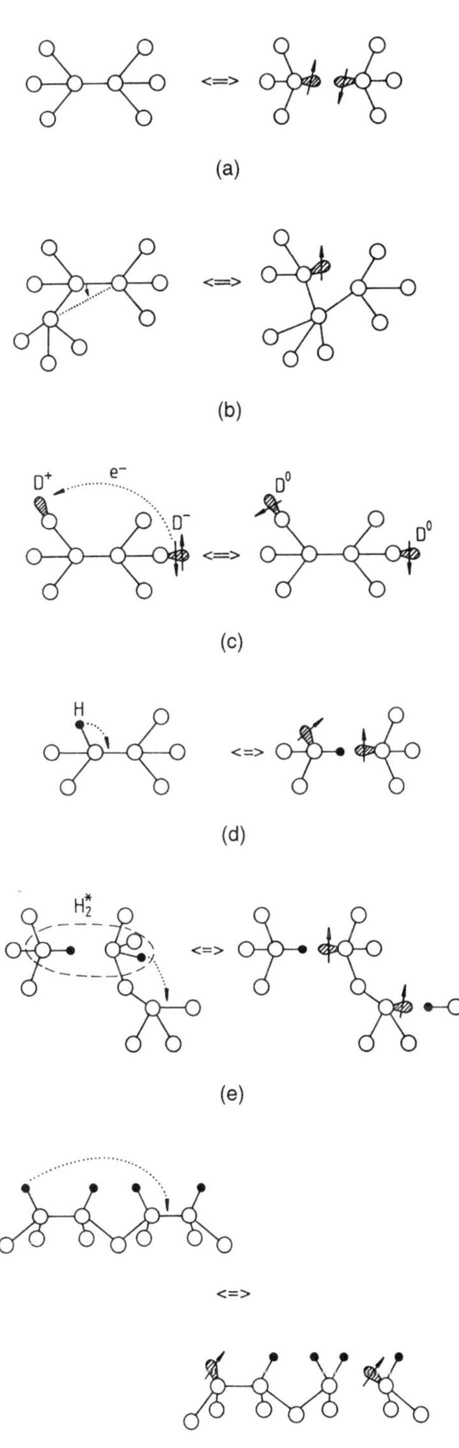

FIG. 31. Atomic models for possible metastability processes: (a) bond breaking, (b) valence alternation, (c) charge exchange, (d) hydrogen bond switching, (e) dissociation of H_2^* complexes, and (f) hydrogen migration on internal surfaces.

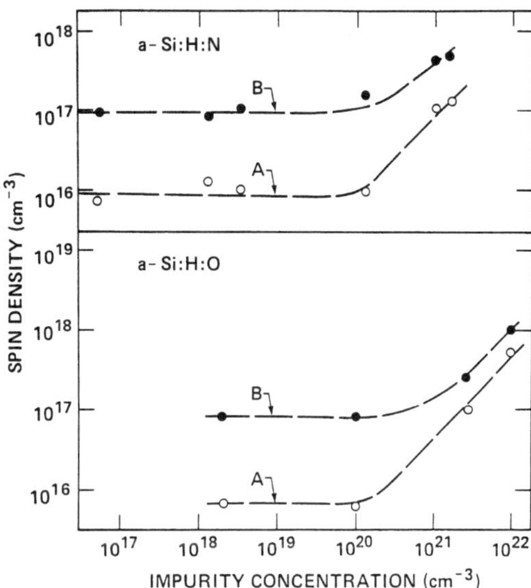

FIG. 32. Dependence of the defect density in the annealed state, curves A, and the light-soaked state, curves B, on the concentration of (a) nitrogen and (b) oxygen impurities.

typically $[P_4]/[P_3] \leq 10^{-1}$ (see Fig. 14). The fourfold dopants are most stable when ionized, and hence an approximately equal density of compensating charged dangling bonds are created (D^- in the case of P, D^+ in the case of B doping).

It is now observed that the densities of gap states and active dopants are only partially defined during the growth of the sample but are also controlled by postdeposition structural relaxation processes, which occur above a certain equilibration temperature T_E (Kakalios and Jackson, 1989). The dopant-defect system is in thermodynamic equilibrium above T_E, which is ≈130 °C for P-doped and $T_E \approx$ 90 °C for B-doped *a*-Si:H. Rapid cooling to temperatures below T_E freezes in the atomic and electronic structure at T_E. The effect of fast versus slow cooling on the conductivity of *n*- and *p*-type *a*-Si:H is shown in Fig. 33. The temperature dependence of the dark conductivity for P- and B-doped *a*-Si:H below T_E is strongly dependent on the cooling rate, which has been varied in the case of Fig. 33 from 0.02 K/s (full lines) to 1 K/s (dashed lines). Rapid thermal quenching from temperatures above T_E (arrows in Fig. 33) results in a higher dark conductivity (i.e., a larger fraction of

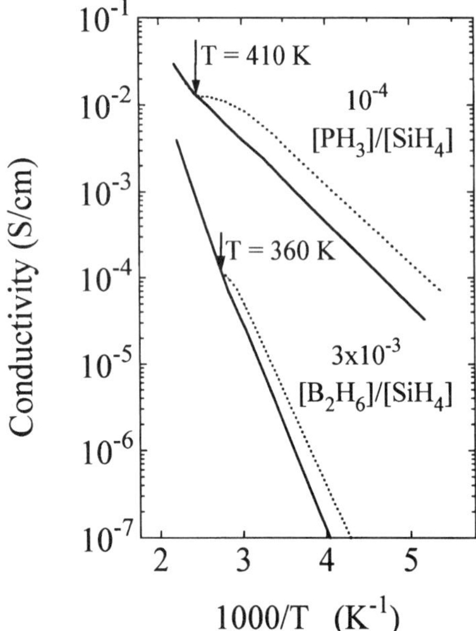

FIG. 33. Dark conductivity of *n*-type and *p*-type amorphous silicon versus temperature after slow cooling (solid curves) and rapid quenching (dashed curves). Arrows at 410 and 360 K indicate equilibration temperatures.

active dopants) than slow cooling, as expected for a thermally enhanced (endothermic) dopant activation reaction:

$$P_3^0 + Si_4^0 + \Delta E \Leftrightarrow P_4^+ + Si_3^-.$$

The frozen-in state obtained by rapid thermal quenching of phosphorus-doped *a*-Si:H below the equilibration temperature T_E is characterized by a higher active dopant concentration $[P_4^+]$. From thermodynamic arguments it is expected that the defect density Si_3^- also increases. However, up to now no significant changes of the defect density could be detected experimentally. Instead, the increase in active dopants leads to an increase of the band-tail carrier density $\Delta n_{BT} \sim \Delta P_4^+$. Analogous arguments are valid for $[B_4^-]$ in *p*-type *a*-Si:H doped with boron.

Experimentally, the density n_{BT} of shallow carriers in band-tail states at a constant temperature $T < T_E$ following rapid thermal quenching is found to decay according to a stretched-exponential relation

$$n_{BT}(t) = n_{BT}(0) \exp[-(t/\tau)^{\beta}].$$

Corresponding decay curves for phosphorus-doped *a*-Si:H ($T_E \approx 130$ °C) are shown in Fig. 34(a). The two parameters characterizing a stretched-exponential relaxation process are the dispersion parameter β and the time constant τ. The dispersion parameter β is given by $\beta = T/T_0$ with $T_0 \approx 580$ K for *n*- and *p*-type *a*-Si:H [Fig. 34(b)]. Physically, stretched-exponential relaxation occurs for a system with an exponential distribution of energy barriers, with kT_0 representing the characteristic decay parameter ($\approx$ spectral width) of the barrier distribution. The time constant τ of the relaxation process is related to the effective barrier height E_{eff} via a Boltzmann term:

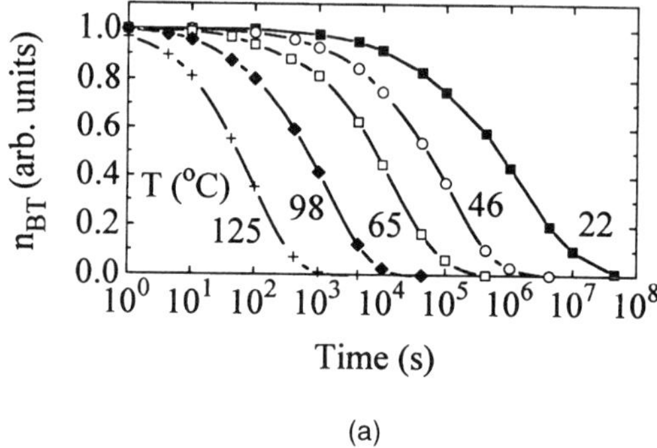

(a)

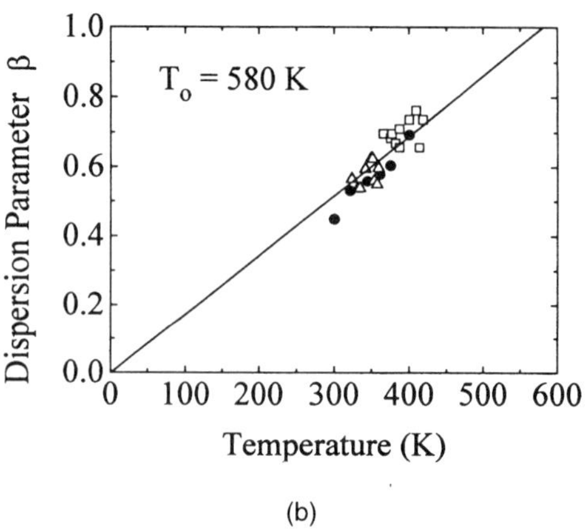

(b)

FIG. 34. (a) Stretched-exponential relaxation of shallow band-tail carriers in *n*-type samples at different annealing temperatures. (b) Temperature dependence of the dispersion parameter β obtained from the relaxation of conductivity in *n*-type and *p*-type samples (open squares and triangles, respectively) and from sweep-out measurements in *n*-type material (solid dots). The solid line indicates the expected behavior for a characteristic temperature T_0 of 580 K.

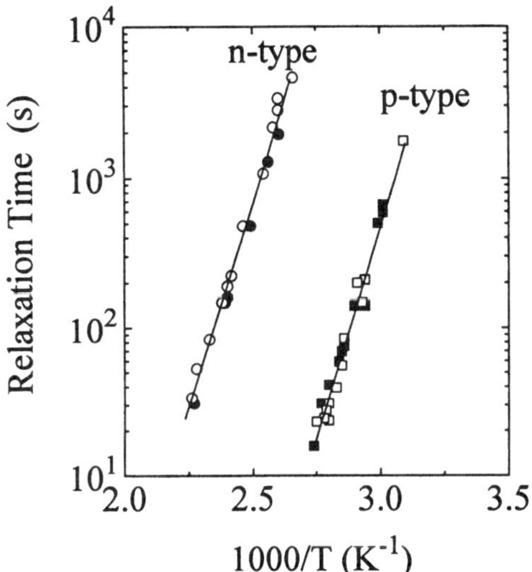

FIG. 35. Arrhenius plot of the relaxation time in doped amorphous silicon samples following rapid quenching.

$$\tau = \tau_0 \exp[E_{eff}/kT].$$

According to Fig. 35, both n- and p-type a-Si:H exhibit a relaxation behavior following thermal quenching, which can be described by an effective barrier of $E_{eff} \approx 1.1$ eV and a prefactor τ_0 between 10^{-12} and 10^{-10} s. Note that these values are very similar to those observed for the annealing of light-induced metastable dangling bonds in undoped a-Si:H described above. It is thus tempting to propose a common structural origin for metastability in both undoped and doped a-Si:H, namely, a reaction between lattice bonds, dangling-bond defects, and dopants that is mediated via the diffusion of hydrogen. Specific processes for doped a-Si:H, which are based on corresponding dopant-hydrogen interactions in crystalline Si, are shown in Fig. 36 (Chang and Chadi, 1988). Thus, substitutional boron is rendered electronically inactive by insertion of a hydrogen atom into one of the four Si–B bonds, while P_4^+ is transformed into the passive state by insertion of H into an antisite of Si ("back bond"). Upon heating, the hydrogen atom can migrate to a nearby weak Si–Si bond, where it becomes trapped by the formation of a Si–H bond and a charged dangling-bond defect. In the same process, the dopant atom becomes activated.

The involvement of hydrogen in a-Si:H metastability phenomena is a recurring theme in the recent literature. As a matter of fact, the participation of hydrogen in structural relaxation of a-Si:H at relatively low temperatures is quite reasonable in view of experimental results concerning hydrogen diffusion in a-Si:H. Hydrogen is found to diffuse in a-Si:H via a dispersive multiple-trapping process with an activation energy of about 1.4 eV and a diffusivity of about 10^{-20}–10^{-18} cm^2 s^{-1} at 400 K. With this diffusivity, hydrogen can move in the amorphous Si network over a distance of up to several tens of angstroms within one hour's time, a distance that is comparable to or less than the average distance between H atoms in the lattice. Thus, it is possible to view the hydrogen atoms as constituting a distinct sublattice in the a-Si:H structure, whose atoms become mobile above a certain temperature T_g. In analogy to network glasses, this temperature T_g has certain characteristics of a typical glass transition temperature: above T_g, the hydrogen atoms are sufficiently mobile to mediate a partial thermal equilibration among particular structural units such

Phosphorus activation

(a) Si H P_3^0

(b) H D^- P_4^+

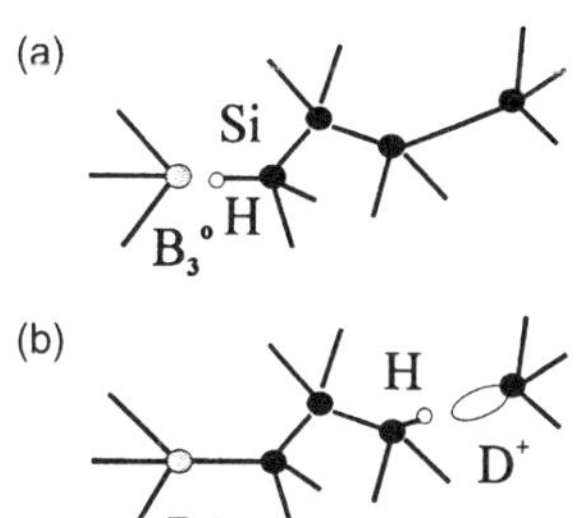

FIG. 36. Microscopic models for thermal activation of dopants via hydrogen migration in phosphorus-doped samples and boron-doped material: (a) describes the dopant passivation by H and (b) the active state of the dopant.

as dopants and defects; below T_g, even the H atoms as the most mobile structural species are frozen into their respective positions, and all relaxation processes come to a halt. In the literature this model is known as the hydrogen glass model (Kakalios and Jackson, 1989). The glass transition temperature T_g is assumed to be equal to the equilibration temperature T_E discussed in connection with Figs. 33 and 34.

3. ALLOYS AND MULTILAYERS

For many applications of a semiconductor it is desirable to have a certain control over the band-gap energy E_g. In the case of *a*-Si:H this is easily achieved by alloying with Ge, C, O, or N. In practice one adds reactive gases, such as GeH_4, CH_4, N_2O, CO_2, NH_3, etc., to the SiH_4 used for preparation of amorphous Si. As already shown in Fig. 32, alloying with nitrogen and oxygen quickly increases the density of structural defects in the material, so that substoichiometric SiO_x and SiN_x alloys play a negligible role for applications. Only wide-gap *a*-SiO_2:H and *a*-Si_3N_4:H close to stoichiometry are used as the dielectric gate insulator in *a*-Si:H thin-film transistors. For smaller variations of the band gap, better results have been obtained with isoelectronic *a*-$SiGe_x$:H and *a*-SiC_x:H alloys. As compiled in Fig. 37, alloying with Ge or C allows a variation of the optical band gap over the range 1 eV $\leq E_g \leq$ 2.5 eV, while maintaining relatively good transport properties, at least in the region 1.5 eV $< E_g <$ 2 eV. The rapid decrease of the photoresponse, σ_{ph}/σ_D, which is evident from Fig. 37 for increasing Ge or C content, is mainly caused by a corresponding increase of deep defects acting as recombination centers. Thus, the dangling-bond density increases from less than 10^{16} cm^{-3} to more than 10^{17} cm^{-3} when going from *a*-Si:H to *a*-Ge:H, and a similar trend holds for alloying with carbon. In addition, the electron mobility in *a*-$SiGe_x$:H and *a*-SiC_x:H is reduced from its value in *a*-Si:H because of additional carrier scattering by chemical disorder [cluster formation, preferential hydrogen bonding to Si or C, etc. (Wang and Schiff, 1993)]. According to Fig. 38, at the same time the hole mobility remains essentially unchanged, indicating that isoelectronic alloying with Ge and C mainly affects the density-of-states distribution close to the conduction-band mobility edge (alloy broadening).

Whereas the optoelectronic properties of pure hydrogenated amorphous silicon have been more or less optimized in the course of the last 20 years, the same is obviously not the case for all possible *a*-Si:H–related alloys. Mixing of reactive gases in a glow-discharge reactor under otherwise unchanged deposition conditions is only the zero-order approach to a very complex problem. In particular, the work on *a*-Ge:H and *a*-C:H has shown that optimized deposition conditions for these amorphous semiconductors are quite different from those used for *a*-Si:H deposition, leaving a lot of room for the improvement of *a*-Si:H–based alloy materials. The need for variable–band-gap material in amorphous silicon solar cells will certainly motivate future work in this interesting area.

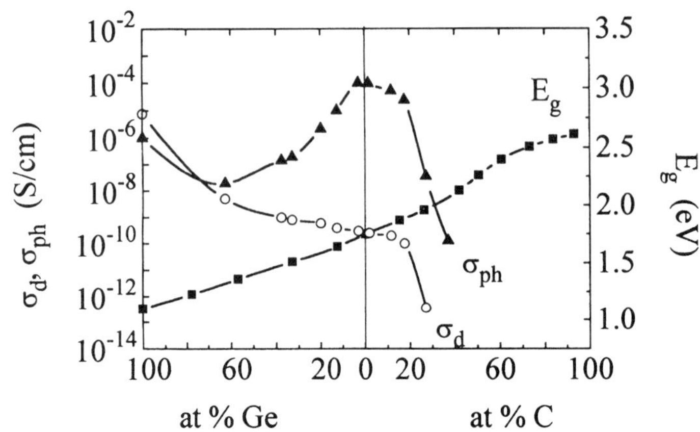

FIG. 37. Dark conductivity (circles), photoconductivity (triangles), and optical band gap (squares, E_g), in silicon–germanium and silicon–carbon alloys as a function of alloy composition.

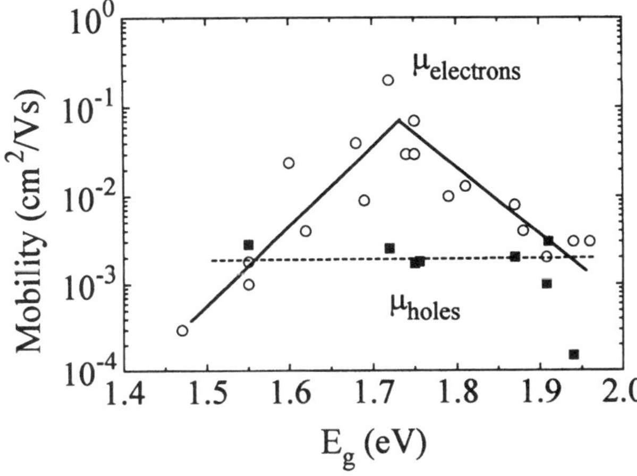

FIG. 38. Dependence of electron and hole mobilities in silicon-related amorphous alloys on the optical band gap, E_g.

In addition to alloying, attempts have also been made to influence the optical and electronic properties of *a*-Si:H–based thin films via the preparation of compositional or doping multilayers. In the field of amorphous semiconductors, the term "multilayer" is preferred over the term "superlattice," since there are no periodic lattices to begin with. Because the deposition of *a*-Si:H and related materials by plasma-enhanced CVD proceeds with a growth rate of typically 1 Å/s, it is fairly easy to produce atomically abrupt interfaces by a corresponding temporal modulation of the reactive-gas partial pressures in the deposition reactor. Indeed, multilayers with periods between 5 and 500 Å and satisfactory long-range periodicity have been routinely produced in this fashion.

The quality of the multilayer structure can be monitored either by direct high-resolution microscopy or by scattering experiments. In the case of compositional superlattices (*a*-Si:H alternating with *a*-$SiGe_x$:H, *a*-SiN_x:H, *a*-SiC_x:H, etc.), combined microscopy and x-ray scattering investigations have shown a high multilayer quality in terms of atomically abrupt interfaces, good periodicity in growth direction, and lateral homogeneity.

Direct evidence for superlattice effects in periodic compositional *a*-Si:H/*a*-SiN_x:H multilayers also comes from Raman-scattering experiments. As a result of the artificial periodicity of the multilayer, the phonon dispersion of long-wavelength acoustic phonons with k vectors parallel to the growth direction is altered by folding back of the approximately linear phonon branches at the superlattice Brillouin-zone boundary $k_z = \pi/L$, where L is the period length of the multilayer. As a consequence, k conservation ($\Delta k \approx 0$) in Raman backscattering is fulfilled not only by the long-wavelength optical phonons near the Brillouin-zone center but also by a ladder of back-folded acoustic-phonon states, ω_n and ω_{-n}, where n indicates the order of folding and the sign describes branches with positive or negative dispersion, respectively ($\partial\omega/\partial k > 0$ or < 0). In the limit of a linear dispersion relation for long-wavelength acoustic phonons, one has $\omega_n \approx \omega_{-n} \approx 2n\nu\pi/L$ (ν is sound velocity), so that the Raman-active $k \approx 0$ acoustic-phonon frequencies should be proportional to $1/L$. The experimental data in Fig. 39 demonstrate that this is indeed the case for $n = 1$.

In contrast to the obvious effects of multilayer formation on the phonon properties of *a*-Si:H–based systems, there is little convincing evidence for pronounced quantization phenomena in the electronic properties of the same kind of multilayers. For example, confinement of electrons and holes in the *a*-Si:H layer of *a*-Si:H/*a*-Si_3N_4:H superlattices should lead to the same quantization condition,

$$E_n = \frac{h^2}{2m^*}\left(\frac{n\pi}{L}\right)^2,$$

for the electron or hole levels as in crystalline quantum wells. Corresponding experimental results in Fig. 40 indeed exhibit a

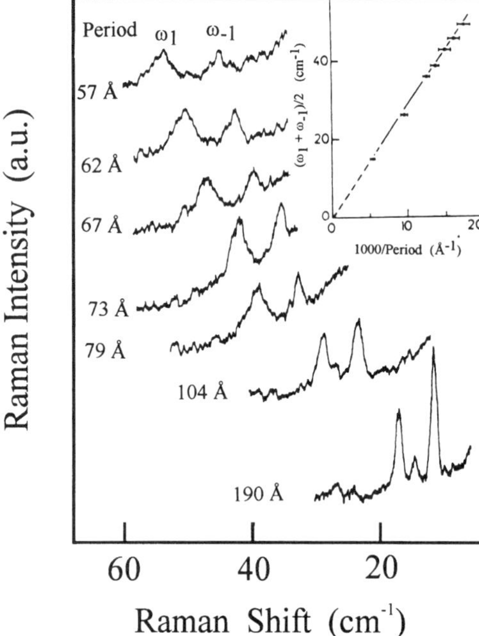

FIG. 39. Raman spectra of a-Si/Si_3N_4:H multilayers with different periods, showing the first and second phonon peaks that become Raman active through back-folding of the acoustic-phonon branches. Inset: The expected linear dependence on inverse layer period. (After Santos *et al.*, 1986.)

trend toward higher band gaps with decreasing L.

On the other hand, the absorption spectra in Fig. 40 remain essentially featureless even for very small well widths, in contrast to the expected steplike density-of-states distribution for two-dimensional systems. It has thus also been proposed that a considerable contribution to the band-gap opening in short-period amorphous multilayers is due to alloying effects and other changes in the a-Si:H–layer composition.

Evidence for quantum confinement effects in short-period a-Si:H–based multilayers and heterostructures has also been claimed on the basis of other experiments, such as resonant tunneling in double-barrier structures or thermally modulated absorption. So far, many of the results claiming evidence for quantum confinement effects lead to conflicting results concerning the effective mass m^* of confined carriers [$m^* \approx (0.3\text{–}0.6)m_0$ for electrons, $m^* \approx m_0$ for holes]. Also, conceptually it is difficult to reconcile quantum confinement effects in 50-Å-period multilayers with the mean free path for carriers in a-Si:H of $\approx$10 Å. Quantum effects require a complete phase coherence of the confined wave function over the entire well width. In a-Si:H this coherence is quickly lost by the strong disorder scattering. Nevertheless, it may be possible that a small fraction of confined carriers experiences less disorder scattering and can actually feel the influence of confining potential barriers.

The second type of multilayers comprises the doping superlattices (n-i-n-i, n-i-p-i, etc.), which are obtained by a periodic modulation of the degree and/or the type of doping during growth. The energy-band diagram of such an n-i-p-i structure in its equilibrium

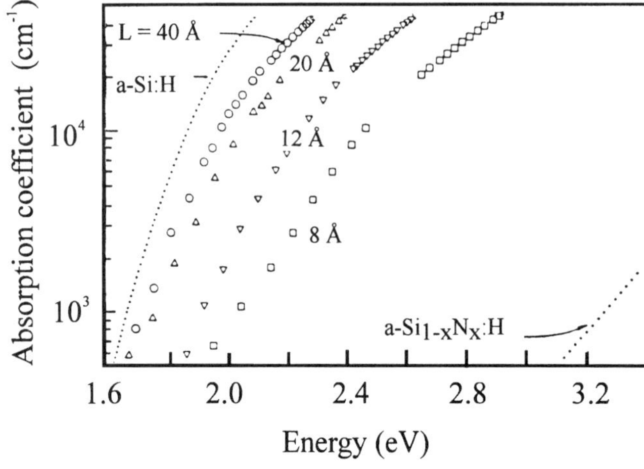

FIG. 40. Absorption coefficients of a-Si/Si_3N_4:H multilayers with small periods L between 40 and 8 Å (after Abeles and Tiedje, 1983).

state is shown in the top part of Fig. 41 by the solid lines. The periodic alternation between n- and p-type layers leads to a corresponding modulation of the band structure in growth direction due to the transfer of electrons from n to p layers. Upon illumination of such a *nipi* structure, photoexcited carriers are separated by the internal fields. Electrons and holes then accumulate in their corresponding majority-carrier regions (electron in n layers, holes in p layers), causing a flattening of the potential modulation as indicated by the dashed curves in the top part of Fig. 41. Recombination between photoexcited carriers is thus hindered by the spatial separation between electrons and holes, leading to a phenomenon known as persistent

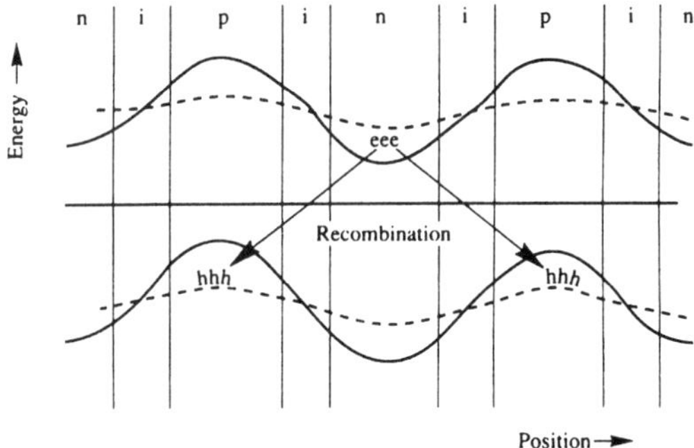

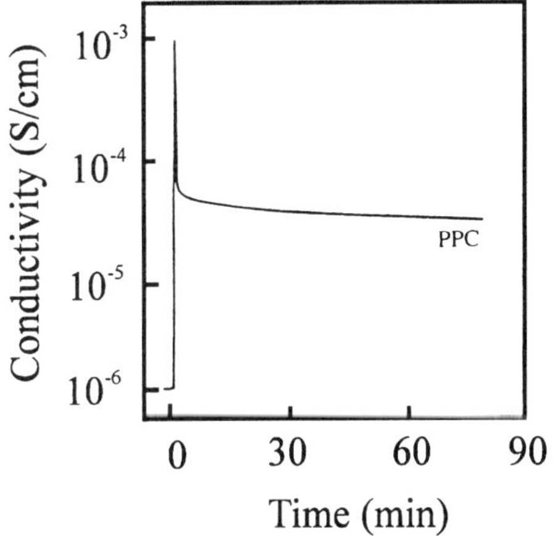

FIG. 41. Schematic band diagram of a *nipi* structure in thermal equilibrium (solid lines) and under illumination (dashed lines). The lower part shows the persistent photoconductivity in such a structure following a short light pulse. (After Kakalios, 1986.)

photoconductivity: because of the increased recombination time of spatially separated electron and hole populations, the photoconductivity perpendicular to the modulation direction is strongly enhanced and may even persist for very long times in long-period multilayers at low temperatures. An example of such a persistent photoconductivity in an a-Si:H doping multilayer is given in the lower part of Fig. 41. Following optical excitation at $t = 0$, the photoconductivity quickly decays at short times but then reaches an almost constant level, far above the dark conductivity, that persists for minutes to hours, depending on temperature and sample structure. In fact, the extremely long persistence of elevated conductivities in a-Si:H doping superlattices has caused a lot of doubt on the applicability of the standard persistent-photoconductivity model outlined above. Even in the absence of thermal excitation, carriers in adjacent n and p layers should recombine rather rapidly via tunneling. An alternative model for the extremely persisting conductivity in a-Si:H *nipi* structures is based on the pronounced structural metastability phenomena in this material (cf. Sec. 2.5). It is known from investigations of bulk doped a-Si:H that illumination leads to a light-induced metastable activation of dopant atoms that, however, has different efficiencies in n- and p-type samples. As a consequence, illumination of a-Si:H doping multilayer structures will cause a different activation of dopants in n and p layers, a situation that may resemble a persistent photoconductivity. In addition, light-induced creation of metastable dangling bonds is known to be differently efficient in n- versus p-type a-Si:H, an effect that may also contribute to a persistent conductivity increase after illumination.

4. DEVICES

4.1 Overview

The importance of state-of-the-art hydrogenated amorphous silicon for device applications is based on a number of advantages compared to other semiconductors. The most relevant advantages are the following:

- low-temperature processing ($\leq$400 °C)

- easy and uniform preparation over large areas
- electronic quality independent of substrate material
- compatibility with crystalline Si technology
- low defect density (comparable to bulk crystalline GaAs)
- high absorption coefficient
- easy transformation in crystalline state with local selectivity
- easy alloying with other atoms for bandgap variation

Of course there are also a number of deficiencies, such as

- low structural stability under electronic excitation
- low doping efficiency
- low carrier mobility

Nevertheless, amorphous Si has been used successfully in a number of electronic devices listed below and partly described in more detail in the following sections. Among *a*-Si:H–based devices one notably finds

- solar cells, photodetectors
- thin-film transistor arrays
- spatial light modulators
- photocopiers, image sensors, scanners
- large-area light-emitting diodes
- sensors
- memory circuits

and others. A good review of these applications can be found in the books edited by Pankove (1984) and Kanicki (1992). Here, we only discuss briefly the first two kinds of devices, namely, solar cells and thin-film transistors, which at present dominate commercially available *a*-Si:H products.

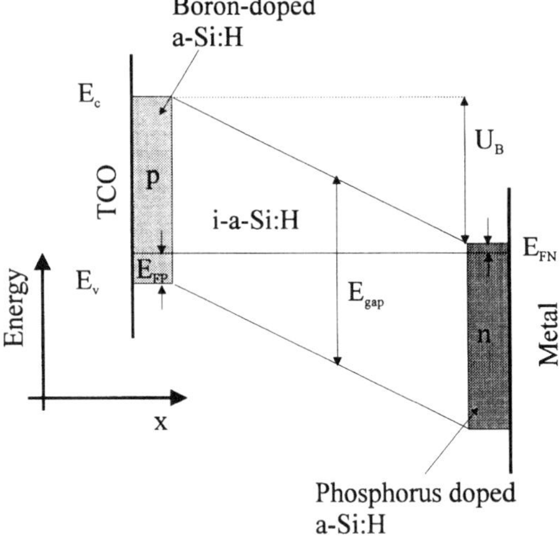

(a)

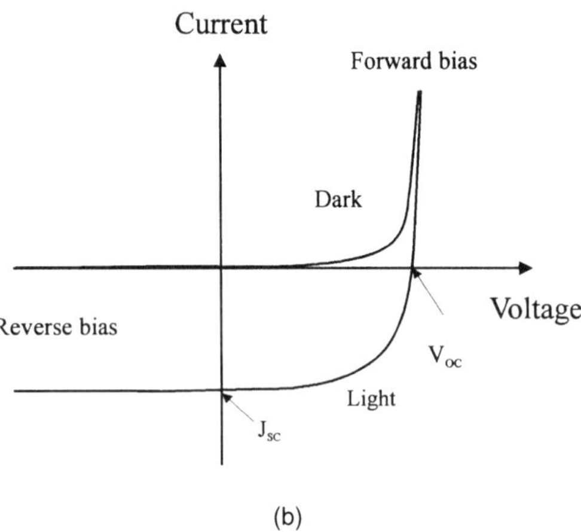

(b)

FIG. 42. (a) Energy band diagram of an *a*-Si:H *p-i-n* photosensor (E_{gap}: optical gap; E_{FP} and E_{FN}: Fermi level of holes and electrons in the *p*-type and *n*-type contact layers; U_B: built-in voltage). (b) Current-voltage characteristics of the *p-i-n* diode in (a) under dark and light conditions (I_{SC}: short circuit current; V_{OC}: open circuit voltage).

4.2 Solar Cells and Photodetectors

The main advantages of amorphous silicon as a solar-cell material are the high optical absorption coefficients in the visible spectral range, allowing the use of thin-film technology (cf. Fig. 8), and the possibility to prepare such films on large-area, low-cost substrates and at relatively low temperatures (see PHOTOVOLTAIC DEVICES; SOLAR ENERGY). A typical solar-cell structure used today is shown in Fig. 42(a). Light is absorbed in an intrinsic (undoped) *a*-Si:H layer of thickness 200–800 nm. The field necessary to separate photoexcited electrons and holes is generated by thin, highly doped *n*- and *p*-type layers (≈10 nm). The whole structure is referred to as a *p-i-n* diode. Electrical contact to the diode is made, for example, by a transparent conducting oxide (TCO) on the *p* side and by a metal such as Al or Ag on the *n* side, which also serves as a back-reflector to increase the effective optical path lengths for weakly absorbed light. In addition, surface texture of the substrate material or the TCO on the micron scale is used for light trapping in the solar-cell device. As substrate material, low-cost lime glass is generally employed, but cells deposited on stainless steel foil or plastic substrates have also been developed.

Because holes in *a*-Si:H have a much lower room-temperature mobility than electrons (see Fig. 12), it is necessary to introduce the light via the *p*-type contact, in order to minimize the distance holes have to drift in the built-in field toward the *p* contact. In more advanced cell designs, one also replaces the *p*-type *a*-Si:H layer by *p*-type *a*-SiC:H alloy layers with a wider optical gap. This decreases the absorption of blue light in the *p* contact. At the same time, the conduction-band offset between *a*-SiC:H and *a*-Si:H serves as a back-diffusion barrier, preventing electrons from recombining in the *p* contact. Further improvement of carrier collection can be achieved by using microcrystalline Si with a higher conductivity than *a*-Si:H as the *n*-type contact.

The electrical characteristics of a *p-i-n* solar-cell structure can be easily understood from the *I-U* curves shown in Fig. 42(b). In the dark, the solar cell behaves like a diode, with a low thermal or leakage current in reverse bias, and a high, exponentially increasing injection current in forward bias. Under illumination, the reverse-bias current increases by the amount of photoexcited and collected carriers. For an incident solar light intensity of 100 mW/cm^2, good *a*-Si:H cells obtain short-circuit currents of $I_{SC} \approx 17$ mA/cm^2. Typical values for the open-circuit voltage V_{OC} in *a*-Si:H solar cells fall between 0.85 and 0.95 V.

For a single *p-i-n* device as shown in Fig. 42(a), the theoretical limit for the conversion efficiency has been estimated to be about 14–15%. In state-of-the-art cells, values of $\eta \approx 12\%$ have been reached, at least in small prototype cells with a total area of about 1 cm^2.

For commercially produced *a*-Si:H solar cells, the simple *p-i-n* structure of Fig. 42(a) has been abandoned in favor of a more complex tandem or triple stack design, where two or three *p-i-n* devices are deposited on top of each other. This has mainly two reasons: For one, stacked cells allow the combination of two or three absorbing layers with different band gaps. Variation of the band gap is achieved by alloying with carbon or germanium (see Sec. 3). The advantage of variable–band-gap stacked cells is a much larger maximum conversion efficiency than for single cells. Estimates indicate that ultimately as much as 25% efficiency could be achieved. An example of such a tandem cell is shown (inverted) in Fig. 43. The device consists of a thin top cell made out of *a*-Si:H and a thicker bottom cell made out of a suitable amorphous SiGe alloy. High-energy photons are absorbed in the top cell, whereas lower-energy photons are absorbed in the bottom cell, which has a smaller optical band gap. The main design parameters of such tandem arrangements are the band gaps and the thicknesses of the two absorber layers. An optimized two-terminal structure has to fulfill the condition of current matching; i.e., the current generated by the top cell has to be equal to that generated by the bottom cell. This is generally not possible for all operation situations encountered in real applications. Another possibility is to isolate top and bottom cells from each other electrically (four-terminal device), but this is too expensive for large-area power generation. On the other hand, four-terminal tandem structures have been used as sensitive color sensors.

A second reason for the preference given to tandem structures compared to single *p-i-n* diodes is the light-induced degradation

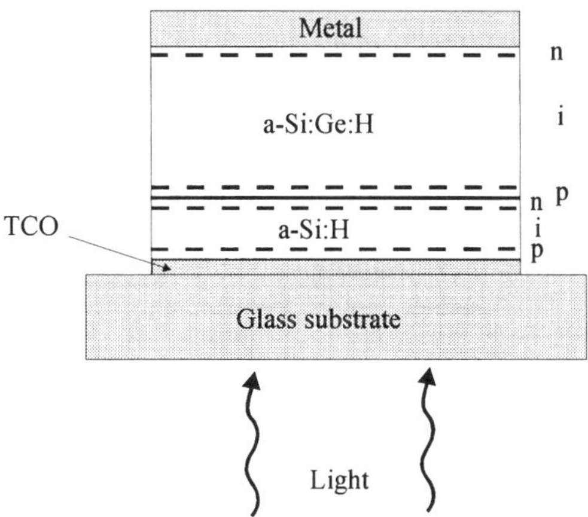

FIG. 43. Layer structure of an *a*-Si:H/*a*-SiGe:H tandem solar cell.

of *a*-Si:H solar cells, which can be linked directly to the light-induced metastable defect creation discussed in Sec. 2.5. The creation of additional defects in the absorber layer causes an increased recombination of photogenerated carriers and a change of internal electric fields. As a result, all three device parameters (U_{OC}, I_{SC}, and *FF*) are decreased during degradation (see Fig. 44), thus strongly affecting the conversion efficiency of the cell. The light-induced degradation of *a*-Si:H solar cells is at present the main hindrance for large-scale applications of such devices, since it decreases the cell efficiency from about 12% to about 8% after long-term exposure to intense sunlight. Since according to Sec. 2.5 the main reason for metastable defect creation in undoped *a*-Si:H is direct nonradiative recombination of photoexcited carriers, photoinduced degradation is particularly severe in devices with a bad collection efficiency. The efficiency of carrier collection is low in regions of the absorber layer with a low electric field or further away from the nearest contact layer than the typical carrier drift length, $\mu\tau F$. As a consequence, cell degradation is much more pronounced in thick cells (*i*-layer thickness > 500 nm) than in thin cells, since in the latter the field $F \approx U_B/d$ (U_B: built-in voltage; *d*: layer thickness) can be maintained at a sufficiently high value to prevent carrier recombination. Thus, a practical way to produce relatively stable solar cells is to divide the total absorber thickness into two separate layers of a *p-i-n-p-i-n* tandem structure. In this way, a compromise between good overall absorbance and good stability can be achieved. Such tandem cells have produced a conversion efficiency around 10% even in their degraded state. Similar efficiencies have also been demonstrated for large-area *a*-Si:H cells (0.1–0.5 m^2), whereas present mass-production modules have lower efficiencies of about 6–8%.

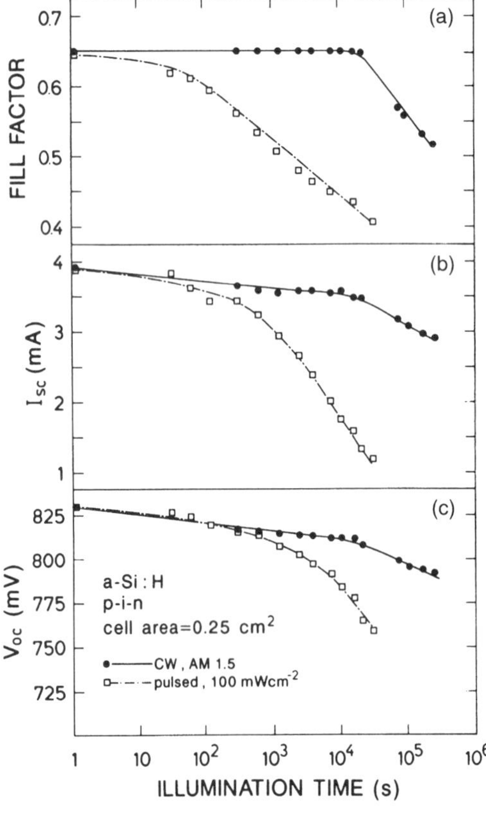

FIG. 44. Light-induced degradation of the main solar-cell parameters for a *p-i-n* cell exposed to 100 mW/cm^2 white light irradiation. Open symbols show accelerated degradation of the same cell using pulsed light from a flash lamp.

The same kind of *p-i-n* structure employed in *a*-Si:H solar cells is also used in various kinds of photodetectors or radiation detectors. Depending on the application desired, the *p-i-n* diodes are either laterally structured as small pixels with dimensions of about 50 × 50 μm^2 for high-resolution scanners or imaging arrays or produced with much larger *i*-layer thicknesses (≈50 μm) for direct detection of ionizing radiation.

4.3 Field-Effect Transistors

In addition to applications as a photodetector, *a*-Si:H is widely used for active thin-film transistor (TFT) devices (see TRANSISTORS, FIELD EFFECT), in particular in the form of active-matrix liquid-crystal displays for portable computers. The basic structure of an *a*-Si:H TFT is shown in Fig. 45. The active layer consists of a thin (<100 nm) undoped *a*-Si:H layer, which is contacted via Ohmic metal/n^+-doped *a*-Si:H layers to form the source and drain contacts. The active layer is isolated from the metal gate by a suitable *a*-Si_3N_4:H or *a*-SiO_2:H dielectric layer produced by plasma-CVD. TFT structures used today have a channel length (source–drain distance) between 10 and 100

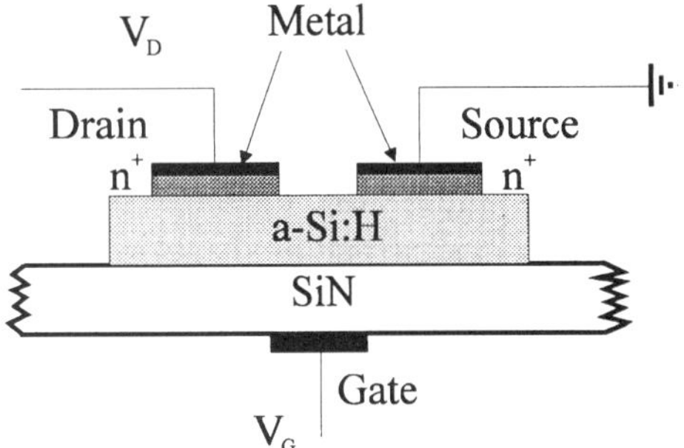

FIG. 45. Schematic design of an *a*-Si:H thin-film field-effect transistor. n^+ indicates heavily phosphorus-doped contact layers; SiN is the silicon–nitride dielectric layer.

μm, depending on the switching speed and maximum source–drain current desired. To a good approximation, the source–drain current can be written as

$$I_{SD} = \mu C_G \frac{W}{L} U_{DS}(U_G - U_{thr} - U_{DS}),$$

where μ is the carrier mobility in the channel, C_G the gate capacitance per unit area, W the gate width, and U_{thr} a threshold voltage, which for *a*-Si:H has a value of approximately 1 V. Typical source–drain currents of *a*-Si:H TFTs are 10^{-11} Å in the "off" state ($U_G = 0$) and 10^{-5} Å in the "on" state ($U_G = 10$ V), both for a source–drain voltage $U_{DS} = 10$ V.

The combination of many such small TFT structures to form a sensor or display array is shown in Fig. 46. Each sensor element or display pixel is addressed by its own TFT. The gate and data (drain) contacts of the TFTs are patterned as indicated schematically in Fig. 46(a) in the form of a gate- and data-line matrix. For displays, the TFTs are turned on to apply a voltage bias to a liquid-crystal element acting as a capacitor in the electric circuit. The TFT has to be able to supply sufficient charge to the electrodes of the liquid-crystal pixel during one frame time to turn the pixel on or off. In sensor arrays, the sensor element (e.g., an *a*-Si:H *p-i-n* or Schottky photodiode) acts as a variable light-dependent resistor that is connected to an external current sensor by turning the TFT on. A micrograph of such a sensor array is shown in Fig. 46(b). Modern TFT array structures for applications as TV screens or computer displays contain typically a million TFTs on an area of several hundred square centimeters and are presently produced at a level of about 100 000 displays per year. Both the size and the production volume of such displays are expected to experience considerable growth in the near future.

5. SUMMARY AND OUTLOOK

In addition to monocrystalline, multicrystalline, microcrystalline, and more recently porous Si, amorphous silicon is a modifica-

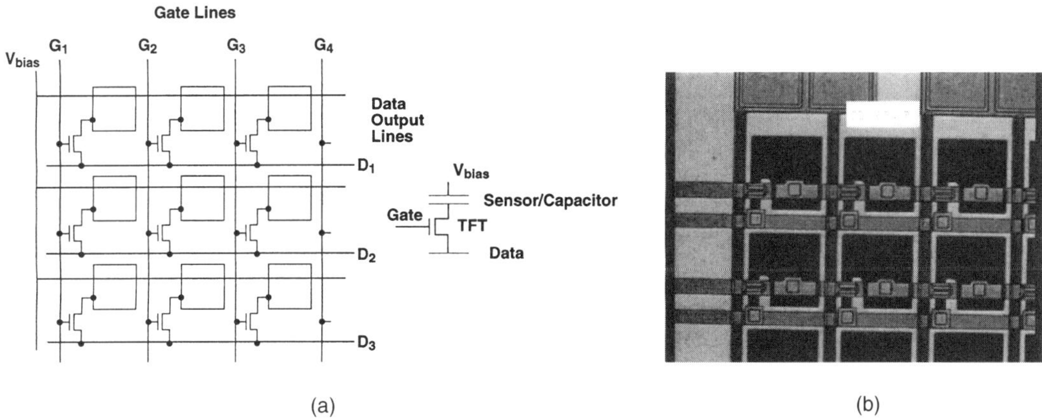

FIG. 46. (a) TFT matrix array used for liquid-crystal displays and for two-dimensional high-resolution image sensors. (b) Electron micrograph of a real device using the basic design of (a). The pixel size is 100 × 100 μm^2.

tion of solid silicon with interesting properties for application as an electronic material. The complete absence of long-range order in amorphous silicon gives rise to a number of specific properties that are qualitatively different from those of any other form of Si. This is particularly noticeable in the optical absorption of visible light, which is stronger than in all other forms of silicon and justifies the interest in *a*-Si:H thin-film photosensors and solar cells. Amorphous Si can be prepared with low defect densities over large areas and on many different types of substrates, which is a prerequisite for most present applications. Despite its large-area capability, amorphous Si can be structured by conventional lithographic techniques, thus providing a unique possibility to combine macroelectronics and microelectronics. Of similar importance for applications is the fact that *a*-Si:H can be doped substitutionally both *n* and *p* type and that the band gap of *a*-Si:H can be tuned over a large range by alloying with Ge, C, O, or N. There are, however, also a number of negative features that very often reduce the potential of *a*-Si:H for high-performance devices. In particular, the low carrier mobility, the low doping efficiency, and the various forms of metastability pose severe limitations on present applications. In many cases it would therefore be desirable to combine the good properties of amorphous Si with those of crystalline Si. Activities in this direction are presently pursued by a number of groups, e.g., using laser recrystallization, preferential etching, or plasma deposition of mixed amorphous–microcrystalline Si thin films.

GLOSSARY

***a*-Si:** Amorphous silicon.

***a*-Si:H:** Hydrogenated amorphous silicon.

***a*-SiGe:H:** Hydrogenated amorphous silicon alloyed with Ge.

***a*-SiC:H:** Hydrogenated amorphous silicon alloyed with C.

***a*-SiN:H:** Hydrogenated amorphous silicon alloyed with N.

***a*-SiO:H:** Hydrogenated amorphous silicon alloyed with O.

***a*-Ge:H:** Hydrogenated amorphous germanium.

***a*-C:H:** Hydrogenated amorphous carbon.

Band-Tail States: Localized states at the band edge of the conduction and valence band.

CB: Conduction band.

Correlation Energy: Relative energy position of the singly and doubly occupied charge states of a defect.

CPM: Constant photocurrent method.

DB: Dangling bond.

Demarcation Energy: Energy that separates states in thermal equilibrium with band states from states that are not yet equilibrated.

Dispersive Transport: Decreasing mobility of carriers with increasing time due to interaction with localized states.

DOS: Density-of-states distribution.

ESR: Electron spin resonance.

Geminate Recombination: Recombination of an electron–hole pair generated by the same photon.

Hyperfine Structure: Splitting of the ESR spectrum due to interaction between electronic spins and nuclear magnetic moments.

IR Absorption: Infrared absorption.

Light Soaking: Prolonged exposure to intense light.

LO(A): Longitudinal optical (acoustic) phonon.

LVM: Local vibrational mode.

Metastability: Spontaneous reversible creation of additional defects and/or dopants under strong electronic excitation.

PDS: Photothermal deflection spectroscopy.

PECVD: Plasma-enhanced chemical vapor deposition.

PL: Photoluminescence.

Photo-CVD: Optical excitation of reactive gas.

p-i-n: Boron-doped (*p*), intrinsic (*i*), and phosphorus (*n*) layer sequence.

SDPC: Spin-dependent photoconductivity.

Staebler–Wronski Effect: Light-induced degradation.

Statistical Shift: Temperature-induced shift of the Fermi level.

TCO: Transparent conductive oxide.

TFT: Thin-film transistor.

Thermalization: Energy relaxation of excited carriers.

Trapping: Capture of electrons or holes into localized band-tail states.

TO(A): Transverse optical (acoustic) phonon.

Urbach Tail: Exponential absorption tail below the band-gap energy.

Urbach Slope: Characteristic energy of the exponential absorption.

Van de Pauw Contact Geometry: Ohmic contacts on the perimeter of a flat sample of arbitrary shape.

VB: Valence band.

Works Cited

Abeles, B., Tiedje, T. (1983), *Phys. Rev. Lett.* **51,** 2003–2007.

Beyer, W. (1985), in: D. Adler, H. Fritzsche (Eds.), *Tetrahedrally Bonded Amorphous Semiconductors,* New York: Plenum, p. 129.

Brandt, M. S., Stutzmann, M. (1991), *Phys. Rev. B* **43,** 5184–5187.

Cardona, M. (1983), *Phys. Status Solidi B* **118,** 463–481.

Chang, K. J., Chadi, D. J. (1988), *Phys. Rev. Lett.* **60,** 1422–1425.

Dersch, H., Schweitzer, L., Stuke, J. (1983), *Phys. Rev. B* **28,** 4678–4687.

Kakalios, J. (1986), *Phil. Mag. B* **54,** 199–218.

Kakalios, J., Jackson, W. B. (1989), in: H. Fritzsche (Ed.), *Amorphous Silicon and Related Materials,* Singapore: World Scientific, p. 207.

Kanicki, J. (Ed.) (1992), *Amorphous and Microcrystalline Semiconductor Devices,* Boston: Artech.

Lannin, J. S. (1984), *Semicond. Semimetals* **21B,** 159–195.

Ley, L. (1984), in: J. D. Joannopoulos, G. Lucovsky (Eds.), *The Physics of Hydrogenated Amorphous Silicon II,* Berlin: Springer, p. 61.

Marianer, S., Shklovskii, B. I. (1992), *Phys. Rev. B* **46,** 13100–13103.

Mott, N. F. (1987), *J. Phys. C* **20,** 3075–3102.

Nebel, C. E., Street, R. A. (1993), *Int. J. Mod. Phys. B* **7,** 1207–1258.

Overhof, H., Thomas, P. (1989), *Electronic Transport in Hydrogenated Amorphous Semiconductors,* Berlin: Springer.

Pankove, J. I. (Ed.) (1984), *Hydrogenated Amorphous Silicon, Part D: Device Applications,* Semicond. & Semimetals 21D, New York: Academic Press.

Santos, P., Hundhausen, M., Ley, L. (1986), *Phys. Rev. B* **33,** 1516–1518.

Spear, W. E. (1977), *Adv. Phys.* **26,** 811–845.

Staebler, D. L., Wronski, C. R. (1980), *J. Appl. Phys.* **51,** 3262–3268.

Street, R. A. (1984), *Semicond. Semimetals* **21B,** 197–244.

Street, R. A. (1985), *J. Non-Cryst. Solids* **77 & 78,** 1–16.

Street, R. A., Knights, J. C., Biegelsen, D. K. (1978), *Phys. Rev. B* **18,** 1880.

Stutzmann, M., Biegelsen, D. K., Street, R. A. (1987), *Phys. Rev. B* **35,** 5666–5701.

Stutzmann, M. (1992), in: J. Kanicki (Ed.), *Amorphous & Microcrystalline Semiconductor Devices* Vol. II, Boston: Artech House, p. 129.

Stutzmann, M., Biegelsen, D. K. (1988), in: H. Fritzsche (Ed.), *Amorphous Silicon and Related Materials,* Singapore: World Scientific, p. 557.

Tauc, J. (1974), in: J. Tauc (Ed.), *Amorphous and Liquid Semiconductors,* Berlin: Springer, p. 225.

Tiedje, T. (1984), in: J. D. Joannopoulos, G. Lucovsky (Eds.), *The Physics of Hydrogenated Amorphous Silicon II,* Berlin: Springer, p. 261.

Wang, Q., Schiff, W. A. (1993), *Mater. Res. Soc. Symp. Proc.* **297,** 419–425.

Further Reading

Mott, N. F., Davis, E. A. (1979), *Electronic Processes in Noncrystalline Materials,* Oxford: University Press.

Street, R. A. (1991), *Hydrogenated Amorphous Silicon,* Cambridge: University Press.

SILICON, CRYSTALLINE

SOKRATES T. PANTELIDES, *Department of Physics and Astronomy, Vanderbilt University, Nashville, Tennessee, U.S.A.*

INTRODUCTION

Silicon is one of the most abundant elements in nature. After oxygen, it is the most abundant element making up the earth's crust (28% by weight) in the form of various compounds. Sand is essentially pure silicon dioxide. Lava from volcanoes consists primarily of molten silicon dioxide. Silicon is also found in many animals and plants. The element silicon (Si, atomic number 14, atomic weight 28.0855) was first isolated by the Swedish chemist Baron Jöns Jacob Berzelius in 1824. In its pure form it is a crystalline solid with a metallic sheen. It also exists in an amorphous form. The melting point of crystalline Si is 1410 °C, and its boiling point is 2355 °C. At 20 °C crystalline Si has a density of 2.33 g/cm^3.

Silicon compounds have long been used in many glasses and ceramics and as ingredients in steel and other alloys. Si is also the main element in synthetic compounds known as silicones, which are a cross between organic and inorganic materials and are used as insulators, lubricants, water repellents, and implants in the human body. Above all, however, *crystalline silicon* is the heart and soul of modern microelectronics that, in turn, is the heart and soul of modern computers, the revolution in communications, information processing, and automation in factories, offices, and homes. This Encyclopedia contains articles on many aspects of silicon-based technology. It also contains articles on amorphous silicon and on several silicon-based compounds and structures. The present article focuses exclusively on crystalline silicon and its physical properties.

Crystalline silicon is a semiconductor, a class of materials that falls somewhere between metals, which have high electrical conductivity, and insulators, which do not conduct electricity. Semiconductors have

3-527-28140-1/96/$5.00 + .50

unique electrical properties that can be customized by the controlled incorporation of certain kinds of impurities.

The electrical properties of crystals were explained by quantum mechanics in the early 1930s. Electrons occupy energy levels that form wide bands. In a metal, the highest-energy band containing electrons is only partially filled, up to an energy called the Fermi energy. Externally applied electric fields can set up electrical currents by pushing electrons near the Fermi energy to adjacent empty energy levels. In insulators, on the other hand, the highest-energy band containing electrons, known as the valence band, is completely filled at $T = 0$ K (absolute zero of the temperature scale) and separated from the next band, known as the conduction band, by a forbidden gap. No conduction is possible when all bands are either completely full or completely empty simply because no energy can be imparted to the electronic system by applying moderate voltages. At finite temperatures some electrons are excited from the valence band into the conduction band so that some conduction is possible. For true insulators at room temperature, such conduction is negligible.

Semiconductors are essentially insulators with fairly small forbidden energy gaps. In pure crystalline forms, semiconductors would be fairly useless materials because they are poor insulators (a significant number of electrons are excited in the conduction band at room temperature) and poor conductors (the number of electrons in the conduction band at room temperature is not large enough to yield conductivity comparable to that of metals).

The technological significance of semiconductors arises primarily from two reasons:

1. Unlike metals, semiconductors exhibit two distinct types of conductivity, namely, by electrons in the conduction band (generally known as *n* type for the fact that electrons carry a negative charge) and by "holes" in the valence band (generally known as *p* type because the holes effectively carry a positive charge).
2. The concentration of electrons in the conduction band and holes in the valence band can be controlled via the introduction of certain types of impurities known as dopants. In this manner, the conductivity can be dominantly *n* type or dominantly *p* type and can be varied over more than ten orders of magnitude.

More importantly, by juxtaposing *n*-type and *p*-type material, one can form a *pn* junction with unique electrical properties. Such junctions act as rectifiers for ac current and were indeed the first solid-state electrical devices, before World War II. The first transistor, invented in 1948 by John Bardeen, Walter Brattain, and William Shockley (Bardeen and Brattain, 1948; Shockley, 1949), consisted of two such junctions. Similarly, semiconductor–metal and semiconductor–insulator interfaces have unique electrical properties. Virtually all microelectronic devices that have been developed since the original transistor exploit the unique electrical properties of *pn* junctions and of semiconductor–metal and semiconductor–insulator interfaces.

The first transistor was fabricated from germanium. Soon after that, however, silicon replaced Ge as the semiconductor of choice. The primary reasons are that Si can be grown in large quantities with much better control than any other semiconductor and that Si has a native oxide (SiO_2) that is stable and has other unique properties, e.g., the quality of the Si–SiO_2 interface. Currently the global sales of Si for use in microelectronics approach \$10 $\times$ 10^9 annually, fueling microelectronics, the world's largest industry, whose global revenues are approaching the 10^{12}-dollar level annually. About 5000 metric tons of single-crystal Si is currently consumed annually.

This article reviews the properties and uses of crystalline Si, with emphasis on its electronic properties and its use in microelectronics. In the first part the article covers atomic arrangements in perfect bulk crystalline Si, at its free surfaces, and at bulk defects. It then describes the vibrational, electronic, transport, thermal, and mechanical properties of Si and finishes with the methods of bulk-crystal and thin-film growth and basic processing steps used in microelectronics.

1. ATOMIC ARRANGEMENTS

1.1 The Bulk Perfect Crystal

In crystalline Si the atoms are arranged in a diamond lattice. This is a face-centered cubic (fcc) lattice, one of the 14 Bravais lattices that are allowed by symmetry considerations in three dimensions, with a basis of two atoms (Burns, 1985). If the origin of coordinates is taken at one of the Si atoms occupying one of the fcc lattice sites, the second atom in the primitive unit cell occupies a site at $(1,1,1)a/4$, where a is the fcc cube edge. For Si, $a = 0.543$ nm. There are eight atoms per primitive cube. Each atom has four nearest neighbors in tetrahedral directions (i.e., the four nearest neighbors are at the apexes of a regular tetrahedron), as seen in Fig. 1. The angle subtended by two neighbors is approximately 109.4°. Each atom has 12 second-nearest neighbors. If one envisions "bonds" between nearest neighbors, these bonds form six-membered rings.

The diamond lattice is a relatively open lattice. If spheres are drawn around atoms so that they just touch, the volume of these spheres is only 34% of the total volume. In contrast, in a monatomic fcc lattice, touching spheres fill 74% of the available volume.

The space group of the diamond lattice is the nonsymmorphic group designated as O_h^7 in the Schoenflies notation and $Fd3m$ in the international crystallographic notation. The point group at an atomic site is T_d. In the open spaces of the diamond lattice, there are interstitial sites that also have full tetrahedral (T_d) symmetry because they are also the centers of regular tetrahedra whose four apexes are occupied by atoms. Another high-symmetry interstitial site is the center of a six-membered ring of atoms. It has full hexagonal symmetry.

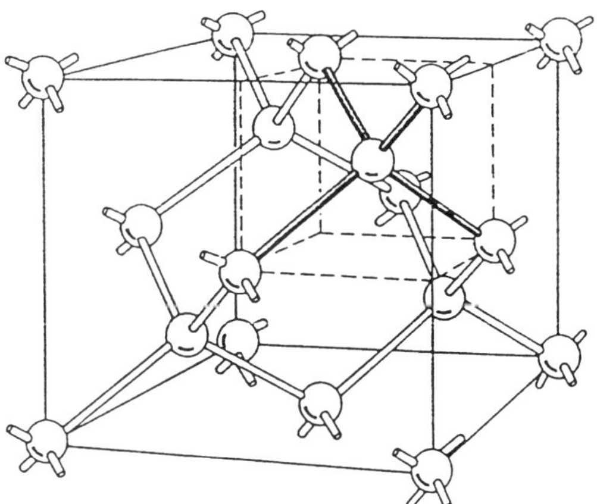

FIG. 1. Schematic showing the tetrahedral coordination of atoms in the diamond lattice.

1.2 Crystalline Surfaces, Interfaces

Since Si is a cubic crystal, the three principal surfaces are the (100), the (110), and the (111). These are shown schematically in Fig. 2 in the case where all atoms are kept at the normal sites they occupy in the perfect crystal. In these "ideal" cases, the surface layers are two-dimensional lattices with primitive translation vectors that we may denote as τ_1 and τ_2. In reality, the surfaces undergo "reconstruction," which means that the symmetry of the surface structure is reduced, leading to a lower-energy state. Equivalently, the surface primitive unit cell is larger, with the primitive translation vectors now being $m\tau_1$ and $n\tau_2$, where m and n are two different integers. One then says that the surface has undergone an $m \times n$ reconstruction.

Low-energy electron diffraction (LEED) is an experimental technique that probes the symmetry of surfaces and can establish the indices $m \times n$. Schlier and Farnsworth (1959) were the first to report a LEED determination of a 2×1 reconstruction of the Si(100) surface. Sometimes at low temperatures a 4×2 reconstruction was also seen. Later, the Si(111) surface was observed to

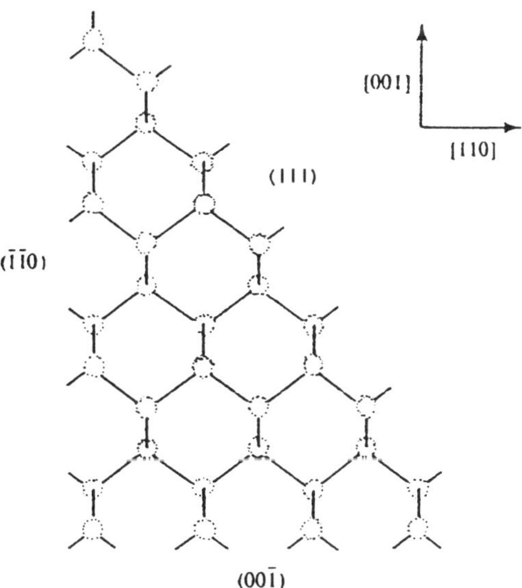

FIG. 2. Schematic showing the atomic arrangements of the principal surfaces of Si in the absence of reconstruction. The crystal is viewed along a $[1\bar{1}1]$ direction. From Harrison (1980).

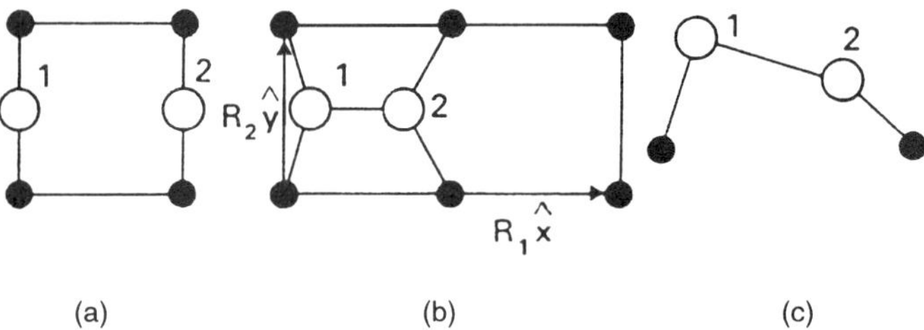

FIG. 3. (a) Top view of the "ideal" Si (100) surface. The open circles depict atoms in the topmost layer, and dark circles depict atoms in the second layer. (b) Top view of the same surface after dimer formation (2 × 1 reconstruction). (c) Side view of the same surface after dimer formation. From Chadi (1979).

undergo several reconstructions as a function of temperature. When cleaved in vacuum at room temperature, the (111) surface exhibits a 2 × 1 reconstruction. If cleaved at very low temperatures (<20 K), a 1 × 1 pattern appears. After annealing to high temperatures (>400 °C), a 7 × 7 reconstruction emerges.

The determination of the actual positions of the surface atoms in the new primitive cell is not an easy task. Intense experimental and theoretical investigations took more than ten years (1970s and early 1980s) to pin down the actual atomic arrangements on Si surfaces. Many experimental and theoretical techniques were developed and contributed to gradual progress. The current consensus is that the surface atoms on the (100) 2 × 1 surface form asymmetric dimers, as illustrated in Fig. 3; the surface atoms on the Si(111) 2 × 1 surface form pi-bonded chains, as illustrated in Fig. 4; and the surface atoms at the (111) 7 × 7 surface form the complex unit cell shown in Fig. 5. The determination of these structures involved the simultaneous elucidation of the surface electronic structure. We therefore postpone further discussion to Sec. 3.6.

The atomic arrangements at Si interfaces depend very much on the second material. For example, it has been determined that a monolayer of hydrogen atoms brings back the surface Si atoms to their "ideal" positions. At the other extreme, the atomic arrangements at the Si–SiO_2 interface are not very regular because SiO_2 is amorphous when grown thermally on Si. In fact it is likely that there exists a thin SiO_x interlayer. Furthermore, it is known that there exist point defects at the interface. The so-called P_b center is believed to be a Si atom that is bonded to only three nearest neighbors, a structure that is commonly referred to as a dangling bond. Many crystalline solids such as transition-metal silicides, CaF_2, etc. form well-ordered interfaces.

1.3 Plane and Line Defects

Stacking faults are the principal plane defects in bulk crystals. An fcc crystal can be thought of as a stacking of planes of close-packed spheres. The planes are stacked in a staggered fashion so that spheres of the second plane sit on top of interstitial sites of the first plane. In this arrangement only half of the interstitial sites have atoms placed on top of them. The third plane of spheres can then go in one of two ways: with its spheres directly over the spheres of the first plane (*ABABA* ... stacking) or with its spheres over the unused interstitial sites of the first plane (*ABCABC* ... stacking). The former gives rise to a hexagonal close-packed (hcp)

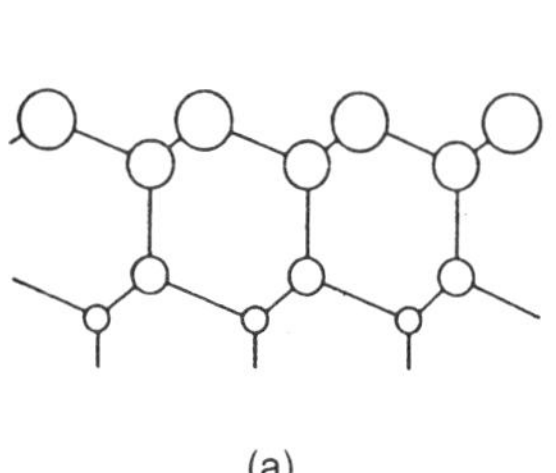

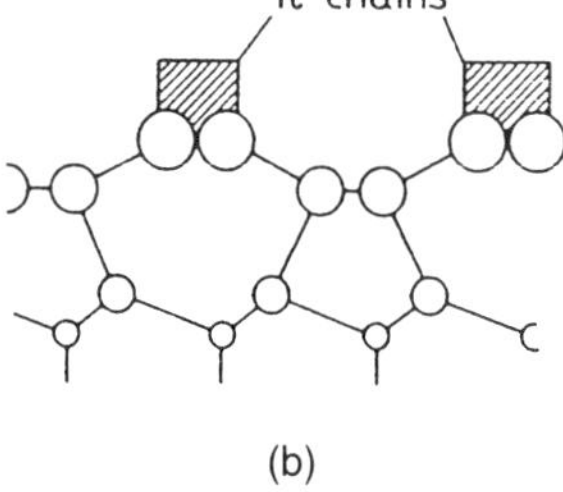

FIG. 4. (a) Side view of a Si(111) "ideal" surface. (b) Side view of the same surface after the formation of pi-bonded chains (2 × 1 reconstruction). From Pandey (1981).

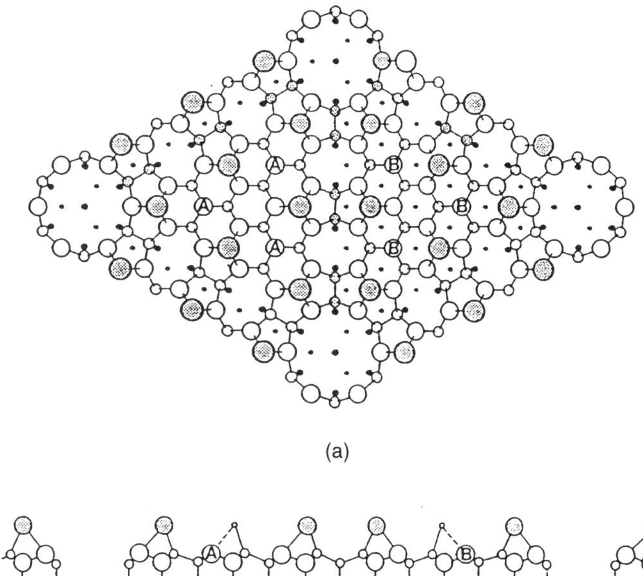

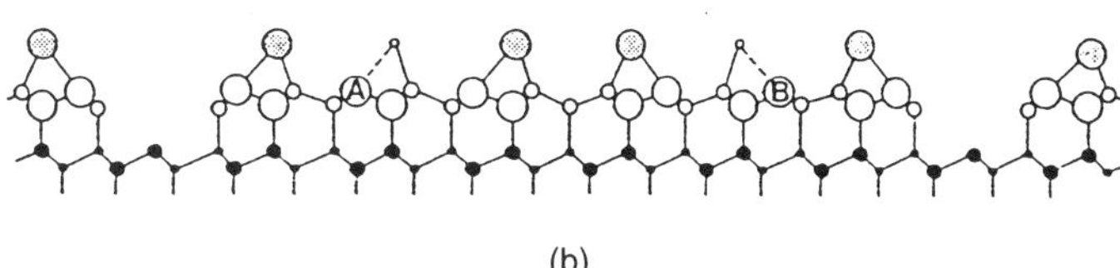

FIG. 5. The complex atomic arrangement of the Si(111) 7 × 7 surface: (a) top view, (b) side view. From Takanayagi *et al.* (1985).

structure while the latter gives rise to an fcc structure. In the fcc case the close-packed planes are {111}. An intrinsic stacking fault corresponds to a missing plane so that the stacking proceeds as *ABCABCBCA*. . . . An extrinsic stacking fault corresponds to an extra plane added in the sequence so that the stacking proceeds as *ABCABCBABC*. . . .

In a diamond lattice, because there are two atoms per unit cell, each fcc {111} plane is actually two planes of atoms, so the stacking sequence should be designated as *AaBbCcAaBb*. . . . In Si, stacking faults correspond to missing or extra *pairs* of planes of atoms. In such faults, the tetrahedral coordination of the atoms is not disturbed so that the total energy of the fault is quite small. Stacking faults where only one plane of Si atoms are missing would disturb the tetrahedral coordination and thus would have a much higher energy.

Dislocations (Hirth and Lothe, 1992) are intrinsic line defects. Edge dislocations correspond to an extra half plane of atoms being inserted in the crystal. The edge of that half plane is the dislocation, and it lies in what is called the slip plane. Screw dislocations correspond to cutting a crystal and shearing it parallel to the plane of the cut by one atom spacing. An important quantity associated with a dislocation is the Burgers vector, defined as follows. One first draws a closed circuit by connecting atoms in a perfect region of the crystal. If the same circuit is drawn around a dislocation, the Burgers vector is defined to be the vector needed to complete the circuit. For edge dislocations the Burgers vector is perpendicular to the dislocation line, whereas for screw dislocations the Burgers vector is parallel to the dislocation line. In the general case of a mixed dislocation, the Burgers vector makes some other angle with the dislocation line. For example, when this angle is 60°, the dislocation is known as a 60° dislocation. For "perfect" dislocations the Burgers vector is exactly equal to a lattice vector. "Imperfect" or partial dislocations are also possible with the Burgers vector being something other than a lattice vector.

As in all crystals, Si may contain edge, screw, or mixed dislocations. Since crystalline Si is fundamentally an fcc crystal, the preferred slip planes for dislocations are the {111} planes, because they have the closest packing of atoms and the maximum interplane separation. This configuration leads to the lowest stress, as calculated using the continuum theory of elasticity. The smallest Burgers vector in the {111} plane, which again minimizes the stress, is in the ⟨110⟩ direction.

In Si it has been found that the prominent dislocations in the early stages of glide are either pure screw or 60° dislocations. There are actually two distinct types of dislocations in Si because of the double-layer atomic arrangement of the diamond lattice. The two types are called dislocations of the glide set and dislocations of the shuffle set. The distinction can be found in any comprehensive book on dislocations.

In Si the perfect dislocations described above dissociate into partial dislocations. Typically, a perfect 60° dislocation dissociates into 30° and 90° partial dislocations. The two partials bound an intrinsic stacking fault.

In dislocations in Si the tetrahedral bonding is disturbed, with some atoms being threefold coordinated. As in surfaces, reconstruction follows to lower the energy of the system. The detailed arrangement of Si atoms in the core of dislocations is still a subject of active research.

1.4 Point Defects

The primitive native point defects in any crystal are the vacancy (missing atom at nominal atomic site) and the self-interstitial (atom at a site that is not a nominal atomic site). Together, they are often called a Frenkel pair. Point defects such as the vacancy occur in different charge states because they have localized energy states within the forbidden band gap. The atomic rearrangements around the point defect may then depend on the charge state. In the case of the vacancy, V^{++} undergoes no lattice reconstruction, i.e., full tetrahedral symmetry is retained (the neighboring atoms only relax in a symmetry-preserving mode that is known as the breathing mode), whereas all other charge states undergo symmetry-reducing reconstructions. These reduced symmetries have been deduced primarily from electron paramagnetic resonance data (Watkins, 1994). The origins of these phenomena are discussed in Sec. 3.6 in the context of the electronic structure of point defects.

The self-interstitial has been more difficult to pin down. Experimentally it has not been observed, apparently because it is highly mobile and reactive. Theoretical calculations that seek to establish its equilibrium configuration by comparing the formation energy of different configurations have found that the lowest-energy configuration depends on the charge state of the defect (see Sec. 3.6). The latest work on the subject finds that the neutral interstitial has a relatively complex structure: it is a form of a split interstitial, namely, a dumbbell of two atoms occupying what is normally an atomic site.

Impurities may occupy substitutional sites, in which case the surrounding lattice usually relaxes a very small amount without lowering the point symmetry. Examples are the dopant impurities (group-III and group-V elements). Other impurities—e.g., Li—occupy the high-symmetry tetrahedral interstitial site. Transition-metal impurities may be either substitutional or interstitial. Oxygen is a rather unique impurity. Its lowest-energy configuration is a low-symmetry interstitial site: oxygen occupies a bridge position between neighboring Si atoms with the Si–O–Si angle being about 145°, mimicking the Si–O–Si bonds in SiO_2. By high-energy electron irradiation, oxygen can be induced into another configuration that is nominally substitutional, but with the oxygen atom shifted off-center, bridging two of its Si neighbors. This center is often referred to as the oxygen-vacancy complex (Watkins and Corbett, 1961). Other impurities—e.g., phosphorus—form true vacancy–impurity pairs; namely, the impurity occupies a substitutional site next to a vacant site. Finally, other complexes are possible—e.g., a boron-hydrogen pair, where the boron is substitutional and the hydrogen occupies a nearby interstitial site.

1.5 Polycrystalline, Amorphous, and Porous Si

For completeness, we mention here briefly forms of Si that are not crystalline.

Polycrystalline Si (see Kamins, 1988; SILICON, POLYCRYSTALLINE), like all polycrystals, consists of crystallites (grains) that meet at grain boundaries (*q.v.*). We do not attempt here a review of grain-boundary structures, which are general for any material. Of interest here would be the particular atomic arrangements in Si grain boundaries. Just like free surfaces, the planes of atoms at a grain boundary undergo relaxations and reconstructions from their "ideal" positions. It is believed that all atoms at Si grain boundaries are fourfold coordinated. Establishing the detailed atomic positions is a demanding task. High-resolution microscopy provides detailed images (Bourret and Rouvier, 1989). Typically one must assume some structure and simulate the image to compare with experiment. Total-energy minimizations also help guide the search for possible structures. Usually, however, more than one structure can be found with similar total energy and similar simulated electron-microscope image.

Real grain boundaries in polycrystalline Si do have point defects that are electrically active and can be passivated by hydrogen. In addition, grain boundaries are low-energy sites for dopant impurities. Since in grain boundaries such impurities are electrically inactive, they are most likely threefold coordinated, which is the low-energy coordination they prefer. It was recently determined that arsenic impurities segregate in Si grain boundaries threefold-coordinated as dimers or ordered chains (Maiti *et al.*, 1996).

Amorphous Si (see SILICON, AMORPHOUS) is a network of fourfold-coordinated Si atoms. For each Si atom the four neighbors are roughly in tetrahedral directions. In contrast to crystalline Si, there is no long-range periodicity. In addition, whereas in crystalline Si the atoms are arranged in buckled six-member rings (where a "member" is a Si–Si bond), in amorphous Si there is a distribution of five-, six-, and seven-member rings.

By viewing "perfect" amorphous Si as a network of fourfold-coordinated Si atoms, one immediately deduces that the native point defects would be threefold-coordinated atoms (generally known as dangling bonds) and fivefold-coordinated atoms. The latter have been labeled floating bonds (Pantelides, 1986). Traditionally, dangling bonds were viewed to be the dominant defect that gives rise to electron-paramagnetic-resonance (EPR) and luminescence signals. The evidence for such an identification, however, is only circumstantial. The concept of floating bonds is of more recent import. They have been found in theoretical simulations of amorphous Si structures. Definitive experimental proof as to the nature of the observed defects and the relative concentrations of dangling and floating bonds is lacking.

Porous Si is a form of disordered Si with a high density of pores. Its claim to fame is that it is the only form of Si known to produce strong luminescence lines in the red region of the optical spectrum. The detailed atomic arrangements that produce luminescence have not been established, though a number of suggestions have been put forth.

2. VIBRATIONAL PROPERTIES—PHONONS

In the previous section, we described the atomic positions in a perfect Si crystal, at surfaces, and at line and point defects. The atoms actually vibrate about these nominal sites. For small vibrations, the allowed frequencies are obtained by a harmonic approximation: the total energy of the system, which is a function of the actual atomic positions, is expanded to second order in the atomic displacements. One then can introduce a set of generalized coordinates (the classical normal vibrational modes) in terms of which the total energy becomes a simple sum of terms, each of which is the Hamiltonian of a simple harmonic oscillator. In the case of a perfect crystal, the normal modes are propagating waves and can be labeled by a wave vector $\mathbf{q}$ and a "branch index" ν. The number of branches is equal to the number of degrees of freedom in the primitive cell, i.e., three for a monatomic solid and six for a diatomic solid like Si. The corresponding frequencies are also labeled by the wave vector $\mathbf{q}$ and the branch index ν. Excitations of these decoupled harmonic-oscillator states are known as phonons.

The phonon spectra of crystalline Si are shown for $\mathbf{q}$ along high-symmetry directions in Fig. 6. The three lower branches are the acoustical branches, with the slope near $\mathbf{q} = 0$ corresponding to the speed of sound in the crystal. In the acoustical branches, the two Si atoms at each lattice site move in tandem. The three upper branches are the optical branches, in which the two Si atoms at each lattice site move in opposite directions. Each set of three branches contains a longitudinal branch—with the atoms oscillating along the direction of wave propagation—and two

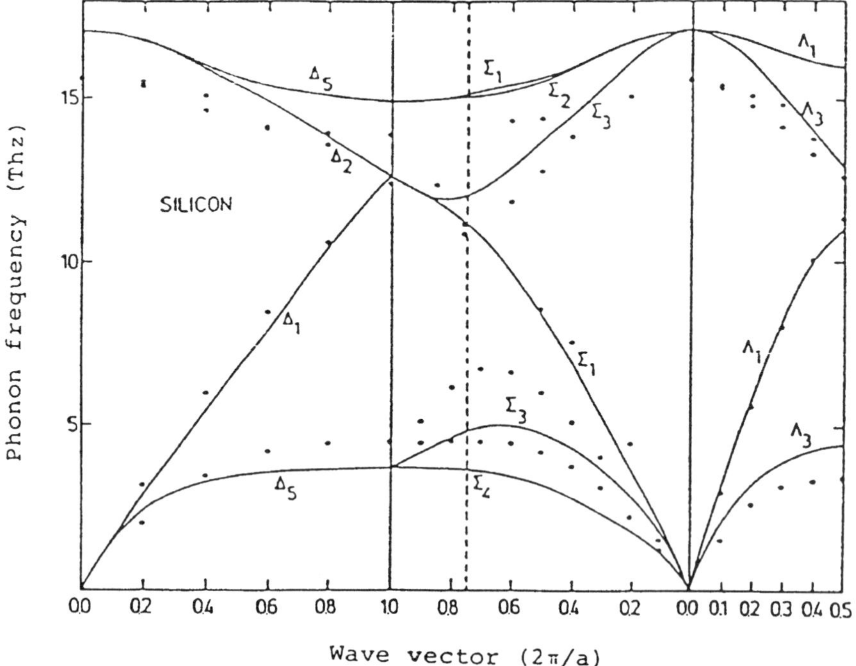

FIG. 6. The phonon spectra of silicon.

transverse branches—with the atoms oscillating perpendicular to the direction of wave propagation.

3. ELECTRONIC STRUCTURE AND PROPERTIES

3.1 The Perfect-Crystal Energy Bands

Electron states in Si are adequately described by the self-consistent-field approximation according to which every electron is acted upon by the same effective periodic potential, arising from all the nuclei and the rest of the electrons. The resulting electron states, characterized by a propagation vector **k** and known as Bloch states, form bands of allowed energies separated by band gaps. The electron energy bands of Si have been calculated by a multitude of techniques, and by now they are thoroughly understood. A plot of the energy bands as a function of the wave vector **k** along high-symmetry directions in the Brillouin zone (i.e., the primitive unit cell in reciprocal space) is shown in Fig. 7. The lowest four bands are completely occupied by the four electrons per unit cell in the Si lattice (two electrons—spin up and spin down—per band at each **k** point) and are known as valence bands. The bands above the forbidden gap are empty, making Si a semiconductor with an energy gap of 1.16 eV at 0 K (1.12 eV at 300 K).

The lowest conduction band has six equivalent minima at $(\pm k_c,0,0)$, $(0,\pm k_c,0)$, and $(0,0,\pm k_c)$, where $k_c \approx 0.82(2\pi/a)$ and a is the primitive cube edge. In the vicinity of each of these minima, the band is parabolic but with different transverse and longitudinal effective masses:

$$E(k) \cong \frac{\hbar^2}{2m_\perp} k_x^2 + \frac{\hbar^2}{2m_\parallel}(k_y^2 + k_z^2) \quad (1)$$

for the minima along the k_x axis. The values of the masses are $m_\perp = 0.19m_e$ and $m_\parallel = 0.97m_e$, where m_e is the mass of a free electron.

The valence bands have a maximum at **k** = 0. At that point, the bands are degenerate. If spin–orbit interactions are ignored, the degeneracy is threefold (times two for spin) with Bloch functions that transform like the

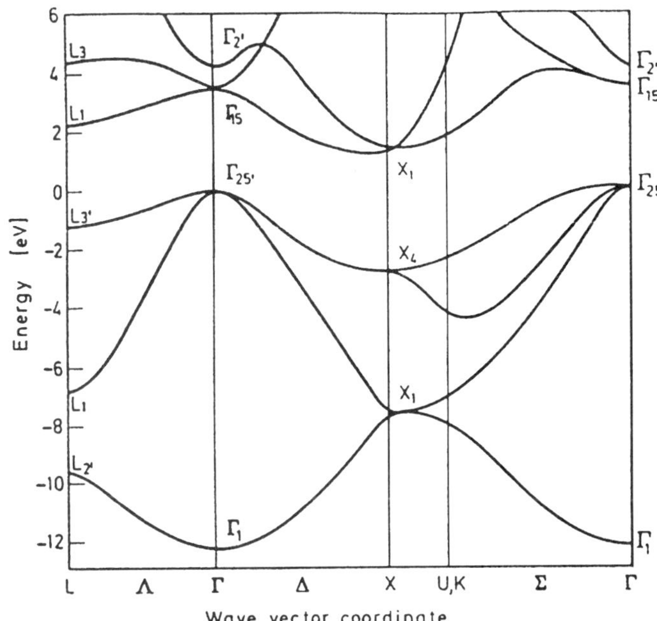

FIG. 7. Energy bands of Si along high-symmetry directions in **k** space. The designations such as L_1, X_4, etc. refer to symmetry properties of the corresponding wave functions. From Chelikowski and Cohen (1976).

functions xy, yz, and zx ($\Gamma_{25'}$ representation of the T_d group). The three bands are the eigenvalues of the following 3×3 matrix developed by $\mathbf{k}\cdot\mathbf{p}$ theory (Kohn, 1957):

$$\begin{pmatrix} Ak_x^2 + B(k_y^2 + k_z^2) & Ck_xk_y & Ck_xk_z \\ Ck_yk_x & Ak_y^2 + B(k_z^2 + k_x^2) & Ck_yk_z \\ Ck_zk_x & Ck_zk_y & Ak_z^2 + (k_x^2 + k_y^2) \end{pmatrix}. \tag{2}$$

When spin–orbit interaction is included, the bands at $\mathbf{k} = 0$ split into a fourfold-degenerate set that transforms like Γ_8 and a twofold-degenerate set that transforms like Γ_7. In the region near $\mathbf{k} = 0$, the top four bands split into the top "heavy-hole" band and the lower "light-hole" band. One can approximate these bands with parabolic bands and assign effective heavy-hole and light-hole effective masses. The values of these effective masses are $m_l = 0.16m_e$ and $m_h = 0.5m_e$. The lowest band is known as the split-off band. At $\mathbf{k} = 0$ the spin–orbit splitting is 44 meV. On the basis of these experiments and $\mathbf{k}\cdot\mathbf{p}$ theory, the structure of the energy bands near the energy gap was established in the 1950s, long before it became possible to calculate the full energy bands from first principles.

Energy-band theory evolved since the 1950s when the first computers became available. The energy bands of Si have probably been calculated with just about every conceivable method. Currently, the most rigorous way to formulate the "one-electron" problem for a crystal is the density-functional theory of Hohenberg and Kohn (1964). In this theory, the total energy E is shown to be a universal functional of the electron density $n(r)$, which in turn is written in terms of one-electron wave functions $\psi_i(r)$:

$$n(r) = \sum_i |\psi_i(r)|^2, \tag{3}$$

where the sum is over occupied states. In the corresponding effective one-electron potential, the exchange-correlation term is not known exactly. The most common approximation is the local-density approximation, where one assumes that the known form of exchange and correlation for the homogeneous electron gas can be used at each r for the local value of $n(r)$. This approximation is often coupled with a pseudopotential approximation, where the core electrons and the strong potential that binds them are replaced by an effective potential whose lowest eigenstates are the valence states. As a result, the oscillatory part of the valence wave functions in the core regions is replaced by a smooth part. It is then possible to expand the crystal wave functions in terms of a practically small set of plane waves.

In density-functional theory, the one-electron wave functions and corresponding one-electron energies are not rigorously wave functions and energy levels. They are merely intermediate quantities that enter the theory, in terms of which the total energy is calculated. Nevertheless the calculated one-electron energies give a good representation of the one-electron energy bands that are measured experimentally. The biggest shortcoming is the fact that the theoretical band gap is only about half of what is measured. In part, the problem is a consequence of the local-density approximation for exchange and correlation. Theories that go beyond these approximations have been developed and yield energy bands that are in full accord with experimental data (Hybertsen and Louie, 1985). Silicon has in fact been the testing ground for such theories.

The structure of the energy bands of Si can be understood in terms of simple arguments based on atomic theory. Si atoms have 1*s*, 2*s*, and 2*p* core electrons. They also have two 3*s* and two 3*p* electrons. One can combine the four 3*s* and 3*p* wave functions into four sp^3 "hybrid orbitals" that point in the four tetrahedral directions. For example, the orbital h_{111} defined by

$$h_{111} = \tfrac{1}{2}(\phi_s + \phi_x + \phi_y + \phi_z) \tag{4}$$

points in the (111) direction. Here ϕ_s is the 3*s* orbital of Si and ϕ_x, ϕ_y, and ϕ_z are the three 3*p* orbitals. The energies of the four hybrid orbitals can then be easily calculated to lie at $\epsilon_h = (\epsilon_s + 3\epsilon_p)/4$, where ϵ_s and ϵ_p are the 3*s* and 3*p* atomic energies, respectively. One can then form bonding and antibonding orbitals by combining hybrids pointing toward each other. If we denote two such hybrids on neighboring atoms by h_1 and h_2, the bonding and antibonding orbitals are defined by

$$b = \tfrac{1}{2}(h_1 + h_2), \tag{5}$$

$$a = \tfrac{1}{2}(h_1 - h_2). \tag{6}$$

If the interaction matrix element $\langle h_1|H|h_2\rangle$, where H is the crystal Hamiltonian, is denoted by $-V_2$, the corresponding energies are $\epsilon_b = \epsilon_h - V_2$ and $\epsilon_a = \epsilon_h + V_2$, and the bonding–antibonding splitting is $2V_2$. The notation used here is that introduced by Walter Harrison of Stanford University, who in the 1970s constructed an elementary theory of semiconductor properties based on these concepts (see Harrison, 1980).

One can construct Bloch functions out of the bonding orbitals and, by using the Hamiltonian matrix elements between neighboring bonding orbitals as parameters, reproduce the known valence bands of Si quite well. The conduction bands arise from the antibonding orbitals.

The densities of one-electron states of Si have been calculated and measured by photoemission techniques. They are shown in Fig. 8. Points of high symmetry are reflected in features of the density of states.

3.2 Optical and Dielectric Response

Optical absorption experiments cause electrons to be excited from valence states to conduction states. The optical reflectivity, which is directly related to the optical absorption, of Si is shown in Fig. 9. The main peak around 4 eV arises primarily from transitions between the highest valence bands and the lowest conduction bands that, as can be seen in Fig. 7, are approximately "parallel." In addition to the band-to-band transitions, electron–hole attraction between the electron and the hole gives rise to excitons (*q.v.*) that lie at energies just below the band gap. The excitons are not resolved in Fig. 9.

X-ray absorption experiments excite electrons from the core states to the conduction bands. Because the core states are flat in **k** space (because of negligible interactions between neighboring core wave functions), the x-ray absorption spectrum should reflect the conduction-band density of states. The observed spectra actually rise much more steeply at the threshold than the corresponding density of states (Fig. 10). The effect has been attributed to electron–hole interactions and the formation of excitons just below the

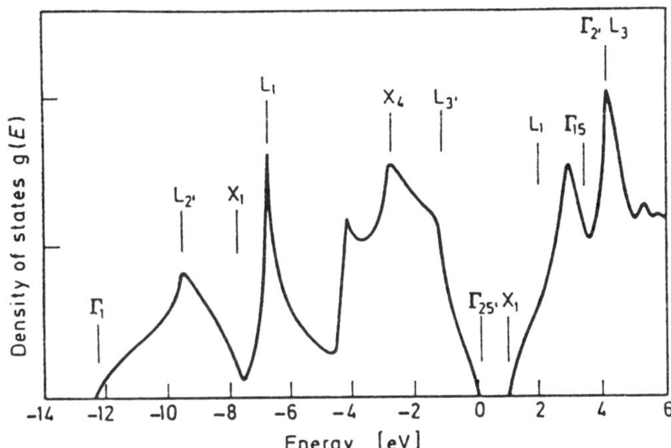

FIG. 8. The density of one-electron states in Si. From Chelikowski *et al.* (1973).

band edge. These excitons have very small binding energies (~30 meV; see Sec. 3.3 for this estimate) and are not resolved.

When an external electrostatic potential $V_{ext}(r)$ is applied to a crystal, the electrons are polarized and give rise to a screening potential. In linear response theory, the total (screened) potential $V(r)$ is given by

$$V(r) = \int \frac{V_{ext}(q)}{\epsilon(q)} e^{-i\mathbf{q}\cdot\mathbf{r}} d^3q, \tag{7}$$

where $V_{ext}(q)$ is the Fourier transform of $V_{ext}(r)$. The dielectric function $\epsilon(q)$ can be calculated in terms of virtual excitations of the core and valence electrons. Results of numerical calculations have been fitted to the following analytic formula (Nara, 1965):

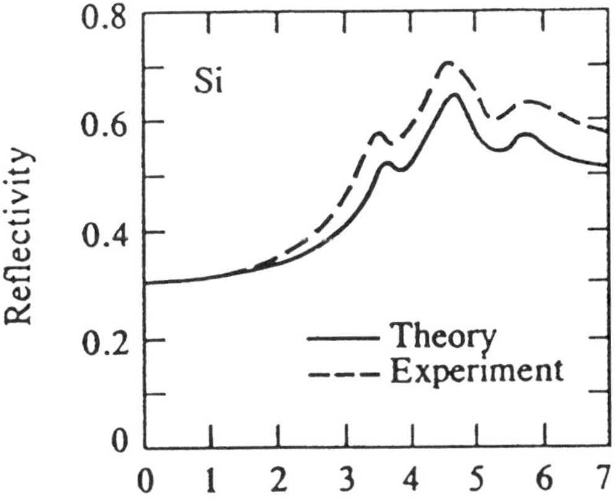

FIG. 9. The optical reflectivity spectrum of Si.

$$\frac{1}{\epsilon(q)} = \frac{Aq^2}{q^2 + \alpha^2} + \frac{Bq^2}{q^2 + \beta^2} + \frac{1}{\epsilon_0} \frac{\gamma^2}{q^2 + \gamma^2}, \tag{8}$$

where $A = 1.175$, $B = 1 - A = -0.175$, $\alpha = 0.7572$, $\beta = 0.3123$, and $\gamma = 2.044$. The constant ϵ_0 is the static dielectric constant $\epsilon(0)$, which for Si is 11.4.

In the simplest model of a dielectric constant, one assumes that all the virtual excitations can be lumped into a single effective excitation at an energy E_0. A well-known sum rule for the oscillator strengths then yields the following expression for ϵ_0:

$$\epsilon_0 = 1 + (\hbar\omega_p)^2/E_0^2, \tag{9}$$

where $\omega_p^2 = 4\pi ne^2/m$ is the so-called plasma frequency for the electron gas whose density is n. Here e is the electron charge and m is the electron mass. Since ϵ_0 and ω_p are known from experiment, this formula allows one to extract a value of the effective dielectric gap E_0, sometimes referred to as the Penn gap (Penn, 1962). This model of the dielectric constant formed the basis of a theory of ionicity for semiconductors by James C. Phillips of Bell Laboratories (see Phillips, 1973). The theory's main thesis is that the dielectric gap E_0 can be written in terms of a homopolar part E_h and an ionic part C in the form

$$E_0^2 = E_h^2 + C^2. \tag{10}$$

In an alternative approach, Harrison (1980) has suggested that the dielectric constant is

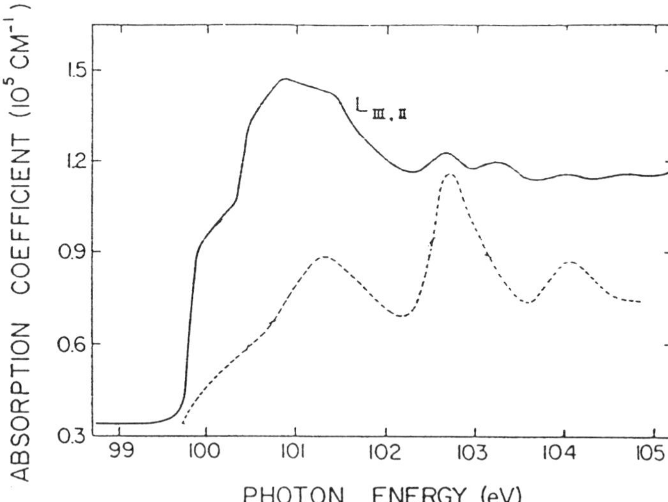

FIG. 10. The x-ray absorption spectrum of Si arising from excitation of $2p$ core electrons. From Brown and Rustgi (1972).

primarily determined by virtual excitations that on average correspond to the bonding–antibonding energy. The solid is then in effect represented by a system of polarizable bonds with a single excitation energy of $2V_2$ in the notation introduced earlier. A formula for the dielectric constant is then obtained that has a different form from Eq. (9) simply because Harrison did not impose the requirement that the oscillator strength of the single average transition be equal to 1. Harrison's theory yields a different ionicity (or polarity) scale for the compound semiconductors.

3.3 Dopant Impurities

When a Si atom is replaced by a group-V element (P,As,Sb), an extra electron is introduced in the crystal and the Si^{+4} core is replaced by a $+5$ core. There is a net increase in the core charge of $+e$. The resulting Coulomb potential is screened, in the simplest approximation, by the dielectric constant ϵ_0:

$$V(r) = -e^2/\epsilon_0 r. \tag{11}$$

The extra electron would normally go to the bottom of the conduction band. If the conduction band had a single minimum with effective mass m^*, the problem would be completely analogous to the free hydrogen atom. A series of hydrogenic bound states appear below the conduction-band edge, but the effective Rydberg is reduced by a factor m^*/ϵ_0^2. If we take round numbers such as $m^* = 0.3$ and $\epsilon_0 = 12$, we find that the effective Rydberg is of order 30 meV, i.e., a small fraction of the Si band gap. At $T = 0$ K, the donor electron would be in the ground state of the hydrogenic series, but at room temperature most donor electrons would be excited into the conduction band where they can contribute to conductivity.

A very analogous situation exists when a Si atom is replaced by a group-III element (B,Al,Ga,In). An electron is now missing, which is equivalent to a hole at the top of the valence band. The net change in the core charge is now $-e$. Hydrogenic states appear above the valence band, where the hole can be bound. Again, at room temperature, most holes are excited into the valence bands.

The simple analysis given above was formulated rigorously by several authors in the 1950s (Kohn and Luttinger, 1955). The resulting theory is known as effective-mass theory, and its range of validity is established. It generally applies when the binding energy is a small fraction of the band gap, and the corresponding hydrogenic wave functions are spread out to encompass several shells of neighbors around the impurity.

In Si, however, the conduction-band minima are anisotropic, as given by Eq. (1). As a result, the p and higher-order states are split. (E.g., the $2p$, $m = 0$ state has a different energy from the $2p$ states with $m = \pm 1$; here m is the azimuthal quantum number. These states are denoted by $2p^0$ and $2p^{\pm}$.) These states were observed by infrared absorption in the 1950s and 1960s. The experimental Rydberg series were in excellent agreement with theoretical effective-mass values.

The Si conduction bands present yet another complication. There are six equivalent minima. Symmetry requires that the bound-state wave functions be constructed as linear combinations of states at all six minima. The coefficients are determined by group theory. The bound-state energies then are determined by intravalley and intervalley terms. For 2*p* and higher-order states, the intervalley terms are negligible. For the 1*s* ground state, however, they are not. As a result, what would have been a sixfold-degenerate 1*s* state splits into a single of A_1 symmetry, a doublet of *E* symmetry, and a triplet of T_2 symmetry (Kohn and Luttinger, 1955). Here the chemical notation for the irreducible representations of the T_d group are used.

If the impurity potential were simply a screened Coulomb potential, all group-V impurities would have the same hydrogenic energy levels. Experimental data show that the 1*s* ground state actually differs from impurity to impurity (Table 1). The reason is because the 1*s* hydrogenic wave function (scaled by m^* and ϵ_0) is sufficiently localized to sample the so-called central-cell region where the actual potential looks more like the difference between the impurity-atom and host-atom pseudopotentials. The deviations of the ground states from the simple hydrogenic result are known as central-cell corrections or chemical shifts.

The valence bands of Si present altogether different complications. The corresponding kinetic energy operator in the hydrogenic equation is now a matrix corresponding to Eq. (2). The resulting hydrogenic spectrum reflects this multiplicity of bands and is quite complex. The spectra have been measured, and detailed calculations have been carried out. Again, however, the ground states of the various impurities exhibit chemical shifts.

Table 1. Donor energy levels in meV, measured from the conduction-band edge (from Aggarwal and Ramdas, 1965).

Impurity	$1s(A_1)$	$1s(T_2)$	$1s(E)$
Phosphorus	45.5	33.9	32.6
Arsenic	53.7	32.5	31.2
Antimony	42.7	32.9	30.6

3.4 Heavy Doping—The Degenerate Limit

When the concentration of shallow impurities exceeds a certain limit, the impurity ground-state wave functions start to overlap, and the discrete energy levels broaden into a band. This is a classic example of the Mott metal-insulator transition. Let us consider a simple hydrogenic 1*s*-like ground state that can take a single electron (a second electron can be bound to form a negatively charged center, but the binding of the second electron, as in the free H atom, is roughly 1/20 of the effective Rydberg). As such levels broaden into a band, there are enough electrons to fill up to half the band. Thus the system is in principle metallic at all impurity concentrations. The system, however, remains insulating up to a critical bandwidth, when delocalization wins. Because of the disorder in the impurity distribution, there are always localized tail states (see ELECTRON STATES, LOCALIZED) at the edge of the band.

Beyond the Mott transition (at about 5×10^{18} cm^{-3} impurities), the impurity band merges with the regular conduction band so that in effect the conduction-band edge is now lower, resulting in a smaller band gap. There are other factors contributing to the band-gap shrinkage in heavily doped Si (Pantelides *et al.*, 1985). The electrons are no longer in discrete states but rather fill a regular band up to the Fermi level. Electron–electron interactions, which are not present in the case of the perfect crystal, have a net effect of lowering the band. The effect is analogous to the usual treatment of electron–electron interactions in a simple metal when one goes beyond the independent electron approximation. The band-gap shrinkage in heavily doped Si is an important effect in Si devices.

3.5 Lattice Point Defects and Deep Impurities

The structure of the electronic states that are localized at a vacancy can be inferred correctly from simple considerations in terms of the sp^3 hybrids that form bonds in the bulk. When an atom is removed and the neighboring atoms are kept frozen at their nominal bulk positions—known as the "ideal" vacancy—we are left with four dangling bonds, i.e., dangling sp^3 hybrids pointing tetrahedrally toward the vacant site. The true localized states need to have the full symmetry of the tetrahedral group and can be represented by the following combinations of the hybrids:

$$\begin{aligned} |A_1\rangle &= \tfrac{1}{2}(h_1 + h_2 + h_3 + h_4), \\ |T_2\rangle_1 &= \tfrac{1}{2}(h_1 - h_2 + h_2 - h_3), \\ |T_2\rangle_2 &= \tfrac{1}{2}(h_1 + h_2 - h_3 - h_4), \\ |T_2\rangle_3 &= \tfrac{1}{2}(h_1 - h_2 - h_3 + h_4). \end{aligned} \tag{12}$$

Here, the h subscripts refer to the four Si atoms neighboring the vacant site and the designations A_1 and T_2 refer to symmetry under the operations of the T_d group, which is the point group of the ideal vacancy. States of A_1 symmetry transform like s orbitals, whereas the T_2 representation is triply degenerate and transforms like p orbitals.

The true localized states are expected to have the same transformation properties as those given above, but the detailed wave functions would of course be more complex. The construction of the states given by Eqs. (12) allows for a simple electron counting. There are only four electrons (one from each of the neighboring Si atoms) to occupy the localized states so that only half of the eight states (4 times 2 for spin) are occupied for the neutral vacancy.

George Watkins at the General Electric Laboratory studied the Si vacancy extensively in the 1960s (see Watkins, 1994). He was able to create and isolate vacancies by high-energy electron bombardment at very low temperatures (4 K). With electron paramagnetic resonance experiments, he was able to determine that the A_1 state is lower in energy and contains two electrons, whereas the T_2 state lies in the gap. The neutral vacancy would thus have two electrons in a sixfold-degenerate state. Such an arrangement triggers the Jahn–Teller effect: any symmetry-lowering distortion splits the T_2 state with a gain in energy that is linear in the distortion amplitude. This linear gain in energy is ultimately balanced by the normal quadratic term in the total energy so that the stable configuration has reduced symmetry. A combination of experiments (EPR, uniaxial stress, optical excitation) led Watkins to conclude that V^+, V^0, and V^- undergo a tetragonal distortion.

The Si vacancy was also the prototype point defect that was studied by the first parameter-free first-principles calculations of the properties of point defects in solids. These techniques were developed in the late 1970s. Baraff and Schluter (1978) at AT&T Bell Laboratories and Bernholc *et al.* (1978) at IBM Research reported the first detailed calculations. The T_2 level is at about the middle of the gap, and the A_1 state is a resonance below the valence-band edge. Electron-density plots for the ideal vacancy are shown in Fig. 11. Note that the wave function of the T_2 state in the gap looks very much like a dangling bond (dangling hybrid) pointing toward the vacant site, with a tail that extends over several neighbors. Note also that, in contrast, the total change in electron density caused by the removal of an atom is very highly localized within roughly an atomic volume.

Further theoretical study of the vacancy by Baraff *et al.* (1979) led to the conclusion that the vacancy is very likely to exhibit so-called "negative-U" behavior. The concept had been introduced earlier by Anderson of Bell Laboratories (1975). Under normal circumstances, if a localized state contains an electron, there is a positive energy cost U to add a second electron to the same level. This energy is often referred to as the Hubbard U. Anderson noted, however, that lattice relaxation around a point defect could in principle lead to enough energy gain so as to overcome the Hubbard U and result in a net negative U. In such a case, the point defect would have either no electrons or two electrons, with the one-electron state being unstable and unobservable. Anderson made the suggestion in the context of chalcogenide glasses to account for the lack of EPR signal from coordination defects. Following the theoretical suggestion for the Si vacancy, Watkins and Troxell (see Watkins, 1994) designed a set of clever experiments that confirmed that the vacancy is indeed a negative-U center.

The self-interstitial in Si has never been isolated. In Watkins's experiments, the electron bombardment creates an equal number of vacancies and interstitials. The signature

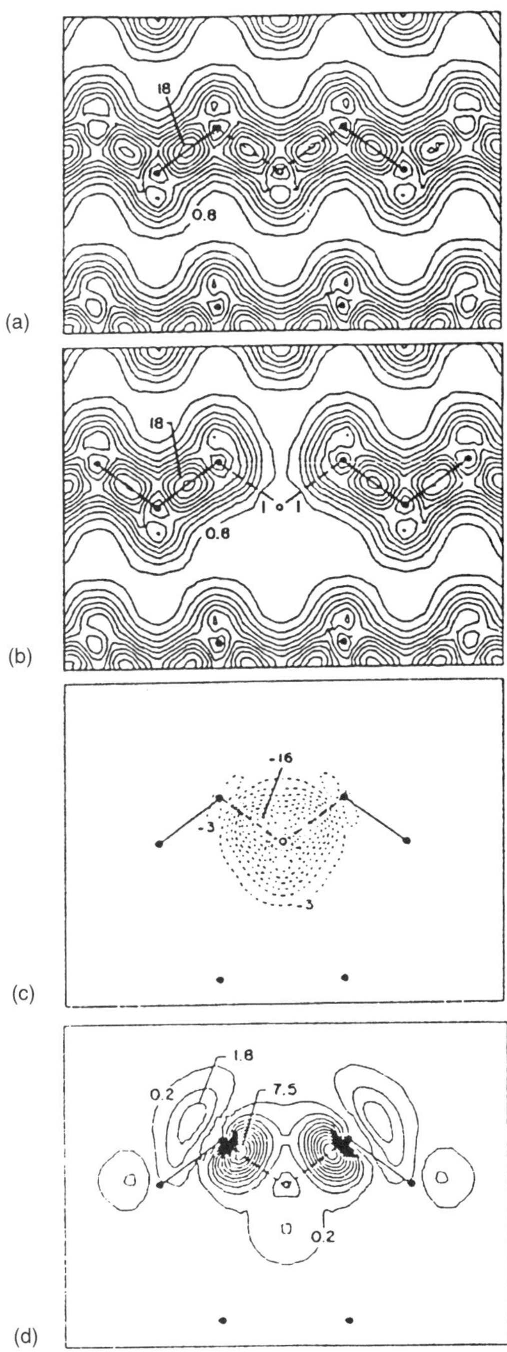

FIG. 11. Electron-density plots in the (110) plane for the "ideal" vacancy in Si. (a) The electron density in a perfect Si crystal; (b) the electron density in a crystal containing a vacancy; (c) the difference between the previous two panels; (d) the electron-density plot for the localized state in the energy gap. From Bernholc *et al.* (1980).

of the interstitial has been indirect. In Al-doped Si, after electron bombardment, the number of vacancies was matched by an equal number of Al interstitials. Since initially all Al atoms are substitutional, the obvious inference is that Si interstitials are extremely mobile even at 4 K, find the substitutional Al atoms, and kick them out. A mechanism for such athermal migration was proposed by Bourgoin and Corbett (1972): Consider an interstitial atom at a site *A* in a positive charge state with a barrier for migration that can be as large as the band gap. Denote the saddle point for this migration by *B*. Now assume that the equilibrium site for the interstitial in the neutral state happens to be the site *B*. The situation is very analogous to what happens at a negative-*U* center where a change of charge state is accompanied by drastic atomic rearrangements. Now consider the interstitial in the presence of free electron–hole pairs. When the interstitial is positively charged at site *A*, it can capture an electron and become neutral. It naturally then rolls over to the nearest *B* site, where it is stable until it captures a free hole to become positive again. Then it rolls over to the nearest *A* site (with some probability that it returns to its original *A* site), and the process continues. In effect, the barrier is overcome by using the energy given up by electron–hole recombination so that a barrier of up to the value of the band gap can be overcome. Detailed first-principles calculations have confirmed that migration along a path involving the tetrahedral and hexagonal sites can proceed athermally. The relevant total energy curves are shown in Fig. 12. The lowest-energy form of the neutral interstitial has since been determined to be a configuration of lower symmetry with two Si atoms forming a dumbbell at a nominal Si site (Blöchl *et al.,* 1993). The athermal path is likely to involve this site and a tetrahedral site.

The simplest deep impurities in Si are the substitutional group-VI elements sulfur, selenium, and tellurium (double donors) and the substitutional group-II elements zinc and cadmium (double acceptors). It is instructive to start with a vacancy and consider the modifications incurred when an impurity atom is added at the vacant site. If a group-I element is added (say copper), one may expect a small interaction between the copper

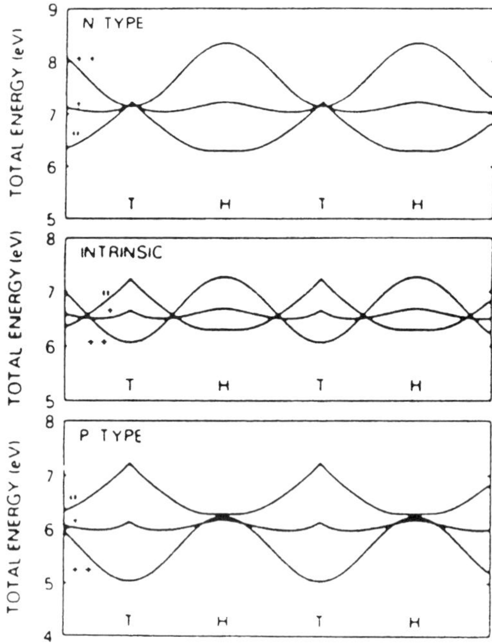

FIG. 12. The total-energy curves for different charge states of the Si interstitial along a path that includes the tetrahedral (T) and the hexagonal (H) sites (from Pantelides, 1994).

3*s* electron and the vacancy A_1 resonance and an even smaller interaction between the copper 3*p* state (which is empty) and the vacancy T_2 state. The net effect should be a T_2 level in the gap containing a total of three electrons in the neutral state. Extending this picture one more step, a group-II impurity should have a T_2 level in the gap with four electrons in the neutral state. Viewed another way, this level in the gap contains two bound holes, which are the two holes one focuses on when treating a Zn impurity as an effective-mass double acceptor. In fact if we extend the theoretical model to a group-III impurity being inserted in a vacant site, we end up with the T_2 state containing five electrons, i.e., only one bound hole, which is the well-known effective-mass hole.

In the model just described, as we proceed from a group-I to a group-II to a group-III impurity, the vacancy T_2 level is successively lowered in energy because the impurity pseudopotential reduces the net repulsion arising from the removal of a Si pseudopotential. Detailed calculations have shown that indeed the level in the gap of group-I and group-II acceptors is virtually identical to that of the vacancy (see Pantelides, 1994).

Group-VI donors have states that are more complex. Four out of the six electrons play the role of Si electrons so that the valence states of the crystal are not seriously perturbed. The two extra electrons again occupy localized states that again have A_1 and T_2 symmetry. Now the A_1 level is in the gap and can take one or two electrons, corresponding to the one or two electrons of effective-mass donors. The T_2 level is a resonance in the conduction bands. The double acceptors have been studied experimentally in great detail by Hermann Grimmeiss's group at the University of Lund in Sweden (see Grimmeiss and Janzen, 1994). It was demonstrated that these deep double donors have full Rydberg excited states just like the single donors. These Rydberg states arise from the Coulombic tail of the impurity potential and should in principle be present in the ionization spectra of any neutral or positively charged double donor. The first and second ionization energies of the group-VI impurities are listed in Table 2.

Substitutional transition-metal impurities can also be understood by starting with a vacancy that has a T_2 level in the energy gap and adding an impurity atom. The relevant atomic energy levels are now the *d* states. The local tetrahedral symmetry splits the five *d* states into a threefold T_2 and a twofold *E* level. Clearly the *E* level does not interact with any of the vacancy localized levels. In contrast, the T_2 level interacts with the vacancy T_2 level. Detailed first-principles calculations have confirmed that the trend of energy levels is consistent with this simple model and provided additional insights. An extensive review of the subject has been given by Zunger (1986). In Fig. 13 we show the results for the series of 3*d* series of ele-

Table 2. First and second ionization energies of group-VI donors in Si (from Grimmeiss and Janzen, 1994).

Impurity	First ionization energy (eV)	Second ionization energy (eV)
Sulfur	0.32	0.61
Selenium	0.31	0.59
Tellurium	0.20	0.41

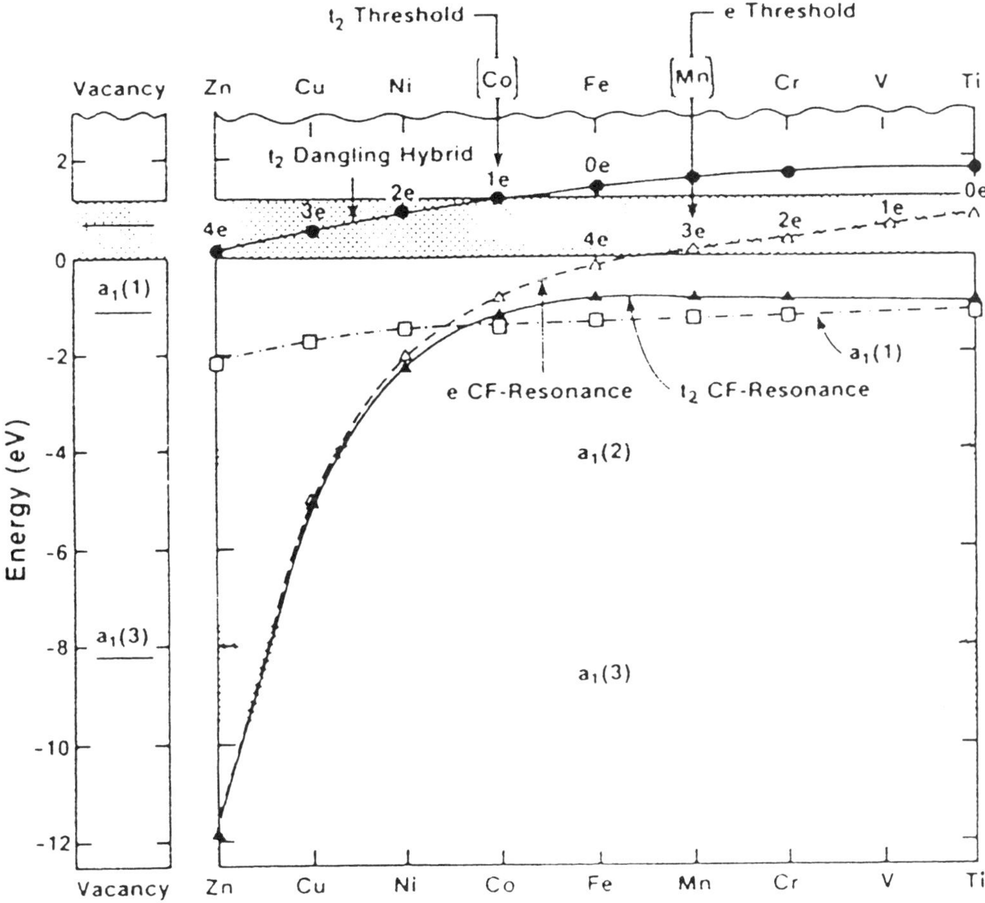

FIG. 13. The localized states in the gap region for transition-metal substitutional impurities in Si (from Zunger, 1986).

ments starting with Zn. As we already noted earlier, Zn, a group-II element, has a T_2 level in the gap that is basically the vacancy level shifted somewhat by the Zn potential. The Zn *d* electrons are near the bottom of the valence bands. Note in Fig. 13 that as we move to Cu, Ni, and so on, the *d*-impurity *d* levels move up in energy and start interacting with the vacancy levels. The T_2 part of the *d* levels interacts with the vacancy T_2 level. They mix and push apart, whereas the E level continues its smooth ascent in energy. It is clear that one needs a detailed calculation to establish the actual ordering and position of the levels relative to the gap. For a complete understanding of transition-metal impurities, one needs to consider further the arrangement of the electron spins and form multiplets. The interested reader may refer to the article by Zunger (1986) and references therein. Finally, transition-metal impurities may also occupy interstitial sites. The same articles contain relevant information.

Another important impurity in Si is oxygen. It is present in bulk Si that is used in the form of wafers in microelectronics and contributes to the mechanical stability of these wafers (oxygen-free Si wafers tend to warp). The lowest-energy configuration of oxygen is a bridge position between two neighboring Si atoms. The resulting Si–O–Si chain is analogous to the Si–O–Si chains that comprise SiO_2. The oxygen 2*p* orbitals form bonds and antibonds with Si sp^3 hybrid orbitals. All the available electrons occupy the bonding orbitals or the nonbonding "pi" oxygen 2*p* orbitals, all of which are low in energy, below the valence-band edge (compare with the fact that the band gap in SiO_2 is of order 10 eV). Thus the center has no electrically active levels in the Si band gap.

After annealing at about 400°C, oxygen atoms form clusters of three to four atoms with electrically active shallow levels. These centers have traditionally been known as "thermal donors," and the belief that they are small oxygen clusters comes from the fact that their concentration scales with some power n of the oxygen concentration, with n being 3–4. These centers have been studied extensively both experimentally and theoretically, but the atomic arrangements that give rise to the donor behavior were a mystery until recently (see Chadi, 1996).

Oxygen also segregates into small SiO_2 particles at the back of wafers. As these particles grow by the capture of additional migrating oxygen atoms, Si interstitials are emitted occasionally for volumetric reasons: as oxygen atoms are inserted in Si–Si bonds, the Si–Si distance stretches from 2.35 Å to about 3 Å (Ourmazd *et al.*, 1984). These SiO_2 particles are effective at capturing unwanted transition-metal impurities such as iron, cleaning the near-surface region of the wafer where devices are fabricated. This phenomenon is called intrinsic gettering.

Under electron bombardment—e.g., the experiments by Watkins described earlier in the context of vacancies—oxygen atoms in the bridge position capture a vacancy and form a vacancy-oxygen pair. The center actually consists of a substitutional oxygen atom that is displaced from the nominal site. The oxygen simply strives to form a bridge configuration with two of the neighboring Si atoms. The electronic structure of this center is simple to understand. The oxygen $2p$ orbitals bond with the two Si dangling hybrids and form Si–O–Si chains that have no levels in the gap. The other two Si dangling hybrids give rise to a state in the gap that takes up the two electrons, making this center a deep double donor.

Hydrogen is a unique impurity that has intriguing behavior in Si. It goes in interstitially and is a fast diffuser so that it has not been observed as an isolated impurity. It has been found to pair with impurities. In particular, it binds strongly with shallow acceptors and deactivates them, i.e., it forms a H–impurity pair that is no longer electrically active as a shallow acceptor (see review by Van de Walle, 1994). Theory has established that the lowest-energy site for the H atom in the case of boron impurities is along one of the B–Si bonds (see Van de Walle, 1994). There is actually a low-energy ring around the boron atom with the barrier for H motion in this ring only about 0.2 eV (see Fig. 14). Hy-

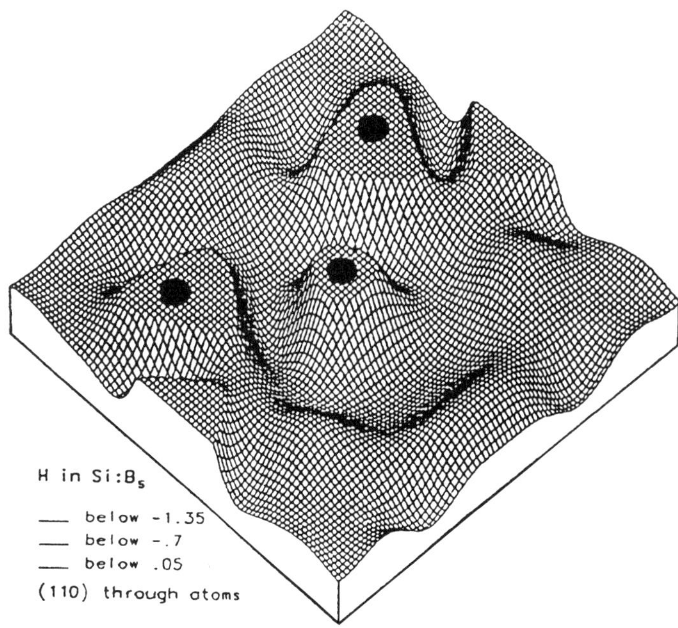

FIG. 14. Total energy for a hydrogen atom in the vicinity of a boron substitutional impurity in Si (from Denteneer *et al.*, 1989).

drogen also forms pairs with donor impurities, but the bonding is much weaker. Finally, H bonds strongly to dangling bonds at vacancies and surfaces forming Si–H bonds.

3.6 Surfaces

In Sec. 1.2 we described how atoms are arranged at the "ideal" primary Si surfaces. If we pursue the simple picture of sp^3 Si orbitals that form bonding and antibonding orbitals in the bulk perfect crystal, we recognize immediately that the creation of an "ideal" surface breaks an array of bonds, leaving the so-called dangling bonds or dangling hybrids. Every such dangling hybrid at a surface contains a single electron whose energy is now at the level of the sp^3 hybrid, much higher than the level of a bonding orbital. Reconstruction occurs because rearrangements of the surface atoms can give rise to some kind of bonding between the dangling sp^3 hybrids. Since the final geometries are not fully tetrahedral, the sp^3 hybridization of course changes, and a more complex description of the electronic states is necessary.

For a given set of atomic positions at a free surface, the electronic structure can be calculated by a variety of techniques. First-principles calculations were pioneered in the early 1970s by Joel Appelbaum and Don Hamann of Bell Laboratories (see Appelbaum and Hamann, 1976) and later by Marvin Cohen's group at Berkeley (see, e.g., Schluter *et al.*, 1975). Semiempirical tight-binding calculations, though less accurate, have the advantage of fast exploration of many structures (see, e.g., Chadi, 1979; Pandey, 1981). Photoemission spectroscopy, pioneered by Dean Eastman at IBM and William Spicer at Stanford (see Eastman and Grobman, 1972, and Wagner and Spicer, 1972, for seminal applications to Si surfaces), was developed as a powerful experimental technique to map out the dispersions of the electron states. Comparison of calculated and measured dispersions is an indirect way to explore surface reconstructions. Similarly, LEED calculations (see, e.g., Duke and Laramore, 1971, 1972) and comparisons with experimental spectra can do the same thing. Scanning tunneling microscopy, developed in the early 1980s by Gerd Binnig and Heine Rohrer of IBM's Zurich Laboratory, is a more direct experimental probe, as it produces atomic-resolution images of surfaces (see Binnig *et al.*, 1983). Finally, first-principles total-energy calculations and dynamical simulations, developed in 1985 by Roberto Car and Michele Parrinello in Trieste, allow direct exploration of reconstructions by computer.

All the above techniques plus many more played a role in unraveling the structure of Si surfaces. In their original LEED article, Schlier and Farnsworth (1959) proposed that the surface atoms of the (100) 2 × 1 surface form dimers. Other competing models were proposed over the years, but Appelbaum and Hamann's calculations showed that the dimer model yielded densities of surface states that were in best agreement with photoemission data. Angle-resolved photoemission data by Himpsel and Eastman (1979), however, found nonmetallic dispersions whereas the theoretical calculations had metallic dispersions. The puzzle was resolved by Chadi (1979) of Xerox with tight-binding calculations. He found that the total energy is lowered and the dispersions become nonmetallic when the dimers are slanted (Fig. 15). These asymmetric dimers account for the 4 × 2

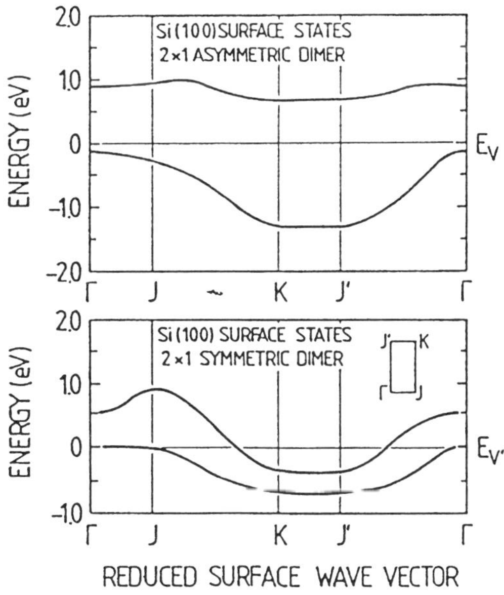

FIG. 15. The surface-state dispersions of a Si(100) 2 × 1 surface for (a) asymmetric dimers and (b) symmetric dimers. The horizontal line in each case marks Fermi energy (all levels below it are occupied by electrons). From Chadi (1979).

structure as well: the dimer slant alternates along a chain of dimers, doubling the size of the unit cell. In the last 15 years, the symmetry of the dimer has been debated, as a number of experiments seemed to find symmetric dimers. It is now well established, however, that the dimers are asymmetric and that some experiments, e.g., scanning tunneling microscopy, may in fact "see" symmetric dimers because of the way they interact with the surface.

The (111) 2 × 1 surface is another interesting story. The ideal surface has an array of dangling bonds sticking out in the ⟨111⟩ direction. Each dangling bond has one electron. In early work by Haneman (1961) it was proposed that a buckling reconstruction would set in: half of the atoms would move up and half would move down. The dehybridization of the sp^3 hybrids would make the down energy level have a lower energy so that both electrons would occupy it. First-principles calculations, however, found that the Coulombic repulsion between the two electrons worked against this picture. Furthermore, calculated dispersions did not agree with photoemission data. The puzzle was resolved by Pandey (1981). Exploring various structures with tight-binding Hamiltonians and comparing the calculated dispersions with experiment, he came upon a reconstruction of pi-bonded chains that worked best. Subsequent total-energy calculations provided confirmation. The atomic rearrangements of the top two layers are significant, but the gain from the pi bonding is sufficient to win over distortion energies.

Finally, the Si(111) 7 × 7 surface has an equally long story. There, experiment pretty much did it alone. First the STM images of Binnig *et al.* (1983) established the existence of adatoms. Subsequent electron-microscopy studies by Takayanagi *et al.* (1985) established the existence of a stacking fault underneath half of the unit cell.

4. TRANSPORT PROPERTIES

Silicon is the prototypical semiconductor and is characterized by n-type or electron conduction and p-type or hole conduction. The conductivity σ is given by

$$\sigma = ne\mu_n + pe\mu_p, \tag{13}$$

where n and p are the electron and hole concentrations, respectively, and the μ's are the corresponding mobilities. The mobility determines the effective average carrier velocity for an externally applied electric field E:

$$\bar{v} = \mu E. \tag{14}$$

In a simple theory, mobilities are given by

$$\mu = e\tau/m^*, \tag{15}$$

where m^* is the carrier's effective mass and τ is a relaxation time, i.e., the average time between collisions by the carrier with whatever scattering centers are present (phonons, impurities, etc.). In the case of electrons in Si, all six equivalent valleys contain on average the same number of carriers. The m^* that appears in Eq. (13) is the so-called conductivity effective mass that can be shown to be

$$\frac{1}{m^*} = \frac{1}{3}\left(\frac{1}{m_\parallel} + \frac{2}{m_\perp}\right). \tag{16}$$

The carrier velocity would increase linearly with the electric field if the mobility were independent of the field. That is true only for weak electric fields, which corresponds to Ohm's law. For such fields, one is interested in the dependence of τ and hence μ on the carrier effective mass and on temperature. In Si the dominant scattering processes are by acoustical phonons and by ionized impurities. The two mean collision times are then combined in the form

$$1/\tau = 1/\tau_1 + 1/\tau_2. \tag{17}$$

The mobility due to scattering by acoustic phonons is given by

$$\mu \sim m^{*-5/2}T^{-3/2}, \tag{18}$$

whereas the mobility due to scattering by ionized impurities is given by

$$\mu \sim m^{*-1/2}N_I^{-1}T^{3/2}, \tag{19}$$

where N_I is the ionized impurity density.

At low electric fields, the carriers maintain a quasi-equilibrium state with the lattice and a Maxwellian velocity distribution. Basically, as the field imparts energy to the carriers, they emit more acoustical phonons than they absorb so that they maintain the same temperature as the lattice. At fairly high fields, the carriers acquire excess energy that they do not share with the lattice very efficiently. They become "hot" with an effective temperature Θ that is higher than the lattice temperature T. The average drift velocity drops below the low-field value $\mu_0 E$ and is given by

$$\bar{v} = \mu_0 E \sqrt{T/\Theta}. \tag{20}$$

At sufficiently high fields, the average drift velocity saturates at about 10^7 cm/s (Fig. 16).

The lifetimes of carriers are determined by the recombination processes that are active. Typically, deep impurities and other defects with localized energy levels in the band gap provide an intermediate step by first capturing a carrier at the deep level and then releasing it for recombination with a complementary carrier in the other band. Such processes can occur in principle by either the emission of light or multiphonon emission.

In the absence of deep impurities and defects, the lifetimes of carriers are primarily limited by Auger recombination; namely, an electron recombines with a hole with the band-gap energy taken up by another carrier that gets excited to a higher energy. The latter carrier subsequently loses the extra energy to the lattice by phonon emission. Experiments to measure these Auger rates are usually carried out in heavily doped but otherwise very pure, dislocation-free single crystals. Theoretical calculations (Laks *et al.*, 1990) have established that in Si, Auger recombination of electrons is a pure electronic process, whereas for holes the process is assisted by phonons. This difference is caused by the details of the energy bands. Both energy and crystal momentum (the wave vector **k**) need to be conserved during the Auger process. For electrons, it is possible to conserve both with pure electronic transitions, but for holes some of the momentum—and hence some of the energy—must be transferred to the lattice by phonon absorption.

Light emission during a recombination process is not generally efficient in Si because it has an indirect band gap. As a result, crystal momentum cannot be conserved during photon emission. In the last several years, visible luminescence has been observed from porous Si, a form of Si with large voids. It appears that this system has an effective direct gap of about 1.7 eV with large cross section for radiative recombination. A number of suggestions have been made about the origins of the effect, but detailed understanding is still incomplete.

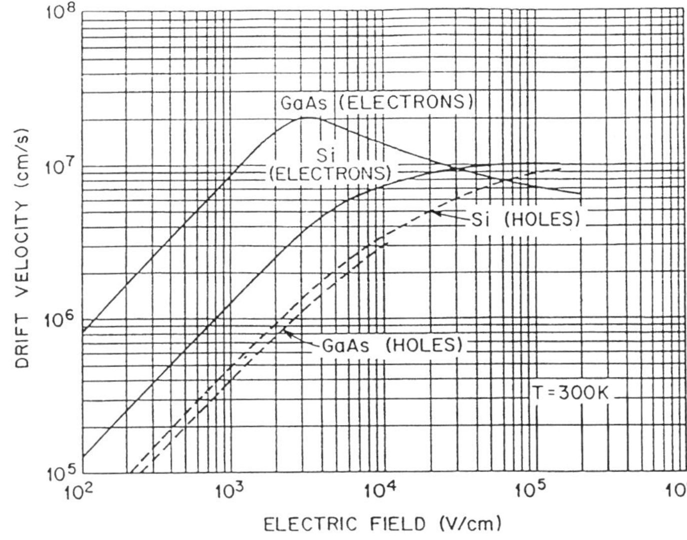

FIG. 16. The drift velocity of electrons and holes in Si and GaS as a function of electric field (from Sze, 1981).

5. THERMAL PROPERTIES

The thermal conductivity of Si has a contribution from the lattice, a contribution from electronic conduction, and a cross term from mixed electron/hole conduction. The direct contribution of conduction carriers to the thermal conductivity is quite small, but the mixed term can be large. The thermal conductivity of Si is shown as a function of temperature in Fig. 17. For comparison, the thermal conductivity of copper is also shown.

A temperature gradient can induce an electric current, and the corresponding linear coefficient P is known as the differential thermoelectric power ($J = -P\partial T/\partial x$). The thermoelectric power is negative for n-type Si and positive for p-type Si.

6. ATOMIC DIFFUSION

Self-diffusion in crystalline Si has been measured by a variety of techniques using radioactive tracers. The diffusion constant has the form

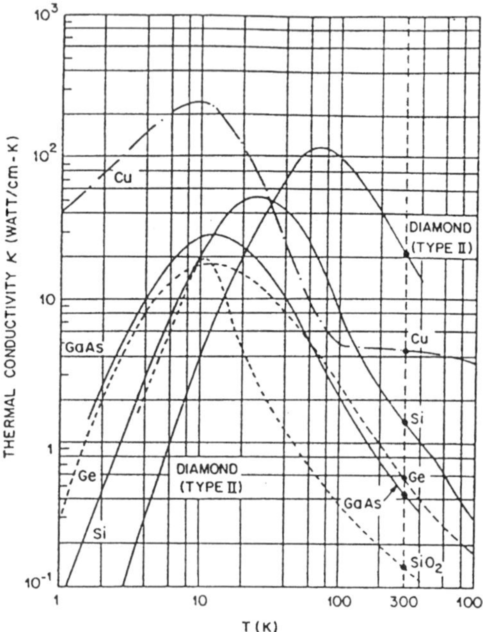

FIG. 17. The thermal conductivity of Si and a few other solids as a function of temperature (from Sze, 1981).

$$D = D_0 \exp(-Q/kT). \tag{21}$$

Measured values of the activation energy Q range from 4 to 5 eV, depending on the range of temperatures that was accessible to the particular experiment. There is no systematic variation of Q, however, so that the range is best viewed as an error bar.

Self-diffusion can be mediated by vacancies or self-interstitials. When a vacancy makes a jump to a neighboring site, a Si atom makes a jump in the opposite direction. A self-interstitial can migrate in the interstitial channels and occasionally kick out a Si atom from a normal atomic site and take its place while the newly created interstitial moves on. In both these mechanisms, the activation energy for self-diffusion is equal to the sum of the defect's formation and migration energies.

Early experiments in the late 1960s and the 1970s led to serious controversies over the role of vacancies and interstitials in Si diffusion. There were strong advocates of one or the other defect as being the dominant or sole mediator of diffusion. First-principles calculations of formation and migration energies were first reported in 1984 by Car *et al.* (see Pantelides, 1994), and the surprise result was that the two defects had comparable formation and migration energies, leading to virtually identical Q's in the range of experimental values. A few years later, Pandey (1986) demonstrated with first-principles calculations that direct exchange between neighboring Si atoms without any defect is possible with an activation energy of about 4.5 eV. In order to determine the relative contributions of the three mechanisms, one would need calculations of the pre-exponentials D_0 or the full diffusion constants. Such calculations were recently carried out (Bloechl *et al.*, 1993), and the results favor the interstitial (Fig. 18).

Dopant diffusion in Si has been measured to have activation energies that are typically about 1 eV smaller than the activation energy of self-diffusion (see Fahey *et al.*, 1989). Dopant impurity diffusion can occur via precisely the same mechanisms as self-diffusion, but at least the defect-mediated mechanisms would have the same activation energy as self-diffusion. Additional mechanisms are possible, however. A vacancy and a substitutional impurity may form a pair, and then

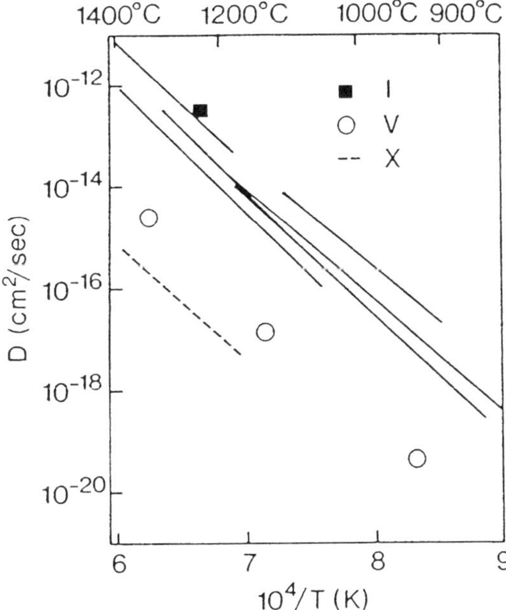

FIG. 18. The self-diffusion constant of Si for the interstitial mechanism (I), the vacancy mechanism (V), and the exchange mechanism (X) compared with experimental data (from Bloechl *et al.*, 1993).

the pair migrates. The process is rather intriguing: The vacancy migrates around a hexagonal ring of atoms and positions itself in "front" of the impurity; the impurity jumps into the vacant site; the vacancy is now at the "back" of the impurity and again migrates to the front via a hexagonal ring, and the process goes on. Calculations have demonstrated that for some impurities the formation plus migration energy of the pair is about 1 eV smaller than the self-diffusion activation energy, in agreement with experiments (Nichols *et al.*, 1989).

An analogous process is possible with a self-interstitial. The self-interstitial can take up the impurity site by pushing the impurity to the next atomic site, where the impurity in turn pushes the resident Si atom out into an interstitial site. The new interstitial can turn around and do the same thing to the impurity and so on. The self-interstitial has yet another possibility: it may simply kick an impurity out into the interstitial region and take its place at the atomic site. The impurity then can migrate in the interstitial region until it kicks back into a lattice site.

Neither experiment nor theory has been able to determine unambiguously the relative roles of vacancies and interstitials. There is sufficient evidence, however, to believe that antimony diffuses predominantly via vacancies, whereas the other impurities have strong interstitial components.

7. MECHANICAL PROPERTIES

The bulk modulus B of a solid is defined by

$$V = -B\frac{dV}{dp}, \tag{22}$$

where V is the volume and p is the pressure. It is an elastic constant that measures the rigidity of the solid against volume changes. The compressibility is defined as the reciprocal of the bulk modulus. For crystalline Si, the bulk modulus is 0.988×10^{12} dyn/cm^2.

At absolute zero, one has $dp/dV = -d^2U/dV^2$, where U is the internal energy, so that

$$B = V\frac{d^2U}{dV^2}. \tag{23}$$

This expression allows the direct evaluation of the bulk modulus from first-principles calculations of the total energy as a function of volume. Theoretical values are within 1% of the experimental value.

For elastic deformations that reduce the symmetry of the solid, one in general needs to introduce 21 independent elastic constants that connect the six independent strain components to the corresponding six stress components:

$$\sigma_i = \sum_{i=1}^{6} c_{ij}\epsilon_j. \tag{24}$$

There are only 21 independent elastic constants because c_{ij} is symmetric. For a cubic solid such as Si, symmetry further reduces the number of independent elastic constants to three. In addition to the bulk modulus discussed earlier, the other two independent elastic constants are c_{12} and c_{44}. Their values for Si are 6.39×10^{11} dyn/cm^2 and 7.95×10^{11} dyn/cm^2, respectively.

Plastic deformation of Si via the generation of dislocations occurs when the material is compressed beyond the elastic limit. The external stress induces dislocations to glide. When dislocations are pinned at the two ends by some crystal defects or impurities, the external stress bows them to the point where new dislocation loops are emitted (Frank-Read sources). In addition to dislocations, large external stresses induce clusters of point defects with a deep electrical level below midgap (about 0.45 eV from the valence-band edge) that pin the Fermi level in lightly doped material. These defects have also been seen by EPR measurements. Annealing data indicate that the level is not directly associated with the dislocations (Kimerling *et al.*, 1981).

8. CRYSTAL GROWTH

8.1 Bulk Crystal

The first step in the production of high-purity single-crystal Si is the reduction of SiO_2 (quartz, common sand) with carbon at high temperatures (in excess of 2000 °C) in a submerged-electrode arc furnace. The oxygen is removed in the form of CO, and the metallurgical grade Si at this point contains 1–2% Fe, Al, and other impurities, including P and B. In the next step, metallurgical grade Si is mixed with anhydrous hydrogen chloride. A number of reactions occur that remove most of the impurities as chloride salts. Silicon is distilled in the form of trichlorosilane ($SiHCl_3$). In a subsequent step, trichlorosilane is decomposed thermally into Si and HCl in the presence of H_2. During the decomposition, Si is deposited in polycrystalline form on a high-purity thin rod of Si known as "slim rod."

The two most important industrial techniques to produce high-grade single-crystal Si are the Czochralski and the float-zone methods (see CRYSTAL GROWTH). In the Czochralski method (Czochralski, 1917), polycrystalline Si is melted in a quartz crucible in argon atmosphere. A single-crystal seed of the desired orientation is dipped into the molten Si and withdrawn from the melt at a controlled rate. The seed and the crucible are usually rotated in opposite directions for uniformity. The diameter of the pulled cylindrical material and hence the resulting wafers was initially 75 mm. Current technology uses 200-mm wafers. Today 300-mm wafers are produced and are expected to be used in microelectronics by the end of the decade.

In the float-zone method no crucible is used. Instead, a rod of polycrystalline Si is converted to single crystal by traversing a molten zone from one end in contact with a single-crystal seed to the other end. The zone melting is done with rf heating. The main difference in the quality of float-zone Si is its low oxygen content. Czochralski Si gets the oxygen and potentially other contaminant impurities from the quartz crucible. By employing vacuum in the growth chamber, float-zone Si can be grown extremely pure.

Czochralski-grown Si is doped as desired by adding appropriate elements or alloys in the crucible so that, when the polysilicon melts, the dopants are distributed uniformly. In the float-zone method, dopants are introduced in the molten zone in the form of gases. During device fabrication, selective doping is done through masks either by in-diffusion or by ion implantation (see ION-BEAM MODIFICATION OF MATERIALS) and subsequent annealing.

8.2 Epitaxial Thin Films

For device fabrication, it is advantageous sometimes to grow epitaxial thin Si films. Two processes are particularly effective. Chemical vapor deposition (CVD) [see CVD (CHEMICAL VAPOR DEPOSITION)] employs a furnace in which many wafers are exposed to a gas of silane. At the Si surface the silane decomposes, and crystalline Si is deposited. In this conventional CVD, used by the microelectronics industry, the temperature is usually high (~1050 °C) in order to keep the surface free of adsorbates. At lower temperatures polycrystalline or amorphous Si is deposited. In recent years Bernard Meyerson of IBM Research (Meyerson, 1986; Meyerson *et al.*, 1988) developed an ultrahigh-vacuum (UHV) version of CVD in which the partial pressures of water vapor and other gases are kept low enough to avoid oxidation and contamination. High-quality crystalline Si films are deposited at ~500 °C. Hydrogen plays an important role in this process by keeping the exposed surface passivated. If the temperature is increased by even 100 °C, hydrogen is removed, and the surface cannot be maintained as clean.

Another method for thin-film deposition is molecular beam epitaxy (*q.v.*) (MBE), where a beam of Si atoms is directed at a Si surface. Low-temperature (~700 °C) crystalline film deposition is again possible. Dopants can be incorporated by simultaneous beam fluxes of atoms or molecules. The drawback of this method is that a film can be deposited on a single wafer at a time by scanning the beam.

Both the UHV-CVD and MBE methods are capable of admixing dopants and of fabricating $Si_{1-x}Ge_x$ films by admixing Ge. For small x the SiGe film can be grown epitaxially on Si with Ge under lateral compression. After a certain thickness, misfit dislocations are induced to relieve the strain. The SiGe system has been studied extensively and can be used to fabricate devices that are much faster than pure-Si devices.

8.3 Silicon-On-Insulator

In microelectronics, Si devices are typically fabricated at the surface of a Si wafer. The Si material underneath the devices is not a good insulator so that devices are not thoroughly electrically isolated from one another. Parasitic transistors through the Si substrate cause a phenomenon called latch-up. Technology designers have found ways to suppress the effect. Ideally, however, one would prefer to have an insulating substrate. Beginning in the early 1960s, the idea of growing thin crystalline Si films on insulating substrates has been investigated extensively. Fabrication of devices and circuits on such Si films would avoid all complications one encounters with regular Si wafers. In addition, such chips are more radiation resistant. For example, circuit failures known as soft errors can be caused by electron–hole pairs generated in the bulk Si substrate by impinging alpha particles. If the substrate is insulating, the effect is minimized. Finally, successful fabrication of circuits on thin Si films grown on insulators would allow three-dimensional stacking of devices.

The technology of silicon-on-insulator (SOI) is quite advanced today and is used for niche applications. The cost of SOI wafers is still significantly higher than regular Si wafers, but new advances may eliminate this differential soon. As the level of integration is approaching the 10^9-devices-per-chip level by the end of the century, SOI technology may play a very important role.

Epitaxial SOI has been grown successfully on sapphire (see, e.g., Vasudev, 1986) even though the thermal properties and crystal structures of Si and sapphire are quite different. The Si films are deposited by a variety of methods, including vacuum evaporation, sputtering (*q.v.*), molecular beam epitaxy, or chemical vapor deposition. The last is the most successful and is used almost exclusively in commercial applications.

Growth of Si on sapphire proceeds initially by the growth of islands that then coalesce to form a film. The nuclei are typically (100)- and (110)-oriented crystallites. The (100) crystallites, however, grow faster and eventually trap and cover the (110) nuclei. As the (100) islands coalesce, stacking faults occur at the island boundaries. As the film grows thicker, the density of defects is reduced significantly, and device-grade films can be grown.

Another approach altogether to SOI is the implantation of oxygen or nitrogen in a regular Si wafer to form a buried oxide or nitride (see Lam, 1986). The process is known as SIMOX (separation by implanted oxide). The implantation is done at about 500 °C, which leads to a good dielectric, but the Si layer at the top is quite defective and thin. After annealing, the Si film improves significantly. Additional layers of Si are then grown epitaxially to achieve the desired thickness for device fabrication.

9. PROCESSING

About ten years after the invention of the transistor, another major breakthrough led to the modern microelectronic revolution: the development of the Si planar technology that allowed the fabrication of entire circuits on the surface of a thin Si wafer. The pioneers of these developments were John Kirtley of Texas Instruments and Robert Noyce of Fairchild Semiconductor. Nowadays entire microprocessors are fabricated on a single piece of a Si wafer (chip). The fabrication of integrated circuits entails a series of processing steps, involving masking for selective diffusion, oxidation, deposition, etching, etc.

The use of photoresists and lithography is a key step for selective processing. The description of all the various processing steps used in Si technology is beyond the scope of this article (see VLSI—INTEGRATED CIRCUITS). We already discussed the deposition of epitaxial Si films in the previous section. In this section we touch briefly on oxidation, etching, and process modeling.

9.1 Oxidation

Oxidation of Si for device fabrication is done by exposing Si to O_2 gas at a high temperature (thermal oxidation; see Katz, 1983). After the initial oxidation, oxygen diffuses through the oxide to the interface where it gets incorporated. As oxygen atoms insert themselves between Si–Si bonds to form Si–O–Si bridges, the Si–Si distance increases from 2.35 Å to about 3 Å. As a result, oxidation leads to the emission of Si self-interstitials (Fahey *et al.*, 1989). The resulting interface is generally of high quality and very flat. There is a small concentration of threefold-coordinated Si atoms (dangling bonds) that are known as P_b centers.

The oxide films used for gate dielectrics in field-effect transistors are very thin and by the end of the decade are expected to be about 40 Å. This requirement is putting new constraints on the quality of the oxide and of the interface. This is a highly active area of research today. It has been found that the introduction of small amounts of nitrogen enhances the quality of the oxide and blocks hydrogen diffusion through the interface, which has detrimental effects.

9.2 Etching

The removal of Si material during device fabrication used to be done with wet chemicals. For increased control, dry etching is now the predominant method (see PLASMA ETCHING). The primary gas in the chamber is fluorine. The atomic-scale processes that result in etching have been studied extensively, both experimentally and theoretically. Initially F ions tie off dangling bonds at the Si surface. Then they penetrate into the subsurface region where they can insert themselves between neighboring Si atoms. As a result, surface Si atoms are gradually surrounded by two, three, and ultimately four F atoms. SiF_2, SiF_3, and SiF_4 molecules are ejected, partly helped by the energy of the impinging ions. F ions may also diffuse interstitially further into the bulk Si, but they can no longer insert themselves between neighboring Si atoms because of the stiffness of the lattice. The near-surface etching is assisted by the presence of holes because the F^- ion needs to get rid of its electron when it inserts itself between neighboring Si atoms (Van de Walle *et al.*, 1988). For this reason, *n*-type material etches at a faster rate because at the surface an inversion layer is formed that traps holes.

9.3 Process Modeling

The equations that describe diffusion in a crystal are well known (Fick's laws). If one assumes a particular mechanism for dopant diffusion (e.g., vacancy–impurity pairs), one can systematically write down the equations that describe the evolution of a dopant profile at high temperatures. These equations need to be solved for the particular geometries of devices. Over the last 20 years, powerful computer programs have been developed that do such modeling in one, two, and three dimensions. As the mechanisms of impurity diffusion have been elucidated over the years, these computer programs have become quite sophisticated. They are used extensively in the design and development of electronic devices and circuits.

Modeling of oxidation has also evolved significantly over the years. The modeling is based mostly on phenomenological models of growth and viscous deformation. The models are used in industrial development. Models for deposition and etching are, however, still at an early stage and are not used extensively by industry. The models are again phenomenological based on simple rate equations but are quite successful in predicting observed profiles.

Integrated circuits are very complex mechanical systems. Stresses can develop as a result of the mismatch of thermal expansion coefficients between the different materials (Si, silicon dioxide, metals). Continuum mechanics successfully models these stresses if the constitutive relations of the relevant materials are known. The latter are usually difficult to establish for thin films and complex structures.

10. SILICON-BASED MICROELECTRONICS

In the last 40 years, Si has been the semiconductor of choice for electronic devices. Though faster devices have been demonstrated using GaAs and other compound semiconductors, they are only used in niche applications. The main advantage of Si is the development of the relevant material growth and processing techniques. Though one can grow dislocation-free single-crystal Si ingots 200 mm in diameter, GaAs material is typically 75 mm in diameter. Process control in Si allows a very high level of integration, whereas GaAs lags behind by one to two orders of magnitude. Thus, even if GaAs devices are faster, that advantage is lost if a single Si chip would be replaced by 20 or more GaAs chips. The interconnects will effectively kill any device advantage.

Process-control improvements proceed at an unrelenting pace as Si technology introduces a new generation of microelectronics every three years or so. The pace is expected to continue for at least three to four more generations. Then, the feature dimensions will be so small that quantum phenomena will set in. In anticipation of this, research in the field of quantum electronics has demonstrated many quantum devices, usually fabricated with III–V compounds. The most likely scenario, however, is that Si technology will gradually adapt to and adopt quantum phenomena, and quantum microelectronics will again be Si based. It is hard to imagine that it will be possible to achieve the necessary manufacturing control with any other material. One might project that even when we get to the point when devices will be fabricated atom by atom using a version of scanning tunneling microscopy, the material is likely to be silicon.

GLOSSARY

Absolute Zero: The lowest achievable temperature where all classical motion ceases.

Acceptor: An impurity that introduces a deficiency of one electron (a hole) in a semiconductor.

Atomic Diffusion: In a solid, the ability of atoms to migrate.

Conduction Band: In a semiconductor, the lowest-energy energy band that, at the absolute zero of temperature, contains no electrons.

Donor: An impurity that incorporates an extra electron in a semiconductor.

Doping: The addition of impurity atoms to a semiconductor to alter the concentration of electrons or holes available for conduction of electricity.

Energy Bands: The bands formed by the allowed energy levels of electrons in crystals.

Epitaxy: The process of growing a crystal layer whose atomic structure is in register with that of the substrate, hence, epitaxial growth.

Hole: An empty energy level in the valence band of a semiconductor; it behaves as a particle carrying a positive charge.

Ionization Energy: In a semiconductor, the energy needed to remove a bound electron or hole to the respective energy-band edge.

Optical: Corresponding to the range of frequencies of visible electromagnetic radiation (light).

Photoemission: An experiment in which a photon is absorbed by a solid and an electron is emitted.

Space Group: A group of mathematical operations describing the translational, rotational, and mixed symmetries of a crystal.

Transistor: A device made by heterogeneous doping of a semiconductor—it acts as a solid-state amplifier or a switch. It is the key device used in microelectronics for logic operations and memory. There are two main types of transistors, known as bipolar and field-effect transistors.

Uniaxial Stress: Stress applied along one axis.

Valence Band: In a semiconductor, the highest-energy energy band that, at the absolute zero of temperature, is fully occupied by electrons.

Works Cited

Aggarwal, R. L., Ramdas, A. K. (1965), *Phys. Rev.* **140,** A1246–A1253.

Anderson, P. W. (1975), *Phys. Rev. Lett.* **34,** 953–955.

Appelbaum, J., Hamann, D. R. (1976), *Rev. Mod. Phys.* **48,** 479–496.

Baraff, G. A., Schluter, M. (1978), *Phys. Rev. Lett.* **41,** 892–895.

Baraff, G. A., Kane, E. O., Schluter, M. (1979), *Phys. Rev. Lett.* **43,** 956–959.

Bardeen, J., Brattain, W. (1948), *Phys. Rev.* **74,** 230–231.

Bernholc, J., Lipari, N. O., Pantelides, S. T. (1978), *Phys. Rev. Lett.* **41,** 895–899.

Binnig, G., Rohrer, H., Gerber, C., Weibel, E. (1983), *Phys. Rev. Lett.* **50,** 120–123.

Blöchl, P., Smargiassi, E., Car, R., Laks, D. B., Andreoni, W., Pantelides, S. T. (1993), *Phys. Rev. Lett.* **70,** 2435–2438.

Bourgoin, J., Corbett, J. W. (1972), *Phys. Lett.* **38A,** 135–137.

Bourret, A., Rouviere, J. L. (1987), in: *Proceedings of the International Symposium on Polycrystalline Semiconductors, Grain Boundaries, and Interfaces,* Berlin: Springer-Verlag, p. 8.

Brown, F. C., Rustgi, O. P. (1972), *Phys. Rev. Lett.* **28,** 497–500.

Burns, G. (1985), *Solid State Physics,* Orlando, FL: Academic.

Car, R., Parrinello, M. (1985), *Phys. Rev. Lett.* **55,** 2471–2474.

Chadi, D. J. (1979), *Phys. Rev. Lett.* **43,** 43–47.

Chadi, D. J. (1996), *Phys. Rev. Lett.* **77,** 861–864.

Chelikowsky, J. R., Chadi, J. D., Cohen, M. L. (1973), *Phys. Rev. B* **8,** 2786–2794.

Chelikowsky, J. R., Cohen, M. L. (1976), *Phys. Rev. B* **14,** 556–582.

Czochralski, J. (1917), *Z. Phys. Chem. Stöchiom. Verwandtschafts.* **92,** 219–221.

Denteneer, P. J. H., Van de Walle, C. G., Pantelides, S. T. (1989), *Phys. Rev. B* **39,** 10809–10824.

Duke, C. B., Laramore, G. E. (1971), *Phys. Rev. B* **2,** 4765–4782.

Duke, C. B., Laramore, G. E. (1972), *Phys. Rev. B* **3,** 3183–3197.

Eastman, D. E., Grobman, W. D. (1972), *Phys. Rev. Lett.* **28,** 1378–1381.

Fahey, P., Griffith, P. B., Plummer, J. D. (1989), *Rev. Mod. Phys.* **61,** 289–384.

Grimmeiss, H. G., Janzen, E. (1994), in: S. T. Pantelides (Ed.), *Deep Centers in Semiconductors,* New York: Gordon & Breach.

Haneman, D. (1961), *Phys. Rev.* **121,** 1093–1100.

Harrison, W. A. (1980), *Electronic Structure and the Properties of Solids,* San Francisco: W. H. Freeman and Co.

Himpsel, F., Eastman, D. E. (1979), *J. Vac. Sci. Technol.* **16,** 1297–1299.

Hirth, J. P., Lothe, J. (1992), *Theory of Dislocations,* Malabar, FL: Krieger Publ.

Hohenberg, P., Kohn, W. (1964), *Phys. Rev.* **136,** B864–B871.

Hybertsen, M., Louie, S. G. (1985), *Phys. Rev. Lett.* **55,** 1418–1421.

Kamins, T. (1988), *Polycrystalline Silicon for Integrated Circuit Applications,* Boston: Kluwer Academic Publ.

Katz, L. E. (1983), in: S. M. Sze (Ed.), *VLSI Technology,* New York: McGraw-Hill.

Kimerling, L. C., Patel, J. R., Benton, J. L., Freeland, P. E. (1981), in: R. R. Hasiguti (Ed.), *Defects and Radiation Effects in Semiconductors 1980,* London: Institute of Physics.

Kohn, W. (1957), in: H. Ehrenreich, F. Seitz, D. Turnbull (Eds.), *Solid State Physics,* Vol. 5, New York: Academic, p. 257.

Kohn, W., Luttinger, J. M. (1955), *Phys. Rev.* **98,** 915–922.

Laks, D. B., Neumark, G. F., Pantelides, S. T. (1990), *Phys. Rev. B* **42,** 5176–5185.

Lam, H. W. (1986), in: B. J. Baliga (Ed.), *Epitaxial Silicon Technology,* Orlando, FL: Academic.

Maiti, A., Chisholm, M., Pennycook, S. J., Pantelides, S. T. (1996), *Phys. Rev. Lett* **77,** 1306–1309.

Meyerson, B. (1986), *Appl. Phys. Lett.* **48,** 797–799.

Meyerson, B. S., Uram, K. J., LeGues, F. K. (1988), *Appl. Phys. Lett.* **53,** 2555–2557.

Nara, H. (1965), *J. Phys. Soc. Jpn.* **20,** 778–784.

Nichols, C. S., Van de Walle, C. G., Pantelides, S. T. (1989), *Phys. Rev. Lett.* **62,** 1049–1052.

Ourmazd, A., Schröter, W., Bourret, A. (1984), *J. Appl. Phys.* **56,** 1670–1681.

Pandey, K. C. (1981), *Phys. Rev. Lett.* **47,** 1913–1917.

Pandey, K. C. (1986), *Phys. Rev. Lett.* **57,** 2287–2290.

Pantelides, S. T. (1986), *Phys. Rev. Lett.* **57,** 2979–2982.

Pantelides, S. T., Selloni, A., Car, R. (1985), *Solid State Electron.* **28,** 17–24.

Pantelides, S. T. (1994), in: S. T. Pantelides (Ed.), *Deep Centers in Semiconductors,* New York: Gordon & Breach.

Penn, D. R. (1962), *Phys. Rev.* **128,** 2093–2097.

Phillips, J. C. (1973), *Bonds and Bands in Semiconductors,* New York: Academic.

Schlier, R. E., Farnsworth, H. E. (1959), *J. Chem. Phys.* **30,** 917–926.

Schlüter, M., Chelikowsky, J. R., Louie, S. G., Cohen, M. L. (1975), *Phys. Rev. B* **12,** 4200–4214.

Shockley, W. (1949), *Bell Syst. Tech. J.* **28,** 435–489.

Sze, S. M. (1981), *Physics of Semiconductor Devices,* 2nd ed. New York: Wiley.

Takanayagi, K., Tanishiro, Y., Takahashi, M., Takahashi, S. (1985), *J. Vac. Sci. Technol. A* **3,** 1502–1506.

Van de Walle, C. G. (1994), in: S. T. Pantelides (Ed.), *Deep Centers in Semiconductors,* New York: Gordon & Breach.

Van de Walle, C. G., McFeeley, F. R., Pantelides, S. T. (1988), *Phys. Rev. Lett.* **61,** 1867–1870.

Vasudev, P. K. (1986), in: B. J. Baliga (Ed.), *Epitaxial Silicon Technology,* Orlando, FL: Academic.

Wagner, L. F., Spicer, W. E. (1972), *Phys. Rev. Lett.* **28,** 1381–1384.

Watkins, G. D. (1994), in: S. T. Pantelides (Ed.), *Deep Centers in Semiconductors,* New York: Gordon & Breach.

Watkins, G. D., Corbett, J. W. (1961), *Phys. Rev.* **121,** 1001–1014.

Zunger, A. (1986), in: H. Ehrenreich, D. Turnbull (Eds.), *Solid State Physics,* Vol. 39, New York: Academic, p. 276.

Further Reading

Böer, K. W. (1990), *Survey of Semiconductor Physics,* New York: Van Nostrand Reinhold.

Harrison, W. A. (1980), *Electronic Structure and the Properties of Solids: The Physics of Chemical Bonds,* San Francisco: Freeman.

Lannoo, M., Bourgoin, J. (1981), *Point Defects in Semiconductors,* Berlin: Springer-Verlag.

Moss, T. S. (1980), *Handbook on Semiconductors,* Amsterdam: North Holland.

Muller, R. S., Kamins, T. I. (1986), *Device Electronics for Integrated Circuits,* 2nd ed. New York: Wiley.

SILICON, POLYCRYSTALLINE

MAURIZIO ARIENZO, *IBM Research Division, T.J. Watson Research Center, Yorktown Heights, New York, U.S.A.*,

WENDY A. ORR-ARIENZO, *IBM Microelectronics Division, East Fishkill, New York, U.S.A.*

INTRODUCTION

Polycrystalline silicon, also referred to as polysilicon, is defined as the state of aggregation of silicon in which a number of single crystals (grains) of various sizes and shapes are intimately networked via grain boundaries. Polysilicon is differentiated from amorphous silicon, since long-range order is present. It is differentiated from single-crystal silicon by the fact that the various grains have different orientations, generating grain boundaries at the crystalline discontinuities. Polysilicon plays a key role in microelectronics. In fact, much of the success of silicon integrated circuits is due to the properties of polysilicon coupled with those of silicon dioxide and silicon nitride. Since it is fully compatible with the high-temperature silicon integrated-circuit processing, it can be deposited during microfabrication on either single-crystal silicon or its conventional dielectrics, silicon dioxide or silicon nitride, creating semiconducting or, by heavily doping, conducting layers. Consequently, it is used as an electrical "contact" to silicon layers, as a capacitor plate on top of oxide [the "gate" of metal-oxide-semiconductor (MOS) transistors], and by running polysilicon lines on top of isolation, either as a local "interconnect" between devices in integrated circuits when doped to low resistivity or as a "resistor" when doped to the required resistivity. Polysilicon is now widely used in CMOS (complementary metal-oxide-semiconductor) technology as the gate, as the contact to both the source and drain, and as the local interconnect between transistors. It is also used to contact both the base and the emitter of bipolar transistors, and as a dopant diffusion source for the extrinsic base and for the single-crystalline emitter. Other applications in microelectronics include its application as a refill and spacer material. In addition to integrated-circuit applications, polysilicon is being employed in solar cells and in thin-film transistors used in flat-panel

3-527-28140-1/96/$5.00 + .50

displays. While most of the considerations in this article apply to the material used in both these applications, the reader is referred to the articles PHOTOVOLTAIC DEVICES, SOLAR ENERGY, and DISPLAY TECHNOLOGY for more detailed descriptions.

1. POLYSILICON DEPOSITION AND MICROSTRUCTURE

Although polycrystalline silicon (polysilicon) can be deposited by sputtering and evaporation, chemical vapor deposition (CVD) is the most common deposition method used in the electronic industry. CVD is the process in which the solid products of a chemical reaction are deposited on a substrate as a thin film. See CVD (CHEMICAL VAPOR DEPOSITION). Silicon CVD is commonly obtained by the pyrolyzation of silane (SiH_4) according to the reaction

$$SiH_4(s) \rightarrow SiO_2(s) + 2H_2(g) \quad (1)$$

or of the silicon halides, primarily dichlorosilane ($SiHCl_2$), trichlorosilane ($SiHCl_3$), and silicon tetrachloride ($SiCl_4$). Pyrolyzation of the halides has been used for years in the production of hyperpure silicon feedstock in the single-crystal silicon wafer manufacturing industry. This process has been particularly successful because of the purity control and its low-cost economics. By pyrolyzing either silane or any of the halides, single-crystal silicon, polysilicon, or amorphous silicon can be obtained, depending on the deposition conditions and on the substrate on which the deposition occurs. Uniform films of high quality and purity can be obtained by CVD, a clear advantage with respect to the films obtained by physical vapor deposition (PVD) methods, such as vacuum evaporation, sputtering, and molecular-beam epitaxy (MBE). PVD is the process in which the material to be deposited is passed into the vapor transport phase by a physical mechanism, like evaporation, sublimation, or ion bombardment. Also, in contrast to the directionality of all PVD techniques, CVD exhibits "conformality," or the ability to deposit a film of uniform thickness over topology or steps. Conformality and reproducibility, key requirements in today's application to device structures, have been the driving forces for the increased application of polysilicon CVD in the microelectronics industry.

1.1 Polysilicon Deposition

Uniquely in CVD, the composition of the transport vapors and of the deposited film are different. Since the chemical reactions involving the transport vapors usually occur at or near the surface, the opportunity for incorporation of contaminants from the gas phase is greatly reduced. In addition, the purity of the gases used for CVD has been steadily increasing over the past decade as a result of technological advances and user demand. The growth rate of the deposited film in a CVD reactor is limited by several kinetic barriers: the gas-phase mass transport of the species to the surface; the adsorption of at least one reactant onto the surface; the reaction on the surface, which may involve several steps each with its own rate; and finally the flux of the reaction by-products away from the surface. Polysilicon is deposited at temperatures between 600 and 650 °C in a low-pressure reactor, in a regime limited by surface reactions rather than adsorption, to obtain uniform nucleation and thus uniform thickness and conformality independently of the material on which it is grown. In this temperature and pressure regime, during the initial phases of the polysilicon growth on any surface, small clusters of adsorbed atoms diffuse on the surface until they reach nucleation sites, where they grow to form polycrystalline films. The amount of surface diffusion is influenced by many parameters, including temperature, pressure, and flow rates. Figure 1 shows the deposition rate as a function of temperature, also called Arrhenius plot, for polysilicon growth for different silane partial pressures (Adams, 1983). As can be observed, the deposition rate increases rapidly as the temperature increases and, for a given temperature, increases, as expected, with increasing silane partial pressure. At lower temperatures, the deposition rate is too slow to be practical, and a film containing microcrystalline silicon embedded in an amorphous network can be obtained. At higher temperatures, gas-phase reactions and silane depletion become significant, and nonuniform films are obtained.

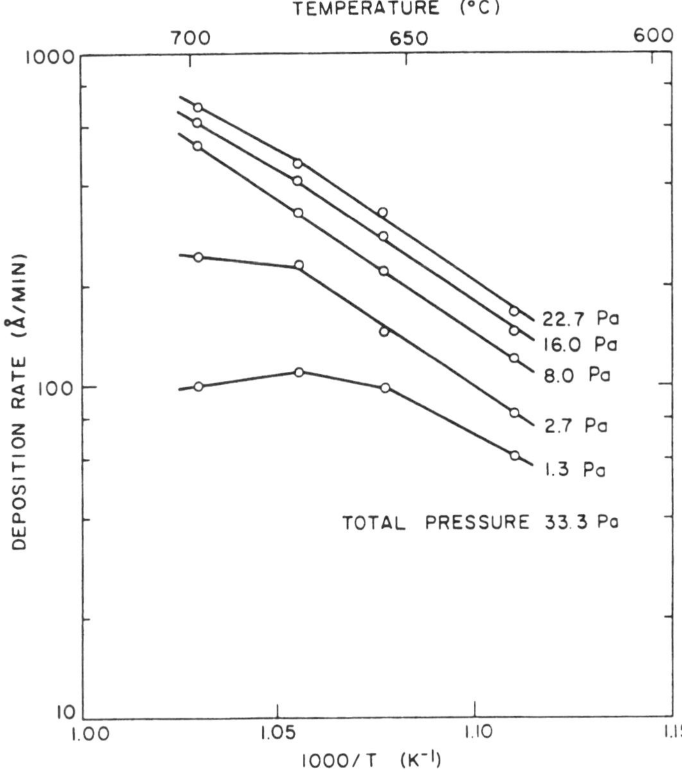

FIG. 1. Arrhenius plot for polysilicon deposition for different silane partial pressures (Adams, 1983).

1.2 Production Reactors

Chemical vapor deposition is performed at low pressures, between 0.1 and 1.0 Torr. Low-pressure CVD (LPCVD) reactors use standard diffusion furnace equipment modified to operate at lower pressures and are described in Sec. 2.2 of the article CVD (CHEMICAL VAPOR DEPOSITION) in this Encyclopedia. The excellent temperature controls present in diffusion and oxidation equipment are used to guarantee good temperature uniformity both across the wafers and from wafer to wafer. These reactors are hot-wall reactors, since the tube walls are at the same temperature as the wafers. As a result, they require periodic cleaning because peeling and particle generation become a concern.

LPCVD is performed at temperatures between 600 and 650 °C, using silane at high inlet reactant-gas mole fraction (little or no carrier gases). These two conditions, together with the lower pressure, cause the rate of the chemical reaction at the wafer surface to be the dominant kinetic mechanism for the growth (Adams, 1983; Rosler, 1977; Kern and Schnable, 1979). It has been demonstrated that virtually no decomposition of the silane source occurs in the gas phase because these reactions are usually surface nucleated (or heterogeneous) given the temperature uniformity inherent in resistance-heated furnaces. The deposited films have superior uniformity and stoichiometry, and very low defect densities. Furthermore, particulates are strongly reduced because of the lack of gas-phase nucleation.

Uniformities in the $\pm 3\%$ range (2σ) are possible both across each wafer and wafer to wafer for a standard LPCVD polysilicon deposition system. Since the interwafer distance is of the same order of magnitude as the molecular mean free path in the gas, while mass transfer is greatly enhanced at reduced pressure, more than 100 wafers of 200-mm diameter can be deposited using standard diffusion quartzware with a growth rate in the 10- to 20-nm/min range.

A typical LPCVD reactor is depicted in Fig. 2. The gases are injected at the system door and exhausted at the rear through a trap using either a mechanical pump or a

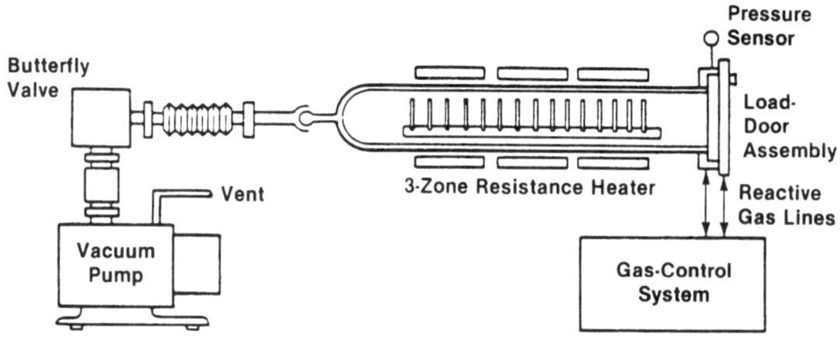

FIG. 2. Schematic of a low-pressure CVD (LPCVD) reactor used for batch deposition of polysilicon in VLSI factories.

booster pump/mechanical pump combination. Pressure control, usually between 0.2 and 1 Torr, requires control of the pumping speed. This can be achieved by introducing a butterfly valve before the mechanical pump, by varying the speed of the booster, or simply by injecting some inert gas (e.g., N_2) before the mechanical pump. However, the latter approach is discouraged because the deposition rate is a strong function of the silane partial pressure, and it is quite difficult to vary the flows and keep the partial pressures constant, especially in the presence of carrier or dopant gases. As discussed earlier, it is the partial pressure of the reactant gases (product of the mole fraction by the total pressure) that is important in determining the deposition rate. The growth rate can be significant at low pressure because of the lack of a carrier gas.

To compensate for silane depletion along the tube caused by the consumption of the reacted gas, most of the processes use a "tilted" temperature profile, with the temperature increasing toward the rear of the tube. This may cause some undesirable effects, such as different grain sizes for wafers at different tube positions, due to the high temperature sensitivity of the growth rate in the temperature range used. Polysilicon deposition using a "flat" temperature profile is now available either by diluting silane, usually 20–30% in nitrogen, increasing the total flow in the system, or by using gas manifolds inside the system. The latter are quartz tubes with appropriately spaced holes to supply reactant gases at different positions inside the tube.

1.3 Safety Considerations

It is well known that silane is explosive in concentrations in excess of 2% in air. To obviate this problem, processes using 2% diluted silane have been developed. If a flow-limiting device is included at the gas cylinder, the risk of explosion in case of a line rupture is reduced. Some cylinder manufacturers now provide such standardized devices on their cylinders. Additional other precautions, needed not only to improve safety but also to reduce the contamination caused by leakage and other sources, have been summarized by Hammond (1980).

The dopant gases used for *in situ* doping (phosphine, arsine, and diborane) are also very toxic, and in LPCVD additional safety problems are associated with the dissolution of the gases in the pump oil. A useful approach is to include a post reactor either in the back portion of the system or as a separate unit in the source cabinet before the pumping stages. Some LPCVD system manufacturers are offering such systems in their units.

1.4 Material Characterization

There are two types of characterization of polysilicon:

1. routine tests that are performed on one or more test wafers on each run (lot monitors) and
2. periodic tests that are performed either in process development or in quality control/process validation.

A standard method to characterize a polysilicon deposition process will include the analysis of the thickness and the resistivity and their uniformity both across the wafer and wafer to wafer across the system load on lot-monitor wafers. This is usually done by using oxide monitors (wafers with approximately 100 nm of silicon dioxide) on which the thickness can be easily measured with either an ellipsometric or an interferometric technique. If the process is *in situ* doped, dopant uniformity is measured by sheet resistance measurement after an appropriate anneal to activate the dopant. This anneal should be as close as possible to the thermal processing that the doped polysilicon will see in the device process since, as will be discussed later, sheet resistance is a complex function of microstructure, dopant distribution, and dopant electrical activation.

Microstructure, dopant profile, and carrier profile are monitored only periodically. By far, the most common characterization technique for polysilicon microstructure is transmission electron microscopy (TEM). A good review of the technique is contained in the book by Thomas and Goringe (1979). TEM is used for grain-size and texture analysis. In a conventional TEM sample prepared by thinning the polysilicon, the various orientations of the grains are made visible through light- and dark-field imaging, highlighting the grain boundaries. By analyzing the diffraction spots, the orientations of the grains can be obtained, and preferential alignment of the grains can be evaluated. In Fig. 3, TEM pictures of a polysilicon film before [Fig. 3(a)] and after [Fig. 3(b)] a thermal treatment are shown. As can be observed, the grains have grown considerably during the thermal treatment (from an average grain size of 30–100 nm for a 90-nm-thick film).

If the TEM sample is prepared in cross section, the presence of grain boundaries in the direction perpendicular to the surface can be verified as well as the presence or absence of voids. Figure 3(c) shows a cross section of the as-deposited film in 3(a). When a sample is viewed in cross section, the thickness of extremely thin films can be accurately determined by use of the directly resolved lattice fringe spacings of the substrate. Using x-ray analysis, more information on the texture and orientation can be obtained, together with, for example, the chemical analysis of inclusions possibly present.

With the introduction of high-resolution TEM and lattice imaging, polysilicon/single-crystal interfaces have been studied (Bravman *et al.*, 1985) to determine the chemical composition of the interface, the presence or absence of an interfacial oxide, the continuity of the interface, and the epitaxial realign-

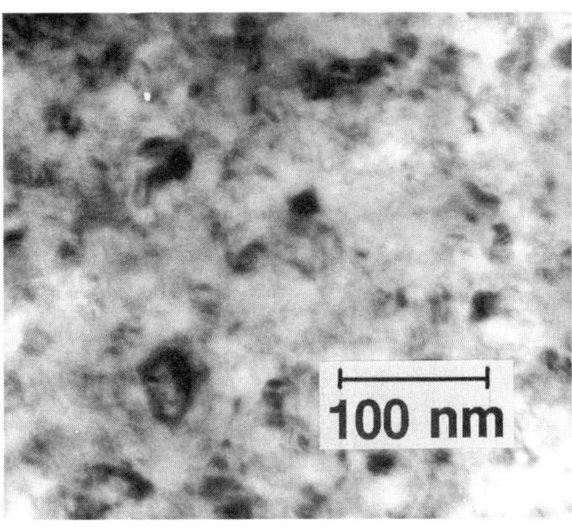

(a)

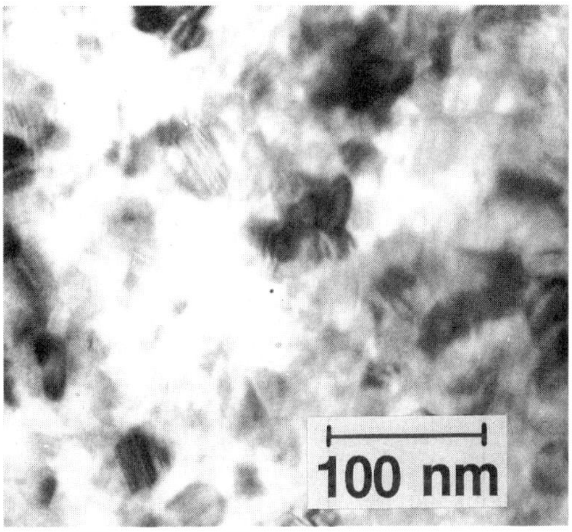

100 nm

(b)

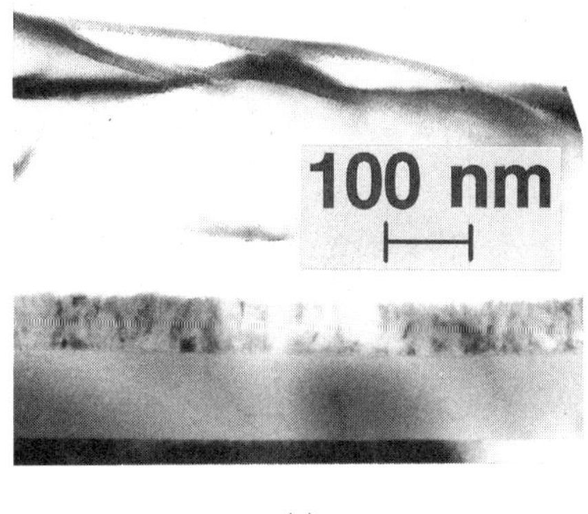

(c)

FIG. 3. Transmission electron micrograph of polysilicon grown on oxide (a) as deposited and (b) after thermal treatment, and (c) cross section as deposited.

ment of the grains at the interface, as we discuss in Sec. 2.6. Finally, combining scanning TEM (STEM) observations with energy-dispersive chemical analysis (EDX) on cross sections, an analysis of the interfacial region is feasible, collecting information on the amount of dopant incorporated there.

Two main dopant profiling techniques are used: Rutherford back-scattering (RBS) and secondary-ion mass spectroscopy (SIMS) (Marcus, 1983). RBS has advantages for dose determination, since it is an absolute measure of the chemical quantity. It cannot be used for boron and lacks sensitivity for arsenic. For doping profiling, RBS lacks resolution where the doping changes by more than one order of magnitude within 20 nm or less. SIMS is the preferred technique for B, As, or P determination, and, since resolutions of 5–10 nm per order of magnitude of dopant change can be achieved, it is also used to study the distribution of dopants in devices where the impurity profiles are very steep.

For carrier profiling, several techniques are used: spreading resistance, junction depletion capacitance, and differential Hall together with the sheet-resistance measurement. Spreading resistance is a well-known method to obtain an electrical depth profile. This technique uses a two-point probe running along a beveled sample edge to identify from the resistivity and type the profile of the active carrier concentration. This method can be used accurately for layers greater than 0.5 μm in depth. However, doping density cannot be accurately inferred in polysilicon because of electrically inactive grain-boundary doping. For junction depletion capacitance, a junction must be formed to determine the active carrier profile on the low-doped side of a unilateral step junction. These measurements can be made using a *p-n* junction, Schottky-barrier diodes, or MOS capacitors formed on the surface of the wafer to observe the depth profile from the surface into the substrate. This method is strongly influenced by the doping level and has limited application because of the junction breakdown.

Hall measurement is the most precise method to determine the mobility and the average active carrier concentration in polysilicon. The two previous methods, in fact, lead to questionable results as a result of the fields present at the grain boundaries and the difficulty of probing single grains. If polysilicon is deposited on thermal oxide, a Hall pattern can be defined by photomasking. A bulk measurement will give information on the bulk resistivity, bulk Hall mobility, and bulk carrier concentration. To find the carrier concentration and mobility profiles within a polysilicon (or single-crystal) deposited film, two methods can be used. In one, various thicknesses can be deposited and the above measurement repeated each time, obtaining the profiles from the data by a single transformation (Kamins, 1971). Another method used (Natsuaki *et al.*, 1983) is to etch layers of the polysilicon film with a dry or wet isotropic etchant and repeat the measurements each time. Again, the profiles can be simply obtained from the measurements. Some of the problems with these two methods are the grain-size variations between films of different thicknesses subjected to the same heat treatment and the difficulty in etching polysilicon isotropically without preferential grain-boundary attack.

2. IMPURITY DOPING, DIFFUSION AND ELECTRICAL PROPERTIES

The first applications of polysilicon in very large-scale integration (VLSI) have been for silicon-gate CMOS transistors and for high-value resistors. In both cases, uniform heavy doping concentration is necessary throughout the polysilicon: heavy in the first case, controlled in the second. More recently, polysilicon has also found a large number of applications as a diffusion source for both bipolar and CMOS processes. By depositing and etching a pattern in a doped polysilicon layer, a self-aligned diffusion can be obtained only in the silicon underneath the patterned polysilicon. The polysilicon itself can also act as the contact to the diffusion and can be used as an interconnection level. Doping of and diffusion in polysilicon have been extensively investigated.

Polysilicon can be doped by diffusion from a solid, liquid, or gaseous source, by ion implantation into the polysilicon, and by the addition of dopant gases during the deposition process (*in situ* doping). The structure of the polysilicon will be influenced by the presence of the dopant, and this, in turn, will influence the dopant diffusion characteristics in the polysilicon, the dopant activation, the dopant segregation, and the carrier trapping at the grain boundary.

None of the three commonly used dopants (B, P, and As) behave in quite the same way. These differences are further enhanced by varying grain-boundary composition and impurity content that are obtained by depositing in various systems and by various techniques. In this section the methods used to incorporate dopants in polysilicon, the thermal diffusion in polysilicon, and the segregation of dopants are discussed, with emphasis placed on the grain-boundary effects. Diffusion throughout the polysilicon/silicon interface is deferred to Sec. 2.6, after the discussion of the interface properties.

2.1 Dopant Incorporation

Gas and solid diffusion sources have been used to incorporate dopants in polysilicon. $POCl_3$ has been used routinely to dope the polysilicon to be used as a gate for CMOS transistors with phosphorus (Kamins, 1971, 1979). Solid sources (such as boron nitride wafers) have been used to simplify the processing steps. However, the most precise method to incorporate dopants in polysilicon is ion implantation. The advantages of ion implantation include the capability to measure accurately the number of ions incorporated (dose) and to adjust the depth at which the ions are implanted by regulating the ion energy (range). For each ion there will be a projected range of penetration, R_P, for every energy used. The depth profile of the implanted ions is approximately the same as in single-crystal silicon, the only difference being the possibility of channeling in aligned grains. However, significant differences in the profiles arise during the subsequent heat treatments. By implantation of arsenic and phosphorus (heavier ions), an amorphous layer is formed within the projected implantation range R_P for doses exceeding the mid-10^{14} cm^{-3} regime. During the subsequent anneal, this amorphous layer will rapidly recrystallize leading, in the case of heavy doping, to large grain growth or even defective crystalline layers.

Variations in polysilicon thickness and the possibility of channeling can affect the dose of dopants incorporated into the polysilicon. Preamorphization of the polysilicon layer or implantation with a tilt angle are also used in polysilicon. Contamination from the implanters (usually heavy metals) can be avoided by implanting through an oxide.

In situ doping can be achieved by the addition of arsine (AsH_3), diborane (B_2H_6), or phosphine (PH_3) during the deposition process. The main advantage with respect to ion implantation is process simplification. In some applications—for example, when polysilicon is used as a plate of the trench capacitors of advanced CMOS circuits (Goto and Inayoshi, 1987) (see Sec. 3.3.2)—*in situ* doping has the distinct advantage of providing the doping inside the trenches. For deep trenches, the only alternatives for doping incorporation are lengthier processes, deep implantation, and long anneals (Arienzo *et al.*, 1986; White *et al.*, 1985).

The addition of dopant sources in the reactors during growth introduces more species that compete for the available adsorption sites. Diborane increases the deposition rate, while the addition of either arsine or phosphine decreases it. For example, in the presence of phosphine, a self-limiting surface layer composed of a layer of adsorbed phosphorus or phosphine has been observed with a variety of UHV probes, reducing the adsorption and decomposition of silane at the surface (Meyerson and Yu, 1984). Diborane, on the other hand, showed no detectable adsorption on the silicon surface.

2.2 Dopant Diffusion

In single-crystal silicon, dopant diffusion proceeds by interaction of the dopant atoms with point defects: vacancies and interstitials. In the case of polysilicon, the grain boundaries play a very significant role in dopant diffusion. This is because grain boundaries provide paths along which dopant atoms can diffuse with less activation energy than in the grains. Dopant atoms can move from within the grains to the grain boundaries, proceed along the grain boundaries, and then diffuse back into the grains. Diffusion kinetics in polycrystalline thin films are discussed at length in Sec. 2 of the article DIFFUSION IN THIN FILMS. Since the electrical activation of the dopant atoms is different in the grains as compared to the boundaries, as discussed in Sec. 2.3, it is im-

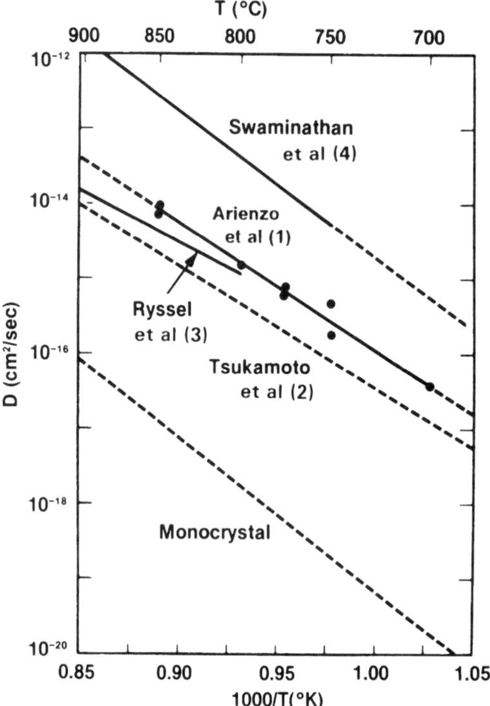

FIG. 4. Diffusion of arsenic in single-crystal and polycrystalline silicon. Experimental data and extrapolations from (curve 1) Arienzo *et al.* (1984), (curve 2) Tsukamoto *et al.* (1978), (curve 3) Ryssel *et al.* (1981), (curve 4) Swaminathan *et al.* (1982).

portant to understand the diffusion coefficients in polysilicon. In the case where polysilicon is used as a diffusion source into single-crystal silicon, the diffusion profile in the single crystal will also be determined by the properties of the polysilicon/silicon interface, as is discussed in Sec. 2.6. Various workers have determined experimentally the diffusion coefficients in polysilicon for arsenic, phosphorus, and boron. The formula governing diffusion is

$$D = D_0 e^{-E_a/kT}, \tag{2}$$

where D_0 is the pre-exponential constant term, E_a is the activation energy of the diffusion process, k is the Boltzmann constant in eV/K, and T is the absolute temperature. The values obtained for both the pre-exponential constant and the activation energy of the diffusion coefficient of a particular dopant species cover a considerable range. Figure 4 is a semilogarithmic plot of the experimental results and extrapolated data to low temperature obtained in the case of arsenic, and Table 1 is a compilation of the data found in the literature (Tsukamoto *et al.*, 1978; Ryssel *et al.*, 1981; Swaminathan *et al.*, 1982) for As-doped and B-doped polysilicon as compared to single-crystalline silicon diffusion values. It can be observed that diffusion in polysilicon is at least a few orders of magnitude faster than in single-crystal or bulk diffusion at any temperature. In the particular case of arsenic, the large scatter in the reported data is partially due to the fact that in most of the experiments, the dopants are introduced by implantation, which amorphizes the surface region. During the subsequent heat treatments, the amorphous layer caused by the implantation damage will recrystallize, resulting in a larger grain size in the implanted region (Michel *et al.*, 1982; Arienzo *et al.*, 1984). These morphology

Table 1. Diffusivity (cm^2/s) of arsenic and boron in single-crystal silicon and polysilicon

Temp (°C)	As crystal	As poly (a)	As poly (b)	As poly (c)	B crystal	B poly
650	1.2×10^{-21}	1.6×10^{-18}	4.4×10^{-17}	9.3×10^{-18}	1.3×10^{-19}	3.5×10^{-18}
700	1.7×10^{-20}	1.3×10^{-17}	5.4×10^{-16}	5.5×10^{-17}	1.2×10^{-18}	2.9×10^{-17}
750	1.9×10^{-19}	8.6×10^{-17}	5.2×10^{-15}	2.7×10^{-16}	8.1×10^{-18}	1.9×10^{-16}
800	1.6×10^{-18}	4.7×10^{-16}	4.1×10^{-14}	1.1×10^{-15}	4.8×10^{-17}	1.1×10^{-15}
850	1.2×10^{-17}	2.2×10^{-15}	2.7×10^{-13}	4.3×10^{-15}	2.4×10^{-16}	5.2×10^{-15}
900	7.0×10^{-17}	9.2×10^{-15}	1.5×10^{-12}	1.4×10^{-14}	1.1×10^{-15}	2.2×10^{-14}
950	3.7×10^{-16}	3.4×10^{-14}	7.3×10^{-12}	4.3×10^{-14}	4.1×10^{-15}	8.1×10^{-14}
1000	1.7×10^{-15}	1.1×10^{-13}	3.1×10^{-11}	1.2×10^{-13}	1.4×10^{-14}	2.7×10^{-13}
1050	6.8×10^{-15}	3.4×10^{-13}	1.2×10^{-10}	3.1×10^{-13}	4.5×10^{-14}	8.4×10^{-13}
1100	2.5×10^{-14}	9.5×10^{-13}	4.2×10^{-10}	7.4×10^{-13}	1.3×10^{-13}	2.4×10^{-12}

Data and extrapolations from (a) Tsukamoto et al., (b) Swaminathan et al., and (c) Ryssel et al. (1981).

changes, as well as the effect produced by residual lattice defects generated by implantation, complicate the dopant redistribution.

Another interesting effect observed during dopant diffusion in polysilicon is grain growth, particularly observed at high doping levels. It has been found experimentally that both phosphorus and arsenic enhance grain growth, while boron does not (Wada and Nishimatsu, 1978; Mei *et al.*, 1982; Smith *et al.*, 1984; Angelucci *et al.*, 1983). The explanation involves the Fermi-level dependence of dislocation motion and the concentration-enhanced diffusivity of arsenic and phosphorus.

2.3 Dopant Segregation and Activation

Unlike single-crystal silicon, the relationship between impurity doping and electrical conductivity is complex. The dopant incorporated in the polysilicon film will either reside interstitially or substitutionally in the crystallites where it will be mostly ionized (active) or segregate in the grain boundary where it will be only partially ionized or not ionized at all (inactive). Since impurity atoms move preferentially in grain boundaries, a substantial amount of impurities are always present in the grain boundaries of polysilicon. The fraction of the dopant atoms segregating to the grain boundaries is proportional to the volume density of grain-boundary segregation sites and, hence, inversely proportional to the size of the grains (Mandurah *et al.*, 1980). The ratio between the electrical concentration and the dopant concentration is called the activation ratio. Generally, the electrical carrier concentration is always lower than the dopant concentration because of the grain-boundary segregation and carrier trapping at states present in the grain boundaries (Solmi *et al.*, 1982). At high dopant concentrations, the trapping sites, generally of the order of 3×10^{12} sites/cm^2 (Baccarani *et al.*, 1978), are saturated with a negligible fraction of the carriers so that only dopant segregation can influence the electrical properties of polysilicon. For arsenic and phosphorus a substantial amount of segregation has been observed by STEM (Rose and Gronski, 1982; Wong *et al.*, 1985). Boron, on the other hand, shows no appreciable segregation: this has been explained (Mandurah *et al.*, 1980) with a model based on the heat of segregation that yields a negative heat of segregation for boron, implying that boron would be depleted from a surface or interface, and a positive heat of segregation for arsenic and phosphorus. In the case of boron, the activation is very high. High concentrations of boron have been observed at the interface (surface) on which polysilicon is deposited. The amount of pile-up at the interface is a function of the interface preparation and has been attributed to the presence of interfacial oxides or contaminants (Michel *et al.*, 1982; Garben *et al.*, 1986).

In order to obtain a film of the desired resistivity, the synergistic effects between grain size, grain growth, dopant diffusion, and segregation must be considered during all thermal processes that the film will undergo. First, as discussed in Sec. 1.4, grain size and dopants and/or contaminants are functions of the deposition system and will strongly affect the film structure both as deposited and during subsequent heat treatments. If an ion implantation is performed, amorphization and defects introduced will further modify the material properties. As discussed earlier, during each heat treatment, dopant will diffuse through and segregate to grain boundaries, and for higher temperatures, the grain size may also increase as a function of the dopant type and amount. Finally, after the last heat treatment, some of the dopant will be ionized, depending on its bonding coordination, and some of the ionized carriers will be trapped. Only the carriers that are not trapped will be participating in electrical conduction.

2.4 Electrical Properties—Resistivity

Dopant segregation, grain size, and grain orientation can all influence the electrical properties of the films. The resistivity of undoped and lightly doped polysilicon is on the order of 10^5 to 10^6 Ω cm, which is several orders of magnitude higher than that of single-crystal silicon, as shown in Fig. 5. In this range the resistivity is nearly constant, and changes in dopant concentration will result in only small resistivity variations. At medium doping levels, the resistivity becomes a strong function of dopant concentration, and a slight increase will cause it to drop sharply. At high doping levels, the resistivity

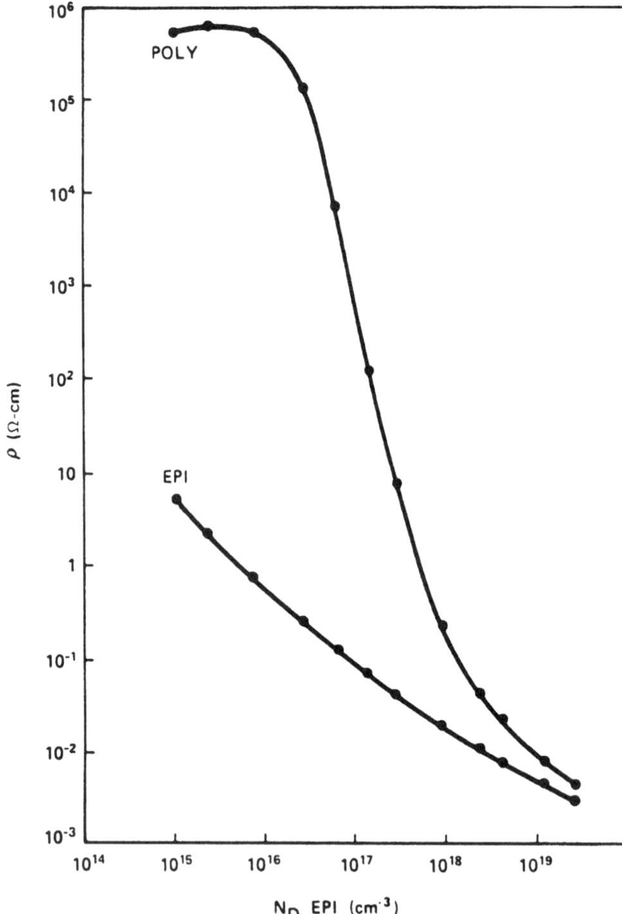

FIG. 5. Resistivity of polysilicon as a function of doping level as compared to that of single-crystal epitaxial silicon (Kamins, 1988).

approaches that of single-crystal silicon, although it will always remain slightly higher (Seto, 1975; Baccarani *et al.,* 1978).

Polysilicon is modeled as an ensemble of small crystallites joined by grain boundaries. Inside each crystallite, the atoms are arranged in a periodic manner forming small single crystals, while the grain boundaries are composed of disordered atoms with incomplete bonding. The high concentration of defects and dangling bonds at the grain boundaries causes trapping states that can immobilize both dopant atoms and charge carriers. Thus, the number of free carriers available for electrical conduction in polysilicon films is always lower than the dopant concentration because of both carrier trapping (Kamins, 1971; Seto, 1975; Baccarani *et al.,* 1978) and dopant-atom segregation at the grain boundaries (Swaminathan *et al.,* 1980; Mandurah *et al.,* 1980).

At low doping levels, the carrier concentration in polysilicon is very close to its intrinsic value in single-crystal silicon. As the doping is increased, the carrier concentration increases slowly and remains very small compared to the average dopant concentration. This is explained by the trapping of most of the carriers at states at or near the grain boundaries, leaving few free to contribute to the conduction; see Fig. 6 (Mandurah *et al.,* 1979). As the doping concentration is increased, the number of trapped carriers will increase in a manner determined by the energy distribution of the traps. Eventually, the traps will approach saturation at a generally accepted concentration in the low 10^{17}-cm^{-3} range. Upon further increase in the doping concentration, the number of trapped carriers will not increase appreciably; however, the potential barrier will decrease and the space-charge region will nar-

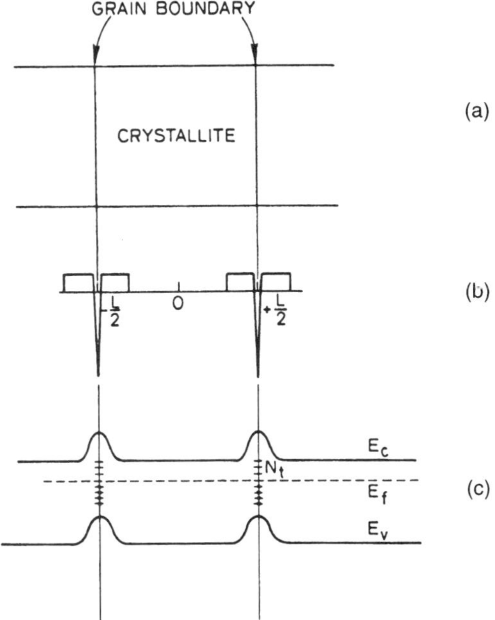

FIG. 6. Model of *n*-type polycrystalline material showing (a) small crystallites surrounded by grain boundaries containing large numbers of traps, (b) resulting space charge, and (c) corresponding energy-band diagram with potential barriers surrounding the grain boundaries (Mandurah *et al.*, 1979).

row. The resistivity will thus decrease sharply with small increases in the dopant concentration as the traps become saturated (Mandurah *et al.*, 1979). Finally, at high dopant concentrations, the region near the grain boundary will no longer limit the conductivity of the samples, and the properties of the material will approach those of the crystallites.

The total dopant concentration is equal to the sum of the active dopant in the crystallites and the dopant segregated at the grain boundaries. Although the dopant atoms in the crystallites have a high probability of being completely ionized, there are several hypotheses on the behavior of the dopant atoms at the grain boundaries. The fraction of the dopant atoms segregated at the grain boundaries is proportional to the volume density of grain-boundary segregation sites and, hence, inversely proportional to the size of the grains. Of the total carriers generated, some ($\sim 10^{17}$) will be trapped at grain-boundary sites, while the remaining n will be finally available for conduction:

$$\sigma = ne\mu, \tag{3}$$

where σ is the conductivity, e is the electron charge, and μ is the mobility (see Sec. 2.5). Figure 7 shows the carrier concentration versus concentration for P- and As-doped films after heat treatment at 1050 °C for 1 h (Solmi *et al.*, 1982). The solid solubilities at the same temperature for P and As in single-crystal silicon are also shown. It can be observed that the electrical carrier concentration is always lower than the dopant concentration, the segregation is more marked for the As-doped films, and as dop-

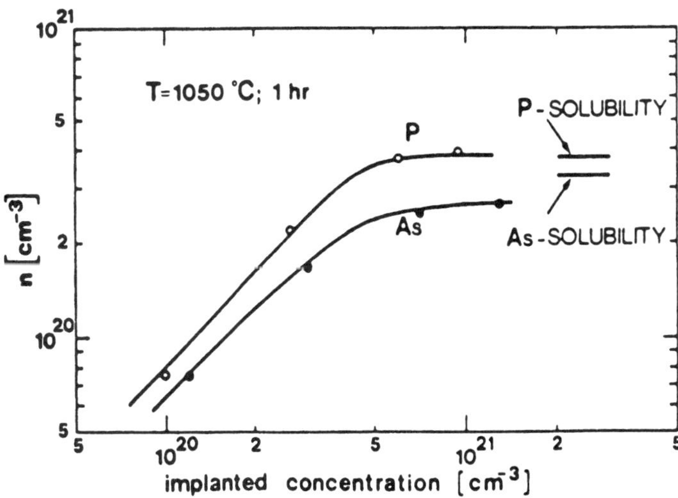

FIG. 7. Carrier concentration vs implanted concentration for P- and As-doped polysilicon films after thermal annealing at 1050 °C for 1 h. Solid solubilities in single-crystal films for P and As are also shown (Solmi *et al.*, 1982).

ing increases, both curves saturate to a value close to the solid solubility at that temperature.

For boron, no measurable changes in carrier concentration were observed, again indicating no appreciable tendency for the boron dopant atoms to segregate at grain boundaries. The boron-doped films are usually activated in the as-deposited form and do not require thermal annealing for activation.

For heavily doped films, the resistivity of the polysilicon is limited by the solid solubility of the dopants in silicon, since we can supersaturate the film with dopant either during deposition or with ion implantation. The resistivity can be reduced only by increasing the mobility, and the dependence of the mobility on the grain size and other polysilicon properties is shown in the next section.

2.5 Electrical Properties—Carrier Mobility

In Eq. (3), the mobility μ describes the ease with which a carrier can move through the polysilicon film. This mobility is influenced by the potential barrier at the grain boundaries and therefore by doping concentration (Seto, 1975). At low doping levels, carrier mobility is on the order of 100 cm^2/V s, approximately one order of magnitude lower than in single-crystal silicon, and is relatively insensitive to changes in dopant concentration. At moderate doping levels, it has been observed to drop to a minimum value, and this occurs over the same range of dopant concentrations where the resistivity is most sensitive to changes in dopant (Seto, 1975). At higher levels, the potential barrier at the boundaries decreases and the mobility increases, approaching that for the single crystal.

The mobility in heavily doped polysilicon thin film is affected by ionized-impurity scattering (μ_i), surface scattering (μ_s), and grain boundary scattering (μ_{gb}), as described by (Solmi *et al.*, 1982)

$$\frac{1}{\mu} = \frac{1}{\mu_i} + \frac{1}{\mu_s} + \frac{1}{\mu_{gb}} . \tag{4}$$

The ionized-impurity scattering contribution is the same as found from analysis of carrier mobility versus electron concentration in heavily doped single-crystal silicon. In single crystals, it was verified that the mobility depends only on carrier concentration and is not affected by electrically inactive dopant. Regarding the surface scattering term μ_s, if the thickness of the film d is comparable to the carrier mean free path λ, the mobility μ in the film differs from the bulk mobility μ_b because of surface scattering. The empirical relationship found is

$$\mu = \frac{\mu_b}{1 + 2\lambda/d} . \tag{5}$$

In the heavy doping range, λ is in the 10–15-nm range, and if the film thickness is in the 100–200-nm range, the mobility will then be about 5–10% lower than μ_b.

The behavior of μ_{gb} in the case of phosphorus doping seems to be well characterized empirically by the grain size only and can be described by the relationship

$$\mu_{gb} = KD^{\alpha}, \tag{6}$$

where $K = 4.5 \times 10^4$ cm$^{2-\alpha}$/V s and $\alpha = 0.59$ are constants and D is the average grain size in centimeters. The direct role of the grain boundary in determining the conductivity in polysilicon films is taken into account in a model where a potential barrier, dependent on the dopant and on processing conditions, is associated with the grain boundary (Mandurah *et al.*, 1981a, 1981b). In the case of arsenic, the mobility in the grain boundary depends strongly on process conditions. The increase of arsenic concentration in the film causes a degradation of the bulk grain crystalline quality (twins and lattice defects) and subsequently a degradation in mobility due to scattering at these defects.

2.6 Diffusion through Polysilicon/Single-Crystal Silicon Interfaces

As discussed earlier, doped polysilicon is often used as a diffusion source for single-crystal silicon during device processing. Through the characterization of the resulting impurity profiles in the single-crystal substrate, it has been found that the interface can profoundly affect the resulting diffusion profiles. More specifically, it has been shown that impurities such as oxygen or carbon that can easily be trapped at the interface during polysilicon deposition can block the diffusion of dopant across the interface. The degree of this effect varies with the dopant and the amount of contaminant. In the case of boron, interface contamination strongly influences the outdiffusion profile in the single crystal. When the interface is free from contaminants, the boron concentration at the interface is equal to the solid solubility at the diffusion temperature. The concentration at the contaminated interface is less than that of the solid solubility limit at the annealing temperature. As the diffusion time increases, the interface concentration increases, gradually approaching the solid solubility limit.

Chemically grown oxides can also substantially retard the diffusion of dopants and have been purposely used to obtain very shallow junctions in silicon (Stork *et al.*, 1985). Studying the oxygen Auger peaks associated with the interface, it has been observed that arsenic diffusion decreases the amount of oxide at the interface. This decrease has been found to be inversely proportional to the initial thickness of the interfacial oxide and the drive-in temperature. These results demonstrate the importance of interface control in achieving reproducible junction profiles.

3. POLYCRYSTALLINE SILICON APPLICATIONS

Polycrystalline silicon has been a key enabling technology in many diverse VLSI applications. It is used extensively in CMOS circuits as a gate contact and for local wiring. Its thermal and chemical compatibility with single-crystal silicon as well as the capability for a wide range of conductivity have made possible the invention of trench (vertical) capacitors and isolation structures. These same properties are utilized to significant advantage in bipolar circuits, increasing switching speeds and decreasing transistor size. Finally, polycrystalline silicon is also used as resistors and resistor contacts.

3.1 Silicon-Gate Technology

Metal-oxide-semiconductor transistors have been used extensively since the 1960s and form the prime building blocks of the well-known *n*-type MOS (NMOS) and CMOS circuits used in memory technology. Through the 1970s, gate electrodes of MOS devices were formed using evaporated aluminum, which was deposited after the source and drain were formed. As circuit densities increased to 1 Mb (1 million transistors/chip), the tight lithographic tolerance required in forming the aluminum gate while ensuring sufficient overlap of source and drain became problematic. This tolerance also drove the need for a large overlap, increasing the parasitic gate-to-drain capacitance, known as the Miller feedback capacitance.

The use of polysilicon as an aluminum-gate replacement is a significant breakthrough. By using polysilicon deposited on the gate electrode as a mask to define both source and drain, a process known as self-alignment, the need for a large overlap of the gate electrode or tight lithographic tolerance is eliminated. Not only does gate polysilicon self-alignment provide miniaturization leverage, it also improves circuit performance by decreasing parametric variability from transistor to transistor. Finally, the precise orientation of gate to drain achieved across all transistors in a circuit by self-alignment considerably reduces and equalizes the Miller feedback capacitance.

Polysilicon gates also provide more suitable threshold voltages for integrated MOS circuits than do aluminum gates. The threshold voltage is the voltage at which a MOS transistor begins to form a conducting channel between source and drain. This voltage is in part determined by the difference in work function between the gate-electrode material and the silicon. While the work function of aluminum is fixed at about 4.1 eV, that of polysilicon can be tailored to the circuit requirement on the basis of doping level and type. For highest electrode conductivity, the polysilicon is normally highly doped by ion implantation. The electron affinity for highly doped *n*-type material, where the Fermi level lies near the conduction band, is close to the work function of aluminum. Therefore, threshold voltages of *n*-gate transistors are similar to those of aluminum-gate transistors. With high *p*-type doping, the Fermi level lies close to the valence band, increasing the work function by nearly 1.1 eV. This makes the threshold voltage of *p*-gate transistors less negative than that of aluminum-gate transistors. The lower threshold voltage is particularly advantageous in CMOS circuits.

3.2 Polysilicon-Emitter Bipolar Transistors

Polysilicon has also provided a number of significant advantages in bipolar circuits. Specifically, its use as an emitter contact has enhanced circuit performance by reducing hole injection from the base even in ultrathin submicron junctions. As a base contact, it has provided a simple means, through the process of self-alignment, to make denser circuits by creating a vertical *npn* bipolar transistor. The process of self-alignment has also enabled miniaturization, eliminating certain critical alignment tolerances.

The most significant early industrial application of polysilicon in bipolar circuits came as an extrinsic base contact in about 1982. The extrinsic base is the conductive region between the lightly doped intrinsic base and the external contact. For high-performance applications, the resistance of the extrinsic base and the capacitance of the doped region must be minimized. The extrinsic base must be separated from the *n*-emitter region to avoid low emitter–base breakdown voltage. In a conventional transistor, this spacing must be at least an alignment tolerance, and this lithographic requirement limited miniaturization.

By using doped polysilicon, the extrinsic base region can be removed from the silicon and placed above an oxide. The polysilicon is highly boron doped and, therefore, very conductive without adding to the capacitance. The intrinsic base contact is achieved by boron out-diffusion from the polysilicon to the single crystal. Furthermore, the emitter can be made to self-align to the extrinsic base so that the two are separated by an oxide sidewall formed on the polysilicon. By using the self-alignment technique, the emitter–base separation can be reduced to far less than an alignment tolerance. Typical sidewall thickness of 250 nm can be formed routinely in production. Therefore, the use of doped polysilicon for the extrinsic base has enabled the reduction of the base resistance and capacitance.

The significant effect of these features on bipolar transistor size is clearly demonstrated in Fig. 8, in which the layouts of two generations of bipolar transistors are compared. The polycrystalline-silicon extrinsic base, coupled with the self-alignment process and advances in photolithography, have enabled the transistor size to be reduced nearly ten times. The speed of polysilicon-base bipolar digital circuits in production has increased from about 4 GHz for conventional transistors to 9 GHz.

A further technical advance was made by using *n*-type polysilicon to contact and form the emitter region. This simplifies the emitter process in two main areas. First, it estab-

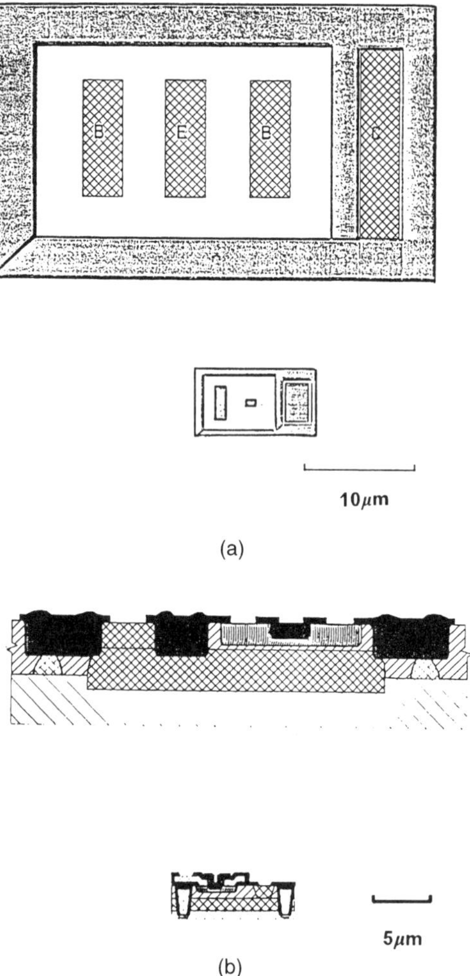

FIG. 8. Layout and cross-section comparison of two generations of IBM bipolar transistors. The transistor (a) is formed with conventional implantation/diffusion, while the transistor (b) is formed using polysilicon.

lishes a low-resistance contact to the single-crystal emitter region without the stringent requirements for metal alloying while avoiding junction penetration. Second, it serves as a source for arsenic dopant for the single crystal, mitigating the requirement for ion implantation. By using polysilicon as an arsenic diffusion source, emitter depths of less than 50 nm can be realized in production without the concerns of implantation damage or contamination. A typical junction profile showing the emitter polysilicon contact, single-crystal emitter, and boron-doped intrinsic base is shown in Fig. 9. With these advances, the circuit speed of digital bipolar circuits in production has exceeded 19 GHz.

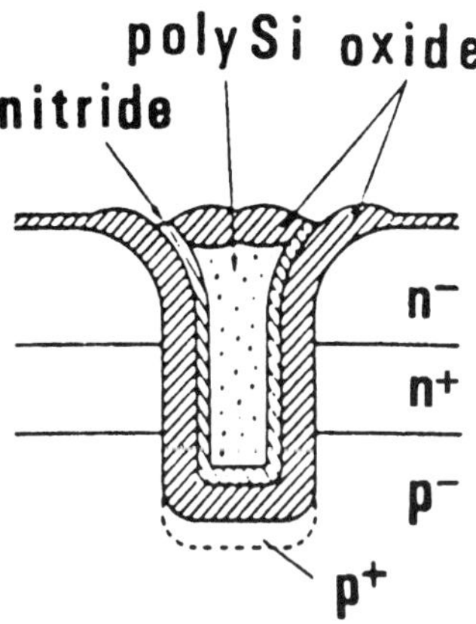

FIG. 10. Schematic of a trench capacitor.

3.3 Other Polysilicon Devices and Applications

3.3.1 Trench Isolation Polysilicon has also provided key leverage in integrated-circuit miniaturization through its use in vertical isolation. Conventional isolation structures formed from thermal oxidation consume considerable area and normally require high-temperature processing. A typical polysilicon-filled trench used for isolation is shown in Fig. 10, and the dimensional advantages of such a structure are clear. Two key technological breakthroughs have enabled the vertical isolation structure to be used successfully in manufacturing. First, reactive ion etching has made possible the formation of deep, vertical grooves in the silicon substrate. Many techniques have been attempted to fill these grooves with suitable dielectrics. Materials such as oxides and polyimide have been employed as fillers, with limited success due to the mismatch in the coefficient of thermal expansion and, in the polyimide case, to the lack of chemical stability. The advent of CVD polysilicon was the second breakthrough in achieving vertical isolation: the deposition conditions are controlled so as to grow uniformly along the walls of the groove. The groove will be filled when a polysilicon layer half as thick as the trench is wide has been deposited. During the trench fill process, polysilicon is also deposited on the surface of the silicon wafer. This layer can be removed using dry etching or by chemical mechanical planarization.

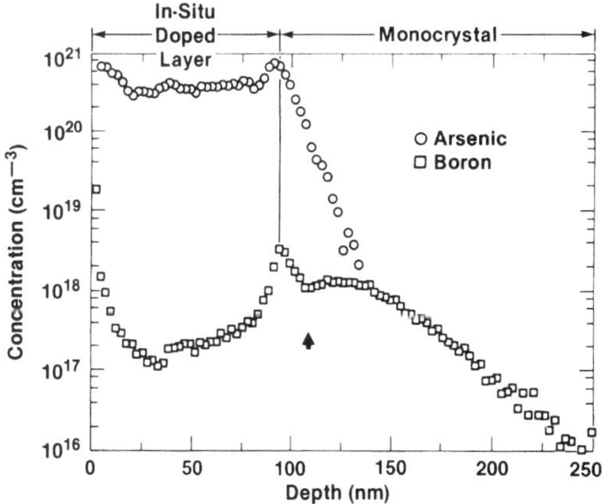

FIG. 9. Typical junction profile of a bipolar transistor formed using *in situ* doped polysilicon.

3.3.2 Trench Capacitors The vertical polysilicon structure of Fig. 10 has been further modified and improved to become a capacitor, providing considerable area reduction in specific MOS circuit applications, especially dynamic random-access memories (DRAMs). In DRAM cells, this new feature permitted the planar storage capacitor to be replaced with the trench capacitor, shrinking the cell size while maintaining the required capacitor value. To form the capacitor, a groove is etched using reactive ion etching into the silicon. The groove is oxidized and additional dielectric deposited as required. Doped polysilicon is deposited to fill the trench forming the counter electrode of the capacitor. The excess polysilicon is removed, and finally contact is made to the trench electrode through a conductive layer of polysilicon deposited later in the process. Charge is stored in the inversion layer formed along the vertical walls of the trench, thus increasing the capacitor's effective storage area without increasing the surface area.

3.3.3 Resistors and Interconnections Given the widespread use of polysilicon in circuit applications and its wide range of resistivity, it is frequently used for local interconnections and for high-value resistors. For example, using polysilicon as a local interconnect may eliminate additional masking layers and metal deposition processing, depending on the specific process sequence. However, since the resistance of polysilicon is higher than that of aluminum, it is generally used only over short distances to avoid signal delay. Polysilicon can be used as a high-value resistor when tight electrical tolerances are not required. A typical application is the load resistor in a static RAM cell where resistors are in the range of 10^9 Ω. Polysilicon can be doped using ion implantation and defined lithographically to achieve the required resistor value. The contact areas require additional dopant to ensure good Ohmic contact. The resistivity is difficult to control with great accuracy since as the film's grains grow during subsequent thermal processing, the resistivity of the polysilicon decreases. The resistivity of lightly doped polysilicon varies with temperature, which also degrades resistor operating tolerances.

4. SUMMARY

In this article the salient features of polycrystalline silicon have been discussed ranging from deposition methods to properties and finally applications. Polycrystalline silicon has been used extensively in the VLSI industry because of its ease of deposition and great range of electrical and physical properties. Polycrystalline silicon has also been extensively studied to understand better the influence of crystalline defects and impurities on semiconductor electrical properties.

GLOSSARY

Bipolar Transistor: Junction transistor that provides current gain due to the proximity of forward- and reverse-biased junctions. These two junctions are formed by doping three regions: the *emitter,* the *base,* and the *collector.*

Carrier Trapping: Capture of a free electron or hole by a dangling bond (or recombination center).

CMOS: Complementary metal-oxide-semiconductor circuit family, utilizing both *p*-type and *n*-type MOS transistors.

CVD: Chemical vapor deposition.

Dopant: A chemical impurity that is introduced in a semiconductor material to alter that material's electrical conductivity.

Fermi Level: Chemical potential of electrons in a solid.

Grain Boundaries: The disordered regions between adjacent crystallites in a polycrystalline solid material.

Interferometric Technique: Optical measurement method that relies on the interference of light at film interfaces to determine thickness.

Ion Implantation: Technique to introduce dopants, in which atoms are separated by mass and are highly accelerated to penetrate the semiconductor.

Ionization: Loss of an electron by an atom.

LPCVD: Low-pressure chemical vapor deposition.

MBE: Molecular-beam epitaxy.

Mobility: Ease by which an electrical carrier may move through a material.

MOS: Metal-oxide-semiconductor transistor utilizing a conductor dielectric gate.

Photolithography: Technique by which circuit images are printed on semiconductor wafers.

Polycrystalline: A solid composed of many crystallites of different crystal orientation.

PVD: Physical vapor deposition.

RBS: Rutherford back-scattering.

SIMS: Secondary-ion mass spectroscopy.

Solar Cell: Semiconductor device in which solar energy is converted into electrical energy.

STEM: Scanning transmission electron microscopy.

TEM: Transmission electron microscopy.

VLSI: Very large-scale integration.

Works Cited

Adams, A. C. (1983), in: S. M. Sze (Ed.), *Dielectric and Polysilicon Film Deposition,* New York: McGraw-Hill, pp. 93–129.

Angelucci, R., Severi, M., Solmi, S. (1983), *Mater. Chem. Phys.* **9,** 235–245.

Arienzo, M., Komem, Y., Michel, A. E. (1984), *J. Appl. Phys.* **55,** 365–369.

Arienzo, M., Megdanis, A. C., Sackles, P. E., Michel, A. E. (1986), *IEEE Trans. Electron Dev.* **ED-33,** 1535–1538.

Baccarani, G., Ricco, B., Spadini, G. (1978), *J. Appl. Phys.* **49,** 5565–5570.

Bravman, J. C., Patton, G. L., Plummer, J. D. (1985), *J. Appl. Phys.* **57,** 2779–2782.

Garben, B., Orr-Arienzo, W., Lever, R. (1986), *J. Electrochem. Soc.* **133,** 2152–2156.

Goto, H., Inayoshi, K. (1987), in: G. Soncini, P. U. Calzolari (Eds.), *ESSDERC '87: 17th European Solid State Device Research Conference,* Bologna: Technoprint, p. 369.

Hammond, M. L. (1980), *Solid State Technol.* **23,** 104–109.

Kamins, T. I. (1971), *J. Appl. Phys.* **42,** 4357–4366.

Kamins, T. I. (1979), *J. Electrochem. Soc.* **126,** 833–837.

Kamins, T. (1988), *Polycrystalline Silicon for Integrated Circuit Applications,* Boston: Kluwer Academic Publishers.

Kern, W., Schnable, G. L. (1979), *IEEE Trans. Electron Devices* **ED-26,** 647–657.

Mandurah, M. M., Saraswat, K. C., Helms, C. R., Kamins, T. I. (1980), *J. Appl. Phys.* **51,** 5755–5763.

Mandurah, M. M., Saraswat, K. C., Kamins, T. I. (1979), *J. Electrochem. Soc.* **126,** 1019–1023.

Mandurah, M. M., Saraswat, K. C., Kamins, T. I. (1981a), *IEEE Trans. Electron Devices* **ED-28,** 1163–1171.

Mandurah, M. M., Saraswat, K. C., Kamins, T. I. (1981b), *IEEE Trans. Electron Devices* **ED-28,** 1171–1176.

Marcus, R. B. (1983), in: S. M. Sze (Ed.), *Diagnostic Techniques,* New York: McGraw-Hill, 524–527.

Mei, L., Rivier, M., Kwart, Y., Dutton, R. W. (1982), *J. Electrochem. Soc.* **129,** 1791–1795.

Meyerson, B. S., Yu, M. L. (1984), *J. Electrochem. Soc.* **131,** 2366–2371.

Michel, A. E., Kastl, R. H., Mader, S. R. (1982), *Nucl. Instrum. Methods Phys. Res.* **209,** 719–724.

Natsuaki, N., Tamura, M., Miyazak, T., Yanagi, Y. (1983), in: *IEDM Technical Digest: International Electron Devices Meeting,* New York: IEEE, pp. 662–665.

Rose, J. H., Gronski, R. (1982), *Appl. Phys. Lett.* **41,** 993–995.

Rosler, R. S. (1977), *Solid State Technol.* **20,** 63–70.

Ryssel, H., Iberl, H., Bleier, M., Prinke, G., Haberger, K., Kranz, H. (1981), *Appl. Phys.* **24,** 197–200.

Seto, J. Y. W. (1975), *J. Appl. Phys.* **46,** 5247–5254.

Smith, D. A., Tan, T. Y., Fontaine, C. (1984), in: *IBM Technical Report RC9361.*

Solmi, S., Severi, M., Angelucci, R., Baldi, L., Bilenchi, R. (1982), *J. Electrochem. Soc.* **129,** 1811–1818.

Stork, J. M. C., Arienzo, M., Wong, C. Y. (1985), *IEEE Trans. Electron Devices* **ED-32,** 1766–1770.

Swaminathan, B., Demoulin, E., Sigmon, T. W., Dutton, R. W., Reif, R. (1980), *J. Electrochem. Soc.* **127,** 2227–2229.

Swaminathan, B., Saraswat, K. C., Dutton, R. W., Kamins, T. I. (1982), *Appl. Phys. Lett.* **40,** 795–798.

Thomas, G., Goringe, M. J. (1979), *Transmission Electron Microscopy of Materials,* New York: Wiley.

Tsukamoto, K., Akasaka, Y., Watari, Y., Kusano, Y., Hirose, Y., Nakamura, G. (1978), *Jpn. J. Appl. Phys. (Suppl.)* **17,** 187–192.

Wada, Y., Nishimatsu, S. (1978), *J. Electrochem. Soc.* **125,** 1499–1504.

White, F., Koburger, C., Abernathey, J., Nesbit, L., Arienzo, M. (1985), in: *IBM Technical Report RC11953.*

Wong, C. Y., Grosvenor, C. R. M., Batson, P. E., Isaac, R. D. (1985), in: *IBM Technical Report RC10802.*

Further Reading

Cullen, G. W. (Ed.) (1989), *Proceedings of the Tenth International Conference on Chemical Vapor Deposition 1987,* Pennington, NJ: The Electrochemical Society, Inc.

Ghandi, S. K., *VLSI Fabrication Principles,* New York: Wiley.

Grove, A. S., *Physics and Technology of Semiconductor Devices,* New York: Wiley.

Kamins, T., *Polycrystalline Silicon for Integrated Circuit Applications,* Boston: Kluwer Academic Publishers.

Robinson, M. D. (Ed.) (1984), *Proceedings of the Ninth International Conference on Chemical Vapor Deposition 1984,* Pennington, NJ: The Electrochemical Society, Inc.

Sze, S. M., *VLSI Technology,* New York: McGraw-Hill.

SILVER, GOLD, AND OTHER NOBLE METALS

MONIKA FIEBERG, *Chimet SpA, Badia al Pino (Arezzo), Italy,*

HERMANN RENNER, *Degussa A.G. (retired), Hanau, Germany,* and

GUENTHER SCHLAMP, *Demetron G.m.b.H. (retired), Hanau, Germany*

3-527-28140-1/96/$5.00 + .50

INTRODUCTION

The term "noble metals" encompasses the eight elements silver, gold, and the six platinum-group metals (PGM) ruthenium, rhodium, palladium, osmium, iridium, and platinum. According to their electronic structures, the precious metals are classified as the end members of the transition metals placed in the periodic table of the elements in groups VIII and IB of the fifth and sixth periods. They are characterized by dense close-packed crystal structures, outstanding resistance against oxidation, high melting temperatures, good to high electrical conductivities, and mechanical and thermal stability. Silver has the highest electrical conductivity of all elements, gold has the highest oxidation resistance of all metals, and osmium and iridium are the elements with the highest densities. According to their individual electronic structures, they show, unique to each, distinctive physical and chemical properties.

Silver and gold are the classical precious metals, used since ancient times for jewelry, tableware, and coins and in dentistry. Besides these still-continuing applications, they are in elemental and in alloyed form of high technical importance as conductor, connector, solder, and braze materials in the fields of electrical engineering and electronics, batteries, and optical and heat-reflecting coatings on architectural glasses. Silver is the most important constituent in photographic emulsions. The main application fields of the platinum metals are catalysis in organic and anorganic chemical reactions, automotive catalysts, electrodes of galvanic and of fuel cells, highly coercive magnetic alloys for recording and memory devices, resistor elements in electronic devices and in the sensor technology, construction materials of corrosion-resistant chemical equipment, and, as chemical compounds, constituents in medical drugs.

1. GENERAL CHARACTERISTICS OF THE NOBLE METALS

1.1 Electronic Structure

The electronic structure of the noble-metal atoms is characterized by the completion of their $4d$ or $5d$ electron levels with increasing nuclear charge, extending for the "light" noble metals (atomic numbers 44–47) from $4d^7 5s^1$ for ruthenium to $4d^{10} 5s^1$ for silver and for the "heavy" noble metals (atomic numbers 76–79) from $5d^6s^2$ for osmium to $5d^{10}6s^1$ for gold. In the condensed metallic state, the electron orbitals overlap, forming energy bands. The electronic structure is described by the density of states (DOS = number of energy levels in a unit energy range) and the Fermi level, which represents the state occupied with maximum electron energy. The state-density curve is almost the same for all noble metals (Yonemitsu, 1991). Figure 1 shows a schematic DOS curve of the noble metals with the positions of their Fermi levels. The different locations of the Fermi levels of each pair of the group homologs relative to the peaks or valleys of the DOS curve account for the differences in their electric conductivities and in their thermoelectric, magnetic, and optical behavior. Because of the very small energy differences between their outer *s*- and *d*-electron levels, *d* electrons can be easily transferred into the valence band. This enables all noble metals to react in several oxidation states, forming numerous anorganic and organic compounds and—with the exception of gold—to perform manifold catalytic activities by intermediate participation in chemical reactions. Structure-related similarities exist between the group homolog elements of the fourth period (iron, cobalt, nickel, copper) in the alloying behavior, in the formation of magnetic alloys, in the formation of complex anorganic and organic compounds, and in catalytic activities.

The strong tendency to complete the 4*d* and 5*d* levels has the effect that for ruthenium, rhodium, silver, platinum, and gold the outer *s* levels are only singly occupied. Palladium has the basic electron configuration $d^{10}s^0$.

The "noble" character of the precious metals is expressed by their positive reduction potential against hydrogen. They share this characteristic with a few other elements such as antimony, copper, mercury, and rhenium. The reduction potential increases along the periods and inside the groups with increasing nuclear charges.

1.2 Physical Properties

The atomic diameters of the transition metals of the second long period are systematically larger than those of their group homologs of the first long period. The differences between the atomic diameters of the homologs of the second and the third long periods are, as a result of the lanthanide contraction, very small. The elements of each of the pairs ruthenium/osmium, rhodium/

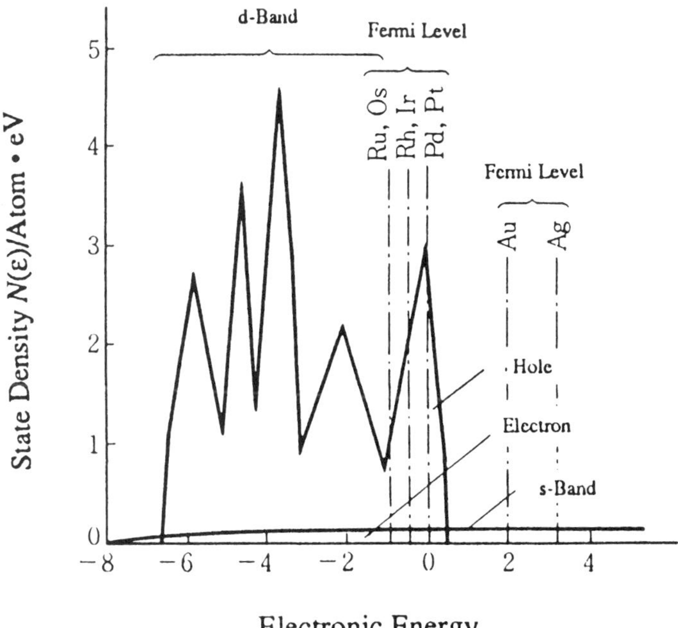

FIG. 1. Schematic DOS curve of the fcc transition metals with location of the Fermi levels of the noble metals relative to the upper state of density (Yonemitsu, 1991, p. 30).

iridium, palladium/platinum, and silver/gold have practically the same atomic diameters.

The crystal structures of the transition metals change with increasing atomic numbers from the more open body-centered-cubic structures to the close-packed structures of highest symmetry and highest coordination numbers. Ruthenium and osmium crystallize hexagonal close packed; rhodium, palladium, silver, iridium, platinum, and gold have face-centered-cubic crystal structure. There are octahedral and tetrahedral holes in the fcc and the hcp structures, which can be occupied by interstitial atoms having maximum atomic radii of $0.41r$ or $0.225r$, respectively (r = atomic radius of the noble metals, if these are considered as close-packed spheres in these lattices).

The densities of all noble metals are above 10 g/cm^3, and those of the the heavy noble metals (osmium, iridium, platinum, gold), above 19 g/cm^3.

The ductility increases in each period with increasing atomic number. Silver, gold, palladium, and platinum exhibit a wide range of ductility; rhodium, iridium, ruthenium, and osmium can only be deformed at elevated temperatures, accompanied by strong work hardening.

The elastic moduli of the single crystals of the fcc noble metals show considerably different values in the ⟨100⟩, ⟨110⟩, and ⟨111⟩ direction, causing anisotropic elastic properties of single crystals and in polycrystalline materials with textures (Table 1).

The high melting temperatures, between 960 °C (silver) and 3045 °C (osmium), are in accordance with the dense packed structures and indicate strong interatomic bonds. The melting points and the recrystallization temperatures decrease in each period with increasing nuclear charge. Likewise, the hardness and the brittleness decrease, while the thermal expansion increases with increasing nuclear charge.

The individual electrical conductivities differ considerably. Silver has the lowest electrical resistivity of all metals. Palladium and platinum have values as large as six times those of silver, and rhodium, iridium, ruthenium, and osmium have intermediate values (Table 2). The higher resistivities compared with those of silver and gold are mainly due to the fact that these elements have smaller numbers of carriers and that the mobility of the current-carrying s electrons is decreased by an increased probability of scattering due to phonons, caused by the higher densities of d states at the Fermi level. The carrier density for silver and gold is 1/atom, while for palladium and platinum, because of d-s interband transitions, it is reduced to 0.6/atom. The total state densities $N(e_F)$ at the Fermi levels are proportional to the electron specific heat coefficients γ:

$$N(e_F)\alpha\gamma \quad (\text{mJ/mol K}^2). \tag{1}$$

The difference of the state densities can be derived from the measured values of γ, which amount for silver and gold to 0.65 and 0.7, respectively, while for palladium and platinum γ is as high as 9.9 and 6.7 (Tanuma, 1991, p. 46). The rule of Wiedemann and Franz of the proportionality of the electrical conductivity and the thermal conductivity is valid for all precious metals. The Lorenz numbers, calculated from the ratios of the measured values of the thermal and their electrical conductivity,

Table 1. *E* moduli in different crystal directions (Jönsson, 1995, p. 204).

Noble metal	Direction	*E* (GPa)
Ag	[100]	44
	[110]	82
	[111]	115
Au	[100]	42
	[110]	81
	[111]	114
Pd	[100]	65
	[110]	129
	[111]	187
Ir	[110]	480
	[111]	660

Table 2. Electrical resistivities of the precious metals (Diehl, 1995, p. 155).

Metal	Resistivity	Measurement error ±
Ru	6.7	—
Rh	4.34	0.01
Pd	9.725	0.015
Ag	1.465	0.005
Os	8.2	0.1
Ir	4.7	0.05
Pt	9.825	0.025
Au	2.03	0.01

Table 3. Superconducting precious-metal alloys (Benner *et al.*, 1991, p. 636).

Alloy	Transition temperature (K)
Nb_3Au	10.9
Nb_3Pt	10.6
ZrRuP	12.3
$Tb_3Os_4Ge_{13}$	14.1
ZRuAs	11.9
YOs_4Ge_{10}	8.6
$ScIr_4Si_{10}$	8.4
$ErRhB_4$	8.5

$$L = \lambda\kappa/T \quad (\mathrm{W}\,\Omega/\mathrm{deg}^2), \tag{2}$$

correspond with values between 2.27 for silver and 2.96 for ruthenium to typical metallic behavior in accordance with the theory of the contribution of the free electrons to the thermal conductivity.

Superconductivity has been found for ruthenium (T_c = 0.49 K), rhodium (T_c = 0.9 K), osmium (T_c = 0.71 K), and iridium (T_c = 0.14 K). Higher transition temperatures are shown by some binary and ternary noble-metal alloys (Table 3) (Khan, 1984; Raub, 1984; Benner *et al.*, 1991, p. 636). The thermoelectric properties of the precious metals are of great practical importance for thermocouples used for measurements at high temperatures. The magnetism of the precious metals is very weak. Silver and gold behave (like copper) diamagnetically, while all platinum-group metals behave paramagnetically.

The optical properties are closely related to the interactions of photons with *s* and *d* electrons. Optical constants give information about electronic band structures, the participation of free and bound electrons, phase formation, ordering processes, and lattice defects. Colors are due to selective light absorption by interband transitions of electrons caused by absorption of photons. Photons can be absorbed by exciting an electron of a filled band just below the Fermi surface. The energy difference between this band and the Fermi surface determines the minimum energy of the photon that can be absorbed by interband transition. For the monovalent metals copper, silver, and gold, these energy differences are 2.1, 4.0, and 2.17 eV, respectively, causing the characteristic brilliant colors of the copper red, silver white, and gold yellow (see also COPPER, p. 269). The reflectivity and the color of the noble metals are of special importance for jewelry and dentistry alloys and for optical and heat-reflecting mirrors. Figure 2 shows the reflectivity of the bulk noble metals in the wavelength range of visible light.

Emissivity values are needed in pyrometrical measurements to calibrate deviations against the emission values of a black body. Table 4 gives some selected values.

1.3 Alloying Behavior

1.3.1 Phase Formation The precious metals form numerous alloys with each other and with many subgroup A and B metals. The alloy systems include types of continuous solid solutions, superlattice structures, miscibility gaps, eutectic systems, and intermetallic phases. They have predominantly fcc, hcp, or bcc crystal structures. Pri-

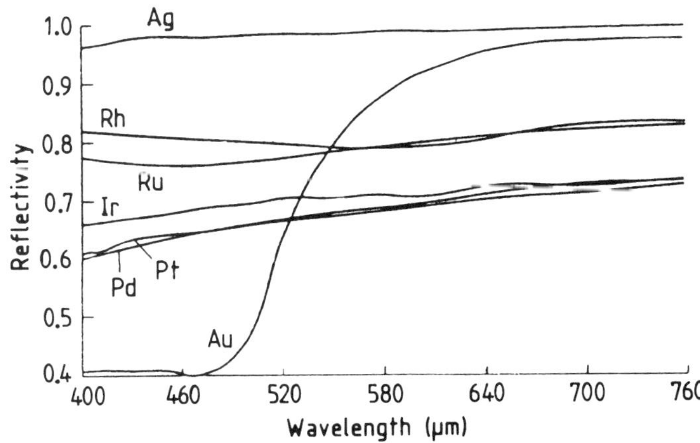

FIG. 2. Optical reflection of the precious metals in the visible light (Völcker, 1995, p. 173).

Table 4. Optical emissitivity of precious metals at 1000 °C[a] (Smithells and Brandes, 1977, p. 1019).

Metal	Emissivity
Ag	0.055[b]
Au	0.16
Rh	0.22
Pt	0.29
Ir	0.36
Pd	0.37
Ru	0.42
Os	0.52

[a]At wavelength 0.65 μm.
[b]At 800 °C.

mary solid solutions in precious-metal alloy systems with metals are generally of the substitutional type, where the solute atoms replace some of the solvent atoms on their lattice positions. Continuous solid solutions are formed between most noble metals, and in the alloy systems of gold, palladium, and platinum with iron, cobalt, and nickel. Interstitial-type solid solutions are formed by palladium and platinum with small atoms like hydrogen and carbon.

Some precious-metal alloys undergo ordering reactions forming superlattices. Their generation is bound to differences of the electrochemical properties and to differences of the atomic sizes of the alloying components. The most frequently formed structures are of the composition types AB or AB_3, which derive from the fcc lattices in the disordered state (Table 5). The transformations are accompanied by changes of physical, mechanical, or chemical properties. These order–disorder transformations are of practical importance for the production of mechanically hard alloys and of high-coercivity magnetic alloys.

1.3.2 Strengthening Strengthening of precious metals and alloys is effected either by solid-solution hardening, by phase trans-

Table 5. Compositions and structures of superlattices in precious-metal alloy systems (Hirabayashi, 1991, p. 105).

Atomic ratio	Composition			Superlattice structure	Fundamental structure
3:1 or 1:3	Pd_3Fe	$PdCu_3$	Au_3Cu $AuCu_3$		
	Pt_3Fe	$PtCu_3$	Au_3Pd		
	Pt_3Ti	$PtNi_3$	Ag_3Pt	$L1_2$	fcc
		$PtMn_3$			
	Rh_3Mo		Ag_3In		
	Rh_3W			$D0_{19}$	
	Ir_3Mo				
	Ir_3W		(Au_3Mn)	$D0_{22}$	fcc
	Pd_3V				
	Pd_3Nb				
	Pt_3V				
			Au_3Cd	$D0_{23}$	fcc
			Au_3Zn		
			Ag_3Mg		
2:1	Pd_2V			Ni_2Cr	fcc
	Pt_2V				
	Pt_2Mo				
1:1	PdFe		AuCuI		
	PtFe			$L1_0$	fcc
	PtV				
	PtNi				
	PtCo				
	PtCu				fcc
	PdCu		AuMn	$L1_1$	bcc
			AgCd	$L2_0$ or B2	
			AgZn		
	RhMo				hcp
	IrMo			B19	
	PtMo				
	IrW				
	PtNb				

formations, by precipitation hardening, by dispersion hardening, or by grain-size refinement.

Dispersion hardening is effected by incorporating finely dispersed, small amounts (0.1–2 vol %) of very small particles of high–melting-point compounds (oxides, carbides), which are insoluble or become soluble only near the melting temperature of the matrix. Special properties are available by composite materials made from noble-metal powders and metallic or nonmetallic additives. They are used for electrical contacts.

Fine grain structures of noble metals and noble-metal alloys improve the working behavior and surface quality. Grain refining and the stabilization of grain sizes to prevent grain growth during annealing operations is performed by alloying with finely dispersed insoluble particles, by segregation of insoluble atoms, or by solute atoms with large dimensional misfit to the host metal. They are applied in gold jewelry and dental alloys, and silver and platinum-metal alloys for spinnerets and seal rings.

1.3.3 Oxygen Noble metals, except gold, form solid and gaseous oxides. Dissolved oxygen affects hardness and working behavior. Silver oxides and platinum-metal oxides have practical importance as catalysts in organic chemical reactions.

1.3.4 Hydrogen All precious metals, except palladium, take only very small amounts of hydrogen in solution. Palladium and certain palladium alloys are unique in their capability to dissolve large amounts of hydrogen.

2. SILVER

Silver, chemical symbol Ag (argentum), atomic number 47, relative atomic mass 107.8682, is in its natural abundance composed of two stable isotopes in the composition of 51.84% ^{107}Ag and 48.16% ^{109}Ag. It is a bright, white metal with the characteristic silver luster. In pure condition it is very soft and malleable. The most important properties are the high electrical conductivity, the oxidation resistance, the optical reflectivity, the alloying behavior, and the light sensitivity of silver halides. The main application areas are film/photography (39%), jewelry and silverware (28%), electronics/electrical technology (14%), and solders and brazes (5%) (quantities in weight percent of the total yearly demand in the western world 1994; Degussa, 1994).

2.1 Electronic Structure

The electronic structure is [Kr] $4d^{10}5s^1$. The band diagrams, describing the electron energy levels, show five lower bands, originating from the fully occupied *d* levels, while one band near the Fermi level corresponds to the *s*-electron level. Figure 3(a) shows the band diagram of silver, and Figure 3(b), the density of states for the *d* and *s* states. Interactions between the 4*d* and the 5*s* states cause the Fermi surface, which is basically spherical for monovalent metals, to intercept the hexagonal faces of the Brillouin zone, forming “necks” in the Fermi surface with ring-shaped intersections around the ⟨111⟩ direction in **k** space [Fig. 3(c)].

The fully occupied 4*d* level at an energy about 4 eV below the singly occupied *s*-like states accounts for the preferred monovalent chemical behavior, for the high electrical conductivity, and for the brilliant reflectivity over the total range of the visible light.

2.2 Special Properties

Selected values of basic physical and chemical properties of silver are given in Table 6.

2.2.1 Mechanical and Thermal Properties The fcc crystal structure with its high symmetry provides many slip planes and is the reason for the good mechanical formability of silver. It has low strength and can easily be cold worked. The hardness of pure, annealed silver is ~25 HV; the work hardening is not very marked. It can be rolled down to foil thickness of ~0.2 μm.

The value of the elastic modulus of single crystals of silver differs in the ⟨100⟩, ⟨110⟩, and ⟨111⟩ directions (Table 1), causing anisotropic elastic properties of single crystals and in polycrystalline material with textures. The recrystallization temperature is $0.28T_m$ (T_m = melting temperature in kelvins) (Jönsson, 1995).

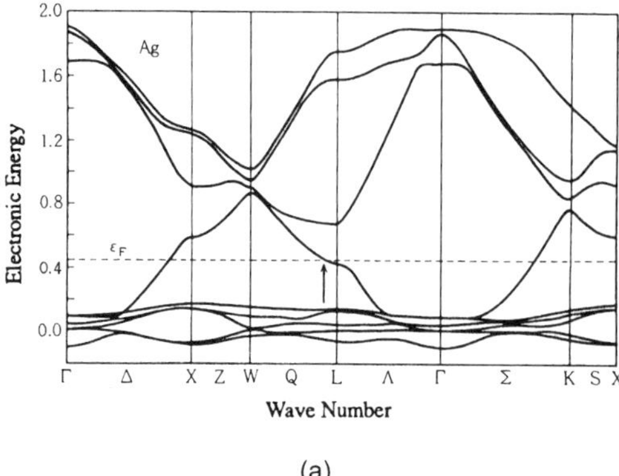

(a)

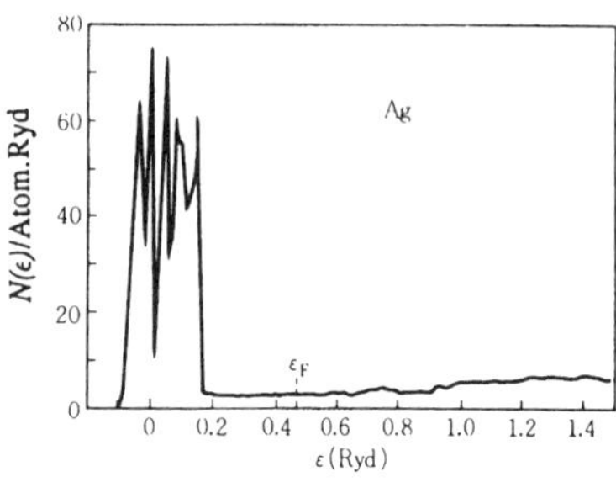

(b)

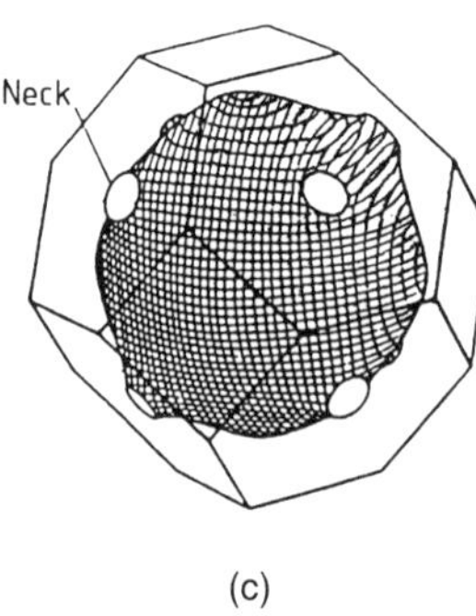

(c)

FIG. 3. Electronic structure of silver (Yonemitsu, 1991, pp. 25–28): (a) band diagram, (b) DOS curve (from Christensen, 1972), (c) Fermi surface.

2.2.2 Electrical Conductivity The electrical conductivity $K = en\mu$ is determined by the number of carriers en and their mobility μ, understood as the mean velocity of the carriers at a given field strength and electrical charge. The number of electrons involved in the Fermi surface of silver is 1 electron/atom, which can be converted by calculation with 1 × crystal density × atomic weight × Avogadro number to 5.8×10^{22} cm^{-3}. The high electrical conductivity of silver arises from its free-electron-like monovalent behavior and the small amplitudes of thermal vibrations due to its dense crystal structure and its low compressibility, which lead to only small electron-phonon scattering.

2.2.3 Magnetic Properties Silver behaves diamagnetically. The diamagnetic susceptibility is independent of the temperature between 0 °C and the melting point. Table 7 gives the susceptibilities at zero and 20 °C. Measurement of magnetic moments is used to determine the oxidation state of elements in alloys and compounds.

2.2.4 Optical Properties In the range of the wavelengths of visible light, silver shows the highest optical reflection of all noble metals. Absorption of light is related to interactions with electrons in filled states below the Fermi level. Photons can be absorbed by excitation of electrons only if the photon energy is higher than the energy difference between the lower filled band and the Fermi level. The difference of about 4 eV corresponds to a wavelength of 3000 Å, which is in the ultraviolet range (Hummel, 1971, p. 161). No visible light is therefore absorbed by electron excitation. Silver coatings are therefore best suited for optical mirrors, reflectors, and transparent heat-reflecting architectural glasses. The high reflectivity of silver is in accordance with the Hagen–Rubens relation (Hagen and Rubens, 1903),

$$R_{opt} = 1 - \sqrt{\nu/\kappa} \qquad (3)$$

where ν = frequency of light and κ = electrical conductivity. This states that at small frequencies, metals with high electrical conductivities are also good optical reflectors.

2.2.5 Chemical Properties Silver is resistant against water and shows excellent resistance to cold and hot concentrated organic acids such as acetic, formic, citric, lactic, fumaric, phthalic, and benzoic acids, fatty acids, and phenol. It is soluble in nitric and sulfuric acid. It is to a much higher degree than most other metals resistant to high-temperature caustic alkalies.

Silver is attacked by low–melting-point metals, such as mercury, sodium, potassium, lead, tin, indium, and bismuth. Reactions with sulfur-containing compounds form

Table 6. Selected values of basic physical and chemical properties of silver and gold (Degussa, 1995).

	Ag	Au
Atomic properties		
Atomic number	47	79
Atomic weight	107.868	196.966
Electron configuration	[Kr]$4d^{10}5s$	[Xe]$4f^{14}5d^{10}6s$
Crystal structure	fcc	fcc
Lattice constant (Å) (10^{-1} nm)	4.086	4.078
Atomic radius (Å) (10^{-1} nm)	1.445	1.442
Mechanical properties		
Density (20 °C) (g/cm^3)	10.49	19.32
Density, liquid at melting point (g/cm)^3	9.35	17.19
Hardness (HV, kg/mm^2)	26	25
Young's modulus (GPa)	82	79
Modulus of rigidity (MPa)	27	28
Poisson's ratio	0.37	0.42
Coefficient of compressibility	0.95	0.58
Thermal properties		
Melting point (mp) (°C)	961.9	1064.4
Vapor pressure at mp (mbar) (1×10^5 N m^{-2})	1.5	0.02
Boiling point (°C)	2212	3080
Thermal conductivity (20 °C) (W·m^{-1} K^{-1})	418	310
Coeff. of linear expansion (0–100 °C) (10^{-6} K)	19.0	14.3
Electrical properties		
Specific electrical resistance (0 °C) ($\mu\Omega$ cm)	1.465	2.0
Temperature coefficient of electrical resistance (0–100 °C) (10^{-3} K)	4.1	4.03
Thermoelectric voltage against Pt (0–100 °C) (mV)	0.740	0.770
Work function (eV)	4.73	5.22
Chemical properties		
Oxidation states in compounds	2, 1	3, 1
Ionization energy (eV)	7.576	9.225
Covalent radius for single bonds (Pauling) (pm)	134	134
Ionic radius (pm) (oxidation number, coordination number)	79 (2+,4) 67 (1+,2)	68 (3+,4) 137 (1+,6)
Electronegativity (Allred and Rochow)	1.4	1.4
Reduction potential (V) with (*n*) electrons	0.800 (1)	1.498 (3)

black silver-sulfide layers. Very thin surface layers of Ag_2O are formed in oxygen or in air below 180 °C. Above this temperature the oxygen partial pressure of the Ag_2O exceeds 760 Torr and the oxide decomposes. Up to temperatures of 800 °C, no visible attack can be seen, though small losses in weight can be measured.

Table 7. Magnetic susceptibilities of precious metals (Diehl, 1995).

Metal	χ(0 K) (10^{-9} m^3 kg^{-1})	χ(20 °C) (10^{-9} m^3 kg^{-1})
Ru	4.85	5.15
Rh	11.56	12.44
Pd	82.68	63.89
Os	0.60	0.69
Ir	1.29	1.51
Pt	13.32	12.19
Ag	−2.5	−2.5
Au	−1.7	−1.76

2.3 Alloying Behavior

2.3.1 Phase Formation Solid solutions are formed with A and B metals in different compositions. Maximum solubilities of B metals with favorable size factor (ratio of atomic diameter of solute to solvent <15%; Hume-Rothery and Raynor, 1956, p. 127) are determined by the electron concentration *e*/*a*, defined by

$$\frac{e}{a} = \frac{v_1(100 - x) + v_2 x}{100},$$

where v_1 = valency of the solvent, v_2 = valency of the solute, and x = atomic percentage of the solute, at the value of e/a = 1.4, these values decreasing with increasing valency of the solute (Raynor, 1976, p. 13).

Intermetallic compounds with B metals with identical or similar crystal structures are formed at compositions corresponding to definite ratios of the number of valence electrons per atom (electron concentration e/a) with values of 3/2, 21/13, and 7/4, ("Hume-Rothery phases"; Hume-Rothery and Raynor, 1956, pp. 201–202). Most of these phases extend over a certain homogeneity range. The observed values of e/a are near the electron concentration at which the Fermi sphere meets the boundary of the first Brillouin zone of the corresponding crystal structures (Kittel, 1980, p. 586). Besides these e/a-determined structures, there exists a huge number of other intermetallic phases, most of them of metallic character with predominantly cubic, hexagonal, or tetrahedric symmetry.

2.3.2 Influence of Alloying Elements

2.3.2.1 Strengthening. Figure 4 shows some examples of the effect of alloy additions on hardening. Silver and gold alloys show the characteristic behavior of solid-solution hardening. Yield stresses depend on the kind and the concentration of solutes. They decrease with increasing temperature to final values ("plateau stress"; Haasen, 1994, p. 325). These are supposed to be caused by remaining interactions of the dislocations with solute-atom groups, which cannot be released by thermal activation. The critical resolved shear stress of the solid solutions is characteristic for each kind of solute.

Precipitation hardening is performed with silver–copper alloys. It is of practical importance for jewelry, dentistry, or industrial apparatus parts and construction components.

Dispersion-strengthened silver and silver alloys have high mechanical strength, corrosion resistance, and electrical conductivity. Figure 5 shows the strength of dispersion-hardened silver compared with pure silver.

2.3.2.2 Grain Refining. The most common used grain-refining additive for silver is nickel, alloyed in concentration of 0.15 wt % ("fine-grain silver"). Gold, zinc, aluminum, antimony, and cadmium increase the recrystallization temperature.

2.3.2.3 Electrical Conductivity. Figure 6 shows some examples of effects of alloy additions on the electrical conductivity. In the low concentration range, the electrical conductivity increases proportionally to the atomic concentration of the solute atom. The absolute values of the resistivity increase for B-metal solutes roughly with the square of the distance of the position of the solute from the solvent in the periodic table (Linde, 1932).

2.3.2.4 Oxygen. Silver forms two oxides, Ag_2O and AgO. Both decompose at higher temperatures (Ag_2O at >150 °C, AgO

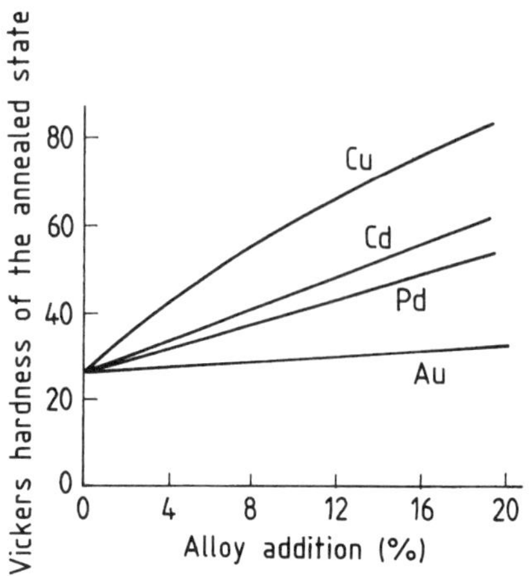

FIG. 4. Influence of alloying elements on the hardness of silver (Butts and Coxe, 1967, p. 111).

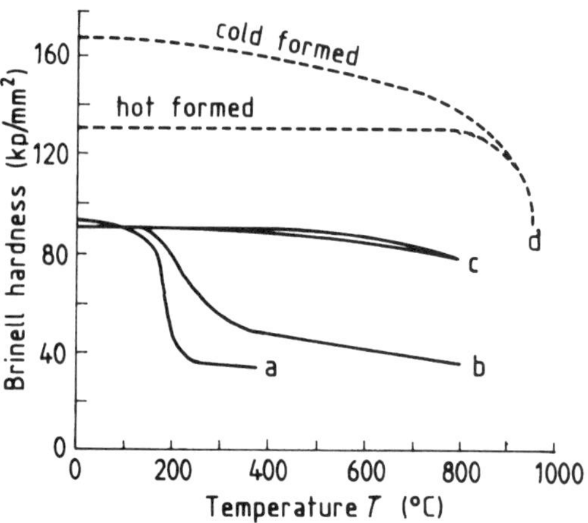

FIG. 5. Brinell hardness of (curve a) fine silver, (curve b) fine-grain silver (99.7%), (curve c) silver (99%) dispersion hardened by oxide-metal powder, and (curve d) silver (99.5%) dispersion hardened by internal oxidation (1 kp/mm^2 ~450 kPa) (Degussa, 1967, p. 36).

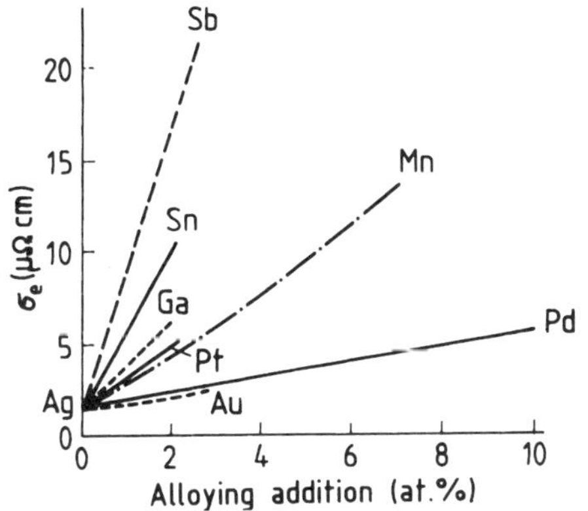

FIG. 6. Influence of alloying elements on the electrical conductivity of silver (Doduco, 1977, p. 20).

at >85 °C). Silver oxides have practical importance as catalysts and are used in considerable quantities in batteries.

Oxygen dissolves in the silver lattice, occupying interstitial lattice sites. The solubility increases with increasing temperature, rises very strongly at the melting temperature by a factor of ~80, and decreases then with further rising temperature (Table 8). The solution of oxygen in liquid silver is favored by the formation of Ag_2O, which dissolves in the liquid silver. The melting temperature of silver decreases with rising oxygen pressure of the surrounding gas atmosphere.

2.4 Abundance, Resources, and Raw Materials

2.4.1 Abundance and Resources The abundance of silver in the region accessible by mining (to a depth of 5 km) amounts to an average of 0.1 to 0.5 ppm, which is at least 100 times higher than that of gold and platinum. The abundance of silver in the oceans is estimated to be ~0.001 ppm (~10^9 t). This enormous reserve cannot, however, be extracted economically. The present known exploitable silver resources are estimated to be ~250 000 t.

Table 8. Oxygen uptake by silver (Degussa, 1967, p. 57).

T (°C)	Solubility (ppm)
200	0.03
400	1.4
600	10.6
800	38.1
973	3050
1000	3000
1100	2700
1200	2500

2.4.2 Raw Materials Native silver occurs in the form of nuggets, dendrites, or lumps. Main sources are silver sulfides in complex ores with lead or lead–zinc found in the United States, Mexico, South America, Africa, and Australia. They account for ~50% of the world silver production. The next most important sources are argentiferous copper ores, with largest reserves in Europe, which contribute ~30% of the world silver demand. Further amounts result from extraction from gold ores (~10%) and tin ores (~2%).

2.5 Production

2.5.1 Primary Production True silver ores are treated by the aqueous cyanide process. The ore is crushed, wet milled, and leached with an aqueous solution of NaCN. Solids are removed by filtering. Silver is then precipitated either by cementation with zinc dust or by electrolytic reduction. By far, most of the silver comes from lead and zinc ores. It is concentrated together with gold and the platinum metals by stirring the lead melt with zinc, precipitating, and separating the formed AgZn crystals. The zinc is distilled, leaving a 50% silver concentrate with predominantly lead (Parkes process). Silver is extracted by blowing air onto the liquid melt. The formed liquid PbO layer dissolves oxides of the base metals and up to 5% silver. It is continuously withdrawn for renewed extraction, leaving highly concentrated silver of about 99% purity (cupellation process). Silver from silver-containing copper ores is collected in the anode slimes of electrolytic copper refining and extracted by hydrometallurgical or pyrometallurgical processes.

Refining is done by electrolysis. Silver is deposited on the cathodes and continuously removed. The remaining slime contains gold and PGMs. Purity comes up with a single run to 99.99, with a double run to 99.999%.

2.5.2 Secondary Production (Recycling) Recycling processes include chemical, pyrometallurgical, hydrometallurgical, and electrochemical processes. The extraction of low-grade scrap is performed by melting with collector metals. These dissolve the major portion of the precious metals. The slags, oxides, and base-metal alloys with lesser amounts of precious metals are separately treated or fed back to the preceding operation. The collector metal for silver and gold is lead; the collector metal for essentially PGM-containing materials is copper. Figure 7 shows schematically the recycling process of silver scrap by lead extraction in the shaft furnace and the concentration by removing base-metal oxides together with PbO in the converter furnace.

2.6 Manufacturing

Fine silver and silver alloys are melted and cast into bars, rods, and granules either discontinuously or in continuous processing under a protecting-gas or reducing-gas atmosphere. Processing to foils and wires is done by extrusion, rolling, or drawing. Pure silver wires can be drawn to very fine diameters without intermediate annealing. Smaller items are produced by investment casting; complex-shaped pieces are made by the "lost-wax method" (see Gold, page 255). Silver powders of different grain sizes and shapes are produced from the melt by water- or gas-spray processes, or by electrolytical precipitation. They serve as base material for powder-metallurgical–produced composites and as components in metallizing preparations. Silver granules are formed by pouring a thin stream of molten silver into water.

Dispersion-hardened silver and silver alloys are produced either by powder-metallurgical methods or by internal oxidation of alloyed base-metal components—e.g., aluminum. In the concentration range of 0.1–0.2 vol % of solute, spherolithic oxide particles of sizes less than 20 nm are formed (Poniatowski and Clasing, 1968). Silver-containing composites are made from mixtures of silver powder with powders of oxides, refractory-metal carbides, or nickel by pressing, sintering, and extrusion. Extruded powder com-

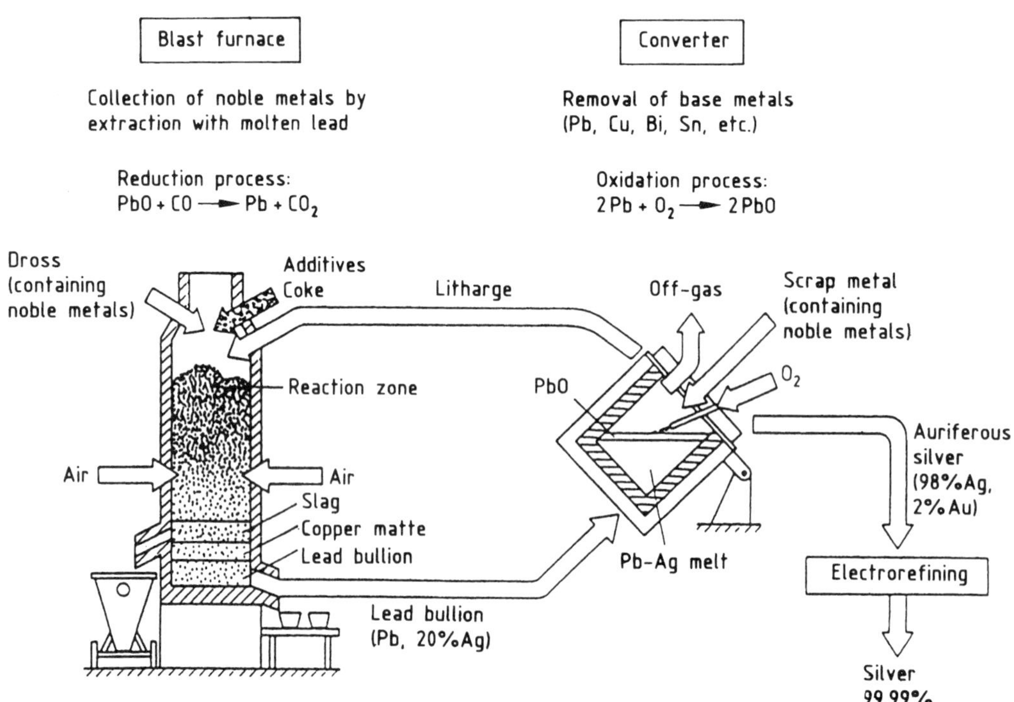

FIG. 7. Schematic view of silver recycling: extraction with lead in the shaft furnace, concentration in the converter (Renner, 1993, p. 125).

pacts show preferred orientation of the added powder grains along the rod axes. Silver composites with continuous length-oriented fibers can be produced starting with thin nickel wires placed inside silver tubes, which are then drawn together down to small diameters. Silver-nickel fiber composites are magnetic. High coercivity results if the nickel-fiber diameters become so small that no Bloch wall can be generated, and therefore only one single domain exists (Stöckel, 1979). Figure 8 shows the dependence of the coercive force on fiber diameter.

Infiltrated composite materials are produced from sintered skeletons of tungsten, molybdenum, or tungsten carbide, into which liquid silver penetrates by capillary forces.

Surface layers are produced

1. from bulk material components by mechanical cladding (rolling, pressing), thermal cladding (soldering, welding), and physical vapor deposition (PVD: evaporation, sputtering), or
2. from chemical compounds or solutions by galvanic deposition, chemical reduction, powder-based preparations (lacquers, pastes), thick-film pastes (screen printing, firing), thermal decomposition (e.g., from resinates), and chemical vapor deposition (CVD).

2.7 Alloys

2.7.1 Special Compositions The alloys AgNi0.15 ("fine-grain silver") (all compositions are in weight percent), AgSi1.5 ("hard silver"), and AgCd5 exhibit excellent corrosion resistance and high electrical conductivity. They are used as construction material for equipment and apparatus in the chemical industry.

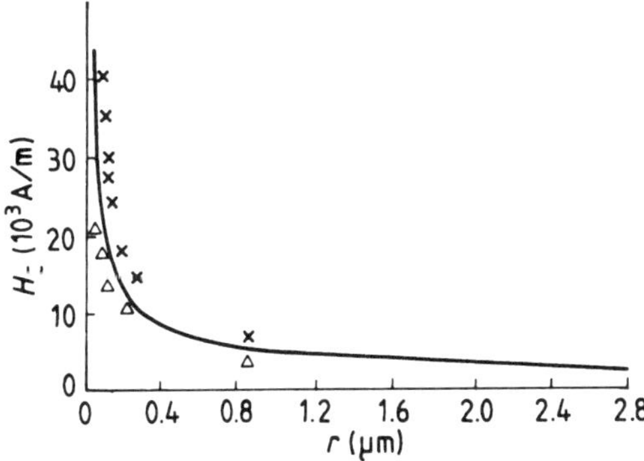

FIG. 8. Coercivity of Ag/Ni wire fiber (AgNi20) composites with small fiber diameters as a function of the fiber diameter. Crosses, 77 K; triangles, 295 K (Stöckel, 1979, p. 238).

Ag–Cu alloys. Copper effects hardening by three different mechanisms: solid solution, improved rate of work hardening, and, as a result of the miscibility gap in the alloy system, hardening by precipitation. The compositions of the majority of industrial important alloys are on the silver-rich side of the eutectic point. Ag–Cu alloys with 3–20 wt % Cu ("hard silver" alloys) are used for contacts, switches, circuit breakers, and motor contactors.

AgCu7.5 ("sterling silver"), AgCu4.2 ("Britannia silver"), and AgCu10 ("coin silver") are used for flat and hollow tableware, jewelry items, coinage, and electrical contacts.

AgCu28 (eutectic) and ternary or multicomponent alloys with Zn, Cd, Sn, Li, and/or Ni find wide application in braze alloys. Ag–Cu–P alloys are braze fillers for flux-free brazing.

AgMg0.25Ni0.2 is used for electrical contacts, and for high–thermal-conductivity spring clips for vacuum tubes, relays, and similar parts. Ag–Mg–Ni alloys harden by heating in air by internal oxidation. Nickel prevents grain growth during annealing treatments. The electrical conductivity is 75% IACS at 20 °C.

Ag–Pd alloys serve as electrical contacts and as conductor material in electronic devices. Alloys with 50–70 wt % silver are applied as electrode layers in capacitors. They are further used for brazing stainless steel, nickel-based, and other heat-resistant alloys. Small tubes of Ag23Pd77 are used in diffusion cells for the production of high-purity hydrogen.

Ag–Hg alloys play an important role in dentistry as material for fillings of cavities.

Silver alloys with tin or lead as binary, ternary, or multicomponent alloys with indium, tin, copper, and/or zinc are used as soft solders for general purpose and in special compositions as solders for electronic devices. The silver additions in the soft-solder matrix increase the mechanical strength and suppress the leaching of silver surfaces during the soldering process.

Ag–Ca alloys are the basis for the production of skeletal silver catalysts. Ag–Cu–Au alloys are used in a great variety of compositions in jewelry and dentistry. AgCd5In15, because of its high thermal-neutron capture cross section and good corrosion resistance against hot water, is used as cladding material of control rods in nuclear fuel reactors.

2.7.2 Solders and Brazes A variety of silver alloy compositions with a wide range of different melting temperatures are used as solders and brazes—*soft solders* (liquidus temperatures below 450 °C), *hard solders or brazes* (liquidus temperatures between 450 °C and 950 °C), and *high-temperature solders* (liquidus temperatures above 950 °C). Table 9 gives some typical compositions. Silver-containing solders based on tin are applied in food canning, in the installation of copper tubes, and in cryotechnology. Silver-containing eutectic binary and ternary alloys of lead and tin are used for the assembling of semiconductor devices.

Silver-based brazes are the biggest group. They are preferably used in industrial applications, electronics, and silverware production. Silver–manganese alloys are especially suited for soldering hard-metal and refractory-metal surfaces (molybdenum, tungsten,

Table 9. Noble-metal–containing solder and braze alloys (Beuers and Krappitz, 1995).

a) Soft solders

Type	Composition	Solidus (°C)	Liquidus (°C)
Sn based	Sn62Pb36Ag2	179 *E*[a]	
	SnAg3.5	221 *E*[a]	
	SnAg25Sb10	228	395
Pb based	Pb92.5Ag2.5Sn5	280 *E*[a]	
	PbAg3	305 *E*[a]	
	Pb92.5Ag2.5In5	307 *E*[a]	
Au based	AuSn20	280 *E*[a]	
	AuGe12	356 *E*[a]	
	AuSi2	370	740

[a]*E* = Eutectic composition.

b) Hard solders (brazes)

Type	Composition	Solidus (°C)	Liquidus (°C)
Ag–Cu	Ag40Cu19Zn21Cd20	595	630
	Ag60Cu27In13	605	710
	Ag44Cu30Zn26	675	735
	Ag72Cu28	779 *E*[a]	
Ag–Mn	Ag85Mn15	950	950
Ag–Pd	Ag68.4Pd5Cu26.6	807	810
	Ag54Pd25Cu21	901	950
	Ag75Pd20Mn5	100	1120
Pd–Cu	Pd18Cu72	1080	1090
Pd–Ni	Pd60Ni40	1237	1237
Au–Cu	Au80Cu20	890	890
	Au33Cu65	990	1010
Au–Ni	Au82Ni18	950	950

[a]*E* = Eutectic composition.

c) Active solders

Alloy	Composition (wt %)						Melting range (°C)	Brazing temperature (°C)
	Ag	Cu	Sn	In	Ti	Pb		
CB 1	72.5	19.5		5	3		730–760	850–950
CB 2	96				4		970	1000–1050
CB 4	70.5	26.5			3		780–805	850–950
CB 5	64	34.5			1.5		770–810	850–950
CB 6	98			1	1		950–960	1030
CS 1	86	10			4		221–300	850–950
CS 2				4	4	92	320–325	850–950

tantalum). Additions of nickel or cobalt increase the strength of the bond. Ag–Pd–Cu alloys are used to join components for high-temperature applications (gas turbines, supersonic aircraft, nuclear technology).

High-purity silver-solder alloys are needed for hard-solder processes under vacuum or in protective-gas or reducing-gas atmospheres. The minimum purity of the components is 99.99%. They must be free of metallic and nonmetallic constituents that have high vapor pressures at the soldering temperature (e.g., zinc, cadmium, arsenic). Palladium additions increase the corrosion resistance and the mechanical strength. These alloys are predominantly applied for solder connections of metallized ceramics with metals, of vacuum electron tubes, and of molybdenum, tungsten, zirconium, and beryllium.

Special compositions for jewelry and dentistry account for requirements of caratage, color, mechanical stability, and corrosion resistance.

Phosphorus-containing copper–silver-solder alloys are self-fluxing on copper, silver, and tin–bronze surfaces. These alloys find application in electrical engineering and in cryotechnology, where solder connections have to withstand temperature differences extending over wide ranges down to −50 °C.

Direct soldering of ceramic is possible with "active solders," containing ~2–4 wt % titanium or zirconium. Bonding is performed by reaction of the refractory metal with the ceramic, forming an adherent bond layer.

2.8 Applications

2.8.1 Photography Silver is used in large quantities as halide compounds, with the main starting material high-purity silver nitrate (see RECORDING, PHOTOGRAPHIC).

2.8.2 Jewelry and Silverware Silver alloys for jewelry and silverware contain copper up to 20%. Alloys with fineness (noble metal content in parts per thousand) of 925/1000 ("sterling silver"), 835/1000, and 800/1000 are the main types. Their low hardness in the soft (quenched) condition (HV 70–85) and the high ductility of sterling silver makes them best suited for sledge, bend, and filigran work. Hardening is performed by annealing at 290 °C, causing segregation of Cu atoms in the grain boundaries of the silver-rich phase. The harder 800/1000 alloy is preferred for the production of cast items with higher resistance against scratching. For enamel work the composition 970/1000 is used to avoid local melting at heating temperature exceeding 780 °C. Silver jewelry has to be protected against corrosion by coating with clear lacquers or by galvanic plating with rhodium.

2.8.3 Dentistry and Biomedical Uses Tooth cavity fillings are made with silver amalgams. They are formed from pasteous mixtures of Ag–Cu–Sn alloy powders and mercury, which harden at room temperature forming a complex matrix of silver amalgam and base-metal compounds. Typical compositions are in the range of 40 wt % silver, 32 wt % tin, 30 wt % copper, 2 wt % zinc, and ~3 wt % mercury.

Colloidal silver sols and organic silver compounds have bactericidal properties. They are used for disinfection. Implants of silver wire, gauzes, and plates have been used in brain surgery. Another area of medical application is acupuncture needles made from high-carat silver and gold alloys.

2.8.4 Electrical Technology Silver, silver alloys, and silver-bearing composites make up the largest amount of *contact materials* as make-break (= switching) contacts, especially in the heavy-current technology. Typical compositions are listed in Table 10.

Fine silver is used as a galvanic coating in thicknesses between 2 and 50 μm to lower the contact resistance and to improve the solderability of base-metal contact pieces. Values of hardness are between 60 and 160 HV, and the electrical conductivity amounts from ~50 to ~80% of that of cast metal.

Laminated composite materials are made by hot pressing, diffusion bonding, hot rolling, cold rolling, or explosive welding. Ag–Ni and Ag–Cu alloys ("fine grain" and "hard-silver" grades), soldered or clad on copper-alloy backings, and Ag–Ni composites are applied for ac and dc devices that work in the low-voltage range and current below 25 A.

Silver–metal-oxide (CdO, SnO_2, ZnO) composites cover the range of 50 to 5000 A in ac switches, especially with high current peaks on makes like low-voltage motor switches, power circuit breakers, and relays in automobiles. Contact faces that have been cut vertical to the axis of the fibers have in-

Table 10. Silver-bearing contact materials (Renner, 1993, p. 156).

Type	Composition (at. %)	Hardness (HV)	Electrical conductivity (m/$\Omega \cdot$mm^2)
Alloys	AgNi(0.15)	100	58
	AgCu(3)	120	52
	↓		
	AgCu(20)	150	49
Composites with:			
Metals	Ag–Ni(10)	90	54
	↓		
	Ag–Ni(40)	115	37
Oxides	Ag–CdO(10)	80	48
	↓		
	Ag–CdO(15)	115	45.5
	Ag–ZnO(88)	95	49
	Ag–SnO_2(8)	92	51
	Ag–SnO_2(12)	100	42
Carbon	Ag–C(2)	40	48
	Ag–C(5)	40	43.5
Refractory metals/compounds	Ag–W(20)[a]	240	26–28
	↓		
	Ag–W(80)[a]	80	42
	Ag–WC(40)	130	24–30
	↓		
	Ag–WC(80)	470	—

[a]Composite made by infiltration of liquid silver into a tungsten skeleton.

creased erosion resistance in make-break contacts compared to composites with random orientation of composite particles. Silver-graphite composites with 2–5 wt % C offer high protection against contact welding under short-circuit currents. Ag–Mo, Ag–W, and Ag–WC composites are applied for power switches that work at very high temperatures. Ag–W composites with tungsten contents above 60% are made by infiltration. They withstand arc erosion and prevent softening of rocket nozzles by the heat that develops by friction during the re-entry into the atmosphere.

Fuses, made from silver or silver alloys in bulk form or clad on copper, are designed to melt, thereby breaking the current flow, if an overload or a short circuit develops.

Silver coatings on copper or aluminum supports are the current-bearing medium of conductor devices in UHF television, radar technology, coaxial cables, and space technology. Surface protection is provided by thin coatings of either gold, palladium, or rhodium.

2.8.5 Electronics Silver is used in bulk material form and as a surface layer, as conductor material, as electrodes, as contact pads in electronic devices, in assembling parts, in thick-film and thin-film circuits, and in printed circuits. Silver-containing soft solders and brazes are widely used for electrical and mechanical connections of electron tubes and semiconductor devices.

Preparations made from silver or silver-palladium powder and organic binders serve for the production of electrode layers and contact faces of capacitors (Fig. 9). Conductor pastes for metallizing of ceramic thick-film circuits (Fig. 10) are made from powders of silver, silver-palladium, and silver-platinum mixtures together with glass frits and organic binders. They are applied by screen printing and firing.

2.8.6 Heat Mirrors Thin layers of silver with thicknesses below ~0.012 μm are transparent for optical light waves in the visible range (380–780 nm) but reflect in the long-wavelength region. These layers, deposited on architectural window glasses by sputtering, reduce energy losses caused by outgoing heat radiation. Additional thin, transparent, and color-neutral oxide coatings serve as antireflex layers and to improve adhesion, corrosion resistance, and mechanical stability. Special sputtering equipment and

FIG. 9. Multilayer ceramic capacitor: cross section, with silver-palladium electrode layers and silver contact edge (Demetron).

target material conditions have been developed for large-scale batch type and continuous processing of coherent layers of homogeneous thickness (Dachselt *et al.*, 1982; Gläser, 1989).

2.8.7 Chemical Technology In chemical equipment, silver is used in bulk form and as surface layers for corrosion-resistant apparatus parts—e.g., crucibles for alkaline fusions or evaporations. For high-temperature applications, dispersion-hardened parts are used. Examples are fittings, bursting disks to protect equipment against damage by overpressure, and wires and rings as sealing materials for oxygen compressors and for ultrahigh-vacuum equipment. Filter elements of spheroidal silver powder are used in fruit presses. Silver-lined reactors find extended use in the food industry as vacuum pans, condensors, and evaporators. Hormones and vitamins generally are prepared in silver equipment.

2.8.8 Catalysis Metallic silver is an excellent catalyst for oxidation processes of organic compounds. It is applied in the form of wire gauzes or as precipitate on γ-Al_2O_3 or silicate carrier. Technically important processes are the production of ethylene oxide and the formation of formaldehyde from methanol.

2.8.9 Batteries The AgO/Zn button battery type is of great practical importance as current source for cameras, watches, hearing aids, and similar small devices. The worldwide yearly demand exceeds 10^9 pieces. Silver oxide forms the cathode; zinc, cadmium, or magnesium forms the anode. AgO batteries have high energy densities (80–250 W·h/dm^2), high discharge current, and high long-life stability. The cell voltage is 1.55 V. Rechargeable units with cadmium anodes reach lifetimes exceeding 1000 cycles (Benner *et al.*, 1991, p. 541).

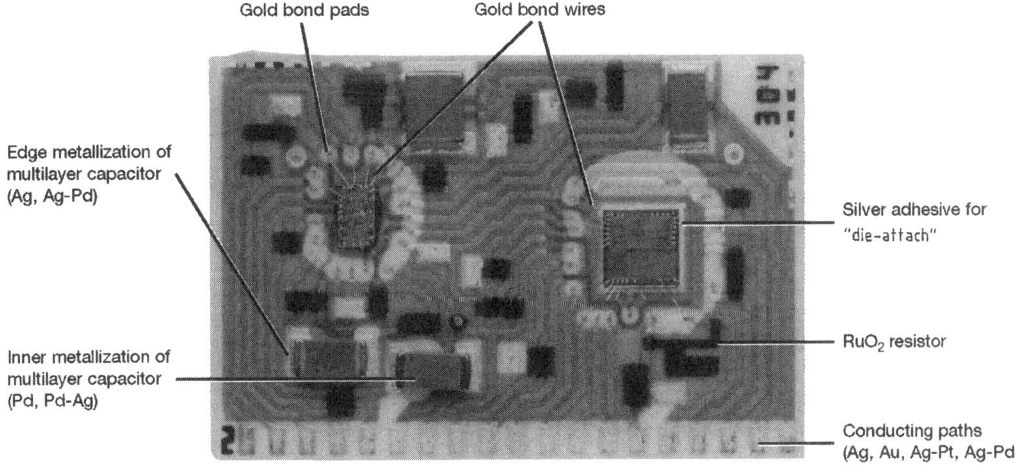

FIG. 10. Thick-film hybrid circuit on ceramic substrate with conductor and resistor paths, bond pads, and assembled semiconductor devices (Dorbath *et al.*, 1991, p. 47).

3. GOLD

Gold, chemical symbol Au (aurum), atomic number 79, relative atomic mass 196.96654 (only one natural isotope), is a soft, brightly shining metal of the characteristic gold-yellow color. In pure condition it is very ductile. The most important properties are its oxidation resistance, its easy workability, its mechanical and thermal stability, and its high electrical conductivity. The main application areas are jewelry (86%), electrical technology/electronics (7%), and dentistry (3%) (quantities in weight percent of the total demand in the western world 1994; Degussa, 1994).

3.1 Electronic Structure

The electronic structure is [Xe] $4f^{14}5d^{10}6s$. Band diagrams for the $5d$ and $6s$ states, the density-of-states curve, and the shape of the Fermi level are very similar to those of silver (Fig. 11). The energy difference between the upper filled d band and the Fermi level is about 2.17 eV.

3.2 Special Properties

Selected values of basic physical and chemical properties of gold are given in Table 6.

3.2.1 Mechanical and Thermal Properties Pure gold is very soft and can easily be cold worked without considerable work hardening. It can be cold deformed more than 99% of its cross section without intermediate annealing. High-purity gold wires can be cold drawn without envelope down to diameters of <10 μm. Very thin beaten gold leaf is hammered in packages to foils as thin as 0.1 μm. The hardness of pure, annealed gold is ~25 HV; the recrystallization temperature is $0.28T_m$ (T_m = melting temperature in kelvins).

3.2.2 Electrical Conductivity The electrical conductivity of pure gold is ~73% of that of silver. The number of electrons involved in the Fermi surface is 5.7×10^{22} cm^{-3}. Minute quantities of magnetic impurities, particularly iron, down to the part-per-billion level, influence the low-temperature electrical transport properties of gold. They cause a minimum in the electrical resistivity, accompanied by anomalies in thermoelectric power, magnetoresistance, and magnetic susceptibility (Kondo effect: Kondo, 1964, 1965). Measurements of the thermoelectric power can be used to determine the content of iron in gold down to ~0.01 ppm (Kopp, 1976).

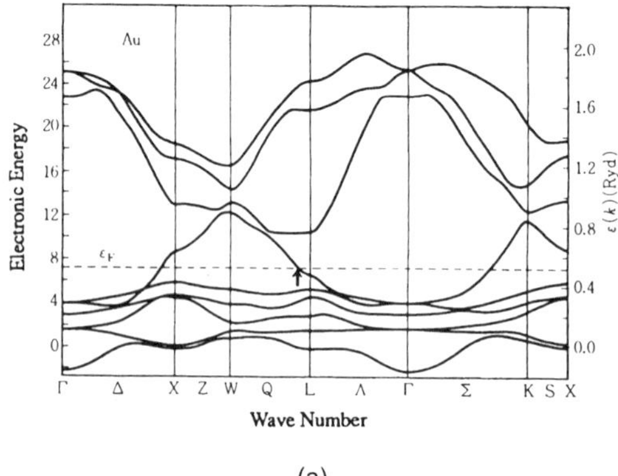

(a)

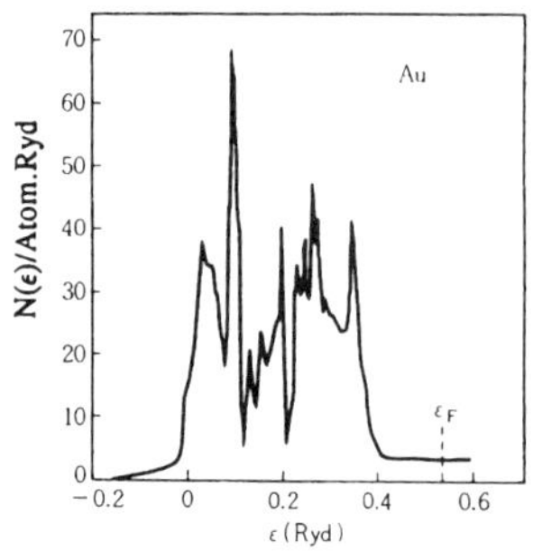

(b)

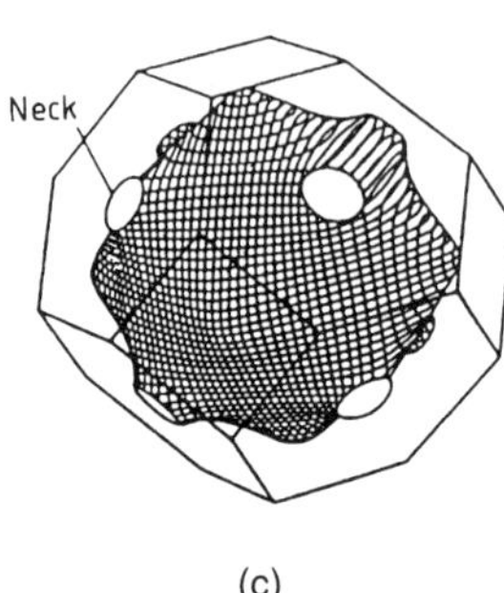

(c)

FIG. 11. Electronic structure of gold (Yonemitsu, 1991, pp. 25–28): (a) band diagram, (b) DOS curve (from Christensen and Saraphin, 1971), (c) Fermi surface.

3.2.3 Magnetic Properties Gold behaves diamagnetically. Its diamagnetic susceptibility is independent of the temperature between 0 °C and the melting point. Table 7 gives susceptibilities at zero and 20 °C.

3.2.4 Optical Properties The reflectivity of gold shows a marked fall at ~550 nm in the visible range, with a minimum of *r* ~0.25 in the near ultraviolet. The *d* bands are, compared with silver, positioned at higher energy levels. Interband transitions occur already at smaller energies (~2.17 eV) in the visible branch. The reflected light contains all wavelengths above 560 nm, which gives the typical gold-yellow color. Because of its high reflectivity in the infrared range, gold films are used in radiant heating and drying devices, for thermal-barrier windows of large buildings, and as reflecting coatings to protect space vehicles and space suits against excessive solar radiation. Ultrafine particles of gold ("gold-black") show selective light absorption. Reflectivity is less than 0.01 in the visible region and less than 0.1 in the infrared. Gold-black forms, therefore, an excellent light-receptive surface coating for radiation detectors and is also suited as selective absorption film for the capture of solar energy (McKenzie, 1978).

Thin films show optical transmission dependent on the wavelength and film thickness. Thin gold films appear green in transmission. Penetration of photons in the metal surfaces is in the range of 10^{-6} cm. Above this thickness, transmission decreases very rapidly.

3.2.5 Chemical Properties Gold does not react with water, dry or humid air, oxygen, ozone, nitrogen, hydrogen, fluorine, iodine, sulfur, or hydrogen sulfide under normal conditions. Gold withstands practically all anorganic and organic acids and aqueous solutions of alkali-metal hydroxides, alkali salts of mineral acids, and alkali-metal oxides. Also, fused caustic alkalies do not attack gold, provided air and other oxidizing agents are excluded. If a hydrohalic acid is combined with an oxidizing agent, such as nitric acid ("aqua regia"), a halogen, hydrogen peroxide, or chromic acid, gold will dissolve in the mixture. Gold reacts readily with dry chlorine. Gold can be easily recovered from solutions of gold-complex compounds by anorganic or organic reducing agents (e.g., iron, tin, or zinc salts, sulfur dioxide, hydrazine, oxalic acid) and by anion exchangers. Colloidal gold forms stable hydrosols of an intensive red or violet color. Colloidal gold is the coloring component of rubin glass.

3.3 Alloying Behavior

3.3.1 Phase Formation Gold forms alloys with most metals. Alloying behavior and phase formation follow principally the same rules as those valid for silver. Gold alloys readily with mercury at room temperature to form an amalgam.

The maximum solubilities of B metals with favorable size factor are reached at *e/a* values of ~1.3. The eutectic alloys of gold with silicon (AuSi3.1), germanium (AuGe12) and tin (AuSn20) show very low melting temperatures (Table 9). In the system gold–copper, superlattice structures are formed at compositions Cu_3Au, CuAu, and $AuCu_3$ by disorder–order transformation from the cubic A1 crystal structure to orthorhombic ($L1_0$ type for CuAu) and cubic ($L1_2$ type for Cu_3Au and $AuCu_3$). Their formation is accompanied by changes of the electrical conductivity and hardness. The superlattice phases form sublattice structures, showing gold and copper on inverted lattice sites. They form coherent domains of nine or five single cells, respectively, separated by the so-called "antiphase boundaries" (APB; Fig. 12).

3.3.2 Influence of Alloying Elements

3.3.2.1 Strengthening. The effect of alloy additions on hardening corresponds principally to the behavior of silver. Figure 13 shows some examples of the effect of alloy additions on hardening; Fig. 14 shows the dependence of the yield stress on the kind

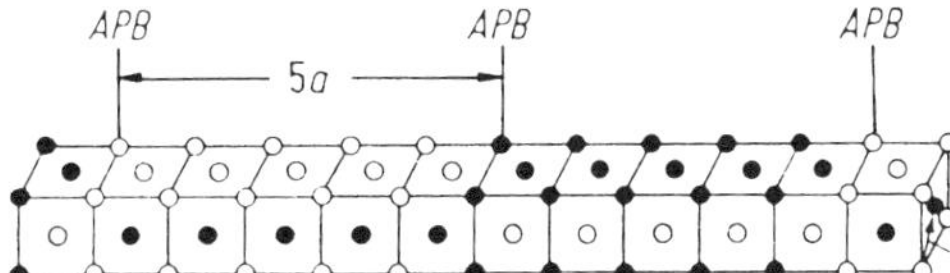

FIG. 12. Antiphase boundaries (APB) in the ordered CuAu superlattice ("CuAuII") (Haasen, 1994, p. 40, copyright Springer Verlag, Heidelberg).

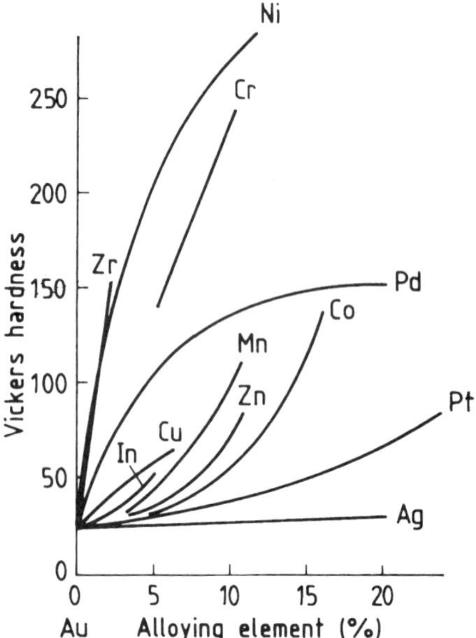

FIG. 13. Influence of alloying elements on the hardening of gold (Wise, 1964, p. 72).

and the concentration of solutes. A miscibility gap in the ternary system Au–Ag–Cu, which broadens with decreasing temperature, enables precipitation hardening in these composition ranges. Additionally, strengthening of the alloys occurs in the composition range >75 wt % Au by the formation of the ordered AuCu phase.

3.3.2.2 Grain Refining. Effective grain-refining additions for castings of gold and gold alloys (Au–Cu, Au–Ag–Cu in 14 and 18 carat) are small amounts of iridium, ruthenium, rhenium, zirconium, barium, tantalum, niobium (0.01%–0.1%), and combinations of Co/Ba, Co/Mo, and Ni/Mo. Zirconium, barium, and cobalt increase the recrystallization temperature by 150–200 °C and suppress grain growth at further annealing. Tin (0.3 wt %) and antimony (0.1 wt %) decrease the recrystallization temperature of 18-ct gold–copper alloys by ~100 °C (Ott and Raub, 1980, 1981, 1982).

3.3.2.3 Electrical Resistivity. Figure 15 shows some examples of the increase of the electrical resistivity by alloying elements. The increase of the resistivity in the low concentration range is proportional to the atomic concentration of the solute. Like in the case of silver, the absolute values of the resistivity for B-metal solutes increase roughly with the square of the distance of the position of the solute from the solvent in the periodic table (Linde, 1932).

3.4 Abundance, Resources, and Raw Materials

3.4.1 Abundance and Resources Gold is distributed very unevenly in the earth's crust. The average abundance is estimated as

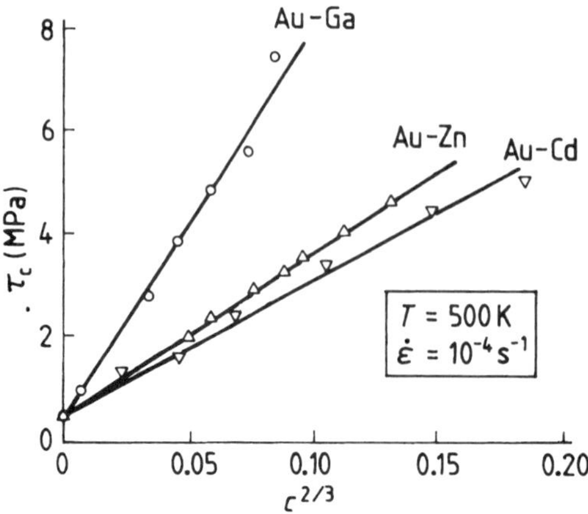

FIG. 14. Critical shear stress of solid-solution hardened gold alloys (Haasen, 1994, p. 326, copyright Springer Verlag, Heidelberg).

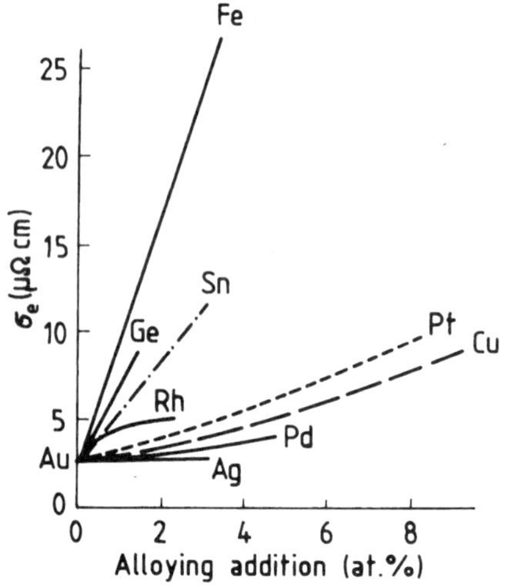

FIG. 15. Influence of alloying elements on the electrical conductivity of gold (Doduco, 1977, p. 50).

~0.005 ppm. The gold content in the oceans varies greatly depending on location. Gold contents of 0.008–4 mg/m^3 have been reported.

World gold reserves for which mining is economically viable are assessed at ~70 000 t.

3.4.2 Raw Materials The main sources contain gold as the metal—generally alloyed with small amounts of silver, copper, or platinum—as particles in quartz veins, in river sands ("placer" deposits), or included in massive rock formations. The average gold contents vary from ~0.5 ppm in placer deposits to ~12 ppm in the rock formations, when separated from the gangue. Most important deposits are in South Africa (Witwatersrand), Russia, the United States, Papua New Guinea (Ok Tedi), and Brazil.

3.5 Production

3.5.1 Primary Production Gold particles are concentrated and separated from the gangue by milling, flotation, or panning. The most important process for extraction is cyanide leaching in the presence of particles of activated carbon ("carbon in pulp" process). The gold-cyanide solution is adsorbed on the surface of the carbon particles. The process of solution and adsorption runs in several stages in countercurrent flow. The loaded carbon particles are removed from the pulp by sieving, and the adsorbed gold cyanide is eluted with hot NaCN solution. Gold is recovered either by precipitation with zinc or by electrolysis with stainless-steel electrodes. New developments are directed to the exchange of carbon against organic resins as adsorbing agent for the AuCN solution ("resin in pulp"). In metallurgical extraction processes, gold and silver follow the same route. Zinc, lead, and copper act as collecting agents for gold through the formation of alloys. Refining is done by leading chlorine gas over the surface of molten gold at about 1100 °C (Miller process), removing silver and base metals as slags and volatile components, and leaving gold of ~99.5% purity. Electrolysis in HCl and $H[AuCl_4]$ in aqueous solution refines to purities of 99.9% to 99.99%. Tetrachloroauric(III) acid can be also extracted from aqueous solution by solvent extraction with organic solvents that form complex gold compounds (Fieberg and Edwards, 1978).

3.5.2 Secondary Production (Recycling) Recovery processes from secondary materials like alloys, sweeps, or surface-coated materials is performed by principally the same kind of processes as applied for the primary production. Gold can be extracted from solutions with very low gold content (<0.1 g/L) by anion exchange resins or activated charcoal.

3.6 Manufacturing

Bars, rods, foils, and wires are manufactured by classical metallurgical processes. For the production of gold leaf, fine gold is rolled down to 20-μm thickness, then placed between sheets of vellum and beaten out with a spring hammer to leaves of thicknesses of ~0.2 μm. Alternatively, those leaves can be produced by cathode sputtering.

Casting of small jewelry items and dental pieces is done by the *lost-wax method*. Wax models of the desired shape are made in a split mold, provided with a joint gate, and embedded in a hardenable paste. After hardening, the wax is melted out and the hollow mold fired at red heat. The molten gold alloy is then poured in. After cooling, the formed pieces are removed, smoothed, and polished.

Gold powders of different grain sizes and shapes are made by precipitation from solutions by means of reducing agents. Thin surface films are produced by galvanic processes or by sputtering; thicker surface layers are produced by applying metallizing pastes made from special gold or gold alloy-powder grades.

3.7 Alloys

3.7.1 Special Compositions AuAg20 (all compositions are in weight percent) is applied for low-voltage electrical contacts, e.g., rivets. AuCo5, AgNi5, and AuAg26Ni3 are used as contact material that is resistant against Ag migration. Au–Ag–Cu alloys ("colored gold alloys") in different compositions are predominantly used in jewelry and dentistry. Melting temperatures are adjusted by additions of zinc or nickel for jewelry alloys and palladium or platinum for dental alloys.

Au–Ni–Cu and Au–Ag–Pd alloys ("white-gold alloys") are jewelry alloys, applied also in different compositions corresponding to requirements for mechanical strength and oxidation resistance. Palladium is used as an alloy component instead of nickel, if nickel may cause allergic reactions. They have higher melting temperatures but are relatively soft and cannot be age hardened by precipitation. They may be made susceptible to precipitation hardening by additions of copper.

Au–Pt alloys are, because of their high corrosion resistance, used for spinnerets in rayon production. AuPt30 serves as a high-melting precious-metal braze with liquidus temperature of 1450 °C and solidus temperature of 1228 °C. The alloys with 40%–65% Au can be effectively age hardened by quenching from 1100 °C and annealing at ~500 °C to tensile strengths of ~1400 N/mm^2, using the existing miscibility in this composition range. Small amounts (~0.5 wt %) of Rh, Rn, or Ir prevent segregation effects during the ingot production. They broaden the miscibility gap and turn the system into the peritectic type by extending the two-phase region up to the solidus.

AuPt10 and AuAg25Pt5 are alloys for electrical contacts working under heavily corrosive conditions. AuCu14Pt9Ag4 is applied for oxide-free slip contacts of high reliability at very low voltage and current—for instance, as a test value transmitter in the form of bent wire pieces or cut foils.

Au–Ni alloys form a continuous series of solid solutions at high temperature. These alloys are used as brazing alloys. The most common composition is AuNi18. Au alloys with small amounts (up to 1–3 wt %) of iron, cobalt, or nickel are electrolytically deposited as hard, wear-resistant surface coatings of electrical contacts.

3.7.2 Solders and Brazes The low-melting-point eutectic alloys with silicon, germanium, and tin have high mechanical strength and are resistant against thermocycling. They are of special interest in the manufacturing of electronic devices.

Gold-based brazing alloys are important for jewelry, dentistry, and high-temperature–resistant solder connections. The compositions of jewelry and dentistry solders must correspond in color, corrosion resistance, and mechanical strength to the materials to be joined. Melting temperatures are adjusted by alloying with zinc and nickel to lower, and with palladium and platinum to raise, the melting temperature. Au–Ni and Au–Cu alloys have high oxidation resistance and good thermal stability up to 800 °C. The production of turbines, jet engines, and components and devices in nuclear and space technology are fields of application (Table 9).

3.8 Applications

3.8.1 Jewelry Gold jewelry alloys are usually classified as colored and as white gold alloys (Table 11). Colored gold alloys are based on the ternary Au–Ag–Cu alloy system. It offers, in different compositions, a variety of colors and the possibility to adjust the mechanical properties according to requirements of production and wear resistance. The colors are described by special parameters (T = tone, S = saturation, D = darkness). Values for different color grades and methods for their determination have been appointed by standards (DIN 5033, 6164, 8238). The binary alloy of 99% gold and 1% Ti shows the bright color of pure gold. Its hardness of 70 HV in soft condition can be raised by work hardening and by subsequent age hardening up to 170 HV, with a fine distribution of the intermetallic $TiAu_4$ phase in the gold matrix (Gafner, 1989). The high caratage and its strength are of interest both for jewelry and for technical applications—e.g., bond wire for semiconductor devices.

White gold alloys contain either nickel or palladium. They are typically used with gold

Table 11. Main components (wt %) of gold-based jewelry alloys (Renner, 1990, p. 526).

	Gold content, wt %	Content of alloy components, wt %
Colored gold		
20 carat	88.3	Ag 0–16.7, Cu 0–16.7
18 carat	75	Ag 0–20, Cu 5–25
14 carat	58.5	Ag 8–34, Cu 7.5–33.5
8 carat	33.3	Ag 8–35, Cu 30–55, Zn 0–20
White gold		
18 carat	76	Cu 4–8, Ni 10–18, Zn 3–6
18 carat	75	Pd 10–20, Cu + Zn 5, rest Ag
14 carat	59	Cu 15–25, Ni 10–16, Zn 5–8

contents of 14 and 18 ct. Nickel white gold may cause allergic reactions. The trend is therefore toward palladium white-gold alloys, which are also superior in their ductility and corrosion resistance.

Very thin (<0.1 μm) layers of gold, silver, platinum, and palladium on ceramic, porcelain, glass, quartz, and mica, predominantly for decoration, but also for technical purposes, are deposited by thermal decomposition of liquid noble-metal preparations. They contain organic noble-metal compounds ("resinates," organic thio compounds, and carboxylates). They are applied by painting, brushing, or screen or offset printing and firing. The adherence on ceramic and porcelain is effected by additions of base-metal oxides (Bi_2O_3, SnO_2) or oxide-forming base-metal compounds. Rhodium-oxide prevents grain growth and agglomeration of the deposited noble-metal pigments (Hunt, 1979). Addition of fine precious-metal powders to the preparations leads to the formation of matte surfaces ("burnished gold," "burnished silver").

3.8.2 Dentistry Pure gold foil and matte gold are used for direct filling of cavities. Gold alloys form the major part of inlays, crowns, multiple-unit bridges, and partial dentures. Special gold alloys are provided for successive surface sealing with ceramic veneers (PFM alloys—porcelain fused to metal). The available hardness of dental alloys ranges from 50 to ~275 HV 5. Alloy additions of silver, palladium, or platinum and copper, tin, indium, zinc, or gallium affect hardness and melting range and improve the adhesion to ceramic (Kempf and Hausselt, 1992).

3.8.3 Electrical Technology Gold surfaces enable the formation of electrical connections with lowest Ohmic resistances. Gold and gold alloys are indispensible for conductor material and low-voltage contacts for communication and information, where even thin oxide layers would cause interruptions or failures in the signal transfer. AuCr2 serves as resistor material in corrosive environments with stable temperature coefficient of resistance (TCR) between -20 and $+40$ °C. Wires of multicomponent alloys based on gold or palladium serve as potentiometer windings. Their specific resistance values range from 15 to 150 $\mu\Omega\cdot$cm with TCR extending from $+10^{-4}$ to -10^{-5} (deg^{-1}). Sliding contacts are made from Pd–Ag or Ag–Au–Cu alloy.

3.8.4 Electronics Gold-plated surfaces serve for connections by pressure, bond, or solder contacts on semiconductor chips and for the joining of assembling parts and package components. The surface coats are deposited by galvanotechnic, by sputtering, or by screen printing of thick-film pastes. Gold powders of different grain sizes and spherical or platelike grain shapes are used in metallizing and thick-film pastes to produce bond pads and conductor layers. The eutectic alloys of gold with silicon, germanium, and tin (Table 9) are used for low-temperature solder connections with high mechanical strength. Gold and gold–silicon alloys are used to connect the semiconductor crystals with the base contact supports ("die-attach") on lead frames, packages, or hybrid circuits. Thin wires ("bond wire", standard diameter 25 μm) made of pure gold, alloyed with small amounts (~7 ppm) of beryllium, serve for the connection of the majority of monolithic integrated semiconductor chips ("IC") with the inner leads of the device (thermocompression bonding). Gold-plated metallic lids and solder preforms of the alloy AuSn20 are used for the hermetic (vacuum tight) sealing of devices.

4. PLATINUM AND THE PLATINUM-GROUP METALS

4.1 Common Characteristics

4.1.1 General Description The six platinum-group elements ruthenium, rhodium, palladium, osmium, iridium, and platinum are high–melting-point, silver-white metals of high density. The most important properties are the high resistance against oxidation, high strength and thermal stability, and their capability to act as catalysts in numerous inorganic and organic chemical reactions.

4.1.2 Electronic Structure Palladium, platinum, rhodium, and iridium have fcc structures. The electronic structures of the isolated atoms are for rhodium [Kr]$4d^85s^1$, for palladium [Kr]$4d^{10}5s^0$, for iridium [Xe]$4f^{14}5d^76s^2$, and for platinum [Xe]-$4f^{14}5d^96s^1$ (see Table 14). They have similar densities of state (DOS) with more than five high d-band peaks that overlap with the broad s band. The Fermi levels of palladium and platinum are only slightly above the d-band peak. The corresponding Fermi surfaces have complicated shapes, forming Fermi hole sheets and Fermi electron sheets (Figs. 16 and 17). The numbers of d electrons of rhodium and iridium are smaller than those of palladium and platinum; the Fermi levels are located at the midslope of the peak of the DOS curve. The band diagrams show an s-like electron sheet around the same symmetry point as for palladium and platinum. The Fermi surfaces have more complicated shapes and consist of several sheets (Figs. 18 and 19).

The crystal structures of ruthenium and osmium are hexagonal closed packed. The electronic structures of the isolated atoms are for ruthenium [Kr]$4d^7 5s^1$ and for osmium [Xe]$4f^{14} 5d^6 6s^2$. The band diagrams and the Fermi surfaces (Figs. 20 and 21) are much more complicated than those of the elements with the fcc structure and show a variety of hole and electron sheets (Yonemitsu, 1991).

4.1.3 Physical Properties

4.1.3.1 Mechanical and Thermal Properties. Palladium and platinum have low strength and can easily be cold worked. The work hardening is not very marked. Rhodium and iridium have much higher strength and have only limited workability at room temperature. They can be easily deformed at higher temperatures (rhodium >200 °C, iridium >600 °C), accompanied by strong work hardening. Ruthenium and osmium exhibit high strength. They can only be deformed at high temperatures and only to a limited degree. Work hardening for these metals is stronger than that for rhodium and iridium. Ruthenium that has been deoxidized by melting with rare-earth metals can be hot rolled.

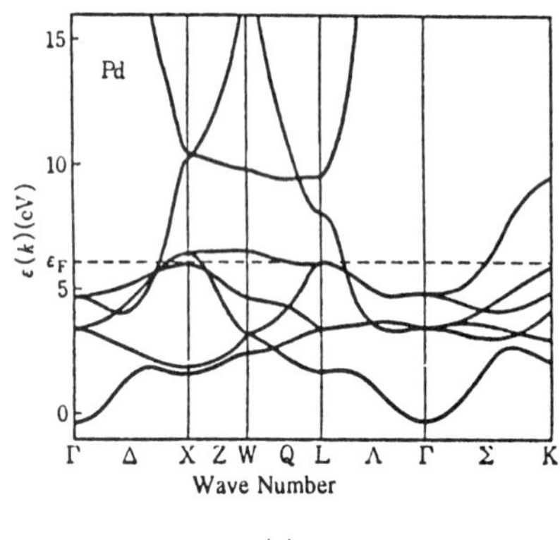

(a)

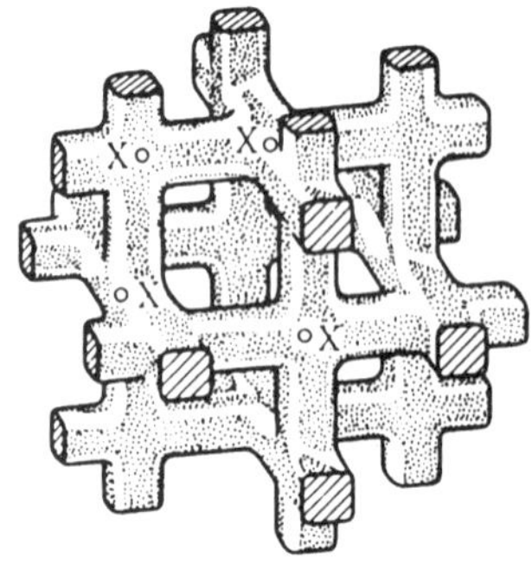

1

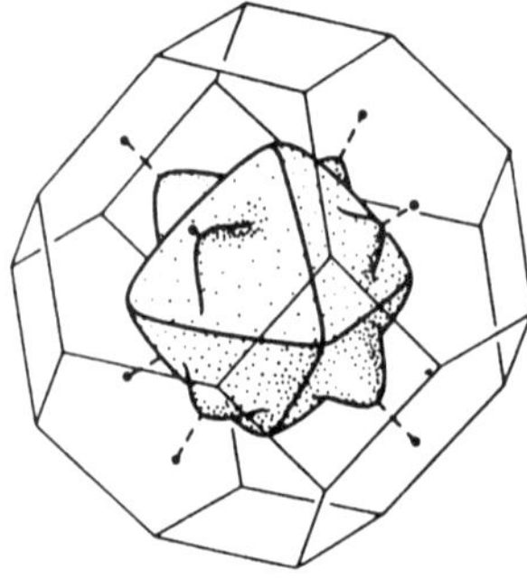

2

(b)

FIG. 16. Electronic structure of palladium: (a) band diagram (from Fong, 1974), (b) Fermi surface: (1) hole sheet, (2) electron sheet (from Mueller *et al.*, 1970).

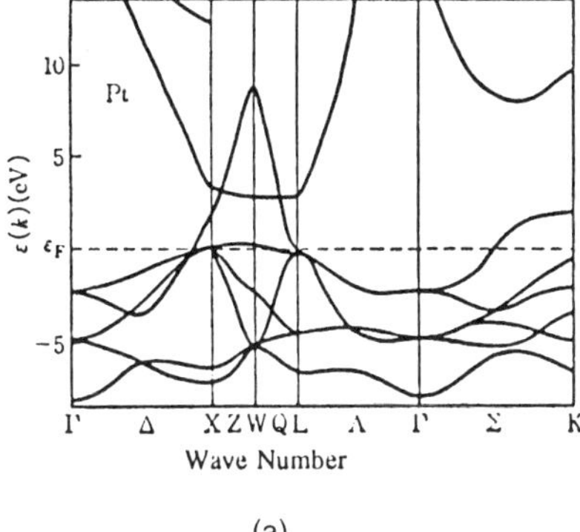

(a)

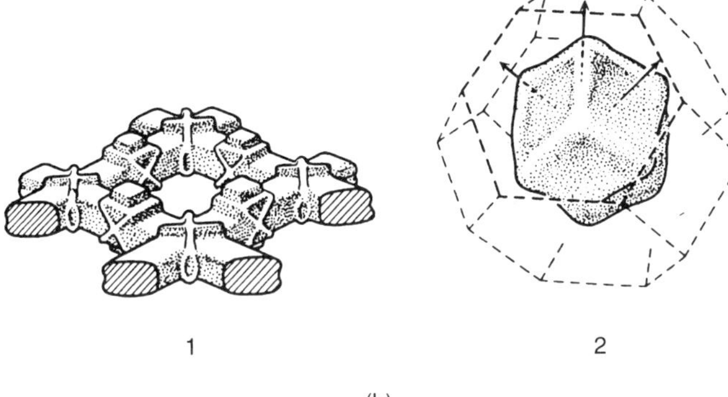

FIG. 17. Electronic structure of platinum (Yonemitsu, 1991, pp. 31–32): (a) band diagram (from Rogers, 1974), (b) Fermi surface: (1) hole sheet, (2) electron sheet (from Ketterson and Windmiller, 1970).

Figure 22 shows the increase of hardness caused by cold deformation. The rate of work hardening increases with increasing modulus of elasticity. There are two separate branches for cubic and hexagonal structures (Fig. 23; Darling, 1973, p. 93). The deformation behavior can be related to the ratio of the bulk modulus to the rigidity modulus (Table 12). Lower values correspond to the harder and brittle metals (Pugh, 1954).

The melting temperatures of the PGMs are between 1554 °C for palladium and 3045 °C for osmium. The recrystallization temperatures in relation to the melting temperatures decrease with increasing nuclear charge (ruthenium $\sim 0.57 T_m$, rhodium $\sim 0.43 T_m$, palladium $\sim 0.41\ T_m$, iridium $\sim 0.55\ T_m$, platinum $\sim 0.31\ T_m$). The melting points, the hardness, and the brittleness likewise decrease. The thermal expansion and the ductility increase with increasing nuclear charge.

4.1.3.2 Electrical Resistivities. The electrical resistivities are between $10.55 \times 10^{-8}\ \Omega \cdot \text{m}$ for palladium and $4.78 \times 10^{-8}\ \Omega \cdot \text{m}$ for rhodium. They decrease in the order palladium > platinum > osmium > ruthenium > iridium > rhodium. (Table 2).

4.1.3.3 Thermoelectric Behavior. Thermocouples made from platinum-group alloys are used to measure high temperatures because of their refractoriness and oxidation resistance. The thermoelectric voltage is defined as the voltage $E_{A,\mathrm{Pt}}$ in mV that a wire made from material A develops in conjunction with a wire made from physically pure platinum. The device forms with two junctions a closed electric circuit, with the two junctions held at different temperatures. One junction is supposed to have a temperature of 0 °C. For metals A and B, one has

$$E_{\mathrm{AB}} = E_{\mathrm{A_{Pt}}} - E_{\mathrm{B_{Pt}}}. \tag{5}$$

The thermoelectric voltage is positive if on the hot junction the current flows from the platinum to material A. The thermoelectric power ϵ,

$$\epsilon_{\mathrm{A_{Pt}}} = \frac{dE_{\mathrm{A_{Pt}}}}{dt} \quad (\mu\mathrm{V/K}), \tag{6}$$

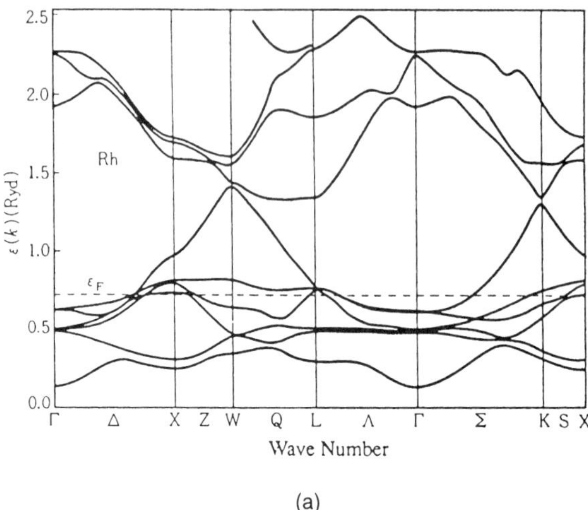

(a)

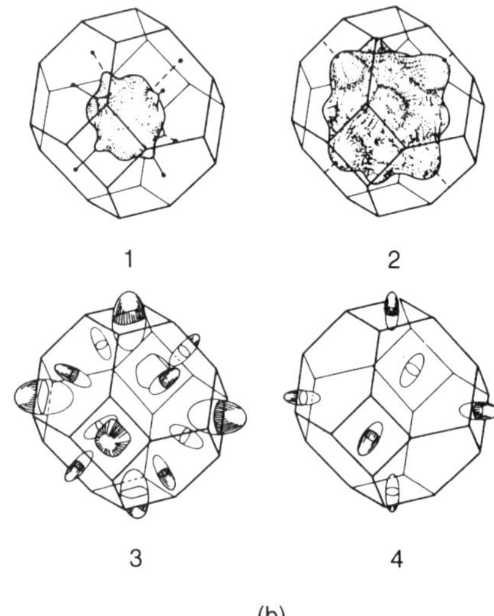

(b)

FIG. 18. Electronic structure of rhodium (Yonemitsu, 1991, pp. 33–34): (a) band diagram (from Christensen, 1973), (b) Fermi surface: (1) and (2) hole sheets in band 1 and 2, (3) and (4) electron sheets in band 3 and 4 (from Coleridge, 1966a, 1996b).

is derived from the thermoelectric voltage as a function of the temperature t by differentiation or by measuring over a small temperature difference. It is a material characteristic and defines the basic thermoelectric behavior of different materials. Figure 24 illustrates the temperature dependence of the thermoelectric power of the precious metals. Metal pairs belonging to the same group in the periodic table show similar values and temperature dependences. The values of the thermoelectric power of palladium and platinum are about five times higher than those of ruthenium and osmium. Copper and silver have positive values and show linear temperature dependence (Tanuma, 1991, p. 53).

For practical purposes, the thermoelectric voltage is the most important parameter. The thermoelectric voltage is changed by alloying elements. Table 13 gives basic values of the thermoelectric voltage of some common thermocouples in millivolts.

4.1.3.4 Magnetic Behavior. The magnetic properties are described by the mass susceptibility χ_m, which is composed of the diamagnetic portion χ_{Dia} and the paramagnetic portion χ_{para} according to

$$\chi_m = \chi_{\mathrm{Dia}} + \chi_{\mathrm{para}} \tag{7}$$

with

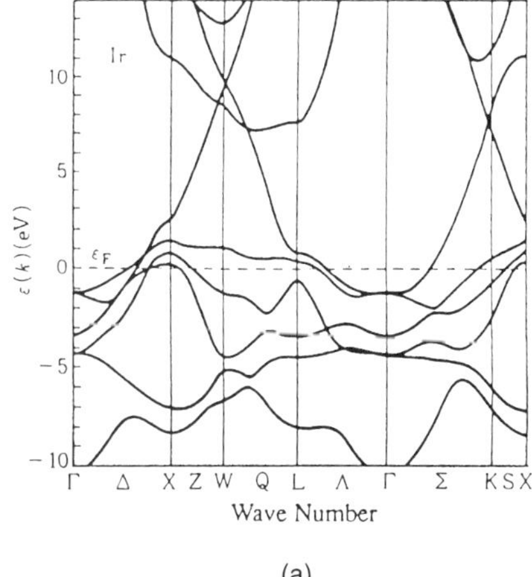

(a)

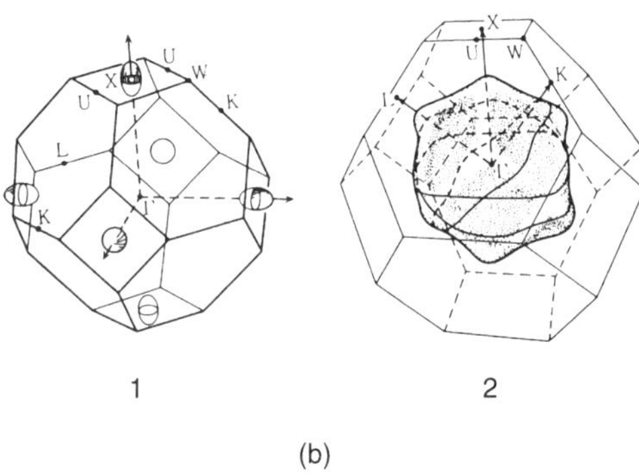

(b)

FIG. 19. Electronic structure of iridium (Yonemitsu, 1991, pp. 33, 34): (a) band diagram (from Smith, 1974), (b) Fermi surface: (1) hole sheet (from Grodski and Dixon, 1969), (2) electron sheet (from Hornfeldt *et al.*, 1973).

$$\chi_{\text{para}} = c/T \tag{8}$$

(Curie's law) and

$$C = M^2/3R, \tag{9}$$

where M is the magnetic moment per mole and R is the gas constant. The platinum-group metals behave paramagnetically. Palladium and platinum exhibit fairly strong paramagnetism ($\chi > 0$), since the Fermi level is located at a peak of the state-density curve in the d band, corresponding to the electronic structure of nickel. The paramagnetic susceptibilities of palladium and platinum decrease with increasing temperature, while in contrast the susceptibilities of rhodium, iridium, ruthenium, and osmium increase with increasing temperature. The change of the susceptibility with the temperature is related to the excitation of the electrons below the Fermi level. The temperature dependence of the magnetic susceptibility of the PGMs (Fig. 25) gives the experimental evidence for the different location of the Fermi levels (Chikazumi, 1991, p. 98). Palladium and platinum tend to become ferromagnetic when they are alloyed with ferromagnetic elements.

PGMs show magnetostriction under the action of a magnetic field:

$$\Delta l/l = s_l H^2, \tag{10}$$

where s_l = specific magnetostriction and l = length of the sample. Based on this and in combination with ordering processes, alloys in the systems Pt–Fe and Pd–Fe show around the Fe_3Pt and Fe_3Pd stoichiometries in the disordered state, like Fe–Ni alloys (e.g., NiFe36, "Invar"), zero or negative coefficients of thermal expansion for similar magnetic reasons (Kussman and von Rittberg, 1950; Kussman and Jessen, 1962).

4.1.4 Alloying Behavior

4.1.4.1 Metals. Compared to silver and gold, the behavior of the PGMs is rather complex both in the range of solid solubilities and in intermetallic-phase formation. Ruthenium, iridium, palladium, and platinum form very stable compounds with the refractory transition elements, especially zirconium and hafnium. Platinum takes up to 18 at. % zirconium into solid solution and reacts even with ZrC and ZrO_2 with formation of the hexagonal compound Pt_3Zr. Relatively high solid solubilities exist for the PGMs in molybdenum and tungsten (Darling, 1973, p. 116).

A high affinity of the PGMs exists for the metals of the rare-earth group. They form numerous intermetallic compounds of different compositions and different structures. In view of the exceptionally high thermal-neutron absorption cross sections of samarium, europium, and gadolinium, these PGM–rare-earth-metal alloys are of special interest for the atomic industry (Funston and McGurty, 1961; Loebich, 1971).

4.1.4.2 Carbon. The platinum-group metals dissolve large quantities of carbon in the molten state, but the solubility in the

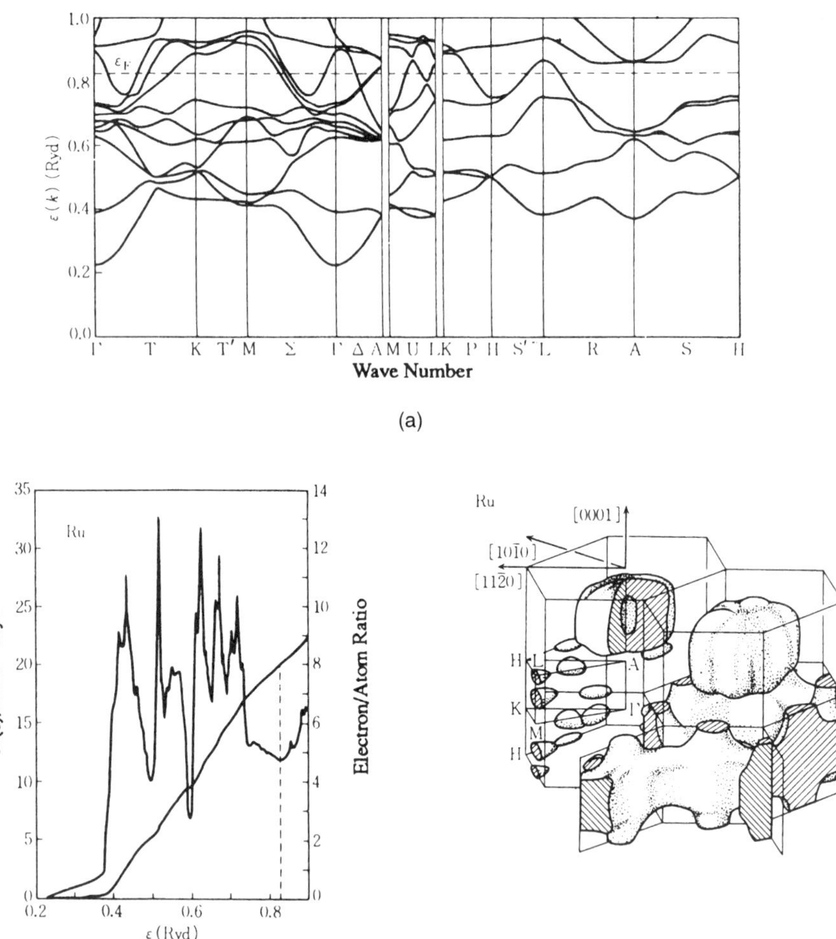

FIG. 20. Electronic structure of ruthenium (Yonemitsu, 1991, pp. 35–37): (a) band diagram, (b) DOS curve (from Jepsen *et al.*, 1975), (c) Fermi surface: connected hole sheets in the seventh and eighth bands, electron sheets in the ninth band (from Alekseev *et al.*, 1979).

solid state is very small. Carbon dissolves interstitially and effects a considerable increase of hardness, accompanied by slightly depressed melting temperatures.

4.1.4.3 Oxygen. The platinum metals form a variety of solid and gaseous oxides in different oxidation states. The solid oxides have tetragonal and hexagonal crystal structures. Rhodium and palladium form stable oxide skins when heated in air. Both also take oxygen into solid solution. The solubility of oxygen increases in the order Rh $<$ Pd $<$ Ag. PdO has a tetragonal crystal structure and is semiconducting; RhO_2 is an electrical conductor; RuO_2 is a semiconductor.

4.1.5 Catalytic Actions The platinum-group metals show pronounced catalytic actions in numerous chemical reactions (e.g., reforming processes in petrochemistry, oxidation of NH_3 to NO, and hydrogenation of hydrocarbon compounds). The catalytic activities are essentially determined by the heat of adsorption of the reactant molecules on the catalyst surface and the bond strength of the adsorbed molecules. The platinum metals provide moderate values of heat of adsorption, which correspond to the dissociation energy of the reactant molecules. The heat of adsorption for each element differs also for different crystal planes; it increases

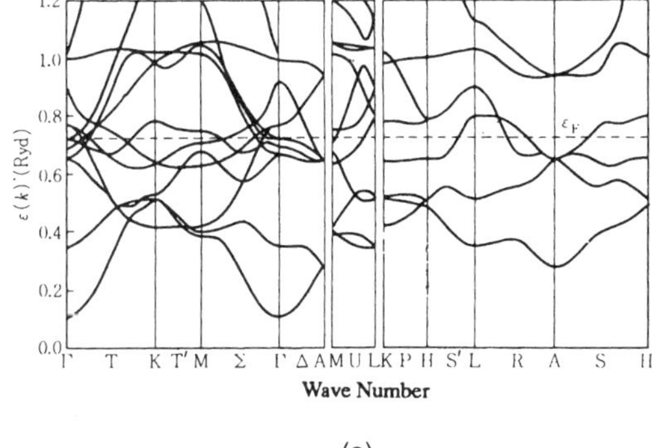

(a)

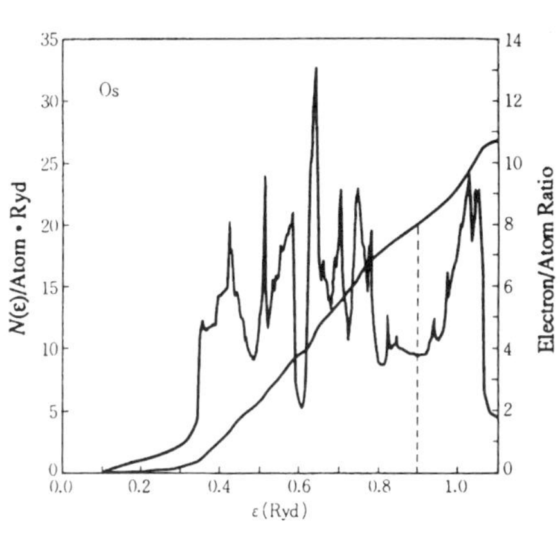

(b)

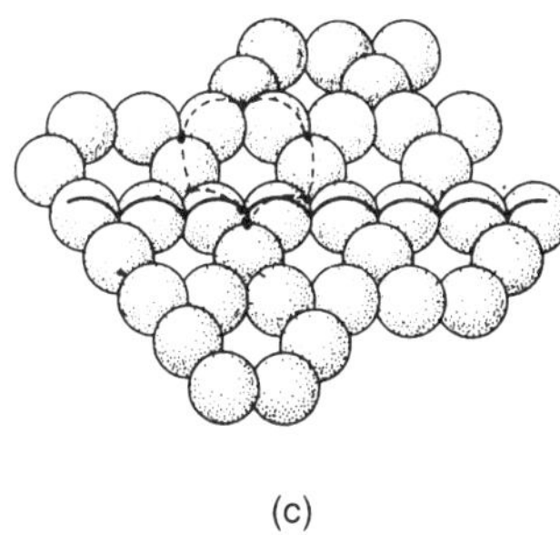

(c)

FIG. 21. Electronic structure of osmium (Yonemitsu, 1991, pp. 35–37): (a) band diagram, (b) DOS curve (from Jepsen *et al.*, 1975), (c) Fermi surface: connected hole sheets of the eighth band (from Plachy and Dixon, 1974).

for fcc crystals with decreasing density of the surface atoms in the order [111] < [100] < [110] (Toyoshiura and Somorjai, 1979; Tamaru and Naito, 1991, p. 266). Catalytic actions are structure sensitive. Correlations exist between the lattice dimensions of catalysts and the dimensions of adsorbed molecules. As a consequence of these influences, the effectiveness of the catalytic action of the individual PGMs differs for different reaction types and follows changing order ranges. They decrease, for example, for the hydrogenation of C=C double bonds in the order Rh > Ru > Pd > Pt, for the oxidation of paraffines Pd > Pt ≫ PdO (Benner

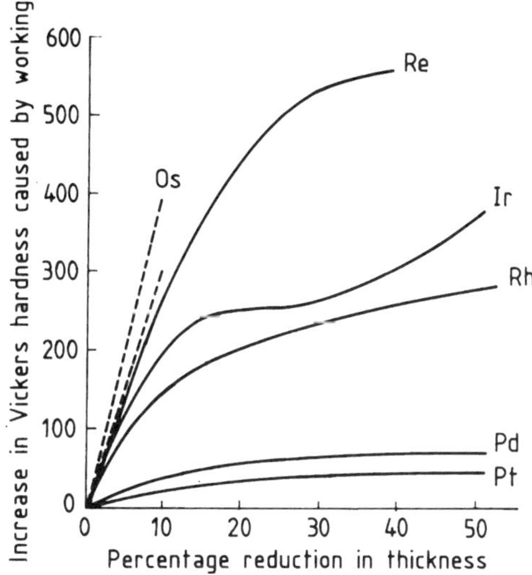

FIG. 22. Work hardening of the PGMs: increase by reduction (Darling, 1973, p. 93).

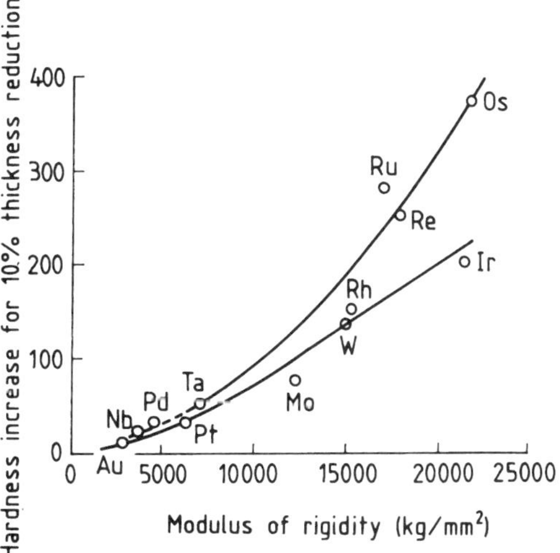

FIG. 23. Work hardening of cubic and of hexagonal metals: hardness increase at different values of the rigidity (Darling, 1973, p. 93).

Table 12. Mechanical properties of PGMs (Darling, 1973, p. 94).

Metal	Crystal structure	Young's modulus (GPa)	Modulus of rigidity, G (GPa)	Bulk modulus, K (GPa)	Poisson's ratio	K/G
Os	hcp	560	220	380	0.25	1.73
Ir	fcc	538	214	378	0.26	1.76
Re	hcp	472	180	340	0.26	1.89
Ru	hcp	430	172	292	0.25	1.71
W	bcc	396	151.4	318.6	0.29	2.11
Rh	fcc	386.4	153	280.1	0.26	1.83
Pd	fcc	128.3	46.1	190.9	0.39	4.13
Pt	fcc	174	62.2	280.9	0.39	4.52
Au	fcc	80.2	28.2	174.6	0.42	6.18

et al., 1991, p. 765), and for the synthesis of methane from CO and H_2O in the order Ru > Rh > Pd > Pt > Ir (Schuit and van Reijen, 1958; Vannice, 1975; Broden *et al.*, 1976). The selectivity of the catalytic actions and the long-time stability of the catalysts can be improved by alloying.

In heterogeneous catalysis the PGMs are applied in wire gauze form or as highly dispersed metal coatings ("platinum black," "palladium black") onto supports such as carbon, γ-Al_2O_3, silica gel, aluminum silicate (zeolite), or $CaCO_3$. The deposition is performed by thermal decomposition or chemical reaction of precious-metal salts. The supporting material can be in the form of powder or granules; the concentration of noble metals varies between 0.5% and 5%. Because of the surface-induced reaction, high specific surfaces (usually in ranges between 100 and 1000 m^2/g) are favorable for higher yields and reaction rates.

In homogeneous catalysis the reactants and the catalyst are both present in liquid, preferably organic, solution. Halide compounds of palladium and complex organic compounds of rhodium, iridium, and ruthenium are important for the stereospecific synthesis of emantiomeric compounds.

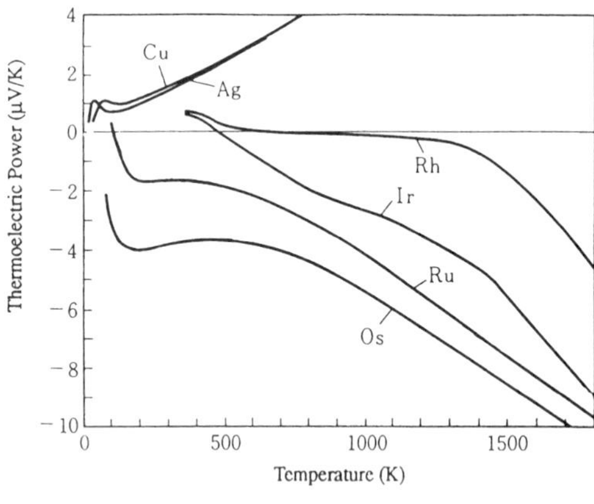

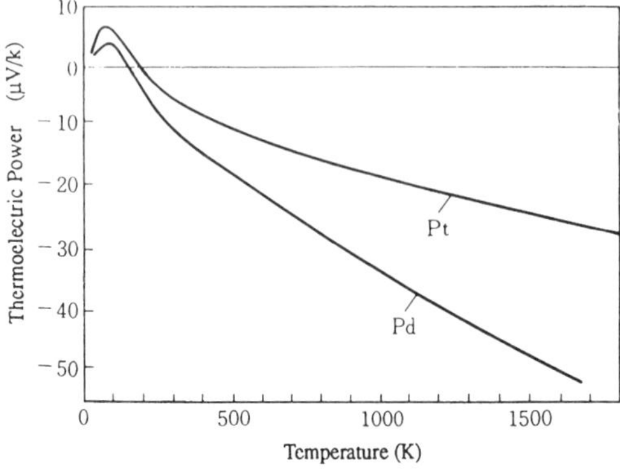

FIG. 24. Thermoelectric power of precious metals (from Vedernikov, 1969).

4.1.6 Abundance, Resources, and Raw Materials

4.1.6.1 Abundance and Resources. The platinum-group metals are among the scarcest of metallic elements. The abundance of platinum and the associated platinum-group metals is estimated as on the order of 10^{-4} ppm, concentrated in planetary regions reaching up to ~30 ppm.

The largest deposits are found in South Africa (Bushveld), in the United States (Stillwater, Oklahoma), in Canada, and in Russia. The latter two are the most important sources for palladium. PGM contents in exploitable minerals and deposits vary from 0.05 ppm (secondary deposits in washable sands) to 3–20 ppm for the most important deposits and up to platinum concentrations of 200 ppm in special ores.

Table 13. Basic values of the thermoelectric voltage[a] of common themocouples (Degussa, 1967, p. 100).

T_1 (°C)	T_2 (°C)	Pt-Rh10/Pt	Pt-Rh20/Pt-Rh5	Rh-Ir60	Pt-el[b]	Pd-or[c]
0	100	0.643	0.074	0.371	3.31	4.6
	500	4.221	1.447	2.562	20.20	27.9
	1000	9.570	4.921	5.495	41.65	59.6
T_1 (K)	T_2 (K)	Au-Co2.1/Cu[d]	Au-Fe0.02/Cu[d]			
4.2	10	0.044	0.093			
	20	0.173	0.208			
	40	0.590	0.423			

[a]In mV.
[b]Pt-el = Platinel, Pd83Pt14Au3/AuPd35.
[c]Pd-or = Pallador, PtIr10/AuPd40.
[d]At. %.

For the platinum-group metals together, the current estimate of workable deposits is ~70 000 tons.

4.1.6.2 Raw Materials. The PGMs occur as native alloys or in mineral compounds in placer deposits, sometimes together with gold, as well as in lode deposits in basic rocks together with nickel and copper. Deposits contain varying compositions in regard to the PGM mix with average distribution of the proportion in percent for platinum, 45 (19–59); palladium, 30 (25–66.5); ruthenium, 5 (1–12); rhodium, 4 (2–7.6); iridium, 1 (<1–2.4); and osmium, <1–1.2.

4.1.7 Production

4.1.7.1 Primary Production. Production steps for the PGMs are schematically shown in Fig. 26. After flotation, the material is melted under alternating reducing and oxidizing conditions, removing iron and SO_2, followed by grinding and then magnetic separation of the enriched PGM-containing Ni–Cu–Fe phase. Solution with HCl/HNO_3 or HCl/Cl_2 removes most of the base metals, leaving a PGM concentrate of above 30%. Osmium and ruthenium are separated by distillation of their volatile oxides from concentrated hydrochloric solutions. The remaining PGMs are separated in aqueous solution by subsequent oxidation, reduction, and precipitation of complex PGM chlorides in the sequence platinum, iridium, palladium, rhodium.

PGMs can also be recovered from solutions by solvent extraction. It works with quantitative yield and is highly selective.

Refining is done by repeated solution, recrystallization, and precipitation of compounds. Conversion of metals to sponge or powder is carried out for platinum, palladium, iridium, and rhodium by thermal decomposition of the ammonium-chloro complexes ("calcination").

Ruthenium is prepared by precipitating RuO_2 from aqueous solution and reduction with hydrogen; osmium is prepared by precipitating potassium osmate (IV) and heating in hydrogen atmosphere or by reduction with $NaBH_4$ in aqueous solution.

4.1.7.2 Secondary Production (Recycling). Recycling of PGM scrap has become of special importance for platinum, palladium, and rhodium with the growing

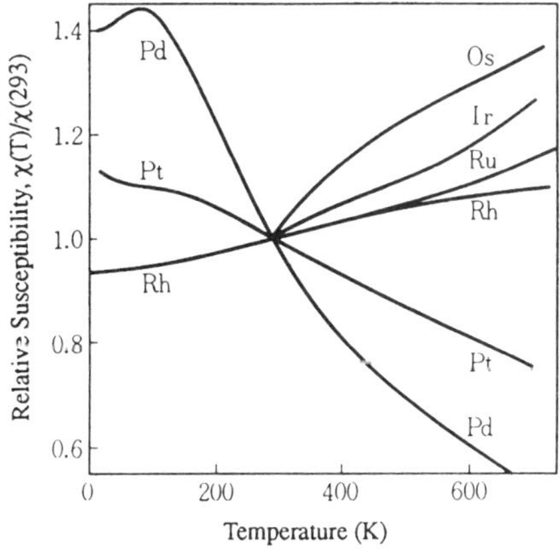

FIG. 25. Temperature dependence of the magnetic susceptibility of the PGMs (from Berkowitz and Kneller, 1969).

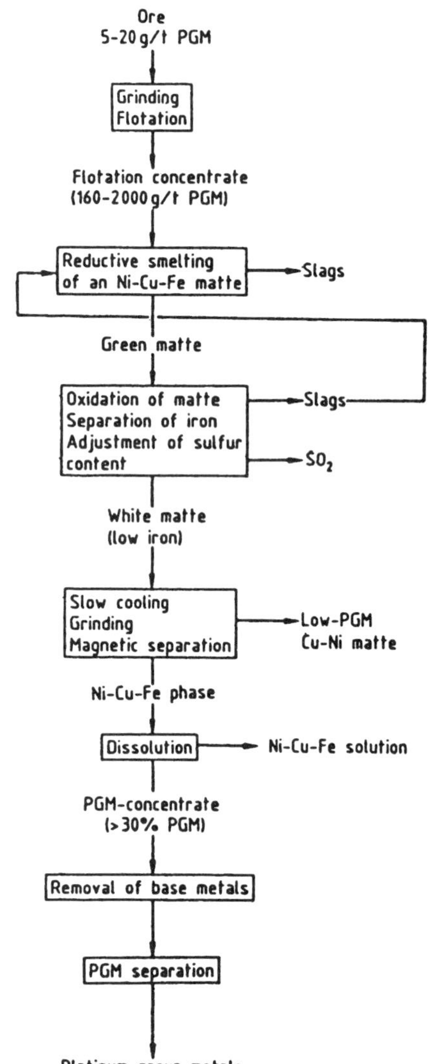

FIG. 26. Production steps for PGMs (Renner, 1992, p. 88).

demand for autocatalysts. Metallic waste is dissolved or treated with chlorine with subsequent reduction and precipitation of PGM powders. Low-grade wastes are recovered by pyrometallurgical processes using lead as collector metal. Supported catalysts and automobile exhaust catalysts are melted in electric furnaces with iron or copper as collector metals, forming alloys with ~20 wt % of PGM (predominantly for platinum, palladium, and rhodium).

4.2 The Elements

4.2.1 Platinum Platinum, atomic number 78, relative atomic mass 195.08, six natural isotopes (between 190 and 198), is a silvery white, bright metal, ductile and malleable, resistant against air and water, soluble only in HNO_3–HCl and molten alkali. The most important properties are oxidation resistance; mechanical, thermal, and electrical stability; and the ability to act as catalyst in numerous different chemical reactions. The main application areas are jewelry (39%), automotive catalytic converters (42%), chemistry (4%), electronics/electrical technology (5%), glass industry (4%), petroleum (6%) (quantities in weight percent of the total yearly demand in the western world 1994; Degussa, 1994).

Selected values of basic physical and chemical properties are given in Table 14.

4.2.1.1 Influence of Alloying Elements. *Metals: Strengthening.* Platinum can be hardened by alloying with other precious metals and base-metal elements. The most effective hardeners are precious metals in the order Ru, Os, Ir, Au, Rh and the base metals in the order Co, Ni, Cu, Fe. Figure 27 shows the increase of hardness by increasing amounts of alloy additions.

Dispersion-strengthened platinum and platinum alloys are produced by coprecipitation of the metals and refractory oxides—for instance, with ZrO_2 (Selman *et al.*, 1974; zirconium-oxide grain-stabilized platinum, "ZGS"®, Johnson Matthey, 1974). Platinum can also be effectively dispersion hardened by incorporating very small amounts (0.04–0.08 wt %) of refractory carbides (TiC, ZrC). Additive strengthening is effected by partial decomposition and reaction of the carbide with the matrix. The low carbide concentration offers the advantage that ductility, electrical conductivity, and the recrystallization behavior are not seriously impaired.

Metals: Electrical resistance. Figure 28 shows the influence of some alloying elements on the electrical conductivity. In the Pt–Mo system, at composition with 30 at. % Pt, a superconducting ordered phase is formed. The partially ordered state shows the maximum effect with the highest critical transition temperature (7.5 K) (Flükiger *et al.*, 1972).

Metals: Magnetic properties. In the Cu–Pt

Table 14. Selected values of basic physical and chemical properties of platinum metals (Degussa, 1995).

	Ru	Rh	Pd	Os	Ir	Pt
Atomic properties						
Atomic number	44	45	46	76	77	78
Atomic weight	101.07	102.905	106.42	190.2	192.22	195.08
Electron configuration	$[Kr]4d^7 5s$	$[Kr]4d^8 5s$	$[Kr]4d^{10}$	$[Xe]4f^{14}$-$5d^6 6s^2$	$[Xe]4f^{14}$-$5d^7 6s^2$	$[Xe]4f^{14}$-$5d^9 6s$
Crystal structure	hcp	fcc	fcc	hcp	fcc	fcc
Lattice constant (Å) (10^{-1} nm)	a = 2.701 c = 4.275 c/a = 1.583	3.803	3.887	a = 2.735 c = 4.319 c/a = 1.579	3.839	3.924
Atomic radius (Å) (10^{-1} nm)	1.325	1.345	1.376	1.377	1.357	1.377
Mechanical properties						
Density (20 °C) (g/cm^3)	12.45	12.41	12.02	22.61	22.65	21.45
Density, liquid at melting point (g/cm)^3	10.90	10.70	10.49	20.1	19.39	18.91
Hardness (HV, kg/mm^2)	250–500[a]	130	50	300–680[a]	200	48
Young's modulus (GPa)	485	386	124	570	538	173
Modulus of rigidity (MPa)	172	153	51	220	214	67
Poisson's ratio	0.29	0.26	0.39	0.25	0.26	0.39
Coefficient of compressibility	0.31	0.36	0.52	0.28	0.28	0.36
Thermal properties						
Melting point (mp) (°C)	2310	1966	1554	3045	2410	1772
Vapor pressure at mp (mbar) (1×10^5 Nm^{-2})	5.5	3.5	20	20	4	0.1
Boiling point (°C)	3900	3700	2970	5000	4130	3827
Thermal conductivity (20 °C) (W·m^{-1} K^{-1})	117	150	75	87	147	73
Coeff. of linear expansion (0–100 °C) (10^{-6} K)	9.5	8.5	11.1	6.5	6.8	9.0
Electrical properties						
Specific electrical resistance (0 °C) ($\mu\Omega$ cm)	6.7	4.34	9.725	8.2	4.7	9.825
Temperature coefficient of electrical resistance (0–100 °C) (10^{-3} K)	4.0	4.6	3.8	4.2	4.3	3.92
Transition temperature of superconductivity	0.47	0.9		0.71	0.11	
Thermoelectric voltage against Pt (0–100 °C) (mV)	0.684	0.70	−0.570		0.660	0
Work function (eV)	4.52	4.8	4.97	4.55	5.40	5.27
Chemical properties						
Oxidation states in compounds	8, 6, 4, 3, 2, 0, −2	5, 4, 3, 1, 2, 0	4, 2, 0	8, 6, 4, 3, 2, 0, −2	6, 4, 3, 2, 1, 0, −1	4, 2, 0
Ionization energy (eV)	7.37	7.46	8.34	8.7	9.1	9.0
Covalent radius for single bonds (Pauling) (pm)	125	125	128	126	127	130
Ionic radius (pm) (oxidation number, coordination number)	62 (4+,6) 68 (3+,6)	55 (5+,6) 67 (3+,6)	64 (2+,4) 86 (2+,6)	39 (8+,4) 63 (4+,6)	63 (4+,6) 68 (3+,6)	63 (4+,6) 60 (2+,4)
Electronegativity (Allred and Rochow)	1.4	1.5	1.4	1.5	1.6	1.4
Reduction potential (V) with (n) electrons	0.455 (2)	0.758 (3)	0.951 (2)	0.85 (8)	1.156 (3)	1.118 (2)

[a]Depending on crystal orientation.

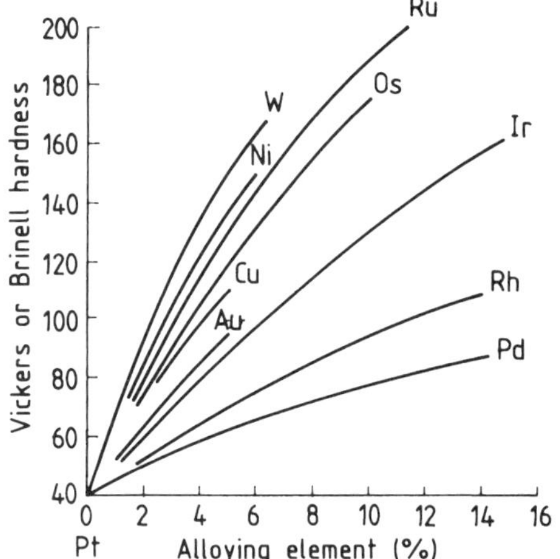

FIG. 27. Influence of alloying elements on the hardening of platinum (Zysk, 1979, p. 689).

system a long-range–ordered rhombohedrical lattice is formed below 812 °C. The equiatomic PtCo alloy is of great importance for permanent magnets. Ordering occurs in such a manner that cobalt and platinum atoms arrange on alternate planes. The strongest increases of hardness and magnetic properties are found at intermediate states where maximum strains exist by the combination of the two different lattice structures. Considerable stresses are caused during the ordering process by coherence of tetragonal platelets on the [110] planes of the cubic lattice. The

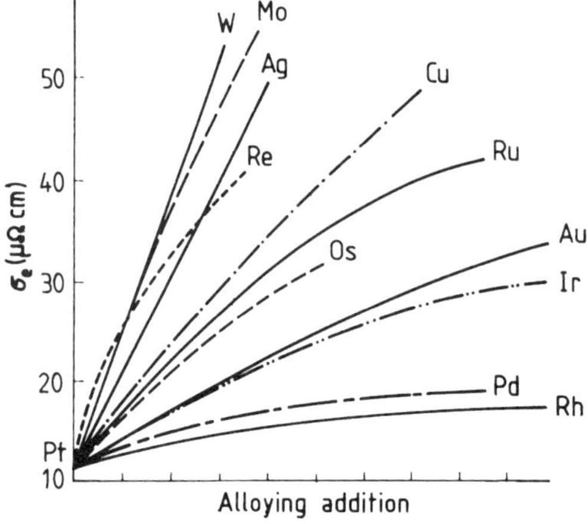

FIG. 28. Influence of alloying elements on the electrical conductivity of platinum (Doduco, 1977, p. 67).

highest magnetic characteristics are developed if the ordered and disordered phases are present in comparable proportions.

Superlattices, formed in Pt–Cr alloys in the composition ranges of 17–65 wt % Cr, are ferromagnetic with the maximum magnetic saturation for the phase PtCr (extending from 43 to 84 wt % Cr) at ~30 at. % Cr. FePt and CoPt with tetragonal crystal symmetries have large magnetic anisotropies. In the alloy systems Fe–Pt and Fe–Pd the saturation moment increases with the addition of noble metals. Such alloys can be used as thin-film head materials for high-density recording devices.

Oxygen. The solubility of oxygen in platinum is very low. Very thin coatings of platinum on reactive metals are therefore an effective protection against oxidation. Platinum forms three solid oxide compounds, which are stable at different temperatures and exhibit different oxygen partial pressures. Solid PtO_2 in the hexagonal modification is used as a catalyst. Volatile oxides form if platinum or platinum–rhodium alloys are heated in oxygen-containing atmospheres. They cause remarkable weight losses in long-term runs. The oxides can be recovered by leading the outcoming gas stream through a gauze of gold-palladium alloy, on which the platinum and rhodium oxides are deposited.

4.2.1.2 Alloys. Platinum–palladium alloys are used in jewelry, for electrical contacts, and for equipment parts for glass production and construction elements in highly corrosive environments.

Platinum–iridium alloys find various applications in electrochemistry, chemical technology, medicine, and jewelry. Iridium contents (in weight percent) are for laboratory ware 0.4%–0.6%, jewelry 5%–15%, medical 10%, electrical contacts 10%–25%, electrodes 10%, and pens, needles, and springs 25%–30%. Maximum hardness (~360 HB) is reached with 30% iridium.

Platinum–rhodium alloys are applied in composition with 10% rhodium as catalyst for the oxidation of NH_3, as a standard element in thermocouples (Pt–Rh of different compositions against Pt), and as heater elements in electrically heated high-temperature furnaces. Pt–Rh alloys with addition of Au are especially resistant against wetting by molten glass.

Ruthenium strengthens platinum very effectively. Ten percent ruthenium raises the hardness to 168 HV. The alloy PtRu5 has the same strength, hardness, and corrosion resistance as the alloy PtIr10 ("hard platinum"). Applications are jewelry items, electrodes for electrochemistry, spark-plug electrodes, electrical contacts, and wear-resistant wires in potentiometers. Higher alloy compositions up to 14% ruthenium and ternary types with 15% rhodium and 6% ruthenium are used for electrical contacts and potentiometer wires intended for heavy-duty applications.

Platinum–nickel alloys have been proven as electrical switches and as high-strength and high-temperature–resistant construction materials in the chemical industry. Compositions reach from trace amounts up to 20% nickel with maximum tensile strength of ~1725 MPa after 90% cold reduction.

Platinum–tungsten alloys, commonly used in compositions with 5–8 wt % Co, have excellent wear resistance. They are used for potentiometers, glow wires in copiers, diode switches, heater elements, and claddings on rods for fuel elements in nuclear reactors.

Platinum–cobalt alloys with compositions near 50 at. % Co form strong permanent magnets with coercivities up to 540 kA/m at room temperature. Typical uses are focusing magnets, hearing aids, miniature motors, medical instruments and implants, and films for magneto-optical recording.

Platinum–gold and platinum–gold–rhodium alloys have higher strength than pure platinum at 1000 °C, finer grain structure, and better resistance against wetting by molten glass.

4.2.1.3 Applications. *Jewelry.* The reflectivity of platinum in the visible-light spectrum of 69% (for comparison: silver 91%, palladium 57%) gives it its characteristic appearance, similar to white gold. This, together with the higher hardness of platinum and platinum alloys compared with gold, has led to a fast-growing platinum demand for jewelry items. Platinum resinates are employed for decoration on porcelain surfaces.

Catalysis. The largest demand for platinum catalysts in the chemical industry comes from the petroleum refinery. Catalysts are Pt–Al_2O_3, Pt–Ir, and Pt–Re. Further uses are hydrorefining, hydrocracking, and hydrogenation processes and synthesis of medical drugs.

Automobile exhaust-gas catalysts take the biggest portion of the total demand of platinum (~42%) and rhodium (~89%). They reduce and remove the gaseous and particulate pollution constituents from exhaust-gas streams by catalytically activated reduction and oxidation reactions. The three-way catalyst for gas engines contains as active part a highly dispersed metallic coat of a mixture of platinum and 20 wt % rhodium and palladium in changing proportions. The metal coat is precipitated on a honeycomb-shaped ceramic support ("monolith"), made from cordierite (Fig. 29). The monolith is coated with a ceria-doped "washcoat" of Al_2O_3, which forms on firing a porous structure with a high specific surface area. The oxidation of hydrocarbons (CH_x) is activated by platinum and palladium, the reduction of nitrogen oxides (NO_x) by rhodium (Amirnazmi and Boudart, 1975; Obuchi *et al.*, 1982, 1983). The reaction has its best yield at an air-to-fuel ratio of 14.7:1. The ratio is measured and maintained by means of two platinum sensors. One is installed in the inlet manifold to monitor the proportion of air in the air-fuel mixture as it enters the combustion chamber; the second forms the tip of the so-called lambda probe (see p. 271), inserted in the exhaust stream ahead of the catalyst to measure the amount of oxygen in the exhaust. Each catalytic converter contains ~2 g of a PGM mixture, either platinum/rhodium or platinum/palladium/rhodium mixture in varying proportions.

Three-way catalysts are also used in thermal power stations.

Electrodes in fuel cells. Fuel cells convert chemical energy directly into electrical energy by catalytic oxidation of hydrogen. The battery-type devices consist of two catalyst-bearing electrodes, separated by an ion-conducting electrolyte. Hydrogen is oxidized at the anode, and oxygen is reduced at the cathode, generating electric current together with some reaction heat. Two fuel-cell types have been developed to practical use: the phosphoric acid fuel cell (PAFC) and the proton-exchange membrane (PEM) fuel cell. The electrodes on the anode side of the PAFC devices consist on the gas side of platinum on coal fiber and on the electrolyte side of a porous structured layer of platinum/carbon and PTFE. Cathode catalysts are Pt/Co/Cr, Pt/Co/Fe, or Pt/Ni alloys. In PEM-

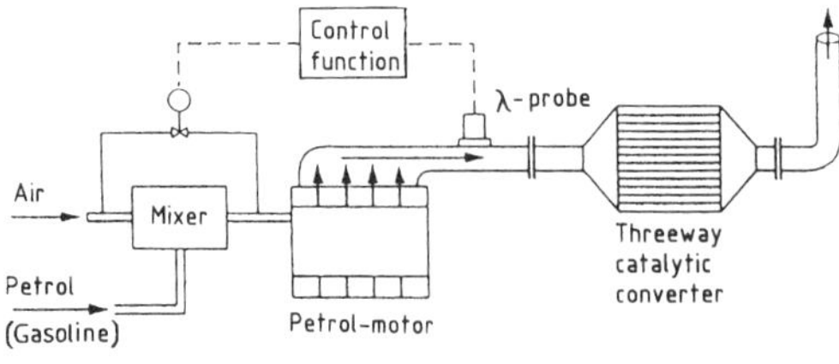

(a)

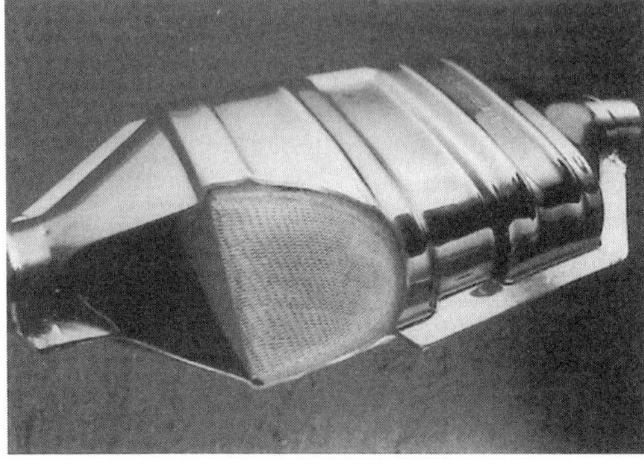

(b)

(c)

FIG. 29. Three-way catalytic converter: (a) schematic view (Prescher, 1995, p. 369), (b) cutaway view (Coombes, 1992), (c) Pt/Rh catalyst surface on the washcoat (Prescher, 1995, p. 366).

type fuel cells the electrodes are directly coated on both sides with porous layers of PTFE loaded with platinum/carbon on the cathode and platinum (+ ruthenium and palladium)/carbon on the anode side (Seibold, 1995).

Sensor devices: Thermocouples. Platinum and Pt–Rh alloys are used for thermocouples to be applied at high temperatures. Standard combinations are the following:

1. "S,"* PtRh10/Pt, up to 1600 °C;
2. "R,"* PtRh13/Pt or
3. "B,"* PtRh30/PtRh6, up to 1800 °C; and
4. IrRh40/Ir up to 2000 °C.

Measurement of very low temperatures can be made with the systems

5. AuFe(0.03 at. % Fe)-Chromel from 4.2 to 273 K,
6. AuCo(2.11 at. % Co)-AuAg(0.37 at. % Ag) or Cu from −240 to 0 °C, and
7. AuFe(0.02 at. % Fe)-Cu from −270 to −230 °C.

Thermocouples for heavily corrosive atmospheres are

8. AuPd40-PdPt14Au3, corresponding to NiCr–Ni;
9. AuPd35-PtPd12,5 corresponding to NiCr–Ni; and
10. AuPd46-PtIr10 corresponding to Fe-Constantan.

Sensors: Resistance thermometers. Platinum resistance thermometers are based on the temperature dependence of the electrical resistivity. The temperature coefficient of electrical resistance of platinum in high-purity condition is 3.927×10^{-3}/K. Basic values (= demanded values for the *R-T* relation) of platinum resistance thermometers are specified in international standards (DIN IEC 751). At temperatures below about 20 K the TCR of pure metals becomes too low in comparison to the absolute value of the resistance and cannot be measured. Alloys of rhodium with 0.5 at. % iron or alloys of gold with 1.15 at. % manganese show by magnetic effects an increasing TCR with decreasing temperature (Kondo effect; see p. 252) to sufficiently high, measurable values.

Film-type resistance thermometers contain thin platinum or iridium deposits of 1–2 μm thickness, deposited on flat ceramic substrates. The required resistance value is adjusted by laser-beam cutting. Iridium layers have values of thermal expansion that are comparable to those of the ceramic. They are therefore applied on devices for temperature measurements at lower temperatures down to ~200 °C.

*The types "S", "R", and "B" correspond to the international standard IEC 584. Chromel is Ni80Cr20. Constantan is Cu40Ni40.

Sensors: Special sensor devices. Platinum-equipped sensor devices are used to detect combustible gases in air or explosive gas mixtures. In the so-called pellistor devices, a platinum wire is placed in a small sphere of Al_2O_3 (diameter ~1.5 mm). The sphere is plated with platinum-black deposit that heats up by catalytically activated reaction of the gas composition in case of the presence of combustible constituents. The increase of temperature is measured by the electrical resistance of the platinum wire.

Platinum contact electrodes are constituents of sensor devices made from semiconducting materials (e.g., SnO_2, ZnO, piezoceramic, NTC resistors) or special ceramics. They serve for the determination of gas concentration, humidity, or pressure load. Oxygen-determining sensors ("lambda probes") use the high electrical conductivity imparted to solid electrolytes (e.g., ZrO_2) by the presence of oxygen ions, compared to their normal conductivity due to electrons. If this ionic conductor acts as a separator between two gas spaces, and if it is provided with platinum electrodes on both sides, a difference in the oxygen partial pressures causes the formation of an electromotive force ("Nernst emf") between the electrodes. Lambda probes are used to control the fuel-to-air ratio in conjunction with automobile exhaust-gas catalysts in Otto engines.

Magnetic storage. Thin-film cobalt-platinum alloys and thin-film cobalt-platinum multilayers play an important role as magnetic media on hard disks in magnetic and magneto-optic storage and recording devices. (see MAGNETIC STORAGE OF INFORMATION).

Electrical technology/electronics. Resistor-heated furnaces working in air at temperature >1200 °C are equipped with heaters made from platinum wires (up to 1600 °C), PtRh40 (up to 1700 °C), or rhodium bands (up to 1800 °C). Pt–Ni, Pt–Rh, and Pt–Ru alloys are used in spark-plug electrodes. Platinum finds applications in electronic devices as barrier layers, as contact pads on semiconductor chips, and as a stabilizing component against silver-ion migration in silver-based metallizing preparations and thick-film conductor pastes for screen printing of hybrid circuits. Platinum resinates are used for the production of conductive coatings on ceramic substrates onto which further layers can be deposited electrolytically.

Chemical technology. Platinum crucibles, dishes, boats, and electrodes for electrogravimetry are basic tools in the laboratory technology. If dimensional stability is required, components are made from the alloys PtIr3, PtAu5, and PtIr0.3.

Platinum coatings serve as corrosion protection on reaction equipment. Porous filters of platinum, palladium, or silver for corrosive media are produced by sintering of spherolithic powders. High-purity optical glass is melted in pure platinum crucibles. Platinum components (skimmers, pouring funnels, protection tubes for thermocouples, level-measuring devices) and dispersion-hardened platinum are used for equipment parts in the glass-working technology. Dispersion-strengthened platinum–rhodium alloys serve as heater elements in high-temperature furnaces. Platinum centrifuges are used for the production of rock wool and slag wool. Spinnerets for the manufacturing of glass fiber and glass wool are made of Pt–Rh alloys. Spinnerets for textile fibers like rayon or acrylic are made from platinum–gold and platinum–rhodium alloys. Platinum–rhodium tubes protect high-temperature thermocouples against oxidation. Platinum-plated wires of iron–nickel or molybdenum, or gold-plated iron–nickel wires, serve as vacuum-tight leads through glass tubes. Platinum and iridium are indispensible crucible materials for the production of high-purity single crystals for optical and laser technology.

Biomedical uses. The stereoisomer dichloro-diammino Pt^{II}-complex compound ("*cis*-platinum") has antineoplastic effects by suppressing the DNA synthesis but not the synthesis of RNA or proteins. This compound and less toxic successor products ("Carboplatin," "Paraplatin") have been proven as highly effective agents against cancer (Rosenberg *et al.*, 1965, 1969). Platinum and platinum–iridium alloys find use for electrodes in pacemakers, in catheters, as marker bands, as guide wires, and as temporary and permanent implants. Irradiated Ir wire, sheated in Pt, can be implanted in the body to deliver controlled doses of radiation for cancer therapy. The radio-opacity of the platinum shields the healthy tissues from radiation, while the tumor is irradiated by the exposed iridium tip. Platinum coils made from thin wire are applied for treatment of aneurysm in the brain, where traditional recourse of surgery is most difficult (Coombes, 1993).

4.2.2 Palladium Palladium, atomic number 46, relative atomic mass 106.42, six natural isotopes (between 102 and 110), is a silvery white, bright metal, malleable and ductile. It is soluble in oxidizing acids and molten alkali. The main application areas are electronics (46%), dentistry (26%), automotive catalytic converters (17%), jewelry and others (10%), and chemistry (7%) (quantities in weight percent of the total yearly demand in the western world 1995; Johnson Matthey, 1995).

Selected values of basic physical and chemical properties are given in Table 14.

4.2.2.1 Influence of Alloying Elements. *Metals: Strengthening.* Figure 30 shows the increase of hardness with increasing amounts of alloy additions. Grain refining of palladium alloys is performed with additions of ruthenium and germanium. Palladium and copper form superlattices in the composition ranges between 10 and 25 wt % palladium (tetragonal $L1_0$ structure) and between 30 and 50 wt % palladium (B2 structure).

Metals: Electrical resistance and magnetic behavior. Figure 31 shows the influence of alloying elements on the electrical resistance.

Palladium alloys with iron and with cobalt show ferromagnetism. On the alloying of palladium with 0.1 wt % cobalt, the palladium atoms surrounding a cobalt atom are strongly magnetized to magnetic moments on the order of 10 Bohr magnetons (M_B), which is ~6 times that of a single cobalt moment.

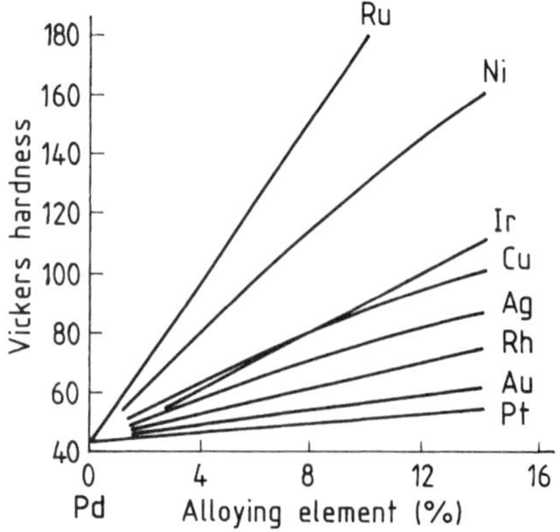

FIG. 30. Influence of alloying elements on the hardening of palladium (Wise and Vinnes, 1979, p. 700).

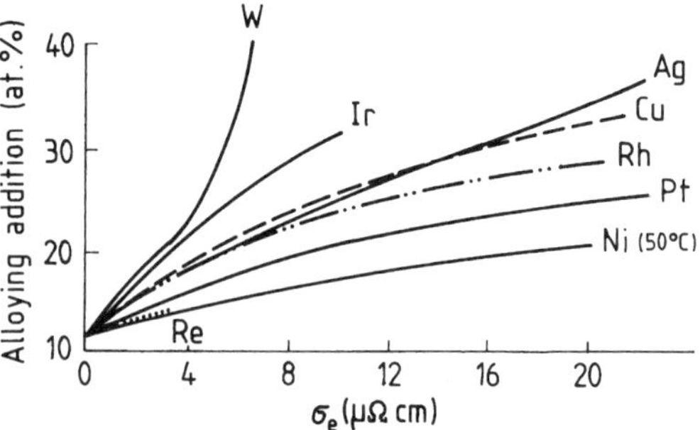

FIG. 31. Influence of alloying elements on the electrical conductivity of palladium (Doduco, 1977, p. 67).

Palladium dissolves in diamagnetic copper, forming diamagnetic alloy phases in the composition ranges up to 50 at. % palladium. The partial dissociation of the metallic palladium according to

$$Pd_{Dia} \leftarrow Pd^{+}_{Para} + e^{-}, \qquad (11)$$

which is responsible for its paramagnetism, is suppressed by the high electron-gas concentration of the metallic copper. The ordered phases in this system have higher diamagnetic susceptibilities than the disordered solid-solution phase. Only disordered Cu–Pd alloys with palladium contents above 50 at. % palladium are paramagnetic. Similar behavior is found when hydrogen is dissolved in palladium, where the paramagnetism of palladium decreases to reach zero at the composition $PdH_{0.66}$ and above, leaving a pure diamagnetic palladium–hydrogen alloy.

Hydrogen. Palladium and certain palladium alloys are unique in their capability to dissolve large amounts of hydrogen, up to ~2800 times their own volume (Moore, 1939; Lewis, 1967, 1994). Hydrogen enters palladium interstitially. The hydrogen atoms are located in the octahedral holes of the fcc lattice. Above 295 °C, palladium and hydrogen form a continuous series of solid solutions with fcc structure. Below that temperature, the phase splits into a fcc palladium-rich α phase and also a fcc hydrogen-rich β phase, forming a miscibility gap that broadens with decreasing temperature. The equilibrium hydrogen pressure at 295 °C amounts to 19.87 atm.; the composition on the critical point corresponds to palladium with 21 at. % hydrogen. The β phase takes up hydrogen up to 1300 times the volume of the palladium, corresponding to a composition of palladium with 50 at. % hydrogen. Further quantities of hydrogen up to a total of 2800 times the volume of palladium can be loaded by cathodic deposition. The two-phase region shows, according to the phase rule, constant vapor pressures at constant temperatures, which amounts to 1 atmosphere at ~140 °C. The occupation of interstitial sites by hydrogen atoms causes expansion of the palladium lattice. The lattice parameters increase with increasing hydrogen content from 3.891 Å for the pure palladium up to 4.06 Å at 75 at. % hydrogen (Nelin, 1971). Thermal cycling of Pd–H alloys in the duplex phase region causes high mechanical stresses due to changes of the lattice dimensions with changing amounts of dissolved hydrogen. Pure palladium, if repeatedly heated and cooled in hydrogen-bearing atmosphere, becomes brittle and cracks. Palladium–silver alloys with 20–25 wt % silver dissolve higher amounts of hydrogen than pure palladium and resist thermal cycling (Lewis, 1982). Ordered A_3B phases of Pd–Mn and Pd–Fe alloys show considerably higher solubility for hydrogen than the corresponding disordered phases. In Pd–Mn alloys, hydrogen supports the ordering process to proceed at lower temperatures (Flanagan and Sakamoto, 1993).

4.2.2.2 Alloys. PdAg23 exhibits extremely high hydrogen permeability and is used for semipermeable membranes for the extraction and purification of hydrogen. The Pd–Ag alloy with 40 at. % Ag has the highest electrical resistivity at room temperature. It finds use for precision resistors in the electrical technology. The specific resistance between 20 and 100 °C is nearly constant and amounts to 42.5 $\Omega \cdot$cm; the specific resistance between ~20 and 800 °C changes from 42.5 to 44 $\Omega \cdot$cm.

Palladium–copper (15–40 wt %) alloys serve as surface material for electrical contacts in the milliampere range and as hard, wear-resistant material for slip rings with high conductivity. Pd–Ag–Cu alloys containing ~45% Pd can be age hardened. These alloys find application in dentistry. Addition of 1% Zn accelerates age hardening. Ternary alloys containing 10 to 25% Pd are used as hard solders for the formation of high-strength brazed joints.

Palladium–gold alloys are the basis of precision resistance alloys. Addition of iron accentuates the formation of long-range ordering. The alloy Au50Pd45Mo5 has the specific resistance of 100 $\mu\Omega \cdot$cm with a temperature coefficient of resistance (TCR) of 0.00012/°C between 0 and 100 °C. The tensile strength reaches values of ~1060 N/mm^2.

Palladium–ruthenium with 4.5 wt % ruthenium is a standard jewelry alloy in the United States. Alloys with ruthenium contents up to 12% serve as material for electrical contacts and as catalysts for the reduction of nitric oxide.

4.2.2.3 Applications. *Electronics:* Palladium is, pure or in combination with silver, the most important component in capacitor electrode layers, in silver conductor paths, and in resistor elements in thick-film circuits.

Dentistry. Palladium-based alloys (palladium content from about 60–80 wt % Pd) contain copper and gallium as hardening elements and tin to lower the melting temperature. The best corrosion resistance is attained with a single-phase structure in the composition range of Cu 4–6 wt %, Ga 5–7 wt %, and Sn 6 wt %.

Chemistry. Hydrogen purification equipment based on the alloy of the composition palladium with 23 wt % silver is used to produce ultrapure hydrogen up to 99.999 99% purity. Palladium-silver permeation membranes are provided in the form of bundled tubes with wall thickness of ~75 μm, sealed on one end. Hydrogen is pressed from outside through the tubes, dissociating and diffusing through the alloy to the inner side of the tubes, leaving most impurities outside the membrane (Hunter, 1960; Knapton, 1977).

The solubilities of the hydrogen isotopes in palladium and palladium–silver alloys differ to an appreciable degree. Diffusion membranes of palladium or palladium–silver alloys can therefore be used for separation and concentration processes of the different hydrogen isotopes. The equilibrium hydrogen/deuterium separation factors vary from ~2.4 at 25 °C to 3.7 at −80 °C (Wicke and Nernst, 1964). For the hydrogen/tritium ratio, a value of 2.8 has been measured at 25 °C (Sicking, 1970).

4.2.3 Rhodium Rhodium, atomic number 45, relative atomic mass 102.905 50, no second natural isotope, is very rare. It is a bright, silvery, and hard metal, oxidation resistant up to 875 K, inert against all acids but attacked by molten alkali. The most important properties are catalytic action in heterogeneous and homogeneous catalysis. The main application areas are automotive catalytic converters (89%), chemistry (3%), electronics/electrical technology (1.8%), and glass (3.3%) (quantities in weight percent of the total yearly demand in the western world 1995; Johnson Matthey, 1995).

Selected values of basic physical and chemical properties are given in Table 14. The characteristic fall of the optical reflectivity in the visible range (Fig. 2) is smallest for rhodium. It has a neutral color; the reflectivity is ~20% lower than that of silver.

4.2.3.1 Applications. *Catalysts.* Rhodium is an important constituent in automotive catalysts and in numerous heterogeneous and homogeneous catalytic reactions. In automotive catalysts, rhodium effects the reduction of nitrogen oxides to nitrogen by intermediate adsorption of NO and CO on the surface followed by dissociation of the NO molecule and reaction of the desorbed oxygen atom with CO to form CO_2 (Madix, 1986; Friend, 1993; Cho, 1994).

Chemical technology. Because of their ability to withstand highly corrosive operating environments, platinum–rhodium alloys find extended use for equipment and apparatus parts in the glass industry, especially for the production of high-quality glass for optical devices, liquid-crystal–display glasses, and screens for television sets and desk-top computers.

Platinum-rhodium alloys find application in net catalysts for the oxidation reaction of ammonia to nitrogen oxide. Complex organic rhodium compounds are of great importance as catalysts for hydroformylation (Osborne *et al.*, 1965; Coffey, 1966) and for the production of enantiomerically pure compounds by stereospecific chemical reactions. Rhodium and iridium are suited as susceptors for medium- and high-frequency fields for inductive heating on air. Iridium crucibles used for induction melting of glasses or ceramics are protected against oxidation by rhodium coatings.

Electrical technology/electronics. Rhodium serves as a hard, wear-resistant, electrically conductive surface coating for slide and pressure contacts. It is a constituent in the most commonly used thermocouples in various compositions (e.g., PtRh10/Pt, PtRh13/Pt, PtRh30/PtRh6) applied for temperature measurements in the medium- and high-temperature range.

Jewelry, mirrors. Silver mirrors and jewelry items made from silver and white-gold alloys are protected against corrosion and wear by thin coatings of rhodium.

4.2.4 Iridium Iridium, atomic number 77, relative atomic mass 192.22, only two natural isotopes (191 and 193), is a bright silvery, hard, high–melting-point metal. Together with rhodium, it belongs to the rarest members of the platinum metals. This is in accordance with the general rule that isotopes with odd atomic numbers are rarer than those with even atomic numbers that are divisible by 4. Iridium is inert against all acids, air, and water, but attacked by molten alkali. The main application areas are chemistry and catalysis (58%), crucibles (5%), and others (37%) (quantities in weight percent of the total yearly demand in the western world 1995; Johnson Matthey, 1995).

Selective values of basic physical and chemical properties are given in Table 14. Iridium has the highest density of all metals. It is an effective hardener and grain-refining agent in PGM alloys.

4.2.4.1 Applications. *Chemistry.* In chlor-alkali plants and in electrolytic cells for the production of sodium chlorate, iridium and iridium–ruthenium are applied as surface coatings on anodes. Iridium crucibles serve for the production of high-purity single crystals for optical and laser technology—e.g., gadolinium–gallium garnets and monocrystalline Al_2O_3 substrates. Special catalytic actions of iridium find application for reforming catalysts in the petroleum refining industry and in the synthesis of pharmaceuticals. Quartz glass fibers are drawn through iridium spinnerets.

Electrical technology. Iridium coatings are used as sensors in layer resistor devices. Thermocouples made from combinations of iridium with iridium–rhodium 10 alloy are suited for temperature measurement at very high temperatures up to 2300 °C. Spark-plug electrodes tipped with iridium-containing alloys are used on helicopters and in heavy-duty automobile spark plugs.

4.2.5 Ruthenium Ruthenium, atomic number 44, relative atomic mass 101.07, seven natural isotopes (between 96 and 104), is a bright, silvery metal, resistant against air, water, and acids, but soluble in molten alkali. The main application areas are electronics/electrical technology (45%), chemistry (45%), and catalysis (10%) (quantities in weight percent of the total yearly demand in the western world 1995; Johnson Matthey, 1995).

Selective basic physical and chemical properties are given in Table 14. Ruthenium is, like iridium, an effective hardener and grain-refining agent for PGM alloys.

4.2.5.1 Applications. *Electronics.* Ruthenium is the most important base element of resistor devices and resistor layers in thick-film circuits. The starting material is RuO_2. Resistor layers are printed with paste preparations made from RuO_2 with additions of PbO and Bi_2O_3. They form by firing complex Ru compounds of high thermal stability and very low temperature coefficients of electrical resistivity. The measurement unit is the so-called "sheet resistivity," defined as the electrical resistance of a square area, measured between opposite edges. Exact resistance values are adjusted by applying small cuts by means of laser beam into the resistor layer ("laser trimming").

Electrochemistry. Ruthenium-coated titanium electrodes are used in the electrolysis of brine to produce chlorine and caustic soda. They lower the overpotential against chlorine, saving energy and giving controlled composition of the chlorine gas with low oxygen concentration.

Catalysis. RuO_2 and complex organic ruthenium compounds are specific catalysts in numerous heterogeneous and homogeneous catalytic reactions. Metal-organic ruthenium compounds are proven as catalysts for the photodissociation of water (Watkins, 1978; Harriman, 1983, 1991).

4.2.6 Osmium Osmium, atomic number 76, relative atomic mass 190.23, seven natural isotopes (between 184 and 192), is a bright, silvery metal, resistant against water, oxygen, and acids, but attacked by molten alkali. Application areas are catalysis, alloy components in hard precious-metal alloys, and electronics.

Selected values of basic physical and chemical properties are given in Table 14. Osmium and iridium are efficient reflectors in the wavelength range below 100 nm (Hass and Hunter, 1974).

4.2.6.1 Applications. Osmium is the least practical used noble metal. There are only a few applications.

Chemistry. Osmium tetroxide is used as a catalyst in special oxidation reactions. Complex organic osmium compounds are used in homogeneous catalytic reactions, especially for drug synthesis.

Optics. Osmium and iridium have been used as mirror coatings for vacuum-UV spectroscopy to study celestial phenomena that cannot be found with Al-based mirrors. Ir-coated mirrors have been used in large-aperture telescopes of space rockets for exploration tasks in enhanced space astronomy (Herzig and Spencer, 1982; Herzig, 1983).

Electronics. Surface layers of osmium and ruthenium on dispenser cathodes lower the electronic work function (Zalm and Van Stratum, 1966). Thin osmium layers are applied on tungsten and molybdenum wires for *thermionic sources.*

5. SUMMARY

Silver, gold, and the platinum metals have the highest oxidation resistance of all metals. They are construction materials in the chemical technology and used as conductor, resistor, and contact materials. The platinum metals and platinum-metal compounds are effective and highly selective catalysts in numerous chemical reactions and in automotive exhaust-gas converters. Platinum and platinum alloys are used in sensor elements and thermocouples. The noble metals are also key materials in the fields of magnetic recording; as energy-saving coatings on architectural glass panes; and as electrodes in fuel cells, electrolytic production equipment, and batteries. New technologies open promising application areas. Progress in the automobile engine technology and the growing demand for multilayer capacitors in electronic devices have caused 35% growth of the palladium demand from 1993 to 1994. The demand for ruthenium grew in the same year by 76%, mainly on account of applications in catalysis, electrodes in the chlor-alkali electrolysis, and alloy improvement. The demand for iridium grew by 36% by new catalytic activities, electrodes, and the need for hard, corrosion-resistant alloys—e.g., for spark plugs in aircraft engines. Because of the limited reserves, the production from lower-grade raw material and the recycling of secondary materials have become of increasing importance.

ACRONYMS

APB: Antiphase boundary
CVD: Chemical vapor deposition
DNA: Deoxyribonucleic acid
DOS: Density of states
fcc: Face centered cubic
hcp: Hexagonal close packed
HV: Vickers hardness
NTC: Negative temperature coefficient
PFM: Porcelain fused to metal
PGM: Platinum-group metal
PTFE: Polytetrafluorethylene
PVD: Physical vapor deposition
RNA: Ribonucleic acid
ZGS: Zirconium-oxide grain stabilized

Works Cited

Alekseev, E. S., Ventsel, V. A., Voronov, O. A., Likhter, A. I., Magnitskaya, N. V. (1979), *Zh. Eksp. Teor. Fiz.* **76,** 215–222 [*Sov. Phys. JETP* **49,** 110–113].

Amirnazmi, A., Boudart, M. (1975), *J. Catal.* **39,** 383–394.

Benner, L. S., Suzuki, T., Meguro, K., Tanaka, S. (Eds.) (1991), *Precious Metals, Science and Technology,* Allentown, PA: International Precious Metals Institute.

Berkowitz, A., Kneller, E. (1969), *Magnetism and Metallurgy,* New York: Academic, p. 259.

Beuers, J., Krappitz, H. (1995), in: Degussa (Ed.), *Edelmetalltaschenbuch,* Heidelberg: Hüthig.

Broden, G., Rhodin, T. N., Brucker, C. (1976), *Surf. Sci.* **59,** 593–611.

Butts, A., Coxe, C. D. (1967), *Silver,* Princeton, NJ: Van Nostrand, p. 111.

Chikazumi, S. (1991), in: L. S. Benner, T. Suzuki, K. Meguro, S. Tanaka (Eds.), *Precious Metals, Science and Technology,* Allentown, PA: International Precious Metals Institute, p. 98.

Cho, B. K. (1994), *J. Catal.* **148,** 697–706.

Christensen, N. E. (1972), *Phys. Status Solidi B* **54,** 551–563.

Christensen, N. E. (1973), *Phys. Status Solidi B* **55,** 117–127.

Christensen, N. E., Seraphin, B. O. (1971), *Phys. Rev. B* **4,** 3321–3344.

Coffey, R. S. (1966), British Patent No. 1,121,642.

Coleridge, P. T. (1966a), *Proc. R. Soc. A* **295,** 458–475.

Coleridge, P. T. (1966b), *Phys. Lett.* **22,** 367–369.

Coombes, J. S. (1992), in: Johnson Matthey Co., *Platinum 1992 (Annual Report),* London: Johnson Matthey Publishers, pp. 37–43.

Coombes, J. S. (1993), in: Johnson Matthey Co., *Platinum 1993 (Annual Report),* London: Johnson Matthey Publishers, pp. 35–38.

Dachselt, W. D., Münz, W. D., Scherer, M. (1982), in: Carl M. Lampert (Ed.), *Optical Coatings for Energy Efficiency,* SPIE Proceedings No. 324, Bellingham, WA: SPIE, pp. 37–43.

Darling, A. S. (1973), *Inst. Met. Rev.* **175,** 91–122.

Degussa (1967), *Edelmetalltaschenbuch 1973,* Frankfurt/Main: Degussa.

Degussa (1994), *Edelmetallmärkte 1994 (Annual Report),* Frankfurt/Main: Degussa A. G.

Degussa (1995), *Edelmetalltaschenbuch,* Heidelberg: Hüthig, cover pages.

Diehl, W. (1995), in: Degussa (Ed.), *Edelmetalltaschenbuch 1995,* Heidelberg: Alfred Hüthig, pp. 155–170.

Doduco (Ed.) (1977), *Doduco Datenbuch,* Pforzheim: Doduco, pp. 19–80.

Dorbath, B., Frischkorn, H. J., Jeske, G., Starz, K. A. (1991), in: Degussa Ressort Metall (Ed.), *Metall Forschung und Entwicklung,* Frankfurt/Main: Degussa A. G., pp. 47–58.

Fieberg, M., Edwards, I. R. (1978), Report No. 1996, National Institut van Metall, Randburg, South Africa.

Flanagan, T. B., Sakamoto, Y. (1993), *Plat. Met. Rev.* **37,** 26–37.

Flükiger, R., Paoli, A., Roggen, R., Yvou, K., Muller, K. J. (1972), *Solid State Commun.* **11,** 61–63.

Fong, C. Y. (1974), *J. Phys. F: Metal Phys.* **4,** 775–782.

Friend, C. M. (1993), *Sci. Am.* **268**(4), 42–47.

Funston, E. S., McGurty, J. A. (1961), *Plat. Met. Rev.* **5,** 56.

Gafner, G. (1989), *Gold Bull.* **22,** 112–122.

Gläser, H. H. (1989), *Glastech. Ber.* **62,** (3), 93–99.

Grodski, J. J., Dixon, A. E. (1969), *Solid State Commun.* **7,** 735–737.

Haasen, P. (1994), *Physikalische Metallkunde,* 3. Aufl., Berlin: Springer.

Hagen, E., Rubens, H. (1903), *Ann. Phys. (Leipzig)* **11,** 873–901.

Harriman, A. (1983), *Plat. Met. Rev.* **27,** 102–107.

Harriman, A. (1991), *Plat. Met. Rev.* **35,** 22–23.

Hass, G., Hunter, W. R. (1974), in: B. J. Thompson, R. R. Shannon (Eds.), *Proceedings of the Ninth International Congress of the International Committee for Optics, Santa Monica, California, October 1972,* Washington, D.C.: National Academy of Sciences, pp. 525–553.

Herzig, H. (1983), *Plat. Met. Rev.* **27,** 108–109.

Herzig, H., Spencer, R. S. (1982), *Appl. Opt.* **21,** 15–16.

Hirabayashi, M. (1991), in: L. S. Benner, T. Suzuki, K. Meguro, S. Tanaka (Eds.), *Precious Metals, Science and Technology,* Allentown, PA: International Precious Metals Institute, pp. 103–115.

Hornfeldt, S. P., Windmiller, L. R., Ketterson, J. B. (1973), *Phys. Rev. B* **7,** 4349–4357.

Hume-Rothery, W., Raynor, G. V. (1956), *The Structure of Metals and Alloys,* London: The Institute of Metals.

Hummel, R. E. (1971), *Optische Eigenschaften von Metallen und Legierungen,* Berlin: Springer, p. 161.

Hunt, L. B. (1979), *Gold Bull.* **12,** 116–127.

Hunter, J. B. (1960), *Plat. Met. Rev.* **4,** 130–131.

Jepsen, O., Andersen, O. K., Mackintosh, A. R. (1975), *Phys. Rev. B* **12,** 3084–3103.

Johnson Matthey Co. (1992), *Platinum 1992 (Annual Report),* London: Johnson Matthey Publishing.

Johnson Matthey Co. (1995), *Platinum 1995 (Annual Report),* London: Johnson Matthey Publishing.

Jönsson, S. (1995), in: Degussa (Ed.), *Edelmetalltaschenbuch,* Heidelberg: Hüthig, pp. 202–223.

Kempf, B., Hausselt, J. (1992), *Interdiscip. Sci. Rev.* **17,** 251–259.

Ketterson, J. B., Windmiller, L. R. (1970), *Phys. Rev. B* **2,** 4813–4838.

Khan, H. R. (1984), *Gold Bull.* **17,** 94–100.

Kittel, C. (1980), *Einführung in die Festkörperphysik,* München: R. Oldenbourg.

Knapton, A. G. (1977), *Plat. Met. Rev.* **21,** 44–50.

Kondo, J. (1964), *Prog. Theor. Phys.* **32,** 37–49.

Kondo, J. (1965), *Prog. Theor. Phys.* **34,** 372–382.

Kopp, J. (1976), *Gold Bull.* **9,** 55–57.

Kussman, A., Jessen, X. (1962), *J. Phys. Soc. Jpn.* **17,** (B1), 136–139.

Kussman, A., von Rittberg, G. (1950), *Ann. Phys. (Leipzig)* **7,** 173–181.

Lewis, F. A. (1967), *The Palladium-Hydrogen System,* New York: Academic.

Lewis, F. A. (1982), *Plat. Met. Rev.* **26,** Pt. I, 70–78; Pt. II, 20–27; Pt. III, 121–128.

Lewis, F. A. (1994), *Plat. Met. Rev.* **38,** 112–118.

Linde, J. O. (1932), *Ann. Phys. (Leipzig)* **4,** 219–248, 353–366.

Loebich, O. (1971), Doctoral Thesis, Technische Hochschule Aachen.

Madix, R. J. (1986), *Science* **233,** 1159–1166.

McKenzie, D. R. (1978), *Gold Bull.* **11,** 49–53.

Moore, C. A. (1939), *Trans. Electrochem. Soc.* **75,** 237–269.

Mueller, F. M., Freeman, A. J., Dimmock, J. O., Fardgna, A. M. (1970), *Phys. Rev. B* **1,** 4617–4635.

Nelin, G. (1971), *Phys. Status Solidi* **45,** 527–536.

Obuchi, A., Naito, S., Onishi, T., Tamaru, K. (1982), *Surf. Sci.* **122,** 235–255.

Obuchi, A., Naito, S., Onishi, T., Tamaru, K. (1983), *Surf. Sci.* **130,** 29–40.

Osborne, J. A., Wilkinson, G., Young, J. F. (1965), *J. Chem. Soc. Chem. Commun.* 131–132.

Ott, D., Raub, Ch. J. (1980), *Metall.* **34,** 629–633.

Ott, D., Raub, Ch. J. (1981), *Metall.* **35,** 543–548, 1005–1011.

Ott, D., Raub, Ch. J. (1982), *Metall.* **36,** 150–156.

Plachy, A. L., Dixon, A. E. (1974), *Can. J. Phys.* **52,** 1295–1303.

Poniatowski, M., Clasing, M. (1968), *Z. Metallk.* **59,** 165–170.

Prescher, N. N. (1995), in: Degussa (Ed.), *Edelmetalltaschenbuch 1995,* Heidelberg: Alfred Hüthig, pp. 351–401.

Pugh, S. F. (1954), *Philos. Mag.* **45,** 823–843.

Raub, Ch. J. (1984), *Plat. Met. Rev.* **28,** 63–75.

Renner, E. (1990), in: W. Gerhartz (Ed.), *Ullmann's Encyclopedia of Industrial Chemistry,* Weinheim: VCH, Vol. A12, pp. 499–533.

Renner, E. (1992), in: W. Gerhartz (Ed.), *Ullmann's Encyclopedia of Industrial Chemistry,* Weinheim: VCH, Vol. A21, pp. 75–127.

Renner, E. (1993), in: W. Gerhartz (Ed.), *Ullmann's Encyclopedia of Industrial Chemistry,* Weinheim: VCH, Vol. A24, pp. 107–163.

Rogers, D. L. (1974), *Phys. Status Solidi B* **66,** K53–K57.

Rosenberg, B., Van Kamp, L., Krigius, T. (1965), *Nature* **203,** 698–699.

Rosenberg, B., Trosko, J. E., Mansour, V. H. (1969), *Nature* **222,** 385–386.

Schuit, G. C. A., van Reijen, L. L. (1958), *Adv. Catal.* **10,** 242–317.

Seibold, K. (1995), in: Degussa (Ed.), *Edelmetalltaschenbuch,* Heidelberg: Hüthig, pp. 414–421.

Selman, G. L., Day, J. G., Bourne, A. A. (1974), *Plat. Met. Rev.* **18,** 46–57.

Sicking, G. (1970), Ph.D. Thesis, Univ. Münster, Germany.

Smith, N. V. (1974), *Phys. Rev. B* **9,** 1365–1376.

Smithells, C. J., Brandes, E. A. (1977), *Metals Reference Book,* 5th ed., London: Butterworth, p. 1019.

Stöckel, D. (1979), *Z. Werkstofftech.* **10,** 230–242.

Tamaru, K., Naito, S. (1991), in: L. S. Benner, T. Suzuki, K. Meguro, S. Tanaka (Eds.), *Precious Metals, Science and Technology,* Allentown, PA: International Precious Metals Institute, pp. 262–307.

Tanuma, S. (1991), in: L. S. Benner, T. Suzuki, K. Meguro, S. Tanaka (Eds.), *Precious Metals, Science and Technology,* Allentown, PA: International Precious Metals Institute, pp. 35–59.

Toyoshima, I., Somorjai, G. A. (1979), *Cat. Rev. Sci. Eng.* **19,** 105–159.

Vannice, M. A. (1975), *J. Catal.* **37,** 449–473.

Vedernikov, M. V. (1969), *Adv. Phys.* **18,** 337–370.

Völcker, A. (1995), in: Degussa (Ed.), *Edelmetalltaschenbuch 1995,* Heidelberg: Alfred Hüthig, pp. 170–173.

Watkins, D. M. (1978), *Plat. Met. Rev.* **22,** 118–125.

Wicke, E., Nernst, G. H. (1964), *Ber. Bunsenges.* **1964,** 224–235.

Wise, E. M. (1964), *Gold,* Princeton, NJ: Van Nostrand, p. 72.

Wise, E. M., Vines, R. F. (1979), in: W. H. Cubberly, H. Baker, D. Benjamin (Eds.), *Metals Handbook,* Vol. 2, *Properties and Selection: Nonferrous Alloys and Pure Metals,* 9th ed., Metals Park, OH: American Society for Metals, pp. 699–705.

Yonemitsu, K. (1991), in: L. S. Benner, T. Suzuki, K. Meguro, S. Tanaka (Eds.), *Precious Metals, Science and Technology,* Allentown, PA: International Precious Metals Institute, pp. 13–35.

Zalm, P., Van Stratum, J. A. (1966), *Philips Tech. Rev.* **27** (3/4), 69–75.

Zysk, E. D. (1979), in: W. H. Cubberly, H. Baker, D. Benjamin (Eds.), *Metals Handbook,* Vol. 2, *Properties and Selection: Nonferrous Alloys and Pure Metals,* 9th ed., Metals Park, OH: American Society for Metals, pp. 688–698.

Further Reading

Bard, J. A. (1979), "Properties of Gold and Gold Alloys," in: W. H. Cubberly, H. Baker, D. Benjamin (Eds.), *Metals Handbook,* Vol. 2, *Properties and Selection: Nonferrous Alloys and Pure Metals,* 9th ed., Metals Park, OH: American Society for Metals, pp. 679–683.

Benner, L. S., Suzuki, T., Meguro, K., Tanaka, S. (Eds.) (1991), *Precious Metals, Science and Technology,* Allentown, PA: International Precious Metals Institute.

Butts, A., Coxe, Ch. D. (1967), *Silver, Economics, Metallurgy, and Use,* Princeton, NJ: Van Nostrand.

Coxe, C. D., McDonald, A. W. (1979), "Properties of Silver and Silver Alloys," in: W. H. Cubberly, H. Baker, D. Benjamin (Eds.), *Metals Handbook,* Vol. 2, *Properties and Selection: Nonferrous Alloys and Pure Metals,* 9th ed., Metals Park, OH: American Society for Metals, pp. 671–678.

Darling, A. S. (1973), *Some Properties and Applications of the Platinum Group Metals,* London: The Institute of Metals, pp. 91–122.

Degussa (1967), *Edelmetalltaschenbuch,* Frankfurt/Main: Degussa.

Degussa (1995), *Edelmetalltaschenbuch,* 2nd ed., Heidelberg: Alfred Hüthig.

Degussa (Ed.) (1991), *Metall Forschung und Entwicklung,* Frankfurt/Main: Degussa Ressort Metall.

Degussa (annual reports), *Edelmetallmärkte,* Frankfurt/Main: Degussa.

Emsley, J. (1991), *The Elements,* 2nd ed., Oxford: Clarendon.

Haasen, P. (1994), *Physikalische Metallkunde,* 3. Aufl., Berlin: Springer.

Hume-Rothery, W., Raynor, G. V. (1956), *The Structure of Metals and Alloys,* London: The Institute of Metals.

Hummel, R. E. (1971), *Optische Eigenschaften von Metallen und Legierungen,* Berlin: Springer.

Johnson Matthey (annual reports), *Platinum,* London: Johnson Matthey Publishing.

Kittel, C. (1980), *Einführung in die Festkörperphysik,* authorized translation of *Introduction to Solid State Physics,* 5th ed., München: R. Oldenbourg.

Kondo, J. (1969), "Theory of dilute magnetic alloys," in: H. Ehrenreich, F. Seitz, D. Turnbull (Eds.), *Solid State Physics,* Vol. 23, New York: Academic, pp. 184–281.

Raub, E. (1940), *Die Edelmetalle und ihre Legierungen,* Berlin: Springer.

Raynor, G. V. (1976), "The Alloying Behavior of Gold," *Gold Bull.* **9,** 12–18, 50–55.

Renner, H. (1992), "Platinum Group Metals and Compounds," in: W. Gerhartz (Ed.), *Ullmann's Encyclopedia of Industrial Chemistry,* Vol. A21, Weinheim: VCH, pp. 75–131.

Renner, H. (1993), "Silver, Silver Compounds, and Silver Alloys," in: W. Gerhartz (Ed.), *Ullmann's Encyclopedia of Industrial Chemistry,* Vol. A24, Weinheim: VCH, pp. 107–163.

Renner, H., Johns, M. W. (1990), "Gold, Gold Alloys, and Gold Compounds," in: W. Gerhartz (Ed.), *Ullmann's Encyclopedia of Industrial Chemistry,* Vol. A12, Weinheim: VCH, pp. 499–533.

Savitskii, E. M., Prince, A. (1989), *Handbook of Precious Metals,* New York: Hemisphere.

Schlamp, G. (1995), "Noble Metals and Their Alloys," in: R. Cahn (Ed.), *Materials Science and Technology,* Vol. 8/9 (K. Matucha, Ed.), *Structure and Properties of Nonferrous Metals,* Weinheim: VCH, pp. 467–587.

Wise, E. M., Vines, R. F. (1979), "Properties of Palladium and Palladium Alloys," in: W. H. Cubberly, H. Baker, D. Benjamin (Eds.), *Metals Handbook,* Vol. 2, *Properties and Selection: Nonferrous Alloys and Pure Metals,* 9th ed., Metals Park, OH: American Society for Metals, pp. 699–705.

Zysk, E. D. (1979), "Precious Metals and Their Uses," in: W. H. Cubberly, H. Baker, D. Benjamin (Eds.), *Metals Handbook,* Vol. 2, *Properties and Selection: Nonferrous Alloys and Pure Metals,* 9th ed., Metals Park, OH: American Society for Metals, p. 659–657.

Zysk, E. D. (1979), "Properties of Platinum and Platinum Alloys," in: W. H. Cubberly, H. Baker, D. Benjamin (Eds.), *Metals Handbook,* Vol. 2, *Properties and Selection: Nonferrous Alloys and Pure Metals,* 9th ed., Metals Park, OH: American Society for Metals, pp. 688–698.

SIMULATION BY MOLECULAR DYNAMICS

BERNI J. ALDER AND ANTHONY J. LADD, *University of California, Lawrence Livermore National Laboratory, Livermore, California, U.S.A.*

INTRODUCTION

Molecular dynamics has come to mean the numerical solution of the classical Newtonian equations of motion for a representative sample of interacting particles. In effect one is solving the 6*N*-dimensional Liouville equation (*N* is the number of particles), from which, without invoking any ergodic hypothesis regarding the accessibility of regions of phase space, one can obtain long-time (equilibrium) average properties (Alder and Wainwright, 1959a). Equilibrium properties can also be obtained by a competitively powerful numerical method called Monte Carlo (Metropolis *et al.*, 1953), which does assume that the ergodic hypothesis is valid, but molecular dynamics is uniquely useful in solving time-dependent many-body problems.

In fact, the computer-based Monte Carlo and molecular-dynamics methods are the only available techniques by which to solve many-particle problems "exactly." It is due to this circumstance that these two techniques have found widespread use in almost all branches of science. Some of the more interesting time-dependent calculations from both the fundamental and applied viewpoints will be illustrated in this article. Before that, what is meant by an exact solution must be

3-527-28140-1/96/$5.00 + .50

clarified. Previously, exact solutions of many-particle systems have only been possible through analytic treatment of idealized systems, such as perfect gases or harmonic oscillators. For more realistic interactions, molecular dynamics can give the exact solution for a finite number of particles by following the equations of motion for a finite time. In most applications, these equations of motion are integrated by a finite-difference scheme, which introduces numerical error (Verlet, 1967). However, for rigid elastic spheres, the trajectories are followed collision by collision, so that the only numerical errors involved are roundoff errors in calculating the time to the next collision (Alder and Wainwright, 1963). The effect of each of these limitations is discussed next.

1. TIME CONSIDERATIONS

The very first attempt at integrating the equations of motion numerically led to the misleading conclusion that statistical mechanical systems do not relax to equilibrium rapidly or that systems very much larger than 100 particles are required to reach equilibrium, that is, to observe ergodic behavior.

1.1 The Fermi–Pasta–Ulam Problem

This first investigation (Fermi *et al.*, 1965; Tuck and Menzel, 1972) involved a linear chain of either 32 or 64 nearly harmonic oscillators with an additional fixed particle at each end. This isolated system of slightly anharmonic particles was started from an initial condition in which all the energy was localized in one of the long-wavelength modes. As the classical equations of motion are integrated in time, one expects that the energy will flow from the long-wavelength modes to the shorter-wavelength modes, until an equipartition of energy among all the modes is reached. However, instead of thermalization, the energy was observed to recur periodically (to a very close approximation) in the original mode after being redistributed in only a few other modes. This lack of ergodicity could not be ascribed to the initial preparation of the system nor to the restricted number of degrees of freedom in the finite system but ultimately was attributed to the weak nonlinearity. As the particles interacted more anharmonically (are more strongly coupled), the system behaved as expected.

1.2 Hard Particles

The validity of the ergodic hypothesis was in fact first demonstrated for hard spheres, the extreme anharmonic interaction. For as few as ten particles it was shown that within statistical errors identical equilibrium results were obtained by Monte Carlo and molecular dynamics. Moreover, a rigorous proof of the ergodic hypothesis has been achieved for four hard spheres (Sinai, 1973). The practical numerical demonstration is, moreover, valid for almost all initial conditions one might randomly choose. Some highly selected initial choices, such as all spheres moving in the same direction with the same speed, would not be ergodic.

1.3 The *H* Theorem

A demonstration that the velocity distribution of a hard-sphere system rapidly and monotonically approaches the Maxwell–Boltzmann equilibrium distribution, from almost any randomly chosen initial velocity distribution $f(\mathbf{v})$, has been carried out by calculating the Boltzmann H function (Alder and Wainwright, 1958),

$$H(t) = \int f(\mathbf{v},t) \ln f(\mathbf{v},t)\, d^3v. \quad (1)$$

For an initial distribution where all the particles had the same speed, but random velocity directions, the H function reached its equilibrium value in less than 200 collisions for a 100-particle system, or less than 4 collisions per particle (see Fig. 1). H-function calculations for particles interacting with an attractive potential, in addition to the hard-sphere repulsive potential, show that thermalization now depends on the rearrangement of particle positions, which can be quite time-consuming at high density, but nevertheless, the H function still approaches the final equilibrium value monotonically.

1.4 Irreversibility

The H theorem attributes a direction in time to a system of particles that obey time-reversible equations of motion. This has led

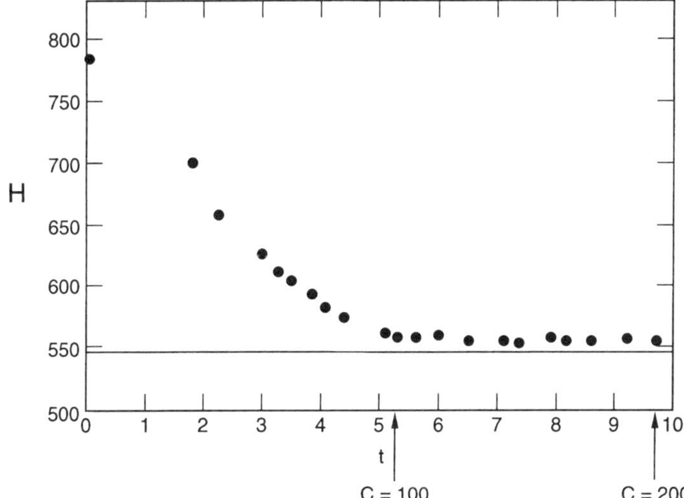

FIG. 1. The behavior of the Boltzmann H function with time t, or collision number C, for 100 hard spheres at low density. The horizontal line represents the equilibrium H value for an infinite system.

to a great deal of discussion in the literature, but the statistical nature of this theorem (Lebowitz, 1993) is perhaps best illustrated by the following computer simulation. Suppose you put four spheres in the left half of a box and integrate the equations of motion forward for some time. Subsequently, you reverse the time, that is, reverse the directions of all velocities; nobody is particularly surprised if, after you integrate for an equal time in the reverse direction, you find the same four particles in their original positions on the left-hand side. On the other hand, if the experiment is repeated for 100 particles, everybody recognizes that upon finding all 100 particles again on the left-hand side, a highly unusual event has occurred. This illustrates not only the irreversible nature of systems as small as 100 particles but also the incredibly long (larger than the age of the universe) Poincaré cycle or the time it takes for all 100 particles to return, without intervention, arbitrarily close to their initial position in phase space. The essential point is that coarse-grained distributions, such as $f(\mathbf{r},\mathbf{v},t)$, evolve monotonically from a low-probability initial state prepared by the experimenter to overwhelmingly more probable equilibrium states.

1.5 Numerical Noise

In the above numerical experiments, the 100 particles only return to their original positions if the time to reversal is less than about ten collisions per particle. For longer times the accumulation of roundoff error in the last of the customary 12 digits with which the positions and velocities are specified causes the reversed trajectory to lead to a different configuration. The cause for this rapid divergence of paths is the dynamical fact that closely spaced trajectories in phase space separate exponentially in time. Roundoff error can be considered as the addition of some random noise to the otherwise completely deterministic dynamics. It has never been observed to cause any problem in calculating any transport property, even if the correlation time involved in determining that property exceeds ten collision times per particle. For equilibrium properties, the noise can be considered equivalent to choosing different members of the microcanonical ensemble. In the process of determining averages over initial states, one has the choice of either running a single long trajectory and chopping it up to establish a series of initial equilibrium states or setting up stochastically a set of initial equilibrium states by, for example, the Monte Carlo method.

Roundoff error can be avoided (Levesque and Verlet, 1993) entirely by algorithmic design as well as eliminated in certain simplified models—for example, lattice-gas models (Frisch *et al.*, 1986), which are discussed later. Experiments with such models show that numerical precision is not a limiting factor. In lattice gases only exact logical operations are required, since they are cellular automata algorithms involving only Boolean arithmetic. These models are exactly reversible microscopically, yet for large systems macroscopic observables evolve monotonically, as expected from the second law of thermodynamics.

1.6 Computational Limitations

Molecular-dynamics calculations for hard spheres make it particularly evident that one is dealing with logically sequential calculations, which cannot take advantage of parallel computers to increase the length of time for which the trajectory can be followed. It is easier to study larger systems on parallel computers than it is to lengthen the simulation time. For particles with continuous interaction potentials, where the parallel aspect of computers can be taken advantage of, for example, in the force-field part of the calculation, up to 100 million particles have been simulated for a few picoseconds of real time (Plimpton, 1995).

Molecular-dynamics calculations for particles with short-range interaction forces typically use periodic boundary conditions so as to minimize boundary effects and thus mimic an infinite system as accurately as possible by surrounding a finite system by replicas of itself. The system can also be viewed as residing on a toroidal surface. Long-range Coulombic interactions can be dealt with by a summation method developed by Ewald (Madelung, 1918) for calculating the total energy of an ionic crystal, since interactions between periodically repeated cells must be taken into account explicitly (Brush *et al.*, 1966). However, more computationally efficient methods have been developed recently (de Leeuw *et al.*, 1980; Greengard, 1994). For the more common short-range–force situation, the periodic cell is further divided into regions of size somewhat larger than the range of the forces, typically three particle diameters, to take advantage of the fact that distant regions do not contribute significantly to the force calculation. This domain decomposition, each cell of which typically contains 100 particles, significantly reduces the calculation time in systems of more than 1000 particles, so that it is then proportional to the number of particles rather than the number of particles squared. The neighbor list in each compartment does not have to be updated until the maximum particle displacement exceeds the difference between the compartment size and the range of the forces. Strips of that size difference occasionally have to be interchanged between nodes of a parallel computer.

1.7 Length of Trajectory

The time-consuming process in a molecular-dynamics simulation is the force calculation necessary to advance all the particles one time step. The step size is usually taken to be the same for all particles and is limited by the requirement that the force does not change too much in the time interval. For simple pairwise interacting particles at liquid densities, that time step typically corresponds to 10^{-14} s in real time. There are many algorithmic choices for updating the particle positions and velocities, but important considerations are simplicity, efficiency, accuracy, stability, and reliability. Another consideration is that only one force calculation per time step is usually practical. Because force can change rapidly with distance, it does not pay to go much beyond third-order accurate integration schemes; this has led to the Verlet leapfrog algorithm as the method of choice (Berendsen and Van Gunsteren, 1986). The algorithm is a simple finite-difference scheme to solve second-order differential equations for the particle coordinates; particle velocities are obtained by differencing coordinates.

In a sample simulation of 10^3 particles, the force calculation requires about 10^4 machine clock cycles per particle per time step or of the order of 10^7 cycles per time step. In a long calculation of 30 h or 10^5 s that runs on a processor with a clock speed of 10^8 cycles/s, a total of 10^{13} cycles can be executed; this means 10^6 of the above time steps or 10^{-8} s of real time. Massively parallel computers allow simulations of much larger numbers of particles, where each processor typically handles a domain consisting of 10^3 particles, but not for longer real times since each processor is subject to the above limitations. Unless clock speeds improve significantly, this limitation will persist. Lattice gases overcome this limitation by reducing both the compartment size and the number of neighbors to the order of ten particles each, as well as significantly reducing the number of cycles per time step. This allows systems to be followed for many orders of magnitude longer in time.

1.8 Rare Events

Because trajectories cannot be followed for macroscopic times, events that occur very infrequently—for example, as a result of a rare fluctuation—cannot be studied without some special preparation of the system that focuses on the unusual happening. In general, dealing with phenomena that occur at vastly different time scales in a given system presents formidable numerical problems unless one can ignore the coupling between the different time scales (Brackbill and Cohen, 1985). Then one can either average over the short time scale or ignore the long time scale.

As an example, consider vacancy diffusion in which a hard sphere occasionally jumps into a neighboring vacancy (Bennett and Alder, 1971). The bottleneck can readily be identified by physical insight, although this is not always the case. One then resorts to the transition-state theory, developed long ago for chemical reactions, which gives the rate of a process as the probability of the system being in the activated state multiplied by the transmission coefficient, which measures whether the reaction will successfully proceed from one side of the barrier to the other, once the particle is at the barrier. The only assumption made in this theory is that of local equilibrium. As a consequence, the jumping sphere is successively confined between two closely spaced walls along the rare-event trajectory, and the system is allowed to equilibrate at each successive position of the two walls (Connick and Alder, 1983). From the pressure difference between the two walls, derived from the collision rate with the walls, the work necessary to drag the sphere to the saddle point can be evaluated (see Fig. 2). Having thus obtained the probability of being at the top of the barrier by an adiabatic process, the two walls are removed, and the transmission coefficient is obtained by finding how often the forward trajectory ends in the hole and the backward trajectory at the original site of the jumping sphere. The transmission coefficient will automatically correct for placing the dividing

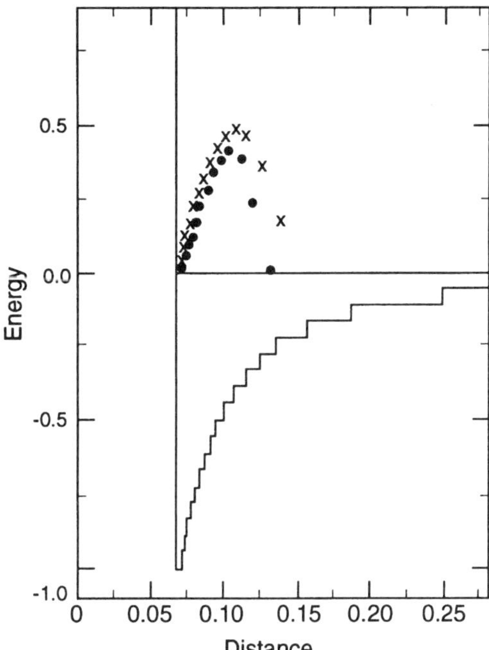

FIG. 2. Energy (dots) and free energy (crosses) of activation for exchanging a sphere in contact with an attracting particle located at the origin with another neighboring sphere, determined by confining the exchanging sphere between successively more distant pairs of walls located at the indicated steps in the attractive potential.

surface somewhere other than at the exact saddle point. There are many variations of this theme, but the key ingredient is the choice of the transition state to be near the bottleneck.

Another example of rare events is the mechanism by which a crystal melts; in this case the bottleneck was not known in advance, and so brute-force long runs were required. The rare events are schematically indicated in Fig. 3. It was shown (Barker *et al.*, 1981) later that this mechanism represents an instability that occurs at densities slightly below melting. However, the only melting criterion that has been rigorously established is the thermodynamic one, namely, that the free energy of the solid equals that of the liquid.

2. EQUILIBRIUM PROPERTIES

Although we do not want to dwell on equilibrium properties, a few key points are worth enumerating insofar as they relate to

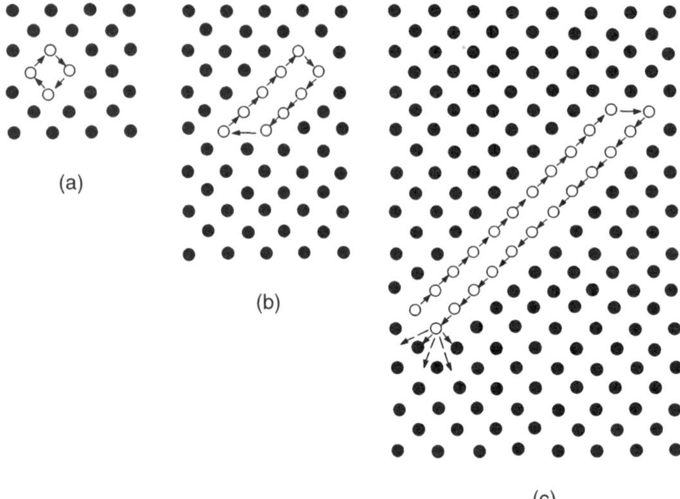

FIG. 3. A schematic diagram of the particle motion (indicated by arrows) in a two-dimensional solid, shown at three successively lower densities near the melting density. (a) First phase of closed-loop motion showing four particles changing position in lattice. (b) As the density is lowered, the loop elongates and contains more particles. (c) Breakup results when the loop becomes so large that the jump of one atom at one end of the loop is no longer correlated with the jump at the other end.

nonequilibrium properties. The first point to consider is the relative efficiency between Monte Carlo and molecular dynamics in determining equilibrium properties. For simple systems such as hard spheres, it is difficult to make a choice (Jacucci and Rahman, 1984). As the properties of more complex molecular systems are studied, Monte Carlo may be most efficient if "smart" or biased Monte Carlo moves can be employed (Rao and Berne, 1979; Siepmann and Frenkel, 1992).

2.1 Phase Transition

The first surprising result that came out of a comparison of Monte Carlo and molecular-dynamics calculations was that, except in regions near a phase transition, a 100-particle system could mimic an infinite system, in that quantitative differences in thermodynamic properties were only of the order $1/N$. Near phase transitions, long-range fluctuations are large, and thus those regions of the phase diagram are greatly distorted in small systems. At a second-order phase transition, such as the critical point between a liquid and a gas, the additional problem of critical slowing down has to be faced, that is, the infrequency with which such collective long-range fluctuations occur. There are special Monte Carlo techniques to deal with that (Swendsen *et al.,* 1992). In first-order phase transitions, such as the liquid–solid transition, the difficulties are the rarity of fluctuations that lead to nucleation of the solid (see Fig. 3) and the large number of particles that are required for the liquid and solid phases to coexist (Alder and Wainwright, 1959a). Free-energy matching of different phases is the most important technique to determine phase boundaries (Frenkel and Ladd, 1984).

In spite of the above difficulties, it proved possible, in very long runs, to demonstrate the coexistence of a solid and fluid phase for two-dimensional systems of hard disks with about 1000 particles (Alder and Wainright, 1959b, 1962) (see Fig. 4). This and complementary simulations of three-dimensional hard-sphere systems settled an important issue, namely, whether purely repulsive forces alone could lead to solid formation. No rigorous analytic proof of this transition is available as yet.

2.2 Ensembles

A further problem in simulating freezing from a liquid is that for small systems the kind of crystal that can form is determined by the boundary conditions, namely, that an integer multiple of unit cells must fit in each dimension in order to form a perfect crystal of a specified structure. This is because most molecular-dynamics calculations use a microcanonical ensemble, where the energy and, in particular, the volume are fixed. However, other ensembles can be formulated

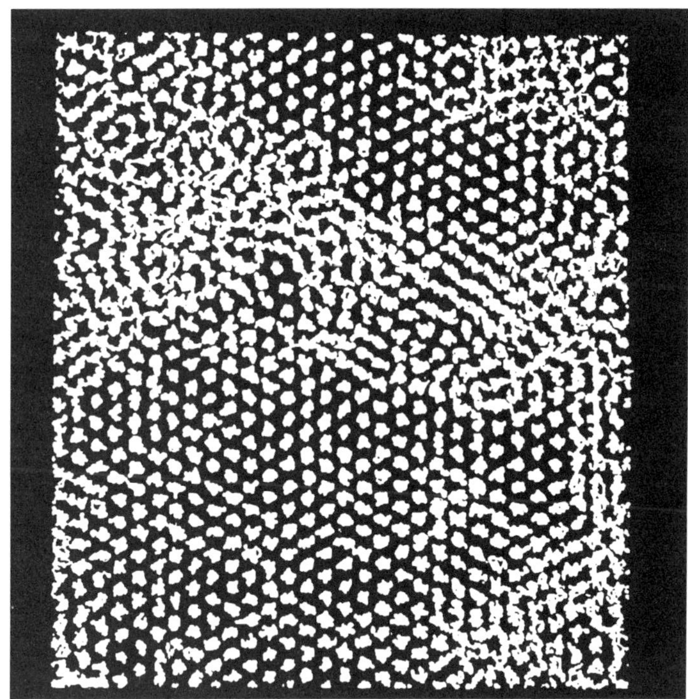

FIG. 4. The trajectories of the centers of particles for a few thousand collisions in the two-phase region. Ordered regions, where particles oscillate about their solid lattice positions, can be clearly seen to coexist with liquid regions where the particle centers trace out short line trajectories as the particles diffuse between lattice positions.

(Anderson, 1980)—in particular, constant-pressure and -temperature ones that correspond more naturally to the freezing process and that allow the system to accept a variety of crystal structures. Such molecular-dynamics ensembles can provide correct equilibrium properties, including the crystal structure (Nosé and Klein, 1983). Correct dynamical properties for such ensembles can only be attained under conditions such that perturbations vary sufficiently slowly that the system is in effect microcanonical for times comparable to a typical relaxation for the process under consideration. A constant-pressure ensemble can also be formulated for Monte Carlo simulations and shows a bimodal density distribution between fluid and solid in the two-phase region (Wood, 1970).

2.3 van der Waals Approximation

The physical significance of the hard-sphere system is that it provides a basis from which to determine the properties of real fluids by treating the attractive potential as a perturbation. The van der Waals equation is an important prototype of this idea; it relates the pressure p, volume V, and temperature T by

$$(p + a/V^2)(V - b) = RT. \tag{2}$$

The pressure is modified by the attractive potential contribution given as a constant a divided by V^2, and the effective volume is reduced by the excluded volume occupied by the spheres b (R is the gas constant). Rewriting this equation as

$$pV/RT = V/(V - b) - a/VRT \tag{3}$$

and replacing the hard-sphere term $V/(V - b)$ by the more accurate computer-generated hard-sphere pressure allows the a/VRT term to be identified as a perturbation due to the attractive forces, namely, the ratio of the potential energy a/V to the kinetic energy RT. This view can be made rigorous if the attractive potential is sufficiently weak and long range (Uhlenbeck *et al.*, 1963), so that no matter how the particles are arranged they are always within each other's range of interaction. Then the mean-field theory represented by the van der Waals equation is rigorously valid; that is, the higher-order attractive perturbation terms $O(T^{-2})$ are negligible. It is also noteworthy that with the establishment of the solid phase for hard spheres, this simple theory contains all three states of matter, that is, solid, liquid, and gas.

The van der Waals theory is the foundation of the modern theory of simple liquids, not only for equilibrium but also for nonequilibrium properties. For transport properties, the van der Waals theory permits the attractive potential to be ignored, since in the mean-field approximation the constant attractive potential never exerts any force on a particle. Indeed, to a good approximation the transport coefficients of hard spheres reproduce those of real monatomic fluids (Dymond and Alder, 1966).

3. COMPLEX MOLECULES

It is fortunate that most gross macroscopic properties are primarily determined by the repulsive forces, since we do not yet know the intermolecular forces very accurately. Furthermore, even for simple systems such as argon, the potential is not, as is often assumed, pairwise additive (Barker, 1986). It is known that the effective pair potential at liquid densities differs by 20% in well depth from the true pair potential (Maitland *et al.*, 1981). Besides repulsion, the next most important force to take into account in more complex molecules is the Coulomb charge distribution. This is often done by empirically assigning partial charges to various atoms in the molecule. Such partial charges can be obtained from quantum mechanical calculations, from which the correct potential energy of interaction must ultimately be obtained. However, this task is at present much too complex to be carried out with sufficient precision. Nevertheless, it is amazing that by merely assigning a charge and size to an atom, one can obtain semiquantitative results, even ignoring the important polarization effects. More delicate questions, such as the conformation of a polymer, require more accurate knowledge of the intermolecular potential. This is because differences in conformation involve tiny energy differences. At present one must resort to effective empirical potentials for complex molecules, obtained (in a building block fashion) by reproducing properties of simpler molecules by molecular dynamics (Lifson *et al.*, 1979).

3.1 Water

It is remarkable how well a distributed-charge model can reproduce the properties of water, even though the hydrogen bond can only be accurately represented quantum mechanically (Rahman and Stillinger, 1971). Studies of water and ionic solutions are, of course, important precursors to the study of biological systems. From molecular-dynamics studies employing simple model potentials for water, dynamic properties such as the residence time of a water molecule near a positively or negatively charged ion can be obtained. Structural properties such as the difference in orientation of a water molecule in the neighborhood of an ion can also be ascertained. Such properties are not very sensitive to the details of the interaction potential.

3.2 Constraint Dynamics

The most commonly used algorithm for molecules, even for water, is constrained molecular dynamics (Ciccotti *et al.*, 1982). The constraint for the water calculation, for example, would be to keep the two oxygen–hydrogen bond lengths fixed at their experimental value. A completely rigid model of water would also keep the angle (the bending angle) between these two bonds fixed. This makes physical sense since the vibrational and bending frequencies are not very different. In general, the motivation for introducing constraints is to eliminate fast modes, such as vibrational modes, which do not significantly affect the macroscopic properties but which do affect the efficiency of the molecular-dynamics calculation. If one had to take into account these fast modes, a considerably smaller integration time step would be necessary. Since vibrational periods are typically at least 100 times less than correlation times for translational or rotational velocities, they can safely be ignored.

In general, complex molecules are modeled by chains of point particles interacting by a potential energy that is split into an intermolecular and an intramolecular part. The latter part is usually written as a sum of various bond-bending or torsional energies, and pair interactions between nonbonded atoms on the same molecule. Most commonly only the direct intermolecular bonds are constrained since they differ significantly in frequency from the others. The simplest and most common implementation of the constrained procedure is to take unconstrained steps in the atomic coordinates, utilizing the leapfrog algorithm, and then apply a resetting mechanism to the positions to satisfy the constraints. The constraints are introduced as a set of Lagrange multipliers to be evaluated by an auxiliary calculation at each time step. This involves solutions to a set of nonlinear algebraic equations that can be obtained iteratively, treating each individual constraint in succession. The Lagrange multipliers represent the forces directed along the bonds required to keep the bond lengths constant. The method works well, even for a very complex network of bond constraints, as demonstrated by simulation of small proteins (Brooks *et al.*, 1988).

3.3 Biological Applications

The principal difficulties in applying molecular dynamics to biological macromolecules are the computational problem of the long time scales on which interesting phenomena occur and the lack of an accurate knowledge of the interatomic forces.

3.3.1 Protein Folding The physical time scales involved in the very important problem of protein folding are of the order of seconds or minutes and impossible to reach computationally (McCammon *et al.*, 1977). The main reason for this slow relaxation rate is the vast dimensionality of configuration space that has to be sampled. The only reasonable approach to the protein-folding problem is to decompose the calculation by investigating marginally stable intermediate secondary structures that can be observed experimentally. Simulations have so far been carried out for folding and unfolding of small peptides and proteins (Bishop and Schulten, 1994) (see Plate 1). The suggestion has been made that such a process might be speeded up by funneling into a singular deep energy minimum (Shakhnovich, 1994). To investigate various folding mechanisms, it is possible to speed up the process by artificially altering the force field along relevant bond angles.

A particular example of the use of molecular dynamics in structural biology is the study of the interaction of proteins with nucleic acids (DNA or RNA). These interactions are the basis of a variety of essential cellular functions, including DNA replication and transcription. The hormone receptor proteins represent a class of proteins that interact with DNA to regulate gene expression. The cocrystal x-ray structure of the DNA binding domain of one member of this class, the glucocorticoid receptor, interacting with a specific sequence of DNA has been determined. However, the observed features in the cocrystal structure are not biologically active. With the use of molecular-dynamics techniques, a complex that represents the biologically active element in this regulatory mechanism was constructed that is more closely related to the physiologic environment [see Plate 1(a)] by placing the system in a drop of water. The simulations indicate that when the protein interacts with DNA, the conformation of the DNA assumes a bent and unwound conformation as compared with uncomplexed DNA [see Plate 1(b)]. These observations are not borne out in the crystallographic environment; however, they are observed on a macroscopic scale by the use of other experimental techniques that are conducted under physiological conditions. The conformational changes that the protein induces in the DNA are essential to a proposed mechanism of gene regulation for this class of proteins by making a particular gene accessible for transcription.

3.3.2 Membranes Fortunately, there are a number of interesting phenomena that occur on much shorter time scales. An example is binding of a small molecule to a macromolecular site, a subject intensively pursued in drug design (Berendsen, 1986). If the rate is too slow, the rare-event algorithm, discussed earlier, can be invoked. Other important studies involve the transport of small molecules across membranes. Bilayer membranes self-assemble by using a molecule that has a polar group at one end that likes to interact with the water solvent (hydrophilic) and a nonpolar group at the other end that does not (hydrophobic). These systems are similar to soap solutions, which can assume a variety of micellar phases, depending on concentration (Smit *et al.*, 1990). In certain concentration ranges a lamellar structure is obtained consisting of bilayers that are permeable to various smaller molecules. Simulations have also been used to study the insertion of lipid molecules into the bilayer.

3.3.3 Electron Transfer Another interesting example where molecular dynamics is making an impact is in the mechanism of electron-transfer rates in photosynthetic reactions (Gehlen *et al.*, 1994). The free electron generated by photon-induced excitations appears to couple electrostatically to the nuclear charges that fluctuate in position in the macromolecule. From molecular dynamics the thermal motion of the charges for a given molecular structure yields the spectral density of the charge fluctuations, which apparently controls the electron transfer rate.

3.3.4 Crystal Structure A final example where molecular dynamics has had an impact is in the determination of crystal structures of proteins from x-ray diffraction. Initial crystal structures can be considerably refined by utilizing additional data that are sensitive to different distance scales and features of the molecular configuration. For example, nuclear magnetic resonance is sensitive to the proximity of different hydrogen atoms within the molecules. Utilizing that information, in addition to x rays, helps define the global structure of the molecules more accurately. Similarly, molecular dynamics, which is sensitive to short-range spatial interactions, can help refine the local protein structure, leading to a considerably more accurate location of the atoms in a crystal structure (Karplus, 1986).

4. TRANSPORT COEFFICIENTS

Molecular-dynamics simulations of single-particle diffusion led to a fundamental disagreement with kinetic theory, namely, that successive collisions in a fluid would be uncorrelated. This "molecular chaos" assumption corresponds mathematically to a Markov relaxation process; it necessarily leads to an exponentially decaying velocity correlation function, contradicting the computer-observed power-law decay (Alder and Wainright, 1970). This unexpected discovery of long-term memory effects through computer experiments is particularly noteworthy since it preceded laboratory experiments (Paul and Pusey, 1981). It led to a profound change in the mathematical structure of kinetic theory (Dorfman and Cohen, 1975). In retrospect this phenomenon could have been predicted, if hydrodynamics had been thought to be applicable on molecular time and distance scales. In fact, quite the contrary was believed, so that the coherent hydrodynamic flows that lead to correlated events were ignored. Surprisingly, the observed correlation effects are quantitatively accounted for hydrodynamically, as is shown in Sec. 4.2. This in turn leads to the intriguing practical consequence that the relatively small systems studied by molecular dynamics for short times can be used as an alternative to continuum mechanics to explain hydrodynamic phenomena. The advantage of this approach is discussed in Sec. 5.

4.1 Autocorrelation Functions

There are two methods to calculate transport coefficients; both have analogies to real experiments. In one case, one studies the decay of spontaneous fluctuations by means of an autocorrelation function (Kubo, 1957). In the alternative method, one sets up a steady nonequilibrium state through an external force (Ashurst and Hoover, 1975); this is discussed below.

Many relaxation processes have been studied by molecular dynamics via correlation functions. For example, dielectric relaxation monitors the average decay in time of fluctuations in the total dipole moment of the system. Other examples involve the decay of orientational components of a molecular system to determine various spectroscopic properties. We concentrate here on the velocity autocorrelation function $\rho(t)$, by which the self-diffusion coefficient D can be determined (Rahman, 1964):

$$\rho(t) = \langle v(0)v(t)\rangle, \tag{4}$$

where $v(0)$ is the velocity of a particle at the origin of time and $v(t)$ is its velocity at some later time t. The diffusion coefficient is the integral of the autocorrelation function:

$$D = \int_0^\infty \rho(t)\, dt. \tag{5}$$

This expression is equivalent to the more familiar Einstein expression

$$D = \lim_{t \to \infty} \langle \Delta r^2(t) \rangle / 6t, \qquad (6)$$

where Δr is the displacement of a particle position in time t. The angular brackets in these expressions refer to averages over many initial equilibrium states. The shear viscosity can be obtained in like manner from an integral of the stress autocorrelation function (Levesque *et al.*, 1973).

4.2 Long-Time Tails

The decay of the velocity autocorrelation function of a fluid of hard disks shows a long positive persistence of velocity that implies that a particle remembers its initial velocity after it has undergone in excess of 100 collisions (Alder and Wainwright, 1969). This long-time tail can be quantitatively accounted for by a hydrodynamic model in which a solid disk is surrounded by a two-dimensional continuum fluid with a shear viscosity equal to that of the hard-disk system. The solid particle is given an initial velocity, which leads to a positive pressure wave in the fluid ahead of the disk and a corresponding rarefaction wave behind. The positive pressure in front and the negative pressure behind equalize by creating a double vortex flow around the particle (see Fig. 5). By comparing the velocity field around the disk calculated by molecular dynamics with the continuum hydrodynamic one under identical conditions, the quantitative validity of hydrodynamics at length scales of three molecular diameters (10 Å) and time scales of 10 mean collision times (10^{-13} s) was established. A straightforward dimensional analysis of the hydrodynamic model leads to the conclusion that the tail decays as $t^{-d/2}$, where d is the dimensionality of the system (see Fig. 6). This power-law decay is to be contrasted with the exponential decay predicted by Markov processes. It is also to be noted that in strictly two-dimensional systems ($d = 2$), the diffusion constant diverges; in fact, in two-dimensional systems none of the transport coefficients exist, requiring a reformulation of continuum hydrodynamics that has not as yet been carried out.

The theoretical prediction of divergent transport coefficients in two-dimensional systems has been confirmed by lattice-gas simulations of the stress autocorrelation function, which shows a long-time t^{-1} tail. The shear viscosity, which is proportional to the integral of the stress autocorrelation function, therefore diverges logarithmically. At liquid densities, near solidification, molecu-

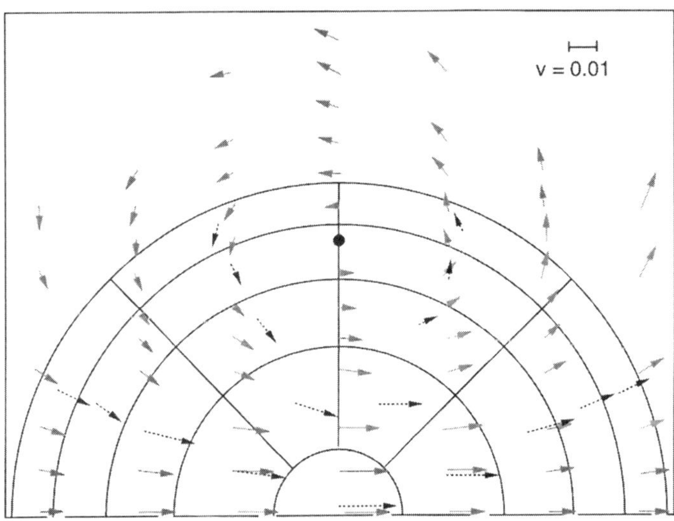

FIG. 5. The velocity correlation between a central particle and its neighboring particles after the central particle has undergone about ten collisions. (Because of symmetry, only the upper half of the diagram is shown; the vectors shown are the average of those of the upper and lower halves.) The size of the central particle is shown by the central half-circle and its velocity by the size of the arrow with a dotted tail near its center. The other such arrows are of a size to indicate the velocity average (as obtained by molecular dynamics) of whatever neighboring particles are found in that section of space. These arrows are to be compared with the light arrows, with line tails, which are obtained from a hydrodynamic solution of the velocity field on a rectangular grid for the identical physical situation. The agreement is striking, confirming the vortex motion.

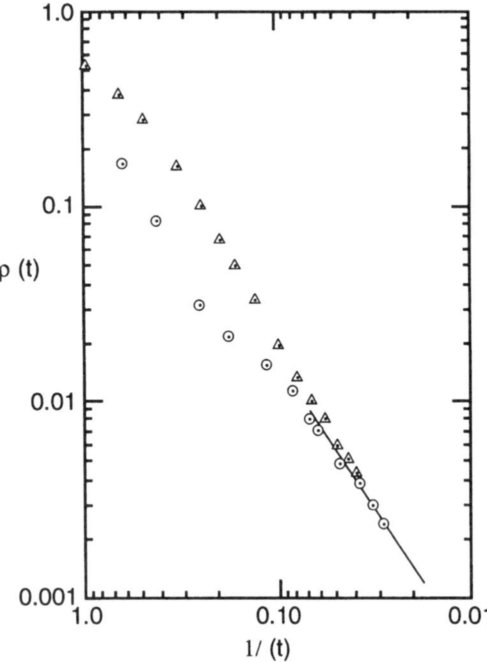

FIG. 6. Comparison of the velocity autocorrelation function $\rho(t)$ as a function of time (in terms of mean collision times) between the hydrodynamic model (circles) and a 500-hard-sphere molecular-dynamics calculation (triangles) at a volume relative to close packing of 3 on a log-log plot. The straight line is drawn with a slope corresponding to $t^{-3/2}$.

lar dynamics shows another type of slow decay for the stress autocorrelation function, characterized by a "stretched" exponential, $e^{-t^{\beta}}$ (β is a constant) (Palmer *et al.*, 1984), similar to experimental observations of viscous and dielectric relaxation. This tail, often called the "molasses" tail, appears to be connected with a slow structural adjustment of the dense fluid in the presence of a shear (Ladd and Alder, 1989). This process is characterized by a series of different exponential relaxations, the totality of which can be expressed as a stretched exponential. The detailed structural relaxation mechanisms in dense liquids have yet to be identified and very likely would lead to a dynamic theory of glass formation.

4.3 Generalized Hydrodynamics

Fluctuations in the hydrodynamic variables (mass, momentum, and energy) also depend on length scale (or wavelength), as well as on amplitude. We forego discussing dependence on amplitude, which leads to nonlinear transport coefficients, because that subject is not well understood. However, we briefly discuss the dependence of the linear transport coefficients on wavelength; this is the subject of generalized hydrodynamics (Alley and Alder, 1983). Introducing a molecular length scale into hydrodynamics permits a quantitative discussion of the corrections to macroscopic hydrodynamics as one approaches atomic dimensions. The only wavelength-dependent fluctuations of atomic dimensions accessible experimentally are density fluctuations, measured by neutron-diffraction studies (see NEUTRON DIFFRACTION). Molecular-dynamics calculation of the neutron-scattering function (Alley and Alder, 1983) have been made, and the experiments compare quantitatively to the prediction of generalized hydrodynamics (see Fig. 7).

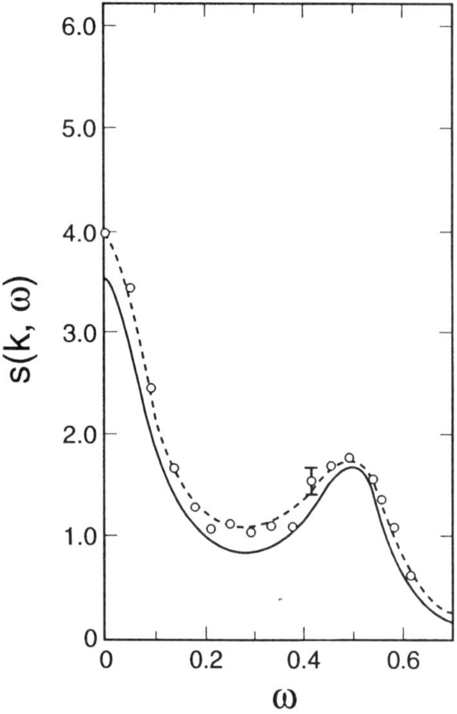

FIG. 7. The neutron-scattering function $S(\kappa, \omega)$ for hard spheres as a function of frequency ω, at a volume three times larger than close packing and a wavelength $(2\pi/k)$ 2.63 times the hard-sphere diameter. The molecular-dynamics results are represented by circles, and the hydrodynamic results, by the dashed curve. The solid curve represents the results of kinetic theory, valid only at lower densities.

As an illustration of generalized hydrodynamics, consider the problem of an infinitely massive sphere of atomic dimensions falling in a fluid, in order to find out what the correction to Stokes's law would be (Alley and Alder, 1980). By analysis of the wavelength dependence of the stress fluctuations, a wavelength-dependent viscosity $\eta(k)$ was deduced by molecular dynamics, with the result that

$$\eta(k)/\eta(0) = (1 + a^2k^2)^{-1}, \tag{7}$$

leading to a correction to Stokes's law of $1 - a/\sigma_{12}$, where σ_{12} is the mean diameter of the falling sphere and the solvent molecules. The crucial distance scale a that is introduced through the generalized viscosity is of the order of molecular size, that is, $a/\sigma_{12} \sim 0.3$ for a dense hard-sphere solvent. This Stokes prediction agrees well with a separate molecular-dynamics calculation of the Stokes friction coefficient and shows that even for atomic dimensions the correction to Stokes's law is only about 30% (see Fig. 8).

4.4 Microscopic Boundaries

The above Stokes problem requires that boundary conditions be imposed on the surface of the sphere to determine the hydrodynamic friction coefficient. Two common boundary conditions correspond to slip or stick; they determine how much of the tangential momentum of a colliding solvent molecule is randomized. From an atomistic view, slip boundary conditions correspond to elastic collisions, whereas stick corresponds to diffusive collisions, whereby the post-collision velocity direction is randomly distributed. The real situation lies somewhere between the above two extremes, depending on the roughness of the surface, and can be ascertained only by a molecular-dynamics simulation. Thus, by forming the falling sphere from smaller spheres representing the size and density of the atoms of the material of which it is constituted, molecular dynamics yields the correct boundary conditions when this aggregate object is suspended in the solvent spheres that collide elastically with the falling sphere (see Fig. 8). Such realistic boundary conditions generated microscopically are important when one studies, for example, fluid flow through microscopic pores (Koplic and Banavar, 1995).

4.5 Nonequilibrium Molecular Dynamics

The linear transport coefficients (diffusion, viscosity, and thermal conduction) can be calculated by monitoring the decay of

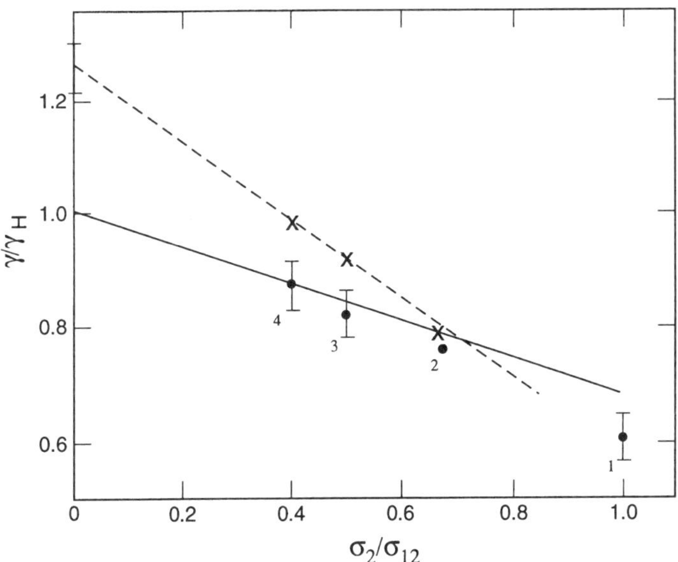

FIG. 8. The ratio of the molecular-dynamics determined friction coefficient γ to that of the hydrodynamic one, γ_H, at various ratios of the diameter of the solvent, σ_2, to that of the average diameter of the sphere and the solvent in which the sphere is immersed, σ_{12}. The four circles with their indicated error bars correspond to the molecular-dynamics results for spheres, respectively, 1, 2, 3, and 4 times the size of the solvent; the solid line represents the generalized hydrodynamic result, extrapolating to the hydrodynamic result for slip boundary conditions, in the large-ratio size limit ($\sigma_2/\sigma_{12} = 0$, $\gamma/\gamma_H = 1$). The crosses correspond to molecular-dynamics results for a big sphere having a rough surface, made up of little spheres of the same size as the solvent. The dashed line shows that in the hydrodynamic limit the results extrapolate to a value in between that for slip boundary conditions (1) and stick boundary conditions (1.5).

fluctuations in an equilibrium ensemble, as discussed above. However, it is also possible to drive a nonequilibrium flow of mass, momentum, or energy and measure transport coefficients directly (Hoover *et al.*, 1980). Moreover, nonequilibrium molecular dynamics (NEMD) is a unique probe of nonlinear transport phenomena, such as occur during the shearing of a polymer solution or the plastic flow of a crystalline solid (Hoover *et al.*, 1982). In such situations, rearrangements of the microstructure in the externally imposed flow field lead to nonlinear responses. Such flows are of considerable practical importance in understanding the rheological properties of complex fluids and solids. NEMD simulations fall into two categories: boundary-driven flows and modifications of the equations of motion.

4.5.1 Boundary-Driven Flows In such simulations flows of momentum or energy are set up by interactions between the particles and some external reservoir, for example, a collection of particles with a preassigned velocity distribution, or a container wall. Two such reservoirs act as a source and sink of momentum and energy and allow nonequilibrium flows to be set up in the system.

Boundary-driven flows are easy to visualize and are closely related to experimental methods. However, in systems of microscopic size, the reservoirs introduce substantial spatial inhomogeneities of heat in NEMD simulations, so that the thermodynamic state of the system is spatially varying. Hence, these methods are not suitable for accurate calculations of transport coefficients.

4.5.2 Modified Equations of Motion For an accurate calculation of transport coefficients, it is necessary to set up a homogeneous perturbation in the distribution function of particle coordinates and velocities, so that the system approximates a small volume of a macroscopic system at constant mass, momentum, and energy but with fluxes of mass, momentum, or energy flowing through it (Hoover *et al.*, 1980; Evans, 1982). To illustrate the principle, consider the simplest case of the diffusion coefficient D of a tagged particle in a system with periodic boundary conditions. The NEMD method consists of applying an external force F to the tagged particle [a compensating force $-F/(N-1)$ must be applied to the $N-1$ other particles] and measuring its mean velocity u. For small enough driving forces, the ratio u/F is equal to D/kT, the self-diffusion coefficient divided by the Boltzmann constant k times the temperature T. The other transport coefficients can also be measured from homogeneous modifications of the equations of motion. Furthermore, it is possible to establish a rigorous connection between the NEMD simulations and the autocorrelation formulas for the linear transport coefficients. NEMD is about as efficient numerically as calculating autocorrelation functions.

Because in NEMD simulations very large gradients are necessary to establish a significant signal-to-noise ratio, large amounts of energy are dissipated, and the systems consequently heat rapidly. Thus, methods of uniformly thermostatting the system had to be developed, and again, suitable modifications to the equations of motion allowed for simulation of a steady state (Nosé, 1984). It should be noted that viscous heating is a second-order effect (in the external driving force), so that the linear transport coefficients are not sensitive to the details of the thermostatting process.

4.5.3 Nearby Trajectories It has been suggested that the signal-to-noise problem in nonequilibrium simulations could be overcome by a "subtraction" technique (Ciccotti *et al.*, 1979). The idea is to start two trajectories from the same initial phase point: an equilibrium trajectory and a weakly perturbed nonequilibrium one. These trajectories might be expected to be highly correlated. If so, the nonequilibrium response could be determined with small statistical noise, by subtracting the fluctuating equilibrium one, which has zero mean, from the nonequilibrium response. This technique also allows the use of small perturbations, resulting in negligible viscous heating. Unfortunately, the chaotic nature of the underlying dynamics, and the resulting exponential divergence of neighboring trajectories, makes this scheme of limited practical value.

5. PARTICLE HYDRODYNAMICS

Encouraged by the hydrodynamic explanation of the long-time tails, one might think of doing other fluid-flow problems by molecular dynamics, instead of solving the continuum Navier–Stokes equations (see Fig. 9) (Meiburg, 1986). There are distinct advantages as well as disadvantages. For the Navier–Stokes equations there is no proof of the existence, convergence, or uniqueness of the numerical solutions, while molecular dynamics, being a particle method, is numerically stable, although computationally very expensive. Navier–Stokes ignores fluctuations while molecular dynamics includes them within the limitations of a finite system. Navier–Stokes is only strictly valid when gradients are small, even though it is often used to calculate shock-wave propagation, while molecular dynamics is valid even when the linear-gradient approximation breaks down. Finally, the continuum approximation of Navier–Stokes is not applicable when the mean free path is comparable to the characteristic length scale of the flow, such as in rarefied gas flows; in such cases, one needs to solve the Boltzmann equation, as discussed in Sec. 5.3.

5.1 Lattice Gas

Although hydrodynamic simulations have been carried out by molecular dynamics (Rapaport, 1988), they are computationally intensive and cannot be pursued for very long times. Thus an alternative, simpler, particle method is highly desirable, particularly since molecular dynamics often produces much more detailed information than is either necessary or digestible. This is the motivation behind lattice-gas models where the particles' positions are restricted to a lattice and the particles have a very limited set of velocities. Collisions between particles still must conserve mass, momentum, and energy. The only other restriction is that of detailed or semidetailed balance, which means that irreversibility is not introduced at the microscopic level or, in other words, into the equations of motion. Detailed balance is satisfied if the probability of a collision event is equal to that of the reverse one, while for semidetailed balance it is true on the average. It can be shown (Frisch *et al.*, 1987) that the conservation laws and semidetailed balance are sufficient to derive the equations of hydrodynamics. Some lattice-gas models violate semidetailed balance to reach a particular goal—for example, to lower the viscosity of the gas so as to achieve higher Reynolds-number flow; such models are nonphysical.

Lattice gases involve only integer algebra and are orders of magnitude faster than molecular dynamics, especially on special-purpose computers. The algorithms are numerically stable and well suited to massively parallel computers. However, because of the limited set of velocities, a number of problems arise that have limited the application of lattice gases. One such problem arises because, at a given lattice point, only distinguishable particles are allowed; that is, all particles must have different combinations of position, speed, and direction. As a consequence, an exclusion effect arises, which leads to a Fermi-like particle velocity distribution rather than the Boltzmann distribution. As a result, special measures must be taken to ensure that translational invariance is obeyed for high-velocity flows. Moreover,

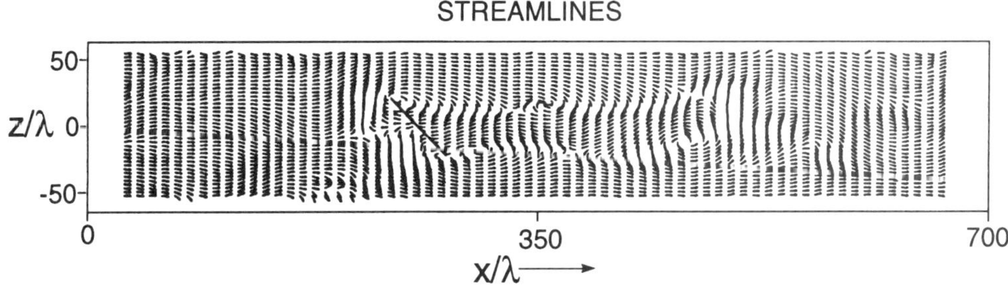

FIG. 9. The velocity of volume elements of a fluid streaming past a plate (shown by the solid line) set at an angle to the flow, in a container 700 mean free paths long, 100 mean free paths wide, and 1 mean free path deep. The molecular-dynamics simulation of some 100 000 spheres shows the alternate vortex-shedding phenomenon behind the plate leading to an observable von Kármán wake.

the lattice must be of sufficiently high symmetry that the viscous stresses are isotropic. Furthermore, lattice gases, being only point particles, are restricted to a perfect-gas equation of state. There is also a computational drawback that in practice long runs are needed to average out the naturally occurring fluctuations. Nevertheless, lattice gases have been used to advantage to gain more detailed knowledge of the behavior of the long-time tails (van der Hoef and Frenkel, 1991). They can also deal with flows in complex geometries, such as porous media and particulate suspensions (Ladd *et al.*, 1988).

5.2 Lattice Boltzmann

Since lattice gases are inherently dilute, one can also solve the corresponding Boltzmann equation and avoid the time-consuming average over the fluctuations (McNamara and Zanetti, 1988). The idea is to solve the Boltzmann equation on a lattice with a discrete set of velocities. In this case one calculates the probability densities of particles that move along lattice links rather than tracking the individual particles. Their interactions at lattice sites are subject to the usual conservation laws. Dealing with the continuous probability density variables avoids the "exclusion" problem of lattice gases as well as their translational invariance difficulties. Furthermore, it proved possible to reduce the viscosity to arbitrarily low values in the lattice Boltzmann scheme, whereas the constraints of lattice gases made that impossible for models consisting of particles with only a few different velocities (Higuera *et al.*, 1989). Thus, lattice versions of the Boltzmann equation can reach high Reynolds-number flows, but at the cost of losing the fast integer-algebra aspect of lattice gases and the absolute numerical stability, since this is no longer a particle method. The method is numerically equivalent to a discretized solution of the Navier–Stokes equations, with similar limits on its numerical stability, although one might have thought that the existence of an H theorem for the Boltzmann equation would have significantly improved stability. These calculations require roughly the same amount of computer time as Navier–Stokes solutions.

5.3 Direct-Simulation Monte Carlo

A particle method has also been developed to solve the Boltzmann equation (Bird, 1976). It has found wide use in rarefied-gas dynamics and has supplemented wind tunnel tests, for example, in airplane design. In this method the particle positions and velocities are described by continuous variables as in molecular dynamics. Collisions are executed stochastically by having particles assigned to cells of approximately a mean free path in size. Collisions among the particles within that cell are selected with the Boltzmann predicted probability, according to the relative velocity of randomly selected pairs and their collision cross section. After the collision, the pair remains in place but with stochastically determined post-collision velocities. The number of collisions is set by the collision rate and the selected time step. After that, all particles are advanced by about one mean free path, after which the collision process within a cell is repeated. In effect this algorithm ignores spatial correlations over a distance of one mean free path, consistent with the assumption of the Boltzmann approach. The procedure is about three orders of magnitude faster than a corresponding molecular-dynamics calculation. By having a full spectrum of velocities, the scheme has none of the limitations of lattice gases or lattice Boltzmann, and even the assumption of an ideal-gas equation of state can be removed by a small modification of this procedure (Alexander *et al.*, 1995). In practice it has not been possible to reach high Reynolds-number flows with this method.

6. MACROSCOPIC PARTICLE DYNAMICS

A wide variety of systems containing macromolecular particles can be studied by molecular dynamics and related methods. Examples of such systems include polymer solutions, suspensions, microemulsions, fluidized beds, grain silos, powder hoppers, and slurries. Particle sizes vary considerably, from less than 0.1 μm in fine-particle suspensions to granular particles of sizes up to a few centimeters. What distinguishes these calculations from molecular dynamics is that collisions between macromolecular particles are inelastic. Energy is absorbed, either by the background fluid or by internal degrees of freedom within the particle itself; in either case the energy dissipated has little effect on the subsequent dynamics, but the loss of kinetic energy leads to complex phenomena. Inelastic collisions can easily promote the formation of transient, but large-scale, clusters of particles. It is this clustering that is the primary cause of the complex dynamics. Three important examples of macroscopic particle dynamics are discussed here: dilute solutions of polymers, dilute to dense suspensions of hydrodynamically interacting spheres, and dense systems of granular particles.

6.1 Polymers and Brownian Dynamics

The simplest model of a polymer chain in dilute solution is the Zimm model, in which the polymer backbone is approximated by a chain of beads connected by harmonic springs. The beads interact with one another via a short-range excluded-volume potential. The model also takes into account the effects of the solvent by a mean-field or "preaveraged" approximation of the hydrodynamic forces acting between beads. The Zimm model successfully accounts for many of the experimentally observed scaling laws for dilute polymer solutions. More realistic models require numerical simulation.

Polymer solutions can be numerically modeled by "Brownian dynamics" in which the beads representing the monomer units of the polymer chain are represented by Brownian particles subject to random forces as well as viscous and intermolecular forces. Brownian dynamics is used to describe the behavior of macromolecules so large and massive that their position and velocity can be considered as frozen during collisional momentum transfer with solvent particles. In other words, the Brownian particle experiences many collisions with solvent particles before it changes its velocity significantly. The dynamics of a Brownian particle are stochastic, as described by the Langevin equation (Ermak and McCammon, 1978). For a single sphere

$$m\dot{u} = -\xi u + R(t) + F, \tag{8}$$

where $\xi = 6\pi\eta a$ is the friction coefficient describing the drag on a sphere of radius a in a solvent of viscosity η, R is the random force representing the effect of collisions between solvent molecules and the Brownian particle, and F is force due to interparticle interactions. The friction constant used is that appropriate for "sticky" collisions, since fluid in contact with a macroscopic solid-particle surface is found to move with the local velocity of the contact point. In a suspension containing many Brownian particles or in a single polymer, the presence of neighboring spheres introduces a many-body hydrodynamic interaction, so that the force on each sphere includes significant contributions from every other sphere (Ladd, 1988).

6.2 Hydrodynamic Interactions

Hydrodynamic interactions are the forces between macroscopic particles arising from their motion; they are transmitted by the fluid in which they are suspended (Happel and Brenner, 1986). If the macroscopic particles have sufficiently small velocities, the fluid flow can be described by low Reynolds-number or "creeping-flow" hydrodynamics. There is then a linear relationship between particle forces and particle velocities,

$$u_i = \sum_{j=i}^{N} \mu_{ij} F_j. \tag{9}$$

The mobility matrix μ_{ij} depends on the configuration of all the spheres in the system. An extension of Eq. (9) to include particle torques and particle rotations can readily be made. Given the complete set of mobilities for translation and rotation, it is then possible to solve for the particle dynamics. However, the mobilities are difficult to calculate in general, so that in many instances hydrodynamic interactions are neglected altogether: this leads to the isolated-sphere results represented by Eq. (8). These are strictly valid only at infinite dilution. Dilute suspensions can be treated fairly simply by using the mobility tensor for an isolated pair of spheres, but even then analytic expressions are only valid at large separations of the pair. When the pair of particles is near contact, the character of the interaction changes; the critical length is then the gap between the neighboring surfaces rather than the interparticle separation. This region is described by lubrication theory.

For denser suspensions it is not possible to enumerate all the many-body hydrodynamic interactions analytically, but direct numerical calculations can determine the mobility matrix for arbitrary configurations of spheres (Brady and Bossis, 1988). This has made it possible to compute the hydrodynamic properties (permeability, sedimentation velocity, self-diffusion coefficient, and viscosity) of random dispersions of rigid spheres, at any concentration, with a high degree of accuracy and reliability (Ladd, 1990). Much less work has been done on the computationally more demanding task of determining the changes in suspension microstructure under imposed external flows. Semiquantitative results have been obtained for small numbers of spheres in Couette flow and channel flow and for bulk sedimentation. Because the above simulations are so time-consuming, it is natural to try the lattice-gas/lattice-Boltzmann models discussed earlier. The simplicity and robustness of the algorithms has made it possible to handle the complex boundary conditions that occur in particle suspensions with relative ease (Ladd, 1994).

6.3 Granular Particle Dynamics

Despite extensive empirical experience, the dynamics of granular flow is poorly understood. There is both a scientific challenge and a technological need to develop an understanding of the mechanisms underlying the flow of grains. In a typical granular flow, such as the emptying of a hopper, there are regions of kinetically excited flows, where the grains have a random "thermal" component to their velocities, and stagnant regions, where the grains are quiescent and in constant contact with one another. Numerical simulation requires a formulation of grain–grain and grain–wall interactions in order to predict these macroscopic flow properties. The interactions between grains or between a grain and a container wall depend on the specific contact mechanics of the two surfaces. The important qualitative feature of granular particle dynamics is that, at each collision between grains, some of the kinetic energy of the colliding pair is lost. This energy is absorbed as heat, by sound waves inside the grains, and by irreversible deformation (plastic flow). Such systems eventually fall into a stagnant rest state with no particle motion, unless energy is supplied from an outside source, for instance, a shearing flow or a gravity flow. More realistic models of grain mechanics allow for distortion of the grains, for rolling and sliding friction between the contact points, and for adhesive surface forces (Walton and Braun, 1986) (see Fig. 10).

7. AB INITIO MOLECULAR DYNAMICS

It is apparent that the fundamental limitation of molecular dynamics is the parametrization of the force field in which the atoms move. In condensed phases many-body forces are not negligible, even for the simplest materials, and must be absorbed into an effective few-body potential. There are two drawbacks to this procedure: first, an effective potential depends on the thermodynamic state, so that different potentials are necessary for dilute and condensed phases; second, this effective potential must be determined by extensive fitting to experimental

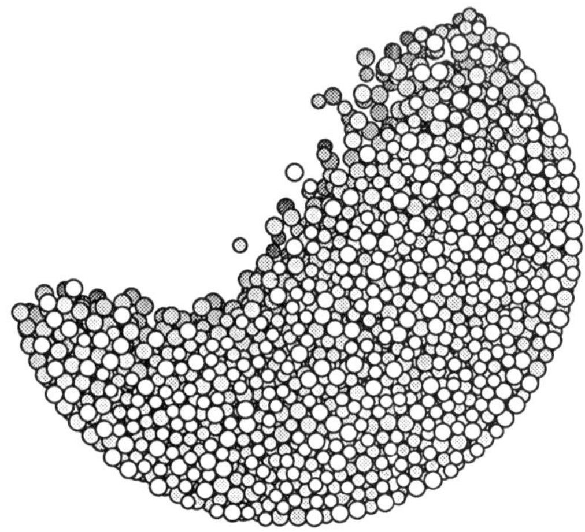

FIG. 10. Simulation of inelastic, frictional spheres in a rapidly rotating horizontal drum (not shown). Coefficient of friction, 0.2; coefficient of restitution for two-body normal impacts, 0.85. The rotation rate is such that at approximately six particle diameters from the outer wall the centrifugal acceleration is equal to the gravitational acceleration. The slice is approximately 5 diameters thick with periodic boundaries along the horizontal axis.

data (Vitek and Srolovitz, 1989). This is a cumbersome, ill-defined, and imprecise prescription. Instead, it would be desirable to introduce an accurate quantum mechanical calculation of the forces, particularly for metals, where pairwise additivity is even qualitatively inadequate. This requires that, for a given instantaneous configuration of the nuclei, the Schrödinger equation be solved to determine the spatial distribution of the electron density. Once the electron density is known, the interatomic forces can be calculated and the (classical) equations of motion of the nuclei solved for a small time step. This approach makes it possible to determine the properties of materials *ab initio,* requiring only the atomic number of the nuclei of the various elements making up the substance.

7.1 The Local-Density Approximation (LDA)

Despite decades of effort, neither the traditional variational methods of molecular quantum mechanics nor the more recent quantum Monte Carlo techniques (Ceperley and Kalos, 1986) have succeeded in developing a practical scheme for solving the Schrödinger equation with sufficient accuracy, for more than a few electrons. For systems with many electrons, the local-density approximation is the most widely used method in *ab initio* molecular dynamics (Car and Parrinello, 1985). It is chosen for its conceptual simplicity and predictive accuracy. Although density-functional theory was originally developed for metals, it can be used to describe covalent bonding as well.

Density-functional theory rests on the principle that the exact ground-state energy is a minimum of the total energy with respect to all possible variations in the electron density. However, the scheme is only computationally viable in the independent-electron or mean-field approximation. In practice this means that the total charge density is expressed in terms of noninteracting wave functions (orbitals). Missing from this description are the effects of electron exchange and electron–electron correlations. However, this exchange-correlation energy has been accurately calculated for a uniform electron gas by quantum Monte Carlo (Ceperley and Alder, 1980). In the LDA approach it is assumed that the exchange-correlation energy at each space point in the inhomogeneous electron system around a nucleus can be replaced by the exchange correlation of the uniform electron gas at the appropriate local electron density. Further corrections caused by the exchange and correlation energy due to the local density gradient are being attempted (Perdew and Wang, 1992). With this approach, systems of 100–1000 atoms can be accurately simulated (Brommer *et al.*, 1992).

7.2 Car–Parrinello Method

The Car–Parrinello method allows explicit treatment of the electronic degrees of freedom. This is accomplished by treating the electronic degrees of freedom as classical particles by incorporating their motion into the Lagrangian. The orbitals of the electronic wave function are assigned a fictitious mass, which serves to make the time scale of the variations in electron density slower than in reality, but still fast compared with nuclear motion (Payne *et al.*, 1992). The wave function thus follows the Born–Oppenheimer (adiabatic) surface staying near the true electronic ground state at all times. Nevertheless, a rather small time step ($\sim 10^{-15}$ s) is still required. More recently the alternative idea of fully relaxing the electronic degrees of freedom to the ground state at every time step has been implemented. The extra cycles necessary to achieve that are compensated for by a less stringent time-step requirement.

The other important ingredient of the Car–Parrinello method is the expansion of the orbitals in terms of a plane-wave set, which requires the use of accurate pseudopotentials to represent the nuclei with the tightly bound core electrons. Otherwise, many basis functions are required to represent the rapid variation of the wave function near the ion core. These core states usually play only a minor role in determining the interatomic forces, and thus most simulations deal only with valence electrons. These interact with one another and with the atomic pseudopotential, which represents the effective interaction due to the nucleus and core electrons. Most importantly, the Car–Parrinello method provides, for the first time, a practical way of coupling rearrangements of the atomic structure and the electronic structure. It is now possible to follow changes in electronic structure with changes in the thermodynamic state as, for example, when a substance melts (Sugino and Car, 1995). Recent applications include properties of amorphous solids, the motion of defects in crystals, and surface reconstruction. Materials studied include covalently bonded structures, semiconductors, and metals.

8. PROSPECTS

Applications of molecular dynamics to condensed systems have mushroomed from its inception, when simple systems were studied, to the investigation of increasingly more complex systems. Simple systems allowed theoretical concepts to be tested quantitatively and, in analogous fashion to real experiments, led either to verification or to modification of theories. Molecular dynamics and closely related methods are still being applied to relatively idealized problems, but in an effort to reach larger distance and time scales. Examples of such problems are dense polymeric systems, glasses, and particle suspensions. Although computer power is increasing rapidly, such calculations usually require algorithmic innovations—for instance, the use of lattice-gas and lattice-Boltzmann models to describe the fluid in two-phase solid–fluid flows. In some cases it may by that there is no substitute for brute-force computing; here parallel computation is often essential.

Another area where large distance and time scales are necessary is in hydrodynamic flow problems, such as rarefied-gas flows, which cannot be described by the Navier–Stokes equations. Here, another variant of molecular dynamics, the Direct Simulation Monte Carlo (DSMC) or Bird method, is useful. It can simulate a wide range of gas flows, on much larger length and time scales than is possible with conventional molecular dynamics. The range of applicability of DSMC and related methods can be enhanced by embedding them in continuum finite-difference or finite-element codes. Thus a Direct Simulation Monte Carlo algorithm (or perhaps even molecular dynamics) could be used to simulate "critical" regions of the problem, such as narrow gaps, less than or comparable to the collisional mean free path, where Navier–Stokes breaks down. The DSMC code can be directly coupled to a continuum code that would be used to simulate the large-scale motions, at enormous computational savings compared with a fully molecular simulation. Such "adaptive" algorithms, where the computational method varies with the local physics, could be very important in simulating a much wider class of fluid flows—for example, flows with very steep gradients or where realistic representations of boundaries are required.

Realistic simulations of complex molecules are even more difficult. In addition to the problems associated with the multiple time scales, there is very little reliable information about the intermolecular forces in real materials. Even if intermolecular forces can be calculated, there is the further difficulty of devising a suitable functional form for the potential; this is especially problematic in systems with large many-body forces. As a result, it seems clear that *ab initio* methods, which calculate the intermolecular interactions *in situ*, will become increasingly important. However, although density-functional methods have been used to calculate some dynamical properties, like diffusion coefficients in liquid metals, much more needs to be done to include electron correlations accurately, and beyond that, to allow electron transport properties, like thermal and electrical conduction, to be determined.

GLOSSARY

Adiabatic: A process where the system is in equilibrium at each stage; thus a reversible process with no change of entropy.

Boltzmann Equation: A kinetic theory of transport that accounts for collisions by the molecular chaos (or Markov) approximation, assumed to be valid for low-density gases.

Cellular Automata: A computational scheme in which neighboring cells of a regular lattice exchange only binary bits of information.

Couette Flow: A flow between a stationary plate and a slowly moving one.

Ensemble: A collection of macroscopically equivalent equilibrium states in which a given set of thermodynamic variables are held fixed while the others are allowed to fluctuate.

Ergodicity: A condition in which time averages are equivalent to ensemble averages.

Markov Process: A process where the probability of an event is independent of preceding events.

Mean Field: An approximation where the effect of the neighbors on a given particle depends only on the average (mean) force they exert.

Mean Free Path: The average distance a particle travels between successive collisions.

Monte Carlo Method: A stochastic method to evaluate ensemble averages.

Navier–Stokes Equation: The equation of macroscopic hydrodynamics, consisting of the conservation laws and the linear dissipation assumption.

Rheology: The study of nonequilibrium properties of highly viscous media where the transport is not linear in the applied gradients.

Reynolds Number: The dimensionless number describing the flow rate, uL/v, where u is the velocity, L is the characteristic dimension of the flow, and v is the kinematic viscosity.

Stokes Drag: The frictional or drag force on a macroscopically large sphere falling freely in a fluid.

Works Cited

Alder, B. J., Wainwright, T. E. (1958), in: I. Prigogine (Ed.), *Transport Processes in Statistical Mechanics,* New York: Interscience, pp. 97–131.

Alder, B. J., Wainwright, T. E. (1959a), *J. Chem. Phys.* **31,** 459–466.

Alder, B. J., Wainwright, T. E. (1959b), *Sci. Am.* **201** (4), 113–123.

Alder, B. J., Wainwright, T. E. (1962), *Phys. Rev.* **127,** 359–361.

Alder, B. J., Wainwright, T. E. (1963), in: *Conference on the Many-Body Problem,* Interscience Publishers, New York, p. 97.

Alder, B. J., Wainwright, T. E. (1969), *J. Phys. Soc. Jpn.* **26,** 267–269.

Alder, B. J., Wainwright, T. E. (1970), *Phys. Rev. A* **1,** 18–21.

Alexander, F. J., Garcia, A. L., Alder, B. J. (1995), *Phys. Rev. Lett.* **74,** 5212–5215.

Alley, W. E., Alder, B. J. (1980), in: M. Balaban (Ed.), *Molecular Structure and Dynamics,* Philadelphia: International Science Services.

Alley, W. E., Alder, B. J. (1983), *Phys. Rev. A* **27,** 3158–3173.

Alley, W. E., Alder, B. J. (1984), *Phys. Today* **37** (1), 56–63.

Andersen, H. C. (1980), *J. Chem. Phys.* **72,** 2384–2393.

Ashurst, W. T., Hoover, W. G. (1975), *Phys. Rev. A* **11,** 658–678.

Barker, J. A. (1986), *Mol. Phys.* **57,** 755–760.

Barker, J. A., Henderson, D., Abraham, F. F. (1981), *Physica* **106A,** 226–238.

Bennet, C., Alder, B. J. (1971), *J. Phys. Chem. Solids* **32,** 2111–2122.

Berendsen, H. J. C. (1986), "Biological Molecules and Membranes," in: G. Ciccotti and W. G. Hoover (Ed.), *Molecular Dynamics Simulation of Statistical Mechanical Systems,* Proceedings of the Enrico Fermi International School of Physics Summer School, Varenna, Bologna: Società Italiana de Fisica, pp. 496–519.

Berendsen, H. J. C., Van Gunsteren, W. F. (1986), in: G. Ciccotti and W. G. Hoover (Ed.), *Molecular Dynamics Simulation of Statistical Mechanical Systems,* Proceedings of the Enrico Fermi Summer School, Varenna, Bologna: Società Italiana de Fisica, pp. 43–65.

Bird, G. A. (1976), *Molecular Gas Dynamics,* London: Oxford University Press.

Bishop, T., Schulten, K. (1994), in: G. Wipff (Ed.), *Computational Approaches in Supramolecular Chemistry,* Boston: Kluwer Academic Publishers.

Brackbill, J. U., Cohen, B. I. (Eds.) (1985), *Multiple Time Scales,* New York: Academic.

Brady, J. F., Bossis, G. (1988), *Annu. Rev. Fluid Mech.* **20,** 111–155.

Brommer, K. D., Needels, M., Larson, B. E., Joannopoulos, J. D. (1992), *Phys. Rev. Lett.* **68,** 1355–1358.

Brooks, C. L., III, Karplus, M., Pettitt, B. M. (1988), *Proteins: A theoretical perspective of dynamics, structure, and thermodynamics,* Advances in Chemical Physics No. 71, New York: Wiley.

Brush, S. G., Sahlin, H. L., Teller, E. (1966), *J. Chem. Phys.* **45,** 2102–2118.

Car, R., Parrinello, M. (1985), *Phys. Rev. Lett.* **55,** 2471–2474.

Ceperley, D., Alder, B. J. (1980), *Phys. Rev. Lett.* **45,** 566–569.

Ceperley, D. M., Kalos, M. H. (1986), in: K. Binder (Ed.), *Monte Carlo Methods in Statistical Physics,* 2nd ed., Topics in Current Physics, Vol. 7, Springer: Berlin, pp. 145–194.

Ciccotti, G., Jacucci, G., McDonald, I. R. (1979), *J. Stat. Phys.* **21,** 1–22.

Ciccotti, G., Ferrario, M., Ryckaert, J.-P. (1982), *Mol. Phys.* **47,** 1253–1264.

Connick, R. E., Alder, B. J. (1983), *J. Phys. Chem.* **87,** 2764–2771.

de Leeuw, S. W., Perram, J. W., Smith, E. R. (1980), *Proc. R. Soc. London, Ser. A* **373,** 57–66.

Dorfman, J. R., Cohen, E. G. D. (1975), *Phys. Rev. A* **12,** 292–316.

Dymond, J., Alder, B. J. (1966), *J. Chem. Phys.* **45,** 2061–2068.

Ermak, D. L., McCammon, J. A. (1978), *J. Chem. Phys.* **69,** 1352–1360.

Evans, D. J. (1982), *Phys. Lett.* **91A,** 457–460.

Fermi, E., Pasta, J. G., Ulam, S. M. (1965), in: E. Amaldi, H. L. Anderson, E. Persico, F. Rasetti, C. S. Smith, A. Wattenburg, and E. Sergré (Eds.), *Collected Works of Enrico Fermi,* Vol. II, Chicago: University of Chicago Press.

Frenkel, D., Ladd, A. J. C. (1984), *J. Chem. Phys.* **81,** 3188–3193.

Frisch, U., Hasslacher, B., Pomeau, Y. (1986), *Phys. Rev. Lett.* **56,** 1505–1508.

Frisch, U., d'Humières, D., Hasslacher, B., Lallemand, P., Pomeau, Y., Rivet, J.-P. (1987), *Complex Syst.* **1,** 649–707.

Gehlen, J. N., Mardin, M., Chandler, D. (1994), *Science* **263,** 499–502.

Greengard, L. (1994), *Science* **265,** 909–914.

Happel, J., Brenner, H. (1986), *Low-Reynolds Number Hydrodynamics,* Dordrecht: Martinus Nijhoff.

Higuera, F., Succi, S., Benzi, R. (1989), *Europhys. Lett.* **9,** 345–349.

Hoover, W. G., Evans, D. J., Hickman, R. B., Ladd, A. J. C., Ashurst, W. T., Moran, B. (1980), *Phys. Rev. A* **22,** 1690–1697.

Hoover, W. G., Ladd, A. J. C., Moran, B. (1982), *Phys. Rev. Lett.* **48,** 1818–1820.

Jacucci, G., Rahman, A. (1984), *Nuovo Cimento* **D4,** 341–356.

Karplus, M. (1986), in: E. Clementi, S. Chin (Eds.), *Structure and Dynamics of Nucleic Acids, Proteins, and Membranes,* New York: Plenum, pp. 113–126.

Koplik, J., Banavar, J. R. (1995), *Annu. Rev. Fluid Mech.* **27,** 257–292.

Kubo, R. (1957), *J. Phys. Soc. Jpn.* **12,** 570–586.

Ladd, A. J. C. (1988), *J. Chem. Phys.* **88,** 5051–5063.

Ladd, A. J. C. (1990), *J. Chem. Phys.* **93,** 3484–3494.

Ladd, A. J. C. (1994), *J. Fluid Mech.* **271,** 311–339.

Ladd, A. J. C., Alder, B. J. (1989), *J. Stat. Phys.* **57,** 473–482.

Ladd, A. J. C., Colvin, M. E., Frenkel, D. (1988), *Phys. Rev. Lett.* **60,** 975–978.

Lebowitz, J. L. (1993), *Physica A* **194,** 1–27.

Levesque, D., Verlet, L. (1993), *J. Stat. Phys.* **72,** 519–537.

Levesque, D., Verlet, L., Kürkijarvi, J. (1973), *Phys. Rev. A* **7,** 1690–1700.

Lifson, S., Hagler, A. T., Dauber, P. (1979), *J. Am. Chem. Soc.* **101,** 5111–5121.

Madelung, E. (1918), *Z. Phys.* **19,** 524–532.

Maitland, G. C., Rigby, M., Smith, E. B., Wakeham, W. A. (1981), *Intermolecular Forces: Their Origin and Determination,* Oxford: Claredon Press.

McCammon, J. A., Gelin, B. R., Karplus, M. (1977), *Nature* **267,** 585–590.

McNamara, G. R., Zanetti, G. (1988), *Phys. Rev. Lett.* **61,** 2332–2335.

Meiburg, E. (1986), *Phys. Fluids* **29,** 3107–3113.

Metropolis, N., Rosenbluth, A. W., Rosenbluth, M. N., Teller, A. H., Teller, E. (1953), *J. Chem. Phys.* **21,** 1087–1092.

Nosé, S. (1984), *J. Chem. Phys.* **81,** 511–519.

Nosé, S., Klein, M. L. (1983), *Mol. Phys.* **50,** 1055–1076.

Palmer, R. G., Stein, D. L., Abrahams, E., Anderson, P. W. (1984), *Phys. Rev. Lett.* **53,** 958–961.

Paul, G. L., Pusey, P. N. (1981), *J. Phys. A* **14,** 3301–3327.

Payne, M. C., Teter, M. P., Allan, D. C., Arias, T. A., Joannopoulos, J. D. (1992), *Rev. Mod. Phys.* **64,** 1045–1097.

Perdew, J. P., Wang, Y. (1992), *Phys. Rev. B* **46,** 12947–12954.

Plimpton, S. (1995), *J. Comp. Phys.* **117,** 1–19.

Rahman, A. (1964), *Phys. Rev. A* **136,** 405–411.

Rahman, A., Stillinger, F. H. (1971), *J. Chem. Phys.* **55,** 3336–3359.

Rao, M., Berne, B. J. (1979), *J. Chem. Phys.* **71,** 129–132.

Rapaport, D. C. (1988), *Phys. Rev. Lett.* **60,** 2480–2483.

Shakhnovich, E. I. (1994), *Phys. Rev. Lett.* **72,** 3907–3910.

Siepmann, J. I., Frenkel, D. (1992), *Mol. Phys.* **75,** 59–70.

Sinai, Ya. G. (1973), in: E. G. D. Cohen, W. Thirring (Eds.), *The Boltzmann Equation,* Vienna: Springer-Verlag.

Smit, B., Hilbers, P. A. J., Esselink, K., Rupert, L. A. M., van Os, N. M., Schlijper, A. G. (1990), *Nature* **348,** 624–625.

Sugino, O., Car, R. (1995), *Phys. Rev. Lett.* **74,** 1823–1826.

Swendsen, R. H., Wang, J. S., Ferrenberg, A. M. (1992), in: K. Binder (Ed.), *The Monte Carlo Method in Condensed Matter Physics,* Berlin: Springer.

Tuck, J. L., Menzel, M. T. (1972), *Adv. Math.* **9,** 399–407.

Uhlenbeck, G. E., Hemmer, P. C., Kac, M. (1963), *J. Math. Phys.* **4,** 229–247.

Vitek, V., Srolovitz, D. J. (Eds.) (1989), *Atomistic Simulations of Materials: Beyond Pair Potentials,* New York: Plenum Press.

van der Hoef, M. A., Frenkel, D. (1991), *Phys. Rev. Lett.* **66,** 1591–1594.

Verlet, L. (1967), *Phys. Rev.* **159,** 98–103.

Walton, O. R., Braun, R. L. (1986), *J. Rheol.* **30,** 949–980.

Wood, W. W. (1970), *J. Chem. Phys.* **52,** 729–741.

Further Reading

Allen, M. P., Tildesley, D. J. (1987), *Computer Simulation of Liquids,* Oxford: Clarendon. A more advanced textbook devoted to the foundation and application of molecular dynamics to liquids.

Brooks, C. L., III, Karplus, M., Pettit, B. M. (1988), *Proteins: A Theoretical Perspective of Dynamics, Structure and Thermodynamics,* Advances in Chemical Physics No. 71, New York: Wiley. A more advanced textbook devoted to proteins.

Cicotti, G., Hoover, W. G. (Eds.) (1986), *Molecular Dynamics Simulation of Statistical Mechanical Systems,* Proceedings of the International School of Physics "Enrico Fermi," Course 97, Amsterdam: North-Holland. Advanced topics in molecular dynamics by some of the principal contributors to the field.

Cicotti, G., Frankel, D., McDonald, I. R. (Eds.) (1987), *Simulation of Liquids and Solids,* Amsterdam: Elsevier. A good collection of pioneering papers in the field.

Gould, H., Tobochnik, J. (1988), *An Introduction to Computer Simulation Methods/Application to Physical Systems,* Pts. 1 and 2, Reading, MA: Addison-Wesley. A general book intended to be used in a course.

Hansen, J. P., McDonald, I. R. (1990), *Theory of Simple Liquids,* 2nd ed., London: Academic. Another more advanced textbook devoted to the foundation and application of molecular dynamics to liquids.

Heermann, D. W. (1990), *Computer Simulation Methods in Theoretical Physics,* 2nd ed., Berlin: Springer. An elementary textbook introduction to molecular dynamics.

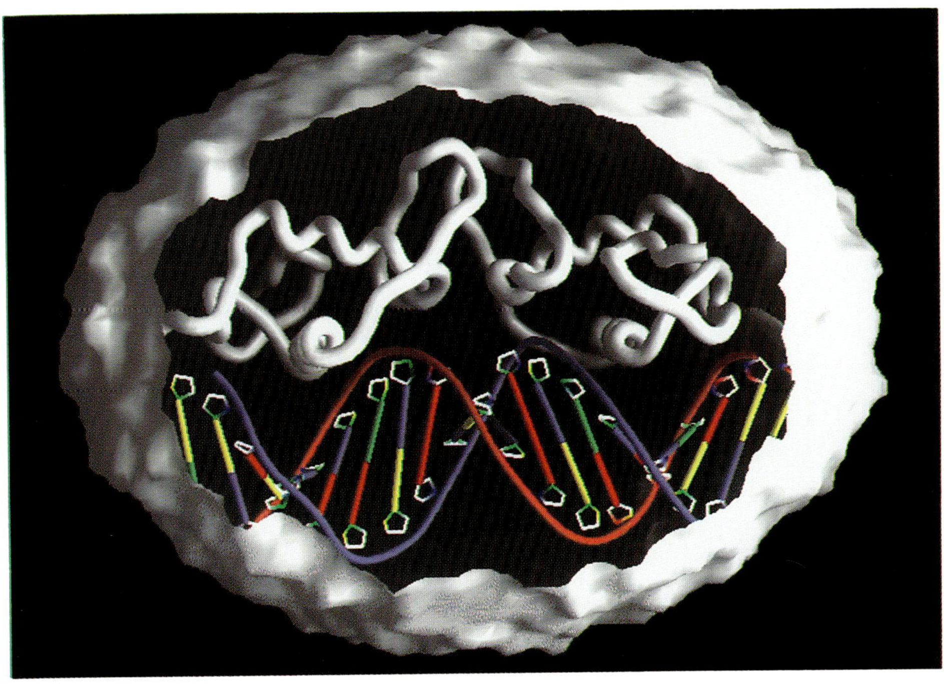

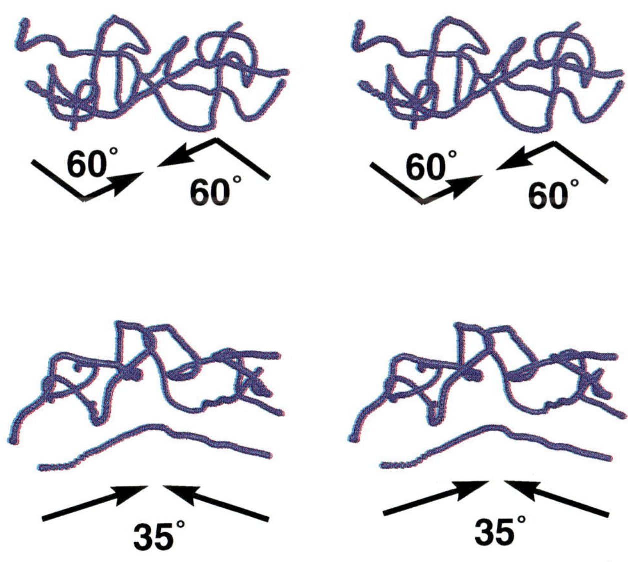

Plate 1. (a) Cutaway view of the solvent (water) surface of the protein–DNA complex. The white "worm" represents two identical proteins (glucocorticoid receptor) facing each other, which bind to a specific sequence of the DNA (color) structure. (b) Line representations, in cross-eyed stereo, of the DNA–protein complex after 70 ps of molecular-dynamics simulation. The system is viewed with the DNA located behind the protein (top) and with the DNA below the protein (bottom). The DNA and the two monomers are represented schematically through a single line, each of which traces the axis of the DNA and the backbone of the protein, respectively. The views illustrate how the DNA is bent through the interaction with the protein dimer; the solid black lines indicate the deviation of the DNA strand from its ideal linear geometry.

SOFTWARE ENGINEERING

HERBERT J. BERNSTEIN, *Bernstein+Sons, Bellport, New York, U.S.A.*

INTRODUCTION

Software engineering is the disciplined use of well-understood techniques to analyze, design, build, repair, or use software. The main focus of software engineering is on building new software systems by analyzing system requirements and then by designing and developing software to satisfy those requirements. As is true of all the engineering professions, software engineering provides the interface between technology and its uses within society.

Much of this article will be devoted to organizational issues and human factors. This reflects the state of the software engineering profession at the time of writing this article. Only a few years before, there had been a widespread belief that automata would rapidly be able to replace people in most of the detailed work of software construction, and the focus of software engineering had been on suites of automated tools that would write software without human intervention. Those approaches failed to provide reproducible results, and most of the effort in software engineering has now turned back toward understanding how people work effectively.

As a young field, software engineering is

3-527-28140-1/96/$5.00 + .50

still defining itself, still learning how to build the tools it needs, learning how to build reliable software. In design and development, the goal of software engineering is to produce software that performs as intended, performs reliably, is delivered on schedule, and is delivered at predictable and reasonable cost. There have been serious problems in achieving these goals. There is an ongoing effort to improve the approaches we now use, but the basic concepts of top-down analysis, bottom-up synthesis, object-oriented representation, and the use of software to process software are, and are likely to remain, the core of software engineering. Approaches to analysis, repair, and operation of existing software systems are strongly influenced by the major approaches used in design and development.

In this article we concentrate on the problems and solutions of interest to physicists. We look at the engineering approaches that are important for producing large scientific production codes and for producing reliable data-acquisition and control systems for experiments. This causes us to look at a much wider range of issues than might be the case in discussing software engineering in the context of business applications. We give less emphasis to any single paradigm, such as object orientation, than to basic engineering approaches.

In this context, the line between engineering approaches needed for software and those needed for hardware blurs. It is, however, important to understand what software is and when it is appropriate. Computers, communications media, detector arrays, and even people are the hardware, the tangible components of information-processing systems. Software is the intangible part, the sets of instructions and lists of data that determine what the tangible components will do. Software is the means by which we organize and adapt appropriate hardware to specific tasks.

Especially in data-acquisition and control systems, it is common to be faced with multiple viable alternative solutions using complex combinations of hardware and software, some relying primarily on hardware, some relying primarily on software, and most somewhere in the middle. How do we choose among such alternatives?

Software is intangible. Hardware is tangible. If we process information with an assembly of physical components, we are using a tangible system for information processing. We can touch the parts, examine them, and determine what the system will do with information. In a laboratory, we might take a signal from a detector, amplify it by a set gain, send it through a low-pass filter, and derive new information from the signal. We use well-understood techniques to select the cables, match the amplifier to the signal, and choose component values for the filter. We apply those techniques with sufficient care and sufficient discipline to have a reasonable assurance that what we build will perform in a predictable manner. Assembling electronic components to process data is an example of electrical engineering.

Now suppose we need a more complex filtration of the signal. Perhaps we need a Fourier transform to help detect certain interesting frequency components. We could feed the signal to a computer via an analog-to-digital converter and write a program to transform the signal. When it is running, the program does not have an obvious tangible expression in the computer. Software is information (a concept with powerful implications, since we are working with information-processing systems). Software changes the computer, making it behave as if it were a different machine than we would deduce from simple examination of the raw hardware. By loading the program into a computer, we have made a virtual machine. Instead of writing a program, we could have built hardware to perform the same information-processing tasks. Though either approach could have been used to solve the problem, using software is often more cost effective. Even though planning the logical sequence of actions to be taken may require the same amount of work for both the real machine and the virtual machine, loading instructions and data into a computer is faster and simpler than wiring a circuit, and it is much easier to modify a program than it is to rewire a circuit. Therefore we choose to build the virtual machine. There is a much better chance that the virtual machine will function reliably if we follow engineering approaches, and discipline ourselves to use well-understood techniques to design the code, instead of compounding the experiment we are performing on the signal with an experiment in programming. For example, we might use a well-tested Fourier-transform code and run it under a proven operating system. We might take care to plan and document the program in a clearly organized and structured manner before writing code. Such disciplined use of well-understood programming techniques is an example of software engineering.

Not all problems can be solved this way. The first time that problems in software design or analysis are faced there may be no well-understood techniques to apply, or the problem may prove to us that techniques we thought we understood we did not understand at all. Then artistry, creativity, and the courage to try and to fail may be needed to chart new territory. Many mistakes may be made, but, usually, with time and patience, the parameters of the problem and the tools needed to solve the problem become well understood. Creating software for similar problems becomes less of a creative exercise, becomes more routine, more predictable, more mechanized, becomes engineering rather than artistry. When we are in the experimental, creative, artistic mode of programming, we may be doing interesting research of great value to future efforts, a necessary prerequisite to future software engineering, but we are not doing software engineering. When we can approach a software problem and know very early in the process how to solve the problem, how long it will take to solve the problem, and how reliable our solution is likely to be, then we are doing software engineering.

In making any estimates of how much effort it may take to solve a particular problem, it helps to step back from the problem itself and first answer some questions. In general, we need to identify precisely what is to be produced. Are we producing a product, a process, or a document? Who will use what we produce? What are the constraints on how to produce it and, once produced, on how it is to be used? Are there standards and codes we should follow? Who is the customer, and when is delivery to be made? Who pays (an often ignored question, but clearly those who pay must be satisfied, or the project has failed), and what are the financial constraints on the project? How do we tell if we have a success, and how do we fix things if there are problems? What happens when we are done?

Such questions sound fussy and mundane, much as if saying we had to go to an architect and make fancy blueprints every time we wished to make a small change in our house. Clearly there are times when just going ahead and remodeling a house by the seat of our pants will work and is the right thing to do, but there are many more times when it might be a disastrous mistake to do so. If it matters that the software we produce work, it is best to produce it in an organized, well-documented manner, to define clearly what we are trying to do and for whom we are trying to do it. Asking such questions encourages us to impose structure on the problem, to reduce the context we have to deal with at any given time to manageable scope, and, most importantly, to keep things simple.

If we plan first and act second, we have a chance to select the right tools to do the job; to select appropriate hardware platforms, appropriate paradigms, languages, and libraries; to consider the human factors involved so that we can match the skills of the developers to the tasks to be done and match the product produced to the needs of its users; to conform to the ever-growing maze of rules and regulations that surround what we do; and to produce a quality product and have some assurance that it actually works.

Among the most powerful tools that have been developed for software engineering are structured programming, which focuses on the top-down analysis—the hierarchical decomposition of complex actions into simpler actions with more limited context—and object-oriented representation, which focuses on interactions among the information structures (objects) involved and state-changing communication among objects. The structured programming approach has particular application to large systems performing numerical calculations, while object-oriented representation lends itself well to handling the kinds of multiple simultaneous interactions encountered in real-time data acquisition and experiment control, as well as in dealing with human control and display interfaces.

Despite the importance of these two approaches, it is not the choice of a particular technique that defines software engineering. We can no more say that software engineering is object-oriented programming or structured programming or any other popular technique than we can say that the only engineering approach to crossing a river is to design a suspension bridge. Just as some river-crossing problems are better solved with ferries than with bridges, some software design problems are better solved with procedural designs than with object-oriented ones. Converting a procedural matrix inversion on a single-processor SIMD (single-instruction, multiple-data, i.e., vector-processor) machine into object-oriented form may be entertaining, but it is not likely to be an optimal choice for solving the problem of inverting a matrix. The wider the range of tools in the software engineer's kit and the better our awareness of the appropriate domain of applicability of each tool, the more likely that the software engineer will find sound solutions to the problems we face. Software engineering is not defined by particular methods but by being methodical.

Because of the limitations of space and the need to focus on issues of methodology, we cannot address a full range of tools and techniques in detail. The tools of software engineering cover a wide range of arts and sciences. In this article we look at examples of two important types of tools: those drawn from the design of communications networks and those drawn from interactive graphics. Applying the mathematical models of queues and graphs that have long been used in the design of telephone systems, highways, and railroads to flows of information helps to provide the means to allocate necessary resources to tasks with minimal waste. Allowing designers and users to work with graphical representations of subsystems of information systems allows for a more natural design paradigm in many cases than does working with commands and lists of numbers.

1. BACKGROUND

A brief history of software engineering can be found under COMPUTER PROGRAMMING LANGUAGES. Here we examine the present state of the field. Why is there a need for a distinct field of software engineering? What problems need to be addressed for software design that differ from those in electrical engineering, digital engineering, or systems engineering? Software engineering has a unique perspective because of the remarkable flexibility of computers. We can change what a computer does much more quickly than we can build the equivalent physical systems. New programs can be loaded into a computer in seconds. This allows the same device to attack a wide range of problems or, more importantly, to attack immense problems one small portion at a time. In comparison to the problems they can solve, computers are simple devices. The problems computers can be used to solve are complex. Software, of necessity, carries the increased level of complexity of the problems being solved. The most challenging issues in software engineering arise because of the need to deal with complexity.

1.1 The Problem

Software engineering is, or at least aspires to be, one of the engineering professions, the one that brings the reliability of engineering approaches to problems that involve computer software. Engineers work at the so-called "socio-technological interface," the interface between science and technology on one hand and the practical needs of society on the other. In simple terms, engineers build things for other people to use. As Herbert Hoover said, "To the engineer falls the job of clothing the bare bones of science with life. . . ." (Wright, 1989, p. 49).

The goal of software engineering is to produce software that performs as intended, performs reliably, is delivered on schedule, and is delivered at predictable and reasonable cost. In order to be able to do that, the software engineer must understand computers and the applications to which they are put. The complexity of the tasks we are able to address with computers makes each of the goals of software engineering difficult to achieve. It is hard to build a reliable system when you do not understand what the system does and when many of the components you will use are themselves systems that are complex and poorly understood. This gives a high priority in software engineering to techniques that help in the management of complexity. The most important issues are psychological. Many people have a difficult time dealing with too much new or unrelated information at one time. They need to have the information context kept to something small—a page or two of text or a handful of distinct concepts—to understand what they are dealing with. Unless we are working on something very familiar, if what we intend to produce goes beyond that context we well may forget parts of the project from the beginning when we are working in the middle and the end, and produce something that, though it works, does not do what we had originally intended. Worse, because of this tendency to lose track of things, we may leave out critical cases and create errors. These conceptual errors and the errors in operation then have to get fixed, causing delays in the project, and making any original estimates of time and cost wrong. These are not abstract concerns. Many more large software projects have failed than have succeeded. The failures are rarely because of a lack of skill in programming, nor because of a lack of powerful tools. The failures are almost universally due to a failure to manage complexity, because the problem took on a life of its own and got away from the people trying to solve it.

However, abstract managerial skills are not sufficient. The engineer must have a solid understanding both of the tools to be used and of the technical details of the problem to be solved. Even more important than understanding what can be done with the resources at hand is to understand what cannot be done with the resources at hand, to have a quantitative understanding of rates and flows and design limitations.

Let us look in more detail at what software is and the environment in which software is produced.

1.2 The Context of Design and Development

Software is information that affects the processing of information. It may be data, instructions, images, etc. It may exist in a computer, on a piece of paper, or in someone's head. It may be a static product or a dynamic process, which is a system able to interact with other systems and change state. It may be a small piece of code created quickly and used by one person for a brief period or a system of millions of lines of code and documentation created by years of effort by teams of people and computers and used by millions of people over decades. It has a life, sometimes the brief life of a butterfly, sometimes the well-ordered life of a child born to great expectations, sometimes the tumultuous vitality of a prolific family, and sometimes a life that never leads anywhere.

The well-ordered life is rare, but easy to understand and illuminating. One begins with a need, a requirement for a system to do something useful with information, say, to perform a certain experiment. Just as the well-ordered personal life tends to arise in the context of generations of similar lives, well-ordered program development tends to arise within well-understood contexts. The need involved may never before have been seen, or never before have been automated, but it has strong family ties to earlier generations of problems or to similar problems that have been automated. The need exists within the context of the systems within which the need was perceived and of similar and related systems for which analysis has been done. We analyze the need within that context, trying to define precisely what it is we are asking for. Perhaps it is to control experimental apparatus, collect data, and try to fit that data by appropriate models. Having clearly defined the need, we carefully design information data structures, flows of information, and flows of control that we believe will meet those needs, and once that design is complete, we translate the abstractions of the design into a concrete implementation as computer code and instruction manuals. We validate and test what we have produced to convince ourselves that we have indeed met the need, and then, if all is well, we put the new system into use. After some period of use, we make some changes to the system, whether because of problems, changing needs, or a changing environment. This happens some number of times until the system somehow feels too old and patched, and, after a new system is built, the old system is laid to rest.

Such a software development life cycle itself defines an interesting information-processing system—one that accepts ideas, produces software, and then applies it. Therefore, it is challenging to attempt to produce software to support this system. As we discuss in Sec. 3.2.6, major portions of this system have been automated using "Computer Assisted Software Engineering" (CASE) tools.

This well-ordered flow from concept to design to construction to use is called the "waterfall model" (see, for example, Pressman, 1992, p. 24), in which analysis and design of the external system to which the software belongs, analysis of the software system, design of the software system, programming the software system, testing of the system, and maintenance of the system are shown as cascading flows from higher to lower levels, on the assumption that it requires considerable effort to make the process flow uphill. If this is the life cycle inherent in the system and in the organizations using it, then the reverse flows do require special care. This is especially true in large bureaucratic organizations with complex hierarchical decision making. The stately procession of steps in the waterfall model (and in the necessary reverse flows to deal with changes) can require a great deal of time.

Many applications, however, require a higher rate of progress or at least a higher rate of change, more in line with the image of a tumultuous family. In this environment new requirements and design concepts are continually being derived from experience with systems and rapidly changing external demands. As with the well-ordered life, the tumultuous family has many generations, but instead of coming neatly, one after the other, they are jumbled up with new family members coming on the scene and usurping the rights of the old before they think themselves old. This is common when there have been several cycles of earlier systems, many of which are still in use, and control over the development process no longer rests with a small group of decision makers but is driven by a democratic market process. Seeing the overall flow of development may be impossible in such chaos, and, in the large, the development environment becomes one in which psychology, group dynamics, and organizational issues dominate the technical issues of software development. Interacting design and implementation decisions may be made independently by many groups of players. We look at some of those issues in Sec. 6. However, in the small, the major fragments of the basic flows seen in the waterfall model remain important, and the related issues are worth considering.

The issues are much the same as in writing a good newspaper article: what, why, who, how, when, and where. At each step in the flow, we need to consider what to produce, why it should be produced, who will produce it for whom, and how, when, and where to produce it. In addition, in the real world, costs and who will pay them always enter the equation, and it is important to define clearly criteria and mechanisms for the customer to decide if we have done what was intended and to bring the software into actual use.

What to produce is a basic issue. In the waterfall model we produce an analysis, then a design, then a program, then a system in use. But what are we actually producing? Are we just producing more and more concrete versions of code or data for a computer? This is rarely the case. Especially when a process is being created, the real demands on the system are likely to be from people rather than computers. The software produced is not just the internal state of a computer. At the conceptual stage of analyzing needs, no code at all may be produced, just documents and conversations about the desired functioning of systems able to process information. Documentation is needed not just of the system in isolation but also as part of the broader systems it supports and with which it interacts.

In order to design software that will meet expectations and to decide if the software is functioning correctly, we need to know why the software is being produced and for whom (or for what). The consideration of why a system is being produced extends beyond understanding the context of systems within which it is to fit, to understanding the goals and purposes of those systems. That requires a clear understanding of who the customers and users are. If output of the system is input to equipment, we only have to consider the interface requirements of the equipment. If the results are intended for people, we have to consider the interface requirements of people, which can be very demanding. The study of people's requirements for use of machines is called ergonomics, and involves consideration of how people work effectively. For example, if we design a system requiring manual entry of certain data, it is important to consider the demands being made on the human operator in order to understand the chances of errors in that data entry process. The choices of keyboards, displays, and even chairs can be software engineering issues.

The user of the system is rarely the only entity imposing constraints on what is to be produced. Most projects exist within a complex context of management, rules, laws, and regulations. Independent of the technical requirements of the project at hand, external standards may be imposed on how code is to be produced and how it is to be used. Depending on the discipline, there may be legally enforceable rules, laws, and regulations that impose the use of certain programming languages and require special features in the system. For example, a government agency might require use of the programming language ADA and, for code controlling certain experiments, require doubly redundant control over safety shutters.

In addition, many projects are paid for by a customer who is not the user. The project is only partially successful if the user is satisfied, but the customer is not happy. The customer may not have the necessary background to decide if the project has done what it is supposed to do, and it often becomes part of the design process to produce necessary information. This layer of management information can be more complex than the project upon which it is providing reports.

Considering for whom the software is being produced constrains how it is to be produced but does not necessarily provide a viable mechanism that will meet those constraints. The constraints may leave no feasible means for production or may be so loose as to leave us without guidance on how to go forward. In the well-ordered environment we can look to the past and similar projects to find mechanisms to resolve this, but, in general, making the compromises on how to go forward and getting them accepted is critical to the success of the project.

Once it is clear what is to be delivered to whom and how it is to be done, the question of schedules arises. Because software has such a reputation for untimely delivery, it becomes important for large projects to track detailed timelines and subproducts and subprocesses. The dependences among tasks have to be considered as part of the project design. If C cannot be delivered until both A and B are done, then, in tracking the project, knowing that either A or B is late gives an early warning that C and everything dependent on C will also be late.

If we can understand the time lines and dependences, then often we can make a crude estimate of costs. Information-processing system costs have hardware and software components. The software costs usually dominate in large systems, and so tracking how many workers will be needed for what steps and organizing for a steady use of people through the project with appropriate contingencies for the inevitable delays is critical to estimating costs.

If the project does succeed, we need some way of convincing ourselves that it is indeed a success. If it fails, it is even more important to be aware of that fact. Testing and validation are important parts of the design process. Deciding who will make the decisions about testing and validation and who will have the authority to approve the product are important decisions in the design process. When the locus of authority to make these decisions is far removed from the locus of the decisions about the actual production of the software, it may become impossible to produce a product that is acceptable.

If the product is found to be flawed, either because of mistakes in design or, more commonly, because of changes in requirements while the design was in progress, then, if the project is not to be scrapped, a mechanism must be found to repair or rebuild the software, which may involve restarting the entire process.

If at the end of all this the product is working and acceptable, we enter the most important and generally least considered part of the life of a software product, the period of its actual operation. Then the user will have access to the software and perhaps do things that the designer never considered. At the very least, the users push systems and discover the limits experimentally. As an abstract pattern of information, software does not degrade with time. Yet it is a common observation that most software does seem to degrade and eventually stop working. This is a result of two phenomena: the environment in which any given software is used keeps changing, and "slight" changes to the software accumulate and corrode it like a layer of rust.

Unlike software, computers and related hardware fail, stop working, and have to be replaced. When they are replaced, it is usually with newer, different machines, which provide at best a slightly different environment for execution. Even when instruction sets are identical, newer machines tend to be faster, which may upset timing assumptions. It is common practice to replace and upgrade operating systems and language compilers during a hardware change. For example, the era of physics on computers has overlapped more than six distinctly different versions of FORTRAN (see COMPUTER PROGRAMMING LANGUAGES), and the transition has in each case disabled some number of programs.

Even when great care is taken to preserve the hardware and operating-system environment, it is a rare software system that does not accumulate maintenance changes, patches, and other improvements. Code that once provided printed lists of numbers for output might have seen a conversion to black and white graphics and then to color graphics. Ensuring that all the changes work with the original software and with one another as intended can be a complex combinatorial problem, eventually making a total redesign simpler than making the next "small" change.

Thus we have extremely complex environments in which we are trying to produce complex systems. We need tools to bring order out of such chaos.

1.3 Tools Used in Solving the Problem

The major tools used in software engineering rest on abstraction and the creation of hierarchies of abstractions to reduce the level of complexity to a manageable level. We save ourselves from having to deal with all issues at once by looking for some common characteristics by which to label groups of issues, hoping to reduce our confusion with new, broader categories. If we choose well, the new labels help. If we choose poorly, the new labels become additional distracting details we must track.

1.3.1 Conceptual Disciplines In the broadest terms, the major simplifying steps we can take are to define clearly the tasks to be performed, to define clearly the client for whom the tasks are being performed, to impose a structure on the tasks and their relationship to the client, to take that structure and reduce the context if it is not sufficiently simple, and, while we are doing this simplification, to try to keep what we are doing simple. This last concept, which in classical engineering is called KISS for "Keep It Simple, Stupid," applies not just to subproblems within the design of a system but to the system as a whole and at all levels in between. The fact that a system is large and may require the efforts of tens, hundreds, or even thousands of people to build is not a reason to give up on KISS. The basic task of software engineering is to take what at first appears to be impossibly complex and beyond the comprehension of any one person and find the tools and techniques to recast the problem in terms one person can understand. At the very least we must make the external appearance of what is produced sufficiently simple for a person to be able use it.

Most large problems require two modes of thought: analysis and synthesis. In analysis, we discover the inner workings of a system. In synthesis, we build a system from components. When we have an existing, working concrete system, analysis is a process of examination. When all we have is a problem requiring us to design a new concrete system, we may have to analyze an abstract system in terms of internal structures that do not yet exist but that we infer from the logical necessities of the abstract system and from our experience. Analysis of components is an essential step in synthesis, since it is difficult to combine components that are not understood, and synthesis can create powerful tools for analysis. We return to this interaction when we discuss top-down analysis and bottom-up synthesis below.

There are a large number of ways to analyze a system—i.e., different, often arbitrary, ways to slice something large into smaller pieces. We can follow the flows of control or the flows of information, or we can try to follow both together. There are always new tools to do this that can be found or synthesized. For each tool, we need to ask if there is sufficient benefit to justify the costs, delays, and complications involved.

1.3.2 Selection of Appropriate Tools In the early days of computing, the designer of a system had a limited range of hardware from which to choose. Picking another platform was a very expensive decision. It is important to get past that mind-set today and understand that the cost of hardware is no longer the dominant factor in system design. Trade-offs between choices of hardware platforms and the time and effort to be invested in using different platforms must be evaluated. The ripple effects of delays caused by changes requiring new documentation and training for experienced programmers and users must also be considered. Similarly, choices have to be made among design paradigms, languages, libraries, software tools, etc., with concern for the impact of delays and costs of any change from that which is familiar. For each tool choice we should attempt to quantify the benefits (estimated savings in time, reduced probability of errors implying a quantitative estimate of reduced rework effort) and costs (estimated expenditures in time and money to acquire or build the tools, train people, etc.). Some tools will show clearer gains than others and will be worth considering further. We look at the tools of software engineering in Sec. 3. There are many among which to choose. It is rare for one tool to be the right choice. In the end what matters is that the tools used be capable of doing the job and be well understood by the people using them. Delaying a project while people are retrained in the latest technology may be worth while for a large project with long time scales, but rarely for a small project. However, one must be aware of the not-invented-here syndrome, in which designers stay with familiar approaches and blind themselves to tools or even complete solutions they might simply pick off the shelf from elsewhere. Minimizing the impact of such biases is one reason why it is advisable to do a quantitative cost/benefit analysis of the impact of the choice of tools, rather than relying solely on qualitative impressions.

1.3.3 Human Factors The element of familiarity is one of a wide range of critical human factors in the design process. Unlike hardware, people are a very expensive resource and need to be used effectively. This is a difficult challenge. People have their own lives and interests, and their interests and skills do not necessarily fit well with the needs of projects. Matching people to tasks and motivating them to work is a subtle and demanding task. Some people work well in highly structured environments; some do not. Some people are natural team players; some work better alone. Encouraging people to work in ways that produce results of consistently high quality is critical to organizations that produce software. The process of making decisions, of selecting one approach over another, depends on human value judgments that are difficult to quantify.

As difficult as it may be to get people to work together, dealing with the clients for software and getting them to state their requirements clearly can be even more difficult. Organized approaches have been developed to deal realistically with people and try to produce a system that actually meets their needs and to some extent even does what they want. We look at these issues in more detail in Sec. 6.2.

1.3.4 Standards and Controls, QA/QC The minimum that is required to obtain quality results is a set of standards against which to measure processes and products. In small groups with innate motivation, these standards may be sufficient, but, in general, full systems are needed for quality assurance and quality control (QA/QC). Quality assurance is the term for the entire set of mechanisms used to assure those who need to be assured (which hopefully includes the clients and users of systems) that the system has an acceptable chance of doing what is needed of it. Quality control is the term to describe the detailed control mechanisms used to ensure that every step of the way in production the quality of the product is being maintained. In software systems, which rarely can be tested for every eventuality, quality control is an essential means of quality assurance. We return to issues of quality in Sec. 7.

2. BASIC CONCEPTS

In the Introduction to this article, we explored the distinction between software and hardware. See COMPUTERS, COMPUTER HARDWARE, and COMPUTER PROGRAMMING LANGUAGES for some of the terminology we will now use.

2.1 What Is Software?

Software is information that affects information-processing systems. From the time of Babbage (see Morrison and Morrison, 1961, pp. 245–295; Hollingdale and Toothill, 1965), the basic elements of software have been operations (actions to be performed), variables (information that may change during a calculation), and constants (numbers that do not change). These concepts spring from the design of computer hardware. A computer typically has a memory, a processor for arithmetic and control, and input-output capabilities. The details of hardware, instruction sets, memory sizes, and device controls now change very rapidly. Software built on such shifting sands must be designed with a certain independence of the particulars of the hardware, or any hardware dependence must be parametrized. Multiple computers may be joined together to form one larger machine. Software must then be able to present and distinguish among various models of machines: multiprocessors based on message passing among processors with independent memories, multiprocessors based on sharing one large memory among many processors in some coordinated manner, or collections of specialized processors working in bucket-brigade fashion to handle large arrays of data efficiently. Systems that provide the ability to manipulate images require software abstractions for graphics and visualization.

From the elementary concepts of operations, variables, and constants, we need to build abstractions that allow us efficiently to map real-world problems onto a wide variety of hardware designs. This is commonly done by working in hardware- and software-independent pseudocode, using graphical representations of systems, and deferring coding to near the end of analysis and design.

2.2 Types of Software

It helps clarify our thinking about software to distinguish some major types of software. Systems software is used to make hardware more manageable. Application development support software provides tools used to build software. Finally, application software is the software solution to a particular problem.

2.2.1 Systems Software Systems software is used to present us with well-controlled reliable interfaces, which isolate us from the specifics of particular hardware and which provide interfaces to the hardware on a level of abstraction more appropriate for use by other software. When a machine will be used by a single task at a time, a simple monitor may be used to provide higher-level interfaces for the execution of jobs. Systems that will be shared by more than one user need operating systems to control the allocation of shared resources. When the sharing will involve multiple threads of independent execution, it can be very useful for the operating system to provide virtual machines, which give each task the illusion of a private computer. Commonly used accesses to system functions can be gathered into collections of routines called application programming interfaces (APIs). The evolving need for graphical user interfaces (GUIs), which allow users to point and click instead of type, and real-time demands such as data acquisition can call for specialized APIs. Services can be provided for process-to-process communications within systems, on local area networks, and on internetworks. Operating system accounting and security, predefined interfaces for the creation of client/server applications, and access locks to allow users to reliably share their own resources without confusion are all useful. Software engineering for systems software and for all software dependent on systems software has been greatly simplified by the widespread acceptance of variants of the UNIX operating system (see Martin *et al.*, 1984). While there are many variants of UNIX and many competing systems, UNIX now provides a common background against which a large portion of software engineering is done, especially in the academic world.

2.2.2 Application Development Support In some cases all that is needed is an operating system and the appropriate application, but, especially for software engineering itself, specific tools are needed to support application development. Assemblers, compilers, interpreters, and revision control systems are an essential minimum. See COMPUTER PROGRAMMING LANGUAGES for a discussion of the first three classes of tools. A revision control system allows us to keep track of multiple versions of documents. As changes are made to a program or design document and it is time to replace an older version, instead of giving the new version of the document a distinct name, or discarding the old document completely, a revision control system allows us to hold both versions and get at either one. In addition, a good revision control system includes access locks, so that in a team development environment, one member of the team can check out a document for changes and be sure that the system will not allow someone to slip in another change out of order.

Perhaps the most important development support tool is a mechanism to maintain commonly accessible libraries of previously developed routines, so that earlier efforts can be reused. A permanent shared file system with library mechanisms is now a given in all software engineering efforts.

2.2.3 Applications In most cases, the ultimate objective of all we have described thus far is the support of applications. There are many types of applications, as many types as there are useful ways to process information. For example, we could have numerical, administrative, laboratory automation, real-time, embedded, GUI, graphical, distributed, client/server, or database applications. Each makes different demands. A numerical application may require extremely efficient access to the processor and libraries of highly optimized linear algebra routines. Administrative applications may require ultrareliable security and development support tools suited to large groups. Real-time applications may require extremely responsive support of multiple threads of execution. Embedded applications may need the same features as real-time, plus specialized device control and the ability to clone applications to diskless environments. Software engineering needs to look at what is common in all of these and make the most of these commonalities while preserving the unique flavor of each application.

3. SOFTWARE ENGINEERING

As we noted in the Introduction, the main focus of software engineering is on building new software systems. However, it is important to see the building of systems as a part of the complex lives of systems and to ensure that the approaches used in construction integrate well with the rest of the design cycle.

3.1 Software Design Cycles

The waterfall model discussed in Sec. 1.2 is one of many possible patterns of life for software. No matter what the actual order of events, however, the software engineer must have the tools to deal with all stages in the life of a project. It is never safe to assume that the process flows one way. The same or similar problems tend to recur, and involvement in earlier or later stages can help to make sense of the current stage. While it may seem desirable to "build it right the first time," this is rarely realistic. Most software products come back for reconstruction, redesign or even reanalysis many times. Nor should we restrict communications to be only between adjacent stages, with analysts communicating only with designers, designers only with programmers, etc. This may lead to serious miscommunications and a failure ever to close on a working product. The safest assumption is that software will flow in a series of nested loops, with all paths possible. Further, as pressures for rapid development become more common, it becomes desirable to skip steps, to make analysis feed directly to code generation, or to allow users to program without intermediaries. This is neither good nor bad, just a fact of the world in which software engineers work. When well-understood techniques are available to achieve such bypasses, an engineer may well use them, but it is important that they be well understood, or the result may be a nonworking system or, worse, a system that appears to work but produces wrong answers. Therefore, let us look at the major design paradigms, understanding that some of them may be applied by automata, rather than by people.

3.2 Software Design Paradigms

There are many ways to look at a system analysis and design problem. None of them is perfect. When faced with a totally new problem, it may be necessary first to invent a design paradigm and then to return to the specific problem. However, most problems of a modest scale or a great deal of regularity (such as many major scientific calculations) are handled more than adequately by some combination of bottom-up synthesis and top-down analysis used together in what is called "outside-in" design.

3.2.1 Bottom-Up, Top-Down, Outside-In Most software analysis and design projects start at one of two places: at the top, considering the highest level of abstraction and working down to understand the necessary details; or at the bottom, starting with existing tools, resources, and capabilities and working up to create more general, abstract tools closer to the level of abstraction of the problem. In most cases, working top down is a process of analysis, discovering the inner structure of the problem, and working bottom up is a process of synthesis, creating the solution to the problem. However, as we shall see, a top-down process can be used as a means of synthesis, and a bottom-up process can be used as a means of analysis.

Consider the flow in Fig. 1. We start with the original problem—for example, the requirement to "do" the experiment mentioned in Sec. 1.2. The overall task is to do the experiment. We can analyze that requirement as consisting of two major subtasks: data acquisition and data processing. We can then further decompose data acquisition into two subtasks: experiment control and data logging. We can also decompose data processing into two subtasks: data reduction and modeling. If we continue this approach to a fine enough level of detail, that detail will map directly onto the necessary code for implementation. Indeed, if we are working in almost any suitable higher-level language, we can make this into a direct code synthesis technique by making each transition from higher-level tasks into lower-level tasks into a sequence of calls to subroutines, which get written after their use has been defined by those calls. That is top-down analysis used as a means of code synthesis.

Now consider the opposite flow in bottom-up synthesis shown in Fig. 2. We start with a collection of simple tools, the software equivalent of knives and screws and levers, and put them together into increasingly more powerful and complex tools until we have created a tool adequate to solve the problem. Just as combining two knives with the fulcrum of a lever produces a pair of scissors, which can do jobs neither knives nor levers can do, combining simple software tools can produce systems of remarkable power and complexity. Returning to the example we used above, we might first build a library of basic digital and analog device-control routines, and mathematical routines for linear algebra, then combine the basic device-control routines into motor-control routines and pulse-acquisition routines, and

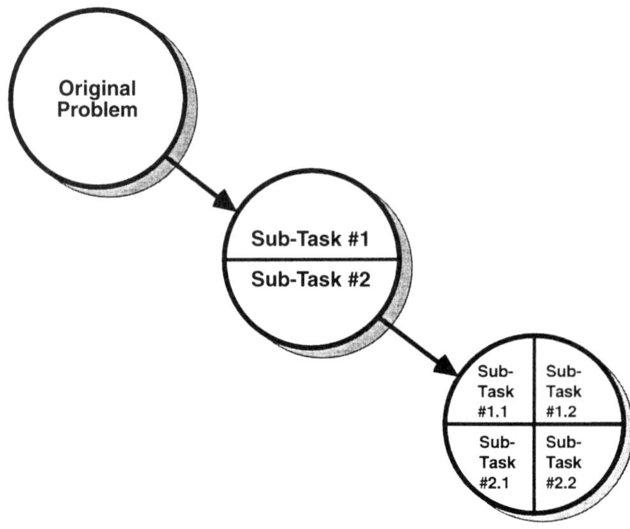

FIG. 1. Top-down analysis.

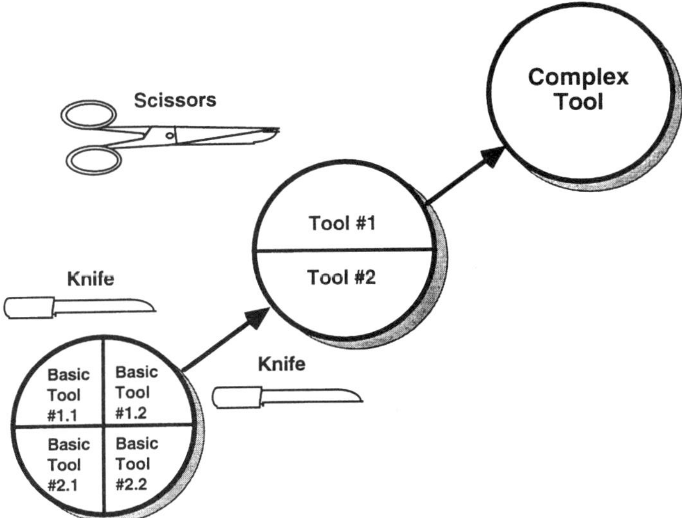

FIG. 2. Bottom-up synthesis.

the linear algebra routines into a least-squares program. That this is a synthesis technique is obvious. It is also useful in analysis, especially when finding errors in existing systems or trying to change them. Following the structure of a system up from the bottom helps to ensure an awareness of the fine details of the actions the system will take.

There are benefits and risks in both of these approaches. One can misuse either technique. In top-down analysis, one could try to skip all intermediate decompositions and go directly to the most elementary subtasks. In bottom-up synthesis, one could try to build the entire system in one pass. In either case, the result is usually a single large tangle of code. Such code is almost impossible to understand and is very prone to errors. However, there is no assurance that hierarchical decompositions used in a top-down analysis will lead to a feasible implementation. When we finally get to the bottom level we may be demanding resources that cannot be delivered. Worse yet, by decomposing each branch independently, we may waste a great deal of effort creating slightly different solutions to similar problems from other branches, making the code unnecessarily complex. Pure bottom-up synthesis suffers from complementary problems. There is no assurance that the structures we build by assembling simple substructures will be combinable into a solution to the overall problem, and we may waste a great deal of effort creating or gathering libraries of substructures that are irrelevant to the problem at hand.

It might appear, however, that both techniques at least reduce complexity by creating hierarchical structure. This is desirable, but it is not an unmixed blessing. Consider the example in Fig. 3. If we did no hierarchical analysis or synthesis, we might build one very large program out of many elementary operations. At any point in that one big program, we have a tremendous width of context, i.e., the number of lines of code, in this case 100 000 lines, all of which must be considered at the same time to understand the operation of the code.

As we recast the large program hierarchically, the width of context is reduced, but the depth increases, and the number of interfaces rises sharply. By the time we have gone four levels deep, no routine demands a width of context of more than 20 lines, but we must deal with over 10 000 routines and over 10 000 interfaces. We have traded one form of complexity for another. Finding the optimum balance point is difficult. The trade-off can be made more economical if there is some significant commonality and reuse among the routines we generate. Bottom-up synthesis provides a bias toward commonality and reuse of code. Therefore, for large projects in which optimizing the depth and width of context and minimizing

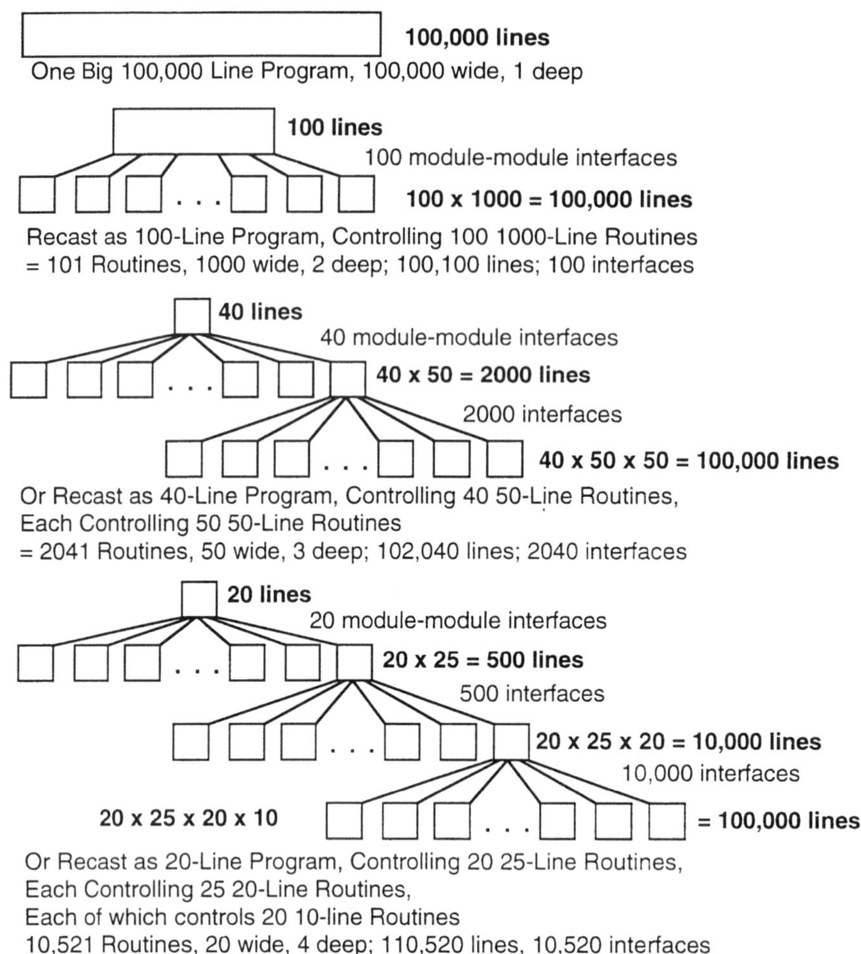

FIG. 3. Width and depth of context.

the number of distinct interfaces and routines can be critical to success, it is common to mix aspects of top-down analysis and bottom-up synthesis, working in on the problem simultaneously from both requirements and resources, i.e., working outside-in. Consider Fig. 4. In dealing with a large problem, it is common to find several pools of existing tools and several areas of broadly specified abstract subtasks. We try to extend the domain of applicability of each pool of tools by building libraries and appropriate application interfaces to serve as virtual tools oriented toward the known subtasks. At the same time, we bring each of the major abstract subtasks closer to the level of the available tools, by further decomposition into simpler and more concrete sub-subtasks, keeping in mind the capabilities of the libraries and interfaces we have. The abstractions and the virtual tools are brought together in a common ground at an intermediate level of abstraction and complexity. It may take many years and many false starts to find the right common ground for a domain of applications, but once found, that common ground can provide the most productive level at which to work.

3.2.2 Functional and Applicative Programming When execution of a piece of code creates a "side effect," a nonobvious change to the state of data structures, the context of that piece of code has been broadened, making it harder to keep track of what it is doing. Functional (or "applicative") pro-

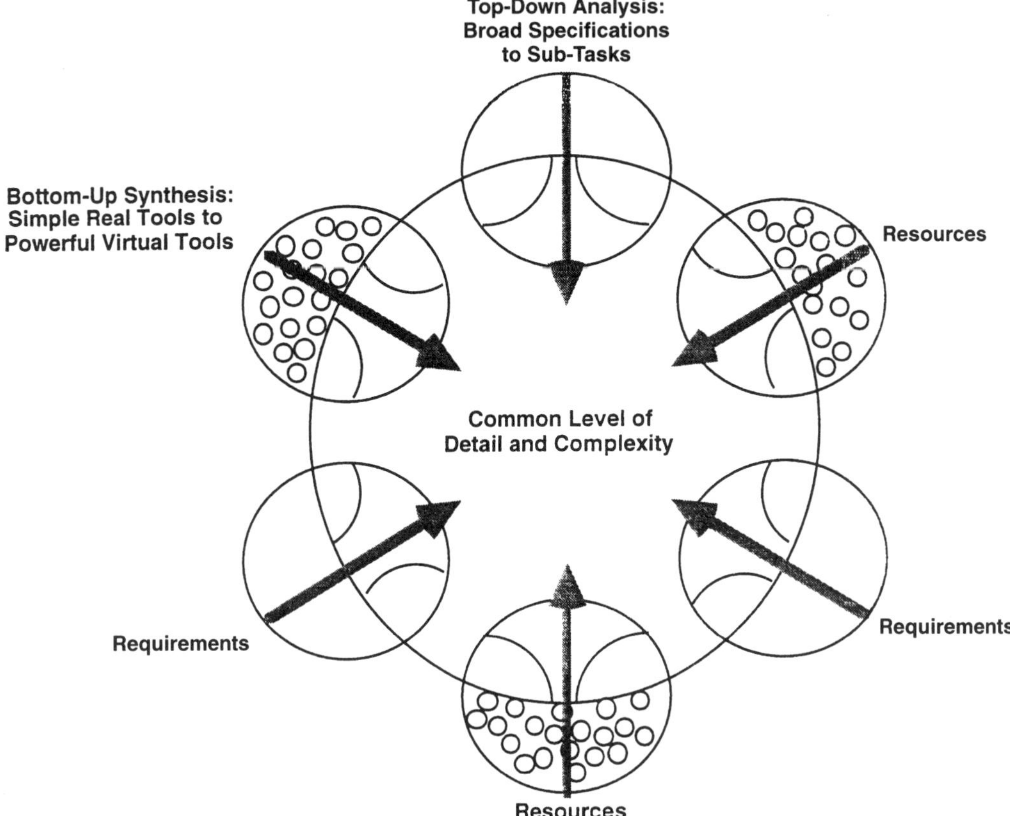

FIG. 4. Working outside-in.

gramming is an approach used to avoid that problem, making it easier to deal with the routine–routine interface problems discussed above. In functional programming, problems are solved by pure function evaluations. Routines have no memory of prior state; they take exactly the same action every time they are called. Thus, no matter how deep a hierarchy we create, at each level and in each routine along that level we need only validate execution locally. The concepts began with work of Church on the "Lambda Calculus" in the 1930s and found a viable programming language expression in McCarthy's LISP. This approach to software design has seen extensive use in natural language processing and symbol manipulation, but the burden of working without side effects can be unacceptable in applications for which side effects are useful, such as device control, operating systems design, and other real-time applications that need to remember where they are as they are interrupted. In such cases other paradigms are better suited. For more background on functional programming, see Hudak (1989).

3.2.3 Structured Programming Structured programming is the application of top-down analysis to the flows of control in an information-processing system and has been used very effectively in systems dominated by considerations of control flow independent of data. Dijkstra (1968) used the technique to design an operating system three levels deep with each routine sufficiently small and well understood that he had the confidence to assert that the system would be "flawless" before having done final testing, and he was essentially correct. To make the technique work, it is important that each module be entered and exited cleanly and used only through its "official" interfaces (a doctrine to which all major software engineering paradigms now adhere). As the technique matured, practitioners tried to ensure clean entries and exits by simply prohibiting the use of direct jumps to particular lines of code, creating so-called "goto-free" code. This extreme rigidity is no longer considered necessary, but certainly unusual changes of flow of control have to be well justified. See Dijkstra (1976).

3.2.4 Proving Correctness Dijkstra attempted to prove the "correctness" of his implementation. Hoare (1969) made this approach rigorous by providing formal proofs of programs. There is a fundamental limitation to this technique. We can only prove an internal consistency between a program and a sufficiently detailed specification. We cannot prove that the program does what was intended by the user. However, removing the possibility of a gap between specification and implementation is highly desirable. While a proof can be attempted after the fact, it is most useful as an intrinsic part of the design and creation of code. By bringing formal methods up into the specifications, the chance of the final system failing to be consistent with those specifications is very small. Such approaches have even been extended to verifying compliance with timing constraints in real-time systems (Felder and Morzenti, 1994).

3.2.5 Software Engineering and Ada Real-time systems are among the most challenging to software designers. Such code is often used for control of devices, and errors can have disastrous consequences. This provides a strong incentive for the application of engineering approaches. Ada (Booch, 1983) was designed to establish a sound engineering environment for the creation of embedded real-time systems, which are information-processing systems embedded into and controlling devices. Ada drew on earlier work in simulation languages and uses "abstract data types" or "classes" that combine data-structure definitions with definitions of the actions that can be taken on those structures. Ada makes use of "strong typing," forcing actions to be consistent with the abstract data-type declarations. Instances of abstract data types coupled with the necessary constants and implementations of the actions make what we now call objects. Only the portions of the specification of an object necessary to determine its interactions with other objects are visible. Private, internal data structures and actions are hidden. This was an important step in assuring that a system could be built reliably, since one could know with certainty which portions of an object could be changed without having to consider interactions with other objects. This "encapsulation," a bottom-up creation of new abstractions, is a powerful design-control mechanism. Ada is still used, but much current work with objects is now done with languages that support additional capabilities such as "inheritance," to which we return in discussing object-oriented programming. Ada provides a complete environment for software development and was an important step in the development of CASE tools.

3.2.6 CASE Tools Computer-aided software engineering (CASE) is a term used to define the use of computer-based tools to automate steps in software development. The original CASE tools are "ordinary" assemblers, compilers, and interpreters (see Computer Programming Languages), which allow software to be defined in human-readable terms. Auxiliary programs that support compilers, such as loaders, linkers, and librarians, allow independent compilation of routines, allow creation and maintenance of libraries of routines for multiple use, and encourage modular development. Interactive context-sensitive editors facilitate the rapid production of both code and documents, and revision-control systems allow the multiple versions created to be tracked. Profilers, monitors, and analyzers allow the software performance to be monitored and, hopefully, improved.

Such tools are often combined into smoothly functioning "seamless" environments, in which the flow from code creation in an editor to compilation to test execution is facilitated. This tool-oriented approach also is used in the administration and control of the development process by means of project management tools. These tools allow time and resources for tasks to be estimated, tasks to be linked showing their dependences, and critical path dependences to be noted. Used interactively, such tools can flag trouble while there is still time to correct the situation. Everything that helps in communication and record keeping becomes a valuable tool in development—databases are used to track the complex dictionaries of entities in object-oriented analysis, electronic mail is used to provide internal project communications among participants, etc. See Fuggetta (1993) for a "Classification of CASE Technology."

Since most people work better with graphical representations than with purely verbal descriptions, there are CASE tools that allow object-oriented software (see below) to be defined graphically. See Armbruster (1995) for a review of such tools. We explore a simple Hypercard® example of partial graphic design of an application in Sec. 8.

3.2.7 Object-Oriented Software As we noted in the discussion of ADA above, at the time this article was being written the most popular software engineering paradigms related to the representation of software systems as collections of communicating objects. The messages that objects send one another may then cause the objects to take some appropriate actions, but the actions are hidden internal mechanisms within the objects, known to other objects only by the outwardly defined communications mechanism. This is precisely how basic computer hardware is designed and presented to software and has long been the realistic approach to design of demanding real-time applications such as operating systems. As general applications are expected to display more real-time interactive capabilities, object-oriented approaches become essential. Object-oriented paradigms have been applied to all stages and aspects of software development, including to software development organizations. The focus on semiautonomous interdependent teams is itself an expression of human software development processes in object-oriented terms.

What is an object? An object is an information-processing system or subsystem or component. In contrast with functional or applicative programming, in object-oriented representations side effects are not only allowed, they are actively encouraged, since an object is allowed to process that same information in different ways at different times by having hidden internal mechanisms; i.e., objects may have a private internal memory of their state that is not accessible to any other object. An object may do anything with the messages it receives that the designer wishes, as long as the object deals with other objects in the ways it is expected to, i.e., that it honors its "contracts" with other objects. This may seem to be a very vague definition. It is. It permits almost anything to be called object oriented. Indeed, we could define an object called "global memory" that accepts the messages "STORE value in location" and "FETCH data from location" and that communicates back the values fetched. We can design all other objects to use global memory to store their internal state and thereby make what had been hidden completely visible. (Such techniques, which evolved in hardware design to permit analysis of internal states of complex logic, have become essential to debugging complex object-oriented systems.)

The main focus of object-oriented representations is on object-oriented programming (OOP), which is done in several languages, the most popular of which is C++ (see, for example, Halladay and Wiebel, 1993). C++, an extension of C, allows abstract data types, encapsulation, inheritance, polymorphism, and dynamic binding.

Let us look at how an experiment control problem might be viewed in terms of objects. Looking at the bottom, we see we have some devices to control, and so we will certainly need some software objects for device control. We are not yet certain of the details of the control we will provide in each case, but it certainly is worthwhile to define an abstract "class" of such objects, each of which would have "methods" to initialize its associated device(s) and its own internal data structures, to send commands and data to its devices, to accept status information and data from its devices, and to shut down the devices and release resources when they are no longer needed. Let us call the class of objects "DeviceController." Now, in looking at an example, we find that we have a four-axis positioner, so we define a particular object "FourAxis," which is a kind of device controller. Therefore we use the general class definition of DeviceController to create the framework we will need and add the specific mechanisms needed for the methods FourAxis will use. When we are building the system, we have incremental encoders for position, but we expect to acquire absolute encoders. We now are facing a design decision whose impact can be carefully controlled through the definitions in FourAxis, with the internals such as encoder type that have limited impact on the processes using FourAxis hidden from view and the internals that do have serious impact, such as the positioning accuracy and timing, made visible through well-defined object interfaces.

In designing the interface to FourAxis, it would improve clarity to allow client objects to work in terms of laboratory coordinate systems for positioning. We find reason to permit polar coordinates in both degrees and radians and Euclidean coordinates in centimeters and inches. So we "overload" the positioning methods of FourAxis to allow the same operation name to work reliably and unambiguously in any of the coordinate systems driven by the type of the position data. FourAxis has become "polymorphic."

That is a small example of bottom-up object-oriented programming. Looking at the same problem top-down, we might start with the view of the entire experiment control system as one object, then decompose it into user-interface, device-control, and data-management subobjects. Clearly the preliminary work we just did on FourAxis fits well with this structure, and the thought about coordinate systems interacts both with the user-interface objects and the data management objects. Hopefully, we will work toward a common meeting ground.

For more on object-oriented software, see Booch (1991).

3.2.8 Usability—Testing and Prototyping A system may work according to its specifications but be difficult to use. It may demand too much of its users or demand things of them that conflict with their training, abilities, interests, or tastes. A system may work for one user and not another. For example, if a manual control input is required and requires use of the right hand, left-handed people may find the system very frustrating. If critical messages are displayed in pure red, those messages may be invisible to the 8% of males who are color blind. Ensuring that a system is usable requires attention at all stages of the design process. Testing and prototyping can be effective techniques to keep a design usable.

We must clearly understand who and/or what will use the system and early in the design process present to them as realistic a simulation as possible of what the system will be like, to see if our understanding of what the system is to do is actually correct. We may have to start with verbal descriptions, but people do not usually absorb the full impact of a system from a verbal description. It is important at least to provide accurate screen images for review and, where possible, to create an adequately functional mockup or prototype for the user to try. Once the system has gotten to the point of having completed modules, testing by using the real modules in place of prototypes is important, both to check for errors in the modules and to see if they meet user expectations. If this is to have any meaning as a test, allowance must be made for the possibility that the software may fail and have to be repaired or redesigned. It is an excellent moral challenge to "build it right the first time" and a goal to which all software engineers aspire, but it is unrealistic in general. It is better to have allowed for extra design cycles and not to have to use them than to fail to allow for those cycles and be left with a useless product.

In many organizations the transition from development to actual use is formalized into a process called a "handoff" or release. The development team is intentionally kept distinct from the operations team so that there will be someone ready to test the product who has not been prejudiced by being involved in developing it. When development thinks it has a product ready for actual use, it prepares a complete handoff package or release kit with all the software and documentation, supposedly ready to stand on its own legs. Operations takes the package, tries to work from it without side communications, and prepares its own tests of usability. If the product has problems, development is given all the details so that they can reproduce the problem, but they are not allowed to tinker with the production version. They have to prepare fully documented patch releases and upgrades, which go through the same handoff procedure. This may seem a cumbersome process, but it has proven so useful that most serious professional developers actively seek parties willing to accept preliminary trial handoffs (so-called "alpha" and "beta" releases) to find remaining problems with their product.

For more on usability, see Nielsen (1992).

3.2.9 Requirements Engineering A system may be usable, yet not usable for the intended purpose, because the intended purpose was not clearly understood or not clearly communicated. Determination of requirements, creating clear, complete, and consistent specifications, is one of the most difficult aspects of software engineering. There are many strongly held opinions on how to address this task, but only partly understood methods with limited records of success. We look at the issues and approaches in requirements engineering involving group efforts, where the designer and the customer (or "stakeholder") are distinct, in Sec. 6. Here let us just consider the abstract issue of defining the question.

A software system exists as part of a larger system. If the system as a whole is not well understood, there is a significant risk that the software will fail to meet the needs of the system. In some areas of software design, the person analyzing the system is untrained and has at best a rudimentary familiarity with the larger systems involved. Fortunately, in the physical sciences, it is not unusual for software designers to be trained and active in the application area. However, even then it can be desirable to put aside any presumptions we may have and come to a new specification question with a completely open mind. We first try to establish a consistent framework within which to pose the questions about the project. We might focus on procedural aspects or information flows or object-oriented representation. Within that framework we ask the question, "what?" That is, we might ask, "what will this data acquisition system do," or "what information do we wish to get out of this system, and what information might we have available to put in," or "what are the major elements of this system, and what services will they provide to one another?" All are valid possible starting points, and in most cases it is worthwhile to pursue all of them and see where they lead, but it is usually less confusing to pursue them one at a time. The answer to the first-level question usually involves sentences about actions taken on information. If we are pursuing a procedural thread, we would then inquire into the detailed meaning of each of the actions, if a data flow thread, into the detailed meaning of each item of information, and if an object-oriented thread, into the detailed structure of each clause. After some amount of iteration and comparison among threads, we are likely to have a trial-balloon document describing requirements.

There is considerable disagreement over the desirability of keeping a wall between the creation of such a requirements document and the creation of code. If we let them merge, then we must impose a strong formality on the specification so that it can be analyzed by software, but we gain the ability to have the software check for consistency and do most of the programming for us. Changes in documentation and changes in code become one and the same thing. On the other hand, the formality we impose may preclude us from clearly stating what we want the system to do, since we can only be as expressive as the formal specification language allows. Arguments that a specification language allows any idea to be expressed, provided we elaborate enough, are unrealistic, since clarity can be lost in the complexity of elaboration.

As of this writing there is an ongoing effort to clarify the choices to be made in requirements engineering and establish the reliable range of utility for the major tools. For a review of the questions, see Siddiqi (1994) and the other articles in the same issue.

3.2.10 Re-engineering Suppose you have a software system that is doing something useful, handling, for example, a diffraction-data acquisition and display for a single detector data stream with four-circle positioning. You want a new system that shows in some ways a strong family resemblance, say an area detector system with an array of data streams and detector roll, pitch, and yaw as additional positioning parameters. Even if the existing system is poorly documented, it is a working prototype for important aspects of the new system and may be worth using as a partial basis for building the new one. The overall process of evolving a system with one functionality into a system with different functionality is called re-engineering. As the value in this reuse of system structure as well as components is recognized, it is becoming the practice of both good designers and good suites of design tools to retain a richly detailed trace of the design flow of a system so that future re-engineering can be facilitated. For older "legacy" systems, an essential first step to permit re-engineering is to reverse engineer the system to discover its inner workings.

Reverse engineering of software is a complex process. Software tools can be used to look for relevant patterns in the code and to attempt to recover the original design information, but often a software system is a complex combination of pieces added by random people at random times in random ways, sometimes without a clear design goal in mind, and often incorrectly. It is rare to reverse engineer a large system and not find previously undiscovered bugs in the process. It is sometimes cheaper and safer to discard the old system completely and start fresh, rather than risk provoking buried bugs in poorly understood sections of code.

If we have (or create) sound documentation of the design of the existing system, it is tempting just to glue on new features. This is a high-risk approach, for the same reasons that starting to extend a house by just knocking down walls and framing new rooms is risky. The whole thing may collapse. For example, our area detector may force a higher interrupt rate on the processor than it can sustain, causing fatal stack overflows. The new system has to be designed before it is built. The old system is there as a resource and helpful guide, not as a straitjacket or a foundation.

In addition to all the standard tools and techniques of software engineering, the use of "wrappers" and "middleware" has proven very effective in re-engineering efforts. A wrapper is a layer of software that presents itself to an existing subsystem as a client with which the subsystem is known to work and that presents itself to outside clients as a server meeting new interface requirements. For example, the positioning subsystem for the single-detector data stream might be a highly efficient work of art with which we have neither the desire nor the need to tinker. However, assume it had messy handling of data structures inconsistent with a good object-oriented design. Rather than modify it, we could create a general positioning layer as a wrapper, providing the necessary data structures and hiding them from the rest of the software, as well as adding the hooks for the area detector roll, pitch, and yaw. If we intended to come back later for a full rewrite, this approach would allow prototyping of the rest of the system quickly and efficiently.

See Winsberg (1995) and Sneed (1995) for more on re-engineering.

4. SECURITY ISSUES

No creation of man is ever completely secure. Imperfections in design, the wear and tear of normal use, environmental insults, deliberate assaults by persons with evil intent, accidental damage by persons who are careless, and simple entropy all conspire against what we create. Information systems are no exception, and, as information systems become increasingly interconnected and interdependent and failures in one system can propagate to many others without any chance for human recognition that anything is wrong, security issues become critical topics in software engineering.

While changes to ensure security are often added to a system after it is built, as with any changes, it would have been better to consider them in the original design. In general, we need to extend the models we use to include all the failure modes the system is likely to encounter. Where possible we have to ensure that failures are detected and reported or corrected. This can increase the complexity of the design, and the tradeoffs of costs and benefits have to be examined.

There are two major categories of failures to consider: those that are due to random events, and those that are due to intentional attacks on the system. For independent random failure following a random probability density distribution p over a domain of events x, with costs c, then the expected cost (usually called the "risk") is

$$\int p(x)c(x)\,dx.$$

We then try to adjust the design to minimize the risk. This can be done by reducing p where c is high or reducing c where p is high. For example, if the cost of losing certain data is high, we can reduce the probability of loss by check sums and replication for that data or reduce the cost of loss by making modules dependent on that particular data able to function from other information.

Caution must be exercised in mechanically using probabilistic risk assessment for securing a system against random failures. Averages can be deceptive. If there is a high-cost, high-probability failure mode, the efforts we put into dealing with that failure mode may blind us to a large number of lower-cost and lower-probability failure modes that make the system performance so annoying as to be unacceptable. This can be viewed either as a misestimation of costs or as a poor choice of measures in the integration.

The more difficult category of failures to model and deal with is that of deliberate attacks on the system. We now must assume an active malign intelligence introducing failures. We can no longer assume a probability distribution as given, but must enter the realm of game theory (*q.v.;* von Neumann and Morgenstern, 1944) in which our opponents will adjust the distribution to our disadvantage. We then have to find a design or operation strategy that will minimize our maximum losses. Thus, even if the compromise of a critical password is estimated by us to be a low-probability event, prudence would argue for building in safeguards to ensure that the damage that can be done by loss of that password be limited to some acceptable level.

In planning methods for dealing with failures—random or intentional—we need accurate models of the behavior of the system. Complex systems built using very abstract models to hide the complexity of the system may also hide critical details of potential failure modes. It does no good to have a system designed with multiple redundant independent data communications paths if the physical implementation puts the data through virtual channels on a common medium. Security modeling has to be done on all levels of abstraction, from the highest to the lowest. For more information on security design approaches, see Baskerville (1993). For considerations in the compartmentalizing of data flows to help protect against intentional system failures and loss of confidential information, see Sandhu (1993).

5. OTHER ISSUES

5.1 Issues in Maintenance

Software should not need maintenance, but it does. While the logical abstractions of code do not wear out, the environment in which software is used changes and necessitates changes in the software. In addition, no matter how carefully software is designed and tested, there is some chance of errors. A person may make a mistake, or a failure in some software or hardware component used to generate or transcribe the software may be reflected in the software. Problems can lie dormant for months or years and then be discovered by users or client systems doing something different or by a change in the environment.

Once we accept the need for change, we must address mechanisms to effect change. Especially when coding is a separate process from analysis, changes can have a corrosive effect, making the system in use drift further and further from its specifications. Suppose we have an experiment with 21-bit encoders for position that we upgrade to 24-bit encoders and just make the minimal changes in the positioner driver software to handle the new encoders. Since we are approaching and sometimes exceeding the accuracy of real arithmetic, failing to reflect this change in the system documentation invites mistakes at a later date. We need mechanisms to hold code and its documentation together. This is perhaps the strongest argument for specifications and other documentation to be made a part of code against which the code is automatically verified or from which the code is generated. Without such automatic tools, tremendous discipline is required of programmers to ensure that they reflect all changes back up to the highest levels of the documentation that might in any way be affected.

5.2 Reuse of Software

A very effective tool for reducing the chance for errors and the need for changes is to reuse existing, appropriate, well-understood, well-tested software. Subject to the constraints stated, there is much to gain from this. The more the same pieces are used in different projects, the better understood they become, and the more certain we can be that all errors have been found and either removed or clearly defined. The time and effort that goes into the initial development gets amortized over a much longer life than would be the case for code used only once. Major libraries of routines are now a staple of software engineering.

There are, however, risks and limitations to this approach. Intellectual property law may give other people rights in software you wish to reuse. The fees to them then become part of the cost equations in doing the project, and potential future fees become a continuing budget item. More importantly, the owners of software sometimes are reluctant to release it in its entirety, providing only external documentation and machine-executable code (known as "binaries") rather than detailed internal documentation and human-readable source code. This limits the degree of understanding of the code that can be achieved and weakens the value of reuse of software as a sound engineering tool.

Even without the intellectual property limitations, and even with every line of source code in front of you, you may fail to understand a software subsystem completely. Subtle errors can lie dormant in the most heavily used routines for decades.

5.3 Decision Support

All of software engineering involves decisions: what design approach to take, which software and hardware to select, what resources to apply to various parts of the problems. The software that is produced may have to make decisions. Making a decision is a complex process, especially because it involves value judgments. Each decision made may create a cascade of further decisions that have to be made. Let us examine some of the issues in making decisions. When decision making is automated, the resulting software provides decision support tools.

In its simplest form, a decision is a choice between two alternatives. Let us call the alternatives *A* and *B*. If we know nothing more about them than their names, there is no way to decide between them. We must know some attributes of *A* and *B* in order to choose between them. Each alternative may have a cost, and each may have a value. The cost is the price we must pay for choosing that alternative, and the value is the benefit we derive. If the costs and values are given in measurable units, the decision would appear to be simple: Choose the alternative for which the value less the cost is greatest. It is not so simple. Suppose alternative *A* costs \$100 and returns a benefit of \$1000. Suppose alternative *B* costs \$1 000 000 and returns a benefit of \$2 000 000. Clearly we have more to gain from alternative *B*. In decision support terms, we say that alternative *B* "dominates" alternative *A*. This leads us to a simple mechanical step in making decisions: Compare values and costs of decision alternatives, and discard the alternatives that are clearly dominated by other alternatives.

However, if we do not have \$1 000 000, we must, with regret, choose alternative *A*. Perhaps we could pursue alternative *B* by borrowing the money and then paying it back when we receive the value of our investment. This sounds attractive but leads up to wonder if we have stated the problem correctly. Are we certain of the value of *B*? Perhaps there is a risk that we will not ever see the \$2 000 000 or that we will see it at a later time.

Let us state the problem more realistically. We need to assign probabilities to the costs and values. Suppose we know that, for each of the two alternatives, there is a 50% chance of the cost being the stated amount and a 50% chance of it doubling. Suppose we know that there is a 90% chance of getting the stated value and a 10% chance of getting nothing. The expected costs and values are then

Alternative	Cost	Value
A	\$150	\$900
B	\$1 500 000	\$1 800 000

Alternative *B* still looks more attractive, as long as we borrow \$1 500 000. We have nine chances out of ten of getting \$2 000 000 and one chance in two of only spending \$1 000 000 of the \$1 500 000 we borrow. But what if we hit the one chance in ten and have to pay back \$1 500 000 with no value in our hands? Alternative *B* could cause ruin. In decision support, the rules of "ruin" and "quasiruin" are applied. If an alternative admits the possibility of a disastrously high net cost, i.e., of ruin, it is rejected; if it admits a net cost that could be absorbed, but that crosses a subjectively set limit of risk, it is rejected as quasiruinous.

Having pruned the alternatives by removing those that are dominated by others and those that may cause ruin or quasiruin, we can compare the remaining quantifiable alternatives by computing appropriately weighted sums of the possible costs and possible values of each alternative. It is tempting simply to apply the probabilities of the various outcomes and use those as the weights, as we did above, but that is rarely sufficient for a sound decision. Each probability must also be multiplied by appropriate subjective measures of the desirability of each outcome, creating what is called the "subjective expected utility" or "SEU." Comparisons of SEUs are at the heart of most decision support tools. However, a word of caution is appropriate. Such methods can only be as good as the accuracy of the values and weights assigned. In many cases it is difficult to assign numeric values accurately, and sometimes it is impossible (see Salvendy, 1987).

6. SUPPORT FOR GROUPS

As Pressman (1995) noted, "People—not technology or process—build software." People working in groups raise some of the most difficult issues in software engineering.

A person developing a completely new program for his own use and on his own initiative has very effective internal lines of communication for the project. He need only consult with himself to define his requirements, approve his specifications, decide on changes, and gain approval from the customer for job completion. Things are not so simple when more than one person is involved as a developer, or if the users, customers, or other decision makers are different people than the developers. Issues of effective communications, team formation, and group interaction become critical to the software development process.

In this respect, software engineering has much in common with any other group engineering activity, and the lessons of classical industrial management apply. Once the job is no longer to be accomplished by an individual working alone, some form of allocation of tasks among people must be made, and some mechanism must be set up to make those allocations of tasks, to monitor progress, and to ensure integration of subtasks into a coherent whole. Except for the use of people instead of computers, this has much in common with the problems presented in organizing a complex program into cooperating modules. The software engineer has the advantage of bringing concepts and tools he routinely uses for software design to the task of designing the process of designing software. There are limits to this, however. People are not computers.

Coordination of tasks being done by different people is a major stumbling block in group development activities. According to Kraut and Streeler (1995), the most valuable and most used coordination technique in software engineering is discussions among peers, especially in the same project. The challenge is to integrate this powerful communication mechanism into the development process in ways that help it fit project goals, rather than in ways that work at cross purposes.

6.1 Organizational Issues

Creating an environment in which people can work effectively on software is difficult, especially when an existing external organizational structure has to be considered. If the organization primarily rewards managerial career paths and not lower-level line activities, then there is a strong disincentive for programmers to stay on the line and code. This has led to the creation of programming management structures that are not related to the technical issues of creating software but that must be considered in the process. If organizational lines of communication are vertical and centralized, the essential interpersonal communications needed for software engineering by groups may be blocked. In some organizational cultures, there is a sharp distinction between analysts who define what has to be created and implementors who create what the analysts have designed. This can be a valuable division of labor, equivalent to that between architects and builders in construction or, as is so often the case in software design, one that breaks down because of blocked or garbled communications. It can work well when the analyst has a deep understanding of both what the client says and what the implementor does, and when the implementor is working on common ground with the analyst, and when implementors have sufficient cross communication with one another and with the analyst to stay on track with one another. In some organizational cultures it is common for those who will use or operate a system to be a distinct group from those who designed it, and perhaps even from those who requested it. A formal process of handoff of a system from development to production happens. Then, when further changes to the system are needed, if smooth paths back into development are not available, conflicts may develop and production may start doing its own programming. As we will see in discussing approaches to ensuring quality in Sec. 7, these traditional data-center structures are under strong competition from more team-oriented, flat hierarchy approaches. For more detail on the classic data-center organization, see Withington (1972).

6.2 Including the Client in the Design Process

People who ask for software from others rarely know exactly what they want when they first ask. Even if they are well versed in both the application domain and the computer technology involved, decisions about very fine details of an information system can seriously affect their perception of that system. As a system develops, it takes on a character of its own and, for good reasons, may drift from the original specifications. Once built, a system may demand too much familiarization time and effort from the client. The obvious way to minimize these problems is to include the client in a significant portion of the design process. As details are refined, the client can react before those details are so deeply embedded in the system design that change is horrendously expensive. As the system evolves, client involvement ensures that mutations are acceptable or at least understood. Being involved in the design gives the client a head start on familiarization. There are many informal and formal approaches to involving clients in software design, ranging from one extreme of presenting the client with *faits accomplis* to take or leave when the project is done to the other, so-called "Scandinavian approach" of having the entire user community be part of the design process throughout the life of the project. Let us look at two approaches. One, called joint application design (JAD), is in the middle of the range, and the other, called participatory design (PD), is an example of full involvement.

JAD began as a way for clients to refine their requirements before presentation to software designers but has evolved into a joint effort on both requirements and design involving clients and software designers. JAD is a very formal method based on meetings. Each meeting is attended by the clients and system designers. The meetings are "facilitated" by a meeting leader. The facilitator must be a neutral party adequately trained to lead the meeting and well versed in system development methods, but not seen as an advocate for either the users or the developers. The facilitator controls the agenda and the discussion and ensures full documentation of the meeting results, usually using a person designated as "scribe" to capture and preserve all the design work done at the meeting. Visual aids, and sometimes highly automated design systems, are used to help present ideas for discussion with maximum clarity.

PD is a less rigid approach with more intimate and continuous user involvement. The focus is brought out of the meeting room and into the workplace where the system will be used. Users and software designers try to teach one another about their working environment and the range of technical options in what is called "mutual reciprocal learning" and do the "design by doing," using hands-on approaches that let the users work with simulations of systems to get a feel for what is being proposed and show the designers what they are doing or want to do with the system. The approach may be very low-tech, with cardboard models, or medium-tech, with simulations on modest personal computers. The demands of such detailed interactions are beyond the capabilities of most existing higher-technology systems.

These approaches both have had some success, but JAD, with its meeting orientation and formality, has had its greatest impact in helping users to clarify their requirements, and its impact tends to fade in the later design stages until prototype systems are available for review. PD maintains a deeper involvement further into the design cycle and, by forcing designers to mock up their ideas before the prototype stage, keeps the development closer to evolving user needs. JAD tends to be a middle-manager methodology, while PD provides more involvement from lower-level line people. Because of that, use of both approaches could be complementary, provided means were found to ensure a commonality of goals between the two groups. For more information on JAD and PD, see Muller and Kuhn (1993) and Carmel *et al.* (1993). For a discussion of a JAD-derived approach for the implementation stages, called the joint implementation process (JIP) involving implementation coordinators (ICs) working in dynamic unstructured environments, see Fallon (1995).

7. QUALITY ASSURANCE AND CONTROL

Quality assurance and quality control begin with a simple goal: to assure ourselves that what we produce will function for its intended customer in its intended manner. In manufacturing industries this effort has been formalized into two related disciplines: quality assurance (QA), which is the overall effort by management to assure itself of the quality of a product, and quality control (QC), which is a control system integrated with the manufacturing process as a mechanism to assist in QA.

Some aspects of QA and QC can be mechanized. One can have inspection and test procedures that are imposed on intermediate and final products. One can use languages and processes that include cross checks for internal consistency and for conformance with machine-readable specifications. But in most respects, software engineering remains a human endeavor and quality can only result when all the people in all the critical paths to creation of a sound product are committed and have the skills and authority to turn that commitment into action. As Demming (1991) notes in commenting on American industrial ills, "measures of productivity do not lead to improvement in productivity [and] . . . new machinery and gadgets are not the answer." When an institutional structure isolates line people from the decision making about their activities with many layers of bureaucracy, tries to treat people as interchangeable parts in a machine, actively encourages them to close their eyes to defects in what they are producing, then quality suffers. There is a growing realization that the quality of production and service efforts can be improved by what amounts to an object-oriented structuring of the organizations that create products and services. This consists of bringing the actions together with the structures upon which they act by creating small functional teams with as much decision-making autonomy as possible and by keeping the cross communications among teams as simple and effective as possible. See the discussion of total quality management by Creech (1994).

There have been serious efforts to bring a sound understanding of QA/QC into software engineering. One very popular effort in the United States has been the capability maturity model (CMM) developed by the Software Engineering Institute (SEI). In Europe efforts have focused on the International Standards Organization (ISO) standards in the 9000 series, particularly ISO 9001.

The SEI CMM focuses on five levels of organizational maturity in the processes of software production. At the lowest level there is no formal structure. On the next level, the organization has built the essential tools to be able to repeat past successes in producing software. It has mechanisms supporting the waterfall model to understand requirements, to plan and track projects, and to verify the quality of the software produced, and it has a software configuration management system. On the third level, the organization-wide scope of software development processes is recognized, creating one organization-wide approach to documentation, standardization, and integration of software. On the fourth level, management of software processes, quantitative measures are used for control of the process. On the fifth and highest level, the organization enters into a phase of continuous improvement using quantitative feedback and anticipating future directions. The view is one of bottom-up growth of quality, with quality spreading into an existing organization and changing the internal culture of the organization.

The ISO 9000 series approach is more

top-down, starting with a top-level organizational commitment to deliver what the organization's customers have asked for. An ISO 9001–compliant organization will have a fully documented quality system integrated into the entire production process, mechanisms to ensure a match between customer requirements and software planning, mechanisms to plan and track projects, mechanisms to fit in externally acquired software whether from third parties or from the customer, mechanisms to trace the life of all software products from conception to creation to installation, mechanisms to plan and control the process of creating software, etc. In most respects, ISO 9000 corresponds to SEI CMM level 4 and does not explicitly address the issue of continuous improvement. However, a lack of commitment to improve and a failure to create mechanisms for continuous improvement would indicate a failure of the top-level organizational commitment on which ISO 9000 is based. See Paulk (1995) for a discussion of the SEI CMM and the ISO 9000 series.

It is interesting to consider the interaction of attempts to achieve quality software with the evolution of organizations into dynamic unstructured environments. Especially in view of the organization-wide demands of the SEI CMM approach and the ISO 9000 approach, one might assume that quality requires imposition of more formal organizational structures. This is generally untrue. Indeed the move to "empower" small teams by giving them autonomy and responsibility has taken place in order to improve quality. Quality is improved by tightened lines of communication for decision making and the emotional commitment of people to a process about which they feel a personal sense of control. But there is also a more subtle reason. Products built by small groups have a better chance of being simple enough to understand. Where it is not possible to handle an entire project within a small cohesive group, quality can often be maintained by having a single facilitator or implementation coordinator who, by having to understand the entire project him/herself, implicitly forces simplification onto the process. KISS is an essential element of systems that produce quality software.

8. EXAMPLES OF TOOLS AND TECHNIQUES

Software engineering draws on tools and techniques from computer science, mathematics, and many other disciplines. A good software engineer will be able to use a variety of programming languages and software libraries. He or she will be proficient in establishing appropriate data structures, be adept at sorting and searching, linear algebra, calculus, the theory of approximation, etc. At this writing, however, two areas of expertise are essential in almost all software engineering problems: an understanding of the basics of communications networks and the use of graphical user interfaces (GUI).

8.1 Communications Networks

The design of information-processing systems requires careful consideration of the interactions among different flows of information. Information from two sources may have to be combined before a critical step can be taken. Portions of an information-processing system may be separated by relatively slow and unreliable communications links. The tools and techniques of network design are of increasing importance in software engineering. We will look at two of the most important concepts from network design, which apply to almost all information-processing systems: that queuing-system response times rise sharply as capacity limits are approached, which teaches us how to allocate sufficient resources to produce a system with good response, and that the flow in a network is limited to the capacity of the narrowest bottleneck, which explains the need to focus design effort in removing the narrowest bottlenecks in a system.

In most cases, each subcomponent of an information-processing system may be viewed as a server of processing requests, acting very much like a teller at a bank. Customers for service arrive in some pattern of arrivals, are placed in some order for eventual processing, and then are served with service times that depend on the nature of their particular requests and the present capabilities of the server.

The study of such systems is called queuing theory, which may be viewed as a subfield of the study of "stochastic processes" (see, for example, Karlin, 1969). Stochastic processes are those governed by the behavior of some family of random variables, $R(t)$, where t is an index variable, often representing time. For a particular value t_0 of t, we do not necessarily have a unique value $R(t_0)$. Rather, we know the probabilities that $R(t_0)$ will assume particular values. By probability, we mean the limiting value over infinitely many trials of thc ratio of the number of times $R(t_0)$ assumes the specified values to the total number of trials.

The model we use in dealing with queues considers customers and servers. We have a supply of customers requiring service. We have servers to provide that service. We do not know when any particular customer may arrive, nor do we know how long a server will take to service a particular customer. We only know the general pattern of arrival of customers and provision of service. For example, we might know that customers arrive at an overall average rate r_c, with no dependence between the arrival of any two customers and that servers serve at an overall average rate r_s, with no dependence between the service time for any two customers.

Such customer and server random variables are said to obey Poisson distributions, with a mean interarrival time of customers of $1/r_c$ and a mean service time of $1/r_s$. If there are m servers, such a model is called an M/M/m system, where M stands for Markov, since Markov chains are an effective tool to study such systems. More generally, if the customer random variables obey a rule X, and the server random variables obey a rule Y, and there are m servers, then the queue model is called an $X/Y/m$ system. In Fig. 5, we see two $X/Y/1$ queues and one $X/Y/3$ queue. For most networks encountered in software engineering problems, using M/M/1 queues provides a sufficiently accurate approximation to the behavior encountered, and we use M/M/1 queues as the primary model.

Analysis of M/M/1 queues shows that the average queue length L is given by

$$L = \rho/(1 - \rho), \quad \rho = r_c/r_s,$$

where ρ, the ratio of the average rate of arrival of customers to the average service rate, is known as the utilization. The average waiting time W can be derived by "Little's Formula" (Little, 1961),

$$L = r_c W,$$

from which we conclude that both the average queue length and the average waiting time rise rapidly toward infinity as the average rate of arrival of customers approaches the average service rate. For example, while we have a queue length of 1 for 50% utilization, the length rises to 4 for 80% utilization

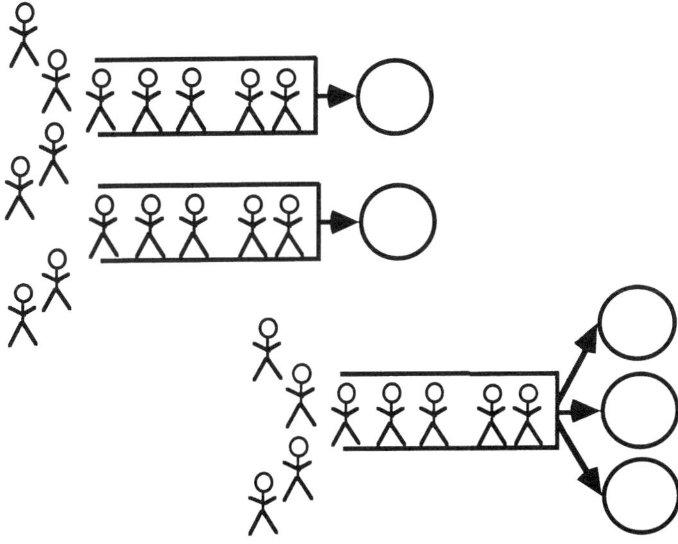

FIG. 5. Two queues with one server each and one queue with three servers.

and to 100 for 99% utilization. Behavior similar to this "soft knee" in an M/M/1 queue is seen in most information-processing systems. For that reason, it is a common rule of thumb to keep at least 20% of the capacity of a system in reserve for almost all systems and a full 50% for systems that must be particularly responsive. For more on the behavior of queues in information-processing systems, see Schwartz (1987).

A full system consists of more than one queue. It will consist of networks of queues, in which the requests serviced by one or more queues become the customers for other queues. We can draw a picture of an information-processing system as a set of processing nodes connected by links along which information flows. The nodes and links joining them form what is called a "graph," which, when we assign a specific direction to each link and forbid links from a node to itself, is called a "network." The study of graphs and networks is called graph theory. For a basic grounding in the subject, see Busacker and Saaty (1965).

When we draw a picture of an information-processing system, we can either draw the precise physical characteristics or abstract only the logical connectivity. If we record the physical locations of the nodes as points in three-space and the paths of the lines as curves in three-space, the sets V of points and E of curves constitute a "geometric graph". While there are design problems that require exact knowledge of node locations and line paths, as in cabling a room full of equipment, much can be learned from the abstract topology of the graph of a system.

Let v and w be nodes, A v-w cut is a set of links, the removal of which disconnects all paths from v to w. A cut is called minimal if restoring any of the links in the cut will reconnect the vertices in question. The capacity of a link is the limit on the rate of flow of information it can carry (e.g., 1000 characters per second). The capacity of a cut is the sum of the capacities of the links in the cut. In Fig. 6, the capacity of each link is given by the number beside it. We have marked a v-w cut of links with total capacity 14, which is less than the capacity of any other cut that could have been used to disconnect v from w.

We can use judiciously chosen minimal cuts to analyze feasible flows between nodes. It is important to realize that we are only looking at the flow between pairs of vertices, not at some generalized total flow in the network.

If we have a feasible flow in which v is the only source and w is the only sink, we say the flow is from v to w. The magnitude of the flow cannot exceed the capacity of a v-w cut (*a fortiori* of a cut of minimal capacity), so the maximal flow from v to w is bounded above by the capacity of a v-w of minimal capacity.

Further, if the maximal flow could not reach the capacity of a cut of minimal ca-

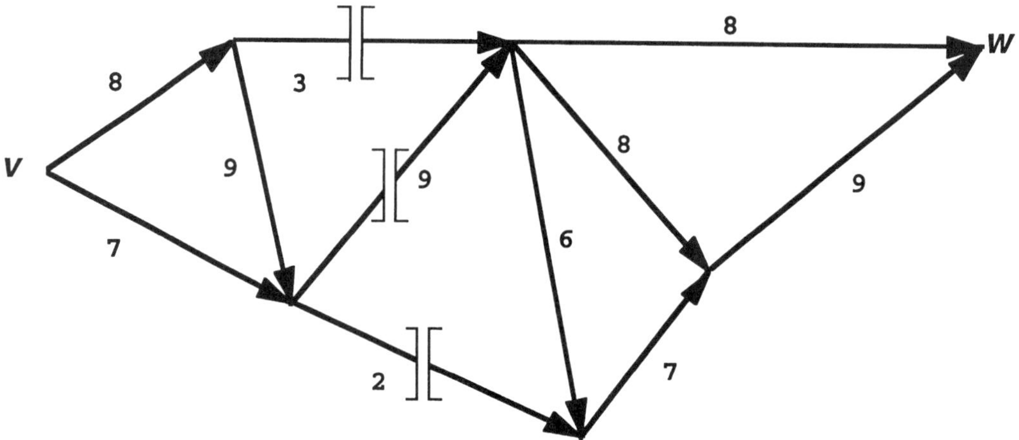

FIG. 6. A network with a *v-w* cut of minimal capacity.

pacity, then the limitation on flow would be due to a cut of still smaller capacity, a contradiction. Thus we have the Ford-Fulkerson "Max-Flow-Min-Cut" theorem:

> The maximal feasible flow from v to w is equal to the capacity of a v-w cut of minimal capacity.

This result is critical to efficient system design. It is saying that we cannot make the flow through an information-processing system be any better than the flow permitted by the capacity of the narrowest bottleneck, that it simply does not pay to put great effort into improving the performance of arbitrary portions of a system. If we want users to see a payback from the effort, we must focus our efforts on those portions of the system where the information flows are constricted. If we combine this with what we have learned about M/M/1 queues, above, we can see that much of the work in software engineering has to be concerned with identifying information flows and capacities and ensuring that we are providing more than adequate capacity at the critical bottlenecks of a system.

8.2 Graphical User Interfaces

A control system for an experiment from the nineteenth century and the first half of the twentieth century would provide its users with a control panel consisting of buttons, switches, dials, and meters. The earliest computers were controlled in a similar manner, but by the 1960s computers and apparatus controlled by computers were managed by typing in commands and looking at lines of characters on paper or on a display screen to see results. For example, to position an axis of a positioner and view related detector output, the user might engage in the following dialog (shown with user responses in bold face):

COMMAND: **POSITION**
WHICH AXIS? **THETA**
THETA = 25.08, NEW THETA? **30.105**
POSITIONING, PLEASE WAIT
POSITIONING, PLEASE WAIT
THETA = 30.105
COMMAND: **DETECTOR**
DECTECTOR = 1.287V
COMMAND:

A lot of good science was done this way, but it is a lot more cumbersome than holding down a switch and watching a position indicator. Graphical user interfaces (GUIs) have allowed for the creation of the modern equivalent of the nineteenth-century control panel on a computer display screen. One can create the necessary software from scratch, but there are now many sound packages that allow rapid and reliable creation of GUI interfaces to information-processing systems. We will look at two of the most popular, Apple Computer's Hypercard® and the so-called "web browsers" based on the hypertext markup language (HTML). Both are examples of hypertext systems, systems that allow documents to be constructed from fragments of text, sound, images, and applications in order to assemble complex assemblies of information with active links among the network of fragments.

Hypercard is an example of a system in which we work to the "WYSIWYG" ("what you see is what you get") paradigm. Hypercard is not the most sophisticated of the GUI-oriented application-building packages, but it was one of the first and is still heavily used, and it illustrates the essential features of such WYSIWYG packages. HTML makes it feasible to build network-based GUI-oriented applications with a high degree of platform independence. In order to provide such independence, there is no assurance that the interface presented on one system will be the same as that presented on another, just that it will work in well-understood and predictable ways. Web browers based on HTML have made distributed graphical access to applications and data commonplace.

If we were to design a control panel for a four-axis positioner, it might look like Fig. 7. The four axes have been labeled with Greek letters, and for each axis, there is an accurate numerical display of the position, buttons to decrease or increase the angular position, and a panel-meter display of the angular position. The detector output is shown as a strip chart. In a full system, we would use a strip chart with more detail, logging the angles in different colors or line styles along with the detector output, but the amount of detail in Fig. 7 should be sufficient for discussion of the graphical user interface. In Fig. 8 we see a partially built con-

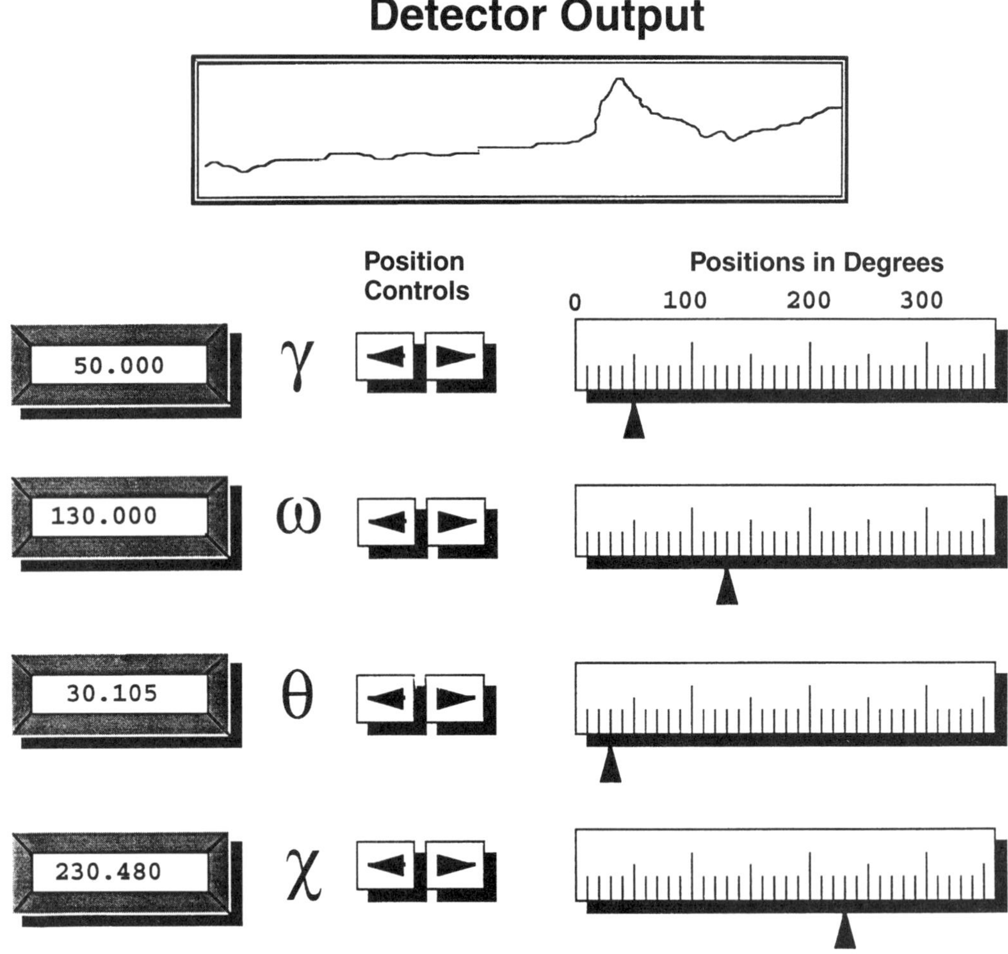

FIG. 7. Design drawing of control panel for four-axis controller.

version of the design drawing into a Hypercard stack. The position controls for the gamma axis have been overlaid with active buttons that can be "pushed" by clicking or holding down the button on a pointing device such as a mouse or trackball. The graphic position indicator on the scale for the gamma axis has been replaced with another button, which can be dragged to the desired angle by placing the pointer into the position indicator and then holding down the button on the pointing device while moving the pointer left or right. The numeric position indicator has been overlaid with a field that displays the current position as the control buttons are pushed or the graphic position indicator is dragged. The user also has the option of typing an exact position into the field. Later in the building of the interface, we would take a section of the display screen for the detector output strip chart and periodically pick up a strip-chart image being maintained by an independent data-logging task.

What speeds the process of creating such useful active images is that they are themselves created by the same sorts of click and drag operations as we just described. For each button, a template of a button was dragged with the mouse to the desired location and then shaped by pushing or pulling on the corners. This saves much tedious measuring with a ruler or the use of graph paper.

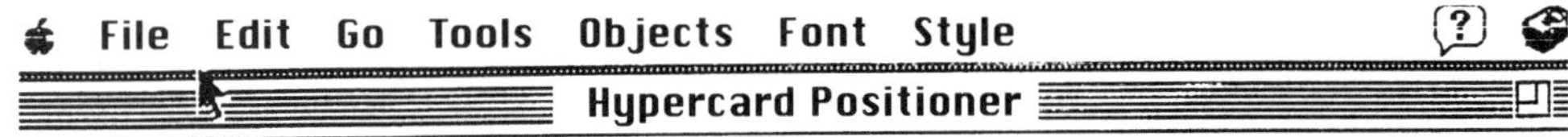

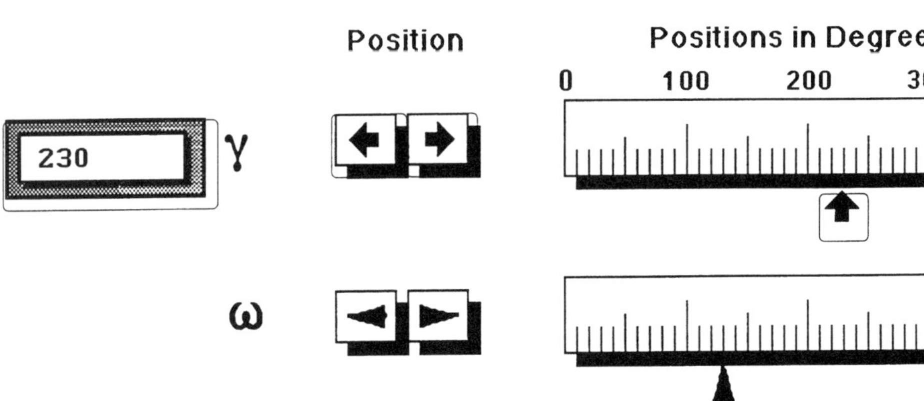

FIG. 8. Screen dump of Hypercard four-axis positioner partially built.

There is much more possible with the use of packages such as Hypercard than just convenient ways to draw images of information-processing systems. Each of the objects in the picture, as well as the entire picture itself, can have an active program, called a "script" in Hypercard, associated with it. Within the scripts are handlers for the messages that may be sent to the object, for example, by pressing the button on a pointing device. Here is part of the script for the positioning button for the gamma axis that decrements the position:

```
on mouseUp
global gammaPos, gammaRate, gammaStart
put .001 into gammaRate
end mouseUp

on mouseDown
   global gammaPos, gammaRate, gammaStart
   put gammaPos into gammaStart
   put .001 into gammaRate
   put gammaPos - gammaRate into xx
   if xx < 0 then put xx+360 into xx
   put xx into gammaPos
   pass mouseDown
end mouseDown

on mouseStillDown
   global gammaPos, gammaRate, gammaStart
   put gammaStart-gammaPos into gd
   if gd < 0 then add 360 to gd
   if gammaRate < .01 then
      if (gd > .01) then
         put trunc (gammaPos*100) into gp
         put gammPos*100-gp into gr
         add .001 to gammaRate
         if gr/100-gammaRate < 0 then
            put gp/100 into gammaStart
            put gp/100 + .01 into gammaPos
            put .01 into gammaRate
         end if
      end if
   else
         .
         . [portion of script omitted]
         .
      end if
   put gammaPos - gammaRate into xx
   if xx < 0 then put xx + 360 into xx
   put xx into gammaPos
   pass mouseStillDown
end mouseStillDown
```

We have provided handlers for three messages in this script, one for the case that the mouse button has just been released, one for the case that it has just been pressed, and one for the case that it is still being held down. Notice that all these handlers do is manage the conversion of information from the state of button presses to two numbers, one held in the global variable gammaPos and one held in the global variable gammaRate. A third global variable gammaStart is used to keep track of when the user started holding down the button. The actual work of displaying the position and managing the actual axis controller is done elsewhere, in the scripts of other objects, which peek at the state of these variables and are also notified of the need for an update by the "pass" commands that cause the messages that were received by these handlers to be passed on to other objects by the rules of inheritance. This incomplete style of programming, where we see only a small part of the effects of an action, this programming by "side effects," is typical of object-oriented designs and appropriate for the design of systems that must handle multiple cooperating autonomous processes. Notice also that the bulk of the code in the script for the case of the button still being held down derives its complexity from human engineering concerns. When we first press the button, we wish the axis to take its highest resolution steps. As we hold it down, we wish the axis to move faster and faster. Therefore we keep increasing gammaRate first in unit steps within each decade and then bump it to the next decade. However, in order to produce a display that causes minimal confusion, it is important to make the transition on rate decades only when the position is on a decade boundary, but also to postpone the transitions long enough to allow for realistic reaction times by human operators. For more on Hypercard and hypertext in general, see Nielsen (1990) and the references therein.

Now suppose we want to allow a similar four-axis control to be operated from a variety of terminal types. We want to design our software just once to allow position control from high-performance workstations, medium-scale graphics-oriented personal computers, and text-oriented terminals. The display station might be directly coupled to the system controlling the position or connected via a wide-area network with long delays. For experimental facilities in the physical sciences with visitors from other institutions, this is a common requirement.

The most popular means of solving such problems is to design the interface for the control application using HTML, the hypertext markup language. This language allows us to specify a hypertext document in broad functional terms, leaving it to specific browser applications on the particular display systems to decide how best to present them. Thus, although HTML documents can be edited with WYSIWYG editors, it is important to realize that the result is a simple text document marked up with minimal instructions for display. For example, the portion of an HTML document to provide control of the gamma position for a four-axis positioner might be as shown in the HTML fragment in Fig. 9. The portion of the image produced on a simple browser by this sec-

```
<!doctype html public "-//IETF//DTD HTML 2.0//EN">
<HTML>
<HEAD><TITLE>Positioner Control Page</TITLE></HEAD>
<BODY> ...

<FORM ACTION="...[positioner server goes here]/cgi-bin/script">
<p> Position Controls:
<H2> Gamma </H2>
Position:<INPUT TYPE="text" NAME="Gamma_Position" VALUE="000.00" SIZE="7">
<menu>
<li><INPUT TYPE="radio" NAME="Gamma_Step" VALUE="up"> up
<li><INPUT TYPE="radio" NAME="Gamma_Step" VALUE="down"> down
<li><INPUT TYPE="checkbox" NAME="Gamma_Rate" VALUE="slow"> slow (.01)
<li><INPUT TYPE="checkbox" NAME="Gamma_Rate" VALUE="medium"> medium (.1)
<li><INPUT TYPE="checkbox" NAME="Gamma_Rate" VALUE="fast"> fast (1)
<li><INPUT TYPE="checkbox" NAME="Gamma_Rate" VALUE="gallop"> gallop (10)
    </menu> ... </FORM>
```

FIG. 9. Fragment of HTML document for one axis of four-axis positioner.

tion of the HTML document is shown in Fig. 10. Note that we have checked the box for "slow" positioning and clicked the "up" positioning button. We could just as well have typed in a new position within the position box. These actions are sent back to the server by the browser, and it will reply with an updated form with an updated position. Different browsers may use different graphics for the check box or the position or other elements, or even do them as straight text, but they all would look the same to the server updating the position. Pictures as good as and better than the examples we used for Hypercard, with portions made active for detection of the position of a pointing device, can be used, as well, but it is a good practice to provide the sort of simple alternative shown here so that even the most primitive display terminals can be used.

HTML and related approaches are evolving rapidly. They have drastically changed perceptions and expectations about what can be done with computers. Coupled with an understanding of the data communications networks upon which they are based, they are becoming the most visible, if not necessarily the most important, of the tools of software engineering. For more on HTML, see Graham (1995).

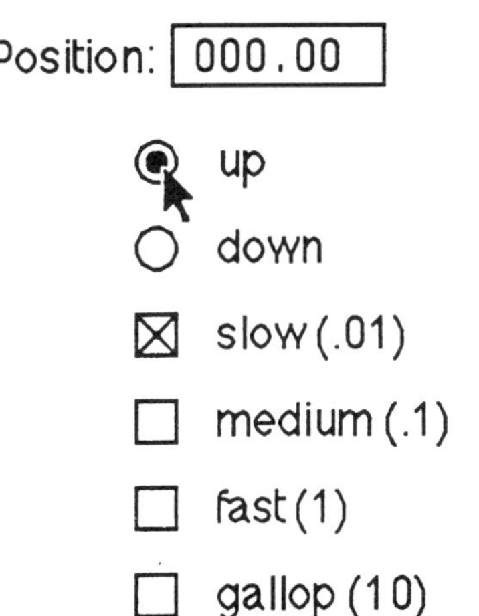

FIG. 10. Browser display for gamma axis.

9. FUTURE DIRECTIONS

Software engineering is a far-from-mature discipline. There is no general agreement on the practical details of how to produce quality software reliably and consistently. Methods that work in one organization may fail completely when transplanted to another. No single method has been a complete success. Top-down analysis in structured programming, side-effect–free applicative programming, using ADA as an environment for software engineering, side-effect–rich object-oriented programming, and bottom-up synthesis to do encapsulation all have advantages and limitations. Where do we go from here?

Software engineering exists within a context of rapidly changing hardware. In the near future we can expect increasing uses of networks of computers as tightly coupled symmetric multiprocessor configurations, as loosely coupled groups intentionally joined to solve cooperatively pieces of a larger problem, and as randomly interacting communities of computers dealing with one another as autonomous sources, transformers, and sinks of information. Software engineering in such environments will have to merge with a more general concept of information systems engineering in which hardware, software, and human-based systems are all components of larger abstract systems for the processing of information and in which the system has to transcend changes in any of these areas. Integrated approaches to hardware, software, and human interactions are needed. See White *et al.* (1993) for a discussion of "Systems Engineering of Computer-Based Systems."

If one were to predict the major impact of such pressures, it would be for an increasing emphasis on the basic tools of top-down analysis, bottom-up synthesis, object-oriented representation, and use of software to process software with an emphasis on simplifying and delayering the processes, as we have already seen in industrial processes. Rather than relying on separate teams specializing in discrete steps in the waterfall model, there will have to be more emphasis on broadly trained people able to follow a project through from beginning to end, able to understand all aspects from the applications to systems software and hardware, and able to follow such processes more efficiently with fewer handoffs.

In the longer term, as distributed processing expands and self-defining processing systems such as neural nets are used, we can expect the need for probabilistic approaches to increase. When one cannot be certain where the "latest" version of information exists, or if it has been lost, delayed, or garbled, when information becomes soft and fuzzy, the best we can do is estimate probabilities that a system is doing what we want and work to make the risk acceptable (see THEORY AND APPLICATIONS OF FUZZY LOGIC, and Graf and Reyneri, 1995). We will learn to build in redundancy, having multiple systems handle the same tasks with as much independence as possible and compare results. Complete uniformity of process will become a source of risk rather than of reliability, since such uniformity would remove the possibility of truly independent checks. The standard accounting practice of one person doing and another checking will have to extend to software systems design.

As always, there will be panaceas presented, perfect methods and systems that will produce perfect software directly from vague statements of requirements without the need for messy, uncooperative human efforts. Some of these systems will work for some problems, but the truly challenging problems in software engineering, the ones that require vision and commitment, will remain a human endeavor.

ACKNOWLEDGMENT

We wish to thank Frances C. Bernstein and Michael E. Bernstein for their helpful comments and suggestions.

GLOSSARY

Abstract Data Type (ADT): Abstract definition of a class of objects giving the data structures, actions, or methods that may be applied and giving the relationships to other ADTs.

Action: Anything an object may do, e.g., by changing its internal state or by sending a message to another object.

Bottom-Up Synthesis: Building a system or product by hierarchically assembling simple tools and resources into more complex and abstract ones.

CASE: Computer-aided software engineering, the use of computer-based tools to automate steps in software development.

Class: An **Abstract Data Type.**

CMM: SEI capability maturity model, an approach to quality assurance based on increasing levels of control over the production process.

Encapsulation: Creation of more abstract classes from more concrete ones, hiding internal methods and data structures not needed for external actions, a bottom-up technique.

GUI: Grapical user interface.

HTML: Hypertext markup language, a language for the creation of hypertext documents that supports text fragments, images, sounds, movies, and application fragments distributed over wide-area networks.

Hypercard®: A registered trademark of Apple Computer Inc., the name of a GUI-based application building system with hypertext capabilities.

Hypertext: A network-oriented document structure providing active links among distributed document fragments.

Inheritance: Definition of a class in terms of another class, so that the new class inherits the data structures and/or methods of the old. Often considered an essential characteristic of object-oriented programming. Inheritance also is used to denote the implied passing of messages up a hierarchy of actual instances of objects created by inheritance among classes.

ISO 9000: An approach to quality assurance promulgated as a suite of standards by the International Standards Organization (ISO). The approach is based on top-level managerial commitment to quality.

Message: Information passed from one entity to another, possibly to itself. For objects, often used interchangeably with **Action.**

Method: Action a class may take.

Multiple Inheritance: Inheritance by a new class from more than one old class.

Object: A specific instance of an **Abstract Data Type,** i.e., an entity in an information-processing system stated in terms of the data it can manipulate, the actions it can take on those data, and its relationships to other objects.

Object-Oriented Analysis (OOA): Analysis of requirements in terms of objects and their relationships. Not necessarily done top-down.

Object-Oriented Programming (OOP): Programming in terms of objects that take actions on the basis of messages passed to them.

Outside-In Analysis and Design: The alternate or simultaneous application of bottom-up synthesis and top-down analysis to a problem, possibly from more than one "top" starting point and from more than one "bottom" set of resources.

Operation: Same as **Action.**

Overloading: Using the same name for different actions in different contexts, e.g., "+" for integer addition, floating-point addition, and logical union.

Polymorphism: The ability to use a common name for multiple classes or methods. See **Overloading.**

Procedure: A **Routine.**

Quality Assurance (QA): Overall management effort to assure the quality of a product, i.e., that the product conforms to the expectations of the customer.

Quality Control (QC): Control system integrated with the manufacturing process as a mechanism to assist in **QA.**

Re-engineering: Using an existing system as a guide in building a new system.

Reverse Engineering: Recreating design documents by analyzing code and external behavior.

Risk: The expected cost of a failure.

Routine: A self-contained portion of code used by other code. It may invoke other routines.

Side Effect: A nonobvious change to the state of data structures resulting from the invocation of a routine.

Software: Information in the form of data and instructions that affects information-processing systems.

Software Engineering: The disciplined use of well-understood techniques to analyze, design, build, repair, or use software.

Structured Programming: Top-down analysis applied to the flow of control of a system.

Top-Down Analysis: Understanding a system or concept by hierarchically decomposing it starting with the most general level of abstraction and working downward toward increasing specificity.

TQM: Total Quality Management, an approach to assuring quality by restructuring organizations into decentralized interdependent semiautonomous teams.

Works Cited

(Some citations appear under Further Reading, below)

Armbruster, J. L. (1995), *Dr. Dobb's J.* **20,** 76–86.

Baskerville, R. (1993), *ACM Comput. Surv.* **25,** 375–414.

Booch, G. (1983), *Software Engineering with Ada,* Menlo Park, CA: Benjamin/Cummings.

Booch, G. (1991), *Object Oriented Design with Applications,* Redwood City, CA: Benjamin/Cummings.

Busacker, R. G., Saaty, T. L. (1965), *Finite Graphs and Networks: An Introduction with Applications,* New York: McGraw-Hill.

Carmel, E., Whitaker, R. D., George, J. F. (1993), *Commun. ACM* **36** (4), 40–48.

Creech, B. (1994), *The Five Pillars of TQM: How to Make Total Quality Management Work For You,* New York: Truman Talley Books/Dutton.

Demming, W. E. (1991), *Out of the Crisis,* Cambridge, MA: Massachusetts Institute of Technology, Center for Advanced Engineering Study.

Dijkstra, E. W. (1968), *Commun. ACM* **11,** 341–346.

Dijkstra, E. W. (1976), *A Discipline of Programming,* Englewood Cliffs, NJ: Prentice-Hall.

Fallon, H. (1995), *How to Implement Information Systems and Live to Tell About It,* New York: Wiley.

Felder, M., Morzenti, A. (1994), *ACM Trans. Software Eng.* **3,** 308–339.

Fuggetta, A. (1993), *Computer* **26** (12), 25–38.

Graf, H. P., Reyneri, L. M. (Eds.) (1995), *IEEE Micro* **15,** 10–11.

Graham, I. S. (1995), *The HTML Sourcebook.* New York: Wiley.

Halladay, S., Wiebel, M. (1993), *Object-Oriented Software Engineering.* Lawrence, KS: R&D Publications, Inc.

Hoare, C. A. R. (1969), *Commun. ACM* **12,** 576–580.

Hollingdale, S. H., Toothill, G. C. (1965), *Electronic Computers,* Harmondsworth, U.K.: Penguin.

Hudak, P. (1989), *ACM Comput. Surv.* **21,** 359–411.

Karlin, S. (1969), *A First Course in Stochastic Processes,* New York: Academic.

Kraut, R. E., Streeler, L. A. (1995), *Commun. ACM* **38** (3), 69–81.

Little, D. (1961), *Operations Res.* **9,** 383–387.

Martin, R. L. *et al.* (1984), *AT&T Bell Laboratories Tech. J.* **63,** 1571–1910.

Morrison, P., Morrison, E. (Eds.) (1961), *Charles Babbage and His Calculating Engines,* New York: Dover.

Muller, M. J., Kuhn, S. (Eds.) (1993), *Communications of the ACM: Special Issue on Participatory Design* Vol. 36, New York: Association for Computing Machinery.

Nielsen, J. (1990), *Hypertext & Hypermedia,* San Diego, CA: Academic.

Nielsen, J. (1992), *Computer* **25** (3), 12–22.

Paulk, M. C. (1995), *IEEE Software* **12,** 74–83.

Pressman, R. (1992), *Software Engineering, A Practitioner's Approach,* 3rd ed., New York: McGraw-Hill.

Pressman, R. (Ed.) (1995), *IEEE Software* **12,** 101–102.

Salvendy, G. (Ed.) (1987), *Handbook of Human Factors,* New York: Wiley.

Sandhu, R. S. (1993), *Computer* **26** (11), 9–19.

Schwartz, M. (1987), *Telecommunications Networks,* Reading, MA: Addison-Wesley.

Siddiqi, J. (1994), *IEEE Software* **11** (2), 18–19.

Sneed, H. M. (1995), *IEEE Software* **12** (1), 25–34.

von Neumann, J., Morgenstern, O. (1944), *Theory of Games and Economic Behavior,* Princeton, NJ: Princeton University Press.

White, S., Alford, M., Holtzman, J., Kuehl, S., McCay, B., Oliver, D., Owens, D., Tully, C., Willey, A. (1993), *Computer* **26,** 54–65.

Winsberg, P. (1995), *Datamation* **41** (9), 36–41.

Withington, F. G. (1972), *The Organization of the Data Processing Function,* New York: Wiley-Interscience.

Wright, P. H. (1989), *Introduction to Engineering,* New York: Wiley.

Further Reading

Blum, B. I. (1992), *Software Engineering, A Holistic View,* New York: Oxford University Press.

Booch, G. (1991), *Object Oriented Design with Applications,* Redwood City, CA: Benjamin/Cummings.

Creech, B. (1994), *The Five Pillars of TQM: How to Make Total Quality Management Work for You,* New York: Truman Talley Books/Dutton.

Dijkstra, E. W. (1976), *A Discipline of Programming,* Englewood Cliffs, NJ: Prentice-Hall.

Muller, M. J., Kuhn, S. (Eds.) (1993), *Communications of the ACM: Special Issue on Participatory Design* Vol. 36, New York: Association for Computing Machinery.

Salvendy, G. (Ed.) (1987), *Handbook of Human Factors,* New York: Wiley.

SOIL POLLUTION

JANICK F. ARTIOLA AND MARK L. BRUSSEAU, *Soil, Water, and Environmental Science Department, University of Arizona, Tucson, Arizona, U.S.A.*

INTRODUCTION

Soil, although often overlooked, is a critical component of the earth and especially for human existence. Soil, directly or indirectly, serves as the major source of food production for humans. All of the structures and dwellings used by humans are built on soil. And, central to the topic of this article, the majority of wastes produced by human activity are purposefully or accidentally released to soil. In this article we define the nature and properties of soils and discuss some of the important interactions between wastes and soil as well as methods for detecting, remediating, and restoring contaminated soil. It should be stressed that the behavior and pollution of water in the subsurface (soil moisture, groundwater) is intimately related to that of soil. Thus, our discussion will, in effect, cover pollution of subsurface water and soil.

1. NATURE OF SOILS

The soil environment is a mass of unconsolidated material that lies between geologic deposits and the atmosphere. Thus, soils cover much of the exposed earth's surface and support many forms of life, including humankind. The soil environment is a highly dynamic system in which a multitude of physical, chemical, and biological processes are constantly operating. Examples of these processes include chemical oxidation/reduction, hydrolysis, physical breakdown of minerals due to temperature changes and mechanical action, and elemental biocycling in organisms that use soils as a growth medium. These energetically diverse processes act on any exposed geologic deposits and transform them into "soil" material.

Soils are formed by the combined contributions of five factors, namely, parent material, climate, topography, biological activity,

3-527-28140-1/96/$5.00 + .50

and time. The geologic deposits provide the parent material from which soils are formed. The composition of the soil will in large part be determined by the parent material. Through their combined actions, climate, topography, and biological factors produce the soil by weathering the parent material. The soil-forming processes operate through the course of time, the fifth factor.

Soil weathering processes decrease rapidly with depth, leading to the formation of soil horizons. These layers, typically parallel to the surface, are studied by soil scientists and provide descriptions of soil-forming processes acting on the horizons of soils. A typical soil profile will have no fewer than three horizons named A, B, and C (see Fig. 1). The A horizon is the layer of soil directly exposed to atmospheric, physical, chemical, and biological processes. This horizon is manipulated to sustain agricultural food production. Horizon B lies directly below A and is considered a zone of accumulation as well as a transitional zone between the exposed A horizon and consolidated geologic material. Horizon C signifies the beginning of unweathered geologic material that may or may not have been the origin of the overlying soil.

Soil is a heterogeneous medium consisting of four major components. The solids component consists of the inorganic materials (e.g., sand, silt, clay) that serve as the support structure of soil. This component is probably what most people think of as soil. However, soil also contains liquids, gases, and biological matter. The liquid component is a solution of water and many dissolved chemicals and is referred to as the soil solution. The gas components that form the soil atmosphere consist of air and fractions of inorganic and sometimes organic gases. The biological component consists of both living (e.g., bacteria) and dead (soil organic matter) material. These components will be discussed in greater detail in forthcoming paragraphs.

Soils are classified first by their physical properties and second by their chemical properties. The commonly accepted triangular textural classification of soil based on percentages of sand, silt, and clay size particles is presented in Fig. 2. It should be noted that clay particles, because of their small diameter, have up to 10 000 times more surface area per unit weight than do sand particles. Thus, clay minerals have unique surface properties that affect many soil physical and chemical processes and will be briefly discussed.

Initial soil classification methods assumed that all clay-size particles are composed of

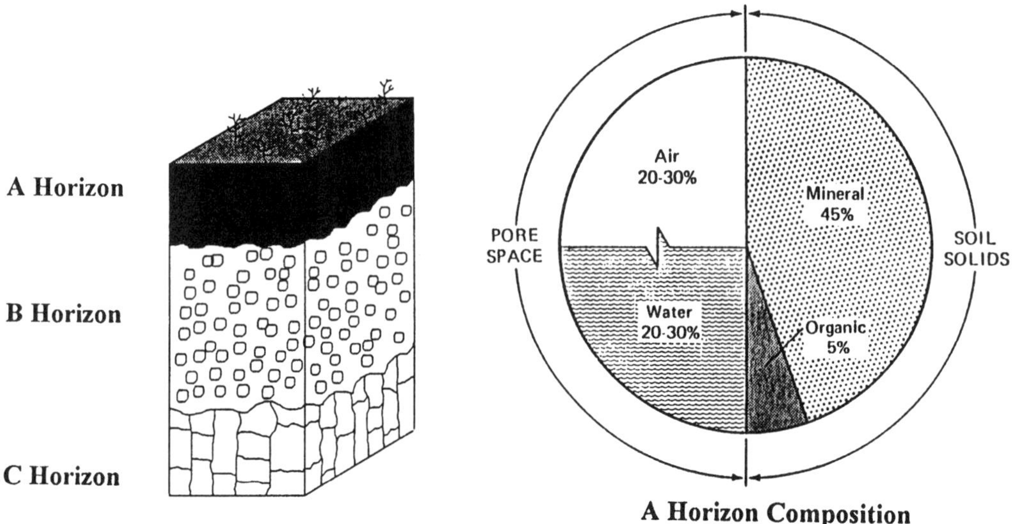

FIG. 1. Soil profile with major soil horizons (A, B, and C) and typical composition (air, water, minerals, and organic matter) of the A horizon (~plow layer). Modified from Buol *et al.* (1973) and Brady (1974).

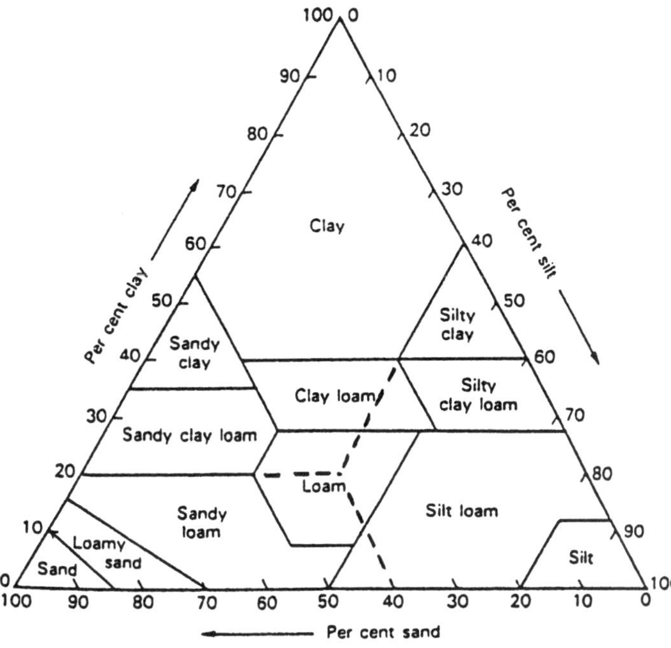

FIG. 2. Textural classes of soils, United States Soil Survey classification. Sand 2.0–0.05 mm, silt 0.05–0.002 mm, and clay less than 0.002 mm in diameter. A typical loam may have about 40% sand and 40% silt particles and about 20% clay. (Buol *et al.*, 1973.)

weathered primary minerals that are called clays, secondary minerals, or phyllosilicates (discussed below). However, it is not uncommon to have in soils significant amount of clay-size particles composed of minerals based on iron, aluminum, and/or manganese oxides, calcium and magnesium carbonates and sulfates, and even organic matter. When these components are found in small quantities in soils (<5%), they act as binding agents among clay, silt, and sand particles. But, when found in larger concentrations, these components can impart irregular and unpredictable physical and chemical properties to soils. Nonetheless, most clay soils have properties distinctly different from sandy soils. Of particular importance is the ability of clay particles to adsorb and to retain water via capillary suction within the micropores of clay particle aggregates. Thus, the water retention capacity of soil decreases rapidly with increasing percentage of sand. Sand particles form large pores that drain easily in response to gravitational forces and do not retain much water.

Clay minerals form during the weathering of primary silicate (silicon based) minerals such as micas and feldspars (Birkeland, 1974). The process of weathering includes selective removal or replacement of elements such as potassium, magnesium, silicon, and calcium from the crystal structures of the primary minerals (aluminosilicates). Clays such as montmorillonite may also be formed by the recrystallization of some minerals such as vermiculite. Substitution of elements such as aluminum(III) for silicon(IV) or iron(II) or magnesium(II) for aluminum(III) results in a gain in net negative crystal charge and gives rise to the negative surface charge of clay minerals. Thus, these minerals have significant cation exchange properties and readily exchange soluble cations with the soil solution. The interactions between negatively charged surfaces of clay and soluble cations have profound implications in soil–plant fertility relationships, as well as the movement and attenuation of inorganic cationic pollutants such as cadmium(II) and lead(II) in the soil environment.

Soils with shrink-swell properties contain dominant amounts of expanding clay minerals, such as montmorillonite (smectite) and vermiculite. Soils with visible surface cracks usually contain significant amounts of these clays. These clay minerals have an expanding interlayer that can gain (absorb) or lose water depending on the amount of water available and thus change crystal dimensions. Kaolinite, a common clay mineral used in ceramics, does not have expanding crystal interlayer spaces and does not exhibit shrink-swell properties during wet-dry cycles.

Organic matter affects the nature and properties of soils in several important ways. The bulk of the organic matter found in soils has its origins in the cycles of plant growth and decay. Dead plant tissue decomposes and undergoes a series of microbially and chemically catalyzed reactions that lead to the formation of humic substances in the soil environment. The mechanisms for the formation of humic substances are not completely understood. But they may include polymerization of small fractions of oxidized organic matter and the restructuring of partially oxidized natural macromolecules such as lignin, the major component of wood. Humic substances have very complex and variable three-dimensional macromolecular structures of polymerized aliphatic and aromatic hydrocarbons (MacCarthy *et al.*, 1990). In general, humic materials have an outer structure with ionizable groups that make them stable in aqueous soil environments. However, the inner structure of these substances appears to be composed of less polar (more hydrophobic) groups. These combined external and internal properties impart to soil humic substances a "reactivity" for organic and inorganic substances. Thus, humic substances play a key role in the fate and transport of pollutants such as metals and of organic chemicals such as pesticides, industrial solvents, and petroleum hydrocarbons.

Humic substances accumulate in the soil environment in response to several environmental factors associated with climate, landscape, and plant species. For example, forested soils in cold climates accumulate several times more organic matter than prairie soils in hot semidesert environments. Humic substances act as binding agents for soil mineral particles and can promote the formation of soil aggregates. These aggregates improve water holding capacity, water movement, nutrient availability, and oxygen-gas transfer properties of soils.

Soils have distinct colorations that extend from black to white and from green to red. The two dominant soil fractions that control color are organic matter, which darkens soils, and iron oxides, which impart reddish-brown tones to otherwise gray silicate minerals. Desert soils that have dominant fractions of calcite and gypsum minerals can have light beige to near-white colorations that may be accentuated by the accumulation of sodic salts. Less common green-colored soils may be found in buried waterlogged soil environments and are due to the formation of reduced iron and manganese minerals.

The soil environment is very rich and diverse in organisms. A list of types and numbers of organisms that inhabit a typical fertile soil is presented in Table 1. Bacteria are by far the most common and diverse group of microorganisms found in soils. In fertile soils, the number of bacteria per gram of soil ranges in the millions. Bacteria are involved in major nutrient recycling processes such as nitrogen fixation, as well as organic-matter decomposition and chemical transformation processes. Soil bacteria are the most active organisms involved in the biotransformation and detoxification of organic wastes and organic pollutants disposed of and/or de-

Table 1. Organisms in the soil environment (from Brown, 1978).

Microorganisms in fertile soil[a]	(10^6 g^{-1})	Animals in mull soil under beech	(10^6 ha^{-1})	% of total animal mass[b]
Bacteria	1–100	Earthworms	1.8	75.1
Actinomycetes	0.1–1	Enchytraeid worms	5.3	1.5
Fungi	0.1–1	Gastropods	1.0	7.0
Algae	0.01–0.1	Millipedes	1.8	10.6
Protozoa	0.01–0.1	Centipedes	0.8	1.8
		Mites and springtails	44.1	0.4
		Other	7.2	3.6

[a]The carbon in soil microorganisms accounts for about 3% of the soil organic carbon; that is, they contain about 1 ton of carbon per hectare in a soil with 1.5% carbon to 15 cm depth.
[b]Total mass of animals was 286 kg/ton of carbon.

posited on land surfaces. Higher organisms such as earthworms are very active not only in organic-matter transformations but also in physical mixing and translocation of minerals in the soil environment.

The nature of soils is very complex and dynamic. Soil scientists must apply fundamental concepts in physics, chemistry, and biology to study and understand the interrelated processes ongoing in the soil environment. Soil systems have a unique capacity to adapt to external stresses. Humankind depends on a healthy soil environment for survival. Thus, humankind must not take the nature of soils for granted by exceeding their ability to cope with external stresses that may damage the soil environment irreversibly.

2. SOURCES OF POLLUTION

There are numerous sources of soil pollution. Some of the sources are the result of planned waste-disposal activities, whereas other sources are the result of accidental releases. Just about all conceivable human activities result directly or indirectly in the production of waste materials. The management, treatment, and disposal of this waste is an important aspect of environmental science and engineering. This aspect is, however, outside the scope of the present article. Brief discussions of the major sources of soil pollution are presented below.

2.1 Controlled Waste Disposal

2.1.1 Landfills Through the course of everyday activity, humans produce enormous quantities of wastes. These wastes consist primarily of paper, plastic, glass, aluminum, and food and plant (e.g., yard) materials. It is estimated that each person in the United States produces approximately 1.8 kg of solid waste each day (Pye *et al.*, 1983). Despite increasing efforts to reduce and recycle this material, much of the waste ends up in landfills. The basic premise of a landfill is to bury the waste in a controlled subsurface environment. The primary means by which landfills can serve as a source of soil pollution is through the generation and escape of landfill leachate.

Landfill leachate consists of water in which are dissolved chemicals that, as a result either of the concentrations involved or of their inherent toxicity, pose a risk to human health or the environment. Landfill leachate is typically high in salts and often contains synthetic organic compounds and heavy metals. Leachate is produced by the infiltration of water through the waste material. As the water moves through the waste, components of the waste dissolve into the water. Current regulations require landfills to be constructed so that leachate production and escape are minimized. However, landfills constructed prior to the regulations were usually not so constructed, and, therefore, many older landfills have problems associated with leachates. Compounding this is the fact that many landfills were located in lowlands near streams overlying porous sediments.

2.1.2 Municipal Sewage Disposal In most larger municipalities, sanitary and storm sewage is processed through waste-water treatment plants. The liquid (treated waste-water effluent) and solid (sewage sludge) wastes produced during treatment are often disposed of by land spreading. This involves placing the wastes on or incorporating the wastes into the soil. Sewage waste can contain synthetic organic compounds, heavy metals, and microbial contaminants (pathogens). Once the waste has been disposed to the soil, rain or irrigation water can dissolve contaminants as it moves through the waste, thereby moving the contaminants deeper into the soil. Some waste components can enter the food chain by plant and animal uptake. Pollutant transport by surface-water runoff and by wind is also a concern.

2.1.3 Residential Septic Systems In smaller municipalities and in most rural areas, sewage waste is disposed through the use of septic systems. A septic system consists of a waste storage tank and distribution piping buried in the soil near a home. Sewage eventually makes its way from the home to the soil. If these systems are improperly designed, they introduce contaminants such as microbial pathogens into the soil.

2.2 Leaking Storage Tanks

Liquids are routinely stored in tanks located above or below the ground. These tanks are used to hold, for example, gasoline, aviation fuel, chemical stocks for manufacturing, and cleaning solvents. An estimated 3 to 5 million underground storage tanks exist in the United States, and it is thought that approximately 10% of these may be leaking (EPA, 1987). Liquids leaking from ruptured tanks or from spills near the tank enter the soil, thereby serving as a source of pollution. Fuel components (e.g., benzene, toluene, xylene) and chlorinated solvents (e.g., tetrachloroethene, trichloroethene) are major subsurface contaminants because of their ubiquitous use. Leaks and spills associated with storage tanks are a major mode of entry into the environment for these compounds. Recently, requirements for emplacing and monitoring storage tanks have been enacted to help reduce such problems.

2.3 Industrial Manufacturing Plants

Waste materials are a byproduct of manufacturing processes, and disposal of these wastes serves as a source of soil pollution. The type of pollution will depend on the item being produced and the method of manufacture. Recently there has been a concerted effort to develop manufacturing processes that minimize waste production and dependence on hazardous materials.

2.4 Commercial Activities

Most commercial activities produce waste materials that find their way into soil. For example, chlorinated organic compounds (e.g., tetrachloroethene, trichloroethene) are used for dry cleaning clothes and for degreasing automobile and airplane engine parts. On-site disposal of cleaning wastes can be a source of soil pollution. Heavy-metal pollution is also associated with commercial activities, such as electroplating and automobile repair.

2.5 Energy and Material Production

Energy-production activities such as electricity generation can be a source of soil pollution. For example, when coal is used for electricity generation, devices are often placed in the plant smoke stack to remove sulfur, metals, and particulates so that they are not released into the atmosphere. The waste materials so collected are often disposed of on site by application to soil. Precipitation infiltrating into the waste may dissolve and transport contaminants deeper into the soil. Mining activities can also serve as a source of pollution, such as surface-water leaching of mining spoils.

2.6 Agricultural Activities

2.6.1 Agrochemicals Fertilizers and pesticides (agrochemicals) are routinely used in current crop production activities in the United States. The use of these agrochemicals has helped to increase crop production capacity greatly. However, the application of these agrochemicals to soil can serve as a source of pollution, especially if their use is not properly managed.

2.6.2 Feedlots Current animal production activities (beef, swine, poultry) make use of feedlots. These feedlots can produce large quantities of animal wastes (manure). The flow of precipitation runoff over the feedlots can wash nutrients such as nitrogen and phosphorus off the site. If the nutrient-laden water enters a surface-water body, it can contribute to the process of eutrophication (see Glossary).

3. MEASURING AND MONITORING POLLUTION

The soil environment is a large repository of anthropogenic and natural pollutants. The soil environment by its nature is able, to a certain extent, to sorb, neutralize, degrade, and retain pollutants and thus protect humankind and other living organisms from deleterious exposure to many kinds of toxic chemicals and pathogens. However, excessive disposal of pollutants into the soil environment usually leads to an imbalance in the processes responsible for this protective effect. The soil then becomes contaminated (overloaded) and is no longer able to cope with the pollutants, therefore releasing them into other environments.

It is worth stressing that soil pollution is a relative term that should be qualified with some examples. For example, a soil that has high concentrations of nitrates, as a result of excessive applications of wastes or fertilizers, may not be damaged nor considered polluted. But this condition usually leads to excessive uptake of N by plants, with negative physiological effects. Also, high levels of nitrates in soils can result in nitrate contamination of surface and groundwater resources. Another example is a soil in which a toxic organic or inorganic chemical has been spilled (such as waste oils, pesticides, solvents, metal plating wastes). Such a spill is likely to influence many chemical and biological processes and therefore induce changes in the soil. This soil is likely to release some of these pollutants into other environments and will need extensive remediation and reclamation. This soil is considered to be severely contaminated.

Soil pollution monitoring has two major objectives. The first is to determine the "health" of the soil itself. The other is to determine the location, amount, and character of any pollutants that may be present in the soil.

Soil pollution can be indirectly estimated by measuring soil properties that are affected by pollutants. These properties include water infiltration, organic matter, biological-microbial activity, plant growth, and biological-species diversity. A good example of this approach to soil organic pollutant monitoring involves changes in the biological activity of some microorganisms when exposed to polluted soils or soil-water extracts. In this case, photoluminescent microorganisms exposed to active soil pollutants die or cease to function, resulting in a decrease in light emission. Since these organisms are sensitive to many common pollutants, their responses do not provide identification of the specific pollutant. New developments in this field have enhanced our ability to identify and quantify soil microbial pathogens (bacteria and viruses). These new soil-monitoring techniques make use of soil microbial population DNA extractions combined with DNA amplification by polymerase chain reaction (PCR).

Other important aspects of soil pollution monitoring involve methods of sample collection and sample preparation, which can be difficult to implement. Typical problems encountered during the collection of soil samples include sample contamination and cross-contamination and pollutant losses such as volatilization in the case of volatile organic chemicals. Soil sample preparation often biases results by contamination, and/or by the exclusion of soil components that may or may not be contaminated. Thus, with all of these aspects considered, the data collected should be sufficient in quality and quantity to be able to quantify with a known degree of precision and accuracy the pollutant(s) status in the soil environment.

Soil pollution is usually measured in the context of a three-dimensional space called the soil environment. Some of the important characteristics of this environment were discussed in Sec. 1 and can be drastically affected by pollutants. Soil pollution measurements require that the extent of the contaminated soil volume or mass be delineated. An example of this approach is finding the extent of pollutant contamination of a soil following a chemical spill. To do this, data on the distribution of pollutant concentrations must be collected. The data can then be plotted in three dimensions using statistically derived interpolation functions that help define a contour of the pollutant in the soil environment. Although this approach may not define the exact extent of the contaminated soil, it is the best tool available to soil scientists to measure distributions of pollutants, short of excavating and analyzing all of the soil.

Temporal measurements of soil pollution are often referred to as soil pollution monitoring. These involve repetitive measurements of soil pollution status over time. The accumulation of pollutants in the soil environment must be monitored over time to ensure that they do not exceed the maximum allowable limits. This practice is becoming more common as more wastes are being disposed of in the soil environment. For instance, federal regulations require that agricultural fields to which municipal solid wastes are applied must be monitored for pollutants such as metals and toxic organic chemicals, as well as for microbial pathogens. In this case a soil-sampling schedule is defined and implemented to help monitor the changes in key pollutants in the soil environment.

There are many methods by which to collect samples that can be analyzed for target pollutants. Solid-phase soil samples are collected using devices as simple as a hand shovel and as sophisticated as a truck-mounted hollow-stem auger. Environmental soil sampling requires that soil samples be collected free of cross-contamination and with minimal physical disturbance. This is usually accomplished by collecting soil core samples. For this purpose a variety of hand hollow-stem and mechanical augers have been devised. The sealed soil cores can be stored and transported to the laboratory for analysis, thereby minimizing field contamination and physical disturbance. Plastic sleeves are used to collect samples for routine soil inorganic parameters, whereas metal sleeves are used when organic contaminants are expected to be present in the soil.

Nondestructive or *in situ* soil monitoring is performed primarily for two soil parameters, these being soil moisture and salinity. Infiltrometers are field devices that are placed on the surface of the soil and allow for the measurement of water infiltration rates and hydraulic conductivity in soils. These two parameters are linked to changes in physical and chemical soil properties, sometimes induced by the presence of pollutants such as soluble salts. There are two devices commonly used for this purpose: the double-ring infiltrometer and the tension-disc infiltrometer. Both devices allow for careful metering of water as it infiltrates into the soil profile.

Neutron-activation probes, time-domain reflectometry (TDR), and electrical-conductivity (EC) soil probes are also used for nondestructive measurements of soil-water content and can be used to measure salinity in soils. However, these devices require some disturbance of the soil environment as an access tube or metal probes must be inserted into the profile. The limitations of these devices include the high cost, the use of a low-energy radioactive source (for neutron probe) or of sophisticated electromagnetic radiation sources and measuring devices (for TDR and neutron probe), and frequent calibrations.

Soil-pore water sampling devices are used to collect water-soluble pollutants from the soil environment. These devices do not directly measure pollutant parameters but allow the *in situ* collection of samples of the soil-pore water, which may then be analyzed for compounds of concern. These devices also require some soil disturbance as they must be inserted into the soil profile.

There are two major types of soil-pore water collection devices: passive or gravity fed and active or suction devices. Passive devices such as glass brick lysimeters are inserted into the soil, and gravitational water is allowed to percolate into them. This water is subsequently withdrawn using an access tube and preserved for analysis. Numerous types of porous cups are used as suction devices to collect soil-pore water. These include porous cups made out of ceramic, Teflon, or stainless steel. These cylindrical devices are inserted into the soil, and vacuum is applied to create suction across the porous membrane in contact with the soil solution. This process draws soil-pore liquid into the cup; the water is subsequently withdrawn via an access tube and subject to chemical analysis.

Methods are also available for collecting samples of the soil gas. This is of special interest for monitoring levels of important gases such as oxygen and carbon dioxide and for determining gas-phase concentrations of volatile organic pollutants (e.g., gasoline components, cleaning solvents). Active gas-sampling devices are conceptually similar to pore-water samples in that volumes of the fluid (in this case gas) are somehow extracted from the soil and then removed to the subsurface for subsequent analysis for the target pollutants. Passive gas-sampling devices are based on emplacing a device in the soil that contains a material to which the gas-phase compound of interest can sorb. After a given time, the device is recovered and the pollutants thermally desorbed and subjected to appropriate analysis.

We have seen that the most common approach to measuring soil pollution is done by measuring the concentrations of pollutants in soil samples or soil-water extracts. This in many instances requires destructive sampling or disruptive sample collection. This activity can potentially expose site workers and sampling personnel to the pollutants. Another concern is that many soil sampling and monitoring programs primarily involve collection and analysis of single-phase samples (e.g., soil-pore water or soil cores). However, many pollutants can reside in several phases associated with the soil environment. In these cases, the determination of total pollutant masses is constrained by several sources of uncertainty when based on single-phase samples. For example, if only soil-pore water samples are available, concentrations and masses of pollutants associated with the solid and gas phases of the soil will have to be estimated by use of distribution coefficients. As is discussed in Sec. 4, the use of these coefficients is subject to and limited by several factors. A complete soil sampling and monitoring program requires collection and analysis of all pertinent phases of the soil.

A further complication to soil environmental sampling is the very limited knowledge about the heterogeneity of the soil environment. Since each soil sample collected is part of a larger population of samples present in the soil, a valid and defensible estimate of soil pollution levels requires

1. *a priori* knowledge of the soil heterogeneity and pollutant variance in the soil system;
2. careful use and implementation of soil sampling equipment, sampling protocols, and analytical techniques; and
3. a rigorous statistical approach to determine the appropriate soil-sampling scheme and to analyze the data.

4. TRANSPORT AND FATE OF POLLUTANTS

An understanding of how contaminants move in the subsurface is required to address environmental problems. For example, such knowledge is needed to evaluate the probability of contaminants associated with a chemical spill reaching an aquifer and contaminating groundwater. Such knowledge is also required to develop and evaluate methods for cleaning up contaminated soils and aquifers. Just as importantly, knowledge of contaminant transport and fate is necessary to design "pollution-prevention" strategies.

Four major processes that control the movement of contaminants in porous media are advection, dispersion, interphase mass transfer, and transformation. Advection, also referred to as convection, is the transport of dissolved matter (solute) by the movement of a fluid responding to a gradient of fluid potential. Water is the primary fluid of concern for the systems in which we are interested. Advection, then, is the resulting movement of a solute or pollutant as it is carried along by the flowing water. It is a primary factor for pollutant transport. Dispersion represents spreading of solute about a mean position, such as the center of mass. Phase transfers, such as sorption, liquid–liquid partitioning, and volatilization, involve the transfer of matter in response to gradients of chemical potential. Transformation reactions include any process by which the physicochemical nature of a contaminant is altered. Examples include biotransformation, radioactive decay, and hydrolysis.

4.1 Quantifying Transport and Fate

The equation that is most often used to describe transport of dissolved contaminants in porous media is (Fetter, 1993; Brusseau, 1994)

$$R\frac{\partial C}{\partial t} = -v\frac{\partial C}{\partial x} + D^*\frac{\partial^2 C}{\partial x^2} - \mu C, \tag{1}$$

where x is a spatial coordinate representing one or more directions, t is time, C is the concentration of contaminant in the fluid (e.g., water), v is the average linear velocity of the fluid in the pores of the medium, D^* is the dispersion coefficient with the asterisk signifying the existence of multiple terms for multiple-dimension systems, μ is a first-order transformation coefficient, and R is the retardation factor, defined as

$$R = 1 + (\rho/\theta)K_d, \tag{2}$$

where ρ is the bulk density of the porous medium, θ is the fractional volumetric fluid content (equal to porosity when pores are completely saturated with the fluid), and K_d is a coefficient representing distribution of the contaminant between the solid and liquid phases (e.g., sorption).

Equation (1), known as the advection–dispersion equation, is by far the most widely used equation for describing the transport of dissolved matter in porous media. It is used in chemical engineering, petroleum engineering, and chromatography, in addition to earth sciences. The term on the left-hand side of Eq. (1) represents the change in contaminant mass occurring at a specified location in response to transport and fate processes. The retardation factor represents the influence of phase transfer (sorption) on transport. The first term on the right-hand side represents advective transport, while the second term represents dispersive transport. The third term represents a loss of contaminant from the solution due to transformation reactions. To develop this equation it must be assumed that the porous medium is homogeneous (e.g., v, D^*, and R are spatially invariant) and that the transfer of contaminant between the solid and liquid phases is instantaneous.

The movement of a contaminant from a point where it is first introduced to a point at which it is being monitored can be described by a breakthrough curve. The measurement and analysis of breakthrough curves is a widely used means of investigating contaminant transport in porous media. An example of a breakthrough curve, generated by solving (1) with $\mu = 0$, is shown in Fig. 3. Note the sharp, symmetrical curve obtained for the transport of a contaminant under ideal conditions (e.g., homogeneous porous medium, instantaneous sorption). However, transport often deviates from this ideal case. A breakthrough curve exhibiting nonideal transport is also shown in Fig. 3. This curve is clearly asymmetrical, exhibiting early breakthrough and tailing. Nonideal transport is often observed in real conditions.

4.2 Dispersion

The dispersion coefficient is usually defined as (Fetter, 1993; Brusseau, 1994)

$$D = \alpha v + D_0/\tau, \tag{3}$$

where α is the dispersivity, D_0 is fluid-phase diffusion coefficient, and τ is a factor accounting for the tortuosity of the porous medium. The first and second terms on the right-hand side of (3) represent the contribution of mechanical mixing and axial diffusion, respectively, to total dispersion. Mechanical mixing results primarily from nonuniform flow, which is induced by nonuniform pore-size distributions. Axial diffu-

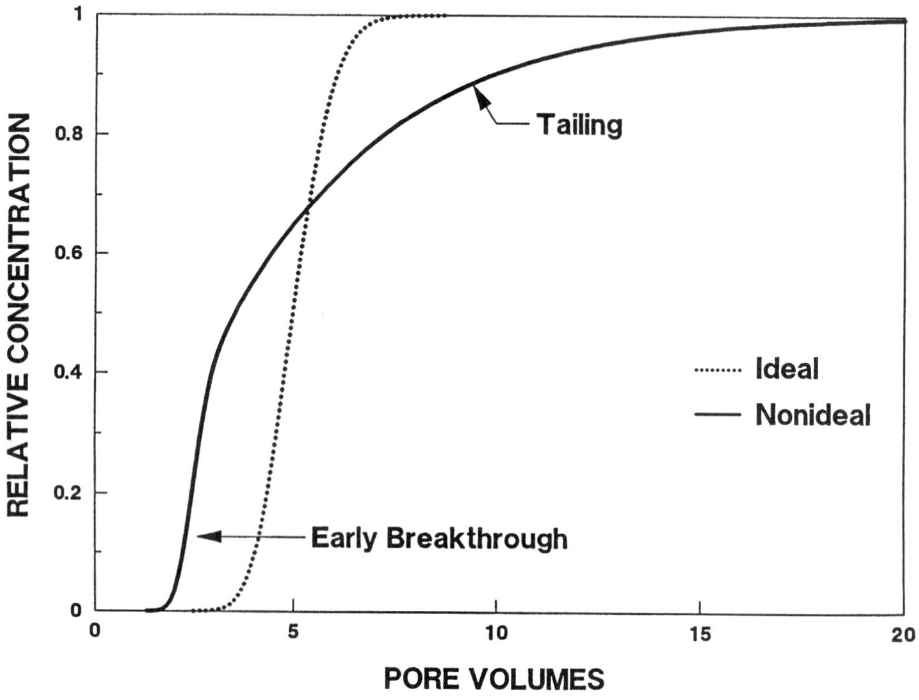

FIG. 3. Ideal and nonideal breakthrough curves. Relative concentration is defined as measured concentration divided by input concentration; pore volume is equivalent to the volume of pumped solution divided by the volume of solution the column can retain.

sion results from natural thermal activity of molecules and usually is described in terms of Fick's law. Generally, axial diffusion is an important component of dispersion only when pore-water velocities are low (e.g., <0.1 cm/h). Such velocities are typical of flow in the vadose zone (the portion of the subsurface between the soil surface and surficial aquifers), in clay layers, and in some consolidated aquifers.

Under ideal conditions, the dispersivity is a function only of the porous medium. A porous medium that has a wide range of particle sizes will also tend to have a wide range of pore-space sizes. Since the flow rate of a fluid is proportional to the size of the pore, a wide range of pore sizes will increase the degree of spreading that occurs during transport. Conversely, a porous medium that is very homogeneous—for example, one that consists of uniform particles—will induce very little spreading.

4.3 Sorption

Sorption or retention is a major process influencing the transport of pollutants in soils. The broadest definition of sorption or retention is the association of pollutant molecules with the solid phase of the soil (soil particles). Properties of the pollutant (e.g., is it nonpolar or polar, organic or inorganic) are key determinants of the degree to which the pollutant will be sorbed or retained by the soil. The specific mechanism by which a particular pollutant is sorbed will depend, in part, on the nature of the pollutant. The physical/chemical properties of the soil are also important.

Sorption is quantified by measuring a sorption isotherm, which is simply a description of the relationship between the concentration of pollutant sorbed by the soil and the concentration in water. Many different forms of isotherms have been proposed and used to describe sorption. The simplest is the linear isotherm (the same idea as Henry's law), which is given by

$$S = KC_w, \tag{4}$$

where S is the concentration of pollutant sorbed by the soil (M/L^3) and K is the sorption coefficient (L^3/M). The larger the value of K, the greater the degree to which a pollutant is sorbed by the soil. A linear isotherm signifies that for all concentrations of pollutant in water, there will always be proportionally the same sorbed concentration. The sorption of many nonpolar organic pollutants is linear. Some organic and many inorganic pollutants exhibit nonlinear sorption. An example of a widely used nonlinear isotherm is the Freundlich isotherm, given by

$$S = K_f C_w^n, \tag{5}$$

where K_f is the Freundlich sorption coefficient and n is, in essence, the power function.

The use of sorption isotherms is based on the existence of equilibrium between the solid and water phases. Research has shown that the rate of sorption of many nonpolar organic pollutants is very slow, taking anywhere from several hours to several months or even years to reach equilibrium. This slow rate of sorption is often due to the slow diffusion of pollutant molecules into spaces in the soil that have very small openings or pores. In addition, pollutants that have been in contact with soil for long times are often difficult to remove from the soil. These factors make it difficult to clean up contaminated soil.

Sorption is such an important process because it reduces the velocity of a pollutant with respect to the movement of water. As long as a pollutant molecule is sorbed by soil, it is prevented from moving with the water. Thus, it will take longer for the pollutant to travel a given distance than it would for water. The effect of sorption on pollutant transport is accounted for by the retardation factor. When the pollutant is not sorbed by the soil (i.e., $K = 0$), the retardation factor R is equal to 1. This means that the pollutant will move at the same velocity as water. When $R = 2$, the pollutant moves at an effective velocity that is half that of water. In other words, the pollutant moves only half as fast as water. When $R = 10$, the pollutant moves at one-tenth the velocity of water. Pollutants that have small retardation factors (<10) are considered to be relatively mobile. They can move rapidly from a spill site and thus quickly contaminate a large area. Conversely, pollutants that have very large retardation factors (>1000) move very slowly with respect to water. Thus, they probably will not create a contaminated zone as extensive as that created by mobile pollutants.

4.4 Transformation Reactions

The transformation of pollutants by microorganisms is referred to as biodegradation. Many organic compounds of interest can be biodegraded by soil microorganisms. Many factors influence the degree to which a compound will be biodegraded in the subsurface. These factors include

1. existence of microbial populations capable of degrading the compound,
2. the availability of the compound to the microbial population (bioavailability), and
3. environmental factors (e.g., *p*H, temperature, nutrient or electron-acceptor availability, etc.).

The bioavailability of a compound is influenced by processes such as mass transfer, rate-limited sorption/desorption, and transport in heterogeneous porous media. Hence, the kinetics of biodegradation can in many cases be controlled by processes other than the physiological activity of the microorganisms.

Some pollutants can be transformed by abiotic (physical/chemical) processes as well. Two major abiotic transformation processes are hydrolysis and radioactive decay. Water is a ubiquitous component of soil systems. As a result, most pollutants will come into contact with water to some extent. It is important, therefore, to understand if and when a pollutant will react with water when they are in contact. Reaction of a pollutant with water is termed hydrolysis. A generalized example of this reaction is given by

$$\mathrm{R}\text{–}X + \mathrm{H_2O} \rightarrow \mathrm{R\text{–}OH} + X^- + \mathrm{H}^+,$$

where R–X is an organic compound with X representing a functional group such as a halide (e.g., Cl). By reacting with water, the original compound (R–X) has been transformed to another compound (R–OH).

The two key factors in hydrolysis reactions are the charge properties of the pollutant molecules and the *p*H. Hydrolysis is essentially an interaction between nucleophiles (substances with excess electrons, such as OH^-) and electrophiles (substances deficient in electrons, such as H^+). Thus, the charge properties of the molecule will govern its reactivity with water. For many compounds, hydrolysis may be catalyzed or enhanced under acidic or basic conditions. This means that the occurrence and rate of hydrolysis is often *p*H dependent. For example, a hydrolysis reaction catalyzed by OH^- would occur more rapidly at higher *p*H values because of larger OH^- concentrations. Hydrolysis can also be influenced by sorption interactions. For example, the *p*H at the surface of many soils is lower than the *p*H of the water surrounding the soil particles. Thus, an acid-catalyzed hydrolysis reaction could be enhanced when the pollutant is associated with the soil.

Radioactive decay is an important transformation reaction for a special class of pollutants: radioactive elements. Radioactive decay is caused by an instability of the nucleus in the atom, whereby either protons and neutrons or electrons are emitted in the form of radiation. This transformation is spontaneous, but the rate at which it occurs varies widely depending on the element concerned.

The rate of radioactive decay and of hydrolysis (for a fixed *p*H) can be described with a first-order equation:

$$\frac{\partial C}{\partial t} = -kC, \tag{6}$$

where k is the rate constant $[1/T]$. A first-order equation can also be used to represent biodegradation under certain conditions (e.g., when the growth of the microbial population is minimal). For transformation reactions it is useful to define a half-life, which is the time required for half of the original pollutant mass to be transformed. For reactions that follow first-order kinetics, the half-life ($T_{1/2}$) is defined as

$$T_{1/2} = 0.693/k_h. \tag{7}$$

This equation can be used to estimate the time required for a pollutant to be transformed, which is related to its persistence in the environment. For example, less than 1% of the original pollutant mass remains after a time period equal to seven half-lives. After ten half-lives, less than 0.1% remains. For a pollutant that has a half-life of 1 day, this would mean that only 0.1% would remain untransformed after 10 days.

4.5 Multiphase Pollution

Some of the most important soil pollutants are immiscible organic liquids. Examples include gasoline, diesel fuel, and aviation fuel associated with various types of fuel stations, chlorinated solvents associated with cleaning facilities, and creosote associated with wood-treatment operations. Separate phases of immiscible liquid may exist in portions of the subsurface at hazardous waste sites contaminated by these liquids. It is well known that when an immiscible liquid moves through a porous medium, a fraction of the liquid can become entrapped, thereby forming a residual saturation. Very large pore-water velocities (i.e., hydraulic gradients) are usually required to displace residual quantities of immiscible liquids. Hence, the primary means of removal will be dissolution into water and volatilization into the soil atmosphere. The residual liquid, therefore, serves as a long-term source of contaminant.

Many organic contaminants of interest are volatile (e.g., gasoline components, chlorinated solvents). For systems that are fully saturated with water (e.g., aquifers), this is of no special concern. However, it is of great significance for the vadose zone, which, because it is usually not completely saturated with water, contains a gas phase. Consequently, a description of the transport of organic contaminants in the vadose zone will be incomplete without consideration of transport in the gas phase. Research in this area has increased greatly with the advent of "soil venting" or "vacuum extraction" as a popular means of remediating vadose-zone contamination.

4.6 Soil Heterogeneity

Soils and geologic materials are inherently nonuniform. This is a direct result of the formation of these materials and their exposure to various events during past geologic times. Thus, properties that affect solute transport can be expected to vary spatially. One of the main reasons for deviations from ideal transport behavior in the field are heterogeneities in the physical, chemical, and biological properties of subsurface material. Any analysis of the transport and fate of pollutants in soil is incomplete without consideration of the nature, magnitude, and impact of heterogeneity.

5. CLEANING UP SOIL AND WATER POLLUTION

The existence of pollutants in soil can pose a risk to human health and to the "health" of the environment. As such, the cleanup of polluted soil and groundwater has become a major component of environmental activity. In the following sections the legislative and technical framework of subsurface cleanup are discussed.

5.1 Legislative Programs for Pollution Cleanup

The first major legislation that dealt with cleanup of contaminated environments was the Water Pollution Control Act (1972). A provision of this act provided authority to fund the cleanup of hazardous substances released into navigable waters. However, there are no provisions for cleanup of contaminated land. The first legislation to directly address cleanup of contaminated land was the Resource Conservation and Recovery Act (1976). This act provided authority to order an operator to cleanup hazardous waste emitted at the site of operation. This act did not cover the many contaminated sites that have been abandoned.

The Superfund program, also known by its official title, CERCLA (Comprehensive Environmental Response, Compensation, and Liability Act—1980) was designed explicitly to address the cleanup of hazardous waste sites. It, along with later amendments (SARA, Superfund Amendments and Reauthorization Act—1986), is the major legislation governing activities associated with the cleanup of environmental pollution. Some states have smaller versions of the federal Superfund program. In addition, the Department of Defense has a separate cleanup program (Installation Restoration Program) for military sites.

Because of its importance, we briefly discuss the major components of Superfund. Its purpose is twofold:

1. to respond to releases of hazardous substances on land and in navigable waters, and
2. to clean up contaminated sites.

The first purpose deals with future releases, whereas the second one deals with sites of existing contamination. There are two types of responses available within Superfund:

1. removal actions: response to immediate threats—e.g., remove leaking drums; and
2. remedial actions: cleanup of hazardous sites.

The Superfund provisions can be used when a hazardous substance is released or when there is a substantial threat of release. They may also be used when there is a release or substantial threat of release of a contaminant that poses imminent and substantial endangerment to public health/welfare.

The first step is to place the potential site in the Superfund Site Inventory. After a preliminary assessment and site inspection, the decision is made as to whether or not the site will be placed on the National Priorities List (NPL). Sites placed on this list are those sites deemed to require a remedial action. There are currently more than 1200 sites on the NPL. The remedial investigation/feasibility study (RI/FS) is the next component. The purpose of the RI/FS is to characterize the nature and extent of risk posed by contamination and to evaluate potential remedial options. The investigation and feasibility study components are normally conducted concurrently and with a "phased" approach. This allows feedback between the two components. The selection of the specific remedial action to be used at a particular site is a very complex process. The goal of the remedial action is that it be protective of human health and the environment, that it maintain protection over time, and that it minimize untreated waste. Specifically, CERCLA mandates that the remedial action must

1. protect human health and the environment,
2. comply with applicable or relevant requirements (ARARs),
3. be cost-effective,
4. use permanent solutions or resource recovery technologies to the maximum extent practicable, and
5. satisfy preference for treatment as a principal element.

After a remedial action has been selected, it is designed and put into action.

An important component of Superfund and all other cleanup programs is the cleanup target. Related to this is the question: How clean is clean? As can be imagined, the stricter the cleanup provision (e.g., lower contaminant concentration), the greater are the attendant cleanup costs. It may require tens to hundreds of millions of dollars to clean large, complex hazardous waste sites completely. In fact, it may be impossible to clean many sites completely. It is very important, therefore, that the physical feasibility of cleanup and the degree of potential risk posed by the contamination be weighed against the economic impact *and* the future use of the site. The consideration of the future use of the site will help focus our scarce resources on those sites that pose the greatest current and future risk.

5.2 Cleanup Technologies

There are three major categories or types of remedial actions:

1. containment—contamination is restricted to a specified domain to prevent further spreading;
2. removal—contaminant is removed from the open environment and transferred to a place of control; and
3. treatment—the contaminant undergoes transformation to a nonhazardous substance.

Since the inherent hazardousness of a contaminant is eliminated only by treatment, the last is the preferred approach of the three. Containment and removal techniques are very important, however, in cases where it is not feasible to treat the contaminant. At many hazardous waste sites, the remedial action in use consists of a combination of containment, removal, and treatment. We focus on each of the three types of remedial actions in turn. Although numerous techniques have been proposed for soil cleanup, only a very few have been tested to any significant extent. Our discussion focuses on those that have been used in the field.

5.2.1 Containment Containment can be accomplished by controlling the flow of the fluid that contains the contaminant, such as controlling the flow of groundwater. This can be done with the use of wells, drains, or physical barriers. The principle of a physical barrier is to install it in front (downgradient) of the contaminated zone, which prevents further spread of the contaminant. An important component of employing physical barriers is the presence of a zone of low permeability into which the physical barrier can be "keyed." Without such a key, the contaminated water could flow underneath the barrier. Usually, physical barriers can be emplaced to depths of about 50 m.

There are three major types of physical barriers: slurry trenches, grout curtains, and sheet piling. Slurry trenches, the most widely used type of barrier, are constructed by filling a trench with clay or a mixture of clay and soil. Grout curtains are formed by injecting into the ground chemicals that are somewhat like cement. In some cases, large sheets of iron (sheet piling) are driven into the ground to form a barrier.

The principle of containment by wells and drains is to manipulate and control water flow by extraction/injection of water. Since contaminant movement is governed largely by water flow, controlling water flow results in contaminant movement being controlled. The key performance factor for this approach is the capture of the contaminant plume. An advantage associated with the use of wells to control contaminant movement is that this is the only containment technique that can be used for deep systems (i.e., >50 m). For this and other reasons, wells are the most widely used method for containment. Some disadvantages associated with wells are the cost of long-term operation and maintenance of the pumps, and the need to store, treat, and dispose of large quantities of contaminated water that is pumped to the surface.

Containment can also be accomplished by immobilizing the contaminant in place. This can be accomplished by cementation, vitrification, freezing, or fixation. Cementation involves solidifying the porous media by injecting cement. A key performance criterion is the long-term stability of the "concrete" block, including the leachability of the contaminant. Vitrification involves "melting" the porous medium to form a "glass" block. This is done by use of an electric current to generate heat (2000–3000 °C). As for cementation, a key performance criterion is the long-term stability of the block, including the leachability of the contaminant. The concept of fixation is to immobilize the contaminant by promoting binding to the solid phase (covalent bond). Key performance factors are delivery of catalyst to target areas and the long-term stability and leachability of the fixated contaminant. Freezing is normally limited to use as a temporary containment technique.

5.2.2 Soil Flushing Soil flushing or "pump and treat" is currently one of the most widely used remedial action techniques. The basic principle of soil flushing is to remove contaminated water from the subsurface by pumping it with one or more wells. Furthermore, clean water brought into the contaminated region by the pumping action will remove additional contamination by inducing desorption from the solid phase (soil particles). The contaminated water pumped from the subsurface is directed to some type of treatment operation. This treatment may consist of air stripping, carbon adsorption, or perhaps an above-ground biological treatment system. There is one major performance criterion when using wells for containment, that being capture of the contaminant. When using wells for contaminant removal (i.e., pump and treat), the effectiveness of contaminant removal is a second major performance criterion in addition to contaminant capture. Recent studies of operating pump-and-treat systems have shown that contaminant removal by pumping is often not very effective. There are many possible factors that can limit the effectiveness of contaminant removal. Some major ones are presence of low-permeability zones, rate-limited desorption, and presence of immiscible liquids.

Because the use of pumping to remove contaminants is a major remedial action technique, methods are being tested to enhance the removal effectiveness. One such approach is to inject into the aquifer a chemical, such as a surfactant, that will promote dissolution and desorption of the contaminant, thus enhancing the removal effectiveness. A key factor controlling the success of this approach is the ability to deliver the chemical to the places that contain the contaminant.

5.2.3 Soil Vapor Extraction The basic principle of soil vapor extraction, or soil venting, is very similar to that of pump and treat: A fluid is pumped through a contaminated domain to enhance removal. In the case of soil venting, however, the fluid is air rather than water. There are two key conditions for using soil venting. First, there must be a gas phase present in the soil, through which the contaminated air can travel. This condition limits the use of soil venting to the vadose zone. Second, soil venting is based on transfer of contaminant from other phases (solid, water, immiscible liquid) to the gas phase. This limits soil venting to volatile contaminants.

Because air is much less viscous than water, much less energy is required to pump air. Thus, it is usually cheaper to use soil venting for removing volatile contaminants from the vadose zone. Once the contaminant is removed from the vadose zone, it is either released to the atmosphere or placed into a treatment system. The major performance factors for soil venting are the same as for water pumping, the effectiveness of capturing and removing the contaminant. The effectiveness of contaminant removal by soil venting can be limited by the same factors that limit removal by water flushing.

5.2.4 *In Situ* Remediation Removing pollutants from soil is often difficult. Hence, techniques have been developed that are meant to treat the pollutant in place. The primary *in situ* treatment technology used to date is *in situ* bioremediation. This usually involves stimulating naturally occurring microorganisms to degrade the pollutant. This stimulation can be effected, for example, by the addition of nutrients or oxygen. The factors previously discussed that control biodegradation (Sec. 4.4) also control the degree to which *in situ* bioremediation will be successful. An additional key factor controlling the efficacy of this technique is the delivery of the nutrients or oxygen to the contaminated domain.

In situ chemical remediation is similar to *in situ* bioremediation in that the contaminant is degraded by promoting a transformation reaction. For chemical remediation techniques, however, this is done with a chemical reaction such as hydrolysis or oxidation/reduction. This approach has been used much less frequently than *in situ* bioremediation.

6. SOIL RECLAMATION

The nature of soils is being constantly challenged by human's activities on the surface of the planet. The reclamation of disturbed and/or contaminated soils is very important to secure a healthy natural soil environment for the future. Soil reclamation implies that the soil must be returned to a state such that natural soil-forming processes (described in Sec. 1) can continue indefinitely. Thus, after soil reclamation activities have been terminated, natural soil physical, chemical, and biological processes and conditions will approximate their original natural state. It is now recognized that soil restoration, that is, returning to exact natural conditions, is often not practically achievable.

Soil systems that often must be reclaimed include

1. agricultural lands that have been excessively cropped, abandoned and mismanaged, resulting in severe physical (erosion), chemical (salt accumulation), and biological (loss of biological diversity) deterioration;
2. lands that have been subjected to extraction of underground resources (mining);
3. lands that have been used to dispose of or stockpile excessive amounts of wastes;
4. lands that have been damaged by natural catastrophes such as flood, fire, and wind erosion; and
5. lands that have been damaged by excessive use or changes in use such as overgrazing, deforestation, and varying urban growth patterns.

Traditional soil reclamation techniques include the physical manipulation of the plow layer (~30 cm) with the addition of a soil amendment. In general, soil amendments re-

Table 2. Bulk density and porosity of soils. Extreme low bulk densities are usually associated with coarse (sandy) soils, whereas extreme high bulk densities are associated with fine textured (clay) soils.

Bulk Density (g/cm^3)	Porosity (%)	Ranking
1–1.3	60–50	Low bulk density, high porosity
1.4–1.7	50–35	Medium bulk density, medium porosity
>1.8	>33	High bulk density, very low porosity

fer to natural materials or chemicals that help restore the natural chemical and physical properties of soils. Early soil reclamation efforts were focused on agricultural lands that had been mismanaged, usually by a combination of poor water quality and poor irrigation practices. These two factors often result in accumulations of salts in the soil profile that severely restrict crop production. Thus, there are several criteria that are used to determine the suitability of soils to sustain plant growth. The two most important physical criteria are permeability and bulk density, and are listed in Tables 2 and 3.

Compacted soils (high bulk density) do not allow proper plant root penetration and severely limit soil–atmospheric gas exchanges. When oxygen diffusion into the soil profile is limited, the soil system can become anaerobic and undergo chemical and biological changes that are detrimental for the sustained growth of most plants. Another way of measuring soil compaction is the soil porosity (the ratio of the total soil volume to the volume of the solids). These two parameters are inversely related (see Table 2).

Table 3. Soil infiltration rankings.

Permeability (cm/h)	Ranking	Texture[a]
<0.15	Very slow	Usually heavy clay soils
0.15–0.5	Slow	Clay soils
0.6–1.5	Moderately slow	Clay loam soils
1.6–5	Moderate, adequate	Loam soils
5–15	Moderately rapid	Sandy loamy soils
15–50	Rapid	Sandy soils
>50	Very rapid	Coarse sandy soils

[a]Other factors besides texture may also affect soil permeability.

Soil-water permeability is very important to insure proper water flow into the soil profile. If water is slow to infiltrate into the soil, it tends to run across the land and create erosion. At the same time, soils with low infiltration rates tend to accumulate salts and may not provide sufficient water for sustained plant growth (see Table 3). The bulk density and infiltration of soils can in general be improved via physical manipulation such as deep plowing. However, such practices offer a temporary remedy at best, as they do not address the more fundamental problems responsible for these deleterious physical changes in the soil.

Changes in soil salinity have a progressive and profound impact on the overall physical and chemical status of soils. Soil salinity is a measure of the soluble salt content of a soil and is measured by electrical conductivity of a soil-water extract. Excessive soil salinity limits plant growth and affects soil microbial activity. Most food production plants are sensitive to salt, and yields are therefore limited when plants are exposed to high salinity. The criteria to evaluate soil salinity are given in Table 4.

Sodium chloride is one of the most deleterious minerals to affect the soil environment. Excessive additions of NaCl from brine spills and/or accumulations due to poor irrigation practices result in several important changes in the chemical and physical structure of clay minerals in soils. Clay minerals have a cation exchange capacity (CEC) or ability to adsorb/exchange solution cations from the soil solution (see Sec. 4). Thus, since NaCl is very soluble in water, the Na^+ ion exchanges onto the clay, causing a change in the positive charge densities of the clay particles. Clay particles that have predominantly monovalent cations such as Na^+ in the exchange sites have a diffuse charge density that makes them repulse other similarly charged clay particles. This change results in the physical dispersion of clay particles in the soil environment. Dispersed clay particles form very small soil pores that severely limit water movement through the soil profile. On the other hand, Ca^{2+} ions do not have this effect on soils. Because clay particles saturated with divalent cations have

Table 4. Soil salinity rankings.

Parameter (mS/cm)	Nonsaline	Slighty saline	Moderately saline	Saline
Electrical conductivity[a]	<4	4–8	8–16	>16

[a]Measured on a water-saturated soil paste extract.

more localized charge densities, they tend to form aggregates with other clay particles. Soil aggregates are of vital importance for a healthy soil environment as they improve porosity, facilitate plant root penetration, improve oxygen and water movement and availability, and provide excellent microenvironments for soil microbes.

The ratio of sodium ions to common divalent ions such as calcium and magnesium provides a good measure of the sodium content in clays. The criterion for evaluating sodium content in clays is called exchangeable sodium percent (ESP) and is defined in Table 5. Alkaline waters used in irrigation of soils can also produce soil conditions similar to those described above. Excessive concentrations of soluble carbonate (CO_3^{2-}) and bicarbonate (HCO_3^-) ions in irrigation water limits availability of soluble calcium ions in the soil solution. This can result in an increase of exchangeable sodium in clay particles, thereby increasing the soil SAR and creating sodic soils (see Table 5).

Saline and sodic soils can be reclaimed by a combination of physical soil manipulations and chemical soil amendments that include the following:

Table 5. Sodium exchangeable percent rankings. These ESP values are generally considered guidelines. Thus, these rankings may vary from soil to soil.

ESP = (Na/CEC) × 100[a]	Ranking	Comments
<10	Adequate	Sandy soils can tolerate ESP values up to 15 without problems
10–15	Borderline	Problems likely to occur in fine texture soils
>15	Inadequate	Infiltration problems

[a]Sodium and cation exchange capacity of clay soils are in milliequivalents per 100 g of soil.

1. Incorporation into the soil plow layer of liberal (>5 tons per acre) amounts of gypsum ($CaSO_4 \cdot 2H_2O$), followed by repeated irrigations.
2. Additions of acidifying materials such as elemental sulfur into the soil, which ultimately produces sulfuric acid, or directly adding sulfuric acid to the irrigation water. This acid reacts with carbonate ions liberating CO_2 gas and forming a more soluble form of calcium, as shown in the following summary reaction:

$$H_2SO_4 \text{ (in water)} + CaCO_3 \text{ (in soil)} \rightarrow CaSO_4 \cdot 2H_2O\text{(gypsum)} + CO_2 \text{ (gas)}.$$

After repeated irrigations the soluble sodium salts are usually leached below the root zone. These treatments restore a more natural chemical and physical balance in the soil environment.

Acid soils develop as a result of continuous fertilization with acid-forming fertilizers. The most common reduced forms of nitrogen fertilizers such as ammonia and urea form nitric acid upon oxidation in the soil environment, thereby lowering the soil *p*H. Soils in the eastern half of the United States tend to become acidic with time, as these soils do not have large reserves of acid-neutralizing minerals such as calcite commonly found in western soils. Acid soils ($pH < 5.5$) quickly lose macro- and micronutrients via leaching and develop aluminum toxicity problems with plants, through the following reaction:

$$Al(OH)_3 \text{ (soil minerals)} + 3H^+ \text{ (in soil solution)} \rightarrow Al^{3+} \text{ (in soil solution, toxic!)} + 3H_2O.$$

Lime and calcite additions to these soils are a common reclamation practice that provide a temporary remedy for the deleterious effects that low *p*H has on both plants and the soil itself. In this case, lime (CaO) or calcium carbonate ($CaCO_3$) additions neutralize

Table 6. Major soil macronutrients.

Element	Desired soil concentration ranges	Desired plant-available concentration and forms
Carbon	0.5–3% in organic matter (O.M.)	C:N:P = 100:10:1
Nitrogen	>0.1% in O.M. optimum C:N = 10:1	>20 mg/Kg as NO_3–N
Phosphorus	>0.01% inorganic form	>10–20 mg/kg as PO_4–P extractable. Variable depending on soil and extraction method
K, Ca	>0.1%; supplies vary depending on mineral form	>100 mg/Kg as water-soluble Ca^{2+} and K^+
Sulfur	>0.1% as gypsum	>100 mg/Kg as water-soluble SO_4^{2-}

excess H^+ ions in the soil, thereby promoting the formation of very insoluble aluminum oxides minerals such as gibbsite.

Soil reclamation may also include the reestablishment of a nutrient balance in the affected soil system. Fertility levels of a soil are usually assessed by measuring the total and plant-available concentrations of macro- and micronutrients. These are listed in Tables 6 and 7. Chemical fertilizers are commonly added to agricultural soils and disturbed lands as a means to restore nutrient balance to the soil system. However, excessive amounts of some fertilizers can produce *p*H and salt problems, as described above.

Organic wastes, such as composted plant residues and municipal solid sludges, are common soil amendments used in the reclamation of soils deficient in organic matter and plant nutrients. Many studies to date indicate that repeated applications to soils in need of organic matter and nutrients have beneficial consequences. However, this practice may not be sustainable (indefinitely) as repeated applications of wastes may lead to increases in toxic metals in the soil environment. Thus, waste applications to the soils require a more rigorous monitoring program to control possible deleterious effects that these materials can have in the soil environment. Nonetheless, recycling organic wastes by using them as soil amendments is very cost-effective and beneficial to the soil environment. This practice is becoming a more acceptable form of waste disposal, as the natural cycle of the major plant and animal nutrients listed in Table 6 is completed in the soil environment.

Following successful soil reclamation activities, plants are usually seeded or planted to provide physical stability to the soil surface and help re-establish the natural biological and nutrient cycle to this environment. To this end, many types of grasses, shrubs, and trees are available (Munshower, 1993). These plant species have been found to be tolerant to the less-than-optimum conditions that recently reclaimed soils have.

Soil reclamation practices are very beneficial and necessary to restore the natural physical, chemical, and biological balance of soil systems. These practices are becoming increasingly more efficient as our understanding of soil science continues to increase. Soil reclamation activities will likely

Table 7. Major soil micronutrients.

Element	Desired soil concentration ranges	Desired plant-available concentration and forms
Boron	>10 mg/kg, variable	>0.5 mg/kg as H_3BO_3–B hot-water extractable
Copper, zinc, iron	>10 mg/kg, variable	>0.5 mg/kg as Cu^{2+} EDTA[a] extractable
Manganese	>100 mg/kg, variable	>5 mg/kg as Mn^{2+} EDTA[a] extractable
Molybdenum	>1 mg/kg, variable	>0.01 mg/kg as MoO_4^{2-}–Mo, but varies depending on presence of other micronutrients and extracting solution used

[a]EDTA is ethylene diamine tetraacetic acid, a metal chelating agent.

increase in the future as more and more soils are being affected by human's activities.

7. SUMMARY

We have briefly discussed the nature and importance of soils and the movement and impact of pollutants in soil and groundwater environments. We have also discussed methods to monitor, remediate, and restore polluted soils and groundwaters. It must be stressed that it is very difficult, expensive, and often impossible fully to remediate and restore polluted subsurface systems. It is clear that preventing pollution from occurring is a much more effective strategy. It is also clear that pollution prevention programs can be fully successful only when the public, industry, academia, and government work together.

GLOSSARY

Advection: The transport of solute by the movement of a fluid; also called **Convection.**

Advection–Dispersion Equation: The most widely used equation for simulating solute transport in porous media.

Aluminosilicates: Minerals composed primarily of aluminum (Al) and silicon (Si) atoms.

Aquifer: A water-bearing formation in the subsurface.

Axial Diffusion: Movement of solute in response to a concentration gradient in the direction of flow.

Breakthrough Curve: A representation of the concentration of solute in a fluid as a function of time at a selected point.

CEC: Cation exchange capacity, a surface property of clay minerals enabling them to exchange cations to and from water solutions containing dissolved ions.

Clays: Secondary minerals. Examples: montmorillonite, vermiculite, and illite.

Convection: A term used interchangeably with **Advection.**

Dispersion: The spreading of solute about a mean position.

Dispersivity: A parameter representing the spreading potential of a solute-porous medium system.

ESP: Exchangeable sodium percent, the percentage of sodium ions that cover the surface of clay particles.

Eutrophication: A process whereby the trophic status of surface-water bodies changes with time, and the type and quantity of biota change as well. This process is mediated by the availability of nutrients and can be deleteriously accelerated by runoff containing high levels of fertilizers.

Facilitated Transport: Enhanced rate of solute transport caused by an increase in apparent solubility and a concomitant decrease in sorption and retardation.

Humic Substances: Complex organic (carbon-based) materials that form from decaying plant and animal tissue in the soil environment.

Hydrodynamic Residence Time: The average time required for fluid particles to traverse the flow domain.

Hydrolysis: Chemical process that breaks a molecule in two by insertion of a water molecule, the hydrogen attaching to one fragment and the hydroxyl to the other.

Ideal Transport: Transport that can be predicted with models based on assumptions of subsurface homogeneity and instantaneous mass transfer.

Local-Scale Dispersion: Spreading caused by axial diffusion and mechanical mixing.

Mechanical Mixing: Pore-scale spreading caused by flow through a macroscopically tortuous and nonuniform porous medium (i.e., one with nonuniform pore-size distribution).

Nonideal Transport: Transport that does not coincide with that predicted from models based on assumptions of subsurface homogeneity and instantaneous mass transfer.

Oxidation and Reduction: Chemical processes that result in the loss or gain of electrons from elements and molecules.

PCR: Polymerase chain reaction, laboratory biochemical process whereby DNA molecules are replicated enabling their detection and characterization.

Plume: A spatial distribution of solute.

Pump and Treat: A method for removing contaminants from porous media based on inducing water flow through the contaminated domain.

Retardation Factor: A parameter in the advection–dispersion equation that accounts for association of solute with the solid phase.

Soil Venting: A method for removing contaminants from porous media, based on inducing gas flow through the contaminated domain.

Solute: Matter dissolved in a fluid.

Sorption: The association of solute with a solid phase.

Sorption Isotherm: A regression of sorbed-phase concentrations against solution-phase concentrations.

TDR: Time-domain reflectometry, electrophysical process that characterizes and measures the propagation of electromagnetic radiation through the soil and relates it to water content.

Vadose Zone: The portion of the subsurface located between the soil surface and surficial aquifers and characterized by conditions of variable water saturation.

Weathering: Chemical and physical soil-forming process that transforms primary minerals into secondary minerals (silt and clay).

Works Cited

Birkeland, P. W. (1974), *Pedology, Weathering, and Geomorphological Research,* New York: Oxford University Press.

Brady, N. C. (1974), *The Nature and Properties of Soils,* 8th ed., New York: MacMillan.

Brown, A. L. (1978), *Ecology of Soil Organisms,* London: Heinemann.

Brusseau, M. L. (1994), *Rev. Geophys.* **32,** 285–314.

Buol, S. W., Hole, F. D., McCracken, R. J. (1973), *Soil Genesis and Classification,* Ames, IA: The Iowa State University Press.

Environmental Protection Agency (1987), Underground Storage Tank Corrective Action Technologies, EPA/625/6-87-01115, Cincinnati, OH.

Fetter, C. W. (1993), *Contaminant Hydrogeology,* New York: Macmillan Press.

MacCarthy, P., Clapp, C. E., Malcom, R. L., Bloom, P. R. (1990), *Humic Substances and Crop Sciences: Selected Readings,* American Society of Agronomy, Madison, WI: Soil Science Society of America.

Munshower, F. K. (1993), *Practical Handbook of Disturbed Land Revegetation,* Boca Raton, FL: Lewis Publishers.

Pye, V. I., Patrick, R., Quarles, J. (1983), *Groundwater Contamination in the United States,* Philadelphia: Univ. of Pennsylvania Press.

Schaller, F. W., Sutton, P. (1978), *Reclamation of Drastically Disturbed Lands,* American Society of Agronomy, Crop Science Society of America, Madison, WI: Soil Science Society of America.

Further Reading

Boulding, J. R. (1994), *Description and Sampling of Contaminated Soils: A Field Guide,* 2nd ed., Boca Raton, FL: Lewis Publishers.

Brady, N. C. (1974), *The Nature and Properties of Soils,* 8th ed. New York: MacMillan.

Brusseau, M. L. (1994), *Rev. Geophys.* **32,** 285–314.

Dixon, J. B., Weed, S. B. (1989), *Minerals in the Soil Environment,* 2nd ed., SSSA Book Series, Madison, WI: Soil Science Society of America.

Klute, A. (1986), *Methods of Soil Analysis—Part 1, Physical and Mineralogical Properties,* 2nd ed. Agronomy No. 9, ASA, Inc., Madison, WI: SSSA, Inc. Publishers, 1188 pp.

Munshower, F. F. (1994), *Practical Handbook of Disturbed Land Revegetation,* Boca Raton, FL: Lewis Publishers.

Page, A. L., Miller, R. H., Keeney, D. R. (1982), *Methods of Soil Analysis—Part 2, Chemical and Microbiological Properties,* 2nd ed. Agronomy No. 9, ASA, Inc., Madison, WI: SSSA, Inc. Publishers, 1159 pp.

Pye, V. I., Patrick, R., Quarles, J. (1983), *Groundwater Contamination in the United States,* Philadelphia: Univ. of Pennsylvania Press.

Wild, A. (1993), *Soils and the Environment,* New York: Cambridge Univ. Press.

Wilson, L. G., Everett, L. G., Cullen, S. J. (1995). *Handbook of Vadoze Zone Characterization & Monitoring,* Boca Raton, FL: Lewis Publishers.

SOLAR CELLS

See PHOTOVOLTAIC DEVICES

SOLAR ENERGY

I. DOSTROVSKY, R. TENNE, AND A. YOGEV, *Weizmann Institute of Science, Rehovot, Israel*

INTRODUCTION

In this article we concentrate on technologies for harvesting solar radiation that reaches the surface of the earth for the benefit of mankind. We do not deal here with those indirect forms of solar energy such as wind energy, hydropower, biomass, or fossil fuels.

The earth intercepts about 5.6×10^6 exajoules (EJ) per year of solar energy (1 EJ = 10^{18} J), about 1.7×10^6 EJ of which are backscattered by the upper atmosphere into space, allowing 3.9×10^6 EJ to reach the surface of the earth or to be absorbed by the lower atmosphere. The bulk of this energy is ultimately reradiated to space in the form of infrared radiation, but a small fraction is converted to the kinetic energy of winds, flowing water, and ocean currents; the potential energy of precipitation; and chemical energy produced by photosynthesis. The latter, biomass, is, of course, the basis of all life on the planet.

Of the total solar radiation reaching the surface of the earth, about 7.9×10^5 EJ/yr are absorbed by the land masses, of which some 9.3×10^4 EJ/yr are in the sunny deserts. If we consider only the latter quantity as available for harvesting, it still amounts to about 300 times the present world demand for energy. At the present time, on a global scale, the proportion of the solar energy harvested directly by mankind is negligible—in

3-527-28140-1/96/$5.00 + .50

spite of the enormous potential of this renewable source and its relatively benign environmental nature. In the present article we review the problems involved in the large-scale utilization of solar energy and the various efforts under way to overcome them.

Solar radiation is not distributed uniformly around the globe, as the map in Plate 1 shows. We note the expected relatively high insolation values for the world's major deserts. Tracts of land that could be devoted to solar energy harvesting without competing unduly with other essential needs (food production, for example) are available in the desert areas. In fact it may be shown that a small fraction of the world's deserts, if devoted to solar energy harvesting, could cover mankind's need "forever." These considerations point to the world's deserts as the future sources of energy.

But Plate 1 also shows the lack of overlap between these potential sources of "unlimited" energy and the major centers of population and industry where this energy is needed. This, perhaps, is one of the major handicaps to the large-scale use of solar energy, for it requires developing technologies for the large-scale storage and transportation of the harvested solar energy. There is, of course, also the inherent problem arising from the intermittent nature of solar radiation (day and night), and its somewhat stochastic behavior (weather dependence). Possible solutions to all these problems are discussed in the following pages.

Solar radiation reaching the earth must be converted to some other forms of energy before it can be put to use by mankind. There are three possible routes for such conversion (see also ENERGY CONVERSION):

1. the thermal route, where the radiation is first absorbed and converted to heat and then to other forms of energy, if needed. (Sec. 1);
2. the direct conversion to electricity (Sec. 2); and
3. the photochemical route, where the radiation induces a chemical reaction generating products of higher-energy content than the starting material (Sec. 3).

Although in nature the last route dominates, it has not so far been possible to emulate it artificially.

1. THERMAL PROCESSES

Solar radiation reaching the earth corresponds to that emitted by a blackbody at 5780 K after some modification by absorption and scattering in the atmosphere. For the thermal processes the exact spectrum of the radiation is unimportant. On reaching the surface a fraction α is absorbed and is converted to heat. Some of this heat, in turn, is lost by various processes, and only the residual amount is available for harvesting. The efficiency of the collection is defined as the ratio of useful heat recovered from the collector to the solar irradiation impinging on an equal area if set to face the sun. Its maximum value, for a given insolation, depends clearly on the magnitude of the losses. In analyzing the various thermal losses of a solar collecting system, one must distinguish between those due to convection and conduction and those due to radiation from the heated surface. The former can, in principle, be reduced to very small values by proper design. The radiative losses, however, cannot be reduced at will. At the very least, the surfaces that are exposed to sunlight (and there may be additional ones) radiate energy back into space. The rate of reradiation is determined by a property of the surface material known as its emissivity, ϵ, and its temperature according to the Stefan-Boltzmann relationship:

$$E = \epsilon\sigma T^4. \tag{1}$$

The value of the constant $\sigma = 5.6705 \times 10^{-8}$ W m^{-2} K^{-4}. The value of ϵ depends on the spectrum (and hence also on the temperature) and may vary from nearly 1 to less than 0.1 in different materials.

Because of this, the thermal efficiency of solar collectors is a somewhat complicated function of the temperature at which it is operated. A very simplified form of this function (where heat losses by conduction and convection and the effect of covers or windows are neglected) will be used to illustrate the relationship between the thermal efficiency of collectors and the temperature:

$$\eta = \alpha - \epsilon\sigma T^4/SI, \tag{2}$$

where ϵ and σ have been defined before, I is the solar flux (in units of W/m^2), and S is degree of concentration of the solar radiation. In Fig. 1 is shown the relation between the efficiency of heat recovery and collector temperature assuming a perfect black surface ($\alpha = 1$ and $\epsilon = 1$, curve *a*). We note the rapid drop in efficiency as the temperature rises toward 100 °C, with useful operational range of 60–70 °C.

A good deal of control of the collector's operating temperature and efficiency is possible by manipulating the parameters of the energy balance equation [Eq. (2)] by proper design and appropriate choice of materials. In this simplified equation there are three parameters that can be manipulated: α, ϵ, and S. In general, each application has its own combination of parameters that optimize the system, and we shall discuss them further for each class. We shall proceed from those involving low temperatures to the highest ones and in each case outline the principles of the design.

1.1 Low-Temperature Applications

These are characterized by temperatures below 100 °C. As is shown in Fig. 2, curve *a*, at this temperature level no special adjustment of the parameters is necessary, and the system will operate with reasonable efficiency of conversion to useful heat with α and ϵ close to 1 (i.e., ordinary black surface) and with no concentration of the solar radiation ($S = 1$). However, a great improvement is possible with relatively little extra cost, through the use of selective surface coatings of the collector. These are surfaces that show very high absorption of the solar spectrum (the absorption coefficient α) while possessing a low emissivity in the infrared region (the coefficient ϵ). Several commercial processes are available that give surfaces with α greater than 0.9 and ϵ of about 0.1. Figure 2, curve *b*, shows the theoretical maximum efficiency curve for a collector fitted with such a selective surface.

1.1.1 Domestic Hot Water The most widespread application of low-temperature solar energy is probably in domestic hot-water supplies. Here the aim is to heat water to about 70 °C and store it for use for about a day. Very simple systems, with no moving parts, have been developed requiring no special maintenance efforts. The heart of such systems is a collector surface of a few square meters area covered by a transparent window (usually ordinary window glass). Many designs are available commercially. Most include a selective coating of the absorber surface. The collector is connected to a storage vessel, and water is circulated through both by the thermo-syphon principle. Such a design is sufficient for locations free from frost. In places where freezing of the water may occur, a somewhat more complicated design is necessary, where a separate loop of an antifreeze medium is used to couple the collector and storage tank.

1.1.2 Solar Ponds Solar ponds are also designed to operate at temperatures below 100 °C. These are aimed at the large-scale

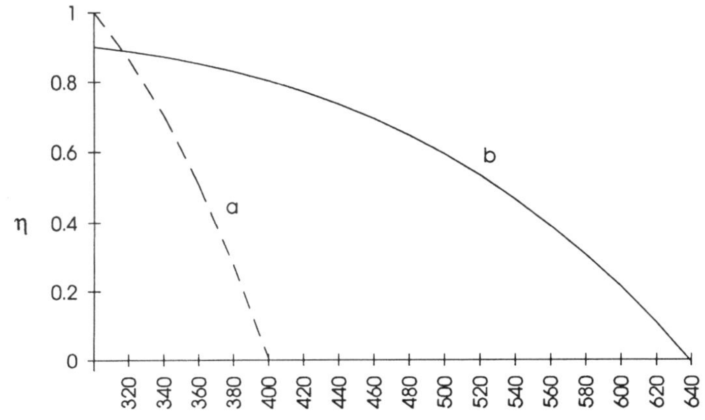

FIG. 1. Efficiency (η) of heat recovery as function of collector temperature (K). Curve *a*, $\alpha = 1$, $\epsilon = 1$, $c = 1$; curve *b*, $\alpha = 0.9$, $\epsilon = 0.1$, $c = 1$.

production and long-term storage of solar hot water for further use either for the generation of electricity or as a source of energy for desalination.

The natural pond or lake does not get heated up much by the sun because convection soon brings any warm water from the lower layers to the surface where it cools. If such convection could be prevented, water heated up by absorption of solar radiation near the bottom would stay there, and its temperature would rise (and could even reach the boiling point). The upper water layers then act as transparent thermal insulation. Convection can be stopped by establishing and maintaining a vertical density gradient in the pond, and this is achieved by creating a salinity gradient. Under such conditions the hot, saline, water near the bottom of the pond will still be heavier than the upper layers, and so no convection will occur (Fig. 2). Systems based on this technology have been under development for several decades in several countries, the largest demonstration plant being in Israel, where this work originated. A 2.5×10^5-m^2 pond, constructed on the north shore of the Dead Sea, supplied hot water to an organic-fluid turbine to generate 2.5 MW of electricity. This plant operated for several years and proved the principle. However, no larger-scale facility has been built anywhere in the world so far. The main handicap arises probably from the low efficiency of the energy conversion to electricity (about 1% overall), which is not completely compensated for by the inherent simplicity of the collection system and its large storage capacity. It is quite possible that as a source of heat for multiple-effect desalination plants specially designed for low–source-temperature operation, solar ponds may become an ideal energy source.

1.2 Medium Temperature

As soon as temperatures above about 200 °C are required, at reasonable recovery efficiency, controlling heat losses by the use of selective surfaces alone becomes insufficient (see Fig. 1, curve *b*). However, the last term of Eq. (2) can be further reduced by increasing the factor *S* in the denominator. This means that the solar flux must be concentrated before impinging on the absorbing surface. The effect of doing this is equivalent to reducing the area of the surface from which heat can be radiated, without at the same time reducing the solar energy input.

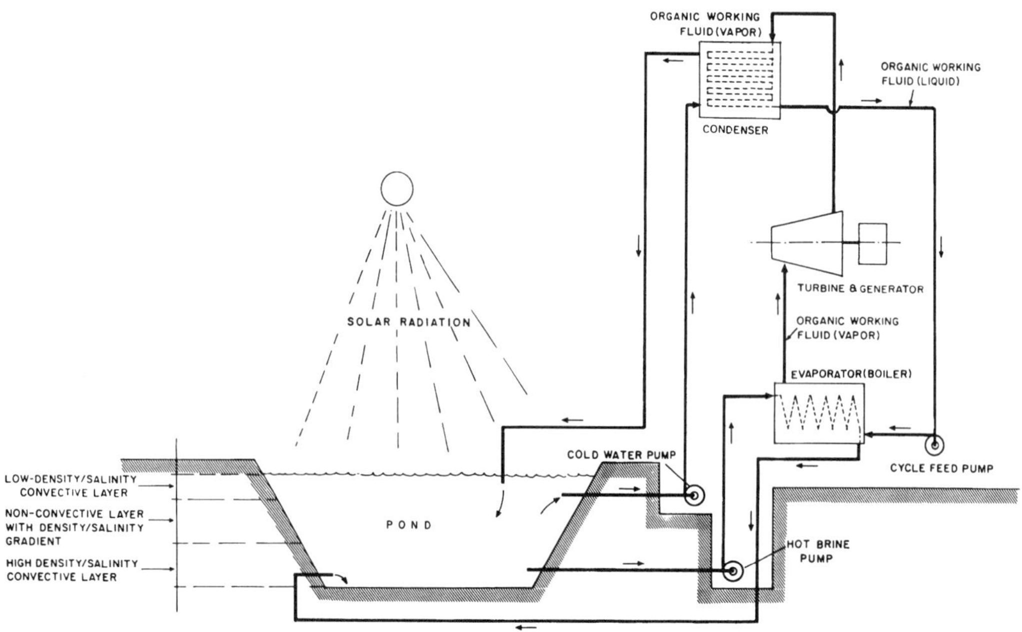

FIG. 2. Schematic diagram of a solar pond power plant.

To achieve this in practice requires a sharp change in the technologies involved and in the economics of the systems.

Once the need for concentration has arisen, the emphasis shifts from the absorber to the concentrating collector. In the low-temperature range, the absorber (which is also the collector) is the extended element, and the cost of the plant is proportional to its size. With concentration, the size of the absorber shrinks drastically, and the extended area is now the collector and not the absorber, which now largely determines the economics of the plant. Ironically, at the very high temperatures—at the other end of the scale—the emphasis shifts back to the design of the absorber (see Sec. 1.4).

The only practical method of concentrating solar radiation over large areas is by reflection, and the cost of the reflecting surface becomes a dominant factor. The degree of concentration required depends very much on the details of the processes that follow. It will be recalled, of course, that in general the higher the source temperature, the higher will be the efficiency of any thermal conversion process (Carnot principle), but so will be the losses. Therefore, for applications where a conversion of solar thermal energy to mechanical energy is involved, there is a conflict in the temperature requirements. For the thermal collection step, the lower the working temperature, the higher will be the efficiency, while for the conversion to mechanical energy, the higher the temperature, the greater will be the efficiency. Obviously an optimum value for the process as whole must exist somewhere. Figure 3 shows the result of the folding of the two efficiency curves. The values shown are for the theoretical ideal efficiency. In reality, of course, the actual values will be much smaller (could be as small as half the theoretical maximum).

1.2.1 Process Heat The temperature range 100–300 °C covers industrial and commercial applications known generically as "process heat." This temperature range is too low for efficient primary electric power generation. At the lower end of this temperature range, fairly simple concentrating devices may be sufficient, and tracking the sun is not essential. At the upper end, shaped mirrors and solar tracking are necessary.

The applications in this temperature range are increasingly moving into the domain of cogeneration (electricity and heat). Because the additional effort of achieving higher solar concentration, and consequently higher temperatures, is not very great, it may become economically advantageous under certain conditions to combine electricity gen-

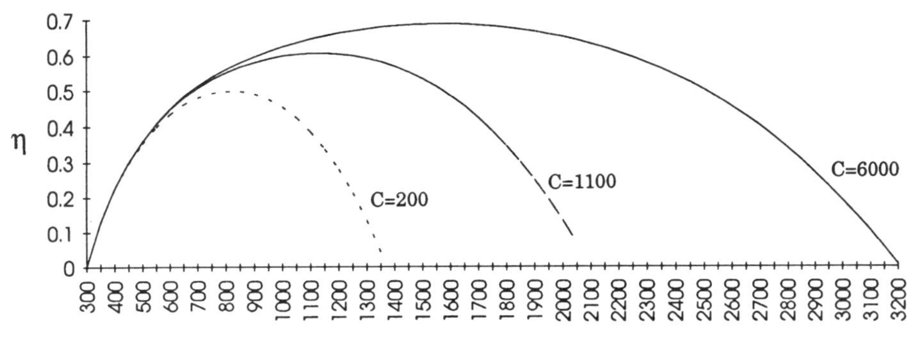

FIG. 3. Efficiency of conversion (η) of solar radiation to mechanical energy as a function of working temperature for various degrees of concentration C. For solar normal irradiation of 1000 W/m^2, ambient temperature of 300 K, $\alpha = 0.9$, and $\epsilon = 0.9$.

eration and process-heat production into one cycle. The high temperature is used to generate electricity, and the rejected heat from this stage, at a lower temperature, serves as process heat. It is likely that in the future the needs for the latter will be covered mostly through such cogeneration plants whenever both electricity and heat are needed in the same site.

1.3 High Temperatures

The temperature range of 300–700 °C is applied mainly to the generation of electric power, using the Rankine cycle. To achieve these temperatures, highly concentrated solar energy is needed (see Fig. 3), and there are several technologies available: parabolic troughs, reflecting dishes, and heliostat fields, all involving tracking the apparent motion of the sun.

1.3.1 Trough Systems This technology is the basis of the largest solar commercial electricity-generating plants erected to date. In southern California, a total of some 350 MW_e (megawatts electric) were in operation as of 1994 using the trough technology. As the name implies, these concentrators consist of long parabolic reflecting troughs, usually of silvered glass, giving a line-focussed solar radiation. Along the focal line a pipe is located that acts as the radiation absorber. The heat generated in it is transferred to a flowing oil stream, which in turn transfers it to a water boiler to generate steam and drive a conventional Rankine-cycle turbine and generator. To reduce heat losses to the minimum, the receiver pipe is coated with a selective surface and enclosed in an evacuated glass jacket. The parabolic trough assembly, usually aligned in a N–S orientation, tracks the sun along one axis during the day using a servo system driven by a sun sensor (Fig. 4). The combination yields a concentration ratio (defined here as the ratio of the area of the aperture collecting the solar radiation to the area from which heat is lost from the absorber tube) of about 26.

The oil exits the solar field at a temperature of 390 °C, and the steam generated by the circulating oil is delivered to the turbine at 371 °C and 100 bar. Overall efficiency (solar radiation to electricity) is reported as

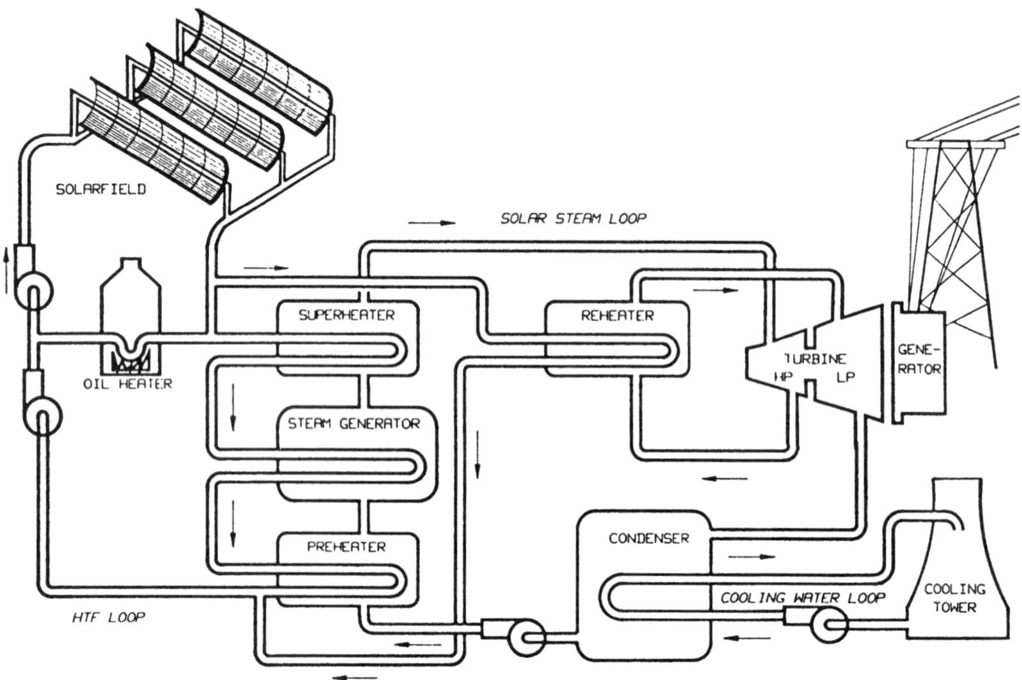

FIG. 4. Schematic diagram of a trough solar power plant (LUZ SEGS VIII).

13.9%. Because of its essentially modular design, there are no special problems in scaling the plant size to meet any demand. This provides the first example of a distributed system, where the heat output of many individual collectors arranged in a "farm" are combined to feed a power block to generate electricity.

In these plants no provision is made for solar energy storage, and instead, natural-gas–burning oil heaters are included in the design as standby. These not only serve to stabilize the operation of the plant against fluctuation in the solar energy flux but also enable it to operate in a fossil-fuel-only mode. They are therefore an example of "hybrid" plants. In California, the maximum permitted fraction of fossil fuel for these plants was 25%, and so the operators have to plan the use of this resource very carefully to maximize their profit. These plants are connected to a large grid and try to feed electricity to it mainly at times of high demand. At their specific locations these match roughly the high-insolation periods.

The design described above is the state of the art of this technology at present. Research and development efforts are under way to eliminate the oil heat-transfer circuit and to raise the steam directly in the absorber tube. If this is successful, it is estimated that considerable economic advantage will result.

1.3.2 Dish Systems The parabolic trough described above delivers a line-focus image with a modest concentration of the solar flux. Higher concentration ratios and, therefore, higher conversion efficiencies can be obtained, in principle, by optical systems that form a "point" image. The simplest of these is the parabolic dish. A single glass mirror, like the ones used in the largest telescopes, would be too heavy and far too expensive to be of practical use in solar energy applications. So other ways were developed by which a reflecting parabolic dish surface could be approximated of as large a size as is consistent with adequate mechanical strength and rigidity and of sufficient optical quality to reach high solar concentration.

It must be remembered that the sun is not a point source, but rather a disc with an angular radius of 4.653 mradians. So even in the ideal case the image will be a disc and not a point. Furthermore, the sun's disc is not uniformly radiant. Scattering processes both in the solar photosphere and in the earth's upper atmosphere cause a spreading of the theoretical "flat top" intensity distribution to a more "bell-like" shape with certain "limb darkening" and spreading beyond the angular radius (the "sunshape"). The actual sunshape has been determined by measurements, and various approximation formulas are available. It should be noted that the sunshape is not independent of atmospheric conditions and for certain very precise applications must be determined locally. These effects will further degrade the image. In general, however, for most practical and quick calculations the idealized disc with uniform distribution may serve as a sufficiently accurate model of the sun.

Many approximations to the ideal configuration of a parabolic dish concentrator have been tried (Fig. 5). They can be classified into the following three types:

1. A reflecting surface assembled from a large number of suitably shaped pieces of a matrix carrying a reflective surface of silvered glass or a metallized plastic film. The resulting reflecting shape can give a reasonable approximation to a paraboloidal surface. Several variants of this technology have been constructed and de-

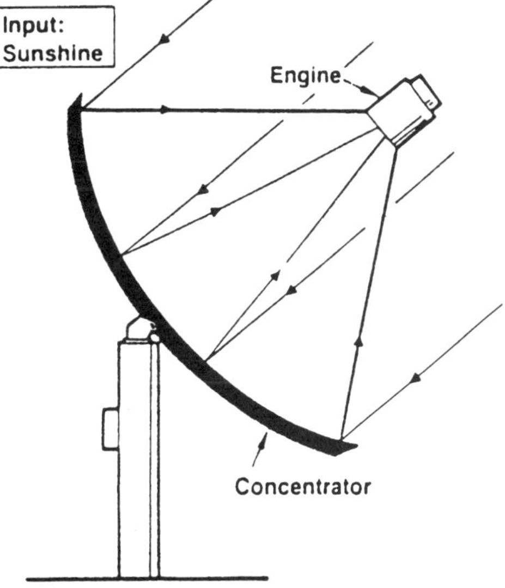

FIG. 5. Parabolic dish collector system.

ployed with reflecting area of up to 400 m^2. Concentrations of a few thousand suns have been obtained.

2. A single stretched reflecting membrane. The best example of this design is that of SBP (Germany). A thin sheet-metal membrane is stretched across a drumlike structure and deformed to the desired shape. For better reflectivity thin second-surface silver glass mirrors are glued onto the metal membrane. The largest dish constructed according to this design has a diameter of 17.5 m and a reflecting area of 227 m^2 (Lampoldshausen, Germany). Smaller units, of 7.5-m diameter, have also been built. They provide a concentration ratio of about 600 and are used to drive Stirling engines (see below) in Almeria, Spain.
3. A multifaceted assembly of reflecting membrane elements. The difficulty of manufacturing large single stretched-membrane dishes has led to the design whereby the reflecting surface is broken up to a number of smaller stretched-membrane units. An example of this approach is the LaJet design, where 24 facets (made of metallized PMMA foil), 1.8 m in diameter each, are assembled on a suitable structure to simulate a large dish of 9.4-m diameter. The curved shape of each facet is achieved by maintaining a slight vacuum behind the membrane. Solar concentration ratio of 225 has been reported.

The difficulties resulting from the need to locate complex energy converting devices at a movable focal point "floating in mid-air" has prompted attempts at designing a fixed-focus dish. A model of such a solar collector is now under development in Germany (Bomin Solar).

1.3.2.1 Dish Farms. The maximum practical size of an individual dish of any of the above designs is limited by mechanical considerations of weight, rigidity, wind load, and other. A size of around 400 m^2 seems to be the limit at present. This means that under good insolation conditions (say 900 W/m^2) there will be some 340 kW of thermal energy at peak time delivered to the absorber. Clearly, therefore, many such units have to be connected together, to provide a meaningful commercial output.

There are two ways of connecting together many dishes into a solar "farm," each requiring a different technology:

1. Each dish can carry only the heat absorbing unit at the (moving) focus, and the hot heat-transfer fluid is then pumped through an extensive piping network to a central generating unit. The reduced aperture of the cavity receivers will reduce the heat losses and thereby improve the efficiency, but on the other hand, the extensive piping network will entail extra heat losses and pumping work. Several dish farms were constructed and operated in the last decade according to this scheme. As examples one can quote the STEP cogeneration plant at Shenandoah, Georgia, U.S.A., where 114 dishes with a combined aperture of 4386 m^2 heated oil to 400 °C to raise steam and produce electricity. A somewhat more recent example is the Solar Plant 1 at Warner Springs, California, U.S.A., where 700 dishes with a combined aperture of 30 261 m^2 generated steam at 371 °C and 41 bar to generate 4.8 MW_e of electricity at peak. In Australia a dish farm uses steam as the heat-transfer medium.
2. In another approach each dish carries a complete electricity-generating unit at its focus, and the output from all of them is combined to give the required plant capacity. For this approach it is necessary to develop a relatively small conversion unit that will accept the highly concentrated radiation at the focus of the dish and convert it to electricity. The unit, together with its heat rejection system, has to be located at the focus and move with the dish as it tracks the sun. This requirement imposes severe constraints on possible designs.

In comparing the two approaches one must remember that in choosing oil as a heat-transfer medium or even steam at "conventional" characteristics, one is back at the relatively poor thermodynamic conditions and must accept the low efficiencies that they entail with little benefit from the higher-temperature capabilities of the dish. It is such considerations that are encouraging the efforts to develop a compact heat engine that will take full advantage of the high solar concentrations available at the focal area of a dish.

One generation system that has been receiving much attention in recent years for this application is the Sterling engine (*q.v.*). This is a compact and self-contained heat engine that can operate at elevated temperatures (>700 °C) with hydrogen or helium as working fluid. The design was conceived in Sweden many decades ago and has been developing slowly. Under the impetus of possible solar energy application, this development has received a boost in recent years. Development programs are under way now in several countries to marry the dish and Sterling engine technologies to produce a compact and reliable solar-power unit. A number of such units are undergoing solar testing at present.

A key component in these units is the absorber that is exposed to the highly concentrated radiation and transfers the high-temperature heat to the operating gas in the engine. Two approaches were demonstrated. In one, the concentrated solar flux impinges directly onto thin tubes through which the gas is circulating. In the second approach sodium is boiled (either in a heat-pipe configuration or in a boiler) and then transfers the heat to the working gas. Efficiencies close to 30% net conversion to electricity have been achieved. The modular nature of these units would permit the combination of any number of them to meet local needs, with the flexibility of a distributed cluster system.

At the present time, apart from the few experimental dish–Sterling rigs, the dish collectors with their high concentration ratios have not achieved much more than the troughs with their very modest concentrations.

1.3.3 Central Receivers A basic issue in all large-scale solar energy harvesting is the need to combine the output from a large number of collecting devices. So far we have described concepts based on combining the outputs of heat-transfer medium from many receivers in a central power-generating unit or combining the electrical outputs from many independent generators. There is, however, another way of achieving the final aim, and it is to combine the solar flux from many concentrating reflectors into one receiver to drive a compact central power-generating unit. This is known as the central receiver (or "Solar Power Tower") concept (Fig. 6).

According to this concept the solar energy is collected by a large number of curved mirrors that track the sun on two axes and direct the image onto a common target, the receiver. These mirrors are known as "heliostats" and are themselves constructed from a number of curved facets, with focal lengths corresponding to their distance from the target. They are mounted on a common tracking structure and each canted in such a way as to approximate a section of a paraboloidal surface.

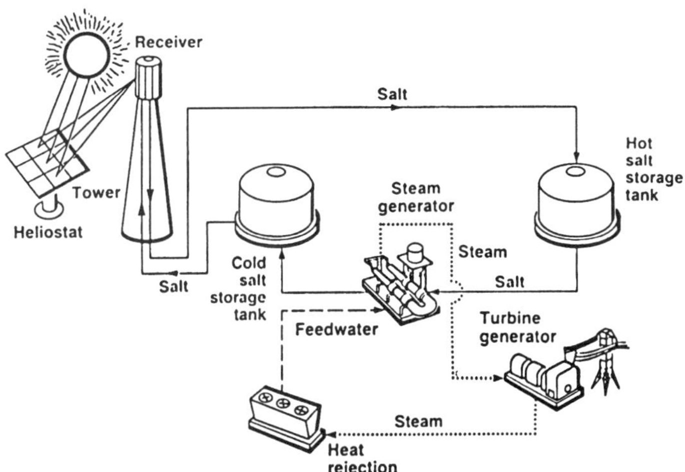

FIG. 6. Schematic diagram of a central receiver plant using molten salt as heat-transfer medium.

For fairly obvious reasons the receiver has to be mounted high enough above the plane of the mirror field to minimize blocking and shadowing of the reflected light from one heliostat by those in front of it. The tower that carries the receiver, which may be over 100 m in height, gives the alternative name to this concept, "solar tower."

Clearly the total amount of solar energy reflected into the receiver is the sum of the light reflected from the individual heliostats. But this varies greatly according to the location of the heliostat in the field and the time of day. The optimization of the layout of the field of heliostats, taking into account the blocking and shadowing by neighbors and the height of the tower, can be made using computer codes that are available for this purpose.

It is easy to see that large amounts of concentrated radiation can be delivered to the receiver to be converted to other forms. The largest field built so far, which collects some 45 MW_t (megawatts thermal) of solar energy, is in Barstow, California, U.S.A. The optimal commercial units would probably be at least ten times larger. Because of the eminent suitability of this concept to large-scale harvesting of concentrated solar radiation, much work has been devoted to the development of the various components.

The key component, in the sense that it determines the economics of the whole system, is formed by the heliostats. They provide the primary energy-collecting and -concentrating function, and their number is large. Many designs have been tried in the last two decades, and work is still continuing. The aim is to reduce cost per square meter and ensure adequate lifetime (some 30 years). At present, the designs seem to converge to heliostats of between 50 and 150 m^2 in area, made of second-surface curved glass facets. Under development are heliostats using metallized plastic films and stretched membranes (similar to those described above for the dishes).

Each heliostat in a field is individually computer controlled. The position of the sun is calculated every few seconds, and for each heliostat the positions of the azimuth and elevation drive motors needed to reflect the beam to its target are determined, and the proper commands are issued. It will be appreciated that the required algorithm here is somewhat more complicated than in the dish case, since the heliostat is not aimed at the sun but along a line bisecting the angle between the sun and the target, as seen from the heliostat. Furthermore, this direction is different for each of the thousands of heliostats in the field and is varying all the time. This may seem a formidable task, and indeed would have been impossible (or prohibitively expensive) a couple of decades ago, but is easily within the capabilities of modern computing technology.

1.3.4 Thermal Receivers The availability of large amounts of concentrated (over 1000 suns) solar energy at the receiver has opened up a new development need—how to handle this intense energy without losing its advantages. These are not at all trivial, and the emphasis of the research and development in recent years has therefore shifted from the problems of collecting and concentrating solar energy to the problem of receiver design at these high flux levels. Fortunately, in the central receiver concept there are fewer physical restraints on the design of the receiver as compared with dishes or troughs.

The mere fact that much energy is absorbed in a small area means that the fraction of the heat lost by all modes (including radiation) is drastically reduced, and the efficiency of heat production will be high. The problem is then how to transfer these high fluxes of energy to the working fluid of the plant. In most of the experimental plants built so far, the working fluid was water, as is customary in conventional power plants. The high-temperature and high-pressure steam was then used to drive a Rankine-cycle power block. Difficulties were encountered with this approach due to the limited heat-transfer capability to water boiling in tubes and the very demanding conditions imposed by the high and rather unstable solar fluxes.

1.3.4.1 Molten-Salt and Sodium Receivers. Because of the limitations of water, much attention was devoted to the search for other heat-transfer media. Two such systems were tried: liquid metal (sodium) and molten salt (a mixture of alkali nitrates). With both of these materials very high heat-transfer rates are possible, leading to efficient solar absorbers.

Both types were tested in Almeria, Spain, in the 1980s, and the expected efficiencies of the receivers were confirmed. Further work on the liquid-sodium receiver was abandoned there because of the perceived hazards of this application.

Molten-salt solar receivers were also developed and tested in France and the United States. An experimental power station designed for 2.4 MW_e (THEMIS) was built and operated for 3 years in Targasonne, France, in the 1980s. Components for molten-salt systems were studied further at Sandia laboratory in the United States. On the basis of the results obtained there, it was decided to modify the Solar One plant at Barstow (California, U.S.A.) (initially designed with a water-in-tube receiver) and change it to a molten-salt system. This modified plant has now (1996) started operation. In all these plants the power block is still based on the conventional Rankine cycle at more or less conventional thermodynamic conditions.

A number of test facilities were built and operated over the last 20 years in several countries ranging in power from about 0.5 to 10 MW_e. They all performed below expectation and showed that none of the technologies tested then was ready for commercial exploitation. Among the problems that surfaced was the great thermal inertia of the systems, which severely reduced the amount of energy supplied.

1.3.4.2 Volumetric Receivers. All the receivers described so far suffer from the disadvantage that large fluxes of heat have to be forced through tube walls and into the working fluid. This constrains the design and limits the possible efficiency of the receivers and the utility of solar energy. Development projects were initiated in several countries aimed at overcoming this difficulty by causing the concentrated solar energy to be absorbed directly in the bulk of the working fluid, usually a gas. Since air and all other working gases are transparent to solar radiation, some means of absorbing it must be devised. Two general approaches were studied. In the first, fine particles of a material suitable for absorbing the solar radiation are suspended in the gas stream, and in the second, some sort of structure is provided that will stop the solar flux and transfer the heat from its surface to the gas stream flowing past it. These approaches led to what are known as the "volumetric receivers." They all involve a cavity, with or without a window depending on the application, illuminated with very concentrated solar radiation. Inside the cavity there is the absorbing structure. Various designs of this structure have been tried: wire-mesh screens, ceramic-fiber screens, specially shaped ceramic forms, and ceramic foams. All these are undergoing laboratory and sun tests. Very encouraging absorption rates (over 2 MW_t/m^2) have been obtained with some of them, leading to physically small receivers of high efficiency and much higher gas temperature (see Fig. 7). For applications such as gas-turbine drives or chemical reactors (Secs. 1.4.1 and 1.4.2), the receivers need a window, and this is possibly the most demanding component. It is required to transmit very high fluxes of solar radiation and to operate hot and under pressure. The size of a module will probably be limited by the maximum size of the window that could be installed. In this connection the application of secondary concentrators (see Sec. 3.2.2) will be crucial, since for a given power rating they will reduce the size of window needed.

Another, and simpler, application is to use a windowless volumetric receiver to heat air (at atmospheric pressure of course) and then use the hot air to raise steam and drive a conventional power block. A 30 MW_e plant using this approach has been designed in Europe but has not been constructed yet.

1.4 Very High Temperatures

There is considerable interest in temperatures above 1000 °C for higher efficiency electricity generation and for a whole range of chemical and materials applications. As can be seen in Fig. 4, the requirement of temperatures above 1000 °C immediately translates to the need for much higher concentrations of the solar energy. And to achieve these, secondary concentrators are required. These devices are discussed fully in connection with solar lasers (see Sec. 3.2.2 and Fig. 18). Basically they are specially configured reflectors, designed according to the principles of nonimaging optics, and known as compound parabolic concentrators (CPCs) and Trumpet-type concentrators. They are placed ahead of the receiver and further con-

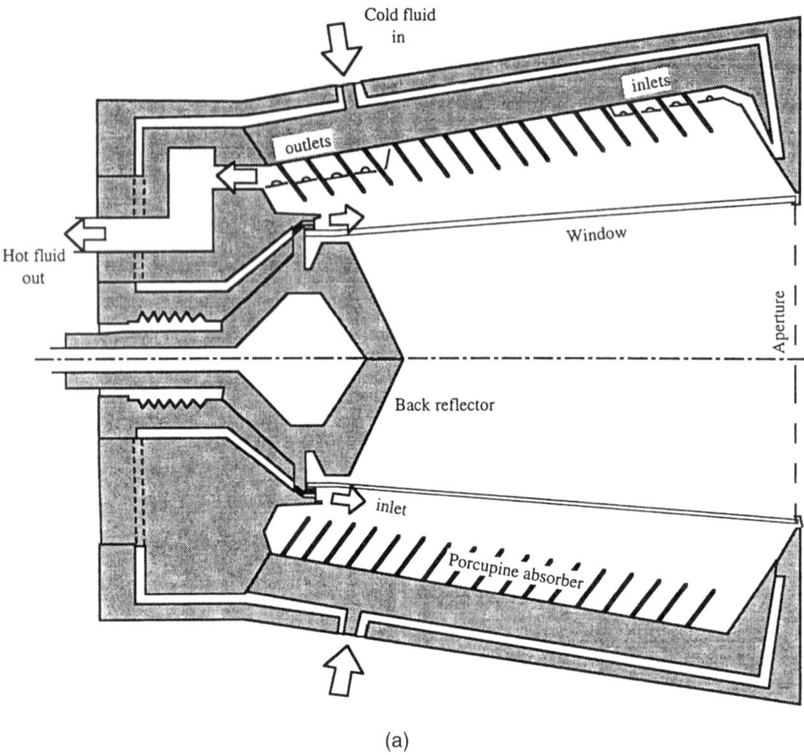

(a)

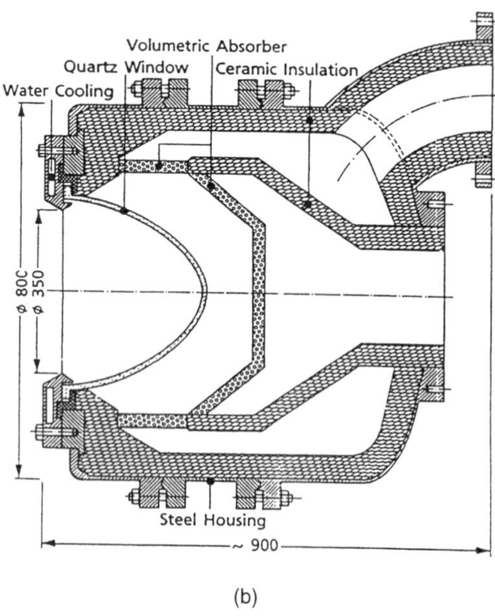

(b)

FIG. 7. Volumetric receivers. (a) DIAPR 50-kW receiver (WIS). Aperture diameter 115 mm. (b) VORBREC receiver (DLR). Dimensions in mm.

centrate the incoming concentrated radiation from a dish or heliostats.

1.4.1 Solar Gas Turbines Although the use of highly concentrated solar radiation is the key to high efficiency of collection in the dish and central receiver technologies, no advantage has been taken so far in these devices (apart from the dish–Sterling combination) of the availability of high temperatures to improve the thermodynamic efficiency of the conversion to electricity. To do so would require a change in both the power block and the receiver that feeds it.

Developments under way today in several countries attempt to adapt the Brayton cycle (gas turbine; see TURBINES, GAS) to solar use (Fig. 8). According to this concept, the working fluid (usually air) is compressed and heated to a high temperature (in a volumetric receiver, for example) and is then allowed to expand through a gas turbine that drives both the compressor and the electricity generator. Commercial gas turbines are available that can operate at 1300 °C, and their conversion efficiency is correspondingly high (over 50% in a combined-cycle mode). The adaptation to solar use requires the development of receivers that can deliver air at the temperature and pressure needed as input to a modern gas turbine (1300 °C and 20 bars).

Early experiments with solar gas turbines, without secondary concentration, were based on a cavity receiver lined with silicon carbide tubes through which air at the required pressure was passed. Temperatures of about 1000 °C were reached and used to drive a hot-air turbine coupled to an electricity generator. The limited heat-transfer capability of the tubes led to a bulky and rather inefficient design. Experiments are now in progress (Karni, 1994) with a volumetric receiver replacing the ceramic-tube one.

1.4.2 Chemical Reactions The very high temperatures that can be reached readily by means of concentrated solar radiation have attracted the attention of material processors and chemists. There are a large number of high-temperature endothermic processes of possible industrial interest, and some of them have been examined as candidate systems for solar energy application. Of particular relevance to the energy field is the search for fuels and for storage systems.

Hydrogen, as an alternative and environmentally benign fuel, has been the center of attention for decades (Fletcher, 1977). It can

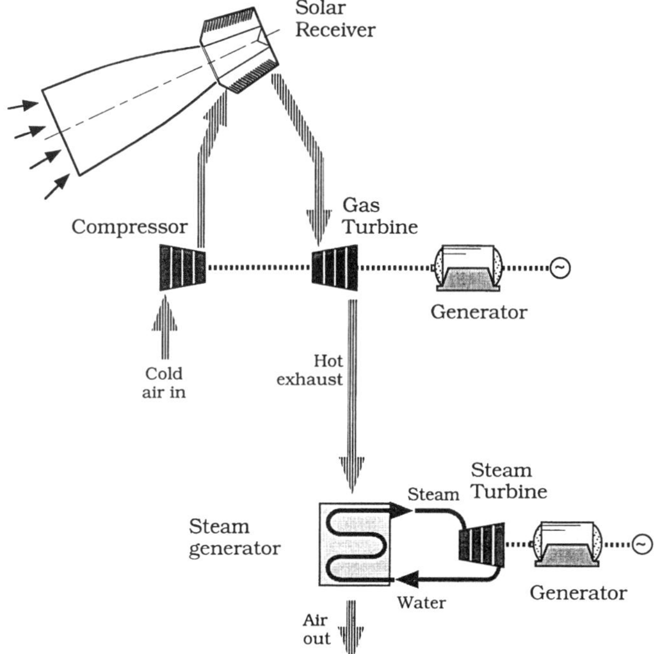

FIG. 8. Solar gas turbine (in combined-cycle mode).

be used not only directly in various engines but also to generate electricity via fuel cells and as a raw material for the synthesis of methanol and other liquid fuels. At temperatures above 2000 °C water is appreciably dissociated to its elements. If a method of separating hydrogen and oxygen, while still at the high temperature, can be developed, there will be a direct route to hydrogen production not involving electricity (Winter, 1988). Research to this end is under way, and some progress has been made.

A way of sidestepping the problem of separating the two elements is to carry out the decomposition of water in two or more steps. Much work was done on such an approach in connection with nuclear energy, but the upper temperature available from the reactors is around 900 °C. This limited the choice to fairly complex chemical systems. The higher temperature available from the concentrated sunlight opens additional, and simpler, reactions. One such reaction scheme involves the decomposition, by intense concentrated sunlight, of magnetite (Fe_3O_4) particles suspended in a gas stream to ferrous oxide and oxygen. The ferrous oxide is then reacted with water vapor to give hydrogen and re-form the magnetite (Kuhn *et al.*, 1994). The thermal decomposition of certain oxides (e.g., zinc) can also make it possible to design similar cycles, and in addition provide a means of energy storage (Fletcher, 1984).

Another direction in the search for fuels is the gasification of biomass (wood, forestry, and agricultural residues) and other carbonaceous matter. Under high temperatures (around 900 °C) such materials (with the addition of water if necessary) form a gaseous mixture containing hydrogen, carbon monoxide, and other components depending on the nature of the feedstock. This mixture can be used as such, or purified to give hydrogen or synthesis gas (a mixture of hydrogen and carbon monoxide), and they, in turn, can be used as starting points for liquid fuels. It should be pointed out here that when the feedstock is biomass, the fuels produced in this way are environmentally benign because their use and production do not add any carbon dioxide to the atmosphere (Fletcher, 1991; Epstein, 1994).

Much of the research described in this section is carried out with the help of a solar furnace. Essentially it is a fixed-dish concentrator illuminated by one or more heliostats. The principal advantage of such a facility is that the experimental zone, at the focus of the dish, is stationary. The largest solar furnace, 1000 kW (thermal), is in Odeillo, France. Many other much smaller solar furnaces, in the tens of kilowatts rating, are operating in research facilities around the world.

Another possible application of solar energy has surfaced in recent years, and it is the detoxification of wastes and the purification of contaminated water and soil. These problems are assuming a major proportion, and considerable efforts are being devoted to solve them. Solar energy can play a part in two ways:

1. by a thermal decomposition of the contaminants at high temperature, and
2. by photochemical decomposition of the contaminants.

The second, and more extensively researched photochemical route, is discussed in Sec. 3.1.

1.5 Storage

No mention has been made so far of a crucial element of solar energy plants, and it is that of energy storage. Energy storage is an important, but not essential, feature of any conventional power system and is used primarily as a means of load leveling and improving the economics.

In the case of solar energy the problem of storage becomes crucial. The sun's radiation is not a steady source of energy either on the short term or on longer terms, diurnal or seasonal. The short-term fluctuations are stochastic in nature and cannot be programmed into the operational procedure, yet may have a profound effect on the performance of the plant. The longer-term storage is particularly important where the solar plant is expected to be the sole, or the major, source of power ("stand-alone" operation).

The technical solutions to these problems vary greatly according to the specific requirements of the system. For the short-term storage, whose purpose is mainly to stabilize the operation of the power-generation equipment, some partial uncoupling (buffering) between the heat-generation part (the solar component) and the power block is the common solution. In systems using a liquid heat-transfer medium—oil, molten salt, or molten metal—a common solution is to provide storage tanks in the heat-transfer loop. In some systems only one tank is provided where mixing of the hot and cold liquid (usually oil) is avoided. This is known as the "thermocline" design. A stratified regime is maintained in the tank with the interface between the hot and cold layers moving up and down according to the charge/discharge status.

More common are the two-tank systems, one cold and one hot. Liquid from the "cold" tank flows to the solar receiver, where it is heated to the working temperature, and from there flows to the hot storage tank. The power-generation part of the plant is fed from the hot tank and is therefore completely decoupled from the solar receiver (Fig. 6). When the heat-transfer fluid is a gas, the same result may be achieved by placing a sensible heat-storage device in the circuit between the receiver and the power block. It may take the form of towers or vessels filled with solids, such as ceramic bricks, metal forms, boulders, gravel, and similar materials.

The various systems have been tried in the experimental plants operated so far and ranged in storage capacity from tens of minutes to a few hours. The shorter–storage-time devices were used mainly as stabilizers to deal with short-term fluctuations of insolation. The longer-duration storage devices were used to improve the economics of the plant by extending the operating periods by a few hours so as to match closer the demand curve. In this case the solar-collecting field must be larger than that needed to provide the nominal output at the design point, with the excess energy collected going to the storage device. The ratio of the size of the actual field to the nominal one is known as the "solar multiple" of the plant.

Another approach to extending the storage capability is to use phase-change systems. In these the heat is stored in the form of latent heat of melting of suitable solids and is released upon solidification. Such devices, therefore, operate by cycling around the melting point of the working substance. The early development of such systems was aimed at the long-term (diurnal) storage of low-temperature heat (for domestic use), and the materials used were various salt hydrates of rather low melting point. More recent advances, aimed at the power industry, developed composite ceramic-salt materials loaded with high-melting salts (Gluck *et al.*, 1992).

The really long-term storage of solar energy is a much more difficult problem. So far only low-temperature seasonal heat-storage capability has been demonstrated, and it is in the solar pond (see Sec. 1.1.2). A more universal system is under development at present and involves the conversion of solar energy into chemical energy, i.e., into suitable chemicals that can be stored indefinitely and release the energy on demand. These systems are discussed in the following section.

The problems associated with storage can be avoided if one is prepared to sacrifice the total dependence on solar energy and adopt the hybrid mode of operation. In this mode the use of fossil fuel together with, or instead of, the sun is possible and permitted. In this case fossil burners are placed in the heat-transfer circuit and supplement or replace the solar radiation. In fact most commercial plants operating today are of this nature (see Fig. 4).

Where cheap natural gas is available, the decisions whether to adopt the hybrid approach, and to what extent, is a matter of economics and environmental considerations.

1.5.1 Chemical Storage The importance of long-term storage of solar energy cannot be overstated. It is the key to the large-scale exploitation of the largest potential source of renewable energy on earth. Without long-range storage ability, solar energy can only supplement existing power networks to a limited degree. A related factor was mentioned early in this article, and it is the mismatch between the regions with ample insolation and regions of intense energy use (see Plate 1). The solution to this problem requires the ability to collect solar energy in the sunny areas, store it, and then deliver it to the centers of demand.

Research and development have been under way for over a decade on a possible solution to both the above challenges. It involves the conversion of solar energy to chemical energy, that is, to the energy stored in energy-rich substances from which they can be recovered. In principle any endothermic, reversible, reacting system could meet such a need, but careful analysis of various possibilities has limited the choice so far to a very few systems. The most widely studied is the reversible reaction between methane and water (or carbon dioxide) to give a mixture of hydrogen and carbon monoxide. This is a well-known reaction in the petrochemical industry where it is known as the "reforming reaction." It proceeds in the presence of suitable catalysts at around 900 °C with the absorption of large amounts of heat. The mixture of the products (hydrogen and carbon monoxide) can be stored at ambient temperature indefinitely. This mixture of gases, also known as "synthesis gas," is a very useful starting material for many chemical compounds (hence the name), including hydrogen and fuels. It can be used in fuel cells to generate electricity, and it can be burnt very effectively in gas turbines and to provide process heat. But in all these applications, most of the energy originates in the methane used in the reforming reaction, and the contribution of solar energy is only 27%.

Nevertheless, this reaction is of great interest to solar energy because of its reversibility. With the help of suitable catalysts, it is possible to reverse the reactions, this time with the evolution of all the solar heat absorbed in the reforming reaction and the regeneration of the starting materials. This makes it possible to use these reactions in a closed-cycle mode (see Fig. 9) to store and transport solar energy without using up any materials and without any emissions to the atmosphere. This cycle has initially been proposed in Germany as a means of transporting nuclear-generated heat over long distances (Harth, 1981). The complete cycle has been demonstrated on about 10 kW (thermal) of solar energy at the Weizmann Institute in Israel, and a scaled-up version (to 500 kW thermal) is now operating there (Levy, 1993). The reversible decomposition of certain metal oxides, mentioned above, being studied at present, is also a possible route to long-term storage of solar energy.

1.6 Economic Considerations

Solar energy is by far the largest renewable energy resource on earth, and moreover it is almost entirely benign from the environmental point of view. And yet it has not been exploited directly on a large scale by humankind. The reason, obviously, is that to date it has not competed successfully with inexpensive fossil sources. The lack of commercial interest, in turn, suppressed research and development efforts. Over the past 30 years, with the increasing awareness of the finite nature of the fossil resources and the adverse environmental effects of continued dependence on them, the attitude has changed somewhat. More effort is being exerted on the international level to develop the commercialization of solar energy, although by no means commensurate with the effort devoted for many decades to the development of nuclear energy. The following paragraphs discuss some of the economic hurdles confronting solar energy.

Since the incoming solar radiation costs nothing, the economics of solar energy is dominated by the capital investment in the plant. Solar radiation is a rather dilute source of energy, in the sense that the flux reaching a unit area on earth is rather low, seldom reaching over 1000 W/m^2. Potential sites for exploitation of solar energy are con-

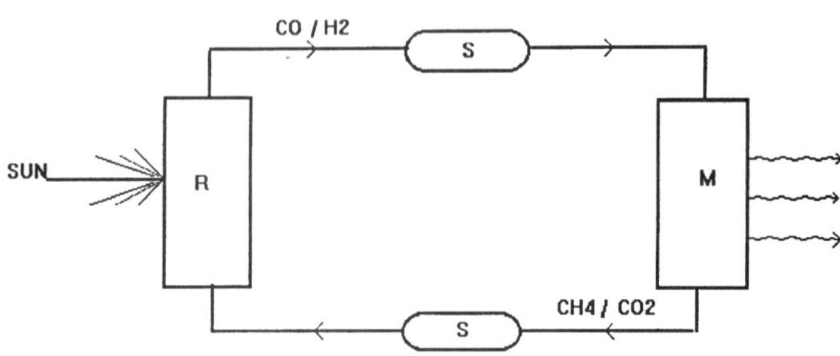

FIG. 9. Solar chemical storage and transport system. R, reformer; M, methanator; S, storage vessels.

sidered very good if the annual irradiation is greater than 2300 kW h/m^2 and good if it is over 2000 kW·h/m^2. This immediately translates to the need for very large areas to collect a meaningful amount of energy. Even if one neglects the cost of the "real estate," the cost of the collecting surface becomes a large fraction of the total cost of the plant.

There are several economic implications associated with the intermittent nature of solar energy. First, of course, is the fact that the solar plant is utilized for a smaller fraction of the day than it could be in a conventional plant. Furthermore, intermittent operation results in additional loss of production due to the thermal inertia of the plant. The need for storage, even for short terms to ensure stability of operation, which is absent in nuclear and conventional power stations, also puts solar energy at an economic disadvantage.

The economics of solar energy depends very much on how it fits into national energy policies. These, in turn, vary greatly from country to country. The parameters involved relate to the level of insolation available, to the availability and economics of other energy resources, to the sensitivity of the nation to environmental considerations, and to the international policies in this regard. Also of importance are questions relating to how solar energy will fit into the power economy. Are the solar stations to be part of a large network, or are they to serve isolated communities or small networks? How large will they be? What are the characteristic load curves?

We have seen already that where fossil fuels are combined with a solar plant, hybrid operation has great advantage. The most economical arrangement is where the solar plant supplies power to a network only at peak load time. This is the time when a utility is prepared to pay most for additional generated electricity. This only applies to locations where the electricity load curve overlaps the insolation curve significantly, a factor that severely limits the possible areas of applicability of solar energy. And it is not an accident that the only example of large commercial solar plants is from such a location (southern California). When the match is not perfect, some storage capacity is required to gain a few working hours, or, alternatively, hybrid operation is adopted. Careful economic analysis is necessary to determine the optimal proportion of the additional storage. In general, small solar plants that feed into a large electric network are relatively free from problems of storage (beyond what is necessary for the smooth operation of the plant).

Modern electricity-generation plants are shifting to a combined-cycle mode of operation, where the efficiency of conversion to electricity is much higher and can reach 60% as compared with about 40% for the "conventional" power station. In the combined-cycle mode, both gas turbines and steam turbines are used. The former use natural gas, operate at high inlet temperature, and exhaust the combustion product gases at temperatures of around 600 °C. These in turn are used to raise steam of a quality suitable to drive a conventional Rankine cycle. The combination leads to the high overall conversion efficiency. A solar component may be integrated into such a combined-cycle system in several ways. At present most such systems assign the solar energy to the "bottoming," i.e., the Rankine steam cycle, and use the fossil fuel to drive the gas turbine. However, with the development of solar-driven gas turbines, it will become possible to use them also in the "topping" cycle (Fig. 10).

Difficult problems arise when the solar plant is required to be of the "stand-alone" type, that is, to be the sole, or major, source of power to the network. This situation may arise when the network being served is small and remote, or when the use of fossil fuels is precluded, by regulation or availability. In such situations the need for large and long-term storage becomes unavoidable. As was mentioned above, the technology required for this purpose is still under development (see Sec. 1.5.1).

2. PHOTOVOLTAIC SYSTEMS

2.1 Introduction

Photovoltaic (PV) cells are solid-state electronic devices that convert electromagnetic radiation to electricity. Full description of the physics of these devices is given in PHOTOVOLTAIC DEVICES, and is not repeated here. Instead we concentrate on the application of

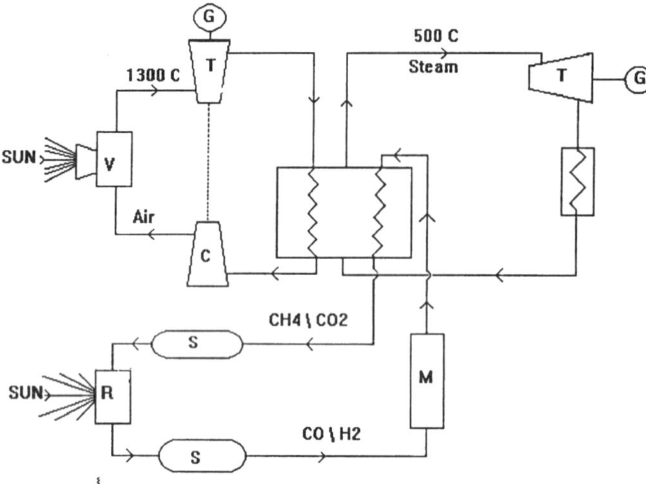

FIG. 10. A combined-cycle solar stand-alone system. V, volumetric receiver; R, reformer; T, turbine; C, compressor; M, methanator; S, storage; G, generator.

such devices to the harvesting of the major renewable energy resource—solar energy.

PV cells are quantum devices in the sense that only photons above a certain energy (the forbidden gap energy) will promote an electron from the valence band to the conduction band of a semiconductor and, if conditions are right, ultimately to an outside circuit to provide useful electric current. PV devices are therefore spectrum sensitive and can only use part of the solar spectrum. In Table 1 (Emery, 1996) we list the major photovoltaic materials relevant to solar energy along with the parameters important for PV operation.

Photons with energy lower than the band gap are not absorbed, while the excess energy of photons with energy higher than the band gap is ultimately converted to heat. Therefore the maximum efficiency is to first order determined by the band gap of the semiconductor from which the PV cell is made. Figure 11 shows these theoretical efficiencies and their dependence on gap energy. In practice the efficiencies are lower than the ideal maximum for a variety of reasons, and much of the research and development in PV is directed to improving it; and in fact the efficiencies have been rising steadily over the last few years.

2.1.1 Single-Crystal Silicon The first Si photovoltaic cell was demonstrated in 1954, a year that marked the birth of the modern photovoltaic cells. This is still the most mature technology today with some

Table 1.

Semiconductor	Band gap (eV)	Type, dopant	PV technology	Efficiency (%)	Type of junction
InP	1.35	*n*: Zn, Cd *p*: Se, Te, Si	Single crystal	21.9	Heterojunction
GaAs	1.45	*n*: Zn, Cd *p*: Se, Te, Si	Single crystal	24.3	Heterojunction
$CuInSe_2$	1.0	*n*: In_{Cu} *p*: Cu_{In}	Thin film	17.7	Heterojunction
CdTe	1.45	*n*: In, Cl *p*: P	Thin film	15.8	Heterojunction
a-Si:H	1.7	*n*: P, Sn *p*: B, Al	Thin film	10	Homo/heterojunction
Si	1.1	*n*: P, Sn *p*: B, Al	Single crystal; polycrystalline thick film	24 18.6	Homojunction

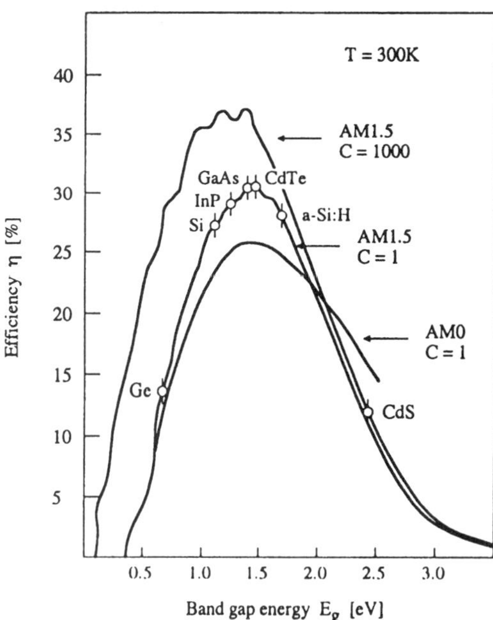

FIG. 11. Ideal solar-cell efficiency as a function of band-gap energy when illuminated under spectral conditions of outer space (AM0) and under terrestrial conditions with the sun at zenith angle of 48° (AM1.5). C = degree of concentration of the solar radiation.

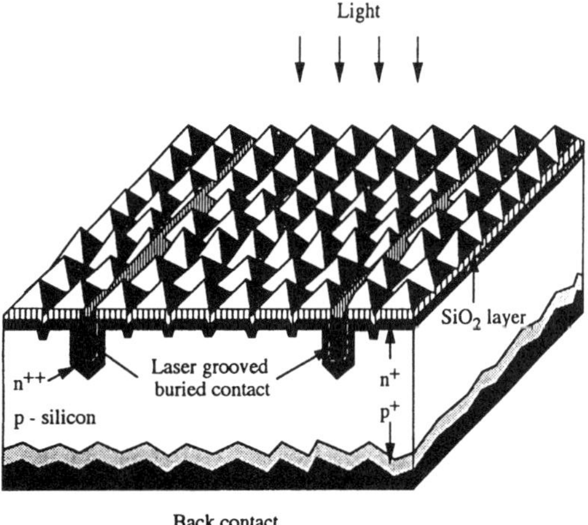

FIG. 12. Schematic representation of high-efficiency design for silicon solar cells.

40% share of the sales of all PV cells. A concise review of this technology has been published by Green (1987). A diagram of one of the most advanced versions of such a cell is shown in Fig. 12. Typical efficiency of laboratory-sized (4 cm^2) advanced PV Si cells is 24% for unconcentrated sunlight, and 28% with concentrated (200 times) sunlight. Commercially available silicon cell modules typically have efficiencies of about 12%. Modules manufactured in a semi-industrial environment have a narrow distribution of efficiencies (19.5% to 21.5%).

The high-performance (>23% efficient) Si cells are made of high-purity silicon prepared by the float-zone or Czochralski technique. They are doped with boron to make them p type. The back surface is made p^+ by Al alloying, which also provides the Ohmic contact and a field to drive away holes from the back surface and thus reduce recombination losses (back surface field). The front (100) surface is texturized by alkali treatment to reduce reflection losses. A further reduction in reflection and front-surface recombination losses is achieved by depositing a layer of SiO_2. Sometimes a second antireflective film of Si_3N_4 or MgF_2 is added.

In some advanced designs the Ohmic contacts are produced by grooving the front and back surfaces by a laser technique and then depositing a series of metals (Ni, Cu, and Ag) in them by an electroless process. These contacts are called "buried contacts" and minimize shadowing without increasing series resistance.

2.1.2 Gallium Arsenide Cells A number of compounds of group-III and -V elements, like GaAs, InP, GaP, etc., are semiconductors. Some of them have useful photovoltaic properties and have been investigated in the hope of producing more efficient or more economic PV cells. Gallium arsenide PV cells are particularly attractive. Single-crystal GaAs has a band gap of 1.41 eV, close to the optimum value on the efficiency–band-gap curve (see Fig. 11). GaAs cells have exhibited efficiencies of 25% with no concentration and 28% efficiency with concentrated sunlight (200 suns).

2.2 Thin-Film Devices

Until the early 1970s, the main applications of photovoltaic devices were in space with a few specialized uses in military and civilian applications. In all these applications, the cost of the PV components was of secondary importance, and emphasis was given mainly to efficiency and reliability. In the last two decades, however, with the increasing importance given to renewable energy sources and environmental issues, attention has been directed to the possibility of large-scale (utility) use of PV. While the currently available single-crystal Si (and GaAs) PV cells show high efficiency and reliability, they suffer from one major disadvantage: they are too expensive, mainly because of the need for very high-purity materials and complex and expensive production methods. Much of the research effort in recent years was directed, therefore, to cost reduction.

The strategies being pursued for cost reduction are aimed at increasing efficiency, reducing the amount of material required, obtaining cheaper materials, and finding simpler, less expensive, production methods. On reflection it will become evident that of the various options available, reducing the amount of material presents the greatest opportunity for cost saving because 100-fold savings are possible, while modifying technologies offers a more modest opportunity for savings. Economic analyses of solar PV plants lead to the conclusion that the efficiency of the less expensive cells must still be high because of the area-proportional value of system costs (balance of system).

2.2.1 Polycrystalline Silicon Polycrystalline silicon is less expensive to produce than single-crystal silicon, and although polycrystalline cells are somewhat less efficient than single-crystal cells, they continue to be a possibly viable PV technology. A number of methods have been used to make polycrystalline silicon for PV use. Large ingots (of up to 100 kg) have been cast from the melt in crucibles of high-purity graphite or silica or silicon nitride. A directional solidification was established by maintaining a temperature gradient in the crucible. The ingot was then sliced into wafers and further processed according to procedures established for the monocrystalline material. The best polycrystalline PV cells are made by this method.

Self-supporting sheets of silicon can be produced by pulling out of a melt or by extrusion from graphite die. Thin films of Si can also be grown on suitable substrates (ceramic, coated steel, and industrial silicon) by one of several methods. These include chemical deposition from solution, vacuum vapor deposition, chemical vapor deposition, and others, all potentially cheaper than high-temperature melt procedures. The thickness of the polycrystalline layers produced by these techniques is of the order of tens of microns, i.e., a factor of 10 lower than in the single-crystal cells.

Since the thickness of Si needed to absorb 90% of the solar photons incident on it is about 200 μm and the minority-carrier diffusion length in polycrystalline Si seldom exceeds 100 μm, thin polycrystalline Si PV cells require design changes. It is necessary to ensure that the light entering the cell is trapped within the active layer and that its effective path within it is much greater than the thickness of the material. A textured front surface (which can be obtained by electrochemical etching—so-called porous silicon), coupled with a mirrorlike back surface, creates a light-trapping configuration that results in a large increase in the absorption coefficient. Efficiency of 17% in 4-cm^2 polycrystalline cells has been achieved.

Even when electronic-grade silicon is used, many undesirable impurities (oxygen, carbon, transition metals), which are minority lifetime killers, may be picked up during processing and must be removed. This is done by heating between 600 and 1000 °C after depositing, by evaporation, a layer of Al to provide gettering action. After the treatment, this film, now partially oxidized, is removed by etching.

A novel approach to the fabrication of Si PV cells has been reported recently (Levine, 1991). Silicon spheres of 0.75 mm diameter are produced by a spray technique. During this process dopants are introduced, and a superficial *p-n* junction is formed. The spheres are embedded in glass matrix and contacts established by metal evaporation. An efficiency of 11.5% has been reported for arrays of this type.

2.2.2 Amorphous Silicon A further major step in the reduction of costs of materials and production has been achieved with the introduction of PV materials of high optical absorption that can be applied in very thin layers, of the order of 1 μm thick, by relatively simple techniques (Zweibel, 1993). We describe here some of the most promising examples.

Amorphous silicon, *a*-Si, is usually prepared by the glow-discharge decomposition of silane gas (SiH_4) in a radio-frequency–generated plasma and the deposition of the resulting silicon on a suitable substrate. The high density of dangling bonds that are formed in the material prepared in this way introduces states within the forbidden gap, which have a deleterious effect on the electronic properties of *a*-Si. An excess of hydrogen (up to about 10%) is introduced to saturate most of the dangling bonds and improve the characteristics of the material (which is then designated as *a*-Si:H).

a-Si:H is unique in that its band gap can be controlled, in the range 1.5 to 1.8 eV, by adjusting the hydrogen content, and even further by alloying with Ge or C. In this way materials may be obtained with band gaps between 1.3 eV (for *a*-Si,Ge:H) and 2.0 eV (for *a*-Si,C:H). This flexibility permits the design of multijunction cells that can be "tuned" for maximum efficiency. Figure 13 shows a diagram of a double-junction *a*-Si:H/*a*-Si,Ge:H cell. *a*-Si:H cells suffer from light-induced degradation. This is particularly severe in single-junction cells. Thus, a 1-cm^2 cell of *a*-Si:H with a 12.5% efficiency (compared with the theoretical maximum of 28%) degrades with time to a stable 6% efficiency.

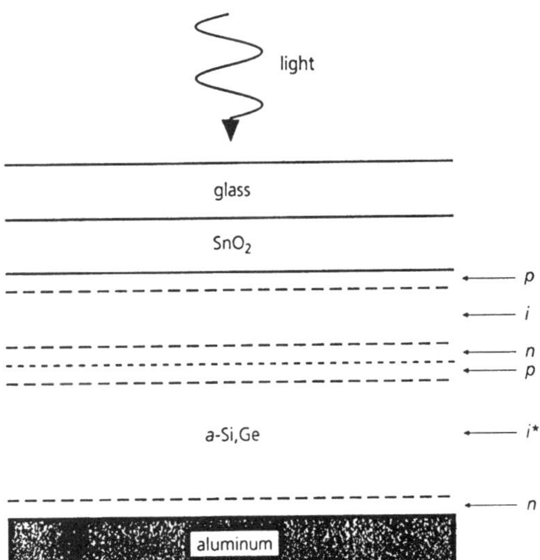

FIG. 13. A schematic diagram of an *a*-Si multijunction cell. The i^* layer of the bottom cell is an amorphous silicon–germanium alloy with a narrow band gap.

A similar three-junction cell starting with 13.7% efficiency degrades to a stable 10% value. Nevertheless, this technology is still very attractive and developing, with modules of 1 m^2 being produced for utility applications. However, unless the efficiency of *a*-Si:H PV panels increases to about 15%, they will not satisfy the requirements for large-scale PV power plants.

2.2.3 Cadmium Telluride Some of the semiconductors of the II–VI family also have properties suitable for thin-film photovoltaic applications. Thus, CdTe has a band gap of 1.45 eV (close to optimum) and a high optical absorption coefficient permitting the use of very thin films (a few microns). In addition it is less sensitive to impurities and lattice defects.

PV cells of this material are formed by vacuum evaporation of CdTe, followed by dipping in $CdCl_2$ solution and annealing at 450 °C. The $CdCl_2$ serves as a flux, which leads to recrystallization of the compound to give large crystallites (>1 μm). The heterojunction is completed by depositing thin CdS film. Somewhat inferior results were obtained with films formed by electrodeposition and screen-printing techniques. HgTe can be used with advantage for the Ohmic back contacts instead of gold. Air annealing of the whole cell at 200 °C for 30 min brings about substantial improvements in these cells. Small-area cells with a solar conversion efficiency of about 16% have been reported (Chu *et al.*, 1992). Large modules (1 m^2) with 8% efficiency are on the market today. As with other technologies where large amounts of Cd are involved, CdTe cells would be subject to strict environmental control, which may impede their utilization on a large scale.

2.2.4 Chalcopyrites Other promising semiconductors for solar applications belong to the group of chalcopyrites, compounds with the formula $CuMX_2$ where M = In,Ga and X = S,Se. The most developed member of this group is the compound copper indium diselenide (CISe) with a band gap of 1.0 eV. Its high absorption coefficient permits the use of very thin films (<0.5 μm). A diagram of a CISe cell is shown in Figure 14. It possesses the highest conversion efficiency of any thin-film cell, 17.7% for a 1-cm^2 cell, 14.6% for a 3.5-cm^2 unit, and 9.7% for a 0.37-m^2 module (Hedstrom *et al.*, 1993). The

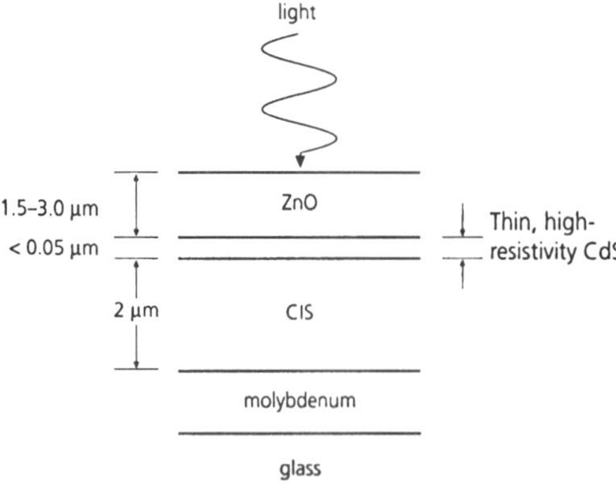

FIG. 14. A copper-indium-diselenide cell.

chalcopyrites can be produced either by vapor deposition or by sputtering of the metals followed by annealing in a selenium-laden gas. The latter mode is the one preferred now. The heterojunction is formed by depositing a 0.05-μm insulating CdS layer followed by a sputtered ZnO transparent conductor a few microns thick.

An intriguing recent discovery is the Cu-deficient ordered vacancy phase (OVP) of *n*-$CuIn_3Se_5$, which forms spontaneously on the last few monolayers of CIS during vapor deposition (Schmid *et al.*, 1993). This new phase has a band gap of 1.5 eV and consequently may serve as the *n*-type layer of an internal heterojunction. Another chalcopyrite, copper indium disulfide, gained attention in recent years. It has the advantage of somewhat higher band gap (1.5 eV) and the replacement of Se by the more benign sulfur. Efficiencies of about 11% and later 12% have been reported for 1-cm^2 cells (Scheer *et al.*, 1993). Dissolution of excess Cu*X*S on the surface of $CuInS_2$ in KCN solution provides the best interface with the overlayer CdS window film (Klenck *et al.*, 1993).

2.2.5 Tandem and Multijunction Cells

In the single-junction PVC, all photons of energy lower than the band gap cannot be absorbed by the semiconductors and are lost. The conversion efficiency of the cells can be increased if layers of two or more photoactive materials are arranged in series in such a way that the photons that were not absorbed in the first layer are absorbed by the following one. Model calculations show (Bennet, 1978) that an AM1 efficiency of 36% can be attained by two junctions and of 40% by three properly tuned junctions. Further increase in the number of junctions leads to only a marginal gain in efficiency. Figure 15 shows a diagram of a typical two-contact arrangement.

The preferred fabrication technique for III–V semiconductor tandem cells is metal-organic chemical vapor deposition. One of the major challenges in fabricating monolithic cells is the Ohmic contact between the two junctions. It is required to be transparent to light, and its lattice must match as far as possible that of the substrate. The preferred arrangement is a layer that acts as a tunnel diode operating under forward bias.

2.3 Concentrators

Up to now we have considered only PV cells that operate under unconcentrated insolation, i.e., not exceeding about 1000 W/m^2. It is clear that if more concentrated sunlight could be used, the area of the PV cells

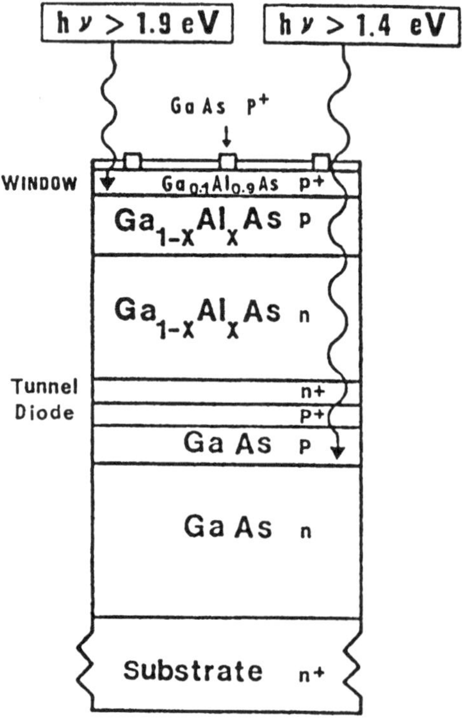

FIG. 15. Schematic diagram of an Al_xGa_{1-x}/GaAs monolithic tandem solar cell with a tunnel diode as intercell Ohmic contact.

required for the same power output would be reduced in inverse proportion. Thus, even a concentration of 100 times, which is considered very modest in solar thermal technologies, will reduce the area of PV cells by a factor of 100 and reduce the cost of cells substantially. However, it will be the cost of the solar collecting and concentrating components that will play a significant role in the economic viability of the plant. Concentrator systems require PV cells that can operate efficiently under intense illumination. Such cells are, indeed, available (Cuevas *et al.*, 1989; Kuryla *et al.*, 1990) at a cost not much greater than cells for unconcentrated sunlight. The economic advantage of operating under concentrated sunlight depends, therefore, on the relative savings on cells as compared to the cost of the solar collector and concentrator components.

A large variety of solar concentrators is available, some designed specifically for PV operation and others through the efforts of the solar thermal development programs. Those designed for PV use are based on Fresnel lenses, while those designed primarily for thermal applications are based on reflectors. The latter are described extensively in Secs. 1 and 3 of this article. With concentrations above about 100, the economic considerations become very similar to those involved in the thermal technology with the replacement of the power-block components used there by photovoltaic power units.

A particularly attractive feature of the use of concentrated sunlight is the possibility of splitting the solar spectrum and illuminating different PV cells with those portions of the spectrum that each cell can use most effectively. A more ambitious proposal involves powering by the same source of concentrated sunlight, at the same time, different conversion technologies, e.g., PV units, lasers, and thermal converters (Krupkin, 1993). Such an arrangement is discussed and illustrated in Sec. 3.

The selection of the various parts of the solar spectrum may be done by means of dichroic filters. These can withstand high light intensities and are relatively inexpensive.

2.4 System Integration

2.4.1 Modules and Arrays The output voltage of an individual cell is low, of the order of 1 V, and the current delivered by it is a few tens of mA/cm^2. A large number of cells have to be connected together in a series and parallel arrangement to obtain the desired output. The hierarchy of these connections starts with the modules that are then connected into arrays that are finally assembled into a system. Because of the large number of units differing slightly in characteristics and operating conditions connected together, and the lack of uniformity in cells of larger area, the efficiency of the assemblies decreases appreciably as we progress along this hierarchy. Thus, the efficiency of a module is considerably lower than that of a laboratory-sized cell, and that of the array smaller still. For this reason we have been careful when quoting efficiencies to mention the size of the unit involved.

Great care has to be exercised in the design and assembly of the final system. In such complex systems provisions have to be made to deal with conditions of off-specification operation of some of the components that may affect seriously the operation of the whole. For example, if one cell of a module loses illumination for some reason, all the string in series with it will be put out of action, but will remain under the voltage of the other modules in parallel with it. This may induce current to flow in the wrong direction through the shadowed cell, heating it ("hot spot") and possibly damaging it permanently. Analogous situations may arise in parallel-connected units. These calamities are avoided by incorporating diodes in strategic points to shunt or prevent the flow of unwanted currents.

2.4.2 Encapsulation and Support A major difference between the laboratory cell and the actual field module is its protection from the weather. The components of an actual photovoltaic power unit are expected to operate, exposed to all weather conditions, for many years, and the economics of the system depends to a large measure on such survival. This is achieved usually through careful encapsulation of the modules.

The assemblies are usually supported on conventional structures, designed, of course, for long outdoor exposures. There is, however, a specific feature of solar assembly support—whether it is stationary or tracking. The simplest design is, of course, the simple fixed support adjusted during construction for maximal solar exposure. Usually this will call for a south-facing (in northern latitudes) or north-facing (in southern latitudes) structure with its normal set at an angle to the vertical equal to the latitude of the location. Then at the equinoxes the panels will receive the maximum insolation, while at other times of the year there will be a reduction of intensity by a factor equal to the cosine of the angle between the sun and the normal to the panel. Such loss can be minimized or avoided entirely if the panel is made to track the apparent motion of the sun. Such double-axis tracking is common in the field of thermal applications of solar energy, and fairly simple and reliable equipment is available. It should be noted that for PV use a far less accurate tracking is sufficient than that needed for the thermal or laser applications (see Secs. 1 and 3). The economic considerations are simple: Is the gain in the annual solar electrical energy produced by a tracking assembly greater than the extra costs of the tracking mechanisms? Concentrating PV systems, of course, always track the sun, and this partly compensates for the fact that they use only the direct normal fraction of the solar radiation.

Obviously the exact relative advantage of one design over the other will be site specific. Comparative studies in various parts of the world showed that, for example, the annual solar radiation per unit area in Rome, Italy, and Denver, Colorado, U.S.A., were the same for the fixed-bed and the concentrating systems, while in Boston, Massachusetts, and Seattle, Washington, U.S.A., the direct normal radiation captured by the concentrating cells was about 80% of the total. Two-axis tracking flat-bed PV assembly will, of course, always receive more total irradiation than the direct normal component.

Another function of the mounting structure is to provide cooling for the PV cells. The efficiency of all PV cells drops rapidly as the operating temperature rises. In fact all the values of efficiencies quoted in this article were those measured at "room" temperature (25 °C). Actual modules operating in the field operate at temperatures slightly higher than ambient, which may often be higher than the nominal value quoted above. This is particularly so with cells operating under concentrated sunlight, and cooling is essential or the cell may even be destroyed. With nonconcentrating cells it is usually sufficient to ensure good thermal contact with the backing structure that may be designed for enhanced radiation. With concentrating cells, forced cooling of the backing must be included in the design, and this, of course, adds to the cost of construction and operation. In advanced designs the hot water obtained from the cooling cycle is used for heating or other household purposes.

2.4.3 Power Conditioning The electrical power produced by PV cells and therefore PV panels and systems is low-voltage direct current. Apart from certain very specific local applications, such power does not match the rest of the local power network and the equipment connected to it. In addition, the raw power produced by the PV assembly is not steady because the solar illumination is not steady even during an apparently clear and cloudless day. So some power conditioning is an essential feature of a PV power supply. The details will, again, depend on the specific site and conditions. The simplest arrangement, for small and independent units, will call for some storage batteries to stabilize the output and an inverter to deliver ac power at the voltage and frequency to match local usage.

For larger units, particularly those connected to networks, more sophisticated controls are necessary. These may include synchronization, various safety interlocks, and power regulators. The last of these are necessary to ensure that the PV units are operated at their maximum power points even when the illumination and load are changing.

2.5 Storage

The problem of energy storage in solar plants is a general one and not specific to photovoltaic systems (see Sec. 1.5). Since the sun does not shine continuously, and in addition disturbances due to weather conditions are unavoidable, solar energy plants suffer a severe handicap. The magnitude of this problem and its possible solutions depend very much on local conditions and policies. This issue is discussed at some length in Sec. 1.5 of this article, and therefore only those aspects more specific to PV systems will be described here.

For small stand-alone units (in the kilowatt range), storage batteries are an adequate solution. For larger systems (in the megawatt range), this solution is not practical. If, however, the solar unit is part of a utility network, then the question of storage depends on the match between the load curve of the network and the insolation curve for the location of the solar plant. If the match is reasonable, then the solar plant can be used as a peaking plant only, supplying power only at times of maximum demand when it is worth most, and storage can be dispensed with. However, if larger-scale utility use or stand-alone operation is contemplated, the problem of storage becomes particularly severe.

The options open to PV systems are more limited than those for thermal systems. The electricity supply industry has been facing this problem for many decades and has evolved some solutions, and these of course are open also to the PV industry and will be discussed further here. However, there is one channel that is particularly suitable to the PV industry. It can take advantage of the fact that the primary product is dc electricity and is thus eminently suitable for electrochemical processes. In particular, the electrolysis of water to produce hydrogen is an attractive route for storage (and use). Hydrogen is a valuable material and is seriously being considered as an environmentally benign replacement of fossil fuel for transportation and other uses. At the same time hydrogen, in a fuel cell, is an excellent source of electricity. Thus, solar energy from PV plants can be stored as hydrogen indefinitely, can be transported in pipe lines, and can be converted back to electricity at will via fuel cells. A joint German–Saudi Arabia 300-kW solar PV-hydrogen experiment (Hysolar) has been running for a number of years (see also Sec. 1.4.2).

As was already discussed in connection with the thermal applications (see Sec. 1.6), the optimal choice of a process and its detailed design depends on many factors in addition to simple cost, most of which are equally valid for photovoltaic systems, and will not be discussed here. They are site specific (in the general sense) and include questions of government policies and environmental regulation.

3. PHOTOCHEMISTRY AND SOLAR LASERS

In this section we deal with the conversion of sunlight to chemical energy without converting it first to heat or electricity. Although solar photochemistry is the basis of all life on earth, and has been studied from the early days of chemistry, up to the present time there have been no major artificial (i.e., nonbiological) applications to industry or the energy economy.

3.1 Applied Solar Photochemistry

Some recent attempts to use solar radiation to detoxify wastes and treat contaminated water may be the first examples of a new trend. The driving force behind this new field is the increasing awareness of the magnitude of environmental damage caused by continuing pollution of the water resources and atmosphere by human activity. The permissible concentrations of most pollutants are so low that the classical technologies are inapplicable or too expensive. The hope that photochemical processes, driven by the plentiful solar radiation, can be used for the degradation of these pollutants has led to a large number of experiments and tests in various parts of the world. Most of these are still in the laboratories (using solar simulators) and are too numerous to be detailed here. A recent review appears in a book by Ollis (1993). Reports of experiments in progress can be found in almost all the international conferences on solar energy, some of which are quoted in the bibliography at the end of this article. In particular we would like to mention the reviews by Larson (1991) and Mills (1993), as well as the field experiment reported by Mehos (1993).

Two different methods are studied. The first method is based on the activation of semiconductors like TiO_2 using the UV fraction of solar radiation and subsequent interaction with organic contaminants leading to destruction of the latter. The main drawback of this method is that only a very small fraction of solar radiation can be utilized.

The other method is based on the excitation of oxygen molecules dissolved in water by energy transfer from an organic dye that absorbs visible solar radiation (Acher, 1990). A more significant fraction of the solar spectrum is involved in this process. The oxygen molecules are excited to their singlet state and can oxidize various organic contaminants present in the solution. For complete detoxification, solar radiation at concentration of more than 1000 suns may be necessary.

The most efficient photochemical processes are those that are induced by monochromatic radiation tuned to the specific molecular species involved. The classical way of obtaining approximately monochromatic light from solar radiation is either by means of absorption filters or by diffraction. A more recently developed method is through the use of solar-driven lasers.

3.2 Solar-Driven Lasers

A detailed description of the physics of lasers is found in LASER PHYSICS and will not be repeated here. A common way of exciting solid-state lasers today is by illumination with a powerful beam of light, usually obtained from an electrically powered source. It is but a conceptually short step to replace the electrically driven light source by intense solar radiation and thus obtain a solar-driven laser. In practice it is somewhat more complicated, and in this section we describe the special features of solar lasers and report on the state of the art today. The motivation for developing such devices is the elimination of the need for a power supply (particularly important in space applications) and the possibility of high overall efficiency of conversion of sunlight to monochromatic and coherent radiation.

3.2.1 Materials for Solar Lasers In Table 2 we list some of the laser materials suitable for solar driving together with their thresholds, spectral absorption bands, and other parameters. In Fig. 16 we show the solar spectrum and note the overlap with the absorption bands of some of the materials listed in Table 2. We note that all of them are garnets possessing high refractive index and high thermal conductivity. They all have absorption bands in the solar spectrum, a high refractive index, and good heat conductance. Yttrium aluminum garnet (YAG) doped with neodymium is a common combination for many laser applications. However, the overlap of the absorption spectrum of the doped crystal with the solar spectrum is rather poor, leading to low efficiency even with chromium as codopant. Gadolinium gallium garnets (GGG) and gadolinium scandium gallium garnets (GSGG) give very high energy-transfer efficiency when codoped with chromium but suffer from poor thermal properties and tendency to fracture. These difficulties may be minimized by spectrally selected illumination and special geometric design of the crystal. It is then possible to double the efficiency as compared with Nd:YAG (Noter *et al.*, 1988).

Table 2. Properties of some laser materials for solar pumping.

Laser material	Center of absorption band (nm)	Lasing threshold (w/cm^2)	Laser wavelength (nm)
NdYAG[a]	540	30	1064
	580		
	740		
	800		
	880		
NdCrGSGG[b]	460	50	1054
	640		
	740		
	810		
	880		
Alexandrite	410	200	750–850

[a]YAG stands for yttrium aluminum garnet.
[b]GSGG stands for gadolinium scandium gallium garnet.

Sapphire doped with Ti^{+3} has an excellent thermal conductivity, but the threshold of brightness for laser pumping is so high that it is practically impossible to remove all the heat that is associated with continuous pumping. It cannot, therefore, be considered a good candidate for a cw solar laser (Krupkin, 1993).

3.2.2 Nonimaging Concentrators As can be seen in Table 2, some of the thresholds for lasing action are within the light intensities easily achievable by the concentration technologies developed for thermal application and are described in Sec. 1. But for some of the more interesting materials, the thresholds are very high and require light intensities that cannot be obtained by these techniques. However, as Winston has shown (Welford, 1989), if in the design of the concentrators the requirement of faithful imag-

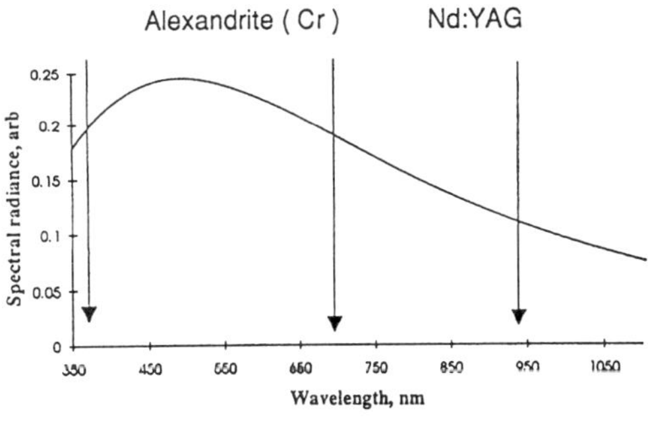

(a)

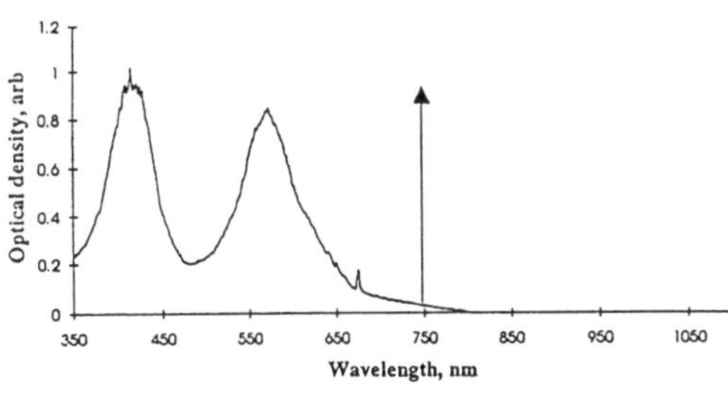

(b)

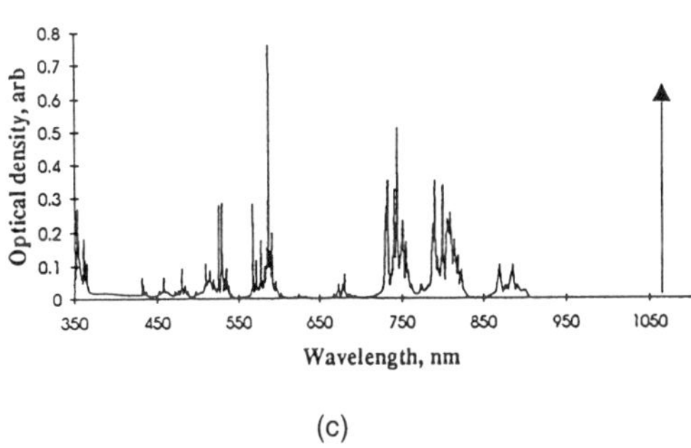

(c)

FIG. 16. Solar spectrum and absorption bands of various laser materials. (a) Solar spectrum outside the atmosphere; (b) absorption spectrum of alexandrite; (c) absorption spectrum of Nd:YAG.

ing is waived, it is possible to attain much higher light intensities.

A very common design of nonimaging concentrators is the compound parabolic concentrator (CPC) shown in Figs. 17 and 18. The three-dimensional shape is obtained by rotating a section of a parabola with its axis tilted an angle θ to the axis of revolution. The degree of concentration provided by this device depends on the ratio of the entrance and exit angular apertures according to the following relations: $C = \sin \alpha'/\sin \alpha$

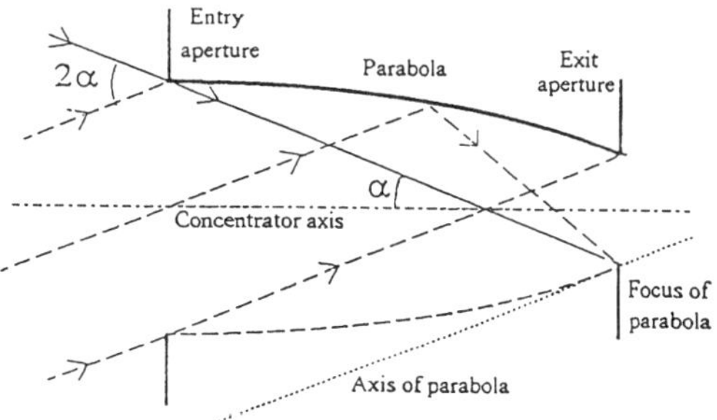

FIG. 17. A compound parabolic concentrator.

for the two-dimensional case (trough), and $C = \sin^2\alpha'/\sin^2\alpha$ for the three-dimensional case, where α is the entrance aperture half-angle and α' is the exit aperture half-angle. One can see that for the value of the angle $\alpha' = 90°$, the concentration is maximal and equals $1/\sin \alpha$ and $1/\sin^2\alpha$ for the two- and three-dimensional cases, respectively. The rays then leave the exit into a 2π solid angle.

For very small entrance angles such as, for example, that of the solar disc (0.266°), the concentration ratio is $46\,200n^2$ (for a 3D device), but the concentrator will be very long and unwieldy. Here n is the refractive index of the medium through which the light propagates within the concentrator. The main use of such CPCs is, however, as secondary concentrators receiving the output of dishes, troughs, or heliostats. In these cases the entrance half-angle will be large and the concentration ratio correspondingly smaller. Thus, for a 45° entrance half-angle the ideal maximum concentration of the CPC will be only 2, but the overall concentration of the assembly can still be several thousand.

When the object to be illuminated is of cylindrical shape (rod or tube), as in most solar laser applications, the configuration shown in Fig. 18 is appropriate (it is a version of 2D concentrators).

When this technology is used with solid-state lasers, the brightness of the pump light can be further increased by refraction. The upper limit can then be increased by n^2 (for a 3D system), where n is the refractive index of the CPC material. In this case, the concentrator itself must be made from transparent dielectric material, relying on total internal reflections for almost lossless operation. Advantage of this circumstance can only be

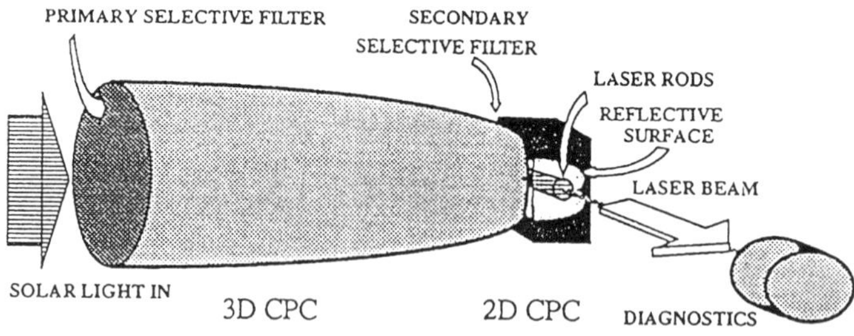

FIG. 18. A solar pumped laser (WIS).

taken if the laser itself is placed in close optical contact with the concentrator and its refractive index is not too different from that of the latter.

3.2.3 Heat Removal and Efficiency Having achieved the high solar fluxes into a solid-state laser, the next major problem is to remove the excess heat generated within it. While about 50% of the light absorbed within the spectral band of the laser material is converted to radiation and leaves the crystal, the remaining light that is absorbed by the solid is converted to heat and has to be removed. Apart from the obvious danger to the crystal by fracture, there is also a very deleterious optical effect. Temperature gradients within the crystal, by causing refractive index variations, can lead to birefrigence and lensing, which interfere with the optics of the laser cavity. Effective cooling is therefore a prerequisite of high-powered solar lasers, with the added complication that the cooling system is part of the optical cavity. In many cases it is the heat-removal capacity that sets the limit to the laser power output. A diagram of a complete solar laser is shown in Fig. 18.

High-power solar lasers can be obtained by bundling together several rod lasers in one solar concentrator (see Fig. 19) in the same optical cavity. Power output of 500 W cw has been obtained from such a device with four 6.3-mm Nd-YAG rods (Thompson, 1992, 1993).

Another way of increasing the power output of lasers, when the rate of heat removal is the limiting factor, is to remove from the illuminating light part of the short-wavelength portion of the spectrum, which contributes only slightly to the lasing. In this way the thermal load is reduced to the minimum. This concept of splitting the solar spectrum and then using each spectral band separately can be generalized to other systems where solar radiation is converted directly to other energy forms. For example, one may consider a more complex system where part of the spectrum is used for lasers, part for photovoltaic conversion, and the rest for thermal applications (Krupkin, 1993). Spectral splitting can also be applied to simultaneous pumping of several lasers with different absorption bands and thus obtain a much higher conversion efficiency. An example of such a pair may be alexandrite and Nd-YAG. See Fig. 20.

The spectral splitting may be obtained with the aid of dielectric filters. These can operate with highly concentrated light without cooling and can be integrated in the optical design of the collection and concentration devices (Krupkin, 1993). Such schemes may be very attractive economically, since 1 m^2 can provide 10 MW of solar light.

3.2.4 The Wavelength of Solar Lasers All available optically pumped lasers are active in the near IR. The holmium laser emits

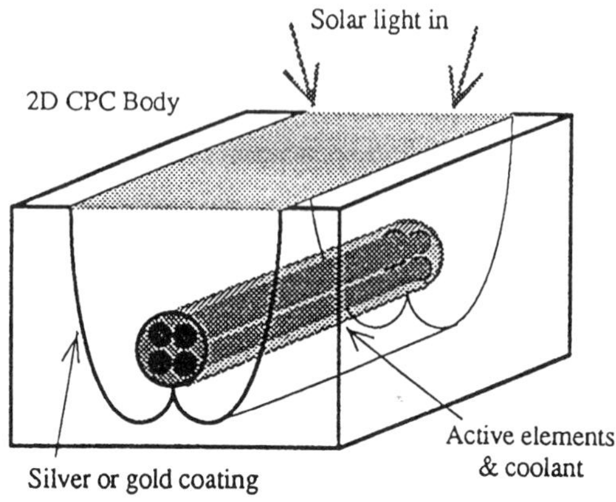

FIG. 19. A cluster of four laser rods in a 2D CPC.

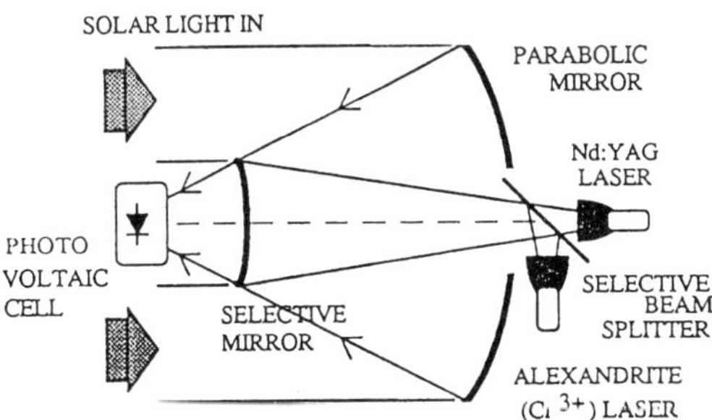

FIG. 20. An example of the use of split solar spectrum to drive two lasers and a photovoltaic cell assembly.

at 2 μm and the neodymium laser at 1.06 μm, and alexandrite is tunable over the range 0.75–0.9 μm. Many applications require shorter wavelength and higher pulse power. Thus, most photochemical reactions require UV light and tunability. Solar lasers are not different from any other continuous-wave lasers, and standard laser techniques for frequency manipulation may be applied to them. Thus, *Q* switching to obtain giant pulses, and the use of nonlinear optical materials to get tunability and frequency up-conversion, have been demonstrated with solar lasers (Arashi, 1993). But since they are not solar-specific techniques, they will not be discussed further here.

3.2.5 Applications of Solar Lasers Solar lasers are still in the early stages of development, and no commercial application has been realized so far. For space applications solar lasers have an obvious advantage due to their independence from other power sources. Even the presently available lasers have enough power for missions such as remote sensing, weather prediction, and communications. Probably the most attractive terrestrial application of powerful tunable solar lasers is in industrial photochemistry, but these have not been developed yet.

Works Cited

Acher, A., Fischer, E., Zelinger, R., Manor, Y. (1990), *Water Res.* **24,** 837–843.

Arashi, H., Kameda, Y. (1993), *Solar Energy* **50,** 447–451.

Bennet, A., Olsen, L. C. (1978), in: *Conference Record, 13th IEEE Photovoltaic Specialists Conference,* New York: IEEE, pp. 860–868.

Buck, R., Abele, M., Bauer, H., Seitz, A., Tamme, R. (1994), in: D. E. Klett, R. E. Hogan, T. Tanaka (Eds.), *Solar Engineering, 1994,* New York: ASME, pp. 73–78.

Cuevas, A., Sinton, R. A., Midkiff, N. E., Swanson, R. M. (1990), *IEEE Electron Devices Lett.* **11,** 6–8.

Chu, T. L., Chu, S. S., Britt, J., Ferekides, C., Wang, C., Wu, C. Q. (1992), *IEEE Electron Devices Lett.* **13,** 303–304.

Emery, K. (1996), in: D. Faiman (Ed.), *Proceedings of the 7th Sede Boqer Symposium on Solar Electricity Production,* Beer Sheva, Israel: Jacob Blaustein Institute for Desert Research, Ben Gurion University of the Negev, pp. 51–63.

Epstein, M., Spiewak, I. (1994), in: *Proceedings of the 7th International Symposium on Solar Thermal Concentrating Technologies, Moscow, September 26–30, 1994,* Moscow: Institute for High Temperature of the Russian Academy of Sciences (IVTAN), pp. 958–971.

Epstein, M., Spiewak, I., Funken, K.-H., Ortner, J. (1994), in: D. E. Klett, R. E. Hogan, T. Tanaka (Eds.), *Solar Engineering, 1994,* New York: ASME, pp. 79–91.

Fletcher, E. A., Moen, R. (1977), *Science* **197,** 1050.

Fletcher, E. A. (1984), *J. Minn. Acad. Sci.* **49,** 30.

Fletcher, E. A. (1991), in: M. E. Arden, S. M. A. Burley, M. Coleman (Eds.), *1991 Solar World Congress,* Tarrytown, NY: Pergamon, p. 790.

Gluck, A., Tamme, R., Streuber, C. (1993), in: M. Macias (Ed.), *Proceedings of the 6th International Symposium on Solar Thermal Concentrating Technologies, Mojacar (1992),* Madrid: Editorial CIEMAT, pp. 445–457.

Harth, R. E., Boltendahl, U. (1981), *Interdisciplinary Sci. Rev.* **6,** 221–228.

Hedström, J., Ohlsen, H., Bodegård, M., Kylmer, A., Stolt, L., Hariskos, D., Ruckh, M., Schock, H. W. (1993), in: R. J. Schwartz (Ed.), *Proceedings of the IEEE 23rd Photovoltaic Specialists Conference, Louisville, Kentucky,* New York: IEEE, p. 364.

Karni, J. (1994), in: *Proceedings of the 7th International Symposium on Solar Thermal Concentrating Technologies, Moscow, September 26–30, 1994,* Moscow: Institute for High Temperature of the Russian Academy of Sciences (IVTAN), pp. 838–848.

Klenck, R., Walther, T., Schock, H. W., Cahen, D. (1993), *Adv. Mater.* **5,** 114.

Kreith, F., Krieder, J. (1978), *Principles of Solar Engineering,* New York: McGraw-Hill.

Krupkin, V., Kagan, Y., Yogev, A. (1993), in: R. L. Holman, R. Winston (Eds.), *Non-Imaging Optics: Maximum Efficiency Light Transfer II,* SPIE Proceedings Vol. 2016, Bellingham, WA: SPIE, pp. 50–60.

Kuhn, P., Enrensberger, K., Frei, A., Oswald,

H. P. (1994), in: *Proceedings of the 7th International Symposium on Solar Thermal Concentrating Technologies, Moscow, September 26–30, 1994,* Moscow: Institute for High Temperature of the Russian Academy of Sciences (IVTAN), pp. 614–626.

Kuryla, M. S., Kaminar, N. R., MacMillan, H. F., Ristow, M. L., Virshup, G. F., Klausmeier-Brown, M. R., Partian, L. D. (1990), in: *Conference Record of the 21st IEEE Photovoltaic Specialists Conference,* Vol 2, New York: IEEE, pp. 1142–1146.

Larson, R. A., Marley, K. A., Schlauch, M. B. (1991), in: D. W. Tedder, F. G. Pohland (Eds.), *Emerging Technologies in Hazardous Waste Management,* ACS Symposium Series No. 468, Washington, DC: American Chemical Society, pp. 66–82.

Levine, J. D., Hotchkiss, G. B., Hammerbacher, M. D. (1991), in: *Conference Record, 22nd IEEE Photovoltaic Specialists Conference,* New York: IEEE, pp. 1045–1048.

Levy, M., Levitan, R., Rosin, H., Rubin, R. (1993), *Solar Energy* **50,** 179–189.

Mehos, M., Turchi, C. (1993), in: M. Macias (Ed.), *Proceedings of the 6th International Symposium on Solar Thermal Concentrating Technologies, Mojacar (1992),* Madrid: Editorial CIEMAT, pp. 1107–1122.

Mills, A., Davies, R. H., Worsley, D. (1993), *Chem. Soc. Rev.* **22,** 417–425.

Noter, Y., Oron, M., Shwartz, J., Weksler, M., Yogev, A. (1990), in: B. P. Gupta, W. H. Traugott (Eds.), *Proceedings of the 4th International Workshop on Solar Thermal Central Receiver Systems,* Bristol, PA: Hemisphere Publishing Corp., pp. 587–596.

Ollis, D. F., Al-Ekabi, H. (1993), *Photocatalytic Purification and Treatment of Water and Air,* Amsterdam: Elsevier.

Scheer, R., Walther, T., Schock, H. W., Fearheiley, M. L., Lewerenz, H. J. (1993), *Appl. Phys. Lett.* **63,** 3294–3296.

Schmid, D., Ruckh, M., Grunwald, F., Schock, H. W. (1993), *J. Appl. Phys.* **73,** 2902–2909.

Thompson, G., Krupkin, V., Yogev, A., Oron, M. (1992), in: M. Macias (Ed.), *Proceedings of the 6th International Symposium on Solar Thermal Concentrating Technologies, Mojacar (1992),* Madrid: Editorial CIEMAT, pp. 553–563.

Thompson, G., Krupkin, V., Oron, M., Yogev, A. (1993), in: M. Oron, I. Shladov, Y. Weissman (Eds.), *8th Meeting on Optical Engineering in Israel: Optical Engineering and Remote Sensing,* SPIE Proceedings Vol. 1971, pp. 400–407.

Welford, W. T., Winston, R. (1989), *High Collection Nonimaging Optics,* New York: Academic.

Zweibel, K. (1993), *Am. Sci.* **81,** 362–369.

Further Reading

Winter, C.-J., Sizmann, R. L., Vant-Hull, L. L. (Eds.) (1991), *Solar Power Plants,* New York: Springer-Verlag.

Möllers, J. (1993), *Semiconductors for Solar Cells,* Boston: Artech House.

Becker, M., Klimas, P. C. (Eds.) (1993), *Second Generation Central Receiver Technologies,* Karlsruhe: Verlag C. F. Muller.

Green, M. A. (1987), *High Efficiency Solar Cells,* Switzerland: Trans Tech Publications.

Winter, C.-J., Nitsch, J. (Eds.) (1988), *Hydrogen as an Energy Carrier,* New York: Springer-Verlag.

At the present time there is an intense international activity in research and development in all areas related to the applications of solar energy. Progress is reported in regular international meetings and symposia (annual or biannual) and published in their proceedings. These publications present a valuable source of updated information on the status of the various topics. Prominent among the organizations sponsoring such conferences are the IEA, IEEE, OSA, and ASME.

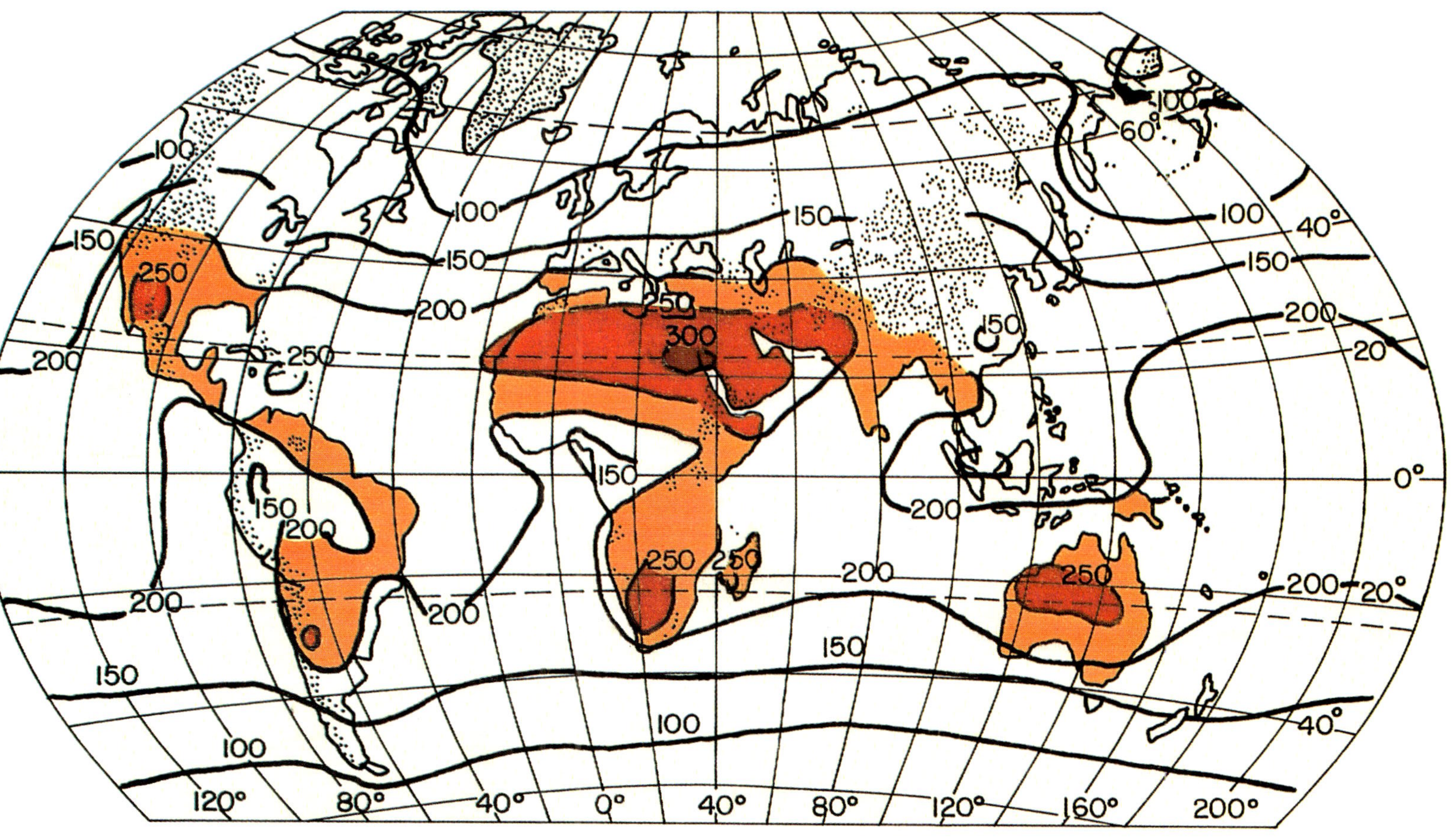

Plate 1. World solar insolation map (W/m² over 24 h, annual average) (source: Kreith and Krieder, 1978).

SOLAR RADIATION

D. F. Smart and M. A. Shea, *Space Physics Division, Geophysics Directorate, Phillips Laboratory, Hanscom Air Force Base, Bedford, Massachusetts, U.S.A.*

3-527-28140-1/96/$5.00 + .50

INTRODUCTION

This article is a general summary of the photonic and particulate emissions from the Sun and describes the effects that these emissions have on our technology. The solar emissions define the character of the entire heliosphere and control the environment that makes life on Earth possible. The general organization of this report starts with the solar structure and electromagnetic radiation, continues with the particulate radiation, and concludes with terrestrial effects resulting from transients in solar emissions.

The Sun is a dynamic star currently having an approximate 11-yr activity cycle. The discussion of the various solar emissions has been separated into emissions from the "quiet Sun" and those from the "active Sun." Quantities that are "well measured" have the briefest description. Quantities that are not well understood nor well measured and are variable have a more detailed discussion in an attempt to define our current understanding.

The effect of solar emissions on the terrestrial environment is worldwide in scope. Research results and technological advances since the advent of the space era in 1957 clearly emphasize the interrelationships of what were once considered separate and independent phenomena. This report discusses both the various interrelationships where they are known and those areas where much is yet to be learned.

1. THE SOLAR STRUCTURE

The Sun is a relatively small, faint, cool star at an average distance of 1.496×10^8 km from the Earth. The composition of the outer layers by mass is approximately 73% H, 25% He, and 2% other elements. This rotating mass ($\sim 1.971 \times 10^{30}$ kg) of hot plasma is in dynamic equilibrium between the gravitational energy of its mass and the energy of its internal fusion heat source. The key word is *dynamic*. The Sun is a very complex and structured body. There are fluctuations in the energy output of the Sun on time scales ranging from seconds to billions of years. The radiative output of the Sun has increased approximately 30% since the Sun began its hydrogen burning sequence over three billion years ago. The Earth's climatic record suggests changes in the radiative output on the time scale of hundreds or thousands of years. The radiative output of the Sun has been observed to vary about 0.3% during a solar cycle since space measurements of the solar irradiance began in 1978.

Models of the Sun are generally organized into various regions where specific processes occur. Each of these regions is often subdivided into smaller specific regions where one particular process dominates. The "standard solar model" (see Cox *et al.*, 1991, for a detailed description) is constantly being updated and improved. For this article the Sun's structure has been simplified and described (see Fig. 1) from the core outward as follows:

1. The core of the Sun has several definitions. The heart, or inner core, where the temperature and density are sufficient to support fusion reactions that are the source of solar energy and the source of the solar neutrino flux, extends to ~0.15 of the solar radius (R_0). The core of the Sun, which contains 96% of the solar mass, extends to $\frac{1}{4}$ of the solar radius.
2. The radiative zone (sometimes called intermediate interior) is the region where radiative transport is the dominant heat-transfer process. This region extends from the core boundary to a poorly defined and model-dependent boundary between about (0.7 and 0.86)R_0. The radiation that begins in the core as high-energy gamma rays is degraded by suppressive absorptions and re-emissions until it finally emerges as visible radiation. Away from the core, the temperature, density, and pressure decrease until free electrons can be trapped by atoms in bound states. This transition in material structure causes an increase in opacity, and so radiation becomes a less efficient method of transporting energy, resulting in a large temperature gradient.
3. The convective zone, where convection becomes the dominant process for transporting heat and magnetic flux, extends from the radiative zone to the photosphere. In the convective zone, there are at least three or four different scales of convection.

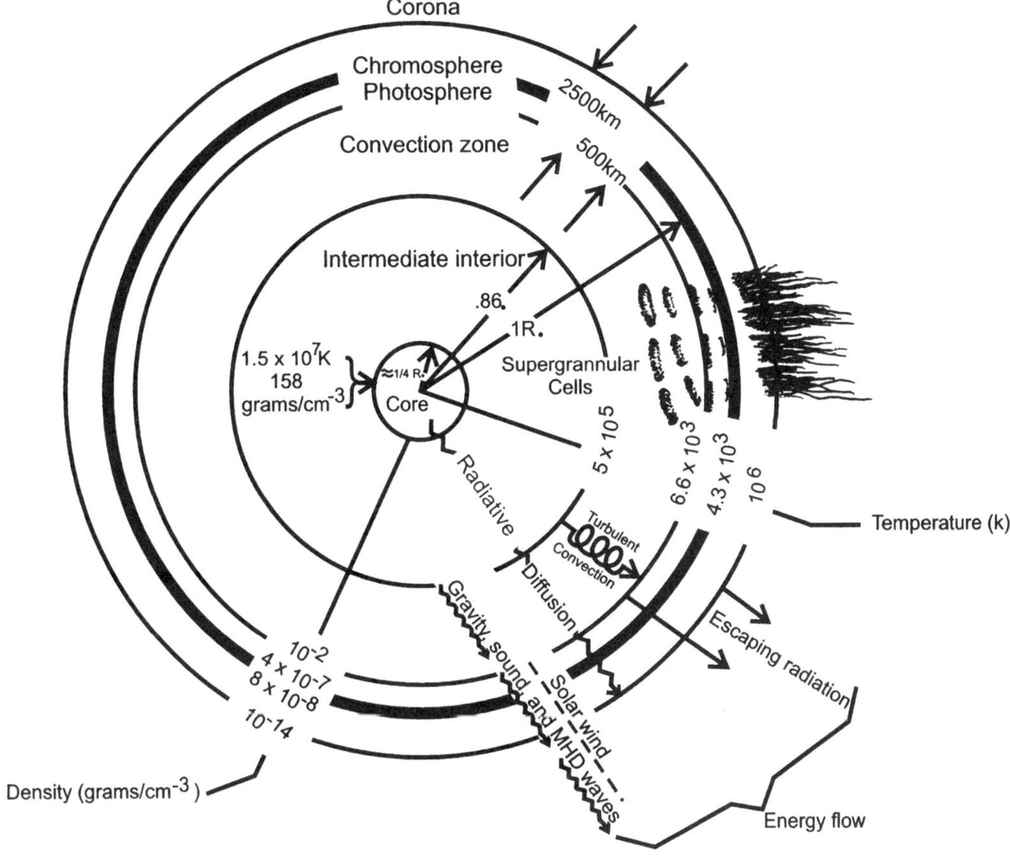

FIG. 1. Idealized general solar structure and modes of energy outflow. The features shown are not to scale.

4. The photosphere, which we observe as the "surface" of the Sun, is where most of the solar energy radiates into space. The photospheric diameter is well defined at the limb of the Sun and is considered to be the diameter of the Sun, having a usually accepted value of 1.391×10^6 km. This photosphere is not a "surface" but actually has a thickness (depending on the wavelength and position) of several hundred kilometers. A formal definition of the extent of the photosphere is the depth in the solar atmosphere from which the probability of a photon escaping is $1/e$. The photosphere "surface" is not uniform but is highly structured, with granulations of the order of 1000 km and supergranulations, which have vertical and horizontal motion apparent in solar photographs. (See Zirin, 1988, for illustrative examples.) Granules have a mean lifetime of about 8 min, while supergranules may have a lifetime of 1–2 days. There are also a number of observable convective motions and oscillation modes. The most prominent oscillation mode has a period of 300 s.
5. The chromosphere is a complex layer of the solar atmosphere above the photosphere and beneath the transition region and the corona. The chromospheric network has a pattern similar to the granulation pattern observable in the photosphere and is an important source of extreme-ultraviolet (EUV) emission. The chromospheric network is also the source of spicules, which appear as brilliant flames of gas that project upward and have lifetimes of 2–5 min. The chromosphere is the source of the strongest lines in the solar spectrum, including the Balmer alpha line of hydrogen (Hα) and the

H and *K* lines of calcium, and is the source of the red (chromium) color often seen during a total solar eclipse. The principal absorption lines in the solar spectrum, called the Fraunhofer lines, originate primarily in the chromosphere. The temperature of the solar atmosphere reaches a minimum in the lowest part of the chromosphere and then increases approximately monotonically with height. The upper boundary of the chromosphere is about 2500 km above the photosphere and is defined as the region where the temperature becomes too high ($\sim 10^4$ K) for hydrogen to give observable line emission. The rise in chromospheric temperature with height is usually explained as being the result of nonradiative energy input via mechanical energy generated in the convection zone. It is presumed that acoustical waves propagate outward and form shocks in the low-density corona, generating collisions between the atoms in the chromosphere. The low densities present also permit the magnetic fields to carry energy in the form of waves.

6. The transition region (an extremely complex region, where the physical processes are poorly understood) is a relatively narrow region ($\sim$100 km) between the chromosphere and the corona where the temperature rapidly increases to about 10^6 K.
7. The corona is a complex, structured region of very hot tenuous gas above the chromosphere extending into space. It has been further subdivided into three components by its emission characteristics: the *E* corona, where most of the line emission occurs; the *K* corona, where the emission is essentially continuum; and the *F* corona, where lightly broadened Fraunhofer lines are the identifying signatures. The corona is dominated by magnetically induced motions, sudden releases of energy, and explosive expansion into space. The corona is heated from below by mechanical, electrical, and magnetic dissipation. Favorite candidates are shock waves produced by the mechanical motion of upward surging granules, propagating waves induced by shocks or associated with the spicules, and Alfvén waves propagating upward from the turbulent photosphere along magnetic lines of force. The average particle's total thermal energy ($\frac{5}{2}kT$) is 0.8 keV, considering fully ionized hydrogen at 2×10^6 K. The poorly understood nonthermal energy input into the corona results in some of the particles acquiring sufficient energy to overcome the gravitational potential energy (2.0 keV per particle at $1R_0$). These particles escape into interplanetary space as the solar wind. The solar corona viewed from space in soft x-ray wavelengths reveals an inner corona whose structure is dominated by the solar magnetic fields.

The solar emissions observable at the Earth reveal many surprises and define the limits of our knowledge and ignorance. The solar neutrino flux measured at the Earth is about a factor of 3 less than expected from our knowledge of fusion reactions and our estimates of the interior of the Sun. The tenuous outer layers of the Sun are much hotter (coronal temperatures range between one and two million degrees) than expected from the photospheric temperature. The majority of the photonic emission of the Sun comes primarily from the photosphere, with absorption and additional emission in the chromosphere and corona. The majority of the radio and x-ray emission comes from the solar corona.

2. SOLAR NEUTRINO EMISSION

Neutrinos are produced by the fusion reactions deduced to occur in the core of the Sun. The standard solar models postulate a proton–proton (*p-p*) chain of reactions, as tabulated in Table 1. The most important neutrino-production processes are the chains noted by (*a*), (b_1), and (b_2).

In the "standard solar model," the product of these *p-p* reactions and the additional neutrino flux from the carbon–nitrogen cycle reactions should result in a neutrino flux at the Earth's orbit of 7.9 ± 0.9 SNU (solar neutrino unit). The SNU definition is a result of using the reaction $^{37}Cl + \nu \rightarrow {}^{37}Ar + e^-$ for neutrino detection where one SNU = 10^{-36} solar neutrino captures per second per target atom. The 20-yr average flux observed by the ^{37}Cl neutrino capture experiment is 2.1 ± 0.2 SNU. The more recent solar neutrino experiments in Europe and Japan designed to measure solar neutrinos at differ-

Table 1. The standard solar model: most important *p-p* reaction chains.

Reaction	Energy (MeV)	
$4^1H \Rightarrow {}^4He + 2e^+ + 2\nu$	26.731	
${}^1H + e^- + {}^1H \Rightarrow {}^2D + \nu$	1.442	$E_\nu = 1.442$ MeV
${}^1H + {}^1H \Rightarrow {}^2D + e^+ + \nu$	1.442	$E_{\nu(max)} = 0.420$ MeV
${}^2D + {}^1H \Rightarrow {}^3He + \gamma$	5.493	
(a) ${}^3He + {}^3He \Rightarrow {}^4He + 2{}^1H$	12.859	
(b) ${}^3He + {}^4He \Rightarrow {}^7Be + \gamma$	1.587	
(b_1) ${}^7Be + e^- \Rightarrow {}^7Li + \nu$	0.862	$E_\nu = 0.862$ MeV (89.5%) $E_\nu = 0.384$ MeV (10.5%)
${}^7Li + {}^1H \Rightarrow 2{}^4He$	17.347	
(b_2) ${}^7Be + {}^1H \Rightarrow {}^8B + \gamma$	0.135	
${}^8B \Rightarrow {}^8Be^* + e^+ + \gamma$	15.075	$E_{\nu(max)} = 14.02$ MeV
${}^8Be^* \Rightarrow 2{}^4He$	2.995	

ent energies are confirming this apparent neutrino deficit. A further detailed description of the solar neutrino measurement techniques can be found in Bahcall (1990) or in references on the solar interior such as Cox *et al.* (1991).

3. SOLAR ELECTROMAGNETIC RADIATION

A crude approximation of the solar visible light emissions is that of a 5770-K "blackbody," varying between 6200 and 5400 K depending on the photospheric depth. The relatively thin photosphere is the primary radiative region above which there are an extended solar atmosphere, the chromosphere, and the corona. These layers of gas above the photosphere complicate the solar emission spectrum by introducing both absorption bands and additional emission from the hot corona. The photonic spectrum received from the Sun is not simple but consists of a very large number of emission and absorption lines. Detailed solar spectral atlases used at solar observatories are very large data files filling computer tapes. A very simplified solar electromagnetic emission spectrum is illustrated in Fig. 2.

3.1 Solar Irradiance from the Quiet Sun

The total irradiance of the Sun received at the Earth is historically called the "solar constant." The value of the total solar irradiance at 1.0 AU adopted for the 1986 solar minimum was 1367 ± 3 W m^{-2} (Lean, 1991). The solar irradiance received at the Earth has an annual modulation of about ±3% due to the elliptical orbit of the Earth about the Sun. Analysis of the solar irradiance variations has identified 11-, 22-, and 80-yr components (Lee *et al.,* 1995). The mean irradiance value between 1984 and 1993 determined from measurements by the Earth Radiation Budget Satellite (ERBS) was 1365.3 ± 0.7 W m^{-2} with an estimated accuracy of 0.2%. Unfortunately, the mean numerical value of equivalent measurements from 1978 to 1993 by the Nimbus 7 satellite was 1372.1. The value from the Solar Maximum Mission Satellite (SMM) was 1367.4; the Upper Atmosphere Research Satellite value was 1365.9; and the Space Shuttle Atmospheric Laboratory for Applications and Science (ATLAS1) Solar Constant (SOLCON) active-cavity radiometer value was 1366.9. These discrepancies are a cause of serious concern since the "solar constant" is a fundamental value used in geophysical models of the Earth's atmosphere and climate (see Foukal, 1990, for a more detailed discussion). The observations are being evaluated and refined by technical specialists with regard to the sensor response over the entire solar spectrum, the sensor calibration, sensor degradation, and measurement precision (see Hudson, 1988, or Donnelly, 1992, for a detailed discussion). There is not yet an ideal device for measuring the solar irradiance, and different space and ground-based experiments yield slightly different values. Figure 3 shows 11 years of unadjusted satellite measurements of the "total solar irradiance," illustrating the solar-cycle trend, the variability of the solar flux, and the discrepancies between different instruments.

The solar electromagnetic emission spec-

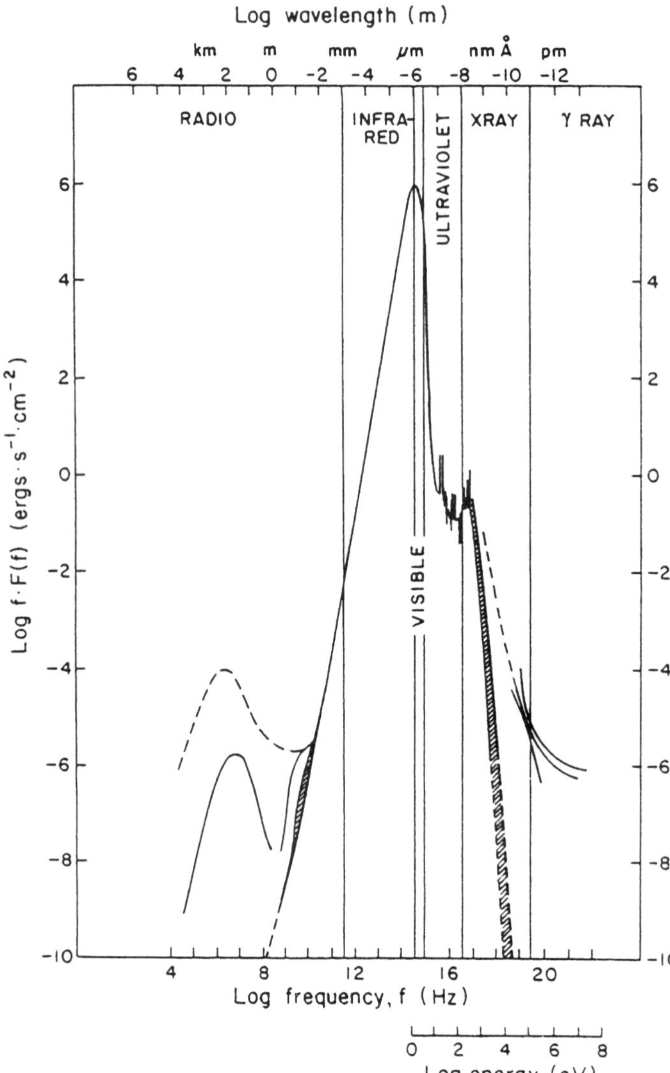

FIG. 2. Overview of the solar radiation spectrum and its variations (adapted from Sawyer *et al.*, 1986). The ordinate is the logarithm of the product of the frequency in hertz by the radiative flux in units of ergs · s^{-1} · cm^{-2} · Hz^{-1}. In the radio region, the solid and dashed lines represent the envelopes expected from "average" and large radio bursts. The shaded areas indicate the solar-cycle variability. The dashed line in the x-ray region indicates the envelope expected for "large" x-ray bursts. The solid lines at the far right indicate gamma-ray bursts.

trum can be divided into regions by wavelength. These regions (see Fig. 2) are detailed in Table 2, adapted from White (1977), and discussed individually in the following sections. Approximately 99% of the total solar radiation is contained in the region between 0.3 and 10 μm. About 2% of the solar irradiance appears at ultraviolet and x-ray wavelengths shorter than 0.32 μm. This radiation is strongly absorbed in the upper atmosphere. More than 70% of the solar irradiance is concentrated in the near-ultraviolet, visible, and near-infrared portions of the spectrum lying between the atmospheric transmission cutoffs near 0.32 and 1.0 μm. Most of this radiation reaches the Earth's surface, and a large portion of this energy enters the lower atmosphere through the evaporation–precipitation cycle of water (see also ATMOSPHERIC PHYSICS; ATMOSPHERIC STRUCTURE; and Figs. 3 and 4 of IONOSPHERE).

3.1.1 The Solar Radio Spectrum The solar radio spectrum ($\lambda > 1$ mm) contains only a very small part of the total solar irradiance; however, the emission in this part of the spectrum is very dynamic. The solar radio emission can be classified as a basic sun component, a slowly varying component, and

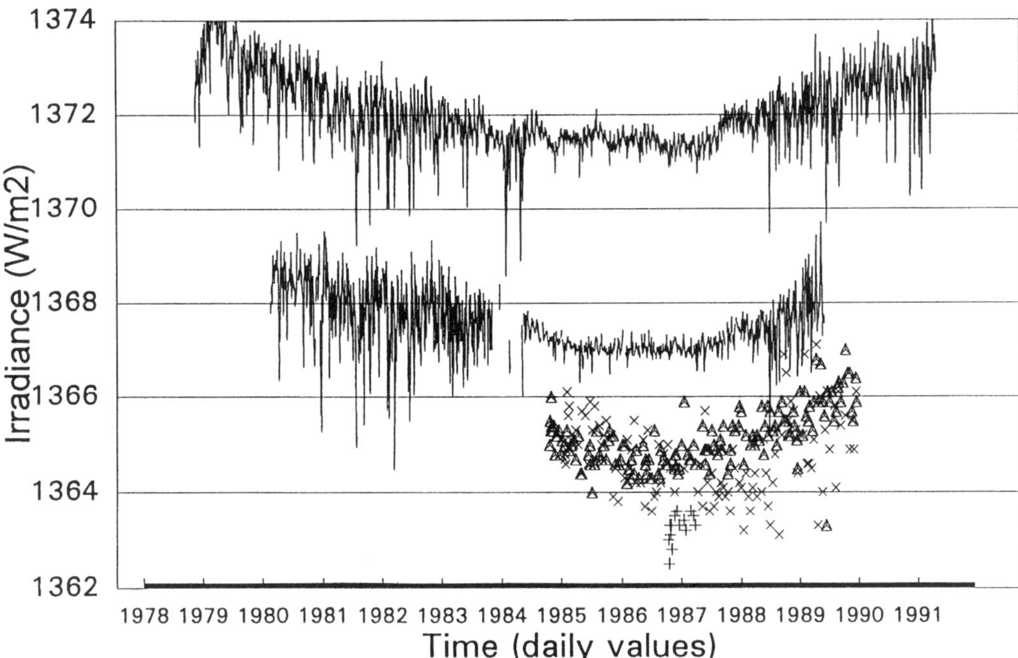

FIG. 3. Unadjusted and uncorrected measurements of the "total solar irradiance" from different instruments on various spacecraft. Data from the NIMBUS 7 spacecraft are at the top of the figure. Data from the Solar Maximum Mission Satellite (SMM) are in the center. Clustered near the bottom are data from NOAA9 (×), NOAA10 (+), and ERBS (Δ), the Earth Radiation Budget Satellite. (Figure courtesy of National Geophysical Data Center.)

a burst component (discussed later in Sec. 5.2.2.1). The solar radio emissions are of interest to radio communicators and radar operators and are diagnostics of solar activity processes. The 10-cm flux is used as a proxy of solar extreme-ultraviolet (EUV) flux.

If the Sun radiated only as a thermal source, the emitted energy density would vary with frequency and temperature according to Planck's radiation law. The following equation gives the equivalent blackbody temperature or brightness temperature:

$$B_f = 2\kappa T f^2 c^{-2} = 2\kappa T \lambda^{-2}. \tag{1}$$

In this equation, B_f is the "brightness" at frequency f in W m^{-2} Hz^{-1} sr^{-1}; the frequency f is in Hz, or cycles per second; the wavelength λ in meters; the temperature T in kelvins; the velocity of light c in meters per second; and Boltzmann's constant κ in joules per kelvin. Figure 4 illustrates the solar radio spectrum and the spectra of a blackbody at various temperatures. In this figure, the solar power-flux density (power per unit area per bandwidth) is plotted against wavelength.

Table 2. Regions of the solar electromagnetic emission spectrum

Spectral region	Wavelength range
Radio	$\lambda > 1$ mm
Far infrared	1 mm $> \lambda >$ 10 μm
Infrared	10 μm $> \lambda >$ 0.75 μm
Visible	0.75 μm $> \lambda >$ 0.3 μm
Ultraviolet (UV)	300 nm $> \lambda >$ 120 nm
Extreme ultraviolet (EUV)	120 nm $> \lambda >$ 10 nm
Soft x rays	10 nm $> \lambda >$ 0.1 nm
Hard x rays	0.1 nm $> \lambda$

3.1.2 The Solar Far-Infrared Spectrum

The far-infrared portion of the solar disk spectrum from 1000 to 10 μm contains an integrated irradiance of about 0.0802 ± 0.026 W m^{-2}, which is 0.057% of the solar constant. This part of the solar emission spectrum is essentially continuous, and there is relatively little solar-cycle change.

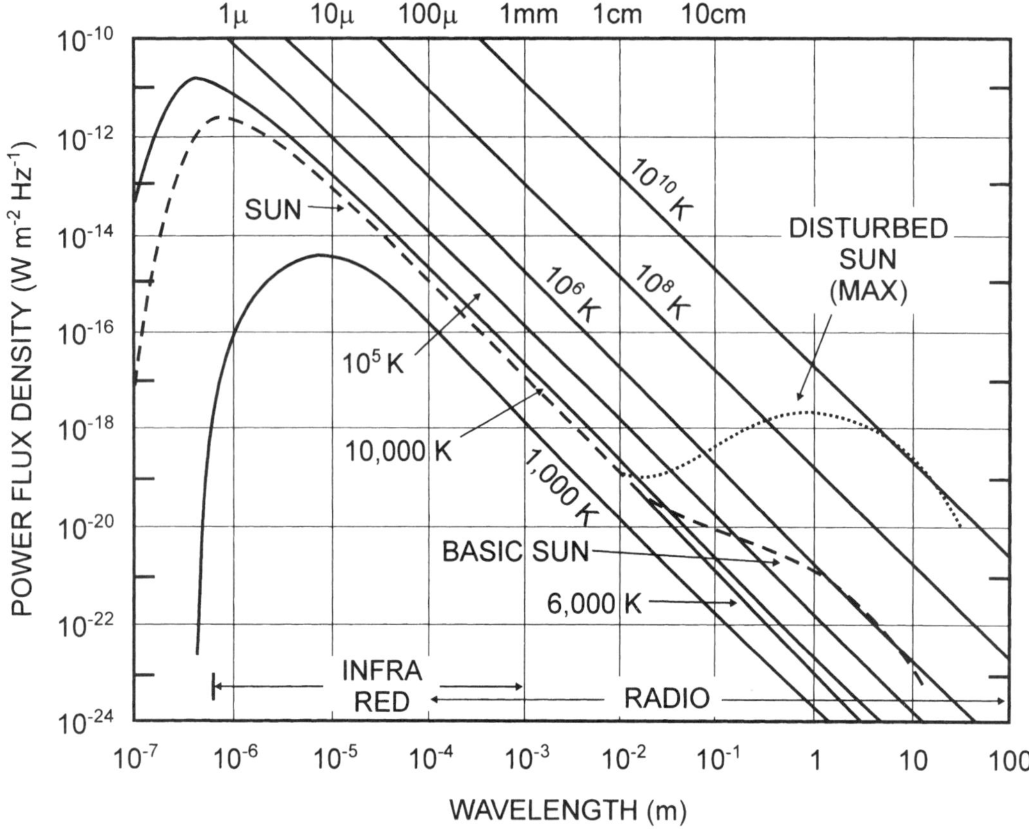

FIG. 4. The solar radio spectrum and the spectra of a blackbody at various temperatures.

3.1.3 The Solar Optical and Infrared Spectrum The optical and infrared portion of the solar spectrum in the wavelength range from 0.3 to 10 nm contains about 99% of the total solar irradiance and is the most important part of the solar constant with respect to the energy input to the Earth. Optical and infrared radiation are usually considered to be "quiet Sun" emissions. There are changes in the emission in the Fraunhofer lines from active regions. The accuracy of the observations remains a point of controversy because of systematic differences between the flux determination found by different measurement methods (see Fig. 3). The absolute accuracy should be in the 1 to 2% range; however, calibration errors of up to 8% have been reported that affect recent measurements of solar irradiance.

3.1.4 The Solar Ultraviolet Spectrum The ultraviolet region of the solar spectrum in the wavelength range from 300 to 120 nm contains only about 1% of the total solar irradiance. This radiation is important because it heats the stratosphere and photodissociates the major molecular constituents of the upper atmosphere, being almost totally absorbed by the ozone and diatomic oxygen molecules (see also ATMOSPHERIC PHYSICS; ATMOSPHERIC STRUCTURES; IONOSPHERE). The spectrum of this region is complex; at about 140 nm, there is a transition from a strong continuum broken by Fraunhofer absorption lines to a weak continuum dominated by emission line spectra. The ultraviolet wavelength region is where there are the greatest technical difficulties in making accurate measurements. The measurement accuracies in this part of the solar spectrum are often about ±10%, sometimes achieving a relative accuracy of 2%. Measurement accuracies of 1% are needed for global modeling of the upper atmosphere, particularly for ozone time variations. Because this portion of the solar spectrum is so complex, reference spectra (see Hinteregger *et al.*, 1981, or Tobiska, 1993) are compiled and updated as more precise measurements become available.

The solar ultraviolet (UV) flux variations contribute between one-fifth and one-third of the solar-cycle enhancement in the total solar irradiance. There are significant changes in the solar ultraviolet due to solar activity. There are 27-day (the average solar rotational period) variations that can be as large at 25% at 120 nm, these variations decreasing approximately exponentially with wavelength to about 1% at 300 nm. The largest changes in the emission in this entire wavelength region occur during major solar flares and can be as great as 16%. Because of the solar-activity dependence, the recent UV and EUV observations are now being published monthly in *Solar-Geophysical Data* by the NOAA National Geophysical Data Center.

3.1.4.1 The Lyman-Alpha Flux. The Lyman-alpha line centered at 121.6 nm contains more radiative energy than all the other shorter wavelengths in the solar spectrum. This "line" is not simple and is actually a complex profile. The Lyman-alpha emission is a principal source of ionization in the upper D region of the Earth's atmosphere at times of low solar activity (see also IONOSPHERE). The integrated Lyman-alpha flux depends on solar activity and can vary between 3.3 and 6.5 erg cm^{-2} s^{-1}. The Lyman-alpha flux variation during a solar rotation can be $\pm 15\%$ and about a factor of 2 between solar maximum and solar minimum.

3.1.5 The Solar Extreme-Ultraviolet Spectrum The solar flux in the extreme ultraviolet region of the solar spectrum is the main ionization source for the Earth's ionosphere (see also IONOSPHERE). It contains a very complex series of resonance lines and continua from the major constituents of the solar atmosphere. The energy flux at wavelengths below 120 nm is small, but it is important because it determines the ionization state of the Earth's ionosphere. At wavelengths greater than 17 nm, the variation during a solar rotation is typically $\pm 15\%$ and can be a factor of 5. The variation over a solar cycle can be two orders of magnitude.

3.1.6 The Solar X-Ray Flux Solar x rays are an important source of ionization in the Earth's ionosphere, with absorption wavelength being a function of depth in the terrestrial atmosphere (see also ATMOSPHERIC PHYSICS; IONOSPHERE). The Lyman-alpha flux and the cosmic radiation are the main sources of ion production in the Earth's ionosphere; however, when the 0.05- to 1-nm flux exceeds 10^{-6} W m^{-2}, the solar x rays become an important, and for large transient events the dominant, source of ionization in the Earth's atmosphere. The variations in the 0.1–1.0-nm flux are so large and unpredictable that the soft x-ray flux is published for general use in *Solar-Geophysical Data.* The 0.1–0.8-nm (1–8-Å) flux varies by orders of magnitude over a solar cycle. Large variations occur during a solar rotation with marked differences from rotation to rotation depending on the number, size, growth rate, and spatial distribution of solar-active regions. During solar-flare events the 0.1–0.8-nm flux can increase by more than three orders of magnitude, and the solar x-ray and gamma-ray bursts are the cause of the "sudden ionospheric disturbance" (SID) troublesome to terrestrial radio communications. For a very large X10-class solar x-ray event, the peak radiative power received at the Earth in the 0.1–0.8-nm region will exceed 1.0×10^{-3} W m^{-2} and will have a duration of several hours; the total energy received in this wavelength band may exceed 1 J. (See Sec. 5.1.1.2 for an explanation of classification of solar x-ray events.)

4. SOLAR PARTICLE EMISSION FROM THE QUIET SUN

4.1 Solar Plasma Emissions

The Sun is constantly emitting plasma, called the solar wind. The existence of the solar wind was hypothesized, and many of its characteristics were theoretically predicted, before it was actually measured by spacecraft. The solar corona is very hot, and as a result of poorly understood nonthermal energy input, some of the particles acquire sufficient energy to overcome the gravitational potential energy (2.0 keV per particle at $1R_0$) and escape into interplanetary space. These particles form a highly ionized plasma in which is embedded the coronal magnetic field. This becomes the interplanetary magnetic field, which is carried out from the Sun by the solar wind. The plasma kinetic energy is generally much larger than the magnetic energy, and the field is said to be "frozen in" the solar-wind flow.

4.1.1 The Solar Wind In the general solar-wind theory devised by Parker (1963) and modified by later researchers, there is a continuous expansive motion of the solar wind, reaching supersonic flow and continuing to accelerate slowly until it approaches an asymptotic value at some great distance from the Sun. In its supersonic flow condition, the solar-wind bulk-motion speed is faster than the speed of sound in the interplanetary medium and is faster than the Alfvén speed at which magnetic wave phenomena propagate. In its radial expansion, the solar-wind density is predicted to decrease as $1/r^2$. Many of the Parker (1963) predictions of the general characteristics have been verified; however, spacecraft measurements show a solar wind that is faster, cooler, and more turbulent, variable, and structured than the predictions of the simple Parker theory. The goal of modern solar-wind theorists is to explain the space observations. However, there is no satisfactory general model of the solar wind. Currently, there are a number of individual theories that attempt to model several aspects of the solar-wind behavior.

Various space probes have sampled the solar wind from distances as close to the Sun as 0.3 AU to as far from the Sun as 70 AU (in 1995), from latitudes in the ecliptic plane to 80° solar latitude. In addition to the limited *in situ* measurement of the solar wind by spacecraft, the interplanetary scintillation radio observation technique can determine the pattern speed of irregularities in the electron density as the plasma moves through the line of sight to various stellar point sources, thus providing information on the structure of the interplanetary plasma. The solar wind is a polytropic plasma with a proton component having an average particle kinetic energy of approximately 1 keV and an electron component having average particle kinetic energy of approximately 100 eV. The solar wind is very structured, and it is customary to refer to solar-wind streams. These solar-wind streams corotate with the Sun and retain their individuality to very large distances from the Sun (they are still observable near the ecliptic plane by our most distant interplanetary spacecraft). Low-speed streams are generally at low solar latitudes and seem to be associated with "magnetically closed" coronal features. The high-speed streams generally originate from "magnetically open" coronal structures called coronal holes. High-speed streams "override" slow-speed streams generating compression density structures and corotating shocks. The result of the stream structure of the solar wind is that the behavior of the individual plasma elements can be approximated as motion radially outward with an essentially constant velocity (except when there are stream–stream interactions) with a slight slowing down of the fast solar-wind streams at large distances. Transient structures propagating in the solar wind are often the result of coronal mass ejections. It is common to consider two different classes of solar wind: the steady or normal solar wind and the transient solar wind. Other classifications often used for the solar wind in the ecliptic plane are slow (<400 km s^{-1}), fast (>600 km s^{-1}), or average.

The general characteristics of the solar wind near the ecliptic plane (normalized to 1 AU) are summarized in Table 3. There is a great variability in the solar wind. Many of the properties of the solar wind are dependent on the flow speed, as shown in this table. However, the momentum flux and the total energy are remarkably independent of solar-wind speed. The HELIOS spacecraft data set covering the period from 1974 to 1982 indicates that the density of the solar wind between 0.3 and 1.0 AU can be modeled as a power law $n(r) = 6.1r^{-2.1}$, where r is the radial distance from the center of the Sun.

4.1.1.1 Solar-Wind Composition. The solar wind is an electrically neutral plasma. The average ionic component consists principally of protons, about 4% alpha particles, and a small number of other positively charged ions. The solar-wind composition and charge state are variable (stream structure and temperature dependent) and different from either the solar photospheric or the solar coronal composition, probably reflecting the ionization and acceleration processes. The measurements of the elemental

Table 3. Solar-wind parameters at 1 AU in the ecliptic plane.

Parameter[a]	Units	Slow solar wind	Fast solar wind	Average solar wind
Proton bulk flow speed	km s^{-1}	335	680	475
Proton density	cm^{-3}	9.5	3.0	6.5
Proton flux density	10^8 cm^{-2} s^{-1}	3.2	1.9	2.7
Proton momentum flux density	10^8 dyn cm^{-2}	2.12	2.16	2.1
Proton temperature	10^5 K	0.5	2.6	1.2
Electron temperature	10^5 K	1.6	1.5	1.4
Proton heat flux	10^{-3} erg cm^{-2} s^{-1}	0.05	0.23	0.13
Electron heat flux	10^{-3} erg cm^{-2} s^{-1}	4.3	5.1	4.3
Alpha temperature	10^5 K	1.4	10.	5.8
Alpha/proton ratio		0.031	0.042	0.04
Total energy flux	erg cm^2 s^{-1}	1.37	1.49	1.45

[a]Parameters averaged from several sources.

abundances of the solar wind (tabulated in Table 4) are still being improved. Some of the most detailed measurements have been made in the magnetosheath of the Earth's magnetosphere.

4.1.2 The Interplanetary Magnetic Field The interplanetary magnetic field (IMF) is convected out of the Sun via the solar wind (a highly ionized conductive plasma). Its basic properties were successfully predicted by Parker (1963); however, spacecraft measurements reveal an IMF that is more turbulent, variable, and structured than the predictions of the simple Parker theory. It is customary to organize IMF data according to the basic Parker theory and to organize variations in the behavior of the IMF by their deviation from the Parker theory predictions.

In the time-stationary spherically symmetric Parker theory, assuming an expanding solar wind at constant velocity at a distance a few solar radii from the Sun, the three components of the interplanetary magnetic field are given by

$$B_r(r,\theta,\varphi) = B(\phi,\theta_0)(b/r)^2, \tag{2}$$

$$B_\theta(r,\theta,\varphi) = 0, \tag{3}$$

$$B_\phi(r,\theta,\varphi) = B(\theta,\varphi_0)(\Omega b^2/vr^2)(r_0 - b)\sin\theta, \tag{4}$$

where b and φ_0 are the distance and longitude at which the inner-boundary radial field $B(\theta,\varphi_0)$ is assumed to exist, v is the solar-wind velocity, and Ω is the solar angular rotation rate. The distances r and b are connected by the equation

$$r/b - 1 - \ln(r/b) = (v/b\Omega)(\varphi - \varphi_0), \tag{5}$$

which for $r \gg b$ reduces to the Archimedean spiral equation. This formulation works well for an average description of the interplanetary magnetic field to the distances traveled by man-made spacecraft.

The basic topology of the interplanetary magnetic field is very important in the understanding of solar energetic-particle events. The outward flow of the solar wind with its "frozen in" magnetic field from a rotating sun leads to an Archimedean spiral geometry. This spiral geometry is easiest to visual-

Table 4. Solar-wind abundances relative to oxygen (Gloeckler and Geiss, 1989).

Element	Relative abundance: Interplanetary	Magnetosheath
H	1900 ± 400	1125 ± 100
He	75 ± 20	45 ± 5
C		0.529 ± 0.013
N		0.129 ± 0.008
O	1.000	1.000
Ne	0.17 ± 0.02	0.114 ± 0.008
Mg		0.106 ± 0.001
Si	0.19 ± 0.04	0.103 ± 0.011
S		0.038 ± 0.009
Ar	0.004 ± 0.001	
Fe	0.19 ± 0.1	0.124 ± 0.004

ize in the ecliptic plane, as illustrated in Fig. 5. The spiral geometry is predicted to be present at all latitudes, and the latest Ulysses solar polar mission results confirm these expectations of the Parker theory.

There are variations in the magnitude of the interplanetary magnetic field observed at all wavelengths. The magnitude fluctuation (field variance σ_B^2) can be approximated by a power law with a slope of -3. Like the solar wind, the interplanetary magnetic field is highly structured. The gross characteristics of the solar magnetic field can be approximated by an inclined dipole (the inclination varying as the solar cycle) with well-defined northern and southern polarities separated by a heliospheric current sheet, as illustrated in Fig. 6. (A useful concept for visualizing this is the ballerina skirt.) A spacecraft in the ecliptic plane would sense changes in the polarity of the interplanetary magnetic field during a solar rotation, as it is sequentially connected to the polarity of the "northern" and then "southern" solar latitudes. This regular variation in the sign of the interplanetary magnetic field was originally called the "sector structure." This structured interplanetary magnetic field exists primarily in the ecliptic plane, and the latitudinal extent of this "sector structure" is called the tilt of the interplanetary current sheet. The inclination angle of the interplanetary current sheet is solar-cycle dependent, minimum at solar minimum and maximum during the solar magnetic polar reversal at solar maximum. Interplanetary spacecraft at heliolatitudes "above" or "below" the current sheet observe a unipolarity interplanetary magnetic field reflecting the solar magnetic cycle polarity of the solar magnetic field. Solar source surface maps of the solar magnetic field modeled and extrapolated to 2.5 solar radii and then projected to 1 AU at the average solar-wind speed have been shown to have a 0.8 correlation with the sign of the interplanetary magnetic field observed at the Earth (Hoeksema and Scherrer, 1986).

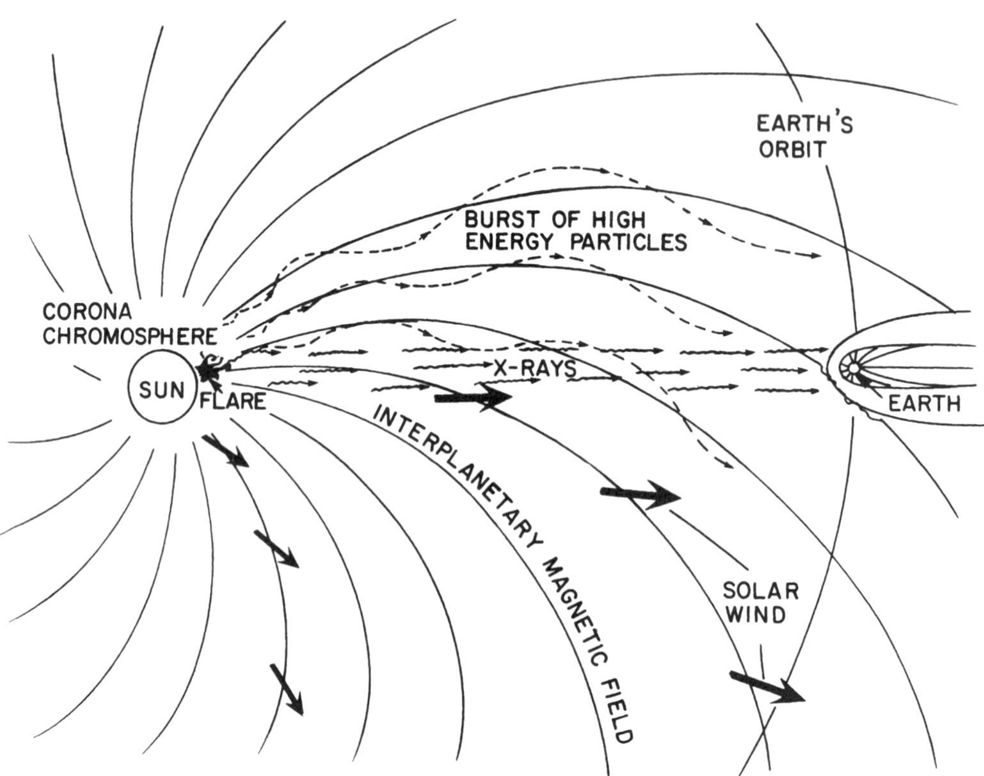

FIG. 5. Conceptual view of the basically Archimedean spiral geometry of the interplanetary magnetic field. The propagation of solar energetic charged particles is controlled and organized by the interplanetary magnetic field.

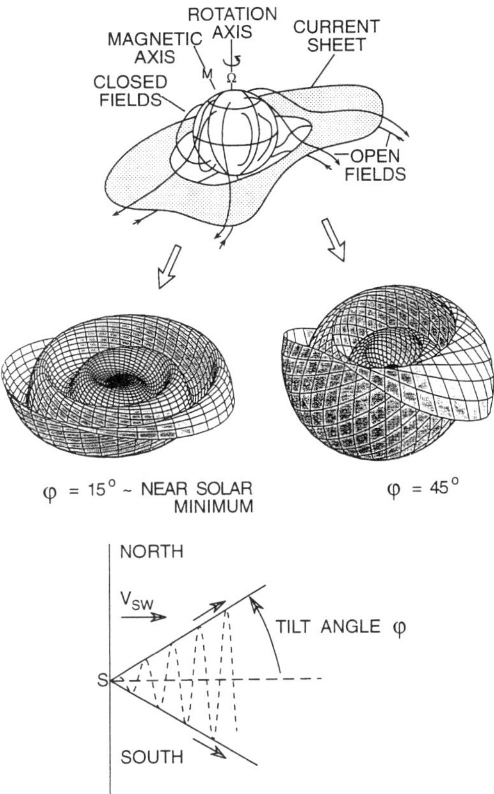

FIG. 6. Conceptual illustration of the heliospheric current sheet during solar minimum and solar maximum. Top: The concept of the heliospheric current sheet. Center: Conceptual view of the amplitude and structure of the heliospheric current sheet near solar minimum (left) and during more active periods of the solar cycle (right). Bottom: Conceptual view of the "wavy" heliospheric current sheet being convected out by the solar wind.

5. SOLAR ELECTROMAGNETIC RADIATION FROM THE ACTIVE SUN

Features visible in the photosphere and chromosphere (and, with space technology, the solar corona) are observed to wax and wane in a cyclic manner with an approximate 11-yr period. The concept of solar activity has been broadened from the classical narrow concept of sunspots and solar flares to embrace an entire range of solar processes, time scales, and energy-release processes. The cyclic behavior affects the electromagnetic output of the Sun; however, the effect is very dependent on the wavelength. The solar-cycle effect is smallest in the visible wavelengths and has a small observable effect on the solar constant (see Fig. 3). The solar-cycle effect is largest at the extreme ends of the electromagnetic spectrum (the long radio wavelengths and the short x-ray wavelengths).

5.1 Solar Activity and Solar Cycles

The cyclic solar activity is the result of variations in the energy in the form of magnetic flux, convected to the solar surface by the action of the solar dynamo operating in the convection zone. The sun has a magnetic cycle with a period of about 22 years, with the average magnetic field observed at the solar poles reversing about every 11 years during the maximum of the solar-activity cycle. This same solar magnetic polarity phase cycle is observed in sunspot pairs where the leader spot and the follower spot have opposite polarities. The sunspots observed on the solar photosphere are manifestations of locally strong magnetic fields. The visible solar flare is the result of the explosive release of energy, stored as magnetic stress, into the solar atmosphere. The number of "flares" on the Sun and the number of magnetic disturbances observed on the Earth are roughly correlated with the solar-activity cycle.

5.1.1 Sunspot Cycles Sunspots and solar flares are the most obvious manifestations of solar activity. The average period of the sunspot cycles for the last 280 years has been 11.1 years with an average rising phase of 4.8 years and an average declining phase of 6.2 years. Figure 7 illustrates the sunspot cyclic behavior since 1900. It is traditional to use the International sunspot number as a proxy to indicate solar activity, and there are many methods of predicting the "next solar cycle." These different methods generate diverse predictions, and there is no universally accepted method of sunspot-number prediction. Some contemporary theories consider the solar cycle to be chaotic and not amenable to precise statistical prediction.

At the beginning of a solar-activity cycle, new sunspots are observed at midlatitudes on the sun, the average sunspot position gradually migrating toward the solar equator as the solar cycle progresses, resulting in a "butterfly diagram" in a map of sunspot heliolatitudes over a solar cycle. Large sunspots

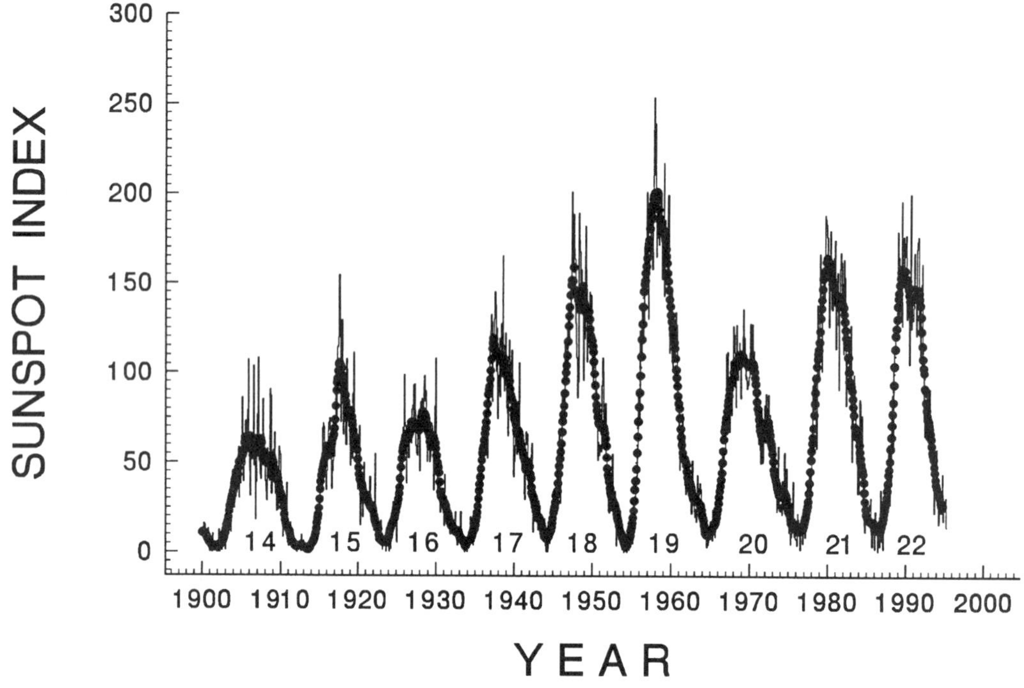

FIG. 7. Sunspot cycles since 1900. The light line is the monthly sunspot index. The large solid dots denote the smoothed sunspot index (a 13-month running mean).

can be observed on the solar surface with an unaided human eye, and the historical record of sunspot observations can be extended for approximately 2000 years. The current cyclic behavior has not always been apparent; during the Maunder minimum and the Spöre minimum, very few sunspots were visible (or reported) for long periods of time (see review by Eddy, 1978). Sunspot numbers extend from 1610 to the present, and the sunspot cycles have been numbered since 1700. Sunspot numbers are considered "reliable" for statistical purposes only from 1848 when the Swiss astronomer Wolf defined rules for counting sunspots and introduced the sunspot index. These numbers are currently called the International sunspot numbers; historically they were called the Zürich sunspot numbers [see McKinnon (1987) and data files maintained by the National Geophysical Data Center]. There are several different types of sunspot numbers, such as the number as determined by the Wolf sunspot index and smoothed sunspot numbers that are a 13-month running mean of the Wolf sunspot index. A less-well-known and little-used sunspot count is the AVSO number (American visual sunspot observations).

5.2 Solar Flares

The English term "solar flare" is the name given to the sudden release of energy (occasionally $>10^{32}$ ergs) in a relatively small volume of the solar atmosphere. The original term referred to a *chromospheric brightening* that is more descriptive of the observations, occasionally seen in white light but usually observed in the hydrogen-alpha line. The term has been expanded to include all of the electromagnetic and particle emissions associated with the energy release. There is a general correlation between the optical importance of a solar flare and its radio and x-ray output, but there are very large variations in the data. In extreme instances, optical subflares are associated with large x-ray emissions, and in other extreme instances very large optical flares have been almost "radio quiet."

The concept of what is a solar flare has undergone considerable evolution. Contemporary concepts based on results from the NASA Solar Maximum Mission and the Japanese YOHKOH satellite favor energy release in loops. As seen from space in ultraviolet and x-ray wavelengths, characteristic looplike structures outline the magnetic fields from active regions. The loop sizes vary considerably, from a few thousand kilometers to more than 100 000 km. The disruption of magnetic stress in these loops seems to result in the solar flare. In a large solar flare, many loops, including entire arcades, are involved. The mechanism that triggers the flare processes is not understood. The impulsive phase of the solar flare is the sudden release of energy either directly or indirectly into electromagnetic and particle radiation. The flaring loop model developed by NASA as a result of the Solar Maximum Mission is shown in Fig. 8.

While there are a number of theories involving processes occurring during a solar flare, there is no general solar flare theory. Current concepts refer to an impulsive phase and a gradual phase. The theoretical concept of how the impulsive phase of the flare proceeds is referred to as the thick-target model. In this model the primary outcome of the magnetic reconnection process is the production of large quantities of energetic electrons and protons. These charged particles will move away from the reconnection site (believed to be in the corona) along the magnetic field lines toward the denser lower portions of the solar atmosphere. As they encounter dense plasma in the chromosphere, they will interact and deposit their energy and further heat the plasma. Energetic electrons will generate gyrosynchrotron radiation as they gyrate along the field lines; they also generate bremsstrahlung (German for braking radiation) as they interact with the denser material. The more energetic ions can cause nuclear interactions, generating gamma rays and knocking neutrons out of the target nuclei. As the tube fills with hot

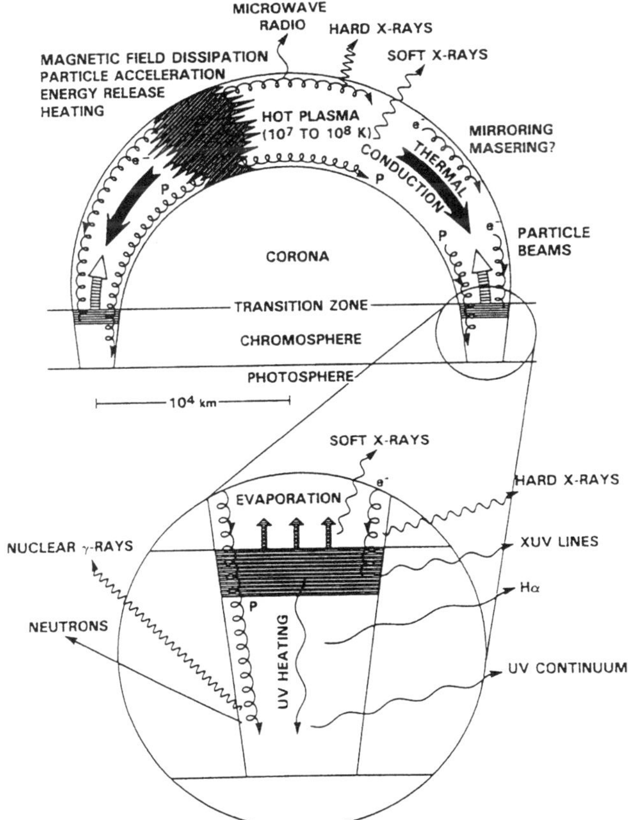

FIG. 8. An idealized flaring loop, illustrating probable sources of the radiation involved (courtesy of NASA).

plasma, it emits ultraviolet and soft x-ray radiation, initiating what is called the gradual phase. Once the heating begins to decrease, the plasma cools by radiation. Figure 9 is a sketch of an idealized solar-flare radiative-emission curve.

The gradual phase is a more time-extended portion of the solar flare. Essentially, explosive evaporation carries large quantities of high-temperature plasma into the magnetic flux tubes where it is observed as ultraviolet and soft x-ray emission. Much of the energy put into the corona by the flare is gradually moved by thermal conduction to the chromosphere where it is radiated away by the strong emission lines of hydrogen and helium.

5.2.1 Solar-Flare Reporting

5.2.1.1 Optical Solar Flares. Optical solar-flare reporting is done in accordance with specifications of the International Astronomical Union. The optical classification of a solar flare is based on its measured optical area (in millionths of the solar disk) corrected for foreshortening and brightness at the time of maximum brightness in the Hα line (centered at 656.3 nm). Optical solar flares are currently (since 1965) ranked in importance as 0 (or *s*), 1, 2, 3, to a maximum of 4. Solar flares are subjectively ranked in brightness as *f* (faint), *n* (normal), or *b* (bright). Prior to 1965, optical solar flares were ranked as *s* (sub) or 1, 2, 3, and 3+. The *Quarterly Bulletin on Solar Activity* is a publication of the International Astronomical Union in which the solar-flare reports have been evaluated and grouped. (Unfortunately, this is a time-consuming process, and the publications are several years "late.") In the United States, the NOAA National Geophysical Data Center collects and maintains files of unevaluated solar-flare reports that are published monthly in *Solar-Geophysical Data*. The statistical occurrence of optical solar flares for the last four solar cycles is given in Table 5.

5.2.1.2 Solar X-Ray Flares. Since about 1969, there have been routine space measurements of the solar x-ray flux by the NOAA Space Environment Laboratory. The NOAA classification of solar-flare soft x-ray emission is generally accepted worldwide. This is a letter–number combination denoting the magnitude of the peak x-ray burst intensity measured at the Earth in the 0.1–0.8-nm band, detailed in Table 6. The letter denotes the class (order of magnitude) of the x-ray emission, and the number describes the magnitude in each class. For example, a

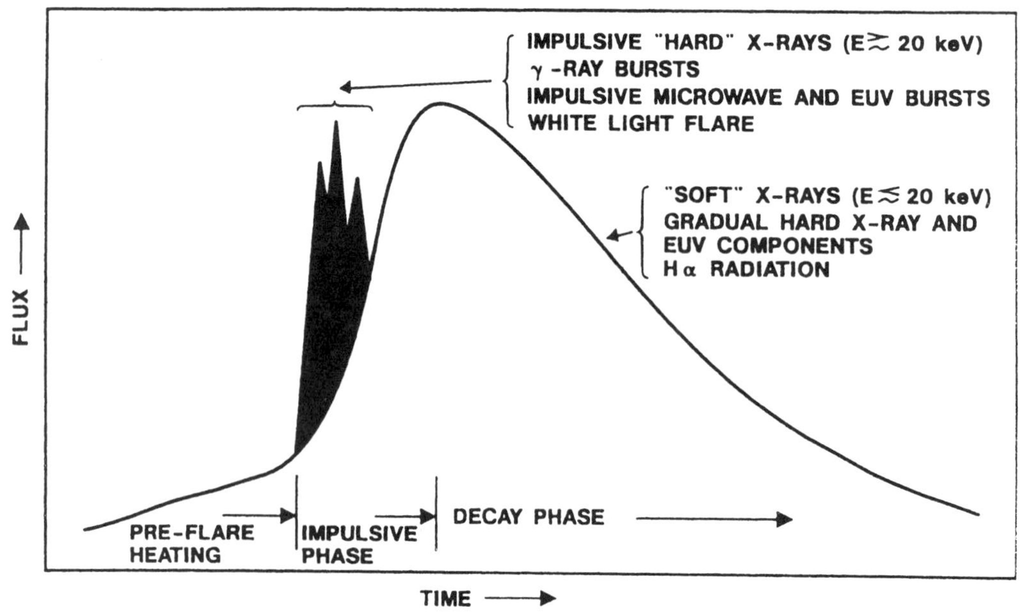

FIG. 9. Idealized solar-flare radiative energy intensity-time curves.

Table 5. Optical solar-flare occurrence frequency.

Optical importance	Approximate flares per year Solar minimum	Approximate flares per year Solar maximum
s[a]	~800	>10 000
1	37	1 100
2	3	120
3	0	10
4	0	2

[a]A curious fact is that there are no rules for the reporting of subflares (individual observatory option), which leads to a skewing of the statistics when all reported optical brightenings are plotted.

Table 6. The NOAA SEL solar x-ray flare classification.

Class	Power in the 0.1–0.8-nm band
B	$<10^{-6}$ W m^{-2}
C	$\geq 10^{-6}$ to $<10^{-5}$ W m^{-2}
M	$\geq 10^{-5}$ to $<10^{-4}$ W m^{-2}
X	$\geq 10^{-4}$ W m^{-2}

peak soft x-ray flux emission of 6.2×10^{-5} W m^{-2} would be classified as an *M* 6.2 event. Figure 10 illustrates the solar flare x-ray frequency observed since 1969.

5.2.2 Solar-Flare Emissions

5.2.2.1 Solar-Flare Radio Emissions. Radio burst emission occurs during the solar flare process; the burst characteristics vary with wavelength. The duration and intensity are extremely variable; the duration may last from seconds to hours, and the intensity may range from small to extremely large. Accelerated electrons are the source of the radio emission, and the principal mechanisms are gyrosynchrotron radiation from electron motion around magnetic field lines or plasma radiation stimulated by electrons penetrating into the preexisting plasma. Another radiation mechanism hypothesized as a result of asymmetric electron distributions in magnetic flux tubes is the cyclotron maser radiation. The electron gyrofrequency is given by

$$\nu_B = eB/2\pi m_e c \approx 2.8 \times 10^6 B. \quad (7)$$

The plasma frequency is given by

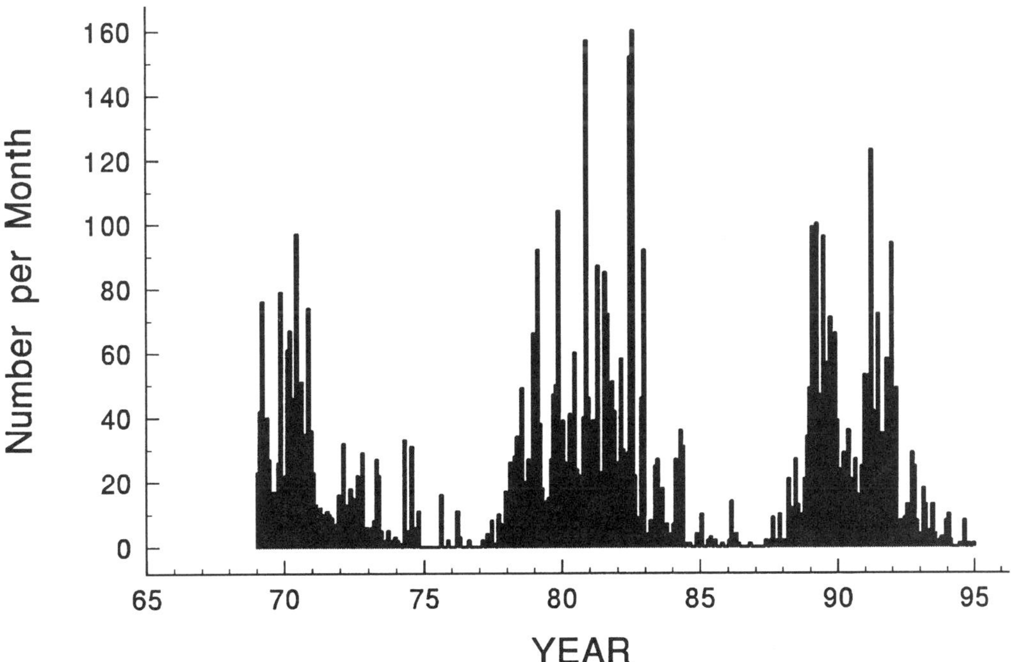

FIG. 10. The frequency of solar flares with x-ray emission $\geq 10^{-5}$ W m^{-2} (Class M1) since 1969 (data courtesy of NOAA Space Environment Laboratory).

$$\nu_p = (n_e e^2/\pi m_e)^{1/2} \approx 9000\sqrt{n_e}. \qquad (8)$$

In these equations e and m_e are the charge and the mass of the electron, B is the magnetic field strength in gauss, n_e is the plasma electron number density, and c is the speed of light.

In the low corona near active regions and/or near sunspots, the magnetic field may be so strong that the electron gyrofrequency is higher than the plasma frequency. Energetic electrons with energy >100 keV that are confined to move around the magnetic field lines emit radiation at frequencies up to 10–100 times their gyrofrequency. In other locations in the solar atmosphere where the magnetic field is weaker and the gyrofrequency is less than the plasma frequency, electrostatic waves (called Langmuir waves) stimulated by anisotropic electron distributions generate plasma radiation.

The intensity of the solar-flare radio burst is generally reported in solar flux units (SFU). One solar flux unit is 10^{-22} W m^{-2} Hz^{-1} or 10^4 Jansky. Moderate solar radio bursts have amplitudes in the hundreds of SFU. Large solar radio events have amplitudes exceeding 1000 SFU. Occasionally a very powerful solar flare may generate a solar radio burst with a peak flux exceeding 10 000 SFU for several minutes. In the United States, the NOAA National Geophysical Data Center (which also functions as World Data Center A for Solar–Terrestrial Physics) collects and maintains files of solar radio bursts, which they publish monthly in *Solar-Geophysical Data*. There is a detailed classification system by spectral type for bursts in the metric wavelength range illustrated in Fig. 11 and described below.

- Type I solar radio emission: A noise storm composed of many short, narrow-band bursts in the meter wavelength range (300–500 MHz), of extremely variable intensity. The storm may last from several hours to several days.
- Type II solar radio emission: Narrow-band emission that begins in the meter range (300 MHz) and sweeps slowly (tens of minutes) toward decameter wavelengths (10 MHz). Harmonics at twice the frequency are often observed. Type II emissions often occur in association with solar flares and fast coronal mass ejections. The metric type II is indicative of a shock wave moving through the solar atmosphere from 1.2 to 2 solar radii.
- Type III solar radio emission: Narrow-band bursts that sweep rapidly (seconds) from decimeter to dekameter wavelengths (500–0.5 MHz). They often occur in groups and are an occasional feature of complex solar-active regions. The fast fre-

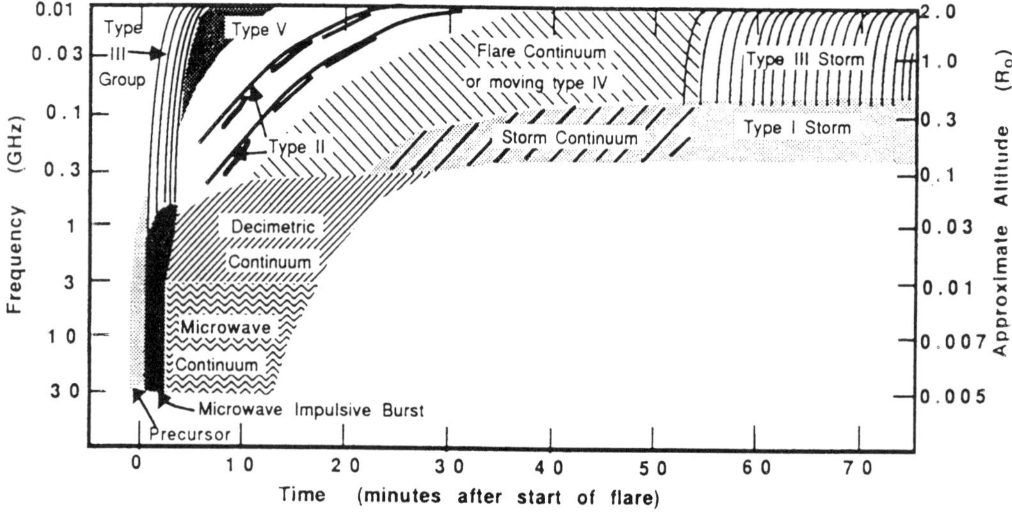

FIG. 11. Schematic diagram illustrating the types of solar radio emission associated with solar flares as a function of dynamic spectra, emission frequencies, and time. The emission altitude is indicated on the right.

quency drift is believed to be caused by fast electrons (electrons traveling about 10% of the speed of light) streaming outward. Type III emission is often associated with solar electron events (see Sec. 6.2.2).

- Type IV solar radio emission: A smooth continuum of broad-band bursts primarily in the meter range (300–30 MHz). These bursts occur with some major flare events; they begin 10–20 min after the flare maximum and can last for hours. Type IV bursts are commonly associated with CMEs (see Sec. 6.1.1).
- Type V solar radio emission: Short-duration (a few minutes) broad-band continuum noise in the dekameter range usually associated with type III bursts.

5.2.2.2 Solar-Flare Gamma-Ray Emission. Ions are simultaneously accelerated by the solar-flare process into the MeV energy range and, in exceptional cases, to the GeV energy range. As these ions travel down the magnetic flux tubes in a coronal loop to the chromosphere or even to the photosphere, there will be nuclear interactions with the ambient material producing nuclear line emission. The fragments of the nuclear interactions are highly radioactive, generating positrons and pi mesons. There is extremely strong emission at 0.511 MeV resulting from electron–positron annihilation. Strong lines are emitted from carbon and oxygen nuclei at 4.44 and 6.13 MeV. Another prominent nuclear line at 2.223 MeV is the result of neutron capture by hydrogen. Theoretical models of the solar-flare high-energy emission are still being developed (Ramaty and Murphy, 1987; Ramaty and Mandzhavidze, 1993). Table 7 lists high-energy–photon production mechanisms.

The gamma-ray emission was initially thought to be limited to the impulsive phase of the solar flare; however, the results from the more sensitive gamma-ray experiments on both the U.S. Gamma Ray Observatory and the Russian MIR space station show that the very powerful solar flares in June 1991 had extended gamma-ray emission for several hours.

The gamma-ray continuum emission is slightly directional, reflecting its high-energy ion-interaction initiation. The high-energy ions traveling down the coronal loops will be magnetically focused and have a high probability of having a velocity component parallel to the solar surface when they interact with the ambient nuclei. The Solar Maximum Mission satellite results noted that gamma-ray emission from the solar flare on the limb of the Sun (as viewed from the Earth) was about an order of magnitude "brighter" than those from flares near the solar central meridian. There is also great variability between the power in the continuum emission and the line emission. The line emission can be observed only if it is stronger than the continuum emission; for some solar flares the electron bremsstrahlung is the dominant emission.

5.2.2.3 Solar-Flare Hard X-Ray Emission. Hard x-ray (ten to hundreds of keV) emissions are most intense during the initial part of the solar flare. Hard x-ray emission is

Table 7. High-energy–photon production mechanisms.

Emissions	Process	Observed photons
Continuum	Electron bremsstrahlung	keV to MeV
Nuclear deexicitation lines	^{4}He (α,n) ^{7}Be*	0.429 MeV
	^{4}He (α,p) ^{7}Li*	0.478 MeV
	^{20}Ne (p,p) ^{20}Ne*	1.634 MeV
	^{12}C (p,p) ^{12}C*	4.438 MeV
	^{16}O (p,p) ^{16}O*	6.129 MeV
Neutron capture line	^{1}H (n,γ) ^{2}H	2.223 MeV
Positron annihilation radiation	$e^+ + e^- \rightarrow 2\gamma$	0.511 MeV
Pion (π^0 and π^+) decay radiation	$\pi^0 \rightarrow 2\gamma$	10 MeV to 3 GeV
	$\pi^\pm \rightarrow \mu^\pm \rightarrow e^\pm$	

extremely impulsive, composed of bursts that range from milliseconds to seconds in duration. The hard x-ray bursts are accompanied by microwave radio and ultraviolet bursts with similar time profiles, suggesting that a common mechanism stimulates the emissions.

The hard x-rays are thought to be the result of interactions between accelerated electrons and ambient ions (bremsstrahlung) as the accelerated electrons travel down the magnetic flux tubes in a coronal loop. The hard x-ray emission energy spectra are often approximated either by a multithermal source with a peak temperature in the several hundred million degree range or by a nonthermal power-law distribution of energies with a spectral index having typical values ranging from -2 to sometimes -5. At low energies the spectrum merges with the low-energy thermal spectrum. There are large variations in the amount of hard x-ray flux emitted from different solar flares. These variations are probably the results of variations in the amount of energy involved in the various acceleration mechanisms. There is not a good correlation between solar-flare optical importance and hard x-ray emission. The correlation between solar-flare hard x-ray emission and the number of energetic particles observed in space is also poor.

5.2.2.4 Solar-Flare EUV and Soft X-Ray Emissions. The hot (multimillion degree) plasma resulting from the solar flare radiates at wavelengths corresponding to the plasma temperature. The measurement technology for soft x-ray measurements is better developed than the EUV measurement techniques, and most of the data reflect measurements in the soft x-ray wavelengths. The x-ray emission at a few kilovolts (soft x rays) increases rapidly to a maximum intensity and then decays on a longer time scale. A general classification of x-ray flares as impulsive or gradual has developed. The x-ray emission from "impulsive" solar flares is generally less than an hour in duration, having characteristic short rise times and rapid decay times. The extended x-ray emission from "gradual" solar flares may persist for several hours. These extended solar x-ray flares are often called long-duration events because of a characteristic long decay time of many hours that is suggestive of extended thermal heating of the plasma, perhaps due to magnetic reconnection as the active regions involved relax to simpler magnetic configurations. The long-duration soft x-ray–emitting flares have a good association with the observations of coronal mass ejections (CMEs) and are considered a good proxy indicator of a probable CME.

6. SOLAR PARTICLE RADIATION FROM THE ACTIVE SUN

The particle emission for the active Sun ranges from enhanced plasma emission to copious fluxes of energetic ions. Both types of emissions have effects on the Earth's atmosphere and human activities. When the solar plasma interacts with the Earth's magnetosphere, the additional energy input causes perturbations that affect communications, navigation, and even our satellites and power grids. The energetic ions affect communications in the Earth's polar regions and are operational constraints on human activities in space.

6.1 Solar Plasma Emissions

It was initially assumed that transient, fast, dense plasma emissions would be associated with the solar-flare activity; however, these associations proved to be tentative at best. Most solar flares (brightenings observable in the Hα emission) have no observable effect in the solar wind. Only the "big flares," those that are conspicuous by emissions in all electromagnetic wavelengths and associated with energetic particles, can be related to transients in the solar wind. Beginning with the Skylab mission in 1972, a new solar phenomenon was identified, the coronal mass ejection. Rather than association with "explosive" energy releases on the Sun, these emissions were found to be related to large-scale changes in solar features, such as the "disappearing filament" or erupting filament and streamer "blow outs." The total energy involved in a very large mass ejection is equivalent to and may exceed the energy released in a large solar flare.

6.1.1 Coronal Mass Ejections There is a complete spectrum of mass motions on the Sun. The small-scale mass motions may contribute to coronal heating and provide additional mass to the solar wind. The large-scale CMEs result in significant restructuring of the solar corona and may involve up to 10^{16} g of material. The large-scale CMEs are self-contained structures of plasma and embedded magnetic field ejected into space, which perturb the structure of the interplanetary medium. Approximately 30% of all CMEs have an internal field topology characteristic of twisted flux ropes (Gosling, 1990, 1993) and are commonly called "magnetic clouds." There is a continuum in the size and speed of coronal mass ejections ranging over several orders of magnitude. The speed of the leading edges of CMEs observed by the Solar Maximum Mission satellite ranged from 10 to 2100 km s^{-1}. Most coronal mass ejections have speeds comparable with the solar-wind speeds. Only about one-third have a speed sufficiently in excess of the solar wind that they can drive an interplanetary shock. Fast coronal mass ejections generate magnetohydrodynamic (MHD) shock waves as they move through the interplanetary medium at speeds faster than the local Alfvén speed. This shock wave imparts energy into the interplanetary medium and is capable of accelerating electrons and ions to higher energies.

Most coronal mass ejections arise from initially large-scale closed structures, with many from preexisting coronal streamers. However, more than one-third of the observed CMEs cannot be associated with any activity observable on the solar surface. The latitudinal distributions of CMEs is similar to that of the streamers and prominences, principally at low latitudes near the "streamer belt" near the solar-activity cycle minimum and becoming very broad near the solar-activity cycle maximum. The occurrence frequency of coronal mass ejections tracks the solar-activity cycle in both phase and amplitude. During solar minimum the average CME rate is about 0.25 per day; during solar maximum the average CME rate is about 2.5 per day (Webb and Howard, 1994; Webb, 1995). It is estimated that during the maximum of the solar-activity cycle, the coronal mass ejections may contribute about 15% of the mass flux in the near-ecliptic solar wind.

Large coronal mass ejections may be observed as density enhancements against the plane of the sky by space-borne instruments that observe the "Thomson scattered" light from density structures. State-of-the-art ground-based coronagraphs are also capable of observing coronal mass ejections but over a much smaller density range than the space-borne instruments. Unfortunately, these space-borne instruments are technically demanding and extremely expensive, and only a few have been orbited. Figure 12 shows a large coronal mass ejection observed on 24 October 1989. This very fast coronal mass ejection was associated with a large solar flare and high-energy solar-particle event.

6.2 Solar Energetic Particles

There has been a dramatic change in perspective regarding the source of energetic-particle events. Since the large transient increases in particle flux could often be associated with solar flares, it was assumed that the solar-flare process was the source of energetic particles observed in space. The names "solar cosmic rays" and "solar energetic particles" were commonly used to describe transient increases in particle flux. Papers written in the 1950s through the 1980s

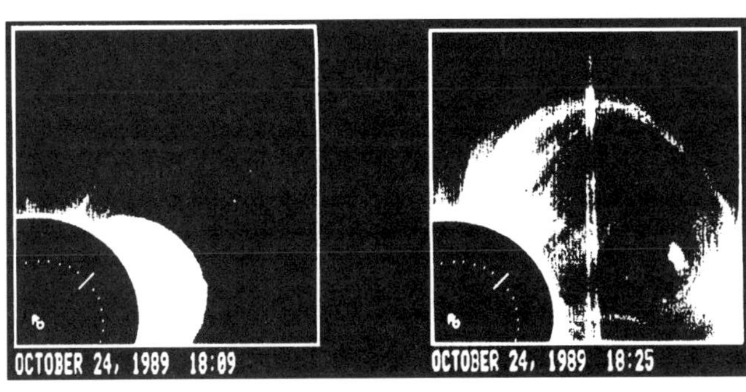

FIG. 12. The coronal mass ejection observed on 24 October 1989 by the SMM coronograph. The dotted circle indicates the visible Sun; the arrow indicates the direction of the solar rotation axis (courtesy of A. Hunthausen, HAO).

reflect the assumption that solar energetic particles were accelerated from the available material in or above the solar-active region in which the flare occurred. As measurement techniques improved, the elemental and isotopic composition of solar particles observed at the various interplanetary spacecraft was found to be consistent with the particles having passed through less than 30 mg cm^{-2} of matter from the acceleration site to their detection location. Thus, the solar photosphere and chromosphere were eliminated as the source region of the particles observed in space since the elemental and isotopic composition of these solar energetic ions did not appear to have undergone fragmentation due to interaction with the significant mass of the solar atmosphere.

A new paradigm suggests that a source of the energetic ions observed in space is the result of acceleration associated with interplanetary shocks generated by fast coronal mass ejections (see Reames, 1995, for a review). It is supposed that a shock wave has accelerated some fraction of the ions in the local solar wind. The diffusive shock acceleration scenario (Lee, 1983), a concept where particles near the shock are scattered back and forth across the shock by waves, gaining energy each time, is favored as a mechanism for continuous acceleration of particles as the shock travels out from the Sun. The accelerated particles stream away from the shock along the interplanetary magnetic field lines. The particle intensity–time profile will depend on how the observer is connected to the shock via the interplanetary magnetic field. If the shock approaches the observer, the maximum low-energy flux will be measured as the shock passes the observer.

Both the solar-flare scenario and the shock-acceleration scenario have enthusiastic advocates. Many of the "large" solar-particle events are associated with "large" solar flares. However, only about $\frac{1}{2}$ of the solar-proton events observed at the Earth can be confidently time associated with specific solar flares. The intensity–time profiles of solar energetic particles leave no doubt that interplanetary shocks also accelerate ions.

Solar-particle events are described by their flux, fluence, and spectra. Solar-particle flux is a measure of particle intensity, which can be either an integral flux above a specified energy level, in units of particles $(cm^2 \cdot s \cdot sr)^{-1}$, or a differential measurement, which specifies the flux at a specific energy in units of particles $(cm^2 \cdot s \cdot sr \cdot MeV)^{-1}$. Omnidirectional particle fluxes are given in units of $(cm^2 \cdot s)^{-1}$. Solar-particle fluence is the total number of particles experienced throughout an individual solar-particle event or episode. Fluence may be given in either directional units of particles $(cm^2 \cdot sr)^{-1}$ or omnidirectional units of particles (cm^{-2}). The fluence is generally of concern for the total radiation exposure. The solar-particle flux (or fluence) as a function of energy is described by the particle spectrum. Many different spectral forms are used. The most commonly used spectral forms are a power law in energy or rigidity (rigidity is momentum per unit charge), an exponential form, or more recently the use of Bessel functions.

Energetic solar-particle events occur in many sizes and on many time scales. Several different classification systems can be used to group different types of events. A relatively recent grouping is to classify them as "impulsive" or "gradual" (a classification of the associated solar flare x-ray activity), as shown in Table 8. The "impulsive" events are generally small particle-flux and fluence events. Some of the "impulsive" events, particularly the ^{3}He- and Fe-rich events, seem to be directly associated with solar flares. Gradual events tend to be larger flux and fluence events with "normal" composition and are associated with coronal mass ejections and/or large, gradual, long-duration x-ray–emitting solar flares.

6.2.1 Propagation Effects in the Interplanetary Medium The transport of solar protons in interplanetary space is controlled by the topology and characteristics of the interplanetary magnetic field. The topology of the magnetic field lines in interplanetary space is controlled by the flow speed of the ionized plasma and the rotation rate of the sun, resulting in the so-called Archimedean-spiral configuration (see Fig. 6). Solar-particle events observed in space resulting from solar activity at the probable "foot point" of an idealized Archimedean-spiral interplanetary magnetic field line leading from the observer back to the Sun are considered to be "well-connected" events. This means that the probable propagation path of the solar par-

Table 8. Properties of impulsive and gradual solar-proton events (adapted from Reames, 1995).

	Impulsive particle event characteristics	Gradual particle event characteristics
Particles	Electron rich	Proton rich
$^3He/^4He$	~1	~0.0005
Fe/O	~1	~0.1
H/He	~10	~100
Charge state of iron	~20	~14
Duration	Hours	Days
Heliolongitude range	<30° (generally "well connected")	~180°
Associated solar radio type	III, V, (II)	II, IV
Associated solar x-ray emission	Impulsive	Gradual
CME association	. . .	96%
Associated interplanetary shock	. . .	Yes
Events per year (at solar maximum)	~1000	~10

ticles was probably along the interplanetary magnetic field lines with minimal scattering due to irregularities in the interplanetary magnetic field. This concept is often referred to as the "favorable propagation path." It is often observed that the highest particle-flux intensities are along this favorable propagation path.

The particle-flux longitudinal gradients in the inner heliosphere are variable, and local interplanetary conditions and structures greatly influence the observed time–intensity profile. The average heliolongitudinal gradient of particle flux is difficult to determine; however, the longitudinal proton-flux gradient used in the U.S. Air Force proton-prediction system is one order of magnitude decrease in flux per radian of heliolongitudinal distance away from the most favorable propagation path (Smart and Shea, 1985).

There is a radial gradient in the spacecraft observation of energetic-particle flux at different distances from the Sun, which is most pronounced in the inner heliosphere. These gradients are the result of an initial near-Sun particle injection and particle propagation away from the Sun along the interplanetary field lines. The most extensive measurements are comparisons between the flux measured by Earth-orbiting spacecraft and interplanetary spacecraft with orbits closer to the Sun than 1 AU (such as the HELIOS spacecraft). There are fewer data available between 1 and 3 AU. The form and magnitude of these radial gradients are the subject of controversy, reflecting the arguments between a solar-flare acceleration source and a CME/shock acceleration source. In the outer heliosphere, at distances of 5 AU and beyond, the particle-flux profiles are dominated by the major interplanetary shock structures.

If it is assumed

1. that the maximum possible prompt solar-proton flux is at the position that is "well connected" to the particle source acceleration region,
2. that diffusion across the interplanetary magnetic field lines is negligible, and
3. that the volume of the magnetic flux tube expands in the manner expected from classical geometry as the distance from the Sun increases,

then a power-law function of the form R^{-3} can be used to extrapolate proton flux at 1 AU to other distances. In the energy range of 10–70 MeV and for radial distances from 1 to 5 AU, the average solar-proton radial gradient derived by Hamilton (1977), considering the probable effects of diffusion, has a functional form of $R^{-3.3}$, where R is the radial distance from the Sun. The limited measurements available suggest solar-proton radial gradient variations ranging from R^{-3} to R^{-4}. For distances less than 1 AU, the proton flux at 1 AU seems to extrapolate with a functional form of R^{-3}. Again, the limited measurements available suggest that radial gradient variations ranging from R^{-3} to R^{-2} should be expected. When extrapolating proton fluence from 1 AU to other distances in the heliosphere, a functional form of $R^{-2.5}$ has been recommended with variations expected that may range from R^{-3} to R^{-2} (Feynman and Gabriel, 1988).

6.2.2 Solar-Electron Events Solar-electron events in space are now recognized to be a common result of solar activity. Solar-electron events were not identified until the beginning of the space era. Systematic identification of solar-electron events began in 1965 when space instrumentation was developed to measure specifically these events. The electron fluxes in space do not constitute a significant source of radiation dose in comparison with the energetic-ion events. Solar-electron events are often used as diagnostics of solar activity or "tracers" of the interplanetary magnetic field topology since their primary motion is essentially confined along the interplanetary magnetic field lines.

Solar-electron events were initially classified into two groups: nonrelativistic solar-electron events, which are the most common, and relativistic solar-electron events. Some events were called "pure" electron events since they did not have a significant ionic component. A more recent grouping is to classify the solar-electron events by the associated soft x-ray emission into impulsive and long-duration (sometimes called gradual) events.

Many of the solar-electron events observed in space do not have an association with visible solar-flare activity; however, there is an excellent correlation between the observation of solar-electron events in space and the occurrence of a type III solar radio burst. There is a very good correlation between large solar x-ray events and the observation of energetic electrons in space at positions that are "well connected" via the interplanetary magnetic field to the solar-flare position on the solar disk.

A comprehensive systematic survey of electron data acquired by the ISEE-3 and HELIOS 1 and 2 spacecraft (Moses *et al.*, 1989) found that the spectral characteristics of solar-electron events could be organized by the type of associated solar-flare soft x-ray event (impulsive or long duration). Solar-electron events associated with long-duration x-ray events (LDE) had a characteristic particle spectrum typified by a power law in momentum with a characteristic spectral index ranging in magnitude between 3 and 4.5. (For electrons, except at low energies, the energy spectrum and the rigidity spectrum are essentially the same.) The solar-electron events associated with impulsive solar x-ray events have a characteristic double power-law spectrum with a spectral index that hardens (the magnitude of the absolute value of the spectral index decreases) with rigidity (or energy). The transition range in these broken power-law spectra generally occurs between 100 and 200 keV, with a spectral index ranging in magnitude from 0.6 to 2 in the low-energy region of the spectra and a spectral index ranging in magnitude between 2.4 and 4.3 in the high-energy range of the spectra.

The characteristics of the electron spectra observed on widely separated spacecraft were found to be similar. After adjustment for flux amplitude, which is a function of relative connection to the solar activity source, the same characteristics in spectral shape are found. The long-duration events have the spectral form of a power law in momentum, and the impulsive events have the double power law even at widely separated heliolongitudinal distances.

The number of solar-electron events in a specific energy range can be described by a power-law distribution. Cliver *et al.* (1991) found that in the interval from 1973 to 1989, the number of solar-electron events in the 3.6–18-MeV energy range and differential flux range of 10^{-2}–10^{2} $(\mathrm{cm}^2 \cdot \mathrm{s} \cdot \mathrm{sr} \cdot \mathrm{MeV})^{-1}$ exceeding a specific threshold could be approximated by a power law with a slope of about -1.3.

6.2.3 Solar-Proton Events It is common to classify energetic solar-proton flux by the observed proton flux and fluence at a specific energy. Proton events in the energy range between 10 and 30 MeV are often used, both for traditional historical reasons and because there are enough statistics to be meaningful. There is a general association of proton-event frequencies observed at the Earth with the solar-activity cycle, but there is no discernible pattern in the proton event occurrence. (See Shea and Smart, 1990, for solar cycles 19–21.) However, the expected solar-proton-event flux and fluence frequency distribution is important for planning space missions, and so there is considerable interest in this type of data. Table 9 gives a summary of the solar-proton fluences observed at the Earth for solar cycles 19–21. Table 10 presents a summary of the largest events that have occurred during solar cycle 22 (not

Table 9. Summary of solar-proton fluence for solar cycles 19–21.

Cycle No.	Start	End	No. of months in cycle	No. of discrete proton[a] events	No. of discrete proton[a] producing regions	Solar cycle integrated solar-proton fluence	
						>10 MeV (cm^{-2})	>30 MeV (cm^{-2})
19	May 54	Oct 64	126	65	47	7.2×10^{10}	1.8×10^{10}
20	Nov 64	Jun 76	140	72	56	2.2×10^{10}	6.9×10^{9}
21	July 76	Sept 86	123	81	57	1.8×10^{10}	2.8×10^{9}

[a]Proton events with ≥10 MeV peak flux ≥10 $(cm^2 \cdot s \cdot sr)^{-1}$.

yet completed as of the preparation of this article).

It had been previously suggested, as a result of the analysis of the solar-proton events observed in solar cycle 20, that there were "ordinary" and "anomalously large" events. However, analysis of a larger data sample of all the solar-proton events in solar cycles 19–21 shows that the proton-event fluence distribution observed at the Earth during the "active years" of the solar cycle is well approximated by a log-normal function (Feynman *et al.*, 1990). Including the solar-proton events of solar cycle 22 improves the statistics and reinforces the utility of the log-normal approximation (Feynman *et al.*, 1993). These data can be used to estimate the probable free-space exposure to solar protons for various space missions beyond the Earth's geomagnetic shield (including geosynchronous orbit). The probable exposure of a space vehicle at 1.0 AU to solar protons with energy >30 MeV during the solar-active years for various mission times is given in Fig. 13.

The number of solar-proton events in a specific energy range can be described by a power-law distribution. The distribution of the most commonly occurring solar-proton events exceeding a specific threshold in the 24–40-MeV energy range can be approximated by a power law with a slope of about −1.1 over the differential flux range between 10^{-3} and 10^{2} $(cm^2 \cdot s \cdot sr \cdot MeV)^{-1}$ (van Hollebeke *et al.*, 1975; Cliver *et al.*, 1991). The occurrence of very large solar-proton events appears to be less frequent than would be expected from an extension of the most commonly occurring event frequency. When large data samples are analyzed, there appears to be a broken power-law distribution. The distribution of the number of solar-proton events with energies ≥10 MeV expected during a solar cycle is given in Fig. 14. These

Table 10. Summary of the largest solar-proton fluence occurrences for solar cycle 22.

Cycle No.	Yr	Begin		End		Event-integrated solar-proton fluence	
		Mo	Dy	Mo	Dy	>10 MeV (cm^2)	>30 MeV (cm^2)
22	89	03	09	03	24	0.12×10^{10}	0.03×10^{9}
	89	08	12	08	18	0.76×10^{10}	1.4×10^{9}
	89	09	29	10	02	0.38×10^{10}	1.4×10^{9}
	89	10	19	10	30	1.9×10^{10}	4.2×10^{9}
	89	11	27	12	03	0.51×10^{10}	0.16×10^{9}
	89	12	30	01	02 (90)	0.21×10^{10}	0.13×10^{9}
	90	05	21	05	31	0.035×10^{10}	0.14×10^{9}
	91	03	22	03	26	0.96×10^{10}	1.8×10^{9}
	91	06	04	06	21	0.32×10^{10}	0.79×10^{9}
	91	07	07	07	12	0.11×10^{10}	0.01×10^{9}
	92	05	09	05	11	0.066×10^{10}	0.02×10^{9}
	92	10	30	11	05	0.35×10^{10}	0.43×10^{9}
	94	02	20	02	22	0.099×10^{10}	0.02×10^{9}
Totals:						5.82×10^{10}	1.0×10^{10}

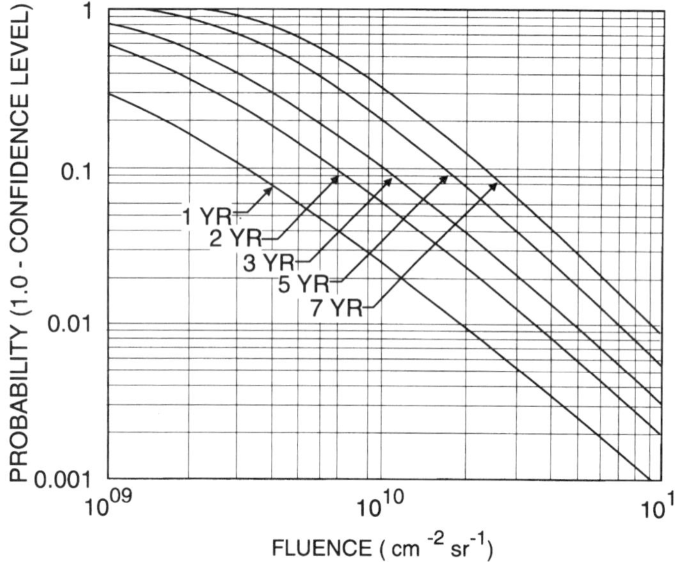

FIG. 13. The probable exposure to protons with energy >30 MeV during the solar-active years for various exposure times in interplanetary space at 1 AU. (Original figure from Feynmann *et al.*, 1993, reproduced by permission of the American Geophysical Union.)

data represent the average of ≥10-MeV proton events with peak flux ≥10 $(cm^2 \cdot s \cdot sr)^{-1}$ measured by Earth-orbiting satellite instruments during solar cycles 20–22. The increasing slope of the curve at very large flux events suggests that there may be limits to the acceleration processes, and extraordinarily large events are very rare. The analysis of total solar-proton fluence data also noted the reduced number of very large events, $>10^{10}$

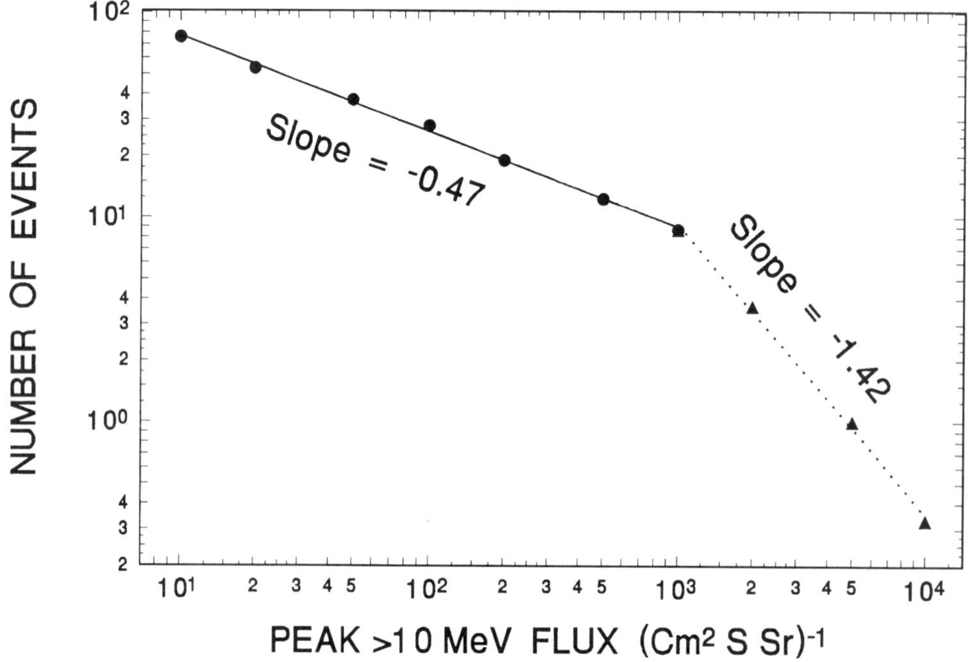

FIG. 14. The expected number of proton events during a solar cycle with protons ≥10 MeV exceeding a specified peak flux threshold. Data from solar cycles 20–22.

cm^{-2}, first noted by Lingenfelter and Hudson (1980) and subsequently confirmed by later investigators.

There is no universally accepted solar-proton classification system. The solar-proton event classification system introduced by Smart and Shea (1971) separated the >10-MeV peak flux in decade groups. A statistical treatment of the ≥30-MeV fluence data by Nymmik (1993) suggests a fluence classification based on groups separated by one standard deviation in the log-normal distribution of fluence (see Table 11) where the average solar-proton-event fluence occurrence is "normal." Stassinopoulos *et al.* (1996) adopted an order-of-magnitude classification system for convenience.

6.2.4 Solar-Particle Event Composition

The composition of "small" solar-particle events is variable and has been separated into distinct composition groups. From an astrophysical point of view, the "iron-rich" and ^{3}He-rich events are extremely interesting, perhaps reflecting processes in the solar corona, but the particle flux in these events is small. The composition of "large" solar-particle events is relatively consistent with an ion selection process based on the first ionization potential of the elements in the solar corona or solar wind.

6.2.4.1 Large Solar-Particle Events. Large solar-particle events tend to have a "normal" composition that is relatively consistent, although there are event-to-event variations, particularly in the helium/hydrogen ratios. The ratio of the abundances of the elemental composition observed in large solar energetic-particle events to the abundances of the elemental composition observed in the photosphere is a function of the first ionization potential. Some researchers contend that large solar energetic-particle events are a reasonable sample of the solar corona (Breneman and Stone, 1985; Mewaldt and Stone, 1989); however, there are relatively large uncertainties in coronal abundance measurements. The elemental composition of large solar-particle events is given in Table 12.

Table 11. The Nymmik solar-proton fluence classification.

Class	$n\sigma$[a]	Omnidirectional ≥30-MeV fluence (cm^{-2})
Small	Average − 1σ	~2 × 10^6
Normal	Average	~9 × 10^6
Large	Average + 1σ	~3 × 10^7
Very large	Average + 2σ	~2 × 10^8
Extremely large	Average + 3σ	~5 × 10^9

[a] σ is one standard deviation of the log-normal frequency distribution.

6.2.4.2 Small and Iron-Rich Solar-Particle Events. When space particle-detector technology developed the capability to determine the composition of "small events," the initial results were regarded as "composition anomalies" since there were distinct differences from "large" solar-particle events. It became common to classify these as ^{3}He-rich or iron-rich events. A more recent classification is to consider the class of particle emission associated with the particle source: "impulsive-flare" events and "gradual events." In general, the fluxes in the impulsive events are orders of magnitude smaller than large or ordinary solar-particle events. The elemental composition of impulsive-flare–associated solar-particle events is detailed in Table 13.

6.2.4.3 Charge State of Solar Particles. The charge state of solar energetic ions provides information on their origin and also affects their propagation through interplanetary space since the charge state determines the magnetic rigidity (momentum per charge) and the particle scattering from magnetic irregularities in space as a function of the particle rigidity. Partially ionized solar energetic ions are more "rigid" (less susceptible to geomagnetic bending) and capable of greater penetration into the Earth's magnetic field, an important effect in radiation-dose calculations. In the early decades of solar particle measurements, it was assumed that the energetic ions would be fully stripped of electrons. As space instrument technology improved and direct ionic-charge measurements became possible, it was found that the elemental nuclei with $Z > 6$ were not fully stripped of electrons. Table 14, adapted from Luhn *et al.* (1985), gives the mean charge states for 12 solar-particle events observed with the ISEE-3 spacecraft between September 1978 and September 1979. These events were specifically selected to exclude fluxes

Table 12. Composition of "normal" solar-particle events (adapted from Reames, 1992).

Element	Z	Mean solar particle (O = 1000)	Mean solar corona (Si = 1000)	Accepted solar photosphere (Si = 1000)
H	1	1 170 000 ± 89 000	7 550 000 ± 570 000	27 900 000
He	2	55 000 ± 3 000	355 000 ± 19 000	2 270 000
C	6	471 ± 14	3 040 ± 90	10 100
N	7	128 ± 4	826 ± 26	3 130
O	8	1 000 ± 30	6 450 ± 190	23 800
Ne	10	151 ± 5	974 ± 32	3 440
Na	11	11.8 ± 1.1	76 ± 7	57.4
Mg	12	203 ± 8	1 310 ± 52	1 074
Al	13	14.0 ± 0.7	90.0 ± 5	84.9
Si	14	155 ± 7	1 000 ± 45	1 000
P	15	0.74 ± 0.11	4.7 ± 0.7	10.4
S	16	35.6 ± 1.1	230 ± 7	515
Cl	17	0.33 ± 0.11	2.1 ± 0.7	5.24
Ar	18	3.3 ± 0.5	21 ± 3	101
K	19	0.53 ± 0.24	3.3 ± 1.5	3.77
Ca	20	12 ± 1.6	77 ± 10	61.1
Ti	22	0.61 ± 0.16	3.8 ± 1.1	2.4
Cr	24	2.3 ± 0.4	15 ± 3	13.5
Fe	26	155 ± 15	1 000 ± 100	900

associated with the passage of interplanetary shocks by the spacecraft.

The charge state of energetic ions is a particularly difficult measurement, and the charge states are generally determined at relatively low energies, typically of the order of 1 MeV or less. The initial measurements have been verified by subsequent independent measurements, and the charge states have been measured at energies of tens of MeV (Oetliker *et al.*, 1995). These higher-energy charge states are generally consistent with the low-energy charge states, except for some unresolved ambiguity in the charge state of iron. The general view is that the charge state of the elements in large events is consistent with coronal temperatures ($\sim 10^6$-K plasma). A recent complex analysis found an ionic charge state of 12 for iron nuclei at ~200 MeV per nucleon. These results were from the analysis of the solid-state nuclear track detectors on the Long Duration Exposure Facility (LDEF) in 28° low altitude Earth orbit during the September and October 1989 large energetic solar-particle events.

The charge state of iron is particularly sensitive to the temperature of the plasma from which the ion is accelerated. In the "gradual" events, the typical charge state of iron seems to be $Q = 14$ and is consistent with the charge state of iron in the solar corona and in the solar wind. In the particle events associated with small impulsive solar flares, the mean charge state of iron is reported to be $Q = 20$, which is consistent with the ions having a source in a hot ($\sim 2 \times 10^7$ K) plasma.

Table 13. Averaged elemental abundances in impulsive-flare associated solar-particle events (adapted from Reames, 1992).

Element	Z	Averaged impulsive-flare (O = 1000)
^{4}He	2	46 000 ± 4 000
C	6	406 ± 18
N	7	172 ± 11
O	8	1000 ± 27
Ne	10	453 ± 18
Mg	12	478 ± 18
Si	14	425 ± 17
S	16	171 ± 11
Ca	20	78 ± 14
Fe	26	1025 ± 30

Table 14. Charge state of "normal" solar-particle events.

Element	Atomic number	Measured mean charge state (Q)
C	6	5.7
N	7	6.3
O	8	7.0
Ne	10	9
Mg	11	10.7
Si	14	11.0
S	16	10.9
Fe	26	14.9

7. TERRESTRIAL EFFECTS FROM TRANSIENT SOLAR EMISSIONS

The consequences of the solar emissions on the Earth have been noted for centuries, although it has not always been recognized that the Sun was the initial source of the energy manifestation. Auroras have been observed in northern Europe throughout recorded history. The effects of magnetic storms on navigation have been noted since the 18th century. As technology grew more complex, the influence of the Sun on this technology became more apparent. In the 19th century, the effects on long telegraph lines were noted. In the 20th century, effects on radio communications, satellite operations, and power grids have been experienced. Figure 15 gives the time scales of solar effects on the Earth's environment.

7.1 Communication Effects

One of the most immediate effects of solar flares is the radio emission that may cause interference throughout the electromagnetic spectrum, effectively jamming radios and radars. The solar x-ray event energy deposition increases the electron density in the ionosphere, which for major perturbations can suddenly disrupt high-frequency (HF) propagation paths (see also IONOSPHERE). These sudden ionospheric disturbances (SIDs) affect HF radio point-to-point communication over the entire sunlit hemisphere, often inconveniencing HF communication users. In some cases, such as aircraft operations, the loss of communication may present control difficulties.

7.1.1 Polar Communication Effects As the solar energetic charged particles initially accelerated as a result of solar activity travel along the interplanetary magnetic field lines and intercept the Earth, additional energy is deposited in the polar atmosphere, increasing the electron density in the polar ionosphere. In the case of major solar-particle events, the dramatic enhancement in the electron density in the *D* region of the

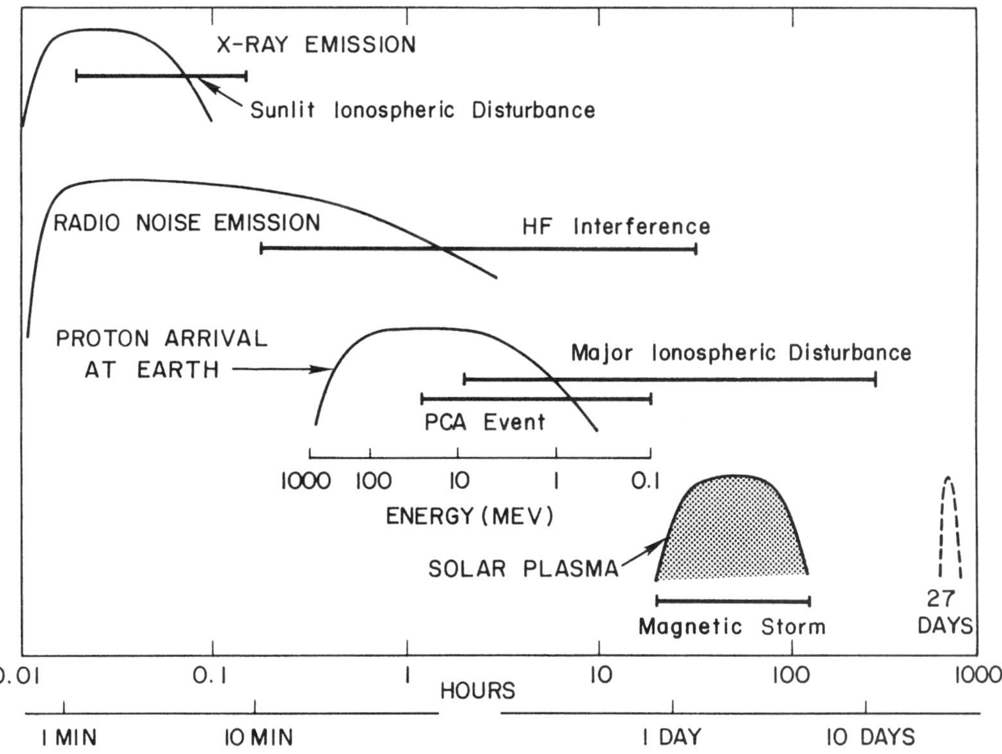

FIG. 15. The time scale of various effects on the Earth. The effects are offset vertically for clarity. (See text for definitions.) The dashed feature at 27 days is indicative of the 27-day recurrence pattern in many solar phenomena affecting the Earth.

Earth's polar ionosphere attenuates radio signals traversing the ionosphere, leading to a phenomena known as polar cap absorption (PCA) events. Analysis of the PCA events recorded in the 1950s, 1960s, and 1970s aided in the determination of the solar-particle radiation levels in these time periods. On rare occasions, the solar-particle flux was so intense that signals from ionosondes used to determine the structure of the Earth's ionosphere were completely absorbed, causing a phenomenon historically known as polar cap blackout. In extreme cases, the polar cap blackout continued for several days during which time point-to-point HF radio communications were essentially completely disrupted. Even with more modern VLF long-range navigation systems, such as the LORAN system, during a solar-particle event (PCA event) the increased ionization in the polar ionosphere modifies the characteristic of the Earth–ionosphere waveguide, resulting in changes in the phase and amplitude of received signals that translate into navigation errors.

7.1.2 Aurora The polar aurora is a consequence of energy initially emitted from the Sun via the solar wind (including the coronal mass ejection) entering the Earth's magnetosphere and, through complex magnetospheric processes, eventually resulting in the local acceleration of electrons that travel down geomagnetic field lines, depositing sufficient energy into the atmosphere so as to stimulate emission from atomic oxygen atoms. (The common auroral emissions are at 557.7 and 630.0 nm, which appear as green and red when the intensity exceeds the color threshold of the human eye.) This locally intense energy deposition again increases the local electron density in the Earth's ionosphere and can interfere with HF point-to-point communication in the high-latitude auroral regions. During large magnetic storms, the aurora moves toward the equator, on occasions reaching the Earth's midlatitudes, transporting what are normally polar-region effects to these latitudes.

7.2 Geomagnetic-Field Variations

The geomagnetic storm, historically named because of variations induced in the Earth's magnetic field, is both a direct effect and an indicator of complex processes in the Earth's magnetosphere resulting from solar activity, such as the coronal mass ejection. The variations induced in the Earth's magnetic field are maximum in the polar regions (up to several thousand nanotesla variation in an average 60 000-nT surface polar magnetic field) and minimum at the Earth's equator (extreme variations to perhaps 400 nT in an average 30 000-nT equatorial surface magnetic field). These variations affect the direction indicated by magnetic compasses, a fact that has been a technological driver leading to the development of other types of navigation systems that might be immune to these effects like the LORAN system and more recent space navigation systems such as the Transit satellites and the Global Positioning satellites. A more recently noticed effect on human activity is the action of induced currents, resulting from extremely intense and rapid magnetic field variations, exceeding the design capacity of power grids, thereby damaging equipment or, in more spectacular cases, causing the collapse of a major power grid such as occurred in Quebec, Canada, in March 1989.

7.3 Atmospheric-Density Effects on Satellite Operations

The density at a specific height in the upper atmosphere changes during a solar cycle as the atmosphere expands and contracts because of variations in the solar EUV flux that heats the upper atmosphere. The upper-atmosphere density controls the lifetime of orbiting satellites, since even these very low densities result in slight drag forces that limit the lifetimes of orbiting objects. Predictions of satellite orbits such as the U.S. space shuttle or the MIR space station are extremely dependent on our ability to specify the atmospheric density. Low-altitude Earth-orbiting satellites must be either periodically reboosted, such as the Russian MIR space station or the U.S. Hubble Space Telescope, or replaced. Failure to reboost a decaying satellite into a higher orbit will result in reentry of the satellite in the Earth's atmosphere, such as happened to the U.S. Skylab spacecraft in July 1979.

Besides the long-term solar-cycle variations, the coronal mass ejection plasma emissions may couple a tremendous amount of energy (in human terms) into the Earth's magnetosphere. The geomagnetic storm is a result of this energy input, which initiates and intensifies current systems flowing in the magnetosphere and ionosphere. In the polar regions, the phenomenon known as Joule heating greatly increases the atmospheric density, thereby changing satellite orbits. During the major geomagnetic storms in March 1989, several hundred orbiting objects were temporarily "lost" to the NORAD catalog of Earth-orbiting objects, because of unanticipated changes in their orbits.

7.4 Solar-Particle Radiation Effects

Energetic-particle ionizing radiation has many effects on today's technology. Besides degrading polar communications as described in Sec. 7.1.1, energetic solar-particle radiation also presents operational constraints on human activities in space.

7.4.1 Radiation-Dose Effects on Solar-Cell Power Systems Many space vehicles are powered by solar cells, which convert a fraction of the Sun's electromagnetic energy into electrical power. Modern spacecraft are designed to last for many years or even decades. The cumulative radiation-dose effect on solar cells results in their becoming gradually less efficient, and eventually their power output drops below the minimum necessary to operate all of the vehicle subsystems. The total fluence from one very large solar proton event can double the cumulative exposure for a solar cycle (see Tables 9 and 10), and the output of solar-cell power systems can be significantly degraded, reducing the operational lifetime of a space vehicle by years (Marvin and Gorney, 1991).

7.4.2 Soft Errors in Electronics Modern semiconductor elements are so small (feature sizes less than a micron) and the currents in an individual element so low that the number of electrons involved in a change of state (from off to on) is of the same order of magnitude as the number of electrons generated by the passage of an energetic ion through semiconductor material. This phenomenon is called "single-event upset" or "soft error" since the device will continue to operate correctly even though the information it contains is now incorrect. Semiconductor devices that are vulnerable to these types of errors are called "soft" devices, and semiconductor elements that are immune to soft errors are called "hard" devices.

There are two primary mechanisms for causing soft errors in space-borne semiconductor electronics. Energetic heavy ions, such as ^{26}Fe, passing through the sensitive volume of a device liberate enough electrons to cause a change of state in a logic element. (Engineers use the term "linear energy transfer" to compute the probability of soft errors.) The other mechanism is that an energetic nucleus has a direct collision with an atom, and in its recoil, the energy imparted to the atom by the collision is converted to enough electrons to cause a change of state in a logic element. Selection of space-qualified parts to minimize the soft-error effect is a major challenge in space-vehicle design.

7.4.3 Solar-Particle Radiation Effects on Human Space Flight The free-space dose that might be experienced by an unshielded astronaut during the largest recorded solar-proton events would be debilitating (exceeding the threshold for inducing nausea) and perhaps potentially lethal. Studies of the probable effects of solar-particle radiation of space mission scenarios to the Moon or Mars recommend that a location with a minimum of 20 $g \cdot cm^{-2}$ shielding be included in the spacecraft to reduce the effects of large energetic solar-particle events to tolerable levels.

The orbits of the Russian MIR space station and the projected International Space Station have been selected partly to minimize the solar-particle radiation risk by taking advantage of the geomagnetic shielding effect offered by the Earth's magnetic field and still permit maximum coverage of the Earth's surface. The Russian MIR space station has the longest actual history of radiation measurements on manned spacecraft during large energetic solar-particle events. The very energetic solar-particle event of 29 September 1989 (evaluated as the third largest in the history of cosmic radiation observations) resulted in a measured radiation exposure to the Cosmonauts in the MIR space station of 0.005 Gy (0.5 REM). Following the extremely intense solar-particle–interplanetary-shock event that occurred on 20 October 1989, during the intense geomagnetic storm that reduced the normal geomagnetic shielding, the Cosmonauts in the MIR space station experienced an additional exposure of 0.015 Gy (1.5 REM).

These large energetic solar-particle events pose no significant hazard to ordinary commercial aviation at altitudes below 40 000 ft. At very high altitudes (in the 50 000–70 000-ft range) of the Concorde SST, there is a possibility that the radiation exposure could exceed the recommended levels for the general public, and these aircraft carry radiation warning devices in case of such an occurrence. Only twice in the operational history of the SST has this occurred, but radiation intensity never approached the hazard level. (On 29 September 1989, and again on 20 October 1989, low-level warnings were registered.)

8. SUMMARY

The Sun is the source of energy that permits life on Earth. It has only been during the past century or so that we have been able to identify some of the more elusive "causes and effects" of solar activity and solar emissions on our lives. Much of our knowledge emanates from our increased technology and our dependence on this technology. With progressive miniaturization of electronic components used in space and on earth, and the increased susceptibility of these components to charged particles and current systems, it is imperative that we understand solar emissions and interplanetary phenomena that perturb our spatial and terrestrial environment.

In the infrared and visible wavelengths, the solar output is relatively constant with small identifiable changes occurring during the solar-activity cycle. However, our current technology is not adequate to provide the precision we desire about the solar constant. The solar electromagnetic radiations are more variable at both ends of the solar electromagnetic emission spectrum. At long wavelengths (radio frequencies) sporadic large solar outbursts occur that are associated with solar-flare activity, the frequency of which is directly related to the solar-activity cycle. At short wavelengths, the frequency of UV and x-ray outbursts that affect the electron density in the Earth's ionosphere and hence its electromagnetic propagation properties are again related to the solar-activity cycle. Unfortunately, the solar-activity cycles have not proven to be very predictable.

It is in the solar plasma and solar-particle emission that the most significant concept changes have occurred. Solar energetic-particle emission constitutes an operational constraint on human space activities. This includes limiting the lifetime of space-vehicle solar-cell power systems and interference with semiconductor logic operations. The occurrence of these solar plasma and particle emissions is sporadic and currently not very predictable.

Research on the solar-wind plasma and embedded magnetic field continues in order to understand better this method of energy transport to the Earth's environment. The energy involved in a coronal mass ejection is of the same order of magnitude as a large solar flare, and those CMEs that are directed toward the Earth impart a large amount of energy into the Earth's magnetosphere, resulting in the terrestrial disturbances once thought to be the result of solar-flare activity. These magnetospheric perturbations, which often are manifested as geomagnetic storms, interfere with navigation systems (particularly those employing magnetic direction-finding systems for orientation) and can temporarily disrupt communication links. The energy input to the Earth's upper atmosphere can change the atmospheric density and affect satellite operations. In extreme cases, a major geomagnetic storm can interfere with and even disrupt electrical power grids.

The very nature of the effect of solar emissions on humans and their environment is worldwide in scope. In reviewing the available information for this article, it was obvious that the research results and technological advances over the past quarter century clearly emphasize the interrelationships of what were once considered separate and independent phenomena. As our technology advances we will continue to learn more about the complexities of solar emissions on the solar–terrestrial system.

GLOSSARY

Active Region (AR): A localized, transient volume of the solar atmosphere in which plages, sunspots, faculae, flares, etc. may be observed. Active regions are the result of enhanced magnetic fields; they are at least bipolar and may be complex if the region contains two or more bipolar groups.

Alfvén Wave: A transverse wave in magnetized plasma characterized by a change of direction of the magnetic field (rather than a change of intensity).

Anisotropy: The ratio of the maximum to the average particle-flux distribution as a function of angle.

Archimedean Spiral: A mathematical curve resulting from a linear angular rotation with increasing distance.

Astronomical Unit (AU): The average distance from the Sun to the Earth, 149.6×10^6 km.

Chromosphere: The layer of the solar atmosphere above the photosphere and beneath the transition region and the corona. It is the source of the strongest lines in the solar spectrum, including the Balmer alpha line of hydrogen (Hα) and the *H* and *K* lines of calcium. It is the source of the red (chromium) color often seen around the rim of the Moon at total solar eclipses.

Chromospheric Brightening: A literal description of the "solar flare" that is optically visible as a brightening of a small area of the solar chromosphere.

Corona: The outermost layer of the solar atmosphere, characterized by low densities ($<10^{-9}$ cm^{-3} or 10^{-15} m^{-3}) and high temperatures ($>10^6$ K).

Coronagraph: An optical device that makes it possible to observe the corona at times other than during an eclipse. An occulting disk blocks light from the solar disk, effectively providing an artificial eclipse.

Coronal Hole: An extended region of the corona, exceptionally low in density and associated with unipolar photospheric regions having "open" magnetic field topology. Coronal holes are largest and most stable at or near the solar poles and are a source of high-speed solar wind. Coronal holes are visible in several wavelengths. They are most notable in solar soft x rays and can be found in the He 1083-nm emission.

Coronal Loops: A typical structure of enhanced corona observed in EUV lines and soft x rays. They are sometimes related to Hα loops. Coronal loops represent "closed" magnetic topology.

Coronal Mass Ejection (CME): A transient outflow of plasma from or through the solar corona. The CMEs are often, but not always, associated with erupting prominences, disappearing solar filaments, and flares.

Differential Particle Flux: The energetic particle flux at energy *E* per unit energy interval, per unit area, per unit time, per unit solid angle of observation, passing through an area perpendicular to the viewing direction. The directional differential flux of solar particles is usually specified in units of $(\mathrm{cm}^2 \cdot \mathrm{s} \cdot \mathrm{sr} \cdot \mathrm{MeV})^{-1}$. The omnidirectional differential flux is usually specified in units of $(\mathrm{cm}^2 \cdot \mathrm{s} \cdot \mathrm{MeV})^{-1}$.

Differential Rotation: The change in solar rotation rate with latitude. Low latitudes rotate at a faster angular rate (~14°/day) than do high latitudes (~12°/day).

Disappearing Solar Filament (DSF): An English term for the sudden disappearance of a solar filament (prominence) on a time scale of minutes to hours. The DSFs are a possible indicator of coronal mass ejections. The proper name of the phenomenon is *disparition brusque* given by French astronomers.

Extreme Ultraviolet (EUV): The portion of the electromagnetic spectrum from approximately 10 to 100 nm.

Facula: White-light plage. A bright region of the photosphere seen in white light, seldom visible, except near the solar limb. Corresponds with concentrated magnetic fields that may precede sunspot formation.

Favorable Propagation Path: A concept suggesting that the Archimedean spiral path from the Earth to the Sun would connect to a specific solar longitude. It is based on the principle that charged particles travel along the interplanetary magnetic field lines that are transported out from the Sun. For an idealized constant-speed solar-wind flow, if the interplanetary magnetic field is "frozen" in the plasma, then the result would form an Archimedean spiral.

Fibril: A linear feature in the Hα chromosphere of the Sun, occurring near strong sunspots and plage or in filament channels. Fibrils parallel strong magnetic fields, as if mapping the field direction.

Filament: A mass of gas suspended over the chromosphere by magnetic fields and seen as a dark ribbon threaded over the solar disk. A filament on the limb of the Sun seen in emission against the dark sky is called a prominence. Filaments occur directly over magnetic-polarity inversion lines, unless they are active.

First Ionization Potential: The energy required to remove the first electron from an electrically neutral atom. (The ionization potential is usually given in electron volts.)

Flare: See Solar Flare.

Fluence: The time-integrated flux throughout a solar-particle event or a specific time period.

Flux: The rate of particle flow through a reference surface. The directional flux is usually specified in units of $(cm^2 \cdot s \cdot sr)^{-1}$. The omnidirectional flux is usually specified in units of $(cm^2 \cdot s)^{-1}$.

Granulation: Cellular structure of the photosphere visible at high spatial resolution. Individual granules, which represent the tops of small convection cells, are 200–2000 km in diameter and have lifetimes of 8–10 min.

Gray (Gy): The new international system unit (SI unit) of absorbed dose of radiation. 1 Gy = 1 $J \cdot kg^{-1}$ = 100 rad.

Hα: The first atomic transition in the hydrogen Balmer series, with wavelength 656.3 nm. This absorption line of neutral hydrogen falls in the red part of the visible spectrum and is convenient for solar observations. The Hα line is universally used for patrol observations of solar flares, filaments, prominences, and the fine structure of active regions.

Heavy Ions: Nuclei of elements such as nitrogen, carbon, oxygen, or iron that are positively charged because some or all of the planetary electrons have been stripped from them.

Heliocentric: A measurement system with its origin at the center of the Sun.

Heliocentric Angular Distance: A measurement of angular distance from the center of the Sun.

Heliographic: Referring to coordinates on the solar surface referenced to the solar rotational axis. (See **Solar Coordinates.**)

Heliosphere: The plasma domain surrounding the Sun and dominated by the solar-wind outflow. The heliospheric boundary, where the solar-wind flow is hypothesized to change from supersonic flow to subsonic flow, is estimated to be at >100 AU.

Integral Particle Flux: The number of particles of energy equal to or greater than E, per unit area, per unit solid angle, per unit time, passing through an area perpendicular to the viewing direction. The directional integral flux is specified in units of $(cm^2 \cdot s \cdot sr)^{-1}$. The omnidirectional integral flux is specified in units of $(cm^2 \cdot s)^{-1}$.

Interplanetary Magnetic Field (IMF): The magnetic field in interplanetary space transported out from the Sun via the solar wind. The average magnitude of the IMF at 1 AU is ~5 nT.

Interplanetary Shock: An abrupt change in velocity or density that is moving faster than the wave-propagation speed in interplanetary space.

Mean Free Path: The distance between scattering events in interplanetary particle propagation.

Photosphere: The lowest visible layer of the solar atmosphere; corresponds to the solar surface viewed in white light. Sunspots and faculae are observed in the photosphere.

Plage: On the Sun, an extended emission feature of an active region that is seen from the time of emergence of the first magnetic flux until the widely scattered remnant magnetic fields merge with the background. Magnetic fields are more intense in plages, and temperatures are higher than in the surrounding, quiescent regions.

Pore: A feature of the photosphere, 1–3 arc sec in extent, usually not much darker than the dark spaces between photospheric granules. It is distinguished from a sunspot by its short lifetime, 10–100 min.

Proton Event: A generic term referring to an enhancement in the energetic-particle flux observed in space, usually as a result of solar activity.

Rigidity: The momentum per unit charge of a charged particle. Rigidity is usually specified in units of MV (10^6 V) or GV (10^9 V).

Sievert (Sv): The new international system unit (SI unit) of biologically effective radiation absorbed dose. 1 Sv = 1 Gy times dose equivalent (Q). 1 Sv = 100 REM.

Shock: A discontinuity in pressure, density, and particle velocity, propagating through a compressible fluid or plasma.

Smoothed Sunspot Number: A statistical 13-month running mean of the sunspot numbers, centered on the month of concern. The 1st and 13th months are given weights of 0.5.

Solar Coordinates: Specifications for a location on the solar surface. The location of a specific feature on the Sun (for example, a sunspot) is complicated by the fact that there is a tilt of 7.25° between the ecliptic plane and the solar equatorial plane as well as a true wobble of the solar rotational axis.

Solar Cycle: The solar cycle refers to the solar-activity cyclic behavior, usually represented by the number of sunspots visible on the solar photosphere. The average length of solar cycles since 1900 is 11.4 yr.

Solar Flare: The name given to the sudden release of energy (often more than 10^{32} ergs) in a relatively small volume of the solar atmosphere. Historically, an optical brightening in the chromosphere, now expanded to cover almost all impulsive radiation from the Sun.

Solar Flare Importance: The classification of a solar flare based on its measured optical area (measured in millionths of the solar disk) and brightness at the time of maximum brightness in the Hα line, by a number–letter combination. Solar flares are ranked in area as importance 0 (or *s*), 1, 2, 3, to a maximum of 4. Solar flares are subjectively ranked in brightness as *f* (faint), *n* (normal), or *b* (bright).

Solar Maximum: The month(s) during the sunspot cycle when the smoothed sunspot number reaches a maximum. The maximum for solar cycle 22 occurred in July 1989.

Solar Minimum: The month(s) during the sunspot cycle when the smoothed sunspot number reaches a minimum. The recent solar minimum between the 21st and 22nd solar cycles occurred in September 1986.

Solar Rotation Rate: Relative to the Earth (synodic), $(13.39° - 2.7° \sin^2\theta)$/day. Relative to an inertial system (sidereal), $(14.38° - 2.7° \sin^2\theta)$/day ($\theta$ = solar latitude). The difference between sidereal and synodic rates is the Earth orbital motion of 0.985°/day.

Solar Wind: The plasma flowing out into space from the solar corona. Typically at 1 AU, the average solar-wind velocity is ~375 $km \cdot s^{-1}$, and the average density is ~5 cm^{-3}.

Spicules: Rapidly changing, predominantly vertical, spikelike structures in the solar chromosphere observed about the limb. Spicules appear to be ejected from the low chromosphere at velocities of 20–30 $km \cdot s^{-1}$, reaching a height of about 9000 km and then falling back or fading. The total lifetime is 5–10 min.

Spray: Luminous material ejected from a solar flare with sufficient velocity to escape the Sun (675 $km \cdot s^{-1}$). There is little evidence that sprays are focused by magnetic fields. Compare with **Surge.**

Subsonic Flow: Plasma flowing slower than the local speed of sound in the medium.

Sunspot: An area seen as a dark spot, in contrast with its surroundings, on the photosphere of the Sun. Sunspots are concentrations of magnetic flux, typically occurring in bipolar clusters or groups. They appear dark because they are cooler than the surrounding photosphere. Larger and darker sunspots sometimes are surrounded (completely or partially) by penumbrae. The dark centers are umbrae. The smallest, immature spots are sometimes called pores.

Sunspot Cycle: The approximately 11 yr quasiperiodic variation in the sunspot number. Other solar phenomena, such as the 10.7-cm solar radio emission, exhibit similar cyclical behavior. The polarity of the solar magnetic field reverses during the maximum phase of each cycle.

Sunspot Number: A quantification of the number of sunspots and sunspot groups visible on the Sun according to the International Astronomical Union (IAU) definitions of sunspot measurement. This number is often used as a proxy of solar activity. The daily sunspot index is identified as *A*. The sunspot index *R* is defined as $R = k(10g + s)$, where s = number of individual spots meeting the IAU criterion, g = number of sunspot groups, and k is an observatory and/or observer factor. The standard sunspot number was for many years derived at Zürich and was referred to as R_z. It is now being derived at Brussels and is called the International sunspot number R_I. Often, the term "sunspot number" is used in reference to the widely distributed smoothed sunspot number.

Supergranulation: A system of large-scale velocity cells that does not vary significantly over the quiet solar surface or with phases of the solar cycle. The cells are presumably convective in origin with weak upward motions in the center, downward motions at the borders, and horizontal motions of typically 0.3 to 0.4 km s^{-1}. Magnetic flux is more intense along the borders of the cells.

Supersonic Flow: Plasma flowing faster than the local speed of sound in the medium.

Surge: A jet of material from active regions that reaches coronal heights and then either fades or returns into the chromosphere along the trajectory of ascent. Surges typically last 10–20 min and tend to recur at a rate of approximately 1/h. Surges are linear and collimated in form, as if highly directed by magnetic fields. Compare with **Spray.**

Ultraviolet (UV): That part of the electromagnetic spectrum between 5 and 400 nm.

Well Connected: A concept suggesting that the Archimedean spiral path from the specified position in space would connect to a specific solar longitude. It is based on the concept that charged particles travel along the interplanetary magnetic field lines that are transported out from the Sun.

X-ray Flare Class: Rank of a solar flare based on its x-ray energy output. Solar x-ray flares are classified by the NOAA Space Environment Services Center by a letter–number combination denoting the magnitude of the peak x-ray burst intensity measured at the Earth in the 0.1–0.8-nm band. See Table 6.

Works Cited

Bahcall, J. N. (1990), *Sci. Am.* **262** (5), 54–61.

Breneman, H. H., Stone, E. C. (1985), *Astrophys. J.* **299,** 959–963.

Cliver, E., Reames, D., Kahler, S., Cane, H. (1991), in: *22nd International Cosmic Ray Conference,* Vol. 3, Dublin: Dublin Inst. for Adv. Studies, pp. 25–28.

Cox, A. N., Livingston, W. C., Matthews, M. S. (Eds.) (1991), *Solar Interior and Atmosphere,* Tucson, AZ: University of Arizona Press.

Donnelly, R. P. (Ed.) (1992), *Proceedings of the Workshop on the Solar Electromagnetic Radiation Study for Solar Cycle 22,* Boulder, CO: ERL, NOAA, DOC.

Eddy, J. A. (Ed.) (1978), *The New Solar Physics,* Boulder, CO: Westview Press for the AAAS.

Feynman, J., Armstrong, T. P., Dao-Gibner, L., Silverman, S. (1990), *J. Spacecraft Rockets* **27** (4), 403–410.

Feynman, J., Gabriel, S. (Eds.) (1988), *Interplanetary Particle Environment,* JPL Publication 88-28, Pasadena, CA: Jet Propulsion Laboratory, California Institute of Technology.

Feynman, J., Spitale, G., Wang, J., Gabriel, S. (1993), *J. Geophys. Res.* **98,** 13281–13294.

Foukal, P. V. (1990), *Sci. Am.* **262** (2), 34–41.

Gloeckler, G., Geiss, J. (1989), in: C. J. Waddington (Ed.), *Cosmic Abundances of Matter,* AIP Conference Proceedings No. 183, New York: American Institute of Physics, p. 49.

Gosling, J. T. (1990), in: C. T. Russell, E. R. Priest, L. C. Lee (Eds.), *Physics of Magnetic Flux Ropes,* Geophysical Monographs No. 58, Washington, DC: American Geophysical Union, p. 343.

Gosling, J. T. (1993), *J. Geophys. Res.* **98,** 18937–18949.

Hamilton, D. C. (1977), *J. Geophys. Res.* **82,** 2159–2169.

Hinteregger, H. E., Fukui, K., Gilson, B. R. (1981), *Geophys. Res. Lett.* **8,** 1147–1150.

Hoeksema, J. T., Scherrer, P. H. (1986), *The Solar Magnetic Field—1976 through 1985,* UAG-94, Boulder, CO: WDC-A, NGDC, NOAA, DOC.

Hudson, H. S. (1988), *Annu. Rev. Astron. Astrophys.* **26,** 473–507.

Lean, J. (1991), *Rev. Geophys.* **29,** 505–535.

Lee, M. A. (1983), *J. Geophys. Res.* **88,** 6109–6120.

Lee, B. L., Gibson, A. M., Wilson, R. S., Thomas, S. (1995), *J. Geophys. Res.* **100,** 1667–1675.

Lingenfelter, R. E., Hudson, H. S. (1980), in: R. O. Pepin, J. A. Eddy, R. B. Merrill (Eds.), *Proceedings of the Conference on the Ancient Sun,* New York: Pergamon Press, pp. 69–79.

Luhn, A., Hovestadt, D., Kleckler, B., Scholer, M., Gloeckler, G., Ipavich, F. M., Galvin, A. B., Fan, C. Y., Fisk, L. A. (1985), in: *19th International Cosmic Ray Conference,* Vol. 4, Greenbelt, MD: NASA Conference Publication 2376, pp. 241–244.

Marvin, D. C., Gorney, D. J. (1991), *J. Spacecraft Rockets* **29,** 713–719.

McKinnon, J. A. (1987), *Sunspot Numbers: 1610–1985,* UAG-95, Boulder, CO: WDC-A, NGDC, NOAA, DOC.

Mewaldt, R. A., Stone, E. C. (1989), *Astrophys. J.* **377,** 959–963.

Moses, D., Dröge, W., Meyer, P., Evenson, P. (1989), *Astrophys. J.* **346,** 523–530.

Nymmik, R. A. (1993), in: *23rd International Cosmic Ray Conference,* Vol. 3, Calgary: Univ. of Calgary, pp. 29–32.

Oetliker, M., Klecker, B., Hovestadt, D., Scholer, M., Blake, J. B., Looper, M., Mewaldt, R. A. (1995), in: *24th International Cosmic Ray Conference,* Vol. 4, Rome: Universitá La Sapienza, pp. 470–474.

Parker, E. N. (1963), *Interplanetary Dynamical Processes,* New York: Wiley-Interscience.

Quarterly Bulletin on Solar Activity, Zurich: Eidgen, Strenwarte.

Ramaty, R., Mandzhavidze, N. (1993), in: J. T.

Ryan, W. T. Vestrand (Eds.), *High Energy Solar Phenomena—A New Era of Spacecraft Measurements,* AIP Conference Proceedings No. 294, New York: Institute of Physics, p. 26.

Ramaty, R., Murphy, R. J. (1987), *Space Sci. Rev.* **45,** 213–268.

Reames, D. V. (1992), in: *Proceedings of the First SOHO Workshop,* ESA Special Publication No. 348, Paris: European Space Agency, pp. 315–323.

Reames, D. V. (1995), *Rev. Geophys. Suppl.* **33,** 585–589.

Sawyer, C., Warwick, J. W., Dennett, J. T. (1986), *Solar Flare Prediction,* Boulder, CO: Colorado Associated Press.

Shea, M. A., Smart, D. F. (1990), *Sol. Phys.* **127,** 297–320.

Smart, D. F., Shea, M. A. (1971), *Sol. Phys.* **16,** 484–487.

Smart, D. F., Shea, M. A. (1985), in: A. S. Jursa (Ed.), *Handbook of Geophysics and the Space Environment,* Bedford, MA: Air Force Geophysics Laboratory, Hanscom AFB, Chap. 6.

Solar-Geophysical Data, Boulder, CO: NGDC, NOAA, DOC.

Stassinopoulos, E. G., Brucker, G. J., Nakamura, D. W., Stauffer, C. A., Gee, G. B., Barth, J. L. (1996), *IEEE Trans. Nucl. Sci.* **43,** 369–382.

Tobiska, W. K. (1993), *J. Geophys. Res.* **98,** 18879–18893.

van Hollebeke, M. A., Ma Sung, L. S., McDonald, F. B. (1975), *Sol. Phys.* **41,** 189–223.

Webb, D. F. (1995), *Rev. Geophys., Suppl.* **33,** 557–583.

Webb, D. F., Howard, R. W. (1994), *J. Geophys. Res.* **99,** 4201–4220.

White, O. R. (Ed.) (1977), *The Solar Output and Its Variations,* Boulder, CO: Colorado Associated University Press.

Zirin, H. (1988), *Astrophysics of the Sun,* London: Cambridge University Press.

Further Reading

Akasofu, S.-I., Chapman, S. (1972), *Solar-Terrestrial Physics,* London: Oxford University Press.

Bruzek, A., Durrant, C. J. (1977), *Illustrated Glossary for Solar and Solar-Terrestrial Physics,* Dordrecht/Boston: D. Reidel.

Durrant, C. J. (1988), *The Atmosphere of the Sun,* Bristol: Adam Hilger.

Foukal, P. (1990), *Solar Astrophysics,* New York: Wiley Interscience.

Hruska, J., Shea, M. A., Smart, D. F., Heckman, G. (Eds.) (1993), *Solar-Terrestrial Predictions-IV,* Boulder, CO: SEL, NOAA, DOC.

Hundhausen, A. J. (1972), *Coronal Expansion and Solar Wind,* New York: Springer-Verlag.

Kundu, M. R., Woodgate, B., Schmahl, E. J. (Eds.)(1989), *Energetic Phenomena on the Sun,* Dordrecht: Kluwer Academic Publishers.

Jursa, A. S. (Ed.) (1985), *Handbook of Geophysics and the Space Environment,* Bedford, MA: Air Force Geophysics Laboratory, Hanscom AFB.

Noyes, R. W. (1982), *The Sun, Our Star,* Cambridge MA: Harvard University Press.

Reviews of Geophysics (1987), *Contributions in Solar-Planetary Relationships,* Vol. 25 Supplement, Washington, DC: American Geophysical Union.

Reviews of Geophysics (1991), *Contributions in Solar-Planetary Relationships,* Vol. 29 Supplement, Washington, DC: American Geophysical Union.

Reviews of Geophysics (1995), *Contributions in Solar-Planetary Relationships,* Vol. 33 Supplement, Washington, DC: American Geophysical Union.

Schwenn, R., Marsch, E. (Eds.) (1990), *Physics of the Inner Heliosphere I,* New York: Springer-Verlag.

SESC Glossary of Solar-Terrestrial Terms (1990), Boulder, CO: SEL, NOAA, DOC.

Solar Cosmic Rays; Proton Fluxes Model (1986), State Standard 25645. 13-86, Moscow.

Stenflo, J. O. (Ed.) (1990), *Solar Photosphere: Structure, Convection and Magnetic Fields,* Dordrecht: Kluwer Academic Publishers.

Sturrock, P. A., Holzer, T. E., Mihalas, D. M., Ulrich, R. K. (1986), *Physics of the Sun,* Dordrecht/Boston: D. Reidel Publishing Co.

Svestka, S. (1976), *Solar Flares,* Dordrecht/Boston: D. Reidel.

Tandberg-Hanssen, E., Emslie, G. (1988), *The Physics of Solar Flares,* London: Cambridge University Press.

Wentzel, D. G. (1985), *The Restless Sun,* Washington, DC: Smithsonian Institution Press.

Wilson, J. W., Townsend, L. W., Schimmerling, W., Khandelwal, G. S., Khan, F., Nealy, J. E., Cucinotta, F. A., Simonsen, L. C., Shinn, J. L., Norbury, J. W. (1991), *Transport Methods and Interactions for Space Radiations,* NASA Reference Publication 1257, Washington: U.S. GPO.

SOLID-STATE DEVICES

See VLSI—INTEGRATED CIRCUITS; MICROELECTRONICS

SOLID-STATE LASERS

See LASERS, SOLID STATE

SOLID, LIQUID, AND GASEOUS ELECTRICAL INSULATION

MARKUS ZAHN, *Laboratory for Electromagnetic and Electronic Systems, Department of Electrical Engineering and Computer Science, Massachusetts Institute of Technology, Cambridge, Massachusetts, U.S.A.*

3-527-28140-1/96/$5.00 + .50

INTRODUCTION

The essential purpose of electrical insulation is to prevent the flow of electricity. An ideal insulator has zero electrical conductivity and is only characterized by its dielectric permittivity ϵ. However, "real" insulators have finite electrical conduction that leads to power dissipation, and to chemical degradation that causes the insulation to age and ultimately to fail. If the electric field exceeds the electrical breakdown strength of the insulator, spark discharges will occur with large values of current often leading to failure.

The presence of matter between voltage-stressed metallic electrodes often modifies the electric field because even though the material is initially uncharged, the electric field causes charge motion, called conduction, and small charge displacements, called polarization. Because of the large number of atoms present—an Avogadro's number of 6.02×10^{23} per gram molecular weight—slight imbalances in the distribution have large effects on the electric field inside and outside the material. We must then self-consistently account for the electric-field effect on charge motion and redistribution, together with the local net charge as a source of electric field.

If the material polarization and conduction constitutive laws are known, the electric-field distribution throughout the insulation can be calculated. However, often materials contain structural defects and impurities. Charge carriers are also generated from electron, light, or nuclear radiation, from high electric-field strengths that cause charge emission from electrodes, by dielectric ionization, or by contact charging of impurities. With imprecise control or knowledge of such physical mechanisms and material constituents, it often becomes impossible to calculate insulation performance, and many tests are necessary to develop statistics of electric breakdown strength and insulation life.

1. IMPORTANT ELECTRICAL PROPERTIES OF INSULATION

To characterize the electrical insulation properties of an insulator, it is necessary to specify the dielectric permittivity ϵ (or relative dielectric constant $\epsilon_r = \epsilon/\epsilon_0$ with $\epsilon_0 \approx 8.854 \times 10^{-12}$ F/m the permittivity of free space), the electrical resistivity ρ (or its reciprocal, the conductivity $\sigma = 1/\rho$), and the dielectric breakdown strength under dc, ac, and impulse voltages. Table 1 lists represen-

Table 1. Typical values of dielectric strength, resistivity, and relative permittivity of common electrically insulating materials.

	Dielectric strength (MV/m)	Resistivity (Ω m)	Relative permittivity
Gases (STP)			
Air	~3		1.0
Nitrogen	~3		1.0
SF_6	~7.5		1.0
Liquids			
Transformer oil	35	$>10^{12}$	2.2
Pure water (25 °C)	~13	$\sim 1.8 \times 10^5$	78
Pure ethylene glycol (25 °C)	~27	$\sim 9 \times 10^6$	38
Solids			
Bakelite	~24	$>10^{10}$	4.9
Cellulose paper	~10	$>10^{13}$	3.8
Ethylene propylene rubber	30–40	$\sim 10^{13}$	2.5–3.0
Mica	~100	$>10^{16}$	5.4
Polycarbonate	~250	$\sim 10^{14}$	3.0
Polymethylmethacrylate	~350	$>10^{13}$	3.5
Polystyrene	~200	$\sim 10^{14}$	2.5
Polyethylene (25 °C)	~570	$>10^{14}$	2.2
Polypropylene	~300	$\sim 10^{16}$	2.2
Porcelain	~10	$>2 \times 10^{11}$	5–7
Glass (Pyrex)	~17	$>2 \times 10^{11}$	5
Barium titanate	~5	$>10^9$	1200

tative values for typical insulating gases, liquids, and solids.

The study of electrical insulation is interdisciplinary, as the insulation properties depend strongly on electrical, thermal, physical, and chemical parameters. The fundamental limit for most high-voltage systems is the insulation strength—that is, the highest voltage that can be safely sustained without spark discharges.

Gaseous and liquid insulation offer the advantage of filling all available space and of often being self-healing after a spark discharge as the debris is dispersed or carried away by a flow. Liquids also increase the rate of heat transfer by their high thermal conductivity and by convection. Transformer oil is about 20–30 times more effective than air in removing the heat generated in the windings and magnetic core of a transformer. A solid offers the advantage of mechanical strength but often has voids that are not self-healing after electrical breakdown, because the insulation is punctured, leaving a carbonized discharge channel that is conducting.

Even vacuum is not a perfect insulator and, in a real system, is really a gas at very low pressure. Electron emission from the cathode due to microscopic particles or protrusions releases dissolved and absorbed gas on the electrodes and causes heating that slightly melts the metal electrode. Eventually, sufficient vapor is produced so that a low-pressure gas discharge can result if sufficient voltage is applied. For a typical ~1-mm gap at a pressure of $\sim 10^{-3}$ Torr, the breakdown field strength is of order E_{bd} ~375 kV/cm, whereas for typical liquids and solids the working breakdown strength is of order 100–200 kV/cm, although thin-film solids in laboratory measurements can have breakdown field strengths that exceed 1 MV/cm.

The initiation of spark discharges is generally by charge injection from the electrodes. Such discharges contribute to thermal, oxidative, radiative, mechanical, and chemical degradation of insulation that lead to aging and loss of life. Thermal failure occurs when the heat generated by Joule heating and by other dielectric losses cannot be easily lost to the surroundings. The temperature increases, which further increases dielectric loss, leading to thermal runaway. For most practical applications it is important to estimate insulation life from sets of measurements at elevated voltage stresses to measure the time to failure under accelerated aging conditions. Various statistical models are used to estimate the probability of in-service insulation failure from laboratory tests.

2. REPRESENTATIVE ANALYTICAL DESCRIPTIONS OF ELECTRICAL INSULATION

2.1 Linear Lossy Dielectrics

2.1.1 Governing Equations In most electrical insulation applications, the frequency or time rates of change of electric field are sufficiently slow that Maxwell's equations are well approximated in their electroquasistatic limit. Then, electrically insulating materials are essentially capacitive with the material characterized by its dielectric permittivity ϵ. Maxwell's equations can then be written as

$$\nabla \times \mathbf{E} = 0 \Rightarrow \mathbf{E} = -\nabla \Phi, \qquad (1)$$

$$\nabla \cdot \mathbf{E} = \rho/\epsilon \Rightarrow \nabla^2 \Phi = -\rho/\epsilon, \qquad (2)$$

$$\nabla \cdot \mathbf{J} + \frac{\partial \rho}{\partial t} = 0, \qquad (3)$$

where for simplicity we take ϵ to be a constant in the region of interest. The irrotational electric field **E** in (1) comes from Faraday's law with negligible magnetic field, which allows the definition of the electric scalar potential Φ. In (2), Gauss's law relates the electric field to the source volume charge density ρ. Together with (1), (2) shows that the potential Φ obeys Poisson's equation. Conservation of charge in (3) relates the current density **J** to the time rate of change of charge density.

If a pair of electrodes at voltage v enclose a dielectric medium, even with ϵ as a function of position, the total energy stored in the capacitor is

$$W = \int_{\text{all space}} \tfrac{1}{2}\,\epsilon |\mathbf{E}|^2 \, dV = \tfrac{1}{2} C v^2, \qquad (4)$$

where C is the capacitance between electrodes.

2.1.2 Ohmic Material For electrical insulation applications, it is usually desired that the current density $\mathbf{J}$ be as small as possible. Many materials are accurately described by an Ohmic constitutive law where σ is the Ohmic conductivity:

$$\mathbf{J} = \sigma \mathbf{E} \Rightarrow v = iR, \tag{5}$$

which is the field form of the familiar circuit Ohm's-law proportionality relating voltage v to current i via resistance R.

A homogeneous lossy dielectric between a pair of electrodes at voltage difference v has a circuit model of resistance R in parallel with capacitance C such that

$$RC = \epsilon/\sigma, \tag{6}$$

where the RC product is independent of voltage, electrode shape, area, or spacing. The power dissipated in the system is

$$P = \int_{\text{all space}} \sigma|\mathbf{E}|^2\, dV = i^2R = v^2/R. \tag{7}$$

2.1.3 Transient Charge Relaxation Substituting Ohm's law of (5) into (3) for constant conductivity σ yields

$$\sigma\nabla\cdot\mathbf{E} + \frac{\partial\rho}{\partial t} = 0 \Rightarrow \frac{\partial\rho}{\partial t} + \frac{\sigma}{\epsilon}\rho = 0, \tag{8}$$

where the right-hand equation eliminates $\nabla\cdot\mathbf{E}$ using (2). The solution to (8) is

$$\rho(x,y,z,t) = \rho(x, y, z, t = 0)e^{-t/\tau}, \tag{9}$$

where $\tau = \epsilon/\sigma$ is called the dielectric relaxation time and equals from (6) the circuit RC time constant. This solution shows that if any region of space has an initial volume charge distribution $\rho(x, y, z, t = 0)$, it will exponentially decay with time constant τ. If an Ohmic region has no initial volume charge so that $\rho(x, y, z, t = 0) = 0$, then the volume charge density remains zero in that region thereafter.

2.1.4 Interfacial Charge

2.1.4.1 Step Transient. Surface charge can accumulate on the interface between two dissimilar lossy dielectric materials. This is illustrated for the series lossy capacitor shown in Fig. 1, which is modeled by a pair of parallel RC circuits connected in series. We consider a step voltage applied at $t = 0$ to an initially unexcited system. Neglecting end effects, the electric field can only be in the x direction and at best vary with that coordinate. In the absence of any volume charge ($\rho = 0$), Eq. (2) for the electric field requires $dE_x/dx = 0$ so that the electric fields in each region, E_1 and E_2, are uniform in space within each lossy dielectric but can vary with time.

The interfacial form of (3) gives the required conservation of charge boundary condition at the interface as

$$J_2 - J_1 + \frac{\partial\sigma_s}{\partial t} = 0; \quad \sigma_s = \epsilon_2E_2 - \epsilon_1E_1, \tag{10}$$

where σ_s is the interfacial surface charge density, which equals the difference in perpendicular displacement fields ($\mathbf{D} = \epsilon\mathbf{E}$) across the interface.

At $t = 0_-$, the interfacial surface charge density is zero and remains continuous at $t = 0_+$ so that $D = \epsilon_1E_1 = \epsilon_2E_2$, while in the dc steady state as $t \to \infty$, Eq. (10) requires that the current density $J = \sigma_1E_1 = \sigma_2E_2$ be continuous across the interface.

The solutions for the electric fields in each region are then

$$E_1 = \frac{\sigma_2 v}{\sigma_2 a + \sigma_1 b}(1 - e^{-t/\tau}) + \frac{\epsilon_2 v}{\epsilon_2 a + \epsilon_1 b}e^{-t/\tau},$$
$$E_2 = \frac{\sigma_1 v}{\sigma_2 a + \sigma_1 b}(1 - e^{-t/\tau}) + \frac{\epsilon_1 v}{\epsilon_2 a + \epsilon_1 b}e^{-t/\tau}, \tag{11}$$

where the time constant

$$\tau = \frac{\epsilon_1 b + \epsilon_2 a}{\sigma_1 b + \sigma_2 a} \tag{12}$$

is a thickness-weighted average of relaxation times in each material. The interfacial surface charge density is

$$\sigma_s = \epsilon_2E_2 - \epsilon_1E_1 = \frac{(\epsilon_2\sigma_1 - \epsilon_1\sigma_2)v}{\sigma_2 a + \sigma_1 b}(1 - e^{-t/\tau}), \tag{13}$$

which is zero at $t = 0$ and increases to a

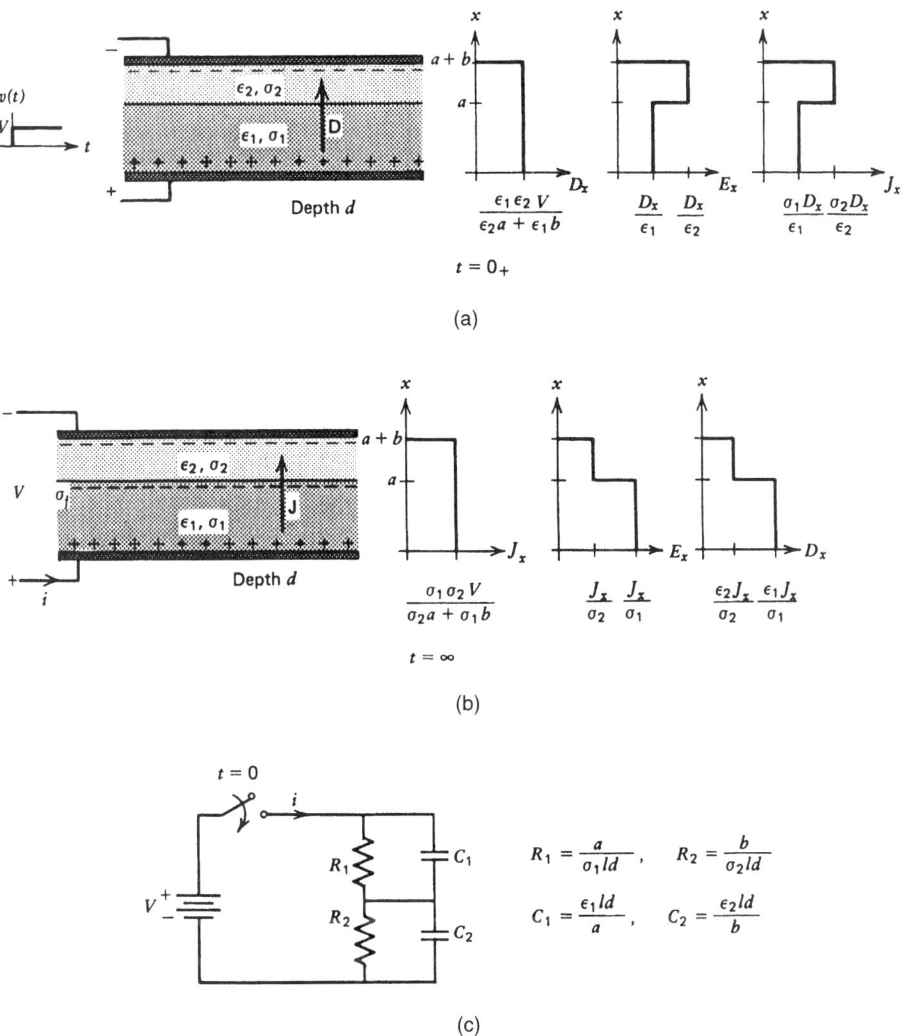

FIG. 1. Two different lossy dielectric materials in series between parallel plate electrodes have permittivities and Ohmic conductivities that change abruptly across the interface. (a) At $t = 0_+$, right after a step voltage is applied, the interface is uncharged so that the displacement field $D = \epsilon_1 E_2 = \epsilon_2 E_2$ is continuous with the solution the same as for two lossless dielectrics in series. (b) Since the current is discontinuous across the boundary between the materials, the interface will charge up. In the dc steady state, the current density $J = \sigma_1 E_1 = \sigma_2 E_2$ is continuous. (c) Each region is circuit equivalent to a resistor and capacitor in parallel.

positive steady-state value if $\epsilon_2/\sigma_2 > \epsilon_1/\sigma_1$, decreases to a negative steady-state value if $\epsilon_2/\sigma_2 < \epsilon_1/\sigma_1$, or remains zero for all time if $\epsilon_2/\sigma_2 = \epsilon_1/\sigma_1$.

2.1.4.2 Sinusoidal Steady State. If, rather than a step voltage, the applied voltage is sinusoidal, the electric fields in each region will also vary sinusoidally with time:

$$v(t) = Re\{\hat{v}e^{j\omega t}\} \Rightarrow E_1(t) = Re\{\hat{E}_1 e^{j\omega t}\}, E_2(t) = Re\{\hat{E}_2 e^{j\omega t}\}. \quad (14)$$

In (14) it is convenient to use complex notation with careted symbols representing complex amplitudes. Then (10) requires that the sum of conduction and displacement current densities be continuous across the interface:

$$(\sigma_2 + j\omega\epsilon_2)\hat{E}_2 = (\sigma_1 + j\omega\epsilon_1)\hat{E}_1. \quad (15)$$

The electric field complex amplitudes are then related as

$$\frac{\hat{E}_2}{(j\omega\epsilon_1 + \sigma_1)} = \frac{\hat{E}_1}{(j\omega\epsilon_2 + \sigma_2)} = \frac{\hat{v}}{[b(\sigma_1 + j\omega\epsilon_1) + a(\sigma_2 + j\omega\epsilon_2)]}, \tag{16}$$

and the interfacial surface charge-density complex amplitude is

$$\hat{\sigma}_s = \epsilon_2\hat{E}_2 - \epsilon_1\hat{E}_1 = \frac{(\epsilon_2\sigma_1 - \epsilon_1\sigma_2)\hat{v}}{[b(\sigma_1 + j\omega\epsilon_1) + a(\sigma_2 + j\omega\epsilon_2)]}. \tag{17}$$

2.2 Charge-Injection Models

2.2.1 Schottky Emission Schottky emission is essentially thermionic emission from a metal electrode into the conduction band of a dielectric taking into account the lowering of the work-function potential barrier by the image force on the injected electron of charge $e = 1.6 \times 10^{-19}$ Coulombs. The injected current density at an absolute temperature T with an electric field E into a dielectric of permittivity ϵ is

$$J = AT^2 \exp\{-[\phi - (e^3E/4\pi\epsilon)^{1/2}]/kT\}, \tag{18}$$

where ϕ is the work function, m is the effective electron mass, $k = 1.38 \times 10^{-23}$ J/K is Boltzmann's constant, $h = 6.6256 \times 10^{-34}$ J s is Planck's constant, and

$$A = 4\pi emk^2/h^3 \sim 120 \text{ A cm}^{-2}\text{ K}^{-2} \tag{19}$$

is the Richardson–Dushman constant of thermionic emission.

2.2.2 Fowler–Nordheim Field Emission Quantum mechanical tunneling of electrons from a metal surface into a dielectric under a strong electric field results in the current density

$$J = \frac{e^3E^2}{8\pi h\phi} \exp[-8\pi\sqrt{2m}\ \phi^{3/2}/3heE]. \tag{20}$$

2.3 Avalanche Modeling

The fundamental model of electric breakdown is the avalanche model, where an injected electron collides with a molecule to release an additional electron. Each of these electrons then continues with more collisions, multiplying the number of electrons and leading to current growth. With α, known as Townsend's first ionization coefficient, denoting the number of ionizing collisions per electron per unit distance, the increase in the number density n of electrons in a distance dx is

$$dn = n\alpha\, dx \Rightarrow n = n_0e^{\alpha x}; \quad \alpha = e^{-\nu/eE\lambda}/\lambda, \tag{21}$$

resulting in the number density of electrons exponentially increasing with distance, with n_0 the original number of electrons at the cathode, $x = 0$. The Townsend coefficient α increases with increasing electric field E, depends strongly on the mean free path λ between collisions, and is characterized by the energy ν required to ionize gas molecules. The conduction current flow through the anode a distance d from the cathode is then

$$I(x = d) = I(x = 0)e^{\alpha d}, \tag{22}$$

where $I(x = 0)$ is the cathode emission current. Such impact ionization by electrons alone is not sufficient to cause electrical breakdown.

As the fast-moving electrons enter the anode, they leave behind the relatively slower positive ions. As these positive ions drift to the cathode, they also collide with neutral molecules and liberate additional electrons with probability γ, known as Townsend's second ionization coefficient. The parameter γ also includes the effects of additional free electrons generated by photoionization processes.

The anode current then becomes

$$I(x = d) = I(x = 0)\frac{e^{\alpha d}}{1 - \gamma(e^{\alpha d} - 1)}. \tag{23}$$

The current grows without bound when the denominator in (23) becomes zero, so that the discharge becomes self-sustaining when

$$\gamma(e^{\alpha d} - 1) = 1. \tag{24}$$

Electronegative gases, such as SF_6, have an electron attachment η, so that Townsend's breakdown criterion of (24) is modified to

$$\frac{\gamma\alpha}{\alpha - \eta}\left[e^{(\alpha-\eta)d} - 1\right] = 1, \tag{25}$$

which requires a larger value of α and thus a larger electric field before the equality is satisfied.

With the mean free path λ inversely proportional to pressure, the Townsend coefficients α and γ are functions of E/p, where E is the electric field and p is the gas pressure. This model is appropriate for gases, but it is also used to describe electrical breakdown in liquids with formation of a gaseous bubble (perhaps due to heating) and in solids that have voids. In a uniform electric field, the curve of gas breakdown voltage versus the pressure–gap (pd) product is known as the Paschen curve, as shown in Fig. 2. For a gap of fixed spacing d, as the pressure decreases, the gas density decreases and the electron mean free path between collisions increases. Thus, each electron makes fewer collisions with gas molecules as it travels to the anode and thus avoids loss of energy due to collisions. A lower electric field can then accelerate electrons to sufficient kinetic energy for ionizing collisions. As the pressure further decreases, the breakdown voltage decreases to a minimum value, v_p, known as the Paschen minimum. As the density is decreased below the Paschen minimum, there are even fewer collisions, so that self-sustaining ionization is maintained only with a corresponding increase in voltage. To the right of the Paschen minimum, there are so many collisions that much of the electron kinetic energy is dissipated so that a large voltage is necessary for ionizing collisions.

2.4 Statistical Methods

To model insulation behavior, three statistical distributions are usually used. The normal distribution, also known as the Gaussian distribution or bell-shaped curve, has the probability $F(x)$ of a value of a measured property x given by

$$F(x) = \frac{1}{\sqrt{2\pi\sigma}} \exp\left[-\frac{(x-\mu)^2}{2\sigma^2}\right], \tag{26}$$

where μ is the mean of x and σ is the standard deviation. The value of μ gives the center of the distribution peak, and σ determines the width of the distribution and describes the scatter of data. This is the simplest statistical method and is typically used to describe such insulation properties as dielectric constant, dissipation factor, insulation resistance, tensile strength, and elongation.

The log normal distribution is described as

$$F(x) = \frac{1}{\sqrt{2\pi\sigma}} \exp\left[-\frac{(\log x-\mu)^2}{2\sigma^2}\right], \tag{27}$$

where μ is now the logarithmic mean and σ is the logarithmic standard deviation. The logarithm can be base e or base 10. This distribution is typically used for such insulation

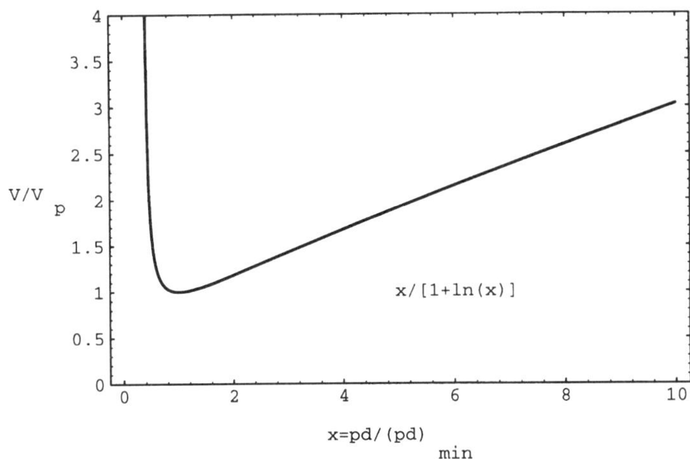

FIG. 2. The Paschen curve describes the voltage breakdown strength in gases as a function of the pressure-gap (pd) product. The nondimensional function given in the plot with voltage and pressure–gap product normalized to the Paschen minimum approximates measured curves, as described in von Hippel (1959).

properties as corrosion, gaseous diffusion, chemical reaction rates, and time to failure.

The Weibull distribution is usually used to describe life and failure distributions, time to failure, and dielectric strength data,

$$F(x) = 1 - \exp\left[-\left(\frac{x - \gamma}{\alpha}\right)^{\beta}\right], \qquad (28)$$

where α is a scale parameter analogous to the mean, β is a shape parameter analogous to the standard deviation, and γ is a location parameter that indicates a threshold below which there is no failure.

2.5 Life-Prediction Techniques

Thermal aging relates aging time to temperature and often assumes a single chemical reaction whose rate depends on an activation energy H and absolute temperature T. The rate of reaction R is usually described by the Arrhenius model,

$$R = A \exp[-H/kT], \qquad (29)$$

where A is a frequency factor. The rate of insulation aging approximately doubles for every 10 °C rise.

To estimate insulation life with accelerated aging by elevating testing voltage v_t above normal service voltage v_s, an inverse-power model is often assumed:

$$t_t/t_s = (v_s/v_t)^n, \qquad (30)$$

where t_t is the testing time, t_s is the time in service, and n is called the life exponent. For many cases, $n \approx 7$. Tests usually take the lowest voltage for v_t to yield failures with a value of $t_t \sim 6$ months. By combining Weibull statistics with (30), an estimate of the probability of in-service insulation failure can be determined from laboratory measurements. Life-prediction techniques are greatly complicated when there are multiple simultaneous stresses, typically from electrical, thermal, mechanical, and environmental factors.

2.6 Space-Charge Effects

2.6.1 Space-Charge Distortion of the Electric Field Most analytical models assume that the electric field is spatially uniform. This motivates many tests to use parallel plane electrodes, but the uniform-field assumption is only true in the central region between electrodes in the absence of net volume charge.

To help understand space-charge effects, often due to charge injection from electrodes as described in Sec. 2.2, we consider the simplest case of planar electrodes with an x-directed electric field E that is distorted by net charge density $\rho(x)$ only dependent on the x coordinate. In such a one-dimensional electrode geometry, Gauss's law of Eq. (2) requires that the slope of the electric-field distribution be proportional to the local charge density:

$$\nabla \cdot \mathbf{E} = \frac{\rho}{\epsilon} \rightarrow \frac{\partial E}{\partial x} = \frac{\rho}{\epsilon}. \qquad (31)$$

For the case of no volume charge, shown in Fig. 3(a), the electric field is uniform, given by $E_0 = v/d$ for a voltage v across a gap d. The electric field drops at a charge-injecting electrode, but its average value E_0 remains constant at v/d. Charge injection thus causes the electric field to increase above the average value at the non-charge-injecting electrode. In Fig. 3(b), unipolar injection has the electric field maximum at the non-charge-injecting electrode, thus possibly leading to electrical breakdown at lower voltages. For bipolar homocharge injection in Fig. 3(c), positive charge is injected at the anode and negative charge is injected at the cathode, so that the electric field is lowered at both electrodes and is largest in the central region. Since electrical breakdown often initiates at the electrode-dielectric interface, this case can allow higher-voltage operation without breakdown, shown by Zahn *et al.* (1986) to be up to ~40% higher in highly purified water. If the voltage suddenly reverses, the electrical field must also instantaneously reverse, but the charge distribution cannot immediately change because it takes some time for the volume charge to migrate. Thus, for early times with voltage reversal, positive charge is near the cathode and negative charge is near the anode, enhancing the fields near the electrodes, as shown in Fig. 3(d). Such a bipolar heterocharge configuration leads to breakdown at lower voltages, as has been found in HVDC polyethylene cables when the voltage is instantaneously reversed

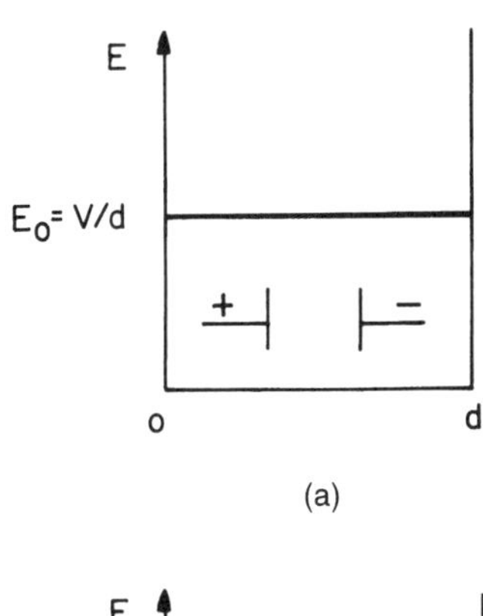

(a)

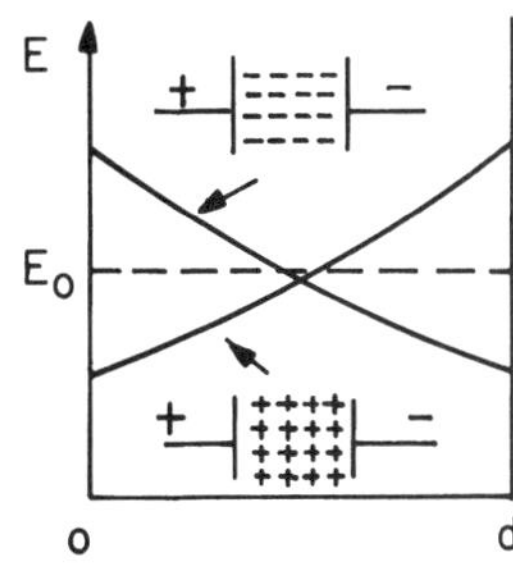

(b)

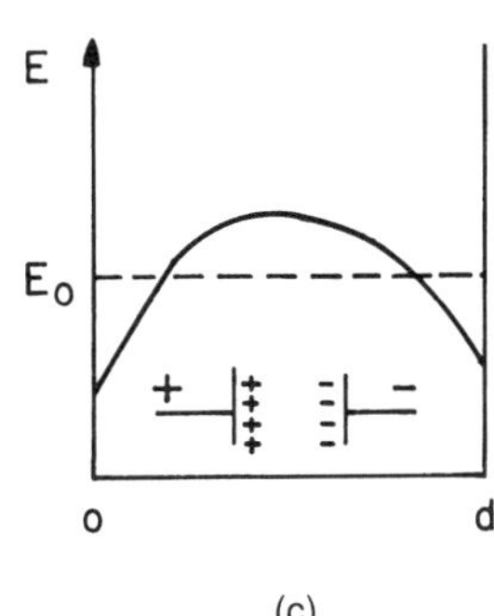

(c)

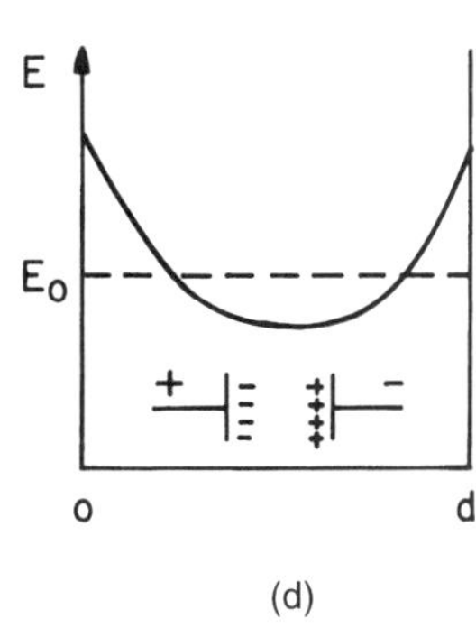

(d)

FIG. 3. Space-charge distortion of the electric-field distribution between parallel-plate electrodes with spacing d at voltage v so that the average electric field is $E_0 = v/d$. (a) No space charge so that the electric field is uniform at E_0; (b) unipolar positive or negative charge injection so that the electric field is reduced at the charge-injecting electrode and enhanced at the non-charge-injecting electrode; (c) bipolar homocharge injection so that the electric field is reduced at both electrodes and enhanced in the central region; (d) bipolar heterocharge distribution where electric field is enhanced at both electrodes and depressed in the central region.

to reverse the direction of power flow. A similar configuration to Fig. 3(d) also occurs when the dielectric is ionized so that free ions are attracted to their image charges on the electrodes.

2.6.2 Governing Equations for Drift-Dominated Unipolar Conduction Because Ohmic conduction involves the presence of equal amounts of positive and negative charge, there can be a net current flow with zero volume charge. However, a volume charge distribution within a dielectric can arise from spatially varying conductivity, irradiation, charge injection from high-field–stressed electrodes as described in Sec. 2.2, dielectric ionization, or contact charging of impurities.

To model net charge transport, we use a drift-dominated unipolar mobility model where the velocity $d\mathbf{r}/dt$ of a positive charge carrier is proportional to the local electric field through the mobility μ, $d\mathbf{r}/dt = \mu\mathbf{E}$, and in turn the electric field is related to the charge density through Gauss's law of Eq. (2). Such a conduction model is often used in the analysis of high-field conduction in solid-state semiconductors when diffusion is negligible.

We consider a one-dimensional model in a medium of thickness d and permittivity ϵ with x-directed electric field E. Without loss of generality, we assume that positive mobile charge with density ρ has a mobility μ. The current-density constitutive law is then

$$\mathbf{J} = \rho \frac{d\mathbf{r}}{dt} = \rho\mu\mathbf{E}. \tag{32}$$

Substituting Eq. (32) into Eq. (3) and integrating with respect to x together with one-dimensional forms of Eqs. (1) and (2) yields the governing equations

$$\epsilon \frac{\partial E}{\partial t} + \rho\mu E = J(t), \tag{33}$$

$$\frac{\partial E}{\partial x} = \frac{\rho}{\epsilon}, \tag{34}$$

$$v = \int_0^d E\, dx, \tag{35}$$

where $J(t)$ is the terminal current per unit electrode area and v is the voltage difference across the electrodes at $x = 0$ and $x = d$. Differentiating (33) with respect to x and using (34) gives an equation for the charge density,

$$\frac{\partial\rho}{\partial t} + \frac{\rho^2\mu}{\epsilon} + \mu E\frac{\partial\rho}{\partial x} = 0. \tag{36}$$

Using (34) in (33), integrating between the electrodes, and then using (35) yields the terminal voltage–current relationship

$$\frac{dv}{dt} + \frac{\mu}{2}[E^2(x = d) - E^2(x = 0)] = \frac{J(t)d}{\epsilon}. \tag{37}$$

The first term is proportional to the familiar capacitive displacement current, while the second term is due to migrating charge.

Equations (33) and (36) are a pair of quasilinear partial differential equations, which can be converted to a set of ordinary differential equations by jumping into the frame of reference of the migrating charge moving at the charge velocity

$$\frac{dx}{dt} = \mu E. \tag{38}$$

In this moving frame of reference, the total time derivatives of electric field and charge density are

$$\frac{dE}{dt} = \frac{\partial E}{\partial t} + \frac{dx}{dt}\frac{\partial E}{\partial x} = \frac{J(t)}{\epsilon}, \tag{39}$$

$$\frac{d\rho}{dt} = \frac{\partial\rho}{\partial t} + \frac{dx}{dt}\frac{\partial\rho}{\partial x} = -\frac{\rho^2\mu}{\epsilon}. \tag{40}$$

The solution to (40) is

$$\rho(t) = \frac{\rho_0(t = t_0)}{[1 + \rho_0(t = t_0)\mu(t - t_0)/\epsilon]} \text{ on } \frac{dx}{dt} = \mu E, \tag{41}$$

where $\rho_0(t = t_0)$ is the initial charge density on a trajectory that starts at $t = t_0$. Physically, this solution of decreasing charge density with time is due to the self-field Coulombic repulsion of charge so that any initial packet of charge expands with time. Even though the total charge within a packet is constant, the density decreases because the volume of charge is increasing. Note that if the initial charge density $\rho_0(t = t_0)$ is zero, the charge density remains zero along the charge trajectory.

If the terminal current density $J(t)$ is known, the electric field along a charge trajectory is obtained by direct time integration of (39), and the position of the charge is found by time integration of the electric field using (38). Often the terminal voltage is constrained so that the terminal current must be found using (37). Because the electric field at both electrodes is generally not yet known, usually a numerical method is needed to solve self-consistently for all variables.

2.6.3 dc Steady-State Distributions

The dc steady-state distributions for electric field and charge density are found by setting the time derivatives in (33) and (36) equal to zero:

$$E_{ss} = \left[\frac{2Jx}{\epsilon\mu} + E_0^2\right]^{1/2}, \tag{42}$$

$$\rho_{ss} = \frac{J}{\mu E_{ss}} = \frac{\epsilon dE_{ss}}{dx} = \frac{J}{\mu}\left[\frac{2Jx}{\epsilon\mu} + E_0^2\right]^{-1/2}, \tag{43}$$

where J is the dc steady-state current density and E_0 is the emitter electric field at $x = 0$, which must be specified as a charge-injection boundary condition, such as from (18) or (20).

The current–voltage relation is found from (35) by integrating (42) between electrodes stressed by dc voltage v:

$$v = \int_0^d E_{ss}\,dx = \frac{\epsilon\mu}{3J}\left\{\left[\frac{2Jd}{\epsilon\mu} + E_0^2\right]^{3/2} - E_0^3\right\}. \tag{44}$$

Simplifying (44) gives solution

$$\frac{16Jd^3}{\epsilon\mu v^2} = \left(9 - 12\left(\frac{E_0 d}{v}\right)^2 + \left\{\left[9 - 12\left(\frac{E_0 d}{v}\right)^2\right]^2 + 192\left(\frac{E_0 d}{v}\right)^3\left[1 - \frac{E_0 d}{v}\right]\right\}^{1/2}\right). \tag{45}$$

For positive charge injection the emitter electric field E_0 must obey the inequality $0 \leq E_0 \leq v/d$. For negative E_0 there is no charge injection as the electric field would push the positive charge back into the electrode. For $E_0 > v/d$, the spatial derivative of E_{ss} could not be positive as required by Gauss's law of

(34) for positive charge injection and simultaneously satisfy (35) of having an average field strength between electrodes of v/d. The positive space-charge distribution distorts the electric field so that the maximum electric field strength occurs at the negative electrode at $x = d$. This field enhancement is important in electric breakdown studies as breakdown initiates where the electric field is largest.

The largest electric field at the non-charge-emitting electrode at $x = d$ is $E(x = d) = 1.5v/d$, which occurs for space-charge limited conditions when $E_0 = 0$ and $\rho_{ss}(x = 0) = \infty$, so that $Jd^3/\epsilon\mu v^2 = 9/8$. As $E_0 d/v$ approaches unity, the electric field distribution approaches a uniform field as the injected charge density and current density become small.

We expect that the injected charge density at $x = 0$, $\rho_{ss}(x = 0) = \rho_0$, would increase as the electric field E_0 at the charge-injecting electrode increases. The simplest relationship that approximates (20) in the high-field limit when the exponential is approximately unity is

$$\rho_0 = AE_0; \quad A \approx \frac{e^3}{8\pi h\phi\mu}, \tag{46}$$

where A is an injection constant. The steady-state current density, which is independent of x, can be particularly evaluated at $x = 0$ as

$$J = \rho\mu E = \rho_0\mu E_0 = \mu A E_0^2. \tag{47}$$

Using (47) in (44) lets us solve for E_0 as

$$\frac{E_0 d}{v} = \frac{3Ad/\epsilon}{(2Ad/\epsilon + 1)^{3/2} - 1}. \tag{48}$$

2.6.4 Space-Charge Effects in Pulsed-Power Technology For inertial-confinement fusion machines, as shown in Fig. 4, and directed-energy devices, high peak power at the terawatt level is needed for short times of the order of 100 ns. Pulsed-power technology collects and stores electric energy at a low input power, typically of order 1 kW, for a long time of order seconds in a capacitor bank, and then delivers this ~1-kJ energy in a much shorter time, ~100 ns, at a much higher power level, ~10^{10} W.

Highly purified water and water/ethylene glycol mixtures have been studied for use as the dielectric in pulse-forming transmission lines because the high relative dielectric constant, $\epsilon_r \approx$ 40–88, and high resistivity, $\rho \approx$ 18–8500 MΩ cm, allow short and efficient low-impedance transmission lines, especially at ~ −35 °C with a mixture of 60% glycol/40% water, as described in Zahn *et al.* (1986).

In testing such pulse-forming lines it was found that at high electric stress, the open-circuit voltage across immersed parallel-plate electrodes decayed more quickly as the initial voltage increased and that the decay rate depended on electrode spacing and was not exponential, as shown in Figs. 5(a) and 5(b)

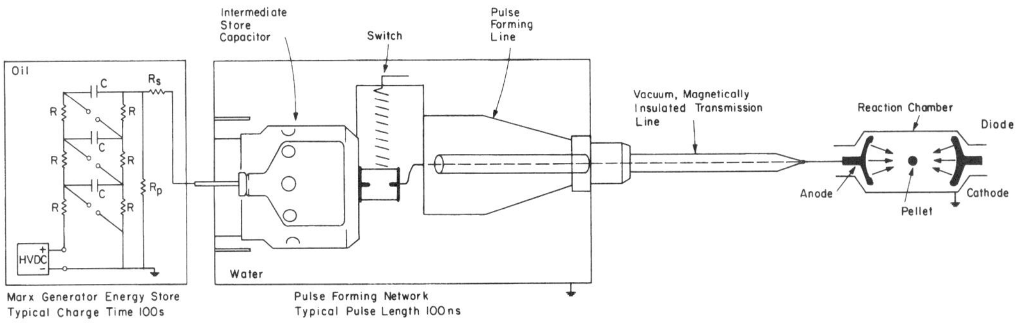

FIG. 4. A representative pulsed-power machine used for inertial-confinement fusion experiments consists of a slowly charged Marx capacitor bank under insulating oil as primary energy storage; a water dielectric capacitor as intermediate energy storage; a switched pulse-forming line that compresses the voltage in time; and a magnetically and vacuum insulated transmission line that delivers the power pulse to a vacuum diode to produce energetic electrons or light ions, which are then accelerated and focused by electric and magnetic fields onto a target, typically a deuterium-tritium pellet.

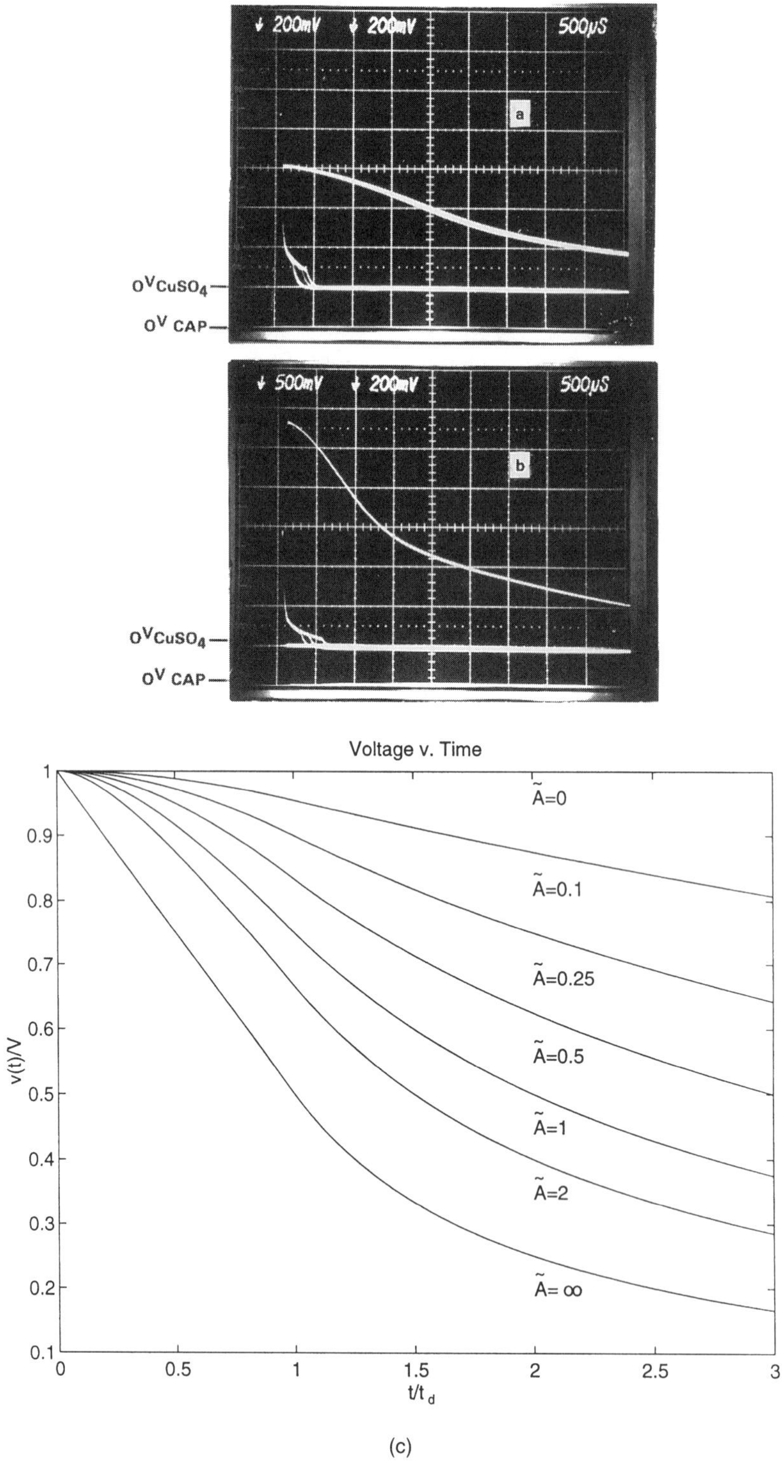

(c)

FIG. 5. (a) Measured open-circuit voltage decay vs. time for an 80% ethylene glycol/20% water mixture at −10 °C with a dielectric relaxation time ~24 ms for parallel-plate stainless-steel electrodes with 5.1-mm gap with initial voltages of (a) 55 kV and (b) 90 kV. Note the change in sign of the curvature of both traces at early time and that by $t \approx 4.5$ ms, both voltages are about the same despite the large difference in initial voltage. (c) Theoretical curves of Eq. (55) using a drift-dominated unipolar conduction model for various values of positive charge injection parameter $\tilde{A} = Ad/\epsilon$ showing nonexponential open-circuit voltage decay.

for two different initial voltages. This disagrees with the simple Ohmic model of a lossy homogeneous capacitor where the open-circuit decay is exponential with decay rate $\tau = RC = \epsilon/\sigma$ independent of capacitor geometry and voltage as given by (6).

To explain the nonlinear behavior, we use the unipolar ion-conduction law of (32) where in a qualitative sense, the quantity $\rho\mu$ is an effective conductivity that varies with time and position. In addition to the dielectric relaxation time of (9), there is a time-of-flight time constant, $t_d = d^2/\mu v$, for injected charge to reach the opposite electrode, which depends on electrode spacing and voltage.

To emphasize space-charge effects, we assume that the Ohmic conductivity is zero and take the capacitor to be initially uncharged in the dielectric volume but with an initial applied voltage v at $t = 0$ and then open circuited thereafter. We find the injection electric field at $x = 0$, $E_0(t)$, using (33) with $J(t) = 0$ and the linear charge-injection law of (46),

$$\frac{dE_0}{dt} + \frac{A\mu}{\epsilon}E_0^2 = 0 \Rightarrow E_0(t) = \frac{v/d}{1 + A\mu vt/\epsilon d}, \tag{49}$$

where the initial condition is $E_0(t = 0) = v/d$.

For those trajectories of injected charge that start at $x = 0$ at time $t = t_0$, Eq. (39) with $J(t) = 0$ shows that the electric field along the trajectory remains constant at the $x = 0$ emitter electric field, $E(t) = E_0(t = t_0)$. Then the trajectory is given from (38) as

$$\begin{aligned}\frac{dx}{dt} &= \mu E_0(t = t_0) \Rightarrow x(t) = \mu E_0(t = t_0)(t - t_0)\\ &= \frac{(\mu v/d)(t - t_0)}{[1 + A\mu vt_0/\epsilon d]}.\end{aligned} \tag{50}$$

In particular, the trajectory that starts at $t_0 = 0$ defines the migrating charge front that reaches $x = d$ at time t_d

$$x_d(t) = \frac{\mu vt}{d};\ t_d = \frac{d^2}{\mu v}. \tag{51}$$

Solving (50) for t_0 gives

$$t_0 = \frac{\mu vt/d - x}{(\mu v/d)(1 + Ax/\epsilon)}, \tag{52}$$

so that the electric field and charge density at time t and position x are

$$E(x,t) = \begin{cases}\dfrac{v(1 + Ax/\epsilon)}{d(1 + Adt/\epsilon t_d)}, & 0 \le x \le x_d(t),\\ v/d, & x_d(t) < x < d;\end{cases} \tag{53}$$

$$\begin{aligned}\rho(x,t) &= \epsilon\frac{\partial E}{\partial x}\\ &= \begin{cases}\dfrac{Av/d}{1 + Adt/\epsilon t_d}, & 0 < x < x_d(t),\\ 0, & x_d(t) < x < d.\end{cases}\end{aligned} \tag{54}$$

The terminal voltage is then computed as

$$\begin{aligned}v(t) &= \int_0^d E dx = \int_0^{x_d(t)} E dx + \int_{x_d(t)}^d E dx;\\ &= \begin{cases} v\left[1 - \dfrac{Adt^2}{2\epsilon t_d^2(1 + Adt/\epsilon t_d)}\right], & 0 \le t \le t_d,\\ \dfrac{v(1 + Ad/2\epsilon)}{1 + Adt/\epsilon t_d}, & t \ge t_d.\end{cases}\end{aligned} \tag{55}$$

Figure 5(c) shows the nonexponential decay of (55) for various values of charge injection constant A, which has similar shape to the measurements shown in Figs. 5(a) and 5(b).

2.7 Drift-Diffusion Conduction

2.7.1 Physical Modeling In such materials as metallic conductors, ionized gases, electrolytic solutions, and semiconductors, the charge carriers can be classically modeled as an ideal-gas plasma. We assume that we have two carriers of equal magnitudes but opposite signs $\pm q$, with respective masses $m_\pm$ and number densities $n_\pm$. These charges can be holes and electrons in a semiconductor, oppositely charged ions in an electrolytic solution, or electrons and essentially stationary nuclei in metals. When an electric field is applied, the positive mobile charges move in the direction of the field, while the negative mobile charges move in the opposite direction. These charges collide with the host medium at respective collision frequencies ν_+ and ν_-. This process then acts as a viscous or frictional dissipation opposing the motion. In addition to electrical and frictional forces, the particles exert a force on their neighbors through a pressure term due to thermal agitation that would be present even if the particles were uncharged. For an ideal gas, the partial pressure is $p = nkT$, where n is the number density of charges, T is the absolute temperature, and $k = 1.38 \times 10^{-23}$ J/K is Boltzmann's constant.

2.7.2 Governing Equations Because the collision frequencies $\nu_\pm$ are typically of the order 10^{13} Hz, the particle inertia is often negligible. Then force equilibrium on each carrier moving at velocities $\mathbf{v}_+$ and $\mathbf{v}_-$ requires

$$0 = q\mathbf{E} - m_+\nu_+\mathbf{v}_+ - \frac{1}{n_+}\nabla(n_+kT),$$
$$0 = -q\mathbf{E} - m_-\nu_-\mathbf{v}_- - \frac{1}{n_-}\nabla(n_-kT), \qquad (56)$$

where we recognize in the last terms that the force per particle due to pressure is given by $(1/n)\nabla p$.

We solve (56) at constant temperature for the velocity of each carrier as

$$\mathbf{v}_+ = \frac{q}{m_+\nu_+}\mathbf{E} - \frac{kT}{m_+\nu_+n_+}\nabla n_+,$$
$$\mathbf{v}_- = -\frac{q}{m_-\nu_-}\mathbf{E} - \frac{kT}{m_-\nu_-n_-}\nabla n_-. \qquad (57)$$

The charge density for each carrier is $\rho_+ = qn_+$ and $\rho_- = -qn_-$. Then the current density for each carrier can be written as

$$\mathbf{J}_+ = \rho_+\mathbf{v}_+ = qn_+\mathbf{v}_+ = \rho_+\mu_+\mathbf{E} - D_+\nabla\rho_+,$$
$$\mathbf{J}_- = \rho_-\mathbf{v}_- = -qn_-\mathbf{v}_- = -\rho_-\mu_-\mathbf{E} - D_-\nabla\rho_-, \qquad (58)$$

where $\mu_\pm$ are the particle mobilities and $D_\pm$ are the particle diffusion coefficients:

$$\mu_\pm = \frac{q}{m_\pm\nu_\pm}, \quad D_\pm = \frac{kT}{m_\pm\nu_\pm}. \qquad (59)$$

We note that the ratio $D_\pm/\mu_\pm$ for each carrier equals the thermal voltage, kT/q (~25 mV at room temperature), the Einstein relation:

$$\frac{D_+}{\mu_+} = \frac{D_-}{\mu_-} = \frac{kT}{q}. \qquad (60)$$

2.7.3 Equilibrium Solutions In equilibrium when the net current density of each carrier is zero, with $\mathbf{E} = -\nabla\Phi$, Eq. (58) shows that the charge densities are related to the potential by Boltzmann distributions:

$$\mathbf{J}_+ = 0 \Rightarrow \rho_+ = \rho_0 e^{-q\Phi/kT},$$
$$\mathbf{J}_- = 0 \Rightarrow \rho_- = -\rho_0 e^{+q\Phi/kT}, \qquad (61)$$

where we use Eq. (60) and $\pm\rho_0$ is the equilibrium charge density of each carrier when $\Phi = 0$.

To find the spatial dependence of ρ and Φ, we use Eq. (61) in the Poisson's-equation form of (2),

$$\nabla^2\Phi = -\frac{\rho_+ + \rho_-}{\epsilon} = -\frac{\rho_0}{\epsilon}(e^{-q\Phi/kt} - e^{q\Phi/kT})$$
$$= \frac{2\rho_0}{\epsilon}\sinh\frac{q\Phi}{kT}, \qquad (62)$$

which is known as the Poisson–Boltzmann equation.

To examine the form of simple solutions to (62), we take the small-potential limit so that $q\Phi/kT \ll 1$:

$$\nabla^2\Phi - \frac{\Phi}{\lambda^2} = 0, \quad \lambda = \left[\frac{\epsilon kT}{2q\rho_0}\right]^{1/2}, \qquad (63)$$

where λ is called the Debye length.

If an electrode is placed at $x = 0$ with potential $\Phi(x = 0) = v_0$ with respect to the potential at $x = \pm\infty$, the potential then decreases exponentially with distance as

$$\Phi(x) = \begin{cases} v_0 e^{-x/\lambda}, & x > 0, \\ v_0 e^{+x/\lambda}, & x < 0. \end{cases} \qquad (64)$$

Similarly, if a point charge Q is inserted into the medium, the solution to (63) is

$$\Phi(r) = (Q/4\pi\epsilon r)e^{-r/\lambda}. \qquad (65)$$

In Eqs. (64) and (65), the potential decays away from the source with characteristic length λ because counter charges of opposite sign are attracted to the source charge, tending to shield the surrounding region.

3. APPLICATIONS OF ELECTRICAL INSULATION

3.1 Gaseous Insulation

3.1.1 Corona An understanding of electrical breakdown in gases can also be applied to a bubble model in liquids or a gaseous cavity model in solids. The fundamental phenomenon is described by the Paschen curve of Fig. 2. Representative voltage (v_P) and pressure-distance products at the Paschen minimum are listed for various gases in Table 2. When the applied electric field exceeds v_P/d, corona onsets, resulting in light, UV radiation, audible noise, radio-frequency interference, nitric acid in the presence of moisture and air, heat generation and power loss, and mechanical erosion of surfaces by ion impacts. Corona partial discharges are localized ionizations of gas forming a plasma of electrons, ions, excited molecules, and free radicals. These form chemically reactive species that often degrade nearby liquid and solid interfaces. Corona is often pulsating, as the buildup of charge near the electrode causes a decrease in the electric field below that necessary to sustain charge emission or dielectric ionization, as illustrated in Fig. 3(b). Corona can be reduced by rounding sharp edges to lower the local electric field. Above the Paschen minimum, an increasing pressure reduces the mean free path between collisions, and higher voltages are necessary for ionization, thereby preventing corona. At a pressure 10^{-4}–10^{-5} Torr, the mean free path is a few meters. Decreasing the pressure to 10^{-6}–10^{-8} Torr increases the mean free path to the order of 50 m.

Table 2. Typical Paschen minimum breakdown voltages and pressure–distance products for various gases [adapted from Meek and Craggs (1978) and Khalifa (1990)].

Gas	v_P (V)	pd at v_P (Pa m)
Air	327	0.754
Argon	137	1.197
Hydrogen (H_2)	273	1.530
Helium (He)	156	5.320
Neon	244	3.987
Carbon dioxide (CO_2)	420	0.678
Nitrogen (N_2)	251	0.891
Nitrous oxide (N_2O)	418	0.665
Oxygen (O_2)	450	0.931
Sulfur dioxide (SO_2)	457	0.439
Hydrogen sulfide (H_2S)	414	0.798
Mercury vapor (Hg)	~425	2.392

In air-insulated overhead power lines, a 350-kV line has corona losses ~4 kW/m in dry weather and ~70 kW/m in wet weather. Raising the voltage to 400 kV increases the corona losses to ~10 kW/m in dry weather and ~180 kW/m in wet weather.

The effects of corona can be reduced by increasing the conductor diameter to lower the electric field. This is most efficiently done using four parallel conductors spaced as a 6-in. square, which approximates the corona characteristics of a 6-in.-diam cable. Two parallel conductors 6–8 in. apart can operate up to ~500 kV.

Some corona characteristics can be applied in useful ways:

1. Xerography—The first step in copying an image is to lay down a uniform layer of static charge from a corona source over the surface of a light-sensitive plate.
2. Electrostatic separation of ores—Corona-charged conducting and nonconducting mineral particles can be separated using the Coulomb force.
3. Electrostatic paint spraying—Corona-charged paint articles will be directed by the electric field onto a grounded surface.
4. Printing on polymer surfaces—Thin films such as polyethylene cannot be wetted by printing inks. Chemical action from rf corona oxidizes the polymer to form surfaces that inks can wet.
5. Ozone generators—Ozone is generated by corona ionization of oxygen.
6. Geiger counter—With a voltage applied to a wire–cylinder electrode geometry to just below corona initiation, a passing cosmic ray causes an avalanche and a measurable current pulse.

Failures in electrically stressed gas-insulated systems usually occur at the interfaces of solid insulators used for mechanical support. Such surface flashovers depend on the materials, surface condition, particle contamination, and the electric fields at the triple junction of electrode, solid insulator, and gaseous insulation.

3.1.2 Streamer Mechanisms With high over-voltages above the Paschen curve in a uniform electric field gap, avalanches grow into irregular filamentary luminous branches known as streamers. According to streamer theory, ionization from an avalanche increases the conductivity, and the emitted light increases photoionization of gas molecules ahead of the streamer. This causes a space-charge distribution that increases the electric field at the tip of the streamer.

The time to breakdown has a statistical lag, t_s, of the average time after voltage is applied before an electron that initiates an avalanche appears in a gap, and a formative time lag, t_f, which is the time the discharge takes to completely develop from the time the initiating electron appears. The statistical time, $t_s \sim 10^{-2}$ s, can be made shorter by ultraviolet radiation of the cathode or large over-voltage of the gap. The formative time is of order $t_f \sim 10^{-4}$–10^{-9} s.

3.1.3 Electronegative Gas—SF_6 Gas-insulated systems (GISs) use sulfur hexafluoride (SF_6), an electronegative gas, as an arc-quenching medium in such high-voltage equipment as circuit breakers, switchgear, and underground cables above 123 kV, electrostatic accelerators, and x-ray equipment. A representative E/p value at the Paschen minimum is 86.2 V/m·Pa for SF_6, versus 23.6 V/m·Pa for air. Mixing a small amount of SF_6 with other less expensive gases such as nitrogen, hydrogen, carbon dioxide, or air increases the breakdown strength of those gases. For example, 20% by partial pressure SF_6 to nitrogen raises the breakdown strength to about 80% of that of pure SF_6. SF_6 is nontoxic but denser than air (so it can cause suffocation in a closed tank), nonflammable, noncorrosive, inert, and nonreactive. It decomposes above 500 °C. Its high molecular weight and low viscosity make it more effective with heat transfer than other gases. Its thermal time constant is about 1000 times shorter than that of air.

3.1.4 Arc Discharges Whenever a circuit breaker or load-break switch is opened while carrying a current, an arc forms between separating contacts. As the contacts move apart, the contact area decreases and the contact resistance goes up. The current becomes highly concentrated and melts metal. The metal vapor between contacts sustains the arc.

3.1.5 Vacuum Insulation Vacuum insulation is limited by Fowler–Nordheim electron emission [Eq. (20)] at microprotrusions on the electrode surface that can typically magnify the local electric field by a factor of 200–500. Electrical breakdown results when the Joule heating from the injected current leads to an explosive growth of plasma that initiates the formation of an arc. Microparticles on the electrode also detach and are accelerated by the electric field to impact explosively on the opposite electrode. The need for vacuum insulation is especially important in designing power systems for space.

3.2 Liquid Insulation

3.2.1 Characteristics Liquids are usually better insulators than solids and gases. Their high density from Paschen's law indicates a higher breakdown strength than gas. Liquid fills the space to be insulated and by thermal conduction and convection helps dissipate thermal losses.

Most insulating liquids used in transformers, cables, capacitors, switches, and circuit breakers are complex mixtures of paraffinic, naphthenic, and aromatic hydrocarbons. The paraffinic base is characterized by the chemical formula $C_{2n}H_{2n+2}$, the naphthenes by $C_{2n}H_{2n}$, and the aromatics by C_nH_n. Naphthenic oils are generally preferred because they are less viscous and have good gas-absorbing properties. Representative physical characteristics of mineral insulating oils for use in transformers, switches, and circuit breakers, with typical ASTM tests, are listed in Table 3.

These oils have relative dielectric constants of about 2, a dissipation factor of about 0.001 at 60 Hz, working electrical breakdown strength of about 120 kV/cm, and a resistivity greater than 10^{12} Ω cm. Their flammability is high, but the hazard from fire is minimal when operating at a temperature far below the flash point.

In transformers, the oil serves the dual function of high-voltage insulation and cooling with oil flow. Analysis of the insulating oil in an operating transformer provides information on the state of health of the system. This monitoring has become more important in recent years because of the trend toward higher voltages and smaller size,

Table 3. Representative physical characteristics and ASTM standards of mineral insulating oils [adapted from Clark (1962) and the ASTM Standards].

Property	
Condition	Clear [D1524]
Viscosity	12 cSt (40 °C) [D445, D88]
Specific gravity	0.91 [D1298]
Color (max)	0.5 [D1500]
Neutralization number	0.03 mg KOH/g [D974, D2440, D1533]
Flash point (min)	145 °C [D92]
Pour point (max)	−40 °C [D97]
Free sulfur	nil [D1275]
Total (fixed) sulfur	0.1% or less [D1275]
Evaporation (8 h/100 °C)	8%
Dielectric strength	30 kV/2.5 mm [D877] 28 kV/1.02 mm [D1816]
Specific heat (30–35 °C)	0.4252
Power factor (100 °C)	0.003 [D924]
Chlorides and sulfates	nil [D878]
Resistivity (100 °C)	$(1–10) \times 10^{12}$ Ω cm [D1169]
Coefficient of expansion	0.00070 [D1903]
Specific optical dispersion	110–115 [D1807]
Thermal conductivity	0.39 cal cm^{-1} sec^{-1}/°C
Refractive index (25 °C)	1.4828
Aniline point	63–78 °C [D611]

thereby increasing the design stress. This has led to the development of on-line hydrocarbon gas monitors, indicators of the degradation of cellulosic insulation, computerized methods to predict temperatures throughout transformer windings by calculating precise oil flow and heat flux, and development of a microprocessor-based transformer temperature monitor to improve monitoring and control of in-service transformers.

Moisture is the most common cause of deterioration, either as tiny droplets mixed with the oil as an emulsion or as a pool at the bottom of the transformer when the droplets coalesce to form larger drops that sink to the bottom. A slower but more damaging deterioration is the formation of acids and sludge, caused by oxidation. Fault conditions such as overheating (hot spots), partial discharges (corona), and continuous arcing lead to accumulation of combustible gases and to chemical degradation of the oil, resulting in changes in the molecular-weight distribution.

In the recent past, mineral-oil flammability was reduced by adding polychlorinated biphenyls (PCBs). However, later research has shown that the nonbiodegradability of PCBs is a health hazard so that current standards do not allow the use of PCB oil. This has resulted in a search for new natural or synthetic oils that are biologically and environmentally safe, yet still have high electric strength, low dielectric dissipation, good chemical stability, low volatility and high flash point, and good arc-quenching properties, and are nonflammable, nontoxic, and inexpensive. Silicone oil is one candidate for replacement for PCB-contaminated oil.

To decrease power dissipation in transformers, oil and pressboard insulation are dried to remove moisture. However, as drying processes have been improved, a new problem has arisen, flow-induced electrification. Here the mobile part of the electrical double layer that forms at the interface of dissimilar materials is swept away by oil flow, leaving the countercharge behind on the interface. The net accumulation of charge on highly insulating surfaces leads to strong electric fields and ultimately to spark discharges that can cause a transformer to fail.

Electrical breakdown in liquids is thought to be primarily due to the following mechanisms:

1. Electronic avalanches as in gases.
2. Bubbles that form near surface irregularities on electrodes that concentrate the electric field and cause Joule heating that boils the liquids to form bubbles. Once a gas region is formed, avalanche theory describes the breakdown phenomena in the gaseous bubble. Figure 6 shows a hydrocarbon liquid stressed by a high-voltage pulse. The electron avalanches within the vapor region grow into treelike structures at the cathode, which can initiate the breakdown streamer shown.
3. Suspended particles that form chains along the electric field to partially short circuit the gap.

3.2.2 Insulation Properties and Standards From the current state of knowledge of oil aging and degradation, the power industry has developed measurement proce-

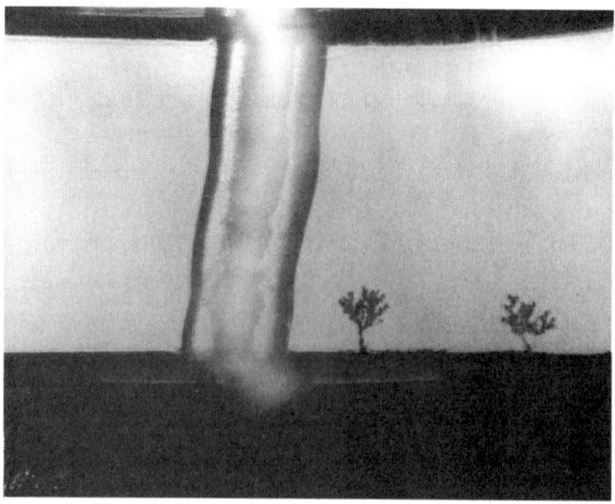

FIG. 6. Parallel-plate electrodes with a hydrocarbon dielectric (toluene) stressed by a 250-kV pulse show electrical trees emanating from the cathode, one of which led to the breakdown streamer that short circuits the 5-mm gap. The photograph was taken about 200 ns after electrical breakdown.

dures, interpretation, and corrective actions for monitoring the electrical, physical, chemical, and thermal properties of oil. The important properties are listed and defined below together with the appropriate ASTM tests.

3.2.2.1 Electrical Properties *Dielectric Breakdown Voltage* [D877, D1816, D3300] is important at 60-Hz power frequency as a measure of the liquid's ability to withstand electric stress at operating and test voltages without failure. Low values indicate the presence of contaminants such as water, dirt, or conducting particles. Impulse testing is used to simulate transient voltage stresses arising from such causes as lightning strokes and high-voltage switching.

Gas Evolution [D2298, D2300] determines the rate of gas evolution or gas absorption under ASTM-specified test conditions of electrical discharge.

Power Factor [D924], the ratio of power dissipation in watts to the product of voltage and current in volt-amperes, equivalent to cos ϕ where ϕ is the phase angle between voltage and current, is a measure of the oil dielectric losses and thus the amount of energy dissipated as heat.

Resistivity [D1169]—High resistivity reflects low content of free ions and ion-forming particles and normally indicates a low concentration of conductive contaminants. Resistivity of less than 15 MΩ cm generally indicates problems.

3.2.2.2 Physical Properties *Aniline Point* [D611]—Testing provides an estimate of the aromatic hydrocarbon content of mixtures and indicates the solvency of the oil for some materials.

Coefficient of Thermal Expansion [D1903] is used to compute the size of the transformer structure to accommodate oil volume over the full temperature range of operation.

Color [D1500]—Oil gets darker with age because of formation of acids, resins, and sludge from insulating materials in the transformer. The color should not progress beyond a light brown, remaining clear and free from turbidity or cloudiness. Black coloring or floating carbon particles indicates severe arcing.

Visual and Qualitative Infrared Absorption Examination [D1524, D2144]—Visual examination of color and homogeneity of the oil permits a determination of whether the oil should be sent to the laboratory for testing. The infrared spectrum indicates the general chemical composition of the oil.

Flash and Fire Point [D92]—The flash and fire points indicate the flammability of an oil. The flash point is the temperature at which vapors above the oil surface first ignite when a small test flame is passed across the surface. The fire point is the temperature at which the oil first ignites and burns for at least 5 s when a small test flame is passed across the surface.

Interfacial Tension [D971, D2285] is a reliable indication of the presence of hydrophilic compounds, which are soluble polar contaminant products of oxidation. A high interfacial tension indicates the absence of undesirable polar contaminants. The test is frequently applied to service-aged oils as an indication of the degree of deterioration.

Molecular Weight [D2224] is a basic quantity required in analysis to convert weight fractions to mole fractions.

Pour Point [D97] is the lowest temperature expressed as a multiple of 5 °F (or 3 °C) at which the oil is observed to flow when cooled without seriously limiting the degree of circulation.

Refractive Index and Specific Optical Dispersion [D1218, D1807]—Changes in refractive index with service indicate changes in composition or the degree of contamination.

Specific optical dispersion is the difference between the refractive indices of light of two different wavelengths at the same temperature, divided by the specific gravity. It serves as a quick index to the amount of unsaturated compounds present in the oil.

Specific Gravity [D287, D1298, D1481]—Insulating oils are usually sold on the basis of volume delivered at 15.6 °C (60 °F), while delivery is usually based on weight. The specific gravity influences the heat-transfer rates. In cold climates, ice may be formed in de-energized transformers below 0 °C, and the maximum specific gravity of the oil should always be less than that of ice, so that ice will settle out at the bottom. When new oil is mixed into apparatus in service, differences in specific gravity may indicate a tendency of the two oils not to mix, which can cause overheating of self-cooled apparatus.

Steam Emulsion Number [D1935] provides an approximate evaluation of the ability of mineral insulating oils to separate from water. The test indicates the presence of contaminants such as metallic soaps, asphaltic materials, etc.

Viscosity [D88, D445, D2261] influences heat-transfer properties and consequently their temperature rise. At low temperatures, the higher viscosity influences the speed of moving parts such as load tap-changer mechanisms and pumps. Viscosity controls mineral insulating oil processing conditions such as dehydration, degasification and filtration, and oil impregnation rates. Viscosity affects pressure drop, oil flow, and cooling rates in circulating oil systems.

3.2.2.3 Chemical Properties *Acidity* [D1534, D1902] in used insulating oils can be compared to that of new oil in detecting contaminants or chemical changes and can be used as a measure for determining when an oil should be replaced to prevent further decomposition and consequent sludging. For a power transformer, 500 kV A and larger, operating at 60 °C maximum, the transformer oil should reach the critical acid number of 0.25 mg KOH/g in about 15 years.

Carbon-Type Composition [D2140]—Percentages of aromatic carbons, naphthenic carbons, and paraffinic carbons relate chemical nature of an oil to other properties related to oil composition.

Characteristic Groups [D2007]—A rapid procedure for classifying mineral insulating oil into hydrocarbon types of polar compounds, aromatics, and saturates.

Compatibility with Construction Material [D3455]—Screening for compatibility of oil with varnishes, dip coatings, core steel, gaskets, and wire enamels.

Copper Content [D2608, D2675]—The presence of copper accelerates the deterioration of oil.

Gas Analysis [D3612] covers the extraction and measurement of gases dissolved in electrical insulating oil having a viscosity of 20 centistokes (100 SUS) or less at 40 °C, and the identification and determination of the individual component gases extracted, including hydrogen (H_2), nitrogen (N_2), carbon monoxide (CO), carbon dioxide (CO_2), oxygen (O_2), methane (CH_4), ethane (C_2H_6), ethylene (C_2H_4), acetylene (C_2H_2), propane (C_3H_8), and propylene (C_3H_6).

Gas Content [D831, D1827, D2945]—Gas bubbles can lead to gaseous ionization in high fields. A low gas content reduces foaming and reduces available oxygen in sealed equipment, increasing service life of the oil.

Inorganic Chlorides and Sulfates [D878]—Presence of inorganic chlorides and sulfates indicates improper refining or contamination affecting corrosivity and oil dielectric properties.

Neutralization Number [D664, D974] determines acidic or basic constituents in oil. The total acid number is the number of milligrams of KOH required to neutralize all acidic constituents in a 1-g sample. A low acid number is necessary to minimize electrical conduction and metal corrosion to maximize the insulation life. In used oils, an increase in acid number from the value for new oil indicates contamination or chemical changes such as oxidation. The acid number should not increase beyond 0.10 mg KOH/g under normal operating conditions.

Oxidation Inhibitor Content, Phenolic [D1473, D2668]—New electrical insulating oil may contain phenolic oxidation inhibitors, 2,6-ditertiary-butyl para-cresal (DBPC) and 2,6-ditertiary-butyl phenol. DBPC reduces oil color, acidity, and sludge in operating transformers.

Oxidation Stability [D1313, D2112, D2440]—Minimizing the development of oil sludge and acidity resulting from oxidation during storage, processing, and long service life reduces electrical conduction and metal corrosion and increases insulation system life, electrical breakdown strength, and heat transfer. Open breathing transformers have faster oxidation deterioration than sealed transformers. The rate of oxidation generally doubles with each 10 °C increase.

Peroxide Number [D1563] magnitude indicates the quantity of oxidizing constituents present. Deterioration of oil results in the formation of peroxides and other oxygen-carrying compounds. The peroxide number measures those compounds that will oxidize potassium iodide.

Saponification Number [D94], the number of milligrams of KOH consumed by 1 g of oil under prescribed conditions. Components that react with KOH may consist of oil oxidation products, added compounds, or contaminants.

Sediment and Soluble Sludge [D1698] determines the amount of sediment and soluble sludge in service-aged oils, including organic and inorganic content. Sediment may deposit on transformer parts and interfere with heat transfer, and may choke oil ducts and thus hinder oil circulation and heat dissipation. Inorganic sediment indicates contamination, and organic sediment indicates deterioration or contamination.

Sulfur, Corrosive [D1275]—Sulfur compounds that severely discolor a copper surface are termed corrosive.

Water Content [D1315, D1533]—In ASTM test D1315, "Test for Water in Insulating Oils by Extraction," moisture is removed from oil by heat, vacuum, and shaking. The moisture is absorbed in a P_2O_5 trap and is determined by the weight gain. ASTM test D1533, "Water in Insulating Liquids," uses the Karl Fischer method, usually in the range of 0–75 ppm, in insulating liquids having a viscosity less than 19 centistokes (100 SUS) at 40 °C. The method is based essentially upon the reduction of iodine by sulfur dioxide in the presence of water. A low water content is necessary to achieve adequate electric strength and low dielectric losses, maximize insulation life, and minimize metal corrosion. A normal power transformer will have a water content under 2% in the paper and 10–20 ppm in the oil.

3.2.3 Analysis of Gas in Transformer Oils

3.2.3.1 Background. A study of gas formation in transformer oil stressed by high fields frequently gives an early indication of abnormal behavior in a transformer and can allow corrective action before the equipment suffers greater damage. Improvements in available instrumentation allow utilities to adopt oil-analysis programs as failure prevention programs rather than just failure analysis programs. The distribution of gases is related to the type of fault, and the rate of gas formation is indicative of the severity of the fault.

3.2.3.2 Gas Cause and Origin. Fault gases are generally due to corona or partial discharge, pyrolysis or thermal heating, and arcing. The hydrocarbon gases methane (CH_4), ethane (C_2H_6), ethylene (C_2H_4), and acetylene (C_2H_2) are produced by mineral oil decomposition. Carbon dioxide (CO_2) and carbon monoxide (CO) arise from thermal decomposition of the cellulosic insulating material. Hydrogen (H_2) is a component of both cellulosic material and any water present in the oil. Oxygen (O_2) and nitrogen (N_2) are not considered fault gases but are usually present.

Arcing causes large amounts of hydrogen and acetylene to be produced, with minor quantities of methane and ethylene. Carbon dioxide and carbon monoxide may also be formed if the fault involves cellulose. The oil may be carbonized.

Low-energy electrical discharges from corona produce hydrogen and methane, with small quantities of ethane and ethylene. Comparable amounts of carbon monoxide and dioxide may result from discharges in cellulose.

Decomposition products from overheated oil include ethylene and methane, together with smaller quantities of hydrogen and ethane. Traces of acetylene may be formed if the fault is severe or involves electrical contacts.

Large amounts of carbon dioxide and carbon monoxide are evolved from overheated cellulose. Hydrocarbon gases, such as methane and ethylene, will be formed if the fault involves an oil-impregnated structure.

All these gases are dissolved in the oil as well as in the gas blanket above the oil. Because of differences in solubilities of these gases in the oil, their distribution in the oil and gas blanket will differ. Equal quantities of hydrogen and acetylene in the gas blanket will result in concentrations in the oil that differ by almost two orders of magnitude because hydrogen is less soluble in oil, while acetylene is more soluble in the oil.

The energy released by any fault, including heat, fragments the oil into smaller hydrocarbon chains and hydrogen. Corona leads to large amounts of hydrogen, sparking gives rise to methane and ethane, severe local heating causes the formation of ethylene, and arcing forms acetylene. These are the key gases for each fault type, but other gases can also be formed. Faults near cellulosic materials give rise to carbon dioxide and carbon monoxide with the ratio of carbon monoxide to carbon dioxide increasing as the severity of the faults increases.

3.2.3.3 Detection Methods. The most widely used method is the determination of the total combustible gases present above the oil. It is rapid and can be continuously monitored, but it does not determine which fault gases are present and therefore cannot indicate either the type or severity of a fault.

Normal oil aging has up to 500 ppm fault gases. Decomposition may be in excess of normal aging with gas levels of 500–2500 ppm, while severe problems are developing for gas levels above 2500 ppm.

A rate of increase of combustible gas generation of 100 ppm or more for a 24-h period, on a continuing basis, with a relatively constant load, indicates a deteriorating condition and requires an assessment of continued operation, especially if acetylene is in excess of 20 ppm.

Dissolved-gas analysis (DGA) takes a sample of oil, and the dissolved gases are extracted and then separated, identified, and quantitatively determined with a gas chromatograph to detect fault gases at the earliest point in time and can ascertain the type and severity of the fault.

Because dissolved hydrogen in oil is of the same order of concentration as the other fault gases, hydrogen monitors using hydrogen diffusion through a Teflon or polyimide membrane have been developed for cheaper and easier measurements that allow continuous monitoring.

3.2.3.4 Interpretation Theories. Faults of thermal or electrical origin can be differentiated by comparing pairs of gases with approximately equal solubilities and diffusion coefficients, as described by Rogers (1978). Thus, an increase in the ratio of ethylene to acetylene above unity indicates an electrical fault, while the ratio of methane to hydrogen exceeding 0.1 suggests a thermal fault and less than 0.1 suggests a corona discharge.

A thermodynamic assessment of the formation of the simple decomposition gaseous hydrocarbons based on equilibrium pressures at various temperatures suggests that the proportion of each gaseous hydrocarbon in comparison with each of the other hydrocarbon gases varies with the temperature of decomposition. This leads to the assumption that the rate of evolution of any particular gaseous hydrocarbon varies with temperature and that at a particular temperature there would be a maximum rate of evolution where each gas attains the maximum rate at a different temperature. This thermodynamic equilibrium suggests that with increasing temperature, the maxima would be in turn methane, ethane, ethylene, and acetylene.

3.3 Solid Insulation

3.3.1 Characteristics Solid insulation is generally used in the form of porcelain, glass, and epoxy resin high-voltage bushings; in oil-impregnated cellulose in transformers; in polymer power cables; in thin-film polymer capacitors; and in mica insulation where high temperature and surface discharges are experienced. The primary advantage of solid insulation is that it also provides mechanical strength, but its primary disadvantage is that solids are not self-healing after electrical discharges, as a conducting channel remains. In carefully controlled laboratory conditions, the breakdown strength in thin-film solids can be as high as 1–15 MV/cm, but in practical working systems the working stress drops to ~200 kV/cm. The primary failure mechanism is electrical discharge across voids. In a free space void with relative dielectric constant of 1 as compared with the typical relative dielectric constant in solids of 2–3, the ac electric field in the void is about 2–3 times that in the solid, often leading to spark discharges.

Thin films usually have many defects of pin holes and irregular thickness, but with thin metallic electrodes they can be self-healing after electrical breakdown as the heat released by the breakdown evaporates the metal, reducing the electric stress across the breakdown channel. The remaining part of the film can be high-voltage stressed.

Glass and ceramic solids can typically work up to ~250 °C. Thermoplastic polymers can be molded and extruded at about 100–120 °C. Heat-curing solids, such as epoxy, have substantial mechanical strength and hardness and prevent the entry of moisture. Epoxy insulators reinforced with glass fibers weigh about one-fourth to one-half as much as corresponding porcelain insulators for the same mechanical and electrical strength.

Many polymers of extremely large molecular size such as polyethylene (C_2H_4), polypropylene (C_3H_6), polyvinylchloride (C_2H_3Cl), polystyrene (C_8H_9), polymethylmethacrylate ($C_5H_9O_2$), and polytetrafluoroethylene (C_2F_4) are highly insulating as a consequence of their covalent bonding. Conduction is usually due to polar or ionic impurities. Cross-linking by chemical or radiation means joins molecular chains at many points to form an extensive, insoluble network. Cross-linked polyethylene (XLPE) has been a replacement for thermoplastic polyethylene in power cables because it can operate at higher temperatures (~90 °C versus ~70 °C for polyethylene) and has improved chemical resistance. It has a higher breakdown strength and longer life, primarily due to its resistance to treeing from acetophenone, which results from a peroxide chemical cross-linking reaction that acts as a tree-retardant additive by absorbing energetic electrons. In the past XLPE was produced by steam curing at 200–220 °C at a pressure of 1.6–2 MPa, but partly in order to minimize absorption of moisture, present processes often use dry curing with hot nitrogen.

Surface failures can occur as a result of surface contamination, cracks, imperfections, and roughness. Insulators should have adequate electrical and mechanical strength with temperature variation, rain, and pollution. When pollution accumulates on insulator surfaces and its soluble ingredients dissolve in moisture from dew or fog, a conducting path is established from the high-voltage conductor to ground. The leakage currents form hot spots that dry up the moisture, forming dry bands. The dry band takes on a major fraction of the voltage drop, causing partial arcs leading to flashover.

3.3.2 Treeing The major degradation process in solids is the formation of electrical and water trees. Treeing is an electrical pre-breakdown phenomena that often initiates at a point of high electric stress and progresses by periodic partial discharges. Treeing lowers the electric breakdown strength and often leads to electric breakdown, and it is the most likely aging mechanism in solid extruded polymers. Electrical trees in solids have a similar appearance to the "trees" shown in Fig. 6 for an electrically stressed hydrocarbon liquid.

Electrical trees are dark, hollow channels resulting from the decomposition of material and are characterized by carbonized branch or bush-type dendrites. Electrical trees initiate from electron injection and mechanical fatigue due to periodic electrical stresses, Joule heating and thermal decomposition, and the electric stress enhancement due to contaminants and voids. Water trees are very fine filamentary paths between small cavities often caused by cracks from pressure through which moisture penetrates driven by an electric field. They are more bushlike or fanlike than electrical trees. Vented electrical or water trees start at an insulation surface, while nonvented (bow-tie) trees grow from contaminants or cavities within the insulation and progress symmetrically outward parallel to the electric field toward both electrodes. Water trees can be made visible by means of methylene blue dye.

Polymer cables are typically used in HVDC power transmission, space and airborne applications, and x-ray machines. HVDC can compete economically with HVAC and offers advantages in long underground, submarine, and long overhead lines, and as interconnections between large ac systems that differ in frequency such as in Japan, which has power networks that operate at 50 or 60 Hz.

The main form of cable deterioration under ac conditions is discharges. At the discharge inception voltage there are four discharges per cycle, which gradually causes erosion and electrical trees that rapidly propagate through the insulation to complete the breakdown. Under dc, the repetition rate of partial discharges is several orders of magnitude less than for similar ac voltages.

4. MEASUREMENT METHODOLOGIES OF ELECTRICAL INSULATION

Electrical insulation applications and research have a venerable history and continue to play an essential role in modern technology. As new materials are developed and as insulation system requirements are increased, the need for improved measurement methods continues to use the most modern tools. In this section current measurement methodologies are briefly described.

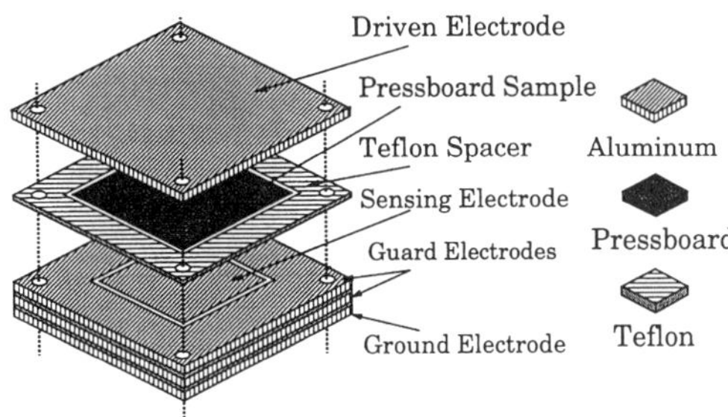

FIG. 7. Structure of the parallel-plate dielectrometry sensor used by Sheiretov and Zahn (1995).

4.1 Dielectrometry Measurements

4.1.1 Sinusoidal Steady State The most direct measurement of the electrical properties of insulators is from terminal voltage–current measurements of electrical impedance. Under ac steady-state operation, every electrical quantity $F(t)$ may be expressed as the real part of a complex quantity varying at radian frequency ω:

$$F(x,y,z,t) = \Re\{\hat{F}(x,y,z)^{j\omega t}\}. \tag{66}$$

If permittivity ϵ and conductivity σ are constant with time, we may write the total current density as a sum of displacement and conduction current densities. In terms of complex amplitudes,

$$\begin{aligned}\hat{J} &= \hat{J}_d + \hat{J}_c = j\omega\epsilon\hat{E} + \sigma\hat{E} \\ &= j\omega\hat{E}(\epsilon - j\sigma/\omega) = j\omega\epsilon^*\hat{E},\end{aligned} \tag{67}$$

where $\epsilon^* = (\epsilon - j\sigma/\omega)$ is the complex permittivity.

The dielectric spectrum of a material is a representation of its complex permittivity $\epsilon^* = \epsilon' - j\epsilon''$ as a function of frequency. The real component, $\epsilon' = \epsilon$, gives the dielectric constant while the imaginary component, $\epsilon'' = \sigma/\omega$, determines the power dissipation (loss) in the material.

In an Ohmic material, ϵ and σ are independent of the frequency or amplitude of the applied electric field, and a plot of log (ϵ''/ϵ_0) versus log ω has a slope -1. In a dispersive material when ϵ'' is plotted against frequency on a log-log scale, it can be characterized by one or more loss peaks.

4.1.2 Parallel-Plate Sensor The simplest measurement is to place the material between a pair of parallel-plate electrodes of known area and separation as described by Sheiretov and Zahn (1995). This test cell can be modeled as a resistor in parallel with a capacitor. The complex admittance of the structure can be measured, and from that its permittivity and conductivity can be calculated.

The parallel-plate sensor shown in Fig. 7 is used to measure the dielectric spectrum of typical transformer insulation, oil-impregnated pressboard, as a function of moisture and temperature. The capacitive structure is composed of the driven electrode, the sensing electrode, and guard electrodes. The input admittance, with which the sensing electrode is loaded, is that of a known parallel *RC* pair, modeled as shown in Fig. 8 with

$$R_T = d/\sigma A, \quad C_T = \epsilon A/d, \tag{68}$$

where d is the plate separation distance, A is the sensing electrode area, and σ and ϵ are the material's conductivity and permittivity, respectively.

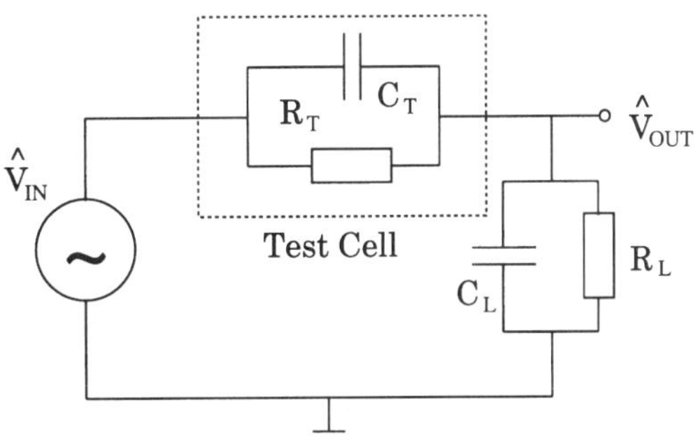

FIG. 8. Equivalent circuit of the test structure for dielectrometry measurements.

For linear time-invariant (LTI) systems we take the standard complex sinusoidal form:

$$v_{IN} = \mathcal{R}\{\hat{v}_{IN}e^{j\omega t}\} \quad v_{OUT} = \mathcal{R}\{\hat{v}_{OUT}e^{j\omega t}\} \qquad (69)$$

with complex voltage amplitudes $\hat{v}_{IN}$ and $\hat{v}_{OUT}$ defined in Fig. 8. A microprocessor-driven controller is responsible for generating the driving voltage and measuring the output voltage. The data that it produces are expressed in terms of a magnitude 20 log(M) [dB] and phase $180\varphi/\pi$ [deg], which are related to the complex amplitudes defined in Fig. 8 as

$$\frac{\hat{v}_{OUT}}{\hat{v}_{IN}} = Me^{j\varphi} \Rightarrow M = \left|\frac{\hat{v}_{OUT}}{\hat{v}_{IN}}\right|,$$
$$\varphi = \angle\left(\frac{\hat{v}_{OUT}}{\hat{v}_{IN}}\right) = \arctan\frac{\mathcal{I}\{\hat{v}_{OUT}/\hat{v}_{IN}\}}{\mathcal{R}\{\hat{v}_{OUT}/\hat{v}_{IN}\}}. \qquad (70)$$

The next step in calculating σ and ϵ is to calculate R_T and C_T from measurements of M and φ, and the known values of R_L and C_L. We define the admittances of the test and load branches as $Y_T = 1/R_T + j\omega C_T$ and $Y_L = 1/R_L + j\omega C_L$, respectively. Then from the voltage-divider relationship we obtain

$$\frac{\hat{v}_{OUT}}{\hat{v}_{IN}} = \frac{Y_T}{Y_T + Y_L} = \frac{1/R_T + j\omega C_T}{1/R_T + 1/R_L + j\omega(C_T + C_L)} = Me^{j\varphi}. \qquad (71)$$

Solving (71) for R_T and C_T in terms of known values for R_L and C_L and measured values for M and φ yields

$$R_T = \frac{1 + M^2 - 2M\cos\varphi}{M[\cos\varphi - M - \omega R_L C_L \sin\varphi]}R_L,$$
$$C_T = \frac{M[\cos\varphi - M + (1/\omega R_L C_L)\sin\varphi]}{1 + M^2 - 2M\cos\varphi}C_L. \qquad (72)$$

Representative measurements in Fig. 9 show that Shell Diala A transformer oil is Ohmic. On a log-log scale, the plot of ϵ'' versus frequency is a straight line of slope -1, which means that σ is independent of the frequency. For linear dielectric materials, ϵ' should also be constant with frequency. The observed rise of ϵ' at low frequencies can be attributed to electrical double-layer formation at the aluminum-oil interfaces.

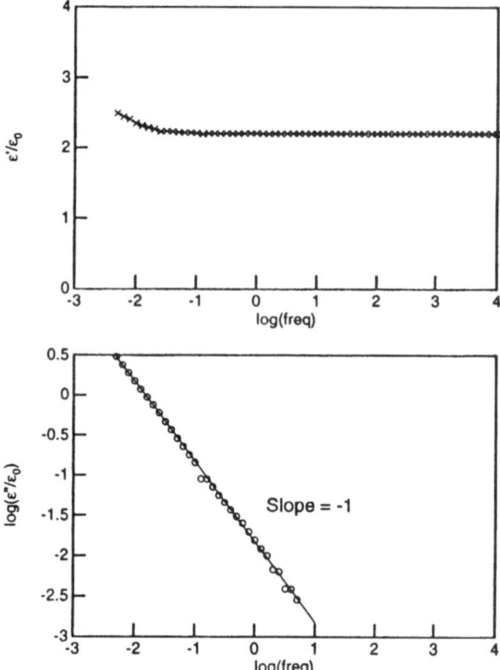

FIG. 9. Complex permittivity of transformer oil measured with the parallel-plate sensor. The -1 slope for the loss measurement indicates that transformer oil is Ohmic. The electrical double layers at the electrode/oil interfaces cause the low-frequency value of ϵ' to increase.

The plot of ϵ'' is not shown for frequencies higher than $10^{0.7}$ Hz. This is because at that high-frequency range the response is dominated by the capacitive element, and the gain/phase measurement is insensitive to the value of σ.

4.1.3 Oil-Impregnated Pressboard

The parallel-plate sensor was also used to study how the dielectric spectrum of oil-impregnated pressboard changes with variations in temperature and moisture content in order to model temperature and moisture dynamics in an operating transformer. First, many samples of pressboard were prepared, each with a different content of water. Then the moisture content of a small piece of each sample was measured, and the remaining sample was placed in the sensor structure, its dielectric spectrum was scanned at five different temperatures, and finally its mois-

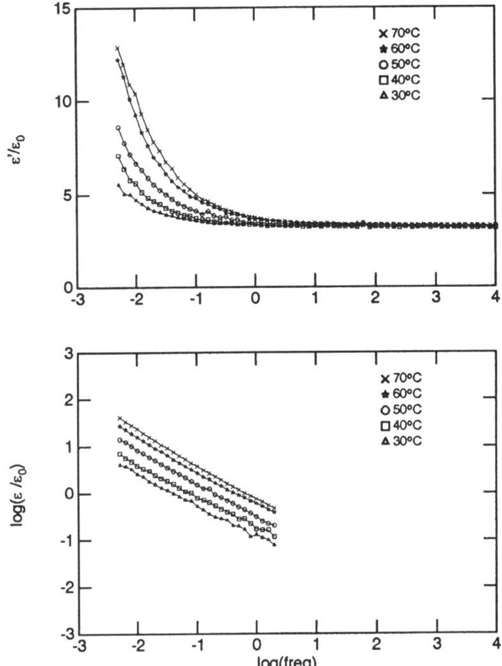

FIG. 10. Dielectric spectra of an oil-impregnated pressboard sample of 2.3% moisture at five temperatures. The slope of the loss characteristic is ~ −0.7, indicating that pressboard is dispersive.

ture content was measured again. Figure 10 shows representative dielectric spectrum measurements as a function of temperature for pressboard with 2.3% moisture. The slope of log ϵ'' versus log ω is -0.7, indicating that pressboard is dispersive, i.e., the conductivity is a function of frequency.

4.1.4 Imposed ω-k Dielectrometry If electrodes are placed side by side in a repeating pattern on one surface of the material, the electric fields will decrease away from the electrodes, and the complex impedance between the electrodes will be most sensitive to the material adjacent to them. Periodic electrode structures with a multitude of interdigitated fingers, as shown in Fig. 11, have been designed. The electric fields are uniform in the z direction and periodic in the y direction with a spatial wavelength of λ, equal to the distance between repeating electrodes. Thus at every surface of constant x, the electric potential is periodic in y and can be expanded as an infinite series of sinusoidal Fourier modes of spatial wavelengths $\lambda_n = \lambda/n$. This is very convenient, because the two-dimensional solutions to Laplace's equation $\nabla^2 \Phi = 0$ in Cartesian geometry are of the form

$$\Phi(x,y) = \sum_{n=1}^{\infty} \Phi_n \,\mathrm{hyp}(k_n x)\, \mathrm{trig}(k_n y), \tag{73}$$

where $k_n = 2\pi/\lambda_n$ is the wave number of the nth mode; hyp(x) stands for any one of the hyperbolic exponential functions sinh(x), cosh(x), e^x, or e^{-x}; and trig(y) stands for one

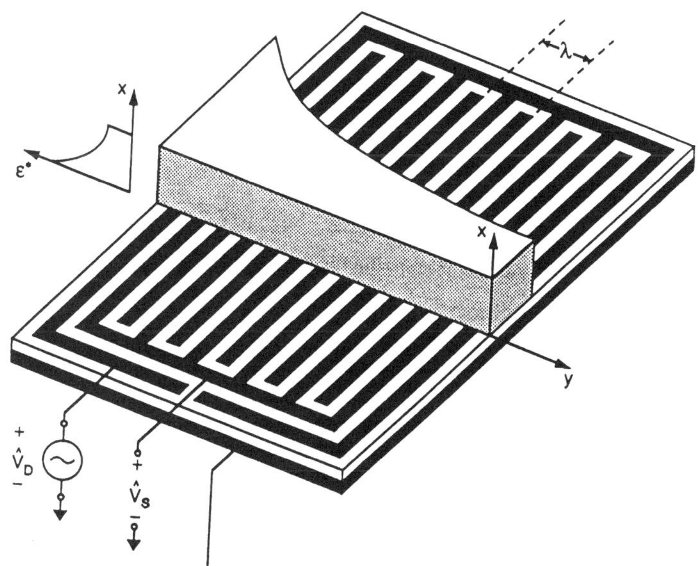

FIG. 11. Interdigital electrode structure for imposed ω-k dielectrometry measurements to measure the complex permittivity ϵ^* of the dielectric material ($x > 0$) above the interdigitated electrodes.

of the trigonometric functions sin(y) or cos(y). For every Fourier mode n, the electric fields decrease with x as $\exp(-2\pi nx/\lambda)$ with the fundamental mode $n = 1$ penetrating farthest into the material.

By designing sensors with various fundamental spatial wavelengths λ, it is possible to measure the dielectric properties of materials at different depths. Combining the results from several such sensors makes it possible to determine the x-dependent spatial profiles of the complex permittivity. The three-wavelength sensor structure shown in Fig. 12 consists of three sets of interdigitated electrodes, deposited on a common flexible Kapton (a polyimide) substrate. Current research is using this sensor to measure the moisture diffusion profiles in transformer pressboard insulation as the temperature changes.

4.2 Flow Electrification

4.2.1 Background to the Problem Electrification has been long recognized as a problem in the flow of petroleum liquids in pipes, through filters, and in charge accumulation in storage tanks. With changes in the chemical makeup of insulating liquids used both to transfer heat and/or to withstand high electric stress, the use of new dielectric materials such as polymers and cellulosic materials, and an increase in the flow speeds for greater cooling, the electrification problem has recently arisen in electric power apparatus and in automotive fuel systems. Polymeric replacements for metal parts in automotive fuel lines, filters, and pumps have caused flow-electrification–initiated discharges that are hazardous.

A number of failures of large oil-cooled power transformers and in a trichlorotrifluorethane-cooled HVDC valve have been attributed to flow-induced electrification. Oil is generally used to cool the windings of a power transformer with the flow by either forced or natural convection through a heat exchanger to dissipate the generated heat. The moving oil provides the transport of entrained charge separated at liquid/solid interfaces by physical, chemical, and electrical phenomena. Moving transformer oil generally entrains positive charge in the charged oil through the transformer structure, with the oil playing the role of the belt of a Van de Graaff generator. Charge separation at interfaces between moving fluid and boundaries with the accumulation of charge on insulators or isolated conductors can lead to high field strengths and electrical discharges. The potential builds up until the rate of charge accumulation equals the rate of charge leakage, or until spark discharges occur.

4.2.2 The Absolute Charge Sensor To overcome measurement uncertainties from charge separation at surfaces of charge-measuring devices, an absolute charge sensor (ACS) using Faraday-cage principles has been developed that provides an absolute measurement of the charge density that is independent of the fluid's electrical properties, the velocity of the fluid, and any electrification process within the instrument. More information about the ACS can be found in the reference by Morin *et al.* (1991a).

A schematic of the ACS is illustrated in Fig. 13. Fluid such as transformer oil or hydrocarbon fuel is drawn into an electrically conductive container as the metal bellows expands. As the ACS is sampling fluid, either the voltage or the current is measured by a high- or low-impedance electrometer, respec-

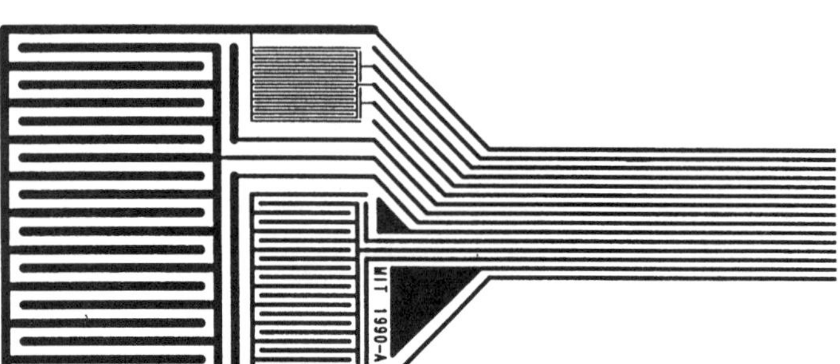

FIG. 12. Structure of a three-wavelength interdigitated dielectrometry sensor to measure spatial distributions of complex permittivity.

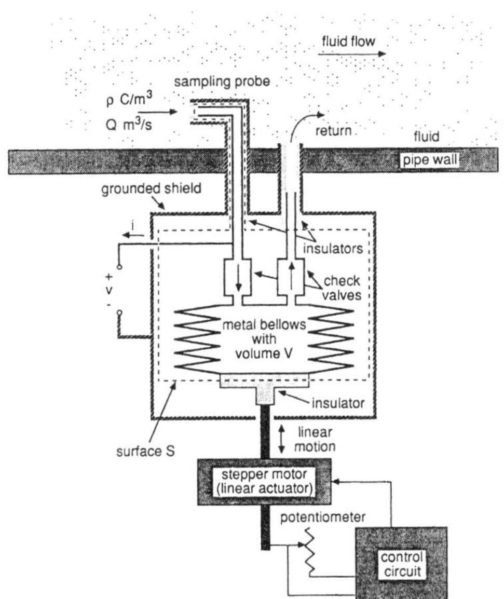

FIG. 13. Electromechanical schematic of the absolute charge sensor (ACS).

tively, connected across the electrical terminals. Once the measurement is complete, the fluid is expelled through the exit check valve and conduit by reversing the direction of the actuator to compress the bellows volume.

A charged body introduced into a perfectly insulating volume surrounded by a conductive container will induce an opposite charge on the interior wall of the container. Thus, without actually being transported to the wall, the net charge induced onto the conductive container inner wall would be equal to but of opposite polarity to that in the fluid. Because the rate of accumulation of the charge so induced on the inner wall of the container constitutes a capacitive electrical current that can be measured externally, it is then possible to measure this net charge by means of an external circuit having a sufficiently low impedance that the potential of the container would remain essentially zero relative to that of the surrounding shield. Because the conductive container forms a Faraday cage, the current measured during the filling cycle is independent of charging processes that occur inside the Faraday cage whether they are due to imaging, charge relaxation, or electrification. The total charge within the ACS with entrained volume charge is balanced by image surface charge on the wall. Thus, as soon as charge enters the ACS inner probe conduit, it is immediately recorded by the electrometer, whether the charge remains entrained in the fluid or is conducted to the wall. The charge density ρ is related to the short-circuit current i as $\rho = i/Q$, where Q is the fluid flow rate into the ACS.

4.2.3 Flow-Loop Measurements A typical measurement flow loop that models oil circulation for cooling in a transformer is shown in Fig. 14. Similar flow loops are also used to model fuel flows in vehicles. In Fig. 14, a pump drives transformer oil through piping connected to a large-volume reservoir (55-gal. oil drum), electrically isolated from the grounded piping by insulated tubing. The system includes a parallel flow loop that is electrically isolated by nylon fittings and contains a paper oil filter that can be valved in or out through valves 2 and 3, a gate valve (valve 1) to control the flow rate through the main flow path, and a flow meter to monitor the total flow rate. The flow speed could also be independently controlled by varying the pump impeller speed from the ac motor controller. The paper oil filter is used as a source of electrification charge that can be controlled by appropriate setting of valves 2 and 3. If all the flow is directed through the filter by fully closing valve 1 and fully opening valves 2 and 3, the maximum flow rate is 18 gal/min and the volume charge density is of order 100–500 $\mu C/m^3$ depending on oil temperature over the range of 65–100 °F. If none of the flow passed through the filter as a result of fully opening valve 1 and fully closing valves 2 and 3, the maximum flow rate is 26 gal/min and the volume charge density is of order 4–10 $\mu C/m^3$.

The electrometer current from the oil drum equals the difference in convection currents entering (measured by ACS 1) and leaving (measured by ACS 2) the oil drum. The ACS-measured charge densities at the inlet, ρ_{in}, and outlet, ρ_{out}, of the oil drum with flow rate Q give the difference in convection currents, i_{acs}, as

$$i_{acs} = (\rho_{in} - \rho_{out})Q. \tag{74}$$

From conservation of charge, the difference in convection currents should equal the total circuit current flowing from the drum

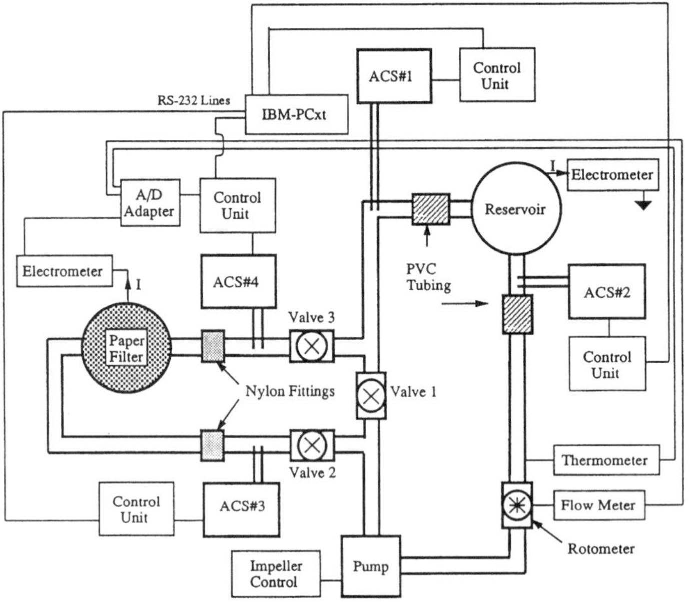

FIG. 14. A pump drives transformer oil through a closed flow loop connected through a large-volume reservoir (55-gal. drum). Two ACSs monitor the charge densities at the inlet (1) and outlet (2) to the electrically isolated reservoir, which is virtually grounded through a current-measuring electrometer. The oil charge density is increased by closing valve 1 and opening valves 2 and 3 to direct the flow through the electrostatically charging paper filter. The charge densities at the inlet and outlet of the paper filter are measured by ACSs 3 and 4. Measured electrometer currents at the paper filter and reservoir should equal the difference in convection currents in and out.

through the electrometer to ground, $i_{acs} = i_E$.

4.2.4 Couette Charger A Couette charger (CC), shown in Fig. 15, simulates flow electrification processes in transformers, where transformer oil fills the annulus between coaxial cylindrical electrodes that can be bare metal or transformer pressboard covered, as described by Morin *et al.* (1991b). The inner cylinder can rotate at speeds giving controlled turbulent flow that brings electric charge to the volume from the electrical double layer at the liquid/solid interfaces. Figure 16 shows the approximate linear charge-density profiles in each hydrodynamic sublayer of thickness δ_1 and δ_2, joining the wall charge-density values ρ_{w1} and ρ_{w2}, to the turbulent core charge density ρ_0. This compact apparatus allows for flexibility in testing various oils and transformer pressboards at controlled temperatures and moisture levels in the oil and pressboard. Flow electrification measurements as a func-

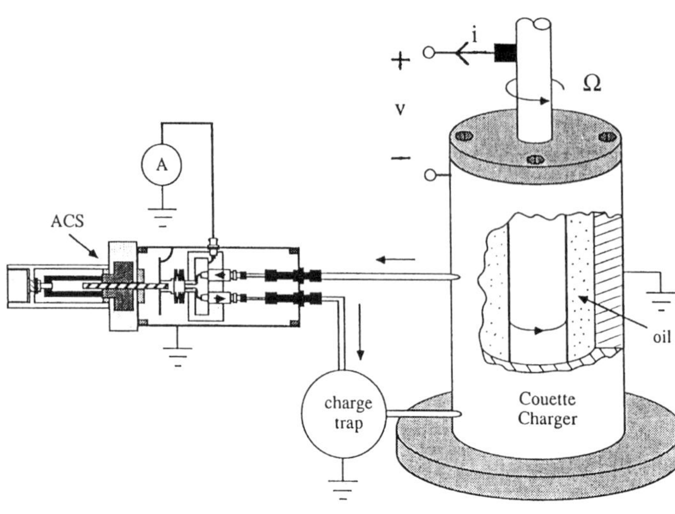

FIG. 15. Couette charger with electrical terminals connecting inner and outer cylinders for measuring open-circuit terminal voltage or short-circuit current or for applying a voltage between the cylinders. The absolute charge sensor is used to measure the charge density in the turbulent core. The charge trap is a large volume where the fluid residence time is much longer than the fluid dielectric relaxation time. It is used so that fluid exiting the ACS into the Couette charger is uncharged.

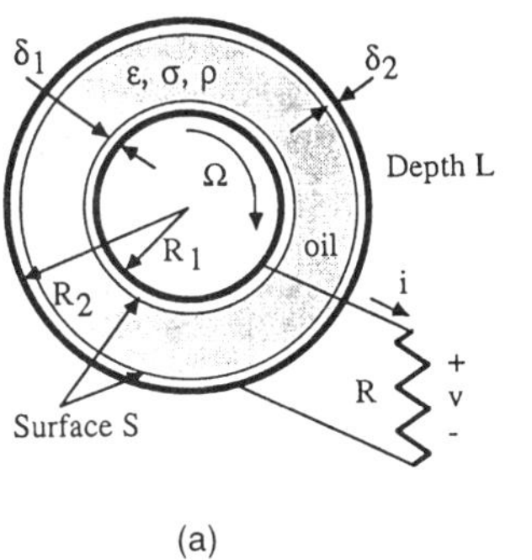

(a)

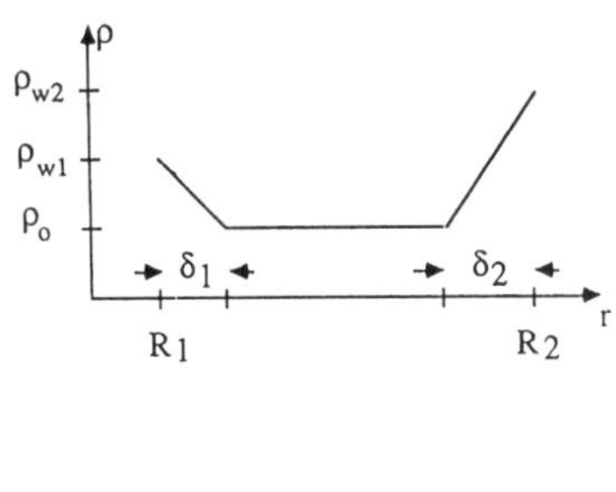

(b)

FIG. 16. (a) Cross section of the Couette charger with load resistance R connected to the terminals. (b) Charge-density distribution between the inner and outer cylinders showing the wall charge densities ρ_{w1} and ρ_{w2} that describe the governing electrokinetics at the oil/cylinder interfaces. Also shown is the approximately linear charge-density profile in each sublayer joining the wall charge-density values to the turbulent core charge density ρ_0 at the edge of each sublayer with thicknesses δ_1 and δ_2.

tion of inner cylinder speed are also performed from open-circuit voltage and short-circuit current terminal measurements, as shown in Fig. 17. Transient measurements with a step change in temperature have shown the charge density to change from an initial value to a new steady-state value, including cases of polarity reversal, both values dependent on the equilibrium moisture levels in oil and pressboard, as shown in Fig. 18.

4.3 Kerr Electro-optic Field-Mapping Measurements

4.3.1 Governing Equations High-voltage–stressed materials are usually birefringent, in which case the refractive indices for light (of free-space wavelength λ) polarized parallel, $n_{\|}$, and perpendicular, $n_{\perp}$, to the local electric field are related by

$$n_{\|} - n_{\perp} = \lambda B E^2, \tag{75}$$

where B is the Kerr constant and E is the magnitude of the applied electric field. The phase shift ϕ between light-field components propagating in the direction perpendicular to the plane of the applied electric field along an electrode length L is

$$\phi = 2\pi B E^2 L = \pi (E/E_m)^2, \quad E_m = (2BL)^{-1/2}. \tag{76}$$

If the birefringent dielectric is placed within a circular polariscope with aligned or crossed polarizers, the transmitted light intensity is

$$\frac{I}{I_0} = \begin{cases} \sin^2 \frac{1}{2}\pi (E/E_m)^2, & \text{crossed polarisers,} \\ \cos^2 \frac{1}{2}\pi (E/E_m)^2, & \text{aligned polarisers.} \end{cases} \tag{77}$$

Field minima and maxima occur when

$$\frac{E}{E_m} = \sqrt{n} \begin{cases} n \text{ odd} \begin{pmatrix} \text{minima} & \text{AP} \\ \text{maxima} & \text{CP} \end{pmatrix}, \\ n \text{ even} \begin{pmatrix} \text{minima} & \text{CP} \\ \text{maxima} & \text{AP} \end{pmatrix}. \end{cases} \tag{78}$$

E_m is thus the field magnitude for first maximum with crossed polarizers and first minimum with aligned polarizers. For test cells of length $L = 1.1$ m with highly purified water at $\lambda = 590$ nm, $B \approx 2.8 \times 10^{-14}$ m/V^2 and $E_m \approx 35$–36 kV/cm. As an example of Kerr measurements of electric field distributions, Fig. 19 shows a circular polariscope experimental apparatus with representative measurement using coaxial cylindrical electrodes.

When a high-voltage pulse is applied, as in Fig. 20, a photomultiplier-tube optical detector goes through the series of maxima and minima related by $\sqrt{n}$ as in (78), as described in Zahn (1988).

4.3.2 Representative Measurements in Highly Purified Water Representative pulsed high-voltage measurements of highly purified water are shown in Fig. 21 with parallel-plate electrodes for different metallic electrodes showing differences in the magnitude and sign of the injected charge as presented by Zahn *et al.* (1986). Stainless-steel electrodes generally inject positive charge; aluminum injects negative charge, while brass can inject either positive or negative

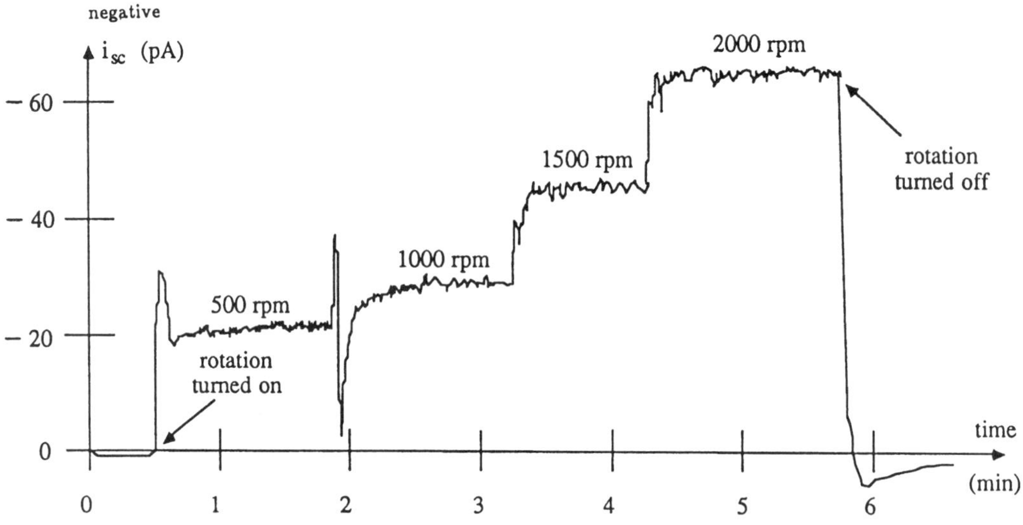

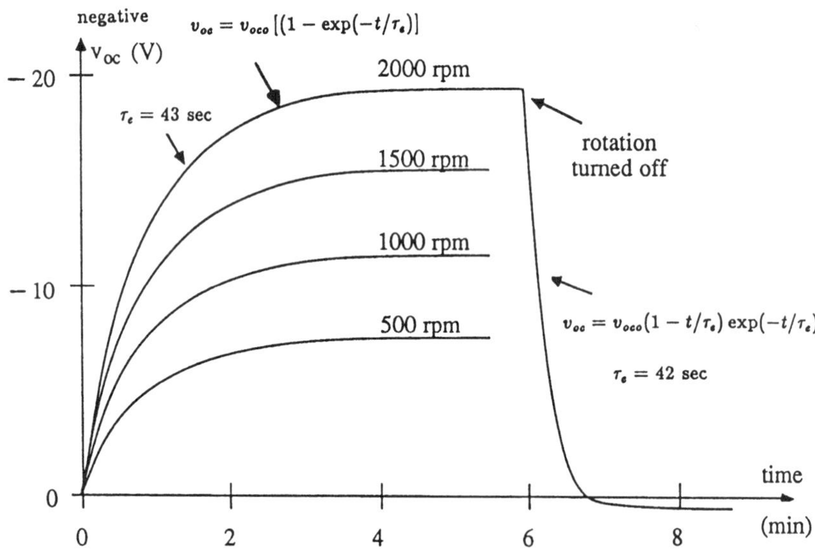

FIG. 17. Measured transients of short-circuit current and open-circuit voltage as functions of time for various inner-cylinder rotational speeds.

charge. Thus by appropriate choice of electrode-material combinations and voltage polarity, it is possible to have uncharged, unipolar charged negative or positive, or bipolar homocharge charged liquid.

Breakdown strengths are generally higher with bipolar injection. For example, with brass–aluminum electrodes, the polarity for bipolar injection had a breakdown strength of ~125–135 kV/cm, while the reverse polarity had negative charge injection with breakdown strength ~90–95 kV/cm. Similarly, stainless steel–aluminum electrodes had a breakdown strength with bipolar injection of ~125–140 kV/cm, while the reverse polarity had no charge injection with a breakdown strength of ~105 kV/cm. This increase in breakdown strength for homocharge distri-

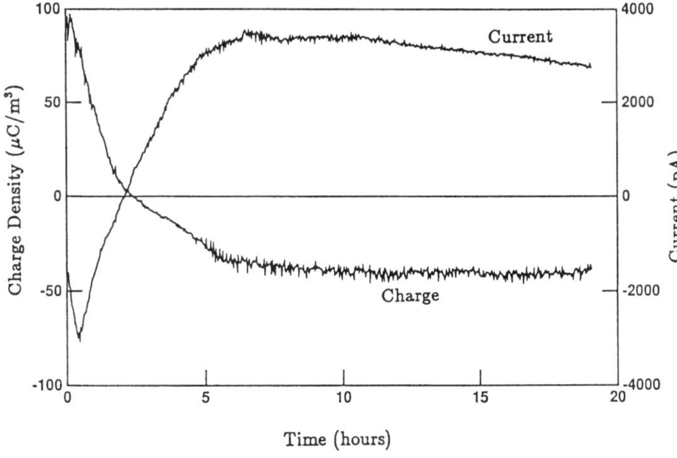

FIG. 18. Representative charge density and short-circuit terminal current responses to a step change in temperature from 40 to 70 °C with pressboard of 1.7% moisture covering inner- and outer-cylinder surfaces. The inner cylinder was rotating at 1000 rpm. Oil moisture started at 21 ppm and ended at a value of 35 ppm, which was approaching equilibrium at 70 °C.

butions is due to the decrease in electric field at both electrodes resulting from the space-charge shielding, as shown in Fig. 3(c). The electric field is increased in the center of the gap, but breakdown does not occur because the intrinsic strength of the dielectric in the volume is larger than at an interface.

4.3.3 Electron-Beam–Irradiated Polymethylmethacrylate Polymethylmethacry-

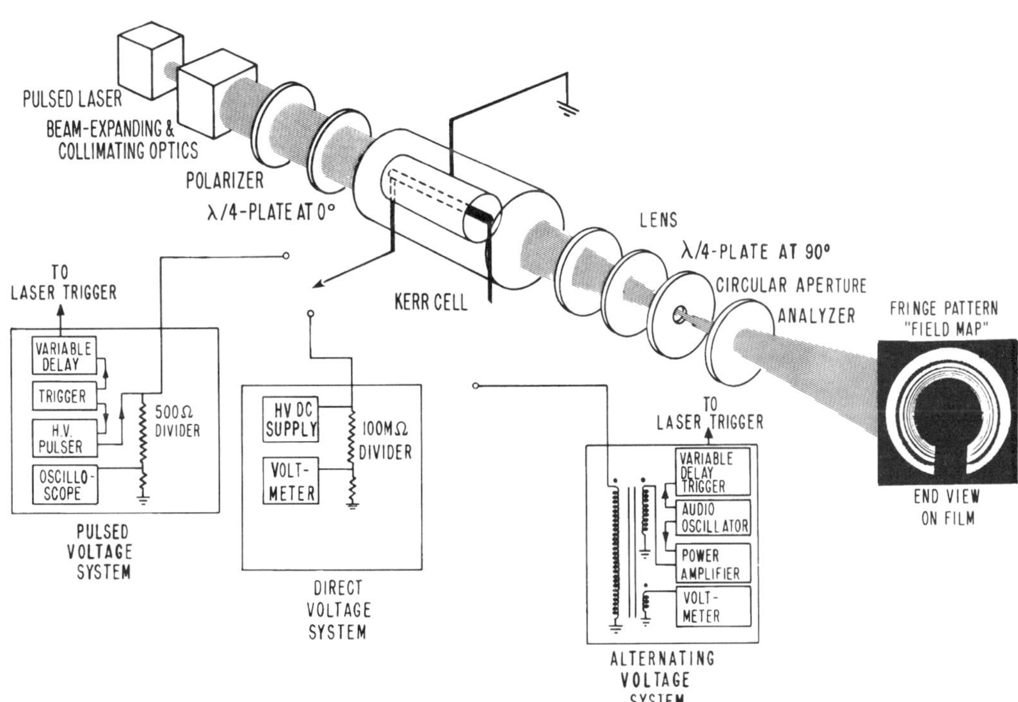

FIG. 19. Circular polariscope experimental configuration shown for Kerr electro-optic field-mapping measurements using coaxial cylindrical electrodes for pulsed, dc, and ac high voltages. The photographic film shows circular light maxima and minima as the nonuniform electric field between cylindrical electrodes satisfies the conditions of Eq. (78) for multiple values of *n*.

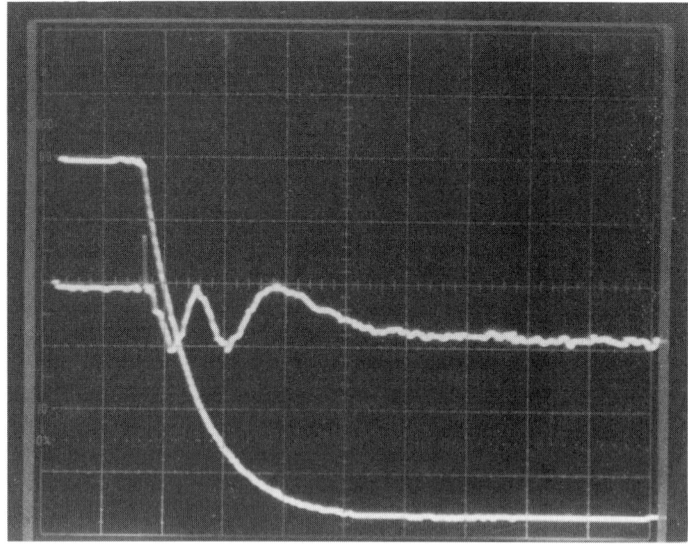

FIG. 20. The output of an optical photomultiplier tube goes through a series of maxima and minima [here n = 1–4 in Eq. (78)] when a 100-kV negative high-voltage pulse is applied, here shown for 77% water, 23% ethylene glycol by weight at 2.7 °C using crossed polarizers (time base 20 μs cm^{-1}). From this measurement, the Kerr constant of the mixture at λ = 633 nm wavelength is $B \approx 2.2 \times 10^{-14}$ m/V^2 with $E_m \approx 45.6$ kV/cm.

late (PMMA) is a convenient highly insulating polymer to work with because it is transparent at room temperature. When an electron beam is used to irradiate PMMA samples, the accumulated trapped charge leads to high internal self electric fields up to 3.5 MV/cm so that for a 10-cm-long PMMA sample, numerous light minima and maxima arise, making photographic and videotape measurements possible, as demonstrated by Zahn *et al.* (1987) and Hikita *et al.* (1988).

With $B \approx 2 \times 10^{-15}$ m/V^2 for PMMA, $E_m \approx 0.5$ MV/cm for L = 10.2 cm and $E_m \approx 0.44$ MV/cm for L = 12.7 cm, for which representative images are shown in the linear polariscope configuration in Fig. 22. Incident light is in the z direction, polarized at $\pm 45°$ to the direction of the electric field E with an analyzing polarizer placed after the sample either crossed or aligned to the incident polarization. The He–Ne laser at 633 nm wavelength has its beam expanded to ≈7.5 cm to allow measurements of the light intensity distribution over the entire sample cross section. Beam splitters allow simultaneous measurements for aligned and crossed polarizers using Polaroid cameras as well as a videotape recording system.

The electron beam is generated by a Van de Graaff generator, exits from the accelerator tube through a thin (76 μm) aluminum window, and passes through ≈50 cm air to the PMMA sample that is short-circuited through current monitors at the top and bottom surfaces. The energy loss in the window and intervening air is about 160 keV. Figure 22 also shows representative data near the cameras for an accelerator beam energy of 2.6 MeV and a current density of 20 nA/cm^2 after ≈60 s of beam irradiation for aligned and crossed polarizers for a $d \approx 1.27$-cm-thick PMMA sample uniformly irradiated over its width, 5.1 cm, and length, 10.2 cm. The sample had $E_m \approx 0.5$ MV/cm. The sample is shown before irradiation to have some mechanical birefringence. Also shown in Fig. 22 is a representative frame of light intensity distribution from a computerized image digitizer for a larger current density of 110 nA/cm^2 after 11 s of irradiation with $E_m \approx 0.44$ MV/cm and L = 12.7 cm.

4.4 Acoustic Measurements of Space-Charge Distributions

In nontransparent materials, acoustic measurements offer an alternative to optical measurements. The two usual approaches for measuring space-charge distributions are a pressure-wave method and a pulsed electroacoustic method. Both methods result in a measured time signal that is a direct image of the spatial profiles of charge density or electric field. In addition to fundamental measurements of conduction properties, trapped charges in solids are useful for microphones, sensors, actuators, and optoelectronics.

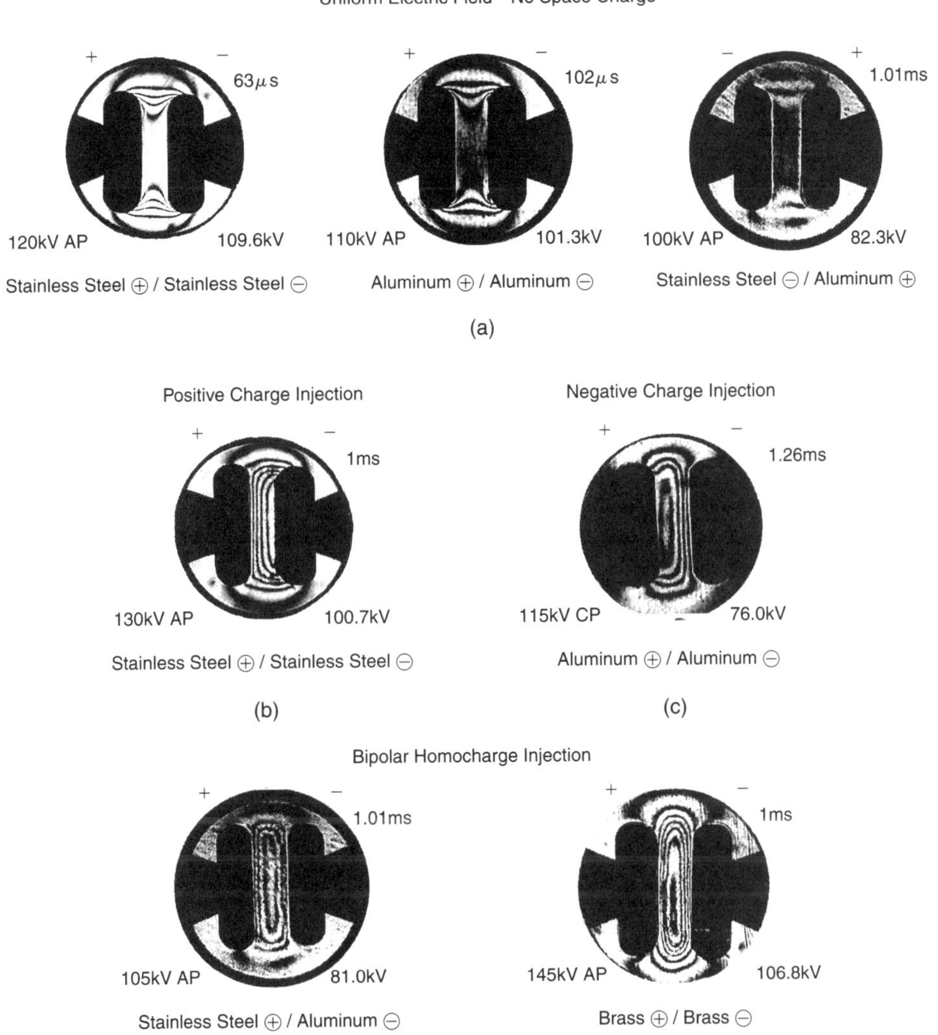

FIG. 21. Representative Kerr-effect measurements with a 1-cm gap between parallel-plate electrodes stressed by a high-voltage step in highly purified water with $E_m \approx$ 35–36 kV/cm. (a) The space-charge–free uniform electric-field distribution so that the Kerr effect shows uniform light transmission in the interelectrode gap region. At times less than about 100 μs after the step was applied, the dielectric is essentially space-charge free for all electrode metal combinations. With stainless steel negative and aluminum positive, the dielectric is space-charge free for all time. (b) Positive charge injection with stainless-steel electrodes so that the electric-field distribution has positive slope with decreased electric field near the positive electrode and increased field at the negative electrode. (c) Negative charge injection with aluminum electrodes so that the electric-field distribution has negative slope with decreased electric field near the negative electrode and increased field at the positive electrode. (d) Bipolar homocharge injection, where the electric field is decreased at both charge-injecting electrodes, with the peak field in the central gap region. In each photograph, the electrode metals and polarity are listed at the bottom; the initial charging voltage and whether aligned (AP) or crossed (CP) polarizers are used are given at the lower left; the instantaneous voltage is given at the lower right; and the time after high voltage is applied is given in the upper right.

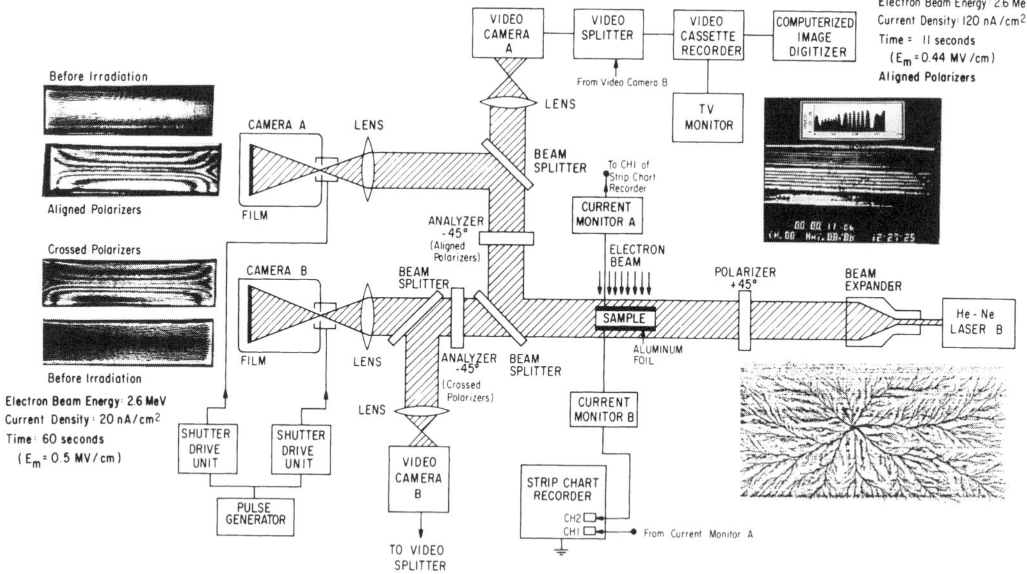

FIG. 22. Apparatus and representative data for Kerr electro-optic field-mapping measurements with simultaneous aligned and crossed polarizers in electron-beam–irradiated samples using photographic film and a computer-interfaced videotape recording system as detectors. In the lower right is a representative electric tree in PMMA after electric breakdown.

4.4.1 Pressure-Wave Method A pressure wave is produced at one end of a sample using a piezoelectric transducer or by the heating from an intense laser pulse, as developed by Lewiner (1986) and Maeno *et al.* (1988). As the pulse propagates through the sample at the acoustic wave speed c, space charge will be displaced with the material causing a displacement current in the sample. The resulting open-circuit voltage is proportional to the electric field, $v(t) = \alpha E(x = ct)$, while the short-circuit current density is proportional to the volume charge density, $j(t) = \alpha c \rho(x = ct)/s$ where s is the sample thickness and α is the proportionality constant related to sample compressibility, pulse pressure, and pulse duration. The shape of measured voltage or current in time is identical to the spatial distribution of electric field E or charge density ρ. Measurements have typically been applied to polymers and ceramics but have also been applied to water trees in polymers and oil-impregnated pressboard. With a typical acoustic speed of $c = 2000$ m/s, a 1-ns pressure pulse has a spatial resolution of 2 μm.

4.4.2 Electroacoustic Method A high voltage is applied, and the Coulomb force on the space charge produces a propagating pressure wave that is detected by a piezoelectric transducer. The transducer signal is proportional to the convolution of the charge density and applied electric field, so that the Fourier transform of the transducer signal in time is proportional to the product of spatial Fourier transforms of charge density and electric field. Recent work has used this method to measure space-charge distributions in XLPE power cables.

4.5 Thermally Stimulated Measurements

A solid sample that is metallized on both surfaces is poled with a high electric field at high temperature, allowing permanent dipoles to orient with the electric field and charge to build up at interfaces and at structural defects. With the electric field maintained, the sample is cooled to trap the charge and dipoles. Then with the sample short-circuited, the sample is reheated at a constant rate of the order of 1 °C/min, and the discharge current is measured with an electrometer, of order 10^{-12}–10^{-10} A. A number of peaks in current as a function of temperature occur whose shape and location are different for each material and characteristic of permanent dipole density, relaxation time, activation energy, and other trapping parameters.

GLOSSARY

Arrhenius Aging Model: An exponential thermal aging model that assumes a single chemical reaction rate dependent on an activation energy and absolute temperature.

Avalanche: A cumulative multiplication of charge carriers due to electric-field–induced impact ionizations.

Corona: Luminous partial discharges in gases.

Debye Length: Approximate thickness of space-charge layer near interfaces for drift/diffusion conduction.

Einstein Relation: Statement that the ratio of molecular diffusivity to charge mobility equals the thermal voltage, $D/\mu = kT/q$.

Electroacoustic Measurements: Electric-field and space-charge measurements by propagating acoustic waves through the dielectric.

Flow Electrification: Electrostatic hazard that arises from entrainment of electrical double-layer charge at interfaces that accumulates on isolated surfaces to cause spark discharges.

Fowler–Nordheim Field Emission: Quantum mechanical tunneling of electrons from a metal into vacuum or a dielectric under a strong electric field.

Kerr Electro-optic Effect: An electric-field–induced birefringence that can be used to optically measure electric-field distributions.

Paschen's Law: In a uniform electric field, the breakdown voltage in a gas is a function only of the pressure-gap product. The minimum voltage for breakdown is called the Paschen minimum.

Poisson–Boltzmann Equation: Poisson's equation for the scalar electric potential under equilibrium drift/diffusion conduction with the charge carriers obeying Boltzmann distributions.

Schottky Emission: The increase in thermionic emission of electrons from a metal due to the electric field.

Streamers: Growth of avalanches into irregular filamentary luminous discharges in gas and liquid dielectrics.

Thermally Stimulated Measurements: Flow of charge as poled dielectric is reheated at constant rate.

Townsend Ionization: Ionizing collisions that can lead to electrical breakdown if the increase in electrons is unbounded.

Treeing: Partial discharge dendritic patterns of hollow or water channels in solids that can lead to electrical breakdown.

Weibull Distribution: A statistical distribution to typically model life, failure, and dielectric strength of dielectrics.

Works Cited

Clark, F. M. (1962), *Insulating Materials for Design and Engineering Practice*, New York: Wiley.

Hikita, M., Zahn, M., Wright, K. A., Cooke, C. M., Brennan, J. (1988), "Kerr Electro-optic Field Mapping Measurements in Electron Beam Irradiated Polymethylmethacrylate," *IEEE Trans. Electrical Insulation*, **EI-23,** 861–880.

Khalifa, M. (1990), *High-Voltage Engineering, Theory and Practice*, New York: Marcel Dekker.

Klinkenberg, A., Van der Minne, J. L. (1958), *Electrostatics in the Petroleum Industry*, Amsterdam: Elsevier.

Lewiner, J. (1986), "Evolution of Experimental Techniques for the Study of the Electrical Properties of Insulating Materials," *IEEE Trans. Electrical Insulation*, **EI-21,** 351–360.

Maeno, T., Futami, T., Kushibe, H., Takada, T., Cooke, C. M. (1988), "Measurement of Spatial Charge Distributions in Thick Dielectrics Using the Pulsed Electroacoustic Method," *IEEE Trans. Electrical Insulation*, **EI-23,** 433–439.

Meek, J. M., Craggs, J. D. (1978), *Electrical Breakdown of Gases*, New York: Wiley.

Morin II, A. J., Zahn, M., Melcher, J. R., Otten, D. M. (1991a), "An Absolute Charge Sensor for Fluid Electrification Measurements," *IEEE Trans. Electrical Insulation*, **26,** 181–199.

Morin II, A. J., Zahn, M., Melcher, J. R. (1991b), "Fluid Electrification Measurements of Transformer Pressboard/Oil Insulation in a Couette Charger," *IEEE Trans. Electrical Insulation*, **26,** 870–901.

Rogers, R. R. (1978), "IEEE and IEC Codes to Interpret Incipient Faults in Transformers, Using Gas in Oil Analysis," *IEEE Trans. Electrical Insulation*, **EI-13,** 349–353.

Sheiretov, Y., Zahn, M. (1995), "Dielectrometry Measurements of Moisture Dynamics and Oil-Impregnated Pressboard," *IEEE Trans. Dielectrics Electrical Insulation*, **2,** 329–351.

von Hippel, A. R. (1959), *Molecular Science and Molecular Engineering*, New York and London: Technology Press of MIT and Wiley.

Zahn, M., Ohki, Y., Fenneman, D. B., Gripshover, R. J., Gehman, Jr., V. J. (1986), "Dielectric Properties of Water and Water/Ethylene Glycol Mixtures for Use in Pulsed Power System Design," *Proc. IEEE,* **74,** 1182–1221.

Zahn , M., Hikita, M., Wright, K. A., Cooke, C. M., Brennan, J. (1987), "Kerr Electro-optic Field Mapping Measurements in Electron Beam Irradiated Polymethylmethacrylate," *IEEE Trans. Electrical Insulation,* **EI-22,** 181–185.

Zahn, M. (1988), "Space Charge Effects in Dielectric Liquids," in: E. E. Kuhnhardt, L. G. Christophorou, L. H. Luessen (Eds.), *The Liquid State and Its Electrical Properties,* New York and London: Plenum Publishing Corp., pp. 367–430.

Further Reading

Insulation Theory, Practice, and Measurement Methods

Annual Book of ASTM Standards; Vol. 10.01, *Electrical Insulation-Solids, Composites and Coatings;* Vol. 10.02, *Electrical Insulation-Wiring and Cable, Heating and Electrical Tests;* Vol. 10.03, *Electrical Insulating Liquids and Gases; Electrical Protective Equipment,* Philadelphia: ASTM.

Cygan, P., Laghari, J. R. (1990), "Models for Insulation Aging under Electrical and Thermal Multistress," *IEEE Trans. Electrical Insulation,* **25,** 923–934.

Gallagher, T. J., Pearmain, A. J. (1983), *High Voltage Measurement, Testing and Design,* New York: Wiley.

Haus, H. A., Melcher, J. R. (1989), *Electromagnetic Fields and Energy,* Englewood Cliffs: Prentice Hall.

Zahn, M. (1979), *Electromagnetic Field Theory, A Problem Solving Approach,* New York: Wiley; also (1987), Malabar: Krieger.

Vacuum

Latham, R. V. (1995), *High Voltage Vacuum Insulation. Basic Concepts and Technological Practice,* London: Academic.

Miller, H. C. (1990), "Electrical Discharges in Vacuum (1877–1979)," *IEEE Trans. Electrical Insulation,* **25,** 765–860.

Miller, H. C. (1991), "Electrical Discharges in Vacuum, 1980–1990," *IEEE Trans. Electrical Insulation,* **26,** 949–1043.

Gases

Bartnikas, R., McMahon, E. J. (1979), *Engineering Dielectrics, Volume I, Corona Measurement and Interpretation,* Philadelphia: ASTM Special Technical Publication 669.

Dakin, T. W. (1962), "Theory of Gas Breakdown," in: *Progress in Dielectrics,* Vol. 4, London: Heywood & Company, pp. 152–198.

Sudarshan, T. S., Dougal, R. A. (1986), "Mechanisms of Surface Flashover along Solid Dielectrics in Compressed Gases: A Review," *IEEE Trans. Electrical Insulation,* **EI-21,** 727–746.

Van Brunt, R. J. (1994), "Physics and Chemistry of Partial Discharge and Corona. Recent Advances and Future Challenges," *IEEE Trans. Dielectrics and Electrical Insulation,* **1,** 761–784.

Liquids

Bartnikas, R. (1994), *Engineering Dielectrics, Volume III, Electrical Insulating Liquids,* Philadelphia: ASTM, Monograph 2.

Erdman, H. G. (1992), *Electrical Insulating Oils,* Philadelphia: ASTM, Special Technical Publication 998.

Gallagher, T. J. (1975), *Simple Dielectric Liquids,* Oxford: Clarendon Press.

Wilson, A. C. M. (1980), *Insulating Liquids: Their Uses, Manufacture and Properties,* Stevenage and New York: Peter Peregrinus, Ltd.

Zahn, M. (guest editor) (1988), "Flow Electrification in Electric Power Apparatus," *IEEE Trans. Electrical Insulation,* **23,** 101–176.

Solids

Bartnikas, R., Eichhorn, R. M. (1983), *Engineering Dielectrics, Volume IIA, Electrical Properties of Solid Insulating Materials: Molecular Structure and Electrical Behavior,* Philadelphia: ASTM, Special Technical Publication 783.

Bartnikas, R. (1987), *Engineering Dielectrics, Volume IIB, Electrical Properties of Solid Insulating Materials: Measurement Techniques,* Philadelphia: ASTM, Special Technical Publication 926.

Blythe, A. R. (1979), *Electrical Properties of Polymers,* Cambridge, England: Cambridge Univ. Press.

Devins, J. C. (1984), "The Physics of Partial Discharge in Solid Dielectrics," *IEEE Trans. Electrical Insulation,* **EI-19,** 475–495.

Gerhard-Multhaupt, R. (1987), "Electrets: Dielectrics with Quasi-Permanent Charge or Polarization," *IEEE Trans. Electrical Insulation,* **EI-22,** 531–554.

Jonscher, A. K. (1983), *Dielectric Relaxation in Solids,* London: Chelsea Dielectrics Press.

Mason, J. H. (1959), "Dielectric Breakdown in Solid Insulation," in: *Progress in Dielectrics,* Vol. 1, New York: Wiley, pp. 3–58.

O'Dwyer, J. J. (1973), *The Theory of Electrical Conduction and Breakdown in Solid Dielectrics,* Oxford: Clarendon Press.

Sessler, G. M. (Ed.) (1980), *Electrets,* Berlin: Springer-Verlag.

Shaw, M. T., Shaw, S. H. (1984), "Water Treeing in Solid Dielectrics," *IEEE Trans. Electrical Insulation,* **EI-19,** 419–452.

Steennis, E. F., Kreuger, F. H. (1990), "Water Treeing in Polyethylene Cables," *IEEE Trans. Electrical Insulation,* **EI-25,** 989–1028.

van Turnhout, J. (1975), *Thermally Stimulated Discharge of Polymer Electrets,* Amsterdam: Elsevier.

SOLIDS, ACOUSTICAL PROPERTIES OF

A. V. GRANATO, *Department of Physics, University of Illinois at Urbana-Champaign, Urbana, Illinois, U.S.A.*

INTRODUCTION

This is a short survey of a very large topic, using selected examples illustrating some of the principal effects found by acoustical methods over the past 40 years. It starts with perfect crystals, or at least acoustically perfect crystals. These are defined as crystals for which the defects are either static, such as dislocations uniformly pinned by impurities, or in low enough concentrations so as not to make their presence known. In perfect crystals, acoustic waves travel without attenuation or dispersion. Measurements of the sound velocity v are most often used to determine elastic constants c through relations of the form $\rho v^2 = c$, where ρ is the density. The literature on elastic constants is old and large, and many excellent reviews and compilations are available (Huntington, 1958; Simmons and Wang, 1971; Fedorov, 1968; Hearmon, 1979). The elastic constants are thermodynamic properties that are functions of applied stress and temperature. Measurements of the change of elastic constants with applied stress have been made in the past few decades to obtain third-order elastic constants, giving the first deviations from linearity in the interatomic force–displacement relations.

Real crystals contain imperfections, which give rise to attenuation and velocity changes. The acoustic response of solids is of two types, relaxational and resonance. These are analogous to the paramagnetic and diamagnetic response of magnetic dipoles to external magnetic fields. Paraelasticity and relaxation are the same. The defect strain results from the competition between the ordering effects produced by an external field on an existing elastic dipole and the disordering effects due to thermal fluctuations. It requires thermal activation over or tunneling through an energy barrier and is normally therefore strongly temperature dependent. Diaelasticity

3-527-28140-1/96/$5.00 + .50

and resonance are also the same. They are the direct response corresponding to an induced elastic dipole under an applied force even where there is no motion over or through a barrier.

Defects break up into two major classes, dislocations and point defects. Little will be said about the Bordoni relaxation peaks arising from dislocation–lattice interactions, but the resonance described by the string model that has been used to explain the dislocation behavior that is seen in a wide variety of situations will be discussed, including the amplitude dependence with its inertial and viscosity effects, as well as the Brownian-motion theory that has been developed for dislocation theory that finds application in many other areas of physics.

Point defects are another major category of defects. Acoustic measurements have been used to determine the configuration and dynamics of self-interstitials in metals and to resolve a long-lasting discussion about what the structure of such defects is. A particularly simple and illustrative example is hydrogen in niobium, of special interest because of the quantum tunneling behavior of the hydrogen. Electrons are another important source of acoustic attenuation. This attenuation proved to be one of the major sources of information for testing and establishing the Bardeen–Cooper–Schreiffer (BCS) (Bardeen *et al.*, 1957) mechanism of superconductivity.

In solid studies, the types of techniques that have been used vary a great deal because the techniques depend on the frequency range, although the basic concepts and ideas are pretty much the same at all frequencies. Measurements have been made at frequencies as low as 10^{-2} Hz and occasionally even lower using stress–strain curves. In the 1-Hz range pendula are used, and in the 100-Hz range bending-beam techniques are often employed. In the kilohertz region, usually a resonant-bar technique is used. The pulse-echo technique is mostly used for the megahertz region from about 10 to 1000 MHz. The available range continues on up into the gigahertz range with phonon spectroscopy techniques available there, and more recently into the terahertz region with again basically pulse techniques, but now the sound is stimulated by laser excitation and usually detected by laser techniques as well. Many of the various measurement techniques available are described by Krautkrämer and Krautkrämer (1983).

Most of these techniques are characterized by a very strong sensitivity, which proves to be a decisive advantage over nonacoustic techniques for many applications. Typically, sensitivity to defects in the parts-per-million range is available. For the pulse-echo technique, one can measure the attenuation of the amplitude and the velocity of the signal. By using interferometer techniques, one can measure velocity changes, for example, at 10 MHz up to about 1 Hz or about one part in 10^7. This is especially important if one wants to know the properties of isolated defects, because typically defects interact strongly through their shear stress fields, and at concentrations of even a thousand parts per million, the interactions become so strong that it is not possible really to study the properties of isolated defects. These interactions severely limit direct techniques such as x-ray, neutron, and phonon spectroscopies for the study of isolated defects.

For the pulse-echo technique, a second property has proved to be particularly important and useful. This is the fact that one can quite easily vary the polarization, for example, by looking at longitudinal and shear waves in different directions. This leads directly to complete symmetry information, useful in determining configurations of defects.

This large field has many international conferences, reports, reviews, and monographs (Nowick and Berry, 1972; De Batist, 1972; Hiki, 1972; Truell *et al.*, 1969; Mason, 1958, 1964; Zener, 1948; Mason, 1964–1968; Mason and Thurston, 1970–1984) available for detailed discussions. An extensive reading list with references to many conference reports has been compiled by Magalas (1996).

1. PERFECT CRYSTALS—ELASTIC CONSTANTS

1.1 Second-Order Elastic Constants of Perfect Crystals

Elastic constants have provided a means for testing various cohesion theories. One approach has been a phenomenological one in which the free energy is expanded in a Taylor series in the strains in a crystal. The first nonzero terms are quadratic in the strains and represented by a fourth-order elastic-constant tensor. The next terms in such a series would be given by a sixth-rank tensor, called the third-order elastic constants (Thurston and Brugger, 1964). The elastic constants, defined as the second derivatives of the free energy against the strains, give the second-order elastic constants, but in general they depend on the strain. These tensors can be written in a reduced notation because of the symmetries involved, and the three familiar elastic constants for the cubic case, C_{11}, C_{12}, and C_{44}, are more usually expressed in a different combination: the Zener (1948) combination of bulk modulus $B = (C_{11} + 2C_{12})/3$ and the two shear constants, C_{44} and $C' = (C_{11} - C_{12})/2$.

1.2 Third-Order Elastic Constants of Perfect Crystals

There are six third-order elastic constants for cubic symmetry, obtainable by measuring the change of the second-order elastic constants with applied stresses. With the sensitivities now available, it is not necessary to apply the kind of stresses (kilobars of pressure) that occurred in most of the older literature. The fact that one can work with extremely low loads means that plastic-deformation–induced defects can be avoided, and this permits measurements under uniaxial strain conditions so that one can get many different kinds of measurements and determine the full set of third-order elastic constants. We mention two important applications for these constants.

1.3 Interatomic Potentials

Third-order elastic constants provide information about interatomic potentials for crystals, as well as useful data for calculations of the properties of defects. Generally speaking, one knows that, at least in the simpler crystals, there is normally a short-range, fast-changing part to the potential and a long-range slow part. For example, in the alkali halides the Coulomb attraction and the closed-core repulsion are the dominant terms. The binding energy is usually determined mostly by the long-range part, typically about 90% from the Coulomb term and perhaps 10% from the short-range repulsive part. When one goes to the lattice constant, this is obtained by a derivative of the potential against distance, so that the short-range part becomes more important, giving more of a 50-50 kind of composition. The second-order elastic constants require another derivative, giving more weight to short-range terms again. The relative values of the third-order elastic constants depend almost completely on the structure and no longer in detail on the potential, and so one can predict relative values fairly well knowing only the geometrical structure (Holder and Granato, 1971).

1.4 Thermal Properties of Crystals

The third-order constants can also be used for descriptions of thermal properties. This was an idea first introduced by Einstein to connect the mechanical and thermal properties of crystals. Using a harmonic oscillator of frequency ω as a model of the solid, he was able to calculate the Einstein temperature ϑ_E (with $\hbar\omega = k\vartheta_E$) by expressing the specific heat in terms of ϑ_E and ϑ_E in terms of the elastic constants. This idea was developed further by Debye, who made a more correct calculation and expressed the specific heat in terms of the normal modes of the crystal, finding the Debye theta Θ_D in terms of the known elastic constants. This works exactly at low temperatures, where in the long-wavelength limit, one can justify use of the continuum elastic theory. This process can be continued. With third-order elastic constants, one can obtain values for the thermal expansion. The same applies for quantities like the temperature dependence of the specific heat and the elastic constants, which, however, require fourth-order elastic constants as well.

2. REAL SOLIDS—ACOUSTIC RESPONSE; ELECTRONS, DISLOCATIONS, AND POINT DEFECTS AS SOURCES OF LOSS

The principal sources of attenuation in metals come from dislocations, point defects, and electrons. Metals not containing defects may be regarded as perfect crystals, but we choose to include them here since the electrons are, like the defects, a source of loss. The losses can be identified by their temperature dependence and their dependence on the point-defect density ϕ, among other things, as illustrated in Fig. 1. The electron loss has a temperature dependence that depends on the electrical conductivity. It is relatively large and constant at low temperatures and falls off quickly at high temperatures. In a superconductor it drops suddenly at the transition temperature T_c [Fig. 1(a)]. The characteristic shape is given very well by the BCS theory. Also, the velocity has a temperature dependence that is T^4 at low temperatures according to theory similar to that for the Debye specific heat (Leibfried and Ludwig, 1961; Garber and Granato, 1975). In a metal there is a contribution to the velocity from the electrons that is T^2, which is normally overwhelmed by the lattice contribution, except at the very lowest temperatures.

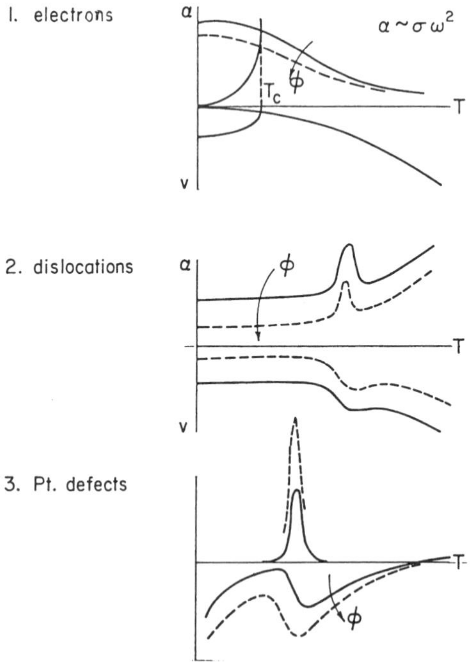

FIG. 1. Schematic temperature and point-defect-density (ϕ) dependences of the principal sources of attenuation. The curves above and below the horizontal line are attenuation and velocity, respectively, as functions of temperature. With increasing point-defect concentration ϕ, the solid curves change to the dashed curves.

In a superconductor there is also a small change in the velocity at the transition temperature. If one irradiates or puts in point defects that change the conductivity, the electronic loss decreases [dashed line, Fig. 1 (a)].

Dislocations provide a second very efficient source of loss, which is felt over the whole available temperature range. For pure and lightly deformed crystals, there is a resonance loss (Granato and Lučke, 1956) that varies approximately linearly with temperature. Point defects immobilize the dislocations and reduce the loss [dashed line, Fig. 1(b)]. For heavily deformed crystals, a relaxation effect known as the Bordoni peak (Bordoni, 1949) occurs, which is less sensitive than the resonance effect to point defects.

Point defects also have a characteristic relaxation loss called the Snoek effect (Snoek, 1941) that has been known for a long time. This gives rise to a loss peak when the frequency matches the rate of motion of the point defect between positions equivalent by symmetry. At the same time, a dispersion in the velocity occurs. More recently appreciated is the fact that there is also a resonance effect for point defects in general (Jäckle *et al.*, 1976; Hunklinger and Arnold, 1976), which gives a negative contribution to the modulus that increases at low temperatures and has a temperature-independent component as well (Granato, 1994a). These are the main mechanisms, and their characteristics are easily identified so that they can be used to sort out multiple sources of loss.

2.1 Electrons

A simple model for the effect of electrons was given by Mason (1955), who supposed that the viscosity of the electron gas produced attenuation in the same way as in the case of acoustic propagation in an ordinary gas or liquid. Mason then applied the standard expressions for sound attenuation due to viscosity, which give, for the amplitude attenuation of transverse waves,

$$\alpha_t = 2\pi^2 f^2 \eta/\rho\vartheta_t^3, \tag{1}$$

where f is the frequency, η is the viscosity, ρ is the density, and ϑ_t is the transverse-wave velocity. For longitudinal waves, a similar expression is obtained with $\alpha_l = (\frac{4}{3})(\vartheta_t/\vartheta_l)^3\alpha_t$, where ϑ_l is the longitudinal-wave velocity. The viscosity η can be expressed in terms of the conductivity from kinetic theory as $\eta = (3\pi^2 N)^{2/3}(\hbar/5e^2)\sigma$, where N is the number of particles per unit volume, $\hbar$ is Planck's constant, e is the electronic charge, and σ is the conductivity. An equivalent result based on a relaxation mechanism was given by Morse (1955), and both are subject to the restrictions $ql_e < 1$, where q is the propagation constant and l_e is the electronic mean free path. Pippard (1955) extended these treatments to arbitrary l_e. The present understanding of ultrasonic attenuation in normal metals has been reviewed by Rayne (1975). An application of the Bardeen–Cooper–Schrieffer (Bardeen *et al.*, 1957) theory of superconductivity by Morse and Bohm (1957) provided a successful explanation for the temperature dependence of the ultrasonic attenuation in superconductors.

2.2 Dislocations

2.2.1 Dislocation Relaxation Effects

The explanation of the Bordoni peak came about mainly through the work of Seeger (1956) and his colleagues at Stuttgart, in terms of kinks on dislocations. That is, dislocations that are nearly aligned along crystallographic directions feel an interaction with the lattice, which provides a periodic potential. For the dislocation to advance, it is necessary to stimulate thermally the creation of a nucleus, a place where the dislocation crosses over into the next valley. That is done by creating a kink pair. There are many models for these kinks, starting with the Frenkel–Kontorova (1939) model. They have a very interesting mathematical structure with particle-type properties. They have a mass, can be accelerated, and have a limiting relativistic velocity given by the sound velocity. The particles can oscillate, radiate, and interact with themselves, with phonons and electrons, and so on. Nowadays the literature is pretty much similar to the older literature, but the kinks are usually referred to as solitons.

The use of the kink-pair model for the explanation of the Bordoni peak has been reviewed recently by Ritchie and Fantozzi (1992). Some questions about the interpretation are still under discussion (Kosugi and Kino, 1993). The relationship of the kink model to the vibrating-string model and to dislocation resonance effects is also reviewed by Ritchie and Fantozzi, and earlier as well by Granato and Lücke (1966).

2.2.2 Dislocation Resonance Effects

For dislocations that lie at a finite angle to the crystallographic directions, the kink density becomes so high that the kinks become irrelevant. That is, although the motion still occurs by kink motion sideways, the displacement actually is very much like that of a simple string. There is an effective tension because a dislocation under stress increases its length or increases its energy. That provides a very strong restoring force. Also phonons and electrons are scattered by a moving dislocation, providing a resistance or viscosity.

One of the characteristic features of the resonant dislocation component to the loss is that it is very amplitude dependent. For example, measurements by Read (1940), shown in Fig. 2, may already be the first strong experimental evidence for the existence of the dislocations in crystals that had been postulated just a few years earlier. The convincing part of the proof is the orientation dependence of the measurements made in single crystals of zinc where the slip planes were known. It takes a shear stress to move a dis-

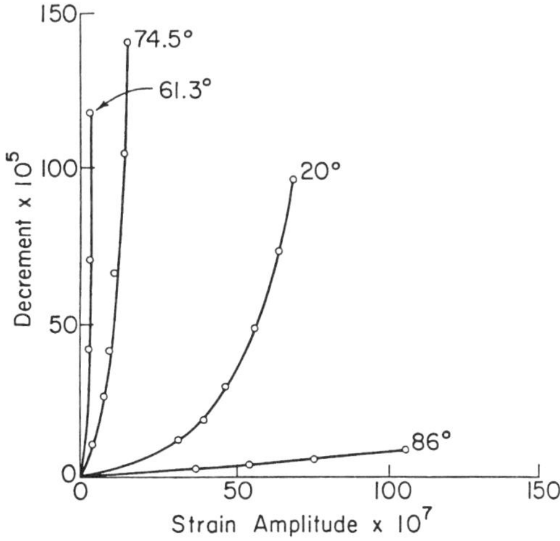

FIG. 2. Variation of decrement with strain amplitude in zinc crystals whose cylinder axes make various angles θ with the hexagonal c axis (after Read, 1940).

location, so that when the slip plane is at 0° or 90° to the tensile axis, there should be no shear stress and no effect. In fact, for crystals where the slip planes were near 0° or 90°, there is very little effect. The maximum effect should and does occur near 45°, providing very strong evidence for the existence of dislocations.

One rather convincing demonstration of the radiation (Eshelby, 1949; Garber and Granato, 1970) that occurs from moving dislocations comes from a simple experiment (Schwenker and Granato, 1969) in which one side of a sample of lithium fluoride is permanently displaced relative to the other so that there is a wall of dislocations on one slip plane. Now, if the dislocations cannot move, then one should simply get the usual echo from the back face of the sample for an incident sound wave. But if the shear stress of the sound wave is polarized so that it can move the dislocations, an additional echo is generated by the wall. Each dislocation radiates a cylindrical wave. By Huyghens's principle, these add up to a macroscopic plane wave that can also be detected. One should expect an echo for the polarization that moves the dislocations and no echo for the polarization at right angles. That is just what happens, as shown in Fig. 3. Also if these

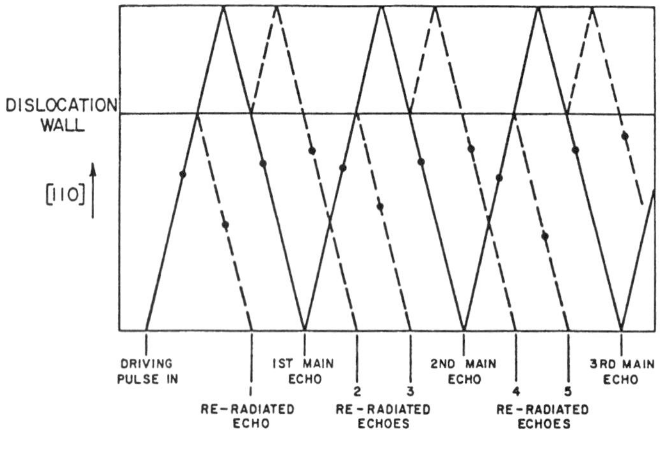

(a)

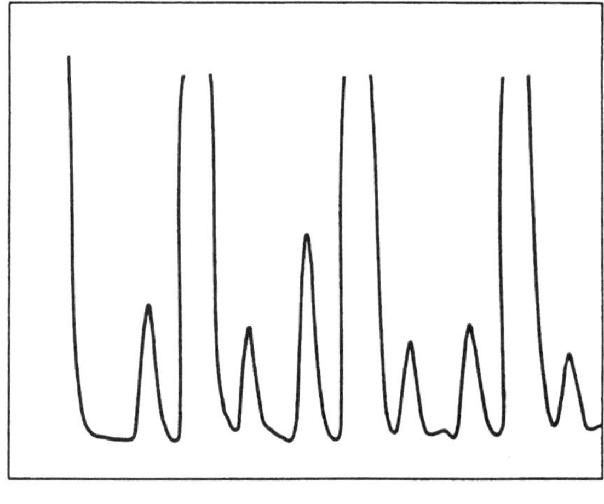

(b)

FIG. 3. (a) Schematic representation of wave positions along the crystal axis vs time for a deformed crystal. (b) The reradiated dislocation wall echoes for a deformed LiF crystal vs display time (after Schwenker and Granato, 1969).

crystals are lightly irradiated so as to pin down the dislocations, the wall echoes go away while the attenuation of the main echoes decreases.

2.2.2.1 Low-Amplitude Dislocation Resonance. For low stress amplitudes, the theory is analogous to dielectric theory, with dislocation displacement taking the place of charge displacement. The elastic shear-constant change $\Delta G/G$ is given by

$$\Delta G/G = \epsilon_d/\epsilon_{el}, \tag{2}$$

where ϵ_d is the dislocation strain and ϵ_{el} is the elastic strain. The dislocation strain is given by the Orowan (1940) relation

$$\epsilon_d = \Lambda b\bar{y}, \tag{3}$$

where Λ is the dislocation density, b is the Burgers vector, and $\bar{y}$ is the average dislocation displacement, obtained from Koehler's (1952) vibrating-string model,

$$A\ddot{y} + B\dot{y} - Cy_{xx} = b\sigma, \tag{4}$$

where A is the dislocation mass per unit length, B is the viscous force per unit velocity, C is the dislocation line tension, and $b\sigma$ is the applied force per unit length. Using Eqs. (4) and (3) in (2), one obtains a damped harmonic oscillator response of magnitude

$$\frac{\Delta G}{G} = \frac{\Omega\Lambda L^2}{6}\frac{1}{(1 - \omega^2/\omega_0^2) + id\omega/\omega_0^2}, \tag{5}$$

where $d = B/A$ and $\omega_0^2 = \pi^2 C/AL^2$, where L is the length of a pinned dislocation segment. In Eq. (5), Ω is an orientation factor taking account of the fact that only the resolved shear-stress component in the slip systems produces dislocation motion.

The real part of Eq. (5) gives the modulus change $\Delta c/c$, and the imaginary part gives the decrement Δ of the vibration, defined in the Glossary. A distribution of dislocation loop lengths L needs to be taken into account for comparisons with experiment.

The first striking evidence for the string model from low-amplitude measurements was that of Thompson and Holmes (1956), who irradiated pure single crystals of copper with neutrons. It is reasonable to assume, with Thompson and Holmes, that the dislocation density does not change in this process, so that observed changes in the damping modulus should be caused only by changes in the average loop length. Thompson and Holmes found that the damping and modulus changes depended upon the fourth and second powers of the average loop length, respectively, as predicted by Eq. (5). In addition, they found that the size of the effect varied considerably from specimen to specimen, all of nominally the same purity. The specimens could be restored to their original condition by a suitable high-temperature annealing treatment. Also, it was found that there was a considerable background damping that was not removed by the radiation. The importance of these results for the further development of the field cannot be overemphasized. They demonstrated that the low-amplitude damping effects follow definite characteristic laws but that it is practically impossible to prepare two different specimens in exactly the same state. Radiation techniques provide a means for continuously varying the "purity" and studying the dependence on loop length in detail. Also the results showed that in order to attempt quantitative analysis of damping measurements, it is necessary to have some means of separating the effects from any unknown background effects. Once these conditions are realized, the method proves a means of extraordinary sensitivity in studying defect interactions.

When pinning occurs without changing the dislocation structure, Eq. (5) can be written in terms of the number of pinners added to a dislocation segment in an irradiation experiment so that a "direct count" scheme for the number of pinners on a dislocation segment is available. Thus, if n_0 is the original number of pinners per unit length of dislocation line, L_0 is the original segment length, n is the number of defect pinners added, and L is the average segment length with the added pinners, then

$$L = \frac{L_0}{1 + n/n_0}. \tag{6}$$

If we define the parameters Y and Z as Thompson and Pare (1960) have done, as the fraction of the modulus defect and decrement remaining after n pinners have been added, it follows that

$$n/n_0 = Z^{-1/4} - 1 = Y^{-1/2} - 1. \tag{7}$$

The ratio n/n_0 may be interpreted directly as the number of pinners added per dislocation segment since, by definition, the number n_0 is the same as the number of dislocation segments in a unit length of dislocation.

Since Y is proportional to L^2 and Z is proportional to L^4, a line of slope unity must result upon plotting Y^2 vs Z on a logarithmic plot if the loop-length expectation of Eq. (5) is realized. Alternatively, plots of $Y^{-1/2}$ and $Z^{-1/4}$ vs radiation time should coincide, as shown in Fig. 4 from the data of Thompson and Pare (1960). The experimental check of the loop-length dependence extends through three orders of magnitude. This is about the normal range of variation found for dislocation damping in high-purity copper crystals. It should be noted that the extent of experimental agreement with theory in this test depends to some extent upon the specimen itself. Deviations are sometimes found. However, these have been interpreted as being caused by the presence of two or more dislocation components that possess different line tensions, and apparently they do not represent a failure of the string model.

The frequency dependence of the damping predicted in Eq. (5) is illustrated in Fig. 5. These are measurements by Stern and Granato (1962) showing the effect of cobalt gamma irradiation on the damping of high-purity copper. Before irradiation, the damping has a maximum at a few megahertz. After 50 h of irradiation in a 6000-Ci cobalt source, the height of the maximum decreased, and the location is increased to about 100 MHz. The damping at low frequencies is much more sensitive to the increased number of pinning points than that at high frequencies, as expected from Eq. (5). More recently, the underdamped dislocation resonance has been observed by Beamish and Franck (1982), in bcc ^{3}He.

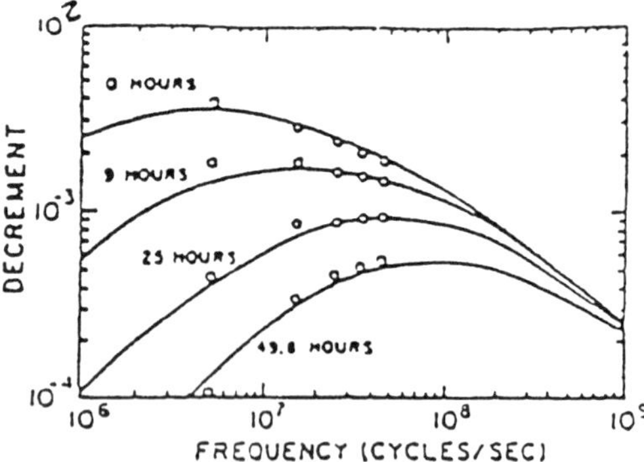

FIG. 5. The dislocation decrement in copper as a function of frequency for several times during cobalt gamma irradiation. The solid curves are theoretical (after Stern and Granato, 1962).

According to Eq. (5), the drag constant B may be measured if the dislocation density is known. Drag constants were determined by Fanti *et al.* (1969) from measurements of ultrasonic attenuation and velocity changes as functions of frequency in specimens of LiF and NaCl for which the dislocation density Λ was determined by etch-pit counts. The values obtained are in good agreement with those to be expected for phonon drag, according to Leibfried (1950). When these values are used to calculate the limiting dislo-

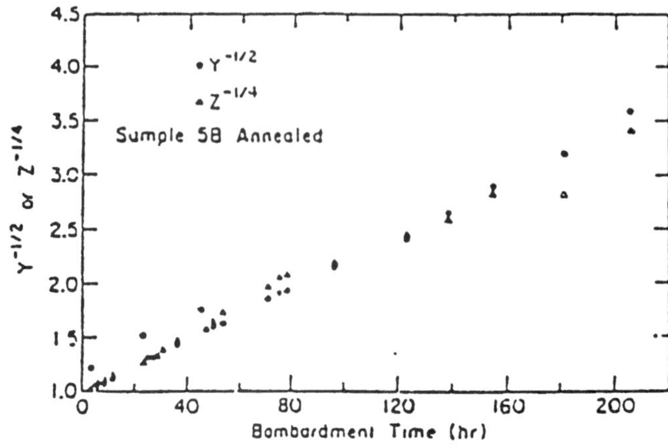

FIG. 4. $Y^{-1/2}$ and $Z^{-1/4}$ vs time at room temperature in copper sample 5*B* (after Thompson and Pare, 1960).

cation velocities expected at high stress at room temperature, according to

$$B_p v = b\tau, \tag{8}$$

the results in Fig. 6 are obtained for NaCl. This figure also shows the velocities at high stresses determined by the etch-pit technique by Gutmanas *et al.* (1963). It may be seen that the etch-pit velocities do not exceed, and appear to be limited by, the velocities predicted by extrapolation of the ultrasonic results from lower stress levels ($\sim 10^2$ cm/s). This is strong evidence that the same drag mechanism is operative on the instantaneous velocities in both stress ranges. As was pointed out by Baker (1962), the fact that velocities determined from ultrasonic measurements are many orders of magnitude higher than those measured by etch-pit techniques at the yield stress and the fact that dislocation displacements as large as 1000*b* are deduced at stress amplitude levels an order of magnitude below the yield stress demonstrate that the Peierls force is ineffective in limiting dislocation motion, at least at room temperature. This result is an example of ways by which acoustic measurements can help to distinguish between various postulated deformation mechanisms.

These results, with others, show that under controlled conditions the predictions of the string model for low strain amplitudes have been verified, and, where contact can be made with results of other measurements, the comparison is favorable. The predictions can therefore be used with some confidence in the study of dislocation–point-defect interactions. The controlled conditions include

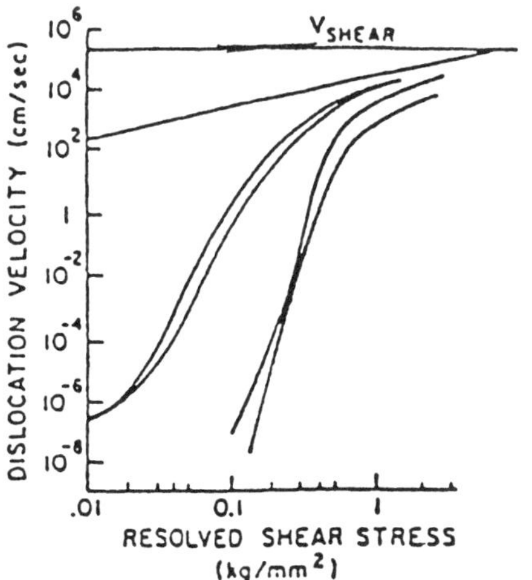

FIG. 6. Dislocation velocities as a function of applied stress in NaCl. The straight line is a linear extrapolation of ultrasonic velocities to higher stresses. The curved lines are the etch-pit measurements of Gutmanas *et al.* (1963) (after Fanti *et al.*, 1969).

1. well-defined experimental arrangements;
2. discrimination of dislocation losses from other losses, for example, as obtained by irradiation; and
3. self-checking arrangements.

For the kilohertz region, the latter includes making both modulus and damping measurements, and for the megahertz region, both modulus and damping measurements as a function of frequency. In experiments, the main complicating feature found over the simple model is the possibility of a distribution of line tensions. This must be assessed by studies of **3** above. The conditions mentioned favor irradiation techniques over plastic-deformation techniques. In the former, low dislocation densities may be used, reducing interaction possibilities, and better estimates of defect concentrations and distributions may be obtained.

2.2.2.2 Amplitude-Dependent Damping Theory. For high enough stress amplitudes, the dislocations may be pulled away from the weak pinning points, as illustrated in Fig. 7. A large increase in strain for no increase in stress then occurs, and for a cyclic stress, the area in the stress–strain curve represents the energy loss per half cycle. This loss is amplitude dependent. It is zero until the stress amplitude equals the mechanical breakaway stress $\sigma_M = U_0/b^2L$, where U_0 is the pinning energy (Cottrell, 1948), and L is the distance between weak pinning points. Beyond this, the decrement, which by definition is the loss per half cycle divided by the maximum stored elastic energy $\sigma^2/2G$, decreases as $1/\sigma_0^2$ and is as shown in Fig. 7(c). In a real crystal, however, there is a distribution of dislocation segments L, which changes the shape of the decrement vs amplitude from that shown in Fig. 7(c) to that in Fig. 7(d).

It might be anticipated that the effect of thermal fluctuations would be to change the breakaway stress from σ_M at $T = 0$ to a lower value σ_T at finite T, as indicated in Fig.

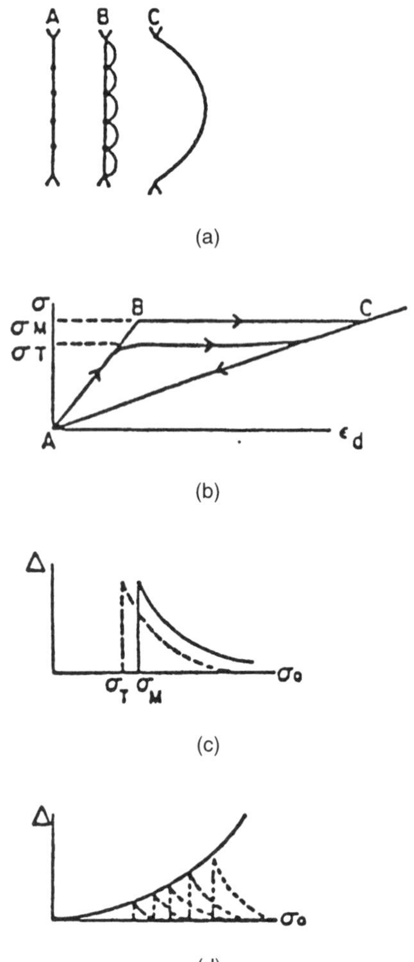

FIG. 7. (a) Dislocation displacement as a function of increasing stress for a dislocation segment with weak interior and strong exterior pinning points. (b) Stress–dislocation-strain curve corresponding to (a) for a cyclic stress. Breakaway occurs at a lower stress σ_T at finite temperatures than the mechanical breakaway stress σ_M. (c) Decrement vs stress amplitude for segments of same length. (d) Decrement vs stress amplitude for a distribution of weak pinning points.

7(b). A summary and physical discussion of the theory is given by Granato and Lücke (1981) and Teutonico *et al.* (1964). For the low-temperature region, the principal results are as follows:

1. The stress amplitude $\sigma_\Delta(T)$ required to produce a constant Δ, at a temperature T, corresponds to a constant fraction of dislocation segments being broken away from point defects and is directly related to calculations for the yield stress of macroscopic flow measurements.
2. For high stress amplitudes, low temperatures, and relatively pure materials, the stress should have the temperature dependence

$$\sigma = \sigma_M - AT^{2/3}, \tag{9}$$

where σ_M is the stress required to overcome the pinning barrier in the absence of thermal fluctuations and A is a constant for constant frequency regardless of the form of the dislocation force–distance law, provided only that the force has a smooth maximum.
3. For low stresses, the activation process no longer depends only on the details of the force–distance law. For such low stresses, breakaway from a single pinning point becomes ineffective because the broken-away state is a relatively high-energy state, and the repinning rate is higher than the unpinning rate. Breakaway to a lower-energy state can now occur only by simultaneous unpinning from more than one pinning point, but the activation energy for this is higher, and the process occurs only at higher temperatures.

Measurements of the critical unpinning stress (stress amplitude for given decrement value) are closely related to macroscopic flow-stress measurements for dilute alloys. An observation of the $T^{2/3}$ law of Eq. (9) is of special significance since this form is required theoretically on very general grounds, and measurements that purport to measure the force–distance profile between a dislocation and a single point defect should meet this requirement as a necessary and crucial test.

Superconducting materials entering the superconducting state suffer a loss of mechanical strength. Since the initial reports by Kojima and Suzuki (1968) and Pustovalov *et al.* (1967), many studies have been made, which have been reviewed by Startsev (1983). This effect was unexpected and unexplainable on the basis of traditional theories of plastic flow, which take no account of viscous or inertial effects.

Internal-friction measurements by Schwarz *et al.* (1977) on pure and dilute-alloy crystals of copper and lead are shown in

Fig. 8. The maximum in the stress of the dilute alloy of copper is similar to a maximum in the flow stress of a similar specimen seen by Kamada and Yoshizawa (1971). For $T <$ 6 K, the data for pure copper follow the strict $T^{2/3}$ dependence of Eq. (9), as had been found earlier for pure aluminum (Schwarz and Granato, 1975) and which was explained in terms of a classical thermally activated process. The measurements in lead are similar to those in copper for the two purities, except that there are additional changes during the *N–S* transition. This change can be used as a key to determine whether the dislocations are overdamped or underdamped in the normal state. In the pure lead, the curve for the *S* state joins smoothly onto the curve for the *N* state below the transition temperature where the drag is still increasing. This shows that the dislocations in pure lead are overdamped. In contrast, dislocations in the dilute alloy are underdamped in the *N* state so that they can become overdamped when the damping is increased when the temperature is raised sufficiently that phonons contribute significantly to the viscous damping. This explains the existence of the anomalous maximum observed for the dilute lead alloy at about 20 K. The fact that the temperature dependence of $\sigma_0(\Delta)$ in the *N* state of the dilute lead alloy is essentially the same as in the dilute copper alloy provides strong experimental evidence that the anomalous maximum in the dilute copper alloy is of inertial origin. Figure 8 also shows that below $T = 3.5$ K, the stress amplitude $\sigma_0(\Delta)$ for the *S* state resumes a linear $T^{2/3}$ dependence with approximately the same slope as in the *N* state. Thus the change in stress vs temperature is predicted well by a purely mechanical model. However, the temperature dependence of the absolute stress is characteristic of a rate process. This suggested to Schwarz *et al.* (1977) that in the presence of inertial effects, dislocation motion is initiated by a thermally activated process and continues by the overcoming of additional obstacles inertially. The amount of plastic deformation in the latter process is determined by the degree of underdamping.

A theory of this process was given by Isaac and Granato (1988), who developed a Brownian-motion rate theory that incorporates viscous and inertial overshoot effects. The theory is an extension to the many degrees of freedom corresponding to dislocation motion of some earlier results found for Brownian motion of a particle in a force field by Kramers (1940), Chandrasekhar (1943), and Wang and Uhlenbeck (1945).

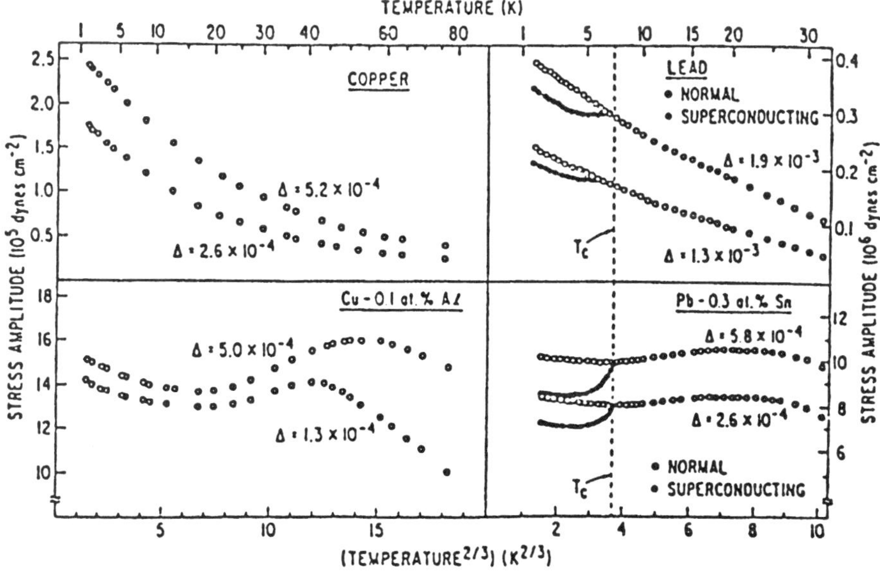

FIG. 8. Temperature dependence of the stress amplitude for constant decrement in internal friction measurements on pure and dilute alloys of copper (left column) and lead (right column) (after Schwarz *et al.*, 1977).

A Langevin equation for a dislocation in a stress and impurity field is solved for a distribution function $p(x,v,t)$, which gives the probability of finding a dislocation at position x with velocity v at time t. The form of this distribution at once explains one of the principal puzzles about many observations of the temperature dependence of the flow stress and its change at the superconducting transition. This is that a purely mechanical inertial theory (Granato, 1971a, 1971b) gives a satisfactory account of the changes in flow stress at given temperature, including the predictions, subsequently verified (Schwarz *et al.*, 1977), that the flow stress change will go to zero below the transition temperature for sufficiently pure materials, and that an anomaly in the temperature dependence of the flow stress above the transition temperature will appear otherwise. On the other hand, this mechanical theory cannot account for the fact that the absolute value of the flow stress has a normal temperature dependence expected from thermal overcoming of barriers in either the normal or superconducting state. This mechanical–thermal duality is explained by the form of the distribution function, which tends to be a cloud centered about the purely mechanical path, which differs for two different viscosities. The path depends on the viscosity while the width of the cloud depends upon the temperature. If one holds the temperature constant and changes the viscosity, the cloud moves, but overall the change is mechanical-like. On the other hand, if one holds the viscosity fixed, and varies the temperature, the cloud position stays fixed, but the cloud width changes, thus emphasizing the fluctuation aspect of the thermal bath. The superconducting transition thus provides a unique opportunity to emphasize either the systematic part of the Brownian-motion effect leading to viscosities or the random part leading to thermal fluctuations.

In fact, these considerations are general, applying for any diffusing "particle" in an external force field. The quantities needed in the theory are the viscosity B and the resonant frequencies of the "particle." The advantages of plastic flow measurements are that dislocations are Brownian particles for which

1. these have been calculated, and mechanisms
2. have been determined,
3. can be measured acoustically, and
4. can be controlled (as in the normal to superconducting transition).

2.3 Point Defects

The diffusion of point defects over obstacles under the influence of an external force is a process that plays a role in many different kinds of physical effects. These have been well studied classically and are normally described by an Arrhenius temperature dependence. More recently, there has been increased interest in quantum mechanical tunneling effects and the effect of viscosity on tunneling rates.

Many different techniques have been applied to such measurements, but ultrasonic measurements have particular advantages. Dipolar point defects interact strongly with ultrasonic shear stress. Measurements of ultrasonic attenuation and velocity as functions of polarization, frequency, temperature, and defect density display both resonance and relaxation behavior. Ultrasonic measurements in the 10–100-MHz range enjoy benefits generally attributable to internal-friction techniques (Robrock *et al.*, 1977). Most notably, they provide information about the symmetry of defects, can often be used at low defect concentration where defect interactions can be neglected, and permit the determination of properties of individual defect species even when several different types of defects are present simultaneously in the sample. In addition ultrasonic measurements can be used to obtain all independent elastic constants and the bulk resistivity from the attenuation in the same specimen. They are particularly well suited for the study of quantum systems. Information about energy gaps and stress coupling strengths is readily obtained from ultrasonic attenuation peaks and velocity changes at low temperature for defects in concentrations of a few parts per million. These peaks would normally occur, if at all, below the readily accessible temperature range for internal-friction measurements at lower frequencies.

Textbooks are available (Nowick and Berry, 1972; De Batist, 1972) for detailed discussions of classical point-defect relaxation effects. Resonant effects for $\omega \ll \omega_0$, where ω_0 is the resonance frequency, are discussed by Granato (1994a). Quantum effects for both relaxation and resonance are reviewed by Hunklinger and Arnold (1976). We outline here the quantum theory, which contains the classical theory as a high-temperature limit.

To illustrate the types of effects that are found, we focus on the system of hydrogen trapped at an impurity in niobium. This is a rather rich system, which has been well studied, with many aspects arising from isotope and multiple-trapping possibilities. We then outline the theory for the simplest system, the two-level tunneling system (TLS), which already contains most of the physical effects of interest, and is easily extended to more complicated systems.

One of the most active areas of condensed matter physics over the past two decades has been the question of the properties of noncrystalline materials, or the structure, dynamics, and physical properties of amorphous and liquid materials. It is known from low-temperature studies of specific heat, thermal conductivity, and ultrasonic attenuation that the properties can be understood in terms of a TLS model, with an unspecified tunneling entity. The same parameters appear in the theories of both the thermal properties and the ultrasonic properties. The system of hydrogen in niobium is a model system for such studies because it is simple, with no need to average over a distribution of barrier heights, as is necessary in the TLS model for amorphous or glassy systems. An additional question of much current interest is the question of the effect of viscosity on tunneling rates. A theory by Caldeira and Leggett (1981) predicts that tunneling rates are reduced by the viscosity provided by interactions of the tunneling system with the environment. There are unique possibilities of studying this effect provided by the possibility of observing the ultrasonic response of tunneling systems in both the normal and superconducting states of superconductors.

The general program for ultrasonic measurements is to measure the temperature, frequency, polarization, and isotope dependence of the attenuation β and velocity v of ultrasonic waves. These represent the out-of-phase and in-phase response of the system to a periodic external force. The attenuation and velocity changes can alternatively be expressed as the quality factor Q^{-1} and modulus change $\Delta C/C$ using $Q^{-1} = \alpha\lambda/\pi$ and $v = \sqrt{C/\rho}$, where λ is the sound wavelength, C is the elastic modulus, and ρ is the density. Q^{-1} and $\Delta C/C$ are given as

$$Q^{-1} = \Delta_R \frac{\omega\tau}{1 + \omega^2\tau^2}$$

and

$$\frac{\Delta C}{C} = \Delta_R \frac{1}{1 + \omega^2\tau^2} + \Delta_s, \tag{10}$$

where Δ_R is the relaxation strength, Δ_s is the resonance strength, τ is the relaxation time, ω is the frequency, and $\omega \ll \omega_0$. For a classical system

$$\Delta_R(T) = \frac{\alpha^2 c}{CkT},$$
$$\Delta_s = \beta c,$$

and

$$\tau(T) = \tau_0 \exp(U/kT), \tag{11}$$

where α is the coupling constant for the defect interaction with the sound wave, β is the diaelastic susceptibility, τ_0 is the prefactor, U is the barrier height for the relaxing point defect, and c is the defect concentration. The same form [Eqs. (10)] holds for quantum systems, but the temperature dependence is different. The relaxation height goes to zero for low enough temperature, and the relaxation time is often weaker than exponential. Quantum effects are generally recognized, though, by analysis of the temperature dependence of Δ and τ. It can happen that the relaxation time is Arrhenius in form but still be quantum in nature. But then the preexponential factor τ_0 is generally much smaller than a characteristic atomic period. By analysis of $\Delta_R(T)$ and $\tau(T)$, the basic parameters of the tunneling system, α and Δ (tunneling gap), are deduced.

2.3.1 Tunneling of Trapped Hydrogen in Niobium Measurements by Sellers *et al.* (1974) of a large isotope effect in the excess specific heat of niobium containing hydrogen or deuterium showed convincingly that hydrogen in niobium formed a quantum tunneling system. More detailed measurements of specific heat by Morkel *et al.* (1978) showed that if there is no oxygen or nitrogen impurity, then there is no excess specific heat. This implies that the tunneling system consists of a complex made up of hydrogen trapped at an oxygen or nitrogen impurity. More specific information has since been obtained from neutron-scattering and ultrasonic measurements. A brief review of these results with references is given by Drescher-Krasicka and Granato (1985).

An example of the type of measurement obtained for the Nb–O/H system is shown in Fig. 9. The specimen contains 700 ppm of oxygen and 100 ppm of hydrogen. In the superconducting state, a small relaxation peak is found for the C' mode at 2.5 K at 10 MHz. The effect is small but can be measured with great accuracy. The relaxation strength behaves quantum mechanically, going to zero at low temperatures. The velocities in these samples show a dispersion at 2.5 K, but they also decrease further at lower temperatures. The latter decrease is a diaelastic effect. In the normal state there is no evidence of a relaxation. Since the hydrogen effect occurs in the C' mode, but not in the C_{44} mode, we obtain immediately important symmetry information useful for discriminating between candidate models for the configuration, which is information not available from specific-heat or neutron-scattering experiments. The relaxation nature of the peak in the superconducting state is established from the shift in the peak position with frequency in the usual way. The remarkable feature of this peak is that it disappears in the normal state, showing a strong and unusual dependence of the relaxation state on the normal electron density. In the normal state, even with the absence of a relaxation, the modulus decreases with decreasing temperature. This is a quantum resonance effect, about which more will be said later.

Measurements as a function of frequency between 10 and 150 MHz permit a determination of the temperature dependence of the relaxation strength and relaxation time. The relaxation time and strength found by Poker *et al.* (1979) are shown in Figs. 10 and 11. The relaxation strength deviates from the classical $1/T$ dependence. The relaxation rate is Arrhenius, but the intercept is too low for a classical process. A classical $1/T$ dependence is shown for reference.

If a specimen containing 700 ppm H and 100 ppm O is rapidly quenched from 240 K, a second much larger peak appears (Huang

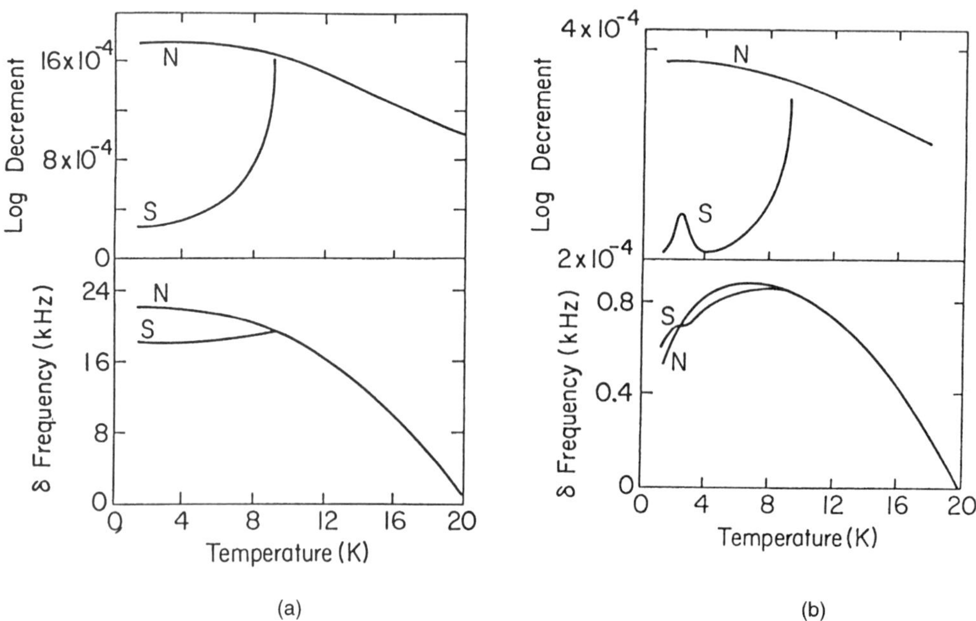

FIG. 9. Decrement (Δ) and velocity change ($\Delta v/v = \Delta f/f$) vs temperature at 10 MHz for Nb containing 100 ppm of oxygen and 700 ppm of hydrogen for the (a) C_{44} mode and (b) C' mode (after Drescher-Krasicka and Granato, 1985).

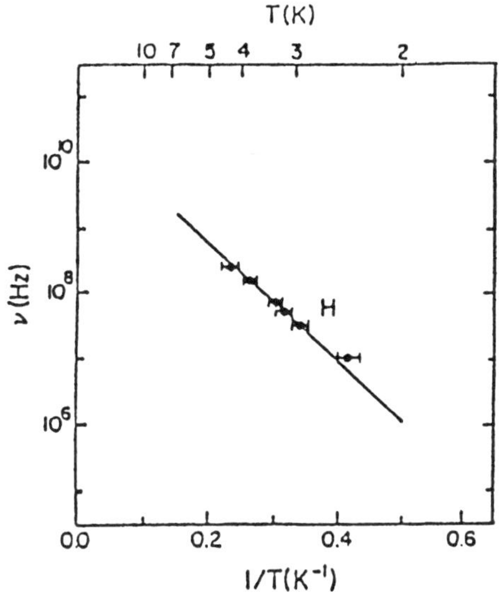

FIG. 10. Logarithm of relaxation frequency as a function of inverse temperature. Straight line is a fit to the data with an activation energy of 1.8 meV.

et al., 1985) near 6.5 K. This peak is unstable against annealing near 90 K, as shown in Fig. 12.

2.3.2 Theory of Two-Level Systems. Relaxation and Resonance Effects The theory of attenuation and elastic-constant changes for TLS in crystals is basically the same as that for amorphous systems, but simpler. For amorphous systems there is not only a distribution of strains but also a distribution of tunneling gaps, extending down to zero. For TLS in crystals, there is a minimum gap, Δ_0, and for ultrasonic measurements $\hbar\omega < \Delta_0$, there is no direct resonant absorption. The theory was given already by Jäckle (1972) and applied immediately to amorphous systems.

The elastic constant C is given by definition as

$$C = \frac{\partial^2 F(\epsilon)}{\partial \epsilon^2}, \tag{12}$$

where F is the free energy per unit volume, ϵ

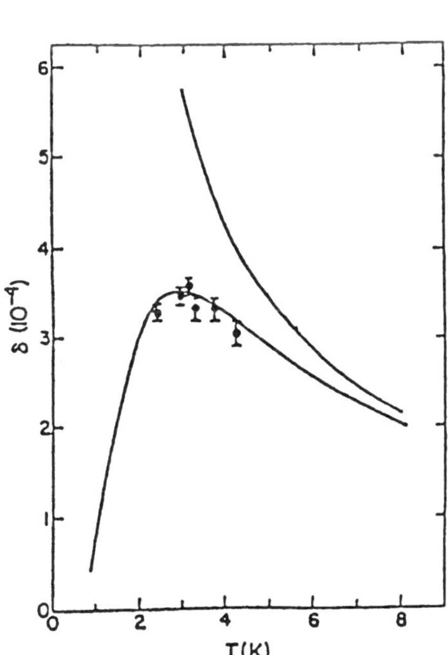

FIG. 11. Relaxation strength of the C' mode vs temperature for hydrogen.

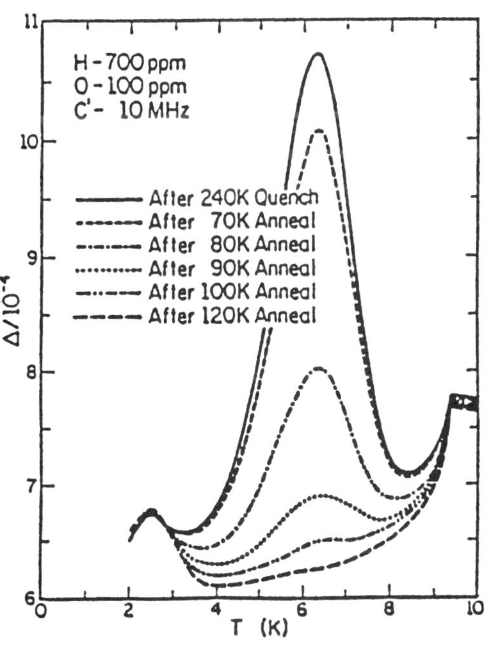

FIG. 12. Decrement vs temperature for a 10-MHz C' mode in a niobium crystal containing about 100 ppm O and 700 ppm H after a rapid cooldown from 240 K. The rapid change of attenuation near 9.2 K is due to the transition from the normal to the superconducting state.

is an elastic strain, and C is the elastic constant belonging to that strain. The free energy is given by

$$F = -kT \ln Z, \tag{13}$$

where Z is the partition function given by

$$Z = \sum_i \exp(-E_i/kT), \tag{14}$$

where $E_i(\epsilon)$ are the states of the system. The relaxation and resonance effects can be illustrated with the simple two-level system shown schematically in Fig. 13. For $\epsilon = 0$, the dipole is supposed to have two equivalent orientations. The ground state is tunnel-split with a gap of 2Δ. If a stress or strain ϵ is applied, one orientation becomes energetically favored. If time is available, transitions will occur, bringing the system back into thermal equilibrium.

For $T \ll \Delta$, only the ground state is populated. For small ultrasonic stress amplitudes, the elastic constant is given by the curvature of the ground state. The dipole moment is induced, proportional to the strain ϵ, and the change is called diaelastic. This response has a resonance-like character. The resonance

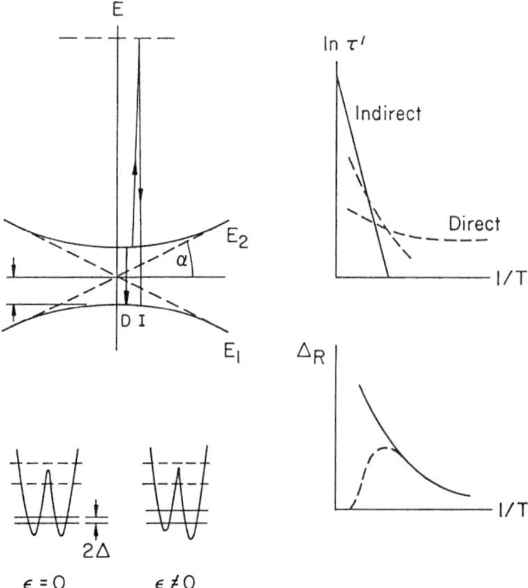

FIG. 13. Schematic illustration of potential, energy levels, transition processes, transition rates, and relaxation strength Δ_R for a two-level system, or a four-level system including dashed lines for E. Classically $\Delta_R = 0$, τ^{-1} is Arrhenius, and $\Delta_R \propto 1/T$.

frequency is normally far above the megahertz range. As a result, no attenuation is seen, but only the low-frequency change of elastic constant is found.

At finite temperature, higher levels become populated, and phonon-induced transitions become possible. For the two-level system shown, however, no relaxation takes place for small-amplitude ultrasonic waves about $\epsilon = 0$ because the energy of the states is flat for small ϵ, maintaining equilibrium population of the states. For $\Delta = 0$, the states are given by the dashed lines. This can represent a classical Snoek system—for example, Fe–C (where one of the levels is doubly degenerate). Now if a strain is applied, equilibrium is disturbed, transitions take place, and the effect is that of a permanent dipole reorienting in a field. This is a paraelastic effect, represented by a relaxation process. The low-frequency (high-temperature) elastic constant is given by Eqs. (12)–(14), and the response for all frequencies, by Eq. (10).

For classical systems Δ_R is inverse with temperature as a result of the competing thermal disordering and external strain-aligning effects. Also Δ_R is proportional to the defect concentration and the square of the strain coupling constant α, given by the slope of $E(\epsilon)$. The relaxation time is Arrhenius with an activation energy U given by the height of the barrier. Transitions may occur by several processes, which act in parallel. Direct transition rates are independent of temperature at the lowest temperature and occur by tunneling, increasing linearly with temperature at higher temperature where they become phonon assisted (Jäckle, 1962). Indirect transitions through other states may also occur, but with a higher temperature dependence. The fastest rate dominates the temperature dependence. At higher temperatures, in the classical regime, transitions through states near the top of the barrier are most effective.

For a somewhat more complicated four-state system, the four levels indicated in Fig. 13 by both the solid and the dashed lines occur. In this system, relaxation paraelastic effects are obtained from the (degenerate at $\epsilon = 0$) first excited states and resonance diaelastic effects from the others. The relaxation disappears, however, for $T < \Delta$ since the excited states become depopulated. The relaxation strength Δ_R then falls below the classical $1/T$ curve at a temperature determined by Δ.

For TLS made asymmetric by a strain ϵ, the gap 2Δ is given by the three parameters Δ_0, α, and ϵ in

$$\Delta_{1,2} = \mp(\Delta_0^2 + \alpha^2\epsilon^2)^{1/2}, \tag{15}$$

where $2\Delta_0$ is the tunnel splitting, $2\alpha\epsilon$ is the strain-induced asymmetry energy, α is a strain coupling coefficient, and ϵ is the external elastic strain. Then

$$Z = 2\cosh(\Delta/kT). \tag{16}$$

Using Eqs. (13)–(16) in (12), one obtains

$$\Delta C/C = \Delta_R + \Delta_S, \tag{17}$$

where

$$\Delta_R = \left(\frac{c\alpha^2}{CkT}\right)\left[\frac{\alpha\epsilon}{\Delta}\right]^2 \operatorname{sech}^2\left(\frac{\Delta}{kT}\right) \tag{18}$$

and

$$\Delta_S = \left[\frac{c\alpha^2}{C\Delta_0}\left(\frac{\Delta_0}{\Delta}\right)^3\right] \tanh\left(\frac{\Delta}{kT}\right) + \gamma c. \tag{19}$$

The factor in parentheses in Eq. (18) is the familiar result for the classical Snoek effect for $E_{1,2} = \mp\alpha\epsilon$. The factor $\operatorname{sech}^2(\Delta/kT)$ cuts off the relaxation at temperatures low enough for the upper state to become depopulated. What is new for a quantum system is the bracketed factor in Eq. (18) for Δ_R and Eq. (19) for Δ_S. The factor $(\alpha\epsilon/\Delta)^2$ in Eq. (18) shows explicitly that for a symmetric potential, there will be no relaxation because there is no energy shift for a small perturbing ultrasonic strain. For large strains $\alpha\epsilon/\Delta \to 1$, the new term Δ_S is the resonance response. Even though there is no resonant absorption for $\hbar\omega \ll \Delta_0$, there is still a response in phase with an applied strain. At zero temperature, where only the ground state is populated, Δ_S is given by the square bracket in Eq. (19), as is readily calculated using $F = E_1$. From Eq. (12), Δ_S is the curvature of this state as a function of strain. This curvature goes to zero for large strains where E_1 becomes linear in strain. At finite temperature, there is some population of E_2 with opposite curvature, reducing the size of the effect by the factor $\tanh(\Delta/kT)$.

For high temperatures,

$$\Delta_R = \frac{\alpha^2 c}{CkT}\frac{\alpha^2\epsilon^2}{\Delta_0^2 + \alpha^2\epsilon^2},$$
$$\Delta_S = \frac{\alpha^2 c}{CkT}\frac{\Delta_0^2}{\Delta_0^2 + \alpha^2\epsilon^2} + \gamma c, \tag{20}$$

and $\Delta C/C$ achieves the classical value

$$\frac{\Delta C}{C} = \Delta_R + \Delta_S = \left(\frac{\alpha^2}{CkT} + \gamma\right)c. \tag{21}$$

For a random distribution of strain, these results must be averaged over a strain distribution function, with the results expressed now in terms of ϵ_0, the average absolute value of the strain. Below the temperature for which $\omega\tau = 1$, the elastic-constant change is given by the resonant response Δ_S. In what follows, we focus mainly on the velocity measurements. For small strain $\beta = \alpha\epsilon_0/\Delta \ll 1$, the resonance is large and relaxation is a small fraction of the total elastic-constant change at high temperatures. For large strain, the resonance is greatly reduced, and the relaxation is a larger fraction of the total change at high temperatures. The total change is the same for both strains at high temperature. For $kT < \Delta_0$, $\Delta C/C$ deviates from the $1/T$ classical high-temperature response and becomes constant for $kT \ll \Delta_0$.

These features provide a means for analyzing data in terms of a TLS formalism. In the superconducting state $\Delta C/C$ changes from $\Delta_R + \Delta_S$ at high temperatures to Δ_S below $\omega\tau = 1$. In the normal state $\Delta C/C$ is given by $\Delta_R + \Delta_S$ over the whole measured temperature range. A deviation from $1/T$ behavior signals the approach of kT to Δ_0. A relaxation that is a small fraction of the total elastic constant change means β is small.

2.3.3 Acoustic Measurements of Tunneling Hydrogen in Niobium The velocity data for a rapidly cooled specimen (Drescher-Krasicka and Granato, 1985) are plotted in Fig. 14. One sees that the resonance is smaller than the relaxation in this case. Independent of the nature of the tunneling system and the strain distribution, the temperature-independent velocity change at low temperature must be a resonance effect.

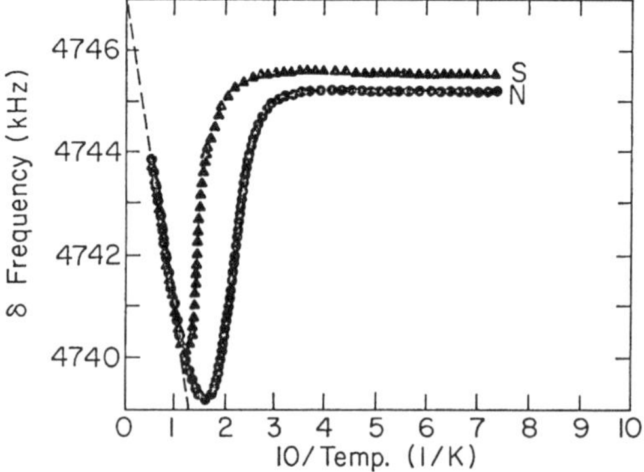

FIG. 14. Frequency change vs 10/T for rapidly cooled specimen. The relaxation strength is large compared to the resonance strength. The velocity change observed at low temperature is larger in the normal state than in the superconducting state (after Drescher-Krasicka and Granato, 1985).

The fact that this is different in the N and S states implies that the tunneling gap must also have changed. This is strong evidence for a correlation of the energy gap with the viscosity of the system.

In Fig. 15, the elastic-constant data, by Maschoff *et al.* (1986), for C' at 10.8 MHz for both the N and S states are plotted vs $1/T$. The two curves, after correction for the electron–phonon interaction, do not come together at the lowest temperatures, as should be expected if Δ_0 does not change. This curve is obtained by subtracting the host and electron–phonon terms from the data. At high temperatures, both the N- and S-state curves vary as α^2/T. This remains true for a strain distribution. Below 2.4 K, in the S state, the relaxation component Δ_R is lost, and only the resonance component Δ_S remains. The deviation from a $1/T$ dependence in the N state occurs for $kT = \Delta_0$, showing directly that $\Delta_0/k \approx 1$ K. The relative size of Δ_R/Δ_S is a measure of $\alpha\epsilon/\Delta_0$, and one sees directly that $\alpha\epsilon < \Delta_0$. Quantitative values for the three parameters of the model, Δ_0, α, and the Lorentzian strain distribution parameter ϵ_0, are obtained by fitting the data as shown by the solid lines in Fig. 15. These are $\Delta_0(S) = 0.094$ meV, $\alpha = 45$ meV, and $\epsilon_0 = 1.1 \times 10^{-3}$. The elastic-constant change, $\delta C(S\text{–}N) = 5.1 \times 10^{-5}$, is 21% and corresponds to a tunneling-parameter change of $\{\Delta_0(S) - \Delta_0(N)\}/\Delta_0(S) = 36\%$. The middle curved line in Fig. 15 is the calculated N-state curve one would expect if Δ_0 does not change, using the parameters Δ_0, α, and ϵ_0 obtained in fitting the S-state data.

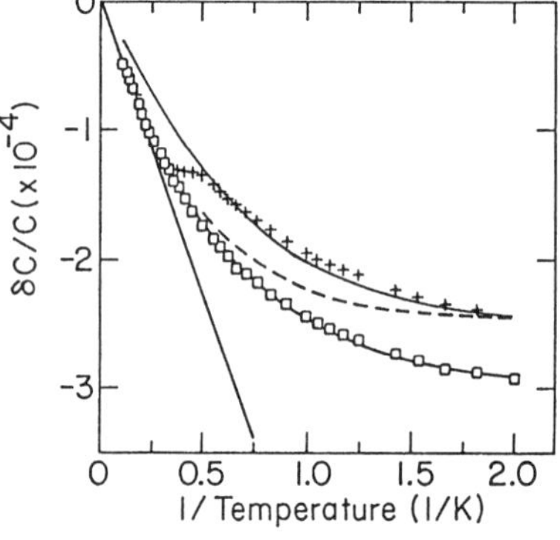

FIG. 15. Relative modulus change vs $1/T$ after background subtraction for N(□) and S(+) Nb–O–H and fits to the two-level system model (solid lines). The middle curve (dashed) is the computed N-state modulus for no change in Δ_0 (after Maschoff *et al.*, 1986).

2.3.4 Classical Point-Defect Resonance Effect

The effect is normally difficult to detect because $\Delta s = \beta c$ in Eq. (20) is both frequency and temperature independent. Although relative changes in $\Delta c/c$ can be measured to great accuracy, absolute changes cannot. Since Δs is normally small, it is only detected when the defect concentration is changed during a measurement, as in a radiation-damage experiment where measurements can be made as a function of radiation flux, as shown in Fig. 16.

This is an example of an important application of acoustic measurements to the determination of the structure of self-interstitials in copper from measurements of the elastic-constant changes measured in thermally radiated copper single crystals (Rehn *et al.*, 1974), where about three parts per million of interstitials were introduced into a sample. One gets changes of the elastic constant that are not very big, in the largest case about 100 ppm, but one can measure that acoustically with very great accuracy. The important features of the results are, first of all, that the effect is linear in defect density and large, about 30 ppm per parts per million of defects for the C_{44} shear constant. The effect is very anisotropic, and this is an important clue for the identification of the structure of the interstitial. Thirty is a very

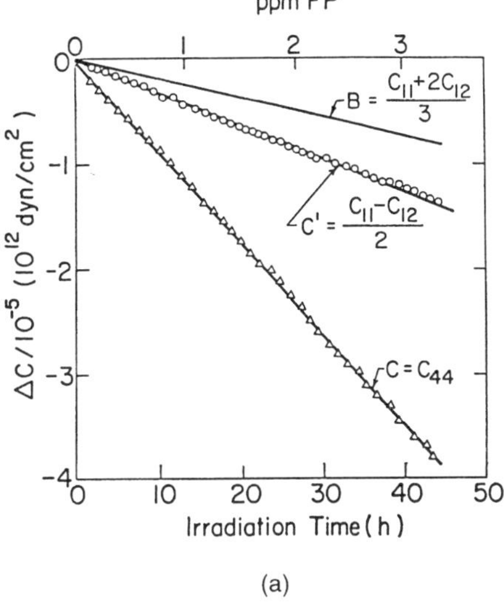

(a)

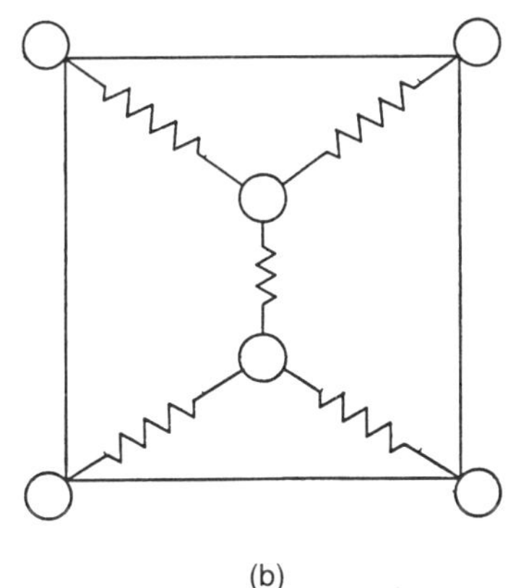

(b)

FIG. 16. (a) Change of the elastic constants of a Cu single crystal after irradiation with thermal neutrons at 4 K (data after Rehn *et al.*, 1974). (b) Split ⟨100⟩ configuration of an interstitial in Cu.

large effect. Normally, property changes such as the volume change or the bulk modulus change per point defect are of order unity. This is more than one order of magnitude larger and that shows that the interstitial produces a very large elastic polarizability in the sample. These clues permit one to identify the structure of the interstitial among several possibilities as one for which two atoms are trying to sit on a normal lattice site, spread out into a dumbbell configuration oriented along ⟨100⟩, as shown in Fig. 16(b), interacting strongly with its neighbors in such a way that it is very susceptible to a shear stress of one type, as compared to the other shear stress, giving a soft spot in the crystal with a resonance mode of the order of $\frac{1}{7}$ of the Debye frequency.

These results form the basis for a new model for liquids and amorphous or glassy solids. According to the interstitialcy model (Granato, 1992), liquids are crystals containing a few percent of interstitials in thermal equilibrium, and glasses are frozen liquids. The strong shear diaelastic polarizability of the interstitial reduces the interstitial formation energy with concentration so that substantial numbers of interstitials exist in thermal equilibrium at high temperature. The properties established for interstitials in fcc metals of

1. a strong shear-modulus softening,
2. a degeneracy in orientation, and
3. low-frequency resonance modes

are those known experimentally to be universal in glasses. Good quantitative agreement has been found for the predictions of the model with the known structure, dynamics, and thermodynamic properties of liquids and glasses (Granato, 1994b).

ACKNOWLEDGMENT

This work was supported by NSF grant DMR 93-19773.

GLOSSARY

Decrement Δ: The ratio of the amplitudes of two successive oscillations of an underdamped harmonic motion. $\Delta = \alpha\lambda = \pi Q^{-1}$, where α is the attenuation, λ is the wavelength, and Q is the quality factor.

Diaelastic Effect: Diaelasticity and resonance are the same. They are the direct response corresponding to an induced elastic dipole under an applied force even when there is no motion over or through a barrier.

Internal Friction: Conversion of mechanical strain energy to heat within a material subjected to fluctuating stress.

Paraelastic Effect: Paraelasticity and relaxation are the same. The defect strain results from the competition between the ordering effects produced by an external field on an existing elastic dipole and the disordering effects due to thermal fluctuations. It requires thermal activation over or tunneling through an energy barrier and is normally therefore strongly temperature dependent.

Relaxation Effect: Same as **Paraelastic Effect.**

Resonance Effect: Same as **Diaelastic Effect.**

Second- (Third-) Order Elastic Constants: Coefficients of the quadratic (cubic) terms in a Taylor series expansion of the free energy in terms of strains.

Works Cited

Baker, G. S. (1962), *J. Appl. Phys.* **33,** 1730–1732.

Bardeen, J., Cooper, L. N., Schrieffer, J. R. (1957), *Phys. Rev.* **108,** 1175–1204.

Beamish, J. R., Franck, J. P. (1982), *Phys. Rev. B* **26,** 6104–6175.

Bordoni, P. G. (1949), *Ric. Sci.* **19,** 851–862.

Caldeira, A. V., Leggett, A. J. (1981), *Phys. Rev. Lett.* **46,** 211–214.

Chandrasekhar, S. (1943), *Rev. Mod. Phys.* **15,** 1–89.

Cottrell, A. H. (1948), in: *Report of a Conference on Strength of Solids, H. H. Wills Physical Lab., University of Bristol, 7–9 July 1947,* London: The Physical Society of London, p. 30.

De Batist, R. (1972), *Internal Friction of Structural Defects in Crystalline Solids,* Amsterdam, London: North Holland Publ. Co., New York: American Elsevier Publ. Co., Inc.

Drescher-Krasicka, E., Granato, A. V. (1985), *J. Phys. (Paris),* Colloq. **C10,** 73–80.

Eshelby, J. D. (1949), *Proc. R. Soc. London, Ser. A* **197,** 396–416.

Fanti, F., Holder, J., Granato, A. V. (1969), *J. Acoust. Soc. Am.* **45,** 1356–1366.

Fedorov, F. I. (1968), *Theory of Elastic Waves in Crystals,* New York: Plenum Press.

Frenkel, J., Kontorova, T. (1939), *J. Phys. Acad. Sci. USSR* **1,** 137–149.

Garber, S. A., Granato, A. V. (1970), *J. Phys. Chem. Solids* **31,** 1863–1867.

Garber, J. A., Granato, A. V. (1975), *Phys. Rev. B* **11,** 3990–3997.

Granato, A. V. (1971a), *Phys. Rev. B* **4,** 2196–2201.

Granato, A. V. (1971b), *Phys. Rev. Lett.* **27,** 660–664.

Granato, A. V. (1992), *Phys. Rev. Lett.* **68,** 974–977.

Granato, A. (1994a), *J. Alloys Compounds* **211/212,** 503–508.

Granato, A. V. (1994b), *J. Phys. Chem. Solids* **55,** 931–939.

Granato, A. V., Lücke, K. (1956), *J. Appl. Phys.* **27,** 789–805.

Granato, A. V., Lücke, K. (1966), in: W. P. Mason, R. N. Thurston (Eds.), *Physical Acoustics,* New York: Academic Press, Vol. IVA, pp. 225–276.

Granato, A. V., Lücke, K. (1981), *J. Appl. Phys.* **52,** 7136–7142.

Gutmanas, E., Nadgornyi, Y., Stephanov, A. V. (1963), *Fiz. Tverd. Tela* **5,** 1021 [*Sov. Phys. Solid State* **5,** 743–747].

Hearmon, R. F. S. (1979), (1) The Elastic Constants of Crystals and Other Anisotropic Materials, p. 1; (2) The Third and Higher Order Elastic Constants, p. 245, in: K.-H. Hellwege, A. M. Hellwege (Eds.), *Landolt-Börnstein Numerical Data and Functional Relationships in Science and Technology,* Group III, *Crystal and Solid State Physics,* Vol. 11, *Elastic, Piezoelectric, Pyroelectric, Piezooptic, Electrooptic Constants, and Nonlinear Dielectric Susceptibilities of Crystals,* Berlin-Heidelberg-New York: Springer-Verlag.

Hiki, Y. (1972), *Elasticity and Anelasticity,* Tokyo: Kyoritsu (in Japanese).

Holder, J., Granato, A. V. (1971), in: W. P. Mason, R. N. Thurston (Eds.), *Physical Acoustics,* Vol. VIII, New York: Academic, pp. 237–277.

Huang, K. F., Granato, A. V., Birnbaum, A. G. (1985), *Phys. Rev. B* **32,** 2178–2183.

Hunklinger, S., Arnold, W. (1976), in: W. P. Mason, R. N. Thurston (Eds.), *Physical Acoustics,* Vol. XII, New York: Academic, pp. 155–215.

Huntington, H. B. (1958), in: F. Seitz, D. Turnbull (Eds.), *Solid State Physics,* New York: Academic, Vol. VII, pp. 214–351.

Isaac, R. D., Granato, A. V. (1988), *Phys. Rev. B* **37,** 9278–9285.

Jäckle, J. (1972), *Z. Phys.* **257,** 212–223.

Jäckle, J., Piehé, L., Arnold, W., Hunklinger, S. (1976), *J. Non-Cryst. Sol.* **20,** 365–391.

Kamada, K., Yoshizawa, I. (1971), *J. Phys. Soc. Jpn.* **31,** 1056–1068.

Koehler, J. S. (1952), in: W. Shockley *et al.* (Eds.), *Imperfections in Nearly Perfect Crystals,* New York: Wiley, pp. 197–216.

Kojima, H., Suzuki, T. (1968), *Phys. Rev. Lett.* **21,** 896–898.

Kosugi, T., Kino, T. (1993), *Mater. Sci. Eng. A* **164,** 368–372.

Kramers, H. A. (1940), *Physica* **7,** 284–304.

Krautkrämer, J., Krautkrämer, A. (1983), *Ultrasonic Testing of Materials,* 3rd ed., Berlin, Heidelberg, New York: Springer-Verlag.

Leibfried, G. (1950), *Z. Phys.* **127,** 344–356.

Leibfried, G., Ludwig, W. (1961), in: F. Seitz, D. Turnbull (Eds.), *Solid State Physics,* Vol. 12, New York: Academic, pp. 275–444.

Magalas, L. (Ed.) (1996), *Mechanical Spectroscopy,* in press.

Maschoff, K. R., Drescher-Krasicka, E., Granato, A. V. (1986), in: A. C. Anderson, J. P. Wolfe, (Eds.), *Phonon Scattering in Condensed Matter V,* New York: Springer Verlag, pp. 64–66.

Mason, W. P. (1955), *Phys. Rev.* **97,** 557–558.

Mason, W. P. (1958), *Physical Acoustics and the Properties of Solids,* Princeton, NJ: Van Nostrand.

Mason, W. P. (Ed.) (1964–1968), *Physical Acoustics,* Vols. I–V, London, New York: Academic.

Mason, W. P. (Ed.) (1964), *Physical Acoustics, Principles and Methods,* New York: Academic.

Mason, W. P., R. N. Thurston (Eds.) (1970–1984), *Physical Acoustics,* Vols. VI–XVIII, London, New York: Academic.

Morkel, C., Wipf, H., Neumaier, K. (1978), *Phys. Rev. Lett.* **40,** 947–950.

Morse, R. W. (1955), *Phys. Rev.* **97,** 1716–1717.

Morse, R. W., Bohm, H. V. (1957), *Phys. Rev.* **108,** 1094–1096.

Nowick, A. S., Berry, B. S. (1972), *Anelastic Relaxation in Crystalline Solids,* New York, London: Academic.

Orowan, E. (1940), *Proc. Phys. Soc. (London)* **52,** 8–22.

Pippard, A. B. (1955), *Philos. Mag.* **46,** 1104–1122.

Poker, D. B., Setser, G. G., Granato, A. V., Birnbaum, H. K. (1979), *Z. Phys. Chem. N.F.* **116,** 39–45.

Pustovalov, V. V., Startsev, V. I., Didenko, D. A., Fomenko, V. S. (1967), *Fiz. Met. Metalloved.* **23,** 312–318.

Rayne, J. A. (1975), in: D. Lenz, K. Lenz (Eds.), *Proceedings of the Fifth International Conference on Internal Friction and Ultrasonic Attenuation in Crystalline Solids, Aachen, Germany, August 1973,* Berlin: Springer, pp. 13–51.

Read, T. A. (1940), *Phys. Rev.* **58,** 371–380.

Rehn, L. E., Holder, J., Granato, A. V., Coltman, R. R., Young, F. W. Jr., (1974), *Phys. Rev. B* **10,** 349–362.

Ritchie, I. G., Fantozzi, G. (1992), in: F. Nabarro (Ed.), *Dislocations in Solids,* Vol. 9, Amsterdam: North Holland, pp. 57–133.

Robrock, K.-H., Rehn, L. E., Spiric, V., Schilling, W. (1977), *Phys. Rev. B* **15,** 680–688.

Schwarz, R. B., Granato, A. V. (1975), *Phys. Rev. Lett.* **34,** 1174–1177.

Schwarz, R. B., Isaac, R. D., Granato, A. V. (1977), *Phys. Rev. Lett.* **38,** 554–557.

Schwenker, R. O., Granato, A. V. (1969), *Phys. Rev. Lett.* **23,** 918–921.

Seeger, A. (1956), *Philos. Mag.* **651;** (1981), *J. Phys. C* **5,** Suppl. 10, 201–228.

Sellers, G. H., Anderson, A. C., Birnbaum, H. K. (1974), *Phys. Rev. B* **10,** 2771–2776.

Simmons, G., Wang, H. (1971), *Single Crystal Elastic Constants and Calculated Aggregate Properties: A Handbook,* 2nd ed., Cambridge, MA: MIT Press.

Snoek, J. L. (1941), *Physica* **8,** 711–733.

Startsev, V. I. (1983), in: F. R. N. Nabarro (Ed.), *Dislocations in Solids,* Vol. 6, New York: North Holland, Ch. 28, pp. 143–233.

Stern, R. M., Granato, A. V. (1962), *Acta Metall.* **10,** 359–381.

Teutonico, L. J., Granato, A. V., Lücke, K. (1964), *J. Appl. Phys.* **35,** 220–234.

Thompson, D. O., Holmes, D. K. (1956), *J. Appl. Phys.* **27,** 713–723.

Thompson, D. O., Pare, V. K. (1960), *J. Appl. Phys.* **31,** 528–535.

Thurston, R. N., Brugger, K. (1964), *Phys. Rev.* **133,** A1604–A1610.

Truell, R., Elbaum, C., Chick, B. B. (1969), *Ultrasonic Methods in Solid State Physics,* New York, London: Academic Press.

Wang, M. C., Uhlenbeck, G. E. (1945), *Rev. Mod. Phys.* **17,** 323–342.

Zener, C. (1948), *Elasticity and Anelasticity of Metals,* Chicago: Univ. Chicago Press.

Further Reading

Bhatia, A. (1967), *Ultrasonic Absorption,* Oxford: Oxford Clarendon Press.

De Batist, R. (1972), *Internal Friction of Structural Defects in Crystalline Solids,* Amsterdam, London: North Holland Publ. Co., New York: American Elsevier Publ. Co., Inc.

Hunklinger, S., Arnold, W. (1976), in: W. P. Mason, R. N. Thurston (Eds.), *Physical Acoustics,* Vol. XII, New York: Academic, p. 155.

Krautkrämer, J., Krautkrämer, A. (1983), *Ultrasonic Testing of Materials,* 3rd ed., Berlin, Heidelberg, New York: Springer-Verlag.

Mason, W. P. (1958), *Physical Acoustics and the Properties of Solids,* Princeton, NJ: Van Nostrand.

Nowick, A. S., Berry, B. S. (1972), *Anelastic Relaxation in Crystalline Solids,* New York, London: Academic.

Schreiber, E., Anderson, O., Soga, N. (1973), *Elastic Constants and Their Measurements,* New York: McGraw-Hill.

Truell, R., Elbaum, C. (1962), "High Frequency Ultrasonic Stress Waves in Solids," in: S. Flügge (Ed.), *Handbuch der Physik,* Vol. X1/2, Berlin: Springer, p. 153.

Truell, R., Elbaum, C., Chick, B. B. (1969), *Ultrasonic Methods in Solid State Physics,* New York, London: Academic.

Zener, C. (1948), *Elasticity and Anelasticity of Metals,* Chicago: Univ. Chicago Press.

SOLIDS, CRYSTALLINE—MECHANICAL PROPERTIES

Rob Phillips, *Division of Engineering, Brown University, Providence, Rhode Island, U.S.A.*

INTRODUCTION

Much of what makes the properties of materials so intriguing is a reflection of their response to external stimuli. In addition to their rich electrical and optical properties, materials reveal a host of remarkable responses when subjected to mechanical loads. One abiding concern is that materials fail in dramatic ways. For example, metals have always been of special interest because of their mechanical properties. During World War II, the *Liberty* ships repeatedly broke in half, sometimes under the presumed shelter of safe harbor. Earthquakes continue to fascinate and horrify us if for no other reason than they show us how illusory is the solidity of our surroundings. By way of contrast, whole industries depend upon the ability to shape and form materials in ways that are almost unbelievable. Metal ingots that begin with dimensions of order meters can, after deformation processing, reach dimensions in excess of kilometers. Thus both failure and intentional deformation form the backdrop for the study of the mechanical properties of materials.

The mechanical properties of materials pose challenges on experimental and theoretical fronts alike. The variety of observed mechanical phenomena is perhaps best typified by a specimen loaded in tension. Under these circumstances, a cylindrical sample, for example, is loaded parallel to the cylinder axis, and the resulting elongation is considered as a function of the applied load. Though the province of the mechanical properties of crystalline solids is wider than evidenced in a tension test, this simple experiment serves to fix some of the key ideas that characterize the type of behavior requiring explanation and that the engineer seeks to control—ideas such as elastic deformation, plastic flow, and eventual catastrophic failure via fracture. These same features are exhibited whether the test is carried out in tension, torsion, or bending.

3-527-28140-1/96/$5.00 + .50

An example of these features is illustrated schematically via the stress–strain curves shown in Fig. 1. Here it is imagined that the tensile specimen has been subjected to increasing loading, which engenders deformation and ultimate failure. Figure 1(a) shows the response of a ductile material to increasing loads and exhibits plastic deformation while Fig. 1(b) illustrates the stress–strain curve for a brittle ceramic, which exhibits no plastic deformation prior to fracture. Elastic deformation is characterized by its reversibility—once the load is removed, the solid returns to some nascent reference configuration that is free of deformation. *Linear* elastic materials are characterized by a linear relation between stress and strain, just the behavior canonized in the law of Hooke

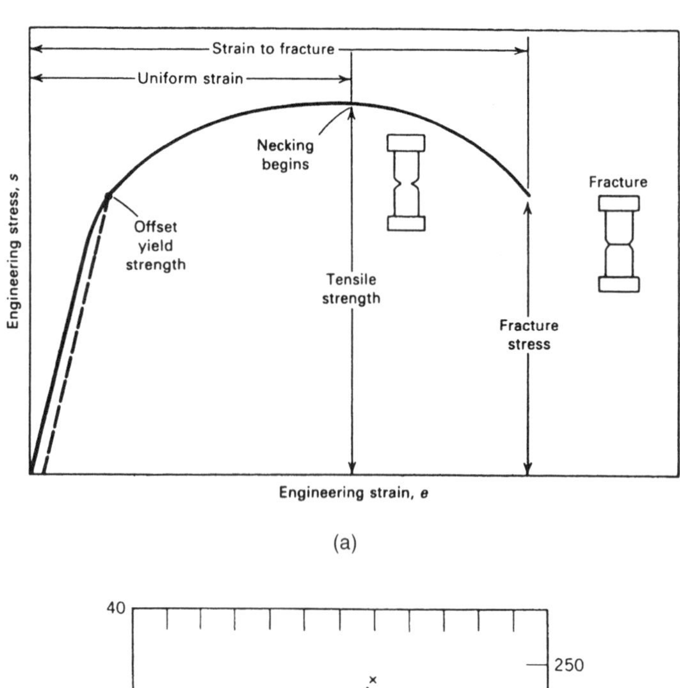

(a)

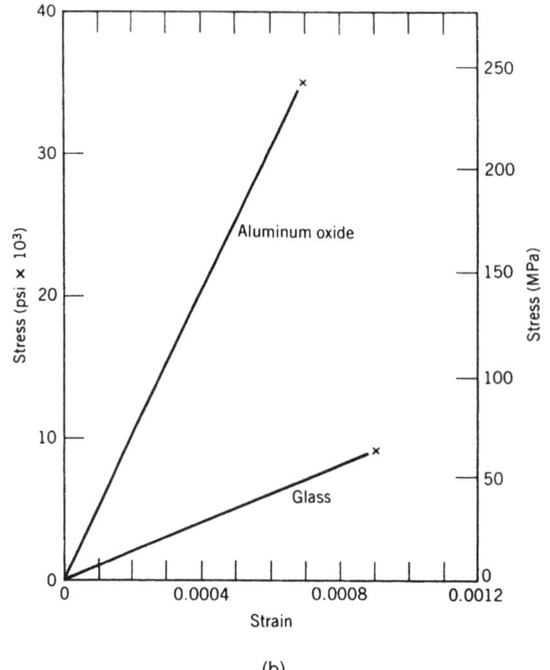

(b)

FIG. 1. Generic stress–strain curves for samples loaded in tension. (a) Stress–strain behavior for ductile material that exhibits plastic deformation. After Boyer (1987). (b) Stress–strain curve for brittle ceramics. After Callister (1994).

that bears his name. This behavior is exhibited in the small-strain region of the figure. At a stress known as the yield stress σ_y, ductile materials undergo permanent deformation. Beyond this point, if the load is removed, the sample does not return to its undeformed state; rather, the specimen exhibits a permanent change of shape. This behavior is familiar to anyone who has bent a clothes hanger and is the basis of technologies such as the sheet forming used in shaping car bodies. For a generic ductile solid, after some range of plastic deformation, the solid fails catastrophically via a fracture process. At this point, the loading part of the experiment has ended.

Our discussion of the mechanical response of a generic crystalline solid has thus far been highly idealized, implying that the mechanical response of a particular material is determined by nothing more than the state of applied stress. However, the observed range of mechanical behavior is determined by a number of competing factors (both external and internal) that we have thus far ignored; factors such as temperature, strain rate, and thermal and mechanical loading history are especially important. Ashby has pioneered a diagrammatic representation of the different observed behaviors known as deformation mechanism maps (Ashby and Jones, 1980), an example of which is shown in Fig. 2. This figure depicts the deformation mechanisms (i.e., elastic response, dislocation-mediated plastic deformation, and creep) that can be realized by a material as simple as elemental copper under different loads and temperatures. For example, at high enough temperatures even for modest loads, it is seen that the solid suffers deformation in the form of creep. Creep deformation refers to the permanent plastic deformation of a material at stresses below the yield stress, usually as a result of heightened temperatures. An additional level of complexity is introduced as a result of the dependence of the dominant mode of deformation on strain rate. Contours of constant strain rate are seen as the thin lines that rib the figure and allow for the data at different strain rates to be collapsed onto a single plot.

Not only are there a variety of mechanisms by which a solid may undergo deformation, but their eventual failure via fracture occurs by a number of mechanisms as well. A fracture mechanism map is illus-

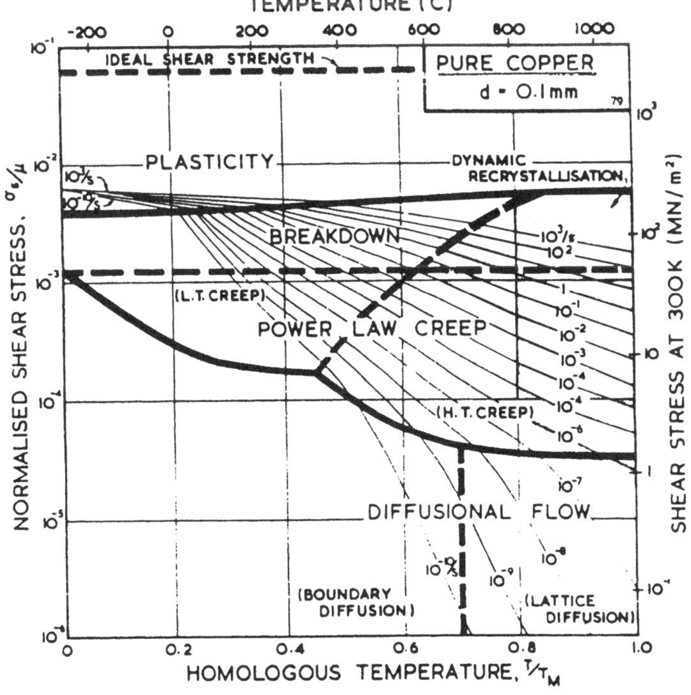

FIG. 2. Deformation mechanism map for elemental Cu with a grain size of 100 μm illustrating various permanent deformation mechanisms available to the solid under thermomechanical loads. Temperature normalized by melting temperature T_m, and stress normalized by shear modulus μ. After Frost and Ashby (1982).

trated in Fig. 3, which demonstrates that at elevated temperatures materials may suffer ductile fracture (fracture preceded by plastic deformation), while at lower temperatures they will fail in a brittle manner via cleavage. One measure of the ductility that is clearly evident in the fracture mechanism map is the reduction in the cross-sectional area of the sample prior to failure.

A second feature that broadens the range of observed mechanical properties is that of various microstructural influences. For example, trace amounts of Au or Al in elemental Cu change the yield stress (stress to plastic deformation) dramatically. This is illustrated in Fig. 4, which depicts the concentration dependence of the yield stress for various impurities in Cu and reveals that the hardening effect is an appreciable fraction of the yield stress itself for a soft metal such as Cu. Similarly, by tuning the grain size of a given material via application of both heat treatment and deformation, the yield strength can again be varied substantially. Thus, while Fig. 1 serves to provide a schematic idea of the types of mechanical behavior, both testing conditions and the material's history should be considered as well.

The geometric description of shape changes in materials and the applied forces needed to maintain them serve as the language of deformation for describing the mechanical behavior of solids. It is most convenient to idealize a solid body as a continuum, with the underlying discrete lattice properties blurred into a set of field variables such as displacement, velocity, strain, and temperature. The deformation of a body is characterized by imagining an initially undeformed state (dubbed the reference configuration) in which the material particles that make up the body are labeled by their Cartesian coordinates **X**. After undergoing some shape change, the material particles find themselves in new positions given by

$$\mathbf{x} = \mathbf{X} + \mathbf{u}(\mathbf{X}), \tag{1}$$

where $\mathbf{u}(\mathbf{X})$ is known as the displacement vector and represents the displacement suffered by the material particle that was initially at position **X**. Such deformations are schematized in Fig. 5, which illustrates that in general the deformation may consist not only of relative motion of material particles but of rotations as well. Though the kinematics of finite deformation as illustrated in Fig. 5 is intriguing in its own right, in this article we emphasize a geometric description of the deformations of interest in terms of the infinitesimal or small-strain tensor, defined by

$$\epsilon_{ij} = \tfrac{1}{2}(u_{i,j} + u_{j,i}). \tag{2}$$

The displacement gradients are given via indicial notation as

$$u_{i,j} = \frac{\partial u_i}{\partial X_j}, \tag{3}$$

and this notation will be used throughout the article. It is important to note that this deformation measure is really only appropriate in the limit of small displacement gradients. In particular, the small-strain tensor, while blind to rigid-body translations, is nonzero in the presence of *finite* rotations. On the other hand, for most purposes, the interesting deformations from the standpoint of the mechanical properties of materials are those involving changes in *relative* position, such as are picked up by the symmetric part of the displacement gradient, more familiar as the strain tensor. The components of the small-strain tensor admit a simple interpretation if we restrict our analysis to the Cartesian directions. The diagonal components of the small-strain tensor determine the stretching of these vectors, while the off-diagonal elements provide a measure of the change in angles between them as a result of the deformation.

While strain serves as the kinematic basis of our discussion of the mechanical properties of materials, kinematics must be supplemented by a description of the forces that result in deformation. In this case, the solid is idealized as a continuum in which forces are transmitted locally and are described in terms of the stress tensor. For example, if we reconsider the tensile specimen described above and imagine a loading in which the tensile axis points along the z direction, the stress tensor is given by

$$\sigma = \begin{pmatrix} 0 & 0 & 0 \\ 0 & 0 & 0 \\ 0 & 0 & \sigma_0 \end{pmatrix}. \tag{4}$$

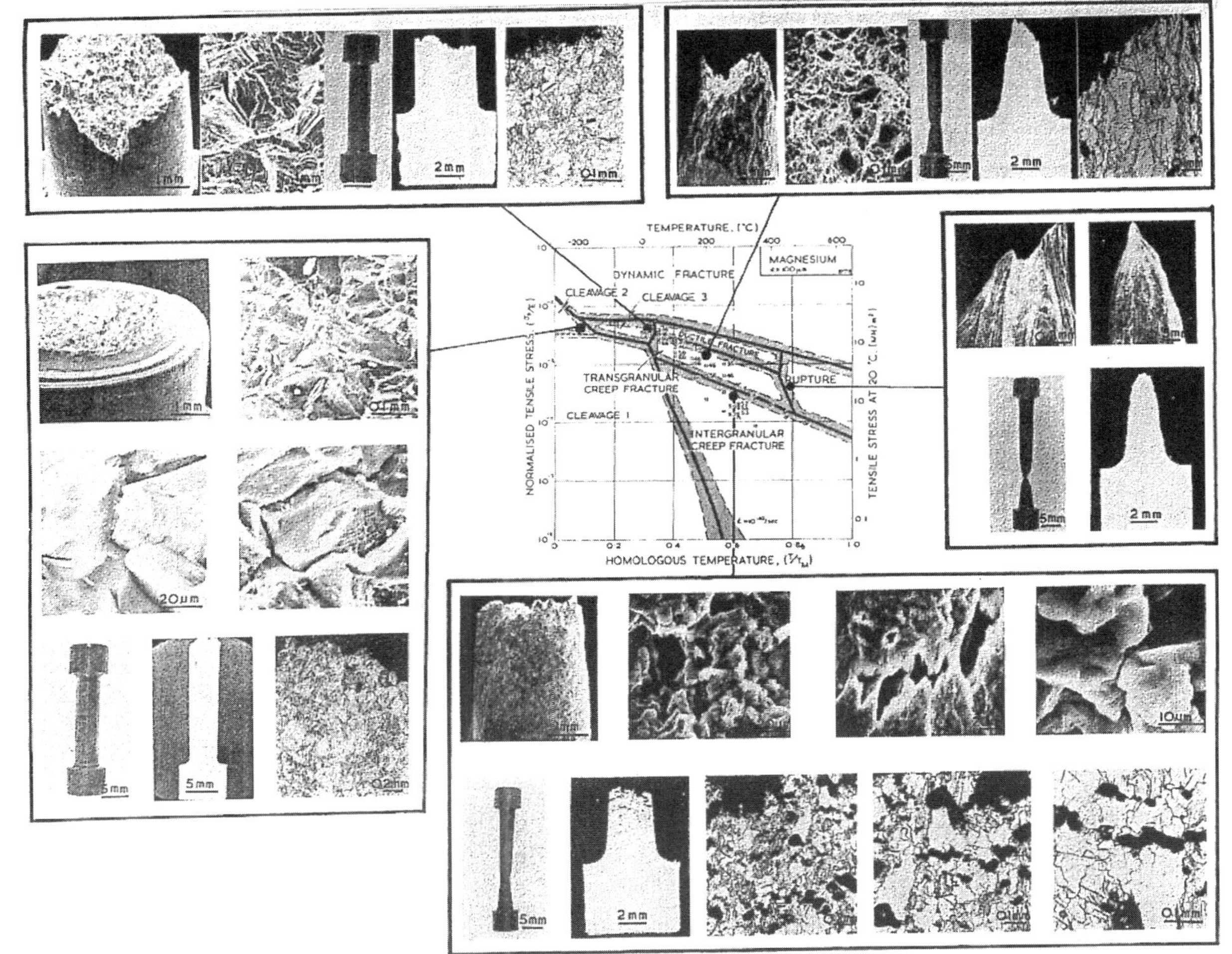

FIG. 3. Fracture mechanism map illustrating various fracture mechanisms available to sol d Mg upon fracture. Central map shows mechanisms for different loads and temperatures, while surrounding figures show fracture surfaces and deformation of sample prior to fracture. After Gandhi and Ashby (1979).

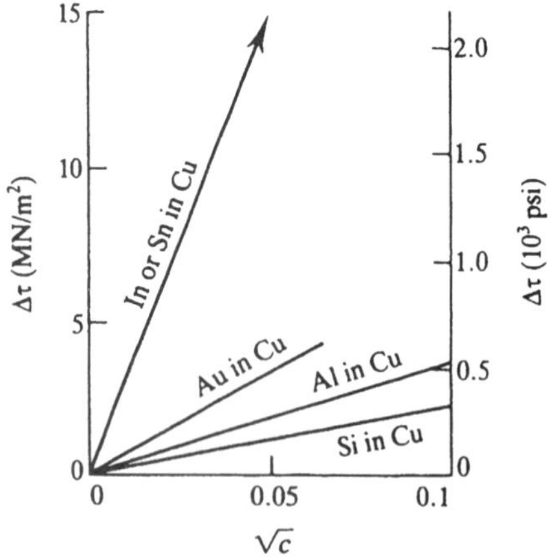

FIG. 4. Dependence of the yield stress of copper on impurity concentration. After Courtney (1990).

Given the stress tensor at a point, the force per unit area acting on a plane with normal **n** passing through that point is determined by

$$t_i^{(\mathbf{n})} = \sigma_{ij} n_j, \tag{5}$$

where $\mathbf{t}^{(\mathbf{n})}$ is known as the stress (or traction) vector. For more complicated states of stress, the off-diagonal components of the stress tensor determine the shear stresses on the planes orthogonal to the Cartesian directions. This can be seen by recourse to Eq. (5) if we consider a plane with normal in the z direction. In this case, σ_{xz} gives the component of the traction vector pointing along the x direction. Hence, these components determine the shearing forces.

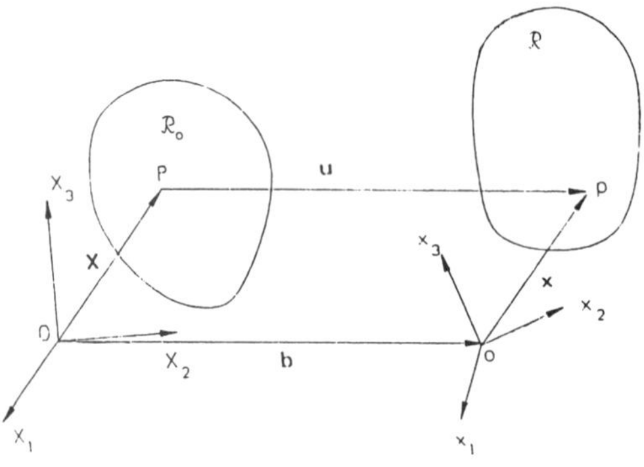

FIG. 5. Schematic of the mapping that carries a reference configuration with material particles labeled by **X** into the deformed configuration, with positions **x** = **X** + **u**(**X**). After Khan and Huang (1995).

The equilibrium of a solid body in the absence of any body forces follows from Newton's first law as

$$\sigma_{ij,j} = 0, \tag{6}$$

where the summation convention, which instructs for the summation over repeated indices, is implied. The symmetry of the stress tensor ($\sigma_{ij} = \sigma_{ji}$) follows from the absence of body moments. The point of our excursion into rudimentary continuum mechanics is that the notions of stress and strain not only have been found as the most powerful tool for describing the experimental basis of the mechanical properties of materials, but they provide the foundation upon which models of elastic and plastic deformation as well as fracture are built.

The plan of the remainder of the article is as follows. The main body of the article begins in Sec. 1 with a discussion of the elastic behavior of crystalline solids and an emphasis on linear elastic response. Section 2 follows by describing the notion of the ideal strength of crystalline solids. The ideal strength concept illustrates the importance of crystalline defects as the agents of mechanical behavior, with dislocations as the primary players in plastic deformation and cracks as the source of fracture at stresses below the ideal tensile strength. Sections 3 and 4 are the centerpiece of the article and consider permanent deformation in the form of plasticity, creep, fracture, and fatigue. Here it is shown how crystal lattice defects mediate the various mechanisms responsible for observed macroscopic mechanical behavior. Section 5 illustrates the subtleties introduced as a consequence of microstructural influences and demonstrates the refinements needed to understand the dependence of mechanical behavior on a material's thermal and mechanical history. At best, our treatment of this vast subject can only be cursory. In particular, the emphasis here is purely on crystalline solids, with metals serving as the basis of the majority of the discussion. The suggestions for further reading gives a list of references that are particularly enlightening.

1. ELASTIC RESPONSE OF MATERIALS

1.1 Phenomenology of Elastic Deformation

Prior to permanent deformation, crystalline solids typically undergo elastic deformation. Such deformation is reversible with the result that the undeformed state may be instantaneously recovered once any applied stresses are removed. The physics of such deformation may be neatly captured via a few simple material parameters known as the elastic moduli. That is, the elastic response of a material subjected to some arbitrarily complicated state of stress is ultimately determined by a set of numbers (up to 21 for crystals of low symmetry) known as the elastic constants. Experimentally, it is often found that there is a linear relation between these strains and the stresses needed to impose them.

The simplest sample for exploring the linear elastic behavior of materials is the tensile specimen. In this case, it is found that there is a linear relation between the dimensionless tensile elongation $\epsilon = (l - l_0)/l_0$ and the stress needed to produce it. This behavior is parameterized via a material constant known as the Young modulus, E, with the result that

$$\sigma_n = E\epsilon, \tag{7}$$

where σ_n is the normal stress. This is Hooke's law in its simplest form and accounts for the initial linear regime depicted in Fig. 1. Hooke's law is perhaps the simplest version of a material constitutive law, a necessary ingredient to supplement the kinematics and balance laws of conventional continuum formulations. Such constitutive laws serve to relate kinematic variables such as the strain or strain rate to the stresses. An accompanying effect of such tensile deformation is a lateral contraction of the sample perpendicular to the applied load that is parameterized via a second elastic material constant known as the Poisson ratio, ν. The strains in this simple case, then, are

$$\epsilon_{xx} = -\nu\epsilon_{zz} = -\frac{\nu\sigma_n}{E}$$

$$\epsilon_{yy} = -\nu\epsilon_{zz} = -\frac{\nu\sigma_n}{E}$$

$$\epsilon_{zz} = \frac{\sigma_n}{E},$$

where we have assumed the tensile axis is in the z direction. For isotropic materials, these two material parameters (E and ν) serve to characterize the elastic material response for loads as complicated as desired as long as the strains are sufficiently small. For larger elastic deformations, this description must be extended to the consideration of higher-order moduli. For many applications, the elastic regime is all that a material is ever hoped to sample—the bending of an airplane's wing while it is being bumped around in turbulence is intended to be reversible.

1.2 Generalized Hooke's Law

Thus far, we have ignored any intrinsic anisotropy induced by the underlying crystal structure of which the sample is composed. However, if we pull on a sample of zinc, the moduli describing both the elongation and the lateral contraction will depend strongly upon the crystal orientation. For example, if we are to pull along the c axis in a zinc crystal with a particular stress, it will lead to one tensile elongation. On the other hand, if we pull along one of the close-packed directions in the hexagonal basal plane with the same stress, the resulting elongation will be different. These anisotropy effects suggest a generalized form of Hooke's law that posits a relation between stress and strain of the form

$$\sigma_{ij} = C_{ijkl}\epsilon_{kl}, \tag{8}$$

where C_{ijkl} is known as the elastic modulus tensor, and the summation convention is in effect. The number of independent elastic moduli is determined by the symmetry of the crystal, with cubic crystals having three independent moduli and hexagonal crystals, five such moduli, and for isotropic solids, the modulus tensor reduces to the two independent material constants (Young's modulus and Poisson ratio) discussed above. For example, in cubic crystals like Al and Cu, the three independent moduli are denoted C_{1111}, C_{1122}, and C_{1212}. A more compact notation is often adopted in which one defines $C_{1111} = C_{11}$, $C_{1122} = C_{12}$, and $C_{1212} = C_{44}$. For the cubic crystal, the use of these constants leads to the conclusion that the stress needed to induce a 1% strain is the same for the (100) and (001) directions, but not for the (111) direction. In cubic crystals, a useful measure of the elastic anisotropy is given by

$$A = 2C_{44}/(C_{11} - C_{12}), \tag{9}$$

a quantity known as the anisotropy ratio. For an isotropic material, this constant is unity.

The observed range of elastic moduli for different materials spans some six orders of magnitude, with diamond at the pinnacle with a Young's modulus of a thousand gigapascals, typical structural materials such as steel coming in with moduli of a few hundred gigapascals, and rubbers and other polymers reaching values of less than 0.1 GPa. These numbers can be put in perspective by considering the stresses needed to induce a 1% strain in a few different materials. To put these stress measures into everyday terms, we first consider the pressure exerted on the floor by one's feet when standing. If we estimate a typical person's weight as 700 N and the contact area between their feet and the floor as 0.04 m^2, this corresponds to a stress of roughly 2×10^4 N/m^2 or 2×10^{-5} GPa. These numbers should be contrasted with the stresses of 10 GPa needed to induce a 1% strain in a cylindrical tensile specimen of diamond and even the stress of 10^{-3} GPa needed to strain a generic rubber specimen by 1%.

A second view of an elastic solid is provided by the notion of a stored energy function. By virtue of its elastic deformation, a solid stores strain energy, which can be recovered upon release of the deformation. The work done in bringing a solid to some state of strain is indicated schematically in Fig. 6. For a linear elastic solid, the strain energy density is given by

$$W(\epsilon) = \tfrac{1}{2} C_{ijkl}\epsilon_{ij}\epsilon_{kl}. \tag{10}$$

As a result, the total strain energy stored in the solid by virtue of its elastic strain is computed by integrating the strain energy density over the entire solid,

$$E_{\text{strain}} = \int W(\epsilon)\, dV. \tag{11}$$

The strain energy provides another window on the significance of the elastic moduli. For example, in the case of a material with cubic symmetry, the strain energy density is given by

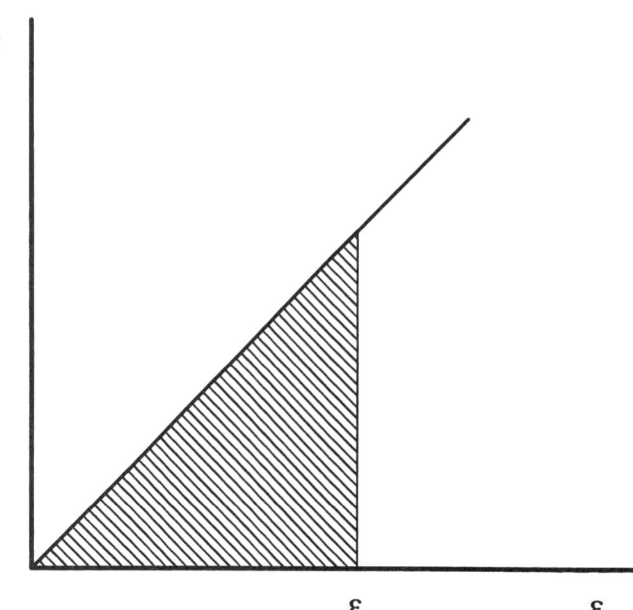

FIG. 6. Stress–strain curve illustrating energy stored in a solid by virtue of elastic deformation. Upon reaching a strain of ϵ_{max}, the strain energy is given by the area of the shaded region.

$$W = \tfrac{1}{2}C_{11}(\epsilon_{xx}^2 + \epsilon_{yy}^2 + \epsilon_{zz}^2) + C_{12}(\epsilon_{xx}\epsilon_{yy} + \epsilon_{xx}\epsilon_{zz} + \epsilon_{yy}\epsilon_{zz}) + 2C_{44}(\epsilon_{xy}^2 + \epsilon_{xz}^2 + \epsilon_{yz}^2). \quad (12)$$

Thus if we consider pure shear for which only the off-diagonal components of the strain tensor are nonzero, the energy cost of such deformations is governed by C_{44}. Alternatively, if we consider purely dilatational strains for which $\epsilon_{xx} = \epsilon_{yy} = \epsilon_{zz}$ and for which $\epsilon_{ij} = 0$ for $i \neq j$, then only C_{11} and C_{12} determine the energy cost. The linear elastic description of such strains breaks down once the strains become sufficiently large. In particular, the energy cost described above, which is quadratic in the strains, is no longer an appropriate description of the strain energy and requires the inclusion of nonlinear effects.

1.3 Microscopic Origins of Elastic Moduli

The use of the type of linear elastic models described above is intriguing for a number of reasons. One important feature of such models is that they replace the underlying lattice degrees of freedom with a continuum whose properties are specified by a few independent material parameters. However, ultimately the values adopted by these material parameters are a direct consequence of the underlying interactions between the atoms that make up the solid. Recent advances in the solution of the Schrödinger equation for electrons in a solid have made the first-principles calculation of material parameters such as the elastic moduli routine. As is shown below, the moduli can be interpreted as the curvature of the potential energy surface for the deformation of interest.

A more transparent microscopic model of the origins of the moduli in terms of bonds between atoms is afforded by considering a pair-potential description of cohesion. In this case, the total energy of the solid may be written as

$$E_{\text{tot}} = \tfrac{1}{2}\sum_{ij}^{N} V(R_{ij}), \quad (13)$$

where the sum is over all of the atoms in the solid and R_{ij} is the distance between the ith and jth atoms. The moduli are computed by seeking an analogy between the microscopic expression for the total energy and that arising from the continuum picture. For a crystal subjected to a homogeneous deformation described by the strain tensor $\boldsymbol{\epsilon}$, the microscopic expression for the total energy may be written as

$$E_{\text{tot}} = \tfrac{1}{2}\sum_{ij}^{N} V((\mathbf{I} + \boldsymbol{\epsilon})R_{ij}). \quad (14)$$

In this expression, $\boldsymbol{\epsilon}$ is the strain tensor and is a measure of the extent to which the crystal is deformed, and hence interatomic distances are altered, and $\mathbf{I}$ is the identity tensor. By expanding the energy to second order in the relevant strains and equating this expression for the energy to that arising from the continuum model, the moduli may be read off as

$$C_{ijkl} = \frac{1}{2\Omega}\sum_{n}\left(V_n'' - \frac{V_n'}{R^n}\right)R_i^nR_j^nR_k^nR_l^n/(R^n)^2. \quad (15)$$

In this expression, the sum is over the neighbors of the atom at the origin, R^n is the distance to the nth neighbor, R_i^n is the ith component of the vector to the nth neighbor, and Ω is the volume per atom in the crystal. The critical insight provided by this model calculation is the relation between the curvature of the interatomic potential (a microscopic quantity) and the elastic moduli (macroscopic observables). Though this model provides useful insights into the microscopic origins of the moduli, for a covalent material such as Si this approach will not capture the bond-bending mechanisms relevant to its moduli, and more sophisticated treatments are required.

1.4 Beyond the Elastic Regime

Earlier we noted that much of the permanent deformation that accounts for the broad variety of observed mechanical properties is mediated by extended defects such as dislocations and cracks. Interestingly, the primary conceptual basis for understanding such defects is the type of linear elastic models elucidated in the preceding section. As we discussed earlier, the phenomenological basis for the continuum description of phenomena such as elastic and plastic response is constitutive models. Thus, despite the fact that the presence of defects such as cracks and dislocations results in macroscopic behavior that cannot be described by linear elastic constitutive models (i.e., Hooke's law no longer applies), the fields surrounding the defects themselves are amenable to a linear elastic treatment.

In subsequent sections our understanding of crack-tip phenomena, for example, will be founded upon an elastic analysis of the crack problem that allows for a characterization of the crack-tip stress fields in terms of an $r^{-1/2}$ singularity, where r denotes the distance from the crack tip. Similarly, elastic analyses of dislocations are characterized by stress fields that decay with a generic r^{-1} singularity and logarithmic interaction energies.

Reference to the stress–strain curve exhibited in Fig. 1 makes it clear that for most materials, the elastic regime constitutes but a small part of the observed mechanical response. A foreshadowing of the qualitative changes that arise beyond the elastic regime is first noted at the so-called proportional limit, the point on the stress–strain curve where linearity is lost. The proportional limit is followed by plastic yield at a stress often referred to as the offset yield strength. The offset yield strength is a stress that will lead to some prespecified permanent strain upon unloading. Once the yield strength is reached, upon unloading there remain permanent strains, and it is the understanding of these permanent strains that occupy us in coming sections.

2. STRENGTH OF CRYSTALLINE SOLIDS

2.1 Ideal Tensile Strength

Early attempts to explain the permanent deformation of crystalline solids centered on the notion of the ideal strength. The ideal strength concept envisions failure as a homogeneous process in which all atoms across a given plane are displaced uniformly, either in shear or in tension. For example, failure in tension was thought to occur as indicated schematically in Fig. 7, with the result that once sufficiently large strains were reached, the atomic bonds could no longer resist the applied tension. This simple model of fracture allows for an atomistically based estimate of both the stress and strain to failure, an estimate found to be in gross disagreement with observed tensile strengths. The ideal strength calculation can overestimate the strength of a solid by as much as a factor of 10^3. This failure of solids to reach their ideal strengths served as an impetus for the development of defect-based models of permanent deformation.

One particularly transparent model of ideal tensile failure approximates the normal stress σ_n that develops in response to the separation of adjacent planes by an opening displacement Δ in excess of their equilibrium spacing as a sinusoidal function

$$\sigma_n = A\sin(2\pi\Delta/a), \tag{16}$$

where a is a length that sets the scale over which cohesion between adjacent planes occurs. The prefactor A is determined by insist-

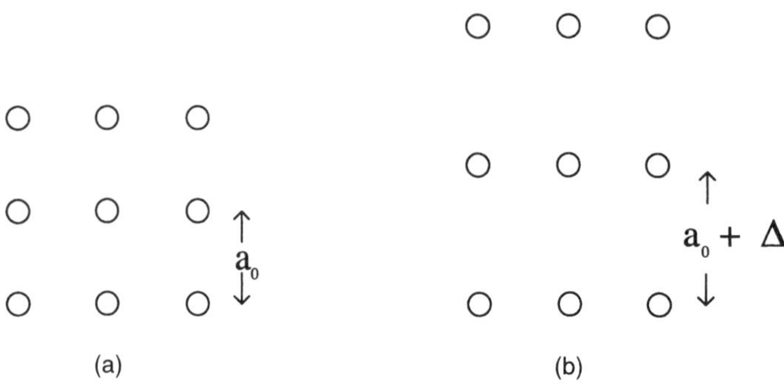

FIG. 7. Fracture mechanism envisioned in calculation of ideal tensile strength. (a) Perfect crystal. (b) Crystal subjected to homogeneous elongation.

ing that in the limit of small opening displacements, the normal stress is linear in the opening displacement, with the proportionality constant given by Young's modulus, E. In particular, it is found that the prefactor A is given by $E/2\pi$. In turn, this model implies that the maximum tensile stress that can be supported by the crystal is $E/2\pi$. While providing a physically simple model of the fracture process, the predictions implied by this model are generally found to be at odds with measured values of both the stress and strain to failure. Our simple model of the normal stresses that develop in response to the normal strains preceding fracture may be challenged as overly facile with the hope that a more realistic law based upon a correct treatment of the atomistic underpinnings of this process would reduce the estimate of the ideal strength. However, such refinements do not suffice to bring this simplified law into correspondence with the experimental values, except perhaps in the case of whiskers that have stress–strain behavior like that shown in Fig. 8, where it is seen that stresses much closer to the ideal stress may be reached. In particular, for Cu with a Young's modulus slightly larger than 100 GPa, the whisker reaches a value of roughly $E/50$. The difference between whiskers and the tensile samples considered until now is that whiskers are nearly defect free, thus lacking the stress-concentration features thought to operate in more macroscopic samples.

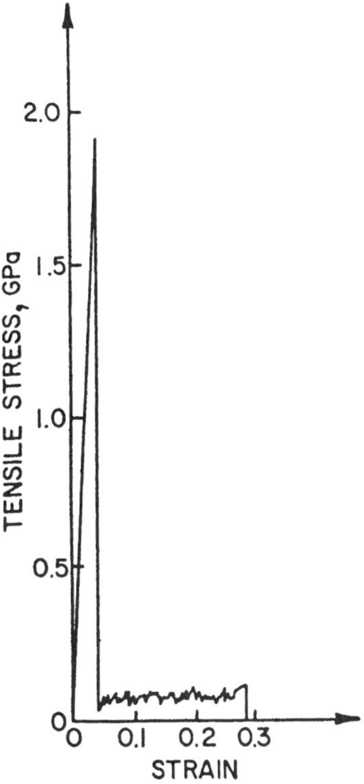

FIG. 8. Stress–strain curve for Cu whisker. Note the stress scale, which is in units of GPa, typical of ideal yield strength, rather than MPa units typical of conventional yield. After Boyer (1987).

2.2 Ideal Shear Strength

Analogous arguments have been advanced for evaluating the strength of solids in shear. In this case, plastic deformation is treated as arising from the shearing of adjacent crystal lattice planes. As in the model of the ideal tensile strength elucidated above, a relation between the shear stresses and slip of adjacent planes is assumed, the simplest of which takes the form

$$\tau = \tau_{\max} \sin(2\pi\delta/a). \tag{17}$$

The prefactor in the stress-displacement relation is again determined by insisting upon a fit to the linear elastic value, which, upon following an argument like that given above, yields the result $\tau_{\max} = \mu/2\pi$, where μ is the shear modulus for the plane of interest. This result provides an estimate of the ideal shear strength of crystalline materials. As might have been expected in light of our experience with the ideal tensile strength, the ideal shear strength is an overestimate of the observed values by as much as a factor of 10^5. The wide discrepancy between the observed and ideal shear strengths of crystalline solids can be gleaned from Fig. 9, which reports the yield stresses observed in wide classes of materials. The scale associated with these strengths is in the megapascal range, while the shear modulus of typical materials is in the gigapascal range. It is noted that the ideal strength is often orders of magnitude higher than the observed strength, with structural materials such as metals only reaching as much as 10^{-5} of their ideal values. Clearly, these simple models of mechanical yielding are missing something crucial.

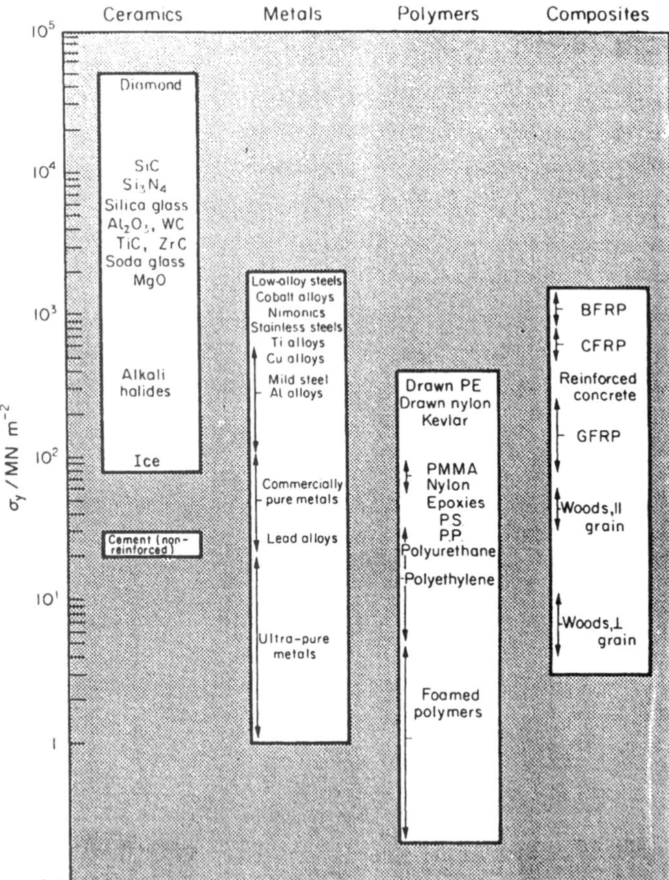

FIG. 9. Illustration depicting range of observed yield stresses for wide classes of materials. Note that the vast majority of materials yield at stresses orders of magnitude lower than estimates based on ideal strength. After Ashby and Jones (1980).

2.3 Defect-Mediated Deformation

The paradox left in the wake of the calculation of both the ideal shear and tensile strengths led to some of the most powerful theoretical insights concerning the mechanical behavior of solids. In particular, these failed models led to the conviction that extended defects such as dislocations and cracks are the carriers of permanent deformation in materials. As will be shown subsequently, cracks can be thought of as stress concentrators, which can in the vicinity of the crack tip ramp up the local stresses to values as high as the ideal tensile strength, while the remote loads remain much lower. It is this ability of cracks to magnify applied stresses locally that resolves the ideal-strength paradox. Similarly, the presence of dislocations reduces the stress needed to initiate plastic deformation. In this case, the presence of these line defects allows for a net translation of one half of a crystal with respect to the other at stresses well below those demanded by the ideal shear-strength calculation. In large measure, then, the interesting deformation of materials is primarily a consequence of the presence of extended defects, most notably dislocations and cracks.

The following sections focus on two central issues in the mechanical behavior of materials, namely, plasticity and fracture. Our treatment of plasticity first considers plastic flow as a rate-independent process that results once a certain critical stress is attained. Subsequently, we consider time-dependent plastic deformation in the form of creep with an emphasis on the mass-transport and dislocation-based mechanisms that make time-dependent permanent deformation possible, even though the stresses are lower than the yield stress. Fracture is then considered both from the standpoint of the stress concentration concept that holds that fracture occurs because of a magnification of stresses due to the presence of flaws such as cracks, and from the perspective of the energetic arguments first introduced by Griffith that assert that fracture is the result of a competition between the release of elastic energy and the creation of new surface. Fatigue is treated briefly, illustrating that fracture can occur (like plastic flow before) at stresses lower than the critical stress when the material is subjected to cyclic loading.

3. PERMANENT DEFORMATION OF MATERIALS

3.1 Phenomenology of Permanent Deformation

Some of the primary criteria for the selection of materials to be used in applications where the material is subjected to mechanical loads center on the competition between a few key material properties, namely, yield strength, weight, and toughness. The yield strength is a measure of the difficulty to induce plastic deformation. For example, among metals, steel is a high–yield-strength material, whereas lead is not. On the other hand, toughness refers to a material's resistance to fracture. Interestingly, much of the emphasis on the permanent deformation of materials focuses on the detrimental consequences of such deformation. On the other hand, deformation processing has existed for as long as man has used metals for the construction of various objects. The use of tungsten filaments in light bulbs relies on the ability to form these metals to extreme plastic strains. Thus, while plastic deformation can have detrimental consequences, in some cases the failure of a material to exhibit sufficient plastic deformation can mean the difference between success and failure in its processing.

3.2 Time-Independent Plastic Deformation

Plastic deformation refers to the irreversible shape change of materials that have been subjected to stresses in excess of their yield strength. Figure 10 shows the stress–strain curve of a generic tensile specimen loaded into the permanent deformation regime. As long as the stress is less than σ_y, upon release of the load the material will traverse the same curve upon unloading that it did upon loading, with the result that the initially undeformed state is recovered. On the other hand, if the sample is loaded to

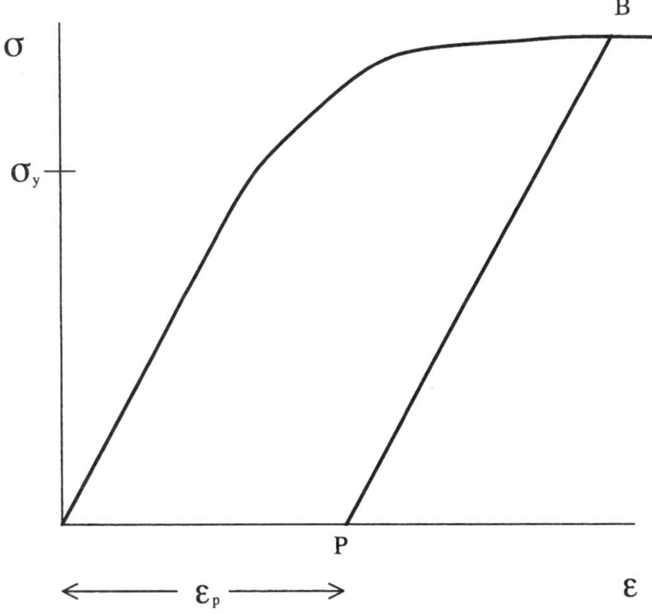

FIG. 10. Stress–strain curve illustrating the characteristics of permanent deformation. Plastic strain ϵ_p is strain that remains after load has been removed. Hardening refers to the increase in flow stress needed to maintain continued plastic deformation once yield has been attained.

point B in the figure, and the loads are then released, the unloading curve will correspond to that labeled BP in the figure. The plastic strain is defined as the strain that remains after unloading and is labeled ϵ_p in the figure. Further, if the sample is now loaded again, to a first approximation, it will load elastically up the curve PB with a new and larger yield stress than was found in the original experiment. The material has hardened, a consequence of the interactions between dislocations.

The simplest microscopic picture of plastic deformation centers upon the motion of dislocations. One immediate task we face is to demonstrate how dislocations can be the carriers of macroscopic plastic strains. In Fig. 11, a schematic of a dislocated crystal shows how a dislocation can be thought of as a one-dimensional boundary between slipped and unslipped regions of the crystal. The planes along which dislocations move are known as slip planes. The unit of slip carried by a particular dislocation is known as the Burgers vector, **b**. In fcc metals, slip typically occurs on {111} planes with Burgers vectors along the ⟨110⟩-type direction. hcp crystals allow for slip on basal planes and the perpendicular pyramidal planes, while in bcc metals, slip typically occurs on {110}- and {112}-type planes with Burgers vector along the ⟨111⟩-type directions.

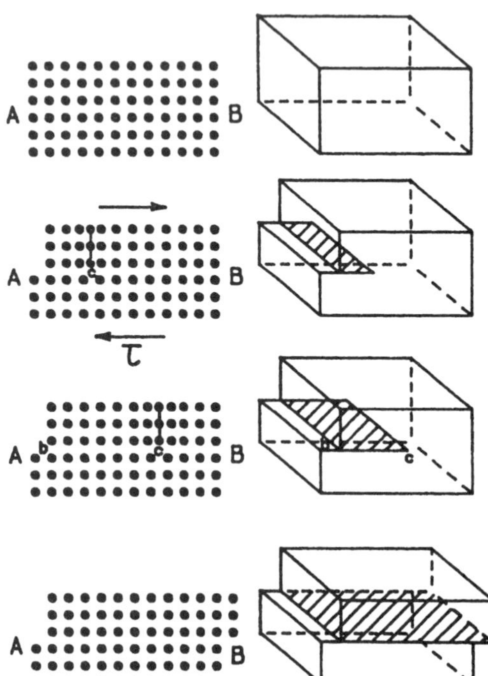

FIG. 11. Dislocated solid, illustrating the way in which dislocation is a boundary between slipped and unslipped regions of crystal. After Khan and Huang (1995).

An argument for the claim that dislocations are the agents of plastic change can be made by imagining the dislocation to move all the way to the opposite boundary of the crystal, as shown in Fig. 11. In this simple two-dimensional model, we note that the average plastic strain carried by such dislocations is given by

$$\Delta\epsilon_p = nb/h, \tag{18}$$

where b is the magnitude of the Burgers vector of the relevant dislocations, h is the height of the region of interest, and n is the number of dislocations. If the dislocations travel with a mean velocity $\bar{v}$, then the time to traverse the distance L is given by

$$\Delta t = L/\bar{v}. \tag{19}$$

In the limit that the number of dislocations is sufficiently large to allow for this average description, the net plastic strain rate can be approximated as

$$\dot{\epsilon}_p \approx \Delta\epsilon_p/\Delta t = \rho b\bar{v}, \tag{20}$$

where $\rho = n/hL$ is the density of mobile dislocations. This result is known as Orowan's equation and is an example of a constitutive law that links a macroscopic quantity (the plastic strain rate) to microscopic quantities (the average dislocation velocity and the density of mobile dislocations). In treating the dislocation density as time independent, we have ignored the effects of dislocation sources, which can enhance the plastic strain rate.

What is the evidence that a dislocation-based mechanism, while plausible, is indeed responsible for plastic deformation in metals? One piece of evidence is that of well-defined slip traces that can be observed on the surfaces of deformed materials. Such slip traces are the points of exit of dislocations at the surface of a crystal, just as indicated schematically in Fig. 11, though for traces to be observed optically, large numbers of dislocations must have exited the crystal. A second crucial outcome of the dislocation model of plastic deformation is that the presence of dislocations must resolve the paradox associated with the failure of materials to reach their ideal shear strength. By virtue of the fact that dislocations respond to an applied stress, it is possible to perform a model calculation (Hirth and Lothe, 1982) of the stress needed to move a dislocation (this is the so-called Peierls stress), which yields a value for the critical shear stress for dislocation motion of

$$\tau_{\text{Peierls}} = \frac{2\mu}{(1-\nu)} \exp\left(\frac{-4\pi\zeta}{b}\right), \tag{21}$$

where ζ is the width of the dislocation and is the length scale over which the slip goes from zero to a full Burgers vector; typical widths range from one to ten lattice spacings. This expression should be contrasted with the ideal shear strength; the Peierls stress is reduced by an exponential factor relative to the ideal shear strength.

A more immediate treatment of the coupling of dislocations to stress fields is provided by the elastic theory of dislocations. This analysis is important not only because it yields insights into the stress needed to induce plastic deformation but also because it is the basis of understanding various hardening mechanisms. For example, solid-solution hardening is a reflection of the interaction between dislocations and the stress fields induced by point defects. Similarly, strain hardening arises because dislocations interact with each other with the result that as the density of dislocations on various competing slip systems increases, dislocation–dislocation interactions inhibit further dislocation motion. In general, for a dislocation with line direction given by the vector $\boldsymbol{\xi}$ and with Burgers vector $\mathbf{b}$ the force on that dislocation per unit length due to a stress $\boldsymbol{\sigma}$ is given by

$$\mathbf{F} = -\boldsymbol{\xi} \times (\boldsymbol{\sigma}\mathbf{b}), \tag{22}$$

a result known as the Peach–Koehler force. As seen below, the Peach–Koehler force presides over phenomena as diverse as conventional plastic yielding and strain hardening arising from dislocation–dislocation interactions.

A critical testing ground for the dislocation-based model of plastic deformation is that of flow in single crystals. It is imagined that the sample is loaded in tension, as shown in Fig. 12. The pioneering observations of Taylor and Schmid (Honeycombe, 1968), among others, led to a number of key insights concerning plastic flow in single crystals. One such insight is the observation that the stress to induce plastic flow is dependent upon the orientation of the underlying crystal lattice with respect to the loading axis. Careful analysis led Schmid to elucidate the law that bears his name, namely, that the stress to induce plastic flow is given by the component of the shear stress in the slip plane that is directed along the slip direction, the so-called resolved shear stress. Plastic flow will initiate when this stress reaches a critical value known as the critical resolved shear stress. For a slip plane with normal $\mathbf{n}$ and slip direction $\mathbf{s}$ as shown in Fig. 12, the resolved shear stress is given by

$$\tau_r = \mathbf{s} \cdot \boldsymbol{\sigma}\mathbf{n}. \tag{23}$$

In terms of the angles ϕ and λ defined in the figure, this leads to the equivalent expression

$$\tau_r = \sigma_0 \cos\phi \cos\lambda, \tag{24}$$

where the geometric term is known as the

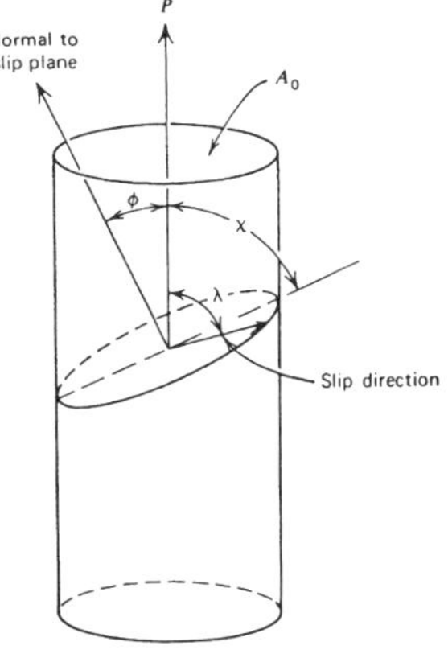

FIG. 12. Cylindrical test specimen that exhibits geometric origins of Schmid factor. Tensile loading occurs along cylinder axis while slip plane is oriented in a direction characterized by angle ϕ, and slip direction is characterized by angle λ. After Hertzberg (1976).

Schmid factor and σ_0 is the remote tensile load (P/A_0 in the figure). Hence there is a characteristic geometric signature for plastic flow in single crystals. In particular, depending upon the orientation of the tensile axis with respect to the crystal axes, the critical resolved shear stress will be reached at different tensile loads as determined by the Schmid factor and shown in Fig. 13. Though the dislocation picture of plastic deformation provides a mechanistic underpinning for such deformation, the phenomenology of permanent deformation via plastic flow requires the motion of dislocations in such large numbers that for many purposes it is preferable to invoke an average description in terms of macroscopic plasticity considerations. In these models, the dislocations are never considered explicitly. Rather, the emphasis is on determining the macroscopic criteria for plastic flow.

We now revisit the stress–strain curve of Fig. 10 with an eye to a phenomenological description of plastic flow. The simplest model of plastic flow holds that a solid undergoes strictly elastic deformation until a critical stress (known as the yield stress) is reached. Of course, so far our analysis has ignored the tensor character of the stress state since we have only considered tensile specimens. In principle, there are six components of the stress tensor that can be specified independently. These more general stress states lead to plastic flow as well. The locus of all points in stress space that result in plastic flow is known as the yield surface. To refine our insights into yield it is necessary to revisit the idea of the stress tensor in order to characterize the conditions under which yield will occur. The yield criterion is most transparent if we choose a coordinate system in which the stress tensor is diagonal. This coordinate system is characterized by the fact that the traction and normal vectors are parallel. This condition can be stated quantitatively in the form of an eigenvalue problem,

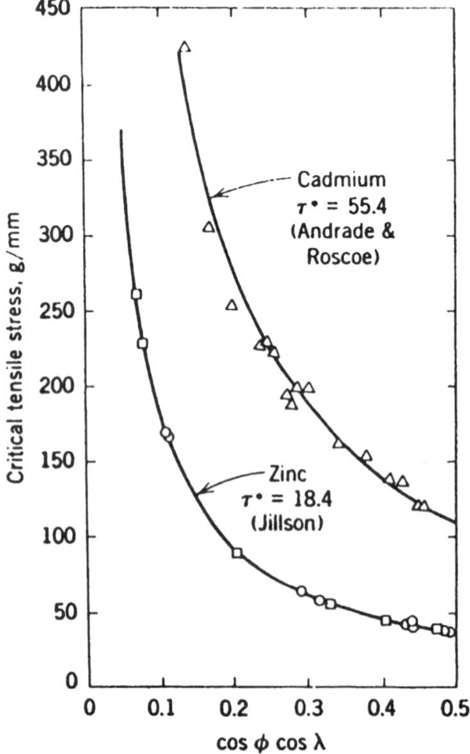

FIG. 13. Experimental data showing correlation between critical tensile load and Schmid factor. After Gilman (1969).

$$\mathbf{t}^{(\mathbf{n})} = \boldsymbol{\sigma}\mathbf{n} = \lambda\mathbf{n}, \tag{25}$$

where $\mathbf{n}$ is known as a principal axis of the stress tensor and λ is a principal stress. A yield condition for plastic flow can be constructed on the basis of the insight that the maximum shear stresses occur on (110) planes with respect to these principal axes, with the value of such shear stresses given by

$$\tau_{\text{shear}} = \tfrac{1}{2}(\sigma_i - \sigma_j), \tag{26}$$

where σ_i and σ_j are the ith and jth principal stresses. The Tresca yield condition holds that plastic yield will occur when the difference between the maximum and minimum principal stresses reaches some critical value. This criterion may be written

$$\tau_{\text{critical}} = \tfrac{1}{2}(\sigma_{\max} - \sigma_{\min}) = k, \tag{27}$$

where k is the critical shear stress for plastic flow. Though this yield condition is consistent with the lessons of the Schmid law since it emphasizes the role of critical shear stresses in inducing plastic flow, it is important to note that this yield criterion is isotropic. In particular, it does not acknowledge the discreteness of the crystal lattice that allows for slip only on certain specific slip planes. Within this criterion, by rotating the state of stress continuously, the plane on which the Tresca condition envisions slip to occur can be rotated continuously as well. On the other hand, isotropic yield is a reasonable representation for polycrystals with their slip systems oriented over a range of directions, and it is this class of behavior that such yield criteria are intended to describe. A second yield condition, known as the von Mises yield criterion, argues that plastic flow will proceed when an averaged version of the shear stresses reaches a critical value. In particular, if the *deviatoric stress* is defined as

$$S'_{ij} = \sigma_{ij} - \delta_{ij}\tfrac{1}{3}tr(\boldsymbol{\sigma}), \tag{28}$$

which amounts to subtracting off the hydrostatic part of the stress, then the *second invariant of the deviatoric stress* is defined as

$$J_2 = \tfrac{1}{2}S'_{ij}S'_{ij}. \tag{29}$$

The von Mises yield criterion contends that plastic flow will initiate when

$$J_2 = k^2. \tag{30}$$

Note that for both of the yield criteria described above, under a state of hydrostatic stress in which $\sigma_1 = \sigma_2 = \sigma_3$, there will be no plastic flow, in keeping with the observations of Schmid and others. Experimental results contrasting these two yield criteria are shown in Fig. 14. These experiments were performed on thin-walled tubes subjected to both torque and tension and show that the initiation of plastic flow occurs for states of stress that are bounded by the Tresca and von Mises values.

Models such as those we have described above are known as models of perfectly plastic solids. These models hold that below some critical stress the deformation will be strictly elastic and that once the critical stress is reached, plastic deformation will proceed. The highly idealized models of plastic flow make no provision for hardening. On the other hand, as illustrated in Fig. 10, plastic deformation is characterized by *hardening* in which it is necessary to increase the stress in order to continue plastic deformation. This hardening is a reflection of the interaction between dislocations, which makes it increasingly difficult to move such dislocations in the presence of the stress fields induced by neighboring dislocations. This topic is taken up in more detail in Sec. 5.

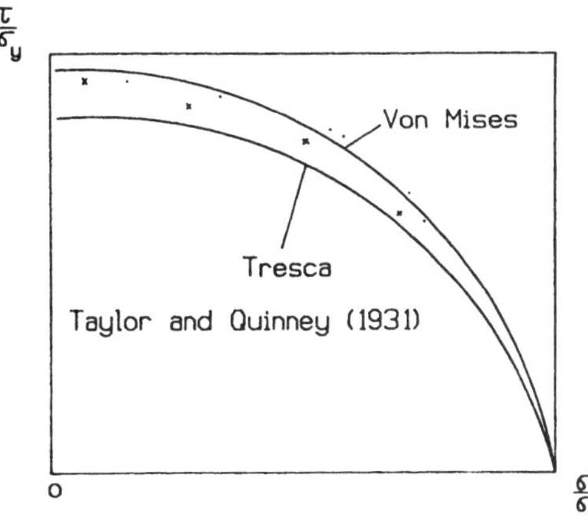

FIG. 14. Experimental comparison of yield stress for different values of shear and normal stress components to those predicted by von Mises and Tresca criteria. After Khan and Huang (1995).

3.3 Creep Phenomena

Our description of permanent deformation thus far implies that for stresses lower than that of the yield strength, a material's deformation will be strictly elastic. This is an idealization of observed behavior. However, once temperatures reach some fraction of the melting temperature, materials exhibit permanent deformation even for stresses below the yield stress. Such processes are generally denoted as creep phenomena, though, as will be seen in the case of fracture, a variety of creep mechanisms are operative there as well.

Besides being of fundamental interest, creep mechanisms are of particular concern in high-temperature material applications, since they pose threats to performance over the life of a component. The two fundamental creep tests that set ideas for these phenomena are those carried out at constant load and those occurring at constant strain. In constant-load creep tests, the sample is loaded, and the subsequent strain is monitored as a function of time, leading to strain vs time profiles of the type shown in Fig. 15.

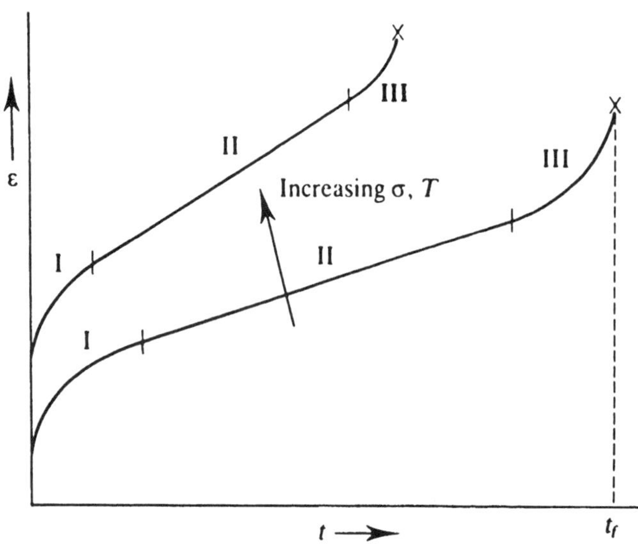

FIG. 15. Generic creep curve exhibiting transient regime (I) followed by steady-state creep regime (II) and tertiary creep regime (III). After Courtney (1990).

Alternatively, the sample can be loaded to a fixed strain with the resulting stress relaxation monitored as a function of time. The underlying mechanisms are defect based and can result from thermally activated dislocation motion, stress-assisted mass-transport mechanisms, and motion of or along grain boundaries.

It is conventional in the analysis of creep curves such as that shown in Fig. 15 to divide the response into three regimes (I, II, and III) known respectively as primary (or transient), secondary, and tertiary creep. The primary regime is characterized by a decreasing creep rate, while the secondary regime exhibits steady-state behavior, as is evidenced in the figure. The third stage reveals an increasing strain rate and precedes failure of the material. Experiments have suggested a number of empirical relations between the strain rate, the applied stress, and the temperature, a typical example of which is

$$\dot{\epsilon} = \dot{\epsilon}_0 \sigma^n \exp(-Q/RT), \tag{31}$$

where Q is the activation energy for creep and n is a phenomenological parameter relating the strain rate to the stress. An interesting correlation between the creep activation energy and that for lattice self-diffusion may be noted, leading to a conjectured relation between creep and mass transport via diffusion. One of the challenges posed by the type of expression just discussed is the construction of microscopically based models that result in such behavior. The two types of thermally activated and stress-assisted creep mechanisms considered here are dislocation-based models and diffusion-based models.

The key idea behind the diffusion-based creep mechanisms is that the presence of a stress field can bias mass transport, leading to a net shape change. Perhaps the simplest such picture is that of Nabarro–Herring creep. Here it is imagined that a sample is loaded in tension in one direction and in compression in an orthogonal direction. Further, it is imagined that the vacancies that mediate bulk diffusion have as their primary source the crystal surface or the boundaries surrounding a grain of size d. The presence of the applied stress state inhibits vacancy formation on the crystal faces in compression while it enhances the vacancy concentration on the faces in tension, leading to a net concentration gradient. As a result of this concentration gradient, a diffusive flux will be set up between these faces, having the net effect of transporting matter to the faces in tension. A simple computation of this diffusive flux leads to the Nabarro–Herring creep rate given by

$$\dot{\epsilon} = A(D_L/d^2)(\sigma\Omega/kT), \tag{32}$$

where D_L is the self-diffusion coefficient for *bulk* diffusion, Ω is the atomic volume, and d

is the average grain size over which such diffusion occurs. A mechanistic variant of this creep mechanism known as Coble creep considers the dominant diffusion to take place along the grain boundaries themselves rather than through the bulk as postulated in the Nabarro–Herring case. In this case, the resulting strain rate is given by

$$\dot{\epsilon} = A(D_{GB}\delta/d^3)(\sigma\Omega/kT), \quad (33)$$

where here the relevant diffusion coefficient is that for grain boundary diffusion (D_{GB}) and δ refers to the grain-boundary thickness.

Reflection on the deformation mechanism map shown in Fig. 2 reveals that in addition to the Nabarro–Herring and Coble creep regimes, there is an additional region of parameter space that is dominated by so-called power-law creep. In this case, it is thought that it is the diffusion-assisted motion of dislocations that carries the plastic deformation. In contrast with the linear stress dependence found for both the Nabarro–Herring and Coble mechanisms, the strain rate in this regime is characterized by a power-law stress dependence ($\dot{\epsilon} \propto \sigma^n$). A mechanistic picture of the processes operating in this regime centers on the idea that pinned dislocation segments are able to surpass the barriers that pin them via a thermally activated climb mechanism. Dislocation climb refers to the motion of a dislocation perpendicular to its slip plane via mass transport, either to or away from the dislocation core.

4. FRACTURE

4.1 Phenomenology of Fracture

As indicated in Fig. 1, at sufficiently large stresses, solids fail via fracture. Taken literally, fracture refers to the breaking of interatomic bonds across some entire cross section of the solid. Though all fractures share certain generic features, an elucidation of fracture at the microscopic level reveals a host of processes that can ultimately result in catastrophic failure.

Fracture that is not accompanied by some prior plastic deformation is known as brittle fracture. Alternatively, fracture can be preceded by massive plastic flow, a phenomenon known as ductile fracture. The variety of fracture mechanisms are captured in the fracture mechanism map shown in Fig. 3. Here it is seen that, depending upon the temperature and the loading, the generic tensile specimen can exhibit a variety of interesting behaviors. Though broadly speaking the division into ductile and brittle fracture is convenient, a more precise characterization of the fracture process must allow for a number of different processes within each of these classes of failure. These mechanisms are summarized pictorially in Fig. 16, which gives an interpretation of the dominant fracture mode in various regimes. Among other things, this figure makes it clear that microstructural influences, particularly grain boundaries, are crucial in determining the dominant fracture mode.

Brittle fracture can occur via either a transgranular or an intergranular mechanism. Intergranular failure points to the important role played by the types of microstructural details normally ignored in a preliminary analysis. Here, the grain boundaries serve as the mechanical weak link, opening up the attendant possibility of toughening mechanisms that center on grain-boundary chemistry. In the ductile regime, it is seen that again a number of different mechanisms are available, with necking and progressive thinning of the cross section as in pulling taffy as the most extreme variant of ductile failure. At elevated stresses, ductile fracture arises from the growth and coalescence of voids under the influence of the concentrated stresses ahead of the crack tip. However, even at loads much lower than the putative critical stress for fracture, creep fracture becomes possible. This type of failure depends upon the operation of mass transport mechanisms that are biased by the presence of the applied and crack-tip stresses. The wide range of mechanisms displayed in the fracture mechanism map makes it clear that a detailed understanding of fracture is a challenging task. On the other hand, certain generic features of fracture can be enumerated by considering both the energetics associated with crack propagation and the stress enhancements caused by the presence of the crack. It is this route that we follow here.

4.2 Stress Concentration

As noted earlier, the conceptual basis for the resolution of the paradox posed by the failure of materials in general to reach their

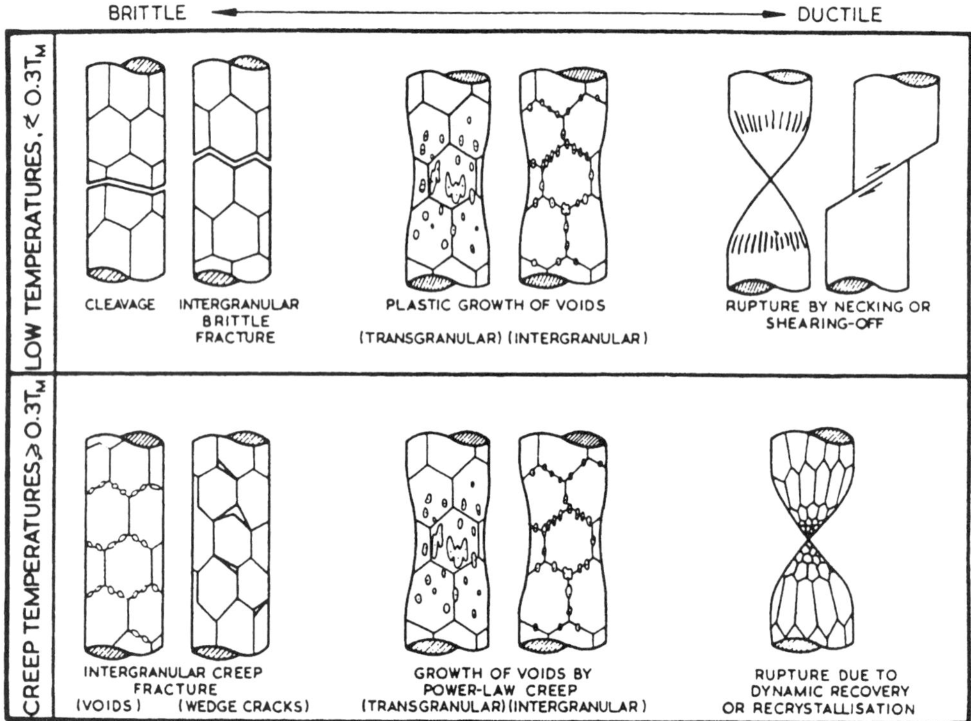

FIG. 16. Pictorial representation of key observed fracture mechanisms for different temperature regimes. After Ashby *et al.* (1979).

ideal tensile strength is the notion of stress enhancement due to the presence of flaws such as cracks, steps, and notches. What is the experimental basis for considering cracks as the agents of fracture? Advances in optical techniques now allow for the direct observation of moving cracks, as is shown in Fig. 17. The idea is to use the nonuniform deformation fields in the near crack-tip region to produce fringe patterns in real time. For the case shown in the figure, the crack is found to travel at somewhat less than half of the shear-wave speed. Post mortem analysis of fracture surfaces also plays a key role in elucidating the underlying fracture mechanism. In Fig. 18, the fracture surfaces from a ductile fracture arising from void coalescence and a brittle intergranular fracture are contrasted. These probes, and many others, provide definitive insights into the role of cracks as the agents of failure via fracture.

With the insight that cracks are the key players in mediating fracture, it is important to systematize the way in which such defects concentrate stresses well in excess of the applied loads. One particularly convenient avenue for examining this is to think of cracks as the singular limit of elliptical holes. A linear elastic analysis of the elliptical hole was first provided by Inglis, who demonstrated that for a geometry such as that shown in Fig. 19, the stress along the continuation of the semimajor axis of the ellipse is given as shown in the figure. This analysis is suggestive in that as the radius of curvature goes to zero (the ellipse becomes increasingly eccentric) the expression for maximum normal stress σ_{yy} that occurs at the termination of the semimajor axis may be expanded to give

$$\sigma_{\max} = \sigma_0(1 + 2\sqrt{a/\rho}). \tag{34}$$

For a given elliptical hole, the stress concentration factor may be defined in terms of the remote load σ_0 as the ratio $\sigma_{\max}/\sigma_0$, which reveals the way in which such geometrical inhomogeneities can magnify stress. Indeed, even a circular hole results in a ramping up of the remote stress by a factor of 3.

The sharp-crack limit can be thought of

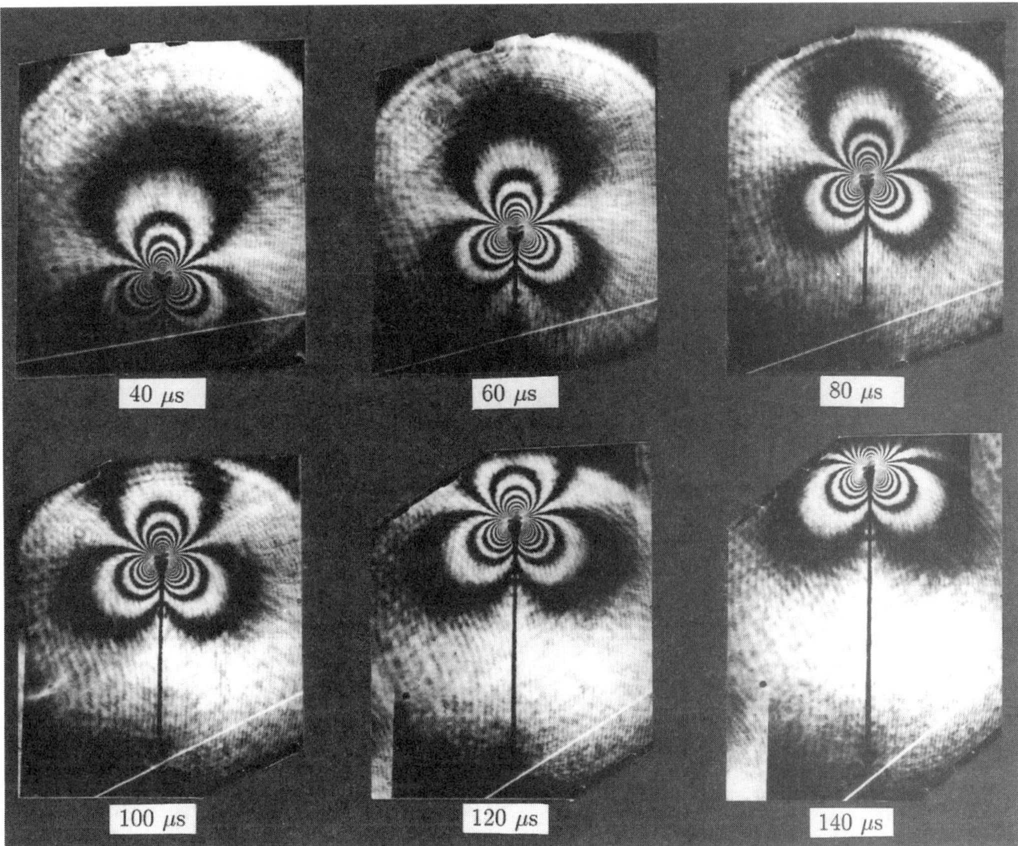

FIG. 17. Time evolution of crack tip observed experimentally. Courtesy of A. Rosakis.

in the context of the elliptical hole discussed above, or it can be solved explicitly. Because the underlying theoretical framework in our analysis is linear elasticity, superposition allows for the consideration of three independent modes of loading that "span" the space of possible loading conditions of a sharp crack, as illustrated in Fig. 20. Recall that the equilibrium equations may be written as $\sigma_{ij,j} = 0$. A solution to the boundary-value problem of the sharp crack in the absence of body forces yields a solution for the nonvanishing components of the stress tensor under mode I loading of the form

$$\sigma_{ij} = \frac{K_I}{\sqrt{2\pi r}} f_{ij}(\theta), \tag{35}$$

where $f_{ij}(\theta)$ is an angular factor [for example, $f_{\theta\theta} = \cos^3(\theta/2)$], K_I is known as the mode I stress intensity factor, and r is the distance from the crack tip.

One of the more beautiful results of the theory of linear elastic fracture mechanics is the way in which this singular solution for the crack problem points the way to a material parameter that characterizes the resistance of a given material to crack propagation. In particular, by performing experiments on specimens with known geometry and with a preexisting flaw, the stress to fracture can be measured. This experiment then defines the critical stress intensity K_{IC}, which serves as a material parameter known as the fracture toughness. Once the fracture toughness is known, the critical load for different geometries can be determined.

4.3 The Griffith Concept

The preceding discussion has highlighted the way in which stresses can be locally enhanced by the presence of stress-concentrating inhomogeneities. A complementary view-

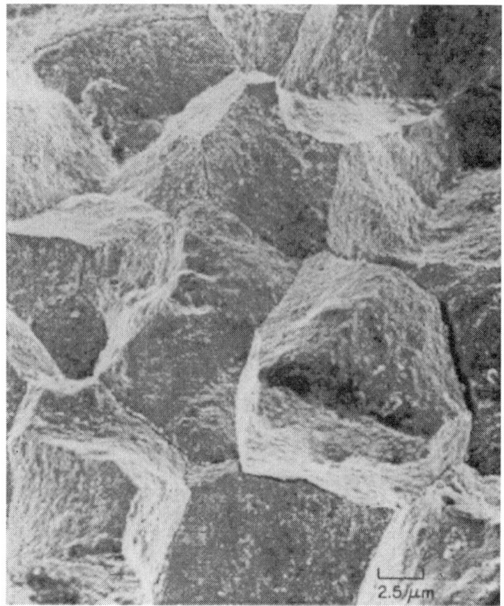

(a)

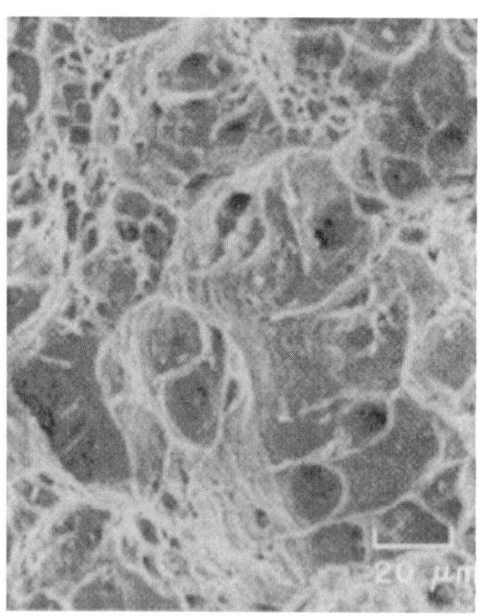

(b)

FIG. 18. Fracture surfaces arising from (a) brittle intergranular fracture and (b) ductile fracture via void coalescence. The intergranular fracture surface clearly exhibits facets from grain structure, while ductile fracture surface shows dimpled features. After Boyer (1974).

point, pioneered by Griffith (1920), evaluates the energetics of crack extension via much the same type of competition between energies familiar from nucleation theory. In particular, the Griffith energy-balance concept argues that the driving force for crack extension is the result of a competition between the elastic energy released upon crack extension and the increasing surface energy attending such extension. If we consider a centered crack of length $2a$ (the sharp limit of the elliptical hole discussed above), on the basis of the elastic stress fields derived above, it can be shown (Lawn, 1993) that the elastic energy release upon extending the crack by amount da is given by

$$dU_{\text{elastic}} = -[\pi\sigma^2 a(1 - \nu^2)/E]\, da, \tag{36}$$

where σ is the applied load, a is the crack half length, E is the Young modulus, and ν is the Poisson ratio. The negative sign indicates that crack extension leads to a reduction in the elastic energy of the system. On the other hand, the energy cost associated with such crack extension in the form of surface energy is

$$dU_{\text{surface}} = 2\gamma\, da, \tag{37}$$

where γ is the surface energy. If we seek the critical condition for which the driving force vanishes, we find that the critical load for crack extension is given by

$$\sigma_c = \sqrt{\frac{2\gamma E}{\pi a(1 - \nu^2)}}\,. \tag{38}$$

This condition is known as the Griffith criterion and implies a fracture toughness of

$$K_{IC} = \sqrt{2\gamma E/(1 - \nu^2)}. \tag{39}$$

For a typical surface energy of 1 J/m^2 and a typical Young's modulus of 100 GPa, this implies a fracture toughness of roughly 0.5 MPa m$^{1/2}$. It is interesting to contrast the Griffith value for the fracture toughness with typical observed values. Only the most brittle materials such as glasses and concrete, for example, have fracture toughnesses as low as those computed from the simple Griffith criterion. On the other hand, the fracture toughness exhibited by materials such as me-

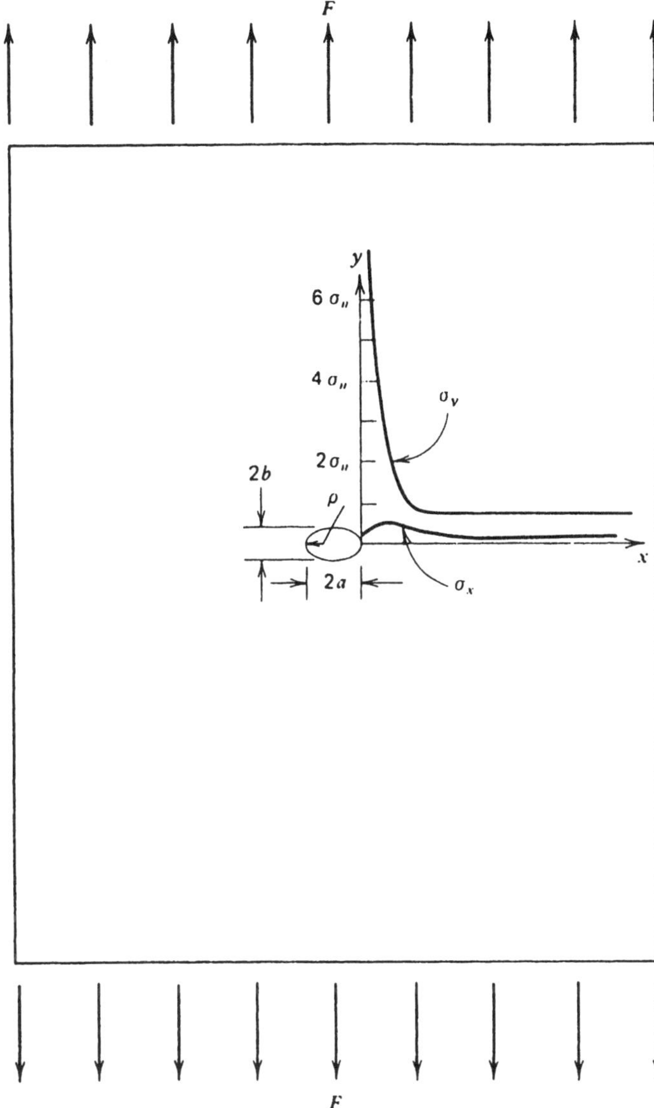

FIG. 19. Geometry considered by Inglis in evaluating stress concentration due to an elliptical hole. Figure shows enhancement of normal stresses (σ_y) relative to remote values (σ_n). Note that notation in this figure is different than in remainder of article. After Collins (1993).

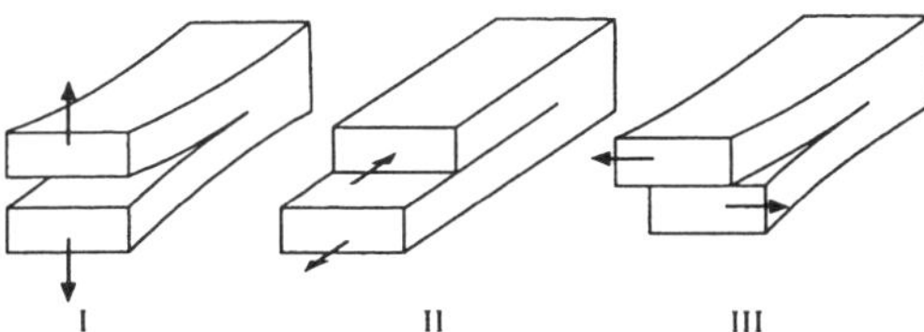

FIG. 20. Three modes of loading that characterize crack-tip response. In mode I, loading is perpendicular to crack faces, while modes II and III involve shears. After Lawn (1993).

tallic alloys easily exceed 10 MPa $m^{1/2}$ and can climb in excess of 100 MPa $m^{1/2}$.

Evidently, the model of fracture toughness described above misses a key feature of most real fracture processes. Irwin (1948) and Orowan (1955) realized that the simple nucleation-theory–type argument formulated above must be augmented by an additional term representing the energy that goes into plastic deformation in the vicinity of the crack tip. At the mechanistic level, one mechanism that engenders such deformation is the emission of dislocations from the crack itself as well as the activation of dislocation sources in the vicinity of the crack tip. We may estimate the size of the region over which plastic deformation is important by recourse to the linear elastic crack-tip stress fields described earlier. The plastic zone is defined as that region over which the near-tip stresses have induced plastic deformation. The size of the plastic zone may be estimated by considering the region over which the elastic stresses have reached the yield strength of the material. That is,

$$\sigma_y = K_I/\sqrt{2\pi r_p}, \tag{40}$$

where r_p is the plastic zone size and is given by

$$r_p = K_I^2/2\pi\sigma_y^2. \tag{41}$$

In its simplest form, the plastic work can be thought of as leading to an effective surface energy of the form

$$\gamma_{\text{eff}} = \gamma_s + \gamma_p, \tag{42}$$

where γ_s is the surface energy and γ_p is the contribution to the effective surface energy arising from the plastic work. Use of this effective surface energy in Eq. (39) rather than its bare value demonstrates qualitatively the way in which plastic dissipation raises the fracture toughness.

A final interesting point concerning the fracture toughness is the notion of the brittle-to-ductile transition. As the temperature of many materials is raised above some threshold value, the fracture mechanism can pass from that of brittle fracture to ductile fracture over a narrow temperature range. Accompanying this change is a significant increase in the material's toughness. An example of this behavior is shown in Fig. 21.

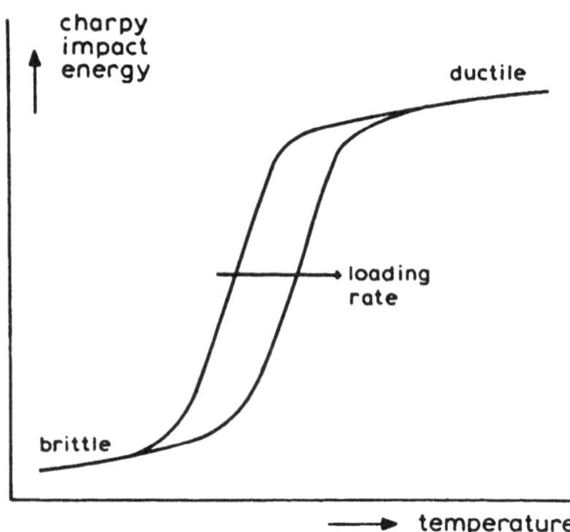

FIG. 21. Brittle–ductile transition as evidenced by energy absorbed in fracture test. High temperatures require more energy to achieve fracture. After Broek (1991).

4.4 Fatigue

A more insidious form of failure is fracture via fatigue. In this case, the material can fail at stresses lower than the stress required for conventional fracture as a result of being subjected to cyclic loading. Hence, even if the service loads have been selected so as to be much lower than the critical stress for fracture, cracks may first be initiated and then slowly grow in response to a cyclic stress field. Many headline structural disasters are ultimately traced to fatigue failure. The detective work of determining the mechanism by which some material failed is aided by consideration of the morphology of the fracture surface. As noted in the previous section, different fracture mechanisms can lead to widely different morphologies in the resulting fracture surfaces. In the case of fatigue failure, the fracture surface reveals one part that reflects the slow–crack-growth regime of fatigue failure and a second part that is a result of the catastrophic fast crack growth that preceded ultimate failure.

One of the major successes in the study of fatigue is the understanding of fatigue crack growth. Here the point is that although the stress amplitude falls short of the stress needed for failure under static loading, subcritical cracks can grow as a result of the cyclic variation in the stresses. Earlier it was shown that the stress intensity factor serves as a governing parameter that presides over fracture processes under constant loads. Remarkably, the variation in the stress intensity has been determined to be a governing parameter in characterizing the slow crack growth found in fatigue. In particular, the rate of crack extension per cycle is given by

$$\frac{da}{dN} = C(\Delta K)^n, \tag{43}$$

where C and n are material constants and $\Delta K = K_{max} - K_{min}$ is the difference between the maximum and minimum stress intensities. An example of this dependence is illustrated in Fig. 22.

This section of the article has, for the most part, treated the various deformation mechanisms as distinct phenomena. However, crack-tip phenomena, for example, should really be seen as the site of many of the mechanisms described above taking place concurrently. Stress concentrations at the crack tip can lead to dislocation nucleation at the tip and activation of sources in the tip neighborhood, emphasizing the coupling between plasticity and fracture.

5. INFLUENCE OF MATERIAL PROPERTIES

So far, our treatment has emphasized the mechanical features common to wide classes of materials without noting the ways in which details such as grain size, chemical composition, and heat treatment dictate these properties. However, from the standpoint of the engineering of materials for specific uses, it is just these details that are manipulated, sometimes with dramatic effect, in order to obtain some desired property. A key tool in the hands of the materials scientist is the ability to control microstructures. Often, such control is carried out via programs of both deformation and annealing that yield variations in the resulting mechanical properties. For example, schoolchildren used to be taught of the bronze age in which the strengthening of copper by additions of tin made possible a new generation of materials. These developments were a consequence of the use of solid-solution hardening to alter dramatically the mechanical properties of a particular class of metals. Chemical effects are important not only because impurities impede the motion of dislocations but also because suitable tuning of the chemistry at grain boundaries can have dramatic consequences as well, resulting in either embrittlement or enhanced ductility.

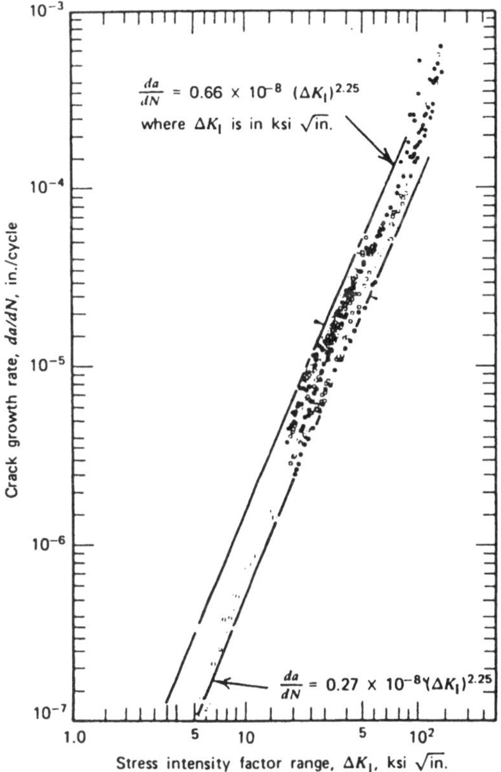

FIG. 22. Crack growth rate as a function of stress intensity difference for steels. Stresses measured in units of kilopounds/in^2 (ksi). After Fuchs and Stephens (1980).

An additional way in which mechanical behavior can be altered beyond that of chemical effects is via programs of heat treatment, which, among other things, have the effect of altering the grain size. It is the purpose of this section to highlight the ways in which the materials designer can tune the mechanical properties of a given material. Our treatment has emphasized the ways in which materials can be altered intentionally. However, anyone who has broken rusted metal knows that nature alters materials as well. Consideration of these effects requires an analysis of the ways in which exposure to various environmental factors can lead to embrittlement and is outside of the scope of this article.

5.1 Grain-Size Effects

Most engineering materials are polycrystalline. That is to say that the microstructure of such materials is built up of a collection of grains, each oriented differently, and bounded by grain boundaries. Interestingly, the geometric character of the grains is surprisingly well described to first approximation in terms of a single parameter—the grain size. This implies that under typical circumstances, the microstructure is composed of a number of grains of similar size. Some examples of microstructures with different grain sizes are shown in Fig. 23. Hall (1951) and Petch (1953) found that the yield strength depends upon this mean grain size according to the simple law

$$\sigma_y = \sigma_0 + k_y d^{-1/2}, \tag{44}$$

where d is the grain size and k_y is a material-specific parameter. This relation implies that larger grain sizes tend to reduce the yield strength of a particular material. An example of this effect is shown in Fig. 24.

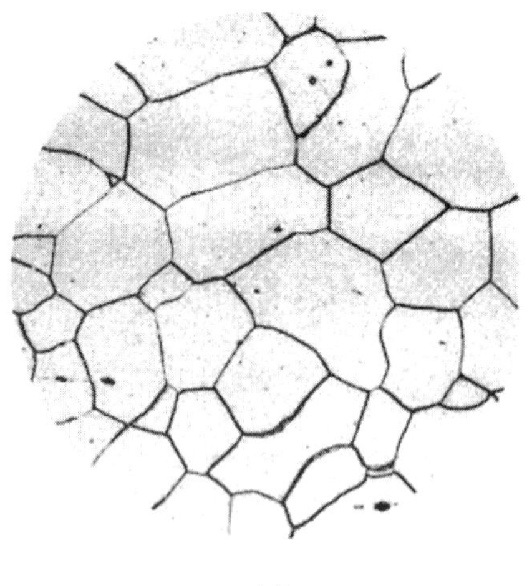

(a)

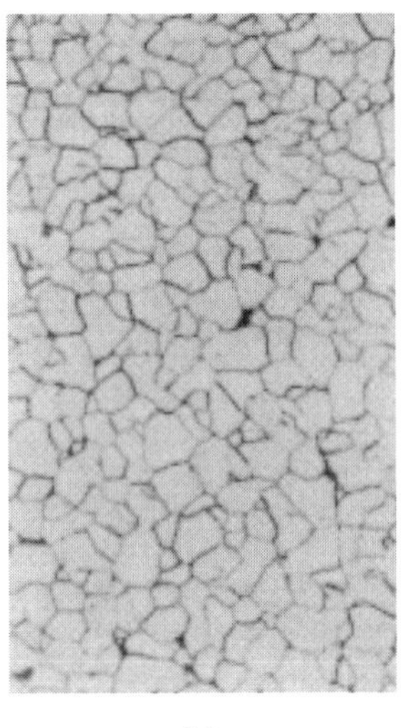

(b)

FIG. 23. Typical microstructures as found in steel. Figures show two different microstructures at same magnification. After Lyman (1972).

As with many of the empirical laws that characterize such mechanical behavior, there are qualitative explanations that serve to indicate the underlying physics of these phenomena. In the case of the Hall–Petch relation described above, the key point is thought to be the role of grain boundaries in inhibiting dislocation transmission from grain to grain. Dislocations are often found to glide until they reach a boundary, which then impedes their motion. As further dislocations accumulate on a given slip plane in the vicinity of the boundary, a dislocation pileup results, which has a stress field of its own. The presence of the dislocation pileup in a given grain enhances the applied stresses that are acting on dislocations in adjacent grains. Macroscopic plastic deformation requires slip to occur in multiple grains, not just those optimally oriented for slip. The stress enhancement due to the pileup in a favorably oriented grain assists the slip in adjacent, less favorably oriented grains. The dependence on grain size enters through the fact that the strength of the dislocation pileup is determined by the grain size, with larger pileups leading to larger stress concentrations in adjacent grains and hence lower yield stresses.

5.2 Work Hardening

The yield stress does not only depend on the grain size but also can be determined by the extent to which a material has been previously deformed. Examination of Fig. 10 reveals that the onset of plastic flow is followed by a regime in which it requires ever higher stresses to maintain plastic deformation. This effect is known as work (or strain) hardening and is attributed to the internal stresses that arise as a result of the dislocation–dislocation interaction. Recall our discussion of the Peach–Koehler force in Sec. 3.2, in which it was shown that a dislocation has an interaction energy with the stress fields that surround it. One source of such stress fields is other dislocations. In this case, if we imagine two parallel edge dislocations, both taken along the x axis for convenience, the x component of the Peach–Koehler force on the dislocation at the origin is given by

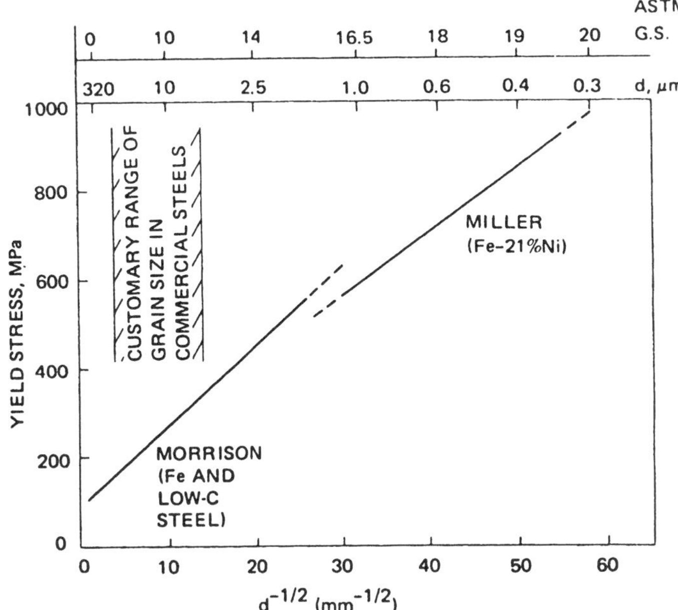

FIG. 24. Relation of yield strength to grain size in steels illustrating the Hall–Petch grain size dependence. After Leslie (1981).

$$F = -\mu b^2/2\pi(1 - \nu)d, \tag{45}$$

where d is the separation between the two dislocations and μ is the shear modulus. The important insight afforded by this result is that the presence of neighboring dislocations can impede the motion of the dislocation of interest.

Taylor formulated a simple but enlightening model of strain hardening that argues that the increase in the flow stress as a consequence of work hardening should scale as $\rho^{1/2}$, where ρ is the dislocation density. Taylor's argument considers an ordered array of edge dislocations with uniform spacing a and interplanar distance d. By appealing to the type of interaction force between neighboring dislocations given above, but by summing the contributions due to all the neighboring dislocations, it can be shown that the critical resolved shear stress to move dislocations scales with the $\rho^{1/2}$ law mentioned.

Though the arguments advanced by Taylor are pleasing, they cannot account for the various stages of hardening revealed in Fig. 25. The Taylor model concentrates on the interactions between dislocations associated with the same slip system. However, once a particular slip system is sufficiently hardened, slip can commence on alternative slip systems. Stage 2 hardening reflects this interaction between dislocations associated with different slip systems. Detailed understanding of the various stages of hardening requires careful analysis of the underlying dislocation processes and resulting dislocation–dislocation interactions.

5.3 Solid-Solution Hardening

As alluded to above, another mechanism leading to an increased yield stress is solid-solution hardening, in which the presence of foreign impurities in the form of point defects can increase the stress needed to move dislocations. As seen in Fig. 4, there is a characteristic scaling of the yield stress with the impurity concentration. However, despite a number of competing models for this behavior, quantitative understanding of this effect remains elusive. One of the difficulties in constructing models of hardening is the role of a number of simultaneous competing mechanisms—mechanisms such as grain-size effects, an initial density of dislocations leading to strain hardening, and the pinning of dislocations by impurities or second-phase particles. Though it is convenient to assume that these affects conspire to alter the flow stress in an additive way, such an assumption may not be warranted.

A simple qualitative model for this hard-

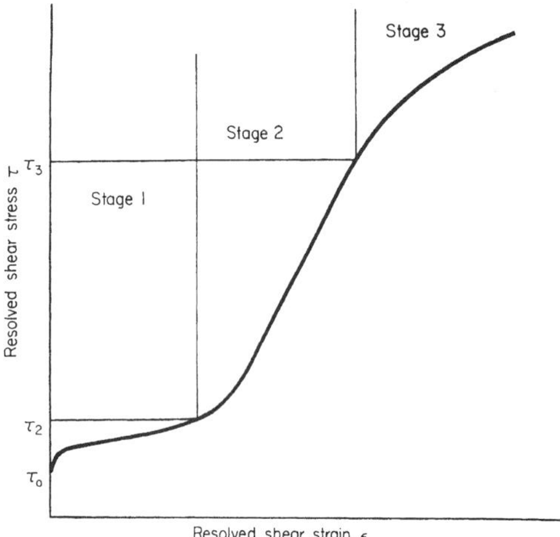

FIG. 25. Schematic of various stages of hardening observed in an fcc metal. After Honeycombe (1968).

ening may be constructed as follows. First, it is necessary to account explicitly for the way in which the stress field of the point defects interacts with the dislocation. One way in which the point defect alters the behavior of the dislocation is through the volume change accompanying the point defect, which is coupled to the hydrostatic component of the stress produced by the dislocation at the position of the point defect. If we consider the geometry depicted in Fig. 26, the hydrostatic pressure at the site of the point defect is given by

$$p(r,\theta) = \frac{(1+\nu)}{(1-\nu)} \frac{\mu b \sin\theta}{3\pi r}, \tag{46}$$

where μ is the shear modulus and, as earlier, ν is the Poisson ratio. For the coupling between the volume change and the hydrostatic stress considered above, there results an interaction energy of the form

$$E_{\text{int}} = A \sin\theta / r, \tag{47}$$

where the constant A depends upon the elastic properties of the medium, the Burgers vector of the dislocation, and the volume change resulting from the presence of the point defect. Differentiation of this expression with respect to the dislocation coordinate x along the slip plane results in an interaction force of the form

$$F = -2Kxd/(x^2 + d^2)^2, \tag{48}$$

where d is the perpendicular distance between the slip plane and the point defect. The interaction force between the point defect and the dislocation is shown in Fig. 27. Note that this result is strictly built around a linear elastic description of the fields due to both the dislocation and the point defect. This implies that for close contact distances between the point defect and the dislocation, this description will break down.

Once the interaction energy between the dislocation and the point defect has been determined, the next step in pushing the model toward a quantitative insight is to obtain an

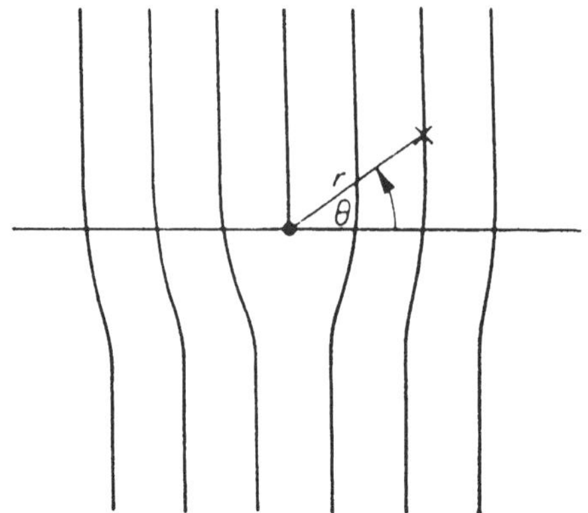

FIG. 26. Schematic of interaction between point defect at (r,θ) and edge dislocation at the origin. After Honeycombe (1968).

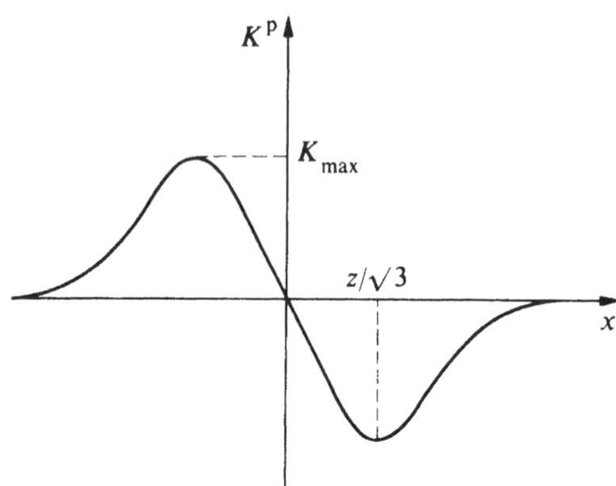

FIG. 27. Force exerted on dislocation due to presence of point defect. After Haasen (1978).

estimate of how a random distribution of such impurities superposes to produce some mean increase in the yield stress. Note that such models presuppose that the distribution of the impurities themselves is not affected by the presence of the dislocation itself, an assumption that breaks down if temperatures are high enough to allow for solute diffusion. In this regime, it is also necessary to consider the breaking away of dislocations from defect "atmospheres."

Theories designed to treat the very dilute limit as well as the higher-concentration regime lead to different predictions as to how the yield strength will scale with impurity concentration. In the Fleischer model (Fleischer, 1963), the impurities are sufficiently dilute so that it can be imagined that the operative process is the breaking away from single pinning points with a force equal to the maximum strength of these impurities. This model leads to a scaling of the yield strength of the form $\sigma_y \propto c^{1/2}$. In the Labusch model (Labusch, 1970), on the other hand, the dislocation is thought to interact with pinning centers of a number of strengths that are spread along the dislocation line with some continuous density. Here the scaling of the yield strength occurs as $\sigma_y \propto c^{2/3}$. On the other hand, in some cases it is experimentally difficult to settle the question of how the yield stress scales with impurity concentration. The key point remains, however, that the point-defect–dislocation interaction is a crucial part of the hardening that is known to occur in real alloy systems.

5.4 Second-Phase Particles

Under appropriate conditions, a particular material can be littered with particles of a second phase, which can significantly alter the mechanical properties of that material. A schematic example of the types of variations in mechanical properties that result from annealing programs is shown in Fig. 28, where the properties of Al–Cu alloys are plotted as a function of aging time. The varying mechanical properties reflect a concomitant change in the disposition of the Cu, which is initially quenched in the Al at concentrations in excess of those allowed by the equilibrium phase diagram at low temperatures. If the alloy is aged at intermediate temperatures, diffusion allows the Cu atoms to cluster into

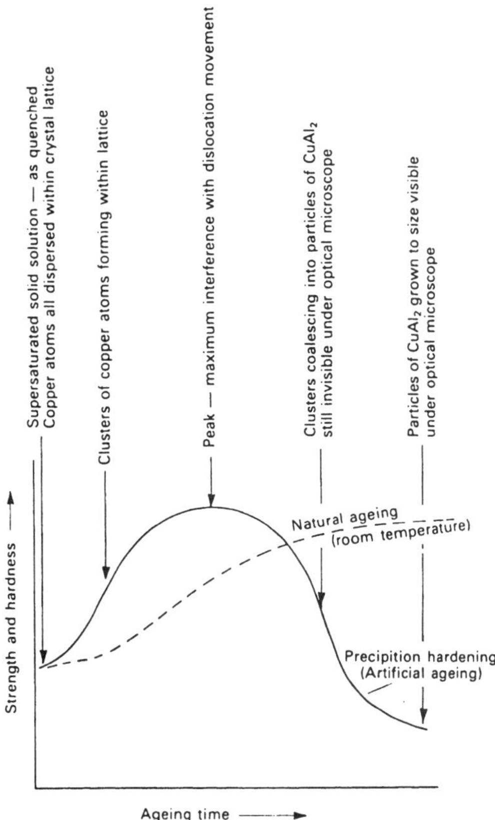

FIG. 28. Diagram illustrating the role of aging in altering the mechanical properties in the Al–Cu system. After Bradbury (1985).

small particles that hinder dislocation motion.

From the standpoint of mechanical properties, the interaction with dislocations is one of the features that makes second-phase particles important. The action of such particles on dislocations has been postulated to lead to two primary mechanisms. In the one extreme, the particles do not allow for the passing of an impinging dislocation. In this case, the dislocation segment between two such particles will bow out until the adjacent bowing segments intersect and pinch off. This mechanism is known as the Orowan process. Alternatively, it is possible that the dislocation can pass through the particle via a cutting mechanism. In this case, the second-phase particles transmit the dislocation with the result that they too suffer slip. These two processes are contrasted in Fig. 29.

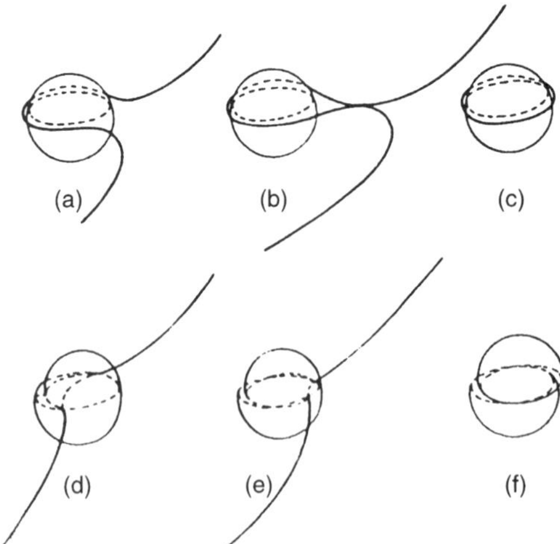

FIG. 29. Illustration of two dislocation mechanisms when second-phase particles are encountered. (a)–(c) Case in which dislocation bows out around particle; (d)–(f) particle cutting mechanism. After Gerold (1979).

Second-phase particles not only are important in dictating the strengthening of some materials but can play a critical role in governing a material's fracture behavior as well. One example of this role is provided by steels, in which it is found that second-phase particles such as carbides and sulfides serve as the sites at which voids are initiated via a mechanism of interfacial debonding (see STEEL). The spacing of these particles, then, in part determines the rate at which these voids coalesce leading in the end to ductile fracture.

6. CONCLUDING REMARKS

The mechanical properties of crystalline solids continue to pose challenges of vital interest to both engineers and physicists. Fundamental questions remain concerning the mechanistic origins of phenomena ranging from ductile fracture via void coalescence to power-law creep. In addition, the ability to optimize materials for both strength and toughness has led to new materials such as composites and lightweight intermetallics. The crucial role of defects in mediating mechanical behavior has been highlighted in this article and will continue to provide problems in applied physics for a long time to come.

ACKNOWLEDGMENTS

I am grateful to N. Bhate, A. Bower, A. Carlsson, R. Miller, V. Shenoy, and E. Tadmor, all of whom made valuable suggestions concerning this article. Thanks also to A. Rosakis for the use of his sequence of figures illustrating dynamic fracture.

GLOSSARY

Brittle Fracture: Fracture in which there is no accompanying plastic deformation.

Brittle–Ductile Transition: Temperature-dependent change in character of fracture mechanism. At low temperatures material fails via brittle fracture, while at high temperatures it fails via ductile fracture mechanisms.

Burgers Vector: The unit of slip carried by a dislocation.

Coble Creep: Diffusion-based creep mechanism in which diffusion takes place along grains.

Creep: Time-dependent plastic deformation of material, usually thought of as occurring at high temperature and small loads.

Ductile Fracture: Fracture preceded by plastic deformation.

Elastic Moduli: Material constants relating stresses to elastic strains.

Fatigue: Failure of material subjected to cyclic loading.

Fracture Toughness: A measure of a material's resistance to fracture.

Hardening: Increase in the stress needed for plastic flow. There are a variety of mechanisms (chemical impurities, second-phase particles, dislocation–dislocation interactions) responsible for such hardening.

Nabarro–Herring Creep: Diffusion-based creep mechanism in which diffusion takes place across grains.

Peierls Stress: The stress needed to move a dislocation against intrinsic lattice resistance.

Resolved Shear Stress: Component of traction vector within slip plane resolved along slip direction.

Slip Plane: Crystal lattice plane along which dislocations move.

Solid-Solution Hardening: Increase in yield stress resulting from chemical impurities.

Strain Energy Density: Elastic energy per unit volume stored in body by virtue of its elastic deformation.

Strain Tensor: Geometric measure of deformation suffered by body. Strain characterizes changes in *relative* position.

Stress Intensity Factor: Numerical constant characterizing the strength of the elastic $r^{-1/2}$ singularity associated with a crack.

Stress Tensor: Tensor carrying information about state of internal forces within continuous body. Components of stress tensor give components of force on given planes.

Yield Stress: Stress above which ductile materials suffer permanent deformation.

Works Cited

Ashby, M. F., Gandhi, C., Taplin, D. M. R. (1979), *Acta Metall.* **27,** 699–729.

Ashby, M. F., Jones, D. R. H. (1980), *Engineering Materials,* Oxford: Pergamon.

Boyer, H. E. (Ed.) (1974), *Metals Handbook,* Vol. 9, *Fractography and Atlas of Fractographs,* Metals Park, OH: American Society for Metals, 8th ed.

Boyer, H. E. (Ed.) (1987), *Atlas of Stress-Strain Curves,* Metals Park, OH: ASM International.

Bradbury, E. J. (Ed.) (1985), *Essential Metallurgy for Engineers,* New York: Van Nostrand Reinhold.

Broek, D. (1991), *Elementary Engineering Fracture Mechanics,* Dordrecht: Kluwer Academic Publishers.

Callister, W. D. (1994), *Materials Science and Engineering,* New York: Wiley.

Collins, J. A. (1993), *Failure of Materials in Mechanical Design,* New York: Wiley.

Courtney, T. H. (1990), *Mechanical Behavior of Materials,* New York: McGraw Hill.

Fleischer, R. L. (1963), *Acta Metall.* **11,** 203–209.

Frost, H. J., Ashby, M. F. (1982), *Deformation Mechanism Maps,* Oxford: Pergamon.

Fuchs, H. O., Stephens, R. I. (1980), *Metal Fatigue in Engineering,* New York: Wiley.

Gandhi, C., Ashby, M. F. (1979), *Acta Metall.* **27,** 1565–1602.

Gerold, V. (1979), in: F. R. N. Nabarro (Ed.), *Dislocations in Solids,* Vol. 4, Amsterdam: North Holland.

Gilman, J. J. (1969), *Micromechanics of Flow in Solids,* New York: McGraw Hill.

Griffith, A. A. (1920), *Philos. Trans. R. Soc.* **221,** 163–198.

Haasen, P. (1978), *Physical Metallurgy,* Cambridge, U.K.: Cambridge University Press.

Hall, E. O. (1951), *Proc. Phys. Soc. London, Ser. B* **64,** 747–753.

Hertzberg, R. W. (1976), *Deformation and Fracture Mechanics of Engineering Materials,* New York: Wiley.

Honeycombe, R. W. K. (1968), *The Plastic Deformation of Metals,* New York: St. Martins Press.

Irwin, G. R. (1948), *Fracturing of Metals,* p. 147, Cleveland, OH: American Society for Metals.

Khan, A. S., Huang, S. (1995), *Continuum Theory of Plasticity,* New York: Wiley.

Labusch, R. (1970), *Phys. Status Solidi* **41,** 659–669.

Lawn, B. (1993), *Fracture of Brittle Solids,* Cambridge, U.K.: Cambridge University Press.

Leslie, W. C. (1981), *The Physical Metallurgy of Steels,* New York: McGraw Hill.

Lyman, T. (Ed.) (1972), *Metals Handbook,* Vol. 7, *Atlas of Microstructures of Industrial Alloys,* Metals Park, OH: American Society for Metals, 7th ed.

Orowan, E. (1955), *Welding J.* **34,** 157–160.

Petch, N. J. (1953), *J. Iron Steel Inst.* **174,** 25–28.

Further Reading

Ashby, M. F., Jones, D. R. H. (1980), *Engineering Materials,* Oxford: Pergamon. A gold mine of insights and phenomenology concerning the mechanical properties of materials; both the best place for a first exposure to topics and a reference for many of the types of figures shown in this article.

Batchelor, G. K. (Ed.) (1958), *The Scientific Papers of Sir Geoffrey Ingram Taylor,* Vol. 1, Cambridge, U.K.: Cambridge University Press. Interesting historical insights into the role played by one of the 20th century's great practitioners of classical physics in the understanding of mechanical behavior.

Boyer, H. E. (Ed.) (1974), *Metals Handbook,* Vol. 9, *Fractography and Atlas of Fractographs,* Metals Park, OH: American Society for Metals, 8th ed. A source of a great variety of pictures illustrating fracture surfaces. An enjoyable afternoon's browsing.

Boyer, H. E. (1987), *Atlas of Stress-Strain Curves,* Metals Park, OH: American Society for Metals. Compendium with wealth of stress–strain behavior for variety of metals. Perusal of this volume leads to insights into elastic behavior, plasticity, and the role of crystal orientation and a number of other topics.

Courtney, T. H. (1990), *Mechanical Behavior of Materials,* New York: McGraw Hill. General textbook with interesting coverage of wide range of topics.

Gleiter, H. (1983), "Fundamentals of Strengthening Mechanisms" in: R. C. Gifkins (Ed.), *Strength of Metals and Alloys,* Oxford: Pergamon. An overview of current understanding of many of the strengthening mechanisms discussed in the article.

Haasen, P. (1979), "Solution Hardening in f.c.c. Metals" in: F. R. N. Nabarro (Ed.), *Dislocations in Solids,* Vol. 4, Amsterdam: North Holland. Review article describing efforts to understand solid solution hardening in metals.

Hirth, J. P., Lothe, J. (1982), *Theory of Dislocations,* New York: Wiley. The classic treatise on the theory of dislocations. Covers elastic theory of dislocations as well as the role played by crystal lattice effects.

Kelly, A., Macmillan, N. H. (1986), *Strong Solids,* Oxford: Clarendon. Excellent account of both ideal strengths of solids in addition to role of defects in mediating mechanical behavior.

Khan, A. S., Huang, S. (1995), *Continuum Theory of Plasticity,* New York: Wiley. Discusses phenomenology of plastic deformation and macroscopic ideas used in explaining the phenomena.

Knott, J. F. (1973), *Fundamentals of Fracture Mechanics,* London: Butterworths. Excellent introductory treatment of fracture mechanics with balanced treatment of theoretical insights and experimental foundations.

Kocks, U. F. (1985), *Metall. Trans.* **16A,** 2109–2129. Discussion of difficulties associated with simplified models of solid solution hardening.

Kocks, U. F., Argon, A. S., Ashby, M. F. (1975), *Thermodynamics and Kinetics of Slip,* Oxford: Pergamon. Intriguing book that shows how dislocation considerations may be used to understand macroscopic constitutive ideas used to describe plastic flow.

Kuhlmann-Wilsdorf, D. (1985), *Metall. Trans.* **16A,** 2091–2108. Discussion of experimental and theoretical background on study of work hardening.

Landau, L. D., Lifshitz, E. M. (1959), *Theory of Elasticity,* Oxford: Pergamon. A terse description of many of the key results and ideas of the theory of elasticity.

Lawn, B. (1993), *Fracture of Brittle Solids,* Cambridge, U.K.: Cambridge University Press. A very readable introduction to fracture mechanics.

Lemaitre, J., Chaboche, J.-L. (1990), *Mechanics of Solid Materials,* Cambridge, U.K.: Cambridge University Press. Encyclopedic view of much of theoretical apparatus used in analysis of solid mechanics.

Lyman, T. (Ed.) (1972), *Metals Handbook,* Vol. 7, *Atlas of Microstructures of Industrial Alloys,* Metals Park, OH: American Society for Metals, 8th ed. This entire series is full of interesting figures and data. The microstructures handbook shows many beautiful and intriguing microstructures.

Martin, J. W. (1980), *Micromechanisms in Particle-Hardened Alloys,* Cambridge, U.K.: Cambridge University Press. An undergraduate text with wide coverage on microstructural hardening mechanisms, especially those arising from second-phase particles within the microstructure.

McClintock, F. A., Argon, A. S. (1966), *Mechanical Behavior of Materials,* Reading, MA: Addison-Wesley. A complete appraisal of the field of mechanical properties as of the mid 1960s. Besides its wide coverage, the book suggests a number of interesting "desk-top experiments" that demonstrate various mechanical effects.

Poirier, J.-P. (1985), *Creep of Crystals,* Cambridge, U.K.: Cambridge University Press. A description of both the phenomenology and mechanisms that govern high-temperature deformation.

Suresh, S. (1991), *Fatigue of Materials,* Cambridge, U.K.: Cambridge University Press. A comprehensive research level monograph on the field of fatigue.

Sutton, A. P., Balluffi, R. W. (1995), *Interfaces in Crystalline Materials,* Oxford: Oxford University Press. This book treats interfaces with the same wealth of information and insights that Hirth and Lothe bring to the subject of dislocations.

SOLIDS, DIFFUSION IN

D. J. YOUNG, *School of Materials Science and Engineering, The University of New South Wales, Sydney, Australia*

INTRODUCTION

Diffusion is the sole mechanism of mass transport in isothermal solids free of external fields (magnetic, mechanical stress, etc.). It is consequently of great practical importance in a number of branches of materials science and engineering. These include semiconductor device fabrication, corrosion and scaling of metals, sintering of ceramics and metals, creep, surface engineering, and a number of phase transformations. The phenomenon can be described without reference to the nature of the solid, or its constituent particles, in the classical approach represented by Fick's (1855) first law

$$\mathbf{J} = -D\nabla C.$$

Here $\mathbf{J}$ is the flux in moles per unit area per second, ∇C is the gradient in molar concentration, and D is the diffusion coefficient, approximated as a constant characteristic of the material. Such an approach lacks any predictive capacity. However, the diffusion coefficient can ultimately be related to the frequency of particle jumps and the lattice constant of a solid, permitting at least qualitative prediction of material behavior.

Section 1 considers the defect species in solids responsible for diffusion, and Sec. 2 uses the thermodynamics of irreversible processes to identify chemical-potential gradients as the driving forces for their diffusion. In Sec. 3, the kinetics of vacancy diffusion is discussed, and expressions are arrived at for self-diffusion, i.e., diffusion in the absence of any gradient other than that of an isotope. This section concludes with a brief examination of experimental data.

In the practical situations of interest where diffusion occurs in a chemical-potential gradient, the diffusion coefficient defined in Fick's law becomes a function of concentration. Irreversible thermodynamics is used in Sec. 4 to produce a more general description, appropriate to this situation. This approach is then applied to alloys in Sec. 5 and to ionic solids in Sec. 6. Finally, the principal methods of measuring diffusion coefficients are described in Sec. 7.

1. POINT DEFECTS

Solid-state diffusion involves the movement of individual particles (molecules, atoms, or ions) that constitute the material.

3-527-28140-1/96/$5.00 + .50

These particles are capable of movement because they vibrate around their mean positions and because the existence of defects in the solid crystal permits an occasional vibration to extend into a translation to an adjacent lattice site. Two common defects are illustrated in Fig. 1 for the case of a pure monatomic solid: a vacancy, or unoccupied lattice site, and an interstitial atom, *i.e.*, one located between normal lattice sites.

A lattice atom can move into an adjacent vacancy, exchanging sites with the defect. Movement via this vacancy mechanism is the most common way in which diffusion occurs. Clearly, the concentration of vacancies present is important in determining the probability of atomic translation occurring. The interstitial species can contribute to diffusion simply by moving into an adjacent interstitial site. This is improbable in pure metals, because the atoms are large, but operates for interstitial impurities such as C, H, N, and O dissolved in metals. The interstitial species can also move onto an adjacent normal lattice site, displacing its occupant into one of the adjacent interstitial sites, as shown in Fig. 2. This is the "interstitialcy" mechanism and is known to be important in the diffusion of silver in AgBr (Friauf, 1957). Whichever the mechanism, the concentration of defects is an important factor in the particle movement rates. The question of defect concentrations is now considered.

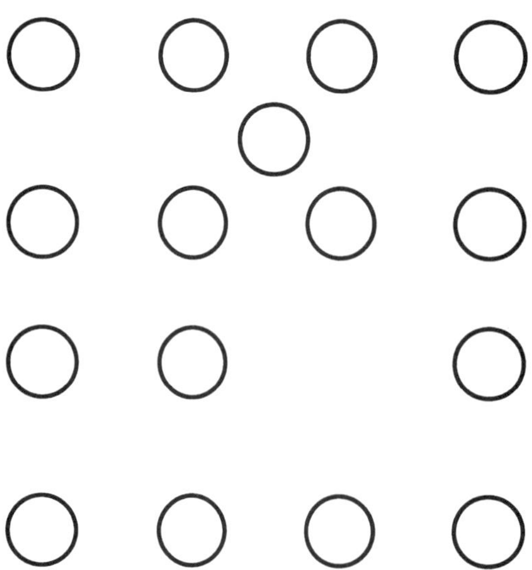

FIG. 1. Schematic view of a solid monatomic crystal containing a vacancy and an interstitial.

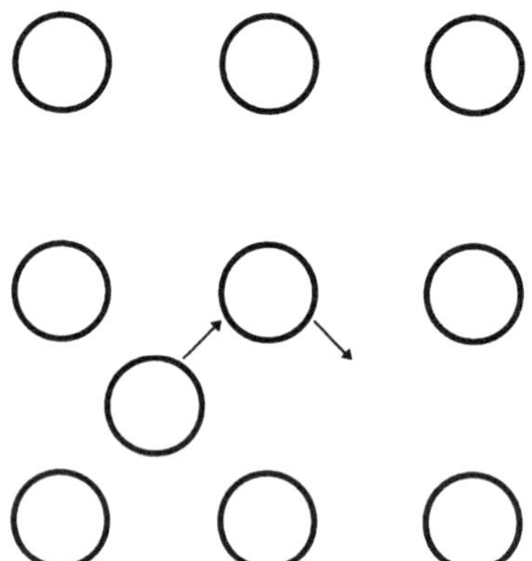

FIG. 2. Interstitialcy diffusion involves simultaneous movement of interstitial and lattice species.

Equilibrium concentrations of point defects in crystals are calculated by the methods of statistical thermodynamics. The application of these methods to crystals has been reviewed in detail by Schottky (1958), and their use in diffusion calculations has been explored by several authors, notably Howard and Lidiard (1964). The Gibbs free energy G for a monatomic crystal containing n_v vacancies and N atoms is

$$G = G_0 + n_v g_v - kT \ln \frac{(N + n_v)!}{N! n_v!}, \tag{1}$$

where g_v is the free energy of formation of a vacancy, k is Boltzmann's constant, and the logarithmic term is recognized as the configurational entropy resulting from the presence of defects. It is this term that makes vacancy formation inevitable at all temperatures above absolute zero. The free-energy minimum representing the equilibrium state of the crystal defines the chemical potential of the vacancies as zero:

$$\left(\frac{\partial G}{\partial n_v}\right)_{T,P,N} = \mu_v = 0. \tag{2}$$

Application of this to Eq. (1), making use of Stirling's approximation ($\ln N! = N \ln N - N$), then yields

$$X_v = n_v/(N + n_v) = \exp(-g_v/kT), \tag{3}$$

where X indicates mole fraction. Thus the equilibrium concentration of vacancies rises rapidly with temperature. A similar expression is obtained for interstitials. These expressions neglect interactions between defects, and are thus suitable only for low defect concentrations.

The same treatment is successful in the case of ionic solids. Vacancy formation in an insulating compound must involve simultaneously both the cation and anion sublattices in order to preserve charge neutrality. In a compound MX of 1:1 stoichiometry, "Schottky pairs" of vacancies result. Similarly, interstitial formation involves the simultaneous formation of a lattice vacancy and an interstitial species, thus again preserving charge neutrality. The same statistical approach applies, and the overall configurational entropy for Schottky defect formation is given by

$$S = kT \ln\left[\frac{(N_m + n_{v_m})!}{N_m! n_{v_m}!} \frac{(N_x + n_{v_x})!}{N_x! n_{v_x}!}\right], \tag{4}$$

where n_{v_m} is the number of cation vacancies and n_{v_x} is the number of anion vacancies. Hence, following the same free-energy minimization procedure, it is found that

$$\mu_{v_m} + \mu_{v_x} = 0, \tag{5}$$

and

$$X_{v_m} X_{v_x} = \exp[-(g_{v_m} + g_{v_x})/kT]. \tag{6}$$

Equations (3) and (6) are recognized as forms of the mass-action law.

2. DRIVING FORCES AND MOBILITIES

Having identified the principal mechanisms of diffusion in crystalline solids, we now seek to calculate the rate of the process. In addition to a knowledge of defect concentrations, evaluation of their mobilities and the driving forces that induce their movement is required. The most general approach to the latter question is provided by the methods of irreversible thermodynamics. The principal concepts were developed by Onsager (1931). More complete accounts are given by deGroot and Mazur (1962) and Prigogine (1967), while the application to solid-state diffusion has also been reviewed (e.g., Howard and Lidiard, 1964; Kirkaldy and Young, 1987).

The procedures of irreversible thermodynamics enable us to calculate the rate of entropy production per unit volume, σ, in terms of the various fluxes flowing within the system. The result is a bilinear expression involving the fluxes themselves and a set of thermodynamic forces. These forces are thereby identified as those responsible for the fluxes, each flux being linearly dependent on all the forces. The description is applicable only to systems that are not far removed from equilibrium and is therefore appropriate to diffusion in a solid within which the local equilibrium state is closely approached.

For isothermal diffusion in a closed, isobaric, field-free system, it is found for a temperature T that

$$T\sigma = -\sum_i \mathbf{J}_i \cdot \nabla \eta_i \tag{7}$$

and hence that

$$\mathbf{J}_i = -\sum_j L_{ij} \nabla \eta_j, \tag{8}$$

where $\mathbf{J}_i$ is the flux of component i, and the coefficients L_{ij} will be discussed shortly. The driving forces are seen to be the gradients $\nabla \eta_j$, known as electrochemical potential gradients. They are defined by

$$\eta = \mu + zF\psi, \tag{9}$$

where z is the effective charge on the particle, F is the Faraday, and ψ is the local electrostatic potential. Gradients in potential constitute fields, but these are internal to the solid, and the conditions for the validity of Eq. (7) are maintained. In the case of a metallic alloy, the constituent atoms have no effective charge, and the driving force is the chemical potential gradient, $\nabla\mu$. This result is intuitively satisfactory in the sense that diffusion is perceived (under the conditions specified earlier) as a process that eliminates differences in chemical potential, thereby achieving equilibration.

Now that the driving forces have been identified, attention is directed to the phenomenological coefficients L, so called because they do not rely on any detailed description of the solid structure or the diffusion mechanism. The method of dealing with the set of equations (8) will be explained and its value illustrated later. For the moment, one-dimensional diffusion in a binary system is examined, and the approximation $L_{12} \simeq 0$ leads to the simple result

$$J_1 = -L_{11}\frac{d\mu_1}{dx}. \tag{10}$$

Methods of evaluating L_{ij} are discussed later.

If we recall that

$$\mu_1 = \mu_1^0 + RT \ln a_1, \tag{11}$$

with μ_1^0 the standard chemical potential and a_1 the activity of component 1, then it is seen that for an *ideal solution* ($a_1 = X_1$),

$$J_1 = -\frac{L_{11}RT}{C_1}\frac{dC_1}{dx}, \tag{12}$$

where C is the molar concentration. This is equivalent to Fick's first law of diffusion,

$$J_1 = -D_1\frac{dC_1}{dx} \tag{13}$$

with

$$D_1 = L_{11}RT/C_1, \tag{14}$$

which is the diffusion coefficient. This coefficient is commonly regarded as a material property. It must be remembered, however, that D_1 varies with the concentration of component 1, and also with the concentrations of all other components. This point is explored further after evaluation of L_{11}.

3. VACANCY AND TRACER (SELF-DIFFUSION) COEFFICIENTS

The measurement of D can be laborious, and considerable effort has been devoted to theoretical calculation of the quantity. Theory does not yet permit *a priori* calculation, but it does provide a number of predictive tools that can be used to minimize experimental effort. The most widely used approach is based on random-walk theories originally due to Einstein (1905, 1906) and Smoluchowski (1908).

3.1 Random-Walk Description

Atom–vacancy exchanges occur about 10^8 times per second in metals just below their melting points. Because the direction of interchange is random and because the number of events is very large, statistical methods work well. In the random progress of a single vacancy over n successive interchanges or jumps, the squared displacement is

$$\mathbf{x}^2 = (\mathbf{x}_1 + \mathbf{x}_2 + \ldots \mathbf{x}_n)^2. \tag{15}$$

Rearrangement to collect diagonal terms yields

$$\mathbf{x}^2 = \sum_{i=1}^{n} \mathbf{x}_i^2 + 2\sum_{j-1}^{n-1}\sum_{k=j+1}^{n} \mathbf{x}_j \cdot \mathbf{x}_k. \tag{16}$$

In a purely random process the terms in the second summation have equal probabilities of being positive or negative. Averaging over the trajectories of many vacancies then yields the mean displacement

$$\langle \mathbf{x}^2 \rangle = \left\langle \sum_{i=1}^{n} \mathbf{x}_i^2 \right\rangle = \sum_{i=1}^{n} \langle \mathbf{x}_i \rangle^2. \tag{17}$$

For cubic lattices all jumps are of the same length, λ. Then

$$\langle \mathbf{x}^2 \rangle = n\lambda^2. \tag{18}$$

The number of jumps in the x direction occurring in time t is related to the isotropic jump frequency ν through

$$n/t = \nu/3 \tag{19}$$

and

$$\langle \mathbf{x}^2 \rangle = \nu\lambda^2 t/3. \tag{20}$$

Since (Manning, 1968)

$$\langle \mathbf{x}^2 \rangle = 2Dt, \tag{21}$$

it follows that

$$D = \nu\lambda^2/6, \tag{22}$$

the Einstein–Smoluchowski relation. However, this diffusion coefficient applies to the vacancy, whereas knowledge of atomic (or ionic) diffusion is usually the objective.

Although the atoms of a single-component solid are continually moving, their indistinguishibility means that the motion is not observable. This difficulty is overcome by observing the diffusional intermixing of different isotopes of the same atom (or ion). In the ideal case, the enthalpy of mixing is zero, and a purely entropic process is observed. The relationship between the diffusion coefficient of the tracer D_{A*} and that of the vacancy D_v is approached by recognizing that the jump frequency of a moving species is proportional to the probability of finding the appropriate species on a nearest-neighbor site. Thus a tracer atom can move only when a vacancy is adjacent (probability X_v), and a vacancy can move when an atom is adjacent (probability $X_A \approx 1$). It follows then that

$$D_{A*} = fD_vX_v, \tag{23}$$

where f is a constant of proportionality. If the tracer movement were completely random, then we would have $f = 1$. However, this cannot be so for diffusion via a vacancy mechanism, and successive jumps are said to be correlated.

The analytic calculation of correlation effects is complicated (Manning, 1968; Kidson, 1978; Murch, 1984) but amenable to Monte-Carlo methods (*q.v.;* see also Murch, 1984). The physical origin of the effects can be easily understood from a consideration of the possibilities confronting a tracer atom immediately after it has interchanged with a vacancy. The most likely event for the tracer is prompt reinterchange with the vacancy. Thus the tracer will travel a shorter distance in a given number of jumps than will a vacancy. This means that the nondiagonal terms in Eq. (16) no longer sum to zero. An approximate estimate of the size of the effect is arrived at (Shewmon, 1963) by noting that for a crystal coordination number of z, the probability that a vacancy will jump back to the tracer position is $1/z$. This will cancel out the first jump, making two tracer jumps ineffective. Thus

$$f \approx 1 - 2/z. \tag{24}$$

Estimates of the correlation factor f made in this way are compared with precise calculations in Table 1. It is to be noted that $f = 1$ for dilute interstitial diffusion because randomness is assured.

The random-walk description informs us that diffusion coefficients can be calculated from a knowledge of defect concentrations and jump frequencies and that the result will vary with direction in anisotropic solids. It also predicts that the correlation factor varies with defect type (values are also available for interstitialcy and divacancy diffusion: Murch, 1991; Philibert, 1991). This fact is made use of in distinguishing diffusion mechanisms in ionic solids, for which the correlation factor can be experimentally accessible (Murch, 1991). In addition, the study of correlation effects has at least the potential to help calculate the diffusion properties of highly disordered solids (Murch, 1984; Sato, 1989). Finally, the methodology has been applied to binary alloys, but with mixed results (Bocquet *et al.*, 1983; Kirkaldy and Young, 1987).

3.2 Absolute-Rate–Theory Description

We turn now to the evaluation of individual jump frequencies, using absolute-rate theory. The first applications to solid-state diffusion were reported by Wert and Zener (1950) and Seitz (1950), and subsequent extensions for various cases have been provided by others (e.g., Lidiard, 1955; Dignam *et al.*, 1973; Young and Kirkaldy, 1984). Alternative approaches exist, notably linear-response theory (Kubo, 1957; Allnatt and Okamura, 1984) and the path-probability method (Kikuchi and Sato, 1969; Sato and Zhang, 1993), but will not be considered here.

When a particle moves from one lattice position to another, it passes through an intermediate state that has a higher energy because adjacent particles must be perturbed

Table 1. Comparison of approximate and exact values of f for vacancies.

Structure	$1 - 2/z$	f
Simple cubic	0.67	0.6555
Body-centered cubic	0.75	0.7215
Face-centered cubic	0.83	0.7814
Diamond	0.5	0.5

from their mean lattice positions in order to accommodate the passage of the moving particle. During this lattice distortion, an activated complex involving the two interchanging species (e.g., particle plus vacancy) is formed. The activity a_{iv} of the complex is described via the equilibrium constant K_{iv} for its formation:

$$\frac{a_{iv}}{a_i a_v} = K_{iv} = \exp\left(-\frac{\Delta H_{iv}}{RT}\right)\exp\left(\frac{\Delta S_{iv}}{R}\right), \quad (25)$$

where ΔH_{iv} is the enthalpy and ΔS_{iv} is the entropy of complex formation.

A profile of the periodic internal-energy surface in a direction parallel to that of diffusion is shown in Fig. 3. An electrostatic field can be externally imposed, or can arise through the movement of the charged species themselves, and will in this case be aligned with the diffusion direction. The height of the energy barrier to diffusion of a charged species is modified by the field, being lower for downfield movement than for upfield movement of an appropriately charged species. It will be assumed that the field does not affect ΔS_{iv}. We may write for the interchange of species i and a vacancy between planes (1) and (2) separated by a distance λ

$$J = m\lambda\nu_{iv}\exp\left(\frac{\Delta S_{iv}}{R}\right) \times \left\{a_i^{(1)}a_v^{(2)}\exp\left[-\frac{\Delta U_{iv} - zF(\psi^{(0)} - \psi^{(1)})}{RT}\right] - a_v^{(1)}a_i^{(2)}\exp\left[-\frac{\Delta U_{iv} - zF(\psi^{(2)} - \psi^{(0)})}{RT}\right]\right\} \times \gamma_{iv}^{-1}, \quad (26)$$

where m is the volume concentration of lattice sites, ν_{iv} is a kinetic frequency term, and γ_{iv} is the activity coefficient for the transition-state complex. Here z is the effective charge of the vacancy, that of the cation being zero. In the more general case of interchange between any two differently charged species, z is replaced by the difference in effective charges of the participating species. Superscripts in parentheses represent the location at which the quantity in question is evaluated.

Equation (26) is cleared of common terms and subjected to Taylor series expansion of the terms $a_v^{(2)}$ and $a_i^{(2)}\exp(-zF\psi^{(2)}/RT)$. Retention of linear terms, in the case where the field is not inordinately high, leads to

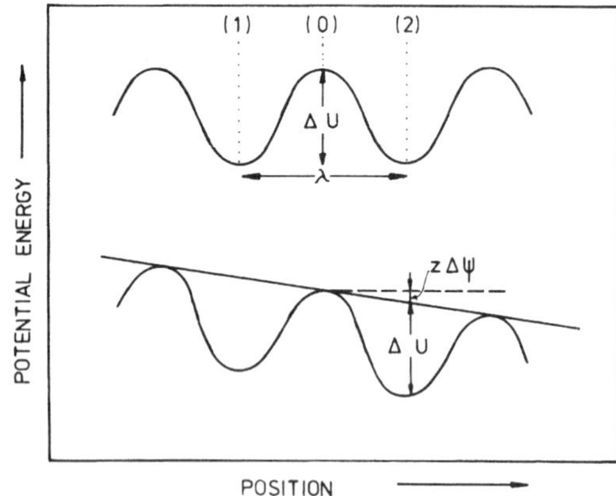

FIG. 3. Potential-energy profile in diffusion direction: upper curve, no electrostatic field; lower curve, showing effect of electrostatic field.

$$J = -\frac{m\lambda^2}{RT}\nu_{iv}K_{iv}a_i a_v\{\nabla\mu_i - \nabla\mu_v - zF\nabla\psi\}, \quad (27)$$

which, upon substituting from Eq. (9), becomes

$$J = -\frac{m\lambda^2}{RT}\nu_{iv}K_{iv}a_i a_v\{\nabla\eta_i - \nabla\eta_v\}. \quad (28)$$

Expressions of this sort always apply to pairwise site exchanges.

For diffusion of noncharged species in a metal or an alloy, $z = 0$, and we obtain

$$J_i = -\frac{m\lambda^2}{RT}\nu_{iv}K_{iv}a_i a_v\{\nabla\mu_i - \nabla\mu_v\}. \quad (29)$$

If the equilibrium condition of Eq. (2) is realized and the off-diagonal terms in Eq. (8) are ignored, this result simplifies to Eq. (10), whence

$$L_{11} = (m\lambda^2/RT)\nu_{iv}K_{iv}a_i a_v,$$

and therefore, in the dilute (ideal) solution approximation, it is found from Eq. (14) that

$$D_1 = \lambda^2\nu_{iv}\exp\left(\frac{-\Delta H_{iv}}{RT}\right)\exp\left(\frac{\Delta S_{iv}}{R}\right)X_v, \quad (30)$$

which is seen to correspond to Eq. (23), except that the correlation coefficient cannot be evaluated from this simple procedure, which omits the off-diagonal L_{ij} terms.

We now combine Eqs. (3) and (30) to determine the temperature dependence of the diffusion coefficient:

$$D = D_0 \exp[-(\Delta H_{iv} + \Delta H_v)/RT], \tag{31}$$

where ΔH_v is the enthalpy (per mole) for vacancy formation, and the remaining constants have been collected in D_0. The expected Arrhenius form is arrived at and is commonly used to interpolate or extrapolate sparse experimental data.

3.3 Self-Diffusion Coefficient Values

The discussion thus far has been concerned with diffusion in a pseudo one-component system ($A + A^*$) and is applicable to both a pure metal and a single sublattice of a pure compound. The measured tracer diffusion coefficient D_{A^*} is also frequently known as the self-diffusion coefficient, because the tracer and solvent species are chemically identical. However, it should be noted that some writers reserve the term self-diffusion coefficient for the quantity D_{A^*}/f. Extensive tabulations of self-diffusion are available (Askill, 1970; Brandes, 1983; Freer, 1980; Mehrer, 1990), and a good general discussion is provided by Philibert (1991). Brown and Ashby (1980) have examined a wide range of data and reconfirmed a number of general correlations.

The self-diffusion coefficient at the melting point, D_{T_m}, is very roughly constant for a given class of materials, as shown in Fig. 4. Using Eq. (31), it is seen that

$$D_{T_m} = D_0 \exp[-(\Delta H_{iv} + \Delta H_v)/RT_m]. \tag{32}$$

The factor D_0 contains the atomic vibration frequency, the lattice constant λ, a lattice-specific factor including the correlation coefficient, and the entropies of vacancy formation and movement. For solids of the same crystal structure and bond type, these quantities are all approximately equal, provided

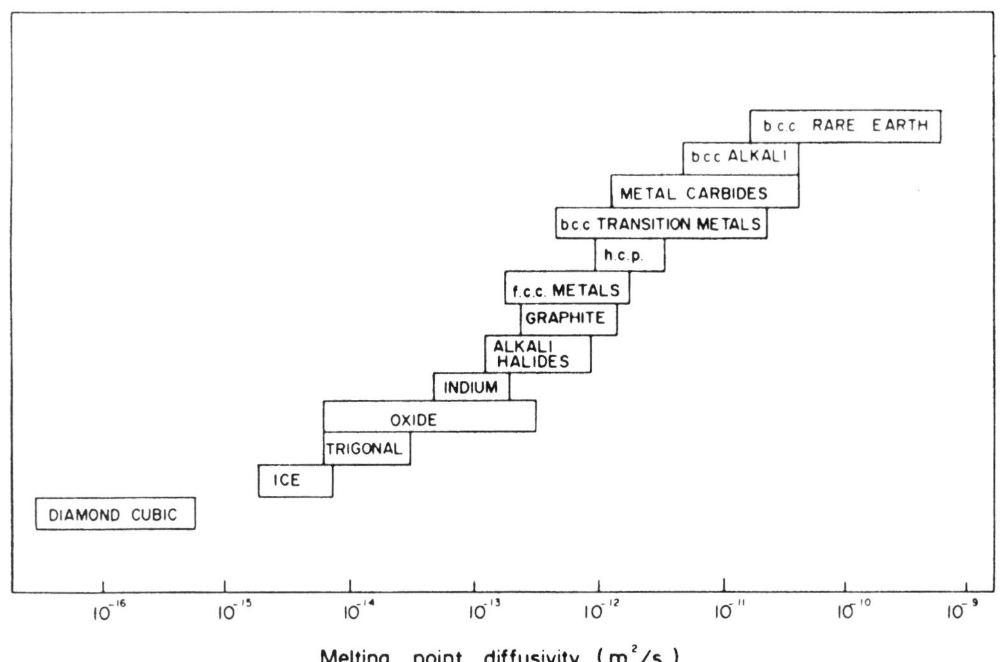

FIG. 4. Comparison of self-diffusion coefficients extrapolated to melting points. Reprinted with permission from *Acta Metall.* **28,** Brown and Ashby, "Correlations for Diffusion Coefficients," pp. 1085–1101, Copyright 1980, with kind permission from Elsevier Science Ltd., The Boulevard, Langford Lane, Kidlington OX5 1GB, UK.

that particle masses (which affect ν) do not vary greatly.

Following classical arguments (e.g., Mott and Gurney, 1964), it is noted that the formation of a vacancy in a constant-mass system involves movement of a particle from the interior of a solid to its surface in the usual event that interstitial formation is impossible. This is energetically equivalent to breaking half the particle's bonds, and the enthalpy of the process would be expected to be proportional to the heat of sublimation, ΔH_s. The movement of a particle into the vacancy during diffusion involves temporary local disordering and is therefore like local melting (Sherby *et al.*, 1973). The enthalpy of movement is expected to be proportional to the heat of fusion, ΔH_m. Thus the activation energy Q appearing in Eq. (32) can be written

$$Q = \alpha \Delta H_s + \beta \Delta H_m, \tag{33}$$

with α and β as constants. At the melting point, phase equilibrium is achieved, and

$$\Delta S_m = \Delta H_m / T_m, \qquad \Delta S_s = \Delta H_s / T_m, \tag{34}$$

with ΔS_m the entropy of fusion and ΔS_s the entropy of sublimation at the melting point. Then we may write

$$Q/RT_m = (\alpha \Delta S_s + \beta \Delta S_m)/R. \tag{35}$$

If these entropy changes are approximately constant for a particular crystal structure and bond type, then so is Q/RT_m, and the constancy of D_{T_m} is explained. Data collected by Brown and Ashby (1980) show the approximate constancy of Q/RT_m for a number of crystal types, as seen in Fig. 5. The data in Figs. 4 and 5 can be used to arrive at order-of-magnitude estimates for self-diffusion at lower temperatures. Other formulas for Q have recently been reviewed by Cahoon and Sherby (1992).

4. USE OF PHENOMENOLOGICAL EQUATIONS

The application of Eq. (8) to a solid is now explored. For an n-component substitutional solid in which diffusion occurs

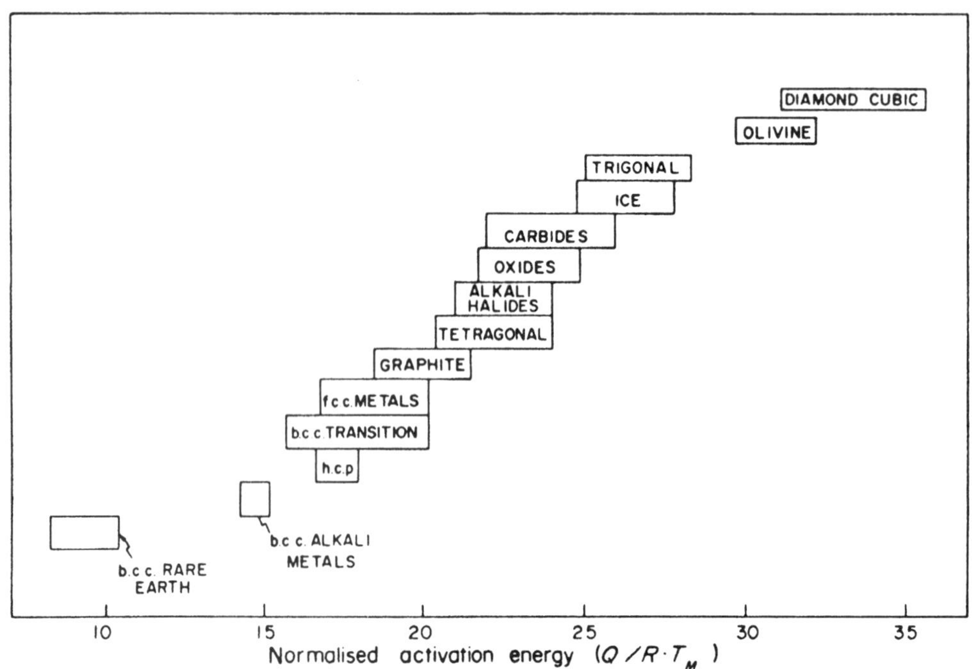

FIG. 5. Normalized activation energies for diffusion in different crystal types. Reprinted with permission from *Acta Metall.* **28,** Brown and Ashby, "Correlations for Diffusion Coefficients," pp. 1085–1101, Copyright 1980, with kind permission from Elsevier Science Ltd., The Boulevard, Langford Lane, Kidlington OX5 1GB, UK.

through pairwise exchanges of lattice-site occupants with vacancies, the transport equations are

$$J_i = -\sum_{j=1}^{n} L_{ij}\nabla\mu_j - L_{iv}\nabla\mu_v, \tag{36}$$

where charge-neutral species are concerned. The equations contain redundancies as a result of site and mass conservation in a lattice-fixed frame of reference:

$$\sum_{i=1}^{n} J_i + J_v = 0, \tag{37}$$

$$\sum_{i=1}^{n} X_i + X_v = 0. \tag{38}$$

The constraints (37) and (38) can be applied to Eqs. (36), reducing them to a linearly independent set. A simpler method is to commence with the entropy-source expression (7), rewritten now as

$$T\sigma = -\sum_{i} J_i\nabla\mu_i - J_v\nabla\mu_v. \tag{39}$$

Application of (37) to (39) then leads to

$$T\sigma = -\sum_{i} J_i\nabla(\mu_i - \mu_v) \tag{40}$$

and its consequence

$$J_i = -\sum_{j=1}^{n} L_{ij}\nabla(\mu_i - \mu_v). \tag{41}$$

Comparison with Eq. (29) shows how the elementary kinetic theory will provide an evaluation of the L_{ij}.

The most important finding of the irreversible-thermodynamics approach is the Onsager reciprocal relationship between off-diagonal terms,

$$L_{ij} = L_{ji}, \tag{42}$$

which applies when Eqs. (41) constitute an independent set. This makes the description of multicomponent diffusion much more efficient and reduces the number of measurements necessary. It is important, wherever possible, to reduce still further the experimental effort required, by arriving at approximations to the L_{ij}.

4.1 Relationship with Correlation Coefficients

The off-diagonal terms describe diffusion driven by gradients other than that of the diffusing species itself. Estimation of these "cross effects" is important. One approach to the problem consists of relating the L_{ij} to correlation effects. A large and difficult literature has accumulated on this subject (see, e.g., Howard and Lidiard, 1964; Manning, 1968; LeClaire, 1975; Allnatt and Allnatt, 1984; Heumann, 1979; Kirkaldy *et al.*, 1987). We consider here only the application to tracer diffusion in the alloy A–A^*–V, i.e., an alloy of A with its tracer isotope A^*, containing vacancies, V. The appropriate independent flux equations are

$$J_A = -L_{AA}\nabla(\mu_A - \mu_v) - L_{AA^*}\nabla(\mu_{A^*} - \mu_v), \tag{43}$$

$$J_{A^*} = -L_{A^*A}\nabla(\mu_A - \mu_v) - L_{A^*A^*}\nabla(\mu_{A^*} - \mu_v), \tag{44}$$

and the correlation factor is related to these coefficients by

$$L_{AA} = \frac{D_vN}{kT}\frac{X_AX_v}{X_A + X_{A^*}}[1 - X_v - X_{A^*}(1 - f)],$$
$$L_{A^*A^*} = \frac{D_vN}{kT}\frac{X_{A^*}X_v}{X_A + X_{A^*}}[1 - X_v - X_A(1 - f)], \tag{45}$$

and

$$L_{AA^*} = L_{A^*A} = \frac{D_vN}{kT}\frac{X_AX_{A^*}X_v}{X_A + X_{A^*}}(1 - f). \tag{46}$$

It is seen that only if tracer jumps are completely uncorrelated ($f = 1$) do the cross terms vanish. Calculations for systems containing more than one chemical component are much more complex. Since the magnitudes of the effects predicted are generally small compared with other factors, they will not be considered further.

In discussing diffusion in real solutions, it becomes impossible to retain the ideal-solution approximation leading from Eq. (10) to

Eqs. (12) and (14). Instead, the activity coefficient γ is introduced,

$$a_i = \gamma_i X_i, \tag{47}$$

and, if off-diagonal terms are neglected, it is found that

$$D_1 = \frac{L_{11}RT}{X_1}\left(1 + \frac{\partial \ln\gamma_1}{\partial \ln X_1}\right), \tag{48}$$

where use has been made of the definition $X_1 = C_1/C$. The magnitude of the thermodynamic effect can be quite significant, as shown in Sec. 5.2.

5. DIFFUSION IN ALLOYS

5.1 Origins of Cross Effects

Equation (8) may be rewritten for atomic diffusion as

$$J_i = -\sum_j L_{ij}\nabla\mu_j, \tag{49}$$

whence it is clear that cross effects can arise through either kinetic interactions, as represented by the Onsager coefficients, or thermodynamic interactions, represented by the variation of chemical potential with composition. For simplicity, we assume equal molar volumes for all constituents and write the local thermodynamic equilibrium statement for the chemical potential of a component in an isothermal, isobaric n-component system:

$$\mu_i = \mu_i(C_1, C_2, \ldots .C_{n-1}). \tag{50}$$

In this case

$$\nabla(\mu_i - \mu_n) = \sum_j^{n-1} \mu_{ij}\nabla C_j, \tag{51}$$

where

$$\mu_{ij} = \frac{\partial(\mu_i - \mu_n)}{\partial C_j}. \tag{52}$$

Substitution into Eq. (41), with μ_v replaced by μ_n, yields

$$J_i = -\sum_j^{n-1} D_{ij}\nabla C_j \tag{53}$$

with

$$D_{ij} = \sum_k^{n-1} L_{ik}\mu_{kj}. \tag{54}$$

Equation (53) is recognized as a generalized Fick equation for the description of diffusion driven by multiple concentration gradients. It is now explicit that the cross effects are due to thermodynamic interactions (μ_{ij}) and kinetic effects (L_{ij}).

The need for this degree of complexity is illustrated by Darken's (1948) famous experiment on carbon diffusion in Fe–Si–C alloys. A bar of homogeneous Fe–0.44%C was joined to one of Fe–0.48%C – 4%Si and diffused at 1050 °C. At this temperature the diffusion couple is fcc austenite, and carbon diffuses rapidly, whereas silicon diffusion is extremely slow. The resulting carbon concentration profile is shown in Fig. 6, where the carbon is seen to have "unmixed," i.e., to have spontaneously adopted a nonhomogeneous concentration distribution. This cannot possibly be predicted or analyzed by use of Fick's law [Eq. (13)]. The reason for the unmixing lies in the thermodynamic effect of silicon on carbon: the addition of silicon to Fe–C alloys increases μ_C. The initial state of the diffusion couple contained equal carbon *concentrations* but a higher *activity* in the half containing silicon. In fact, the system shown in Fig. 6 is moving toward the equilibrium state in which μ_C would be everywhere the same.

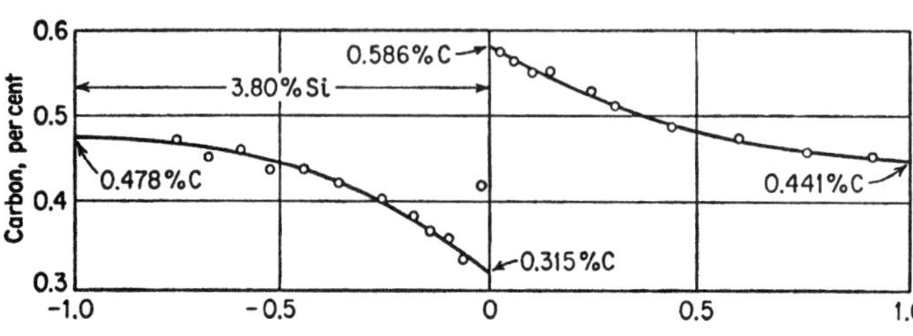

FIG. 6. Distribution of carbon in Fe–Si–C infinite diffusion couple, after 13 days at 1050 °C. Abscissa is fractional position. After Darken. Reprinted with permission from *Trans. Metall. Soc.* **180,** p. 43 (1949).

Solid-solution thermodynamic models are usually expressed in terms of mole fractions, and Eq. (52) is therefore transformed into

$$\bar{\mu}_{ij} = \frac{\partial(\mu_i - \mu_j)}{\partial X_j} = C\mu_{ij}. \tag{55}$$

In the case of equal partial molar volumes, it can be shown (Miller *et al.*, 1986; Kirkaldy and Purdy, 1969) that the D matrix, as determined on a mole fraction scale, is given by the matrix product

$$C[D] = [L][\bar{\mu}] \tag{56}$$

where the $\bar{\mu}$ matrix is symmetric. A useful model for dilute solution (solvent component n) was provided by Wagner (1952):

$$\ln \gamma_i = \ln \gamma_i^0 + \sum_{j=1}^{n-1} \epsilon_{ij} X_j, \tag{57}$$

where γ_i^0 is the value of the activity coefficient at infinite dilution and ϵ_{ij} are the "interaction parameters." Combination of Eqs. (47), (55), and (57) leads to

$$\bar{\mu}_{ij} = RT\epsilon_{ij} \qquad (i \neq j), \qquad \epsilon_{ij} = \epsilon_{ji}; \tag{58}$$

$$\bar{\mu}_{ii} = RT[1/X_i + \epsilon_{ii}]. \tag{59}$$

For a ternary system, the two necessary flux equations are found in Eq. (53), the coefficients of which are evaluated from Eqs. (56) and (59) as

$$\begin{aligned}
D_{11} &= RT\{L_{11}(\epsilon_{11} + 1/X_1) + L_{12}\epsilon_{21}\},\\
D_{12} &= RT\{L_{11}\epsilon_{12} + L_{12}(\epsilon_{22} + 1/X_2)\},\\
D_{21} &= RT\{L_{22}\epsilon_{21} + L_{21}(\epsilon_{11} + 1/X_1)\},\\
D_{22} &= RT\{L_{22}(\epsilon_{22} + 1/X_2) + L_{21}\epsilon_{12}\}.
\end{aligned} \tag{60}$$

In an ideal solution all $\epsilon_{ij} = 0$, and the D_{ij} reduce to the purely kinetic form

$$D_{ij} = RTL_{ij}/X_j. \tag{61}$$

For real solutions, if no kinetic cross effects occur, i.e., $L_{ij}(i \neq j) = 0$, it is clear that the diffusional cross terms are nonzero. In this case the dilute-solution limit (X_1, $X_2 \to 0$) may be described by

$$D_{12}/D_{11} = X_1\epsilon_{12}/(1 + \epsilon_{11}X_1), \tag{62}$$

and

$$D_{21}/D_{22} = X_2\epsilon_{21}/(1 + \epsilon_{22}X_2). \tag{63}$$

Thus the ternary coefficients are determined uniquely by the binary ones in dilute solutions.

5.2 Interstitial and Substitutional Diffusion

For interstitial diffusion there are negligible correlations between sublattices so that the approximation $L_{ij}(i \neq j) = 0$ is valid. Practical examples are steels Fe–C–M in which M is a substitutional metal (Si, Mn, Ni, Cr, Mo, Co) and the carbon is interstitial. The variation of D_{CM}/D_{CC} with carbon concentration is shown in Fig. 7, compared with the predictions of Eq. (62) using independently measured interaction parameters (Brown and Kirkaldy, 1964). Agreement is quite good, and the error of up to about 10% could be due to inaccurate values of the interaction parameters.

In considering diffusion in substitutional alloys, it is normal to use a volume-fixed frame of reference. If the partial molar volumes can be adequately approximated as

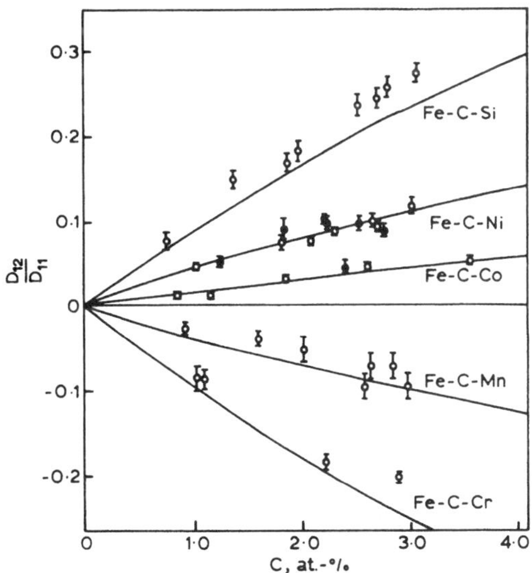

FIG. 7. Variation of D_{12}/D_{11} in Fe–C–M alloys. The solid lines show the thermodynamic prediction. After Brown and Kirkaldy. Reprinted with permission from *Trans. Metall. Soc.* **230**, p. 223 (1964), a publication of the Minerals, Metals & Materials Society, Warrendale, PA.

equal, then the frame of reference is simply defined by

$$\sum_{i=1}^{n} J_i = 0. \tag{64}$$

Physically, this corresponds to recognition of the fact that diffusion of a particular species from a solid region does not leave void space behind: rather another species will diffuse into it. It is recognized that this conservation law leads directly to a diffusional cross effect: movement of one component leads necessarily to the movement of others. The process can be visualized as shown in Fig. 8, where it is seen that while vacancies are necessarily involved, they function simply as intermediaries. The kinetics of this process is described using absolute-rate theory applied to the interchange of any two species i and j, thereby arriving at estimates of L_{ij}.

The applicability of this description to experimental data for ternary alloys has been discussed by Kirkaldy and Young (1987), Purdy (1990), and Dayananda (1991, 1992). However, to understand measured data, it is necessary to relate the experimental (laboratory) reference frame with the lattice frame used in theoretical discussions. The physical manifestation of the difference between the two frames is known as the Kirkendall effect.

5.3 Kirkendall Effect

The discussion commences with a substitutional binary alloy AB in which diffusion occurs via atom-vacancy exchanges, and Eq. (29) applies to both A and B. It is seen that these individual fluxes are in general not equal and opposite. If, for example, $D_A > D_B$ in a sample initially A rich on the left, there will be an excess flux of A from left to right over B atoms moving to the left. Thus the diffusion zone as a whole drifts to the left, compensating for the accumulation of matter and hydrostatic pressure that would otherwise occur on the right. Since the diffusion zone is generally a small part of a larger sample, measurements of position that are referred to the end of the sample (the laboratory reference frame) are affected by this drift, and so, in consequence, is the estimate of diffusion rate. This is dealt with by relating the two frames of reference.

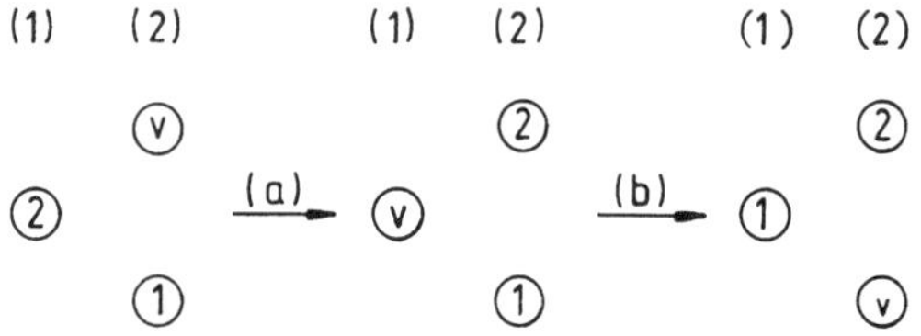

FIG. 8. Site-exchange model for diffusion in $(A,B)O$ with fixed anion sublattice. Species shown are the two different cations and a vacancy.

In the laboratory frame, Eq. (64) applies. However, in the lattice frame [to which Eq. (29) refers] it does not. We therefore denote this latter frame by primes, observing that

$$J'_A + J'_B = -J'_v. \tag{65}$$

If the lattice frame moves with respect to the laboratory frame with a velocity v, then

$$J_i = J'_i + C_i v, \qquad i = A,B, \tag{66}$$

where the nonprimed fluxes refer to the laboratory frame. These equations are solved using (64) to obtain

$$v = -\frac{J'_A + J'_B}{C_A + C_B}, \tag{67}$$

or, upon resubstitution,

$$J_A = -J_B = X_B J'_A - X_A J'_B. \tag{68}$$

In the simple situation in which the off-diagonal Onsager coefficients are set equal to zero, and local equilibrium applies ($\nabla\mu_v = 0$), Eq. (13) applies. Since, moreover, for an isobaric system in which partial molar volumes $\bar{V}$ are equal,

$$\bar{V}(C_A + C_B) = 1, \tag{69}$$

combination of Eqs. (67), (68), (69), and (13) yields

$$v = \bar{V}(D_A - D_B)\nabla C_A, \tag{70}$$

$$J_A = -(X_B D_A + X_A D_B)\nabla C_A. \tag{71}$$

The value of v can be measured using inert markers.

The first demonstration of lattice drift was performed by Smigelskas and Kirkendall (1947) using the diffusion arrangement shown schematically in Fig. 9. Annealing caused rapid outward diffusion of the zinc and inward drift of the molybdenum markers. The effect is quite general and is widely used in diffusion measurements.

For an infinite diffusion couple (sample much larger than the diffusion zone), it can be shown that

$$C_A = C_A(\lambda), \qquad \lambda = x/t^{1/2}, \tag{72}$$

and hence

$$v = \frac{D_A - D_B}{t^{1/2}}\frac{dC_A}{d\lambda}. \tag{73}$$

Because the markers are located at a point of fixed composition and therefore at a fixed value of $dC_A/d\lambda$, Eq. (73) integrates immediately to yield

$$x_m = 2(D_A - D_B)\frac{dC_A}{d\lambda}t^{1/2} \tag{74}$$

for the marker displacement.

The quantities D_A, D_B are known as the *intrinsic diffusion coefficients* because they refer to diffusion with respect to the lattice planes in the presence of an activity gradient. It is necessary now to relate these to the measured *tracer* coefficients, D_{A^*}, D_{B^*}. Recognizing that the tracer coefficient refers to an ideal solution for which Eq. (14) applies and that the intrinsic coefficient refers to a solid solution for which Eq. (48) applies, it is found that

$$D_A = D_{A^*}\left(1 + \frac{d\ln\gamma_A}{d\ln X_A}\right). \tag{75}$$

Since from the Gibbs–Duhem equation for equilibrium in a solution

$$\sum X_i d\mu_i = 0, \tag{76}$$

we may write

$$1 + \frac{d\ln\gamma_A}{d\ln X_A} = 1 + \frac{d\ln\gamma_B}{d\ln X_B}; \tag{77}$$

then Eq. (71) becomes

$$J_A = -\tilde{D}\nabla X_A, \tag{78}$$

with

$$\tilde{D} = (X_B D_{A^*} + X_A D_{B^*})\left(1 + \frac{d\ln\gamma}{d\ln X}\right). \tag{79}$$

This is the Darken–Hartley–Crank equation, and $\tilde{D}$ is the *chemical diffusion coefficient* (Darken, 1948; Hartley and Crank, 1949). The quantity $\tilde{D}$ is obtained from a diffusion-couple measurement, and if markers are used values of D_A, D_B may also be obtained. If the thermodynamic factor in Eq. (77) is known, then the validity of Eq. (79) can be verified. It has been found to be rigorously true for dilute solution but slightly inaccurate for rich solutions in which the correlation effects suggested by Manning (1968) may have some importance.

The above analysis has been extended to multicomponent systems (see, e.g., Kirkaldy and Young, 1987). The lack of balance among the intrinsic diffusive flows always leads to a compensating mass flow of material. That is to say, diffusional cross effects arise even in the absence of kinetic or thermodynamic correlations. Thus even a component with a negligible intrinsic mobility will move. The simple form of Fick's law fails, and the generalized form (53) must be used.

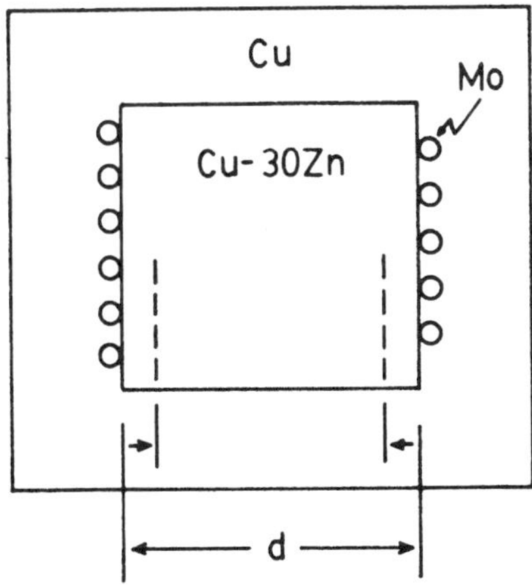

FIG. 9. Schematic representation of diffusion couple used to demonstrate lattice drift.

6. DIFFUSION IN IONIC SOLIDS

Self-diffusion in ionic solids is discussed elsewhere in DIFFUSION AND IONIC CONDUCTION IN SOLIDS. Attention will be confined here to chemical diffusion in the presence of activity gradients. This is technologically important, e.g., in the manufacture of ceramic materials and in the growth of oxide scales on alloys. Discussion is facilitated by the use of relative building units (Kroger *et al.*, 1959). These are used to provide a link between macroscopic thermodynamic/kinetic properties and the point-defect structure, thus allowing access to the phenomenological equations of irreversible thermodynamics.

6.1 Relative Building Units

We employ the defect notation of Kroger and Vink (1956) in which a lattice species is represented by the symbol S_m^*. Here the subscript represents normal occupancy in a perfect crystal, the principal symbol represents the species actually present, and the superscript represents the charge of the species relative to normal site occupancy: a prime for negative, a dot for positive, and a cross for zero charge. In a ternary substitutional oxide (A,B)O containing fully ionized cation vacancies and positive holes as the only defects, the species constituting the solid are

$$A_m^{\times}, B_m^{\times}, V_m'', \mathrm{O}_0^{\times}, h^{\bullet}, \tag{80}$$

and thus outnumber the thermodynamic compositional variables (two independent variables for a three-component compound).

The electrochemical potential of each species is defined through Eq. (9). However, this is not operationally useful as neither η nor ψ can be measured locally. This is overcome by grouping species in charge-neutral sets. If these sets are chosen so as to preserve the cation/anion ratio characteristic of the solid, they are building units. Obvious units for (A,B)O are $\{A_m^{\times} + \mathrm{O}_0^{\times}\}$, $\{\mathrm{O}_0^{\times} + V_m'' + 2h^{\bullet}\}$, etc. A subset of building units is made up of relative building units that consist of the difference between two lattice species. Thus they represent compositional change arising from the replacement of one species with another and may therefore be used to describe diffusion (Young and Kirkaldy, 1984). A flux of units $\{A_m^{\times} + V_m^{\times}\}$ corresponds to diffusion of A via a vacancy mechanism; a flux of $\{A_m^{\times} - B_m^{\times}\}$ corresponds to interdiffusion.

Since relative building units conserve sites and charge, their movement necessarily satisfies the flux constraints appropriate to diffusion in the absence of an external field: Eq. (37) together with charge conservation

$$2J_{V''} = J_{h^{\bullet}}, \tag{81}$$

where the fluxes are measured in a solvent-fixed reference frame provided by an immobile anion sublattice. The constraints (37) and (81) serve both to reduce the transport equation set (8) and to eliminate dependences among the species set (80). Thus, for example, movement of a vacancy must be accompanied by movement of positive holes and an opposing cation flux. The relative building units that describe these exchanges are

$$\{A_m^{\times} - V_m'' - 2h^{\bullet}\}, \{B_m^{\times} - V_m'' - 2h^{\bullet}\}, \{A_m^{\times} - B_m^{\times}\},$$

and the third unit is merely a linear combination of the first two. The oxygen sublattice is assumed to be immobile but can still play a chemical role. The appropriate unit is $\{\mathrm{O}_0^{\times} + V_m'' + 2h^{\bullet}\}$. It is seen that combination in the appropriate proportions of the above units yields a solid $(A,B)_{1-\delta}$O of any desired degree of substitution or nonstoichiometry.

Individual lattice species lack thermodynamic significance, but a precise meaning attaches to the building units that they constitute. This is seen from the reactions leading to the introduction of defects:

$$\tfrac{1}{2}\mathrm{O}_2(g) = \mathrm{O}_0^{\times} + V_m'' + 2h^{\bullet}, \tag{82}$$

$$A(g) + V_m'' + 2h^{\bullet} = A_m^{\times}, \tag{83}$$

$$B(g) + V_m'' + 2h^{\bullet} = B_m^{\times}, \tag{84}$$

the Gibbs equilibrium equations for which yield

$$\tfrac{1}{2}\mu_{\mathrm{O}_2} = \eta(\mathrm{O}_0^{\times}) + \eta(V_m'') + 2\eta(h^{\bullet}), \tag{85}$$

$$\mu_A = \eta(A_m^{\times}) - \eta(V_m'') - 2\eta(h^{\bullet}), \tag{86}$$

$$\mu_B = \eta(B_m^{\times}) - \eta(V_m'') - 2\eta(h^{\bullet}). \tag{87}$$

Thus the relative building units have potentials corresponding to the chemical potentials of the constituent elements. Appropriate grouping of the individual species (80)—for which kinetic models can be developed—has dealt with the redundancy of the description while preserving the detailed microscopic description.

6.2 Reduction of Transport Equations

Reduced sets of phenomenological equations are found from the entropy-source expression

$$-T\sigma = J_A\nabla\eta_A + J_B\nabla\eta_B + J_v\nabla\eta_v + J_h\nabla\eta_h \tag{88}$$

by applying the constraints (37) and (81). Inspection of the latter reveals that five distinguishable pairs of species can be chosen for elimination, leading to the results

$$\begin{aligned}-T\sigma &= J_A\nabla\{\eta_A - \eta_v - 2\eta_h\} \\ &\quad + J_B\nabla\{\eta_B - \eta_v - 2\eta_h\}; \\ &= J_i\nabla\{\eta_i - \eta_j\} + J_h\nabla\{\tfrac{1}{2}\eta_v + \eta_h - \tfrac{1}{2}\eta_j\}, \\ &\qquad\qquad i,j = A,B; \\ &= J_i\nabla\{\eta_i - \eta_j\} + J_v\nabla\{\eta_v + 2\eta_h - \eta_j\}, \\ &\qquad\qquad i,j = A,B.\end{aligned} \tag{89}$$

The entropy source term has a unique value, and these expressions are equivalent. Using the defect equilibria (86), (87), and the flux constraint equations, it is readily demonstrated that each of Eqs. (89) is equivalent to

$$-T\sigma = J_A\nabla\mu_A + J_B\nabla\mu_B, \tag{90}$$

where now the chemical potentials are those of the elements.

The above procedure removes the redundancy in the species set (80). The flux equations arising from Eq. (90) are

$$\begin{aligned}J_A &= -L_{AA}\nabla\mu_A - L_{AB}\nabla\mu_B, \\ J_B &= -L_{BA}\nabla\mu_A - L_{BB}\nabla\mu_B.\end{aligned} \tag{91}$$

Because isothermal diffusion experiments conducted in the absence of an external field disclose nothing but mass transfer rates, the description is complete. It is seen that the use of relative building units is consistent with the fact that the thermodynamics and diffusion kinetics of ionic crystals can always be described in terms of elemental chemical potentials. Information on correlations among the diffusing species can be recovered from further consideration of (89).

The relative building units involving A revealed in Eqs. (89) are

$$\{A_m^\times - B_m^\times\}, \qquad \{A_m^\times + V_m'' - 2h^\bullet\}.$$

The former unit corresponds to diffusion via an exchange mechanism involving vacancies as intermediaries, but with $J_v = 0$, as illustrated in Fig. 8. A strict correlation arises through the fact that the ions share the use of the same vacancies. The second unit describes counter-current flow of material and vacancies and expresses the strict correlation with free-carrier movement. This mechanism can operate only when a gradient of vacancies is available, the effect upon which of an oxygen potential is explicit through Eq. (82). A linear combination of the fluxes arising from the movement of these units represents a complete accounting for diffusion of A, and similarly for B:

$$\begin{aligned}J_A &= -l_1\nabla(\mu_A - \mu_B) - l_2\nabla\mu_A, \\ J_B &= -\lambda_1\nabla(\mu_B - \mu_A) - \lambda_2\nabla\mu_B.\end{aligned} \tag{92}$$

Rearrangement and comparison with (91) yields

$$\begin{aligned}L_{AA} &= (l_1 + l_2), \qquad L_{BB} = \lambda_1 + \lambda_2, \\ L_{AB} &= l_1, \qquad L_{BA} = \lambda_1.\end{aligned} \tag{93}$$

Because l_1 and λ_1 refer to the exchanges $\{A^*–B^*\}$ and $\{B^*–A^*\}$, they must be identical, and symmetry of $[L]$ is assured. Similar treatments of more complex examples have been presented by Young *et al.* (1979, 1990) and Kirkaldy and Young (1987).

A practical example of the application of this methodology is the description of solid-solution oxide scale growth on alloys (Young and Gesmundo, 1988). The rate of scale growth supported by cation diffusion is expressed in terms of scale thickness, x_s:

$$\frac{dx_s}{dt} = V\sum_i J_i, \tag{94}$$

where V is the volume of oxide formed per mole of metal. Oxide diffusion data have been deduced from a number of binary-alloy oxidation experiments (Narita *et al.*, 1982). The type of data and the degree of success are illustrated in Fig. 10.

The principal value of this approach lies in equations of the type shown in (93), whereby microscopic kinetic evaluations of the l_i lead to estimates of L_{ij} and hence the D_{ij}. In this way it can be shown that for substitutional cation interdiffusion in (A,B)O when $\nabla\mu_0 = 0$,

$$\tilde{D} = \frac{D_{A^*}D_{B^*}}{C_A D_{A^*} + C_B D_{B^*}}(C_A + C_B) \qquad (95)$$

for the case of ideal or Henrian (see Glossary) solution. The description is easily extended to the quaternary compound (A,B,C)O. An analogous expression (79) was found for binary alloys to result from the differing intrinsic mobilities, which led to compensating bulk material flow. No such effect is possible in ionic solids, in which the anion lattice is immobile. It is instead the electrostatic potential or field (developed by charge separation within the solid) that brings into balance the fluxes of charged species having different mobilities.

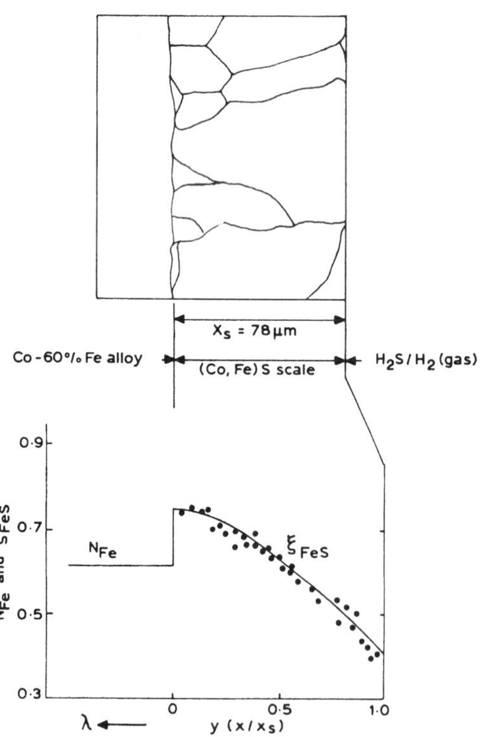

FIG. 10. (Co,Fe)S scale formed on Co–60Fe alloy at 600 °C and $p_{S2} = 2 \times 10^{-6}$ atm in 1 h. Compositions measured by electron probe microanalysis (ξ_{FeS} denotes mole fraction of FeS), and continuous line shows best fit to diffusion model. After Young *et al.* (1980). Reprinted with permission from the Electrochemical Society.

7. EXPERIMENTAL METHODS

In the most common diffusion measurements, the movement of a system toward homogeneity is observed and compared with the predictions of the diffusion equations. These equations, together with appropriate boundary conditions, yield solutions for the one-dimensional case of the general form

$$C_i = C_i(x,t,D). \qquad (96)$$

Thus D is evaluated by fitting the expressions to experimental data $C_i = C_i(x,t)$. In these diffusion couple experiments, two different mixtures are brought into contact at a planar interface, and diffusion is observed along the direction normal to that interface. The method is appropriate to diffusion in the presence of a concentration gradient (chemical diffusion) and is also used for self-diffusion studies where the only gradient is that of a radiotracer.

Measurements can also be carried out on homogeneous materials by observing the interaction of the solid's constituent particles with a transient perturbation, or their subsequent relaxation to a lower energy state. Interaction or relaxations that occur via translational motion (or are thereby modified) display a time dependence determined by particle mobilities, and their measurement leads to a determination of D. Because individual particle events occur over very short ranges, these techniques are valuable for measuring very slow self-diffusion. However, because they are not sensitive to position within a sample, they cannot reveal gradient effects and will not be described here. A general review of these techniques is provided by Philibert (1991). For a more extensive review of the important nuclear magnetic resonance technique applied to solids, see Stokes (1984).

Two types of diffusion couple are important. If sample dimensions and the period of diffusion are such that concentrations at the ends of the sample do not change, then the experiment is described as an *infinite* diffusion couple, which is used to measure chemical diffusion. In a tracer diffusion measurement, the couple consists of a thin film of isotopically labeled material on the end of a long bar of unlabeled but compositionally identical material. In both cases the experiment involves preparation of the couple (while avoiding diffusion), an annealing period at constant temperature in a controlled environment, subsequent analysis of the concentration profiles, and calculation of the diffusion coefficients. Experimental aspects have been reviewed in some detail by Dunlop *et al.* (1992).

7.1 Calculation of Diffusion Coefficients

Predicted profiles of the form (96) are obtained from Fick's law [Eq. (13)], which is subject to the continuity condition

$$\frac{\partial C_i}{\partial t} = -\frac{\partial J_i}{\partial x}, \tag{97}$$

leading to Fick's second law

$$\frac{\partial C_i}{\partial t} = D\frac{\partial^2 C_i}{\partial x^2}, \tag{98}$$

where D has been approximated as constant. The solution of (98) is required for the appropriate boundary and initial conditions. Methods and a number of solutions are available from Carslaw and Jaeger (1959) and Crank (1970).

The thin-film solution applies to the tracer diffusion experiment:

$$C(x,t) = \frac{M\exp(-x^2/4D_{A*}t)}{2(\pi D_{A*}t)^{1/2}}, \tag{99}$$

where M is the amount of labeled material. With knowledge of the annealing time t, D_{A*} is evaluated from a logarithmic plot according to this equation.

For an infinite diffusion couple consisting initially of one half containing a uniform concentration C_0 and the other a concentration C_1, after diffusion time t we have

$$\frac{C(x,t) - C_0}{C_1 - C_0} = \tfrac{1}{2}[1 - \operatorname{erf}(x/2\sqrt{Dt})], \tag{100}$$

where erf is the Gaussian error function,

$$\operatorname{erf} z = \frac{2}{\sqrt{\pi}}\int_0^z \exp(-u^2)\,du. \tag{101}$$

Corresponding solutions are available for ternary systems (Kirkaldy 1958a,b). These solutions rely on D being constant, an improbable event in the presence of a concentration gradient. Instead, the analysis of Boltzmann (1894) and Matano (1933) must be used. Here the new variable $\lambda = x/\sqrt{t}$ is introduced, and the final solution is

$$\tilde{D}(c') = -\tfrac{1}{2}\left(\frac{d\lambda}{dc}\right)\int_0^{c'} \lambda\,dc, \tag{102}$$

with

$$\int_{C_0}^{C_1} x\,dc = 0 \tag{103}$$

defining the origin of coordinates. Graphical or numerical evaluations of the differential and the integral in (102) are used to evaluate $\tilde{D}(c')$. Errors are particularly large as $C \rightarrow C_0$ or C_1. For this reason, the method is best suited to diffusion zones of substantial compositional width. For narrow compositional widths, it is often better instead to approximate $\tilde{D}$ as constant and use error function solutions (see, e.g., Jan *et al.*, 1991).

For multicomponent systems where the diffusion equations (53) hold, the Matano analysis is extended (Kirkaldy, 1957) for the infinite diffusion couple:

$$\frac{1}{2}\int_{C_i^0}^{C_i} \lambda\,dC_i = -\sum_j^{n-1} D_{ij}\left.\frac{dC_j}{d\lambda}\right|_{C'}, \tag{104}$$

with the origin defined by

$$\int_{C_i^0}^{C_i^1} x\,dC_i = 0, \tag{105}$$

the latter condition being simultaneously satisfied for all components. Data on diffusion in ternary metallic solutions have been collected and reviewed by Kirkaldy and Young (1987) and Dayananda (1991). The difficulty of constructing couples that will satisfy Eq. (105) has prevented extension to higher-order systems.

This discussion has been based on the assumption that diffusion occurs by a random process. In real materials, certain paths such as dislocations or grain boundaries may be favored, and this will be apparent in distortions of the concentration profile. A good discussion is provided by Kaur and Gust (1989).

GLOSSARY

Building Unit: Group of lattice species that, when added to or taken from a crystal, preserve its cation/anion site ratio and its electroneutrality.

Chemical Potential: Partial molar free energy.

Correlation Coefficient: A number reflecting the degree of correlation (or departure from the purely random) between successive diffusive jumps of a tracer atom or ion.

Cross Effect: Effect of the potential gradient of one component on the diffusive flux of another.

Henrian Solution: Dilute solution in which the activity coefficient of solute is independent of concentration.

Interaction Parameter: The numerical value of the differential of the logarithm of an activity coefficient with respect to the concentration of a different component.

Mobility: Velocity of a particle moving under the influence of a field having unit magnitude.

Nonstoichiometry: Deviation in the composition of a compound from the ideal stoichiometric ratio of components, the deviation being accommodated by the presence of vacancies or interstitials. For example, $Fe_{1-\delta}O$ contains high concentrations of cation vacancies.

Off-Diagonal Terms: Terms in phenomenological diffusion equations expressing dependence of a given component flux on the potential gradients of other components. Also known as *cross terms*.

Phenomenological Equation: Equation stating linear dependence of each component flux on the chemical potential gradients of *all* components.

Point Defects: Lattice sites that differ in their occupancy or charge level from the situation in a perfect crystal.

Tracer: Radioactive isotope used to enable observation of particle movement and hence diffusion.

Works Cited

Allnatt, A. R., Allnatt, E. L. (1984), *Philos. Mag.* **A 49,** 625.

Allnatt, A. R., Okamura, Y. (1984), in: G. E. Murch, H. K. Birnbaum, J. R. Cost (Eds.), *Non-traditional Methods in Diffusion,* Warrendale, PA: AIME.

Askill, J. (1970), *Tracer Diffusion Data for Metals, Alloys and Simple Oxides,* New York: IFI/Plenum.

Bocquet, J. L., Brebec, G., Limoge, Y. (1983), in: R. W. Cahn, P. Haasen (Eds.), *Physical Metallurgy,* Part 1, Amsterdam: North Holland.

Boltzmann, L. (1894), *Ann. Phys. (Leipzig)* **53,** 90.

Brandes, E. A. (Ed.) (1983), *Smithells Metal Reference Book,* 6th ed., London: Butterworths.

Brown, A. M., Ashby, M. F. (1980), *Acta Metall.* **28,** 1085.

Brown, L. C., Kirkaldy, J. S. (1964), *Trans. AIME* **230,** 223.

Cahoon, J. R., Sherby, O. D. (1992), *Metall. Trans. A,* **23A,** 2491.

Carslaw, H. S., Jaeger, J. C. (1959), *Conduction of Heat in Solids,* Oxford: Clarendon Press.

Crank, J. (1970), *The Mathematics of Diffusion,* Oxford: University Press.

Darken, L. S. (1948), *Trans. AIME* **175,** 184.

Darken, L. S. (1949), *Trans. AIME* **180,** 430.

Dayananda, M. A. (1991), *Diffusion in Metals and Alloys,* Landolt-Bornstein, Berlin: Springer.

Dayananda, M. A. (1992), *Defect Diffus. Forum* **83,** 73.

de Groot, S. R., Mazur, P. (1962), *Non-Equilibrium Thermodynamics,* Amsterdam: North Holland.

Dignam, M. J., Young, D. J., Goad, D. W. G. (1973), *J. Phys. Chem. Solids* **34,** 1227.

Dunlop, P. J., Harris, K. R., Young, D. J. (1992), in: B. W. Rossiter, R. C. Baetzold (Eds.), *Physical Methods of Chemistry,* 2nd ed., Vol. 6, New York: Wiley.

Einstein, A. (1905), *Ann. Phys. (Leipzig)* **17,** 549.

Einstein, A. (1906), *Ann. Phys. (Leipzig)* **19,** 371.

Fick, A. E. (1855), *Ann. Phys. Chem.* **94,** 59.

Freer, R. (1980), *J. Mater. Sci.* **15,** 803.

Friauf, R. J. (1957), *Phys. Rev.* **105,** 843.

Hartley, G. S., Crank, J. (1949), *Trans. Faraday Soc.* **45,** 801.

Heumann, T. (1979), *J. Phys. F* **9,** 1997.

Howard, R. E., Lidiard, A. B. (1964), *Rep. Prog. Phys.* **27,** 161.

Jan, C.-H., Svenson, D., Zheng, X.-Y., Lin, J.-C., Chang, Y. A. (1991), *Acta Metall.* **39,** 303.

Kaur, I., Gust, W. (1989), *Fundamentals of Grain and Interphase Boundary Diffusion,* 2nd ed., Stuttgart: Ziegler Press.

Kidson, G. V. (1978), *Philos. Mag.* **37,** 305.

Kikuchi, R., Sato, H. (1969), *J. Chem. Phys.* **51,** 161.

Kirkaldy, J. S. (1957), *Can. J. Phys.* **35,** 435.

Kirkaldy, J. S. (1958a), *Can. J. Phys.* **36,** 899.

Kirkaldy, J. S. (1958b), *Can. J. Phys.* **36,** 907.

Kirkaldy, J. S., Purdy, G. R. (1969), *Can. J. Phys.* **47,** 865.

Kirkaldy, J. S., Young, D. J. (1987), *Diffusion in the Condensed State,* London: Institute of Metals.

Kirkaldy, J. S., Young, D. J., Lane, J. E. (1987), *Acta Metall.* **35,** 1273.

Kroger, F. A., Stieltjes, F. H., Vink, H. J. (1959), *Philips Res. Rep.* **14,** 557.

Kroger, F. A., Vink, H. J. (1956), *Solid State Phys.* **3,** 307.

Kubo, R. (1957), *J. Phys. Soc. Jpn.* **12,** 570.

LeClaire, A. D. (1975), in: N. B. Hannay (Ed.), *Treatise in Solid-State Chemistry,* Vol. 4, *Reactivity of Solids,* New York: Plenum.

Lidiard, A. B. (1955), *Philos. Mag.* **46,** 1218.

Manning, J. R. (1968), *Diffusion Kinetics for Atoms in Crystals,* Princeton: Van Nostrand.

Matano, C. (1933), *Jpn. J. Phys.* **8,** 109.

Mehrer, H. (1990), *Diffusion in Metals and Alloys,* Landolt-Bornstein New Series, Berlin: Springer.

Miller, D. G., Vitagliano, V., Sartorio, R. (1986), *J. Phys. Chem.* **90,** 1509.

Mott, N. F., Gurney, R. W. (1964), *Electronic Processes in Ionic Crystals,* New York: Dover.

Murch, G. E. (1984), in: G. E. Murch, A. S. Nowick (Eds.), *Diffusion in Crystalline Solids,* Orlando, FL: Academic Press.

Murch, G. E. (1991), in: R. W. Cahn, P. Haasen, E. J. Kramer (Eds.), *Materials Science and Technology,* Vol. 5, *Phase Transformations in Materials,* New York: VCH.

Narita, T., Nishida, K., Smeltzer, W. W. (1982), *J. Electrochem. Soc.* **129,** 209.

Onsager, L. (1931), *Phys. Rev.* **37,** 405; **38,** 2265.

Philibert, J. (1991), *Atom Movements Diffusion and Mass Transport in Solids,* Courtaboeuf, France: Les Editions de Physique.

Prigogine, I. (1967), *Introduction to the Thermodynamics of Irreversible Processes,* New York: Wiley-Interscience.

Purdy, G. (Ed.) (1990), *Fundamentals and Applications of Ternary Diffusion,* New York: Pergamon.

Sato, H. (1989), in: A. R. Laskar, S. Chandra (Eds.), *Superionic Conductors,* New York: Academic Press.

Sato, H., Zhang, H. (1993), in: B. Fultz, R. W. Cahn, D. Gupta (Eds.), *Diffusion in Ordered Alloys,* Warrendale, PA: TMS.

Schottky, W. (1958), *Halbleiterprobleme* **4,** 235.

Seitz, F. (1950), *Acta Crystallogr.* **3,** 355.

Sherby, O. D., Robbins, J. L., Goldberg, A. (1973), *J. Phys. Chem. Solids* **34,** 1025.

Shewmon, P. G. (1963), *Diffusion in Solids,* New York: McGraw-Hill.

Smigelskas, A., Kirkendall, E. (1947), *Trans. AIME* **171,** 130.

Smoluchowski, M. von (1908), *Ann. Phys. (Leipzig)* **25,** 205.

Stokes, H. T. (1984) in: G. E. Murch, H. K. Birnbaum, J. R. Cost (Eds.), *Nontraditional Methods in Diffusion,* Warrendale, PA: AIME.

Wagner, C. (1952), *Thermodynamics of Alloys,* Reading, MA: Addison-Wesley.

Wert, C., Zener, C. (1950), *J. Appl. Phys.* **21,** 5.

Young, D. J., Gesmundo, F. (1988), *Oxid. Met.* **29,** 169.

Young, D. J., Kirkaldy, J. S. (1984), *J. Phys. Chem. Solids* **45,** 781.

Young, D. J., Delamotte, E., Kirkaldy, J. S. (1979), *Can. J. Phys.* **57,** 722.

Young, D. J., Narita, T., Smeltzer, W. W. (1980), *J. Electrochem. Soc.* **127,** 679.

Young, D. J., Dorward, R. C., Kirkaldy, J. S. (1990), in: G. Purdy (Ed.), *Fundamentals and Applications of Ternary Diffusion,* New York: Pergamon.

Further Reading

Barrer, R. M. (1951), *Diffusion in and through Solids,* Cambridge, U.K.: Cambridge Univ. Press.

Crank, J. (1970), *The Mathematics of Diffusion,* Oxford: Oxford Univ. Press.

Denbigh, K. G. (1951), *The Thermodynamics of the Steady State,* London: Methuen.

Flynn, C. P. (1972), *Point Defects and Diffusion,* Oxford: Clarendon Press.

Kirkaldy, J. S., Young, D. J. (1987), *Diffusion in the Condensed State,* London: Institute of Metals.

Kofstad, P. (1972), *Nonstoichiometry, Diffusion and Electrical Conductivity in Binary Metal Oxides,* New York: Wiley-Interscience.

Kroger, F. A. (1973), *The Chemistry of Imperfect Crystals,* 2nd ed., Amsterdam: North Holland.

Manning, J. R. (1968), *Diffusion Kinetics for Atoms in Crystals,* Princeton, NJ: Van Nostrand.

Mrowec, S. (1980), *Defects and Diffusion in Solids,* Amsterdam: Elsevier.

Philibert, J. (1991), *Atom Movements Diffusion and Mass Transport in Solids,* Courtaboeuf, France: Les Editions de Physique.

Porter, D. A., Easterling, K. E. (1981), *Phase Transformations in Metals and Alloys,* Wokingham, Berks: Van Nostrand Reinhold.

Prigogine, I. (1955), *Introduction to the Thermodynamics of Irreversible Processes,* Springfield, IL: Charles C. Thomas.

Shewmon, P. G. (1963), *Diffusion in Solids,* New York: McGraw-Hill.

SOLIDS, DYNAMIC HIGH-PRESSURE EFFECTS IN

W. J. NELLIS, *Condensed Matter Physics Division and Institute of Geophysics and Planetary Physics, Lawrence Livermore National Laboratory, University of California, Livermore, California, U.S.A.*

INTRODUCTION

Dynamic high pressures are applied to materials to increase density and temperature, to alter crystal structure and microstructure, and to change physical and chemical properties. These effects are achieved at high pressures, and many are retained on release from high pressures. This article treats high pressures achieved dynamically by shock compression. In fact, the terms dynamic and shock are used interchangeably to describe high pressures. Shock compression is achieved by high-velocity impact, laser pulses, and explosives, for examples. Typically, shock pressure pulses have a duration of ~1 μs, although other pulse lengths can be achieved depending on how the dynamic pressure is achieved. Because of their fast adiabatic nature, dynamic high pressures cause heating of the specimen, both homogeneous and heterogeneous depending on

3-527-28140-1/96/$5.00 + .50

the nature of the specimen and the magnitude of the pressure. The rate of application of pressure is very high, which causes very high strain rates and the generation of very high densities of lattice defects. Shock pressures up to several hundred GPa (1 Mbar) and temperatures up to several thousand K are common. On release of pressure, very high quench rates up to 10^{12} bar/s and 10^9 K/s are achieved, which can cause retention of high-pressure phases at ambient conditions. These quench rates are the maxima in macroscopic specimens because they occur at the speed of sound. Materials made using dynamic high pressures are fine grained because of the fast time scale. Dynamic high pressures can be applied to virtually arbitrary volume, depending on the method employed.

Thus, dynamic methods

1. achieve very high pressures, densities, temperatures, strain rates, and quench rates;
2. affect crystal structure, microstructure, and properties, often uniquely;
3. treat specimen volumes of arbitrary size;
4. are used both to investigate states of matter at high pressures and to produce unusual materials after recovery from high pressures; and
5. are used to make materials that are produced commercially.

A well-known example of the last is the synthesis of fine-grained diamond powder using explosives (Cowan *et al.*, 1968).

1. SCOPE AND BACKGROUND

1.1 General

The purpose of this article is to give the reader an overview of dynamic high-pressure techniques and examples of materials that are made using dynamic high pressures. Both are important in their own right, and they are interrelated. For example, determining what occurs at dynamic high pressures can guide experiments to recover materials from dynamic high pressures.

1.2 History

The outbreak of World War II stimulated the development of shock-compression research. The primary purpose was to obtain equation-of-state data to generate material models to design devices for national defense. These same equations of state and computer codes are now used to design and interpret a wide variety of interdisciplinary high-pressure experiments, which extend well beyond the initial goals. Examples include general effects of shock waves, the nature of condensed matter at extreme pressures and temperatures, and the structure and properties of materials recovered from high shock pressures. Computations are essential for interpreting and guiding experiments. As one example of multidisciplinary activity, dynamic high-pressure experiments provide laboratory simulations to understand the deep interiors of the Earth and the outer giant planets, which are at high internal pressures and temperatures, as well as the cratered surfaces of bodies of the solar system, which were impacted at high velocities (Syono and Manghnani, 1992).

1.3 Relation to Static High Pressures

Static high pressures are achieved with presses of various types and sizes. Static compression takes seconds or longer to achieve and lasts for longer times. The difference in time duration between static and dynamic methods causes substantial differences in temperature and microstructure. For example, static high pressures typically cause larger grain size and fewer lattice defects than dynamic pressures. Static high-pressure methods, including methods to heat specimens, are described in HIGH-PRESSURE TECHNIQUES and SOLIDS, STATIC HIGH-PRESSURE EFFECTS IN; descriptions are also given by Jayaraman (1986) and by Bundy (1988).

Briefly, for small-scale scientific research at extremely high pressures, a pair of diamond anvils forced against each other is used with suitable gasketing to confine the specimen. Sample dimensions are typically 50 to a few hundred μm in diameter and up to a few tens of μm thick. Pressures range from 1 to a few hundred GPa at room temperature. Specimens are heated either by raising the temperature of the entire cell or by laser heating the specimen locally in a spot about 10 μm in diameter. Much higher pressures are achieved with laser heating that ranges up to 5000 K. Large-volume multianvil presses use much larger samples for scientific research and for commercial applications. Sample dimensions are typically a millimeter up to a centimeter or more on edge. Maximum pressures range up to about 20 GPa at room temperature and decrease as the temperature of the apparatus is increased because strength decreases with increasing temperature. If the specimens are in bulk thermodynamic equilibrium at both dynamic and static high pressures, data from the two types of experiments are related by a thermal model. Because of this, primary pressure scales at static high pressures above about 7 GPa are derived from shock-wave data.

1.4 Scalability

Results obtained using small specimens in the laboratory can be scaled up in size using different dynamic high-pressure systems. If rate effects are not important, pressures and temperatures determined with a diamond cell could also be obtained in shock experiments with larger samples. Scaling to larger specimens using dynamic high pressures can be done using existing computational and experimental shock-wave technology.

2. METHODS TO GENERATE SHOCK PRESSURES

A shock is essentially a discontinuity in flow and thermodynamic variables that propagates in a specimen at supersonic velocity (see SHOCK WAVES). For a strong shock in dense matter, the discontinuous rise in pressure can occur on a subnanosecond time scale. For a shock amplitude comparable to material strength or in a porous specimen, the rise in pressure can be tens of nanoseconds or longer and have a multiwave structure, depending on material strength, phase transitions, and porosity. Since the shock pulse is applied for short times, dynamic pressures can only cause structural changes that occur during the time of the pressure pulse.

Dynamic high pressures are generated typically by hypervelocity impact, by rapid energy release from high explosives, or by rapid energy deposition by lasers. The impact method is illustrated in Fig. 1, in which a projectile is accelerated to velocities in the range 0.1 to 8 km/s with a two-stage light-gas gun. The corresponding pressures range from 0.1 GPa to a few 100 GPa. For velocities below about 2 km/s, a single-stage gun is sufficient. The pressure pulse is generated on impact of the projectile onto the target. Much higher pressures and temperatures in the plasma range can be generated with lasers than with the other methods. Explosive and laser-generated shock generators are described in SHOCK WAVES.

Typical shock samples are tens of millimeters in diameter and a few millimeters thick for gun and explosive-driven experiments. Specimens have been as small as 100 μm on a side and 20 μm thick in gas-gun and laser experiments, as well as up to a few meters in length in explosively driven production experiments. Typical durations are

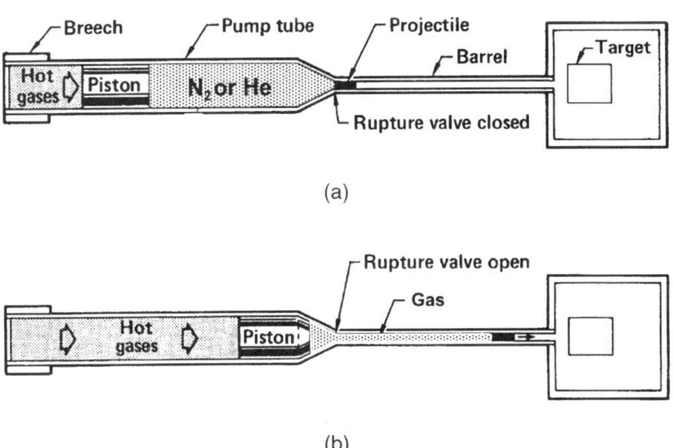

FIG. 1. Schematic of a two-stage light-gas gun. The first stage is the pair consisting of the driving gases produced by the burned gunpowder and the piston; the second stage is the pair consisting of the light gas and the projectile. The first stage compresses the light gas, which breaks the rupture valve at a predetermined pressure. Light gases have a high sound speed, which causes high projectile velocity, which in turn causes high pressure on impact of the projectile with the target. Light gases are N_2, He, and H_2, which produce maximum impact velocities of about 2, 5, and 8 km/s, respectively.

50 to 500 ns for gas-gun experiments, a few microseconds for explosives experiments, and a few nanoseconds for lasers. Small dimensions are often used for investigating effects, while large dimensions are often used for applications. Explosives are used for large-volume samples; lasers are used for surface layers.

The temperatures achieved in shock compression are determined by choice of shock pressure, specimen, and geometry. For example, shock temperatures are influenced by choice of specimen density and morphology, initial temperature, and shock-wave structure. Porous samples have lower initial and shock densities and higher shock temperatures than samples initially at crystal density. Lower initial temperatures cause lower shock temperatures. The highest temperature at a given pressure is achieved by a single shock to that pressure; a pulse with several step rises in pressure causes a lower final shock temperature than a single shock to the same final pressure.

3. SHOCK COMPRESSION

Dynamic high-pressure experiments have short time durations. Thus, fast electronic, optical, and x-ray techniques have been developed to diagnose specimen response. Since the shock process is so fast, the location of the shock wave must be synchronized with the turn-on of the recording system. Modern instrumentation makes the triggering and diagnostics possible with remarkable accuracy. Shock experiments are often performed together with theoretical calculations and have stimulated the development of condensed matter theory at high pressures (see EQUATIONS OF STATE), molecular-dynamics and Monte Carlo techniques, simulations of the dynamic response of powder particles, and hydrodynamic computer calculations. Articles in these areas are presented elsewhere (Davison and Graham, 1979; Schmidt *et al.*, 1992, 1994).

3.1 Simple Shock Waves

A simple, one-dimensional shock front typically involves an effective sharp step rise in shock pressure and in other variables in going from the unshocked to the shocked state. A simple shock front is illustrated in Fig. 2, where P_H is shock pressure, V_H is specific shock volume, E_H is specific internal shock energy, u_s is shock velocity, u_p is mass velocity behind the shock front, and x is distance. The zero-subscripted variables are initial values ahead of the shock front. The spatial width of the discontinuity of a strong shock in dense solids is several lattice spacings, which corresponds to a rise time $\ll 1$ ns. Material is shocked directly along the Rayleigh line from (P_0,V_0) to (P_H,V_H), as shown in Fig. 3. A plot of the locus of any two shock parameters is called the Hugoniot, shock adiabat, or shock-compression curve. Typical representations are u_s vs u_p, which are determined in an experiment, and P_H vs V_H, which are derived from u_s vs u_p.

The example in Fig. 2 represents a simple wave in which all effects except the equation of state are said to be overdriven and so rep-

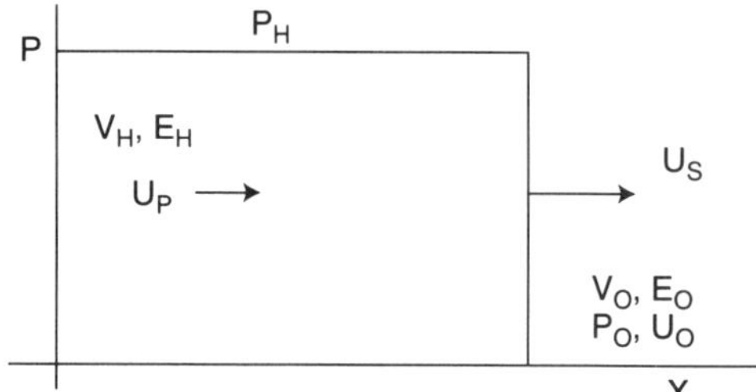

FIG. 2. Schematic of a simple shock wave traveling at velocity u_s in the x direction. This figure illustrates that the shock is essentially a traveling discontinuity. (Note: u_p and u_s are velocity vectors.)

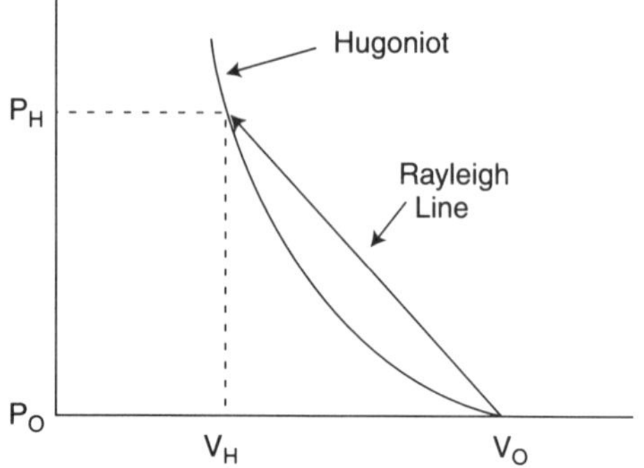

FIG. 3. Illustration in P-V space of the shock compression of a material initially at V_0 to a P-V point on the Hugoniot curve. The loading is along the Rayleigh line.

resents a relatively high-pressure wave. Figure 2 could also represent a purely elastic wave below the Hugoniot elastic limit (HEL). P_{HEL} is the maximum shock pressure at which the material acts as an elastic solid in shock compression. In Fig. 4(a) P_B is the minimum pressure required to overdrive the HEL, that is, to have a single plastic wave with no sign of the HEL in the wave profile. At intermediate pressures between the HEL and P_B, a two-wave structure exists, as shown in Fig. 4(b). In this case, a shock wave at intermediate pressure P_C splits into two waves. The leading wave at P_{HEL} is called the longitudinal elastic precursor because its wave speed is higher than that of the plastic wave at a pressure above P_{HEL}.

3.2 Rankine–Hugoniot Relations and Equations of State

The experimental design of shock-wave equation-of-state experiments is based on the Rankine–Hugoniot equations, which are derived from conservation of momentum, mass, and energy across the front of a shock wave (Fig. 2). These equations are given by

$$P_H - P_0 = \rho_0(u_s - u_0)(u_p - u_0), \quad (1a)$$

$$V_H = V_0[1 - (u_p - u_0)/(u_s - u_0)], \quad (1b)$$

$$E_H - E_0 = \tfrac{1}{2}(P_H + P_0)(V_0 - V_H), \quad (1c)$$

where $\rho_H = 1/V_H$ is mass density and the other variables are defined in Sec. 3.1. For a single wave, u_s is often measured; u_p is measured or is calculated using the shock-impedance-match principle, which states that P_H and u_p are continuous across an interface. Shock impedance is defined to be $\rho_0 u_s$, the slope of P_H vs u_p from Eq. (1a), where $P_0 = u_0 = 0$. For the case of a symmetric impact in which an impactor of a given material is accelerated to velocity u_I and strikes a target of the same material, $u_p = u_I/2$ by symmetry. Equations 1 define all primary pressure scales above about 7 GPa, both static and dynamic.

For many solids in the absence of phase transitions, experimental Hugoniot data follow the relation

$$u_s = C + S u_p, \quad (2)$$

where C is bulk sound speed at $P = 0$ and S is the slope of the linear relation between u_s and u_p. The constant S is a function of the pressure derivative of the isentropic bulk modulus (Ruoff, 1967). This result can be used to calculate P_H, V_H, and E_H from Eqs. (1). For comparison, the experimental Hugoniot curve and the theoretical 0-K isotherm and isentrope of Al are plotted in Fig. 5 (McMahan, 1976). For strong shocks, the total specific internal energy of Eq. (1c) becomes $E = \frac{1}{2}P_H(V_0 - V_H)$. As shown in Fig. 5, the total specific internal energy is represented by the area of the triangle under the Rayleigh line. Internal energy caused by isentropic compression to the same volume is reversible and is represented by the area under the isentrope. Reversible internal energy is removed from the sample when pressure is released. Irreversible internal energy is, thus, represented by the hatched area in Fig. 5. This irreversible internal energy is the reason for shock-induced heating and is the reason why shock pressures cause higher thermal and total pressures than isothermal,

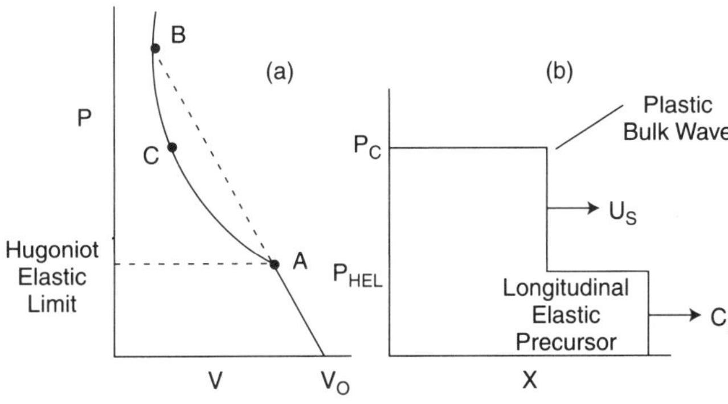

FIG. 4. (a) Hugoniot of a solid with a Hugoniot elastic limit, $P_A = P_{\mathrm{HEL}}$. P_B is the minimum shock pressure required to overdrive the HEL, that is, to achieve a single shock wave. (b) At intermediate pressure P_C, a two-wave structure propagates. The elastic precursor travels at the longitudinal sound speed C_L, followed by a plastic shock traveling at shock velocity U_S.

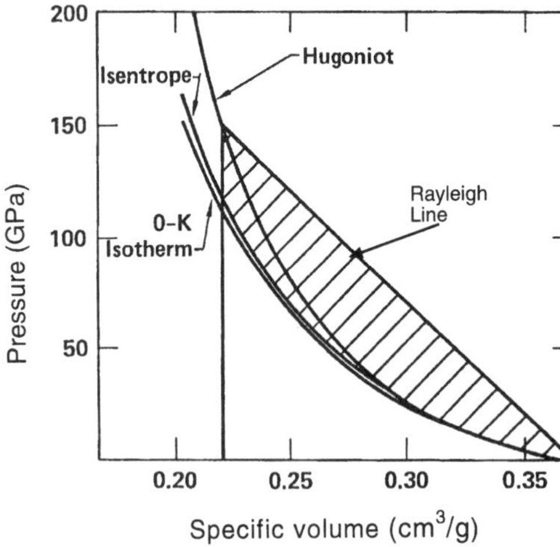

FIG. 5. Illustration of irreversible internal energy (shaded region) achieved in shock compression. This example is for Al shocked to 150 GPa and shows the measured Hugoniot and calculated 0-K isotherm and the isentrope from crystal volume (after McMahan, 1976).

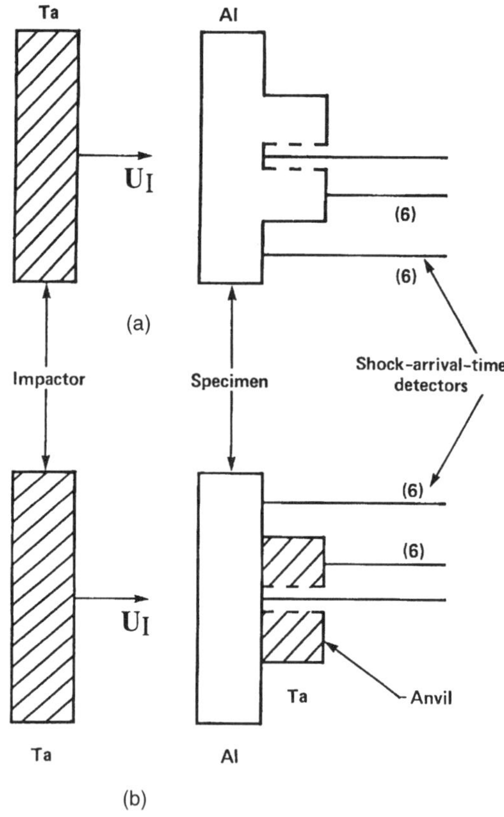

FIG. 6. Illustration of (a) single-shock and (b) double-shock equation-of-state experiments.

static compression to the same volume. In addition to heating materials, irreversible shock energy is also available for inducing phase transitions, changing microstructures, causing chemical reactions, and absorption by other internal degrees of freedom.

Single-shock equation-of-state experiments are illustrated in Fig. 6(a), in which a Ta impactor strikes a stepped Al target. Impactor velocity u_I and shock velocity u_s in the target are obtained by measuring the time between flash x rays of the impactor and the shock-transit time across the measured step height. u_I and u_s and the shock-impedance-match principle determine (P_H, V_H, E_H) in the target with $u_0 = 0$ in Eqs. (1). Double-shock equation-of-state experiments are illustrated in Fig. 6(b), in which a Ta impactor strikes a soft Al target backed by a stiff Ta anvil. After the first shock traverses the Al, it reflects backward off the stiffer Ta anvil, double shocking the Al and sending a shock forward into the Ta anvil. By taking the first-shock state as the zero-subscripted variables in Eqs. (1), measuring u_I, and measuring u_s in the anvil, the second-shock state can be obtained from Eqs. (1). Double shocking to a given volume produces lower pressures and temperatures than a single shock to the same volume.

Hugoniot equation-of-state data for Al, Cu, Mo, Ta, and Pt, for examples, have been measured up to a few hundred GPa using a two-stage light-gas gun at higher pressures and a planar explosive system at lower pressures (Marsh, 1979; van Thiel, 1977). Maximum pressures in these experiments are 170, 330, 470, 560, and 660 GPa, respectively (Mitchell and Nellis, 1981; Holmes *et al.*, 1989).

3.3 Off-Hugoniot Equations of State: Thermal Equation of State

Shock-compression equation-of-state data provide a reference curve for calculating off-Hugoniot states, that is, states that are not on the Hugoniot. Below about 100 GPa, thermal pressure at a given volume can be calculated from the Grüneisen model, which assumes that $(\partial P/\partial E)_V = \gamma/V$ is a constant, where γ is the Grüneisen parameter, dependent only on V. The Grüneisen model assumes that thermal energy is absorbed by the lattice. γ is obtained by measuring various discrete (P,E) states at the same V and approximating differentials with differences. The latter can be obtained by single- and double-shock data to the same volume using both solid and porous specimens. Porous specimens, with lower initial densities than solid ones, are more compressible and thus have larger internal energies and thermal pressures at a given volume than do shock-compressed crystal-density specimens. Equation-of-state data for porous and solid specimens of Al, Cu, and Fe suggested that γ/V is constant (Carter *et al.*, 1971), which is a common form used for γ for most materials. Also, γ has been determined from sound speed measurements, described in Sec. 3.7. Heat capacities are estimated to calculate temperatures from internal energies.

The latest approach to theoretical calculations of the equation of state of metals at high pressures and temperatures has been developed by Moriarty (Mitchell *et al.*, 1991). Electron band theory is used to calculate the 0-K isotherm in P-V space, to which are added electron-thermal and ion-thermal contributions to the pressure as functions of V and T.

3.4 Pressure Scales and Standards

Shock data are the primary data for static pressure scales above about 7 GPa. At higher static pressures the optical shift of ruby fluorescence and x-ray diffraction densities are calibrated vs shock-wave–reduced P-V isotherms. A shock-wave–reduced isotherm is that in which internal shock energies are taken into account to derive a 300-K isotherm from the Hugoniot curve, as described in Sec. 3.3.

Since static-pressure diamond-anvil-cell experiments are now up to the 500-GPa range (Ruoff *et al.*, 1990), accurate 300-K and higher-temperature P-V isotherms are needed for static pressure scales to ultrahigh pressures. Common x-ray standards for static high-pressure experiments are Pt and Mo because of their high atomic numbers and strong x-ray scattering. Powders of these metals are x rayed at very high static pressures to obtain sample volume. The corresponding pressure is derived from shock-reduced isotherms by the methods in Sec. 3.3. The optical ruby-fluorescence shift is a pressure scale calibrated by this x-ray diffraction method (see SOLIDS, STATIC HIGH-PRESSURE EFFECTS IN).

3.5 Shock Temperatures and Optical Emission Spectra

Shock temperatures have been measured in shocked transparent solid specimens by fitting the spectral dependence of radiation emitted from the shock front to a blackbody or graybody emission spectrum. If the specimen is transparent, the optical emission is observed for several times 100 ns while the shock front transits the specimen. The effective emission temperature can be taken in many cases as the shock temperature of the specimen. Results for transparent CsI are an example (Radousky *et al.*, 1985). Because of their thin optical depths of only about 100 Å and $<10^{-12}$ s, which precludes accurate measurements, shock temperatures of opaque solids are calculated as in Sec. 3.3. In one case, shock temperatures of metallic Fe have been measured by observing optical radiation from partially released Fe through a transparent diamond window, taking into account shock-wave reflections and thermal transport at the Fe–window interface (Yoo *et al.*, 1993). Shock temperatures of powders have also been measured (Boslough, 1992).

3.6 Shock-Wave Profiles

In shocked solids at pressures just above the HEL or, similarly, just above phase-transition pressures, a shock can have a multiple-wave structure (Graham and Asay, 1978). By measuring time-resolved shock-wave profiles, equations of state, mechanical constitutive properties and phase transition pressures under dynamic loading are derived.

3.6.1 Elastic–Plastic Flow and Material Strength Ideal elastic–plastic flow is illustrated in Fig. 4. For $P_A = P_{\mathrm{HEL}} < P_C < P_B$, a two-wave structure propagates. Because elastic wave speeds are greater than plastic ones, the elastic wave propagates as a precursor ahead of the plastic wave. The HEL is typically about a factor of 2 greater than the yield strength. The HELs of Cu, hardened steel, alumina, and B_4C are about 0.3, 2, 10, and 15 GPa, respectively. HELs are directionally dependent in anisotropic crystals.

A solid material with strength supports shear stress. As a result, the Hugoniot lies above the hydrostat in *P-V* space for a strong solid, stress is a more accurate representation than pressure, and stress is different in different directions. The principal stresses σ_i and the hydrostatic pressure P are given by

$$\sigma_i = -P + s_i, \quad i = 1, 2, 3 \tag{3}$$

and

$$-P = \tfrac{1}{3}(\sigma_1 + \sigma_2 + \sigma_3), \tag{4}$$

where the s_i are the stress deviators in the principle directions i and $\Sigma s_i = 0$. The yield strength Y^0 is such that

$$s_1^2 + s_2^2 + s_3^2 \leq (Y^0)^2. \tag{5}$$

For a perfectly plastic material, Y^0 is a constant independent of pressure. For many materials, called elastic plastic, work hardening increases yield strength Y for shock stresses above the HEL. That is, as shock pressure increases the amount of plastic work and associated shock temperature, Y increases to a maximum and then decreases until $Y = 0$ at melting, where individual stress deviators $s_i = 0$ in the fluid state. For Cu, Y^0 is about 0.1 GPa, and Y_{max} is about 2.5 GPa at 70-GPa shock stress and 1000-K shock temperature (Morris and Fritz, 1980). Cu melts at a shock pressure of about 230 GPa (Moriarty, 1986).

3.6.2 Shock-Induced Phase Transitions

Phase transitions can be induced by dynamic high pressures and temperatures. If crystal volume changes in the transition, then pressure and material velocity also change. By measuring time-resolved pressure or mass-velocity histories, phase changes can be detected. The graphite-to-diamond transition is an example. The *P-V* curve and wave structure are similar to those in Fig. 4, except that in this case P_A is the onset phase transition pressure, rather than the HEL. The measured velocity of an interface between shocked graphite and a LiF window is shown in Fig. 7. The first wave, or velocity jump, is graphite in the shock-compressed state at 21 GPa. The second wave is caused by the transformation to diamond, which occurs in 10 ns in the highly ordered pyrolytic graphite specimen, oriented with the *c* axis of the hexagonal-graphite crystal structure parallel to the shock direction. The fast speed of the phase transition is the signature of a martensitic transformation from hexagonal graphite probably to hexagonal diamond (Erskine and Nellis, 1992). The kinetics of this transformation are quite sensitive to the degree of crystalline order and to the relative orientation of the specimen with respect to the direction of shock propagation. The velocity profile of Fig. 7 was measured using a fast optical interferometer called a VISAR, or velocity interferometer for a surface of any reflector (Barker and Hollenbach, 1972). Shock-wave profiles can also be measured with a Fabry–Pérot interferometer or with piezoelectric gauges.

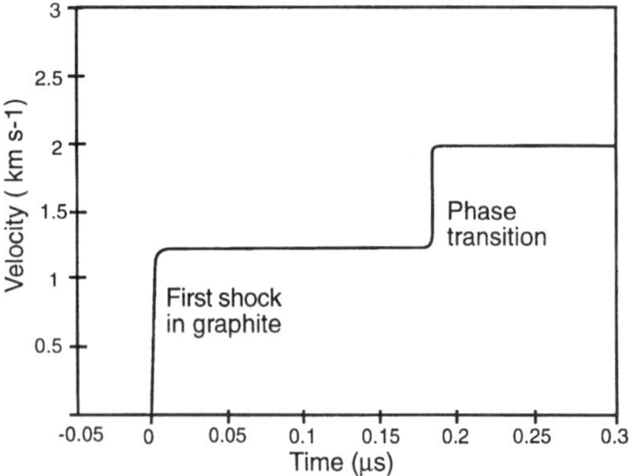

FIG. 7. Measured wave profile in highly oriented pyrolytic graphite shocked into the region of the phase transition to diamond. In this case, the first knee in the curve corresponds to the onset of the phase transition, rather than to the HEL illustrated in Fig. 4 (after Erskine and Nellis, 1992).

Melting is also induced by high shock pressures and temperatures. Melting temperatures increase with pressure. However, shock temperatures also increase with pressure and eventually exceed the melting temperature at sufficiently high pressures (Moriarty, 1986).

3.7 Shock Release

A calculated temporal wave profile of a strong planar shock in Cu is shown in Fig. 8 (Neumeier *et al.*, 1989). This shock is in-

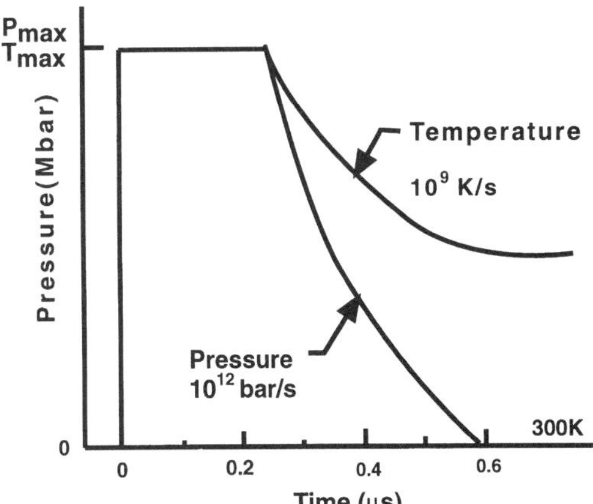

FIG. 8. Shock and release profiles calculated in planar geometry for Cu shocked to high pressure. The pressure releases to zero, but the temperature does not because of the irreversible internal shock energy shown in Fig. 5 (after Neumeuier *et al.*, 1989).

duced by impact; a shock is also induced in the impactor and travels back toward the rear surface of the impactor (Fig. 6). On reaching the rear surface of the impactor, pressure drops, and a release wave is transmitted forward into the impactor, eventually overtaking and releasing the original shock wave as it travels in the target. At overtake or release, both pressure and temperature decrease. The release rates are about 10^{12} bar/s and 10^9 K/s. While pressure releases to zero quickly, temperature releases to a residual value, which is higher than the initial value because of irreversible shock heating (Fig. 5). The residual temperature approaches the ambient value by thermal conduction.

3.7.1 Speed-of-Sound Measurements The high-pressure release wave described in Sec. 3.7 travels longitudinally at the speed of sound c in the material, which is traveling at mass velocity u_p in the laboratory frame. Thus, the release wave travels at velocity $u_p + c$, which is greater than u_s. The longitudinal speed of sound C_L of a high-pressure shock state is determined by measuring the depth in the target at which the release wave just overtakes the shock, provided u_s and u_p are known (Brown and McQueen, 1986).

3.7.2. Grüneisen Parameter and Phase Transitions For the case in which shear strength is negligible, the bulk thermodynamic speed of sound C_B is given by

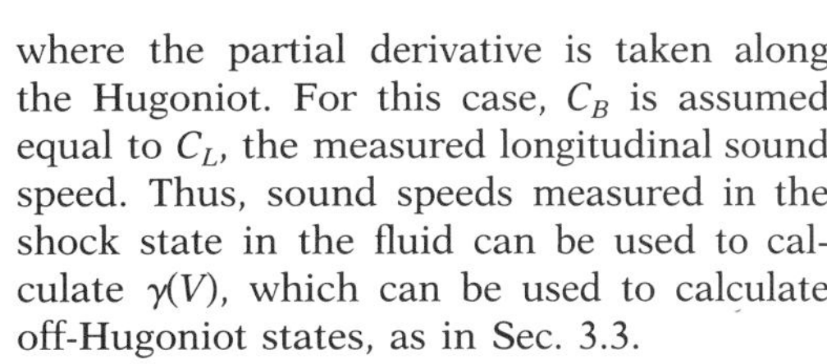

$$C_B = V_H\left\{\left(\frac{\partial P}{\partial V}\right)_H\left[(V_0 - V_H)\left(\frac{\gamma}{2V_H}\right) - 1\right] + P_H\left(\frac{\gamma}{2V_H}\right)\right\}^{1/2}, \quad (6)$$

where the partial derivative is taken along the Hugoniot. For this case, C_B is assumed equal to C_L, the measured longitudinal sound speed. Thus, sound speeds measured in the shock state in the fluid can be used to calculate $\gamma(V)$, which can be used to calculate off-Hugoniot states, as in Sec. 3.3.

For the case in which a solid has shear strength, we have

$$C_B^2 = C_L^2 - \tfrac{4}{3}C_T^2, \quad (7)$$

where C_T is the sound speed of the shear wave. Melting is often taken to occur at the shock pressure at which $C_L = C_B$, where C_L is measured and C_B is calculated from Eq. (6). Because the speed of sound is related to a derivative of pressure with respect to volume, sound speed is usually more sensitive to phase transitions than Hugoniot P-V relations.

3.8 Electrical Resistivity

Electrical resistivities are measured up to the 100-GPa range. The resistivity of FeO at 100 GPa shock pressure is about 100 $\mu\Omega$-cm (Knittle *et al.*, 1986), which is in the range for metallic Fe at 100 GPa.

3.9 Raman Spectroscopy

Raman spectroscopy has been used to observe shifts in the vibrational frequency of diamond with increasing shock pressure (Gupta *et al.*, 1989).

3.10 Flash X-Ray Diffraction

Flash x-ray diffraction has been performed on graphite, BN, and LiF single crystals at shock pressures up to 100 GPa. The observation of diffraction shows that the material has long-range crystalline order, despite high densities of shock-induced lattice defects. The experimental x-ray density is observed to be equal to the Hugoniot density [Eq. (1b)] up to 100 GPa shock pressures for LiF (Johnson and Mitchell, 1980).

4. COMPUTATIONAL SIMULATIONS

The experimental data described above are often used to develop material databases, such as equations of state and constitutive properties, for implementation into computer codes. These codes are often called hydrodynamic, even for solids that have strength. Hydrodynamic codes are used to simulate, design, and interpret a variety of dynamic high-pressure experiments. In addition, molecular-dynamics, Monte Carlo, and electron-energy-band techniques are now used to understand dynamic high-pressure phenomena on an atomic or molecular scale (Schmidt *et al.*, 1992).

5. MATERIALS

Dynamic pressures discussed above are diagnosed and simulated computationally in real time. In this section, we discuss how dynamic high pressures are used to produce novel crystal structures, microstructures, and associated properties. That is, specimens are subjected to dynamic pressures and recovered intact for characterization. Pressure is applied at strain rates up to 10^8/s and higher, and materials are shock compressed to pressures as high as 100 GPa and temperatures as high as a few times 1000 K, held at pressure for times on the order of microseconds, quenched at rates up to 10^{12} bar/s and 10^9 K/s (Fig. 8), and recovered for characterizations of structure and properties. These quench rates occur at the physical limit for macroscopic samples because they occur in bulk at the speed of sound. Even higher quench rates are achieved by thermal diffusion in micron-thin layers at interfaces. These high quench rates mean that phases and structures synthesized at high shock pressures and temperatures might be quenched at ambient conditions. Materials treated this way often have unique structures and tailored properties.

Effects can be investigated in small samples using, for example, a small gas gun or a laser-heated diamond anvil cell. Larger specimens can be produced by using larger dynamic systems, such as explosives or large lasers (see SHOCK WAVES; Cowan *et al.*, 1968).

5.1 Experimental Methods

Materials are recovered from dynamic high pressures using a variety of methods. As examples, we discuss diamond powders made with explosives and a method to investigate effects in small specimens using a gas gun.

5.1.1 Explosive Systems to Synthesize Diamond Probably the best known example of materials produced commercially with dynamic pressure is the synthesis of diamond powders for abrasives (Cowan *et al.*, 1968). This system is about 4 m long and about 1 m in diameter. A mixture of carbon and metal powders is placed inside a cylindrical explosive system, which drives the carbon into the pressure–temperature stability field of diamond. Polycrystalline diamond powders are synthesized from the carbon, which is quenched thermally by the metal powders before the pressure is released. Because the process is fast, polycrystalline diamond is produced with a range of particle sizes of about 0.1 to 40 μm.

Nanocrystalline diamond powders are produced by the nucleation and growth of diamond within the reaction products of detonated explosives. Carbon is a common component, or carbon additives are mixed with explosives. Detonation takes the reaction products to the pressure-temperature stability field of diamond, which forms on phase separation of carbon from the other reaction products. The fast pressure-release path quenches the diamond formed at high pressure. Typical grain size is about 4 nm, small because of the microsecond time scale (Lyamkin *et al.*, 1988). These grains agglomerate into particles about 0.1 μm across. This work was based on the theoretical analysis of the Hugoniot data of hydrocarbons, which assumed that hydrocarbons decompose into diamond-like carbon and molecular hydrogen at high shock pressures and temperatures (Ree, 1979).

5.1.2 Two-Stage Light-Gas Gun Effects of dynamic high pressures in small specimens are investigated using a two-stage light-gas gun (Fig. 1). A shock is generated on impact of a projectile with a target capsule in which a specimen is embedded. Typical dimensions of a gun for this purpose are a length of 6.5 m and a bore diameter of 20 mm. Maximum impactor velocity is 4 km/s with He driving gas, which produces 120 GPa (1.2 Mbar) for a Cu plate impacting a Cu target. Velocities up to 7 km/s could be achieved with H_2 driving gas. Specimens are typically 10 mm in diameter and 0.001 to 1 mm thick. Initial temperatures can be varied in the range 100–1300 K. A representative configuration is illustrated in Fig. 9.

5.2 Specimens

Various types of specimens are subjected to dynamic high pressures and characterized. A few are described below, most of which are shocked with a two-stage gun.

5.2.1 Films Films are used because a wide variety of materials can be made with this method, mixing is on a fine scale for fast reactions during fast pressure pulses, and maximum quench rates are achieved in thin layers that can retain shock-induced structures and properties. To illustrate that films can be recovered successfully from high dynamic pressures, Nb films 1–10 μm thick were recovered from 100 GPa shock pressure (Koch *et al.*, 1990).

5.2.2 Hard Materials Potentially new hard materials can be synthesized and recovered from dynamic high pressures. For example, shock-compressed mixtures of C_{60} fullerenes and Cu powders produced fine-grained diamond (Sekine, 1992). Nanocrystalline metastable diamond films have been produced by shock compression of 2-μm-thick C_{60} fullerene films (Yoo *et al.*, 1992).

5.2.3 Bond Strengths between Film and Substrate The strength of bonding between a film made by conventional means and its substrate is measured by launching a short-pulse planar shock into a substrate, which then transits the film. The shock reflects as a tensile wave from the free surface of the film and travels back in the direction from which the original shock came. The tensile stress required to delaminate the film from the substrate is determined from the shape of the time-resolved free-surface velocity history of the film, which is measured with a fast optical velocimeter similar to the design of Barker and Hollenbach (1972). Short-pulse incident shock waves are generated by impact of thin metal foils (Nutt, 1992) or by a laser (Gupta *et al.*, 1993).

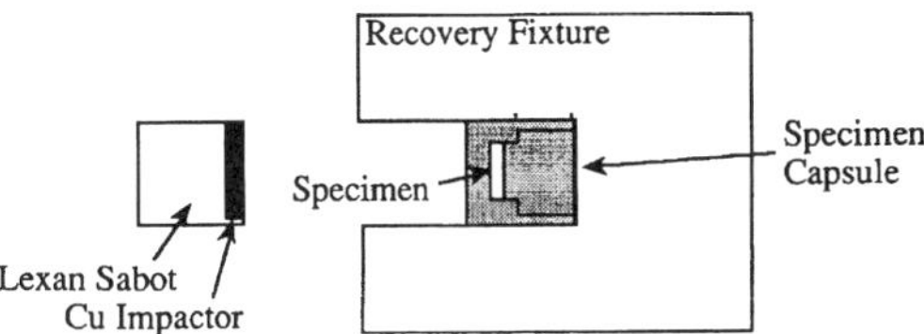

FIG. 9. Illustration of an experiment to recover a specimen shocked to high pressure. The projectile is launched by the gun in Fig. 1.

5.2.4 Single Crystals Three-phase SiO_2 has been synthesized above 20 GPa by shocking single-crystal quartz, which fractures into grains. The three phases are lamellar glass and quartz within grains and melt glass between grains (Gratz *et al.*, 1992). Above about 40 GPa, only amorphous SiO_2 is found.

5.2.5 Highly Defected Superconducting $YBa_2Cu_3O_{7-x}$ Shock deformation in ceramics induces dislocations and stacking faults, which enhance magnetic hysteresis and magnetic levitation force in superconducting $YBa_2Cu_3O_{7-x}$. By using highly ordered melt-textured specimens and orienting them with respect to the shock direction, these brittle oxide specimens can be shocked without macroscopic fracture. After annealing in oxygen, the magnetic hysteresis at 1 kOe and 70 K is about 20% greater than the value before shock at 7 GPa (Nellis *et al.*, 1994).

5.2.6 Powder Consolidation Dynamic compaction rapidly consolidates powders by depositing compressive energy on particle surfaces, which are heated heterogeneously, often bond together, and then quench thermally to the interiors of the particles, forming a dense compact. Advantages include

1. focussing energy on particle surfaces so that whole systems do not need to be heated for compaction,
2. consolidating nanocrystalline powders so quickly that grains do not have time to grow, and
3. bonding powders of composite materials.

A common problem with this method is cracking that occurs on release of pressure. Single-piece disks of both conventional and rapidly solidified alumina–zirconia ceramics have recently been consolidated by a reverberating shock wave (Freim *et al.*, 1994). Diamond particles have been shock consolidated. For example, good results have been obtained by choice of diamond particle size and a relatively small admixture of graphite particles (Potter and Ahrens, 1988). A computational model is being developed for dynamic compaction that could be applied to a wide variety of materials, pressures, and consolidation rates and, thus, address key issues computationally (Benson and Nellis, 1994).

5.2.7 Shock-Induced Chemical Reactions Dynamic high pressures and temperatures cause chemical reactions between powder particles and, thus, offer a way to synthesize novel materials. Recent results are described in Schmidt *et al.* (1994). Shock-induced defects also enhance chemical reactivity and catalysis (Graham *et al.*, 1986; Thadhani, 1993).

6. SUMMARY

High dynamic pressures achieve unusual states of matter through changes in density and temperature, which cause changes in properties at high pressures. In recovered specimens, high dynamic pressures, densities, temperatures, and strain rates can cause changes in microstructures and phases, which in turn can change properties of materials. An extensive array of fast diagnostics have been developed to probe matter at high dynamic pressures. Computational and theoretical methods interpret results and are used to design dynamic high-pressure experiments. The fast real-time experiments are important for validating computational models and for determining materials that can be recovered from dynamic high pressures. A wide variety of materials have been recovered intact for characterization of structure and properties. Material produced with dynamic high pressures are produced commercially, such as synthetic diamond powders.

ACKNOWLEDGMENT

This work was performed under the auspices of the U.S. Department of Energy by the Lawrence Livermore National Laboratory under Contract No. W-7405-Eng-48.

APPENDIX A: UNITS AND CONVERSION FACTORS

The SI system provides five basic units used in high-pressure research: meter (m), kilogram (kg), second (s), ampere (A), kelvin (K), and mole (mol). Pressure is the derived quantity of force per unit area. Many kinds of units of pressure are used in engineering and science according to the particular traditions. In this article we use the units GPa (Gigapascal) and kilobar (kbar); 1 GPa = 10 kbar, where 1 GPa = 1 N m^{-2} × 10^9 and 1 kbar is the pressure of about 1000 atmospheres. Units of density are 1 g cm^{-3} = 1 kg m^{-3}.

GLOSSARY

Bond Strength: Strength of the bond between a film and its substrate. One technique to measure this is to measure the strength of the tensile wave, induced by a shock pulse, required to break this bond.

Dynamic Powder Compaction: Compaction of a powder specimen by a dynamic pressure pulse.

Dynamic Strength: The strength of a material at dynamic high pressure. Strength is generally higher for short pressure pulses than for static pressures.

Elastic–Plastic Flow: Flow in which a material first compresses elastically to the Hugoniot elastic limit and plastically at higher shock pressures.

Fabry–Pérot Interferometer: A velocity interferometer that measures the time dependence of the velocity of a surface.

Flash X-Ray Diffraction: Pulsed x-ray diffraction from a material at high shock pressure.

High Strain Rate: The strain rate $\dot{V}V^{-1}$ achieved by shock compression ranges up to $>10^8$ s^{-1}, the physical limit in macroscopic bodies.

Hugoniot: Locus of states reached by shock compression from a given initial condition. Common representations include pressure vs volume, shock velocity vs mass velocity, and pressure vs mass velocity.

Nanocrystalline Diamond: Grains of diamond 1–20 nm across, produced by high shock pressures. The brief time duration (of order microseconds) restricts particle growth to small dimensions.

Pulsed Raman Spectroscopy: Light scattering by vibrational modes at high shock pressures.

Quench Rates: Rates at which pressure and temperature decrease when shock pressure releases. These rates, about 10^{12} bar/s and 10^9 K/s, are the physical limit in macroscopic bodies.

Shock-Induced Chemical Reactions: Chemical reactions caused by high shock pressures, densities, temperatures, and defect concentrations. Synthesis of new phases from elemental powders is an example.

Shock-Induced Chemical Reactivity: Chemical reactivity induced by shock-induced defects. Enhanced catalytic activity is an example.

Shock-Induced Defects: Lattice defects, usually dislocations and stacking faults, induced in a lattice by the high–strain-rate deformation produced by shock compression.

Shock Wave: A discontinuity in flow and thermodynamic variables that propagates at supersonic velocity.

Single-Stage Gun: A gun consisting of a projectile and a breech in which the projectile is accelerated by compressed gas or by burned gunpowder in the breech.

Two-Stage Light-Gas Gun: A gun consisting of two stages of driving gas and piston. The first stage consists of the products of burned gunpowder and a heavy piston; the second stage consists of light gas and a light projectile. The light gas and the projectile are separated by a nozzle that amplifies the gas velocity and by a valve that opens when the gas pressure exceeds a given value. The light gas has a high sound speed and, thus, produces a high velocity when it accelerates the projectile.

VISAR: A velocity interferometer for a surface of any reflector, which measures the time dependence of the velocity of the surface.

Works Cited

Barker, L. M., Hollenbach, R. E. (1972), *J. Appl. Phys.* **43,** 4669–4675.

Benson, D. J., Nellis, W. J. (1994), *Appl. Phys. Lett.* **65,** 418–420.

Boslough, M. B. (1992), in: S. C. Schmidt, R. D. Dick, J. W. Forbes, D. G. Tasker, (Eds.), *Shock Waves in Condensed Matter 1991,* Amsterdam: Elsevier Press, pp. 617–620.

Brown, J. M., McQueen, R. G. (1986), *J.Geophys. Res.* **91,** 7485-7494.

Bundy, F. P. (1988), *Phys. Rep.* **167,** 618–630.

Carter, W. J., Marsh, S. P., Fritz, J. N., McQueen, R. G. (1971), *Accurate Characterization of the High-Pressure Environment,* Washington, DC: National Bureau of Standards, pp. 147–158.

Cowan, G. R., Dunnington, B. W., Holtzman, H. A. (1968), U.S. Patent No. 3,401,019.

Davison, L., Graham, R. A. (1979), *Phys. Rep.* **55,** 255–379.

Erskine, D. J., Nellis, W. J. (1992), *J. Appl. Phys.* **71,** 4882–4886.

Freim, J., McKittrick, J., Nellis, W. J. (1994), in: S. C. Schmidt, J. W. Shaner, G. A. Samara, M. Ross, (Eds.), *High-Pressure Science and Technology—1993,* New York: American Institute of Physics Press, pp. 1263–1266.

Graham, R. A., Asay, J. R. (1978), *High Temperatures—High Pressures* **10,** 355–390.

Graham, R. A., Morosin, B., Venturini, E. L., Carr, M. J. (1986), *Annu. Rev. Mater. Sci.* **16,** 315–341.

Gratz, A. J., Nellis, W. J., Christie, J. M., Brocious, W., Swegle, J., Cordier, P. (1992), *Phys. Chem. Minerals* **19,** 267–288.

Gupta, Y. M., Horn, P. D., Yoo, C. S. (1989), *Appl. Phys. Lett.* **55,** 33–35.

Gupta, V., Yuan, J., Martinez, D. (1993), *J. Am. Ceram. Soc.* **76,** 305–315.

Holmes, N. C., Moriarty, J. A., Gathers, R. G., Nellis, W. J. (1989), *J. Appl. Phys.* **66,** 2962–2967.

Jayaraman, A. (1986), *Rev. Sci. Instrum.* **57,** 1013–1031.

Johnson, Q., Mitchell, A. C. (1980), *High Pressure Science and Technology,* Oxford: Pergamon, pp. 977–978.

Knittle, E., Jeanloz, R., Mitchell, A. C., Nellis, W. J. (1986), *Solid State Comm.* **59,** 513–515.

Koch, R., Nellis, W. J., Hunter, J. W., Davidson, H., Geballe, T. H. (1990), *Pract. Met.* **27,** 391–405.

Lyamkin, A. I., Petrov, E. A., Ershov, A. P., Sakovich, G. V., Staver, A. M., Titov, V. M. (1988), *Dokl. Akad. Nauk SSSR* **302,** 611–620 (in Russian).

Marsh, S. P. (1979), *Los Alamos Shock Hugoniot Data,* Berkeley: University of California Press.

McMahan, A. K. (1976), *Bull. Am. Phys. Soc.* **21,** 1303.

Mitchell, A. C., Nellis, W. J. (1981), *J. Appl. Phys.* **52,** 3363–3374.

Mitchell, A. C., Nellis, W. J., Moriarty, J. A., Heinle, R. A., Holmes, N. C., Tipton, R. E. (1991), *J. Appl. Phys.* **69,** 2981–2986.

Moriarty, J. A. (1986), *Shock Waves in Condensed Matter,* New York: Plenum, pp. 101–106.

Morris, C. E., Fritz, J. N. (1980), *J. Appl. Phys.* **51,** 1244–1246.

Nellis, W. J., Weir, S. T., Hinsey, N. A., Balachandran, U., Kramer, M. J., Raman R. (1994), in:

S. C. Schmidt, J. W. Shaner, G. A. Samara, M. Ross, (Eds.) *High-Pressure Science and Technology—1993,* New York: American Institute of Physics Press, pp. 695–697.

Neumeier, J. J., Nellis, W. J., Maple, M. B., Torikachvili, M. S., Yang, K. N., Ferreira, J. M., Summers, L. T., Miller, J. I., Sales, B. C. (1989), *High Pressure Res.* **1,** 267–289.

Nutt, G. L. (1992), *J. Mater. Res.* **7,** 203–213.

Potter, D. K., Ahrens, T. J. (1988), *J. Appl. Phys.* **63,** 910–914.

Radousky, H. B., Ross, M., Mitchell, A. C., Nellis, W. J. (1985), *Phys. Rev. B* **31,** 1457–1462.

Ree, F. H. (1979), *J. Chem. Phys.* **70,** 974–983.

Ruoff, A. L. (1967), *J. Appl. Phys.* **38,** 4976–4980.

Ruoff, A. L., Xia, H., Luo, H., Vohra, Y. K. (1990), *Rev. Sci. Instrum.* **61,** 3830–3833.

Schmidt, S. C., Dick, R. D., Forbes, J. W., Tasker, D. G. (1992), *Shock Compression of Condensed Matter—1991,* Amsterdam: Elsevier Press.

Schmidt, S. C., Shaner, J. W., Samara, G. A., Ross, M. (1994), *High-Pressure Science and Technology—1993,* New York: American Institute of Physics Press.

Sekine, T. (1992), *Proc. Jpn. Acad. Ser. B.* **68,** 95–99.

Syono, Y., Manghnani, M. H. (1992), *High-Pressure Research: Applications to Earth and Planetary Sciences,* Tokyo: Terra Scientific Publishing.

Thadhani, N. N. (1993), *Prog. Mater. Sci.* **37,** 117–226.

van Thiel, M. (1977), *Compendium of Shock Wave Data,* Livermore: Lawrence Livermore Laboratory Report UCRL-50108.

Yoo, C. S., Nellis, W. J., Sattler, M. L., Musket, R. G. (1992), *Appl. Phys. Lett.* **61,** 273–275.

Yoo, C. S., Holmes, N. C., Ross, M., Webb, D. J., Pike, C. (1993), *Phys. Rev. Lett.* **70,** 3931–3934.

Further Reading

Asay, J. R., Shahinpoor, M. (Eds.) (1993), *High Pressure Shock Compression of Solids,* New York: Springer-Verlag.

Bushman, A. V., Fortov, V. E. (1987), in: A. E. Scheindlin, V. E. Fortov (Eds.), *Thermal Physics Reviews,* London: Harwood Academic Publishers, pp. 220–336.

Courant, R., Friedrichs, K. O. (1976), *Supersonic Flow and Shock Waves,* New York: Springer-Verlag.

Duvall, G. E., Graham, R. A. (1977), *Rev. Mod. Phys.* **49,** 523–579.

Eliezer, S., Ghatak, A., Hora, H. (1986), *An Introduction to Equation of State: Theory and Applications,* Cambridge: Cambridge Univ. Press.

Graham, R. A. (1993), *Solids Under High-Pressure Shock Compression,* New York: Springer-Verlag.

McQueen, R. G., Marsh, S. P., Taylor, J. W., Fritz, J. N. (1970), in: R. Kinslow (Ed.), *High-Velocity Impact Phenomena,* New York: Academic Press.

Ross, M., Young, D. A. (1993), *Annu. Rev. Phys. Chem.* **44,** 61–87.

Schmidt, S. C., Shaner, J. W., Samara, G. A., Ross, M. (Eds.) (1994), *High-Pressure Science and Technology—1993,* New York: American Institute of Physics Press.

Zel'dovich, Ya. B., Raizer, Yu. P. (1966), *Physics of Shock Waves and High-Temperature Hydrodynamic Phenomena, Volumes I and II,* New York: Academic Press.

SOLIDS, STATIC HIGH-PRESSURE EFFECTS IN

RUSSELL J. HEMLEY AND HO-KWANG MAO, *Geophysical Laboratory and Center for High-Pressure Research, Carnegie Institution of Washington, Washington, D.C., U.S.A.*

INTRODUCTION

Pressure is firmly established as a key variable in many areas of the natural sciences. When pressure is increased on solids, interactions among atoms increase, in some cases radically altering the physical and chemical properties of the material. In contrast, changing interatomic distances by varying composition is restricted by the discrete nature of chemical elements, and tuning by temperature is complicated by thermal excitations (i.e., vibrations) and ultimately the melting and boiling points of the material. Measurements of physical properties tuned over extended pressure ranges now provide crucial experimental bases for testing and developing theories of condensed matter. In addition to its fundamental significance, the effects of pressure on solids have important implications for the syntheses of materials. New materials, with novel physical and chemical properties, have been discovered, and the newly accessible pressure domain provides the prospect of greatly enlarging the number of known materials. High-pressure experiments on geological and planetary materials are providing a unique window on the interiors of the Earth and other planets.

In the past decade, high-pressure research has been revolutionized, principally as a result of breakthroughs in diamond-cell technology. Static pressures of several hundred GPa (gigapascal, 1 GPa = 10 kbar = 10^4 bar) can be reached in the laboratory (Fig. 1). Even more significantly, physical properties of materials can be characterized *in situ* under these conditions, and the accuracy of many high-pressure measurements is now approaching those on samples under ambi-

3-527-28140-1/96/$5.00 + .50

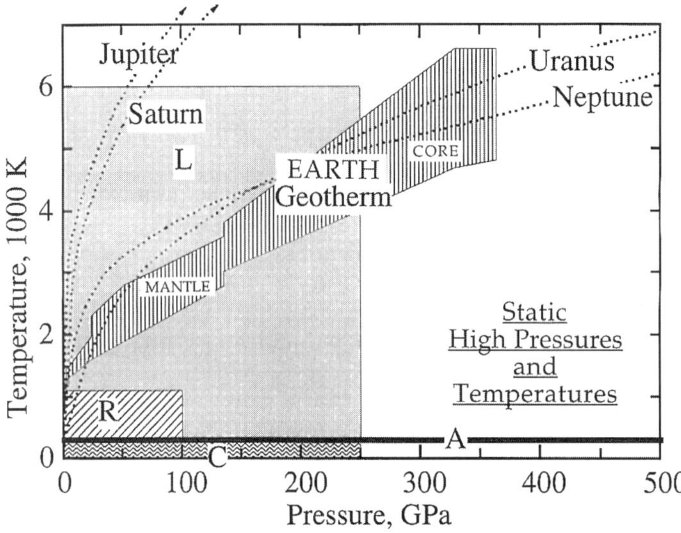

FIG. 1. Pressure–temperature range for static high-pressure experiments superimposed upon the pressure–temperature profiles (geotherms) of the interiors of the Earth and other major planets. Most experiments, including the highest-pressure records, have been performed at the ambient temperature (trajectory "A"). Region "C" denotes the cryogenic range, which has been particularly useful for fundamental studies of physics and chemistry of the materials. The range of simultaneous high temperatures and high pressures achieved by external-resistance heating and/or laser heating are denoted by *R* and *L*, respectively. These ranges represent the reported maxima rather than the intrinsic limitation of the diamond cell. The regions of the Earth's mantle (to 135 GPa) and core (135–364 GPa) are also indicated in the figure.

ent conditions. Typically, a number of high-pressure transitions are observed in each solid in this wide range of pressures. In addition, theoretical work has been growing at a steady rate in all aspects of high-pressure work. Also, high-pressure materials theory, including large-scale computations, approximate methods, and analytical theory, has advanced significantly and plays an essential role in the interpretation of experimental results and in making useful predictions.

1. HIGH-PRESSURE TECHNIQUES

Nearly all static pressure techniques are based on a few conceptually simple principles in which pressure (defined as force/area) is maximized, generally along with tradeoffs in one or more other variables, such as sample size, temperature range, or ability to measure specific physical or chemical properties. High-pressure techniques are reviewed in the article by that title. Bridgman articulated many of the basic principles of high-pressure techniques in use to this day. This included the discovery of the "principle of unsupported area" and the invention of the Bridgman anvil, based on the "principle of massive support." Although the pressure range of laboratory measurements has increased by nearly two orders of magnitude since this early work, devices in use today may be considered descendants of many designs outlined in the early work of Bridgman and others. This advance has come about as a result of continued optimization of many of these early designs, and the fabrication and/or implementation of high-strength materials, such as metal carbides, sintered diamond, and single-crystal diamond.

1.1 Large-Volume Devices

Developments until the 1960s consisted mainly of what are now considered "large-volume" devices. These included the belt or girdle apparatus, the solid-media piston-cylinder apparatus, the split-sphere multianvil (or multipiston) device, the tetrahedral press, and the uniaxial Bridgman (or Drickamer) apparatus (see Sec. 3.2.1 and HIGH-PRESSURE TECHNIQUES). Such large-volume devices create a relatively uniform sample environment at high pressure and temperature, while maintaining the chemical and physical integrity of the samples. Experiments with such devices, while limited to lower pressures (<40 GPa), can produce well-characterized samples for phase equilibria, spectroscopy, calorimetry, x-ray diffraction, and ultrasonic investigations on recovered samples. Studies to temperatures of order 3000 K have been performed. Other modern devices permit a number of *in situ* measurements, including x-ray diffraction. Such experiments have sample dimensions of order several millimeters. New developments include the use of sintered diamond for anvils, instead of tungsten carbide, for reaching higher pressures. Diamond synthesis continues to be an active area of research with these devices. Particularly significant is the growth of chemically and isotopically pure large diamond single crystals.

1.2 Diamond-Anvil Cells

Much of today's knowledge of the behavior of solids at very high pressures can be attributed to the use of diamonds to compress materials in a diamond-anvil cell. The cell is based upon the original Bridgman anvil design, in which the sample is placed between the polished culets of two diamonds and is contained on the sides by a metal gasket. Since pressure is force divided by area, enormous pressures can be generated by opposing anvils with a very small tip (culet), providing that the anvil material is strong enough to withstand the pressure without yielding (Jayaraman, 1983). Flawless single-crystal diamond is chosen for its strength. The anvils exert pressure on a thin metallic gasket within which a circular sample chamber is confined (Fig. 2). Samples immersed in pressure-transmitting media are sealed in the chamber. The gasket also functions as a supporting belt without which the anvils will fail at a maximum pressure of only 35 GPa.

Diamond is transparent to much of the electromagnetic spectrum—including high-energy x and γ radiation above 10 keV and low-energy ultraviolet–visible–infrared radiation below 5 eV. Numerous diffraction, scattering, and absorption probes can be employed for determination of electronic, dynamic, and structural properties of samples under pressure. One of the main constraints on diamond-cell technology is the small sample volume, which in turn is determined by the limited anvil size (~0.1–0.4 carat; 1 ct = 0.2 g). At a constant anvil size, higher pressures are reached with a reduction in sample volume: there is a tradeoff between the maximum pressure and the maximum volume of the compressed sample. For instance, diamonds with 1-mm large culet diameter are generally limited to pressures below 5 GPa; diamonds with 20-μm small culet diameter are needed for pressures above 300–400 GPa. The diameter of the sample chamber is approximately one-half of the culet diameter, and the thickness, one-tenth. Studying such minute samples requires development of microsampling techniques.

1.3 Pressure Calibration

In lower-pressure work, the pressure may be calculated from the applied force and the geometry of the apparatus. Such calculations become increasingly inaccurate at higher pressures because of the deformation of materials of the apparatus. The limit of the accurate range of pressures determined by force per unit area is about 7 GPa. Because of this, an internal standard is required. A fair amount of work has gone into the problem of pressure calibration at very high pressures. If a physical property of a material is pressure dependent, it can be calibrated to provide a pressure scale. One type of calibrant is a fixed point, which is a discontinuous change in a sample property that has independently been determined to occur at a specific pressure and temperature. An example is a phase transformation, which can be observed following quenching to ambient conditions or observed *in situ* at high pressures (and variable temperature), for example, as a change in crystal structure or electrical resistance. Another class of calibrants exploits a systematically varying pressure–property relation; for this, highest precision obtains when the property exhibits a large pressure effect and can be readily measured. Examples are unit-cell volume of simple metals (*P-V* equations of state) determined by x-ray diffraction. Pb and Pt equations of state have been determined by shock-wave compression to 600 GPa for use as x-ray standards for static pressure experiments (Holmes *et al.*, 1989) (see SOLIDS, DYNAMIC HIGH-PRESSURE EFFECTS IN). Another class comprises spectroscopic techniques, the most popular being ruby fluorescence, which has been calibrated to 180 GPa against equations of state of simple metals (Bell *et al.*, 1986).

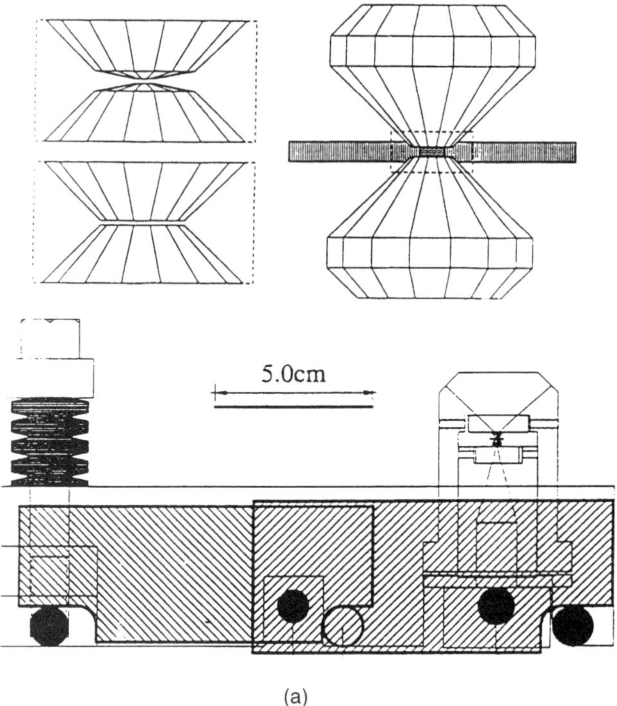

(a)

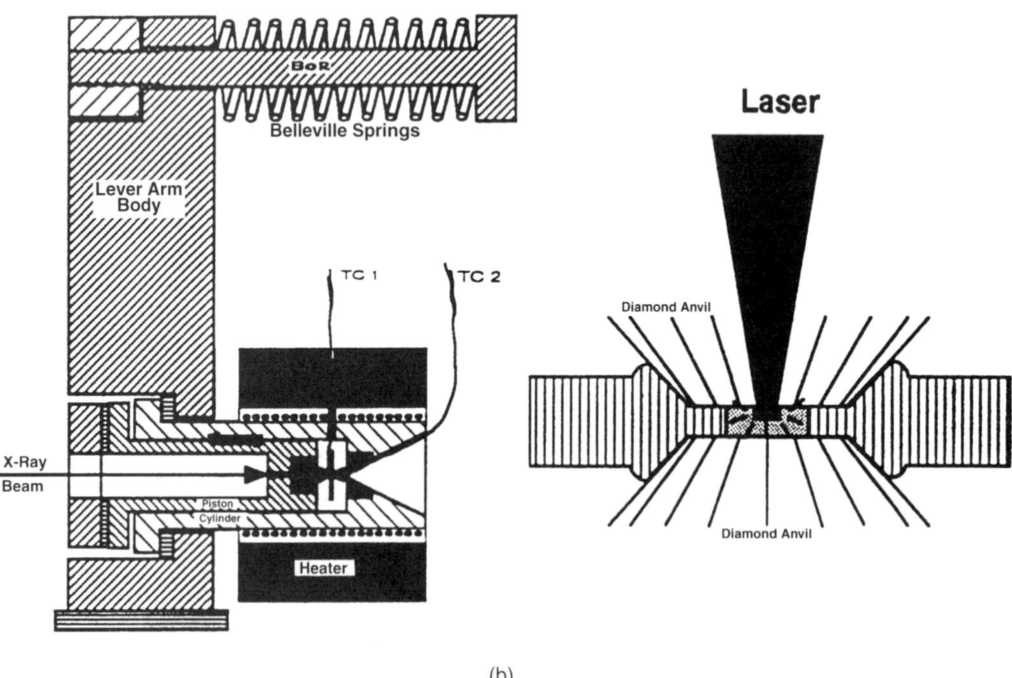

(b)

FIG. 2. (a) Diamond-anvil cell design. Top left: Beveled anvils used for reaching multimegabar pressures, and flat anvils used for lower-pressure studies. Top right: Opposed brilliant-cut diamond anvils and gasket. Bottom: Multimegabar diamond-anvil cell with piston-cylinder and two-stage lever arm. (b) High-temperature/high-pressure methods with the diamond-anvil cell. Left: Resistively (externally) heating technique. Right: Laser (internal) heating technique.

1.4 Variable Temperature

1.4.1 Cryogenic Methods In principle, there are no limits on the lowest temperatures for high-pressure studies. A variety of lower-pressure devices (defined here as those achieving maximum pressures of less than 7 GPa), including piston cylinders, have been used to liquid helium temperatures. For practical purposes, however, again the small size and therefore low thermal load of anvil devices are ideally suited for low-temperature studies with flow and bath cryostats as well as refrigerators (Jayaraman, 1983). Diamond-anvil cell experiments into the megabar range (>100 GPa) have been performed with cryostats at liquid helium temperatures. There have also been a few studies at lower pressures (~1 GPa) to temperatures below 0.1 K.

1.4.2 Resistive Heating A variety of resistance techniques can be exploited for high-*P-T* studies, including both external and internal methods [Fig. 2(b)]. External furnaces provide accurate and controlled temperatures to ~1300 K, the primary limitation of which is the graphitization of diamond (i.e., it is unstable relative to graphite at low pressures). These heaters include large coils surrounding the entire assembly (which provide more uniform temperatures but typically lower maximum pressures) or smaller heating elements located in the immediate vicinity of the cell (providing higher temperatures but requiring accurate characterization of axial temperature gradients). Recently, the advantages of both types of furnaces have been realized by implementing both in a two-stage heating technique. Alternatively, the sample can be heated within the sample chamber with a miniature heater, which circumvents the problem of breakdown of diamond because the sample is heated locally and the hottest portion of the diamond is in its stability field.

1.4.3 Laser Heating Higher temperatures (>3000 K) can be reached in diamond-anvil cell experiments by heating samples within the cell with lasers. Crucial to this is absorbance of the sample at the appropriate wavelengths [Fig. 2(b)]. For this purpose, far-infrared CO_2 lasers operating at 10.6 μm and Nd-YAG (yttrium aluminum garnet) and Nd-YLF (yttrium lithium fluoride) lasers at 1.06 μm are used (Boehler, 1996; Jeanloz, 1990; Ming and Bassett, 1974). New developments include the use of laser-heating techniques in combination with synchrotron x-ray methods for simultaneous high-*P-T* diffraction measurements (Fiquet *et al.*, 1996; Yoo *et al.*, 1995).

2. HIGH-PRESSURE MEASUREMENTS

One of the significant features of current high-pressure research is the development of an increasingly wide range of *in situ* techniques that are being applied to the study of static pressure effects in solids. Table 1 lists some of the major categories of analytical techniques used in these studies. Because different techniques require different sample sizes, there is a wide variation in the pressure range possible for different classes of experiments with current technology. In general, great progress in microsampling techniques for probing materials at the highest pressures (>100 GPa) has resulted from recent advances on two fronts: radiation sources and the detector systems. Lasers and synchrotrons provide collimated beams of electromagnetic radiation many orders of magnitude more brilliant than those from conventional light sources. In addition, new spallation sources provide an unparalleled source of neutrons. Multichannel detector technology—such as energy-dispersive x-ray detectors, two-dimensional position-sensitive x-ray detectors, storage-phosphor x-ray imaging plates, image intensifiers, intensified diode-array detectors, charge-coupled detectors, and new Fourier-transform interferometer techniques—allows simultaneous detection in thousands of channels, thus providing sufficient measurement sensitivity for studying ultrasmall samples at ultrahigh pressures. These developments have also resulted in marked improvement in measurement accuracy and sensitivity at high pressures.

2.1 Diffraction

A variety of diffraction techniques are typically used to probe the structures of materials at high pressure (Jayaraman, 1983). These include x rays, neutrons, and optical diffraction experiments. The use of these re-

Table 1. Examples of various measurement techniques and materials studied at high pressures.

Technique	Materials (P_{max})
Diffraction	
X-ray	
Powder	CsI, Fe, W, Re, Pt, Pb (>300 GPa)[a,b]
Single crystal	SiO_2 (70 GPa),[c] $Ar(H_2)_2$ (150 GPa)[d]
Amorphous	SiO_2 glass (42 GPa)[e]
Neutron	
Crystalline	D_2 (30 GPa),[f] Fe (35 GPa)[g]
Amorphous	Liquid D_2 (1 GPa)[h]
Spectroscopy	
X-ray (EXAFS, XANES)	Br_2 (58 GPa)[i]
Optical	
Absorption	H_2 (~300 GPa),[j] Diamond (420 GPa)[k]
Luminescence	Ruby (>300 GPa)[l]
Reflectivity	W, Pt, Pb (>300 GPa)[k]
Multiphoton	KI, RbI (1 GPa)[m,n]
Raman	
Spontaneous Raman	Diamond (400 GPa)[k]
Coherent anti-Stokes Raman	H_2 (37 GPa)[n]
Hyper-Raman	$SrTiO_3$ (5 GPa)[o]
Infrared	H_2 (250 GPa)[j,p]
Acoustic	
Brillouin scattering	SiO_2 (58 GPa)[q]
Impulsive stimulated scattering	$(Mg,Fe)_2SiO_4$ (12 GPa)[r]
Ultrasonic interferometry	Mg_2SiO_4 (12 GPa)[s]
Acoustic emission	Si, Ge (70 GPa)[t]
Inelastic neuron scattering	H_2 (1 GPa)[u]
Mössbauer effect	FeO (100 GPa)[v]
Nuclear magnetic resonance	$FeSiF_6 \cdot H_2O$ (7 GPa),[w] H_2 (7 GPa)[x]
Nuclear quadrupole resonance	Cu_2O (2 GPa)[y]
Electron paramagnetic resonance	$CdCr_2Se_4$ (8 GPa)[z]
Transport	
Electrical	
Resistivity	$(Mg,Fe)SiO_3$ (100 GPa)[aa,bb]
Superconductivity	$HgBa_2Ca_{n-1}Cu_nO_{2n+2+d}$ (30 GPa),[cc] Se (60 GPa)[dd]
Thermoelectric power	LiH (50 GPa)[ee]
Magnetic	
Susceptibility	Sr (50 GPa)[ff]
Thermal conductivity	$(Mg,Fe)_2SiO_4$ (12 GPa)[s]
Macroscopic techniques	
Deformation and strength	Diamond (170 GPa),[gg,hh,ii] SiO_2 glass (80 GPa)[jj]
Macroscopic length	SiO_2 glass (10 GPa)[kk]

[a]Mao *et al.* (1989). [b]Ruoff *et al.* (1992). [c]Mao *et al.* (1994). [d]Loubeyre, (1996). [e]Meade *et al.* (1992). [f]Glazkov *et al.* (1988). [g]Shil'shtein *et al.* (1984). [h]Ishmaev *et al.* (1988). [i]Itie *et al.* (1986). [j]Mao and Hemley (1994). [k]Vohra (1992). [l]Xu *et al.* (1986). [m]Lipp *et al.* (1993). [n]Daniels *et al.* (1994). [o]Yamanaka and Inoue (1989). [p]Hanfland *et al.* (1993). [q]Zha *et al.* (1994). [r]Zaug *et al.* (1992). [s]Li *et al.* (1995). [t]Meade and Jeanloz (1989). [u]Kobelev *et al.* (1989). [v]Pasternak *et al.* (1993). [w]Vaughan *et al.* (1971). [x]Lee *et al.* (1989). [y]Reyes *et al.* (1992). [z]Sakai and Pifer (1985). [aa]Peyronneau and Poirier (1989). [bb]Li and Jeanloz (1991). [cc]Gao *et al.* (1994). [dd]Akahama *et al.* (1993). [ee]Babushkin *et al.* (1993). [ff]Bireckoven and Wittig (1988). [gg]Mao *et al.* (1979). [hh]Mao and Hemley (1991). [ii]Weidner *et al.* (1994). [jj]Meade and Jeanloz (1988). [kk]Meade and Jeanloz (1987).

quires a sufficiently transparent material either as the gasket or as the anvil. The transparency of diamond to hard x rays, neutrons, and optical wavelengths makes the diamond-anvil cell an ideal device for a wide range of diffraction studies over the broadest pressure regime. The development of techniques utilizing synchrotron radiation to probe materials has revolutionized many areas of science. Synchrotron radiation is intense, polarized, and highly collimated light emitted over a wide continuum of the spectrum by electrons accelerated in circular orbits to the velocities approaching that of light. In addition, the radiation is produced with a precise time structure that can be exploited for time-resolved and kinetic studies of samples under pressure (Skelton *et al.,* 1991). By far, the most commonly used radiation is the hard and soft x-ray region for diffraction, scattering, and spectroscopic applications.

2.2 Spectroscopy

High-pressure optical spectroscopic methods provide crucial information on behavior of solids at very high pressures (Hemley *et al.,* 1987; Jayaraman, 1983). Vibrational spectroscopy provides detailed information on bonding properties, phase transitions, and thermodynamic properties. Electronic spectroscopy includes absorption, reflectivity, and various luminescence techniques. Again, the development and application of intense synchrotron sources for spectroscopy has been highly significant for high-pressure studies—for example, in the lower energy range, from near- to far-infrared wavelengths. Laser techniques using visible radiation are particularly useful for studies at very high pressures because the laser beam can in principle be focused down to 1 μm, or near the diffraction limit of the radiation, to provide high spatial resolution within a sample [e.g., at ultrahigh pressures above 300 GPa (Hemley *et al.,* 1987)]. In recent years, there have been growing applications of nonlinear optical techniques to study electronic transitions, acoustic velocities, and thermal conductivity at high pressure (Daniels *et al.,* 1994; Zaug *et al.,* 1992). High-pressure microspectroscopic methods have been extended to infrared wavelengths with the development of synchrotron infrared techniques (Hanfland *et al.,* 1993). Other spectroscopic methods include Mössbauer spectroscopy, which has measured to pressures near 100 GPa (Pasternak *et al.,* 1993), and EXAFS, measured to as high as 58 GPa (Itie, 1988).

2.3 Transport

The effect of pressure on the electrical resistance of solids was first studied systematically by Bridgman using a variety of large-volume apparatus. Important subsequent measurements with large-volume apparatus include the resistivity studies of the graphite-to-diamond transition and the phase diagram of carbon (Bundy, 1963; Togaya *et al.,* 1994). Investigations at the highest pressures typically use the four-lead technique with the diamond-anvil cell. The technique was first used in the 10-GPa range by Mao and Bell (1981) in a study of the electrical conductivity of (Mg,Fe)O to 30 GPa. More recently, measurements have been carried out at high temperature and pressure using simultaneous resistance heating or laser-heating techniques (Knittle and Jeanloz, 1986; Li and Jeanloz, 1991; Peyronneau and Poirier, 1989). Direct resistance techniques for determining the pressure dependence of superconductivity have been developed at low temperatures (Akahama *et al.,* 1993; Erskine *et al.,* 1986). Recently, the technique has been used to investigate the pressure dependence of superconductivity in high-temperature superconductors (Gao *et al.,* 1994). At the highest pressures, the materials in the sample chamber deform, causing the leads and contacts to the sample to break or insulation to be damaged. To overcome these problems, and to reach still higher pressures, microlithographic techniques have been used to construct microcircuits on the tips of diamond anvils (Hemmes *et al.,* 1989).

A number of techniques have been developed for the study of the effect of pressure on magnetic properties—for example, variations in magnetic moment with pressure, paramagnetic–diamagnetic transitions, and superconductivity. Magnetic methods can thus be used to study changes in electrical properties (e.g., superconductivity), thereby obviating the need for introducing electrical leads into the sample chamber (Timofeev, 1992). The use of inductive techniques to detect superconductivity through the Meissner effect is considered the most reliable. Such experiments can utilize superconducting flux transformers, multiple modulation techniques, and SQUIDs (Bireckoven and Wittig, 1988; Wijngaarden *et al.,* 1991).

2.4 Macroscopic

Another class of techniques is that involving measurements and observations of the entire sample at very high pressures (or stresses). Griggs first showed that solid materials deform and undergo diffusive creep under pressure (see Griggs and Handin,

1960). More recently, such measurements of deformation and material strength have been extended to higher pressures with diamond-anvil cell techniques (e.g., Meade and Jeanloz, 1988; Sung *et al.*, 1977). The capability of such cells for *in situ* observations also permits direct observations of compressibility from changes in macroscopic sample dimensions (Meade and Jeanloz, 1987). In addition, the pressure dependence of phase transitions such as melting and crystallization can be directly observed (e.g., Lu *et al.*, 1991; Williams *et al.*, 1987).

3. EXAMPLES OF HIGH-PRESSURE STUDIES

Below we give a few examples of the application of new high-pressure techniques to scientific problems. These applications typically involve the use of a number of the techniques described above.

3.1 Fundamental Physics

Pressure of several hundred gigapascals causes the density in a solid to increase as much as 50% for the most incompressible materials (e.g., ceramics and many metals), and 1000% for compressible ones (e.g., solidified gases). A variety of pressure-induced transitions typically occur over this range of compressions. These transitions include crystallographic phase transitions (both discontinuous and continuous), electronic and magnetic transitions, and novel transitions involving localization of excitations (both vibrational and electronic).

3.1.1 Insulator–Metal Transitions With increasing density in a material, electrons in general become less stable localized in bonds and near atomic cores. From the early days of quantum theory it was predicted that above a critical density electrons in all materials must delocalize into conduction states, thus forming a metal (Wigner and Huntington, 1935). The metallization process could involve, for example, a change in molecular or crystallographic structure, or an overlap of electronic valence and conduction bands. The insulator–metal transition in iodine at 16 GPa is well established (Balchan and Drickamer, 1961) and has become a classic example of such a pressure-induced insulator–metal transition. Recently, phase transitions, crystal structures, and electronic properties of iodine have been characterized well into the metallic state at multimegabar pressures (276 GPa) by optical and x-ray diffraction measurements and band-structure calculations (Reichlin *et al.*, 1994). Another example of interesting high-pressure behavior is provided by the pressure-induced transitions in Xe, a rare-gas solid with van der Waals bonding, and CsI, an alkali halide with ionic bonding. Despite their very different bonding properties at ambient pressure, both materials transform to metals above approximately 150 GPa (Goettel *et al.*, 1989; Reichlin *et al.*, 1989, 1987). Moreover, in addition to the convergence in bonding properties, both materials transform to an hcp-like structure and have identical equations of state at these pressures (Jephcoat *et al.*, 1987; Mao *et al.*, 1989).

3.1.2 Hydrogen The behavior of hydrogen at very high pressure is a fundamental problem in condensed matter physics. At low pressures and temperatures, hydrogen is an insulating molecular solid. However, it has been predicted theoretically that at extreme pressures, the molecules will dissociate to form a monatomic metal (Wigner and Huntington, 1935). Moreover, this hypothetical monatomic phase of dense hydrogen was subsequently predicted to exhibit several unusual properties, including very high-temperature superconductivity (Ashcroft, 1968) and a liquidlike ground state (Brovman *et al.*, 1972). The characterization of hydrogen at very high densities has become a major goal of high-pressure research. There has been marked progress in the effort in recent years (Mao and Hemley, 1994). Experiments demonstrate that the molecular bond is stable to at least ~250 GPa (Hemley and Mao, 1988), but several intriguing changes in bonding and physical properties are observed at intermediate pressures (Fig. 3). The changes can be understood in terms of a series of symmetry-breaking transitions with increasing pressure and decreasing temperature (to phases II and III). At higher pressures, there is a drop in the intramolecular stretching mode frequency and a large increase in its absorption strength (phase III) (Hanfland *et al.*, 1993; Hemley and Mao, 1988). Notably,

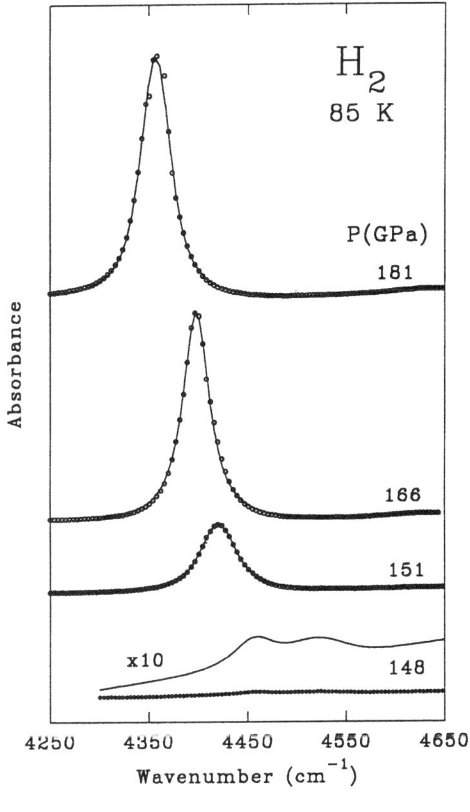

(a)

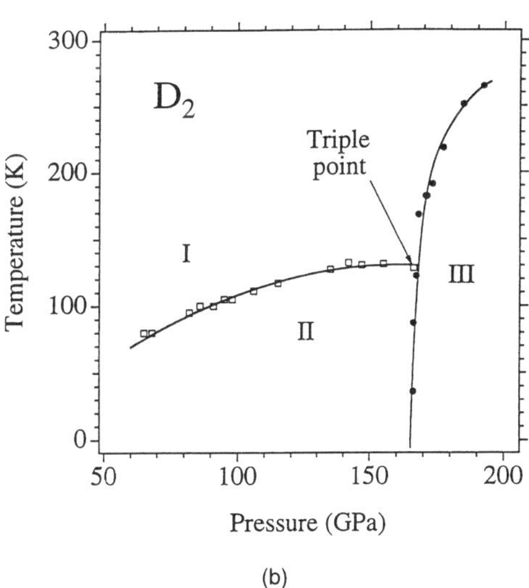

(b)

FIG. 3. High-pressure studies of solid hydrogen isotopes (Mao and Hemley, 1994). (a) Synchrotron infrared absorption spectroscopy of hydrogen showing the intensity enhancement at the transition at 150 GPa to phase III. (b) Phase diagram of deuterium at 100–200 GPa. The boundary between I and III terminates at a higher-temperature invariant point, above which the transition becomes continuous.

the material remains transparent in the visible and midinfrared spectrum (Eggert *et al.*, 1991; Hemley *et al.*, 1996), indicating that metallization occurs at higher pressures.

Despite its apparent simplicity, accurate theoretical calculations for dense hydrogen are particularly difficult and pose a difficult challenge to modern condensed-matter theory. This difficulty arises from the significant quantum character of the atomic nuclei (zero-point energy) associated with their low mass, which requires the proton and electron dynamics to be treated on the same theoretical footing (Ashcroft, 1993). This has been done in recent quantum Monte Carlo calculations for both the atomic and molecular phases at high pressure (Natoli *et al.*, 1993). The large zero-point motion can play a significant role in stabilizing structures [e.g., the diamond structure for the atomic phase (Natoli *et al.*, 1993)]. In addition, the unusual dynamics of the dense solid may give rise to new phenomena such as new types of pairing interactions and high-T_c superconductivity in the theoretically predicted molecular metallic state (Moulopoulos and Ashcroft, 1991).

3.1.3 Superconductivity Physical properties of fundamental and technological interest, such as superconductivity, can also be favorably tuned or enhanced by pressure. Understanding pressure effects on T_c played a major role in the discovery of the first 90-K cuprate superconductors (Wu *et al.*, 1987). The recently discovered mercury-based oxides $HgBa_2Ca_{n-1}Cu_nO_{2n+2+d}$ ($n = 1, 2, 3$) have superconducting onset temperatures (T_c) of 94, 128, and 135 K, respectively (Gao *et al.*, 1994). Their T_c's were found to increase with pressure. When studied in a diamond cell with the four-lead electrical conductance technique (maximum pressure of 45 GPa), record high T_c's of 118, 154, and 164 K were reached (Fig. 4).

3.1.4 Semiconductors There are also interesting high-pressure effects in semiconductors, in part due to processes associated with changing the band gap, including closure of the gap at moderate pressures and crystallographic transitions to metallic states (see Yu *et al.*, 1995). The prototype example of the latter is the transformation in Si from the semiconducting diamond structure to the metallic β-Sn structure at 12 GPa (Jayara-

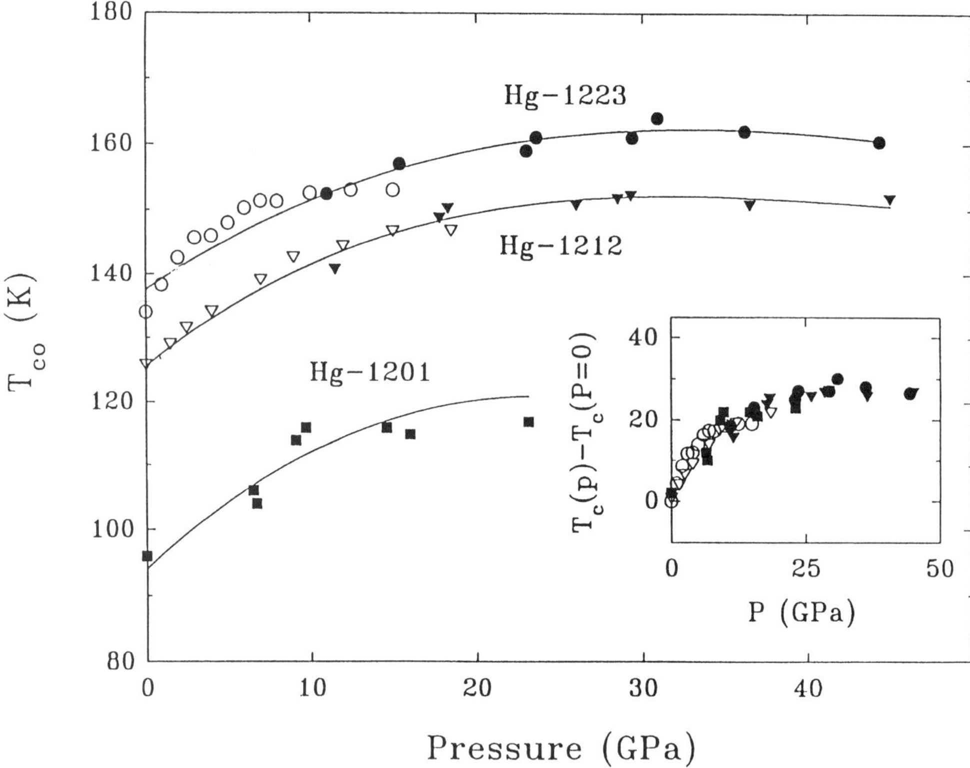

FIG. 4. Pressure dependence of the superconducting onset temperature T_{co} for $HgBa_2Ca_{n-1}Cu_nO_{2n+2+d}$ (Gao *et al.*, 1994).

man, 1983). Recent diffraction studies reveal that a number of the high-pressure phases in semiconductors have structures that are more complex than previously thought (McMahon *et al.*, 1994). Recent studies have also revealed unusual pressure effects on deep-level impurity centers and electronic levels in multiple quantum well and superlattice structures (Paul *et al.*, 1994).

3.1.5 Melting After centuries of study, a quantitative and complete microscopic understanding of melting is still not in hand. Pressure studies provide important tests of theories of melting. The form (slope) of the melting curve under pressure provides information on the structural and thermodynamic changes in the solid and liquid as a function of pressure. There have been numerous debates about the melting behavior of specific materials at very high pressure, including melting of materials that constitute interiors of the planets (Boehler, 1996). Whether or not a melting line will in general terminate in a critical point like the vaporization curve has been a recurring question in high-pressure physics. Experimental work to this day has failed to find evidence for such solid–liquid critical behavior. However, if sufficiently extreme conditions are reached, new phenomena associated with atomic breakdown are expected. The plasma phase transition in fluid hydrogen has been predicted to occur at 8000 K and 190 GPa and has a negative *P-T* slope (Saumon and Chabrier, 1989). The nature of such transitions, including the associated critical phenomena, is an important question addressed in current research.

3.1.6 Pressure Amorphization When a crystalline material is compressed to pressures well outside of its stability field, it may convert to an amorphous material if the temperature is too low for it to recrystallize to the equilibrium high-pressure crystalline phase. This unusual transformation was first documented in ice I; instead of transforming to the equilibrium crystalline phase, ice I transforms to a dense glass when compressed above 1 GPa at 77 K (Mishima *et al.*, 1984). Such pressure-induced amorphization transformations have been observed in a growing number of materials during the past ten years (Hemley *et al.*, 1994). These transformations may access unusual metastable states with varying degrees of disorder and possibly unique properties.

3.1.7 Phonon Localization Transitions

Pressure can also be used to study unusual transitions associated with the localization and delocalization of excitations. Most solids can be described by the harmonic approximation; in some others, anharmonicity prevails. In the harmonic case, noninteracting phonons couple strongly and form a broad continuum band (Kimball *et al.*, 1981). In the anharmonic case, strong interaction of phonons leads to the formation of bound states only weakly coupled in the crystal. For example, the bound state in the vibrational spectrum forms a sharp peak separated from the origin of the broad two-phonon continuum band by the quantity of its anharmonicity. If the relative magnitudes of the continuum band width and the anharmonicity could be varied so that the former would exceed the latter, it was postulated, the bound two-phonon bound state would disappear into the continuum band. Such a transition was observed for the first time by varying pressure in solid deuterium (Fig. 5) (Eggert *et al.*, 1993). At ambient condition, the two-phonon bound state is well separated from the continuum, but at high pressures the continuum bandwidth, which is a measure of the intermolecular coupling, increases sharply. The two become equal at 36 GPa, and indeed, the two-phonon band vanishes into the continuum.

3.2 Chemistry and Materials Science

High-pressure methods up to several gigapascals have been used for years for synthesis of materials—for example, in the growth of high-quality single crystals using hydrothermal and piston–cylinder techniques. However, at higher pressures, bonding properties in materials can be altered considerably, in some cases drastically, as the relative energies of the valence electrons change in response to compression. Thus, chemical af-

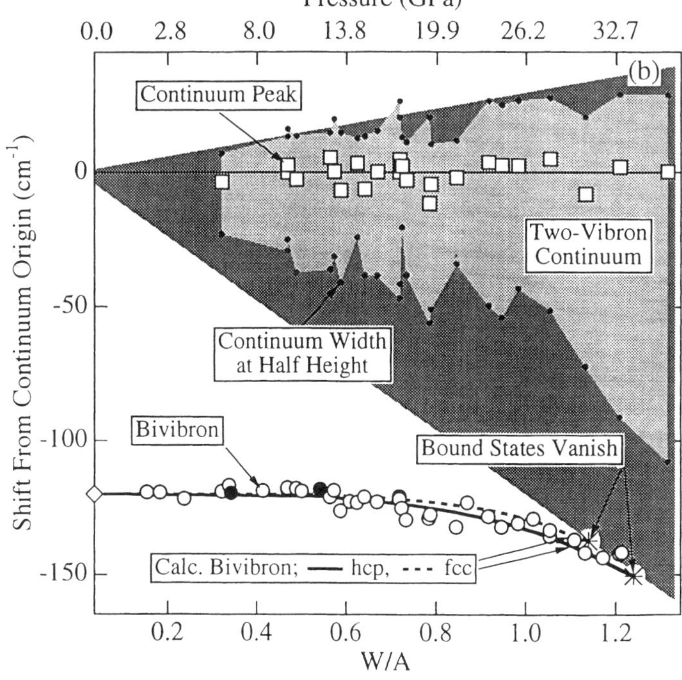

FIG. 5. Pressure-induced two-phonon ("bivibron") bound–unbound transition in solid deuterium. The shaded area shows the continuum bandwidth, *W*. *A* is anharmonicity (Eggert *et al.*, 1993).

finities of the elements can become very different from those at ambient pressure. Thus, new metastable materials with novel or otherwise useful physical and chemical properties can be tailored by appropriate high-pressure synthesis. In this regard, it is important to examine the mechanical stability on decompression of solid phases that are metastable at ambient pressure.

3.2.1 Diamond and Superhard Materials Probably the most celebrated example of high-pressure synthesis was the production of diamond by General Electric Co. in 1955 (see Bundy *et al.,* 1996). This advance was important for demonstrating the economic and technological utility of high-pressure research as well as for initiating new and innovative static high-pressure techniques in general. [Subsequently, diamond was synthesized by shock-wave methods in 1961 (DeCarli and Jamieson, 1961).] Together with the development of chemical vapor deposition techniques for diamond growth, high-pressure diamond synthesis remains a major industry today. The primary product of this industry is diamond powders for abrasives. A highly significant advance in diamond synthesis in the 1990s is the growth of large, perfect single-crystal (gem quality) diamonds, including those that are close to being isotopically pure and having unique physical properties, such as exceptionally high thermal conductivity (Wei *et al.,* 1993) and strength (Vohra *et al.,* 1994). This breakthrough is turning out to be very important for high–heat-load optics applications (e.g., synchrotron radiation research) as well as for anvils in high-pressure experiments. Diamond is predicted to transform to a metallic phase at pressures above 1.2 TPa (12 Mbar) (Biswas *et al.,* 1984; Yin and Cohen, 1983). So far, these pressures have not been reached in static experiments; diamond-cell experiments using natural diamonds do reveal signs of transformations and plastic flow under the extreme nonhydrostatic stresses produced at the tips of the anvils (Mao and Hemley, 1991).

Unusual transformations in graphite to a diamond-like phase have been documented under pressure [pressure-induced metal-to-insulator transition (Bundy *et al.,* 1996)]. The high-pressure phase has the characteristics of a superhard material like diamond. New superhard phases have also been produced from subjecting fullerenes to high pressures (Nunez-Regueiro *et al.,* 1995). The theoretical prediction of superhard carbon nitrides, in particular β-C_3N_4 (Liu and Cohen, 1989), has inspired a large number of experimental and theoretical investigations. Recently, a new class of such materials having a bulk modulus (incompressibility) greater than that of diamond has been predicted theoretically (Teter and Hemley, 1996).

3.2.2 New Low-Z Compounds Inert-gas molecules interact weakly at low density through van der Waals forces. They also do not readily form stoichiometric compounds at ordinary pressure. However, such solid compounds can be made at high pressures, as was first demonstrated with the synthesis of $He(N_2)_{11}$ (Vos *et al.,* 1992). Since then, a number of such compounds of this novel type, $NeHe_2$, $Ar(H_2)_2$, and several in the CH_4–H_2 system, have been observed at high pressures (Loubeyre, 1996). Their crystal structures and stoichiometry can be understood in terms of molecular packing efficiency. Clathrate hydrates consist of hydrogen-bonded networks of cages containing larger guest molecules typically bound by van der Waals forces. They are unstable at moderate pressures, as the open networks break down under compression. In a high-pressure study of the H_2–H_2O binary system, a novel type of clathrate with 1:1 ratio was discovered (Vos *et al.,* 1993) (Fig. 6). In this high-pressure clathrate, H_2O and H_2 form two interlocking networks, both with the diamond structure. With the efficient packing of molecules afforded by the structure, the clathrate is stable to at least 30 GPa.

3.2.3 New Amorphous Materials A number of amorphous materials, including high-silica glasses, can be irreversibly compacted by pressure; that is, a densified form can be retrieved on quenching to ambient conditions (Hemley *et al.,* 1994). This provides materials with a range of optical properties, densities, and chemical properties that can be continuously tuned by changing synthesis conditions. New quenchable nitride glasses with high refractive indices have been produced in recent high-pressure studies (Grande *et al.,* 1994).

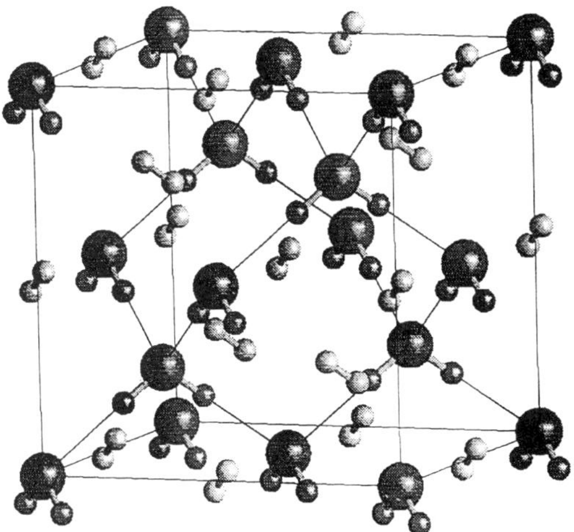

FIG. 6. Crystal structure of the high-density clathrate. The dumbbells are idealized representations of the rotationally disordered hydrogen molecules; the large spheres denote the oxygens of the H_2O molecules (Vos *et al.*, 1993).

3.2.4 Biological Materials There are interesting effects of pressure on biological materials, and most of these changes occur at comparatively low pressures (e.g., below 1 GPa). These include pressure-induced denaturation of proteins, an effect first observed by Bridgman in 1914 in egg albumen. Pressure is also used for food purification (similar to heat treatment or pasteurization) (Hayashi, 1991). The stability of micro-organisms at high temperatures and moderate pressures (<0.1 GPa) is important for understanding the extraordinary biological activity associated with volcanic seafloor vents (e.g., "black smokers").

3.3 Earth and Planetary Science

Laboratory experiments utilizing high pressure and/or high temperature offer the only means to examine directly the chemistry and physical properties of materials that constitute deep planetary interiors.

3.3.1 Earth's Crust Applications of the new techniques described above are providing new insights into a number of high-pressure phenomena associated with the properties and processes of deeper portions of the Earth's crust. Recently, large enhancements of solubility of minerals, such as quartz in aqueous solutions, have been observed under lower crustal pressures (i.e., 0.5–2 GPa) (Manning, 1994). Pressure- and temperature-induced dehydration reactions of low-pressure hydrous minerals have been observed directly, including observations of the kinetics of such transformations using synchrotron x-ray diffraction. Diamond-anvil studies have also been used to determine the equations of state of aqueous fluids to deep crustal conditions (Shen *et al.,* 1992). In addition, the methods provide direct observation of crystal growth of crustal minerals.

3.3.2 Earth's Mantle Studies of pressure-induced polymorphism in minerals have been particularly important for developing progressively more refined models of the mineralogy and composition of the Earth's deep mantle (Liu and Bassett, 1986). The general role of phase transitions in the Earth's deep interior, and how these may give rise to the well-known seismic wave discontinuities, was articulated by Birch (1952). The olivine–spinel transformation was discovered in silicates and investigated in detail in the 1960s, and it remains the leading hypothesis for explaining the 400-km discontinuity. Magnesium metasilicate (enstatite, $MgSiO_3$) transforms to the perovskite-type phase (Fig. 7) (Liu, 1975) at pressures corresponding to the second major seismic discontinuity in the Earth's mantle at 660 km (or 23 GPa). Moreover, silicate minerals in general transform to perovskite-dominated assemblages at lower mantle pressures. This has given rise to the view that (Mg, Fe)SiO_3 perovskite is the dominant mineral in the Earth's lower mantle, and as a result of the large volume of this region, it is likely to be the most abundant mineral in the planet as a whole.

Pressure-induced polymorphism of SiO_2—in particular, the transition from quartz to the high-pressure phases coesite (Coes, 1953) and stishovite (Stishov and Popova, 1961)—had a significant impact on understanding of the deep solid Earth. The higher-pressure form is a tetragonal structure with the rutile-type structure and involves a change in coordination of Si by O from fourfold (the coordination found in crustal silicates) to sixfold (the typical coordination found in the deep mantle). Recent diamond-cell experiments demonstrate that stishovite transforms

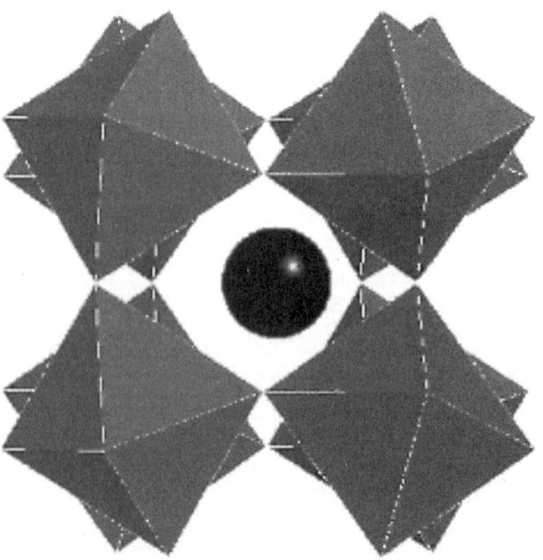

FIG. 7. Crystal structure of orthorhombic (Mg, Fe)SiO_3 perovskite, the dominant phase of the lower mantle of the Earth and most abundant mineral in the planet. The structure consists of corner-sharing SiO_6 groups (octahedra) with Mg and Fe ions (spheres) in the larger crystallographic sites.

to an orthorhombic form near 50 GPa (Kingma *et al.,* 1995). In all of this work, theory has played an important role by providing both fundamental understanding of observed mineral behavior and predictions to be tested subsequently by experiment (Bukowinski, 1994; Cohen, 1994).

3.3.3 Earth's Core *In situ* diamond-anvil cell studies have been instrumental in the characterization of polymorphism and melting of iron to core conditions (Boehler, 1996; Mao *et al.,* 1990; Williams *et al.,* 1987). Although theories for the dominant role of metallic hydrogen in the Earth's core were eventually discredited on the basis of the low density predicted for metallic hydrogen relative to that of the core, the effect of alloying hydrogen with iron (an effect that does not occur at low pressures) was not ruled out. The latter idea gained some favor beginning in the 1980s when iron hydride was shown to be stable under pressure (Badding *et al.,* 1991).

3.3.4 Other Planets High-pressure studies are also providing new insights into the nature of the interiors of other planets in the solar system. Experiments on silicates, oxides, and sulfides yield information relevant to other terrestrial planets such as Mars. The large gaseous planets are believed to consist of approximately 90% hydrogen, with most of this material existing at ultrahigh pressures in a fluid metallic state. Together with shock-wave studies, static high-pressure investigations of hydrogen and helium, together with other low-Z materials (e.g., the C–H–O–N system), thus provide an experimental basis for understanding the structure, composition, dynamics, and evolution of the large gaseous planets. In addition, studies of low-Z molecular materials (e.g., CH_4, NH_3, CO_2, H_2O) are relevant to the moons that orbit these bodies (Johnson and Nicol, 1986). Of particular interest is the observation of new solid compounds in mixtures of simple molecular materials (Vos *et al.,* 1993).

4. OUTLOOK

4.1 Higher Pressures

The upper limit of static pressure techniques is currently unknown. Efforts to reach the highest static pressures continue (Mao *et al.,* 1989; Ruoff *et al.,* 1992; Xu *et al.,* 1986). In principle, the diamond cell is limited by possible ultrahigh-pressure transformations in diamond. Studies to >300 GPa reveal no evidence for metallization of diamond; as discussed above, theory predicts that this should occur at significantly higher hydrostatic pressure. However, even this may not represent a practical limit since the lower-pressure insulating diamond may be metastable above the transition (just as diamond is metastable relative to graphite at ambient pressure). Stress-induced transition and plastic flow has been documented in regions of high stress concentration localized at the anvil tips, but this may be circumvented by further refinements of cell designs. Further advances may result from the discovery and synthesis of new superhard materials for anvils. These advances may also be useful for the related goal of enlarging sample volumes in order to improve the accuracy and precision of measurements as well as permitting a still broader spectrum of techniques to be used to probe materials under extreme pressure conditions.

4.2 New Analytical Techniques

High-pressure research is still in the early stages of utilizing a number of new analytical techniques. These include, for example, the construction of ultrabright synchrotron sources (such as the European Synchrotron Radiation Facility in France, the Advanced Photon Source in the United States, and SPring-8 in Japan), which will make new classes of diffraction, spectroscopic, and imaging experiments possible with high-pressure devices. Also, new spectroscopic measurements will become possible with the continued development of infrared and ultraviolet synchrotron sources, including those utilizing coherent radiation (free-electron lasers). There are efforts to develop lithographic techniques for producing microcircuits on anvil surfaces [perhaps in conjunction with CVD diamond (McCauley and Vohra, 1995)] for electrical and magnetic measurements at multimegabar pressures. There is also the prospect of observing new phenomena at ultralow (millikelvin) temperatures and megabar pressures, currently uncharted territory.

5. SUMMARY

With recent breakthroughs in high-pressure technology, a wide range of measurements can now be conducted *in situ* on compressed solids over an unprecedented range of pressures. In the small number of systems investigated to date, a variety of pressure-induced transitions have been observed in solids under ultrahigh pressures. It is conceivable that novel physical phenomena may emerge with detailed study of matter in the new pressure dimension made possible by the development of new techniques. Analogy could be drawn to the discoveries of superfluidity and superconductivity. These properties were unpredicted and unexpected and could only be found with the application of extreme temperature conditions and careful experimental observations. Searching for novel physical phenomena that are unique to extreme pressure conditions requires similar efforts.

Works Cited

Akahama, Y., Kobayashi, M., Kawamura, H., Endo, S. (1993), *Rev. Sci. Instrum.* **64,** 1979–1983.

Ashcroft, N. W. (1968), *Phys. Rev. Lett.* **21,** 1748–1750.

Ashcroft, N. W. (1993), *J. Non-Cryst. Solids* **156–158,** 621–630.

Babushkin, A. N., Pilipenko, G. I., Gavrilov, F. F. (1993), *J. Phys. F: Condensed Matter* **5,** 8659–8664.

Badding, J. V., Hemley, R. J., Mao, H. K. (1991), *Science* **253,** 421–424.

Balchan, A. S., Drickamer, H. G. (1961), *J. Chem. Phys.* **34,** 1948–1949.

Bell, P. M., Xu, J., Mao, H. K. (1986), in: Y. Gupta (Ed.), *Shock Waves in Condensed Matter,* New York: Plenum, pp. 125–130.

Birch, F. (1952), *J. Geophys. Res.* **57,** 227–286.

Bireckoven, B., Wittig, J. (1988), *J. Phys. E.: Sci. Instrum.* **21,** 841–848.

Biswas, R., Martin, R. M., Needs, R. J., Nielsen, O. H. (1984), *Phys. Rev. B* **30,** 3210–3213.

Boehler, R. (1996), *Philos. Trans. R. Soc. London A* **354,** 1265–1278.

Brovman, E. G., Kagan, Y., Kholas, A. (1972), *Sov. Phys. JETP* **34,** 1300–1315.

Bukowinski, M. S. T. (1994), *Ann. Rev. Earth Planet. Sci.* **22,** 167–205.

Bundy, F. P. (1963), *J. Chem. Phys.* **38,** 618–630.

Bundy, F. P., Weathers, M. A., Bassett, W. A., Hemley, R. J., Mao, H. K., Goncharov, A. F. (1996), *Carbon* **34,** 141–153.

Coes, L. (1953), *Science* **118,** 131–132.

Cohen, R. E. (1994), in: P. Heaney, G. V. Gibbs, C. T. Prewitt (Eds.), *Reviews of Mineralogy,* Washington, D.C.: Mineralogical Society of America, pp. 369–402.

Daniels, W. B., Lipp, M., Strachan, D., Yoo, C., Zhang, H. M., Yu, Z. (1994), in: S. C. Schmidt, J. W. Shaner, G. Samara, M. Ross (Eds.), *High-Pressure Science and Technology—1993,* New York: American Institute of Physics, pp. 1499–1502.

DeCarli, P. S., Jamieson, J. C. (1961), *Science* **133,** 1821–1822.

Eggert, J. H., Moshary, F., Evans, W. J., Lorenzana, H. E., Goettel, K. A., Silvera, I. F., Moss, W. C. (1991), *Phys. Rev. Lett.* **66,** 193–196.

Eggert, J. H., Mao, H. K., Hemley, R. J. (1993), *Phys. Rev. Lett.* **70,** 2301–2304.

Erskine, D., Yu, P. Y., Chang, K. J., Cohen, M. L. (1986), *Phys. Rev. Lett.* **57,** 2741–2744.

Fiquet, G., Andrault, D., Itie, J. P., Gillet, P., Richet, P. (1996), *Phys. Earth Planet. Sci.,* in press.

Gao, L., Xue, Y. Y., Chen, F., Xiong, Q., Meng, R. L., Ramirez, D., Chu, C. W., Eggert, J. H., Mao, H. K. (1994), *Phys. Rev. B* **50,** 4260–4263.

Glazkov, V. P., Besedin, S. P., Goncharenko, I. N., Irodova, A. V., Makarenko, I. N., Somenkov, V. A., Stishov, S. M., Shil'shteyn, S. S. (1988), *JETP Lett.* **47,** 661–664.

Goettel, K. A., Eggert, J. H., Silvera, I. F., Moss, W. C. (1989), *Phys. Rev. Lett.* **62,** 665–668.

Grande, T., Holloway, J. R., McMillan, P. F., Angell, C. A. (1994), *Nature* **369,** 43–45.

Griggs, D., Handin, J. (1960), in: D. Griggs, J. Handin (Eds.), *Rock Deformation,* Memoir 79, New York: Geological Society of America, pp. 347–364.

Hanfland, M., Hemley, R. J., Mao, H. K. (1993), *Phys. Rev. Lett.* **70,** 3760–3763.

Hayashi, R. (1991), *High Press Res.* **7,** 15–21.

Hemley, R. J., Mao, H. K. (1988), *Phys. Rev. Lett.* **61,** 857–860.

Hemley, R. J., Bell, P. M., Mao, H. K. (1987), *Science* **237,** 605–612.

Hemley, R. J., Prewitt, C. T., Kingma, K. J. (1994), in: P. Heaney, G. V. Gibbs, C. T. Prewitt (Eds.), *Reviews of Mineralogy,* Washington, D.C.: Mineralogical Society of America, pp. 41–81.

Hemley, R. J., Mao, H. K., Goncharov, A. F., Hanfland, M., Struzhkin, V. V. (1996), *Phys. Rev. Lett.* **76,** 1667–1670.

Hemmes, H., Driessen, A., Kos, J., Mul, F. A., Griessen, R., Caro, J., Radelaar, S. (1989), *Rev. Sci. Instrum.* **60,** 474–480.

Holmes, N. C., Moriarty, J. A., Gathers, G. R., Nellis, W. (1989), *J. Appl. Phys.* **66,** 2962–2967.

Ishmaev, S. I., Sadikov, I. P., Chernoyshov, A. A., Isakov, S. L., Vindryaeskii, B. A., Kobelev, G. V., Sukhoparov, V. A., Telepnev, A. S. (1988), *Sov. Phys. JETP* **68,** 1403–1406.

Itie, J. P., Jean-Louis, M., Dartyge, E., Fontaine, A., Jucha, A. (1986), *J. Phys. (Paris)* **47,** 897–900.

Jayaraman, A. (1983), *Rev. Mod. Phys.* **55,** 65–108.

Jeanloz, R. (1990), *Annu. Rev. Phys. Chem.* **40,** 237–259.

Jephcoat, A. P., Mao, H. K., Finger, L. W., Cox, D. E., Hemley, R. J., Zha, C. S. (1987), *Phys. Rev. Lett.* **59,** 2670–2673.

Johnson, M. L., Nicol, M. (1986), *J. Geophys. Res.* **92,** 6339–6349.

Kimball, J. C., Fong, C. Y., Shen, Y. R. (1981), *Phys. Rev. B* **23,** 4946–4959.

Kingma, K. J., Cohen, R. E., Hemley, R. J., Mao, H. K. (1995), *Nature* **374,** 243–245.

Knittle, E., Jeanloz, R. (1986), *Solid State Comm.* **13,** 1541–1545.

Kobelev, G. V., Sukhoparov, A. A., Telepnev, A. S., Vindryaevskyi, B. A., Ishmaev, S. N., Sadikov, I. P., Chernychov, A. A. (1989), in: N. V. Novikov, Y. M. Chistyakov (Eds.), *High Pressure Science and Technology, Proceedings of the XIth AIRAPT Conference,* Kiev: Naukova Dumka, pp. 294–298.

Lee, S. H., Conradi, M. S., Norberg, R. E. (1989), *Phys. Rev. B* **40,** 12492–12498.

Li, B., Liebermann, R. C., Gwanmesia, G. D. (1995), *Eos Trans. Am. Geophys. Union (Fall Meeting)* **76,** F619.

Li, X., Jeanloz, R. (1991), *J. Geophys. Res.* **96,** 6113–6120.

Lipp, M. J., Blechschmidt, J., Daniels, W. B. (1993), *Phys. Rev. B* **48,** 687–694.

Liu, A., Cohen, M. L. (1989), *Science* **245,** 841.

Liu, L.-g. (1975), *Geophys. Res. Lett.* **2,** 417–419.

Loubeyre, P. (1996), *Proceedings of the XV AIRAPT Conference,* in press.

Lu, G. Q., Nygren, E., Aziz, M. J. (1991), *J. Appl. Phys.* **70,** 5323–5345.

Manning, C. E. (1994), *Geochem. Cosmochem. Acta* **58,** 4831–4839.

Mao, H. K., Bell, P. M. (1981), *Rev. Sci. Instrum.* **52,** 615.

Mao, H. K., Hemley, R. J. (1991), *Nature* **351,** 721–724.

Mao, H. K., Hemley, R. J. (1994), *Rev. Mod. Phys.* **66,** 671–692.

Mao, H. K., Bell, P. M., Dunn, K. J., Chrenko, R. M., DeVries, R. C. (1979), *Rev. Sci. Instrum.* **50,** 1002–1009.

Mao, H. K., Wu, Y., Hemley, R. J., Chen, L. C., Shu, J. F., Finger, L. W. (1989), *Science* **246,** 649–651.

Mao, H. K., Wu, Y., Chen, L. C., Shu, J. F., Jephcoat, A. P. (1990), *J. Geophys. Res.* **95,** 21737–21742.

Mao, H. K., Shu, J., Hu, J., Hemley, R. J. (1994), *Eos Trans. Am. Geophys. Union (Fall Meeting)* **75,** 662.

McCauley, T. S., Vohra, Y. K. (1995), *Appl. Phys. Lett.* **66,** 1486–1488.

McMahon, M. I., Nelmes, R. J., Wright, N. G., Allan, D. R. (1994), in: S. C. Schmidt, J. W. Shaner, G. A. Samara, M. Ross (Eds.), *High-Pressure Science and Technology—1993,* New York: American Institute of Physics, pp. 629–636.

Meade, C., Jeanloz, R. (1987), *Phys. Rev. B* **35,** 236–244.

Meade, C., Jeanloz, R. (1988), *Science* **241,** 1072–1074.

Meade, C., Jeanloz, R. (1989), *Nature* **339,** 616–618.

Meade, C., Hemley, R. J., Mao, H. K. (1992), *Phys. Rev. Lett.* **69,** 1387–1390.

Ming, L. C., Bassett, W. A. (1974), *Rev. Sci. Instrum.* **45,** 1115–1118.

Mishima, O., Calvert, L. D., Whalley, E. (1984), *Nature* **310,** 393–395.

Moulopoulos, K., Ashcroft, N. W. (1991), *Phys. Rev. Lett.* **66,** 2915–2918.

Natoli, V., Martin, R. M., Ceperley, D. M. (1993), *Phys. Rev. Lett.* **70,** 1952–1955.

Nunez-Regueiro, M., Marques, L., Hodeau, J. L., Bethoux, O., Perroux, M. (1995), *Phys. Rev. Lett.* **74,** 3483–3486.

Pasternak, M., Taylor, R. D., Dillon, T., Jeanloz, R. (1993), *Eos Trans. Am. Geophys. Union (Fall Meeting Suppl.)* **74,** 416.

Paul, W., Burnett, J. H., Cheong, H. M. (1994), in: S. C. Schmidt, J. W. Shaner, G. Samara, M. Ross (Eds.), *High-Pressure Science and Technol-*

ogy—1993, New York: American Institute of Physics, pp. 545–548.

Peyronneau, J., Poirier, J. P. (1989), *Nature* **342,** 537–539.

Reichlin, R., Ross, M., Martin, S., Goettel, K. A. (1987), *Phys. Rev. Lett.* **56,** 2858–2861.

Reichlin, R., Brister, K. E., McMahan, A. K., Ross, M., Martin, S., Vohra, Y. K., Ruoff, A. L. (1989), *Phys. Rev. Lett.* **62,** 669–672.

Reichlin, R., McMahan, A. K., Ross, M., Martin, S., Hu, J., Hemley, R. J., Mao, H.-K., Wu, Y. (1994), *Phys. Rev. B* **49,** 3725–3733.

Reyes, A. P., Ahrens, E. T., Heffner, R. H., Hammel, P. C., Thompson, J. D. (1992), *Rev. Sci. Instrum.* **63,** 3120–3122.

Ruoff, A. L., Xia, H., Xia, Q. (1992), *Rev. Sci. Instrum.* **63,** 4342–4348.

Sakai, N., Pifer, J. M. (1985), *Rev. Sci. Instrum.* **56,** 726–731.

Saumon, D., Chabrier, G. (1989), *Phys. Rev. Lett.* **62,** 2397–2400.

Shen, A. H., Bassett, W. A., Chou, I.-M. (1992), in: Y. Syono, M. H. Manghnani (Eds.), *High-Pressure Research: Application to Earth and Planetary Sciences,* Tokyo: Terra Scientific; Washington, D.C.: American Geophysical Union, pp. 61–68.

Shil'shtein, S. S., Glazkov, V. P., Makarenko, I. N., Somenkov, V. A., Stishov, S. M. (1984), *Sov. Phys. Solid State* **25,** 1907–1909.

Skelton, E. F., Ayers, J. D., Qadri, S. B., Moulton, N. E., Cooper, K. P., Finger, L. W., Mao, H. K., Hu, Z. (1991), *Science* **253,** 1123–1125.

Stishov, S. M., Popova, S. V. (1961), *Geochemistry (USSR)* **10,** 923–926.

Sung, C.-M., Goetze, C., Mao, H. K. (1977), *Rev. Sci. Instrum.* **48,** 1386–1391.

Teter, D. M., Hemley, R. J. (1996), *Science* **271,** 53–55.

Timofeev, Y. A. (1992), *Prib. Tekh. Eksper.* **5,** 186–189.

Togaya, M., Sugiyama, S., Mizuhara, E. (1994), in: S. C. Schmidt, J. W. Shaner, G. Samara, M. Ross (Eds.), *High-Pressure Science and Technology—1993,* New York: American Institute of Physics, pp. 255–258.

Vaughan, R. W., Lai, C. F., Elleman, D. D. (1971), *Rev. Sci. Instrum.* **42,** 626–629.

Vohra, Y. K. (1992), in: A. K. Singh (Ed.), *Recent Trends in High Pressure Research,* New York: International Science, pp. 349–358.

Vohra, Y. K., McCauley, T. S., Gu, G., Vagarali, S. S. (1994), in: S. C. Schmidt, J. W. Shaner, G. A. Samara, M. Ross (Eds.), *High-Pressure Science and Technology—1993,* New York: American Institute of Physics, pp. 515–518.

Vos, W., Finger, L. W., Hemley, R. J., Hu, J. Z., Mao, H. K., Schouten, J. A. (1992), *Nature* **358,** 46–48.

Vos, W. L., Finger, L. W., Hemley, R. J., Mao, H. K. (1993), *Phys. Rev. Lett.* **71,** 3150–3153.

Wei, L., Kuo, P. K., Thomas, R. L., Anthony, T. R., Banholzer, W. F. (1993), *Phys. Rev. Lett.* **70,** 3764–3767.

Weidner, D. J., Wang, Y., Vaughan, M. T. (1994), *Science* **266,** 419–422.

Wigner, E., Huntington, H. B. (1935), *J. Chem. Phys.* **3,** 764–770.

Wijngaarden, R. J., Ennige, E. N., Scholtz, J. J., Hemmes, H. K., Griessen, R. (1991), in: R. Pucci, G. Piccitto (Eds.), *Molecular Systems under High Pressure,* Amsterdam: Elsevier, pp. 157–180.

Williams, Q., Jeanloz, R., Bass, J., Svendsen, B., Ahrens, T. J. (1987), *Science* **236,** 181–182.

Wu, M. K., Ashburn, J. R., Torng, C. J., Hor, P. H., Meng, R. L., Gao, L., Huang, Z. J., Wang, Y. Q., Chu, C. W. (1987), *Phys. Rev. Lett.* **58,** 908–910.

Xu, J., Mao, H. K., Bell, P. M. (1986), *Science* **232,** 1404–1406.

Yamanaka, A., Inoue, K. (1989), *Rev. Sci. Instrum.* **60,** 3253–3257.

Yin, M. T., Cohen, M. L. (1983), *Phys. Rev. Lett.* **50,** 2006–2009.

Yoo, C. S., Akella, J., Campbell, A. J., Mao, H. K., Hemley, R. J. (1995), *Science* 1473–1475.

Yu, P. Y., Paul, W., Weinstein, B. A. (Eds.) (1995), "Proceedings of the 6th International Conference on High Pressure Semiconductor Physics," *J. Phys. Chem. Solids* **56,** 311–672.

Zaug, J., Abramson, E. H., Brown, J. M., Slutsky, L. J. (1992), in: Y. Syono, M. H. Manghnani (Eds.), *High-Pressure Research: Application to Earth and Planetary Sciences,* Tokyo: Terra Scientific; Washington, D.C.: American Geophysical Union, pp. 157–166.

Zha, C. S., Hemley, R. J., Mao, H. K., Duffy, T. S., Meade, C. (1994), *Phys. Rev. B* **50,** 13105–13112.

Further Reading

Bridgman, P. W. (1949), *The Physics of High Pressure,* London: G. Bell and Sons.

Hazen, R. M. (1993), *The New Alchemists,* New York: Random House.

Ichimaru, S., Ogata, S. (Eds.) (1995), *Elementary Processes in Dense Plasmas,* New York: Addison-Welsey.

Liu, L.-g., Bassett, W. A. (1986), *Elements, Oxides, Silicates,* New York: Oxford Univ. Press.

Schmidt, S. C., Shaner, J. W., Samara, G. A., Ross, M. (Eds.) (1994), *High-Pressure Science and Technology—1993,* New York: American Institute of Physics.

Wentorf, R. H. (Ed.) (1962), *Modern Very High Pressure Techniques,* Washington, D.C.: Butterworths.

SOLIDS

See CRYSTALLINE STATE; ELECTRON STRUCTURE OF SOLIDS; ELECTRONIC TRANSPORT IN SOLIDS; HEAT-PULSE PROPAGATION IN SOLIDS; INTERACTION OF SOLIDS WITH PARTICLES AND RADIATION; IONIC CONDUCTION AND DIFFUSION IN SOLIDS; MAGNETIC ORDERING IN SOLIDS; MATERIALS PREPARATION, SOLIDS; MECHANICAL AND ELASTIC WAVES IN SOLIDS; MECHANICAL PROPERTIES OF SOLIDS; PHASE TRANSFORMATIONS, STRUCTURAL; POINT DEFECTS AND EXTENDED DEFECTS IN CRYSTALS; RADIATION DAMAGE IN CRYSTALS; SOLIDS, ACOUSTIC PROPERTIES OF; SOLIDS, CRYSTALLINE, MECHANICAL PROPERTIES OF; SOLIDS, DIFFUSION IN; SOLIDS, DYNAMIC HIGH-PRESSURE EFFECTS IN; SOLIDS, STATIC HIGH-PRESSURE EFFECTS IN; SURFACES AND INTERFACES OF SOLIDS, STRUCTURE OF; THERMAL CONDUCTION IN SOLIDS

SOLUBILITY AND MIXING IN FLUIDS

Robert Hołyst, *Institute of Physical Chemistry, Polish Academy of Sciences, Warsaw, Poland*

INTRODUCTION

The standard definition (Mc-Graw Hill Editors, 1984) of solubility is as follows: It is the ability of the substance to form a solution with another substance. The solution is a single homogeneous liquid, solid, or gas phase that is a mixture in which the components (liquid, gas, solid, or the combination thereof) are uniformly distributed throughout the mixture. Finally, miscibility denotes the tendency or capacity of two or more liquids to form a uniform blend, that is, to dissolve in each other. For historical reasons the substance less abundant in the mixture is called a solute, while the more abundant one is called a solvent. Here we shall concentrate on nonelectrolytes, i.e., solutions in which none of the components is in the form of free ions.

Traditionally, solubility and mixing belong to the realm of chemistry and material sciences, and many standard textbooks on physical chemistry (Nernst, 1904; Atkins, 1994) or chemistry (Grant and Higuchi, 1990; James, 1986) treat this problem, usually within the scope of thermodynamics. Here, apart from thermodynamics, we discuss other issues such as the statistical mechanical theories of mixtures, the relation between intermolecular interactions and demixing, the coupling between ordering and demixing, and the kinetics of demixing, which includes spinodal decomposition in binary liquid mixtures. The following special examples will be discussed: polymer blends, diblock copolymers, liquid crystals, and ternary mixtures including surfactants (amphiphiles). Within the scope of the physics of solubility and mixing, one can study such diverse phenomena as mixing of two simple liquids, collapse of the polymer chain in a solvent, flocculation of colloidal particles upon the addition of polymer chains, mixing of two polymer components in a liquid state, ordering of copolymers, and ternary mixtures of oil, water, and surfactant or formation of micelles in aqueous solutions. In reality, very rarely do we have completely pure substances; thus the properties of mixtures and the phenomena of mixing and demixing are of prime importance for science and technology.

An old alchemist's maxim, "similia similibus solvuntur" ("like dissolves like"), is the oldest rule of solubility. This rule can be a

3-527-28140-1/96/$5.00 + .50

very good guide in the study of mixing and solubility, provided one can precisely define what in a given case is the degree of likeness. For example, two simple liquids of low molecular mass can easily mix, say, at room temperature, while after polymerization they demix well above the room temperature.

1. THERMODYNAMICS

Solubility also means the maximum amount of the solute that can be solubilized in the solvent under given thermodynamic conditions. Within thermodynamics one formulates simple rules for solubility of gases and solids in liquids (Hildebrand and Scott, 1950):

1. The solubility of a gas is proportional to its partial vapor pressure (Henry's law).
2. The gas with the higher critical temperature and boiling point is more soluble than one with a lower critical temperature.
3. The solubility of a gas diminishes with increasing temperature.
4. The solubility of a solid increases with increasing temperature.
5. A solid having a higher melting point is less soluble at a given temperature than one having a lower melting point, provided that the enthalpies of melting are comparable.
6. A solid with large enthalpy of melting is less soluble at a given temperature than one with small enthalpy of melting, provided that the melting temperatures are comparable.
7. If rule **1** is satisfied for a solute in two immiscible solvents in contact, then the ratio of its concentration in these solvents is, for a given pressure and temperature, a constant (Nernst distribution law).

These rules are not always satisfied, sometimes only for very dilute solutions. Nonetheless, in many cases they provide very valuable information about solubilities of different substances in different solvents based on a few thermodynamic properties of pure substances. All these rules can be expressed in simple mathematical forms (Atkins, 1994; Hildebrant and Scott, 1950); e.g., rules **4–6** can be expressed as follows:

$$\ln x = \frac{-\Delta H^s_{\text{solute}}}{R}\left(\frac{1}{T} - \frac{1}{T_m}\right), \tag{1}$$

where x is the mole fraction of the solute in the saturated solution (solution in equilibrium with the solute solid), $\Delta H^s_{\text{solute}}$ is the enthalpy of melting for the solute, T_m is its melting temperature, R is the gas constant, and T is the temperature.

1.1 Colligative Properties

The elevation of the boiling point, the depression of the freezing point in a solution, and osmosis are properties that in the first approximation do not depend on the specific nature of the solute, but only on its amount in the solvent. Such properties are called colligative properties. For a dilute solution we find that the addition of the solute into the pure solvent raises its boiling temperature by

$$\Delta T = \left(\frac{RT_b^2}{\Delta H^v_{\text{solvent}}}\right)x, \tag{2}$$

where T_b is the boiling temperature of the pure solvent and $\Delta H^v_{\text{solvent}}$ is the enthalpy of vaporization for the pure solvent. Similarly the addition of a small amount of solute into the solvent lowers its freezing temperature by

$$\Delta T = -\left(\frac{RT_m^2}{\Delta H^s_{\text{solvent}}}\right)x, \tag{3}$$

where T_m is the freezing temperature of the pure solvent.

The phenomenon of osmosis (*q.v.*) takes place when a solution is separated from the pure solvent by a semipermeable membrane, which allows the flow of solvent from the pure phase to the solution. The pressure that has to be applied to stop this flow is called the osmotic pressure Π. For a dilute solution one gets the van't Hoff equation relating this pressure, the volume of the solution, V, and the number of moles of the solute in the solution n:

$$\Pi V = nRT. \tag{4}$$

Since n/V is given by the total mass of the solute divided by the molar mass M and the volume, combination of a measurement of the osmotic pressure and the van't Hoff formula allows one to determine the molar mass of the solute. This method is particularly useful for determining the mass of macromolecules; however, in this case it is often necessary to include in Eq. (4) the next term in the virial expansion of the osmotic pressure.

1.2 Binary Mixtures: Liquid–Vapor Coexistence

In an ideal mixture the components mix in all proportions without any change of volume or enthalpy. In the ideal binary mixture, the partial vapor pressure of component A (or B) coexisting with the ideal liquid mixture is given by $P_A = x_A(g)P$ [$P_B = x_B(g)P$]. The total pressure P is related to the (Raoult law) vapor pressures above pure A (P_A^*) and pure B (P_B^*) liquids by

$$P = P_A + P_B = x_A(l)P_A^* + x_B(l)P_B^*. \tag{5}$$

These two equations define the vapor [$x_A(g) = 1 - x_B(g)$] and the liquid [$x_A(l) = 1 - x_B(l)$] compositions at coexistence. In Fig. 1 the composition–pressure diagram is shown; the region between the two curves (given by the above equations) is the two-phase region, where the relative amounts of vapor and liquid are given by the lever rule (Atkins, 1994).

This type of diagram (now temperature and composition, Fig. 2) is the basis of fractional distillation. We start with the A, B mixture at point 1 on the diagram and evaporate it (1–2 dashed line). Next we condense it again until we reach point 3. The resulting liquid mixture is now richer in component B. Repeating these steps (evaporate it along the 3–4 dashed line, etc.), we can obtain almost pure B phase. The analogous process is used to purify solids and is called zone refining. In many mixtures the process of distillation is stopped at a certain point where the composition of the vapor is the same as the composition of the coexisting liquid (Figs. 3 and 4). This point is called the azeotrope. Since at this point the liquid boils without changing its composition, the fractional distillation proceeds only until the positive azeotrope is met (Fig. 4). In the example shown in Fig. 3, we reach a negative azeotrope by straight distillation, i.e., by continuously removing the vapor from the vessel of the boiling liquid mixture. For example, the mixture of ethanol and water has the azeotrope at T = 78 °C and at 4% of water (by mass); the mixture of water and nitric acid has the azeotrope at T = 122.4 °C and 60% of HNO_3 (mole fraction) (at 1 atm); that of

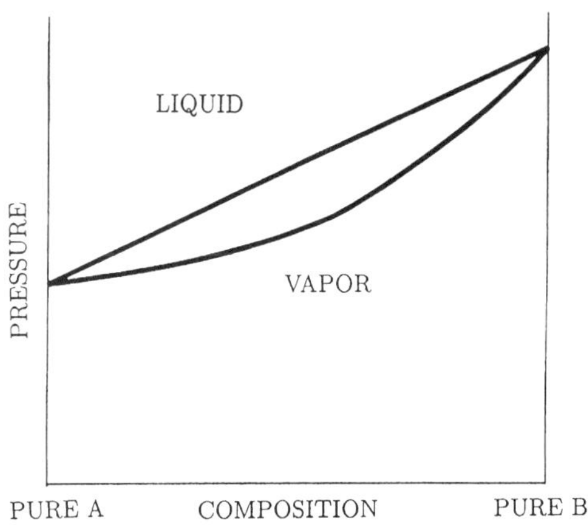

FIG. 1. The pressure–composition phase diagram for an ideal binary mixture (A,B components) [see Eq. (5)]. The region between the solid lines is the two-phase region where vapor and liquid coexist. Along the upper curve, evaporation of the liquid starts, and along the lower curve, condensation of the vapor starts.

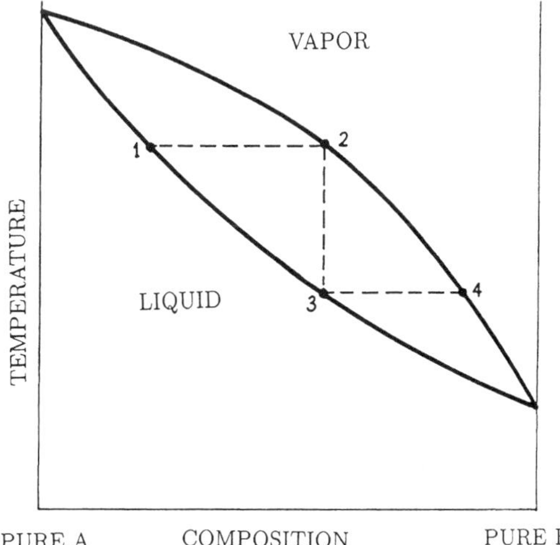

FIG. 2. The temperature–composition phase diagram for the binary mixture. The dashed line shows how the process of fractional distillation proceeds. In this way we can obtain almost pure (B-rich) liquid.

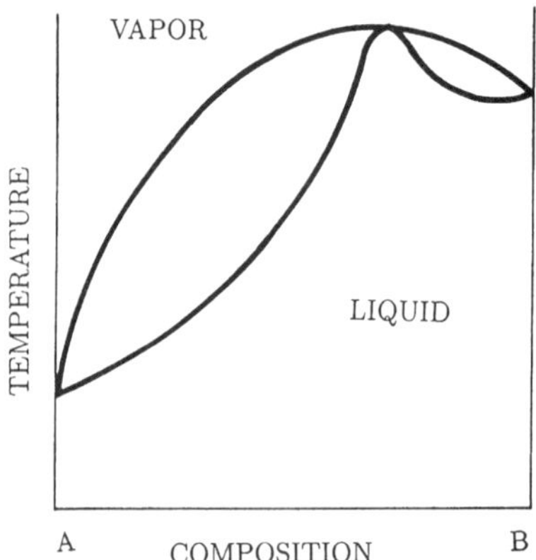

FIG. 3. A phase diagram with a high-boiling or negative azeotrope (maximum on the temperature–composition diagram). At this point liquid boils without changing its composition. This type of diagram occurs, for example, for water/nitric acid, chloroform/acetone, and hydrochloric acid/water.

hydrogen chloride and water has the azeotrope at $T = 108$ °C and 20% (by mass) of HCl (at 1 atm). Of course, the location of the azeotrope changes with pressure. The following rules can help us in deciding when we can expect an azeotrope in a liquid mixture (Rowlinson and Swinton, 1982):

1. The closer the vapor pressures of the pure components of the mixture, the more likely is azeotropy. It is inevitable at any temperature at which they are the same. It also means that when the vapor pressures are very close, then even small departures from ideality of the mixture can produce an azeotrope.
2. The closer the vapor pressures of the pure components, the more rapidly does the azeotropic composition change with temperature.
3. An increase of temperature and of vapor pressure in a positive azeotrope (maximum on the p-x diagram and minimum on the T-x diagram, Fig. 4) increases the mole fraction of the component whose vapor pressure increases more rapidly with temperature. The converse holds for a negative azeotrope (Fig. 3).

These rules are not without exceptions, but usually can be a very useful guide.

1.3 Partially Miscible Liquids

Mixtures are often far from being ideal, and consequently under suitable conditions we can expect demixing in the liquid state. In some systems demixing can take place as we lower the temperature (Fig. 5) or as we raise the temperature (Fig. 6). More complicated cases are shown in Figs. 7–10 (Atkins, 1994; Landau and Lifshitz, 1980). The maximum (minimum) of the curve in Fig. 5 (Fig. 6) is called the upper (lower) critical (or consolute) point. Gases usually mix very well at normal pressures; however, at very high

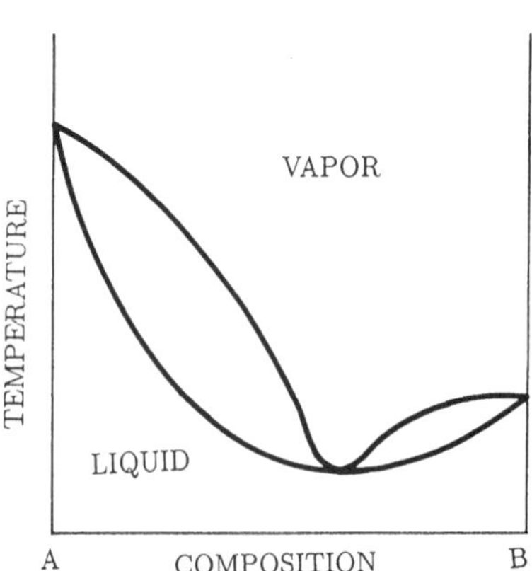

FIG. 4. A phase diagram with a low-boiling or positive azeotrope (minimum on the temperature–composition diagram). This type of diagram can occur for water/ethanol and dioxane/water.

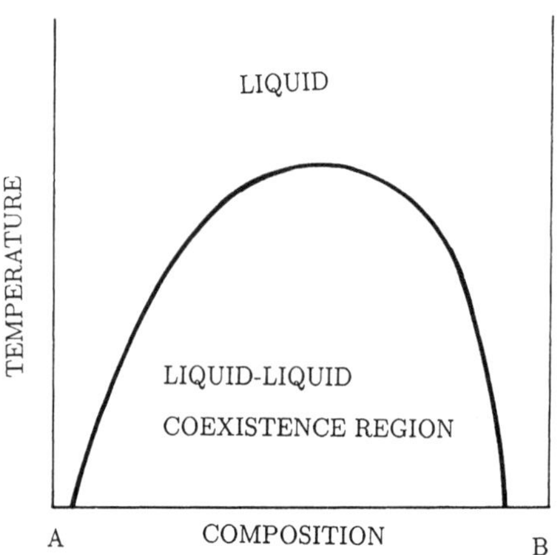

FIG. 5. Typical phase diagram for partially miscible liquids with an upper critical point (maximum on the coexistence curve) (e.g., methanol and hydrocarbon liquids).

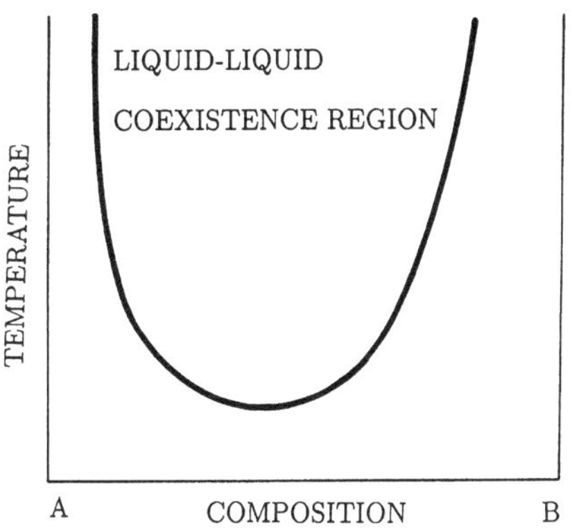

FIG. 6. Typical phase diagram for partially miscible liquids with a lower critical point (minimum on the coexistence curve) (e.g., water and diethylamine or triethylamine).

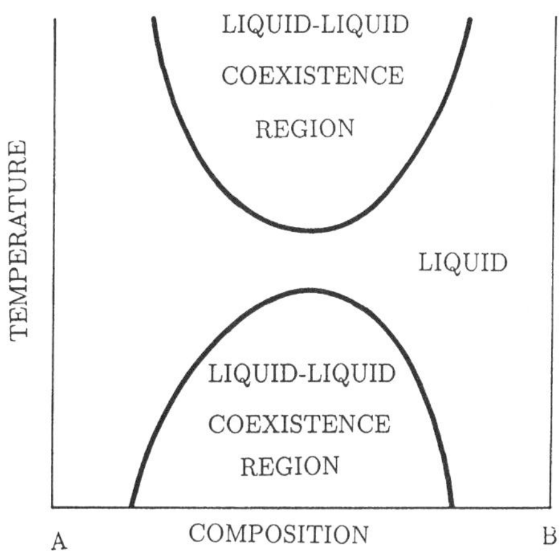

FIG. 8. The phase diagram with a lower and an upper critical point and a miscibility gap between these point (e.g., acetone and polystyrene).

pressures they can demix; then sometimes the phrase gas–gas immiscibility is used (Rowlinson and Swinton, 1982). The densities of these fluid phases at high pressures are comparable to the density of the liquid phases at "ordinary" low-pressure critical points.

Figure 10 is very similar to the liquid–solid phase diagram for a binary mixture (instead of vapor we have liquid, and instead of liquid we have solid). The lowest temperature of the liquid mixture at freezing in this case is obtained for the eutectic composition (point "E" on the diagram).

For ternary mixtures the phase diagram is often represented in the form of the Gibbs triangle. The Gibbs phase rule states that in a system of r components and M coexistent phases it is possible arbitrarily to preassign $r - M + 2$ variables from the set T, P, x_j^i, $i = 1, \ldots, M$, $j = 1, \ldots, r - 1$ (Callen, 1960). In particular, it means that the maximum number of coexistent phases in the system is $r + 2$. One should note here that if the system spontaneously orders in an arbitrarily small external field (magnetic or electric), then the number of coexistent phases can be larger if we include these fields in the set of thermodynamic variables.

So far we have considered liquid mixtures

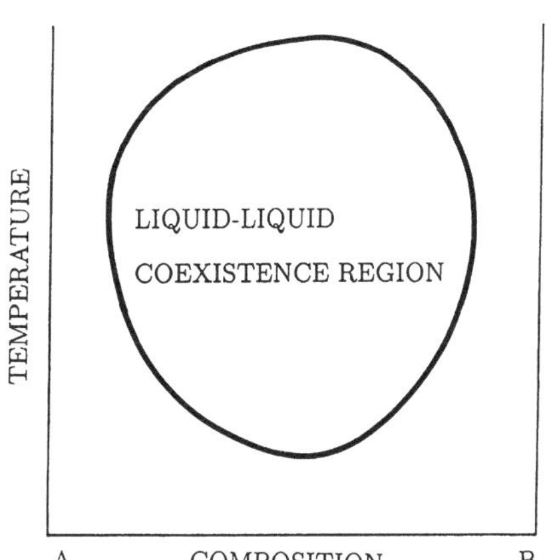

FIG. 7. The phase diagram with upper and lower critical points (e.g., nicotine and water).

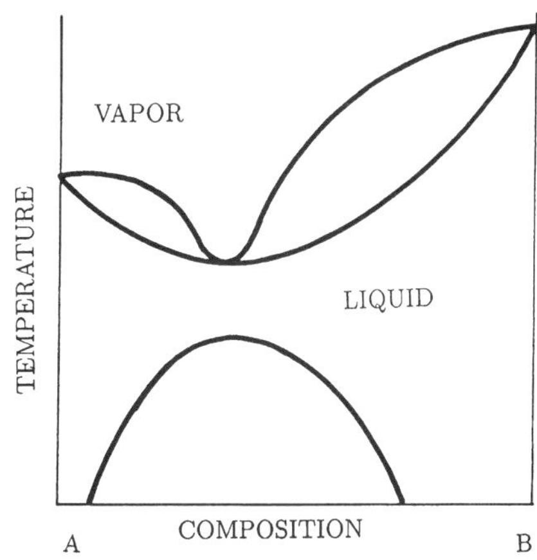

FIG. 9. The phase diagram for a mixture where the upper critical point is below the boiling curve.

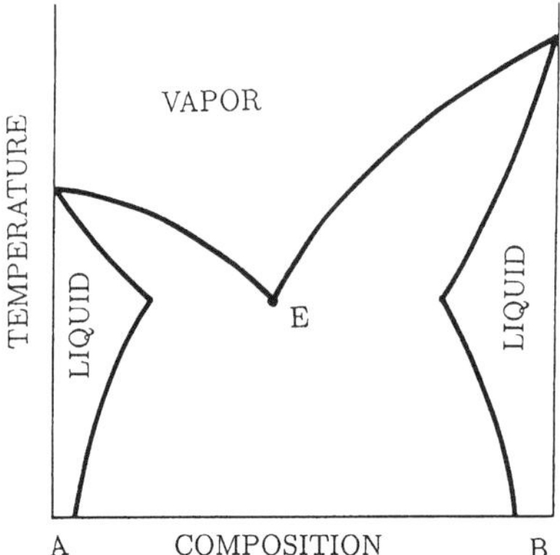

FIG. 10. The phase diagram for a mixture not fully miscible in the liquid state (e.g., water/diethyl ether, water/chloroform, and high hydrocarbons and methanol).

in a bulk system. The confinement of the fluid in a capillary also affects mixing and solubility. The detailed discussion of the thermodynamics of confined mixtures can be found in Evans and Marconi (1987). However, the theoretical problem of whether confinement increases or decreases solubility and mixing is still not resolved. The problem is especially important for narrow pores of the size of few molecular diameters.

2. STATISTICAL MECHANICAL THEORIES OF MIXTURES

The modern statistical mechanical approach to inhomogeneous and/or homogeneous fluids and fluid mixtures is based on density-functional theory (DFT) (Evans, 1979). The central quantity of interest in DFT is the Helmholtz free energy $F[\rho_1 \cdots \rho_m]$, which is a unique functional of the local densities $\rho_i(\mathbf{r})$ ($i = 1 \cdots m$) in an m-component simple atomic mixture. In a homogeneous system, $\rho_i = N_i/V$, where N_i is the number of molecules of type i and V is the volume. The Helmholtz free energy as a functional of the densities completely specifies its statistical and thermodynamical properties. The equilibrium state of the system corresponds to the global minimum of the grand thermodynamic potential, obtained from F by a Legendre transform.

In the case of anisotropic liquids (e.g., liquid crystals) consisting of elongated rigid molecules, the one-particle distribution function $\rho(\mathbf{r},\omega)$, which depends on the position of the center of mass $\mathbf{r}$ of the molecule and its orientation in space ω, represented by the three Euler angles, describes the structure of the system. Consequently, the free energy of liquid crystals is a functional of $\rho(\mathbf{r},\omega)$. More complicated cases are also possible, depending on the structure and flexibility of molecules.

For simplicity we confine further discussion to simple atomic fluids. The free energy can be split into an ideal part and an excess part as follows:

$$F[\rho_1 \cdots \rho_m] = F_{id}[\rho_1 \cdots \rho_m] + F_{ex}[\rho_1 \cdots \rho_m]. \tag{6}$$

For the noninteracting system the excess part is zero, and the ideal part can be easily calculated from the partition function:

$$F_{id} = k_{\mathrm{B}}T \sum_{i=1}^{m} \int d^3r\rho_i(\mathbf{r})\{\ln[\rho_i(\mathbf{r})\lambda_1^3] - 1\}, \tag{7}$$

where λ_i is the thermal de Broglie wavelength of the ith component in the mixture. The integrals are performed over the whole volume of the system. The excess part, arising from the interactions, can be written in the following general form:

$$\begin{aligned} F_{ex}[\rho_1 \cdots \rho_m] = \sum_{i,j=1}^{m} \int d^3r_1 d^3r_2 \rho(\mathbf{r}_1)\rho(\mathbf{r}_2) \\ \times \int_0^1 dl(l-1) \\ \times c_{ij}(\mathbf{r}_1\mathbf{r}_2; [l\rho_1, \cdots, l\rho_m]). \end{aligned} \tag{8}$$

The functions c_{ij} are the Ornstein–Zernicke (OZ) direct correlation functions (Hansen and McDonald, 1986). Although this relation is formally exact, the form of the direct correlation functions is in general unknown for inhomogeneous and even for homogeneous liquids. Various approximations to the exact DFT functional have been proposed: the weighted density approximation (WDA) (Tarazona, 1985; Curtin and Ashcroft, 1985) and the modified weighted density approximation (MWDA) (Denton and Ashcroft, 1989) for pure systems, later applied to binary mixtures of hard spheres (Denton and Ashcroft, 1990, 1991) (for a different approach to mixtures, see Xu and Baus, 1987, 1992;

for a review, Baus, 1990). Density functional theories have been also applied to the description of structural phase transitions in liquid crystals (Poniewierski and Holyst, 1988, 1990; Allen *et al.*, 1993). Recent reviews on this subject are given by Löwen (1994) and Allen *et al.* (1993). In general, the theory has been aimed at explaining the properties of inhomogeneous systems from the known properties of homogeneous systems. Usually, the form of the direct correlation functions for homogeneous systems are required as the input for the theory. For the homogeneous mixture, these functions are defined via the Ornstein–Zernicke equation:

$$h_{ij}(|\mathbf{r}_1 - \mathbf{r}_2|) = c_{ij}(|\mathbf{r}_1 - \mathbf{r}_2|) + \sum_{k=1}^{m} x_k\rho \int d^3r_3 c_{ik}(|\mathbf{r}_1 - \mathbf{r}_3|) \times h_{kj}(|\mathbf{r}_3 - \mathbf{r}_2|), \quad (9)$$

where $\rho = \Sigma_{i=1}^{m} N_i/V$ is the total density and h_{ij} is the two-point correlation function. This set of equations with suitable closure [e.g., Percus–Yevick (PY) for hard spheres; Henderson and Leonard, 1971] can be solved, and the functions c_{ij} can be calculated. Now combining these results together with MWDA completely specifies the properties of the inhomogeneous and/or homogeneous mixture.

In general, solving the OZ equation for mixtures interacting through a complicated potential is very difficult. One can then resort to some old methods known in the theory of mixtures as the one-fluid approximation in the more general scheme known as conformal theories of mixtures (Sec. 2.3) (Rowlinson and Swinton, 1982; Hansen and McDonald, 1986).

2.1 Homogeneous Systems and Ideal Mixtures

In homogeneous mixtures, one defines mixing functions; i.e., for any thermodynamic function $A(N_i)$, the mixing function is defined as

$$A^{\text{mix}} = A(N_1 \cdots N_m) - \sum_{i=1}^{m} A_i(N), \quad (10)$$

where

$$N = \sum_{i=1}^{m} N_i$$

is the total number of particles and A_i is the thermodynamic function of interest for a pure system of the ith component. The mixing function vanishes for a pure system. If the interactions between particles are the same, then the form of the thermodynamic potentials is very simple. The Helmholtz free energy (same for Gibbs free energy) is then [Eqs. (6–8)]

$$F^{\text{mix}} = Nk_{\text{B}}T \sum_{i=1}^{m} x_i \ln x_i. \quad (11)$$

Strictly speaking, real mixtures are nonideal, except in very special cases when we have mixtures of isotopes of low molecular mass and when quantum effects are negligible. However, often the ideal solution is a very good starting point in the description of many properties of mixtures—e.g., gas–liquid coexistence.

2.2 Homogeneous Systems and Excess Mixing Functions

The difference between the actual value of the thermodynamic mixing function and its ideal value [Eq. (11)], for the same temperature, composition, and volume (or pressure), is called the excess function {the same name is used in DFT [Eq. (8)], but its meaning there is different}. The simplest form of the mixing function for a binary mixture is given by the Guggenheim quadratic form (Henderson and Leonard, 1971; Rowlinson and Swinton, 1982),

$$F^E = N\chi x_1 x_2, \quad (12)$$

obtained in the simplest lattice approximation. In the lattice approximation both the excess Gibbs free energy and the excess Helmholtz free energy are the same, and $\chi/\kappa_{\text{B}}T$ is the interaction parameter independent of temperature and pressure (volume). Lattice models are often used for the description of certain properties of mixtures (Furman *et al.* 1977; Walker and Vause, 1983; Carneiro and Schick, 1988). By comparing Eqs. (10)–(12) and Eqs. (6)–(8), it follows that in general χ is a complicated integral of the direct correlation function. Only in the limit of low density is χ independent of x_i. In this limit $c_{ij} = f_{ij} = \exp(-v_{ij}/k_B T) - 1$, where $v_{ij}(|\mathbf{r}|)$ is the two-body interaction potential between molecules of components i and j. In this approximation χ can be explicitly calculated from Eq. (8).

Usually in thermodynamics one defines by Eq. (12) the Gibbs free energy G^E, assuming that χ is a function of temperature and pressure. Then the heat of mixing (enthalpy) is

$$H^E = N\left(\chi - T\frac{\partial \chi}{\partial T}\right)x_1 x_2, \tag{13}$$

and the change of volume in the system upon mixing is

$$V^E = N x_1 x_2 \frac{\partial \chi}{\partial P}. \tag{14}$$

Most mixtures of simple liquids (e.g., argon, krypton) have positive heat of mixing; i.e., heat is absorbed, but the excess volume V^E can be of either sign. The excess quantities are usually small; e.g., V^E per mole in a 50% solution (in mole fraction) of tetrachloroethene in cyclopentane at 25 °C is -8 mm^3 (the mixture contracts) (total volume is ~10 cm^3) and the excess heat of mixing H^E is 800 J/mol.

2.3 Conformal Fluid Theories and the One-Fluid Approximation

One of the oldest approaches to the theory of homogeneous mixtures dates back to van der Waals and is known as the one-fluid approximation. In this approach the excess properties of the mixture are expressed in terms of the quantities of a hypothetical pure fluid. It applies to a mixture in which the interaction potential $v_{ij}(r)$ satisfies the scaling relation

$$v_{ij}(r) = \epsilon_{ij} v(r/\sigma_{ij}) \tag{15}$$

for all i, j. Often σ_{ij} and ϵ_{ij} (for $i \neq j$) are related to pure components. The most frequently used combination rules are as follows: $\sigma_{ij} = (\sigma_{ii} + \sigma_{jj})/2$ and $\epsilon_{ij} = \sqrt{\epsilon_{ii}\epsilon_{jj}}$, known as the Lorentz and Berthelot rules, respectively. The hypothetical fluid is characterized by the potential given by Eq. (15) with some energy parameter ϵ and length parameter σ. In order to specify these parameters, one makes the approximation

$$h_{ij}(r) = h(r/\sigma), \tag{16}$$

and sets

$$\sigma^3 = \sum_{i,j=1}^{m} x_i x_j \sigma_{ij}^3. \tag{17}$$

The interaction parameter for this hypothetical one-component fluid reads

$$\epsilon = \sum_{i,j=1}^{m} x_i x_j \epsilon_{ij} \frac{\sigma_{ij}^3}{\sigma^3}, \tag{18}$$

and the interaction potential is $\epsilon v(r/\sigma)$. Now the excess mixing properties of the mixture are related to the excess properties of this fluid. This approximation is known as the van der Waals one-fluid approximation (Henderson and Leonard, 1971). The comparison with computer simulations for the model binary mixture of Lennard–Jones fluids shows a very good agreement with the theory (Hansen and McDonald, 1986). The approximation breaks down when the molecules of different components are very disparate in sizes and have very different energy parameters. Another of its drawbacks is that it always leads to a negative volume change V^E upon mixing. More complicated approximations are also possible but usually lead to worse agreement with computer simulations. For real systems, the simple Lennard–Jones mixture constitutes a rather poor approximation (Hansen and McDonald, 1986).

One has to remember that excess mixing quantities are rather small and that small deviations of the model interaction potential from the real potential between the molecules can change even the sign of the excess quantity (e.g., V^E).

3. INTERACTION POTENTIAL AND MIXING

The intermolecular forces have their origin in quantum mechanics and classical electrostatics. According to the Hellman–Feynman theorem, the interactions between the molecules can be calculated according to the laws of electrostatics, once the distribution of electrons has been determined from the Schrödinger equation. From the simple Coulomb potential originate various important interactions at the molecular level. The dispersion van der Waals attractive interactions are responsible for boiling (gas–liquid transition). It is now known that the reduction of the range of attraction in the system results in the complete disappearance of the liquid phase even if the attraction is strong at short distances (Tejero *et al.*, 1994). It implies, for example, that C_{60}, where interactions are of the short range, is a substance that should not have a liquid phase (Hagen *et al.*, 1993). Molecules, when brought together, strongly repel each other, preventing the overlap of electronic orbitals. This interaction gives the molecule its size and shape and is known as hard-core repulsion or steric interaction. Steric interactions are responsible for the freezing transition (Alder and Wainwright, 1957), the structure of liquids (Hansen and McDonald, 1986), and structural transitions in liquid crystals (Onsager, 1949; Frenkel, 1991). Van der Waals attraction and steric repulsion are very important since they are present in every molecular system. Apart from them there are many specific interactions. One interesting example considered here is hydrogen bonding occurring between, e.g., water molecules. This interaction is responsible for the facts that ice has lower density than liquid water and that the highest density of water occurs at 4 °C. The extremely high boiling and freezing points of water are also due to this interaction and the tetrahedral structure formed by water molecules. Here we try to answer partially the question of how these various interactions affect mixing.

3.1 Van der Waals Interactions

If we apply the Berthelot rule, which holds reasonably well for van der Waals forces (Israelachvili, 1985), in the Guggenheim model [Eq. (12)] for the binary mixture, we find

$$\chi \sim (\sqrt{\epsilon_{11}} - \sqrt{\epsilon_{22}})^2 > 0. \qquad (19)$$

Since χ is always greater than zero, we may conclude that van der Waals interactions promote demixing. At sufficiently low temperatures, the excess part of the Gibbs free energy outweighs the ideal part [Eq. (11)], and demixing occurs. With this simple model we obtain the type of phase diagram shown in Fig. 5. In the first approximation the square roots of the energies ϵ_{ii} are proportional to the polarizability of the molecules of ith component.

The van der Waals interactions for any two bodies in vacuum are always attractive. Also they are attractive for identical bodies in a solvent. However, one may have repulsive van der Waals interactions between two different solute molecules in a solvent. This happens whenever the index of refraction (related to polarizabilities) of the solvent is intermediate between those of two different solute molecules (Israelachvili, 1985). It happens in many mixtures of organic solvents and different organic polymers (Van Oss *et al.*, 1980).

3.2 Steric Interactions

It has been believed until recently that steric interactions alone cannot induce demixing. This belief has been based on the fact that the ideal entropy of mixing decreases if the demixing transition takes place. This decrease of the entropy S can be compensated by the decrease of the energy U so that the free energy $F = U - TS$ decreases upon mixing. Since in the hard-core system the interaction energy is zero, one has not expected the demixing transition in such a system. This conclusion is supported also by the explicit calculation of the parameter χ [Eq. (12)] for a binary fluid mixture of hard spheres in the low-density approximation. One finds $\chi < 0$ in this limit and concludes that mixing is favored in the system.

Recent computer simulations have shown that this is not the case and that in a binary mixture of a model system of hard-core molecules differing in size only, the demixing transition driven by the increase of entropy occurs. Although the ideal entropy is the largest in a homogeneous mixture, the entropy associated with the free volume is larger in the demixed fluid (Biben and Hansen, 1991; Frenkel and Louis, 1992; Van Duijneveldt and Lekkerkerker, 1993; Dijkstra and Frenkel, 1994; Dijkstra *et al.*, 1994; Frenkel, 1994). At a certain concentration of solute molecules, the latter outweighs the former, and demixing occurs. The role of steric interactions in real mixtures has not been resolved yet. Clearly, it is a many-body effect.

3.3 Hydrogen Bonding

The unique properties of water follow from its ability to form a tetrahedral structure in the liquid, induced by hydrogen bonding (Israelachvili, 1985). The interaction is electrostatic, but its strength (10–40 kJ/mol) is one order of magnitude larger than the strength of the van der Waals energy (~1 kJ/mol). This interaction depends on the orientation of molecules and in general is not pairwise additive.

The solubilization of molecules that do not form hydrogen bonds (e.g., alkanes and other hydrocarbons) in a hydrogen-bonding solvent (methane, water, acetic acid) involves forming a clathrate cage around the solute molecule (Fig. 11). The solvent molecules adapt such an orientation close to the non-bonding solute as to saturate its hydrogen bonds with other solvent molecules. It means that solvent molecules around the solute molecules are more ordered than the molecules in the bulk solvent. It leads to the decrease of entropy in the process of solubilization, which for this reason becomes unfavorable. For n-butane in water the heat of mixing is $\Delta H = -4.2$ kJ/mol, while the change of the entropy $-T\Delta S$ (at $T = 25$ °C) is 28.7 kJ/mol. Thus, the change of entropy contributes 85% to the change of the Gibbs free energy $\Delta G = \Delta H - T\Delta S$. For longer hydrocarbons the contribution of entropy to the total change of the Gibbs free energy is even larger (Israelachvili, 1985). The low solubility of non-hydrogen-bonding molecules in water is entropic in nature and is known as a hydrophobic effect. The solubility of hydrocarbons in a hydrogen-bonding solvent (e.g., methane) decreases with the length of the hydrocarbon; e.g., the critical temperature for n-pentane is $T_c = 287$ °C, while for n-hexane it becomes $T_c = 307$ °C (at normal pressure) (Rowlinson and Swinton, 1982).

Methane and water mix very well in all proportions at room temperature, and this is due to the fact that both form hydrogen bonds. We expect that molecules of the right geometry, containing electronegative atoms (oxygen atoms in alcohols, nitrogen atoms in amines), are capable of forming hydrogen bonds. At low temperatures the hydrogen bonds forming between the solute and solvent molecules enhance mixture ($\chi < 0$). If we raise the temperature, the bonds are broken and the liquids demix above the lower critical temperature as a result of van der Waals forces ($\chi > 0$). Thus, hydrogen bonds are responsible for enhanced mixing at low temperatures and consequently for the existence of the lower critical point (Fig. 6) (Walker and Vause, 1983). We expect that this phenomenon should strongly depend on the molecule geometry.

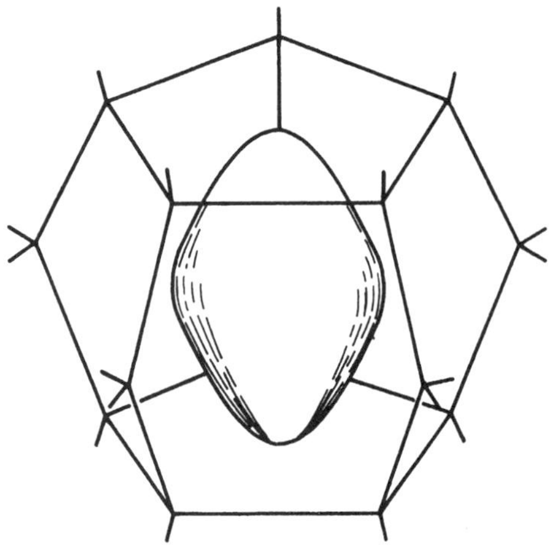

FIG. 11. The clathrate cage around a non-hydrogen-bonding solute molecule. The lines indicate the hydrogen bonds between the solvent molecules.

4. POLYMER BLENDS AND POLYMERS IN SOLUTIONS

The simple lattice model presented in Sec. 2 [Eq. (12)] has been applied to polymer blends and polymers in solutions by Flory and Huggins (Flory, 1953). The free energy for the binary mixture of two homopolymers A and B consisting of N_A and N_B monomers, respectively, is given by the Flory–Huggins expression:

$$\frac{F^{\mathrm{mix}}}{Mk_{\mathrm{B}}T} = \frac{x_A}{N_A}\ln x_A + \frac{x_B}{N_B}\ln x_B + \frac{\chi}{k_{\mathrm{B}}T}x_Ax_B. \tag{20}$$

Here $x_A = n_AN_A/(n_AN_A + n_BN_B)$ is the mole

fraction of A monomers in the mixture; n_i is the number of polymer molecules of component i (A or B); $M = n_A N_A + n_B N_B$ is the total number of monomers in the system, and χ, in polymer physics and chemistry, is called the Flory–Huggins interaction parameter. In the Flory–Huggins theory, incompressibility is assumed, just as in the Guggenheim model of simple mixtures. The critical temperature is

$$T_c = \frac{2N_A N_B}{k_B \chi(\sqrt{N_A} + \sqrt{N_B})^2}, \tag{21}$$

and at the critical composition

$$x_A^c = \frac{\sqrt{N_B}}{\sqrt{N_A} + \sqrt{N_B}}. \tag{22}$$

The ideal mixing entropy is reduced by a factor N_A and/or N_B, and, thus, the critical temperature is very large for polymer mixtures (N_A, $N_B \gg 1$). Indeed, polymers separate very easily. Even small differences in the interaction potential get magnified by a factor of N_A, N_B, and, thus, the critical temperatures are often much higher than the temperature at which the polymer molecule breaks through the breakage of chemical bonds linking the monomers. Even a binary mixture of isotope polymers (one deuterated) can undergo a demixing transition at room temperature if N_A, N_B are sufficiently large (Gehlsen *et al.*, 1992).

A mesoscopic theory of polymer chains has been formulated by Edwards (1966) and applied to polymer blends, polymers in solutions, membranes in solutions, etc.

4.1 Polymer Blends

Despite the fact that the Flory–Huggins theory has been used for more than four decades, only recently has it been carefully checked experimentally (Bates *et al.*, 1988; Gehlsen *et al.*, 1992) and in computer simulations (Deutsch and Binder, 1992). It follows from computer simulations that χ parameter has the following form:

$$\chi/k_B T = \alpha/T + \beta, \tag{23}$$

where α and β are constants independent of temperature or composition. In the first approximation α is related to the attractive van der Waals forces, while β is related to steric interactions (called sometimes excluded-volume interactions). Sometimes, in experimental works, it is assumed that α depends linearly on the concentration x (Roe and Zin, 1984). Typically, we find $\alpha \sim 1$ and $\beta \sim 10^{-3}$–10^{-4}. From the previous section we know that the effective interaction χ is related to the integrals of the direct correlation functions (Schweizer and Curro, 1988), and thus χ can be, in principle, a complicated function of temperature, density, and concentrations. In general, χ should not depend on the global architecture of the polymer chain.

In polymer systems the quantity of interest is the radius of gyration, i.e., the linear size of the region occupied by a single polymer molecule consisting of N monomers of length l. It is defined as follows (Doi and Edwards, 1986):

$$R^2 = \frac{1}{N^2} \sum_{i=1}^{N} \sum_{j=1}^{N} \langle (\mathbf{r}_i - \mathbf{r}_j)^2 \rangle. \tag{24}$$

Here $\mathbf{r}_i$ gives the location of the ith monomer in the polymer chain and $\langle \cdots \rangle$ denotes the statistical average. For the polymer blend the dependence of R on N is (the leading term in N) as follows:

$$R = \sqrt{Nl^2/6}, \tag{25}$$

characteristic for the noninteracting chain. It means that interactions between monomers in the same polymer chain are screened by the presence of other chains (Doi and Edwards, 1986; De Gennes, 1979) or a solvent (Edwards, 1975). As the temperature is lowered, the scaling of R with N does not change. However, the chains shrink progressively (R decreases) as the critical temperature is approached (Hołyst and Vilgis, 1994).

One can deduce the structure of a polymer blend from scattering experiments. The scattering intensity $S(q)$ has the following form (De Gennes, 1979):

$$S^{-1}(q) = \frac{1}{x_A g(N_A,q)} + \frac{1}{x_B g(N_B,q)} - 2\frac{\chi}{k_B T}, \tag{26}$$

where $g(N,q) = 2N(y + e^{-y} - 1)/y^2$ and $y =$

$Nq^2l^2/6$. The contrast for neutron scattering is achieved by deuteration of one of the components. This approximation is known as the random-phase approximation (RPA). The characteristic length in the problem is given by the radius of gyration R [Eq. (25)].

From the theoretical point of view, it is interesting to note that although the critical point in a binary mixture of polymers belongs to the universality class of the Ising model (see UNIVERSALITY; PHASE TRANSITIONS: RENORMALIZATION AND SCALING), in the limit $N \to \infty$ the mean-field theory becomes exact. In particular, the critical exponents measured for the system have the mean-field values for $|T - T_c|/T_c \geq 1/N$ (De Gennes, 1977; Hołyst and Vilgis, 1993). This behavior has been confirmed in experiments (Schwahm *et al.*, 1987).

Polymer blends are compressible mixtures (Floudas *et al.*, 1993), but the role of this factor in the Flory–Huggins theory is still under theoretical study (Lifschitz *et al.*, 1994).

4.2 Polymers in Solutions

In the theory and experiment one often discusses three types of solvents: good, bad, and theta solvents. Since for $N_A \gg N_B = 1$ the critical concentration is very low ($x_A^c \sim \sqrt{1/N_A}$), we can expand F^{mix} in x_A and find for this dilute solution

$$\frac{F^{\text{mix}}}{Mk_{\text{B}}T} = \frac{x_A}{N_A} \ln x_A + Bx_A^2 + \tfrac{1}{6}x_A^3 \cdots, \qquad (27)$$

where $B = \frac{1}{2} - \chi/k_{\text{B}}T$ is the osmotic virial coefficient (Atkins, 1994; De Gennes, 1979). For $B > 0$ the solution is classified as bad, for $B < 0$ it is classified as good, and for $B = 0$ it is classified as a theta or Flory solvent. The nature of the solvent depends on temperature. For each solvent there is a unique temperature (called the theta temperature) when $B = 0$ and the solution becomes nearly ideal. In a theta solvent the chains behave, in the first approximation, according to Eq. (25). In a good solvent the polymer molecule swells and

$$R \sim N_A^{\nu} \qquad (28)$$

with the Flory exponent $\nu \approx \frac{3}{5}$, which should be compared with $\nu = \frac{1}{2}$ [Eq. (22)] for a polymer chain in a blend and a polymer chain in a theta solution. In a bad solvent the chains shrink and $\nu \leq \frac{1}{2}$. In fact, close enough to the critical temperature each solvent becomes a bad solvent. The Flory–Huggins theory predicts correctly the location of the critical point (De Gennes, 1979; Hołyst and Vilgis, 1993) but not the shape of the coexistence curve close to the critical point. More information on polymers in solutions can be obtained in a recent monograph (Des Cloizeaux and Jannik, 1990).

It is often assumed that the theta temperature must reflect the competition between the attractive van der Waals forces and the short-range steric interactions. However, it does not have to be the case. As shown explicitly by Frenkel and Louis (1992), in a simple lattice model of a polymer chain consisting of monomers interacting through steric interactions only and solvent consisting of small molecules interacting through the same steric interaction, the Flory–Huggins parameter is

$$\chi/k_{\text{B}}T = \tfrac{1}{2}c \ln(1 + z_s) > 0. \qquad (29)$$

Here c is the coordination number of the lattice and z_s is the fugacity of the solvent molecules. Thus, in this system with purely steric interactions, we may have a bad, good, or theta solvent by changing the chemical potential of the solvent (Dijkstra *et al.*, 1994). The mechanism of demixing in such a mixture is clear. The fraction of the volume accessible to small solvent particles increases when the large particles cluster. Of course, in this model system, demixing is a purely entropic effect. Also the coagulation (clustering) of colloidal particles in a polymer solution is an entropic effect (Meijer and Frenkel, 1994). The clustering of colloidal particles increases the total number of accessible polymer configurations. The effective attraction between the colloidal particles, responsible for coagulation, is not pairwise additive (Shaw and Thirumalai, 1991), and that is why its quantitative theoretical description is so difficult. The role of steric interactions in polymer solutions is not fully understood.

We note that in the process of preparation of a polymer blend, one first prepares a solution of A and B polymers in a solvent and then evaporates the solvent. The comparison of the theory and experiment is often plagued by the inevitable polydispersity of the polymer masses. And finally, we point out that often instead of using mole fraction one uses either volume fraction or simply the density of monomers in the description of the free energy of mixing in polymer solutions (Atkins, 1994; De Gennes, 1979).

As we see, polymers are not easily mixed even at high temperatures. Usually chemical means are used to induce mixing. One of them is a chemical modification of one of the components in such a way as to induce hydrogen bonds between the different components. For example, PSD (deuterated polystyrene) and PBMA [poly(butyl methacrylate)] are not miscible. However, modifying PSD by attaching at random OH groups along the chain makes the PSD(OH)/ PBMA blend miscible. This is because OH groups from PSD(OH) form hydrogen bonds with CO groups on any segment of PBMA (Hobbie *et al.*, 1994). Another way to induce miscibility is even more direct. One can simply join one polymer chain with the other by chemical methods. In this way diblock and multiblock copolymers are formed.

5. ORDERING AND DEMIXING

There are many examples of demixing induced by ordering and vice versa. These effects occur in complex fluids such as diblock copolymers (Bates, 1991; Bates and Fredrickson, 1990), amphiphilic systems (Gompper and Schick, 1994; Laughlin, 1994), and mixtures of liquid crystals (De Gennes and Prost, 1993).

5.1 Mixtures of Liquid Crystals

For commercial applications in the display industry, only mixtures of liquid crystals are used. By mixing two nematic liquid crystals, one can lower the freezing point (eutectic composition; see Sec 1) without strongly affecting the isotropic–nematic phase transition point. Such a mixture exhibits anisotropic properties in the liquid state in a much wider temperature range than the pure substances. Usually the point at which the nematic–isotropic transition occurs is called the clearing point.

Miscibility is used as a first test for new phases in liquid crystals. If the two materials give the same texture and are miscible in all proportions maintaining this texture, they must have the same symmetry. However, we have learned that demixing does not have to reflect differences in the symmetry of phases. Differences in the shape and size of molecules are sufficient to induce demixing. Nonetheless, the miscibility test is still very useful, since it is quick and simple.

NMR allows us to measure internuclear distances of solute molecules if the solvent is nematic. Elongated isomers are more readily solubilized by nematic solvents, and this effect is used in chromatography.

Doping the nematic solvent with a chiral solute induces a cholesteric (twisted) ordering in the system, with a very large pitch (the wavelength of the twist). Changing the composition of this mixture allows us to vary the pitch continuously. The pitch, in a dilute solution, is inversely proportional to the concentration of the solute molecules. In chiral, ferroelectric smectic C^*, one can increase the pitch by dissolving in the system a nonchiral solute.

In a homogeneous solution of small monomeric units and nematic solvent, the fast polymerization reaction between the monomers results in demixing and leads to the formation of polymer-dispersed liquid crystals (PDLC), which are used as switchable windows, light shutters, and displays. Demixing is induced here since the length of the polymeric component of the mixture grows in the polymerization reaction, and according to Eq. (21) the critical temperature gets higher for longer chains. If the reaction is fast, droplets of the nematic solvent, which form in the process of demixing, are homogeneously distributed in the solid polymeric matrix. Usually the reactions require more components such as curing agents, catalysts, *etc.* (Doane, 1990).

The extremely rich phase behavior of liquid crystals and liquid-crystalline polymers has been described theoretically in several models: the lattice spin model (Sivardière, 1980), the Onsager model for elongated molecules (Deblieck and Lekkerkerker, 1980; Van Roij and Mulder, 1994), and a combination of the Flory–Huggins model for mixtures and the Maier–Saupe model for liquid crystals (Brochard *et al.*, 1984; Maïssa and Sixou, 1989; Hołyst and Schick, 1992a). The following results follow from these theoretical studies. The nematic–isotropic phase transition in mixtures is always accompanied by partial demixing. The nematic phase is always richer in the component that orders more easily. The azeotrope at the nematic–isotropic phase transition is found in binary mixtures where both components have very similar isotropic–nematic transition temperatures in pure systems. The azeotrope concentration is very close to the critical concentration for demixing inside the nematic phase. All the diagrams shown in Figs. 1–10 are found in binary mixtures of liquid crystals if instead of vapor we have the isotropic phase and instead of liquid the nematic phase. The topology of many diagrams is even more complicated (Brochard *et al.*, 1984). The nematic order parameter (De Gennes and Prost, 1993) always couples to the concentration. The demixing at the isotropic–nematic phase transition is driven by ordering; however, inside the nematic phase we can also observe isotropic demixing (Figs. 5 and 9), which changes the nematic order parameter. The demixing inside the ordered nematic mixture has the same origin (isotropic van der Waals or steric interactions) as the demixing shown in Fig. 5 for a simple isotropic liquid mixture. The roles of the anisotropic steric (shape) interactions and attractive interactions in the process are not fully understood. Demixing in an ordered nematic liquid mixture (Casagrande *et al.*, 1982; Dorgan and Soane, 1990) or ordered smectic liquid mixture (Sigaud *et al.*, 1990) has been observed in experiments, confirming many of the above-stated predictions.

5.2 Diblock Copolymers

Polymers are usually not miscible (Sec. 3), but at the same time for practical applications we need homogeneous mixtures. One way to prevent a demixing transition is the chemical bonding of the *A* polymer to the *B* polymer. In this way diblock and multiblock copolymers are formed (Bates and Fredrickson, 1990). Although phase separation is now prohibited on the macroscopic scale, the system can undergo so-called microphase separation. The "demixing" in the *AB* diblock copolymer system takes place at the scale given by the size of the radius of gyration and at much lower temperatures than the demixing in the polymer *AB* blend. The system forms many ordered phases in a liquid state: lamellar, gyroid (Matsen and Schick, 1994), diamond, cubic, and hexagonal. The microscopic interactions responsible for the formation of these phases are the same as those responsible for the demixing transition in the *AB* polymer blend (Leibler, 1980), and the same Flory–Huggins parameter χ is used in the description of both systems. The *AB* diblock copolymer molecules form *A*-rich and *B*-rich domains, and chemical bonds joining *A* and *B* molecules reside at the interface between the domains (Fig. 12). The molecules are stretched in the ordered phase at low temperature, and the characteristic period of the ordered structures scales with the number of monomers in a molecule, N, as $N^{2/3}$. The mixing of the *A* and *B* homopolymers can be enhanced by the addition of the *AB* diblock copolymer. The latter has a similar effect on the *AB* polymer blend as an amphiphilic molecule on the mixture of oil and water, although amphiphiles can be more effective in enhancing mixing between immiscible liquids (Hołyst and Schick, 1992b).

5.3 Amphiphilic Systems

Amphiphilic molecules consist of two parts: a polar head that can form hydrogen bonds and a hydrocarbon tail. Such mole-

FIG. 12. Lamellar phase (periodic in one direction) of the *AB* diblock copolymer. The dots denote the position of the points where an *A* block (dashed line) is chemically joined to a *B* block (solid line).

cules are not easily solubilized in water because of their long tails. At a certain, usually low, concentration called critical micelle concentration, the amphiphilic molecules aggregate and form closed structures called micelles. The interior of the micelle consists of the hydrocarbon tails and has the structure and density of the hydrocarbon liquid. The polar heads reside on the surface of the micelle, shielding solvent molecules (water) from the interior hydrocarbon liquid. The geometrical shape of the micelles depends on the maximum length of the hydrocarbon chain, l_c, the cross-sectional area occupied by the polar head, a_0, and the hydrocarbon volume, v (Israelashvili, 1985; Degiorgio, 1985). For example, spherical micelles can form for $v/a_0 l_c < \frac{1}{3}$, nonspherical micelles (in particular cylindrical) for $\frac{1}{3} < v/a_0 l_c < \frac{1}{2}$, and flat bilayers or vesicles for $\frac{1}{2} < v/a_0 l_c < 1$. They are usually polydisperse in size. As we see, the hydrogen bonding and hydrophobic effect are responsible for the formation of these structures. When the concentration of amphiphilic molecules becomes comparable to the concentration of the water solvent, certain ordered, bicontinuous phases are formed such as simple cubic (Fig. 13), diamond (Fig. 14), and gyroid structures (Fig. 15). The internal surfaces of these structures are formed by the polar heads. The structures are periodic in three dimensions and the internal surfaces may assume the configuration of minimal surfaces. The minimal surface is characterized by zero mean curvature at every point; each point on this surface is a saddle point.

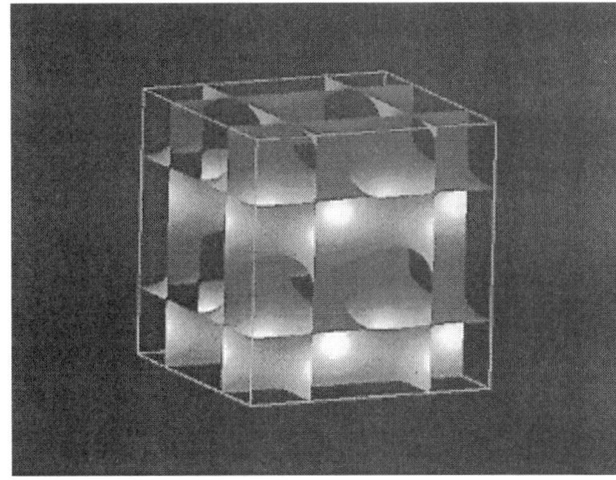

FIG. 14. The cubic bicontinuous phase in the amphiphilic system. Symmetry $F\bar{4}3m$. Known as the Schwartz *D* triply periodic minimal surface. Legend as in Fig. 13.

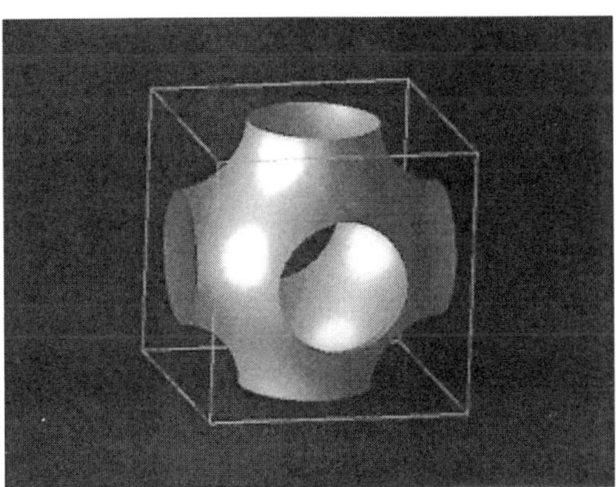

FIG. 13. The cubic bicontinuous phase in the amphiphilic system. Symmetry *Pm*3*m*. Known as the Schwartz *P* triply periodic minimal surface. Here only the unit cell is shown. The internal interfaces are shown only. In the binary mixture of water and amphiphiles, water is on both sides of the surface. In the ternary mixture of oil, water, and amphiphile, we have oil on one side of the surface and water on the other. The interface is formed by the amphiphilic molecules. Water (oil) forms interconnected channels that span the whole volume.

Amphiphilic molecules strongly reduce the surface tension between hydrocarbon liquid and water and enhance mixing between these two immiscible liquids. One characteristic structure formed in the ternary mixture of oil, amphiphiles, and water is the microemulsion. This homogeneous phase is characterized experimentally by enhanced scat-

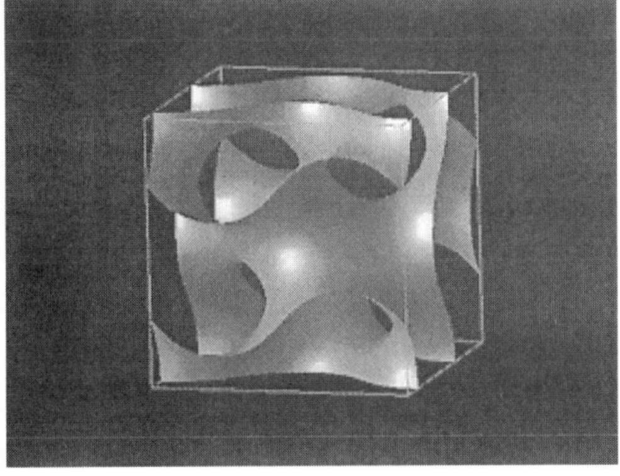

FIG. 15. The gyroid structure. Known as the Schoen *G* surface. Symmetry *Ia*3*d*. Legend as in Fig. 13.

tering at a certain nonzero **q** vector. The form of the water–water scattering intensity

$$S(\mathbf{q}) = (a + g\mathbf{q}^2 + c\mathbf{q}^4)^{-1} \tag{30}$$

with negative g fits the data from scattering experiments very well (Teubner and Strey, 1987). It reveals the internal structure of the microemulsion with a characteristic length of order of 400 Å (the size of the amphiphile molecule is about 30 Å). Inside the microemulsion, water-rich regions of the size of the characteristic length are separated by the amphiphiles from oil-rich regions of a similar size. On the macroscopic scale, the system can be characterized by the large amount of these internal water–oil interfaces. Various surface and bulk properties of microemulsions are described by Ciach (1992) and Gompper and Schick (1994). The simplest Landau model that can be used for the description of bulk and surface properties of microemulsion is given by the following functional:

$$F[x(\mathbf{r})] = \int d^3r[(\Delta x)^2 + g(x)(\nabla x)^2 + f(x)], \tag{31}$$

where Δ denotes the Laplacian and ∇, the gradient,

$$g(x) = g_2x^2 - g_0, \tag{32}$$

and

$$f(x) = x^2(x - 1)^2(x + 1)^2. \tag{33}$$

Here x is the local concentration difference between oil and water, $f(x)$ is the simplest Landau bulk free energy for the three phases (pure water, pure oil, and microemulsion) at coexistence, and g_2 and g_0 are constants. Interestingly, the bicontinuous phases shown in Figs. 13–15 correspond to the local minima of this functional.

6. KINETICS OF DEMIXING

When a homogeneous binary AB mixture above its critical point is suddenly cooled (quenched) below its critical temperature, it ceases to be in thermodynamical equilibrium [for a review on kinetics, see Langer (1992)]. The homogeneous state can be now either a metastable or an unstable state. In the case of a metastable state, the process of demixing requires, in the first place, nucleation of droplets of the minority phase, say the A-rich phase. Then the droplets start to grow. At first they grow independently, and their radius $L(t)$ changes with time t according to the formula

$$L(t) \sim \sqrt{t}. \tag{34}$$

This behavior has been observed in binary polymer blends (Cumming *et al.*, 1990). At a later stage a significant fraction of A molecules disappear from the homogeneous mixture (most of them form droplets), and competitive growth starts. Small droplets decrease in size, and the A atoms diffuse toward large droplets, which grow. This mechanism is known as the "evaporation–condensation" mechanism and is described by the Lifshitz–Slyozov–Wagner (LSW) law for the growth rate of large droplets:

$$L(t) \sim t^{1/3}. \tag{35}$$

When the system is quenched into the thermodynamically unstable region, the demixing proceeds via the spinodal decomposition mechanism. Early stages of this process are described by the Cahn–Hilliard theory. According to the theory the system becomes unstable with respect to small fluctuations, of wave vector **q** smaller than some value $\mathbf{q}_0$. The key prediction of the theory is the exponential growth of the scattering intensity $S(\mathbf{q},t)$ in time, with a well-defined maximum at $\mathbf{q}_{max} = \mathbf{q}_0/\sqrt{2}$. Interpenetrating A-rich and B-rich domains of the size of $L \sim 1/\mathbf{q}_{max}$ are formed. The mixtures, of low viscosity, undergoing spinodal decomposition do not remain for a long time in their unstable early configuration, in contrast to the high–molecular-weight polymer blends. The latter are very viscous liquids, and the whole kinetics of phase separation is very slow, allowing a detailed experimental observation of the process (Bates and Wiltzius, 1989). In the late stages of spinodal decomposition, the scattering intensity can be represented by the following scaling form:

$$S(\mathbf{q},t) \approx L^d(t)Y(\mathbf{q}L(t)) \times \text{const}, \tag{36}$$

where Y is a scaling function, d is the dimension of space, and $L(t)$ is the time-dependent length characterizing the size of the interpenetrating A-rich and B-rich regions when the volume fractions are equal. $L(t)$ grows in time, the pattern coarsens, and the interfacial area, whose energy drives the coarsening, decreases. In the case of unequal volume fractions, $L(t)$ is the size of A (minority phase) droplets [Eq. (35)]. The late-stage configuration depends on the initial volume fractions, but not on the early-stage mechanism of demixing. In the late stage we have either a dilute gas of droplets of the minority phase in the sea of the majority phase, a dense system of the droplets of one of the phases in a sea of the another, or an interpenetrating network of A-rich and B-rich domains, which coarsen in time. The growth rate and the growth mechanism for the first case is described by the LSW law [Eq. (35)]. In the second case a Binder–Stauffer mechanism of collisions and coalescence of droplets is valid. It gives $L(t) \sim t^{1/3}$, similarly as in the first case. Finally, in the last case, if the coarsening proceeds via the flow induced by the surface tension of interfaces, the scaling law $L(t) \sim t$ holds. The scaling function Y and the late stage of spinodal decomposition have been studied in computer simulations (Koga and Kawasaki, 1993). Although the field of kinetics of demixing or, in general, of first-order phase transitions is rather old, new experimental results suggest that the problem is far from being understood. First of all, a new mechanism of coarsening called "collision-induced collision" has been observed (Tanaka, 1994). It has been also shown that surface effects and confinement play an important role in spinodal decomposition (Tanaka, 1993; Jones *et al.*, 1991; Wiltzius and Cumming, 1991). Finally, the problem of heat release during demixing has been addressed in experimental studies (Bailey and Cannell, 1993). Concluding: the kinetics of demixing is still an active field of research.

GLOSSARY

Azeotrope: The point on the phase diagram where two different phases of same composition coexist. Usually, in simple liquids the two phases are vapor and liquid.

Eutectic Composition: The liquid mixture composition with the lowest freezing point.

Excess Thermodynamic Quantities: The difference between actual mixing functions and ideal mixing functions.

Ideal Mixture: Mixture in which all the components mix in all proportions without change of volume or enthalpy. Interactions between various components in the mixture are the same [see also Eq. (11)].

Miscibility: The tendency or capacity of two or more liquids to form a uniform blend, that is, to dissolve in each other.

Mixing Function: The difference between a function defined for the mixture and the sum of these functions for pure systems. The mixing function vanishes for the pure system [Eq. (10)].

Solubility: The ability of the substance to form a solution with another substance. It also denotes the maximum amount of the solute that can be solubilized in the solvent at the given thermodynamic conditions.

Solute: The substance less abundant in the mixture.

Solution: A single homogeneous liquid, solid, or gas phase that is a mixture in which the components (liquid, gas, solid, or the combination thereof) are uniformly distributed throughout the mixture.

Solvent: The substance most abundant in a mixture.

Spinodal Decomposition: Process of phase separation of the mixture in the region (of the phase diagram) of thermodynamic instability of the mixture.

Upper (Lower) Consolute (or Critical) Temperature for a Liquid Mixture: Temperature, for the binary mixture of A and B components, above (below) which A and B components mix in all proportions; i.e., they exhibit full miscibility.

Works Cited

Alder, B. J., Wainwright, T. E. (1957), *J. Chem. Phys.* **27,** 1208–1209.

Allen, M. P., Evans, G. T., Frenkel, D., Mulder, B. M. (1993), in: I. Prigogine, S. A. Rice (Eds.), *Advances in Chemical Physics,* Vol. 36, New York: Wiley, pp. 1–166.

Atkins, P. W. (1994), *Physical Chemistry,* Oxford: Oxford Univ. Press.

Bailey, A. E., Cannell, D. S. (1993), *Phys. Rev. Lett.* **70,** 2110–2113.

Bates, F. S. (1991), *Science* **251,** 898–905.

Bates, F. S., Fredrickson, G. H. (1990), *Annu. Rev. Phys. Chem.* **41,** 525–557.

Bates, F. S., Wiltzius, P. (1989), *J. Chem. Phys.* **91,** 3258–3274.

Bates, F. S., Muthukumar, M., Wignall, G. D., Fetters, L. J. (1988), *J. Chem. Phys.* **89,** 535–544.

Baus, M. (1990), *J. Phys. Cond. Matter* **2,** 2111–2126.

Biben, T., Hansen, J. P. (1991), *Phys. Rev. Lett.* **66,** 2215–2218.

Brochard, F., Jouffroy, J., Levinson, P. (1984), *J. Phys. (Paris)* **45,** 1125–1136.

Callen, H. B. (1960), *Thermodynamics,* New York: Wiley.

Carneiro, G. M., Schick, M. (1988), *J. Chem. Phys.* **89,** 4368–4373.

Casagrande, C., Veyssié, M., Finkelmann, H. (1982), *J. Phys. (Paris) Lett.* **43,** L671–L675.

Ciach, A. (1992), *Polish J. Chem.* **66,** 1347–1381.

Cumming, A., Wiltzius, P., Bates, F. S. (1990), *Phys. Rev. Lett.* **65,** 863–866.

Curtin, A., Ashcroft, N. W. (1985), *Phys. Rev. A* **32,** 2909–2919.

Deblieck, R., Lekkerkerker, H. N. W. (1980), *J. Phys. (Paris) Lett.* **41,** L351–L355.

De Gennes, P. G. (1977), *J. Phys. Lett.* **38,** L441–L443.

De Gennes, P. G. (1979), *Scaling Concept in Polymer Physics,* Ithaca, New York: Cornell Univ. Press.

De Gennes, P. G., Prost, J. (1993), *The Physics of Liquid Crystals,* Oxford: Clarendon Press.

Degiorgio, V. (Ed.) (1985), *Physics of Amphiphiles: Micelles, Vesicles and Microemulsions,* Proceedings of the International School of Physics "Enrico Fermi," Course XC, Amsterdam: North Holland.

Denton, A. R., Ashcroft, N. W. (1989), *Phys. Rev. A* **39,** 4701–4708.

Denton, A. R., Ashcroft, N. W. (1990), *Phys. Rev. A* **42,** 7312–7329.

Denton, A. R., Ashcroft, N. W. (1991), *Phys. Rev. A* **44,** 8242–8248.

Des Cloizeaux, J., Jannik, G. (1990), *Polymers in Solutions: Their Modelling and Structure,* Oxford: Clarendon Press.

Deutsch, H.-P., Binder, K. (1992), *Macromolecules* **25,** 6214–6230.

Dijkstra, M., Frenkel, D. (1994), *Phys. Rev. Lett.* **72,** 298–300.

Dijkstra, M., Frenkel, D., Hansen, J. P. (1994), *J. Chem. Phys.* **101,** 3179–3189.

Doane, J. W. (1990), in: B. Bahadur (Ed.), *Liquid Crystals: Applications and Uses,* Singapore: World Scientific, pp. 362–396.

Doi, M., Edwards, S. F. (1986), *The Theory of Polymer Dynamics,* Oxford: Clarendon Press.

Dorgan, J. R., Soane, D. S. (1990), *Mol. Cryst. Liq. Cryst.* **188,** 129–146.

Edwards, S. F. (1966), *Proc. Phys. Soc.* **88,** 265–280.

Edwards, S. F. (1975), *J. Phys. A* **8,** 1670–1680.

Evans, R. (1979), *Adv. Phys.* **28,** 143–200.

Evans, R., Marconi, M. B. (1987), *J. Chem. Phys.* **86,** 7138–7148.

Flory, P. (1953), *Principles of Polymer Chemistry,* Ithaca, NY: Cornell Univ. Press.

Floudas, G., Pakula, T., Stamm, M., Fischer, E. W. (1993), *Macromolecules* **26,** 1671–1675.

Frenkel, D. (1991), in: J. P. Hansen, D. Levesque, J. Zinn-Justin (Eds.), *Liquids, Freezing and Glass Transition,* Ecole d'été de physique théorique, Session LI, Amsterdam: North Holland, pp. 689–762.

Frenkel, D. (1994), *J. Phys. Cond. Matter* **6,** A71–A78.

Frenkel, D., Louis, A. A. (1992), *Phys. Rev. Lett.* **68,** 3363–3365.

Furman, D., Dattagupta, S., Griffiths, R. B. (1977), *Phys. Rev. B* **15,** 441–464.

Gehlsen, M., Rosendale, J. H., Bates, F. S., Wignall, G. D., Lotte, H., Almdal, K. (1992), *Phys. Rev. Lett.* **68,** 2452–2455.

Gompper, G., Schick, M. (1994), in: C. Domb, J. L. Lebowitz (Eds.), *Self-Assembling Amphiphilic Systems,* Vol. 16, *Phase Transitions and Critical Phenomena,* New York: Academic, pp. 1–181.

Grant, D. J. W., Higuchi, T. (1990), *Solubility Behavior of Organic Compounds,* New York: Wiley.

Hagen, M. H. J., Meijer, E. J., Mooij, G. C. A. M., Frenkel, D., Lekkerkerker, H. N. W. (1993), *Nature* **365,** 425–426.

Hansen, J. P., McDonald, I. R. (1986), *Theory of Simple Liquids,* New York: Academic.

Henderson, D., Leonard, P. J. (1971), in: D. Henderson (Ed.), *Physical Chemistry (an Advanced Treatise),* Vol. VIIIB, New York: Academic, pp. 414–510.

Hildebrand, J. H., Scott, R. L. (1950), *Solubility of Nonelectrolytes,* 3rd ed., New York: Reinhold Publishing Corp.

Hobbie, E. K., Bauer, B. J., Han, C. C. (1994), *Phys. Rev. Lett.* **72,** 1830–1833.

Hołyst, R., Schick, M. (1992a), *J. Chem. Phys.* **96,** 721–729.

Hołyst, R., Schick, M. (1992b), *J. Chem. Phys.* **96,** 7728–7737.

Hołyst, R., Vilgis, T. A. (1993), *J. Chem. Phys.* **99,** 4835–4844.

Hołyst, R., Vilgis, T. A. (1994), *Phys. Rev. E* **50,** 2087–2092.

Israelachvili, J. N. (1985), *Intermolecular Interactions and Surface Forces,* New York: Academic.

James, K. C. (1986), *Solubility and Related Properties,* New York: Dekker.

Jones, R. A. L., Norton, L. J., Kramer, E. J., Bates, F. S., Wiltzius, P. (1991), *Phys. Rev. Lett.* **66,** 1326–1329.

Koga, T., Kawasaki, K. (1993), *Physica A* **196,** 389–415.

Landau, L. D., Lifshitz, E. M. (1980), *Statistical Physics,* 3rd ed., Pt. 1, Oxford: Pergamon Press.

Langer, J. S. (1992), in: C. Godrèche (Ed.), *Solids Far from Equilibrium,* Cambridge, U.K.: Cambridge Univ. Press, pp. 298–362.

Laughlin, R. G. (1994), *The Aqueous Phase Behavior of Surfactants,* New York: Academic.

Leibler, L. (1980), *Macromolecules* **13,** 1602–1617.

Lifschitz, M., Dudowicz, J., Freed, K. F. (1994), *J. Chem. Phys.* 3957–3978.

Löwen, H. (1994), *Phys. Rep.* **237,** 251–324.

Maïssa, P., Sixou, P. (1989), *Liq. Cryst.* **5,** 1861–1873.

Matsen, M. W., Schick, M. (1994), *Phys. Rev. Lett.* **72,** 2660–2663.

Mc-Graw Hill Editors (Eds.) (1984), Dictionary of Chemical Terms, 3rd ed., New York: McGraw-Hill.

Meijer, E. J., Frenkel, D. (1994), *J. Chem. Phys.* **100,** 6873–6887.

Nernst, W. (1904), *Theoretical Chemistry,* New York: MacMillan.

Onsager, L. (1949), *Proc. NY Acad. Sci.* **51,** 627–655.

Poniewierski, A., Hołyst, R. (1988), *Phys. Rev. Lett.* **61,** 2461–2464.

Poniewierski, A., and Hołyst, R. (1990), *Phys. Rev. A* **41,** 6871–6880.

Roe, R.-J., Zin, W.-C. (1984), *Macromolecules* **17,** 189–194.

Rowlinson, J. S., Swinton, F. L. (1982), *Liquids and Liquid Mixtures,* Guilford, UK: Butterworth Scientific.

Schwahm, D., Mortensen, K., Yee-Madeira, H. (1987), *Phys. Rev. Lett.* **58,** 1544–1546.

Schweizer, K. S., Curro, J. G. (1988), *Phys. Rev. Lett.* **60,** 809–812.

Shaw, M. R., Thirumalai, D. (1991), *Phys. Rev. A* **44,** R4797–R4800.

Sigaud, G., Nguyen, H. T., Achard, M. F., Twieg, R. J. (1990), *Phys. Rev. Lett.* **65,** 2796–2799.

Sivardiére, J. (1980), *J. Phys.* **41,** 1081–1089.

Tanaka, H. (1993), *Phys. Rev. Lett.* **70,** 2770–2773, 3524.

Tanaka, H. (1994), *Phys. Rev. Lett.* **72,** 1702–1705.

Tarazona, P. (1985), *Phys. Rev. A* **31,** 2672–2679.

Tejero, C. F., Daanoum, A., Lekkerkerker, H. N. W., Baus, M. (1994), *Phys. Rev. Lett.* **73,** 752–755.

Teubner, M., Strey, R. (1987), *J. Chem. Phys.* **87,** 3195–3200.

Van Duijneveldt, J. S., Lekkerkerker, H. N. W. (1993), *Phys. Rev. Lett.* **71,** 4264–4266.

Van Oss, C. J., Absolom, D. R., Neumann, A. W. (1980), *Colloid. Surf.* **1,** 45–56.

Van Roij, R., Mulder, B. (1994), *J. Phys. II France* **4,** 1763–1769.

Walker, J. S., Vause, C. A. (1983), *J. Chem. Phys.* **79,** 2660–2676.

Wiltzius, P., Cumming, A. (1991), *Phys. Rev. Lett.* **66,** 3000–3003.

Xu, H., Baus, M. (1987), *J. Phys. C* **20,** L373–L380.

Xu, H., Baus, M. (1992), *J. Phys. Cond. Matter* **4,** L663–L668.

SOLUBILITY AND SEGREGATION IN ALLOYS

CHARLES WERT, *Department of Materials Science and Engineering, University of Illinois at Urbana-Champaign, Urbana, Illinois, U.S.A.*

INTRODUCTION

The recorded use of materials dates back more than 200 centuries, and so the sciences of physics and chemistry are only recently on the scene. Thus, it is instructive to begin the discussion by noting how ancient craftsmen made useful materials, especially of metals. We can see how both solutions of elements and preparation of segregated phases produced desirable products. To be sure, pottery and glass were also important, but the inability of craftsmen to alter their chemistry and internal structure appreciably made their form and surface design more important. Thus, major eras have been named the Bronze Age and the Iron Age, not the pottery or glass age. Three examples of metallic alloys illustrate how more complex products were developed as understanding increased of the chemistry and the internal arrangement of atoms: gold, copper, and iron.

Gold and copper (along with silver) were first acquired in nearly pure form. At first, they could be altered only by pounding (to harden them) or by heating (to soften them). Gold in nearly pure form was hammered into thin sheets to stretch its usefulness. It was later alloyed with mercury to form an amalgam that could be spread in thin layers on copper, brass, or bronze. A thin coating of gold was formed after evaporation of the mercury, a coating whose thinness and quality rivals modern electroplated coatings. Silver and copper are often found associated with native gold, and further, both form complete solid solutions with gold; craftsmen around the world developed alloys comparable to the 10- to 22-carat gold alloys presently used for jewelry and for coins.

Copper was also used in pure form, hammered and annealed to produce a desired strength and hardness. It was alloyed with tin—found also in pure form, but also widely present in cassiterite, which could be easily reduced by a wood fire—to provide a bronze with properties comparable to modern bronzes. Brass was also produced by heating copper in the presence of zinc carbonate, whose dissociation provided zinc. Finally, middle eastern craftsmen and much later, Peruvian Indians, made copper–arsenic alloys with good casting properties.

Iron has an even more complex history. It was first obtained in metallic form from iron-rich meteorites, in which it is always alloyed with nickel. Iron oxides can be reduced in a wood fire only with a forced air stream, and only gradually was a means found to provide a spongy iron-slag product from iron oxides. It could then be transformed into more useful materials by heating—particularly to cast irons and later to steels. Craft practice far predated an understanding of the role of carbon in strengthen-

3-527-28140-1/96/$5.00 + .50

ing steels—the tortuous path of this understanding is well described in the books of Smith (1962, 1981) and of Aitchison (1960). To note how recently modern steels have developed, recall that the beams of the Eiffel Tower are made from a rolled iron containing strings of slag—wrought iron.

These examples—and many others that might be cited—describe ancient alloys that are either solid solutions or mixtures of solutions and segregated phases. That they resulted in useful products is a remarkable testimony to the acumen of early craftsmen.

Materials development still provides a vast frontier of opportunity. Within the past 50 years, countless new materials with novel combinations of solutions and segregated phases have entered the marketplace. Tiny "shear pins" hold a filling of my tooth in place; a light-cured epoxy fills in a chipped spot on an incisor, color-matched to the adjacent enamel. Small additions of impurities to silicon make the computer possible. Silica glass fibers of extraordinary purity transmit voice or data signals with low loss using infrared radiation—the small loss inherently present can be made up by laser amplification every 50 km in short lengths of fiber containing Er in solution. Super-bright light-emitting diodes (LEDs) are bright enough to serve as warning lights in the back windows of automobiles; they are highly efficient and have a lifetime of perhaps 100 years. The bearings and bearing races in my 50-year-old bicycle seem unworn—even after some 60 000 miles. My father could expect to get 5000 miles from a set of tires in 1925; now I am unhappy if I don't get 50 000. A modern jet aircraft engine could make about 200 trips around the world at the equator between major overhauls.

These and a thousand other possible examples describe the enormous gains that have been made in materials science and engineering in the past 50 years. Steady gains were made through the centuries, of course, but a chief factor in the recent extraordinary progress has been the avalanche of new techniques of examination of materials—especially the microscopic, spectroscopic, and diffraction techniques that have come from physics. This section describes the nature of solutions and of materials containing segregated phases and shows how applications from science have aided in their specification.

Part of our understanding of solutions and compounds is based on chemistry, part on physics. Yet practitioners of both of these sciences have great difficulty in predicting a precise combination of elements that will produce a useful product. Practical alloy development is still largely the domain of the technologist, and most useful alloys are an exquisite combination of elements heated in a particular way to satisfy a specific need. Reading the vast literature on materials demonstrates forcefully, though, that the science and technology are merging into a continuum of practitioners.

1. THERMODYNAMICS OF ALLOYS

Most materials used technologically for products are in a metastable state, not in a thermodynamically stable configuration. Nevertheless, it is useful to start with the *thermodynamics* of the stable state.

A metal like copper can dissolve other elements by substituting them on sites in the host crystal; the extent of solubility may be calculated by application of the laws of thermodynamics. Let the incremental change in Gibbs free energy, ΔG, in substitution of atoms of B for the host atoms A be expressed by two terms, an incremental change in enthalpy, ΔH, and temperature times a change in entropy, $T\Delta S$:

$$\Delta G = \Delta H - T\Delta S. \tag{1a}$$

The change in entropy may be divided into two terms, an entropy of mixing and an excess entropy, ΔS_{excess}, occasioned by the possible alteration of the vibrational states of atoms in the crystal by the alloying addition. The first may be calculated from statistics of atom placement, and so we may write

$$\Delta G = \Delta H + RT(X_a \ln X_a + X_b \ln X_b) - T\Delta S_{\text{excess}}. \tag{1b}$$

Here X_a is the atom fraction of element A; X_b is that of element B. Several possibilities exist in specification of the minimum in ΔG, depending on the relative values of the several terms. A few special cases are described in the following paragraphs.

The enthalpy change ΔH may be zero, or it may be so small that its effect on ΔG is outweighed by the entropy term for the temperature range of interest. Then the components of Eq. (1b) may yield a monotonic negative value of ΔG for all compositions, and complete solubility exists. An example of this condition for the terms of Eq. (1b) (with ΔS_{excess} set equal to zero) is shown in Fig. 1(a), and the corresponding phase diagram for the binary alloy AuAg (which satisfies this condition) is shown in Fig. 1(b). In both the liquid and the solid alloys, complete miscibility exists for all compositions. Upon melting, the two-phase region demanded by thermodynamics is remarkably narrow, as can be seen; this results from similarity of electronic structure and nearly equal atomic radii of the two elements. This complete solid miscibility results in the presence of naturally appearing solutions of various composition valuable to early craftsmen—historians lump them together under the term *electrum*.

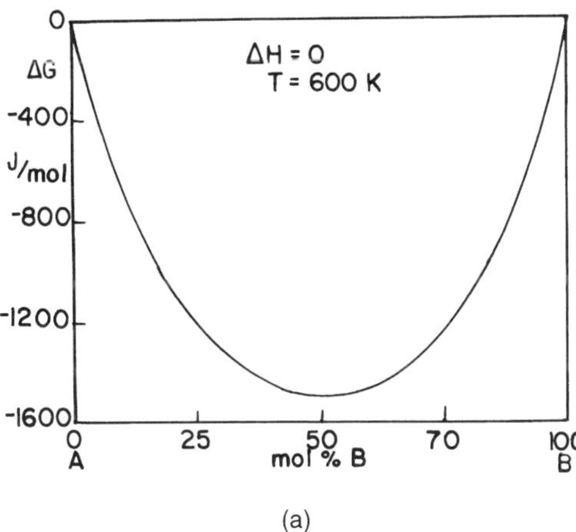

(a)

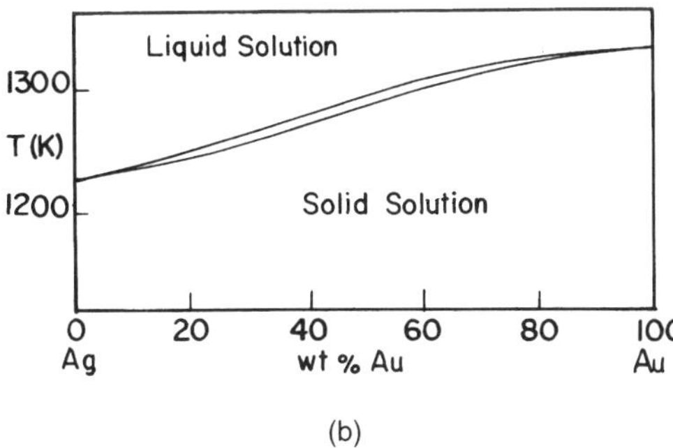

(b)

FIG. 1. (a) Free energy vs concentration for $\Delta H = 0$ and $T = 600$ K. (b) Phase diagram for the Au–Ag system.

Not many alloy systems have such a simple phase diagram. A more common case is one in which ΔH is large enough for concentrated solutions so that a mixed-phase region exists over a large range of the solid alloys. The thermodynamic parameters for that case are sketched in the upper part of Fig. 2. The enthalpy increase on addition of the alloying element at either end of the concentration range is large enough that ΔG goes through a minimum at specific concentrations and then increases. For alloys of intermediate concentration, a solid solution is energetically unfavorable with respect to separation into two phases whose concentrations are given by the common tangent line to the curve of ΔG vs concentration. Those concentrations c_1 and c_2 are functions of temperature—this sketch is appropriate for a temperature of about 340 K for the values of ΔH assumed. The large entropy contribution to ΔG for liquid alloys causes a depression of the freezing point for alloys of intermediate concentration. A typical phase diagram for such a system is that drawn for Pb–Sn, shown in the lower half of Fig. 2. This alloy system includes the technologically impor-

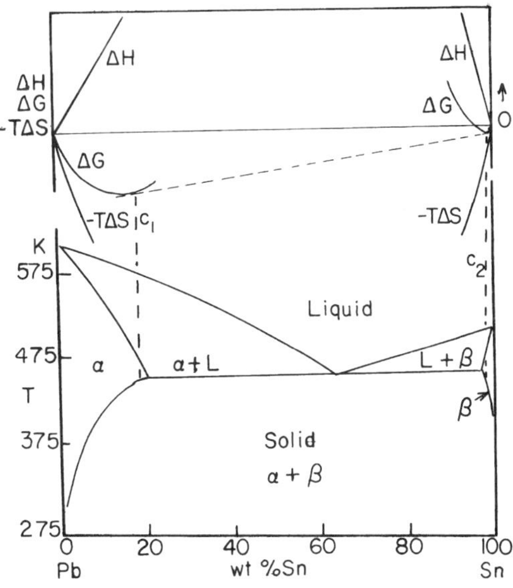

FIG. 2. Phase diagram for Pb–Sn. A eutectic. The upper part of the figure shows the components of Eq. (1) at the extremes of the concentration range. For dilute enough alloys, solid solutions exist. But for concentrated alloys, a mixed-phase region is energetically favorable.

tant solders. The lowest-melting alloy at the eutectic composition 60% Sn, 40% Pb is the common soft solder; compositions away from that point remain mushy during a large part of their solidification. Such phase diagrams also exist for ceramic combinations; complete solubility exists for the system NiO–MgO, Fig. 3(a), and a eutectic for the system MgO–CaO, Fig. 3(b). Observe that such portrayals for ceramic combinations are usually expressed as relationships between compounds, not between the constituent elements.

The "eutectic alloy system," which we (in cold climates) perhaps use more widely than any other, though, is that of ice and salt (NaCl). At one atmosphere of pressure, ice melts at 32 °F, NaCl at 1474 °F. When salt is put in contact with ice, a solution results, which lowers the melting point as a salt-water solution forms. Finally, when the water solution reaches 23 wt % salt, the temperature has fallen to −6 °F. This is the eutectic point. Adding more salt will not lower the water temperature further: the minimum is −6 °F (at a concentration of 23% salt). We all know that, at least in general, for we are aware that salting our sidewalks and streets does not help if the temperature is "too cold."

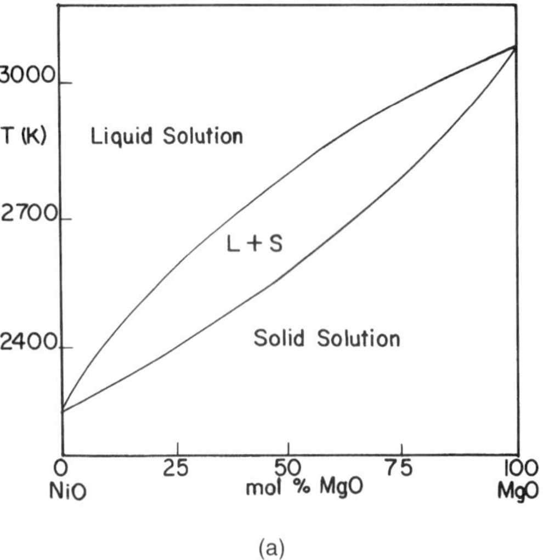

(a)

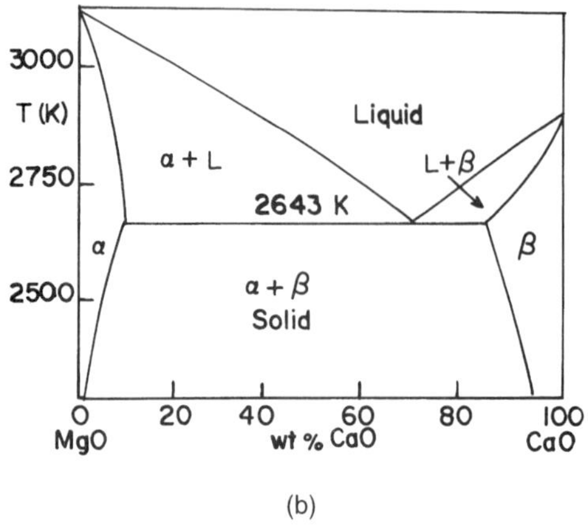

(b)

FIG. 3. Oxide phase diagrams: (a) MgO–NiO, (b) MgO–CaO.

Countless examples of binary and multicomponent phase diagrams (and others with complex sets of intermediate compounds) have been determined and published, mostly by metallurgists and ceramists. First-principles calculations have correctly predicted a few phase diagrams, but their number is so small compared with the vast number of possible combinations of alloying elements that such successes are not much more than demonstrations of the correctness of the physical models. In addition, virtually all commercial materials are multicomponent combinations not easily amenable to modeling, and so lines in binary diagrams (and surfaces in multiconstituent systems) have usually been determined experimentally. Again, though, microscopy and the diffraction and spectroscopic techniques of physics are of inestimable value in determination of the character of phases present in alloy systems.

The extremes of such phase diagrams are often technologically important. For dilute solution of B in A, X_a is about unity, and X_b may be expressed as the ratio of the number of B atoms to the total number of atoms in the solid, N. If the enthalpy required to introduce an atom B is a constant δh, then, at equilibrium,

$$X_b = n/N = \exp(-\delta h/RT) \tag{2a}$$

or in logarithmic form,

$$\ln(n/N) = -\delta h/RT. \tag{2b}$$

Thus, the limiting solubility of B in A is exponential in $1/T$ and δh is the unit enthalpy of solution. Countless important phase diagrams have limiting solid solutions that fit that equation. Not only do the phase boundaries of Pb in Sn and Sn in Pb fit that expression, but so also do such alloys such as Cu, Zn, and many other elements dissolved in Al, alloys that provide the important hardened duralumins. Beryllium dissolved in Cu has a limiting solubility line obeying this equation; proper heat treatments provide a strong spring alloy, as well as nonmagnetic tools, useful around large magnets. They are also nonsparking tools, so that they are useful in potentially explosive gas atmospheres. Carbon, nitrogen, and oxygen dissolved in such metals as Fe, Ta, Nb, and V obey this equation. Examples of alloys whose free energy obey Eq. (2) are nearly endless.

Equation (2) also describes the solubility of vacancies and other point defects dissolved in metals, insulators, and semiconductors. Investigations of δh for these point defects—usually in the range 0.5–3 eV per defect—have been an important part of diffusion studies in solids. See the data for the concentration of single vacancies in Cu, plotted according to Eq. (2b) in Fig. 4; δh is about 1.3 eV/vacancy.

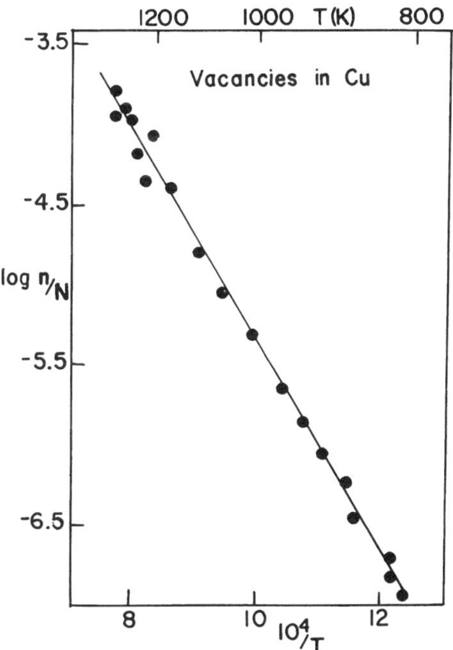

FIG. 4. Equilibrium concentration of vacancies in Cu. The data points are a small fraction of reported values. The slope of the line is the energy of formation, 1.3 eV per vacancy.

An important factor of Eq. (1) is the way in which ΔG approaches zero as the concentration of element B approaches zero. We can see the key features with the aid of the upper curves of Fig. 2. Since ΔH approaches zero linearly with X_b and the entropy term approaches zero with infinite negative slope, a minimum must always exist in ΔG for some finite concentration, no matter how large is the term ΔH. Thus, impurities always exist in thermodynamic equilibrium in solids, and costly purification schemes must often be employed to purify them—as for silicon, whose use in electronic devices is made possible only by zone refining.

The semiconductors, silicon and germanium, present a special case of purification and subsequent alloying. The enthalpy of solution of elements in these semiconductors may be very large; under these circumstances the solubility line above the eutectic may not show the usual linear behavior as is seen in Fig. 1 for Pb in Sn. Rather the line may loop around in a behavior called *retrograde* (moving backward); see an example for Sb in Ge, Fig. 5. Such retrograde behavior for a wide range of alloying elements in Ge is shown in Fig. 17 of the article GERMANIUM. Such lines for all impurities are significant in the purification of Si and Ge by zone refining.

Measurement of the components of the Gibbs free energy of multicomponent systems is an immense task, but it continues methodically. Also difficult is extrapolation of fragmentary values to other temperatures and determining the "best" set of values when data conflict. Finding minima in multicomponent alloys is not always easy, and making sure that these are not metastable minima is difficult. Many calculational and smoothing computer programs are now in wide use. For binary and ternary systems, programs as simple as macro-driven spreadsheet files may be appropriate; for more complex systems, more sophisticated programs are required. The journal *CALPHAD* and the annual CALPHAD conferences provide perhaps the best source of such information. Data bases of thermochemical values are being assembled, some on line, others on diskettes or CD-ROMS, still others in analytical form. Handbooks of tables of data, once common, are becoming passe. A consortium of organizations worldwide is engaged in

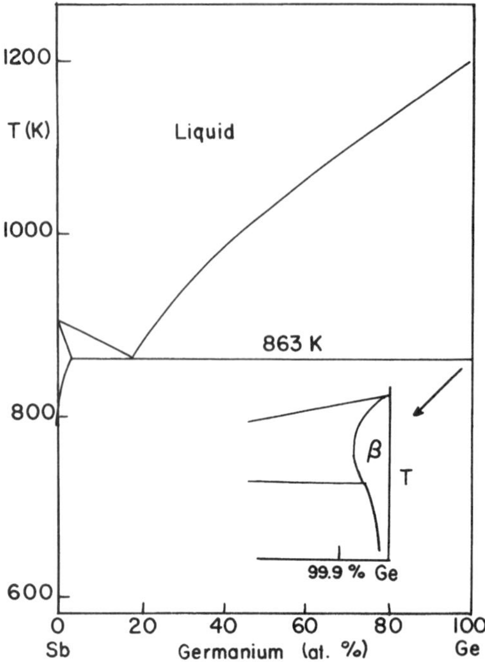

FIG. 5. The Ge–Sb eutectic. The solubility of Sb above the eutectic temperature shows retrograde behavior. The narrow region labeled β is the region of doping. After Wert and Thomson (1970).

collecting and publishing both phase diagrams and supporting thermochemical data on both binary and multicomponent systems (that consortium was assembled under the leadership of ASM International).

2. CONCENTRATED SOLUTIONS

Some alloys of technological importance display great solubility, among them the brasses. The reason for this large solubility was examined by the metallurgist Hume-Rothery (1948), not only for the alloys based on copper but also for those based on the companion metals gold and silver. He and his colleagues (using straightforward metallurgical techniques) assembled a huge amount of data on phase equilibria of alloys of these noble metals, analysis of which led them to propose that the extent of solubility depended on a number of rather simple "rules." The most significant are the following:

1. Large solubility is favored if the elements are not far apart in electronegativity.
2. It is also favored if the atom sizes are not too different (the difference in atomic radii should be less than 15%).
3. These two factors being favorable, the solubility (for these face-centered-cubic alloys) still does not exceed that value corresponding to an average electron to atom ratio (*e*/*a*) of about 1.4.

The results for copper alloyed with several multivalent elements are shown in Fig. 6. One sees that the limit of solubility occurs for a value of *e*/*a* in the range 1.3 to 1.4. A two-phase mixture exists for larger values of *e*/*a*, but other phases (having different crystal structures) exist at higher *e*/*a* ratios: 1.5, 1.6, and 1.7.

Hume-Rothery and his colleagues assembled their many experimental observations in the 1920s and 1930s, just when the band theory of solids was being developed by physicists. Using ideas from that theory, the group proposed that the limitations placed on the *e*/*a* ratio were the result of the reduction in electronic density of states when the expanding Fermi surface impinged on spe-

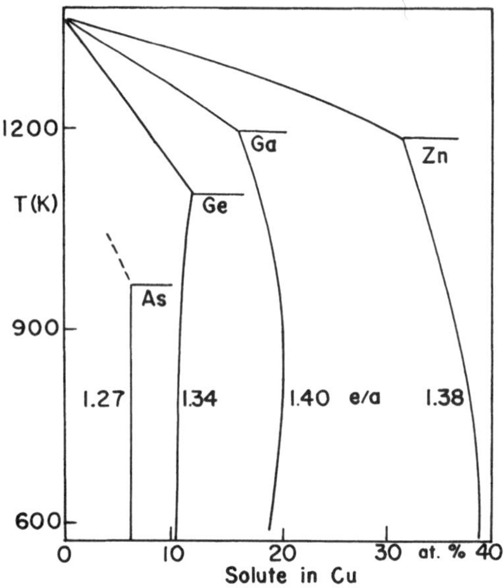

FIG. 6. Multivalent impurities in Cu have decreasing solubility as their valence difference with Cu becomes larger. But the *e*/*a* ratio at maximum solubility is about the same—between 1.3 and 1.4. After Barrett and Massalski (1966).

cific boundaries of the Brillouin zones as the concentration of multivalent elements was increased. For the face-centered-cubic copper alloys, this was assumed to be impingement of a spherical Fermi surface on octahedral faces of the first zone boundary. The critical *e/a* ratios for stable phases at higher concentrations result from impingement of the Fermi surface on boundaries of zones that become successively more nearly spherical.

The experimental evidence of the Hume-Rothery group is not to be doubted; an excellent summary is found in the book of Barrett and Massalski (1966). The simple interpretation initially proposed provoked much discussion among physicists, in particular of the "magic" *e/a* ratios. That the initial explanations were too simple is certain; experiments in the late 1950s showed that the Fermi surface already is in contact with the zone boundary for pure copper. Thus the limiting *e/a* ratio of 1.4 could not be explained by the initial ideas of Hume-Rothery and his colleagues. An alternative model based on more refined principles was proposed shortly thereafter in a series of articles by Blandin and colleagues at Orsay (see, for example, Blandin *et al.,* 1966). Other solutions were later offered by Heine and Weaire (1970) and by several authors in the more recent books *The Physics of New Materials* (Fujita, 1994) and *Electron Theory in Alloy Design* (Pettifor and Cottrell, 1992).

One can appreciate the difficulty of making accurate enough calculations by noting that the enthalpy difference associated with changes in phase is usually very small compared with the total enthalpy of either phase. Consider a really simple case—the enthalpy change for pure iron when it transforms at 1183 K from the low-temperature bcc phase to the high-temperature fcc phase is only about 6% of the enthalpy of the bcc phase just before the transformation. Even in this simple one-element case, one must calculate a relatively small difference between two large numbers. How much more difficult is the task for an alloy of two (or more) elements! These difficulties do not prevent physicists and materials scientists from attempting calculations of many types. Such attempts are described in many of the references given in the list of Further Reading.

Many combinations of elements other than the noble metals form technologically important solid solutions, of course—perhaps none more important than the stainless steels, combinations principally of Fe, Ni, and Cr. These elements have about the same atomic radii—about 0.125 nm. They are in the 3*d* series, so that their electronic similarities are favorable for large solubility. Iron (below 1183 K) and chromium (at all temperatures up to its melting point) have the same crystal structure (body-centered cubic), and so they are miscible in any ratio below 1183 K (except for the intrusion of a troublesome low-temperature σ phase at a 50:50 Fe–Cr ratio). Nickel has a different crystal structure (face-centered cubic), and so a change in crystal type must occur at some ratio for either Fe–Ni or Cr–Ni alloys. That limitation is not insurmountable, in technological practice.

Thousands of combinations of the three have been produced, of which the following are of primary importance. A ternary solid solution of 74% Fe, 18% Cr, and 8% Ni is of special value. It has good strength, yet can be rolled, forged, or pressed into complex shapes. It is resistant to corrosion and oxidation. Its commercial designation is type 304—but it is known in the trade as "18-8," and that symbol often appears inconspicuously on products such as flatware or serving dishes. Its corrosion resistance comes from a thin surface layer of the oxide CrO_2, which forms instantly on the surface—and which re-forms if the surface is abraded or scratched. Because of the high Cr content, 18-8 is not ferromagnetic. This alloy and other ternary combinations can be strengthened further by addition of a small amount of carbon, which forms a segregated carbide with appropriate heat treatment. Two binary higher-nickel alloys—55% Fe, 45% Ni and 22% Fe, 78% Ni—are both ferromagnetic with excellent permeability. Indeed, substitution of a little Mo in the latter alloy produces Supermalloy, an alloy with the highest permeability known—a rod becomes fully saturated, magnetically, when it is placed parallel to the Earth's field.

An enormous number of other alloys of Fe are also important. At one extreme are the "plain-carbon" steels, the cheapest steel, used for a myriad of products. At the other extreme are the complex multicomponent steels alloyed with small amounts of Ni, Cr,

Mo, W, Nb, . . . along with carbon. After a hard landing of an aircraft on a runway, I always give a private word of thanks for "4340," a multicomponent high-strength steel with good ductility, used for such crucial applications as aircraft landing gear. Compilations of compositions, properties, and applications are found in many handbooks and commercial catalogs.

One interesting group of stainless steels are the iron meteorites. The iron always is alloyed with nickel—present in the range 7% to more than 15%. This is the concentration range for which phase separation occurs into an Fe-rich bcc phase (called kamacite) and a higher-Ni fcc phase (called taenite). Because of the billions of years of their existence, the iron–nickel alloy of most iron-based meteorites has separated into bands of kamacite and taenite of massive size easily visible with the unaided eye. The pattern is known as the "Widmannstätten structure" after the Viennese scientist who first identified it in the early 1800s. Many pictures of such bands are to be found in the handbook authored by Buchwald (1975)—that work is both an excellent compilation of the astronomical nature of such meteorites and the metallurgy of their constituents. Past technological use of "iron" of meteoritic origin has been discussed by Aitchison (1960) and by Smith (1960, 1962, 1968, 1981). The reader may find amusing the historical uncertainty that has arisen over "first rights of discovery" of the Widmannstätten pattern, for another observer, Thomsen, may have observed it somewhat earlier (see Smith, 1962). The varied paths of melting, freezing, segregation, and redistribution of minerals in the 50 or so parent bodies of meteorites are well described by Dodd (1986).

3. GAS–METAL EQUILIBRIUM

Equilibrium between an atmosphere of a gas and atoms of that gas dissolved in a metal has been studied extensively. Such characterization is important because of the gradual dissolution of compressed gases in the walls of metal containment cylinders and because of reactions of atmospheric gases with hot metals and alloys. An excellent review of the thermodynamics and kinetics of such interactions and a huge compilation of data are found in the book *Gase und Kohlenstoff in Metallen* written by the group at the Max-Planck-Institute in Stuttgart under the leadership of E. Gebhardt (Fromm and Gebhardt, 1976). They also carried on much work themselves, especially for the diatomic gases dissolved in the bcc metals. One example is given here from the work of Cost (1963) to show the principles involved, the pressure-temperature-composition equilibrium in the Nb–N system.

First, the solution of a diatomic gas dissolved atomically in a metal should obey Sievert's law—the dissolved concentration should vary as the square root of the gas pressure. Measurements for such equilibrium, Fig. 7, show that law to be obeyed; the lines through the data points are drawn with a slope of 2.

Next, the partial molar heat of solution of nitrogen in Nb [for dilute solutions, this is ΔH of Eq. (1)] should be a constant with both temperature and composition. That this

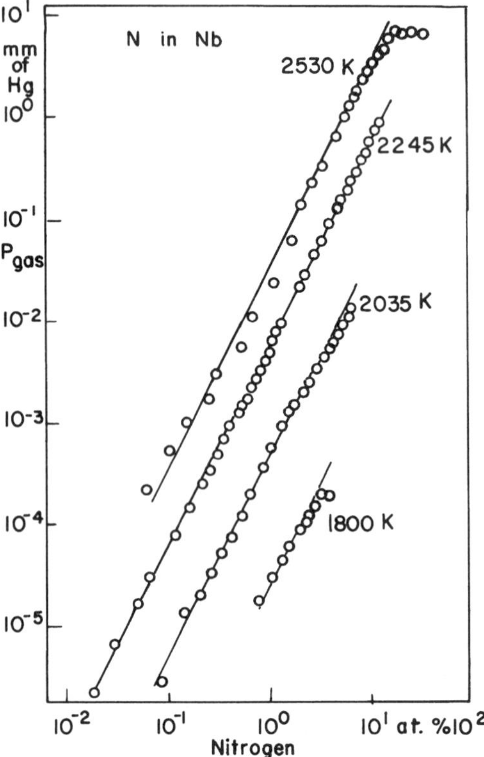

FIG. 7. The concentration of N atoms in Nb as a function of the gas pressure of nitrogen at constant temperature. After Cost and Wert (1963).

law is obeyed can be shown from the data plotted in Fig. 8. The straight lines on the plot of log(p) vs $1/T$ yield a value for ΔH of about -190 kJ/g-atom of nitrogen. The limiting line of Fig. 8 (delineated by the filled circles) yields a value of about -280 kJ/g-atom of nitrogen for the standard heat of formation of the equilibrium solid compound Nb_2N. From these data, the standard molar entropy of solution of nitrogen in Nb may be calculated to be -62 J/K·g-atom of nitrogen, a number just about that expected from a random distribution, with little excess entropy, i.e., change in vibrational character of the metal crystal. Finally, the limiting terminal solubility of N in Nb with respect to the nitride Nb_2N is displayed in Fig. 9; the standard heat of solution of the nitride in Nb at high temperatures is about 88 kJ/g-atom of nitrogen.

These measurements specify completely both the equilibrium between the gas N_2 and N atoms dissolved in the metal and the equilibrium between the nitride Nb_2N and N atoms dissolved in the metal. From the derived parameters, the pressure-temperature-composition phase diagram may be drawn. That construct is shown in Fig. 10 (in coordinates log p, $1/T$, and log c). The phase diagram is a eutectic with eutectic composition at point G. At these low pressures, the composition of the nitride is invariant with either pressure or temperature. The limits of the solubility of the terminal solid-solution phase, labeled α, are planes in these coordinates whose slopes and lines of intersection are given by

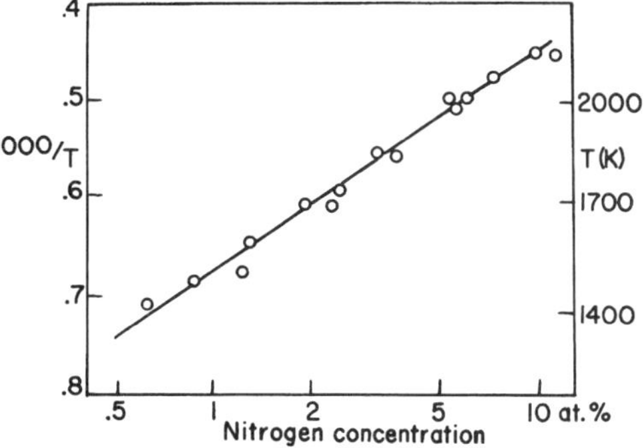

FIG. 9. The concentration of N atoms in solid solution in equilibrium with the stable nitride, Nb_2N, as a function of temperature. After Cost and Wert (1963).

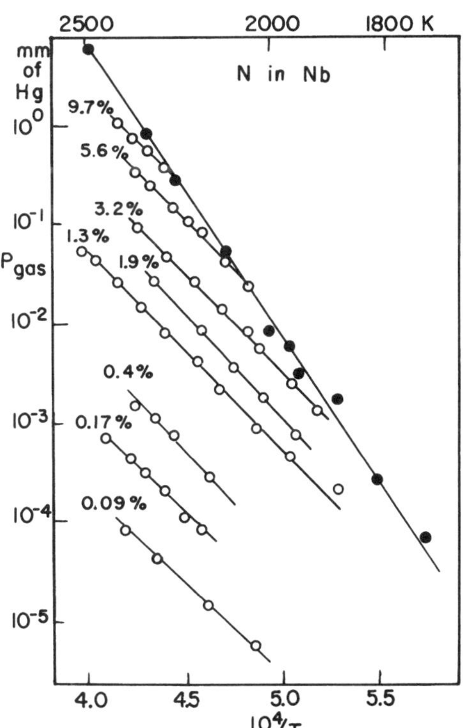

FIG. 8. The pressure-temperature equilibrium at constant concentration of N atoms in Nb. After Cost and Wert (1963).

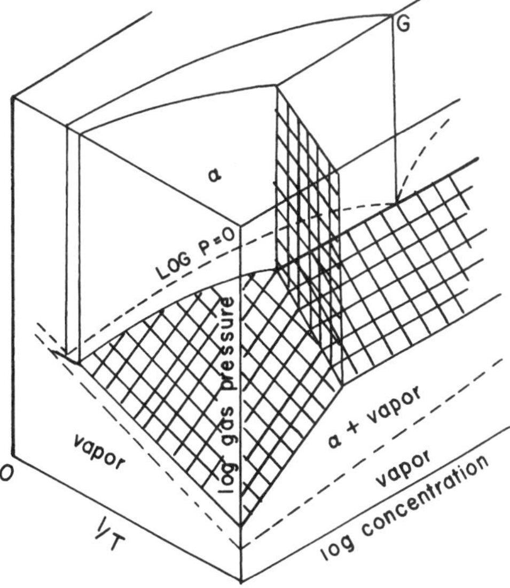

FIG. 10. A part of the phase diagram of N in Nb as a function of log pressure, $1/T$, and log concentration. After Cost and Wert (1963).

the values of the constants derived from the data embodied in Figs. 7–9. In addition, the concentration of dissolved nitrogen may be expressed in terms of temperature and gas pressure p (in pascals) as

$$c(\text{atom }\%) = 7 \times 10^{-3}\{p\}^{1/2} \exp(190\,000/RT).$$

From data in the book of Fromm and Gebhardt (1976), similar evaluations can be made for many gas–metal systems.

4. COMPOUNDS

Few alloy systems display large solubility. Most are dominated by compounds that exist as second phases. An example is shown in Fig. 11 for the system Cu–Zn. Although Zn is dissolved in the fcc Cu to about 40%, above that concentration several phases exist at higher compositions. Each of these phases exists over a range of several atom %; their compositions are not sharply set. These phases are called the Hume-Rothery phases; the pattern also exists for other binary alloys, such as Ag and Au with these alloying elements (see Chap. 11 of Barrett and Massalski, 1966). Their stability results from conduction-electron interactions—the phases themselves are still metallic conductors. An immense number of important alloy systems have their stability controlled by such phases.

The second set of important phases are the ionic crystals—formed from elements with large differences in electronegativity, such as the metal oxides. A typical example is the double eutectic system SiO_2–Al_2O_3 dominated by an oxide of Al and Si, mullite, Fig. 12. This oxide has a large energy of formation; the large value of ΔH overwhelms the entropy term of Eq. (1), and the solubility of mullite in both SiO_2 and Al_2O_3 in equilibrium with the oxide is small. All of these compounds, SiO_2, Al_2O_3, and mullite, are of large technological value. Many metallic and ceramic alloy systems are controlled by the high stability of such ionic compounds.

A large number of metallic and ceramic systems fall between these two extremes, especially for multicomponent combinations of elements. The Superalloy metallic phases based on Ni have been exquisitely combined to give a mixture of solid solutions and second phases to provide properties that make possible the turbine blades in jet engines. Iron alloys range from the solid-solution stainless steels to the cast irons (with their segregated graphite) to multiphase steels used for aircraft landing gear. The ionic oxides of copper combined with other metallic additions are under intense scrutiny for high-temperature superconductivity (see SUPERCONDUCTIVITY, HIGH-TEMPERATURE).

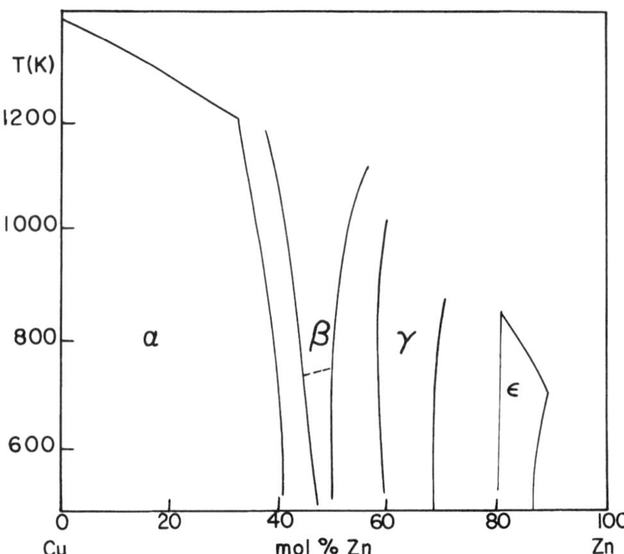

FIG. 11. The Hume-Rothery phases in Cu–Zn.

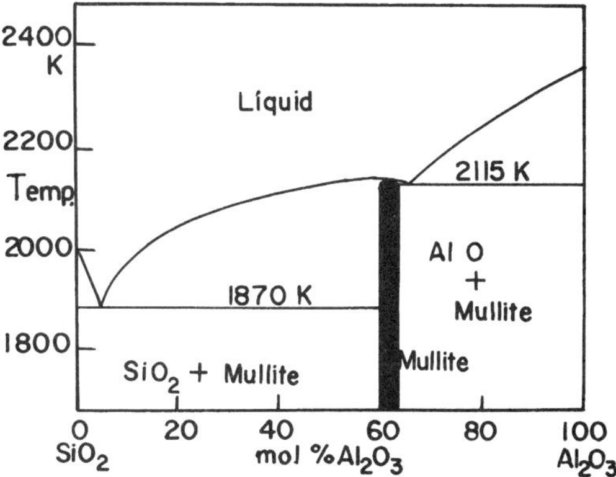

FIG. 12. Phase diagram dominated by an intermediate compound with large stability.

5. ORDERING

The thermodynamics of solid solutions described above is predicated on randomness of atom placement in the crystal. For many alloys, randomness is nearly achieved, but slight departures occur in most instances. One such departure is short-range order, in which atoms of one type have a preference of atoms of the second type for nearest neighbors. Two types of measurements of physical origin are commonly used to investigate the changes in this ordering—changes in electrical resistivity accompanying changes in temperature and x-ray scattering at small angles.

In some instances, the changes may be of long range with predictable atom placement over large distances. An example occurs for the phase 50% Cu–50% Zn; see the location of this phase in Fig. 11. The phase, called β brass, has the bcc crystal structure above 540 °C with nearly random placement of Cu and Zn in the crystal. This randomness gradually disappears with decreasing temperature, and below 540 °C (the short-dashed line in Fig. 11) both the Cu and Zn have only atoms of the other as nearest neighbors.

6. PRECIPITATION OF SECOND PHASES

Precipitation of second phases may occur when a temperature change makes a solid solution unstable. Several distinct reactions may occur. Chief among them is a process called *nucleation and growth.* Nuclei of the second phase (initially unstable) form within the solution by thermal fluctuations. These nuclei then grow by diffusion to form massive quantities of the second phase. An example of such a process may be described for nitrides forming in bcc iron (a choice based not so much on the technological importance of this alloy system as for the opportunity it affords to see how increasingly sophisticated techniques from physics aid in understanding the precipitation process). A thermodynamically stable solid solution of nitrogen in iron may be prepared at some high temperature (900 K is appropriate). Upon rapid cooling to some lower temperature (500 K is appropriate), a stable phase, Fe_4N, nucleates and grows as plates on the (210) planes of the iron. This phase is in equilibrium with a solid solution of lower concentration.

But this is not all. In this instance another nitride exists, $Fe_{16}N_2$, whose free energy in bcc iron is higher than that of Fe_4N, but which forms first at temperatures around 500 K, perhaps because of easier nucleation. The compound forms first as platelets on the (100) planes of the bcc iron and reaches a temporary equilibrium with a solid solution of N in the bcc iron, an equilibrium higher in N in the bcc iron than is true for N in equilibrium with the stable Fe_4N phase. The relationship of these phases and the thermodynamic parameters of this edge of the phase diagram may be seen in Fig. 13. The limiting solubility lines are straight lines in these coordinates, log c vs $1/T$ [in accordance with Eq. 2(b)]. The equations of the lines are put in as labels in the figure. The two sets of measurements, made by independent observations of Dijkstra (1949) and of Fast and Verrijp (1955), agree well. (Both measurements were made using mechanical relaxation techniques described in the last section.) The insets to this figure show optical micrographs of both the metastable phase $Fe_{16}N_2$ and the stable phase Fe_4N, which form in sequence. Observe that the solubility line in equilibrium with the metastable phase $Fe_{16}N_2$ occurs for all temperatures at a higher N concentration than that for the stable phase Fe_4N. This thermodynamic relationship always must exist in phase equilibria where metastable phases occur.

Optical micrographs such as those inset

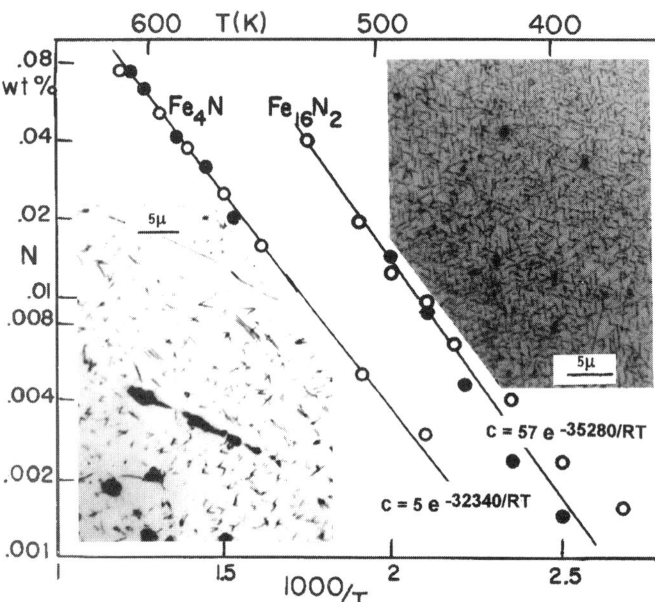

FIG. 13. A part of the phase diagram for Fe–N showing equilibrium for both the stable phase Fe_4N and the metastable phase $Fe_{16}N_2$. The open circles are taken from Dijkstra (1949), the filled circles from Fast and Verrijp (1955). The adjacent insets show precipitates: $Fe_{16}N_2$, upper right; Fe_4N, lower left. The large black blobs are iron oxide, initially present in the iron.

into Fig. 13 clearly show that $Fe_{16}N_2$ forms on the (100) habit plane of the host iron. Later calculations of elastic energy indicated that the preferred habit plane of such a tetragonal precipitate in a cubic host might not be precisely the (100) plane. See Wen *et al.* (1981). This possibility was found to be the case in this instance; electron microscope imaging by Ferguson *et al.* (1984) showed that the "plates" in the inset of Fig. 13 are, in fact, puckered rosettes of platelets, segments that are, indeed, tilted from the (100) plane by about 10°. The average habit plane, though, is (100).

No unusual magnetic characteristic of the compound $Fe_{16}N_2$ seemed evident when it was first observed as a metastable nitride in iron. However, crystals of this phase produce an unusual magnetic effect when grown epitaxially on the surface of an In–Ga–As wafer, whose chemistry was selected to produce a lattice constant close to that of $Fe_{16}N_2$. It possesses a "giant" magnetic moment per iron atom—more than 3 Bohr magnetons, a value much higher than should be expected.

Almost all commercially valuable alloys are heat treated to produce second phases to enhance some desirable combination of properties: strength, ductility, corrosion resistance, magnetic or electrical characteristics, ... In nearly all instances the appropriate final product is not a mixture of thermodynamically stable phases but a combination of metastable phases that experience shows give the best product. The steels, aluminum alloys, brasses, permanent magnets, bronze, semiconductor chips, . . . , all are metastable combinations. Processing is carefully formulated to enhance the desirable features; many failures of products occur because of improperly designed processes or because of errors in carrying out specified processes. In addition, the necessity of making appropriate metastable combinations is a major reason that first-principles calculations of physics are of small importance in formulating multicomponent alloys with useful properties. Again, though, the experimental techniques of physics—especially the electron microscope, together with spectroscopic and diffraction techniques—are of inestimable value in characterization of useful combinations.

Another mode of phase separation exists when the concentration of the alloy lies within the *spinodal* lines of the free energy plot. Understanding this reaction may be aided by the plots of Fig. 14. This is an extension of the plot of ΔG vs concentration (of Fig. 2) over the entire range. At high temperatures, the free-energy curve may be negative and monotonic in concentration; see

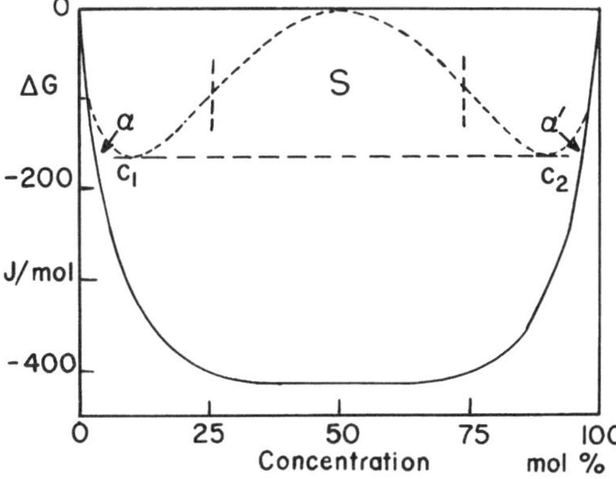

FIG. 14. Free-energy plot for an alloy system showing the influence of the value of ΔH on the extent of solubility. The solid curve shows the case where ΔH is small enough that ΔG varies monotonically with concentration over the entire range. With ΔH large enough, the curve takes the form drawn with the short-dashed line. At either extreme, regions of solid solubility exist. Between the points of common tangency, segregated phases of composition C_1 and C_2 coexist. However, within a central region, labeled S, segregated phases can form with no requirement for nucleation.

the solid line. A complete solid solution exists for this condition. At lower temperatures, where the entropy term is reduced, a curve like the short-dashed line may exist. For alloy compositions in the regions marked α and α', solid solutions exist. For alloys with intermediate composition, two equilibrium phases exist whose compositions C_1 and C_2 are defined by the common tangent line. But between these concentrations, another region S exists, within which $d^2\Delta G/dc^2$ is negative. For alloys with composition within this region, the solution is unstable with respect to small compositional fluctuations within itself. Such fluctuations are commonly bands that grow in spacing and in concentration differences as time goes on. Ultimately, the equilibrium phases C_1 and C_2 should result. This mode of phase separation, called *spinodal decomposition*, is attractive because no nucleation barrier exists.

An excellent example of this effect is decomposition of an oxide of Co and Fe, $(Co,Fe)_3O_4$, which had been deposited on a Si(111) surface. After being rapidly cooled from a high temperature where it exists as a solid solution (Hirano *et al.*, 1993), two phases formed spinodally upon aging at 975 K, one phase rich in Co and the other rich in Fe. The progress of the band formation with time is clearly shown in Fig. 15 by the splitting of the [310] diffraction line of Cu $K\alpha$ into two adjacent peaks. A photograph by the authors shows that the alternate bands of Co- and Fe-rich phases are some 5 to 10 nm broad after some 45 minutes of aging. A great many similar examples of this continuous precipitation have been reported in which such continuous banding occurs: solid metal alloys, solid and liquid ceramic combinations (see GLASSES), and liquid polymeric mixtures.

A striking example of this type of decomposition are bands that have been seen in the olivine-rich achondrite Divnoe (Petaev and Brearly, 1994). Using electron-microprobe microanalysis, the authors observed alternate bands of iron-poor and iron-rich phases on a scale of 1 or 2 μm. They propose that the bands in the olivine phase of this meteorite formed at relatively low temperatures (around 500 K) during many millions of years.

The spinodal concept goes back to Gibbs (1928), but many others have contributed to

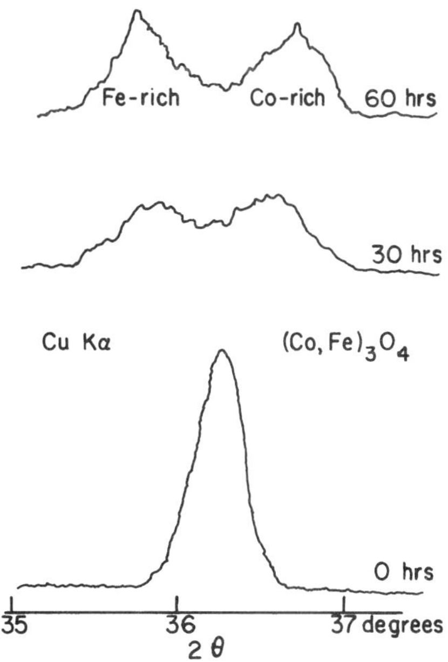

FIG. 15. Formation of the spinodal phases in decomposition of a mixed oxide of Co and Fe. Successive x-ray patterns show formation of alternating Co-rich and Fe-rich bands. After Hirano *et al.* (1993).

its development. Quantitative details of the initial stages of this process have been described more recently by Cahn and Hilliard (1958) in terms of a nonlinear diffusion equation; later stages are not well described. Whether the many reported examples of such phase separation are all spinodal phase changes fitting the Cahn–Hilliard formulation (or one of the many modifications proposed) is not clear, since competing nucleation-and-growth modes could also occur. Both experimental and theoretical investigations provide a fertile area for both materials science and applied physics research.

The strengthening of aluminum alloys containing several percent of such elements as Cu, Li, and Zn displays a sequence of several such metastable phases before the stable phase $CuAl_2$ forms. The commercially important alloys are treated to produce a banded structure resembling the spinodally produced phases. These first phases are thin plates, only a few atoms thick, which are enriched in the alloying constituent. They strengthen the alloy by greatly impeding the motion of dislocations. With time, other metastable phases form, and finally the phase $CuAl_2$; at this point the alloy is much softer than that containing the earlier metastable phases. First found in a laboratory in the German city Düren, the phases have been given the name Guinier–Preston zones in honor of the two French and English scientists who first described the details of their structure. The "overaging" of the aluminum alloys, with consequent softening, requires use of other alloys for high-performance, supersonic aircraft.

The historical account of the development of hardened aluminum alloys is fascinating. In part it shows initial applications of a craftlike nature, with improvements that come later with basic understanding of the geometry of the metastable precipitated phases. The Wright brothers, for example, fabricated a light-weight, but strong, aluminum–copper alloy for the crankcase of their early aircraft engine in 1903. Recent examination by Gayle and Goodway (1994) of the microstructure of the microstituents from specimens cut from the original crankcase shows the presence of thin plates, Fig. 16. Some 8 years after the flight of their plane, the *American Flyer,* Wilm (1911) showed the hardening that takes place upon a certain

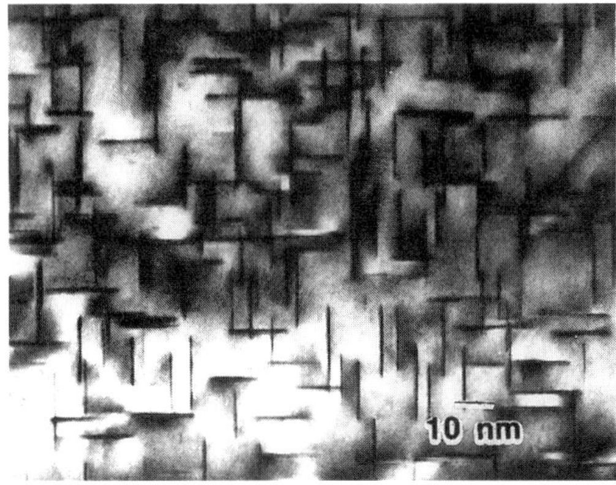

FIG. 16. Thin plates of Cu–Al zones in the aluminum alloy from the crankcase of the aircraft engine from the first Wright brothers' aircraft. Reprinted with permission from Gayle and Goodway (1994). Copyright 1994 American Association for the Advancement of Science.

heat treatment of such alloys. Gradually, the deduction was made that the hardening came from formation of a precipitate during this heat treatment, a precipitate postulated by some to be $CuAl_2$; see Merica *et al.* (1920). Some years later, Guinier and Preston independently showed, by x-ray analysis, that the precipitate was not $CuAl_2$, but consisted of thin segregated platelets of some metastable, Cu-rich preprecipitate phase, now called Guinier–Preston zones [see Guinier (1938); Preston (1938)]. Still later, electron-microscope examination gave direct visual evidence of the geometry of the segregated phases, such as shown in the example of Fig. 16.

Other modes of formation of segregated phases may occur. One, of importance, is decomposition of a single supersaturated phase along a reaction front to produce two new phases. The most famous example is the decomposition of austenite (a face-centered-cubic solid solution of carbon in iron) into two new phases arranged in a lamellar pattern—the new phases being a nearly pure iron (a body-centered-cubic phase) and the compound Fe_3C. The product phases are called *pearlite;* a typical example is shown in Fig. 17. This steel is hard, but not brittle—hence it is useful for such applications as a hammer, which is hard enough not to be dented by a nail, yet not so brittle that chips of steel would break off during the impact.

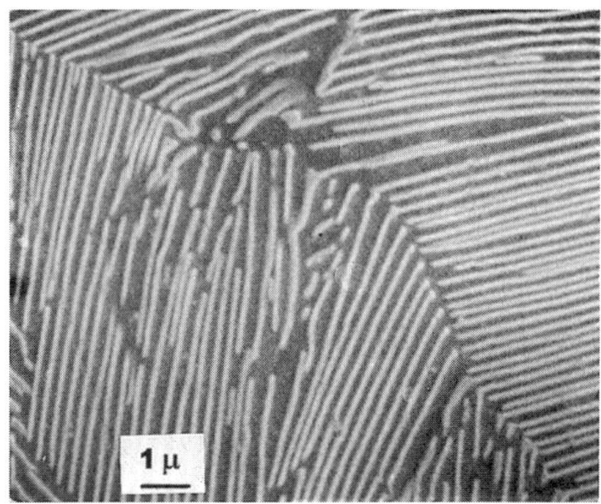

FIG. 17. Micrograph of a typical pearlite in a eutectoid steel. This layered structure shows of alternate bands of ferrite (nearly pure iron) and the carbide Fe_3C. Courtesy of R. Bohl.

7. RAPIDLY SOLIDIFIED ALLOYS

The metals used by humans for at least 8000 years were always aggregations of crystals. Although inorganic glasses were also used during much of that time, no way had been found to make their metallic counterpart; solidification of molten alloys always resulted in a crystalline product. Then in 1958, Prof. Pol Duwez and his colleagues in California produced, by rapid quenching, samples of a series of *amorphous* solid *alloys* in the eutectic system Cu–Ag (Duwez, 1960; Klement *et al.* 1960). Soon (as always happens) anyone could make such alloys—hundreds of alloy systems have been found to yield amorphous material with fast enough quenching. The first samples were measured in grams, but now many thousands of tons/year of amorphous magnetic materials alone are being sold.

Amorphous film production by rapid quenching has been supplemented by other methods. Milling of fine powders can produce amorphous material. Fragmentation of the structure of certain crystalline alloys by hydrogen absorption also may produce an amorphous product. Amorphous surface layers may be produced by laser heating or by electroplating. Irradiation may induce amorphization. Also interdiffusion of ultrathin layers of certain combinations can produce amorphous materials. See Greer (1995).

Rapidly quenched alloys are generally solid solutions; rapid quenching inhibits phase separation and compound formation. Their composition, though, may be well outside the solid-solution range of slowly cooled alloys. The full array of diffraction techniques using x rays, electrons, and neutrons has been employed to determine atom configurations. They show that the amorphous state is not a unique state; processing variables and alloy differences make the product somewhat variable. However, a geometrical short-range order of atom positioning is observed that is highly similar to that in the liquid state. A chemical short-range order is often observed—traces of the chemical affinity characteristic of the crystalline state may remain.

Magnetically soft iron-based amorphous ribbon has remarkably low core loss, about $\frac{1}{3}$ that of grain-oriented silicon iron. It is therefore attractive for power transformers from an engineering perspective. Its higher material cost offsets part of that economic gain, but some 10 000 transformer units are being installed in the United States each year, many additional units in Japan.

Not all rapid quenching of a liquid alloy produces an amorphous product—crystalline phases sometimes occur or can be formed by devitrification. One such phase is the magnetically hard compound $Nd_2Fe_{14}B$. Rapid quenching from the liquid state prevents formation of other equilibrium phases, which would form for that composition upon slow cooling. This rapidly cooled compound (and other rare-earth combinations) has a magnetic energy product in excess of 200 kJ/m^3, a value far exceeding that of any other material. In just a few years it is rapidly replacing the manufacture of Alnico and $SmCo_5$ magnets, the previously common high-energy-product materials. Magnets may be produced from the rapidly solidified ribbons by stacking them and pressing or by use of a binder material. Sintering of appropriately prepared powders may also provide a suitable fabrication technique for this material.

Yet other novel materials—the quasicrystals—may be produced by rapid solidification of particular alloys. These alloys have very large atom ratios compared to the relatively small ratios for most compounds; a particular quasicrystal of aluminum, copper, and iron has the atom ratio $Al_{65}Cu_{23}Fe_{12}$. X-ray Bragg reflection patterns show sharp spots.

However, the patterns show icosahedral (fivefold) symmetry, so that the usual crystallographic space-filling mode by linear translation of the unit cell does not exist. Further, the quasicrystal "cell" is so large—in the range 500–2500 nm—that it is indeterminate, and detailed placement of specific atoms cannot be made. Although the diffraction spots are sharp, careful examination shows a slight irregularity that is related to small improper atom placement, a defect called a phason.

Many quasicrystals are rapidly solidified (ternary, or higher) aluminum alloys, for which the rapid cooling extends the normal solubility range. The materials are generally poor conductors, a condition assumed to result from an extremely low density of states at the Fermi level. The *e/a* ratio is about 2.2 electrons per atom, a number generally considered to be the location of a "Hume-Rothery pseudogap", a designation somewhat of a turnaround from the skepticism of the earlier Hume-Rothery *e/a* concept for crystalline alloys.

Quasicrystals have been observed as a stable feature of a few phase diagrams at high temperature. See, for example, a cut through the Al–Fe–Cu ternary at 680 °C in Fig. 18 (Bradley and Goldschmidt, 1939). Other such phases have apparently been observed in the Al–Cr and Al–Mn binary systems. The reader may see an excellent review of the subject in a paper by Kelton (1993). No practical application has been found for these materials; they are an interesting laboratory curiosity. Undoubtedly, though, use will be found for them (see CRYSTALLINE STATE; QUASICRYSTALS).

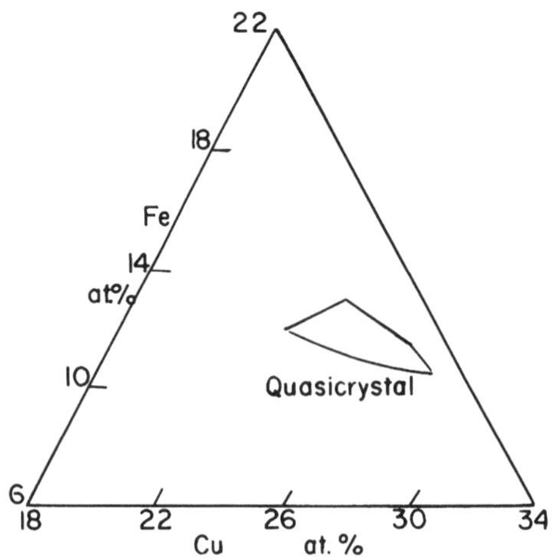

FIG. 18. A cut at 950 K through the ternary AlFeCu phase diagram showing the range of compositions of the quasicrystal phase. For simplicity, the equilibrium relationships with the other crystalline binary and ternary phases are not shown.

8. MECHANICAL AND DIELECTRIC SPECTROSCOPY

Previous paragraphs have emphasized the important role played by the spectroscopic techniques in specification of properties of solutions and segregated phases. In this final section, examples of two spectroscopies, mechanical and dielectric loss, are made to demonstrate their importance for specifying atomic configurations and kinetics of rate processes for metals, ceramics, and polymers. The principles can be best introduced using a formalism defined by Zener in which Hooke's law is modified to take account of both the usual interatomic force constants and a contribution from additional time-dependent relaxations that may produce additional strain. In the simplest case the relationship between stress and strain is modified by addition of only their first time derivatives:

$$\sigma + \tau\dot{\sigma} = M_R\epsilon + M_U\tau\dot{\epsilon}. \qquad (3)$$

Here, the parameter M_R is the usual static elastic constant, and M_U is the elastic constant at infinite frequency. The time constant τ is called the relaxation time; it is that time required for the relaxation to achieve $1/e$ of its completion value. Strictly, three relaxation times have been defined: one at constant stress, τ_s, one at constant strain, τ_e, and a third, τ, defined as the square root of the product of these two, $\tau = (\tau_s\tau_e)^{1/2}$. For most cases, there is little numerical difference between the three, so I use, for these examples, the designation τ.

This expression describes cases for constant applied stress, constant strain, or a time-dependent stress and strain. An enormous number of solids have been shown to possess recoverable atomic rearrangements described by this expression (see Zener, 1948; Nowick and Berry, 1972; McCrum *et al.*, 1967). The usual case is that for which a sinusoidal stress is applied at some angular frequency ω; the strain, of course, possesses the same frequency ω, but it is not in phase with the stress. The elastic modulus has a real and an imaginary part, therefore. These components may be expressed in the following way:

$$M_{\text{real}} = M_U - (M_U - M_R)/(1 + \omega^2\tau^2), \quad (4a)$$

$$M_{\text{imag}} = (M_U - M_R)\omega\tau/(1 + \omega^2\tau^2). \quad (4b)$$

These expressions are called the Debye equations, since they were first derived by Debye (1928, 1945) for the dielectric case, described later. The first equation describes the change in the real part of the elastic constant as the frequency decreases. The second gives the energy loss as a function of frequency, a loss termed the internal friction. In principle, either of these equations can be used for interpretation of measurements; but the second is customarily used. In succeeding paragraphs, examples are used to illustrate the applications to solutions and compounds. These examples depend on changes of atom positions upon application of a stress, i.e., on diffusion. Since thermally activated motion varies with temperature according to an Arrhenius expression $e^{-H/RT}$, τ is assumed to be given by an equation $\tau = \tau_0 e^{H/RT}$, where τ_0 is the time constant at infinite temperature. Then, one has the choice of making measurements as a function of temperature at constant angular frequency ω, or as a function of frequency at constant temperature. For experimental simplicity, the first is commonly used. For simplicity of nomenclature, Eq. (4b) is usually rewritten in terms of the phase angle δ between stress and strain as

$$\tan\delta = A\omega\tau_0 e^{H/RT}/(1 + \omega^2\tau_0^2 e^{2H/RT}). \quad (5)$$

All spectroscopic techniques have two significant parameters; this type is no exception. The first is a *rate constant*, here the material parameter τ. The second is the *amplitude factor A*; it describes the strength of the relaxation. Although both τ and A have been used extensively to characterize solutions and compounds, most of the examples described here refer to the first, the rate constant.

The first example is for one of the oldest metallic solutions used by mankind, a Cu–Zn brass. Zener showed that a plot of $\tan\delta$ vs T has the form shown in Fig. 19 for a measuring frequency of 620 Hz. Zener proposed that this effect is caused by reorientation under stress of Zn–Zn dipoles in the brass, the rate being determined by local diffusion of the zinc atoms. At low temperatures, the rate of jumping is slow, i.e., τ is large, so that $\omega\tau$ in Eq. (4b) in both the numerator and denominator is large compared with 1, and the energy loss given by Eq. (4b) is low. At high temperature, the value of τ is small, so that $\omega\tau$ is small, and the energy loss is again small. At that temperature for which $\omega\tau$ is about unity, the energy loss is at a maximum.

This measurement has been repeated for many such substitutional alloys (see Nowick and Berry, 1972), and the following features have been observed:

1. The effect is indeed a reorientation of strain dipoles under stress. The strength of the effect, i.e., the magnitude of tan δ, for dilute alloys is proportional to the square of the atomic concentration, as would be expected for random placement of alloying atoms among the crystal sites.

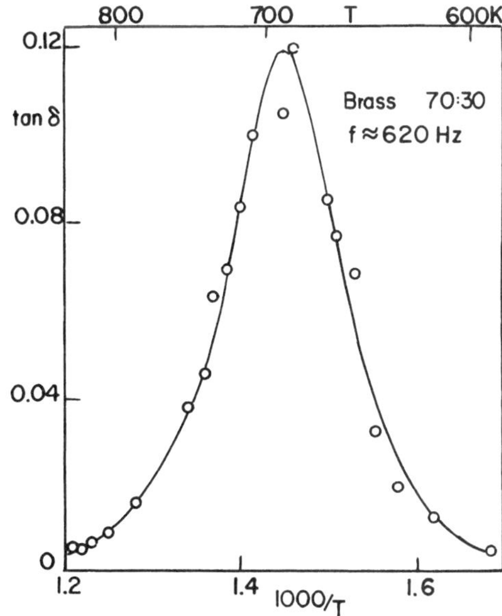

FIG. 19. Zener's first measurement of internal friction caused by spatial reordering under a mechanical stress of Zn–Zn dipoles in a 30% brass. After Zener (1948).

2. Measurements on single crystals give the proper orientation effects for nearest-neighbor pairs.
3. The activation enthalpy [H in Eq. (5)] is closely the same as that measured by other means for diffusion of the alloying elements.

A second example, first examined by the physicist Snoek and his colleagues (Snoek, 1941) and later developed in detail by Zener and his colleagues, is for interstitial alloys. It is typified by the solutions described earlier: nitrogen in bcc iron and nitrogen in niobium (also bcc); but it also is characteristic of other small atoms, carbon, oxygen, and hydrogen, dissolved in the bcc metals. These elements do not replace atoms of the host metal upon alloying; rather, they fit into interstices in the host crystal. If the host is fcc, the dissolved atoms fit into sites also with cubic symmetry, and the local distortion is merely a volumetric change. But if the host crystal is bcc, the distortion has a tetragonal component, and strain dipoles thus exist that can reorient under stress with a rate constant characteristic of local diffusion.

One of the most complete sets of measurements of diffusion in such is for carbon atoms in bcc iron. A set of measurements at several frequencies near 1 Hz is shown in Fig. 20 (Wert and Zener, 1949). Gradually, as experimental techniques developed to allow measurement over a much broader range of frequency, determination of the diffusion coefficient was extended over a range of some 10^{15} [see Fig. 21, adapted from an article of Lord and Beshers (1966)]. The wide range of the measured values allows the activation energy H to be determined with high reliability—the value is 81 470 J/mol. Extrapolation of the data to high temperatures shows a nonlinearity in this semilogarithmic plot, implying some additional enhancement of the diffusion rate at high temperatures, a phenomenon not yet explained.

Interstitials in the bcc metals reorient under the application of an applied stress because the tetragonal symmetry of such an interstitial impurity is lower than the cubic symmetry of the cubic host—the interstitial is thus a strain dipole. Isolated interstitials in fcc hosts (such as the stainless steels) do not produce a mechanical energy loss because they fit into interstices also of cubic symmetry, and thus they produce no strain dipole. But a strain dipole could exist for close H–H pairs or close clusters of H with a substitutional alloying atom, such as Ni or Cr. In such a case, an internal friction peak would be observed if the dipole strength were large enough and the clusters were present in suf-

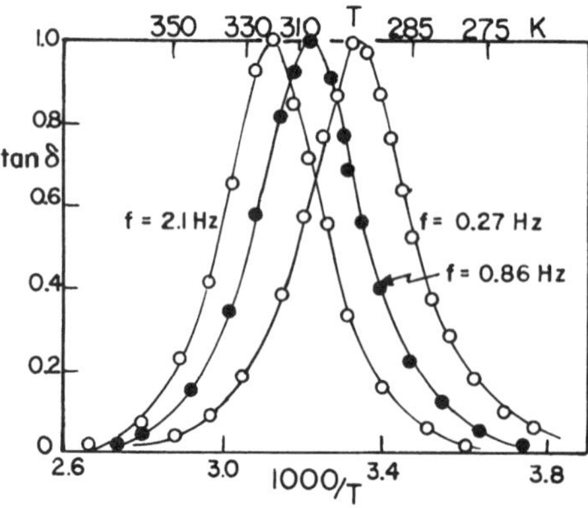

FIG. 20. Measurements of the internal friction Snoek peak of carbon in iron at several frequencies of measurement. The peak shifts in temperature as the frequency of the applied stress is changed. After Wert and Zener (1949).

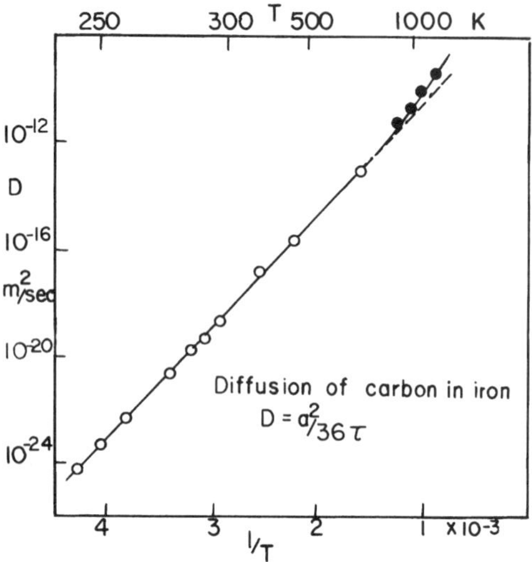

FIG. 21. Plot of diffusion coefficients for carbon in iron as a function of $1/T$. Only a minute fraction of the many reported values are shown. This line represents measurements of a physical parameter over as wide a range of values as for nearly any physical parameter ever measured. After Lord and Beshers (1966).

ficiently large concentration. Many such measurements have been reported for hydrogen in stainless steels of many compositions. Loss peaks have been observed in the temperature range 200–300 K for frequencies from 200 to 2000 Hz. See a measurement for hydrogen in three Fe–Ni binary alloys plotted in Fig. 22 (Nishino *et al.*, 1987); a damping peak is seen at 240 K for a frequency of 700 Hz. These authors supposed that the loss mechanism is indeed the reorientation of H–H pairs in the alternating stress field. Such an interpretation would require that the magnitude of the peak, i.e., the constant A in Eq. (5), should increase as the square of the dissolved H concentration. A later study by Gavriljuk *et al.* (1993) showed that A increases linearly as the H concentration increases, not as the square, and so it appears that a Ni–H cluster is favored. Variation of the rate constant τ with temperature shows that the enthalpy of motion, about 36 kJ/mol, is about right for hydrogen-ion diffusion in the stainless steels.

The measurements described in the preceding paragraphs relate to local diffusion of atoms. However, energy loss may occur by relaxation effects associated with long-range diffusion of atoms for proper specimen geometry. Let a rod be flexed so that a linear stress gradient exists across the rod. Atoms in the compression region possess a higher potential energy than those in the dilated region and will diffuse to reduce their energy. If the stress is periodic with angular frequency ω, this diffusion may result in an alternating strain relaxation that is out of phase with the stress—energy loss occurs. The relaxation time has been shown by Gorsky (1935) to be given in terms of the diameter of the rod, d, and the diffusion coefficient, D, by the expression $\tau = d^2/\pi^2 D$. The energy loss is a maximum, as before, when $\tau = 1/\omega$. This technique has been especially useful for measurement of hydrogen and deuterium in the bcc metals; see the example in Fig. 23 for diffusion of these isotopes in Nb, Ta, and V. Perhaps no measurement of diffusion has ever been made more frequently than this one, since the measurement has been for many years part of a laboratory exercise in the physics program of the Technical University of Munich, in Garching.

The data plotted in Fig. 23 allow us to examine the isotope effect on diffusion. The classical ratio of the diffusion coefficients for hydrogen and deterium, D_H/D_D, at any temperature should be 1.4 (i.e., $2^{1/2}$). This is so since the vibrational frequency of an atom in a crystal should vary as the reciprocal of the square root of the mass of the atom, $1/m^{1/2}$. The data of the figure show a much larger ratio than that for all three metals V, Nb,

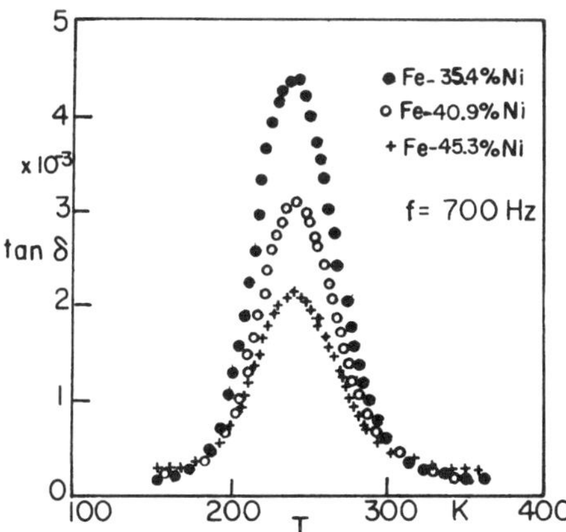

FIG. 22. Internal-friction peak for binary Fe–Ni alloys containing dissolved hydrogen. For these three alloys, the peak stays constant in temperature, i.e., τ is a constant, although the magnitude of the peak changes, i.e., the constant A changes. Nishino *et al.* propose that the peak is caused by relaxation of H–H clusters, but later investigations suggest that the relevant cluster is Ni–H. Curves after Nishino *et al.* (1987).

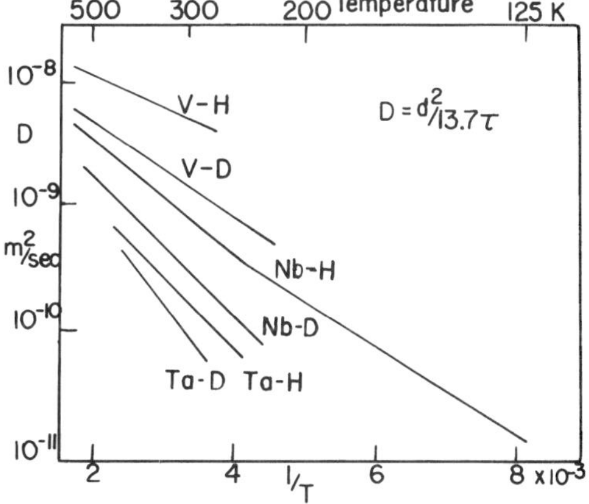

FIG. 23. Measurements of the diffusion coefficient for isotopes of Hydrogen in the bcc metals using the Gorsky effect. After Schaumann *et al.* (1970).

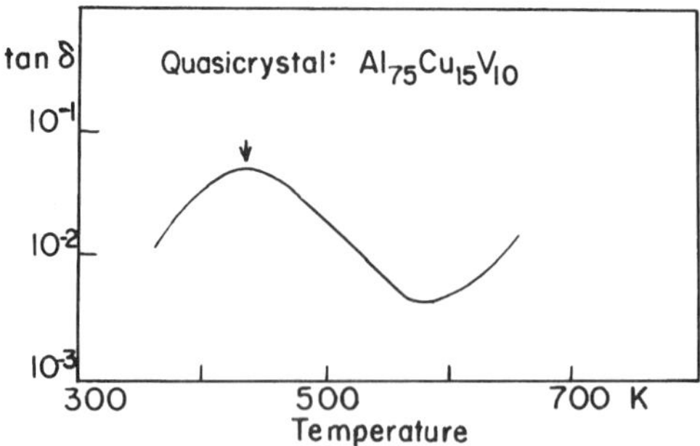

FIG. 24. Internal friction peak resulting from atom reorientation under a mechanical stress for a quasicrystal of Al, Cu, and Fe. After Okumura *et al.* (1994).

and Ta, and so the classical model is not correct in this instance.

Similar loss measurements exist for rapidly solidified alloys—both for amorphous alloys and for quasicrystals. One such measurement is shown in Fig. 24 for the quasicrystal FeAlCu (Okumura *et al.,* 1994). The origin of the energy dissipation is thought by the authors possibly to be caused by an atomic rearrangement associated with phasons.

The "plastic" nature of polymers provides ample opportunity for atomic rearrangements to produce internal friction loss. Energy loss occurs both for polymers formed from a single monomer and for polymer blends. In general, three distinct peaks are seen, as is seen in Fig. 25 for nylon. That at the lowest temperature is called the γ peak. It is caused by reorientation under stress of segments of the main chain. The next peak, the β peak, is caused by reorientation of side chains. That at the highest temperature, termed the α peak, is caused by a structural change at the glass transition temperature. The applications of internal friction to polymer structure occurred during the same time frame as was true for the metals, mainly after 1950. Even so, the two paths of investigation have remained almost separate; few investigators have worked in both. The atomic arrangements that produce these mechanical relaxations in polymers may also be characterized by neutron scattering (Frick and Richter, 1995).

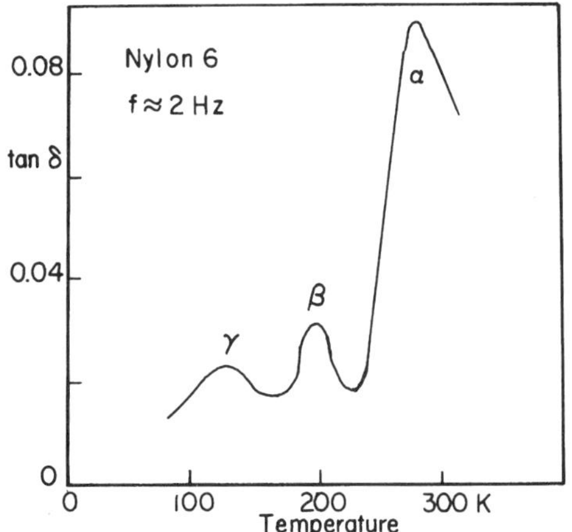

FIG. 25. Internal friction loss peaks for nylon-6.

Ceramic materials have a special attribute that has provided an additional technique of investigation. Many are good insulators, and they thus may show dielectric loss. That application depends on a set of principles first enunciated by Debye (1928, 1945). For a lossy material, the electric polarization **P** may not be in time phase with the field **E**; the controlling expression has the same functional relationship as for the mechanical case,

$$\mathbf{P} + \tau\dot{\mathbf{P}} = \epsilon_0(\chi_R\mathbf{E} + \tau\chi_U\dot{\mathbf{E}}). \tag{6}$$

Here, ϵ_0 is the permittivity of a vacuum and χ_R and χ_U are the susceptibilities in the relaxed and unrelaxed states, respectively. A real and an imaginary part of the dielectric capacitivity exist, and the dielectric loss in an ac field **E** has the form

$$\tan\phi = A'^*\omega\tau_0 e^{H/RT}/(1 + \omega^2\tau_0^2 e^{2H/RT}).$$

Defects in insulators that have both a mechanical strain dipole and an electric moment may show both mechanical loss and dielectric loss. This joint behavior may allow additional features of the defect to be determined. An example of such a determination is shown in Fig. 26 for both $\tan\delta$ and $\tan\phi$ using tetragonal ZrO_2 containing 3 mol% Y_2O_3. A loss peak appears in the same temperature region for both (Weller, 1994), and so the authors conclude that the same defect is responsible for both the mechanical and the dielectric loss. They supposed that this defect is an associated yttrium-vacancy complex that reorients under the influence of

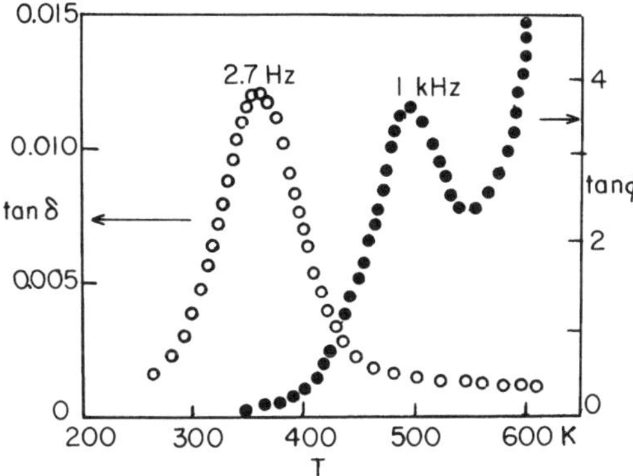

FIG. 26. Mechanical and dielectric loss spectra for ZrO_2 containing 3 mol% of Y_2O_3. The peak is caused by a spatial reorientation, under either an electric or a mechanical stress, of an yttrium-vacancy pair. It is displaced between the two measurements because of the frequency difference. This difference, in fact, allows the activation energy for the reorientation to be determined: 0.9 eV. After Weller (1994).

both a mechanical stress and an electric field. Measurements at several frequencies show that both have the same activation enthalpy, *H*, of about 90 kJ/mol. Weller and his colleagues have made many measurements on other ionic crystals, as have Wachtmann (1963) and Nowick and Heller (1963). A huge number of observations have been reported by a myriad of investigators for the many high-T_c superconductors.

9. CONCLUSIONS

Materials have been used by mankind for tools and weapons for hundreds of thousands of years. Modification of metals and ceramic materials by heat treatment and alloying began perhaps 200 centuries ago. The first metallic alloys were fabricated about 7000 B.C., gold and copper alloys in the beginning and later alloys of iron. The role of physics came, therefore, rather late—only in the last two centuries. Development of materials for mechanical applications has profited in the past century more by the application of physical measurement techniques than by theory; theory has been more after-the-fact explanatory rather than predictive. Development of materials for electrical applications—especially for modern electronics—though, has been aided immensely by both theory and measurement techniques of physics. Polymers, whose production in synthetic form has occurred only in the past century, have been characterized almost from the beginning by the physical and chemical sciences.

GLOSSARY

Alnico: A permanent-magnet alloy developed in the 1930s. It contained aluminum, nickel, and cobalt (and small quantities of other elements) as well as iron. It was a big improvement over the carbide-hardened steels previously used.

Bohr Magneton: The magnetic moment associated with the electron spin. Its value is 9.27×10^{-24} A-m^2.

Brillouin Zones: Boundaries in momentum space at which the energy–momentum relation suffers a discontinuity.

Eutectic: Point of lowest melting temperature as the composition of an alloy is changed.

Habit Plane: The plane of the host crystal along which a precipitate forms.

High-Strength Alloy Steels: Steels alloyed with small amounts of metallic elements (commonly totaling less than 5%). These improve the ability of the alloy to be hardened by influencing the nature of the carbide precipitate. A widely used alloy is "4340." Its composition is about the following: Ni: 1.8%; Cr: 0.8%; Mo: 0.25%; Mn: 0.7%; Si: 0.3%; C: 0.4%. When properly heat treated, its tensile strength at room temperature is about 250 000 psi (about 1750 MPa). It will deform in tension about 10% to fracture.

Kamacite: The low-Ni phase of the Fe–Ni constituent of the iron meteorites.

Phase: A physically homogeneous part of a material's system.

Phase Diagram: A graphical construct in which the phases present are plotted as a function of composition, temperature, and pressure.

Sigma (σ) Phase: An intermediate Fe–Cr phase with composition nearly 50:50. It forms at temperatures below 800 K. It is a brittle phase, usually not desired in alloy steels.

Spinode: Dictionary definition—cusp. In our context, it refers to a maximum in the slope of the line of G vs composition. At that point, $d\Delta G/dc$ goes through zero with increasing composition. Between the spinodal points $d^2\Delta G/dt^2$ is negative.

Taenite: The high-Ni phase of the Fe–Ni constituent of the iron meteorites.

Vacancy: A vacant lattice site in a crystal.

Widmannstätten Pattern: A pattern of precipitates of two phases of iron–nickel solid solutions that form in the iron meteorites over billions of years. The most striking pattern occurs when plates of an iron-rich phase (kamacite) precipitate on (111) planes of the remaining iron–nickel phase (taenite). The pertinent figure shows that pattern for one of the most famous iron meteorites (Gibeon) that fell in the early 1800s in Southwest Africa (now Namibia). The white bands are the kamacite—they have formed on planes of the taenite having octahedral symmetry. Meteorites that show this pattern are called, therefore, *octahedrites.*

Works Cited

Aitchison, Leslie (1960), *A History of Metals,* New York: Interscience Publishers.

Barrett, Charles S., Massalski, T. B. (1966), *Structure of Metals,* 3rd ed., New York: McGraw-Hill.

Blandin, A., Friedel, J., Saada, G. (1966), *J. Phys. (Paris)* Colloque **27,** C3, 128.

Bradley, A. J., Goldschmidt, H. S. (1939), *J. Inst. Met.* **65,** 389–418.

Buchwald, Vagn (1975), *Handbook of Iron Meteorites,* Vols. 1, 2, and 3, Berkeley: University of California Press.

Cahn, J. W., Hilliard, J. E. (1958), *J. Chem. Phys.* **28,** 258–267.

Cost, J. R., Wert, C. A. (1963), *Acta Metall.* **11,** 231–242.

Debye, P. (1928, 1945), *Polar Molecules,* New York: Dover.

Dijkstra, L. (1949), *Trans. AIME* **185,** 252–260.

Dodd, Robert T. (1986), *Thunderstones and Shooting Stars, The Meaning of Meteorites,* Cambridge, MA: Harvard Univ. Press.

Duwez, Pol, Willens, R. H., Klement, W., Jr. (1960), *J. Appl. Phys.* **31,** 1136–1137.

Fast, J. D., Verrijp, M. B. (1955), *J. Iron Steel Inst.* **180,** 337–343.

Ferguson, P., Dahmen, U., Westmacott, K. H. (1984), *Scr. Metall.* **18,** 57–60.

Frick, B., Richter, D. (1995), *Science* **267,** 1939–1945.

Fromm, E., Gebhardt, E. (1976), *Gase und Kohlenstoffe im Metallen,* Berlin: Springer.

Fujita, F. (1994), *The Physics of New Materials,* Berlin: Springer.

Gavriljuk, V. G., Hänninen, Yu, Tarasenko, A. V., Tähtinen, S., Ullako, K. (1993), *Scr. Met. Mater.* **28,** 900–906.

Gayle, Frank W., Goodway, Martha (1994), *Science* **266,** 1015–1017.

Gibbs, J. Willard (1928), *The Collected Works of J. Willard Gibbs,* New York: Longsmans Green.

Gorsky, W. S. (1935), *Phys. Z. Sowjetunion* **8,** 457–465.

Greer, A. Lindsay (1995), *Science* **267,** 1947–1953.

Guinier, A. (1938), *C. R. Acad. Sci.* **206,** 1641–1643.

Heine, Volker, Weaire, D. (1970), in: H. Ehrenreich, F. Seitz, D. Turnbull (Eds.), *Solid State Physics,* vol. 24, New York: Academic Press, 249–462.

Hirano, S., Yogo, T., Kikuta, K., Asai, E., Sugiyama, K., Yamamoto, H. (1993), *J. Am. Cer. Soc.* **76,** 1788–1792.

Hume-Rothery, W. (1948), *Atomic Theory for Students of Metallurgy,* London: Institute of Metals.

Kelton, K. F. (1993), *Int. Mater. Rev.* **38,** 105–137.

Klement, K., Willens, R. H., Duwez, P. (1960), *Nature* **187,** 869.

Lord, A. E., Beshers, D. N. (1966), *Acta Metall.* **14,** 1659–1672.

McCrum, N. G., Read, B. E., Williams, G. (1967), *Anelastic and Dielectric Effects in Polymeric Solids,* New York: Wiley.

Merica, P. P., Waltenberg, H. G., Freeman, J. R. (1920), *Trans. AIME* **64,** 3–25.

Nishino, Y., Kato, T., Tamaoka, S., Asano, S. (1987), *Scr. Metall.* **21,** 1235–1239.

Nowick, A., Berry, B. (1972), *Anelastic Relaxation in Crystalline Solids,* New York: Academic Press.

Nowick, A. S., Heller, W. R. (1963), *Adv. Phys.* **12,** 251–298.

Okumura, H., Tsai, A. P., Inoue, A., Masumoto, T. (1994), *Mat. Sci. Eng. A* **182,** 781–1792.

Petaev, Michail I., Brearley, Adrian J. (1994), *Science* **266,** 1545–1547.

Pettifor, D., Cottrell, A. H. (1992), *Electron Theory in Alloy Design*, London: Inst. of Metals.

Preston, G. D. (1938), *Proc. R. Soc.* **A167,** 526–538.

Schaumann, G., Völkl, J., Alefeld, G. (1970), *Phys. Stat. Sol.* **42,** 401–413.

Smith, C. S. (1960), *A History of Metallography,* Chicago: University of Chicago Press.

Smith, C. S. (1962), *Geochim. Cosmochim. Acta* **26,** 971–972.

Smith, C. S. (1968), *Sources for the History of the Science of Steel,* Jointly Published by the Society for the History of Technology and the MIT Press. Cambridge, MA and London, England.

Smith, C. S. (1981), *A Search for Structure,* Cambridge, MA: The MIT Press.

Snoek, J. L. (1941), *Physica* **8,** 711–733.

Wachtman, J. B., Jr. (1963), *Phys. Rev.* **131,** 517–527.

Weller, M. (1994), *Abschlussbericht DFG—Kolloquium Keramische Hochleitungswerkstoffe* (to be published by VCH, Weinheim).

Wen, S. H., Kostlan, E., Hong, M., Khachaturyan, A. G., Morris, J. W. (1981), *Acta Metall.* **29,** 1247–1254.

Wert, C. A., Thomson, R. M. (1970), *Physics of Solids,* New York: McGraw-Hill.

Wert, C. A., Zener, C. (1949), *Phys. Rev.* **76,** 1169–1175.

Wilm, A. (1911), *Metallurgie* **8,** 125–130.

Zener, C. (1948), *Elasticity and Anelasticity of Metals,* Chicago: University of Chicago Press.

Further Reading

Alper, Allen M. (Ed.) (1970–1978), *Phase Diagrams: Materials Science and Technology,* Vols. I–VII, San Diego: Academic Press. Excellent articles by competent writers.

Alper, Allen M. (Ed.) (1995), *Phase Diagrams in Advanced Ceramics,* San Diego: Academic Press.

Baker, H. (Ed.) (1992), *Handbook of Phase Diagrams,* ASM International, OH: Materials Park.

Bennett, L. H. (Ed.) (1980), *Theory of Alloy Formation,* Pittsburgh: TMS-AIME.

Bergeron, C. G., Risbud, S. H. (1984), *Introduction to Phase Equilibria in Ceramics,* Columbus, OH: Amer. Cer. Soc.

Blythe, A. R. (1979), *Electrical Properties of Polymers,* Cambridge, U.K.: Cambridge Univ. Press.

CALPHAD: Computer Coupling of Phase Diagrams and Thermochemistry, Oxford: Pergamon Press. Includes publication of a summary of the Proceedings of the annual CALPHAD conferences.

Gaskell, David R. (1981), *Introduction to Metallurgical Thermodynamics,* Washington, DC: Hemisphere.

Grimwald, Mark (1992), The Metallurgy of Gold, *Interdisciplinary Sci. Rev.* **17,** 371–381.

Guinier, A. (1959), "Heterogeneities in Solid Solutions," in: H. Ehrenreich, D. Turnbull (Eds.), *Solid State Physics,* Vol. 9, New York: Academic, pp. 293–398.

Haasen, Peter (1978), *Physical Metallurgy,* Cambridge, U.K.: Cambridge Univ. Press.

Massalski, T. B., Okamota, H., Subramanian, P. R., Kacprzk, L. (Eds.) (1990), *Binary Alloy Phase Diagrams,* 2nd ed., Materials Park, OH: ASM International.

Matsumiya, Tooru, Nogami, Atsushi, Sawada, Hideaki (1995), "Monte Carlo Simulations of Intermetallic Compounds," in: *Advanced Materials and Processes,* Materials Park, OH: ASM International, 51–53.

Nambu, Shinji, Sato, Akini, Sagala, D. A. (1992), "Computer Simulation of Spinodal Decomposition in the Tetragonal TiO_2–SnO_2 System," *J. Am. Ceram. Soc.* **79,** 1906–1913.

Paneth, F. A. (1960), "The Discovery and Earliest Reproductions of the Widmanstätten Figures," *Geochim. Cosmochim. Acta* **18,** 176–182.

Quasicrystals: a. (1991), *Reports on Progress in Physics,* Vol. **54**, London: Institute of Physics, 1373–1425. b. (1993), *J. Non-Cryst. Solids* **153–154,** Elsevier, Amsterdam.

Sakurai, Yoshifumi (Ed.) (1993), *Current Topics in Amorphous Materials,* Amsterdam: Elsevier.

Seitz, Fred (1992), *The Science Matrix,* Berlin: Springer. A review of the development of physical thought from ancient time to the present.

Stosks, G. M., Turchi, P. E. A. (1994), *Alloy Modeling and Design,* Pittsburgh: TMS-AIME.

Vickers, Charles (1923), *Metals and Their Alloys,* New York: Henry Cary Baird and Co. An excellent account of the state of knowledge of metallic alloys in the pre-X-ray diffraction, pre-electron microscopy, pre-Band Theory era.

Villers, Pierre, Prince, Alan, Okamoto, Hiraoki (Eds.) (1995), *Handbook of Ternary Alloy Phase Diagrams,* Vols. 1–10, Materials Park, OH: ASM International. The largest compilation of ternary phase diagrams yet published.

SONIC NOISE

GEORGE C. MALING, JR., *Noise Control Foundation, Poughkeepsie, New York, U.S.A.*

INTRODUCTION

Noise is defined as *undesired sound.* The classical *system problem* approach (Bolt and Ingard, 1957) to any problem related to the control of noise is to treat the problem in three parts, the *source,* the *path,* and the *receiver.* In this approach, each component of the problem can be examined separately, and appropriate action can be taken. Even when noise *control* is not the objective, the system approach is useful for describing sonic noise.

From a physical viewpoint, noise is equivalent to sound, and the physics of noise generation and transmission over a path to a receiver have little to do with the term *noise.* When the sound reaches a receiver, a human being or an animal, however, the extent to which the sound is undesired has to be considered. The important effects are primarily psychological (i.e., loudness or annoyance) and physiological (i.e., damage to the hearing mechanism). It is the combination of the physical aspects of generation and propagation of sound, its effects on receivers, and the need to find engineering techniques for noise abatement that makes noise and its control an important professional discipline for noise-control engineers.

There is a wide variety of noise sources in modern society. Sources in and around the

3-527-28140-1/96/$5.00 + .50

home include gardening equipment, lawn-maintenance equipment, kitchen and home-cleaning appliances, and furnaces. Sources at work include industrial machinery, ventilating-system noise, and office equipment. Sources that intrude upon the community include surface-transportation noise—generated by automobiles, trucks, trains, and outdoor recreational vehicles—and aircraft—commercial, general aviation, military, and helicopters.

It is not possible to detail all of the characteristics of all of these sources in this article. The term *emission* is used when the acoustical characteristics of a source are being described. The sound-power level of the source or the sound-pressure level measured under carefully prescribed environmental conditions are common descriptors of noise emission.

The path over which the source reaches the receiver generally provides some attenuation of noise, although, in some cases, the noise may be amplified. Indoor and outdoor sound propagation must both be considered. The characteristics of silencers, mufflers, and similar devices placed between a source and a receiver are considered to be part of the path. Characteristics of sound-absorptive materials used to change the attenuation over the path are an important part of noise control.

Finally, the effects of noise on the receiver must be considered. Important topics include methods for rating the effects of noise, both psychological and physiological, on humans and animals; criteria for control of noise; and regulatory aspects of noise—the setting and enforcement of limits on noise. The sound-pressure level is the commonly used measure of noise incident on a receiver, and the term *noise immission* is sometimes used to describe noise that reaches a receiver.

1. MEASURES OF NOISE IMMISSION AND NOISE EMISSION

1.1 Sound Pressure and Sound-Pressure Level

The immission of noise upon a receiver is usually described in terms of sound-pressure level. Sound consists of the very small variations in pressure p above and below atmospheric pressure, which propagate with a speed c, the speed of sound. A propagating plane wave has the form

$$p(x,t) = p(x - ct), \tag{1}$$

where x is distance and t is time. Although air is fundamentally a nonlinear medium, the pressure fluctuations are, in most situations, so small relative to the atmospheric pressure that a linear approximation may be used. Exceptions include propagation of sonic booms and high-intensity sound generation from engines.

Sound pressure at some location is conventionally defined as a root mean square (rms) pressure averaged over a stated time duration. The level of the sound pressure is determined from the mean square pressure $\overline{p^2}$ and a reference sound pressure p_0, defined to be 20 μPa. Thus, the sound-pressure level may be expressed as

$$L_p = 10 \log(\overline{p^2}/p_0^2). \tag{2}$$

As an example, a sound-pressure level of 60 dB corresponds to an rms sound pressure of 0.02 Pa. This is small compared with atmospheric pressure, which is approximately 10^5 Pa.

The particle velocity u associated with a sound wave is obtained from the linear (one-dimensional) momentum equation given by

$$\rho \frac{\partial u}{\partial t} = -\frac{\partial p}{\partial x}, \tag{3}$$

where p and u are instantaneous values of the pressure and particle velocity at some location and ρ is the density of the medium (air).

1.2 Sound Power, Sound-Power Level, and Sound Intensity

The sound-pressure level defined above may, in some cases, be used as a measure of the noise *emission* of a sound source. An example is specification of the noise emissions of road vehicles. The sound pressure measured under well-defined environmental conditions is also sometimes used to specify the noise emission of machinery and equipment (*emission sound-pressure level*).

The specification of the noise emissions of sources is important because governmental bodies are beginning to require the use of national and international standards to determine noise emissions in terms of sound-power level and emission sound-pressure level, and to require the publication of noise-emission values in specifications for industrial machinery and consumer products.

For stationary noise sources, the sound power and sound-power level produced by the source are most satisfactory as descriptors of noise emission. The sound power W emitted by a source is expressed (Beranek and Vér, 1992, Chap. 4) in terms of the surface integral

$$W = \int_S \mathbf{I} \cdot d\mathbf{S}, \tag{4}$$

where W = sound power (W), $\mathbf{I}$ = sound intensity (W/m^2), $d\mathbf{S}$ = element of surface area (m), and S = surface that surrounds the source. The sound intensity $\mathbf{I}$ is the average rate of sound energy transmitted in the specified direction per unit area at a given position. The integral of Eq. (4) is over a surface that surrounds the source (the *measurement surface*), and the dot product of the two vectors indicates that the component of sound intensity normal to the surface is taken. The integral may be taken over any closed surface surrounding the source (hemispherical, spherical, or irregular), but in practice, rectangular or hemispherical measurement surfaces are most often used.

Most noise sources of practical interest are *directive,* which means that the sound intensity varies over the measurement surface. In practice, the surface integral of Eq. (4) is approximated by dividing the measurement surface into segments, denoted by S_i, and by replacing the integral by a sum over the segments:

$$W = \sum_{i=1}^{N} I_i S_i. \tag{5}$$

Here, I_i is the magnitude of the sound intensity normal to the ith surface segment, and S_i is the area of the ith surface segment.

When the sound power W has been determined, the sound-power level L_W, expressed here in bels, is determined relative to a reference sound-power level W_0 of 10^{-12} W by

$$L_W = \log(W/W_0). \tag{6}$$

The sound-power level may also be expressed in decibels. One decibel is one-tenth of a bel. When the decibel is used as the unit of both sound-power level and sound-pressure level, confusion between the quantities may arise.

1.3 Determination of Sound Power and Sound-Power Levels

Sound intensity and therefore sound power can be measured directly with modern acoustical instruments. If the sound pressure at each of two closely spaced microphones is measured, the difference between these two pressures divided by the microphone spacing is an approximation to the pressure gradient on the right-hand side of Eq. (3), and the particle velocity in the sound wave can be determined by time integration. Standard methods have been developed (ISO, 1993, 1996) for selecting the number of surface segments used in Eq. (5) and determining the accuracy of the calculated sound-power level.

Determination of sound power by means of sound-pressure level measurements is a more widely used technique. Two test environments are in common use: anechoic (or hemi-anechoic) rooms and reverberation rooms.

1.3.1 Determination of Sound Power in Hemi-Anechoic Rooms The sound field produced by a source in a hemi-anechoic room approximates a free field over a reflecting plane. The sound intensity normal to a surface that surrounds the source is approximated by

$$\mathbf{I} = \overline{p^2}/\rho c, \tag{7}$$

where $\overline{p^2}$ is the mean square sound pressure at a point, ρ is the density of air, and c is the speed of sound. Using nominal values for ρ and c and standard reference quantities for sound pressure and sound power, the sound-power level (in bels) produced by the source is given approximately by

$$L_W = 0.1[L_p + 10 \log(S/S_0)], \tag{8}$$

where L_p is the average sound-pressure level

in decibels over the measurement surface, S is the area of the measurement surface in square meters, and S_0 is 1 m^2.

1.3.2 Determination of Sound Power in a Reverberant Room A reverberation room is a room in which all surfaces are highly reflective for sound waves. The sound-pressure level produced by a source is approximately independent of the field position within the room, and the space-average sound-pressure level can be determined from measurements at a few microphone positions or by means of a moving microphone. The sound-power level of a source can be calculated from the space-time–averaged sound-pressure level and a knowledge of the reverberation time (see ACOUSTICS, ARCHITECTURAL) in the room. A more commonly used method is to compare the average mean square sound pressures in the room produced by the source under test and by a source of known sound-power output (a *reference sound source*). In this *comparison method*, the sound-power level of the source under test, L_{WE} (in bels), is given by

$$L_{WE} = L_{WR} + 0.1(L_{pE} - L_{pR}), \tag{9}$$

where L_{WR} = sound-power level of the reference sound source, L_{pR} = average sound-pressure level in the room produced by the reference sound source, and L_{pE} = space-time–averaged sound-pressure level produced by the sound source under test.

An overview of standard methods of determining sound-power level from sound-pressure level measurements is given in ISO (1980).

2. SOUND SPECTRA

The sound pressure at a point is a function of time and, therefore, may also be described in the frequency domain. Three descriptions of noise spectra are in common use. The frequency-domain descriptions below may also be used for sound-power levels.

2.1 Broad-Band Sound-Pressure Levels

A single-number rating for a noise spectrum is obtained by determination of the level of the mean square pressure over a wide frequency range such as 20 Hz to 20 kHz—depending on the characteristics of the spectrum and the frequency response of the microphone used for the measurements. For sound-level meters, the standard frequency weighting C provides a good approximation to a broad-band sound level. This frequency weighting C is not widely used, except for determination of peak levels of noise of impulsive character.

Another standard frequency weighting is A-weighting. Figure 1 illustrates the design goal for the response of frequency weighting A. The response for frequency weighting A decreases especially at frequencies less than 1 kHz and approximates the frequency response of the human ear at low sound-pressure levels. Sound-pressure levels measured with use of this weighting are called *A-weighted sound levels.* Although it is the *level* that is weighted, and not the *unit of level,* the decibel (dB), the designation dB(A) or even dBA is sometimes used to indicate the A-weighted sound-pressure level in decibels. When A-weighting is applied to a sound-power spectrum, the result is defined to be the *noise-power emission level,* i.e., the A-weighted sound-power level in bels.

2.2 Octave and One-Third Octave Band Levels

While the A-weighted sound-pressure level frequently is adequate for evaluating human response to noise, a more detailed description of the spectrum is required for the specification of the noise emission of sources, for selection of noise-control measures, for diagnosis of noise problems, for determination of sound transmission through structures, and often for calculation of single-number ratings related to human response to noise (such as loudness). A description of noise spectra in terms of *octave* or *one-third octave bands* is common. The transmission characteristics of octave and one-third-octave band filters have been standardized. An octave band or one-third octave band is usually specified by its nominal midband frequency. The lower pass-band frequency of an octave-band filter is one-half the upper pass-band frequency. There are three one-third octave bands in an octave band. For most noise spectra, an adequate description of the spectrum is given by octave-band levels having

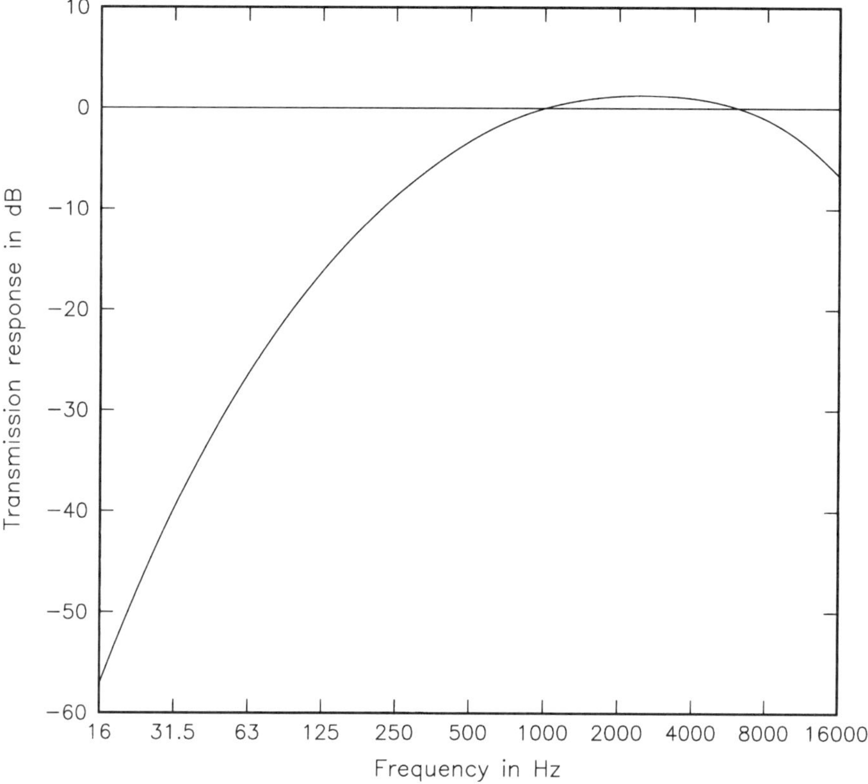

FIG. 1. Relative response vs frequency for A-weighting.

nominal midband frequencies of 125, 250, 500, 1000, 2000, 4000, and 8000 Hz.

2.3 Narrow-Band Levels

There are some situations where a description of a spectrum in narrow bands is required. Examples include the spectra of some axial-flow fans, propellers, and some rotating electrical machinery. Nearly all available narrow-band spectrum analyzers perform narrow-band analysis by using the fast Fourier transform; the characteristics of these instruments are not standardized; however, 4096-line analyses are common. Constant-percentage narrow-band spectrum analyzers are also available and provide the narrow-band levels over a wide range of frequencies in "real time."

3. INSTRUMENTATION FOR NOISE MEASUREMENT

3.1 The Sound-Level Meter

The basic instrument for measurement of noise levels is the *sound-level meter.* The two basic measurements obtained with a sound-level meter are exponential-time–averaged sound levels measured with standardized frequency weightings and time-averaged sound levels determined without exponential time weighting. The exponential time weightings are designated *fast* (or F) and *slow* (or S). The former corresponds to a time constant of 0.125 s, and the latter corresponds to a time constant of 1 s. All conventional sound-level meters are required to have frequency weighting A and time weighting F. Using these time weightings, one can estimate an average level over a specified time interval or determine the maximum sound level with a given time weighting. An example of a time-averaged sound level is an 8-h average sound level that one might measure over the duration of an 8-h work period. The measurement microphone for a sound-level meter is typically a capacitor, electret, or crystal microphone.

A block diagram of a typical sound-level meter is shown in Fig. 2. The signals are amplified, filtered (frequency weighted), and averaged using rms averaging. The output indicator may be either analog or digital, or a combination of digital and quasi-analog. The electroacoustical characteristics of sound-level meters and integrating-averaging sound-level meters have been standardized (IEC, 1979, 1985).

3.2 Spectrum Analyzers

In sound-level meters, the standard frequency weightings are usually implemented by means of analog circuit elements. Design-goal frequency weightings may also be applied to constant-percentage bandwidth sound-pressure levels or to the output of a fast Fourier-transform analyzer.

The sound-level meter may be a simple hand-held analyzer or part of a complicated data acquisition system in which frequency-weighted sound-pressure signals from several microphones are digitized, and the results are stored in computers for later analysis or transmitted directly to a central computer for real-time processing. Some of the most complex systems in use are for monitoring of noise levels around airports (Timmerman, 1993; Nitsche, 1995).

4. MECHANISMS OF NOISE GENERATION

There are four fundamental mechanisms for the generation of sound. These are usually described as *source terms* in the wave equation. The basic fluid flow equations (see ACOUSTICS, LINEAR; ACOUSTICS, NONLINEAR) are usually put in the form of a wave equation for a medium at rest with source terms on the right-hand side of the equation. The sound-pressure levels are assumed to be small enough to avoid nonlinear effects. The mathematical details are in the articles referenced above. For the purposes of this chapter, the acoustic wave equation may be written

$$\nabla^2 p - \frac{1}{c^2}\frac{\partial^2 p}{\partial t^2} = S_m + S_d + S_q + S_h. \tag{10}$$

The left-hand side of Eq. (10) is a wave equation for a medium at rest. The first of the four source terms on the right-hand side of Eq. (10), S_m, represents generation of noise by fluctuating *volume* (monopole) sources. A classical example is the baffled loudspeaker where the loudspeaker diaphragm "pumps" air and generates sound. The second term, S_d, represents fluctuating forces in the medium. For example, air turbulence can cause fluctuating lift forces on an airfoil, and sound is generated by these forces. This mechanism is responsible for the noise generated by low-speed axial-flow fans and centrifugal blowers. Such sources are often referred to as *dipole* sources because there is no net introduction of mass into the medium. A classical example is two closely spaced monopole sources that operate out of phase. Dipole sources radiate sound less efficiently than monopole sources because of the partial cancellation of the sound waves. This principle is used in many methods for "cancellation of noise by noise," or "active noise control," a subject of intense current research. See Sec. 9 for a discussion on active noise control.

The third source term, S_q, represents a source term for generation of sound by aerodynamic sources (quadrupole sources) and was first understood by Lighthill (1954). The

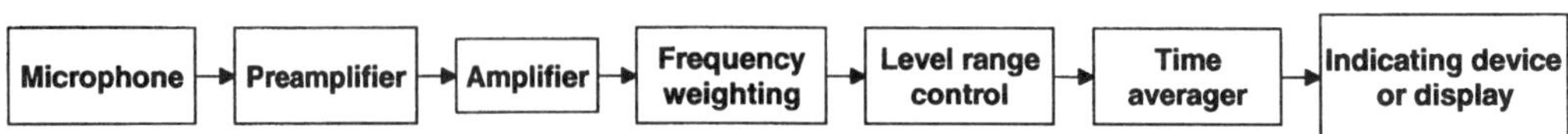

FIG. 2. Block diagram of the principal components of a sound-level meter. The time averager can include exponential weighting for sound-level meters. The frequency response of the microphone is sometimes included in the frequency weighting.

basic mechanism is momentum fluctuations across fixed surfaces in a fluid; sound generation takes place without the influence of solid surfaces. This mechanism is responsible for sound generation by turbojet engines where the major portion of the radiated sound is generated in the air exhaust downstream of the engine itself. A classical description of quadrupole sources is in terms of two closely spaced dipoles where the dipole moments are in opposite directions. Quadrupole sources are less efficient radiators of sound than dipole sources because of the cancellation that takes place between the two dipoles.

The fourth source term, S_h, represents the generation of sound by the introduction of fluctuating heat source into a medium. It is a less common source than those described above. One example is the *Rijke phenomenon,* which occurs when heated filaments are introduced into the lower half of a vertical tube. A strong tone is generated by a feedback mechanism in which the particle velocity in the air induces fluctuating heat flow from the heated filament and acts as a source of sound.

5. STATIONARY NOISE SOURCES

Noise emissions from stationary sources are usually described by the sound-power level emitted by the source. Typical sources include machinery in factories, large air-moving devices used to circulate cooling air in buildings, air-conditioning units, and business equipment used in offices.

The general approach to reduce the noise from stationary sources is to determine the relative strength of the noise emitted from various parts of the source, and to apply appropriate noise-control measures. Sound intensity is one useful tool for identifying noise sources. Another tool is the selective covering or wrapping of portions of a machine with a heavy material such as lead-loaded vinyl in combination with sound-absorptive materials.

Noise-control measures may include replacement of solid surfaces such as protective covers with perforated metal or wire screens, reduction of the size of vibrating surfaces, reduction of the speed at which impacts occur, and reduction of the area of impacting surfaces, damping machine surfaces with materials that absorb energy when flexed, enclosing areas of high noise emission, lowering the speed of rotating equipment, mechanical isolation of equipment so that it does not transmit sound energy into structures, and reduction of the time rate of change of force and pressure.

6. MOVING NOISE SOURCES

Aircraft and road vehicles are the two most common examples of noise generated by moving sources. Although there have been attempts to use sound-power level as the measure of noise emission of such sources, simplified measurement procedures have been adopted for practical reasons. A review of transportation noise in general has been given by Nelson (1987).

6.1 Aircraft Noise

For jet-powered and propeller-driven airplanes, the engines are the principal sources of noise during ground roll, climbout, and landing. Sound produced by vortices shed by the extended landing gear and by wing leading- and trailing-edge devices (flaps) contributes to the noise level under landing-approach flight paths and may be an important source of noise in the future as engine noise is reduced.

The methodology of determining the noise emission of aircraft for certification purposes is well developed and will be covered in this section. Certification noise levels are not intended to represent noise levels that are measured in communities around airports; methods for predicting noise levels around airports are described in Sec. 10.2.

Certification noise limits are specified at three measurement points—known as *lateral* (or sideline), *flyover* (or takeoff), and *landing approach.* The limits are a function of takeoff gross weight and apply to both jet-powered and propeller-driven airplanes. The limits apply to new-design transport-category civil airplanes and to retrofit modifications of old-design transport-category civil airplanes.

Perceived noise levels are determined from 500-ms–average one-third octave band sound-pressure levels, at 500-ms intervals, and are adjusted for the additional annoyance caused by prominent tones, if present. Data measured during a noise-certification compliance test are adjusted to reference atmospheric conditions, reference flight paths, and reference engine-power settings.

Certification limits are set by Part 36 of the Federal Aviation Regulations in the United States (FAR 36) and by Annex 16 to the Convention on International Civil Aviation (ICAO, 1993). The effective perceived noise level (EPNL), a quantity that is regulated by the above documents, is determined from the time integral of tone-corrected perceived noise level.

The first issue of FAR Part 36 became effective in 1969; ICAO Annex 16 was first issued in 1971. The main objective of those regulations was to ensure that noise produced by new aircraft would be less than the noise produced by earlier designs. In the years since the initial issue, the certification noise-level limits were gradually reduced in stages. Stage 1 airplanes are the noisiest and include the original Boeing 707 and McDonnell Douglas DC-8. Stage 2 airplanes have lower noise levels than stage 1 airplanes, and the noise levels of airplanes that comply with stage 3 are even lower. For example, the lateral (sideline) effective perceived noise-level limit at a takeoff gross weight of 200 000 lb ($\approx$91 Mg) is 104.8 dB for a stage 2 airplane with two or three engines, and 97.5 dB for a stage 3 airplane—independent of the number of engines. An airplane that complies with the stage 3 requirements of FAR Part 36 also complies with the Chapter 3 requirements of ICAO Annex 16.

A summary of engine-noise–control designs is available (Hubbard, 1991). A summary of retrofit applications intended to allow stage 2 (Chapter 2) airplanes to meet a phaseout deadline of the year 2000 for stage 3 (Chapter 3) compliance has been published by Marsh (1994). Eldred (1992) has given an overall review of the aircraft noise problem.

6.2 Road-Vehicle Noise

The main sources of road-vehicle noise are engine noise, exhaust noise, and tire–pavement interaction noise. Which source predominates depends on the speed of the vehicle, the condition of the muffler, the design of the tires, and the road surface. Typical noise-reduction measures for vehicles include the use of sound-absorptive materials, optimization of the exhaust-system muffler, optimal design of the cooling fan, and control of gear and transmission noise.

Authorities in most industrialized countries have introduced legislation to control road-vehicle noise by establishing noise-emission limits. New vehicles in many countries are subject to "type approval," and older vehicles may, in some countries, be required to meet noise limits for vehicles in use. The most stringent noise regulations for new road vehicles are currently (1996) in effect in the European Union (Directive 92/97/EEC) where, for example, automobiles must comply with a limit of 74 dB and heavy trucks with a limit of 80 dB. The measurement procedures are described below.

Nationally and internationally standardized noise-measurement procedures for road vehicles differ in some respects, but ISO 362 (ISO, 1994) is widely recognized. This document specifies that the test vehicle is driven over a designated test area with a standardized road surface. Fixed microphones are located 7.5 m from the vehicle path center line. (In the United States, the microphone distance is 15 m, and the relevant standards are SAE J986 and SAE J366.) Vehicles approach the test area at a constant speed, usually 20 to 50 km/h, but the throttle is fully open when the vehicle is within ± 10 m of the microphone positions. The maximum A-weighted fast sound level during the acceleration is measured.

A general review of road-vehicle noise is available (Sandberg, 1995). This review covers the effects of regulations on road-vehicle noise emissions on the noise levels measured in communities.

7. SOUND PROPAGATION

Equation (1) describes the propagation of a plane wave; there is no attenuation of sound with distance. Since sound generally propagates as a spherical wave or into a hemisphere above a reflecting plane, a $1/r$ dependence must be added to Eq. (1) to determine the rms sound pressure at a location at a distance r from the source. The sound-pressure level decreases at a rate of 6 dB per doubling of distance from the source. The sound intensity varies inversely as the square of the distance from the source (inverse-square spreading). This spreading must be taken into account for sound propagation indoors and outdoors. Other effects that influence the variation of sound level with distance from the source are different for sound propagation indoors and propagation outdoors and will be described separately.

7.1 Sound Propagation Indoors

Indoors, the most important effect that modifies the $1/r$ dependence of sound pressure with distance from the source is the reflection of sound from fixed room surfaces. The amount of sound reflected depends on the sound-absorptive characteristics of the surface, usually described in terms of the *absorption coefficient*. There is a variety of methods used to describe absorption coefficients (see ACOUSTICS, ARCHITECTURAL), but for the purposes of this article, the absorption coefficient α is defined as

$$\alpha = 1 - |R|^2, \tag{11}$$

where $|R|^2$ is the squared magnitude of the reflection coefficient, p_r/p_i. Here p_r is the rms amplitude of the reflected sound and p_i is the rms amplitude of the sound incident on the surface.

The classical variation of sound level with distance in a room occurs when reflections from all room surfaces contribute equally to the level of the reflected sound in the room. In this case, the sound field is conveniently described in terms of *direct sound* and *reflected sound*. Near the source, the $1/r$ term predominates, and far from the source, sound reflected from the room surfaces predominates. The total mean square pressure is obtained by adding component mean square sound pressures, since, at least for broad-band sounds, the direct and reflected sounds are essentially uncorrelated.

Under these conditions, the variation of sound level with distance is (see ACOUSTICS, ARCHITECTURAL)

$$L_p(r) = L_W + 10 \log\left(\frac{Q}{4\pi r^2} + \frac{4}{R}\right), \tag{12}$$

where all terms are in decibels and L_W is the sound-power level of the source. Here, r is the distance from the source, Q is a source directivity factor (Beranek and Vér, 1992 Chap. 4), and R is the room constant, which is related to an average Sabine absorption coefficient α_s (neglecting air absorption) by $R = \alpha_s/(1 - \alpha_s)$. At large distances from the source, the level is uniform within the room, and one is in the *reverberant field*. The room constant R is not equal to the reflection coefficient R in Eq. (11).

In most situations of interest in noise control, the classical relationship of Eq. (12) rarely holds because the reflections from all of the room surfaces usually do not contribute equally to the sound-pressure level at a point. Many rooms of practical interest have low ceilings that are partially sound absorptive, and the uniform level in the reverberant field does not occur. In addition, any room of interest contains scattering objects, and the scattering from these objects must be accounted for in any calculation of sound-pressure level vs distance from a source.

Methods for prediction of noise levels in factories have been described by Hodgson (1986). One useful method for describing the variation of sound level with distance indoors has been published by the European Computer Manufacturers Association (ECMA, 1985).

For a "flat room" having a length and width $>3h$ where h is the room height, the variation of sound-pressure level L_p with distance d from a source having a sound-power output L_W may be calculated from

$$L_p = L_W - 20 \log(h/h_0) + \Delta L_f \tag{13}$$

(taken from the ECMA report), where h_0 is 1 m and ΔL_f is given as a function of d/H in Fig. 3. The two parameters in the figure are q, the density factor of the reflecting objects in the room ($q = 0.2$ in the figure), and the average absorption coefficient, $\bar{\alpha}$. The density factor is given by

$$q = (1/4S_f)\Sigma S_e, \tag{14}$$

where S_f is the surface area of the room floor in square meters and ΣS_e is the total area of the scatterers in the room in square meters. The average absorption coefficient is the average of the Sabine absorption coefficients of the ceiling and floor of the room at

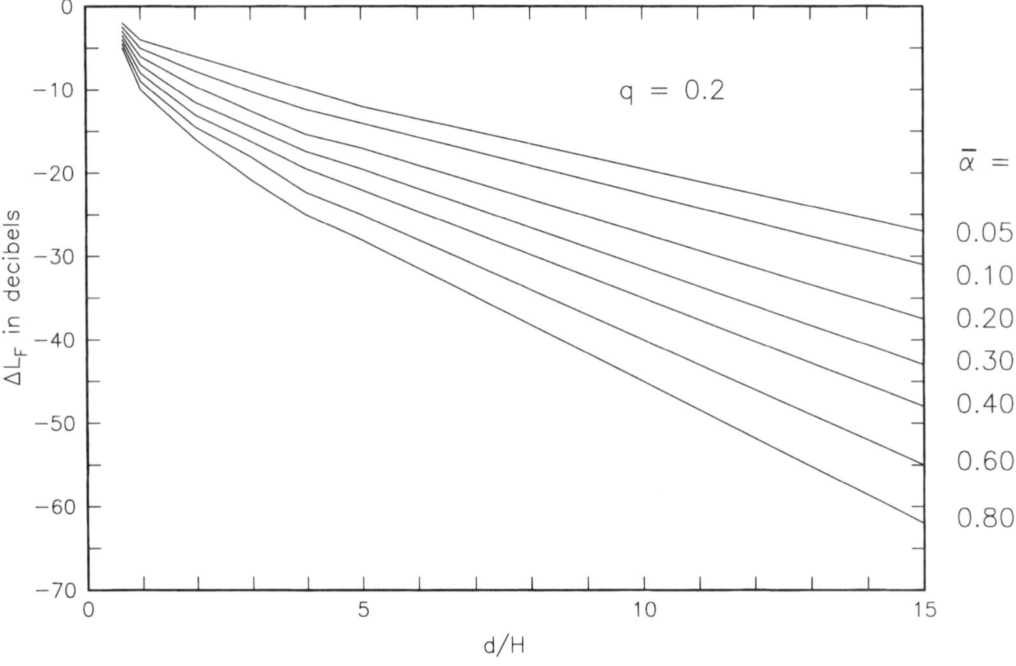

FIG. 3. Correction, ΔL_F, for calculating the decrease with distance of the sound-pressure level from a source for a density factor $q = 0.2$.

frequencies of 250, 500, 1000, and 2000 Hz. For Eq. (13), unlike the classical reverberant-room theory described by Eq. (12), the sound-absorptive properties of the walls in the room are not important.

A second effect that changes the variation of sound-pressure level with distance in rooms is *air absorption.* Sound is absorbed in air because of viscous and heat-conduction effects and (mainly) by oxygen and nitrogen vibrational relaxation phenomena (ANSI, 1995b). This mechanism affects sound levels in large spaces such as concert halls, and is very important in rooms that have very highly reflective room surfaces or for small-scale models, but does not usually play a major role in indoor noise problems.

7.2 Sound Propagation Outdoors

Inverse-square spreading is the most important effect that reduces the sound-pressure level as distance from a source is increased. There are, however, several other factors that influence outdoor sound propagation:

1. Atmospheric absorption. At long distances and at high frequencies, the effects of atmospheric absorption are significant. These effects are usually described by an attenuation coefficient in decibels per meter. Atmospheric absorption is greatly dependent on air temperature and water-vapor content (ANSI, 1995b). As an example, at 4000 Hz, at a temperature of 20 °C and a relative humidity of 50%, it is 29 dB/km (Harris, 1991).
2. Ground effects. When the sound source is located above a ground surface, sound waves that reflect from the ground will constructively and destructively interfere with those propagating directly from the source. In general, the ground is partially reflecting; the reflected wave is modified in amplitude and phase by its interaction with the ground surface. The amount of attenuation attributable to this ground interaction and its variation with frequency depend on the surface characteristics, the source and receiver heights, and their separation. The effects of the ground are largest for intermediate frequencies (~500 Hz) when the source is above the ground

(1 m or more). If the source is close to the ground at all frequencies, all frequencies above about 500 Hz are highly attenuated.

3. Temperature gradients and wind-speed gradients. The speed of sound is proportional to the square root of the absolute temperature of the medium. The normal temperature lapse with height above the ground means that a sound wave front moves more rapidly near the ground surface. This causes the wave front to bend upward, creating a "shadow zone" of low sound-pressure level near the surface. Wind-velocity gradients have a similar effect since the speed of the wave is the wind speed plus the speed of sound. Since the wind speed is usually less close to the ground, propagation upwind tends to bend the wave front upward, creating a shadow zone similar to that produced by a normal temperature lapse. For sound propagation downwind, the opposite effect occurs, and the wave front is bent toward the ground; there is no significant attenuation of the sound wave relative to the attenuation in the absence of wind.
4. Turbulence. Since turbulence always accompanies wind outdoors, the effects of atmospheric turbulence must be included in any analysis of sound propagation outdoors. There are three major effects of turbulence. First, at single frequencies, atmospheric turbulence affects the coherence between waves that propagate directly from the source and waves that are reflected from the ground surface. This results in sound-pressure levels that are higher than would be expected in the absence of turbulence at those points where destructive interference between the waves occurs. Second, atmospheric turbulence produces scattering into the shadow zones caused by air-temperature and wind-speed gradients, tending to raise the sound-pressure level in these areas. Third, in a highly directive sound beam, the presence of turbulence results in an excess attenuation produced by sound scattering out of the beam. Turbulence is an important *source* of sound only in high-speed flows (Lighthill, 1954).

The basic equation for sound propagation outdoors is (Beranek and Vér, 1992, Chap. 5)

$$L_p(r) = L_W - 20\log(r/r_0) + 10\log Q - \log(\Omega/4\pi) + \text{constant} - A, \qquad (15)$$

where all terms are in decibels. Here r_0 is a reference distance of 1 m, Q is a directivity factor (Beranek and Vér, 1992, Chap. 4), the $\Omega/4\pi$ term is related to the solid angle into which the source propagates, the constant depends on the units used, and A is an attenuation that depends on all of the effects described above. Several reviews of sound propagation outdoors are available (Embleton, 1982, 1989; Embleton and Daigle, 1991; Ingard, 1953).

8. SOUND-ATTENUATING ELEMENTS

In the previous section, the reduction of noise over the path from the source to the receiver was discussed for the case where no attempt was made to erect a barrier between the source and the receiver. Such barriers take many forms. Outdoors, noise barriers are being installed along highways to protect residents from the effects of traffic noise. Indoors, walls and other structures form barriers that attenuate sound. In the design of building air-conditioning systems, sound-absorptive ducts are commonly used to reduce the noise level produced by ventilating fans in rooms. Sound-absorptive materials are also used in rooms and in factories to reduce sound levels. Machinery noise reduction is frequently accomplished through partial or full enclosures around machines.

8.1 Outdoor Noise Barriers

The actual performance of barriers designed, for example, to reduce traffic noise depends greatly on traffic flow, since the noise source is basically a line source. It also depends on the height of the source above ground so that the predominant noise sources must be identified. The source heights will be different for slow-moving trucks and for high-speed automobile traffic.

Computational methods for the prediction of barrier noise attenuation are changing rapidly. One set of programs used by noise-control engineers is the STAMINA 2.0/OPTIMA model (Bowlby *et al.*, 1982).

8.2 Transmission Loss of Walls and Other Structures

Indoors, walls and other structures act as barriers to noise. The motion of a wall is controlled by its mass and stiffness, the wall being driven by the pressure of the sound wave incident on one side of a wall. Sound is reradiated on the other side, usually at a much lower level, by the motion of the wall. The sound attenuation of a wall is generally described by a *mass law* (see ACOUSTICS, ARCHITECTURAL) modified by effects such as flanking transmission and coincidence effects. The latter occur when the bending-wave speed in the wall is equal to the speed of sound along the wall.

The sound attenuation of composite structures such as walls with windows is of interest when protecting residents near highways and airports from transportation noise. Sound intensity has proven to be a useful tool for measuring the transmission loss of both solid walls and composite structures. A review of progress in building acoustics and modern methods for the analysis of sound transmission paths is available (Cops and Vermeir, 1995).

8.3 Sound-Absorptive Materials

For a material to absorb sound, it must have a *flow resistance;* that is, the ratio of the pressure drop across the material, Δp, in pascals and the air flow velocity u in meters per second through it must be finite. The flow resistance ζ is usually normalized by the product of air density ρ in kilograms per cubic meter and the speed of sound c at standard sea-level air pressure and an air temperature of 15 °C:

$$\zeta = (\Delta p/u)/\rho c. \quad (16)$$

Materials that have a normalized flow resistance in the range 1–2 are normally good sound absorbers.

A second requirement is that the particle velocity in the sound wave within the material must not be too small in order to achieve high viscous losses from friction. As an example, a thin material with a finite flow resistance is placed close to a solid wall. An incident sound wave is reflected from the wall; the boundary condition is that the particle velocity in the sound wave normal to the (solid) wall is zero, and the incident pressure is doubled at the wall. The material will be ineffective as a sound absorber because of the low (near zero) particle velocity through the material. The same material placed in a region of high particle velocity may be a very effective sound absorber.

Sound-absorptive materials may be bulk materials such as glass-fiber materials or other porous materials. Sound is also absorbed by nonporous materials such as screens and perforated metals placed over partitioned backing materials (i.e., honeycomb structures). In many noise-control situations, a composite structure is used; one example is a bulk material protected by a perforated metal facing. Resonance effects in such structures often enhance the sound absorption at particular frequencies.

A complete description of sound-absorption technology, and computer programs that may be used to estimate sound attenuation by composite structures, sound attenuation in ducts, resonance effects, and sound attenuation in porous materials, is available (Ingard, 1994).

9. ACTIVE NOISE CONTROL

A sound field can be described in space and time by a sound pressure field $p(\mathbf{x},t)$ where $\mathbf{x}$ is a position vector and t is time. To apply active control to this field, one may use a secondary sound source that produces a field closely approximating $-p(\mathbf{x},t)$ within some region of space. Since, in a linear approximation, the two pressure fields add, there is a net reduction of sound pressure in the region. Active noise control is basically an interference phenomenon. Active control can also in theory be applied to the boundaries of an enclosed space as well as to vibrating structures.

The first patent on active noise control was obtained by Lueg (1934), and an early demonstration of how active control can be applied was given by Olson and May (1953). Since that time, improvements in transducers and the development of electronic controllers that use digital signal processing techniques have made active control of sound practical, at least at low frequencies, in many cases (Sommerfeldt and Hamada,

1995; Tichy, 1995; Elliott and Nelson, 1994; Nelson and Elliott, 1992).

Four application areas are of practical interest:

Noise-canceling headsets. Use of a microphone, electronic controller, and loudspeaker mounted in a headset to generate an out-of-phase signal in the ear canal has proved to be effective in reducing noise exposure for persons wearing the headsets (Bartholomae and Stein, 1991).

Global reduction of sound. Placing an out-of-phase secondary source in the vicinity of primary source effectively converts a monopole to a dipole and reduces the radiation efficiency of the primary source. The important parameter is kd, where k is the wave number and d is the distance between the primary and secondary sources. For effective cancellation, $kd \ll 1$. This technique has been demonstrated by Neise and Koopman (1988) in connection with the cancellation of blade-passage frequencies and harmonics in centrifugal fans.

Cancellation of sound in ducts. Use of sound-absorptive materials is an effective way of attenuating sound in ducts such as those used in buildings for heating and ventilating systems. At low frequencies, sound-absorptive materials are less effective, and active techniques may be used to reduce low-frequency sound. A *feed-forward* system is illustrated in Fig. 4. An input microphone detects the sound in the duct, and a loudspeaker is used as a secondary source. The electronic controller calculates the signal wave form to be sent to the loudspeaker in order to minimize, usually in a least-squares sense, the sound pressure at the error microphone. This technique is most useful at low frequencies when only a plane wave can propagate in the duct. See Beranek and Vér (1992, Chap. 15).

Spatial reduction of sound pressure. Effective reduction of the sound-pressure level in a given volume remote from a source is only possible at low frequencies, and it generally requires multiple microphones to sense the field, multiple loudspeakers to act as secondary sources, and a multichannel electronic controller to generate the optimal loudspeaker inputs. Such a system has been described by Peterson *et al.* (1994), and active control of road noise inside vehicles has been described by Sutton *et al.* (1994).

10. EFFECTS OF NOISE ON PEOPLE

As mentioned in the Introduction, receivers of noise are usually human beings; the effects can be broadly divided into two categories, physiological and psychological. A summary of the effects of noise on humans has been published for the World Health Organization (Bergland and Lindvall, 1995).

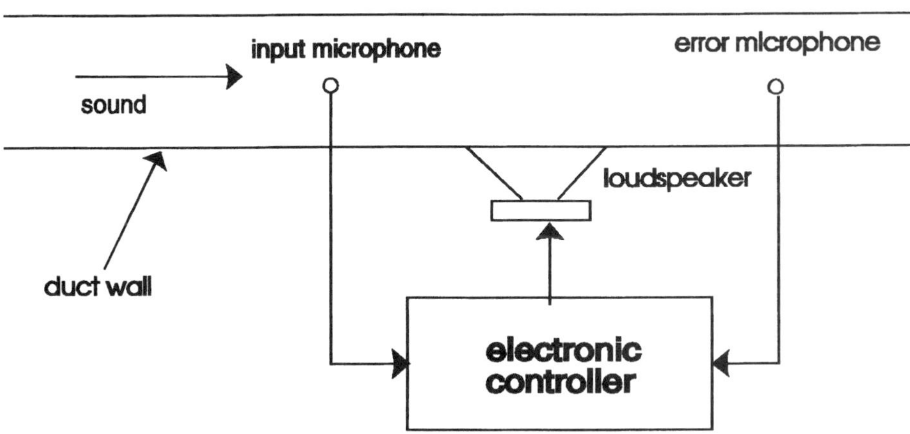

FIG. 4. Concept for a system to cancel sound in a duct.

10.1 Physiological Effects

The most important physiological effect of noise is hearing loss caused by long-term exposure to high levels of noise. This hearing loss may occur through occupational exposure—for example, workers in factories—or through noise exposure through leisure activities or activities around the home. The amount of hearing loss depends on both sound-pressure level and exposure time. It is widely believed (Suter, 1993) that cumulative noise exposure should be calculated on an "equal-energy" basis; that is, a 3-dB increase in sound level accompanied by a decrease by one-half in the exposure time results in the same noise exposure and produces the same amount of hearing loss. This position was adopted by the American Conference of Governmental Industrial Hygienists. The U.S. Department of Labor uses different criteria.

In the 1970s, the U.S. Environmental Protection Agency published a study of the effects of noise (EPA, 1974) and concluded that exposure to an A-weighted sound level of 70 dB averaged over a 24-h period is "safe" for the most noise-sensitive persons, at the frequencies of sound that cause the greatest damage, and when the exposure to noise is for long periods of time. For an exposure of 8 h/d, the "safe" level increases to 75 dB, according to the equal-energy theory.

In industrial situations, it is common to begin monitoring hearing acuity when 8-h exposure levels are 80 dB or greater and to provide hearing protection or engineering controls when 8-h exposure levels exceed 85 dB. The relationship between hearing loss, sound level, and exposure time has been standardized (ISO, 1990), and a description of worldwide requirements on exposure to hazardous noises and recommendations for future action is available (Embleton, 1994).

10.2 Psychological Effects

The primary psychological effect of noise is annoyance. Annoyance depends not only upon the physical characteristics of noise but also on the time of day, background sound levels, personal attitudes, and a number of other considerations. The most serious problems with outdoor noise in today's society are the effects of noise in communities around airports and exposure to road-traffic noise. The former are being addressed by careful control of takeoff and landing patterns of airplanes, by land-use planning around airports, and, in some cases, by sound insulation of homes near airports. A-weighted sound levels in a number of forms are used to describe noise levels around airports. These include the community noise equivalent level (CNEL) and the day–night average sound level (DNL). The day–night average sound level is described in the next paragraph.

Cumulative exposure to noise is often specified through the day–night average sound level L_{dn}, which is a 24-h average A-weighted level determined with the sound exposures during the hours of 24.00–07.00 and 23.00–24.00 increased by a factor of 10. In the 1970s, the U.S. Environmental Protection Agency identified 55 dB as the day–night average sound level "required to protect the public health and welfare with an adequate margin of safety" (EPA, 1974).

The psychological effects of single events are more difficult to quantify. One measure of the noise from a single event, such as an aircraft flyover or automobile passby, is the A-weighted *sound exposure level*. The sound exposure level is given by

$$L_{AE} = 10 \log\left[\frac{\int_0^T p_A^2(t)\,dt}{p_0^2 t_0}\right]. \tag{17}$$

Here, p_A^2 is the squared instantaneous A-weighted sound pressure. The reference sound exposure in the denominator of Eq. (17) is given by the square of the reference pressure, $p_0 = 20\ \mu\text{Pa}$, and the reference time, $t_0 = 1$ s.

As an aid to land-use planning near airports, the Federal Aviation Administration has developed (FAA, 1996) a new version of the Integrated Noise Model (INM, Version 5.1) that enables noise levels in communities around airports to be calculated considering such variables as type of aircraft, number of operations, and flight paths. The main enhancements in Version 5.1 are the addition of military aircraft, the noise-power–distance database from the U.S. Air Force NOISEMAP model, and support for the Microsoft Windows 95® operating system.

As an aid to land-use planning near highways, the U.S. Federal Highway Administration has developed a new (1996) highway noise prediction model, the Traffic Noise Model (Menge, 1996), to replace the STAMINA 2.0/OPTIMA model (Bowlby *et al.*, 1982). The Traffic Noise Model (TNM) incorporates variable ground impedance, reflection from sound-absorptive walls, multiple diffraction, and shielding from barriers, rows of buildings, and trees.

In the United States, near federally funded highways, traffic-noise reduction is being addressed through federal regulations (FHWA, 1976) and, in many cases, the construction of noise barriers along highways. Attempts are also being made to control road-traffic noise through noise-emission standards as described in Sec. 6.2. The relationship between community noise exposure and regulatory limits is being studied. It has been found (Sandberg, 1995) that imposition of limits does not always reduce the noise levels to which communities are exposed. A summary of the effects of noise on people has been given by Eldred and von Gierke (1993).

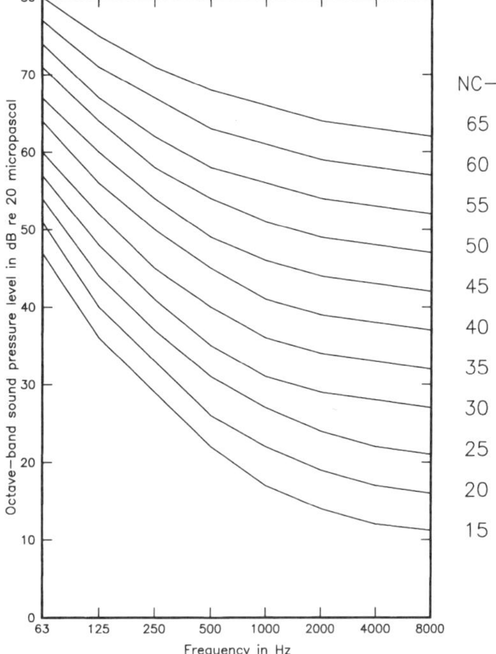

FIG. 5. Noise criterion (NC) curves.

11. CRITERIA FOR NOISE CONTROL

11.1 Indoors

Criteria for control in the workplace for purposes of hearing conservation have been discussed above. Methods of rating noise for the purpose of providing a satisfactory working environment either use an A-weighted level or use a method that requires an octave-band (or one-third octave band) analysis. In the latter case, a set of *noise criterion* (NC) curves may be used to convert an octave-band spectrum into a single-number rating (Beranek, 1990). Alternatively, *balanced-noise criterion* (NCB) curves or *room criterion* (RC) curves may be used to determine when a given noise specification is met (Beranek, 1992; ANSI, 1995a).

For use of the NC curves, shown in Fig. 5, the octave-band spectrum is plotted on the graph, and the highest point of penetration into the set of NC curves determines an NC value for the spectrum. This is a "curve-tangent" method.

Use of the NCB or RC curves, illustrated in Figs. 6 and 7, is more complex but has

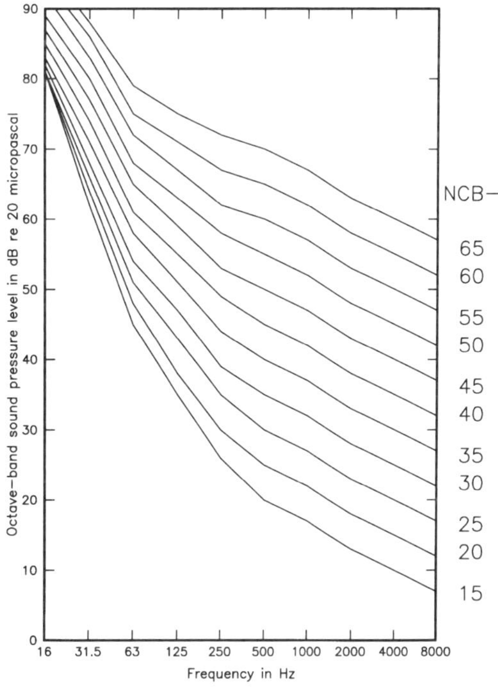

FIG. 6. Balanced-noise criterion (NCB) curves.

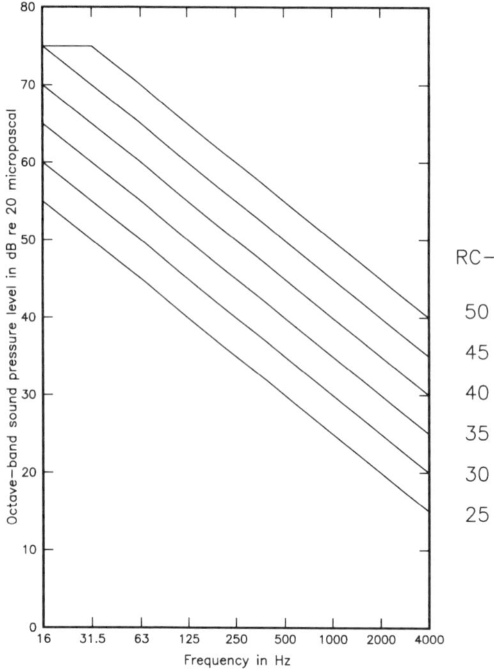

FIG. 7. Room noise criterion (RC) curves developed from measurements of noise from heating, ventilating, and air-conditioning equipment noises in unoccupied offices.

the advantage that the quality of the sound is evaluated and that *rumble* (low-frequency noise) and *hiss* (high-frequency sound) can be explicitly taken into account. If a room noise specification is given in terms of a NCB value (NCB–*NN*, where *NN* is an integer), a given octave-band spectrum complies with this specification if all of the following three criteria are met:

1. The average of the octave-band levels in the 500-, 1000-, 2000-, and 4000-Hz octave bands does not exceed *NN*.
2. The levels in the octave bands from 16 to 500 Hz do not exceed the average calculated in step **1** plus 3 dB.
3. The octave-band levels in the bands from 1000 to 8000 Hz do not exceed an $\mathrm{NCB}_{\mathrm{avg}}$ curve, where this average curve is calculated by the following procedure (ANSI, 1995a). Find the three NCB curves that have the same levels in the 125-, 250-, and 500-Hz bands as the measured sound-pressure levels in those bands, respectively. Call the three curves so determined $\mathrm{NCB}_h(f)$, and calculate the average, $\mathrm{NCB}_{\mathrm{avg}}$.

In the above procedure, step **1** is speech-interference level compliance, step **2** is rumble compliance, and step **3** is hiss compliance. An example of an NCB specification for an occupied activity area is that in a large conference room, an NCB curve in the range 25–30 is suitable (ANSI, 1995a).

If a specification is given in terms of an RC value (RC–*NN*, where *NN* is an integer), a given octave-band spectrum complies with this value if both of the following conditions are met (ANSI, 1995a):

1. The average of the octave-band levels in the 500-, 1000-, and 2000-Hz octave bands does not exceed *NN*.
2. Any level in the octave bands below 500 Hz does not exceed the RC–*NN* curve by more than 5 dB, and any level in the octave bands at 1000 Hz and above does not exceed the RC–*NN* curve by more than 3 dB.

The first criterion relates to average midband levels, and the second relates to rumble and hiss. An example of an RC criterion for an unoccupied space is that for conference rooms. An RC curve in the range 25–30 is a suitable specification (ANSI, 1995a).

Room criteria for noise from heating, ventilating, and air-conditioning systems have been discussed by Blazier (1995) and by Kingsbury (1995).

11.2 Outdoors

There has been a wide variety of noise surveys that attempt to define the relationship between community annoyance and some measure of noise emission. Two such measures are the sound level averaged over some time interval (i.e., the level of the rms sound pressure measured over a given time interval), and the day–night average sound level defined in Sec. 10.2. The results from several studies have been summarized by Eldred and von Gierke (1993) with a graph, shown in Fig. 8, for the best estimate of the percentage of the population highly annoyed vs the day–night average sound level. As an example, it is estimated that more than 30% of the population is highly annoyed by aircraft noise when the day–night average

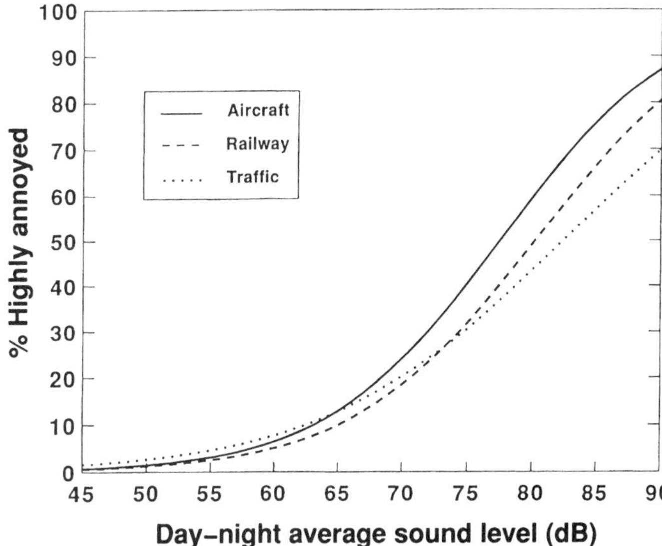

FIG. 8. Percentage highly annoyed vs day–night average sound level from aircraft, road traffic, and railway noise. Based on data from Fidell *et al.* (Eldred and von Gierke, 1993).

sound level exceeds 73 dB. Similar data have been summarized by Finegold *et al.* (1994).

12. REGULATORY ASPECTS OF NOISE: STANDARDS

12.1 Regulations

Noise regulations can be divided into two general areas, regulations on noise *emission* and regulations on noise *immission*. In the United States, major emphasis is placed on regulation of the noise emissions of aircraft through Federal Aviation Administration (FAA) regulations, and these regulations are generally in harmony with international requirements (ICAO). In the 1980s, the European Union began a program to regulate the noise emissions of specific noise sources, but the modern approach is to issue general regulations applicable to a wide variety of equipment and to prepare measurement standards for specific sources. A summary of European noise regulations has been given by Higginson *et al.* (1994). There is a variety of noise-immission regulations, many issued by state and local authorities. A summary of requirements was compiled by Beranek and Vér (1992, Chap. 17).

12.2 Standards

Several organizations produce standards related to noise and its measurement and control. Internationally, standards for measuring instruments are developed by Technical Committee 29 of the International Electrotechnical Commission (IEC). These standards include requirements and specifications for microphones and sound-level meters, sound calibrators, octave- and one-third–octave-band filter sets, and sound-intensity analyzers.

International standards on noise are developed by Technical Committee 43 of the International Organization for Standardization (ISO). Subcommittee 1, Noise, develops standards on determination of the noise emission of sources and other related topics. Subcommittee 2 develops standards on building acoustics, including sound transmission and determination of the properties of acoustical materials.

In the United States, there is a federated standards system coordinated by the American National Standards Institute (ANSI). Several national organizations produce noise standards in their specific areas of interest.

The Acoustical Society of America coordinates the activities of Accredited Standards Committees S1, S2, S3, and S12. Committee S1 develops standards on physical acoustics, including terminology, microphones, sound-level meters, and filter sets. S2 develops standards on mechanical vibration and shock, condition monitoring, and diagnostics. S3 develops standards on bioacoustics—standards that relate to safety, tolerance, and comfort. S12 develops noise standards, including noise-emission standards and standards for the description of environmental sound, for hearing conservation, and for occupational noise exposure.

There are several other national organizations that have significant activities related to noise and its control. The American Society for Testing and Materials (ASTM) develops standards related to building acoustics and acoustical materials, the Society of Automotive Engineers (SAE) develops standards for measurement of the noise of moving sources such as road vehicles and aircraft, and the American Society of Heating, Refrigerating, Airconditioning and Ventilating Engineers (ASHRAE) develops noise standards related to air-conditioning equipment.

There are several other professional and trade associations that develop standards related to specific areas of noise and its control.

ACKNOWLEDGMENTS

The author would like to thank T. F. W. Embleton, G. A. Daigle, A. S. Harris, R. K. Hillquist, K. U. Ingard, and U. Sandberg for reading portions of the manuscript and contributing many helpful suggestions for changes and additions to the text. A. H. Marsh read the entire manuscript and suggested many helpful changes and additions.

GLOSSARY

This glossary has been adapted from American National Standard acoustical terminology (ANSI, 1994) and a noise-control engineering glossary (Anon., 1995).

A-Weighted Sound-Pressure Level: Ten times the logarithm of the ratio of A-weighted squared sound pressure to the squared reference sound pressure of 20 μPa, the squared sound pressure being obtained with fast (F) (125-ms) exponentially weighted time averaging. Alternatively, slow (S) (1000-ms) exponentially weighted time averaging may be specified. Unit, decibel (dB); symbol, L_A.

Anechoic Room: Test room whose surfaces absorb essentially all of the incident sound energy over a frequency range of interest, thereby affording nearly free-field conditions over a measurement surface.

Average Sound-Pressure Level in a Room: Ten times the logarithm of the ratio of the space and time averages of squared sound pressure to the squared reference sound pressure, the space average being taken over the total volume of the room, except for the regions of the room where the direct field of the source and the near field of the boundaries are of significance. Unit, decibel (dB).

Bel: Unit of level when the base of the logarithm is ten, and the quantities concerned are proportional to power. Symbol, B.

Continuous Spectrum: Spectrum of a wave whose frequency components are continuously distributed over a frequency region.

Decibel: Unit of level equal to one-tenth of a bel. Symbol, dB.

Discrete Tone: A discrete tone can be either a sinusoidal variation (see **Pure Tone**), in which case the frequency spectrum would show a single component at the sinusoidal frequency, or, more typically, a nonsinusoidal variation, in which case the spectrum would show a component at the fundamental frequency and other components at harmonics of the fundamental.

Filter Nominal Midband Frequency: For a set of contiguous one-third octave bandpass filters, frequency of a specified series such as the preferred frequency series that includes 100, 125, 160, 200, 250, 315, 400, 500, 630, 800, and 1000 Hz, extended by successive multiplication or division by 10. For a set of contiguous octave bandpass filters, the nominal midband frequency is one of a series such as the preferred frequency series that includes 16, 31.5, 63, 125, 250, 500, 1000, 2000, 4000, and 8000 Hz, extended by successive multiplication or division by 10. Unit, hertz (Hz).

Hemi-anechoic Room: Test room with a hard, reflecting floor whose other surfaces absorb essentially all the incident sound energy over the frequency range of interest, thereby affording nominal free-field conditions above a reflecting plane.

Integrating–Averaging Sound-Level Meter: Device for measuring the level of the ratio of a mean square frequency-weighted sound pressure during a stated time interval to the square of the reference sound pressure.

Level: In acoustics, logarithm of the ratio of a quantity to a reference quantity of the same kind. The base of the logarithm (always 10, unless otherwise specified), the reference quantity, and the *kind* of level shall be explicitly specified. Examples of kinds of levels are sound-pressure level, sound-power level, and sound-intensity level. For common logarithms on the base 10, the symbol for logarithm, for the purposes of this article, is "log."

Noise: Undesired sound.

Noise Emission: Airborne sound radiated by a well-defined noise source (e.g., a machine under test).

Noise Immission: Airborne sound received at the ear of an observer, being a composite of all sounds in the vicinity of the observer.

Noise-Power Emission Level: A-weighted sound-power level. Unit, bel (B).

Particle Velocity: In a sound field, the velocity caused by a sound wave propagating through a given infinitesimal part of the medium, with reference to the medium as a whole. Unit, meter per second (m/s); symbol, u.

Porous Sound Absorber: Material with interconnected voids that presents resistance to airflow through the material.

Pure Tone: (a) Line spectrum consisting of a signal at a single frequency. (b) Sound wave, the instantaneous sound pressure of which is a simple sinusoidal function of time. Also referred to as a **Discrete Tone.**

Reverberation Room: Room having a long reverberation time, specially designed to make the sound field therein as diffuse as possible.

Sabine Absorption: Sound absorption defined by the Sabine reverberation-time equation. Unit, metric sabin. In a reverberant room of volume V, speed of sound c, and decay rate d, Sabine absorption is $A = 0.921Vd/c$.

Sound: (a) Oscillation in pressure, stress, particle displacement, particle velocity, etc., in a medium with internal forces (e.g., elastic or viscous), or the superposition of such propagated oscillations. (b) Auditory sensation evoked by the oscillation described above.

Sound Exposure Level: Ten times the logarithm of the ratio of a given time integral of squared instantaneous A-weighted sound pressure, over a stated time interval or event, to the reference sound pressure determined from the product of the squared reference sound pressure of 20 μPa and reference duration of 1 s. The frequency weighting may be otherwise if stated explicitly. Unit, decibel (dB); symbol, L_{AE}.

Sound Intensity: Average rate of sound energy transmitted in a specified direction at a point through a unit area normal to this direction at the point considered. Unit, watt per square meter (W/m^2); symbol, I.

Sound-Level Meter: Device used to measure sound-pressure level with a standardized frequency weighting and indicated exponential time weighting for measurements of sound-pressure level. Without time weighting, measurement of time-average sound-pressure level is achieved by use of an integrating–averaging sound-level meter. Sound exposure level is determined using an integrating sound-level meter.

Sound Power: Sound energy radiated by a source per unit of time. Unit, watt (W); symbol, W.

Sound-Power Level: Logarithm of the ratio of a given sound power in a stated frequency band or with a stated frequency weighting, to the reference power of one picowatt (1 pW). Unit, bel (B) or decibel (dB); symbol, L_W.

Sound-Pressure Level: Ten times the logarithm of the ratio of the mean square pressure of a sound, in a stated frequency band or with a stated frequency weighting, to the square of the reference sound pressure of 20 μPa. Unit, decibel (dB); symbol, L_p.

Spectrum: (a) Description, for a function of time, of the resolution of a signal into components, each of different frequency and (usually) different amplitude and phase. (b) "Spectrum" is also used to signify a continuous range of components, usually wide in extent, within which waves have some specified common characteristic, e.g., "audio-frequency spectrum."

Spectrum Analyzer: Device used to determine the frequency spectrum of a sound.

Speed of Sound: At a temperature of 15 °C, the speed of sound in air at sea level is approximately 341 m/s and is proportional to the square root of absolute temperature; symbol, c.

Time-Average Equivalent Continuous *A*-Weighted Sound-Pressure Level: Ten times the logarithm of the ratio of the *A*-weighted mean square sound pressure to the square of the reference sound pressure. The duration of the integration time must be explicitly stated. Unit, decibel (dB); symbol, L_{AT} or L_{AeqT}.

Works Cited

Anon. (1995), *Noise/News International* **3,** 161–168.

ANSI (1994), Standard S1.1, New York: Acoustical Society of America.

ANSI (1995a), Standard S12.2, New York: Acoustical Society of America.

ANSI (1995b), Standard S1.26, New York: Acoustical Society of America.

Bartholomae, R. C., Stein, R. R. (1991), in: *Proceedings of NOISE-CON 91,* 239–244, Poughkeepsie, NY: Noise Control Foundation.

Beranek, L. L., Vér, I. (Eds.) (1992), *Noise and Vibration Control Engineering: Principles and Applications,* New York: Wiley.

Berglund, B., Lindvall, T. (Eds.) (1995), *Community Noise,* Stockholm: Stockholm University and Karolinska Institute.

Blazier, W. E. Jr. (1995), *Noise Control Eng. J.* **43,** 53–63.

Bolt, R. H., Ingard, K. U. (1957), in: C. Harris (Ed.), *Handbook of Noise Control,* 1st ed., Chap. 22, New York: McGraw-Hill.

Bowlby, W., Higgins, J., Reagen, J. (Eds.) (1982), Report FHWA-DP-58-1, Washington, DC: Federal Highway Administration.

Cops, A., Vermeir, G. (1995), *Noise/News International* **3,** 10–25.

ECMA (1985), Report No. TR27, Switzerland: European Computer Manufacturers Association (ECMA).

EPA (1974), EPA 550/9-74-004, Washington, DC: U.S. Environmental Protection Agency.

Eldred, K. M. (1992), in: *Proceedings of INTER-NOISE 92,* 3–12, Poughkeepsie, New York: Noise Control Foundation.

Eldred, K. M., von Gierke, H. E. (1993), *Noise/News International* **1,** 67–89.

Elliott, S. J., Nelson, P. A. (1994), *Noise/News International* **2,** 75–98.

Embleton, T. F. W. (1982), *Noise Control Eng. J.* **18,** 30–39.

Embleton, T. F. W. (1989), in: *Proceedings of the Institute of Acoustics,* 11, 5, 1–16: United Kingdom.

Embleton, T. F. W. (1994), *Noise/News International* **2,** 227–237.

Embleton, T. F. W., Daigle, G. A. (1991) in: H. H. Hubbard (Ed.), *Aeroacoustics of Flight Vehicles,* NASA Reference Publication No. 1258, Chap. 12, Hampton, VA: National Aeronautics and Space Administration (republished, 1995, New York: Acoustical Society of America).

FAA (1996), *Integrated Noise Model, Version 5.1,* Washington, DC: Federal Aviation Administration.

FHWA (1976), 23 CFR, Chap. 1, Part 772, Washington, DC: See also *Federal Register,* 23 April 1976.

Finegold, L. S., Harris, C. S., von Gierke, H. E. (1994), *Noise Control Eng. J.* **42,** 25–30.

Harris, C. M. (Ed.) (1991), *Handbook of Acoustical Measurements and Noise Control,* 3rd ed., Chap. 3, New York: McGraw-Hill.

Higginson, R. F., Jacques, J., Lang, W. W. (1994), *Noise/News International* **2,** 156–184.

Hodgson, M. R. (1986), in: *Proceedings of INTER-NOISE 86,* 2, 1319–1322, Poughkeepsie, New York: Noise Control Foundation.

Hubbard, H. H. (Ed.) (1991), *Aeroacoustics of Flight Vehicles,* NASA Reference Publication No. 1258, Vols. 1 and 2, Hampton, VA: National Aeronautics and Space Administration.

ICAO (1993), *International Standards and Recommended Practices, Environmental Protection,* Volume 1: *Aircraft Noise,* Annex 16 to the Convention on International Civil Aviation, 3rd ed., Montréal: International Civil Aviation Organization.

IEC (1979), Standard No. 651, Switzerland: International Electrotechnical Commission.

IEC (1985), Standard No. 804, Switzerland: International Electrotechnical Commission.

Ingard, K. U. (1953), in: *Proceedings of the Fourth Annual National Noise Abatement Symposium,* 4, 1–15.

Ingard, K. U. (1994), *Sound Absorption Technology,* New York: Noise Control Foundation.

ISO (1994), Standard No. 362:1994(E), Geneva, Switzerland: International Organization for Standardization.

ISO (1980), Standard No. 3740, Geneva, Switzerland: International Organization for Standardization.

ISO (1990), Standard No. 1999, Geneva, Switzerland: International Organization for Standardization.

ISO (1993), Standard No. 9614-1, Geneva, Switzerland: International Organization for Standardization.

ISO (1996), Standard No. 9614-2, Geneva, Switzerland: International Organization for Standardization.

Kingsbury, H. F. (1995), *Noise Control Eng. J.* **43,** 64–72.

Lighthill, M. J. (1954), in: *Proceedings of the Royal Society of London, Ser. A* **211A,** 564–587; **222A,** 1–32.

Lueg, P. (1934), U.S. Patent No. 2,043,416.

Marsh, A. H. (1994), in: *Proceedings of INTER-*

NOISE 94, Vol. 1, Poughkeepsie, New York: Noise Control Foundation, pp. 199–204.

Menge, C. (1996), in: *Proceedings of INTER-NOISE 96,* Vol. 6, St. Albans, Herts, U.K.: Institute of Acoustics, pp. 3121–3126.

Neise, W., Koopman, G. H. (1988), in: *Proceedings of INTER-NOISE 88,* Vol. 2, Poughkeepsie, New York: Noise Control Foundation, pp. 801–804.

Nelson, P. A., Elliott, S. J. (1992), *Active Control of Sound,* New York: Academic.

Nelson, P. M. (1987), *Transportation Noise Reference Book,* London: Butterworths.

Nitsche, V. (1995), in: *Proceedings of INTER-NOISE 95,* Vol. 2, Poughkeepsie, New York: Noise Control Foundation, pp. 755–762.

Olson, H. F., May, E. G. (1953), *J. Acoust. Soc. Am.* 1130–1136.

Peterson, D. K., Weeks, W. A., Nowlin, W. C. (1994), in: *Proceedings of NOISE-CON 94,* 315–320, Poughkeepsie, New York: Noise Control Foundation.

Sandberg, U. (1995), *Noise/News International* **3,** 82–113.

Sommerfeldt, S., Hamada, H. (Eds.) (1995), in: *Proceedings of ACTIVE 95, The 1995 International Symposium on Active Control of Sound and Vibration,* Poughkeepsie, New York: Noise Control Foundation.

Suter, A. H. (1993), *Noise/News International* **1,** 131–151.

Sutton, T. J., Elliott, S. J., McDonald, A. M., Saunders, T. J. (1994), *Noise Control Eng. J.* **42,** 137–147.

Tichy, J. (1995), in: *Proceedings of INTER-NOISE 95,* 1, 17–28, Poughkeepsie, New York: Noise Control Foundation.

Timmerman, N. S. (1993), in: *Proceedings of NOISE-CON 93,* 425–428, Poughkeepsie, New York: Noise Control Foundation.

Further Reading

Anon., Home Page of the Institute of Noise Control Engineering, http://users.aol.com/inceusa/ince.html.

Bell, L. H., Bell, D. H. (1994), *Industrial Noise Control,* New York: Marcel Dekker.

Beranek, L. L. (Ed.) (1988), *Noise and Vibration Control,* revised ed., Poughkeepsie, New York: Institute of Noise Control Engineering.

Beranek, L. L., Vér, I. (Eds.) (1992), *Noise and Vibration Control Engineering: Principles and Applications,* New York: Wiley.

Harris, C. M. (Ed.) (1991), *Handbook of Acoustical Measurements and Noise Control,* 3rd ed., New York: McGraw-Hill.

Noise Control Engineering Journal and *Noise/News International,* published by the Institute of Noise Control Engineering, P.O. Box 3206 Arlington Branch, Poughkeepsie, NY 12603, USA.

CONTENTS OF PREVIOUS VOLUMES

Volume 1

Volume 2

Volume 3

Volume 4

Volume 5

Volume 6

Volume 7

Volume 8

Volume 9

Volume 10

Volume 11

Volume 12

Volume 13

Volume 14

Volume 15

Volume 16

Volume 17

LIST OF RECOMMENDED UNITS AND SYMBOLS

RECOMMENDED UNITS AND CONVERSION FACTORS

The SI system provides six basic units: meter m, kilogram kg, second s, ampere A, kelvin K, candela cd, and mole mol.

Some important derived units are also allowed and bear special names, e.g.:

1 N (newton) = 1 kg m s^{-2}
1 J (joule) = 1 N m = 1 kg m^2 s^{-2}
1 W (watt) = 1 J s^{-1} = 1 kg m^2 s^{-3}
1 Pa (pascal) = 1 N m^{-2} = 1 kg m^{-1} s^{-2}

For mass, the gram g, or the metric ton t which equals 1000 kg, may be used instead of kilogram kg.

The so-called "long ton" (UK) and "short ton" (US) have been abandoned and will not be used in the *Encyclopedia of Applied Physics*.

For pressure, the bar (name and symbol alike) may be used, and for temperature, the degree celsius °C.

From all of these, decimal multiples or fractions can be derived:

Power of ten	Prefix	Symbol	Power of ten	Prefix	Symbol
10	deca	da	10^{-1}	deci	d
10^2	hecto	h	10^{-2}	centi	c
10^3	kilo	k	10^{-3}	milli	m
10^6	mega	M	10^{-6}	micro	μ
10^9	giga	G	10^{-9}	nano	n
10^{12}	tera	T	10^{-12}	pico	p
10^{15}	peta	P	10^{-15}	femto	f
10^{18}	exa	E	10^{-18}	atto	a

For mass, multiples or fractions are derived from g, not from kg (since the latter already contains a prefix), e.g., mg.

Units with the prefixes are considered as one entity and can, therefore, be raised to any power, e.g., cm^3.

The liter is now considered as synonymous with dm^3 (which does not hold in the older literature!). Please use the capital letter L as unit symbol (following a recent IUPAC recommendation). The use of the Ångström unit is discouraged; it should be replaced by fractional meters (1 Å = 100 pm = 0.1 nm).

SELECTED QUANTITIES, UNITS, AND SYMBOLS

Name of quantity[a]	Symbol[b]	SI unit[c]	Name of unit	Other units
Space and time				
length*	l	m	meter[d]	
breadth, width	b	m		
height	h	m		
radius	r	m		
thickness	d	m		
area	A, S	m^2		a (are) h (hectare)
volume	V	m^3		L, l(liter)[e]
plane angle	$\alpha, \beta, \gamma, \vartheta, \varphi$	1, rad	radian	° (degree) ′ (minute) ″ (second)
solid angle	ω, Ω	1, sr	steradian	
wavelength	λ	m		
wave number	σ, ν	m^{-1}		
time*	t	s	second	min (minute) h (hour) d (day)
frequency	ν, f	s^{-1}		Hz (hertz)[f]
relaxation time	τ	s		
velocity	u, v	$m\ s^{-1}$		km/h
acceleration	a	$m\ s^{-2}$		
Mechanics				
mass*	m	kg	kilogram	g (gram) t (tonne)[g]
(mass) density	ρ	$kg\ m^{-3}$		g/cm^3
momentum	$\mathbf{p}$	$kg\ m\ s^{-1}$		
angular momentum	$\mathbf{L}$	$kg\ m^2\ s^{-1}$		
force	$\mathbf{F}$	N	newton[f]	
moment of force	$\mathbf{M}$	N m		
weight	G, W	N		
pressure	p	Pa	pascal	bar (bar)[h]
energy	E, W	J	joule	W h (watt hour)[i] eV (electron volt)[j]
work	W, A	J		
power	P	W	watt	J/s, V A[k]
Molecular Physics and Thermodynamics				
thermodynamic temperature*	T	K	kelvin	°C (degrees Celsius)
Celsius temperature	ϑ, t			°C
number of entities	N			
Avogadro constant[l]	N_A, L	mol^{-1}	(particles) per mole	
Boltzmann constant[m]	k, k_B	$J\ K^{-1}$		
Planck constant[n]	h	J s		
(molar) gas constant[o]	R	$J\ mol^{-1}\ K^{-1}$		
(quantity of) heat	Q	J		
entropy[p]	S	$J\ K^{-1}$		
internal energy[p]	U	J		
Helmholtz function,[p] (Helmholtz) free energy, Helmholtz energy	F, A	J		
enthalpy[p]	H	J		
Gibbs function,[p] (Gibbs) free energy, Gibbs energy	G	J		
heat capacity[p]	C_p, C_V	$J\ K^{-1}$		

Name of quantity[a]	Symbol[b]	SI unit[c]	Name of unit	Other units
Chemical Physics				
amount of substance*	n	mol	mole	
relative atomic mass	A_r	1		
relative molecular mass	M_r	1		
atomic mass constant	m_u	kg		u (atomic mass unit)[q]
mass of a portion (of substance B)	m_B,m(B)	kg		g (gram)
molar mass (of substance B)	M_B,M(B)	kg mol^{-1}		
concentration (of substance B)	c_B,c(B)	mol m^{-3}		mol/L
mole fraction[r] (of substance B)	κ_B,κ(B)	1		
mass fraction (of substance B)	ω_B,ω(B)	1		%,‰,ppm,ppb
volume fraction (of substance B)	φ_B,φ(B) ϕ_B,ϕ(B)	1		%,‰,ppm,ppb
mass concentration	ρ	kg m^{-3}		g/L
molality		mol kg^{-1}		mmol/kg
volume concentration[s]	σ	1		
molar volume	V_m	m^3 mol^{-1}		L/mol
molar heat capacity	C_m	J mol^{-1} K^{-1}		
molar conductivity	Λ_m	S m^2 mol^{-1}	(S: siemens)	
Faraday constant[t]	F	C mol^{-1}		
Electricity and Magnetism				
quantity of electricity[u]	Q	C	coulomb	
charge density	ρ	C m^{-3}		
electric potential	ϕ,V	V	volt	
electric potential difference, voltage	U,$\Delta\phi$,ΔV	V		
electric dipole moment	$\mathbf{p}$, $\mathbf{p}_e$	C m		
electric current*	I	A	ampere	
electric current density	j	A m^{-2}		
electric field strength	$\mathbf{E}$	V m^{-1}		
electric displacement	$\mathbf{D}$	C m^{-2}		
capacitance	C	F	farad	
permittivity	ε	F m^{-1}		
relative permittivity	ε	1		
dielectric polarization	$\mathbf{P}$	C m^{-2}		
electric susceptibility	χ_e	1		
polarization (of a particle)	α	m^2 C V^{-1}		
magnetic flux	Φ	Wb	weber	
magnetic flux density	$\mathbf{B}$	T	tesla	
magnetic field strength	$\mathbf{H}$	A m^{-1}		
permeability	μ	H m^{-1}, N A^{-2}	(H: henry)	
relative permeability	μ_r	1		
magnetization	M	A m^{-1}		
magnetic susceptibility	χ	1		
molar magnetic susceptibility	χ_m	m^3 mol^{-1}		
(electrical) resistance	R	Ω	ohm	
(electrical) conductance	G	S	siemens	
(electrical) resistivity	ρ	Ω m		
(electrical) conductivity	κ,o	S m^{-1}		
self-inductance	L	H	henry	
Radiation				
radiant energy	Q, W,Q_e	J		
luminous intensity*	I	cd	candela	
radiant intensity	I_e	W sr^{-1}, W		
emissivity, emittance	ε	1		

Name of quantity[a]	Symbol[b]	SI unit[c]	Name of unit	Other units
absorptance	α	1		
reflectance	ρ, R	1		
transmittance	τ	1		
absorption coefficient:				
linear (decadic)	a	m^{-1}		
molar (decadic)	ε	$m^2\ mol^{-1}$		
refractive index	n	1		
molar refraction	R_m	$m^3\ mol^{-1}$		
angle of optical rotation	α	1, rad		
Transport Properties				
flux of quantity X	J_X, J	(varies)		
mass flow rate	$q_m, \dot{m}$	$kg\ s^{-1}$		
volume flow rate	$q_V, \dot{V}$	$m^3\ s^{-1}$		
heat flow rate	Φ	W		
thermal conductivity	κ, k, λ	$W\ m^{-1}\ K^{-1}$		
coefficient of heat transfer	h	$W\ m^{-2}\ K^{-1}$		
thermal diffusivity	a	$m^2\ s^{-1}$		
diffusion coefficient	D	$m^2\ s^{-1}$		
thermal diffusion coefficient	D_T	$m^2\ s^{-1}$		
viscosity	η, μ	Pa s		
kinematic viscosity	ν	$m^2\ s^{-1}$		

[a]SI base quantities are marked by asterisks (*)
[b]Recommended by IUPAC
[c]SI base units as well as derived and supplementary units are listed; all are to be used with prefixes as needed
[d]do not use "metre"
[e]do not use "litre"; $1\ L = 10^{-3}\ m^3$
[f]$1\ Hz = 1\ s^{-1}$; $1\ N = 1\ kg\ m\ s^{-2}$
[g]Formerly metric ton; $1\ t = 10^3\ kg$
[h]$1\ bar = 10^5\ Pa$
[i]$1\ W\ h = 3.6 \times 10^3\ J$
[j]$1\ eV = 1.602\ 189 \times 10^{-19}\ J$
[k]$1\ W = 1\ J/s = 1\ VA$
[l]$N_A = 6.022\ 136\ 7 \times 10^{23}\ mol^{-1}$
[m]$k = 1.380\ 658 \times 10^{-23}\ J\ K^{-1}$
[n]$h = 6.626\ 075\ 5 \times 10^{-34}\ J\ s$
[o]$R = 8.314\ 510\ J\ mol^{-1}\ K^{-1}$
[p]Molar quantities can be distinguished from the quantity of a system by adding the subscript m; e.g., molar internal energy U_m, in $J\ mol^{-1}$
[q]$1\ u = 1.660\ 565\ 5 \times 10^{-27}\ kg$
[r]A more accurate, but rather uncommon, name is "amount-of-substance fraction"
[s]σ refers to the total volume of a mixture, whereas the volume fraction φ relates the volume of a substance to the volume of several components before mixing
[t]$F = 9.648\ 530\ 9 \times 10^4\ C\ mol^{-1}$
[u]Also called electric charge